9th International Symposium on Hydraulic Structures 2022

Roorkee, India
24 - 27 October 2022

ISBN: 978-1-7138-9977-8

Printed from e-media with permission by:

Curran Associates, Inc.
57 Morehouse Lane
Red Hook, NY 12571

Some format issues inherent in the e-media version may also appear in this print version.

International Association for Hydro-Environment Engineering & Research (IAHR)
Paseo Bajo Virgen del Puerto, 3
28005 Madrid
Spain

Email: iahr@iahr.org

Additional copies of this publication are available from:

Curran Associates, Inc.
57 Morehouse Lane
Red Hook, NY 12571 USA
Phone: 845-758-0400
Fax: 845-758-2633
Email: curran@proceedings.com
Web: www.proceedings.com

Table of Contents

Oct 24th, 12:00 AM

Full Proceedings: Hydraulic Structures

International Association for Hydro-Environment Engineering and Research

Recommended Citation

Palermo M., Ahmad Z., Crookston B., and Erpicum, S. (2022). "9th IAHR International Symposium on Hydraulic Structures (9th ISHS)". *Proceedings of the 9th IAHR International Symposium on Hydraulic Structures – 9th ISHS*, 24-27 October 2022, IIT Roorkee, Roorkee, India. Utah State University, Logan, Utah, USA. 573 pages (ISBN 978-1-958416-07-5).

Foreword to Proceedings of the 9th International Symposium on Hydraulic Structures

M. Palermo[1], Z. Ahmad[2], B. M. Crookston[3], and S. Erpicum[4]
[1] DESTEC-Department of Energy, System, Territory and Construction Engineering,
University of Pisa
ITALY
[2] Department of Civil Engineering,
Indian Institute of Technology Roorkee
INDIA
[3] Civil & Environmental Engineering Dept., Utah Water Research Laboratory
Utah State University
Logan UT 84321
UNITED STATES
[4] Urban and Environmental Engineering Research Unit - HECE Research Group,
Liege University
BELGIUM
E-mail: michele.palermo@unipi.it

Abstract: *The Proceedings of the 9th International Symposium on Hydraulic Structures (ISHS 2022) of the International Association of Hydro-environmental Engineering and Research (IAHR) covered many aspects of hydraulic structures engineering, among others, energy dissipators, intakes and outlets, non-linear weirs, physical and numerical modelling, scour around hydraulic structures, sediment transport, spillways, and training and protection structures. The event organisation motivated the sharing of information and experience between researchers and engineers with different origins and backgrounds. This paper presents an event summary: the proceedings peer-review process, the proceedings content as well as the event organisation.*

Keywords: *Experience mixing; Hydraulic structures; Peer-review process; Proceedings.*

1. INTRODUCTION

Floods, droughts, acute temporal and spatial variations in the water resources, as well as aging infrastructure pose serious challenges to hydraulic engineers (Erpicum et al., 2021). Water continues as a precious resource and populations throughout the globe must store, divert and regulate flows via hydraulic structures to meet sustainable societal needs and objectives. Recent advancements in the analysis and design of hydraulic structures have furthered the knowledge and capabilities of engineers and scientists to provide and maintain economical, eco-friendly and sustainable structures while making improvements where needed for historical structures.

The aim of the 9th International Symposium on Hydraulic Structures (ISHS2022) is to provide a distinctive opportunity for those working in this area to present their works and to share the recent advancements in the field of hydraulic structures.

The 9th Symposium, ISHS 2022 was held in Roorkee, India during 24-27 October 2022 and continues the successful ISHS series with prior events held at Santiago de Chile, Chile (8th ISHS 2020), Aachen, Germany (7th ISHS 2018), Portland, USA (6th ISHS 2016), Brisbane, Australia (5th ISHS 2014), Porto, Portugal (4th ISHS 2012), Nanjing, China (3rd ISHS 2008), Ciudad Guayana, Venezuela (2nd, ISHS 2006), and Theran, Iran (1st ISHS2004).

This short paper summarizes the review process of the ISHS 2022 Proceedings, which have been published open access with a Digital Object Identifier and ISBN by Digital Commons Utah State University. It also lists the proceedings content and symposium organisation.

2. ISHS 2022 PROCEEDINGS

2.1. Peer review process and topics

Papers published in the Proceedings of the 9th IAHR International Symposium on Hydraulic Structures have been peer-reviewed (single blind) for technical content through a formal and rigorous process, as detailed below. The proceedings are hosted by the University Library at Utah State University publication. Each paper was allocated a direct object identifier (DOI) and is accessible open access at the Utah State University Institutional Repository DigitalCommons@USU {https://digitalcommons.usu.edu/}. Each work is available to users through DigitalCommons@USU pursuant to a Creative Commons Attribution CC BY 4.0 License.

The Scientific Committee (86 members) was led by Dr. Michele Palermo with close collaboration with the IAHR Hydraulic Structures Leadership Team and support from the Local Organizing Committee. The Panel of Reviewers was drawn from the Scientific Committee and other international experts in the field, for a total of 80 members. After an initial review of abstracts, the Scientific Committee received 73 full draft papers. These submitted papers were peer-reviewed (single blind) by two independent reviewers, according to an evaluation sheet whose criteria (typical of academic journal paper reviews) were established by the Scientific Committee. Authors were then requested to revise their manuscript in accordance with the reviewers' comments and editorial recommendations, and to submit a revised version with a rebuttal letter responding point-by-point to all reviewers' comments for final review before inclusion in the proceedings. The final number of papers accepted for publication was 60. The symposium proceedings were edited by Michele Palermo (Chair of the Scientific Committee, University of Pisa, Italy), Zulfequar Ahmad (Chair of the Symposium, IIT Roorkee, India), Brian Crookston (Chair of IAHR Hydraulic Structures Technical Committee, Utah State University, USA) and Sebastien Erpicum (Past Chair of IAHR Hydraulic Structures Technical Committee, University of Liege, Belgium).

The scope of the ISHS 2022 proceedings is broad, with five papers on spillways, ten on energy dissipators, five on intakes and outlets, five on non-linear weirs, five on sediment transport, ten on physical and numerical modelling, ten on scour around hydraulic structures, five on training and protection structures, and five focusing on other themes pertaining to hydraulic structures. The papers are based on field investigations, numerical modelling, or experimental investigations. For sure, these proceedings showcase the wide and growing scope of hydraulics structures engineering challenges, methodologies, and solutions.

2.2. Referencing

The papers published in the proceedings are available open access at DigitalCommons@USU {https://digitalcommons.usu.edu/}. DigitalCommons@USU is the institutional repository of the Utah State University.

The full bibliographic reference of the 9th ISHS proceedings is:

Palermo M., Ahmad Z., Crookston B., and Erpicum, S. (2022). "9th IAHR International Symposium on Hydraulic Structures (9th ISHS)". *Proceedings of the 9th IAHR International Symposium on Hydraulic Structures – 9th ISHS*, 24-27 October 2022, IIT Roorkee, Roorkee, India. Utah State University, Logan, Utah, USA. (ISBN 978-1-958416-07-5).

Each paper of the proceedings must be referenced as noted by the University Library at Utah State University, for example:

Mulligan S., and Pope, S. (2022). "A Novel Dual Inlet Tangential Vortex Drop Shaft: An Analytical and Numerical Study" in "9th IAHR International Symposium on Hydraulic Structures (9th ISHS)". *Proceedings of the 9th IAHR International Symposium on Hydraulic Structures – 9th ISHS*, 24-27 October 2022, IIT Roorkee, Roorkee, India. Palermo, Ahmad, Crookston, and Erpicum Editors. Utah State University, Logan, Utah, USA, 10 pages (DOI: 10.26077/7031-ff72) (ISBN 978-1-958416-07-5).

2.3. Statistical summary

60 peer-reviewed papers were written by 176 authors from 18 countries and 5 continents, and 6 keynote lectures were presented during the symposium.

3. SYMPOSIUM ORGANISATION AND REVIEWERS

Because of the exceptional circumstances worldwide induced by the Covid-19 virus pandemic, the 9th IAHR International Symposium on Hydraulic Structures was held in hybrid mode at Indian Institute of Technology (IIT) Roorkee, India, during 24th -27th October, 2022.

3.1. Local organizing committee

- Zulfequar Ahmad (Chair), IIT Roorkee, Roorkee, India
- Ajay K. Singh, NTPC, India
- Ajay Pradhan, Cetus Consulting Solution Services Private Limited, India
- Arun Kumar, IIT Roorkee, Roorkee, India
- Ashish Pandey, IIT Roorkee, Roorkee, India
- C S P Ojha, IIT Roorkee, Roorkee, India
- K S Hariprasad, IIT Roorkee, Roorkee, India
- N K Goel, IIT Roorkee, Roorkee, India
- P K Sharma, IIT Roorkee, Roorkee, India
- R Vinnarasi, IIT Roorkee, Roorkee, India

3.2. International scientific committee

- Michele Palermo (Chair), University of Pisa, Italy
- Abdorreza Kabiri-Samani, Isfahan University of Technology, Iran
- Ajay K. Singh, NTPC, India
- Ajay Pradhan, Cetus Consulting Solution Services Private Limited, India
- Akbar Safarzadeh, University of Mohaghegh Ardabili, Iran
- Alessandro Valiani, University of Ferrara, Italy
- Anton J. Schleiss, EPFL, Switzerland
- Arturo Marcano, CVG EDELCA, Venezuela
- Arun Kumar, IIT Roorkee, India
- Banjamin Dewals, Liege University, Belgium
- Benjamin Hohermuth, ETH Zurich, Switzerland
- Blake Tullis, Utah State University, USA
- Brian Crookston, Utah State University, USA
- Bruce Melville, Auckland University, New Zealand
- Cecilia Lopardo, Istituto National del Agua, Argentina
- Chandrashekhar Ojha, IIT Roorkee, India
- Claudio Fattor, Istituto National del Agua, Argentina
- Cristián Escauriaza, Universidad Católica de Chile, Chile
- Daniel B. Bung, Aachen University of Applied Sciences, Germany
- Daniel Valero Huerta, IHE Delft, The Netherlands
- David Zhu, University of Alberta, Canada
- Deep Roy, University of Pisa, Italy
- Dipankar Roy, Madanapalle Institute of Technology and Science, India
- Eric Lesleighter, Lesleighter Consulting, Australia
- Fabian Bombardelli, University of California, Davis, USA
- Francesco Granata, Università degli studi di Cassino e del Lazio Meridionale, Italy
- Giuseppe Oliveto, Università degli studi della Basilicata, Italy

- Hazi Azamathulla, University of the West Indies, Trinidad
- Hubert Chanson, University of Queensland, Australia
- Iacopo Carnacina, Liverpool John Moores University, UK
- Isabella Schalko, ETH Zurich, USA
- James Yang, Royal Institute of Technology, Sweden
- Jesse Brown, U.S. Army Corps of Engineers, USA
- Jorge Matos, University of Lisbon, Portugal
- Jorge D. Bacchiega, Istituto National del Agua, Argentina
- Jose M. Adriasola, Bechtel, Chile
- José M. Carrillo, Universidad Politecnica de Cartagena, Spain
- Joshua Mortensen, U.S. Bureau of Reclamation, USA
- Juan Cabrera, Universidad Nacional de Ingeneria, Portugal
- Juan Pablo Toro Labbe, Universidad Andres Bello, Chile
- Keil Neff, Stantec, USA
- Kerry Mazurek, University of Saskatchewan, Canada
- Kevin Ruswick, Schnabel Engineering, USA
- KS Hariprasad, IIT Roorkee, India
- Laura Feimster, Schnabel Engineering, USA
- Luis G. Castillo, Universidad Politecnica de Cartagena, Spain
- Mario Oertel, Helmut-Schmidt-University Hamburg, Germany
- Marty Teal, West Consultants, USA
- Masoud Ghodsian, Tarbiat Modares University, Iran
- Matthias Kramer, UNSW Canberra, Australia
- Michael Phillips, U.S. Army Corps of Engineers, USA
- Michael Pfister, University of Applied Sciences and Arts Western Switzerland HES-SO, Switzerland
- Mohanad Khodier, Yarmouk University, Jordan
- Narendra K. Goel, IIT Roorkee, India
- Nils Ruther, NTNU, Norway
- Óscar Link, Universidad de Concepción, Chile
- Pablo Spalletti, Istituto National del Agua, Argentina
- Patricio A. Moreno-Casas, University of the Andes, Chile
- Pramod K. Sharma, IIT Roorkee, India
- Prem L. Patel, SVNIT Surat, India
- Rafael Moran, Universidad Politecnica de Madrid, Spain
- Rajib Das, Jadavpur University, India
- Rita Carvalho, University of Coimbra, Portugal
- Robert Boes, ETH Zurich, Switzerland
- Robert Janssen, Bechtel, Australia
- Rollin Hotchkiss, Brigham Young University, USA
- Roshni Thendiyath, National Institute of Technology Patna, India
- Sahameddin Mahmoudi Kurdistani, IA.ING – Hydraulic Division, Italy
- Sanjay Giri, Deltares, The Netherlands
- Seyed Haji, South Florida Water Management District, USA
- Sean Mulligan, National University of Ireland, Galway, Ireland
- Sebastien Erpicum, Liege University, Belgium
- Silke Wieprecht, University of Stuttgart, Germany
- Silvia Meniconi, Università degli Studi di Perugia, Italy
- Simone Pagliara , ETH Zurich, Switzerland
- SK Zeeshan Ali, IIT Hyderabad, India
- Stefan Felder, University of New South Wales, Australia
- Stefano Pagliara, Universty of Pisa, Italy
- Subhasish Das, Jadavpur University, India
- Subhasish Dey, IIT Kharagpur, India
- Thanos Papanicolaou, Northwest hydraulic consultants, USA
- Tony Wahl, U.S. Bureau of Reclamation, USA
- Valentin Heller, The University of Nottingham, UK

- Vito Ferro, University of Palermo, Italy
- Youichi Yasuda, Nihon University, Japan
- Zulfequar Ahmad, IIT Roorkee, India

3.3. Reviewers

- Abdorreza Kabiri-Samani, Isfahan University of Technology, Iran
- Ajay K. Singh, NTPC, India
- Ajay Pradhan, Cetus Consulting Solution Services Private Limited, India
- Alessandro Valiani, University of Ferrara, Italy
- Anton J. Schleiss, EPFL, Switzerland
- Arturo Marcano, CVG EDELCA, Venezuela
- Banjamin Dewals, Liege University, Belgium
- Benjamin Hohermuth, ETH Zurich, Switzerland
- Blake Tullis, Utah State University, USA
- Brian Crookston, Utah State University, USA
- Bruce Melville, Auckland University, New Zealand
- Cecilia Lopardo, Istituto National del Agua, Argentina
- Chandrashekhar Ojha, IIT Roorkee, India
- Claudio Fattor, Istituto National del Agua, Argentina
- Daniel B. Bung, Aachen University of Applied Sciences, Germany
- Daniel Valero Huerta, IHE Delft, The Netherlands
- David Zhu, University of Alberta, Canada
- Deep Roy, University of Pisa, Italy
- Dipankar Roy, Madanapalle Institute of Technology and Science, India
- Eric Lesleighter, Lesleighter Consulting, Australia
- Fabian Bombardelli, University of California, Davis, USA
- Francesco Granata, Università degli studi di Cassino e del Lazio Meridionale, Italy
- Giuseppe Oliveto, Università degli studi della Basilicata, Italy
- Hazi Azamathulla, University of the West Indies, Trinidad
- Hubert Chanson, University of Queensland, Australia
- Iacopo Carnacina, Liverpool John Moores University, UK
- Isabella Schalko, ETH Zurich, USA
- James Yang, Royal Institute of Technology, Sweden
- Jesse Brown, U.S. Army Corps of Engineers, USA
- Jessica Di Nardi, University of Pisa, Italy
- Jorge Matos, University of Lisbon, Portugal
- Jorge D. Bacchiega, Istituto National del Agua, Argentina
- Jose M. Adriasola, Bechtel, Chile
- José M. Carrillo, Universidad Politecnica de Cartagena, Spain
- Juan Cabrera, Universidad Nacional de Ingeneria, Portugal
- Juan Pablo Toro Labbe, Universidad Andres Bello, Chile
- Keil Neff, Stantec, USA
- Kerry Mazurek, University of Saskatchewan, Canada
- Kevin Ruswick, Schnabel Engineering, USA
- KS Hariprasad, IIT Roorkee, India
- Laura Feimster, Schnabel Engineering, USA
- Luis G. Castillo, Universidad Politecnica de Cartagena, Spain
- Mario Oertel, Helmut-Schmidt-University Hamburg, Germany
- Marty Teal, West Consultants, USA
- Masoud Ghodsian, Tarbiat Modares University, Iran
- Matthias Kramer, UNSW Canberra, Australia
- Michael Phillips, U.S. Army Corps of Engineers, USA
- Michael Pfister, University of Applied Sciences and Arts Western Switzerland HES-SO, Switzerland
- Michele Palermo, University of Pisa, Italy

- Mohanad Khodier, Yarmouk University, Jordan
- Narendra K. Goel, IIT Roorkee, India
- Nils Ruther, NTNU, Norway
- Óscar Link, Universidad de Concepción, Chile
- Pablo Spalletti, Istituto National del Agua, Argentina
- Patricio A. Moreno-Casas, University of the Andes, Chile
- Pramod K. Sharma, IIT Roorkee, India
- Rafael Moran, Universidad Politecnica de Madrid, Spain
- Rajib Das, Jadavpur University, India
- Rita Carvalho, University of Coimbra, Portugal
- Robert Boes, ETH Zurich, Switzerland
- Robert Janssen, Bechtel, Australia
- Roshni Thendiyath, National Institute of Technology Patna, India
- Sahameddin Mahmoudi Kurdistani, IA.ING – Hydraulic Division, Italy
- Sanjay Giri, Deltares, The Netherlands
- Sean Mulligan, National University of Ireland, Galway, Ireland
- Sebastien Erpicum, Liege University, Belgium
- Silke Wieprecht, University of Stuttgart, Germany
- Silvia Meniconi, Università degli Studi di Perugia, Italy
- Simone Pagliara, ETH Zurich, Switzerland
- SK Zeeshan Ali, IIT Hyderabad, India
- Stefan Felder, University of New South Wales, Australia
- Stefano Pagliara, Universty of Pisa, Italy
- Subhasish Das, Jadavpur University, India
- Subhasish Dey, IIT Kharagpur, India
- Thanos Papanicolaou, Northwest hydraulic consultants, USA
- Tony Wahl, U.S. Bureau of Reclamation, USA
- Valentin Heller, The University of Nottingham, UK
- Vito Ferro, University of Palermo, Italy
- Youichi Yasuda, Nihon University, Japan
- Zulfequar Ahmad, IIT Roorkee, India

4. PHILIP H. BURGI BEST PAPER AWARD

The Philip H. Burgi Best Paper Award was instituted at ISHS 2018 in Aachen, Germany to honor the first chair of the IAHR Hydraulic Structures Committee. The aim of the Philip H. Burgi Best Paper Award is to reward the best technical paper and presentation from the symposium. The following two papers were selected for the award during the 9th IAHR International Symposium on Hydraulic Structures:

Angulo M.A., Rivetti A., Lucino C.V., Liscia S.O. (2022) "Anti vortex device to operate pump intakes below the minimum submergence" in "9th IAHR International Symposium on Hydraulic Structures (9th ISHS)". *Proceedings of the 9th IAHR International Symposium on Hydraulic Structures – 9th ISHS*, 24-27 October 2022, IIT Roorkee, Roorkee, India. Palermo, Ahmad, Crookston, and Erpicum Editors. Utah State University, Logan, Utah, USA, 9 pages (DOI: 10.26077/3fa1-7f82) (ISBN 978-1-958416-07-5).

and

Jaiswal A., Ahmad Z., Mishra S.K. (2022). "Effect of Diameter and Inlet-depth on Hydro-suction performance of a Suction pipe" in "9th IAHR International Symposium on Hydraulic Structures (9th ISHS)". *Proceedings of the 9th IAHR International Symposium on Hydraulic Structures – 9th ISHS*, 24-27 October 2022, IIT Roorkee, Roorkee, India. Palermo, Ahmad, Crookston, and Erpicum Editors. Utah State University, Logan, Utah, USA, 8 pages (DOI: 10.26077/aa4f-76f0) (ISBN 978-1-958416-07-5).

5. ACKNOWLEDGEMENTS

The organization of the workshop would not have been possible without the active involvement of all the members of the local organizing committee and the international scientific committee. We would also like to thank all the reviewers for their valuable support in reviewing the manuscripts and all the Authors for their contributions. We also acknowledge and give special thanks for the support of the University Library of Utah State University for the production and publication of the proceedings.

We would also like to acknowledge the support of our sponsors and knowledge partners, Central Board of Irrigation and Power (CBIP), Central Water Commission (CWC), Cetus Consulting Solutions Services Pvt Ltd, Consulting Engineers Association of India (CEAI), Indian Society for Hydraulics (ISH), Inland Waterways Authority of India, NCC Limited, SIGNET, SMP – Syama Prasad Mookerjee Prot, Kolkata, and THDC India Limited. Finally, financial support provided by the IIT Roorkee under its Programme of Celebration of 175th year is greatly acknowledged.

6. REFERENCES

Erpicum, S., Crookston, B., Bombardelli, F., Bung, D.B., Felder, S., Mulligan, S., Oertel, M. and Palermo, M. (2021). Hydraulic structures engineering: An evolving science in a changing world. WIREs Water. 8(2), e1505.

DOI: https://doi.org/10.26077/7031-ff72

A Novel Dual Inlet Tangential Vortex Drop Shaft: An Analytical and Numerical Study

S. Mulligan[1] & S. Pope[2]
[1] *VorTech Water Solutions Ltd, Galway, Ireland*
[2] *Armatec Environmental Ltd, New Plymouth, New Zealand*
E-mail: sean.mulligan@vortechws.com

Abstract: *Urban drainage systems are undergoing substantial upgrades to enhance resilience to increased urbanisation, population growth and more intense rainfall events. Typically, in these hydraulic structures, the challenge of dropping water flows over significant elevation differences is overcome using vortex drop shaft structures. Many options are available for transferring flows from a single upstream channel to a lower level using independent vortex drop shafts. However, there are many cases where a manhole is to convey discharge from multiple flow inlets. In this work, a novel dual inlet tangential vortex drop shaft was developed during a full-scale commercial project proposal. The paper presents analytical and multiphase numerical modelling studies of various geometries of the structure. Analytical methods were used to derive conditions to maximize the free drainage discharge of inlets and were tested using a three-dimensional multiphase numerical model. The performance of the structure was evaluated for two drop tube diameters with scenarios including (1) equal flows in symmetric inlets, (2) unequal flows in asymmetric inlets and (3) independent inlet flows. The results demonstrate that the analytical model overestimates the free-drainage discharge. Nonetheless, interaction between flows and opposing inlets in the design did not appear to pose any significant stability or performance issues. Moreover, when tested at extreme flow rates up to flooding conditions, the numerical results demonstrate that the vortex air core remains intact. To conclude, a cost-benefit analysis of this type of configuration is presented which provides an economic argument for consideration of the dual inlet drop shaft in future schemes.*

Keywords: *Energy Dissipation, Multiphase Modelling, Numerical Modelling, Urban Infrastructure, Vortex Drop Shafts.*

1. Introduction

Across the world, urban drainage systems are undergoing substantial upgrades to enhance resilience to increased urbanization, population growth and more intense rainfall events, and to improve environmental protection. Such hydraulic infrastructure includes the installation of combined sewer overflow intercept structures and deep sewer conveyance systems. Typically, in these hydraulic structures, dropping water flows over significant elevation differences is overcome using vortex drop shaft structures (see Figure 1). These structures provide efficient energy dissipation whilst reducing air entrainment, turbulence, and odours, thus increasing the sustainability and lifespan of the infrastructure (Mulligan et al. 2019; Knauss 1987).

There are several proven options available for transferring flows from a single upstream channel to a lower level using independent vortex drop shafts. The choice of option depends on (1) if the approach flow is subcritical or supercritical, (2) the available footprint, (3) the requirements to pass large solids or debris or (4) sensitivity to upstream hydraulic grade line changes (Knauss, 1987; Vischer and Hager 2018; Mulligan et al. 2019). Types of configurations most commonly available are scroll, tangential and spiral which are described in detail in previous sources of literature along with examples of physical and numerical modelling studies (Knauss 1987; Mulligan et al. 2019).

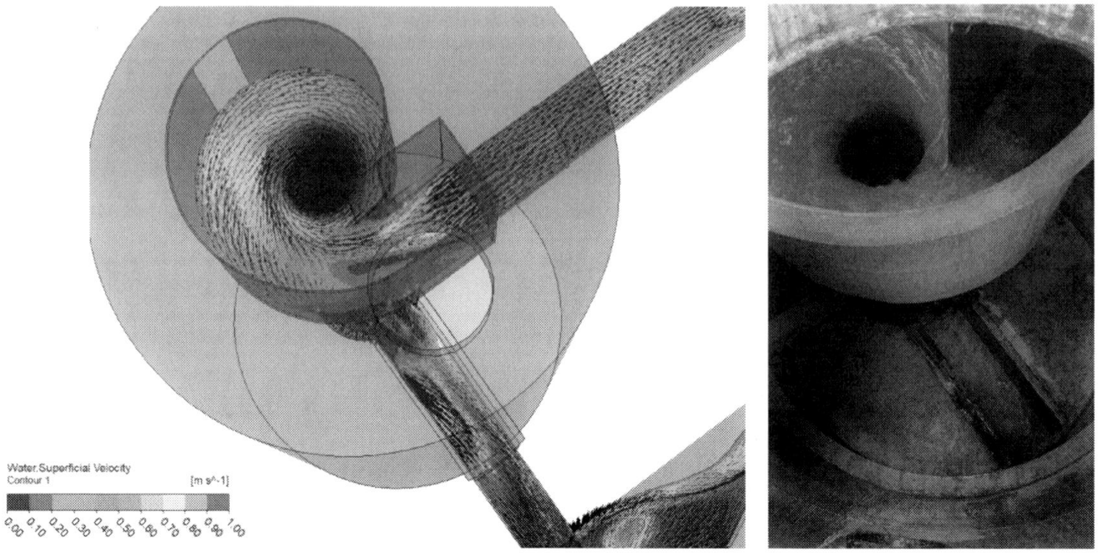

Figure 1. Example comparison between a multiphase numerical model data and prototype free-surface distribution in a scroll type vortex generator project in New Zealand. Project as part of an industry design and fabrication collaboration between VorTech Water Solutions (Ireland) and Armatec Environmental (New Zealand).

In recent years, there has been increased demand for energy dissipation structures in urban drainage systems, due to the construction of major urban drainage infrastructure across cities such as Hong Kong, London, Dubai and Abu Dhabi, among others. The growing demand poses new challenges in cases, for example, where footprint is limited underneath existing urban areas, large flows and elevation differences limit efficient designs, and where there are requirements to connect large diameter tunnels carrying small discharges to small drop structures. There is also the pressing requirement to maintain economy and sustainability of such hydraulic structures amongst stakeholders. Another common challenge is the need to combine multiple inlets in one shaft or manhole. In such cases, multiple independent drop shafts can be installed. However, there may be imperatives instead transfer water in a single drop shaft with multiple inlets, which includes the need to reduce available footprint, material, installation costs, operational costs and indeed, total lifecycle costs.

Some examples of vortex drop structures receiving multiple inlets include the structures installed in Cossonay and in Zürich (see description in Pfister et al. 2018). In these structures, the inflow branches entered at different elevations but connected into a primary trunk channel to convey to a spiral styled vortex drop shaft. A drop structure in Bern (also described by Pfister et al. 2018) connected two separate and spiralling channels to a spiral styled vortex generator. In the study of Pfister et al. (2018), the authors designed a new vortex drop shaft for Cossonay where the original structure had reached its design capacity. In the new design, they had the challenge of conveying the flow from four inlets at four different elevations into the drop shaft. This was solved using a junction chamber into which the four inflow branches would deliver their discharge prior to conveyance to the spiral vortex drop shaft. The physical study carried out demonstrated that such an approach was feasible. Weiß et al. (2010), described another configuration where, in order to avoid the construction of a junction chamber, two feed pipes were fed into a scroll or circular styled vortex drop structure.

In the available examples of vortex drop structures with multiple inlets, so far only scroll and spiral types are described, often necessitating additional junction boxes to deliver the flow. As far as the authors are aware, there is no case available yet in the literature where a tangential style vortex drop structure was adopted with multiple inlets. Tangential types offer advantages over spiral and scroll types due to their smaller footprint and lower upstream head requirements. However, in the case of multiple inlets, the tangential inlet designs require special considerations to ensure stability of the flow is maintained for various conditions, particularly due to the "wrap around interference" (Yu and Lee 2009; Plant and Crawford 2016) expected between the two jets.

Therefore, a study was conducted by an industry collaboration between VorTech Water Solutions and Armatec Environmental to propose and develop the basis of a novel design for a tangential vortex drop shaft with dual inlets. Motivation for the study was derived from a full-scale design proposal for a section of an urban drainage system in the Dubai which required the conveyance of unequal flows from two pipes in a single manhole (See Section 2.1). This study provides an early design basis for the dual inlet tangential vortex drop shaft design and reports on the computational fluid dynamics simulations performed as a proof of concept.

2. Case Study and Design

2.1. Overview

As part of an urban drainage infrastructure project in Dubai, vortex drop structures were required in a manhole to transfer water from two 2000 mm sewer pipe sections A and B which deliver design flows Q_A = 929 l/s (from left to right in Figure 1) and Q_B = 1,176 l/s from right to left respectively. Both pipes align coaxially and meet at the shaft at the position outlined in Figure 2. At this position, the challenge was to safely transfer both flows into the lower level positioned approximately 6 m below the pipe inverts.

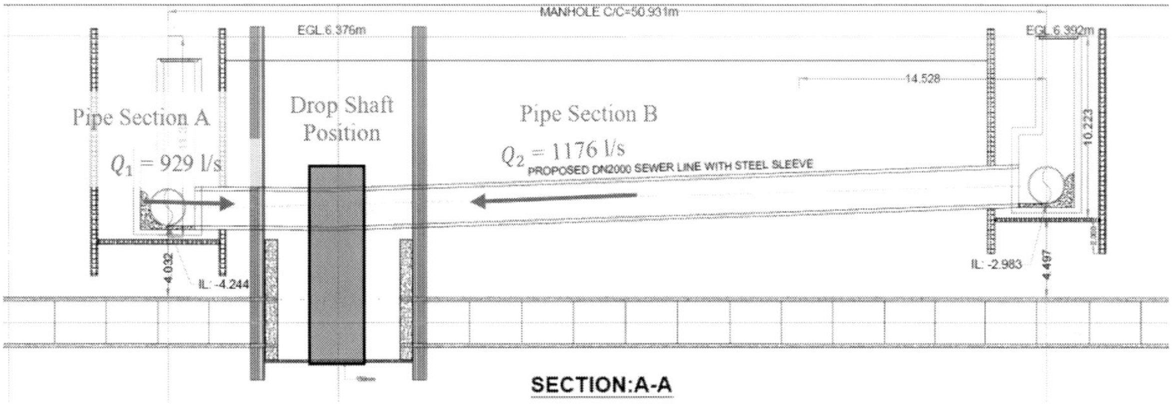

Figure 2. Cross sectional schematic of the pipe and shaft scheme.

Due to the available footprint and drivers to economise the scheme without compromising on performance, a novel dual (double) inlet tangential vortex drop shaft structure was proposed. The structure was designed to receive the total design flow 2.105 m³/s and there was a requirement to ensure:

- A full air core is maintained throughout the vortex drop structure.
- The interaction between both flows is to be smooth and the flow is to be stable.
- That for independent flow conditions, a back up into the non-operational inlet is not experienced.

2.2. Design Equations

To achieve the hydraulic requirements set out in Section 2.1, the following design considerations were initially derived to ensure capacity and to minimize interreference between inflow jets during the design. For a given design total discharge Q, the minimum drop shaft diameter D can be sized based on the empirical relationship provided by Vischer and Hager (1995) as follows:

$$D = k(Q^2/g)^{1/5} \tag{1}$$

where g is the gravitational constant. In the case of a dual inlet, Q can be treated as the total combined flowrate of both inlets ($Q = Q_A + Q_B$) because the discharge is related to the shaft diameter (Vischer and Hager 1995) and thus independent of the number of inlets upstream. For conservative design purposes, a value of k =1.2 can be adopted.

The control shift discharge is defined as the discharge which causes the critical section to shift from the approach flow channel crest to the tangential inlet (Yu and Lee 2009). Similar to a standard tangential inlet, the following expression for the control shift discharge Q_c can then be derived for each inlet independently like in the study of Yu and Lee (2009):

$$Q_c = \frac{\sqrt{g}e(2z/3)^{3/2}}{\left(Cos^{3/2}\beta - (e/B)^{2/3}\right)^{3/2}} \tag{2}$$

where e is the junction width, z is the vertical height from the approach flow invert to the junction invert, β is the channel slope, B is the approach flow channel width. The depth and critical depth at each junction when $Q < Q_c$ can also be obtained from the following equation:

$$y = y_{ca} = \sqrt[3]{\left(\frac{Q}{B}\right)^2 \frac{1}{g}} \tag{3}$$

Typically, designs are based on ensuring that a minimum drop tube diameter is maintained according to Eq. (1) and the design discharges are less than the critical discharge of the inlet $Q < Q_c$. In the case of the dual inlet, this latter condition can be checked for each inlet.

An additional criterion related to maintaining a free-drainage condition was developed by Yu and Lee (2009). Their experiments have shown that under some conditions in a single inlet tangential vortex generator, the supercritical inlet discharge creates an unstable flow, even when $Q < Q_c$. This is due to the condition that occurs after the flow turns through 360° and interacts with the parallel inflow jet leading to a swell, backflow or even a hydraulic jump. This condition is expected to be more severe when two tangential inlets are combined in a single drop shaft, where jet interaction can occur after the flow turns through 180° where it meets the opposing inlet.

Similar to the study of Yu and Lee, (2009), the limiting condition of "free drainage" for dual symmetric inlets can be obtained by a first-order approximation through:

$$\Delta z = v_z \Delta t \tag{4}$$

Where v_z is the axial velocity Δz is the drop height of the water surface after the flow turns 180° in the drop shaft and where Δt is the time it takes for the flow to turn through 180° and meet the second inlet. Using the flow tangential velocity at the inlet, v_t, Δt can be replaced in Eq. 1 as follows:

$$\Delta z = v_z \left(\frac{\pi D}{2 v_t}\right) \tag{5}$$

and subsequently, (by following the derivation outlined in Yu and Lee (2009)), can be defined as follows:

$$\Delta z = tan\beta \frac{\pi D}{2-2e/D} \tag{6}$$

where $\pi D/2$ is the circumference between inlets. Assuming the inlet jet flows are fully symmetric (i.e. the critical depth y_c and flows Q_c are identical for each inlet) the free-drainage discharge for dual inlets can then be determined in terms of the inlet geometry as follows:

$$Q < Q_f = \left(tan\beta \frac{\pi D}{2-2e/D}\right)^{3/2} \sqrt{ge}\, cos^2\beta \tag{7}$$

where Q_f is the free-drainage discharge for the inlet to be evaluated. In the case of the dual inlet design, it is proposed that $Q_f > Q$ for each inlet if flows are symmetric to avoid any flow instability.

In the case of asymmetric flow conditions (which will be more common in practice), an additional measure that should be ensured is that the free-drainage condition is obtained at the second inlet. This can be ascertained through the height difference H_A between Inlet A and Inlet B which should not be less than:

$$H_A = tan\beta_A \frac{\pi D}{2-\frac{2e_A}{D}} - \sqrt[3]{\left(\frac{Q_A}{B_A}\right)^2 \frac{1}{g}} \tag{8}$$

The height difference H_B between the Inlet B and Inlet A can be derived in a similar fashion.

2.3. Design

Based on the total design discharge (Q = 2.105 m^3/s), using Eq.(1), two drop tube diameters were chosen, D_1 =1.0 m (which is just on the limit of minimum diameter for k = 1.2) and D_2 =1.2 m. Two types of dual inlets were also tested (1) symmetric inlets and (2) asymmetric inlets, to account for the larger flowrate in one of the approach flows. The geometric parameters of the vortex generator are outlined in Table 1 which generally aligns with accepted parameters (Plant and Crawford 2016).

In Model 1 and Model 3, $Q < Q_f$ was maintained according to Eq. (7). In Model 2, $Q > Q_f$ determined analytically suggested that the flow from the first inlet will interfere with the second inlet (see figures in **bold**).

Table 1. List of vortex drop shaft geometric parameters.

Parameter	Model 1 – Symmetric		Model 2 – Asymmetric		Model 3 – Asymmetric	
	Inlet A	**Inlet B**	**Inlet A**	**Inlet B**	**Inlet A**	**Inlet B**
D	1.2 m		1.0 m		1.2 m	
B/D	1.00	1.00	1.00	1.00	1.00	1.00
e/D	0.3	0.3	0.3	0.3	0.3	0.3
L_1/D	2.2	2.2	1.65	2.2	1.65	2.2
z/D	1.263	1.263	0.801	1.263	0.801	1.263
β	30.00	30.00	30.00	30.00	30.00	30.00
Q	0.929 m^3/s	0.929 m^3/s	**0.929 m^3/s**	**1.176 m^3/s**	0.929 m^3/s	1.176 m^3/s
Q_c	2.99 m^3/s	2.99 m^3/s	1.441 m^3/s	6.790 m^3/s	14.080 m^3/s	2.989 m^3/s
Q_f	1.586 m^3/s	1.586 m^3/s	**0.765 m^3/s**	**0.765 m^3/s**	1.586 m^3/s	1.586 m^3/s
Δz	1.295 m	1.295 m	1.295 m	1.295 m	1.295 m	1.295 m

3. Numerical Modelling Methodology

3.1. 3D Model

The model of the design (Figure 2 and 3) was integrated into ANSYS SpaceClaim. The numerical modelling domain consists of only short sections of Pipe A and Pipe B (with no gradient included) upstream of the shaft and a 6 m length of drop shaft below the pipe inverts. Energy dissipation was not of interest in this study and therefore downstream detail was omitted (flow exits the domain through the drop shaft only).

3.2. Mesh Generation

An unstructured tetrahedral mesh was adopted as outlined in Figure 3. 15 inflation layers were applied to all walls and surfaces to capture and resolve boundary layers. This was particularly important to resolve the free-surface jet flows in the drop shaft (see Plant and Crawford 2016; Carty et al. 2019). Face sizing varied between 0.030 m and 0.1 m. A total number of 1.8 million elements was used throughout the structures.

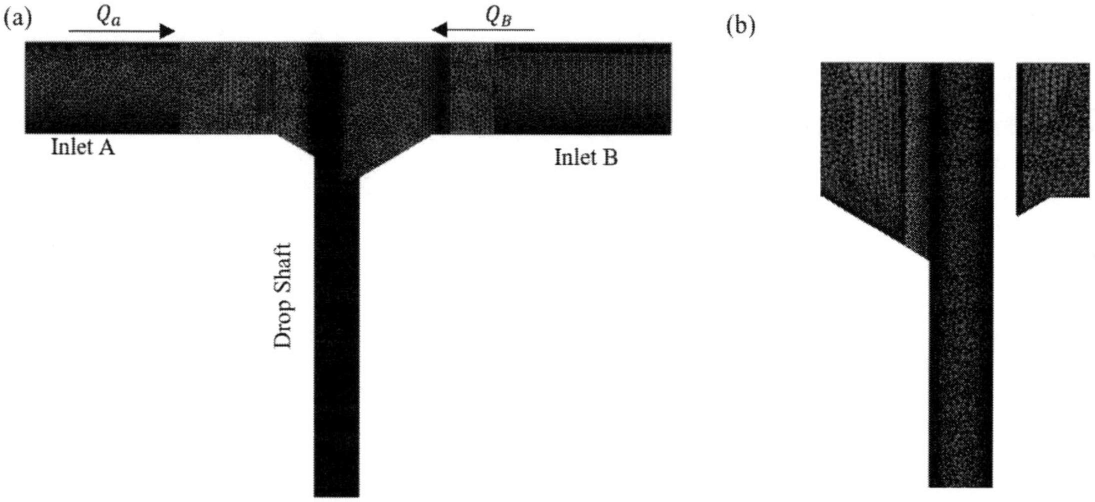

(c)

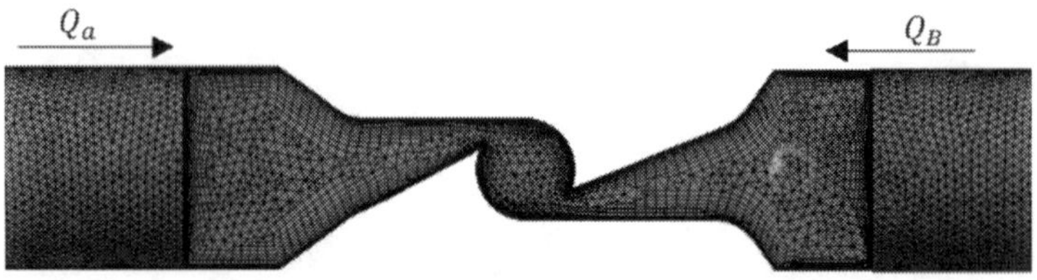

Figure 3. Overview of mesh for Model 2 (a) elevation, (b) end view and (c) plan.

3.3. Model Setup

3D multiphase (air-water) simulations were conducted using ANSYS CFX. The boundary conditions assigned in the model were as follows:

- Mass flow normal to inlet was adopted for the inlet boundary condition.
- An opening boundary condition subject to atmospheric conditions was assigned to the outlet at the bottom of the drop pipe.
- The domain top was treated as an opening subject to atmospheric conditions.
- All walls were treated as no-slip boundary conditions.
- The k-Omega SST turbulence model with curvature correction was adopted.
- Two water phases (of identical liquid properties) were conveyed into the domain at Inlets A and B. The domain consisted of 100% air phase as an initial condition. The liquid phases were treated as continuous phases.

These settings were adopted from best practice guidelines verified with good accuracy in previous physical-numerical modelling studies of tangential vortex drop shafts (Plant and Crawford 2016; Carty et al 2019). Ten scenarios were investigated for the three models as outlined in Table 2:

Table 2: Scenarios investigated

Model 1 – Symmetric ($Q < Q_f$)	**Pipe Section A**	**Pipe Section B**
Scenario 1 – Combined	929 l/s	929 l/s
Scenario 1 – Individual 1	929 l/s	0 l/s
Model 2 – Asymmetric ($Q > Q_f$)	**Pipe Section A**	**Pipe Section B**
Scenario 1 – Combined	929 l/s	1,176 l/s
Scenario 2 – Individual 1	929 l/s	0 l/s
Scenario 3 – Individual 2	0 l/s	1,176 l/s
Model 3 – Asymmetric ($Q < Q_f$)	**Pipe Section A**	**Pipe Section B**
Scenario 1 – Combined	929 l/s	1,176 l/s
Scenario 2 – Individual 1	929 l/s	0 l/s
Scenario 3 – Individual 2	0 l/s	1,176 l/s
Scenario 4 – Increased Flow	2,000 l/s	2,500 l/s
Scenario 5 – Flooding Condition	6,000 l/s	6,000 l/s

4. Results and Discussion

4.1. Free-Surface Interactions

Figure 4, 5 and 6 outlines the free surface distribution across the overall structure for both flow scenarios where green represents flow stream from Inlet A and blue represents flow stream from Inlet B. In almost all cases, the interaction between jets first occurs at the inlets (see Figure 4), despite in most cases the condition of $Q_f > Q$ being maintained analytically. Notwithstanding, the free surfaces from both jets interact smoothly to form a single and stable annular jet extending down the extent of the drop pipe modelled. As outlined in Figure 5 and 6, this jet covers the full circumference of the drop tube from the vortex generator to the outlet (where in a single inlet, the drop tube perimeter is often only partially covered – see Carty et al (2019)). This suggests that greater energy dissipation efficiencies can be achieved in a dual inlet due to the increased drop tube surface coverage.

In the symmetric dual inlets (Model 1), the interaction between the jets causes a negligible effect on upstream hydraulics and an almost fully symmetrical combined jet formation is observed in the downstream pipe.

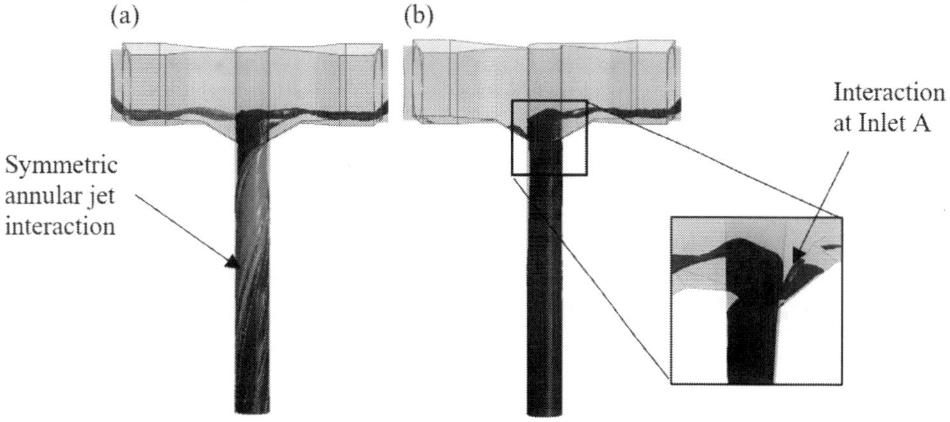

Figure 4. Symmetrical dual inlet structure (Model 1) for (a) Q_1 = 929 l/s and (b) Q_2 = 929 l/s only.

In Model 2, where $Q_f < Q$, the jet interaction causes a slight increase in the upstream channel water depth. In the case where Inlet B is operating independently, the flow causes a back up which extends up the ramp and into the upstream channel of Inlet A. This is overcome in Model 3 ($Q_f > Q$) where the free surface of the inlet jet after turning 180 degrees is aligned almost parallel to the Inlet A ramp (however significantly higher than what was predicted by Eq. (5) and (8)). Despite the jet interaction and increases in upstream head in some cases, all Models appeared to perform generally well with good stability of the vortex and annular jet interactions for each inlet.

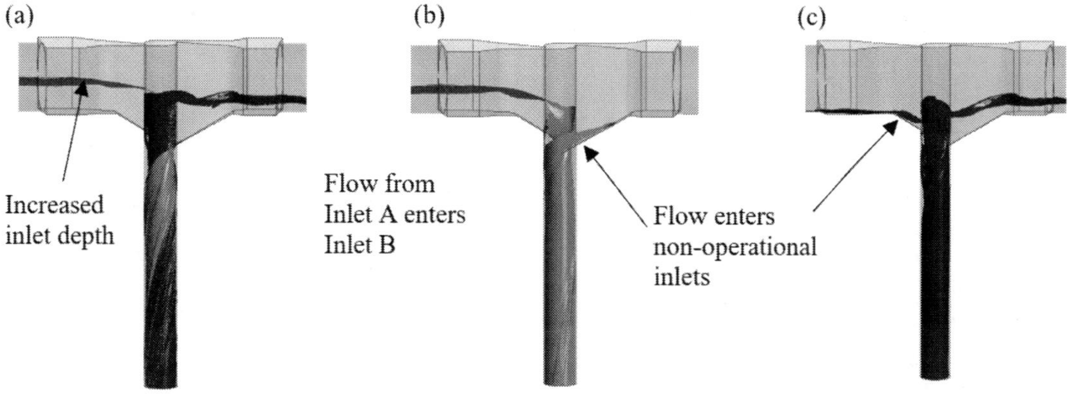

Figure 5. Asymmetrical dual inlet structure (Model 2) for (a) Q_1 = 929 l/s and Q_2 = 1176 l/s, (b) Q_1 = 929 l/s only and (c) Q_2 = 1176 l/s only.

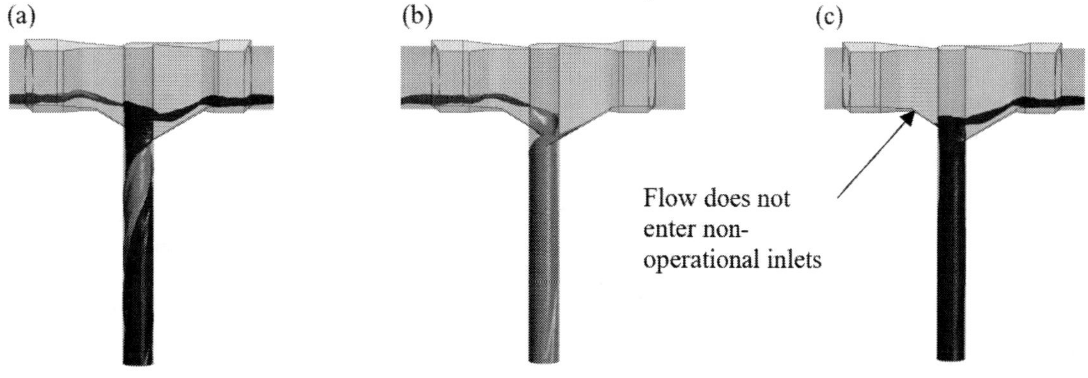

Figure 6. Asymmetrical dual inlet structure (Model 3) for (a) Q_1 = 929 l/s and Q_2 = 1176 l/s, (b) Q_1 = 929 l/s only and (c) Q_2 = 1176 l/s only.

4.2. Wrap Around Interference

As described previously, the Δz and Q_f values were overestimated considerably by Equation (5) and (8). Table 3 outlines the quantitative differences between the analytical and numerical values for Δz (in Model 2, a negative Δz was observed due to the standing wave of the inflow jet). Similar overestimations of the Δz or free-drainage discharge have been identified in the past by Plant and Crawford (2016) where their physical models experienced Q_f at significantly lower flows than were calculated analytically. They propose that the difference is attributed to the equation not considering the length of the intake or the width of the approach channel, both of which have the potential to influence the orientation of the flow entering the drop shaft. There is also a question of numerical model uncertainty since the dual inlet numerical model was not validated using physical model data.

Table 3: Scenarios investigated

Parameter	Model 1 – Symmetric $Q_f > Q$		Model 2 – Asymmetric $Q_f < Q$		Model 3 – Asymmetric $Q_f > Q$	
	Inlet A	Inlet B	Inlet A	Inlet B	Inlet A	Inlet B
Δz (Analytical)	1.295 m	1.295 m	1.295 m	1.295 m	1.295 m	1.295 m
Δz (Numerical)	0.318 m	0.318 m	0.520 m	-0.217 m	0.541	0.142 m

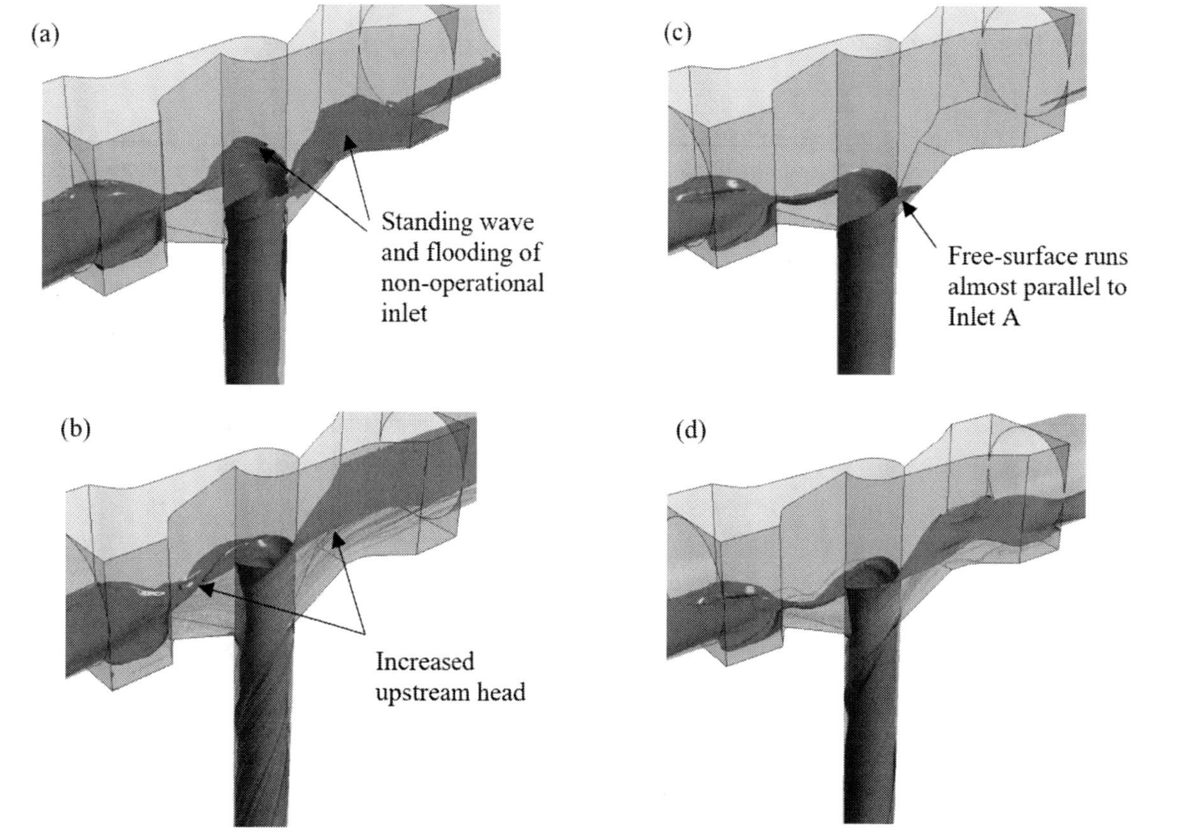

Figure 7. Wrap around inference for Model 2 (a) Inlet B active only, (b) both inlets active and for Model 3, (c) Inlet B active only, (d) both inlets active and for Model 3, outlining the effects on upstream water levels.

4.3. Model 3 – Flooding Condition

Finally, it was noted that flow interactions appeared to be smooth across all models for the design flows which resulted in stable flows down the vortex drop shaft. As an additional evaluation of hydraulic robustness, two increased flow conditions were evaluated, $Q = 4.5$ m^3/s and $Q = 12$ m^3/s (flooding/overtopping condition). The results of the free-surface profiles through the structure for each case are outlined in Figure 8 and despite overtopping at $Q = 12$ m^3/s, the vortex air core for the dual annular jets remains intact for the extent of the drop shaft examined.

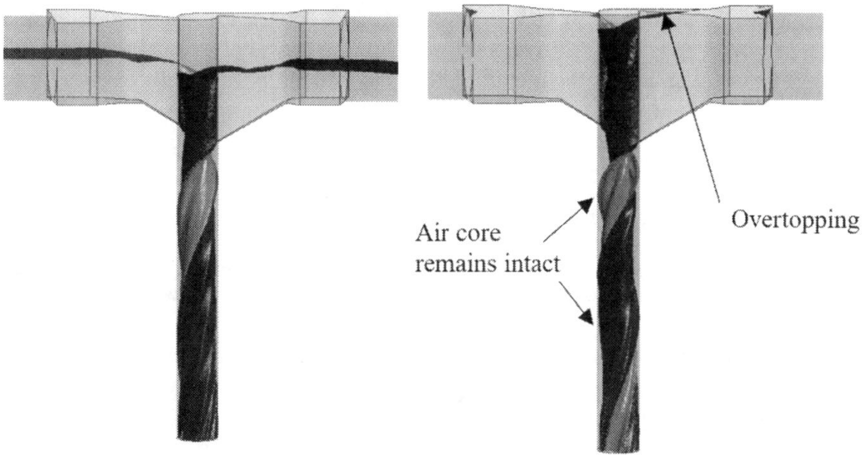

Figure 8. Model 3 (a) for Q_1 = 2000 l/s and Q_2 = 2500 l/s and (b) for Q_1 = 7000 l/s and Q_2 = 7000 l/s.

5. Cost Benefit Analysis

Although multiphase numerical models of tangential vortex drop shafts have provided accurate results in past literature, there is a need to verify the numerical data on jet interaction posed in this new design using a physical model or prototype data sets. Notwithstanding this uncertainty, this section provides brief insights into the expected commercial impact of considering a dual inlet over two single inlets. The example outlined in this study for a 1.0 m drop tube is evaluated and costed for two different drop shaft lengths as outlined in Table 4. Costs were provided by Armatec Environmental for GRP fabrication and considers the costs of the drop shaft product only (excludes connections, installation costs and civil works etc.).

Table 4: Cost comparison between two single and one dual inlets of different drop shaft lengths.

	2 x Single Inlets (760 mm diameter)		**Dual Inlet (1050 mm diameter)**	
	Cost Per Unit ($)	**Total Cost ($)**	**Cost Per Unit ($)**	**Total Cost ($)**
Vortex Generator	10,000	20,000	20,000	20,000
Drop Pipe (6 m)	7,500	15,000	15,000	15,000
	Total Cost	**35,000**		**35,000**
Vortex Generator	10,000	20,000	20,000	20,000
Drop Pipe (30 m)	38,000	76,000	50,000	50,000
	Total Cost	**96,000**	**Total Cost**	**70,000**

When considering the drop shafts alone, it can be understood that for small drops, the cost for two single structures may be similar to the cost of a dual inlet structure. However, for a larger drop shaft, economy can be realized through the reduced price per unit length of drop shaft. Here, a saving of 27% is demonstrated in the drop shaft structure alone.

Regardless of the length of the drop structure, additional savings (not quantified in this paper) can be obtained through the following aspects:

- Reduced installation time and costs.
- Fewer pipe supports or mounting brackets.
- Savings in construction of energy dissipation chamber/stilling basin.
- Smaller access shaft.
- Reduced excavation requirements.

Reduced maintenance costs are also expected for a single shaft where routine inspection time can be reduced. A single unit also poses a lower risk of blockage or damage. In aggregate, the dual inlet solution can provide significantly lower total lifecycle costs for the project where applicable in such applications.

6. Conclusions

In this study, a novel dual inlet tangential vortex drop shaft is proposed to solve design challenges as part of a full scale drop structure project proposal. With an understanding of the flow interaction between two jets and flow stability throughout the structure not available in the literature, this paper focused on developing the fundamentals and insights into the dual inlet design, based on existing analytical models supported by a 3D multiphase numerical model. Analytical models to assess the free-drainage condition for dual inlets were developed. Three designs were produced based on two different drop shaft diameters for symmetric and asymmetric approach flow conditions (where one of the designs was to fail the criteria for free drainage according to the analytical model). All simulations indicated that the free-drainage discharge was significantly overestimated which resulted in jet interaction at opposing inlets for flowrates significantly lower than expected. Nonetheless, despite the jet interaction observed, the simulation results of the free surface demonstrated that the flow remained smooth with a full air core observed in all cases. Moreover, when the limits of the vortex drop shaft were pushed to cause an overtopping of the vortex chambers, the air core still remained intact, exhibiting the robustness of performance within the hydraulic structure. Although multiphase numerical models of tangential vortex drop shafts have provided accurate results in past literature, there is a need to verify the numerical data on jet interaction posed in this new design using a physical model or prototype data sets. Notwithstanding this uncertainty, a cost comparison between two single inlets and one dual inlet structure identified that significant savings could be obtained in total lifecycle costs. It is recommended that future work should focus on physically modelling the dual inlet design and verifying the numerical model, resolving discrepancies in the free-drainage analytical problem, and using the results to develop guidelines to minimize interaction between the jet and thus further improve hydraulic efficiency of the structure.

7. ACKOWLEDGMENTS

The authors would like to express thanks to VorTech Water Solutions Ltd and Armatec Environmental Ltd for providing resources to carry out this project.

8. REFERENCES

Carty, A., O'Neill, C., Nash, S., Clifford, E. and Mulligan, S., (2019). "Hydrodynamic modelling approaches to assess mechanisms affecting the structural performance and maintenance of vortex drops shaft structures". *Journal of Structural Integrity and Maintenance,* 4(3), 162-178.

Knauss, J. E. (1987) Swirling flow problems at intakes. AA Balkema.

Mulligan, S., Plant, J., Nash, S. and Clifford, E., (2019), "Vortex drop shaft structures: state-of-the-art and future trends". *In Proc., 38th IAHR World Congress–Water Connecting the World,* IAHR. Panama. 3860-3869.

Plant, J. and Crawford, D., (2016). "Pushing The Limits Of Tangential Vortex Intakes: Is Higher Capacity And Flow Measurement Possible In A Smaller Footprint?". *Proceedings of the Water Environment Federation, 2016*(12), 4108-4136.

Pfister, M., Crispino, G., Fuchsmann, T., Ribi, J.M. and Gisonni, C., (2018). "Multiple Inflow Branches at Supercritical-Type Vortex Drop Shaft". *Journal of Hydraulic Engineering*, 144(11), 05018008.

Vischer, D.L. and Hager, W.H., (2018). "Vortex drops. In Energy dissipators", 167-181. Routledge.

Weiss, G., Brombach, H. and Hohl, E., (2010). "Hydraulic model tests on a stormwater vortex drop shaft: Verification of special conditions". *NOVATECH 2010.*

Yu, D. and Lee, J.H., (2009). "Hydraulics of tangential vortex intake for urban drainage". *Journal of hydraulic engineering*, 135(3), pp.164-174.

DOI: https://doi.org/10.26077/64f1-4c4b

Angle Effects of LIDAR Measurements on a Flat Surface and in High-velocity Spillway Flows

R. Li[1], K.D. Splinter[1], M. Kramer[2] & S. Felder[1]
[1] *Water Research Laboratory, School of Civil and Environmental Engineering, UNSW Sydney, Australia*
[2] *UNSW Canberra, School of Engineering and Information Technology (SEIT), Canberra, ACT 2610, Australia*
E-mail: s.felder@unsw.edu.au

Abstract: *Measurements in aerated high-velocity spillway flows have been traditionally conducted with point-source instruments. LIDAR is a promising remote sensing technology that may be able to provide more detailed insights into the free-surface properties. So far only one laboratory study has investigated the use of LIDAR technology in air-water flows on a spillway model showing that free-surface data in the aerated part of the flow can be recorded with high spatial and temporal resolution. However, the spillway model, as well as the aerated flow region, were quite short and further research is needed to test the performance of LIDAR technology on a longer spillway. Herein, the present study tested a LIDAR in an 8.6 m long laboratory spillway with strip roughness. To ensure consistency of results along the spillway, the performance of the LIDAR was first assessed without flowing water along a straight wall as well as in the spillway model without water. The results showed that the LIDAR was affected by viewing angle that can be corrected using a scaling factor. The LIDAR was subsequently used to measure the free-surface properties of skimming flows. Comparative results for three different LIDAR measurement positions showed effects on the free-surface elevations and standard deviations linked with the different grazing angle of the LIDAR relative to the free-surface and with the waviness of the free-surface. A potential correction scheme for this is discussed.*

Keywords: *Air-water flows, remote sensing, physical modelling, instrumentation, hydraulic structures, signal processing.*

1. Introduction

High-velocity air-water flows on spillways occur commonly in flow conveyance systems such as spillways and chutes. The flows are non-aerated at the upstream end but become strongly aerated downstream of the inception point of free-surface aeration (Chanson 1996). Highly aerated flows in spillways are characterized by entrapped and entrained air (Killen 1968) where the entrained air penetrates deep into the flow column, while the entrapped air is characterized by free-surface instabilities and waves (Valero and Bung 2016). The air entrainment and entrapment vary strongly in space and time, which makes measurements of the flow depth and other aerated free-surface parameters challenging. Accurate measurements of the free-surface elevations provide important information for the calculation of design parameters such as rate of energy dissipation and flow resistance (Chanson 1996; Wood 1991). In non-aerated flows, pointer gauges and acoustic displacement meters (ADMs) are commonly used for flow depth measurements, while in air-water flows, the most suited instrument are phase-detection intrusive probes (Felder and Chanson 2015, Kramer et al. 2020a). A key limitation of all these instruments is that they provide data at a fixed measurement point and only very limited spatial and temporal information of the flow processes.

Previous studies used ADMs to measure time-varying free-surface features of aerated flows at a single point, revealing free-surface motions with broad frequency spectrum (Zhang et al. 2018). While ADMs appeared to be a suitable instrument for aerated high-velocity flows (Bung 2013; Felder and Chanson 2014), on second look, there is a strong disagreement between characteristic air-water flow elevations measured with a phase-detection probe and the mean elevations measured with ADMs, with variations in void fractions between 50% and 90% (Bung 2013; Felder and Chanson 2014; Kramer and Chanson 2018; Zhang et al. 2018). Several factors may affect the ADM measurements including the ADM type, the size of the ADM beam and the nature of the air-water flows (Li et al. 2021a). In contrast, phase-detection probes are considered very reliable instruments in high-velocity air-water flows on spillways, and have been successfully used for understanding of fundamental flow processes (e.g. Chanson and Toombes 2002; Felder and Chanson 2015) and spillway design (e.g. Felder and Chanson 2016; Hunt et al. 2014; Ohtsu et al. 2004). However, phase-detection intrusive probes only collect information at one location, or if multiple probes are used simultaneously at several locations (Hohermuth et al. 2021). In any case, the calculation of the air-water flow properties, of the characteristic elevations and associated design parameters relies on time and depth averaging.

Other instrumentation may be more promising for the measurement of more spatially- and temporally varying observations of the free-surface. Thorwarth (2008) observed highly non-stationary free-surface motions in pooled stepped spillways with a video camera, while more recent studies used high-speed cameras to record the air-water properties at the vicinity of the sidewall (Bung and Valero 2016; Zhang et al. 2018). The flow properties next to the sidewall may differ from the channel centerline (Kramer and Chanson 2019), and sidewall measurements are also not a viable option for field measurements. More recently, application of top-view video cameras has been explored in hydraulic jumps (Bung et al. 2020) and spillways (Kramer and Felder 2021) showing promising prospects for the recording of design parameters in the laboratory and in large-scale prototype spillways (Chanson 2022; Kramer and Felder 2021).

Another type of remote sensing instrumentation is LIDAR technology that has been explored across a range of air-water flows, including classical hydraulic jumps (Felder et al. 2021; Li et al. 2021a; Montano et al. 2018), hydraulic jumps in stilling basins (Li et al. 2021b), stepped spillway flows (Kramer et al. 2020b), and a water confluence (Rak et al. 2020). These studies have shown that LIDAR can measure the free-surface of highly aerated flows continuously with high spatial and temporal resolution, allowing a detailed documentation of the dynamic free-surface processes. The application of LIDAR technology in high-velocity flows has been limited to a single investigation in a relatively short, stepped spillway with a short air-water flow region (Kramer et al. 2020b). Kramer et al. (2020b) confirmed the applicability of LIDAR in such flows providing measurements of free-surface elevations, fluctuations, as well as characteristic free-surface time and length scales. A comparison with data recorded with a phase-detection intrusive probe suggested that the LIDAR may be able to predict the void fraction distributions. However, further investigations of LIDAR performance in longer spillways are needed to better understand the LIDAR's ability for potential prototype measurements. Herein, this study applied LIDAR technology in a long spillway model with strip roughness. Following tests of the LIDAR on solid surfaces without flows, this paper presents LIDAR measurements of a skimming flow regime using three different LIDAR positions along the spillway. The results suggested a strong effect of the LIDAR position on the free-surface measurements and a possible correction scheme is discussed.

2. Experiments and Data Processing

New experiments were conducted at the UNSW Water Research Laboratory using a 2D industrial LIDAR (SICK LMS511) with a wavelength of 905 nm and a laser beam width < 20 mm for measurement distances < 10 m (SICK 2015). Simple tests with spacers suggested a smaller beam width of approximately 5 mm in the laboratory (Li et al. 2021a). The same LIDAR was previously used in a range of applications including classical hydraulic jumps (Felder et al. 2021; Montano et al. 2018) and hydraulic jumps in a stilling basin (Li et al. 2021b). For all experiments in the present study, the LIDAR was sampled at 35 Hz with an angular resolution of 0.25°.

In the first experiment, the LIDAR was tested in dry conditions along a concrete flat surface with a length of 12 m (Figure 1). For this experiment, the LIDAR was positioned 1.6 m above the bed. The LIDAR continuously scanned the concrete surface for 60 seconds. The grazing angle between the LIDAR beam and the horizontal surface was indicated by θ (Figure 1).

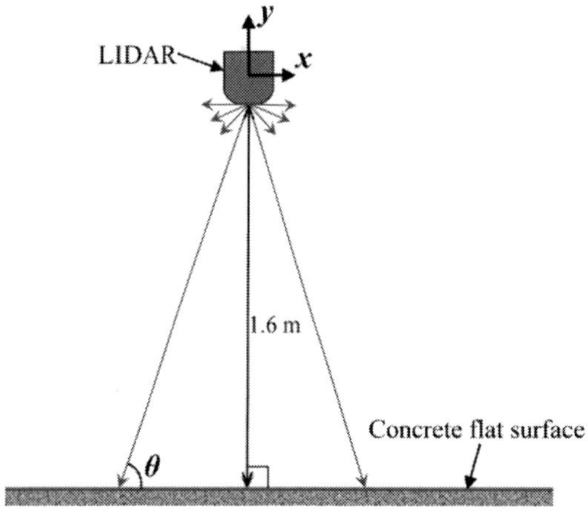

Figure 1. Setup of LIDAR test on long flat surface.

A second set of experiments was conducted in a large-scale spillway model (Figure 2). The spillway had a length of 8.6 m, a width of 0.8 m and a slope of 10.8°. At the spillway's upstream end, a large header tank and a broad-crested weir with an upstream rounded corner provided steady and smooth inflows. Twenty-four identical wooden strip blocks of 0.07 m width and 0.045m height were installed along the spillway model. The first strip block was positioned 0.24 m from the upstream end to allow smooth transition of flows from the broad-crested weir into the spillway chute. The consistent spacing between blocks was 0.28 m along the spillway (Figure 2c). A flow rate q = 0.288 m^2/s was investigated, which corresponded to a Reynolds number (defined with hydraulic diameter) of 8.4×10^5. The flows resembled skimming flows with a non-aerated flow region at the upstream end, and strongly aerated flows downstream of the inception point of free-surface aeration. The inception point was visually determined and occurred at approximately x = 2.5 m between the 7th and 8th block (Figure 2a). The depth-averaged mean flow aeration was approximately 0.13 at the inception point and approximately 0.2 in the downstream uniform flow region. The free-surface of the flows were parallel to the channel bed skimming over the strip roughness elements while strong fluctuations of the free-surface were consistently observed downstream of the inception point. The flow motions between the strips were characterized by stable recirculation patterns with irregular ejection processes between the cavity and the skimming flows above. For the spillway experiments, the LIDAR was inclined at the same plane as the spillway bed so that θ = 90° was perpendicular to the bed (Figure 2b). The LIDAR was positioned at an elevation of 1.5 m above the spillway bed at three longitudinal locations X_{LIDAR} = 2.7, 5.2 and 7.4 m relative to the spillway's upstream end (Figure 2b). These LIDAR positions provided a wide range of grazing angles 10° ≤ θ ≤ 90° along the spillway. For each LIDAR position, the LIDAR continuously measured the dry bed for 5 minutes and measured the flows for 30 minutes along the centerline of the spillway.

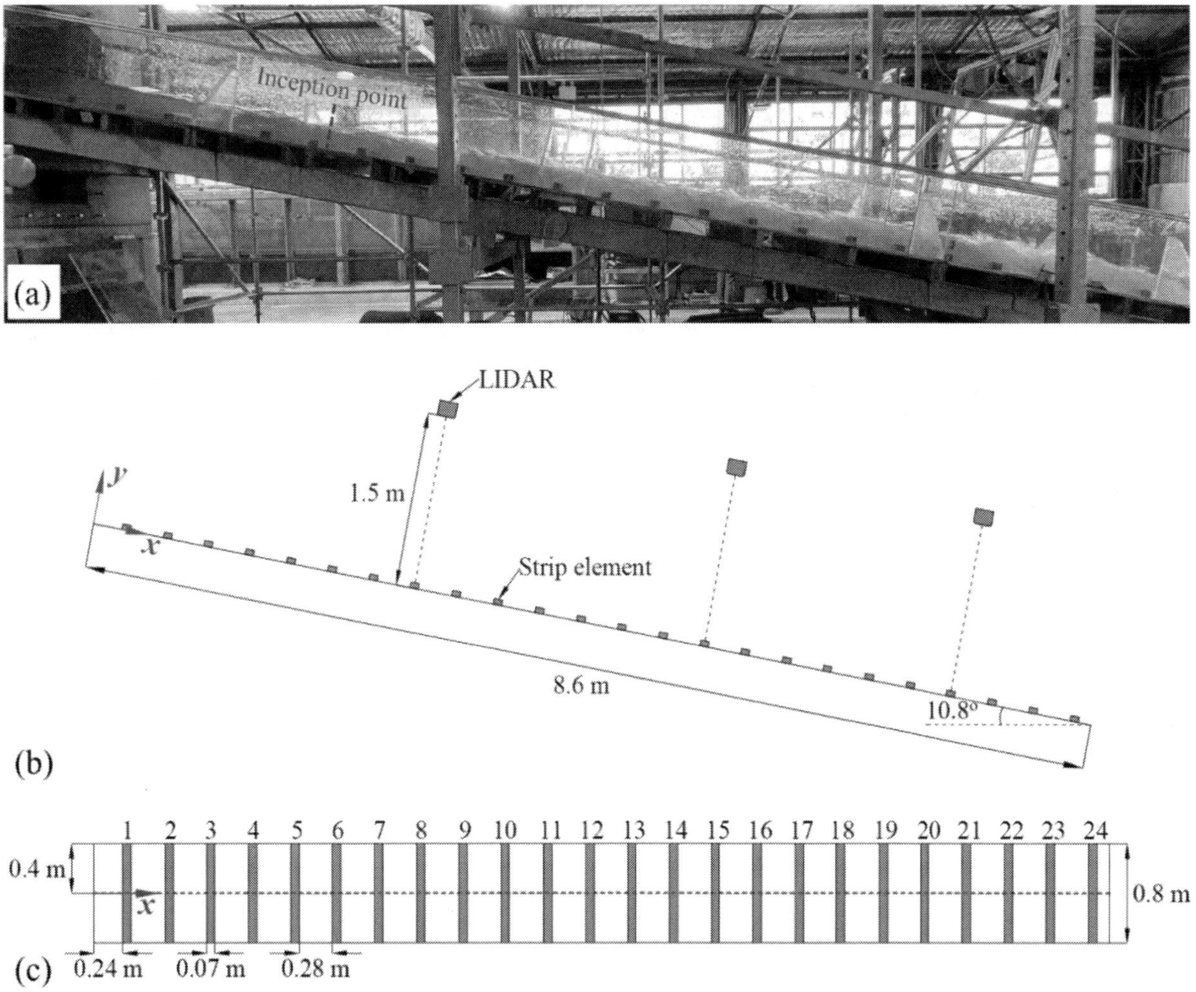

Figure 2. Spillway model in the present study; (a) photo of spillway model and skimming flow patterns for q = 0.288 m^2/s; (b) side-view sketch of the 24 strip roughness elements along the spillway and 3 positions of LIDAR; (c) top-view sketch with the dimensions of the spillway.

For all experiments, the raw signal of distances and angles were recorded using Sopas ET software of SICK. The data processing was conducted in MATLAB following the general procedure presented in Li et al. (2021a). The LIDAR data were first translated into a Cartesian coordinate system. Then all data were interpolated onto a constant x-axis grid based on the average of all recorded frames. Finally, statistical calculations including mean elevations and standard deviations were conducted using the interpolated data. Note that for the spillway experiments, meaningless signals below the channel bed were replaced with NaNs, and no signal filtering was applied in the present study.

3. Results (1): LIDAR Tests in Dry Conditions

Figure 3 shows the mean profiles of the flat concrete surface measured by the LIDAR and corresponding grazing angles. Note that the figure was not scaled to clearly show the measurement distortion on a long flat surface. It appeared that the LIDAR slightly overestimated the surface elevation with increasing distances and decreasing grazing angles resulting in a slight distortion curvature. This effect was also observed in previous studies using LIDAR to measure long straight surfaces (Martins 2017; Streicher et al. 2013). Martins (2017) indicated that this could be due to the energy distribution over the elliptical footprint. According to Katzenbeisser (2003), this distortion might be linked with the scaling of viewing angle and can be corrected using a scaling factor. Different correction factors were tested for the present case and a factor of -0.0027 performed the best for the present LIDAR to adjust the distortion. This means that the grazing angle θ was adjusted as:

$$\theta = 0.9973 \times \theta_{raw} + 0.243 \qquad (1)$$

where θ_{raw} is the grazing angle recorded by the LIDAR before correction.

After the distortion correction with Eq. (1), the uncertainty of mean LIDAR elevations on a flat surface was less than 10 mm for a measurement range up to 12 m with $10° \leq \theta \leq 90°$ (Figure 3a). Since Eq. (1) linearly scaled the grazing angles, the standard deviations y' of LIDAR measurements using raw and corrected LIDAR signals were similar and are shown in Figure 3b. The standard deviations were lower than 10 mm for measurement distances up to 12 m with the sampling time of 60 s (> 2100 data points). For grazing angles larger than 40°, y' was approximately 4 mm, while y' increased for lower grazing angles. This can be linked with longer measurement distances and larger elliptical footprint of the LIDAR signal (SICK 2015; Soudarisssanane et al. 2011). In the present study, all LIDAR measurements on the large-scale spillway were corrected using Eq. (1) to eliminate the distortion due to angle at large distances. Note that the distortion effect was only observed in the long spillway and not in the much shorter air-water flow phenomena (e.g., hydraulic jumps) investigated in previous studies (e.g. Li et al. 2021b; Montano et al. 2018).

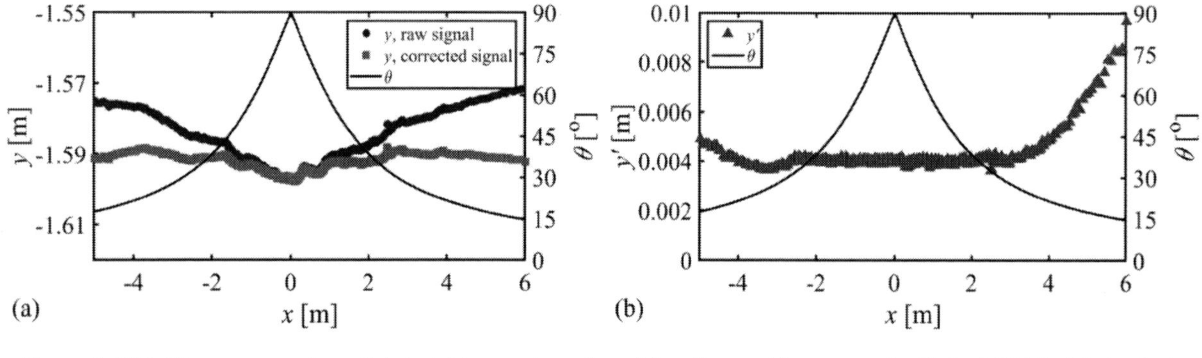

Figure 3. LIDAR measurements on flat straight concrete surface: (a) surface profile measured with the LIDAR; (b) standard deviation of the LIDAR signal.

Figure 4 shows the mean bed profiles measured with the LIDAR positioned at X_{LIDAR} = 5.2 m before and after the correction using the angle scaling factor described above (Eq. (1)). The bed profile after angle scaling compared well with the measured bed profile with an average difference of 3 mm for all data points along the spillway. The standard deviation of the bed measurements was less than 6 mm along the spillway. In contrast to the flat bed tests presented above, the strip roughness had an additional source of error due to signal blocking induced by low grazing angles. This was most apparent in the cavity between strip elements at locations further away from the LIDAR (i.e., smaller θ at the upstream and downstream end) where the elevated strip blocked the signal from resolving the vertical edge and subsequent cavity. This is visible in Figure 4, particularly for $x < 2.4$ m, where less data points along the bed are resolved compared to closer to the LIDAR position.

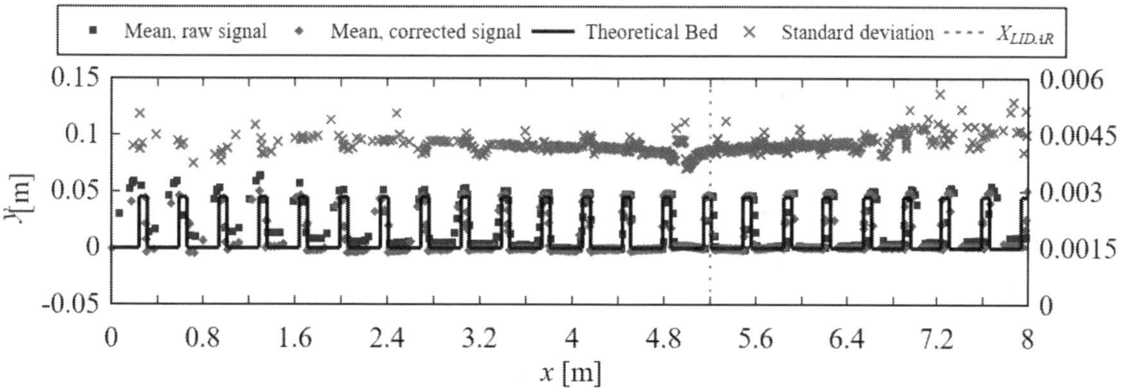

Figure 4. Comparison between the spillway bed measured with the LIDAR before and after distortion correction; standard deviation of the LIDAR measurements ($X_{LIDAR} = 5.2$ m, $Y_{LIDAR} = 1.5$ m).

4. Results (2): Free-surface Elevations in Skimming Flows along Spillway with Strip Roughness

Now that a correction method has been established for the LIDAR data to account for angle effects, measurements of free-surface flows are presented. Figure 5 shows typical instantaneous free-surface profiles from the upstream end to the aerated region recorded by the LIDAR for the three LIDAR positions. The measurements with high temporal and spatial resolution enhanced the visualization of the fast free-surface fluctuations and advection of free-surface waves along the spillway. Notably, additional videos generated using raw LIDAR signals (not shown here) suggested that the LIDAR was able to continuously track the free-surface waves propagating along the spillway.

As shown in Figure 5, there is clear differences in the instantaneous free-surface profiles measured for the three different LIDAR positions. This was linked with both the flow aeration levels and the LIDAR positioning effects. The LIDAR was not able to consistently record the free-surface in the flows with low flow aeration since LIDAR signals tended to penetrate into the flow (Allis et al. 2011; Montano et al. 2018; Rak et al. 2017). This resulted in the very low LIDAR elevations close to the bed at locations upstream of and close to the inception point and was most noticeable for the LIDAR positioned near the inception point ($X_{LIDAR} = 2.7$ m). In the downstream aerated region, the instantaneous free-surface profiles were consistent with visual observations showing strong free-surface fluctuations. However, it can be observed that the LIDAR position also influenced the overall signal with stronger free-surface fluctuations recorded near the LIDAR for all three positions. As such, an upstream positioned LIDAR tended to measure higher free-surface elevations in the downstream aerated region (Figure 5a), while the LIDAR positioned at the downstream end recorded lower signals in the water column (Figure 5c). This was linked with the effect of grazing angles, since LIDAR signals with small grazing angles might be blocked by the extensive water splashes and free-surface perturbations in the air-water flow and is further discussed in Section 5.

Figure 6 shows the mean free-surface elevations d recorded by the LIDAR for the three LIDAR positions. The free-surface standard deviations d' recorded by the LIDAR were added as whiskers (Figure 6). The LIDAR was not able to record the free-surface profile in the non-aerated region upstream of the inception point ($x < 2.5$ m), which was expected due to LIDAR signal penetration into the flow. In the region just downstream of the inception point ($2.5 < x < 5$ m), measurements with a LIDAR position close to the measurement locations ($X_{LIDAR} = 2.7$ m) were still affected by the low flow aeration leading to low mean free-surface profiles (Figure 6a). In the downstream strongly aerated region ($x > 5$ m), differences up to 30% were observed between the mean free-surface elevations recorded with different LIDAR positions (Figure 6a). The free-surface standard deviations d' recorded with different LIDAR positions showed a relative difference up to 50% in the aerated flow (Figure 6b). As noted above, the large differences in free-surface properties recorded with different LIDAR positions are due to different grazing angles of the LIDAR signal and the wavy nature of the free-surface. These effects and a potential method to interpret LIDAR measurements in aerated spillway flows are further discussed in Section 5.

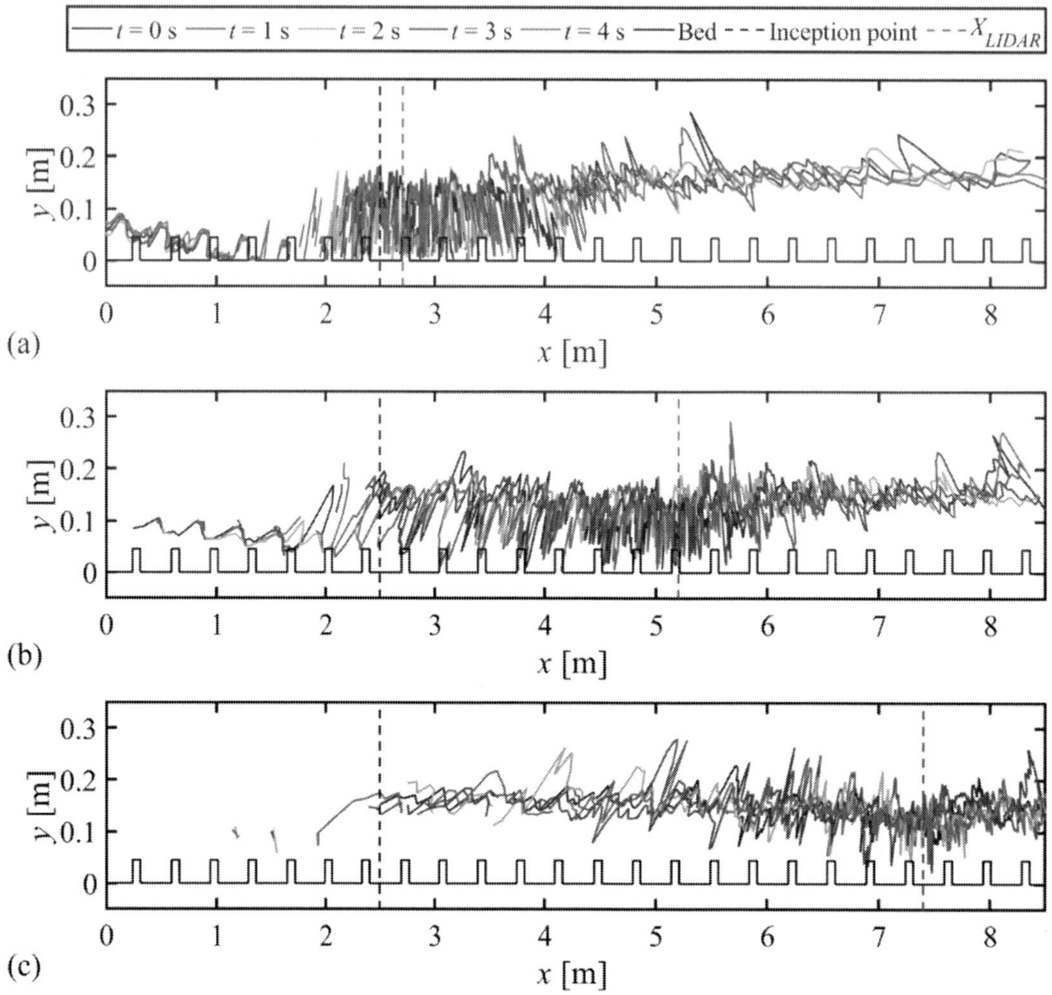

Figure 5. Instantaneous free-surface profiles using corrected LIDAR data for 3 LIDAR positions at 5 consecutive seconds in the skimming flows on the spillway: (a) X_{LIDAR} = 2.7 m; (b) X_{LIDAR} = 5.2 m; (c) X_{LIDAR} = 7.4 m.

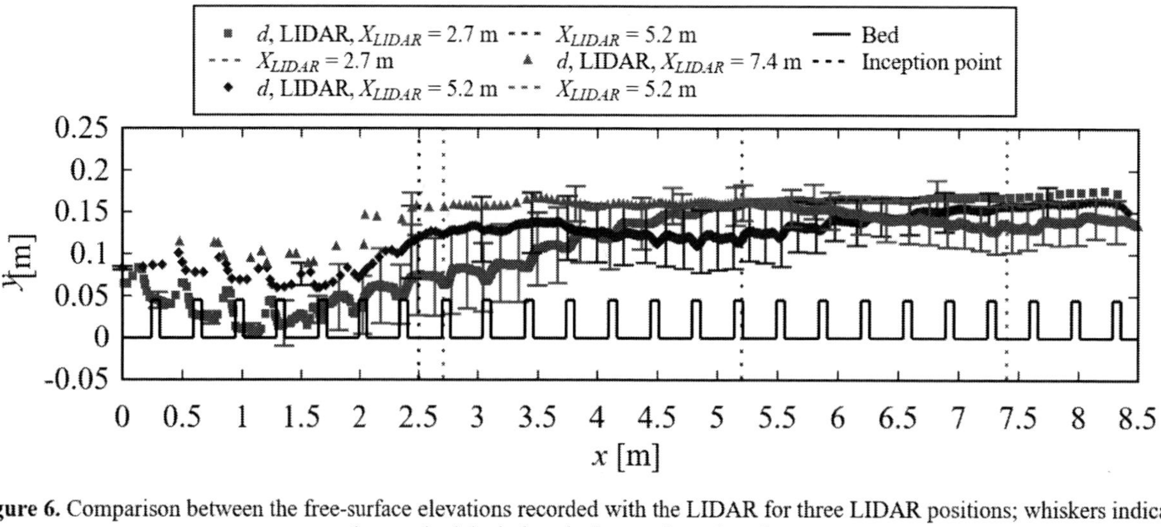

Figure 6. Comparison between the free-surface elevations recorded with the LIDAR for three LIDAR positions; whiskers indicate the standard deviations in free-surface elevations.

5. Discussion

Differences in air-water flow free-surface properties recorded with different LIDAR positions were previously observed in hydraulic jumps (Li et al. 2021a). Li et al. (2021a) suggested LIDAR positions upstream of the jump toe to avoid signal blocking due to free-surface waves and water ejections in the jump roller. These issues have also been noted by researchers studying breaking waves and bores in the surf zone (Blenkinsopp et al. 2010; Martins et al. 2016). In the present study, LIDAR tests in the dry bed conditions on the spillway with strip roughness showed that small grazing angles resulted in signal blocking by strip roughness elements at certain locations. Similarly, the large differences in free-surface properties measured with different LIDAR positions in the high-velocity air-water flows were linked with the angle effects and the nature of the free-surface. While some of these angle effects are due to internal properties of the LIDAR and can be corrected for using Eq. (1), other factors are a function of the aerated flows themselves. Figure 7 presents a sketch to further describe the angle effects in the aerated skimming flows. When grazing angles are close to 90°, i.e. the LIDAR is directly above the point of interest, there is a higher probability of the signal penetrating deeper into the air-water flows. In the inception region with little flow aeration, a LIDAR positioned close to the measurement point tended to penetrate deeper into the flow before hitting a bubble to reflect off, while a LIDAR positioned far from the point penetrated less into the flow before hitting a bubble (Figure 7b). This was also due to internal properties of the LIDAR where longer measurement distances caused larger and elliptical footprints of the LIDAR beam, as well as lower signal strength (SICK 2015; Soudarissanane et al. 2011) resulting in less penetration through the surface in the non-aerated region. In the downstream aerated region with strong free-surface perturbations, an oblique LIDAR signal with small grazing angle was not able to record the troughs of free-surface waves or the entrained bubbles, since the LIDAR signal was blocked by the peaks of the free-surface waves and water splashes. This effect resulted in elevated free-surface profile measured by an oblique LIDAR with small grazing angles (Figure 7c).

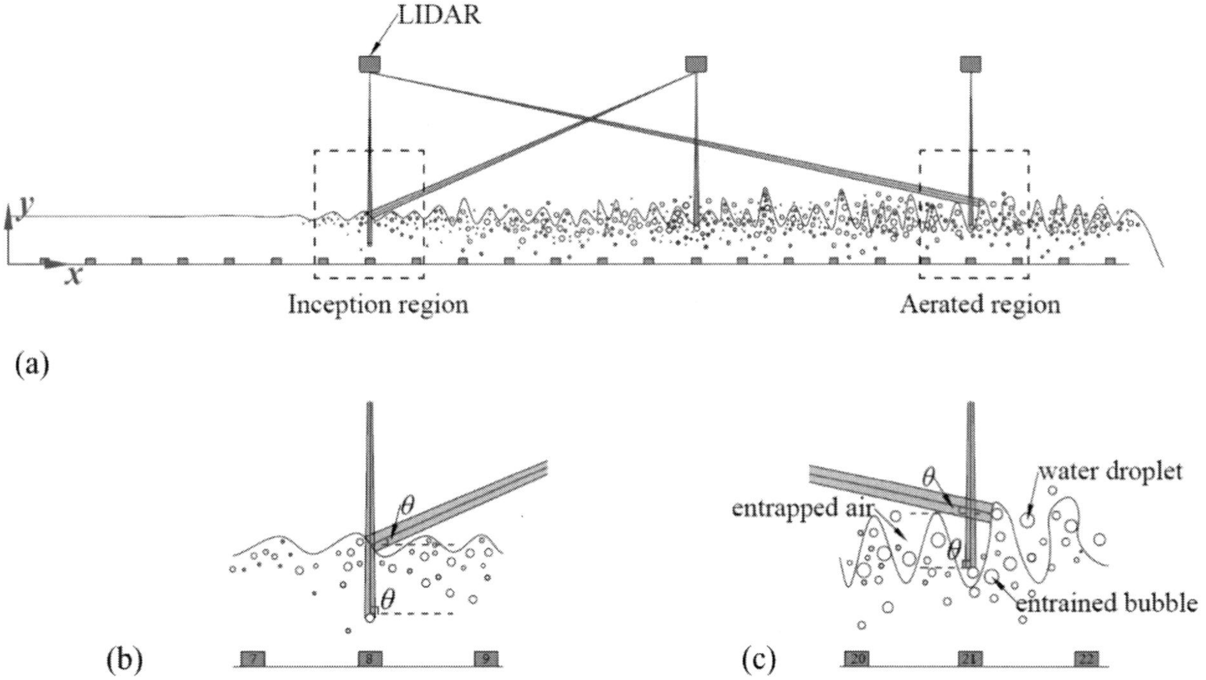

Figure 7. Sketch of the angle effect on the free-surface measurements in skimming flows: (a) angle effect along the spillway; (b) angle effect in the inception region with low flow aeration; (c) angle effect in the aerated flow region.

Conceptually, the link between entrained bubbles and LIDAR have been presented by several authors. Montano (2019) and Li et al. (2021b) showed that the mean LIDAR elevations in aerated hydraulic jumps were close to the elevations with maximum bubble count rate. Additionally, Kramer et al. (2020b) showed that the LIDAR signal distribution was comparable to the distribution of bubble count rate in the downstream aerated region and proposed that the LIDAR signal was able to measure the bubbles in aerated flows. Preliminary analysis of the present spillway experiments also observed similar distributions of LIDAR signal and bubble count rate irrespective of the LIDAR position (not shown here). The comparison between LIDAR measurements and bubble count rate pointed out a promising direction for the correction of the angle effects presented in this manuscript. With this correction method it is believed that all signals can be aligned irrespective of the LIDAR position, enabling the use of LIDAR measurements to understand internal air-water flow properties. Current research is therefore combining measurements of LIDAR with intrusive air-water flow conductivity probes.

6. Conclusion

The present study tested the performance of LIDAR technology on a long flat concrete surface as well as in high-velocity skimming flow in a spillway with strip roughness. The mean elevations of the flat concrete surface measured with the LIDAR showed slight distortion due to scaling of viewing angles. A correction equation was proposed to correct the distortion of all LIDAR measurements with low grazing angles. In the high-velocity spillway flow, the LIDAR was not able to measure the free-surface in the non-aerated region, while the LIDAR captured distinct free-surface waves propagating along the spillway and strong free-surface fluctuations in the aerated region. The LIDAR measurements with different LIDAR positions showed differences up to 30% in mean free-surface elevations and differences up to 50% in standard deviations in the aerated flow. These differences were linked with different grazing angles and the nature of the aerated flows with free-surface waves and perturbations. Comparison between LIDAR signal distribution and bubble count rate distribution may provide opportunities to correct LIDAR signals for the prediction of internal air-water flow properties and is a topic of current research.

7. ACKNOWLEDGMENTS

The authors thank Rob Jenkins (Water Research Laboratory, UNSW Sydney) for his technical help.

8. REFERENCES

Allis, M.J., Peirson, W.L., and Banner, M.L. (2011). "Application of LiDAR as a measurement tool for waves." *Proc., International Offshore and Polar Engineering Conference*, Hawaii, USA, 8, 373–379.

Blenkinsopp, C.E., Mole, M.A., Turner, I.L., and Peirson, W.L. (2010). "Measurements of the time-varying free-surface profile across the swash zone obtained using an industrial LIDAR." *Coast. Eng.*, 57(11–12), 1059–1065.

Bung, D.B. (2013). "Non-intrusive detection of air-water surface roughness in self-aerated chute flows." *J. Hydraul. Res.*, 51(3), 322–329.

Bung, D.B., Crookston, B.M., and Valero, D. (2020). "Turbulent free-surface monitoring with an RGB-D sensor: the hydraulic jump case." *J. Hydraul. Res.*, 59(5), 779–790.

Bung, D.B., and Valero, D. (2016). "Optical flow estimation in aerated flows." *J. Hydraul. Res.*, 54(5), 575–580.

Chanson, H. (1996). Air bubble entrainment in free-surface turbulent shear flows, Academic Press Inc, San Diego, California, USA.

Chanson, H. (2022). "On Air Entrapment Onset and Surface Velocity in High-Speed Turbulent Prototype Flows." *Flow Meas. Instrum.*, 83, Paper 102122.

Chanson, H., and Toombes, L. (2002). "Air–water flows down stepped chutes: turbulence and flow structure observations." *Int. J. Multiph. Flow*, 28(11), 1737–1761.

Felder, S., and Chanson, H. (2014). "Air-water flows and free-surface profiles on a non-uniform stepped chute." *J. Hydraul. Res.*, 52(2), 253–263.

Felder, S., and Chanson, H. (2015). "Phase-detection probe measurements in high-velocity free-surface flows including a discussion of key sampling parameters." *Exp. Therm. Fluid Sci.*, 61, 66–78.

Felder, S., and Chanson, H. (2016). "Simple Design Criterion for Residual Energy on Embankment Dam Stepped Spillways." *J. Hydraul. Eng.*, 142(4), 04015062.

Felder, S., Montano, L., Cui, H., Peirson, W., and Kramer, M. (2021). "Effect of inflow conditions on the free-surface properties of hydraulic jumps." *J. Hydraul. Res.*, 59(6), 1004–1017.

Hohermuth, B., Boes, R.M., and Felder, S. (2021). "High-velocity air-water flow measurements in a prototype tunnel chute – scaling of void fraction and interfacial velocity." *J. Hydraul. Eng.*, 147(11), 04021044.

Hunt, S.L., Kadavy, K.C., and Hanson, G.J. (2014). "Simplistic Design Methods for Moderate-Sloped Stepped Chutes." *J. Hydraul. Eng.*, 140(12), 04014062.

Katzenbeisser, R. (2003). "On the calibration of lidar sensors." *Proc., ISPRS 3-D Workshop Reconstruction from Airborne Laser-Scanner and InSAR Data*, Dresden, Germany, 59–64.

Killen, J. M. 1968. "The Surface Characteristics of Self Aerated Flow in Steep Channels." Ph.D Thesis, University of Minnesota, Minneapolis, USA.

Kramer, M., and Chanson, H. (2018). "Free-surface instabilities in high-velocity air-water flows down stepped chutes." *Proc., 7th International Symposium on Hydraulic Structures*, Aachen, Germany.

Kramer, M., and Chanson, H. (2019). "Optical flow estimations in aerated spillway flows: Filtering and discussion on sampling parameters." *Exp. Therm. Fluid Sci.*, 103, 318–328.

Kramer, M., Hohermuth, B., Valero, B., and Felder, S. (2020a). "Best practices for velocity estimations in highly aerated flows with dual-tip phase-detection probes." *International Journal of Multiphase Flow*, 126, 103228.

Kramer, M., Chanson, H., and Felder, S. (2020b). "Can we improve the non-intrusive characterization of high-velocity air–water flows? Application of LIDAR technology to stepped spillways." *J. Hydraul. Res.*, 58(2), 350–362.

Kramer, M., and Felder, S. (2021). "Remote sensing of aerated flows at large dams: proof of concept." *Remote Sens.*, 13(14), 2836.

Li, R., Splinter, K.D., and Felder, S. (2021a). "Aligning free surface properties in time-varying hydraulic jumps." *Exp. Therm. Fluid Sci.*, 126, 110392.

Li, R., Splinter, K.D., and Felder, S. (2021b). "LIDAR Scanning as an Advanced Technology in Physical Hydraulic Modelling: The Stilling Basin Example." *Remote Sens.*, 13(18), 3599.

Martins, K. (2017). "Wave transformation in the surf-zone." Ph.D Thesis, University of Bath, UK.

Martins, K., Blenkinsopp, C.E., and Zang, J. (2016). "Monitoring Individual Wave Characteristics in the Inner Surf with a 2-Dimensional Laser Scanner (LiDAR)." *J. Sensors*, 2016, 1–11.

Montano, L. (2019). "An experimental study of free surface dynamics and internal motions in fully aerated hydraulic jumps." Ph.D. thesis, School of Civil and Environmental Engineering, University of New South Wales, Sydney, Australia.

Montano, L., Li, R., and Felder, S. (2018). "Continuous measurements of time-varying free-surface profiles in aerated hydraulic jumps with a LIDAR." *Exp. Therm. Fluid Sci.*, 93, 379–397.

Ohtsu, I., Yasuda, Y., and Takahashi, M. (2004). "Flow Characteristics of Skimming Flows in Stepped Channels." *J. Hydraul. Eng.*, 130(9), 860–869.

Rak, G., Hočevar, M., and Steinman, F. (2017). "Measuring water surface topography using laser scanning." *Flow Meas. Instrum.*, 56, 35–44.

Rak, G., Steinman, F., Hočevar, M., Dular, M., Jezeršek, M., and Pavlovčič, U. (2020). "Laser ranging measurements of turbulent water surfaces." *Eur. J. Mech. B/Fluids*, 81, 165–172.

SICK. (2015). LMS5xx Laser Measurement Sensors: Operating Instructions. Germany.

Soudarissanane, S., Lindenbergh, R., Menenti, M., and Teunissen, P. (2011). "Scanning geometry: Influencing factor on the quality of terrestrial laser scanning points." *ISPRS J. Photogramm. Remote Sens.*, 66(4), 389–399.

Streicher, M., Hofland, B., and Lindenbergh, R.C. (2013). "Laser Ranging For Monitoring Water Waves In The New Deltares Delta Flume." *Proc., ISPRS Workshop Laser Scanning*, Antalya, Turkey, 271–276.

Thorwarth, J. (2008). "Hydraulisches Verhalten der Treppengerinne mit eingetieften Stufen – Selbstinduzierte Abflussinstationaritäten und Energiedissipation. (Hydraulics of Pooled Stepped Spillways – Self-induced Unsteady Flow and Energy Dissipation.)." Ph.D Thesis, University of Aachen, Germany (in German).

Valero, D., and Bung, D.B. (2016). "Development of the interfacial air layer in the non-aerated region of high-velocity spillway flows. Instabilities growth, entrapped air and influence on the self-aeration onset." *Int. J. Multiph. Flow*, 84, 66–74.

Wood, I.R. (1991). "Free surface air entrainment on spillways." Air Entrainment in Free-surface Flows, I.R. Wood, ed., IAHR, Hydraulic Structures Design Manual 4, Hydraulic Design Considerations, A.A. Balkema, Rotterdam, the Netherlands, 55–84.

Zhang, G., Valero, D., Bung, D.B., and Chanson, H. (2018). "On the estimation of free-surface turbulence using ultrasonic sensors." *Flow Meas. Instrum.*, 60, 171–184.

DOI: https://doi.org/10.26077/3fa1-7f82

Anti vortex device to operate pump intakes below the minimum submergence

M.A. Angulo[1], A. Rivetti[1], C.V. Lucino[1] & S.O. Liscia[1]
[1]*Laboratory of Hydromechanics, UNLP, La Plata, Argentina*
E-mail: mauricio.angulo@ing.unlp.edu.ar

Abstract: *Thermal power plants require water cooling pump stations to inject fresh water into the condenser. Generally, these pump stations have vertical mixed-flow pumps with large flows and low-to-medium heads that take water from rivers or reservoirs. Under certain natural conditions, water sources can reach extraordinarily low levels, thus affecting pump operation as the minimum recommended submergence might not be met, causing the power plant to be powered off. This causes both a very high loss of profit for the company and a decrease in the available power for the electrical supply system. Therefore, expanding the range of submergence under which pumps could be operated safely for special, transitory, and emergencies is an alternative worth exploring. Pump operation below the minimum level gives rise to unacceptable vortices on the free surface, air entrainment inside the circuit, higher mechanical vibrations, and lower efficiencies due to higher pre-swirl at the suction bell. Because of this, an anti-vortex device is proposed to prevent free surface vortex formation to allow the operation of the pump over a wider range of submergence. The device has a grating shape and was designed and optimized based on scale model tests performed on a typical pump bay following the ANSI standard intake design. Flow analyses through particle image velocimetry (PIV), acoustic Doppler velocimetry (ADV) velocity profiles, and swirl-meter measurements were carried out for a range of levels under the minimum and then compared with the device installed under the same conditions. Experiments proved the effectiveness of the solution to reduce free surface vortex formation and pre-swirl, allowing pumps to operate down to half of the recommended submergence.*

Keywords: *Pump station, submergence, free surface vortex.*

1. Introduction

Large pump stations are required to provide cold water for thermoelectric power plants. Water is extracted from rivers or lakes and then pumped to the gas turbine condenser to cool vapor. As this process is critical for continuous operation, pumps should never stop. Usually, vertical mixed-flow pumps are installed at these pump stations (Kushwaha 2015), able to provide high flows at medium-to-low heads. Under certain natural conditions, water sources may cause the free surface level to descend below the guaranteed submergence. Such operating condition is not recommended as surface vortices may appear from the free surface level and grow enough to let air enter the pump sump, injecting air inside the cooling system and inducing vibrations, noise, and efficiency loss at pumps. In such cases, the pump station has to be shut down, thus affecting the electrical supply system. The power plant owner/operator has to deal with power shortage and thus loss of profit.

To operate under the minimum submergence, different solutions were evaluated. For example, deepening the pump bay locally at the pump suction bell and/or placing the pump suction at the lowest possible, but both alternatives have high civil and mechanical costs and long schedules to implement the solutions. Besides, general guidelines concerning geometry and flow characteristics of pump intake structures design, ANSI/HI 9.8 (2018) standard proposes several devices to prevent or reduce submerged vortices formation employing splitters or fillets and suggests a few devices for superficial vorticities, such as vertical or tilted curtains and horizontal gratings; Harun et al. (2020) tested a floor splitter plate that reduces floor vortices and decreases the intensity of swirling at the inlet flow; Norizan et al. (2019) studied the optimization of floor splitters while Sherestha et al. (2021) optimized the bell mouth shape to suppress vorticity at the entrance. Just a few references were found of structures designed to prevent free surface vortices formation for submergible pumps, most of them related to sumps with lateral intakes, such as the "vortex breaker", mentioned by Jones et al. (2008), a structure which consists on bars parallel to flow direction, and an angle with the free surface, and rafts located above lateral intakes and Park et al. (2018), who propose a floating anti-vortex device which consists of four legs and a floating body surrounding the intake pipe.

The problem caused by superficial vortices is also addressed at intake structures of hydropower plants, where sometimes it is necessary to install vortex suppressors designed to increase the distance between the water surface and the intake by installing, for example, a floating plate in the path of the vortex (Taghvaei 2012).

Regarding CFD simulations, Constantinescu et al. (1998) simulated a pump bay; Arocena et al. (2021) implemented simulations of the bay including the pump itself to measure the impact of vortices on the efficiency. However, as numerical simulations must capture a small enough vortices scale, meshes grow increasingly finer and, as a result, the computational costs are high when trying to simulate many scenarios combining different free surface levels and the geometry changes at the bay.

We propose an antivortex device, a floating grid (FG), that allows the operation of pumps safely under the guaranteed submergences and also below the submergences recommended on the reference code ANSI/HI 9.8. The proposed device is partially submerged and surrounds the vertical pump geometry that reduces significantly the free surface vortices type 3 to 6 reducing them to lower types 1 and 2. The tests carried out are aimed at comparing the operation of a pump station below the minimum submergence with and without the antivortex device. The pump bay design is based on standard guidelines and has no fillets, splitters, or any other geometry element to prevent vortex formation. Photographs, videos, PIV measurements, ADV profiles, and swirl-meter measurements were performed to contribute to the comparison.

2. Materials and methods

2.1. Experimental set-up

The project is based on a real-scale pump station. A physical model of one bay of the pump station was built (scale 1:11.875) and installed inside a test channel of the Laboratory of Hydraulics of the UNLP. The channel consists of a closed-loop circuit that provides the inflow of water from a constant level reservoir. The pump is simulated using a siphon pipe and the pump case geometry corresponds to the commercial vertical pump Torishima SPV 1500 with a nominal flow of 25,000 m³/h and 43.5 m head. The suction bell diameter is 2.28 m. In Figure 1, a detail of bay geometry and dimensions is shown. The prototype scale has a corner, back wall, and side wall fillets, also a center splitter to reduce sub-surface vortices. The test reported in this paper has none of these structures, as the target was to test a more general bay design.

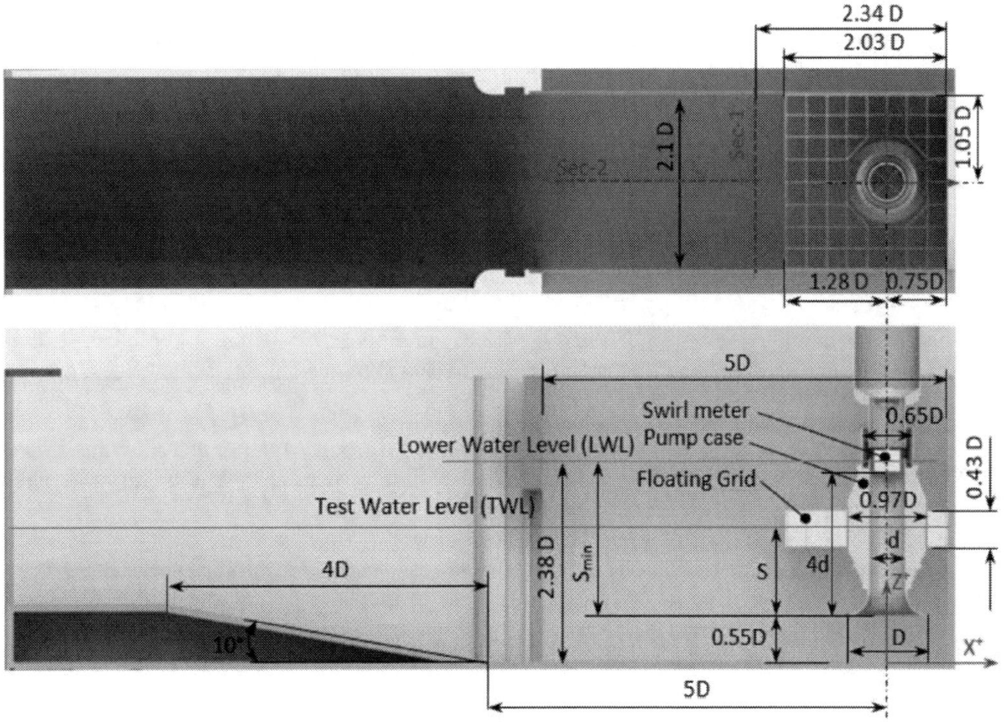

Figure 1. Plan view and cross-section of the pump bay, its dimensions are parametrized by the pump suction bell diameter (D). The swirl meter position is indicated in the cross-section, its diameter is 0.75d and the height 0.6d.

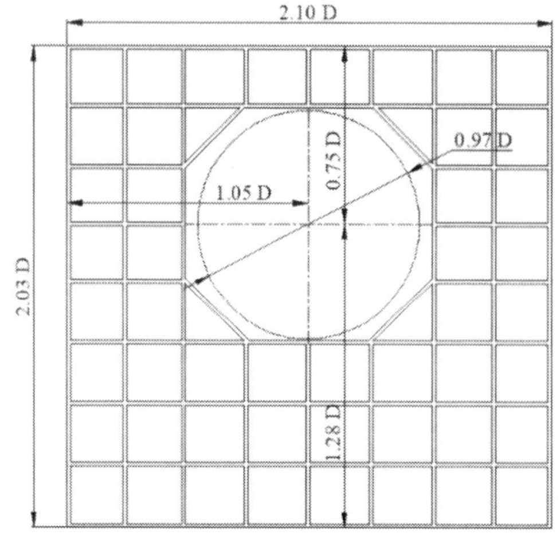

Figure 2. Plan view of the floating grid, it has 8x8 cells with a hole to avoid the pump case contact (dashed circle). The grid is orthogonal but is not uniform and should be adjusted to the pump case diameter.

Concerning the proposed antivortex device (Figure 2), the design concept is aimed at reducing the size of the free surface vortices by splitting them into higher numbers. As the available energy of each vortex decreases, so does its ability to depress the free surface. The grid is neither in contact with the pump body nor with the bay walls. It is submerged at 70 % of its height (0.3 *D*), and as it is floating, it follows the surface level permanently. The grid is orthogonal and has 8 columns by 8 rows, the cells are not of the same dimensions but the size of the sides of the squares is between 0.23 *D* and 0.27 *D*. The height is 0.43 *D* (1 m at prototype) and the grid is extended 1.28 *D* from the pump shaft axis.

The tests were done for a flow of 1.28 Q_n equal to 0.0183 m^3/s (32.000 m^3/h at prototype). This higher flow is under the code recommendations (ANSI/HI 9.8, section 9.8.5.3) to test the final design at 1.5 times the Froude scaled flow to increase circulation and produce stronger vortices resulting in a conservative prediction even when no scale effects are probable. The bay-free surface level is expressed by the relative submergence, *S/D*. The submergence *S/D* = 1.83 is the minimum free water level (*MWL*) at the bay recommended by the pump designer. This value is under ANSI 9.8/HI recommendations expressed by the equation [1].

$$S/D = 1 + 2.3F_D \quad (1)$$

$$F_D = \frac{V}{\sqrt{gD}} \quad (2)$$

Where *S* is the submergence, *D* is the outer diameter of the suction bell, F_D is the Froude number, *V* is the mean velocity at the suction bell, and *g* is the acceleration of gravity.

Reynolds number at the model is *Re* = 135,080 and the Weber number is *W* = 1,050; in both cases, they verified the recommended limits of *Re* > 110,000 and *W* > 720 (Odgaard 1986) to account for scale effects due to friction and surface tension.

The inlet flow at the bay was measured with an orifice plate with corner tapping following ISO 5157-2:2003 guidelines. The free surface level at the bay was measured with a vertical ruler at the entrance. The free surface fluctuation was negligible for all tests.

Table 1. Performed tests. Water levels at the bay are expressed as a ratio with the suction bell diameter and also as the ratio between the descent of water level (relative to the *LWL*) and the minimum submergence recommended by the pump designer.

S/D [-]	(*LWL-TWL*)/S_{min} [%]	Without FG	With FG
1.83	0	1	1-FG
1.61	12	2	2-FG
1.30	29	3	3-FG
1.17	36	4	4-FG
0.95	51	5	5-FG
0.82	58	6	6-FG

In order to detect the presence of vortices at the free surface and to quantify the intensity, three approaches were implemented: the PIV measurements technique, velocity profiles with ADV measurements, and circulation estimations by means of swirl meter measurements. The ANSI 9.8/HI classification of the free surface vortices was adopted to describe vortex observations. Small Polypropylene balls were used as tracers and classify vortices between types 3 and 4 which can pull particles but not air.

It was assumed for all tests that the grid does not affect the development of subsurface vortices. Therefore, the swirl angle variations between the cases with and without a floating grid are exclusively due to its influence.

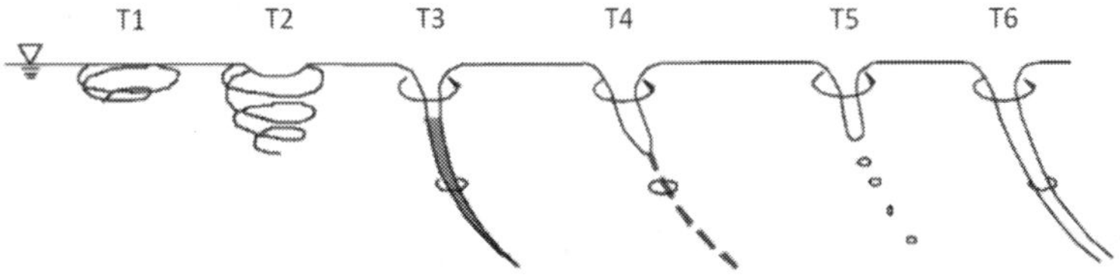

Figure 3. Classification of free vortex type according to ANSI 9.8/HI

Figure 4. For a) and b) side views, *S/D* = 1.17 compared with and without FG. For c) and d) side view, *S/D* = 0.95 compared with and without FG. In c) a vortex core type 4 is pointed with a dashed line which is not visible for d).

2.2. PIV measurements

Two cameras were located above the pump bay, one in front of and the other behind the pump case, in order to obtain a zenithal view suitable to capturing the free surface vortices. Videos were recorded simultaneously and then merged in a unique frame. As the vortices occurrence is unsteady, the video is 180-second long. Small Polypropylene particles were seeded at the bay as surface flow tracers. The time lapse where vortices appeared was analyzed utilizing the PIVlab tool.

2.3. ADV measurements

Two velocity profiles were measured for all tests, one cross-section (Figure 1, Sec-1) and a longitudinal section (Figure 1, Sec-2). Each section was divided into a grid with 35 points. Measurements were made with an ADV Vectrino Plus (Nortek). 3D velocities were measured for 5 minutes at each point and industrial powder was used to increase water turbidity to get a signal-to-noise ratio, $SNR > 15$ dB, and correlation factors > 70 %. Raw data was post-processed with WinADV 2.03 (USBR) filtering spikes to guarantee good velocity profiles.

2.4. Swirl meter measurements

To measure the intensity of flow rotation in the suction pipe (θ) a swirl meter was installed following ANSI/HI 9.8 recommendations. The number of revolutions is accounted for during a lapse of 10 minutes for long-term evaluation and 30 seconds for the short term. This velocity is used to calculate de swirl angle utilizing the following equation:

$$\theta = tan^{-1}(\pi\, d\, n/u) \tag{3}$$

Where d is the diameter of the pipe at the swirl meter (83 mm), n represents the revolutions per second of the swirl meter and u is the average velocity at the swirl meter.

3. Results

3.1. Free surface velocity and vorticity (PIV)

In Figure 5, PIV results are shown in terms of instant (Ω_z) and average ($\bar{\Omega}_z$) vorticity in vertical direction Z. Without FG, average vorticity shows that superficial vortices are located around the left area of the suction pipe. Close to the pump case, vortices are negative (clockwise rotation), and close to the side wall, they are positive (counterclockwise rotation). This result is consistent with the shear wall effect in the superficial flow. The highest vortex intensity was obtained for a submergence of 0.95 *S/D*. At 0.82 *S/D*, the free surface is more stable in terms of vortex formation. In Figure 5.b, instant vorticity shows the same pattern. The frame chosen shows an individual vortex formation type IV. Figure 5.c shows the results with the FG. Average vorticity in the Z direction is not analyzed since all the values are close to zero. Instant vorticity shows vortex formation in the area located upstream of the FG. The maximum intensity is for a submergence equal to1.17 *S/D*. Inside the grid cells, no vortex formation is visible in this analysis.

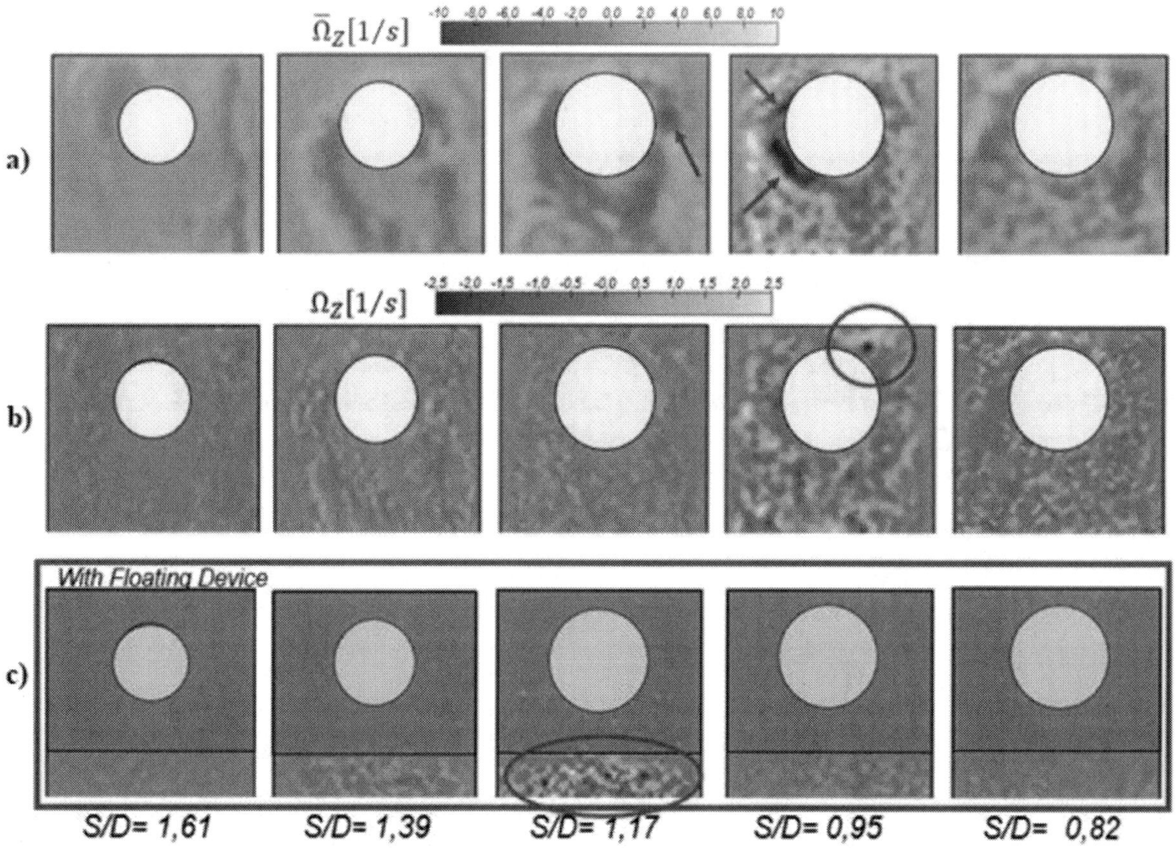

Figure 5. Free surface vorticity in the Z direction was obtained with PIV measurements at different submergence levels, ranging from 1.61 to 0.82 *S/D*. a) Average vorticity ($\bar{\Omega}_z$) for 2 min window. Maximum vortex development is seen at *S/D* between 1.17 and 0.95 close to the suction pipe (red arrows). b) Instant vorticity (Ω_z) frame without FG case. A free surface vortex is identified for *S/D =0.95* (red circle). c) Instant vorticity (Ω_z) frame for FG case. Vorticity inside the grid cells is almost cero. For *S/D* between 1.39 and 0.95, a random vorticity pattern is detected inside the cells (red circle).

3.2. Velocity profile upstream the pump intake (ADV)

Figures 6 and 7 show average velocity profiles and standard deviation of velocity for a window size of 300 s. For a submergence of 1.83 *S/D*, the flow distribution remains mostly undisturbed by the presence of the FG. As submergence decreases, the flow distribution appears to be more uniform and stable with the presence of the FG. Without the FG, the flow tends to concentrate more on the left side of the channel. The standard deviation of the velocity module shows that, for submergences below 1.61 *S/D*, velocity fluctuation increases between 15% to 40%.

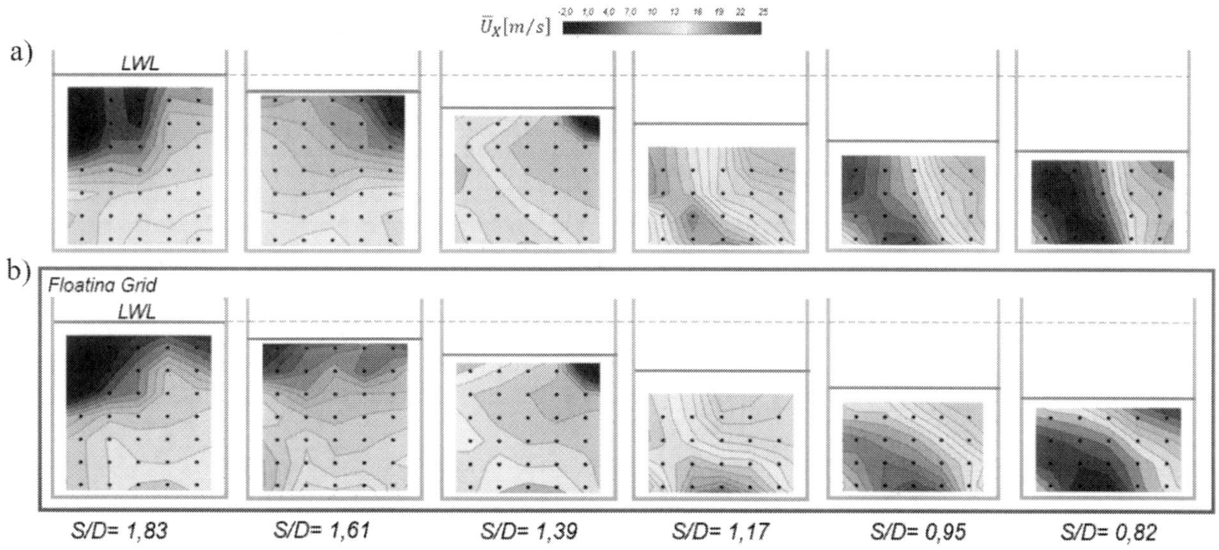

Figure 6. Average X-velocity ($\overline{U_x}$) profiles for a cross-section at a distance of 0.31D upstream of the FG. Submergence levels decrease from left to right ranging from 1.83 to 0.82 *S/D*. Measured probe's locations are indicated with black dots. a) Without FG, the flow tends to concentrate to the left for submergences below 1.39 *S/D*. b) With FG, the flow is more uniform with a small tendency to the left for submergences between 0.95 and 0.82 *S/D*.

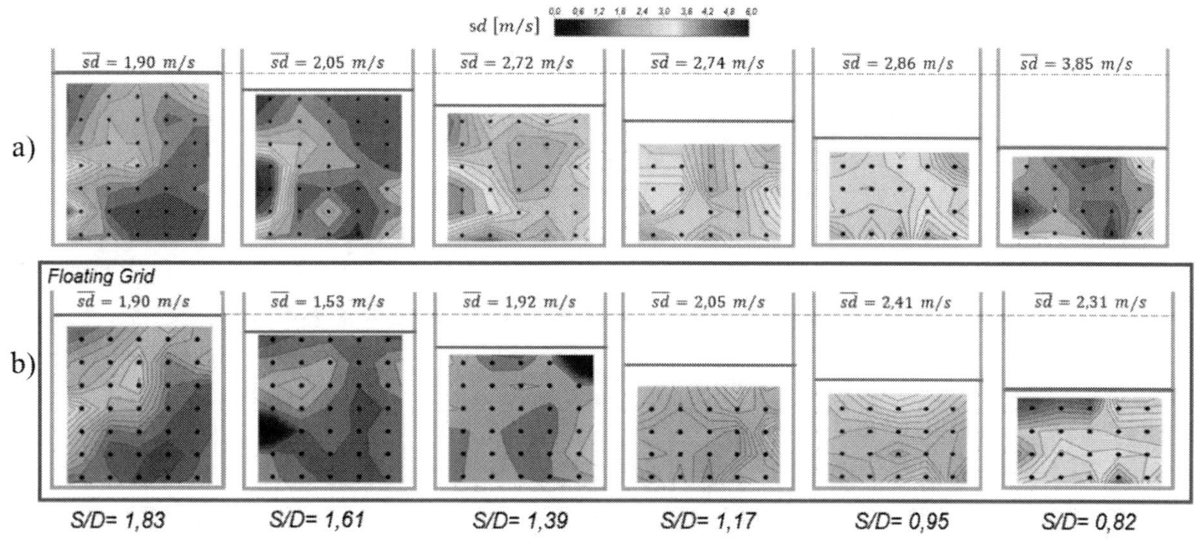

Figure 7. Standard deviation (sd) profiles of velocity module for a cross-section at a distance of 0.31D upstream of the FG. Submergence levels decrease from left to right ranging from 1.83 to 0.82 *S/D*. Measured probe's locations are indicated with black dots. a) Without FG, the flow is more unstable for smaller submergence levels. Mean deviation values decrease from 1.9 m/s to 3.85 m/s going as submergence spans from its maximum to its minimum). b) With FG, the standard deviation distribution remains more uniform as the submergence level decreases. Mean deviation values are 15% to 40% smaller as compared to a).

3.3. Swirl in the suction pipe (Swirl meter)

Table 2 and Figure 8 show the analysis of the flow rotation and the inlet of the pump. For the minimum submergence allowed by the ANSI standard, S/D = 1.83, the acceptance criteria of maximum rotation are achieved for both cases (with and without FG). The long-term swirl average angle is below 5° and the short-term swirl average angle is below 7° (Figure 8.a). For submergence levels below 1.39, without the FG, the acceptance criterion is no longer achieved. For S/D = 1.61, the swirl angle is nearly zero since the direction of rotation changes from negative to positive from

S/D = 1.83 to *S/D* = 1.39. This result is in accordance with the flow distribution analyzed in Figure 6, where the flow changes the tendency to be concentrated on the right of the channel for *S/D* = 1.83 to the left for *S/D* below 1.39. Figure 8.b shows that swirl fluctuation is larger without the FG.

Table 2. Swirl in the suction pipe, with and without the floating grid at different submergence levels.

Case		Swirl in the suction pipe S/D = 1.83	S/D = 1.61	S/D = 1.39	S/D = 1.17	S/D = 0.95	S/D = 0.82
Without Flotating Grid	Long-term Swirl angle [°]	-5.2	0.4	9.1	8.5	13.1	14.3
	Short-term Swirl angle [°]	-5.9	0.4	11.2	10.8	18.8	17.9
	Swirl angle Desviation [°]	0.4	0.0	0.8	1.0	1.2	1.3
With Flotating Grid	Long-term Swirl angle [°]	0.0	1.2	1.6	2.6	3.6	6.5
	Short-term Swirl angle [°]	0.0	1.2	1.5	2.6	3.8	4.2
	Swirl angle Desviation [°]	0.0	0.1	0.6	0.2	0.4	0.8

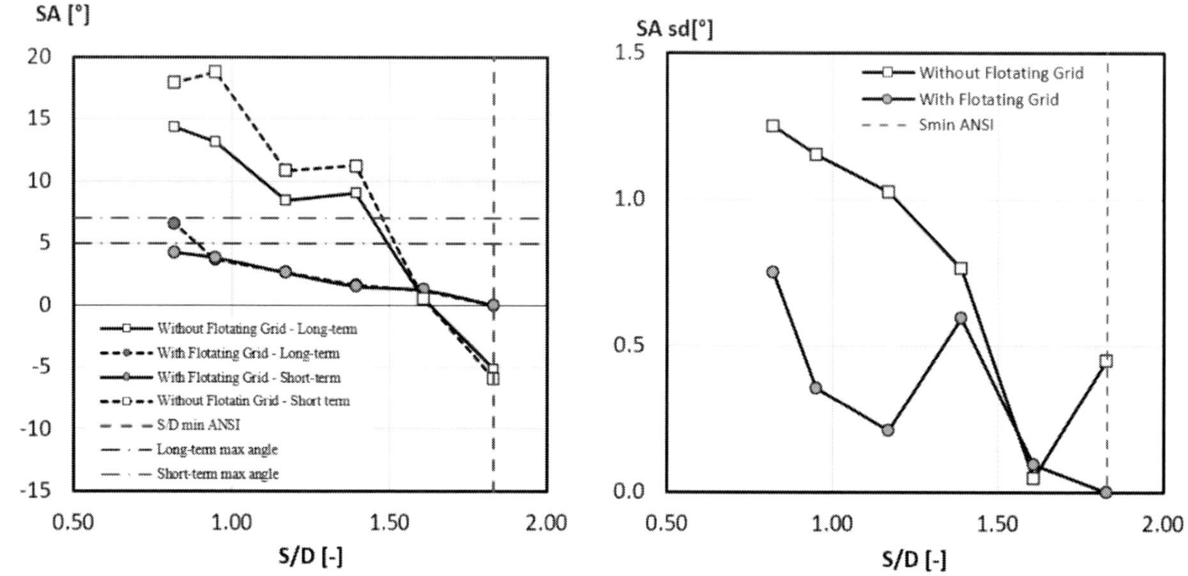

Figure 8. Swirl in the suction pipe, with and without the FG for a submergence level between 0.82 to 1.83 *S/D*. Left, Short-term, and the long-term average value of swirl angle. Without the FG, swirl angle values do not accomplish ANSI limits for *S/D* below 1.61. Right, Standard Deviation for swirl angle. Without the FG, swirl fluctuation is larger.

3.4. Flow visualization

Figure 9 shows an interpretation of superficial flow visualization for all the tested cases. Submergence levels of 1.39, 1.17, 0.95, and 0.82 as *S/D* ratios are included in the same scheme since they have the same superficial pattern. The main two differences for the cases with FG (Figure 9.b) are that the superficial vortices develop upstream of the FG, since the flow inside the grid cells has no movement and the vortices strength is much smaller for *S/D* < 1.61 (Table 3), and that the flow pattern as a function of submergence is the same with and without FG; for *S/D* = 1.83, the flow concentrates on the right side, the swirl is positive inside the suction pipe, and for superficial vortex. For *S/D* = 1.61 the flow and vortex formation are symmetric. Swirl inside the suction pipe is zero. For *S/D* < 1.39, the flow configuration is the opposite of *S/D* = 1.83, the flow concentrates on the left side and the swirl is negative inside the suction pipe and for the superficial vortex.

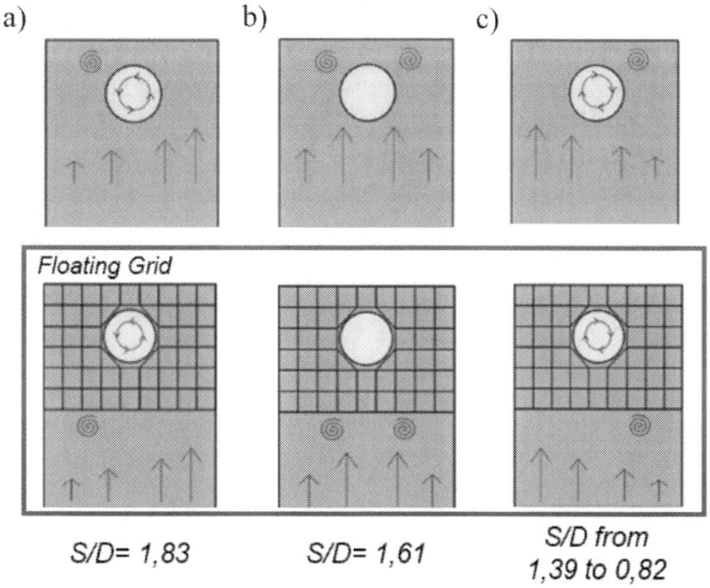

Figure 9 – Schematic interpretation of superficial flow visualization. a) Without FG: for *S/D* = 1.83, the flow concentrates on the right side and the swirl is positive inside the suction pipe and for the superficial vortex. For *S/D* = 1.61 the flow and vortex formation are symmetric. Swirl inside the suction pipe is zero. For *S/D* < 1.39, flow configuration is the opposite of *S/D* = 1.83. b) With FG: flow configuration remains the same as without FG for all the cases, superficial vortex location now is located upstream the FG.

Table 3. Vortex type characterization for every tested case according to ANSI standard (figure 3).

	Vortex type visualization					
Case	**1.83 S/D**	**1.61 S/D**	**1.39 S/D**	**1.17 S/D**	**0.95 S/D**	**0.82 S/D**
Without Floating Grid	T1	T1	T2	T3	T4	T5
With Floating Grid	T1	T1	T1	T1	T3	T4

4. Discussion and conclusions

Physical model tests were performed to evaluate the capability of the design named Floating Grid, to reduce vortex formation at the pump bay for a vertical submerged pump. The work aims to test the effectiveness of the device but not to optimize its geometry. Tests were carried out for a large flow of 1.3 times the design flow in order to achieve a conservative prediction. The design of the pump bay corresponds to a standard bay without fillets or splitter. The surface water level at the bay started from the minimum recommended by the pump designer (in coincidence with the ANSI code guidelines) *S/D* = 1.83, to a lower water level of *S/D* = 0.82, which can also be expressed as the 60 % of the minimum submergence.

Different techniques to quantify and describe the flow and the swirl formation were applied, such as PIV analysis, ADV velocity profiles, swirl angle measurements and vortex visualization and classification.

Based on these results, the most appropriate technique to evaluate the Floating Grid efficiency was the swirl angle measurements. The FG reduces the swirl angle at the entrance of the pump letting the same pump bay design achieve a swirl angle under 5° (as ANSI/HI code recommends) for an *S/D* = 0.82 instead of *S/D* = 1.83. The rest of the flow measurements prove that the flow with the presence of the FG is more stable, keeps vortices far from the suction bell and the pump case, and also reduces the vortices intensity, as shown in Table 3. Results have shown that the FG reduced the contribution of free surface vortices to the swirl angle as was expected, but part of the remaining circulation could be attributed to subsurface vortices.

Regarding the grid design, the minimum cell dimension to meet modular design guidelines would be 0.25 *D* x 0.25 *D*, but as it depends on the pump body diameter and the presence of bottom fillets, cell dimensions can be slightly adjusted to fit the bay geometry. A grid extending upstream will contribute to keeping upstream free surface far from the suction bell, but it is not recommended to reduce this dimension. Also, a grid with smaller cell dimensions will help to reduce vortex intensity inside the cells, but a denser and longer grid will result in a more expensive solution.

It is expected that the efficiency of the grid will decline with higher flows so the 0.82 *S/D* submergence could not be achieved without a free vortex surface.

The key to good efficiency is that the FG can track the free surface level preventing big vortices from arising or grouping, and keeping them off the pump. Designers and pump station owners can implement this solution without the need to carry out a study on a physical model, as the main dimensions of the floating grid are indicated in this work.

Future work will focus on the comparison between different device designs and extend the test area to higher flows and lower levels.

5. ACKNOWLEDGMENTS

This work was possible with the collaboration of J. F. Varvasino and F. Balzacca both undergraduate students of Hydraulic Engineering and the support of Central Térmica Vuelta de Obligado company (CVOSA).

6. REFERENCES

ANSI/HI 9.8 (2018). American National Standard for Pump Intake Design, Hydraulic Institute, New Jersey.

Arocena, V.M., Abuan, B.E., Reyes, J.G.T., Rodgers, P.L and Danao, L.A.M. (2021) "Numerical investigation of the performance of a submersible pump: prediction of recirculation, vortex formation, and swirl resulting from off-design operating conditions." *Energies 2021*, 14, 5082.

Constantinescu G.S. and Patel, V.C. (1998). "Numerical model for simulation of pump-intake flow and vortices." *J. Hydraul. Eng.* 124:123-134.

Harun, Z. (2020). Vortex Dynamics Theories and Applications, IntechOpen, London.

Jones, G.M., Sanks, R.L., Tchobanoglous, G. and Bosserman, B.E. (2006). Pumping Station Design, Elsevier, USA.

Kushwaha, T.N. (2015). "CW pump fluid induced vibration troubleshooting methodology". *Procedia Engineering*, 144, 274-282.

Norizan, T.A., Harun, Z., Abdullah, S. and Mohtar, W.H. (2019). "Effects of floor splitter height on the effectiveness of swirl angle reduction in pump intake." *J. Adv. Res. Fluid Mech. Therm. Sci.,* 57, 1, 32-39.

Odgaard, J.A. (1986). "Free-surface air core Vortex". *J. Hydraul. Eng.*, 112(7), 610-620.

Park, I., Kim, H.-J., Seong, H. and Rhee, D.S. (2018). "Experimental studies on surface vortex mitigation using the floating anti-vortex device in sump pumps". *Water*, 10(4), 441.

Shrestha, U. and Choi, Y.-D. (2021). "Bellmouth shape optimization for the suppression of flow instability in a pump sump model." KSFM *J. Fluid Mach.*, 24(5), 49-57.

Taghvaei, S.M., Roshan, R., Safavi, K. and Sarkardeh, H. (2012). "Anti-vortex structures at hydropower dams." *Int. J. Phys. Sci.,* 7(28), 5069-5077.

DOI: https://doi.org/10.26077/8d2a-290f

Effect of deficient tailwater on the performance of slotted roller bucket of Indira Sagar Dam, Madhya Pradesh

Vankayalapati S. Ramarao[1] Amit Kulhare[2]
[1&2] *Scientist -C, Central Water and Power Research Station, Khadakwasla, Pune-411 024. India*
Email: sramaraov@gmail.com

Abstract: *Slotted roller bucket type energy dissipators are provided for spillways when the tailwater depth is 1.2 times the sequent depth required for the formation of a hydraulic jump. In slotted roller buckets, the dissipation of energy occurs by the lateral spreading of jet passing through bucket slots in addition to the formation of two complementary rollers as in the solid bucket. These are provided for spillways where the downstream river bed is of sound rock. However, when the tailwater levels are not realised during the release of floods over the spillway and surface and ground rollers are not formed, it results in a possible sweep of the flood out of the bucket. Deficient tailwater levels during the initial period of operation of the bucket leading to ski action, generation of high hydrodynamic pressures during roller formation due to very high incoming velocities and occurrence of negative pressures on the bucket teeth, are some of the hydraulic parameters that lead to damage to the slotted roller bucket. Indira Sagar Dam is 653 m long and 92 m in height and was built on the River Narmada in Madhya Pradesh. The main and auxiliary spillway comprises 12 and 8 spans respectively of size each 20 m x 17 m and is designed to dispose of a Probable Maximum Flood of 83,400 m^3/s. The energy dissipator provided was original in the form of a slotted roller bucket. The project is in operation from 2004-2005. Discharges of the magnitude 30,000 m^3/s were released during the monsoon of 2013 and then it was observed that the entire slotted roller bucket in front of spans 6 to 12 was washed away. The teeth of the roller bucket were overturned and thrown away by the flood waters. Theoretical analysis for actual discharges passed over the spillway from 2006 to 2012 indicated that the roller action was not forming for these conditions and instead ski action was taking place due to deficient tailwater levels. There was also a transition from ski action to roller action for increasing discharges till the tailwater build was realised. In fact, this is the inherent shortcoming of roller bucket type of energy dissipators and damages to the bucket have been reported elsewhere also. In this paper, the authors account for the causes that damaged the slotted roller bucket of Indira Sagar dam, M.P and recommendations suggested for improving the performance of energy dissipation arrangement.*

***Keywords**: Slotted roller bucket, tailwater level, bucket teeth, apron, ground roller, surface roller, ski jump bucket*

1. Introduction

Slotted roller buckets are provided as energy dissipators for spillways. In the slotted roller buckets, a part of the flow passes through the slots, spreads laterally and is lifted away from the channel bottom by a short apron at the downstream end of the bucket and the flow is depressed and distributed over a great area providing less violent flow concentrations compared to those in a solid roller bucket. The velocity distribution just downstream of the bucket is more akin to that in a natural stream, that is high velocities at the surface and lower velocities at the bottom (IS 7365, 2010). However, it must be ensured sufficient raise of tailwater levels while designing the slotted roller buckets otherwise it would affect the functioning of the same, causing erosion of the bucket and its downstream. While designing of slotted roller bucket, for a high head spillway exceeding the total head of 50 m or so, specific care should be taken especially for the design of the teeth, to ensure that the teeth will perform cavitation free. After positive assurance of tailwater level within operating range between minimum and maximum tailwater depth limits and availability of rocky bed, the slotted roller buckets are required to be designed in such a way that, incoming, velocity is less than 15 m/s and the bucket lip is set above the river bed (IS 7365, 2010). The slotted bucket is self-cleaning, provides protection against excessive scour and undermining of the bucket structure and is shown to be of minimum size consistent with good performance.

Indira Sagar Project (ISP) is situated on River Narmada, 10 km from village Punasa in Khandwa district of Madhya Pradesh. ISP is a multipurpose project with an installed capacity of 1000 Mega Watt and provides irrigation benefits to about 1,23,000 hectares. ISP is the mother project for the downstream projects in Narmada Basin with the largest reservoir in India, having 12.22 x 10^9 m^3 storage. Indira Sagar dam is a 653 m long and 92 m high gravity dam, with a curved dam axis having a radius of 880 m. Main and auxiliary spillways comprise 12 and 8 spans respectively of size 20 m x 17 m and are designed to dispose of a PMF of 83,400 m^3/s. The original energy dissipator was in the form of a slotted roller bucket with different bucket invert levels for both the spillways. The surface Power House is situated on the right bank of the river to house 8 Francis turbines of 125 Mega Watt each. The water conductor system consists of Head Race Channel with carrying capacity of 2200 m^3/s and the water is discharged back into the river Narmada through 850 m long Tail Race Channel after power generation. Slotted roller bucket of Indira Sagar spillway was designed to handle the design head of 74.65 m (Head difference between Maximum Water Level (MWL) El. 263.35 m and bucket invert El. 189 m). Figure 1 shows the original cross section of the main spillway.

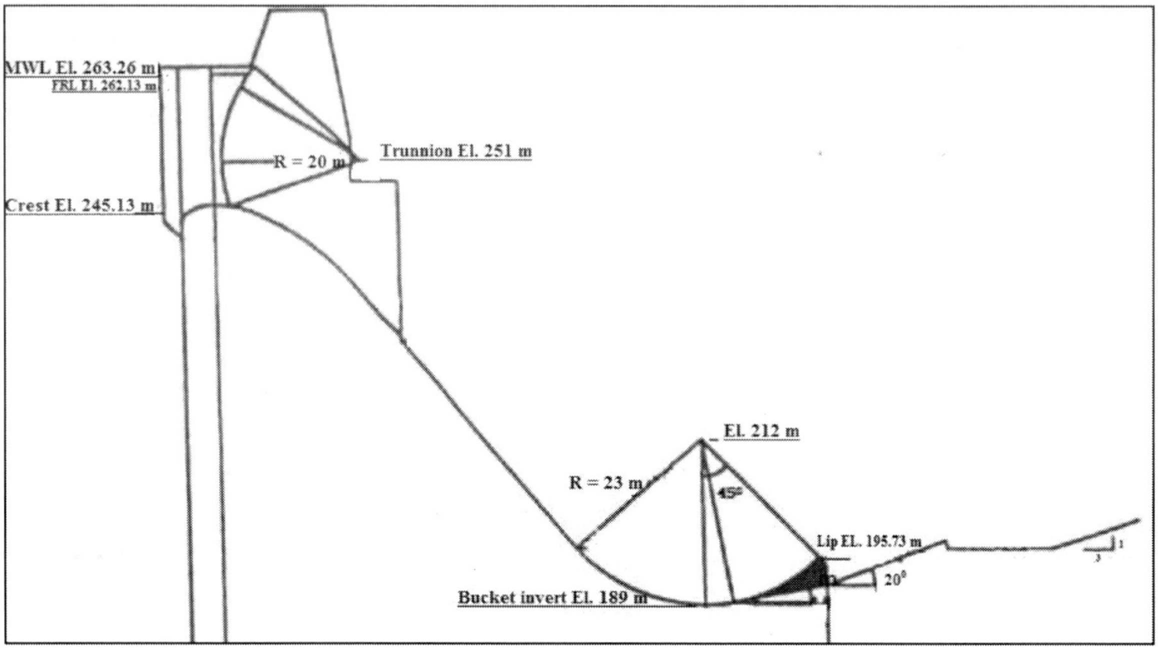

Figure 1. Cross section of main spillway of Indira Sagar dam, Madhya Pradesh

2. Performance of slotted roller bucket of main spillway

The Indira Sagar project is in operation from 2004-2005. According to Project authorities, it was observed that discharges of the magnitude 30,000 m^3/s were released for a considerable time during the monsoon of 2013 and it was observed that the entire slotted roller bucket in front of spans 6 to 12 was washed away. It was necessary to assess the possible reasons for the damage to the energy dissipator and recommendation of remedial measures. It was observed that the slotted roller bucket of the Indira Sagar dam spillway was damaged remarkably. Almost all the teeth of the bucket along with the apron downstream of the lip were overturned and thrown away by the flood waters. Out of the total 71 teeth of the main spillway only about 33 were seen intact, the remaining teeth were seen at a distance of about 100-150 m from bucket invert. Lip and teeth of the bucket in bays 6-12 have got dislodged from the main dam. The spillway was in operation with high discharges and it was opined that the damage to the energy dissipation arrangement is due to both hydraulic as well as structural reasons. Deficient tailwater levels during the initial period of operation leading to ski action, generation of high hydrodynamic pressures during roller formation due to very high incoming velocities of the order of 30 m/s and negative pressures on the teeth, were few of the hydraulic parameters leading to damage to the bucket. Damages to surface concrete were observed at various locations of the area of lip/ downstream apron. The surface of the teeth intact was found damaged due to abrasion/ erosion of concrete. These damages may have failed to boost up required tailwater levels resulting progressive erosion of bucket downstream. Photos 1 to 4 show the eroded bucket portion of spillway and severity of damage occurred to bucket. While designing slotted roller buckets, specific model tests should therefore be conducted to verify the pressures on the teeth and bucket invert should accordingly be fixed at such an elevation as to restrict the sub-atmospheric pressures to the permissible magnitude. As per calculations the velocities of flow for the period under consideration (2006-12) were about 30 m/s. It was opined that energy dissipator blocks got separated from the spillway due to untreated lift joints. These lift joints were formed possibly due to the placement of concrete at the lift joint after considerable time leading to separation due to shrinkage and no cold joint treatment. These construction joints facilitated ingress of water and high hydrodynamic pressures as a result major part of the energy dissipation arrangement got dislodged. The dislodged blocks were carried by the flow and dumped at about 100-150 m downstream forming an endsill type structure. This facilitated rise in tailwater besides the building of tailwater in due course of time and formation of the hydraulic jump at the toe of the spillway, which may prevent further damage to the spillway toe.

Photo 1. Damage to bucket portion of spillway

Photo 2. The eroded apron of bucket

Photo 3. The detachment of bucket teeth

Photo 4. Exposed reinforcement of teeth

2.1 Analysis of data (CWPRS TR No. 5223, 2014)

The provision of slotted roller buckets is based on the availability of high tailwater levels and good rock downstream of bucket. The performance of a slotted roller bucket depends upon the following factors:

1) Magnitude and duration of floods passed over the spillway and corresponding reservoir water levels.
2) Schedule of operation of spillway gates while passing the above floods
3) Tailwater levels realized downstream of slotted roller bucket corresponding to various floods.
4) Tailwater rating curve adopted while designing energy dissipation arrangement.
5) Hydraulic parameters viz. intensity of discharge, velocity, Froude Number, Bucket radius factor, minimum, maximum and sweep-out tailwater levels for different design discharge conditions.

The discharges released through the main spillway and the corresponding tailwater levels observed are tabulated in Table 1. The analysis of data indicated that the observed tailwater levels were below or marginally above the design tailwater levels for most of the flood conditions. Building up of tailwater may take considerable time with heavy flood releases with considerable high velocities. The analysis indicated that the roller action was not forming for most of these conditions and instead ski action might be taking place due to deficient tailwater levels. There was also a transition from ski action to roller action for increasing discharges till the tailwater build was realised. The Froude number was found to be of the order of 9. In fact, this is the inherent shortcoming of roller bucket type of energy dissipators and damages to the bucket have been reported elsewhere also (CBIP Publication No. 247, 1998). From the data, it is known that the tailwater available was marginally above or sometimes lesser than the sweep out depth for the floods under review. Fig 2 shows the plot between the tailwater depth and sweep-out depth for various floods under review. Slotted roller bucket type energy dissipators are provided for spillways when the tailwater depth is 1.2 times the sequent depth required for the formation of a hydraulic jump. But, for most of the flood releases, the observed tailwater levels were marginally above the jump height levels and also lesser than the designed tailwater rating curve for which the slotted roller bucket was designed. Fig 3 shows the plot of designed tailwater depth, observed tailwater depths and jump height for the discharges under review.

Table 1. Tailwater depth realised corresponding to the outflow flood (IS 7395, 2010)

S. No	Disch. observed, Q	Width of spillway (m)	Intensity of disch., q	Res. Water Level	Tail water level	FRL-TWL (m)	FRL - crest (m)	Theo Velocity, V_1	Actual Velocity of flow entering the bucket, Va	Initial depth, d_1	Froude Number, F	Available tailwater Depth, = Max.TWL- 189 (m)	Sweep out depth (m)
1	9140	258.46	35.36	254.66	204.0	50.66	9.53	31.53	29.95	1.18	8.80	15.00	17.59
2	9254	258.46	35.80	259.13	206.5	52.63	14.0	32.13	30.53	1.17	9.00	17.50	17.59
3	9576	258.46	37.05	259.94	206.5	53.44	14.8	32.38	30.76	1.20	8.95	17.50	18.07
4	13272	258.46	51.35	259.95	207.5	52.45	14.8	32.08	30.48	1.68	7.50	18.50	19.38
5	14370	258.46	55.60	254.74	208.5	46.24	9.61	30.12	28.61	1.94	6.55	19.50	19.43
6	14843	258.46	57.43	254.76	209.7	45.06	9.63	29.73	28.25	2.03	6.32	20.70	19.31
7	15372	258.46	59.48	259.68	210.6	49.08	14.5	31.03	29.48	2.02	6.63	21.60	20.17
8	15588	258.46	60.31	259.94	210.5	49.44	14.8	31.15	29.59	2.04	6.62	21.50	20.18
9	16692	258.46	64.58	259.94	210.8	49.14	14.8	31.05	29.50	2.19	6.36	21.80	20.80
10	17772	258.46	68.76	259.94	210.8	49.14	14.8	31.05	29.50	2.33	6.17	21.80	21.45
11	23100	258.46	89.38	259.70	211.5	48.20	14.5	30.75	29.21	3.06	5.33	22.50	23.25

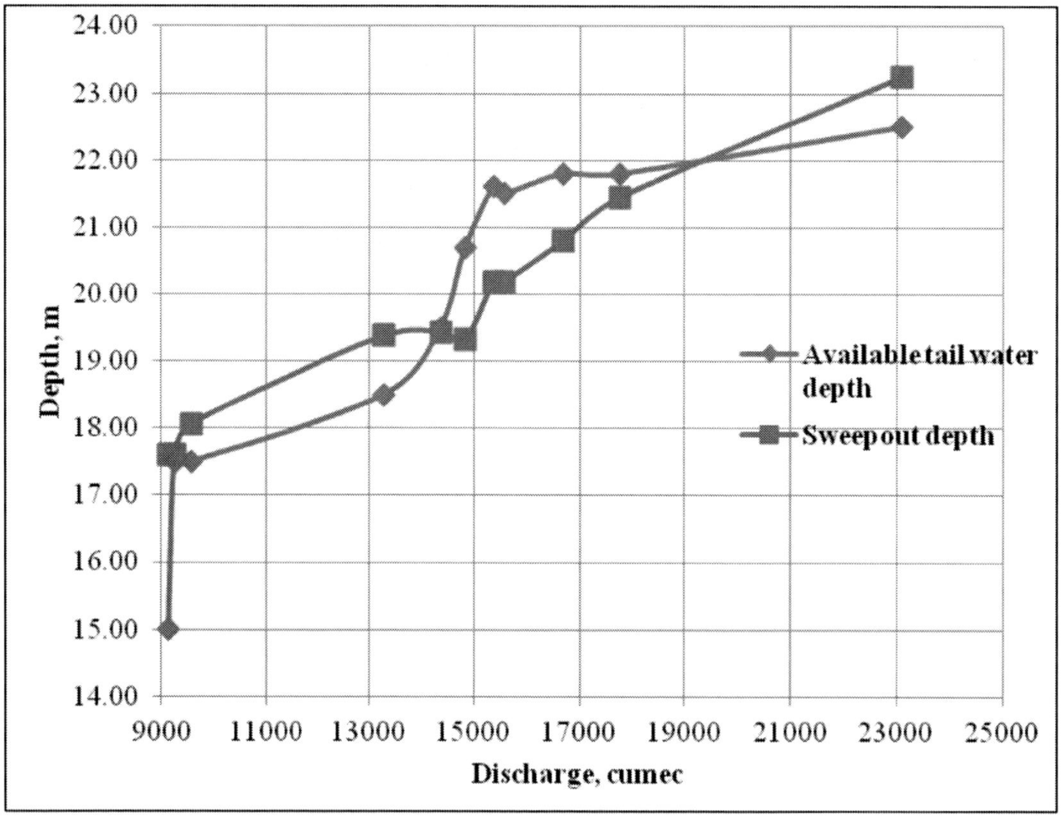

Figure 2. Tailwater depth realised versus sweep out depth

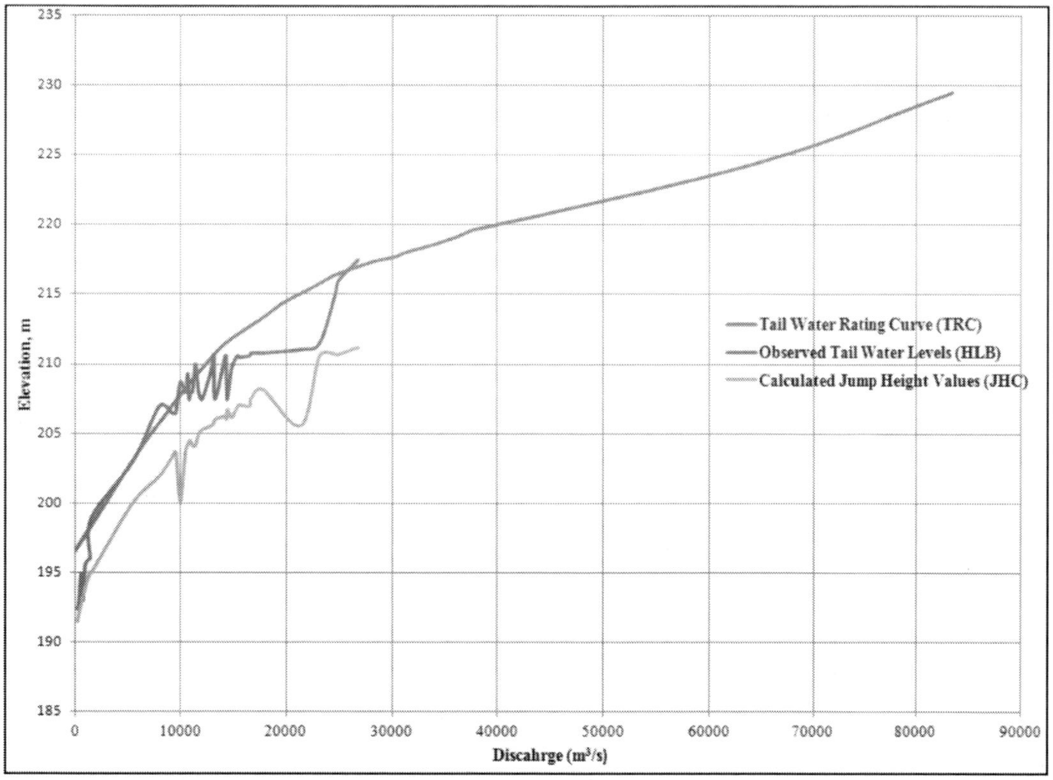

Figure 3. Plot showing tailwater depth and sweep out depth for floods occured

3. Remedial measures suggested

As it was experienced that slotted the roller bucket does not function satisfactorily during the last few years of operation due to its inherent limitations, it was felt that the ski jump bucket is a better long term solution. It has the advantage that it will have the same footprint as the roller bucket and also downstream apron's footprint will remain the same. This will enable the construction of ski jump bucket raising the already existing downstream coffer dam constructed during dam construction.

4. Modified energy dissipator in the form of ski jump bucket

Ski jump bucket is a widely adopted form of energy dissipator which is an upturn solid bucket that is recommended when the tailwater depth is insufficient for the formation of roller action/ hydraulic jump, or when the downstream river bed comprises of sound rock and capable of resisting the high velocity scouring jets released from the bucket. In this form, the flow released from spillway glacis is thrown away from the toe of the dam to a considerable distance downstream as a free discharging jet that falls into the plunge pool, avoiding excess scour caused to the downstream bed. The energy of incoming flow is dissipated in this formation by internal friction within the jet, interaction between the jet and the surrounding air, the diffusion of jet in the tailwater and the impact of jet in the plunge pool. The hydraulic design of ski jump bucket consists of determining bucket shape, its invert elevation, its radius, lip elevation and exit angle, trajectory length and elimination of scour downstream of spillway (IS 7365, 2010).

For the Indira Sagar spillway, the original energy dissipator in the form of slotted roller bucket with invert at El. 189 m and lip at El. 195.73 m was modified to the form of ski jump bucket with a radius of 50 m, invert at El. 199.75 m, lip at El. 210 m and the lip angle is 37° which was recommended by CWC in consultation with CWPRS. The radius of bucket was increased from 23 m to 50 m to accommodate the ski jump bucket profile. Fig 4 shows the cross section of ski jump bucket superimposed on a slotted roller bucket type energy dissipator. Photo 5 shows downstream view of spillway with ski jump bucket.

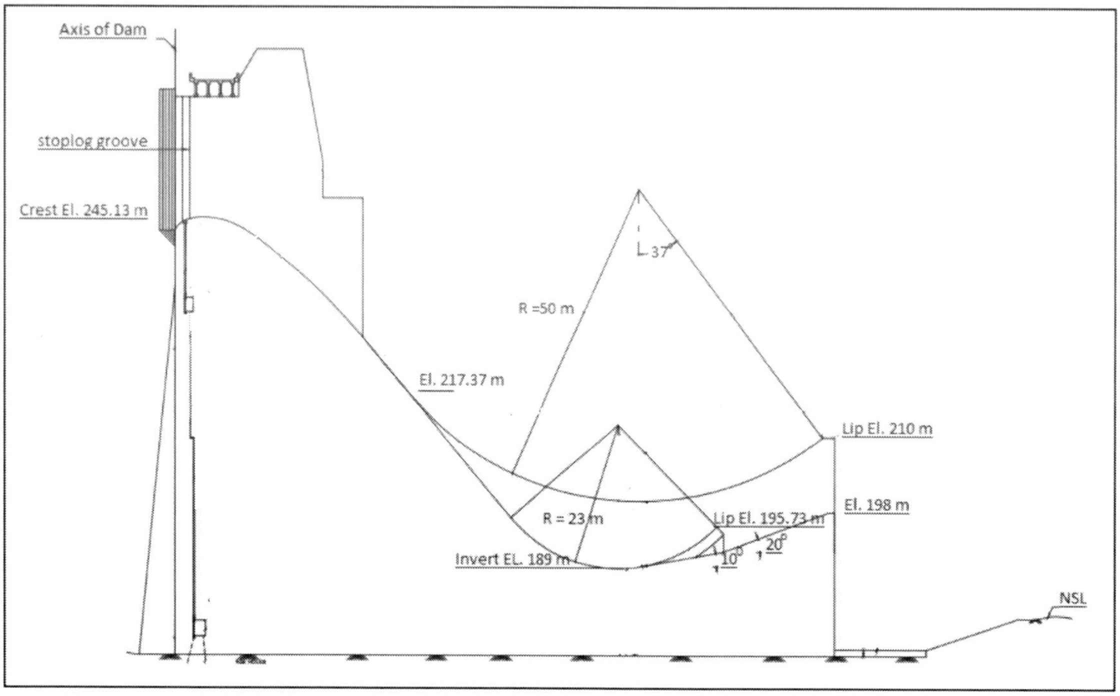

Figure 4. Cross section of modified energy dissipator

Photo 5. Downstream view of spillway with ski jump bucket

5. Performance of ski-jump bucket of main spillway (CWPRS TR No. 5579, 2018)

Hydraulic model studies were carried out on a 1: 130 scale geometrically similar 3-D comprehensive model for assessing the performance of ski-jump bucket for various operating conditions of spillway. Studies indicated that the discharges of 55,727 m^3/s and 62,566 m^3/s could be passed with the ungated operation of main and auxiliary spillways keeping the reservoir water level at Full Reservoir Level (FRL) at El. 262.13 m and MWL El. 263.35 respectively and at Full Reservoir Level (FRL) El. 262.13 m, the discharge of about 32872 m^3/s could be passed through the main spillway. Performance of ski-jump bucket of the main spillway was satisfactory as the clear ski action was seen forming in the bucket. The throw distance was found to be about 86 m from the lip of the bucket. Photo 6 shows the performance of ski jump bucket while passing discharge of 32,872 m^3/s at Full Reservoir Level (FRL) El. 262.13 m with all 12 gates fully open.

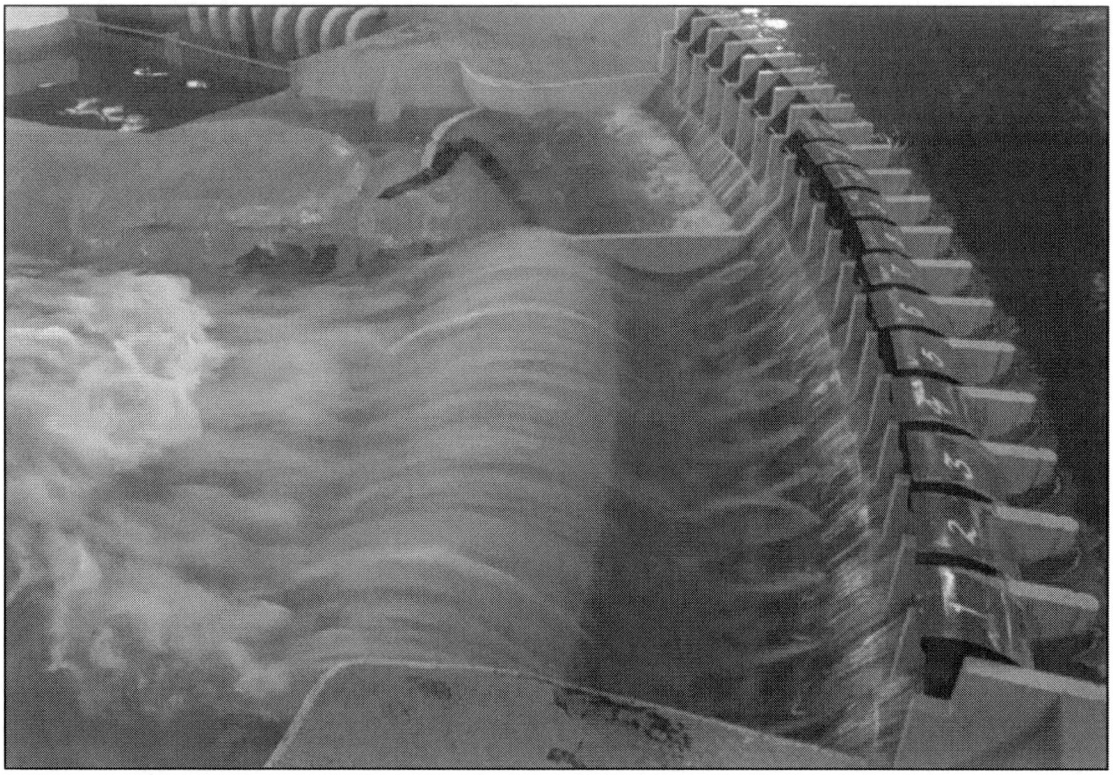

Photo 6. Performance of ski jump bucket, Q= 32872 m^3/s at Full Reservoir Level (FRL) El. 262.13 m

6. Conclusions

The building up of tailwater levels plays a crucial role in the effective functioning of slotted roller buckets. If the tailwater levels are insufficient to form roller action in the bucket, there is a possibility of sweeping out of the jump causing cavitation of bucket surface and teeth and erosion of the same. For Indira Sagar dam spillway, a major portion of bucket and teeth of energy dissipator has been remarkably eroded/ uprooted due to unsatisfactory functioning of slotted roller bucket for insufficient building up of tailwater levels. From the analysis and site observations it could be concluded that the damage to the energy dissipation arrangement was due to both hydraulic as well as structural reasons. It was opined that roller bucket may not be a suitable arrangement for energy dissipation for the spillway of Indira Sagar dam and must be replaced with ski jump bucket type of energy dissipator for main spillway where major damage has occurred. Hydraulic model studies were carried out in CWPRS for the modified energy dissipator as ski jump type bucket and found satisfactory. The project was completed and the ski jump bucket has been performing satisfactorily.

ACKNOWLEDGEMENT

The authors are grateful to Dr. R. S. Kankara, Director, CWPRS for his continued support in writing technical papers. Authors are also grateful to project authorities of NHDC Ltd for their support in providing required data during studies. The support from staff of SED Division in writing this paper is also highly acknowledged.

REFERENCES

1. Beichley G. L., Peterka A. J.(1959). "The hydraulic design of slotted spillway bucket energy dissipators." *Proceedings of The ASCE-Los Angeles Convention*, Los Angeles, California, February 9-13, 1959
2. CBIP Publication No. 247 (1998). Performance Evaluation of Slotted Roller Bucket.
3. CWPRS Technical Report No.5223 (2014). "Hindcasting Studies for assessing the performance of Spillway and Energy Dissipator, Indira Sagar Dam, Madhya Pradesh, October 2014".
4. CWPRS Technical Report No.5579 (2018). "Hydraulic Model Studies for Indira Sagar Dam Spillway, Madhya Pradesh, March 2018"
5. IS 7365 (2010). "Criteria for Hydraulic Design of Bucket type Energy Dissipators", Bureau of Indian Standards, New Delhi- 110002.

DOI: https://doi.org/10.26077/2d70-9137

Energy dissipation and flow regime downstream of trapezoidal piano key weirs

X. Shen[1] and M. Oertel[2]
[1,2]*Hydraulic Engineering Section, Helmut-Schmidt-University, Hamburg, Germany*
[1]*E-Mail: xiaoyang.shen@hsu-hh.de*
[2]*E-Mail: mario.oertel@hsu-hh.de*

***Abstract**: In the past two decades, increasing numbers of piano key weirs (PKW) have been constructed due to their sufficient hydraulic performance and optimized space requirements. Extensive experiments and numerical studies have been carried out to identify the optimal geometry and maximize the discharge capacity. However, less attention has been paid to address energy dissipation processes and the flow structure downstream, especially for in-channel applications. Within the present study, symmetrical trapezoidal piano key weir models were fabricated in two model sizes with different cycle numbers via a 3D printing technique. Both models were tested under equivalent hydraulic conditions to investigate the potential influence of the cycle number and model size regarding energy dissipation processes and the flow structure downstream of the weir. Data show identical downstream residual energy for both investigated weirs, indicating that a model with 30 cm weir height and single cycle is able to deliver reasonable results in terms of total energy dissipation. However, the flow structure downstream of the weir was observed to differ with actual weir size and cycle number, indicating that small-scaled models with an insufficient cycle number may not be able to represent a similar outflow characteristic as the prototype. Sufficient cycle number was found to be beneficial for a more stable downstream flow condition. To facilitate practical design of geometrically similar weirs, observational insights of downstream flow condition development have been provided and analytical equations have been proposed for localization of aerated flow region as well as residual energy estimation.*

***Keywords:** Cycle number, Energy dissipation, Flow characteristics, Piano key weir, Scale effect*

1. Introduction

The piano key weir (PKW) was firstly introduced by Hydrocoop (Blanc and Lempérière 2001; Lempérière and Ouamane 2003) as a further development of labyrinth weirs. With sufficient hydraulic performance and reduced structural footprint, this innovative type of a non-linear weir has been introduced as a favorable up-to-date solution against the changing conditions on waterways induced by climate change. In the last two decades, an increasing number of PKWs have been constructed worldwide (Crookston et al. 2019). Extensive studies (e.g. Leite Ribeiro et al. 2012; Erpicum et al. 2014; Cicero et al. 2016) have contributed to identify the primary and secondary geometric parameters that can significantly influence the hydraulic performance, such as magnification ratio (L/W), weir height (P) and key width ratio (W_i/W_o). Remarkable progress has been made in optimizing the geometry and maximizing the discharge capacity for this kind of weir. As a further step, more and more attention has been recently paid to the flow regime and energy dissipation downstream of similar non-linear weirs. Despite the studies focused on spillway applications, there are only a few previous studies that have addressed the flow characteristics downstream of a PKW or labyrinth weir for channelized application (Oertel 2018). For example, Merkel et al. (2018) tested rectangular labyrinth weirs with varying tailwater conditions. Lopes et al. (2011) investigated flow properties and residual energy downstream of labyrinth weirs with different side wall angles. More recently, four laboratory-scaled type-A rectangular piano key weirs with different key width ratios were studied in Eslinger and Crookston (2020) with the focus on quantification of energy dissipation. These studies show that unlike the relatively uniform outflow downstream of regular linear weirs, highly turbulent and aerated tailwater with the presence of shock wave intersection was observed and the energy dissipation was found to be greatest at lower heads and decrease with increasing upstream head for both labyrinth and piano key weirs. According to Pfister et al. (2013), the number of cycles does not affect the unit hydraulic behavior of a PKW in terms of head-discharge relationship, which means that a single-cycle PKW is representative for multi-cycle models thanks to the symmetrical plan form. However, considering the complex outflow trajectory over such folded crest arrangements, notable differences of downstream flow structures could be expected between single- and multi-cycle models. Further research is thus required to address the correlation between the PKW's cycle-number and downstream flow condition. Furthermore, scale effects induced by force ratios that are not similar between laboratory-scaled models and prototype sizes have always been a major topic for hydraulic experimental research. Previous studies have proposed many useful recommendations to limit scale effects. For labyrinth and PKWs, although different suggestions exist (Oertel 2019; Tullis et al. 2020; Shen and Oertel 2021), it has been largely accepted that a minimum upstream head of 3 cm suggested by Erpicum et al. (2016) is sufficient to avoid significant scale effects when examining the discharge capacity. Up to date, it has been rarely discussed whether the downstream flow condition and the energy dissipation would also differ between models with varying sizes. To address these research gaps, symmetrical trapezoidal piano key weirs in two size-scales with two different cycle-numbers were tested in the present study.

2. Model setup

Physical experiments were conducted in two horizontal rectangular flumes. The first flume has the dimension $L_{f,1}$ = 10 m (length), $H_{f,1}$ = 0.6 m (height) and $W_{f,1}$ = 0.3 m (width); the second flume is $L_{f,2}$ = 20 m (length), $H_{f,2}$ = 1.0 m (height) and $W_{f,2}$ = 1.0 m (width). Both flumes feature transparent side walls to facilitate flow pattern observations. Discharges were controlled by magnetic-inductive flowmeters (Krohne Optiflux, accuracy: ±0.3%) installed in the supply pipelines. For both channels, flow straighteners were additionally installed at flow entrance to produce a more uniform inflow and to compensate water surface fluctuations. Flow depths were collected via ultrasonic sensors (Microsonic Mic-130, resolution: 0.025 mm, accuracy: ±1%) at 20×P (P denotes the weir height) downstream of the weir model for residual energy estimation. The position of the hydraulic jump was measured with a measuring tape in longitudinal direction. For the current study, 23 model runs were performed with laboratory discharges up to 0.33 $m^3/(s \cdot m)$ for trapezoidal piano key weirs in two model sizes (P = 30 cm and 40 cm) with two different cycle numbers (N = 1 and 2.5). All model runs were performed without artificial interference on tailwater elevation. Tested model geometries and main parameters are presented in Table 1 and Figure 1. All models were fabricated via 3D printing technique using polylactic acid (PLA) material. To facilitate the comparison between models with different scales, dimensionless parameter $H_t P^{-1}$ will be used to denote the relative upstream head, with P as the weir height and H_t calculated via $h_t + v_m^2(2g)^{-1}$ (h_t represents the upstream flow depth measured relative to weir crest and v_m is the mean upstream velocity averaged over the total flow depth upstream $P + h_t$).

Table 1. Main geometric parameters of tested PKW models (mm)

Models	Crest shape	Cycle	P	B	L	W	W_i/W_o	T_s	R
$P_{30,qrup}$	Quarter-round upstream	1.0	300	480	1103	300	1	30	15
$P_{40,qrup}$	Quarter-round upstream	2.5	400	640	3677	1000	1	40	20

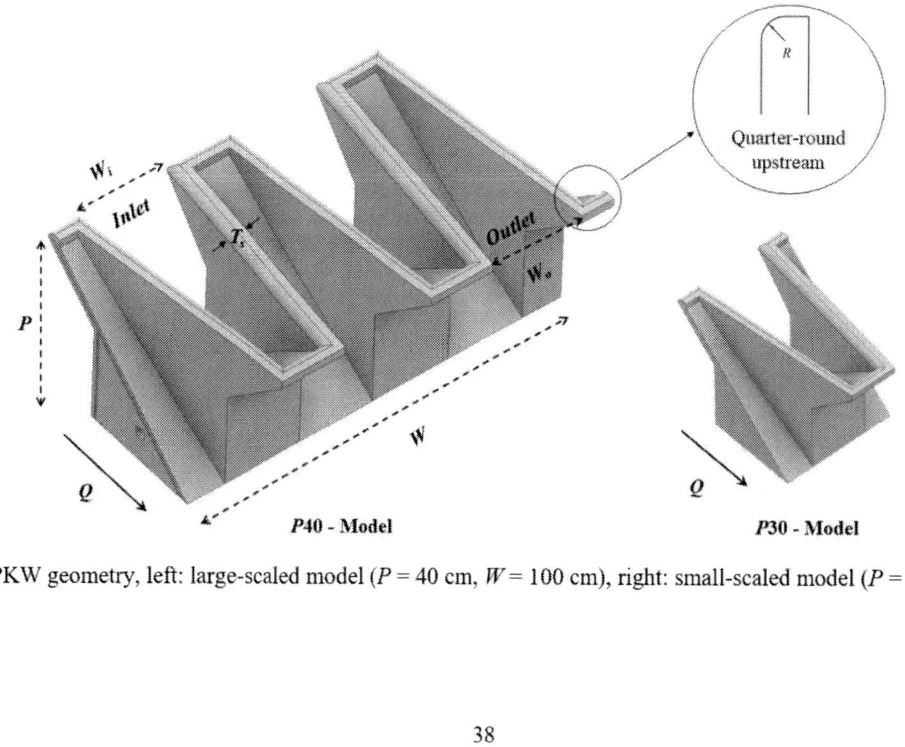

Figure 1. PKW geometry, left: large-scaled model (P = 40 cm, W = 100 cm), right: small-scaled model (P = 30 cm, W = 30 cm)

3. Experimental results

3.1. General downstream flow characteristics

Figure 2 presents the evolution of the downstream flow structure for the multi-cycled $P = 40$ cm model with increasing relative upstream heads. In general, for all heads tested herein, the flow field at the weir toe was observed to be very complex and is generated by a combination of multiple types of outflow (see Fig. 2a): (1) free-falling nappes over downstream apex of inlet keys, (2) inclined jets along the outlet key ramps, and (3) mixed side nappes over side walls which will lead to a transition into submerged condition under high heads. For low heads with $H_tP^{-1} < 0.12$, it can be seen in Fig. 2(a) that the free-falling nappe did not interfere with the spreading inclined jets flowing down from the ramp, which are likely mainly responsible for shock wave intersections. The vertical nappes fall on the channel bed, generating splash and dissipating a large portion of energy. With increasing upstream head and approach velocity, as shown in Fig. 2(b), significant turbulent mixing and air bubble entrainment occurs. At this stage, the nappes within outlet sections start to interact; the free-falling nappes begin to interfere with the inclined jets from outlet ramp and generate circulating flow regions behind the nappe. As upstream heads further grow ($H_tP^{-1} \geq 0.30$), local submergence occurs between the outlet sections and the inclined ramp jets tend to have similar trajectory as the free nappe over the inlet key downstream apex. As a consequence, a critical boundary region, where hydraulic jumps form, starts to be visible, as can be seen in Fig. 2(c). Although relative upstream heads (H_tP^{-1}) greater than 0.5 was not performed in the present study, it can be expected that for larger heads with outlet sections fully submerged, the downstream flow structure will eventually perform a transition into the typical outflow downstream of a linear weir with fully developed hydraulic jump and quasi uniform inflow. Nonetheless, it should also be noted that PKW discharge will likely always be more three-dimensional than a linear weir.

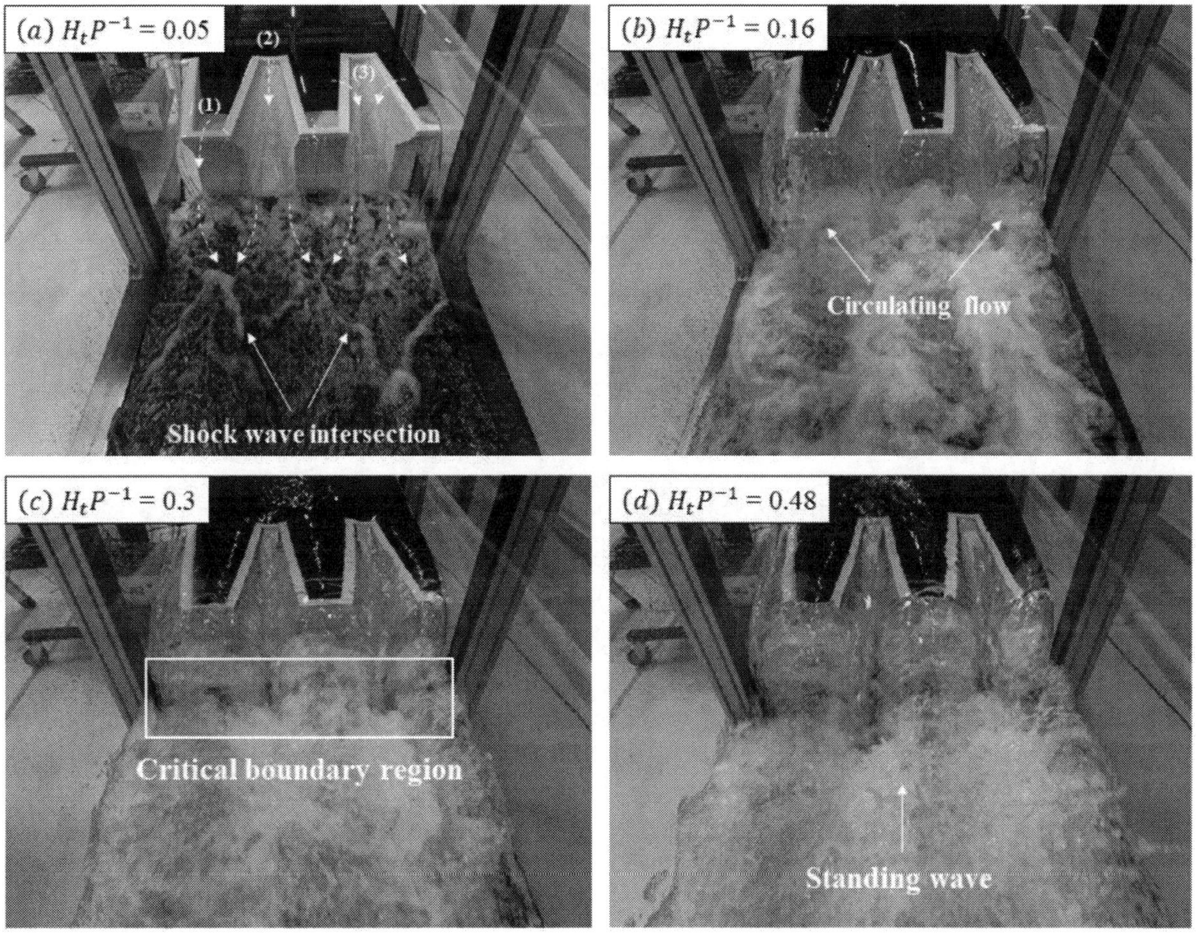

Figure 2. Downstream flow structure for $P_{40,qrup}$ model with upstream relative heads H_tP^{-1} ranging from 0.05 to 0.48

To facilitate the estimation of downstream flow conditions for similar weirs configurations, a preliminary classification of observed flow conditions is provided in Fig. 3 using relative upstream heads (H_tP^{-1}) and relative Froude number ($\mathsf{F}_{relative}$) as parameter. The relative Froude number can be calculated via $\mathsf{F}_{relative} = \mathsf{F}_{h1}\mathsf{F}_{ht}^{-1}$, where F_{h1} denotes the calculated Froude-number in downstream and F_{ht} represents the Froude-number in upstream, calculated with the upstream flow depth measured relative to the weir crest (h_t). Note that these results only apply to similar PKW configurations for horizontal channels without tailwater control.

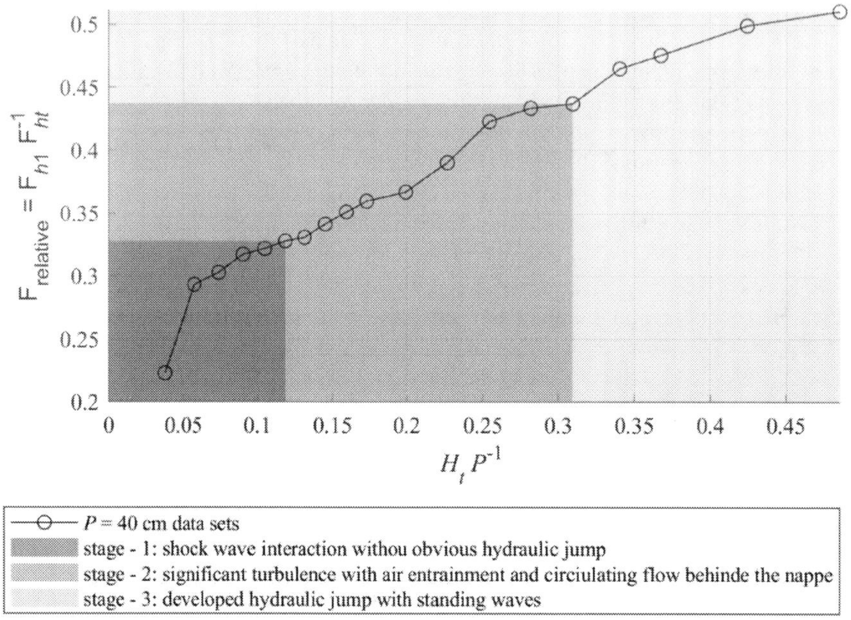

Figure 3. Classification of downstream flow conditions using $\mathsf{F}_{relative}$ and H_tP^{-1}

3.2. Influence of cycle number and model size

Based on laboratory observations, differences were identified between a single- and multi-cycle PKW in terms of resulting downstream flow regimes. A detailed side-view comparison is shown in Fig. 4. For very small heads such as $H_tP^{-1} = 0.05$, it can be clearly seen that the downstream flow regime of the single-cycle model deviates from the subcritical condition downstream of the multi-cycle model by having a weak hydraulic jump. The most likely reason is surface tension which cannot be scaled. For the same upstream relative head, the outflow of the down-scaled $P = 30$ cm model (Fig. 4, left) was still hindered by significant scale effects. In this case, the nappe collision is less sufficient and the inclined ramp jet becomes the main current that maintains its supercritical condition until the hydraulic jump forms further downstream. The cycle number influence at this stage is not identifiable due to significant scale effects of the outflow. As the upstream head increases, scale effects become negligible and the $P = 30$ cm single-cycle model starts to show identical flow condition as the $P = 40$ cm multi-cycle model (see Fig. 4, $H_tP^{-1} = 0.10$). In general, relative to the model with multiple cycles, the tailwater condition of the single-cycle model seems to be more unstable. One possible cause could be the lack of sufficient shock wave intersection between the cycles, which promotes the energy dissipation and stabilizes the surface fluctuation. As an example, relative to the multi-cycled model, more significant surface undulations can be seen at $H_tP^{-1} = 0.18$ for the single-cycle model. Furthermore, for increased upstream heads with $H_tP^{-1} > 0.35$, the flow condition downstream of the $P = 30$ cm model was observed to be more supercritical with a lower flow depth and hydraulic jump formation at farther downstream. In contrast, for the $P = 40$ cm model the hydraulic jump was formed more closer to the free-falling nappe. This observation could be an indicator that a lager cycle number could be beneficial for a more stable downstream flow field and potentially with less river bed erosion.

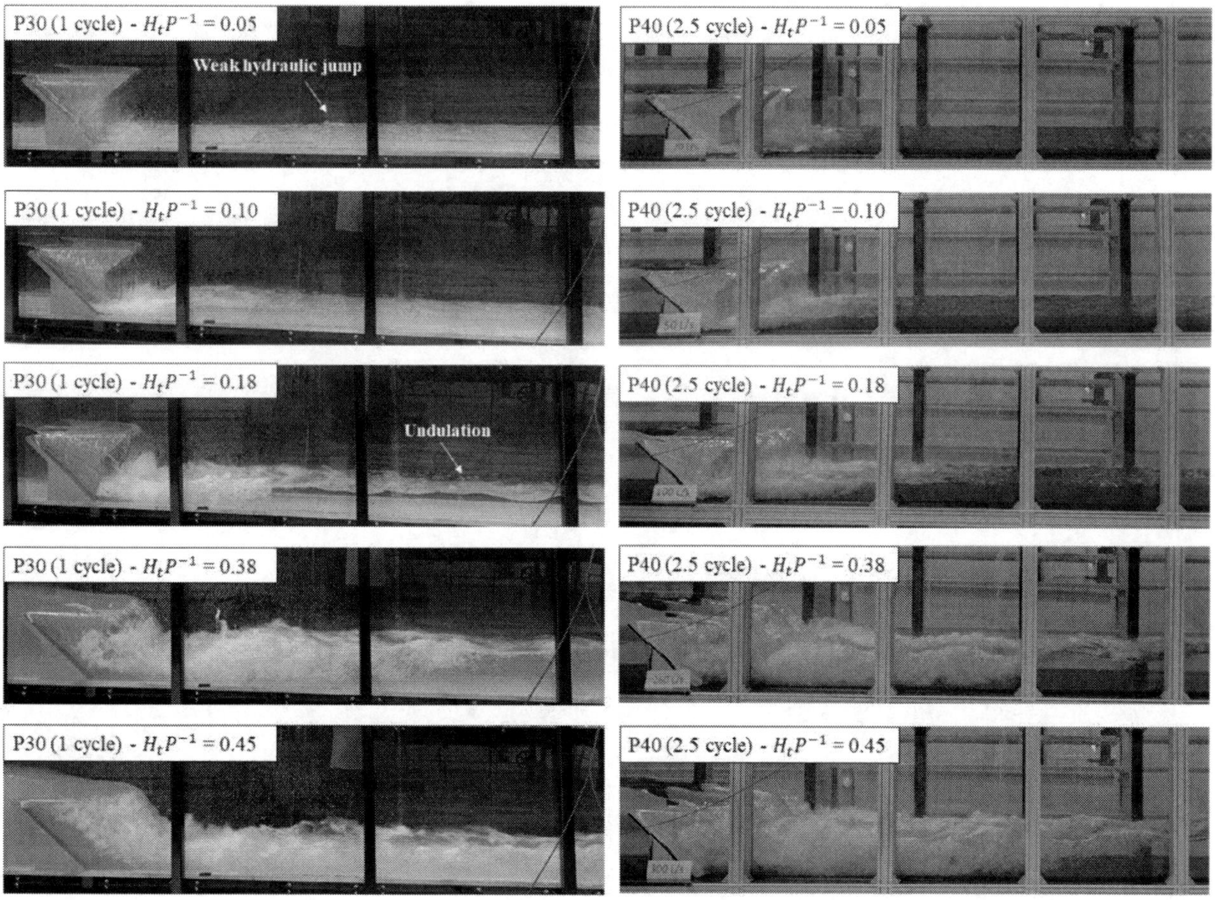

Figure 4. Visual comparison of downstream flow regimes between single-cycle and multi-cycle model, left: $P = 30$ cm weir height, right: $P = 40$ cm weir height

3.3. Energy dissipation

Figure 5 demonstrates the evaluated relative length of the aerated flow region downstream of the weir model for increasing relative upstream heads H_tP^{-1}. As Fig. 4 shows, the flow condition at weir toe was highly turbulent and aerated with surface fluctuations extending farther downstream for most of the heads. The position of flow depth measurement downstream of the weir should thus avoid the aerated flow region in order to obtain a reasonable representative flow depth for residual energy estimation. Therefore, the length of the aeration region downstream was visually determined for the 40 cm model for all investigated discharges. As shown in Fig. 5, the length of the aeration region generally increases with increasing upstream heads and tend to follow a linear trend: $y = ax+b$, with $a = 17.38$ and $b = 0.1874$. For all tested discharges, the maximum length of the aeration region was found to be approximately 8 times the weir height. Flow depth (h_1) measurements were thus performed at $20{\times}P$ downstream of the weir model to avoid additional surface undulation influences and to guarantee plausible results.

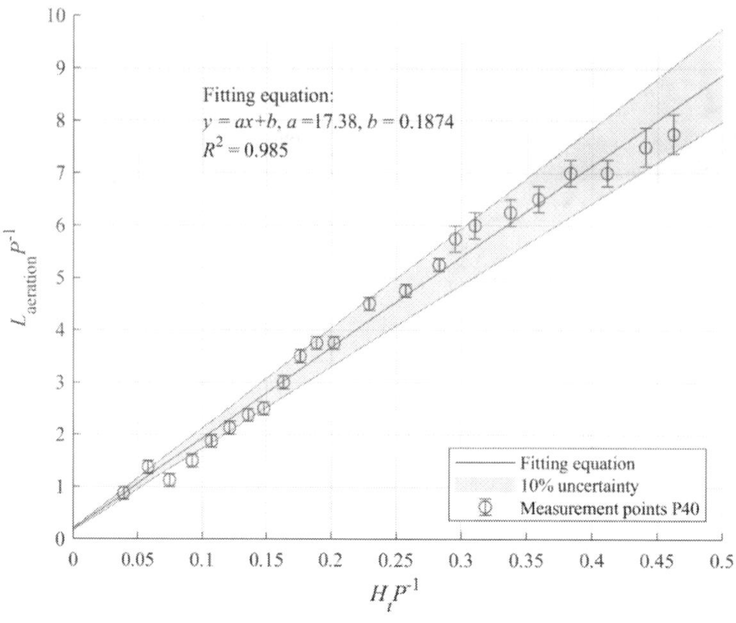

Figure 5. Relative length of aerated flow region downstream of the PKW

The downstream specific residual energy can be computed via:

$$H_1 = h_1 + \frac{v_1^2}{2g} \tag{1}$$

Where H_1 indicates the actual residual energy downstream of the weir; v_1 is the corresponding mean velocity at the same section ($v_1 = QA_1^{-1}$, $A = Wh_1$). According to Eslinger and Crookston (2020), this estimated mean velocity is representative for the actual flow velocity distribution with negligible relative error of less than 5% since the flow at this section has been calmed down into quasi two-dimensional conditions. Figure 6 compares the relative residual energy ($H_1H_0^{-1}$) downstream of both tested weir configurations, in which H_0 denotes the upstream total head (calculated via the same approach as downstream). As the results show, no appreciable deviations between the P30 and P40 model can be identified. This indicates that a 30 cm model with a single-cycle seems to be able to reproduce an identical final residual energy, although the downstream flow regimes were not quite similar as shown in Fig. 4, where the P = 40 cm model with multi-cycles was observed to have a more sufficient turbulent mixing and air entrainment. In summary, although the single-cycle 30 cm model produces deviated downstream flow regimes relative to the P = 40 cm model with multiple cycles due to scale effects and the lack of flow interactions between weir cycles, the final specific residual energy is comparable for both models as long as the measurement is performed downstream of the hydraulic jump.

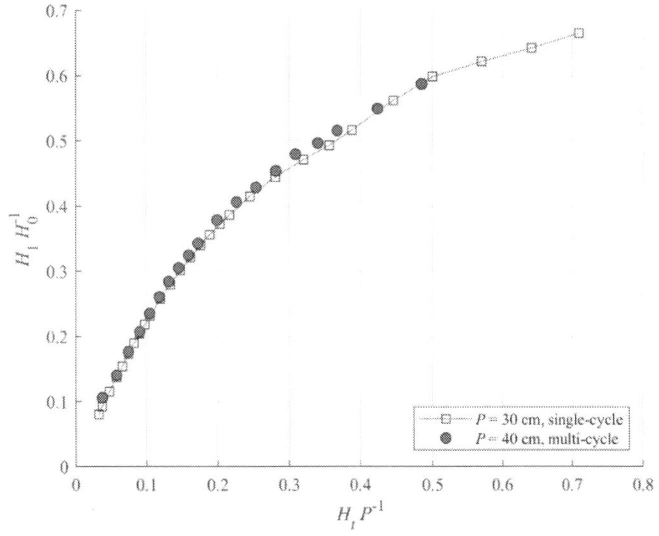

Figure 6. Residual energy downstream of the PKW

In addition, based on the principle of minimum energy heads, the minimum required energy for a rectangular channel H_{min} can be calculated in accordance with the critical flow depth h_c as follows:

$$H_{min} = 1.5h_c = 1.5\sqrt[3]{\frac{Q^2}{gW^2}} \tag{2}$$

Where H_{min} presents the minimum possible residual energy; h_c is the critical flow depth for a Froude number of $\mathsf{F} = 1$; W is the total channel width. Finally, the actual measured residual energy can be plotted against the minimum possible residual energy $H_{min} = 1.5h_c$, as shown in Fig. 7. It can be seen that the actual residual energy downstream of the PKW was always slightly larger than the minimum value, indicating a good efficiency in terms of energy dissipation. For practical applications, a linear best-fitting equation was developed to estimate the final residual energy:

$$H_1 = 1.533 \cdot h_c + 0.004492 = 1.533 \cdot \sqrt[3]{\frac{Q^2}{gW^2}} + 0.004492 \tag{3}$$

As illustrated in Fig. 7, good agreement between Eq. 3 and laboratory data with a relative error less than 5% can be determined. Nonetheless, it should be noted that the energy additionally dissipated by hydraulic jump or shock wave intersection was not included in the present study due to the complex inflow condition at the weir toe.

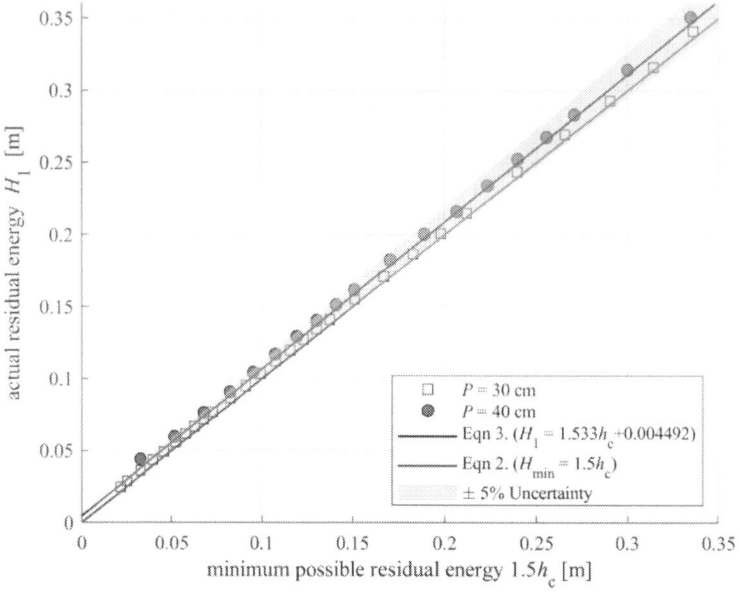

Figure 7. Comparison of final residual energy at 20×P downstream of the weir model

4. Conclusion

Within the present study, the evolution of downstream flow structures and energy dissipation were investigated by testing trapezoidal piano key weir models with different sizes and cycle numbers. In general, the flow condition downstream of PKWs for in-channel applications without any artificial tailwater control was observed to be very complex with massive turbulent mixing, air bubble entrainment and shock wave intersection. With increasing upstream heads, significant nappe splash and jet collision result in an intensive circulating flow region at the weir toe. Relative to an upscaled model (P = 40 cm) with multiple cycles, the single-cycle model (P = 30 cm) was found to have deviated downstream flow regimes due to scale effects and the lack of flow interaction between the cycles. Especially for relative upstream heads larger than 0.35 where significant scale effect diminishes, the single-cycle P = 30 cm model still tended to have a more supercritical condition near the weir toe with intensified surface fluctuation and hydraulic jump formation farther downstream. A sufficient cycle number is thus believed to be beneficial for a more stable downstream flow condition. Nonetheless, collected results show that although deviated flow regime was produced for both tested models, the single-cycle model with P = 30 cm weir height was able to deliver an identical final residual energy and can thus be considered as reasonable choice for estimation of energy dissipation as long as suitable measuring positions are ensured. Follow-up studies will be further conducted in Helmut-Schmidt-University's hydraulics laboratory with similar model geometries in other sizes to expand the measurement campaign and provide a more comprehensive comparison.

5. REFERENCES

Blanc, P., and Lempérière, F. (2001). "Labyrinth spillways have a promising future" *International Journal of Hydropower and Dams,* 8(4), 129-131.

Cicero, G. M., and Vermeulen, J., and Laugier, F. (2016). "Influence of Some Geometrical Parameters on Piano Key Weir Discharge Efficiency" *6th International Symposium on Hydraulic Structures*, Utah State University. Portland, Oregon.

Crookston, B., and Erpicum, S., and Tullis, B. P., and Laugier, F. (2019). "Hydraulics of Labyrinth and Piano Key Weirs: 100 Years of Prototype Structures, Advancements, and Future Research Needs" *Journal of Hydraulic Engineering,* 145(12), 1-7.

Erpicum, S., and Archambeau, P., and Pirotton, M., and Dewals, B. J. (2014). "Geometric parameters influence on Piano Key Weir hydraulic performances" *5th International Symposium on Hydraulic Structures*, Routh. Brisbane, Austrilia.

Erpicum, S., and Tullis, B. P., and Lodomez, M., and Archambeau, P., and Dewals B. J., and Pirotton, M. (2016). "Scale effects in physical piano key weirs models" *Journal of Hydraulic Research,* 54(6), 692-698.

Lempérière, F., and Ouamane, A. (2003). "The piano keys weir: a new cost-effective solution for spillways" *International Journal of Hydropower and Dams,* 10(5), 144-149.

Lopes, R., and Matos, J. (2011). "Flow properties and residual energy downstream of labyrinth weirs" *Labyrinth and Piano Key Weirs,* 97-104, Taylor & Francis, Liege, Belgium.

Leite Ribeiro, M., and Pfister, M., and Schleiss, A. J., and Boillat, J. (2012). "Hydraulic design of A-type Piano Key Weirs" *Journal of Hydraulic Research,* 50(4), 400-408.

Merkel, J., and Belzner, F., and Gebhardt, M., and Thorenz, C. (2018). "Energy Dissipation Downstream of Labyrinth Weirs" *7th International Symposium on Hydraulic Structures*, Utah States University. Aachen, Germany.

Oertel, M. (2018). Piano Key Weir research: state-of-the-art and future challenges, *7th IAHR International Symposium on Hydraulic Structures*, ISHS2018, Aachen, Germany.

Oertel, M. (2019). "Size-scale Effects of an A-type Piano Key Weir, *Proc. 38th IAHR World Congress*, Panama City, Panama, doi:10.3850/38WC092019-0954.

Pfister, M., and Boillat, J. L., and Schleiss, A., and Laugier, F., and Leite Ribeiro, M. (2012). "Piano Key Weirs as efficient spillway structure" *24th Congress of CIGB –ICOLD,* EPFL, Kyoto, Japan.

Shen, X., and Oertel, M. (2021). "Comparative Study of Nonsymmetrical Trapezoidal and Rectangular Piano Key Weirs with Varying Key Width Ratios" *Journal of Hydraulic Engineering,* 147(11), DOI: 10.1061/(ASCE)HY.1943-7900.0001942.

Tullis, B. P., and Crookston, B. M., and Young, N. (2020). "Scale Effects in Free-Flow Nonlinear Weir Head-Discharge" *Journal of Hydraulic Engineering,* 146(2), 1-9.

DOI: https://doi.org/10.26077/817c-13b1

Experimental and Numerical Investigation on Critical Submergence for Square Water Intakes

B. Das[1], Z. Ahmad[1] & P. K. Sharma[1]
[1]*Department of Civil Engineering, Indian Institute of Technology Roorkee, Roorkee, India*
E-mail: bdas@ce.iitr.ac.in

***Abstract**: Critical submergence for square water intakes in an open channel flow has been investigated experimentally and numerically in this paper. The depth of water at which the tail of a free surface vortex core just reaches the tip of an intake causing air entrainment is referred to as critical submergence. The formation of an air-entraining vortex in the vicinity of an intake is considered to be a severe problem for pumps. Experiments were conducted in a concrete flume of 9.47 m length, 0.5 m wide, and 0.6 m deep with an intake of size 0.04 m × 0.04 m under uniform flow for different flow conditions. A three-dimensional Multiphase CFD Model was also developed for simulating critical submergence for the intakes. SST k-ω turbulence model together with the volume of fluid (VOF) two-phase (water-air) model were used to simulate the flow field in the water intake system. In the CFD simulation study, critical submergence was identified using phase volume fraction analysis and free surface streamline analysis. From both experimental and numerical studies, it is found that the approach flow Froude number and intake flow Froude number play an important role for estimating critical submergence depth. The critical submergence increases with an increase in intake Froude number and decreases with an increase in approach Froude number for a square Intake. The outcomes of the CFD simulation were validated with experimental data.*

***Keywords:** Critical Submergence, Square Water Intakes, Open Channel Flow, Vortex Formation, CFD, ANSYS FLUENT.*

1. Introduction

Water intake is one of the most common diversion structures which are applied for the safe withdrawal of water from open channels. In some cases, the goal is to maximize the head over the intake to ensure energy requirements are minimized, but in other cases the aim is to minimize the head over the intake which can be a challenge with regards to critical submergence. Intake structures can be classified based on the source of water, their geometry, positioning, orientation, etc. and can be provided with transitions for minimizing the head losses. Vortex formation near intake structures is a serious concern and it requires a vast expanse of research and investigations. Insufficient water cover at an intake can cause vortex formation. These vortices lengthening to the intakes from the free surface can cause air entrainment which is not appropriate for the smooth and efficient functioning of intakes. Figure 1 shows a side view of free surface air-entraining vortex resulting in critical submergence (S_C) in the present experimental study.

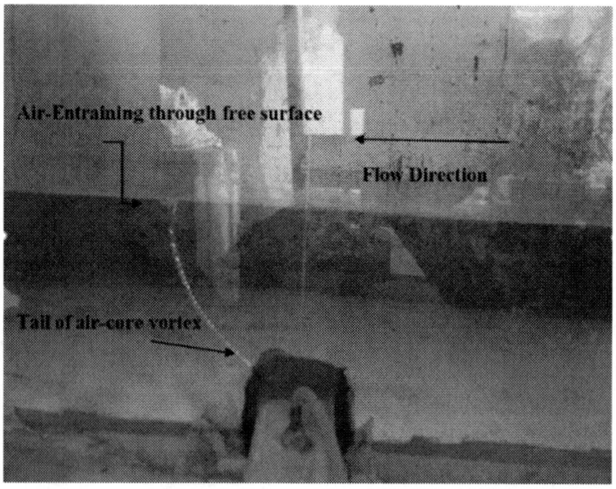

Figure 1. Lateral view of free surface air-entraining vortex resulting critical submergence at square Intake 0.04 m×0.04 m

The depth of water at which the tail of a free surface vortex core just reaches the tip of intake causing air entrainment is referred to as critical submergence for that intake (Yildirim et al., 2000). Numerous studies on S_C have been performed from the mid of 20th century and empirical equations were also developed for the calculation of S_C in which Denny (1956), Gordon (1970), Jain (1977), Anwar and Amphlett (1980), Hite and Mih (1994), Yildirim (2004), Ahmad et al. (2008), etc. are the main study among them. Most of these studies were conducted on circular intakes configuration. Sparse studies have been reported related to square water intakes for S_C. Ahmad et al. (2008) calculated S_C for hydraulic intakes in an open channel flow using potential theory by considering intakes as point sink and placed under rectilinear flow. Guo et al., (2017) pump intake air-entrainment was investigated through numerical simulation using VOF multiphase model.

Most of the studies have been simulated considering a single phase for critical submergence estimation, which is not enough for free surface interaction. Critical submergence was significantly affected by the intake direction at dual-pipe intakes. Khanarmuei et al. (2019) studied the critical submergence by changing the angle of pipe intakes from vertical to horizontal. Hashid et al. (2020) studied a side circular intake with and without bell-mouth transition. They found that the critical submergence for a horizontally placed side circular intakes decreases under bell-mouth transition. The present study focused on the experimental investigation of critical submergence for single square water intake under uniform flow conditions in an open channel. The most significant parameters have been determined which affect the S_C and are also responsible for vortex formation in the vicinity of the intake. A three-dimensional Multiphase CFD Model was also developed for simulating critical submergence for the intakes. CFD simulation outcomes were validated with experimental data.

2. Experimental Setup and Measurements

All the experiments were conducted at the Hydraulic Engineering Laboratory of IIT Roorkee. An open channel with a length of 9.47 m, a width of 0.5 m, and a height of 0.6 m were used for this study which is shown in Figure 2. Water was supplied to the main channel through an approach channel by a pipe fitted with a valve. Splitter plates and flow wave suppressors were used to straighten the flow and to dampen the surface disturbance, respectively. Main channel length upstream of the intake was sufficient to form a developed flow at the intake section. The intake was placed at a distance of 3.47 m from the downstream end. Intake pipe of size 0.04 m×0.04 m were used for this study. Intake pipe was positioned at a clearance with $C = 0$ and placed in flush with the left wall of the concrete flume. A sluice gate was provided at the end of main channel in order to maintain the desired flow depth by adjusting the gate. The intake pipe was of length 0.75m and attached to an outlet pipe which was connected to a 15HP centrifugal pump. A magnetic flow meter was installed on the outlet pipe for measuring intake discharge as shown in the Figure 2.

Discharge in the intake pipe was controlled using a valve fitted in the intake pipe. The water drawn from the intake was discharged into a sump. Similarly, water discharging from the tail gate was passed over a pre-calibrated sharp crested rectangular weir to measure the discharge. This discharge was also passed to the sump. The water from the sump was then recirculated into the experimental setup for further experimentation. Experimental study on critical submergence was carried out for different flow conditions. An initial flow was maintained in the open channel and intake system, and looking for any sign on the water surface for vortex formations. If there was no vortex formation, flow depth was decreased using the downstream sluice gate. This was continued until a free surface vortex was observed at intakes. Critical submergence is the depth of flow at which the tail of first air-core vortex reaches the intake. Approximately 1 to 2-hour time was taken until the vortex was detected. If there was no vortex observed, the submergence was decreased by lifting the sluice gate available in the downstream end. When the air entraining vortex developed, the measurements related to main channel discharge Q_m, intake discharge Q_i, critical submergence S_C, and uniform flow depth Y_m were noted. Experiments were performed for different flow condition having constant intake discharges for varying main channel discharges and constant main channel discharges with different intake discharges.

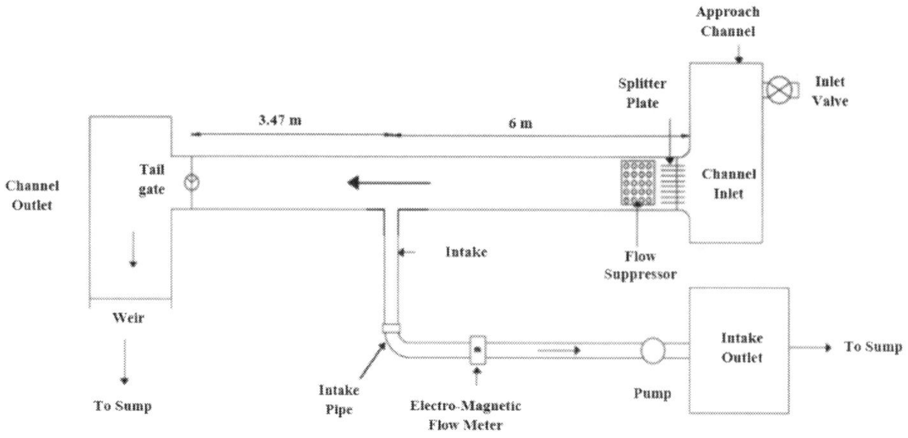

Figure 2. Schematic Diagram of Present Experimental Setup

3. Numerical Model

The computational fluid dynamics (CFD) model has been used to simulate air-water flows for critical submergence. The simulations have been done according to the logic behind the experimental setup with seeking air-core vortex to detect the critical submergence. Critical submergence has been estimated by solving a Multiphase 3D turbulence model grouped in Reynolds-Averaged Navier-Stokes (RANS) Equation. The volume of fluid (VOF) model (Hirt and Nichols 1981) is used to simulate the water-air interactions. The governing equations involved in the present 3D Multiphase VOF model study include the conservation equations of mass, momentum, and energy. Pressure based boundary conditions have been adopted for channel inlet, outlet, and intake pipe outlet. A steady-state time dependency simulation was adopted in the present CFD Simulations study for identifying the critical submergence. There were some trial simulation runs required for detecting the precise critical submergence at which the tip of the air-core vortex just reached the intake. A workstation computer having 64GB RAM and dual Intel Xeon parallel processors have been used for simulating the present problems. It was taken several hours to finish the simulation. CFD codes solve the partial differential equations in each control volume of the fluid domain with different methods. The ANSYS FLUENT program (version 2020R) was adopted in order to test its accuracy in solving present problem. The governing equations involved in the present 3D Multiphase VOF model study include the conservation equations of mass and momentum. These equations may be written as:

$$\frac{\partial \rho}{\partial t} + \nabla.(\rho u) = 0 \quad \text{(Conservation of mass)} \tag{1}$$

For incompressible fluid,

$$\frac{D\rho}{Dt} = 0 \tag{2a}$$

$$\frac{\partial(\rho u)}{\partial t} + \nabla.(\rho u^2) = -\nabla p + \nabla\tau + F_{st} \quad \text{(Conservation of momentum)} \tag{2b}$$

Also, the multiphase VOF model by Hirt and Nichols (1981) is used for solving present problem, in which volume fraction can be written as

$$\frac{\partial \rho}{\partial t}(\rho\alpha_m) + \nabla.(\rho\alpha_m u) = 0 \tag{3}$$

Here the total volume fraction (α_m) of two phases, water and air should be equal to unity.

$$\alpha_m = \alpha_w + \alpha_a = 1 \tag{4}$$

ANSYS Fluent 2020R2 uses finite volume method for solving these governing partial differential equations.

The Reynolds-Averaged Navier–Stokes (RANS) equations coupled with K-omega (k-ω) turbulence model Wilcox (2008) has been used. It is a two-equation transport model solving for kinetic energy (k) and turbulent frequency (ω). The resulting equations written as:

$$\frac{\partial\rho}{\partial t}(\rho k) + \frac{\partial}{\partial x}(\rho k u_i) = \frac{\partial}{\partial x_j}\left(\Gamma_k \frac{\partial k}{x_j}\right) - \rho u'_i u'_j \frac{\partial u_j}{x_i} - \rho\beta^* f_{\beta^*} k\omega \tag{5}$$

$$\frac{\partial\rho}{\partial t}(\rho\omega) + \frac{\partial}{\partial x_i}(\rho\omega u_i) = \frac{\partial}{\partial x_j}\left(\Gamma_\omega \frac{\partial\omega}{x_j}\right) - \alpha\frac{\omega}{k}\left(\rho u'_i u'_j \frac{\partial u_j}{x_i}\right) - \rho\beta f_\beta \omega^2 \tag{6}$$

Effective diffusivity of k,

$$\Gamma_k = \mu + \frac{\mu_t}{\sigma_k} \tag{7}$$

Effective diffusivity of ω,

$$\Gamma_\omega = \mu + \frac{\mu_t}{\sigma_\omega} \tag{8}$$

Turbulent eddy viscosity,

$$\mu_t = \frac{\rho k}{\omega} \tag{9}$$

where k is the turbulence kinetic energy, ω is the turbulence frequency, σ_k and σ_ω are the turbulent Prandtl number for k and ω , x is spatial variable, t is time, β and β^* are the model coefficients, u is flow velocity, i, j is the variable number and ρ is the fluid density.

$$\text{Turbulence production is represented by, } G_k = -\overline{\rho u'_i u'_j}\frac{\partial u_i}{\partial x_j} \tag{10}$$

Whereas production for turbulence frequency is represented in Eq.5 as,

$$G_\omega = -\alpha\frac{\omega}{k}\overline{\rho u'_i u'_j}\frac{\partial u_i}{\partial x_j} \tag{11}$$

Turbulence production can be also expresses in another form using Boussinesq's approximations as

$$G_k = -\sqrt{2}\, u_t S_{ij} \tag{12}$$

$$\text{Strain rate magnitude, } S_{ij} = \frac{1}{2}\left(\frac{\partial u_i}{\partial x_j} + \frac{\partial u_j}{\partial x_i}\right) G_\omega = -\alpha\frac{\omega}{k}\overline{\rho u'_i u'_j}\frac{\partial u_i}{\partial x_j} \tag{13}$$

Production Limiter was applied to avoid the excessive build-up of k at stagnation point as shown in the form,

$G_k = \min(G_k, C_{lim,}\varepsilon)$,

Here the coefficient C_{lim} is having a default value of 10.

$$\text{Dissipation of } k \text{ is given as, } Y_k = \rho\beta^* f_{\beta^*} k\omega \tag{14}$$

$$\text{Similarly, Dissipation of } \omega \text{ is given as, } Y_\omega = \rho\beta f_\beta \omega^2 \tag{15}$$

The model constants used in the present study are $\alpha^*_\infty = 1$, $\alpha_\infty = 0.52$, $\beta^*_\infty = 0.09$ and $\beta_i = 0.072$

3.1 Geometrical Modeling and Computational Mesh

A three-dimensional computational domain for modeling the experimental setup of square water intake models were created in the CAD software namely Ansys Space Claim. The dimensions were kept the same as in the experimental setup, only 2.25 m length of the flume was considered to reduce the calculation efforts. The cross-sectional dimension of the flume of 0.5 m x 0.5 m was created. The distance between upstream end of the domain and intake section was kept 1.5m to allow for uniform flow development. Figure 3 shows the geometry drawing of the 3D numerical model implemented for the simulation of critical submergence depth.

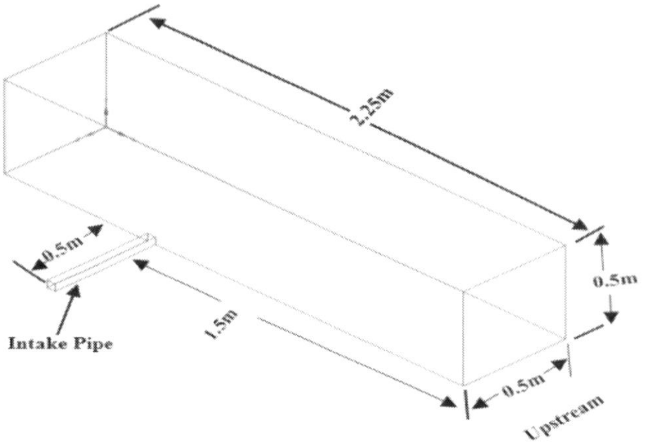

Figure 3. Model geometry of the flume used for numerical study

The meshing for computational domain were done in Ansys Fluent Meshing available on Ansys Workbench. A Poly-hexcore meshing was adopted for the discretization of the computational domain. Three different mesh sizes were adopted and the part of geometry like intake pipe and vicinity of the intakes section was finely meshed while the remaining portion of the geometry was kept element size coarser. During meshing process, the watertight CAD model of domain was imported in the meshing module, then domain was divided into polyhedral and hexahedral cells using poly-hexcore meshing technique. A mesh independence study was conducted for making computational domain mesh independence by adopting three different meshes (1489790,753586 and 349463). This study was performed as per the guidelines of ASME (Celik et al., 2008) using Grid Convergence Index (GCI) for confirming the mesh density which does not affect the discrete solution. The approach velocity (V_u *in the x-direction*) near the vicinity of the intake, was selected for mesh independence study. Thus in present study medium size mesh (753586) were used without compromising the quality and accuracy of the solution. Figure 4 shows the isometric view of meshed geometry of the computational domain and figure 5 exhibits the polyhedral and hexahedral cells.

Figure 4. Isometric view of Poly-hex core meshed geometry of computational domain

Figure 5. Poly-hex core meshed geometry with cutting plane in Y- direction showing polyhedral and hexahedral cells

The hexahedral cell core yields greatly regular cells which decreases the computational effort compared to tetrahedral counterparts. Also, there is a clear main flow direction in our numerical problem which makes implementation of hexahedral cells more efficient and desirable. Polyhedral cells are present near the boundary regions to capture the complex geometry of intake system. These cells also have more number of neighbours which allows for increased accuracy of computations.

3.2 Boundary Condition

Pressure based boundary condition have been adopted in the present simulation study. In channel inlet, the pressure inlet boundary conditions were provided with certain velocity and flow depth as a dynamic pressure and static pressure respectively. The channel outlet was provided pressure outlet boundary condition with flow depth as a static pressure. No-slip wall boundary conditions were used at the solid walls and the intake pipe wall surfaces. The upper boundary of the air phase was kept open to the atmosphere by assigning atmospheric pressure condition by giving pressure inlet boundary condition. In intake outlet, pressure outlet boundary condition was provided by putting appropriate withdrawal pressure. The illustration of boundary condition is shown in figure 6 as follows.

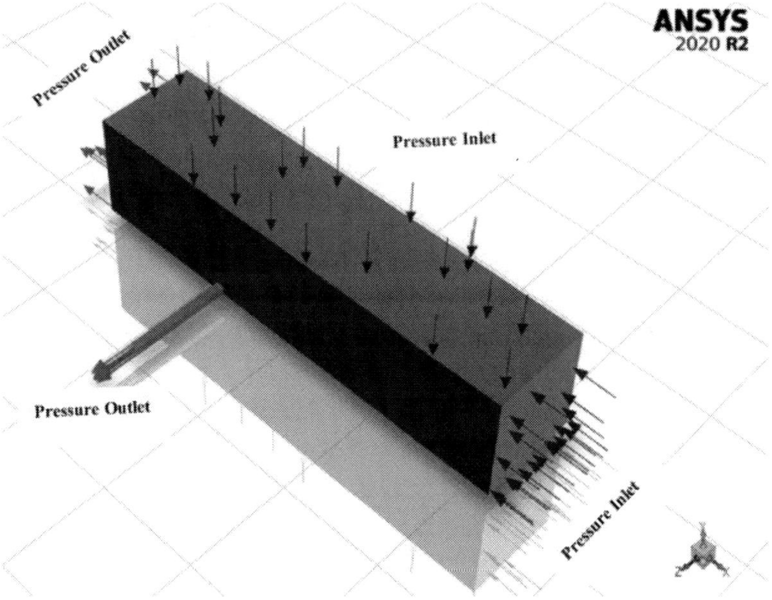

Figure 6. Illustration of the boundary conditions of computational Domain.

4. Result and Discussion

For the simulation of air-water interaction, the VOF Model helped to study the vortex formation at critical submergence at square water intakes. Steady state time dependency simulation was performed to identify air-entrainment in flow at critical submergence depth. In the CFD simulation study, critical submergence was identified using phase volume fraction analysis and free surface streamline analysis. Air-entrainment clearly identified in a VOF Model multiphase simulation (Guo et al., 2017). The air and water phase having different densities thus results to easily visualize air-water interaction at critical submergence condition during vortex formation. In numerical study the critical submergence is considered, when the air phase starts to enters the intake pipe. Computation of critical submergence in CFD study is controlled by intake outlet withdrawal pressure and flow depth in the domain. These are the two key parameters which helps to observe vortex formation for critical submergence. Thus physics in the numerical model is provided in such a way that experimental results may validate the numerical model. In present CFD simulation study, it was observed that an iso-surface of 6% Air Volume Fraction (AVF) gives the similar structure of vortex formation as visualize in experimental program. Thus iso-surface of 6% AVF consider the threshold value for vortex formation for evaluating critical submergence. Similarly, spiral patterns of free surface streamline entraining towards the intakes is also considered the case of air entrainment for critical submergence determination. The *SST k-ω* turbulence model were used for solving the present problem.

There are many ways in present numerical study to identify vortex formation for S_C like- free surface streamlines, free surface velocity vectors, iso-surface of AVF, contour of air volume fractions and volume rendered images etc. All these methods were adopted in present CFD study for determining critical submergence. Figures 7a. and 7b. show the iso-surface of AVF and Contour of AVF in vertical plane (z = 10 mm) of vortex formation at critical submergence. Figure 8. shows free surface streamlines patterns resulting free surface vortex at S_C, with approach flow Froude number (Fr) = 0.4 and Intake flow Froude number (Fri)= 4.15. These two parameters are greatly responsible for vortex formation and critical submergence depth findings. Similarly, Figures 9a. and 9b. are showing volume rendered image of AVF and free surface velocity vectors, respectively resulting free surface vortex at S_C, with Fr = 0.243 and Fri = 4.55.

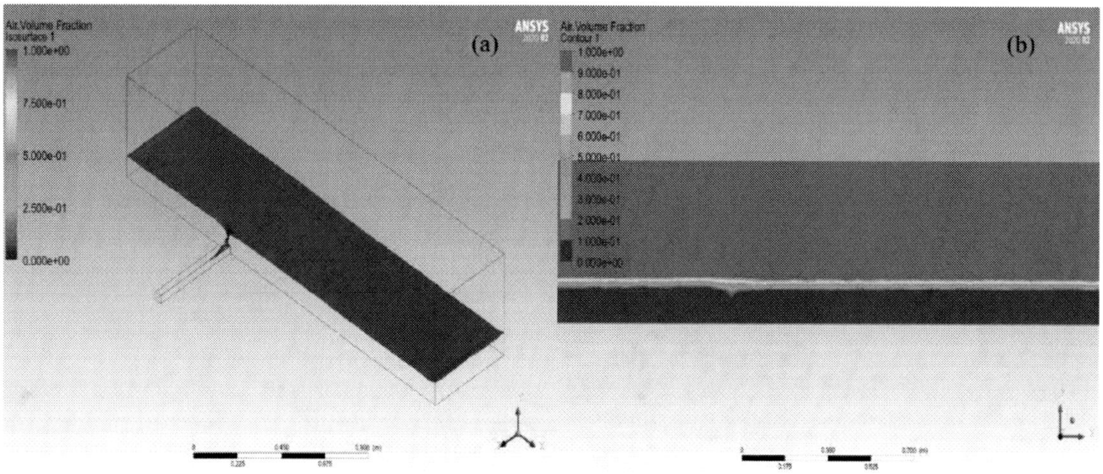

S_C = 0.105 m $\quad$ Fr = 0.4 $\quad$ Fri = 4.15

Figure 7. Computational domain displaying (a) Isosurface of free surface vortex core at S_c (b) Contour of air-water phase fraction along Vertical plane (z = 10mm) of flow direction resulting S_C

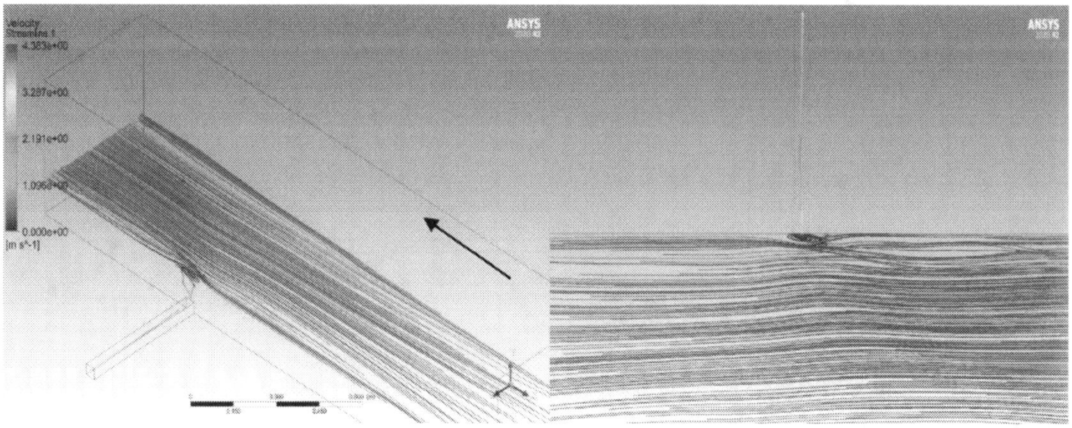

Fr = 0.4 *Fri* = 4.15

Figure 8. Free surface streamlines at S_C showing free surface vortex at two different planes

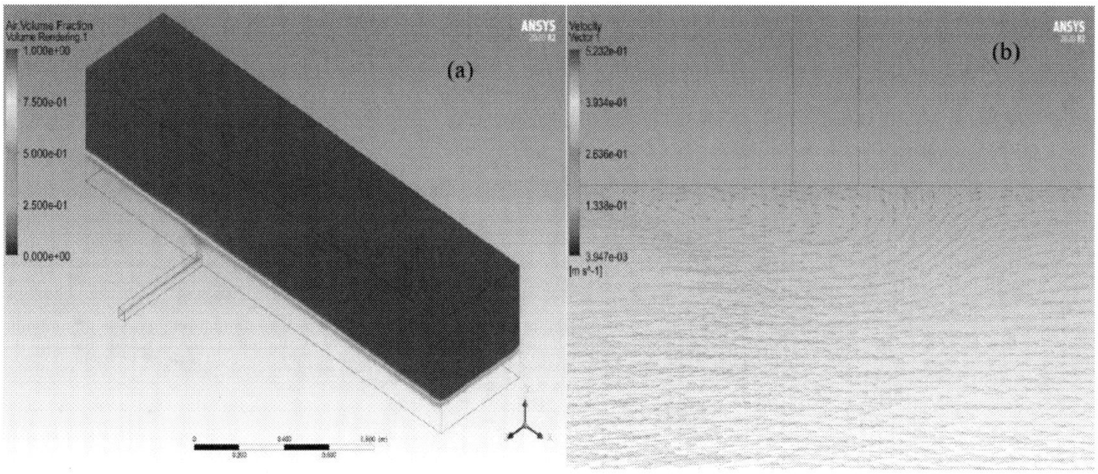

$S_C = 0.125$ $Fr = 0.243$ $Fri = 4.55$

Figure 9. Computational domain showing (a) Volume rendered image of AVF at S_C (b) Velocity vectors displays free surface vortex at S_C

From the analyses of results, it was observed that both experimental and CFD studies are able to find critical submergence under uniform approach flow conditions in open channel flow. Fr and Fri play a great role to estimate critical submergence in CFD study and experimental consideration. The critical submergence increases with an increase in intake Froude number and decreases with an increased approach Froude number for square Intake. Also, CFD simulation outcomes were validated with experimental data with an average error of ±7%. This much error is acceptable as vortex formation is visualized by means of eye judgement in the experimental setup for calculating critical submergence. Figures 10 and 11 shows the validation of experimental results with CFD study.

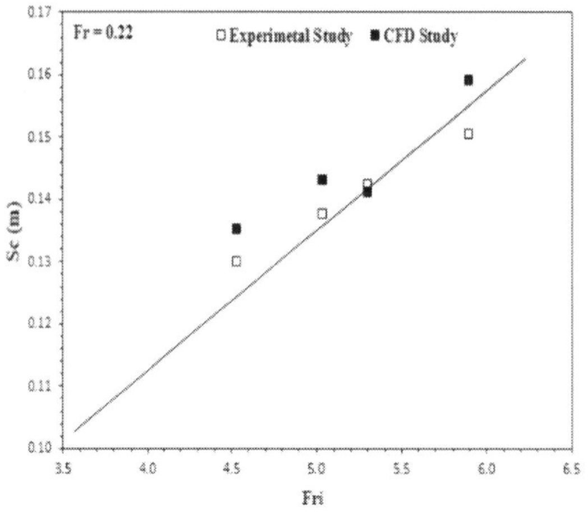

Figure 10. Comparison of experimental and CFD results of critical submergence (Sc) with intake Froude number (Fri)

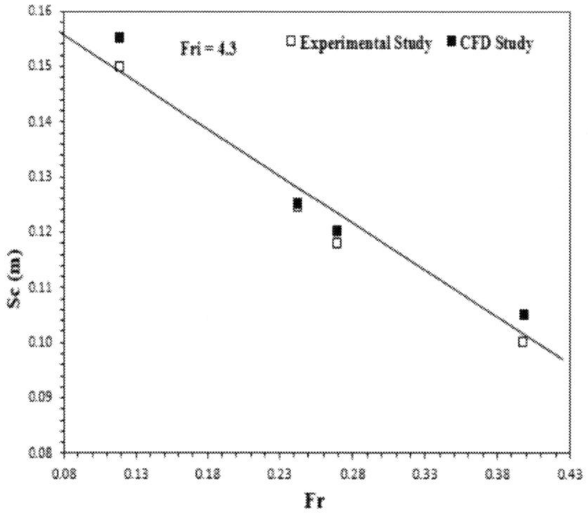

Figure 11. Comparison of experimental and CFD results of critical submergence (Sc) with with approach flow Froude number (Fr)

5. Conclusion

In this present study, critical submergence for square water intakes in an open channel flow has been investigated experimentally and numerically using the commercially available computational fluid dynamics (CFD) software ANSYS FLUENT, with the k-ω based Shear-Stress Transport (SST) model. In the CFD simulation study, critical submergence was identified using phase volume fraction analysis and free surface streamline analysis. The vortex formation in the experimental approach which is visualized by eye judgement on basis of that critical submergence is calculated. From both experimental and numerical studies, it is found that the approach flow Froude number and intake flow Froude number play an important role for estimating critical submergence depth. The critical submergence increases with an increase in intake Froude number and decreases with an increased approach Froude number for square Intake. Overall, the numerical model was found to capture reasonably well critical submergence depth. Also, the modelling technique was validated with experimental data with good agreements. Future work could be done with other turbulence models to see what benefits it could bring compared to computational costs.

6. References

Ahmad, Z et al. (2008). "Critical submergence for horizontal intakes in open channel flows." *Dam Engineering.*, 19, 71–90.

Ahmad, Z. and Hashid, M. (2020). "Experimental and numerical study of critical submergence for dual intakes." In *Proceedings of 8th IAHR International Symposium on Hydraulic Structures.*, Santiago, Chile

Anwar, H. O., and Amphlett, M. B. (1980). "Vortices at vertically inverted intake." J. Hydraul. Res.,18, 123–34.

Celik, I. B et al. (2008). "Procedure for estimation and reporting of uncertainty due to discretization in CFD applications," Transactions of the ASME. *J Fluids Eng.*, 130 / 078001, 1-4.

Denny, D. F. (1956). "An experiment study of air-entraining vortices at pump sumps." *Proc Institution of Mechanical Engineers.*, 170(2), 106-116.

Gordon, J. L. (1970). "Vortices at intake structures." *Water Power.*, 4, 137-138.

Guo, Z. W et al. (2017). "Three-dimensional simulation of air entrainment in a sump pump." *J. Hydraul. Eng.*, 143, 1–9.

Hashid, M et al. (2015) "Discharge characteristics of lateral circular intakes in open channel flow." *Flow Meas Instrum.*, *46*, pp.87-92.

Hashid, M et al. (2021) "Critical submergence for side circular intake in an open channel flow." J. Hydraul. Res., *59*(1), pp.136-147.

Hirt, C. W., and Nichols, B. D. (1981). "Volume of fluid (VOF) method for the dynamics of free boundaries."

J. Comput. Phys., 39, 201–225.

Hite, E. J., and Mih, C. W. (1994). "Velocity of air-core vortices at hydraulic intakes." *J. Hydraul. Eng.*, 120, 284–297.

Jain, A. K. (1977). "Vortex formation at vertical pipe intakes." Ph.D. Thesis presented to the *University of Roorkee.,* Roorkee, India.

Khanarmuei, M et al. (2019) "Effect of dual intake direction on critical submergence and vortex strength." J. Hydraul. Res., 57(2), 272-279.

Yildirim, N. (2004). "Critical submergence for a rectangular intake." *J Eng Mech.*, 130(1), 1195–1210.

Yildirim, N et al. (2000). "Flow-boundary effects on critical submergence of intake." *J. Hydraul. Eng.*, 126, 288–297.

DOI: https://doi.org/10.26077/4151-67ed

Feasibility Assessment of the Application of Delft3D Model to Simulate Flow over a Spillway in a Dam: Study Case of Hampaturi Dam in Bolivia

F.D. Oliveira[1] & M.W. Heredia[1]
[1]*Hydraulics Laboratory, Universidad Mayor de San Simón, Cochabamba, Bolivia*
E-mail: marceloheredia.g@fcyt.umss.edu.bo

Abstract: *Numerical modelling is an attractive alternative over physical modelling. However, sometimes both kind of techniques are coupled. This is the case of Hampaturi Dam in La Paz, Bolivia. A physical model was built for the optimization of the spillway. Hence, a numerical model was implemented with Delft3D computational model. Delft3D is a model based on two-dimensional flow equations (hydrostatic equations), and full three-dimensional equations (hydrostatic equations). Both modes were tested for the flow over the spillway, which includes a steep discharge channel that poses an additional difficulty. The goal of this research is to test the model (in both modes) on these flow conditions in order to determine the feasibility in applying Delft3D on this kind of structure. The model was calibrated and validated for both modes, hydrostatic and non-hydrostatic, by the variation of Manning's roughness coefficient and eddy viscosity, reproducing the observed flow velocities and water depths. Following, a sensitivity analysis was carried out to determine the influence of different parameters in both modes, like: number of vertical layers, bed roughness and eddy viscosity. This analysis was executed by using an independent mesh solution. Results show a significant influence of the tested parameters. Then, the influence of the channel steepness on the performance of the model was tested. Results show that non-hydrostatic mode produces smooth flow profiles. Finally, it was observed that both modes can operate up to inclination angles of 60° for the discharge channel, setting an operation limit for Delft3D in this kind of structure.*

Keywords: *Spillway dam, Delft3D model, three-dimensional model, hydrostatic, non-hydrostatic, sensitivity analysis.*

1. Introduction

In this paper are presented the results of the simulation of the flow over a spillway in dams, and the aim of this research is to show the benefits and opportunities for using the Delft 3D model to simulate flow on this kind of structure.

The comprehension of fluids behavior in structures, such as spillways, can be improved by the application of physical models. Although there are limitations with this kind of modelling, it is possible to use complementary tools, such as numerical models, which allow the setting and evaluation of different parameters according to the needs of the project.

The case study of this research was the physical modelling of the "La Paz – Hampaturi Alto" dam, carried out by the Hydraulics Laboratory of the Universidad Mayor de San Simón in Bolivia (LHUMSS), with the purpose of endorsing the performance of the Delft 3D computational model, by developing simulations of the flow over the spillway and the steep discharge channel, in two and three dimensions, for hydrostatic and non-hydrostatic modes, in order to be able to establish the feasibility of the application for the simulation of this kind of flow, the advantages in the study of the design, and the easiness for the optimization of this type of hydraulic structure.

The evaluation of the application of this numerical model involved the examination of its performance, from the calibration of the model to the sensitivity analysis and the variation of the structure geometry, for the hydrostatic and non- hydrostatic modes, through fit coefficients, such as Nash Sutcliffe, the correlation coefficient R^2, and the RMSE coefficient. The calibration process was executed to obtain parameters, which provided representative and first-rate results. Afterwards, different values were assigned to physical parameters, in order to quantify the influence of each on the results. Subsequently, with the same goal, the structure geometry was modified; the angle of inclination of the discharge channel was increased, as were the dimensions of the basin's shock absorbers.

In this paper, first the background and the case study were introduced, then the numerical model is described and finally, the modeling results are discussed.

2. Project Background

2.1. Assessment of the Spillway Performance of Hampaturi Dam

Due to water scarcity in La Paz, Bolivia, the Ministry of Environment and Water proposed the construction of new dams to store and supply drinking water like the Hampaturi Dam with a height of 32.5 m. The local water supply company commissioned the design of the dam to the INYPSA-CGA Engineering Company. The design task included the assessment of the spillway performance through the application of physical modelling, which was executed by the Hydraulics Laboratory of Universidad Mayor de San Simón in Bolivia.

2.2. Physical Model of Hampaturi Dam

In 2016, the Hydraulics Laboratory of the Universidad Mayor de San Simón presented the results and details of the simulations in the "Final Report of the Physical Modeling La Paz - Hampaturi Alto". The whole physical model includes a feeding and recirculation system of water, an upstream flow calming reservoir, the dam spillway, the front apron, the three buffer pits and the restitution channel to the river. Its length and width were of 20 m and 3 m, respectively. The length scale defined for this modeling was 1:20. The model was a fixed bed, not distorted and without the inclusion of sediments, with a concrete and uniform sand surface roughness, with an adopted scaled Manning value of 0.010 $m^{-1/3}s$ for the physical model (see Figure 1).

Figure 1. General view and layout of the physical model built in LHUMSS

Two series of tests were carried out in the model. The first test aimed to measure water depths at the spillway crest and the second test aimed to measure water depths and flow velocities at the dissipation structures in order to assess their performance. Six discharges were applied to run the tests (see Table 1). All the analysis is based on the runs executed using a discharge of 20 L/s, which is the most relevant for physical modelling study purposes, because it was the design discharge of the prototype for a decamillenary return period.

Table 1. Series of discharges applied to the physical model (LHUMSS, 2016)

Prototype (m^3/s)	20	25	30	35.3	40	45
Physical model (L/s)	11	14	17	20	22	25

The Manning roughness value adopted by LHUMSS for the physical modeling was 0.010 $m^{-1/3}s$ based on an inverse calculation from the Bresse equation for open channel flow. Then, control points were defined throughout the model for the measurements of water depth and flow velocity, which are depicted in Figure 2.

It was important to validate the results by using a complementary numerical method to the physical model due to the difficulties that arise when taking measures of some parameters such as the structure geometry and the lack of instruments of precision, among others. The CFD-type models were the first option as a complementary tool. However, they require a great deal of computational effort, so it was imperative to look for an alternative, such as Delft3D model.

As no studies or analyses were found that validate the use of the Delft3D program for its use in modeling this type of structure and flow, an opportunity was seen to test this tool.

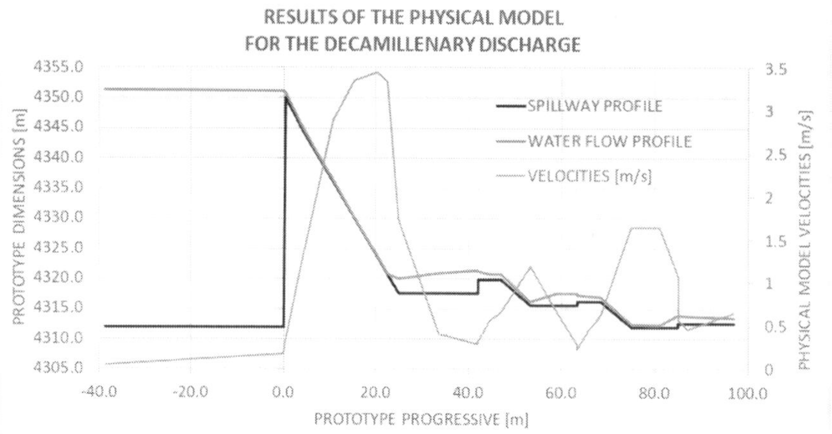

Figure 2. Measured profiles for water depths and flow velocities in the physical model

3. Numerical Modelling

The computational model Delft3D is designed to simulate hydrodynamic and morphological processes, mostly in coastal areas, rivers, and estuaries by using the hydrodynamic and sediment transport equations. The pre and post processing of the simulation is prepared through specific modules, such as Delft-RGFGRID, Delft-QUICKIN, Delft-FLOW and QUICKPLOT, that allow the simulation to be managed according to the requirements of the project, and define a suitable time step, a simulation time, control points, control sections and others.

3.1. Calibration of the numerical model

The physical model provided the data, so the numerical simulation could be properly calibrated in terms of the flow levels and velocities. It was necessary to increase the initial entrance conduit to the spillway and the restitution channel of the physical modeling length to avoid instabilities and the formation of backwaters due to the changes in slope in short sections (see Figure 3).

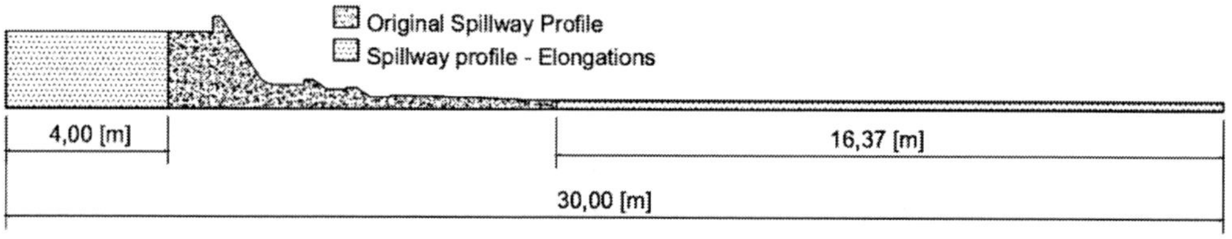

Figure 3. Extensions in the profile of the original spillway

The model was calibrated in 2 dimensions and 3 dimensions (one and three vertical layers, respectively), using the parameters of Table 2:

Table 2. Parameters used in the calibration

Parameter	Values		
Discharge	20 L/s		
Manning Roughness Coefficient	0.010 $m^{-1/3}s$	0.012 $m^{-1/3}s$	**0.017** $m^{-1/3}s$
Eddy viscosity	0.1 $m^{-1/3}s$	0.01 $m^{-1/3}s$	**0.001** $m^{-1/3}s$

The bold values from the Table 2, displayed better results.

3.2. Numerical simulation scenarios

To validate the use of the numerical model Delft3D, several scenarios were carried out, utilizing different values for certain parameters to evaluate the performance of the software in response to the variation of those values.

The evaluation of the results obtained from the numerical simulations made with the Delft3D model were compared with ones from the physical model, making use of the Nash-Sutcliffe coefficient of efficiency and other methods, to evaluate the impact of the sizes of the cells used to build the numerical meshes, the time step, the number of layers for a vertical analysis, and the constant related to the fluid state, such as viscosity and the channel roughness.

The precision of the bathymetry is given by the size of the elements chosen in the Delft3D-QUICKIN module, according to the purpose of each simulation. At each vertex of the grid, a depth value relative to a reference level is assigned, and the surface of the desired structure is formed through the interpolation of the loaded information. For this case study, it was decided to evaluate different cell values in the longitudinal direction (M), keeping the cells constant in the transverse direction (N), in order to also analyze the quality of the results obtained (Ochoa, 2014).

The analysis was performed in 2D and 3D using a discharge of 20 L/s and the most optimal parameters obtained from the model calibration, where the eddy viscosity value is $0.001 m^2/s$ and the Manning roughness value is $0.017\ m^{-1/3}/s$. Thus, hydrostatic mode simulations were carried out with 7 different sizes of elements or cells in the longitudinal direction and with a fixed size in the transverse direction of Δy=0.25 m. Table 3 lists the different grid sizes and mesh resolutions used in the sensitivity analysis (LHUMSS, 2016):

Table 3. Grid sizes and mesh resolutions used in the sensitivity analysis

Δx (m)	0.25	0.20	0.15	0.10	0.05	0.03	0.02
Mesh resolution M×N	120×3	150×3	200×3	300×3	600×3	1000×3	1500×3

The Non-Hydrostatics simulations were carried out in 3D using 2, 3 and 4 layers, which are equivalent to the number of cells in the vertical direction. The following table shows the different sizes for the elements of the grid.

3.3. Timestep and simulation time selection

The stability of the simulations depends mainly on the accurate selection of the time step. According to Palacio (2002), the quality of the results could be higher by using a smaller time step value, making this factor highly influential in the simulations. Determining an adequate coefficient was imperative for obtaining good results, or more to the contrary, its value could modify the results of the model and make it unstable.

The default time step proposed by the Delft3D-FLOW module is 1 minute. Having a relatively wide range to choose the appropriate time step, it was convenient to carry out simulations with different values and then analyze the results and their stability.

On the other hand, the simulation time refers to the period of time that is sought to be represented and analyzed. This factor can also affect the stability of the simulations, since they require a "spin-up time", which consists of solving the Continuity equation and Navier – Stokes equation, until finding constant or homogeneous values.

3.4. Physical parameters effect

The physical parameters analyzed throughout the simulations were the roughness and the eddy viscosity because they were challenging to recreate with total accuracy during the physical simulations.

The chosen roughness values, corresponding to the prototype and scaled roughness Manning values of the concrete, $0.010\ m^{-1/3}s$ and $0.017\ m^{-1/3}s$ respectively, were evaluated in the simulations to determine its influence on the results after being used under the same conditions.

The eddy viscosity is related to the flow speed, and may directly affect the results. The values used in the simulations were determined by trial and error, since there was no reference value for the physical model. For the non-hydrostatic simulations, horizontal and vertical values of the eddy viscosity were needed and determined using the same method, and followed the criteria of the Delft3D manual which established that the horizontal value must be greater than the vertical value.

3.5. Geometry and stability

Based on the calibrated model of the Hampaturi dam and the necessary changes made in the structure to avoid instabilities, waves, and other phenomena that may alter the behavior of the flow, some other structural characteristics were modified as part of the research in order to analyze the output data of the flow depths and velocities, and to test the limitations of the numerical model stability with these small changes.

In the physical model, the angle of the discharge channel that connects the spillway and the buffer pits was 55° with respect to the horizontal plane. This angle was increased to 60°, 70° and 80°, so as to observe the impact on the stability of the results with this slope increase.

Subsequently, the basin's shock absorbers were lengthened to twice their original length in order to evaluate the impact that each of these could have and how it would influence the behavior of the flow over the structure.

4. Discussion of Results

4.1. Calibration of the model

As a first step in the assessment of the computational model, a calibration process was carried out. Delft3D model was calibrated based on data and measurements from the physical model, since it was possible to find values for certain parameters such as the efficiency coefficients obtained for the simulations and both of the flow levels and velocities. These allowed the calibration of these results to be validated. The model was calibrated using a discharge of 20 L/s, in 2D and 3D (one and three vertical layers, respectively). Three different Manning coefficient values were used: 0.010 $m^{-1/3}/s$, 0.012 $m^{-1/3}s$ and 0.017 $m^{-1/3}s$. Also, three different eddy viscosities were used: 0.1 m^2/s, 0.01 m^2/s and 0.001 m^2/s. Furthermore, the best results obtained for a Manning coefficient was 0.017 $m^{-1/3}s$ and for an Eddy viscosity was 0.001 m^2/s.

The calibration results were assessed using a goodness-of-fit analysis. Regarding the water levels in 2D mode, the Nash-Sutcliffe Efficiency reaches values above 93% of efficiency and correlation coefficients above 95%. Regarding the flow velocities in 2D mode, the Nash-Sutcliffe Efficiency presents negative values close to zero, correlation coefficients reach values up to 35% and RMSE values up to 1.4. Nevertheless, in 3D mode the goodness-of-fit for water levels improves, showing a Nash-Sutcliffe efficiency improvement of 3% compared to 2D mode. Similarly, the goodness-of-fit for flow velocities improves, with a Nash-Sutcliffe efficiency of about 65%.

The results from the simulations for the calibration process and those from the physical model, are presented in the next figures:

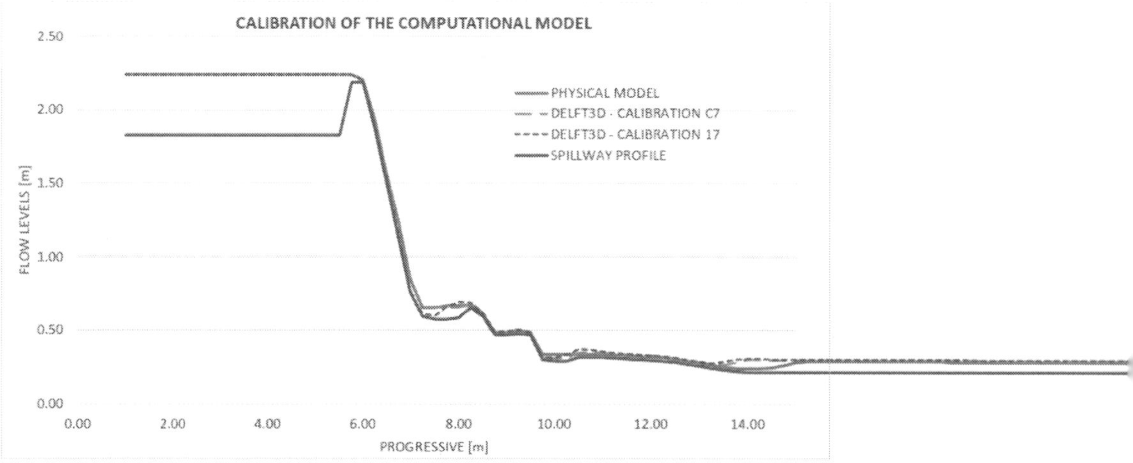

Figure 4. Comparison of flow levels - Physical model vs. Calibration C7 ($v=0.001[m^2/s]$; $n=0.017\ m^{-1/3}s$) vs. Calibration C17 ($v=0.01[m^2/s]$; $n=0.017\ m^{-1/3}s$) – Profile view

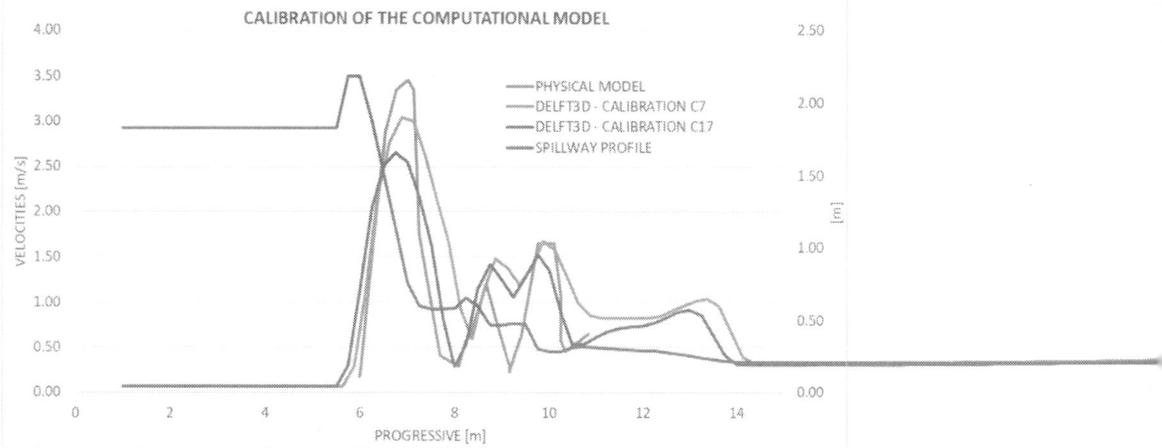

Figure 5. Comparison of flow velocities - Physical model vs. C7 (v=0.001[m^2/s]; n=0.017 $m^{-1/3}s$) vs. C17 (v=0.01[m^2/s]; n=0.017 $m^{-1/3}s$)

4.2. Sensitivity analysis for hydrostatic and non-mode

The assessment of the Delft3D computational model involves the execution of a sensitivity analysis in order to determine the influence of several model parameters on the model's performance and results. All of this analysis can only be carried out in a numerical mesh-independent manner, which involves obtaining results that are independent of the grid size. Therefore, the first step for the sensitivity analysis is the generation of results that are independent of the grid size through the consecutive refinement of the numerical mesh.

4.2.1. Analysis of grid size influence

The simulation results were assessed by the Nash-Sutcliffe Efficiency Coefficient. Figure 6 and Figure 7 depict the evolution of the Coefficient for the mesh resolutions previously defined, in hydrostatic and non-hydrostatic modes respectively. The figures below show an improvement of the model results when refined meshes are used.

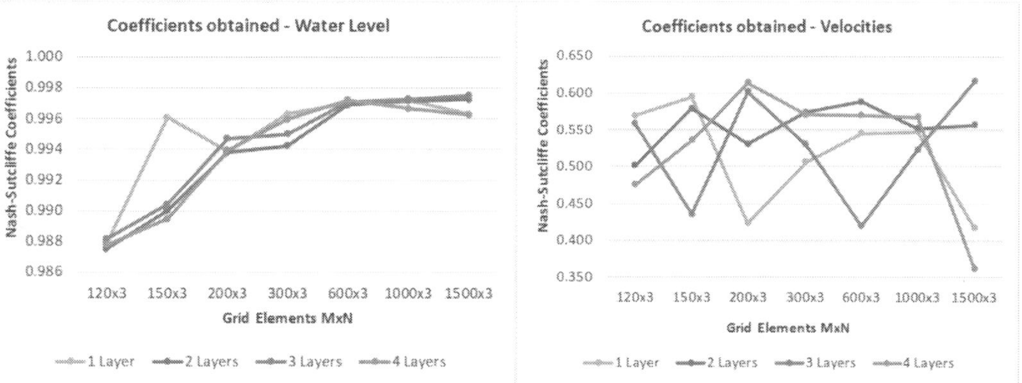

Figure 6. Nash-Sutcliffe Efficiency coefficients obtained for water levels and flow velocities in hydrostatic mode

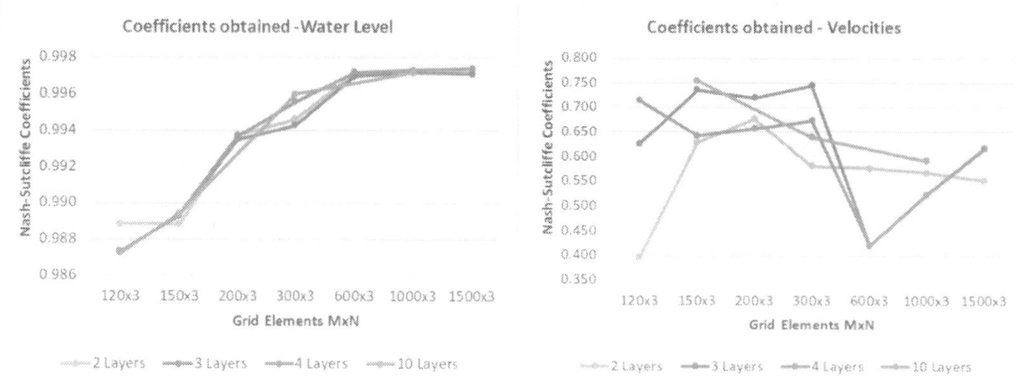

Figure 7. Nash-Sutcliffe Efficiency coefficients obtained for water levels and flow velocities in non-hydrostatic mode

4.2.2. Analysis of bed roughness influence

The goal for this analysis was to assess the influence of the bed roughness on the model results. Hence, two values were prescribed: 0.010 $m^{-1/3}s$ and 0.017 $m^{-1/3}s$, corresponding to Manning values for concrete surfaces, where the roughness coefficient of 0.017 $m^{-1/3}s$ presents more "realistic" results by the reproduction of a hydraulic jump observed in the physical model. Figure 7 and Figure 8 depict the efficiency coefficients calculated for the model result, reaching an efficiency of 99.8%.

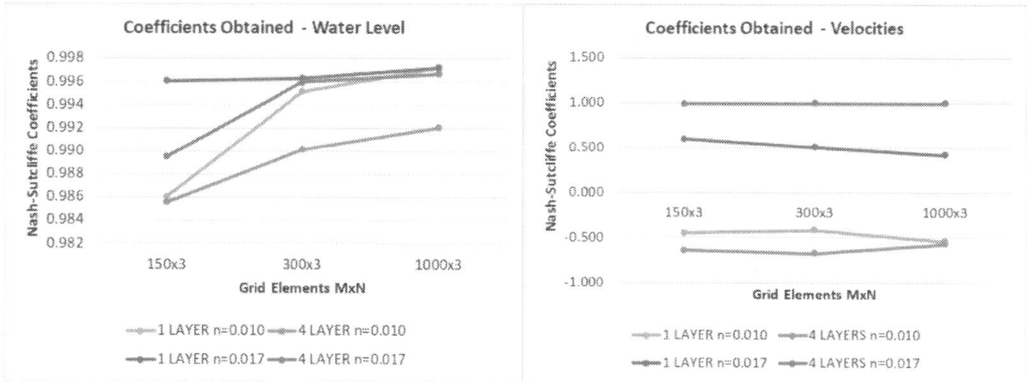

Figure 8. Nash-Sutcliffe Efficiency coefficients obtained for water levels and flow velocities in hydrostatic mode with roughness coefficients of n=0.01 $m^{-1/3}s$ and n=0.017 $m^{-1/3}s$

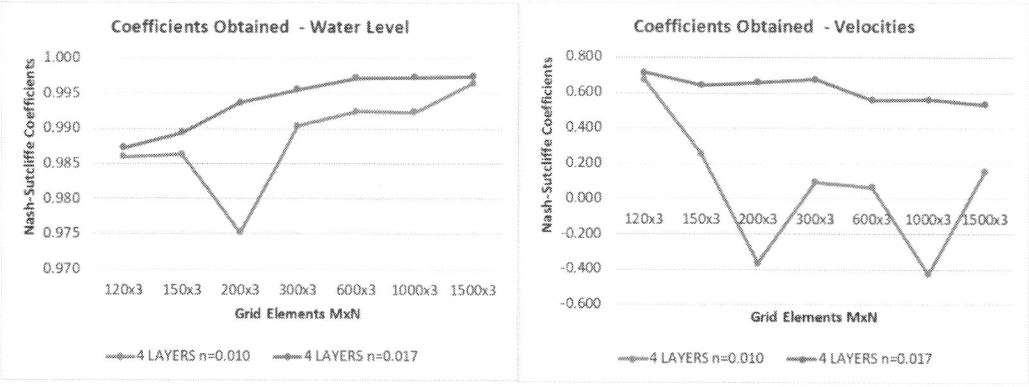

Figure 9. Nash-Sutcliffe Efficiency coefficients obtained for water levels and flow velocities in non-hydrostatic mode with roughness coefficients of n=0.01 $m^{-1/3}s$ and n=0.017 $m^{-1/3}s$

4.2.3. Analysis of eddy viscosity coefficient influence

Two viscosities were used, a horizontal one for 2D mode and a vertical one for 3D mode. In order to analyze the impact of the viscosity values, the results of the Nash-Sutcliffe coefficients for 3D simulations are presented below. From these results, it was observed that by bringing the vertical viscosity closer to the value chosen for the horizontal viscosity, better results are obtained; however, it is important to take into consideration the condition that these two values must satisfy, Vh>Vv.

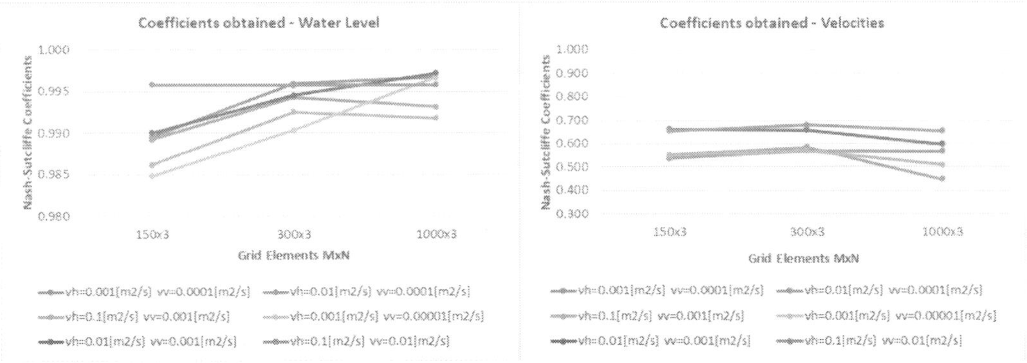

Figure 10. Nash-Sutcliffe Efficiency coefficients obtained for the analysis of the eddy viscosity influence in hydrostatic mode

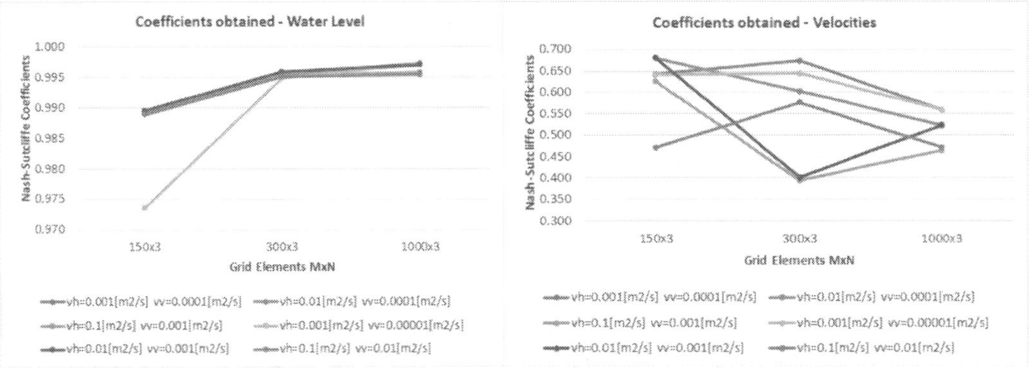

Figure 11. Nash-Sutcliffe Efficiency coefficients obtained for the analysis of the eddy viscosity influence in non-hydrostatic mode

4.2.4. Analysis of the geometry influence

The last step in the assessment work was the analysis of the influence of the geometry on the model results and performance. The slope of the spillway discharge channel was modified, where three slopes were tested: 60°, 70° and 80° using the calibrated roughness and viscosity. The slopes were implemented on three mesh resolutions of 300, 600 and 1000 grid cells in the longitudinal direction. The dissipation structures were extended to: 2 m, 1 m, and 1 m. respectively.

The results obtained from these simulations will be compared with those of a CFD-type simulation at a later stage, to analyze the flow behavior when the channel slope is increased up to 70°, since the water levels decrease up to very low levels, although the model was calibrated and the corresponding parameters were adjusted, such as the numerical parameter known as "threshold depth", the results are inaccurate leading to numerical instabilities. Flow velocities also present some stability problems when the slope is increased, especially for slopes greater than 60°.

4.2.5. Comparison of results

When applying hydrostatic and non-hydrostatic modes, the formation of hydraulic jumps in the basins, located at the end of the discharge channel when a change in slope occurs, is observed. The main difference observed in the profiles of water levels refers to how they change, from profiles without many vertices to profiles with straight lines, especially when jump phenomena appear, showing an unnatural appearance. It can be concluded that the non-hydrostatic mode does not provide significantly better results in terms of velocities. However, when simulating in hydrostatic mode, good quality results can be obtained.

5. Conclusions

The study had the purpose of validating the feasibility of using Delft3D model to simulate flow in a spillway and the energy dissipation structures that are implemented downstream of a dam. This analysis was based on the physical model of the "La Paz" – Hampaturi Alto dam, performed by LHUMSS. The model was calibrated based on the information provided by LHUMSS (2016). From the calibration, it was observed that time step, viscosity and roughness are important and decisive factors in the simulations.

2D and 3D mode were tested. For 3D simulations, a maximum of 4 vertical layers were used, which are equivalent to the number of cells in the vertical direction. It was observed that flow levels improve by 5% and flow rates by 10% in almost all the simulations from the use of the second layer.

During the sensitivity analysis, a study of the impact due to the modification of the geometry was carried out, as well as the effect that the variation of the slope of the discharge channel and the dimensions of the dissipators could have. With this evaluation, it is concluded that it is not advisable to carry out simulations that present angles higher than 60° in their bathymetry, since numerical instabilities could appear, which pose a limitation to the application of the model.

Tests carried out for non-hydrostatic mode have shown no significative improvement in the results compared to the hydrostatic mode, especially considering the fact that the non-hydrostatic mode demands big computational effort. Then, it could be concluded that the hydrostatic mode might be enough for the simulation of flow in spillways.

Bearing in mind that this research corresponds to a first phase of evaluation of the applicability of the Delft3D computational program in the simulation of the flow over spillways, a detailed comparative analysis with a CFD-type model is planned ahead.

Finally, it can be noted that Delft3D computational model is a tool that can adequately simulate flow over spillways, taking into consideration factors such as the size of the numerical grid and the impact that each parameter has on the behavior of the flow, without neglecting the influence of geometry.

6. Acknowledgments

This work was carried out in collaboration with the personnel of the Hydraulics Laboratory of Universidad Mayor de San Simón. A special acknowledgment for our colleague Galo Muñoz who kindly provided all the information of the physical modelling project.

7. References

Chow, V.T., (1959). Open-channel hydraulics: New York, McGraw-Hill, 680 p.

Deltares (2022). "Delft3D-Flow, Simulation of multi-dimensional hydrodynamics flows and transport phenomena including sediments". User Manual, Deltares, Delft

Donigian A.S. (2000). "HSPF Training Workshop Handbook and CD, Lecture #19, Calibration and Verification Issues". Slide #L19-22. EPA Headquarters, Washington Information Center, 10-14 January 2000, Presented and prepared for USEPA, Office of Water, Office for Science and Technology, Washington DC.

Heredia M. W., (2003), Assessment of the efficiency of stabilization works for the confluence of Desaguadero and Jachajahuira rivers by the application of numerical modelling, Master Thesis, Universidad Mayor de San Simón, Cochabamba, Bolivia.

Lesser, G. R., J. A. Roelvink, J. A. T. M. van Kester, and G. S. Stelling (2004), "Development and validation of a three-dimensional morphological model". Coastal Eng., 51(8–9), 883–915.

LHUMSS, (2016), Physical model of Hampaturi Dam, La Paz-Bolivia, Technical Report, Hydraulics Laboratory, Universidad Mayor de San Simón, Cochabamba, Bolivia (in Spanish).

Nash, J., Sutcliffe, J. (1970). "River flow forecasting through conceptual models part I - A discussion of Principles". Journal of Hydrology, 10, 282–290.

Palacio T. C. A., (2002), "Metodología para la validación de modelos hidrodinámicos utilizando amplia información de campo: aplicación a la bahía Meldorf en la costa del mar del norte alemán", Universidad Nacional de Colombia, Colombia.

Romero M., Heredia M.W. (2005), "Assessment of the effect of stabilization works at the river confluence Desaguadero-Jachajahuira, Bolivia", Proceedings of the 31st International Congress of the International Association of Hydraulic Research - IAHR, Seoul, Korea.

Sutherland J. and Barfuss S.L., (2011); "Composite Modelling: Combining Physical and Numerical Models", 34th IAHR World Congress, Brisbane, Australia.

DOI: https://doi.org/10.26077/5b39-665f

Hydraulic Behavior of Spill channel of Polavaram dam spillway without left embankment

Vankayalapati S. Ramarao[1] Yogendra Nath Srivastava[2]
[1] *Scientist -C, Central Water and Power Research Station, Khadakwasla, Pune-411 024. India*
[2] *Additional Director, Central Water and Power Research Station, Khadakwasla, Pune-411 024, India*
Email: sramaraov@gmail.com

Abstract: *Spill channel is a flood disposing channel excavated on the downstream of spillway which joins the main river course. The purpose of spill channel is to divert the spillway released flood safely to the downstream river course. The alignment of the spill channel may be straight of curved or partially straight and partially curved. It may be in steep gradient or in flat gradient and subjected to supercritical flows or subcritical flows. While disposing of the flood, the flow conditions in the spill channel would be straight without cross waves. Water surface profiles in the convex and concave banks of the spill channel must be confined within the top elevation of the embankments of channel to avoid spilling over the embankments. Thus, the hydraulic design of the spill channel involves keeping the velocities in the channel to allowable limits to prevent erosion of embankments and also confining the water levels to within the top elevation of the embankments. Suitable foundations are necessary to keep the embankments of spill channel stable. If a suitable foundation is not available for embankments, the embankment sometimes may have to be removed, if it would not cause problem in disposing of the desired quantum of flood. Not much literature is available relating to these kinds of studies carried out on large spill channels carrying more than 1.41 lakh m³/s discharge and without one embankment. The general procedure for selection of optimum channel parameters is of trial and error by considering number of alternatives and the hydraulic model studies play a key role in optimising the design of spill channel. Hydraulic model studies of Polavaram Irrigation Project, Andhra Pradesh were carried out in CWPRS, Pune on 1:140 scale 3-D comprehensive model to assess the subcritical flow conditions in spill channel with and without the provision of left embankment for various magnitudes of flood ranging upto 141583 m³/s. Studies enabled optimisation of layout of spill channel to pass the design flood with uniform flow conditions in the spill channel. In this paper, the various aspects of optimization of the hydraulic design of spill channel in disposing of the flood safely to downstream, based on the results of hydraulic model studies are discussed.*

Keywords*: confluence, flood, guide bund, hydraulic design, spill channel, spillway.*

1. Introduction

Spillway is designed to dispose off excess flood safely to downstream and the energy of flood is dissipated by the energy dissipator and then it is released to the main river course through the spill channel. Spill channel is an excavated flood channel which carries the spillway releases safely to downstream. The spill channel should be stable both structurally and hydraulically to dispose of the entire flood to downstream by confining flood within the top elevations of embankments. The embankments must be stable enough to carry the high velocity flood and should not cause erosion of downstream. Modifications need to be taken up to suite the design of these components to the dam safety criteria.

Presently, the procedure for selecting optimum channel parameters like alignment, shape, size, and bottom slope, lining is not available because each and every site has unique features and requires special considerations and is effectively designed by trial and error by considering a number of alternatives (Hanif Chaudhry, 2008). Hydraulic model studies play crucial role in finalizing the effective layout of channel.

Polavaram Irrigation Project is a multipurpose national project being constructed on the Godavari River near Ramayyapeta, in West Godavari district, Andhra Pradesh. Water from the project is utilized to meet the demands of irrigation, industries, drinking water and power generation. The project envisages irrigation benefits to 1618 sq. km in East Godavari and Visakhapatnam districts under Left Main Canal by diverting 2.2 x 10^9 m³ and to 1295 sq. km in West Godavari and Krishna districts under Right Main Canal by diverting 2.2 x 10^9 m³. In addition to irrigation benefits, generation of 960 Megawatt of hydroelectric power and water supply for industries in Visakhapatnam besides drinking water supply to villages and towns enroute is also envisaged under the project. Project design flood is estimated as 101941 m³/s and Probable Maximum Flood (PMF) is estimated at 141583 m³/s, which is required to be passed through spillway comprising of 48 spans of size 16 m (wide) x 20 m (height) and through spill channel. The Full Reservoir Level (FRL) of reservoir is at El. 45.72 m and gross storage capacity at FRL is 5.5 x 10^9 m³ with a live storage capacity of 2.1 x 10^9 m³. The other components of project include Earth Cum Rock Fill (ECRF) dam with plastic concrete diaphragm wall, Approach channel, Guide bund and Spill channel.

2. Spill channel downstream of Polavaram dam spillway

The spill channel of Polavaram dam spillway takes off from the end sill of stilling basin at Chainage, Ch. 142 m and continues for a length of about 2778 m (up to Chainage Ch. 2920 m) and is followed by a pilot channel that discharges flood into the main stream of Godavari River at about Chainage, Ch. 3570 m. The width of spill channel at its start is 1043.40 m and is gradually reduces to 1000 m in a length of about 215 m. The width remains constant from thereafter and up to its junction with pilot channel at Chainage Ch. 2920 m. A curve is provided in spill channel at Chainage Ch. 705 m and ends at Chainage Ch. 1473 m. The radius of curvature of the bend is 1000 m. The spill channel design parameters, the cross sectional area, bed slope, side slopes were so designed as to contain the designed flood water levels up to the design discharge. The Manning's roughness coefficient adopted for design was 0.025 $s/m^{1/3}$. The spill channel side slopes are 1 H: 2 V in hill cutting zone and are in 1.5 H: 1 V in earth embankments. In order to contain the flood levels within the spill channel, it is proposed to construct an embankment having maximum top elevation of El. 35.50 m, wherever required. Construction of embankment was originally proposed on both the left and right banks for a significant length of spill channel downstream of Chainage Ch. 800 m. A pilot channel was originally conceived, having a total length of about 650 m, at the downstream end of the spill channel to convey the discharges from spillway into Godavari River. The bed width of pilot channel at its junction with spill channel is 1000 m and is gradually reduced to 660 m in a length of 200 m. The bed level at the start of spill channel is El. 8.8 m and the elevation at the exit is El. 8.26 m. The spill channel is proposed to be lined with cement concrete blocks.

3. Hydraulic model (CWPRS TR No. 5576)

A geometrically similar model of 1:140 scale comprising of the spillway and the spill channel, was constructed in CWPRS, Pune. Entire river topography was incorporating by reproducing the river cross sections up to 4800 m along the upstream and up to 3700 m along the downstream of the dam axis. Entire river bed was cement plastered, after consolidating the moorum laid beneath. The original layout comprising of spill channel of length 2780 m, pilot channel of length 660 m were also reproduced with cement plaster and painted with the enamel paint. The model was provided with suitable water conductor system along with Rehbock and V- notch weirs for the measurement of discharge. Arrangements were made for measurement of water levels and velocities as and where required in the model. Digital water levels were used for the measurement of water levels and propeller type digital current meters were used for the measurement of velocities along the spill channel for various operating conditions. Figure 1 shows the aerial view of 1:140 scale Polavaram model (original layout) and Figure 2 shows upstream view of spillway with spill channel.

Figure 1. Aerial view of 1:140 scale Polavaram model (original layout)

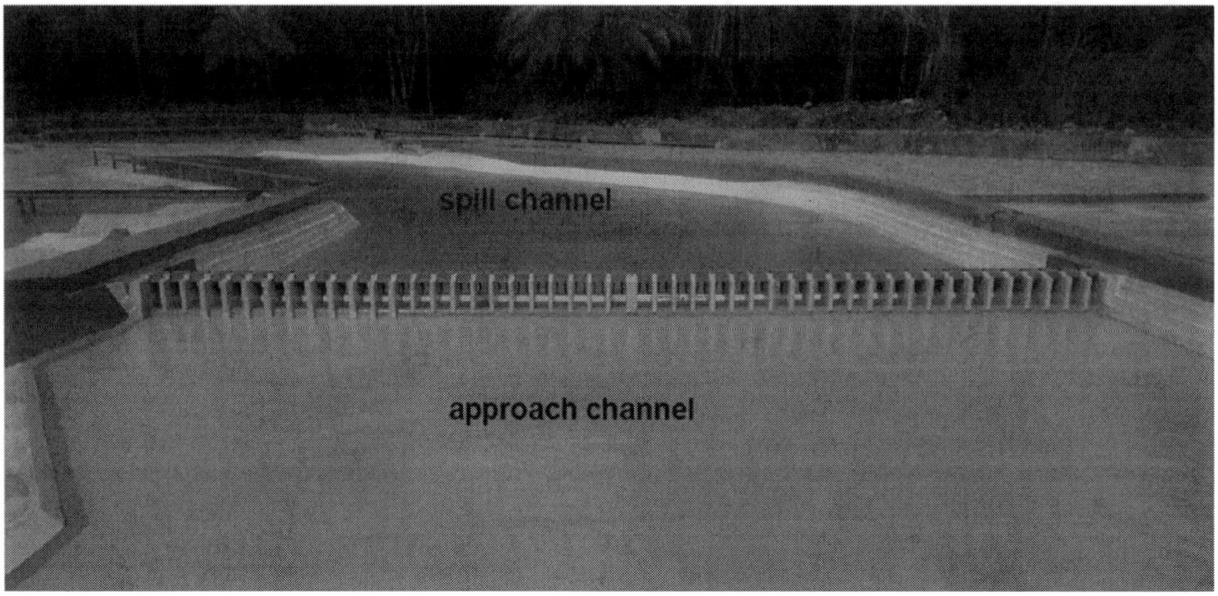

Figure 2. Upstream view of model with spillway and spill channel

The accepted relationship of hydraulic similitude, based on Froudian criteria were used to express the mathematical relation between the dimension and hydraulic quantities of the model and the prototype. The general relations expressed in terms of model scale are given in Table 1.

Table 1. Model Scale Relation for Various Dimensions

Dimension	Scale Relation
Length	1:140
Area	1:19600
Velocity	1:11.832
Discharge	1:231910.32
Time	1:11.832
Manning's `n'	1: 2.2787

Hydraulic model studies were carried out on the 1:140 scale 3-D comprehensive model for optimising the hydraulic design of spill channel for various operating conditions for the flood releases up to the design discharge of 141583 m^3/s (PMF) and reservoir water levels up to FRL El. 45.72 m.

4. Hydraulic behavior of spill channel with left and right embankments

Initially, the hydraulic model studies were conducted for the original design of spill channel with proposed both the left and right embankments. This spill channel had a constricted exit in the form of pilot channel. Studies conducted for flood releases indicated that, for the flood releases up to 75% of PMF, the flood was confined within the top elevation of both the embankments of spill channel and the water surface was well below the top elevation of embankments. However, for the discharge of PMF, there was observed a possibility of both embankments getting submerged due to restricted exit for the flow through the spill channel. This enabled the re-alignment of the spill channel with the removal of constriction at its exit.

To reduce the further obstruction to the passage of flood through spill channel and to improve the flow conditions downstream of pilot channel, the widening of spill channel was done up to Ch. 3570 m and dressed further another

300 m to enable to join the existing river bed level. Figure 3 shows the aerial view of model with realigned spill channel after dressing the pilot channel.

For this modified layout of spill channel, flow conditions were improved and the water surface levels were well below the top elevation of spill channel embankments, even for the discharge of PMF. The velocities that were observed along spill channel were of the order of 1-7.1 m/s for various flood releases and of the order of 141583 m^3/s, for the ungated operation of the spillway. These velocities were necessary in designing the size of cement concrete blocks for the lining of the spill channel, to resist the instantaneous hydrodynamic forces caused by flood releases in spill channel.

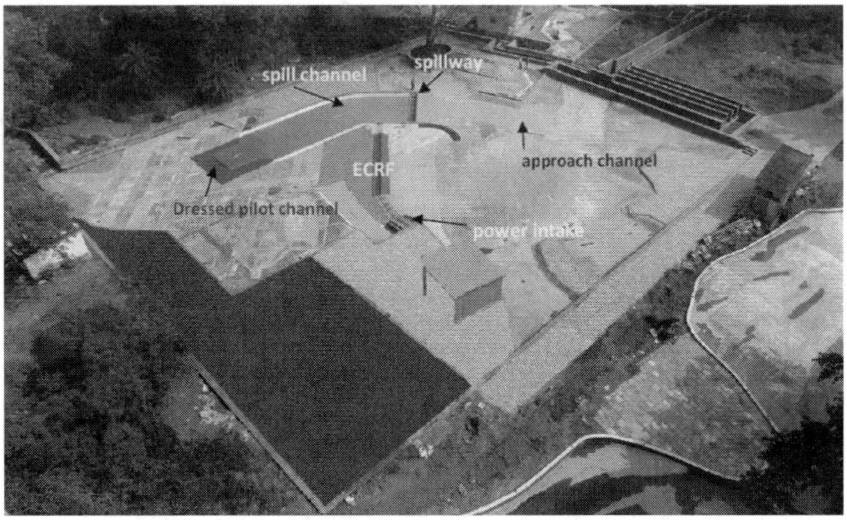

Figure 3. Aerial view of model with realigned spill channel with dressed pilot channel.

4.1 Flow conditions at the exit of spill channel

With the removal of pilot channel and thereby widening of the spill channel to 1000 m and extension of the spill channel by another 300 m with the same bed level of El. 8.26 m, the flow conditions were improved at the exit of spill channel and its confluence with the main river. The velocities of the order of 1 - 3 m/s were observed on left, 1.4 – 6.1 m/s along centre and 2.1 – 7.4 m/s on right side of the exit of spill channel. These velocities further reduced when the flow moves further downstream and hit left bank of the river, due to formation of stagnant pool along left bank. These observed velocities are important for designing the river bank protection works along left flood bank downstream of spill channel.

5. Modifications proposed to spill channel- the channel without left embankment (CWPRS TR No. 5949)

Subsequently, it was proposed to study the performance of spill channel without left side embankment, in the model, as there was observed non-availability of suitable foundation in the river bed for the construction of left embankment. Thus studies were carried out on the physical model by removing the existing left side embankment of the spill channel. Figures 4 and 5 show the dry model of the spill channel without its left side embankment.

Figure 4. Aerial view of model of spill channel showing left embankment to be removed

Figure 5. Downstream view of modified Spill channel

5.1 Hydraulic model studies for optimising layout of spill channel without left side embankment

The non- provision of the left embankment of the spill channel was replicated in the model and studies were carried out to optimise the layout of the spill channel. The experiments were carried out for the representative floods and the flow conditions along spill channel were observed for 60% of PMF (84,950 m^3/s) and PMF (141583 m^3/s) for ungated operation of spillway. Studies indicated that for the discharge of 84950 m^3/s with ungated operation of spillway, the flow conditions in spill channel were satisfactory and subcritical and straight forward flows were observed along spill channel. There were no dominant lateral flows moving towards ECRF dam from the spill channel and their velocities were not significant. Even for the discharge of 141583 m^3/s, with ungated operation of spillway, flow was predominantly straight and satisfactory without left embankment. No dominant lateral flows were observed moving towards ECRF dam and the observed velocities were marginal between left bank of spill channel and ECRF dam. Moreover, maximum observed velocities in the spill channel along left embankment, following the removal of left embankment were reduced drastically from 2.1 m/s to 6.9 m/s to 1 m/s to 5.1 m/s for PMF along the length of the channel and along right side the velocities were marginally increased. Table 2 shows velocities in spill channel without left embankment, while passing a discharge of 141583 m^3/s with ungated operation. Overall, the flow conditions along spill channel even without left embankment were found to be satisfactory. Figure 6 shows flow conditions in spill channel while passing PMF through spillway for ungated operation.

From the Table 2 it is known that along left side of the spill channel, the velocities were reduced without left embankment than those of with the left embankment. This may be due to nearly stagnant pool formation along the downstream of the ECRF dam. Also along the centreline of spill channel, the velocities were reduced without left embankment. Maximum observed velocities along centreline following removal of left embankment reduced from 7.1 m/s to 6.9 m/s for PMF at Ch. 2700 m. However, along right side of the spill channel, the velocities were also increased without left embankment than with it. Maximum observed velocities along right embankment following removal of left embankment increased from 6.8 m/s to 7.5 m/s for PMF at Ch. 2400 m. Moreover, the velocities downstream of spill channel and in the river confluence, the velocities are mostly remaining unchanged without left embankment. Figures 7, 8 and 9 and show the velocities observed while passing PMF along left, centre and right of spill channel without left embankment respectively.

Figure 6. Flow conditions in spill channel
(Q= 141583 m^3/s, ungated operation of spillway)

Table 2. Velocities observed along spill channel with and without left embankment for PMF (m/s)

	Left		Centre		Right	
Chainage	Modified layout	Original Layout	Modified layout	Original Layout	Modified layout	Original Layout
140	2.9	6.1	6.5	8.6	7.6	8.6
300	1.6	4.5	5.0	5.9	6.1	5.2
600	2.6	5.4	5.1	5.9	5.7	4.7
900	4.9	6.9	5.6	5.4	4.2	4.1
1200	5.1	6.5	5.9	6.0	4.6	4.2
1500	1.7	6.1	6.1	6.1	6.1	5.3
1800	1.0	3.6	6.5	6.6	6.4	5.9
2100	1.0	2.4	6.6	6.4	6.6	6.4
2400	1.0	2.3	6.7	6.9	7.5	6.8
2700	2.2	2.6	6.9	7.1	7.3	6.7
2920	2.1	2.7	6.6	6.3	6.6	6.3
3570	2.7	2.5	6.7	6.4	7.5	7.5
3900	2.0	2.1	6.7	6.1	7.9	7.4

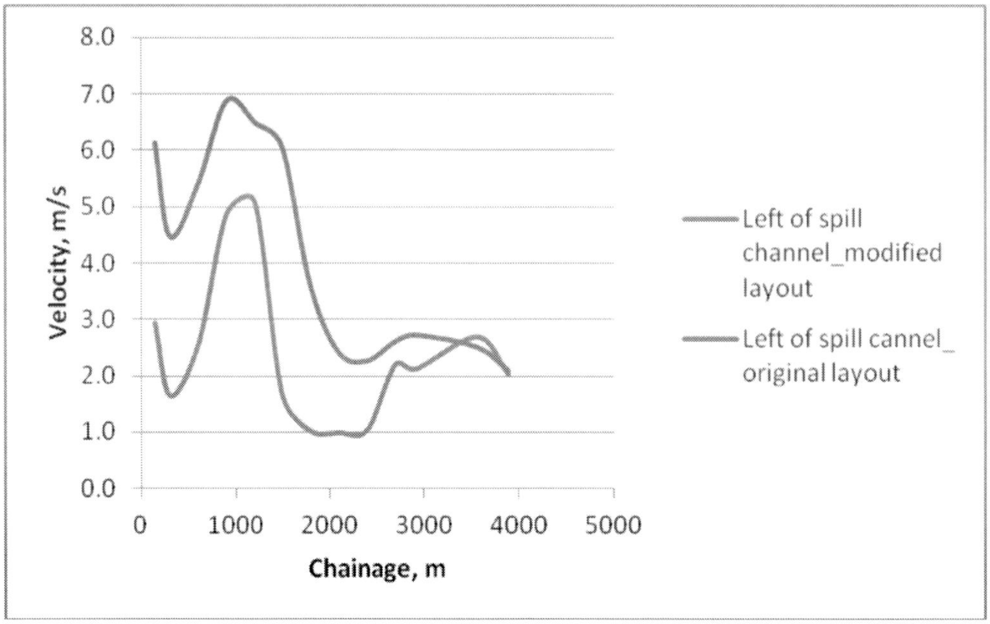

Figure 7. Velocities along left bank of spill channel with and without left embankment for PMF

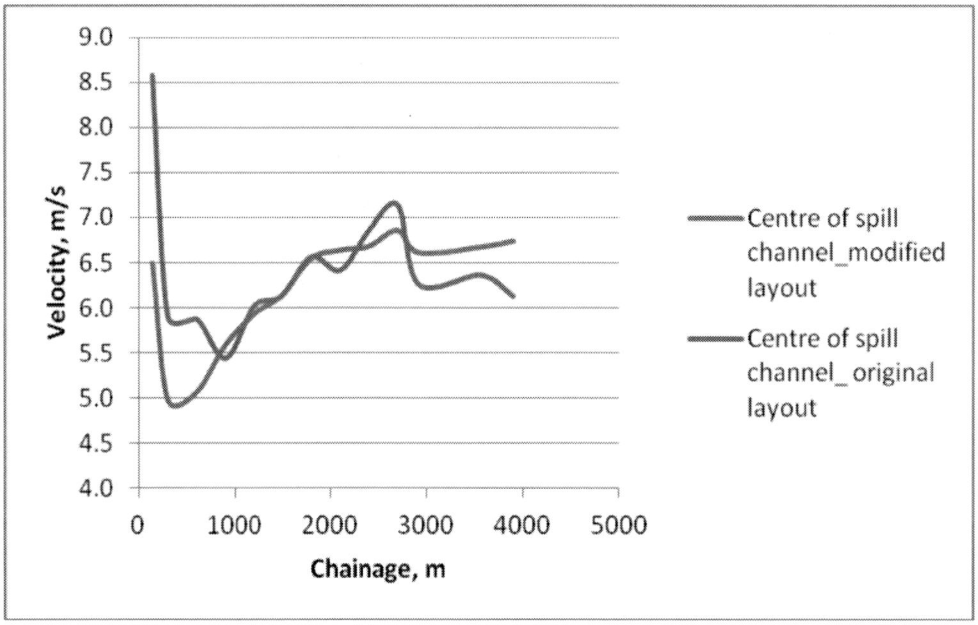

Figure 8. Velocities along centerline of spill channel with and without left embankment for PMF

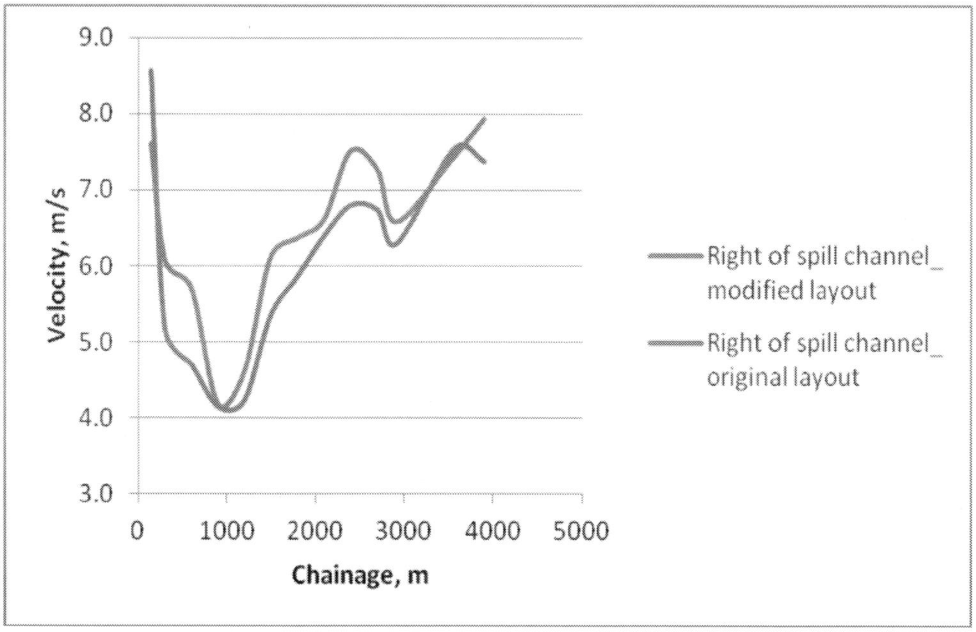

Figure 9. Velocities along right bank of spill channel with and without left embankment for PMF

6. Conclusions

Spill channels are provided downstream of spillways to dispose of the flood safely to the main river course. But the passage of flow in the spill channel should be straight forward without predominant lateral flows and the water surface profiles in the convex and concave banks of the spill channel must be confined within the top elevation of the embankments of channel to avoid spill over. If a suitable foundation is not available for embankments, the embankment sometimes may not be provided, but it should not cause serious problem in disposing of the flood. Thus, the hydraulic design of the spill channel involves keeping the velocities in the channel to allowable limits to prevent erosion of embankments and also confining the water levels to within the top elevation of the embankments. Hydraulic model studies play a key role in optimising the design of spill channel, with such special types of arrangements, which are not available in literature. One of such study was carried out for the spill channel of Polavaram irrigation project, Andhra Pradesh, where modifications were required to improve the performance of spill channel in disposing off the flood. The spill channel was modified by removing the constriction of pilot channel to improve the flow conditions. Further the studies were carried out for various operating conditions by removing the left embankment of the spill channel. From the studies, it was observed that the spill channel performance was satisfactory even without the left embankment. Moreover, the velocities along left bank and centreline of spill channel were reduced due to removal of left embankment without left embankment. However, along right side of the spill channel, the velocities were increased without left embankment. The velocities downstream of spill channel and in the river confluence were remaining more or less the same even without the left embankment. The recommendations of CWPRS based on hydraulic model studies are being implemented and the project is under construction.

7. ACKNOWLEDGEMENT

The authors are grateful to Dr. R. S. Kankara, Director, CWPRS for his continued support in writing technical papers. The authors are also grateful to the project authorities of Andhra Pradesh Water Resources Department (APWRD) for their support in providing required data during model studies. The support from staff of SED Division in writing this paper is also highly acknowledged.

8. REFERENCES

CWPRS Technical Report No. 5576 (2018). "*Hydraulic model studies for spillway and power intake of Polavaram Irrigation Project, Andhra Pradesh, 1:140 scale 3-D comprehensive model.*"

CWPRS Technical Report No. 5949 (2021). "*Additional hydraulic model studies for modified approach channel with upstream guide bund and modified spill channel of Polavaram Irrigation Project, Andhra Pradesh, 1:140 scale 3-D comprehensive model.*"

Chaudhry H. M. (2008). "*Open Channel Flow*", Springer New York, NY, 2008.

DOI: https://doi.org/10.26077/4afc-9644

Hydrodynamic behavior of stilling basins with deflectors to reduce supersaturation of total dissolved gases: analysis of Yacyretá and Jorge Cepernic dams

C.A. Fattor[1], J.D. Bacchiega[1], F. Romero[1] & C. Fernández[1]
[1]*Hydraulic Structures Division, Hydraulics Laboratory, National Institute of Water, Ezeiza, Argentina*
E-mail: cfattor@ina.gob.ar

***Abstract**: High air entrainment to the hydraulic jump is one of the inherent aspects of this macroturbulent flow, extending its presence from the free surface to the bottom. Depending on the variety of fishes in this natural habitat, important values of supersaturation of total dissolved gases (TDG) could be reached immediately downstream of the energy dissipation structure, putting under risk the life of fishes. This article deals with the influence of deflectors designed to reduce TDG on the hydrodynamic behavior of energy dissipators of two important dams. The first one is Yacyretá dam (Paraná River, Argentina – Paraguay) where horizontal deflectors were installed after an environmental incident with important mortality of fishes due to supersaturation of TDG. The second case corresponds to Jorge Cepernic dam (Santa Cruz River, Argentina), which is currently under construction. Hydrodynamic evaluation of the stilling basins under initial design conditions and those derived from the installation of the deflectors was carried out by means of two-dimensional physical models of each of the mentioned projects. The analyses showed a decrease of the energy dissipation rate for Yacyretá energy dissipator, with a hydraulic jump length that exceeds the length of the stilling basin, aspects that demand a continuous supervision of the works. On the other hand, the evaluation carried out for Jorge Cepernic dam does not exhibit significant changes of the hydrodynamic behavior of the stilling basin, except in a region downstream of the deflectors.*

***Keywords:** Physical modeling, stilling basins, total dissolved gases.*

1. Introduction

Stilling basins are hydraulic structures designed to achieve strong energy dissipation downstream of spillways by means of the formation of a hydraulic jump. The project of these structures has a wide variety of possibilities, depending on factors as difference between reservoir and tailwater levels, unit discharge flow, Froude number, submergence, among others, besides geological, constructive and economic issues.

High air entrainment to the hydraulic jump is one of the inherent aspects of this macroturbulent flow, extending its presence from the free surface to the bottom. The mutual convergence of high pressures and air entrainment would reach that air bubbles become dissolved with certain pressure level. Under these circumstances, important values of supersaturation of total dissolved gases (TDG) could be reached immediately downstream of the energy dissipation structure, putting eventually under risk the life of fishes in their natural habitat.

The inclusion of horizontal deflectors in the proper elevation is one possible structural measure to reduce the supersaturation of TDG downstream of the stilling basin, by means of the change of the direction of the flow entering to the energy dissipator. In this way, the flow direction is superficially re-oriented for low discharge flows (frequent discharges) reducing air entrainment due the modification of the initial conditions established in the design of the stilling basin. Under these conditions, energy dissipation rate is even effective because hydraulic design is related to the highest flows.

On the other hand, high flow discharges are less effective in the reduction of air entrainment to the hydraulic jump and it would eventually reduce the efficiency of the stilling basin. This issue is specially considered in the present article taking into account the experience of two multipurpose dam projects.

In effect, the hydrodynamic behavior of two stilling basins related to dam projects where TDG was or would eventually be a problem were analyzed by means of two and three-dimensional physical models.

The current article deals with the evaluation and the analysis of hydrodynamic behavior of stilling basins of Yacyretá dam and Jorge Cepernic dam, where deflectors were considered to mitigate supersaturation of TDG and initial aspects considered in the hydraulic design of these structures were modified. The relevance of these consequences is analyzed hereinafter.

2. Description of the structures

2.1. Yacyretá dam (Paraná River, Argentina – Paraguay)

Yacyretá is a multipurpose dam located on Paraná River, on the border between Argentina and Paraguay. It is a 67 km long earthen dam, with a maximum high around 30 meters, which contains a hydroelectric powerplant, a navigation lock, fishes' elevator and two spillways, each one of them located in the two branches of the river at this location.

Both spillways were designed for a period of return of 10,000 years. The maximum flow discharge of Principal spillway is 55,000 m^3/s while Aña Cuá spillway allows a flow discharge of 40,000 m^3/s, with a total discharge of 95,000 m^3/s.

Principal spillway, analyzed in this article, has 18 spans, with the spillway crest at 63.50 m level and aeration step (anticavitatory purpose) at 61.70 m level, being 82.00 m the highest water reservoir level. The energy dissipation is achieved by means of a 342 m wide and 94 m long stilling basin, with bottom elevation going from 41.00 to 39.00 m at the end of this structure, before reaching the bottom riverbed with an end inverse slope (Figure 1).

This structure was operating without problems concerning supersaturation of total dissolved gases until August 1994, when a singular operative condition of the spillway provoked an increase of TDG and fishes' mortality downstream of the stilling basin. As a consequence of this, immediate evaluations were carried out in order to understand this serious environmental issue and to implement those necessary measures to avoid similar events in the future.

After this environmental incident, multiple researches were developed, including physical understanding of the processes, evaluation of similar incidents in other projects, field measurements of TDG towards downstream of the spillway and a set of hydraulic studies to reach an environmental and engineering solution.

Physically speaking, high levels of total dissolved gas in the river lead to the development of gas bubble disease (GBD) in smolts, as well as other aquatic life. This condition involves the formation of bubbles in the fish's organs, tissues, and vascular system. Even though there is uncertainty as to the significance of GBD-induced mortality at low levels of supersaturation (<110%), it is clear in all studies that as the amount of supersaturation increases (> 110%) the rate of mortality increases significantly. The transition between low levels of generally sublethal effects to the higher level lethal condition involves a shift in the bubble-related mechanisms that lead to death. Specifically, at levels of supersaturation below the threshold fish are more susceptible to death related to infection and stress while above the threshold fish experience death from large intravascular bubbles (White, 1991).

In this project, TDG as high as 145% was measured downstream of the spillways, decaying to TDG around 110% at around 90 km from the dam, making evidence the elevated and risky values of TDG. The research of environmental incidents like this on other dams found relevant information for some of them located on Snake and Columbia rivers, USA, were fishes' mortality due to TDG was observed. In those structures, deflectors to reduce TDG downstream of their stilling basins were considered, built and operated.

In fact, the environmental incident of Yacyretá spillway demanded an integral answer to this issue. In this context, National Institute of Water developed studies to analyze and optimize deflectors to reduce supersaturation of TDG (Angelaccio et. al 1997, Bacchiega et al 1999). These measures were implemented in prototype.

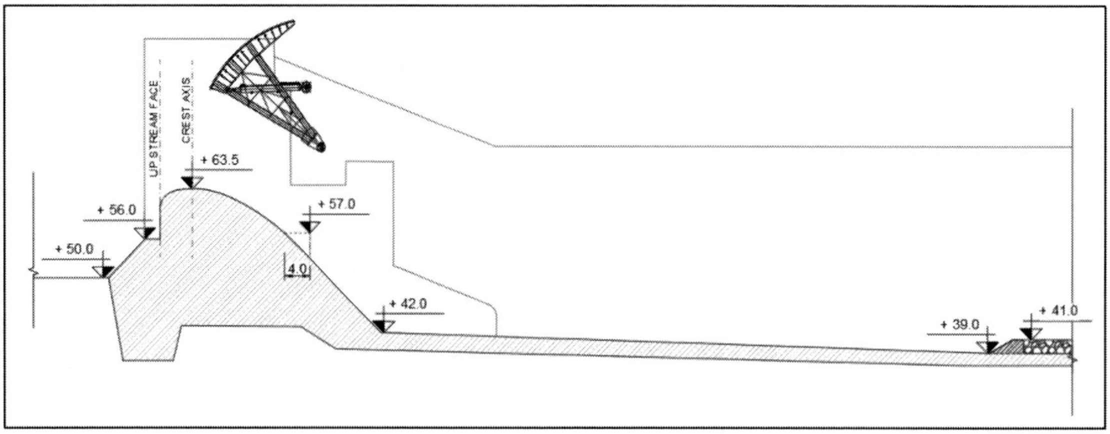

Figure 1. Spillway and energy dissipator Yacyretá – Lateral view

This environmental incident involving massive fishes' mortality was very important for future considerations of supersaturation of total dissolved gases in dam projects in the context of their development in Argentina, being Jorge Cepernic dam an example of this learning.

2.2. Jorge Cepernic dam (Santa Cruz River, Argentina)

Jorge Cepernic is a hydroelectric dam purpose currently under construction on Santa Cruz River, at the south of Argentina. Its spillway was designed for a period of return of 10,000 years, corresponding a flow discharge of 4,163 m^3/s, having 5 spans with spillway crest at 105.00 m level, maximum water surface elevation at 115.00 m and a 72 m wide and 85 m long stilling basin with the bottom at 66.30 m level. The exit sector of the stilling basin has 12 equally spaced blocks located on the vertical transition (1v:2h) until reaching the bottom bed of the tailwater channel (Figure 2).

In this project, structural measures to mitigate supersaturation of TDG was not initially considered but they were included in a second stage of the design, during the construction phase.

Studies concerning the risk of mortality for fishes on Santa Cruz River, leaded to finally consider the disposition of deflectors to diminish total dissolved gases downstream of the stilling basin of Jorge Cepernic dam, following the criteria exposed to Yacyretá spillway.

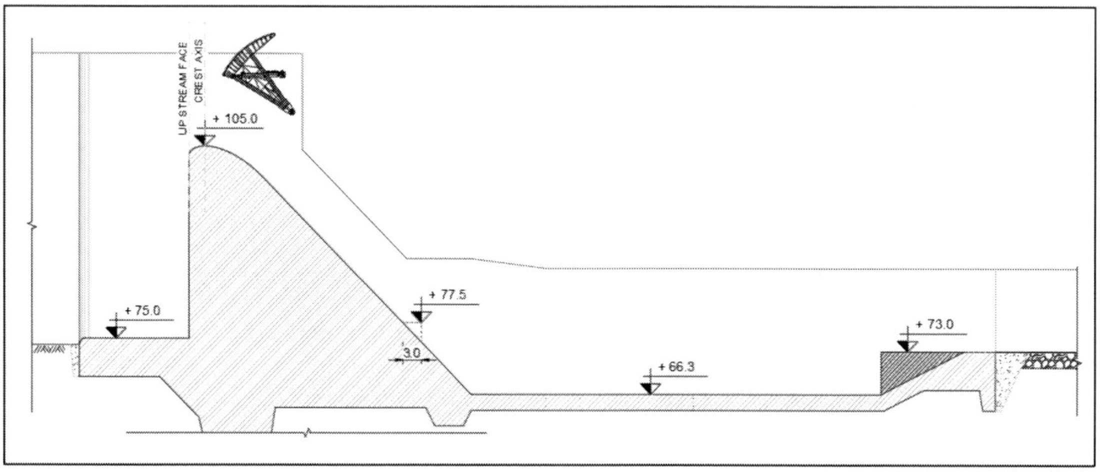

Figure 2. Spillway and stilling basin - Jorge Cepernic dam – Lateral view

Besides that, the bottom outlets are located in the same structure of the spillway, in correspondence with the second and the fourth span. As a consequence of that, horizontal deflectors (Length: 3.0 m; Elevation: 77.50 m) would be disposed on extreme spans and the central span.

3. Experimental methodology

The evaluation of spillways with the problems of supersaturation of TDG downstream of the stilling basin, implies a series of analyses that involve the hydrodynamic evaluation of the stilling basin according to the original project and the evaluation of alternatives to mitigate supersaturation of TDG.

The evaluation of deflectors, which is not specifically comprehended in this article, considers the definition of a frequent flow discharge, as well as deflector location and dimensions. The objective of the deflectors is to deflect the water jet towards the surface of the stilling basin flow so that entrained air is not experiencing the very high-pressure zone close to the stilling basin bottom.

Dimensions and location of deflectors are related to previous definitions of design flow discharges and tailwater levels. Elevation of the deflector must ensure adequate performance for the widest range of flow discharge, taking especially into account frequent operations and discharges. On the other hand, the length of the deflectors must guarantee the change of the flow direction, reducing as high as possible the air entrainment into the hydraulic structure of the hydraulic jump.

The definition of an optimal solution also involves the analysis of spillway operation as well as simultaneous operation of spillway and hydroelectric power-plant. In this last situation, tailwater level is higher downstream of the stilling basin if flow discharged by the spillway is the same compared with the first evaluation. This complex matrix of possibilities leaded to a wide range of possibilities of deflectors to be analyzed by physical modelling, specially concerning their elevation. Hydraulic variables include flow discharge, reservoir elevation and tailwater elevation, being necessary to observe and/or to measure during the tests the elevation of the aerated zone, flow pattern, velocity profiles at different locations of the stilling basin, among the most important parameters.

From the viewpoint of the evaluation of the hydrodynamic behavior of the stilling basins of the dams, both two-dimensional and three-dimensional physical models were developed for each project. As regards the two-dimensional physical models, both of them were built at scale 1:50, following a Froude similarity law and reaching Reynolds number higher than 10^5 that allows proper representation of macroturbulent flow of the hydraulic jump. Even though three-dimensional physical models are not treated in this article, they were many important to analyze effects coming from upstream, from different gates operation and local scour downstream of the energy dissipator, among other aspects.

To study the two spillways (in both projects), the first one according to the project and the second one including a horizontal deflector, is necessary to measure parameters like water levels, mean pressures, fluctuating pressures and velocities. Mean and fluctuating pressures were measured in points located at the most important locations of the stilling basin as well as the downstream face of the deflectors.

The fluctuating pressures records were carried out through instrumentation including pressure transducers, acquisition boards and digital filters. The sample time was 0,01 sec, in order to include macroturbulent phenomenon that are characterized for the low peak frequency. The duration of the sampling is almost three minutes in order to guarantee ergodicity and stationery, because these conditions are necessary to allow a right statistical analysis.

The analysis of the hydrodynamic behavior of a conventional spillway and a modified spillway including a deflector tested and finally selected to diminish supersaturation of TDG downstream of the plunge pool, was done taking into account the flow pattern and pressure measurements (static and fluctuating terms).

In relation with the pressure fluctuations, the record of this parameter allows to characterize both important parameters:

- the intensity of turbulence in different areas of the stilling basin, by means of the root mean square of the fluctuating pressures (RMS). This is a good index to analyze turbulence intensity, widely used for this purpose in stilling basins.
- the total pressure $p/\gamma_{0.1\%}$, which is a parameter of the tendency to cavitation due to the fluctuating pressures (Lopardo et. al 1982), being $p/\gamma_{0.1\%}$ the results of adding the mean pressure (p/γ) and the semi amplitude of fluctuating pressures with a non-exceedance occurrence of 0,1% ($p'/\gamma_{0.1\%}$).

Both relationships can be analyzed as a function of the location in the stilling basin, among other variables.

For the present article, the scenarios as listed in Table 1 were considered to the analysis of hydrodynamic behavior of both structures.

Table 1. Flow discharges by the spillways

Dam spillway	Q_{MAX} (m^3/s)	Reservoir elevation (m)	Tailwater elevation (m)
Yacyretá (Principal)	55,000 m^3/s	84.00	66.40
Jorge Cepernic	4,100 m^3/s	115.00	82.50

Consequently, considering the results obtained of the studies of both hydraulic structures, it is possible to carry out a comparison of the hydrodynamic behavior.

4. Analysis of results

The hydrodynamic changes of the behavior of the stilling basins produced by the disposition of a deflector on the spillway profile are analyzed on the basis of the interpretation of the flow pattern and the evaluation of pressure fluctuations on the bottom.

These evaluations are presented for each project, especially considering the pressures field related to the maximum discharge.

4.1. Yacyretá dam – Principal spillway

Even though the core of this article is related to hydrodynamic conditions of the stilling basin before and after the disposition of a deflector to reduce supersaturation of TDG, it is important to understand the flow pattern not only for high discharges but more frequent discharges.

Figure 3 shows the flow pattern for a condition given for the maximum water reservoir elevation (WSL) and small sector gate opening (GO), without the deflector (original design) and with deflector. As can be seen, the flow conditions on the deflector gets an almost horizontal exit of the jet with very low effects as regards air entrainment towards the bottom of the dissipator. Deflectors are 4 m long and are located at elevation 57.00 m, being disposed in all the spans.

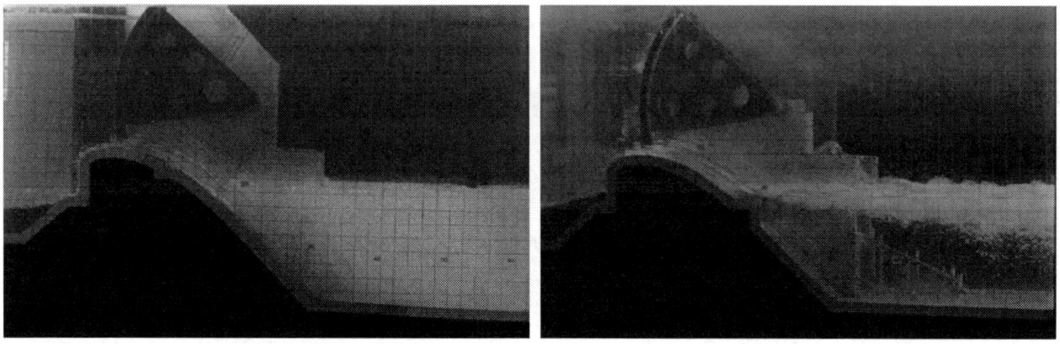

Figure 3. Yacyretá dam–Spillway–Flow condition without and with deflector (WSL=82.0m, GO=2.0m,Q=6,640 m^3/s)

An important aspect of Principal spillway of Yacyretá dam is that the air-entrainment devices disposed in each span to reduce risks of cavitation were eliminated jointly with the construction of the selected deflectors, showing explicitly the importance assigned to the resolution of the issue of fishes' mortality.

Under these conditions, which allow the evaluation and final selection of the best alternative of deflector considering the minimum deepening of the jet, the stilling basin was analyzed from the hydrodynamic viewpoint with especial emphasis of the highest flows because these conditions represent the most important ones from the viewpoint of the turbulence intensity and cavitation risks.

Figure 4 shows the flow pattern for the maximum flow discharge and the condition with deflector, being able to observe the presence of a recirculation flow generated behind the deflector and a displacement of the hydraulic jump towards downstream. This displacement of the initial section of the hydraulic jump would implies that the length of the stilling basin would not be enough to reach most of the turbulence energy being dissipated inside the dissipator, being eventually necessary to evaluate local scour downstream of the end sill.

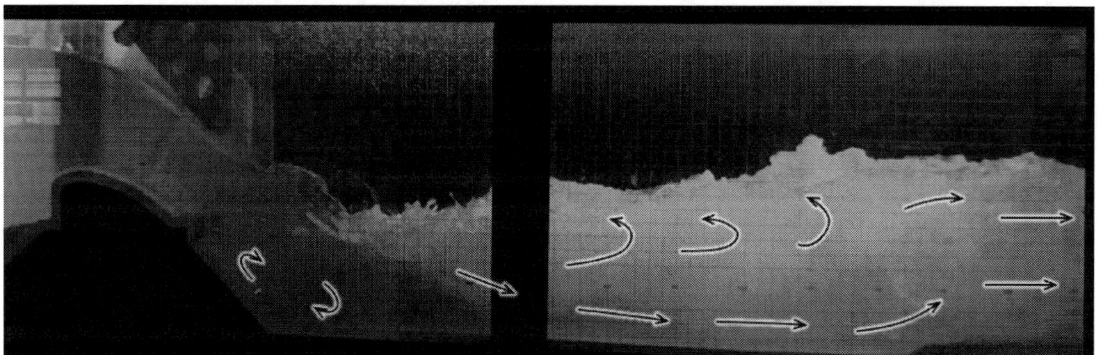

Figure 4. Yacyretá dam–Principal spillway – Flow conditions with deflector (WSL=82.00 m, GO=Total, Q=55,000 m^3/s)

On the other hand, Figures 5 and 6 allow observing the longitudinal variation of RMS and $p/\gamma_{0.1\%}$ on the centerline of the bottom.

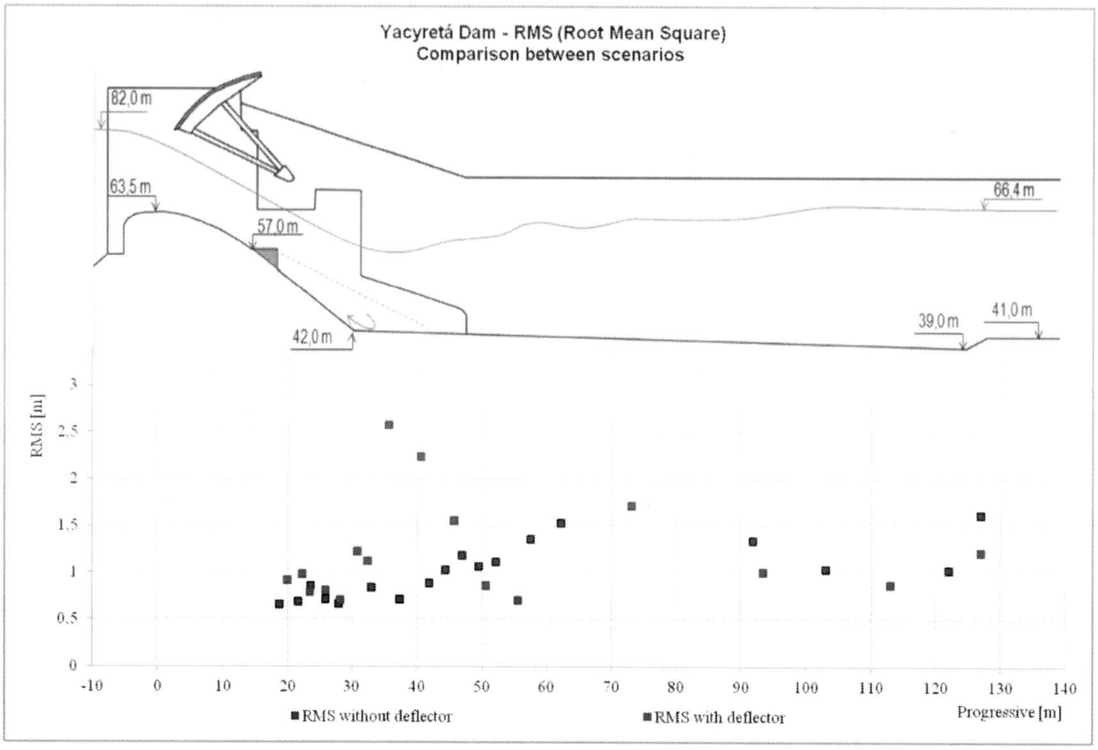

Figure 5. Yacyretá dam – RMS variation without and with deflector

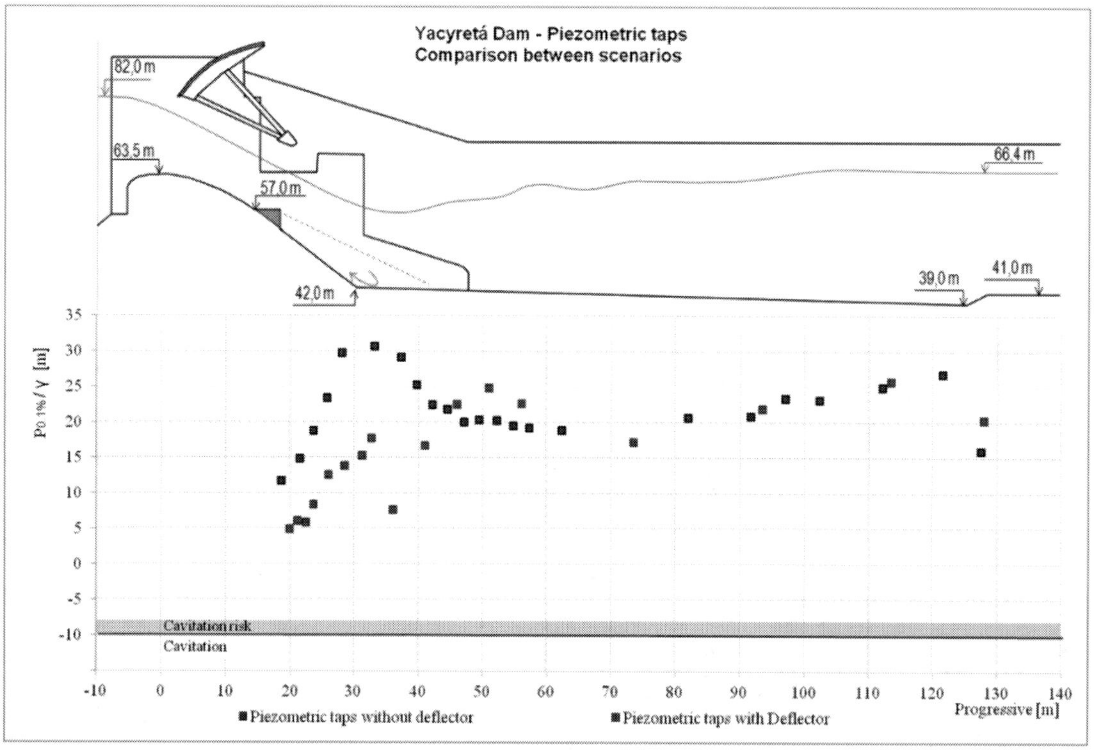

Figure 6. Yacyretá dam – $p/\gamma_{0.1\%}$ variation without and with deflector

4.2. Jorge Cepernic dam

Following the same concepts expressed for Yacyretá dam, Figure 7 shows the flow pattern for a condition given for the maximum water reservoir elevation (WSL) and small sector gate opening (GO), without and with deflector respectively.

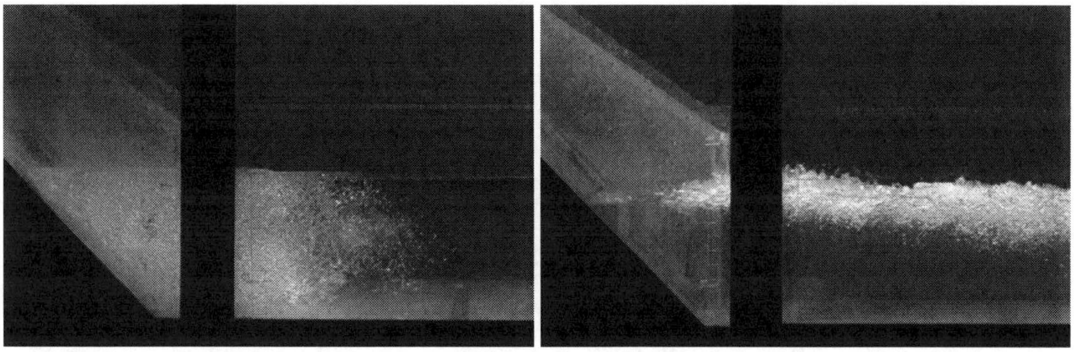

Figure 7. Jorge Cepernic spillway–Flow conditions without and with deflector (WSL=114.00 m, GO=1.0 m,Q=300 m^3/s)

As can be seen, the deflector leads to an almost horizontal exit of the jet with minimum effects as regards air entrainment towards the bottom of the dissipator. It must be said that deflectors are 4 m long and are located at elevation 77.50 m, being disposed in those spans without outlet bottoms.

Figure 8 shows the flow pattern for the maximum flow discharge and the condition without deflector, while Figure 9 illustrates the behavior with the deflector.

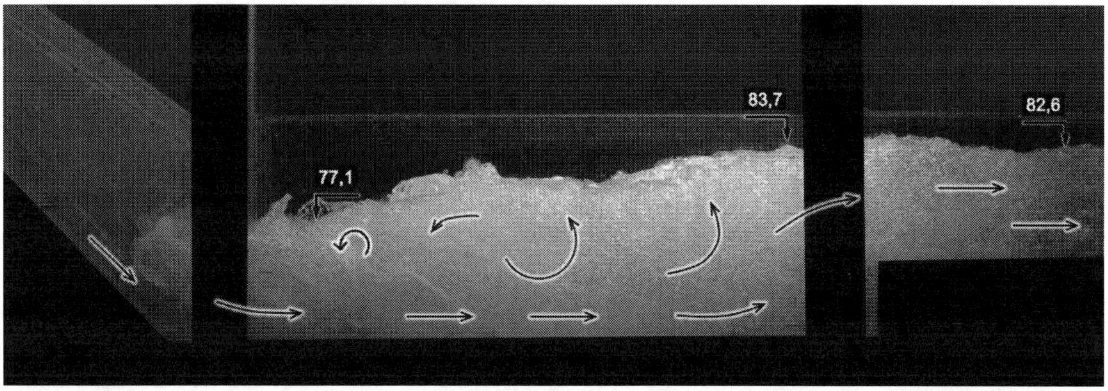

Figure 8. Jorge Cepernic spillway – Flow conditions without deflector (WSL=115.00 m, GO=Total, Q=4,100 m^3/s)

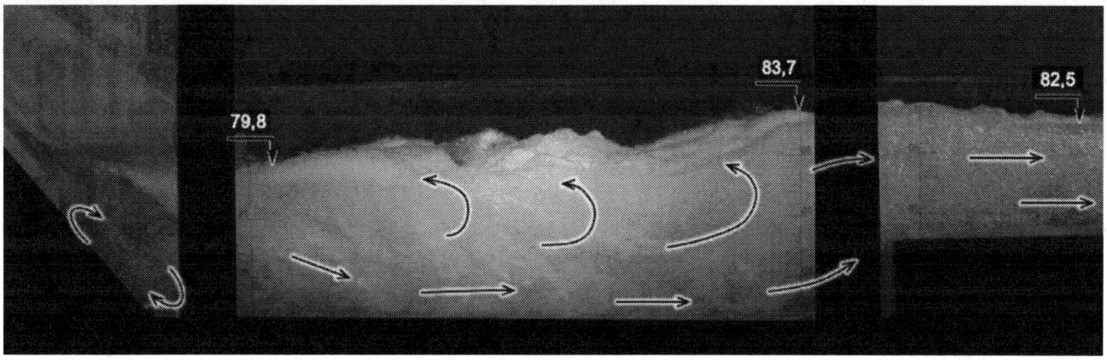

Figure 9. Jorge Cepernic spillway – Flow conditions with deflector (WSL=115.00 m, GO=Total, Q=4,100 m^3/s)

Figure 8 and 9 show a frequent and an extraordinary condition respectively, being able to see that the effect of the deflector for high flows is the displacement of the initial section of the hydraulic jump, exactly as it were previously mentioned for Yacyretá spillway.

On the other hand, Figures 10 and 11 allows observing the longitudinal variation of RMS and $p/\gamma_{0.1\%}$ on the centerline of the stilling basin, comparing conditions without deflector and with the selected deflector.

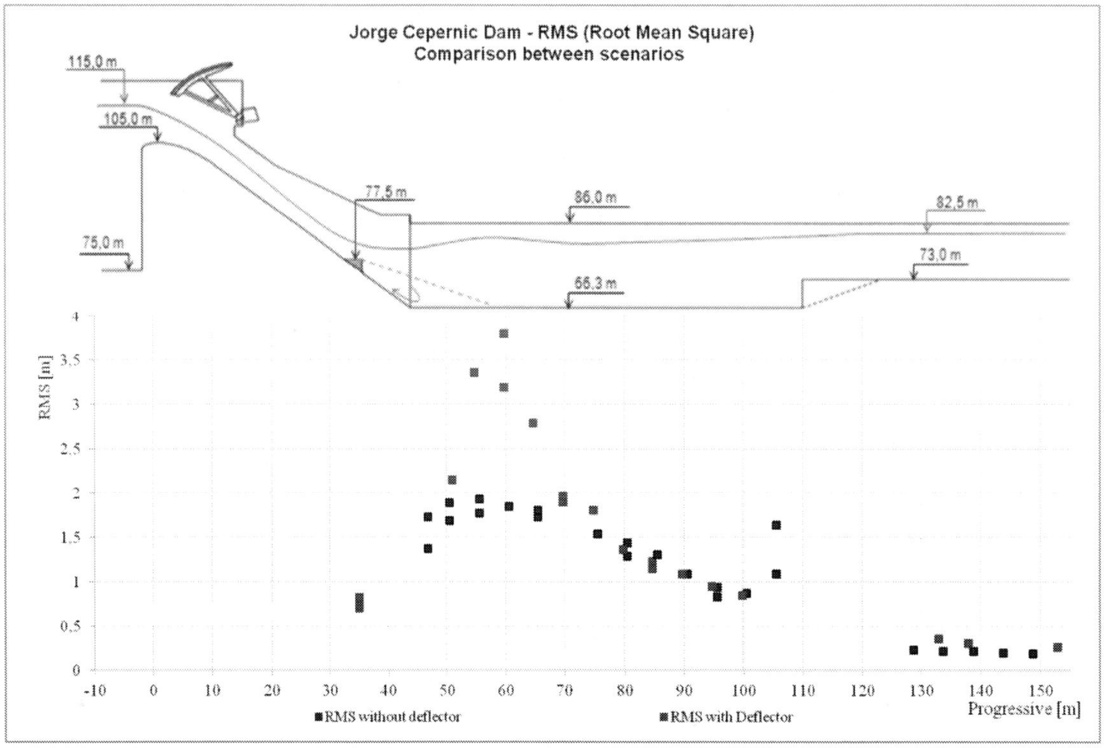

Figure 10. Jorge Cepernic spillway – RMS variation without and with deflector

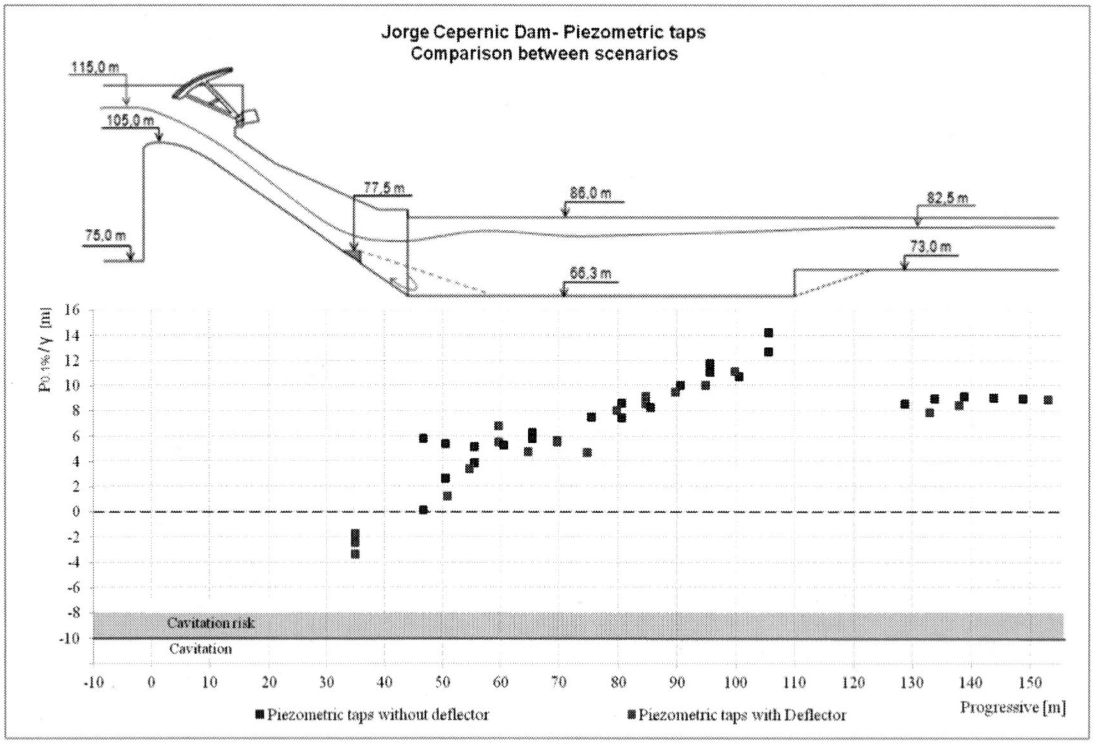

Figure 11. Jorge Cepernic spillway – $p/\gamma_{0.1\%}$ variation without and with deflector

5. Evaluation of results and discussion

Main results of the evaluation of Yacyretá dam's Principal spillway and Jorge Cepernic spillway were previously presented as regards the changes in the flow pattern and pressure fluctuations distribution on the stilling basin due to the incorporation of a deflector to reduce the supersaturation of TDG.

Even though the spillways and their stilling basins are certainly different in terms of unit discharge, difference of elevation between reservoir and tailwater, among others, there are common aspects observed in the flow pattern as well as the parameters RMS and $p/\gamma_{0.1\%}$.

In fact, the analysis of the results allows expressing the following statements:

- The location of the deflector is effective to deviate the direction of the flow for low discharges (frequent operation of the spillway) and strongly reduce the air entrainment towards the bottom. Under these conditions, the jet moves at the surface and establishes a recirculation flow with opposite direction with respect to that generated by the roller of the hydraulic jump.
- Spillways operation with high flows generate a deepening of the jet towards the bottom of the stilling basin, being clearly observed a separation flow behind the deflector.
- Measurements of pressure fluctuations exhibit that the presence of the deflector changes the location where turbulence intensity is higher, with maximum values of RMS displaced towards downstream and more intensity with the presence of the deflector. However, these differences of RMS values are less significant at the end of the stilling basin.
- In relation to the parameter that defines cavitation risk, the values of $p/\gamma_{0.1\%}$ decay in the separation zone behind the deflector due to the reduction of mean pressures and an increase of the pressure fluctuation.
- Despite of the reduction of $p/\gamma_{0.1\%}$ there are not cavitation risks.

6. Conclusions

After comparing the hydrodynamic behavior of the spillway for the current configuration and that resulting from the presence of the deflector, it is deduced that there is a change in the configuration of the flow and a consequent change in the records of fluctuating pressures detected in the spillway profile and in the stilling basin.

This change is observed by means of the presence of a recirculation zone as a consequence of the action of the deflector on the flow and the displacement of the beginning of the hydraulic jump, which produces an increase in the levels of turbulence in this zone.

Finally, it is important to mention that implementation of structural measures, like deflectors to reduce supersaturation of TDG, should be considered since the first phases of the project in order to avoid eventual environmental incidents or hydrodynamic conditions out of the optimum condition when measures are implemented after the completion of the dam and all their complementary structures.

7. ACKNOWLEDGMENTS

The authors want to express their acknowledgement to technical staff of Hydraulics Laboratory of National Institute of Water involved in the measurements carried out in the physical models.

8. REFERENCES

Angelaccio, C., Bacchiega, J., and Fattor, C.A. (1997). "Effects of the Spillways Operation on the Fishes Habitat: Study of Solutions", *XXVII IAHR Congress, August 1997*, San Francisco, U.S.A.

Bacchiega, J., Fattor, C.A., and Barrionuevo, H. (1999). "Environmental Considerations in the Design of Spillways: Analysis of Hydrodynamic Behavior". *XXVIII IAHR Congress, August 1999*, Graz, Austria.

Lopardo, R.A., De Lío, J.C., and Vernet, G.F. (1982). "Physical modelling on cavitation tendency for macroturbulence of hydraulic jump", *International Conference on the Hydraulic Modelling of Civil Engineering Structures*, Coventry, 1982, pp 109-121.

White, P.G. "Effects of supersaturation of dissolved gases on the fishery of the Bighorn River downstream of the Yellowtail afterbay Dam", U.S. Department of the Interior Bureau of Reclamation, Missouri Basin, Region 6, 1991.

DOI: https://doi.org/10.26077/8b70-3ba4

Investigation of energy attenuation, flow resistance and impending motion of downstream bed material in rock ramps

Vishal Singh Rawat[1], Thendiyath Roshni[1], Michele Palermo[2], Simone Pagliara[3] & Deep Roy[4]
[1]*Department of Civil Engineering - National Institute of Technology, Patna, India*
[2]*Department of Energy, Systems, Territory and Construction Engineering, University of Pisa, Pisa, Italy*
[3]*Laboratory of Hydraulics, Hydrology and Glaciology (VAW), ETH Zurich, Zurich, Switzerland.*
[4]*Department of Civil Engineering – National Institute of Technology, Tiruchirappalli, India,*
E-mail: roshni@nitp.ac.in

Abstract: *Block ramps are ecofriendly drop structures, which ensure stable downstream river bed, peculiar to flows over macro roughness elements. It uniquely serves an essential paradigm in riverine management, to encounter deliveries in an ecologically sound manner. It permits safe fish passage, stabilizes stream banks and bed profiles and creates habitat diversity. Study of flows over block ramps are quite extensive and are associated with many intricacies. In this paper, flume experiments were conducted in the hydraulic laboratory of National Institute of Technology Patna. The particle densimetric Froude number (F^*) was calculated for the mobile bed and chosen to predict the hydraulic conditions for incipient sediment motion in the uniform beds and it is compared with the established literature formulations to estimate the stability of the beds under large-scale roughness conditions. Scour volumes from each experimental run is quantified and the intensity of sediment motion was determined. Experimental data analysis allowed describing the mechanism of incipient motion of the mobile bed in the stilling basin, energy dissipation, flow resistance and in the interstitial flow over rock ramps and intensity of block movement for various flow regimes. It was found that the dimensionless shields stress increases with ramp slope and the intensity of ramp motion decreases with the shear stress. The results indicated that friction resistance increases with relative submergence for the tested range of experiments. A comparison with previous literature with uniform sediment transport indicates that relative roughness of block is responsible for increasing the dimensionless Shields stress. The results of physical testing can be used to assess and predict the effective dissipation of energy and its impact on the stability of rock structures.*

Keywords: *Densimetric Froude number, energy dissipation, frictional resistance, incipient motion, intensity, rock ramps.*

1. Introduction

The restoration of longitudinal connectivity in riverine systems is one of the major requirements in river rehabilitation. Rock ramps or block ramps, also known as fish ramps, have notably replaced many anthropogenic barriers such as dams and weirs under reasonable costs. Not only had they satisfied the hydraulic criteria for fish migrations by stabilizing stream bed and limit the channel scour as well. Rock ramps are generally small section of a slopy channel which spans the entire river width, generating large-scale roughness in the form of boulders. They are generally used in mountainous terrains and differ by their bed material, size of boulders (mean diameter D ranging from 0.3 m to 1.5 m) and their arrangement. In lateral hydraulic structures, a huge quantity of kinetic energy gets dissipated by generating hydraulic jump at the toe. But a significant amount of energy (30-50%) still remains in the flow after the jump. This remaining energy, if not taken care, has high potential to further erode the bed. Rock ramps results in high energy dissipation due to confined variation of slope and presence of large roughness elements. Several investigations have been done to correlate the amount of energy dissipation and various ramp characteristics. Platzer (1983) performed experiments on block ramps with slope S equal to 0.1; energy dissipation on stepped channels for skimming flow conditions was explored by Christodoulou (1993). Empirical equations for energy dissipation at nappe and skimming flow conditions was given by Chanson (1994). Pagliara and Chiavaccini (2006a and 2006b) carried out a systematic study on energy dissipation in reinforced rock ramps. Ghare et al. (2010) proposed a dimensionless mathematical model correlating the bed material and step height of the rock ramp. Later, Pagliara et al. (2015) studied energy dissipation characteristics of both transition and uniform flow conditions in a tapered channel over large-scale roughness condition.

Rock ramps also provide high flow resistance due to the presence of traction, wake vortices, localized hydraulic jump and jetting flow bounded by every rough element. Extensive studies on flow resistance in block ramps have been done so far. Bathurst (1978) distinguished three roughness conditions, i.e., large-scale roughness $h/d_{84}<1.2$, intermediate-scale roughness $1.2< h/d_{84}<4$ and small-scale roughness $4 < h/d_{84}$, where h is the flow depth and d_{84} is the characteristic diameter of the bed. Morris (1961) and Baiamonte and Ferro (1997) investigated the outcome of large-scale roughness on frictional resistance. An experimental study on hilly sloped channels with angular rip rap was conducted by Abt et al. (1988) to determine the effect of slope on frictional resistance. Bathurst (2002) proposed the semi-logarithmic resistance relationship from in-situ site investigations. Frictional resistance for steep slope beds was investigated by Rice et al. (1998). In current research, Pagliara and Chiavaccini (2006c) examined frictional resistance for reinforced block ramps having projected boulders. Dey and Raikar (2007) evaluated rough bed friction for homogeneous bed for slope range of $0.28 < S < 0.80$. Pagliara et al. (2010) suggested that Darcy-Weisbach friction factor for rock chutes is a function of relative equivalent depth and chute inclination. Oertel and Bung (2015) analyzed scour and bed stability in cross bar block ramps. Although several investigations on rough inclined beds on in-situ and field situations have been performed, the effect of flows over block ramps in large-roughness conditions with steep slopes is yet to be explored.

Similarly, particle stability in the downstream stilling basin is also a salient feature for the design of block ramps. The bed material resting at the downstream of rock ramps must remain steady unless the shear stress exceeds the critical counterpart. Hence a critical flow parameter must be identified and investigated. Particle stability is generally estimated by comparing acting shear stress with the critical counterpart (Aguirre Pe et al. 2003). The most notable work on the initial movement of sediment particles has been performed by Shields (1936). Shields (1936) empirically established a functional relationship between critical shear stress τ_c^* and grain Reynolds number R_e^*, using different flume data. The original Shields diagram has been reproduced and modified by many researchers. Some significant deviations of the observed critical shear stress from those predicted by Shields curve can be observed. For instance, Knoroz (1953) obtained a threshold curve from experiments with sand without the characteristic dip in the transitional zone. Furthermore, Shields curve has been modified using extensive flume studies by Miller et al. (1977). The revised threshold curve was drawn for rough, turbulent flows with a similar shape as original Shields graph but with a lower value of $R_e^*= 0.045$. Similarly, Yalin and Karahan (1979) derived the modified graph using empirical evidence. Overall, the compiled data sets by both investigators exhibited significant scatter for rough, turbulent flow conditions.

Profuse studies have been done based on particle motion so far. Rajaratnam and Macdougall (1993) performed flume experiments for homogeneous sand beds and disclosed the dependency of densimetric Froude number on maximum scour depth. A comprehensive work on the motion of bed sediment in a trapezoidal flume channel was carried out by Kovacs and Parker (1994). Later, Aguirre Pe et al. (2003) determined particle movement based on critical particle densimetric Froude number (F_c^*) and relative depth ($d/D<10$), with d indicating the flow depth and D the characteristic diameter of the bed material. The outcomes of study indicated that F_c^* can be a better interpreter for particle motion in steeper slopes ($0.005<S<0.10$) and large-roughness conditions ($0.2<d/D<10$). Oertel and Bung (2015) also investigated the stability of cross bars in terms of critical particle densimetric Froude number. Rawat et al. (2021) conducted flume investigations for a range of flow rates and ramp slopes, involving two homogeneous stilling basin materials and an additional arrangement with rock sills. In doing so, authors analyzed the effect of d/D on the incipient motion at various bed configurations.

Although many studies have been carried out to determine the application of rock ramps with steeper gradient for dissipation of energy and to maintain reasonable riverine stability, a more comprehensive research is needed to improve current design approaches of rock ramps. The present experimental study investigates energy dissipation and frictional resistance on the ramp. Furthermore, stilling basin material stability is also studied. Large-roughness conditions were reproduced, and the hydraulic process was analyzed.

2. Experimental programme

Experiments were carried out in hydraulics engineering laboratory of Department of Civil Engineering, National Institute of Technology Patna, Bihar, India. Experiments were planned to explore the flows over rock ramps and its characteristics in the downstream stilling basin. Tests were conducted for different ramp slopes in a 0.8 m wide, 0.50 m high and 7 m long flume (Fig. 1). The walls of the flume were made by glass to facilitate visualization during the experiments. The recirculating flume includes a downstream plunge pool to receive water and send it back upstream through an excavated channel in the ground connecting both sides of the flume. The ramp was placed at a distance of 2 m from the outlet to guarantee a well distributed flow across the flume. A point gauge is mounted on the flume for measurement. The flume is kept horizontal during the experiments and the slope of block ramps were changed at the beginning of each test. All experiments were conducted with a fixed ramp length of 1.5 m; whereas the height of the ramp varied to adjust the slope S that ranged from 0.1 to 0.2. Table 1 reports the ranges of variation of tested parameters for this present work.

Table 1. Ranges of experimental data

Hydraulic Parameters	Present study
Ramp slope, S (%)	10-20
Discharge, Q ($m^3 s^{-1}$)	0.02-0.04
F_c^*	2.41-3.89

Figure 1. Photograph of experimental setup with ramp (top right side) and bed material (bottom right side)

3. Results and discussion

3.1. Energy dissipation

In the present study, energy dissipation mechanism over rock ramps was analyzed. The relative energy dissipation (ΔE_r) between the inlet and the toe of rock ramps was evaluated for various slopes (i.e., $0.1 \leq S \leq 0.2$) and plotted as function of Y_c/H (Pagliara and Chiavaccini 2006a), where Y_c is the critical depth and H is the height of the ramp. Figure 2 represents the variation of relative energy dissipation with Y_c/H derived from present study and previous investigators. Results show that relative energy dissipation $\Delta E_r=(E_0-E_2)/E_0$ (with E_0 and E_2 indicating the total energy at the beginning and the toe of the ramp, respectively) decreases with Y_c/H for similar flow rates, ramp inclinations and rock ramp length. Consequently, the amount of energy dissipation depends on ramp height, all other parameters being constant. Moreover, the present experimental data lie in the large-scale roughness condition which tends to dissipate more energy compared to small-scale and intermediate-scale roughness counterpart. The present experimental values were contrasted against empirical expression for base block ramps developed by Pagliara and Chiavaccini (2006a).

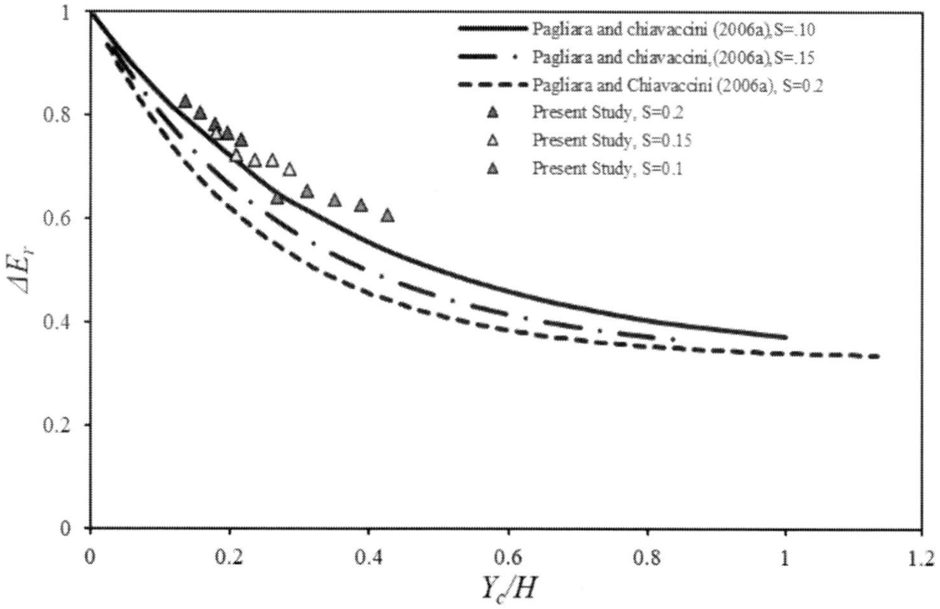

Figure 2. Variation of relative energy dissipation with Y_c/H for flows over rock ramps

3.2. Flow resistance

In the present study, experimental runs have been performed in rough slopy beds to estimate the effect of ramp roughness on flow resistance. Figure 3 depicts the variation of friction factor with relative roughness (Y_c/D_{50}), where D_{50} is the characteristic bed diameter of the ramp. Following the classification of roughness scales made by Pagliara and Chiavaccini (2006), present tests (with $0.5 \leq Y_c/D_{50} \leq 0.8$) pertain to large-scale roughness condition (i.e., $Y_c/D_{50} < 2.5$). Overall, Figure 3 indicates that the Darcy-Weisbach friction factor $(8/f)^{0.5}$ (estimated using the equation proposed by Rice et al. (1998)) is a function of the relative roughness (Y_c/D_{50}) and the ramp slope. Namely, the parameter $(8/f)^{0.5}$ is a monotonic decreasing function of Y_c/D_{50} and S., i.e., lesser the ramp slope, more will be the interaction between the flow surface and ramp material, and, consequently, higher will be the drag resistance.

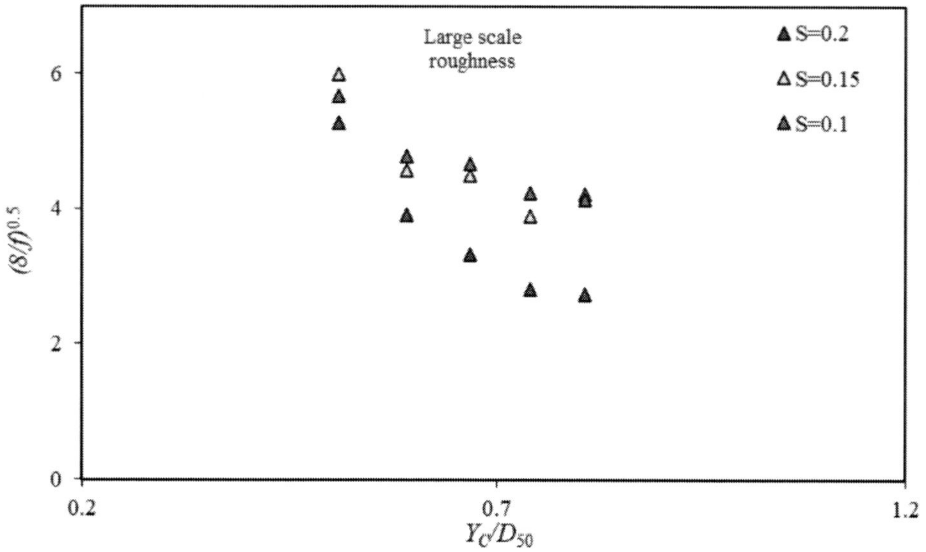

Figure 3. Variation of friction factor $(8/f)^{0.5}$ with Y_c/D_{50} for large-scale roughness condition.

Flow resistance as a function of Froude number at the toe of the ramp is also shown in Fig. 4. Data from previous literature (Abt et al. 1988, Rice et al. 1998, and Pagliara and Chiavaccini 2006) were also included for comparison. Overall, present results appear to be consistent with those of former investigations. Namely, the flow resistance increases with ramp slope. In addition, for large-scale roughness, Froude number also affects flow resistance around ramp roughness elements (via the drag force).

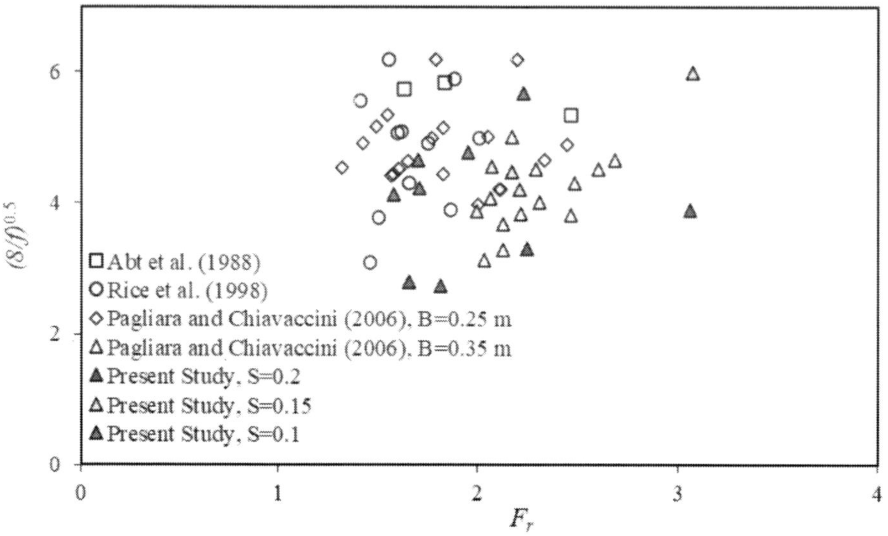

Figure 4. Variation of Friction factor $(8/f)^{0.5}$ with Froude number F_r

3.3. Mobilization of stilling basin material

As for the stilling basin sediment stability, the densimetric Froude number (F^*) at the toe of the ramp was calculated. The densimetric Froude number is defind as $F^* = U/[(s\text{-}1)\,gD]^{1/2}$ (where U = approaching flow velocity to the stilling basin, s=ratio of sediment and fluid densities, g=acceleration due to gravity, and D=characteristic diameter of stilling basin bed particle). Figure 5 shows experimental data from the existing study, which falls in the range of large-scale roughness 0.54<d/D<1.63 (Bathurst et al.1983). It can be observed that present values of densimetric Froude number vary between 2.1 and 3.89 for 2.44 <d/D<6.18. Figure 5 also shows dependency of densimetric Froude number on relative depth d/D (where d is the flow depth) and compared with the critical densimetric Froude number of the previous studies. The previous studies include equations proposed by Aguirre-Pe et al. (2003) and Oertel and Bung (2015) and data from literature with similar experimental approaches (e.g., Lischtvan and Libediev 1959, Bathurst et al. 1983, Bathurst et al. 1984, Neil 1967, Ashida and Bayazit 1973, Aguirre Pe 1975, Olivero 1984 and Rawat et al. 2021). This plot infers that if the ramp bed is made with similar bed material of that of stilling basin, it would have been mobilized corresponding to the incoming flow.

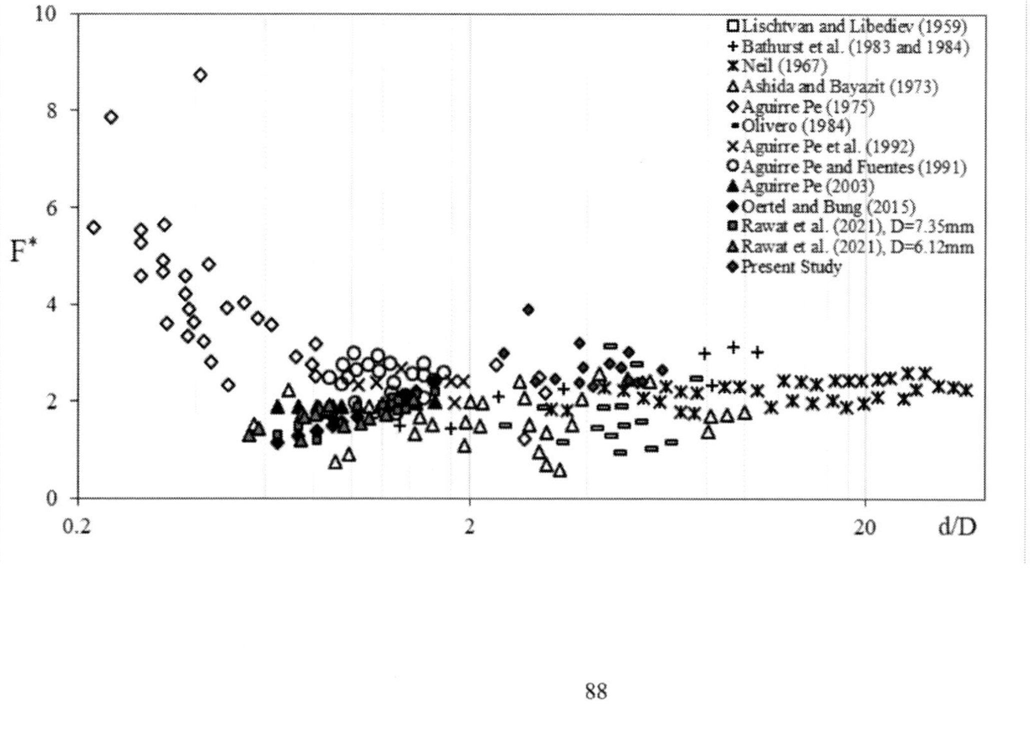

Figure 5. Variation of critical particle densimetric Froude numbers with relative roughness

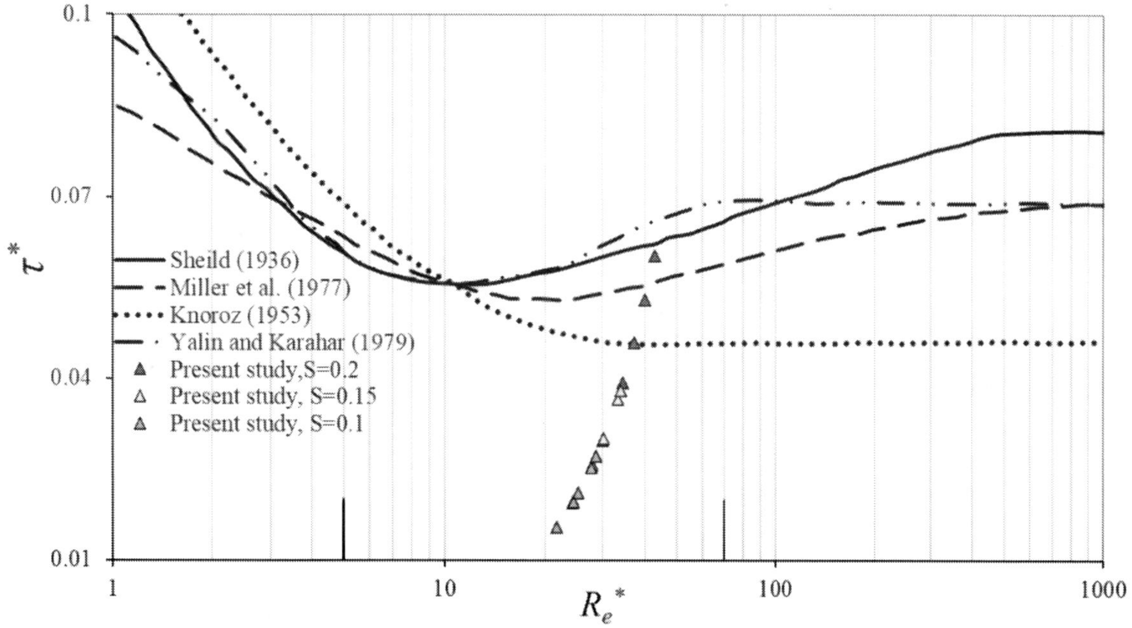

Figure 6. Dimensionless shear stress versus grain particle Reynolds number

To assess the mobility of stilling basin bed material, dimensionless shear stress (τ^*) was evaluated and compared with the critical shear stress or threshold shear stress from Shields diagram. Dimensionless shear stress is calculated as $\tau^* = \frac{\tau_b}{(\rho_s - \rho) g D_{50}}$, where τ^*= Dimensionless shear stress, τ_b = bed shear stress of the ramp, g = gravitational acceleration, D_{50} = median diameter of stilling basin, ρ_s=sediment density of stilling basin bed material, ρ = fluid density. Figure 6 depicts the variation of dimensionless shear stress for different grain particle Reynolds number ($R_e^* = \frac{V^* D}{\upsilon}$), where $V^*$is the shear velocity, and υ is the kinematic viscosity of water. Several observations were made for different flow rates, ramp slopes and bed shear stress (Shield 1936). Namely, in Fig. 6, curves relative to incipient motion conditions (obtained by several researchers for horizontal beds) are contrasted against current experimental data. Based on the traditional interpretation of Shields diagram, sediment motion is only possible if shear stress exceeds the critical counterpart (i.e., experimental points are located above transition curves). However, for flows over inclined ramps ($0.1 \leq S \leq 0.2$), mobilization of bed sediments occurred for all ranges tested in the present study. This is consistent with the findings of Bolhassani et al. (2015), who argued that, for flows over different bed slopes, the threshold shear stress cannot be evaluated via the Shields curve, which is valid only for flows over horizontal beds. In fact, present results reveal that also data below the Shield curve pertain to test characterized by particle transport. Consequently, further analyses of the effect of slope on incipient motion condition are needed.

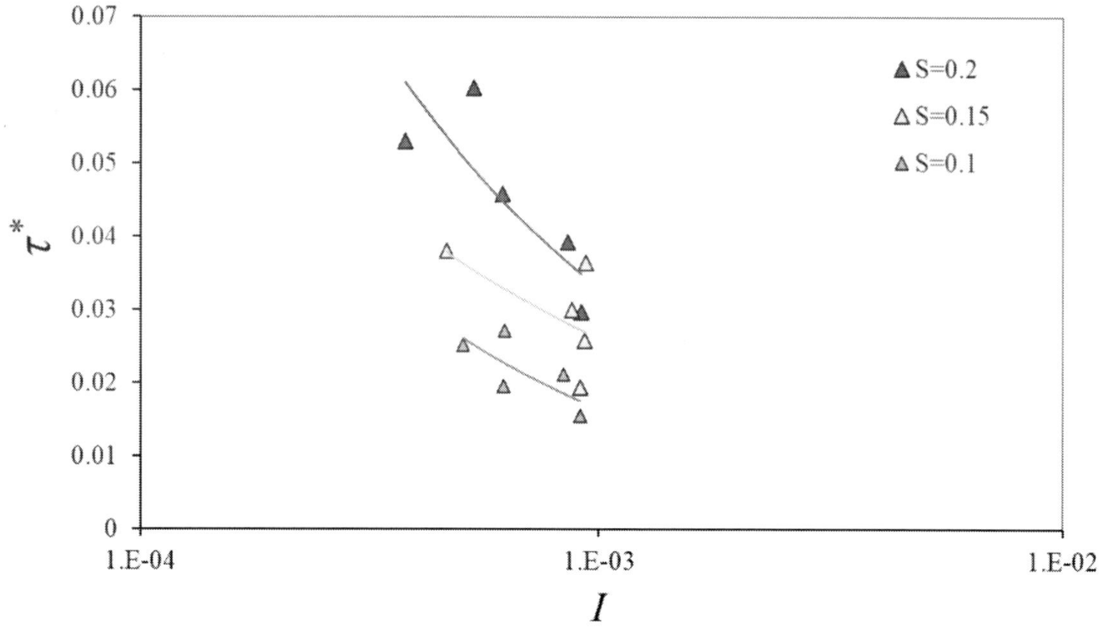

Figure 7. Variation of intensity of sediments motion with dimensionless Shields stress

Likewise, Figure 7 shows the variation of intensity of sediments motion (I) with dimensionless shields stress for flows over inclined block ramps. Intensity is a fraction of all particles in the bed surface moved every second given by $I = \frac{\mathrm{m}}{\mathrm{NT}}$, where m is the number of particle displacements observed during the time interval T on the area of the surface of a mobile bed, and N is the number of surface particles in the area A. Shvidchenko and Pender (2000) defined the intensity I as the relative number of bed sediments moving away from the ramp during each experimental run. Overall, Figure 7 evidences that, for current tests, I ranged from 10^{-4} s^{-1} (one in 10,000 surface particles in motion per second) to 10^{-3} s^{-1} (one in 1000 surface particles in motion per second). Furthermore, as it should be expected, the dimensionless shields stress decreases with the intensity of sediment motion and ramp slope.

4. Conclusions

Rock ramps are huge, macro-roughened, eco-friendly components, randomly dropped on the riverbed. The present experimental study aims to provide further understanding on energy dissipation, frictional resistance and incipient motion occurring in the flows over rock ramps. The nature of flows over rock ramps and the dissipative mechanism under large-scale roughness conditions have been investigated. The dependence of energy dissipation on the dimensionless parameter Y_c/H revealed that ΔE_r decreases with the ramp slope. Experimental evidence revealed that the Darcy-Weisbach friction factor is a function of relative roughness (Y_c/D_{50}), ramp inclination and incoming Froude number. Experimental data were compared with those derived from previous literature to corroborate present analysis. Namely, incipient motion of bed sediment for rock ramps and the effects of ramp slopes on the dimensionless shields stress (τ^*) were also analysed. To describe the flows over rock ramps, the densimetric Froude number (F^*) was calculated and plotted as a function of relative depth (d/D). It was found that the dimensionless shields stress (τ^*) decreases with the intensity of sediment motion. Conversely, an opposite trend occur for increasing values of ramp slope and R_e^*.

5. ACKNOWLEDGMENTS

The first two authors (V.S. Rawat and T. Roshni) received funding from DST-SERB [Grant no. ECR/2017/000984].

6. REFERENCES

Abt, S. R., Wittier, R. J., Ruff, J. F., and Khattak, M. S. (1988). "Resistance to flow over riprap in steep channels." *Journal of the American Water Resources Association (JAWRA)*, 24(6), 1193-1200.

Aguirre Pe, J. (1975). "Incipient erosion in high gradient open channel flow with artificial roughness elements." *Proc., 16th Congr. Int. Assoc. Hydraul. Res.* San Paulo Brazil.2, 137-180.

Aguirre Pe, J., Olivero, M.I.A.L., and Moncada, A.T. (2003). "Particle densimetric Froude number for estimating sediment transport." *Journal of Hydraulic Engineering.*, 29(6), 428-437.

Ashida, K., and Bayazit, M. (1973). "Initiation of motion and roughness of flows in steep channels." *Proc., Int. Assoc. Hydraul. Res., Istanbul*, Turkey, 475–484(1).

Baiamonte, G., and Ferro, V. (1997). "The influence of roughness geometry and Shields parameter on flow resistance in gravel-bed channels." *Earth Surface Processes and Landforms*, 22(8), 759-772.

Bathurst, J. C. (2002). "At-a-site variation and minimum flow resistance for Mountain Rivers." *Journal of Hydrology*, 269(1-2), 11-26.

Bathurst, J. C., Graf, W. H., and Cao, H. H. (1983). "Initiation of sediment transport in steep channels with coarse bed material." In B. M. Summer and A. Muller (Ed.). *Mechanics of sediment transport* (pp. 207-213). Rotterdam, The Netherlands.

Bathurst, J.C. (1978). "Flow resistance of large-scale roughness." *Journal of the Hydraulics Division*, 104(12), 1587-1603.

Bolhassani, R., Afzalimehr, H., and Dey, S. (2015). "Effects of relative submergence and bed slope on sediment incipient motion under decelerating flows". *Journal of Hydrology and Hydromechanics*, 63(4), 295.

Christodoulou, G. C. (1993). "Energy dissipation on stepped spillways". *Journal of Hydraulic Engineering*, 119(5), 644-650.

Dey, S., and Raikar, R.V. (2007). "Characteristics of loose rough boundary streams at near-threshold." *Journal of Hydraulic Engineering*, 133(3), 288-304.

Ghare, A.D., Ingle, R.N., Porey, P.D., and Gokhale, S.S. (2010). "Block ramp design for efficient energy dissipation." *Journal of Energy Engineering*, ASCE, 136(1), 01–05.

Knoroz, V. S. (1953). "Non-eroding velocity for fine sand." *Hydrotech. Canstr., Engl. Transl.*, 8, 21-24 (in Russian).

Kovacs, A., and Parker, G. (1994). "A new vectorial bedload formulation and its application to the time evolution of straight river channels." *Journal of fluid Mechanics*, 267,153-183

Lischtvan, L.L. and Lebediev, V.V., (1959). "Gidrologia i gidraulika v mostovom doroshnom, straitielvie". *Hydrology and Hydraulics in Bridge and Road Building.*

Miller, M. C., I. N. McCave, and Komar, P. D. (1977). "Threshold of sediment motion under unidirectional currents." *Sedimentology*, 24 (4), 507-527.

Morris, H.M. (1961). "Design methods for flow in rough conduits". *Transactions of the American Society of Civil Engineers*, 126(1),.454-473.

Neill, C. R. (1967). "Mean-velocity criterion for scour of coarse uniform bed-material." *Proc., 12th Congress Int. Assoc. Hydraul. Res.,* 3, 46–54.

Oertel, M., and Schlenkhoff, A. (2012). "Crossbar block ramps: flow regimes, energy dissipation, friction factors, and drag forces." *Journal of Hydraulic Engineering,* 138 (5), 440–448.

Oertel, M., and Bung, D.B. (2015). "Stability and scour development of bed material on crossbar block ramps". *International Journal of Sediment Research*, 30(4), 344-350.

Olivero, M. L. (1984), "Movimiento Inciiente de Particulas en Flujo Torren-cial", University of Los Andes, Meridad, Venezuela. 169.

Pagliara, S., and Chiavaccini, P. (2006a). "Energy dissipation on block ramps." *Journal of Hydraulic Engineering*, 132 (1), 41–48.

Pagliara, S., and Chiavaccini, P. (2006b). "Energy dissipation on reinforced block ramps." *Journal of Irrigation and Drainage Engineering*, 132 (3), 293–297.

Pagliara, S. and Chiavaccini, P. (2006c). "Flow resistance of rock chutes with protruding boulders." *Journal of Hydraulic Engineering*, 132 (6), 545–552.

Pagliara, S., Carnacina, I. and Roshni, T., (2010). "Self-aeration and friction over rock chutes in uniform flow conditions". *Journal of Hydraulic Engineering*, 136(11), 959-964.

Pagliara, S., Roshni, T., and Palermo, M. (2015). "Energy dissipation over large-scale roughness for both transition and uniform flow conditions." *International Journal of Civil Engineering*, 3(3), 341-346.

Platzer, G. (1983). "Die hydraulik der breiten blocksteinrampe—eine Bemessungsgrundlage." Bundesanstalt für Wasserbauversuche und hydrometrische Prüfung, Vienna, Austria (in German). gr.-8°, 81 p., [60] Bl.

Rajaratnam, N., and Macdougall, R.K., (1983). "Erosion by plane wall jets with minimum tailwater." *Journal of Hydraulic Engineering*, 109(7), 1061-1064.

Rawat, V.S, Roshni, T., and Carnacina, I. (2021). Experimental study of flows over block ramps on stability of non-cohesive beds. *Arabian Journal for Science and Engineering*. 46, 10333–10339 DOI : 10.1007/s13369-021-05348-6

Rice, C. E., Kadavy, K. C., and Robinson, K. M. (1998). "Roughness of loose rock riprap on steep slopes." *Journal of Hydraulic Engineering*, 124(2), 179-185.

Shields, A.(1936). "Anwendung der Aehnlichkeitsmechanik und der Turbulenzforschung auf die Geschiebebewegung". PhD Thesis Technical University Berlin.

Shvidchenko, A., and Pender, G. (2000). "Flume study of the effect of relative depth on the incipient motion of coarse uniform sediments." *Water Resour. Res*, 36(2), 619-628.

Yalin, M.S. and Karahan, E., (1979). "Inception of sediment transport." *Journal of the hydraulics division*, 105(11), 1433-1443.

DOI: https://doi.org/10.26077/9da6-8da5

Model Studies for Desilting Basin for Teesta-VI H.E. Project, Sikkim – A case study

M.Z. Qamar[1], M.K. Verma[2], A.P. Meshram[3] & Dr. Neena Isaac[4]
[1]*Scientist 'C', Central Water and Power Research Station, Pune, India*
[2]*Scientist 'D', Central Water and Power Research Station, Pune, India*
[3]*Assistant Research Officer, Central Water and Power Research Station, Pune, India*
[4]*Scientist 'E', Central Water and Power Research Station, Pune, India*
E-mail: ziaul_qamar@rediffmail.com

Abstract: *Desilting basins play an important role in run-of-river hydro power projects on Himalayan Rivers. These rivers carry huge amount of sediment with them, due to steep slopes and fragile geology of the region. The suspended sediment enters through the power intake and ultimately the power house. This causes heavy damage to the turbines and other under water parts. Therefore, desilting basins are provided to eliminate suspended sediment from the water conductor system. The design of desilting basin is verified on a physical model for 90% removal of suspended sediment coarser than 0.2 or 0.3 mm and efficacy of flushing tunnel below in transporting the settled sediment. Central Water and Power Research Station, Pune has conducted physical model studies for desilting basin for various hydro power projects in India and neighbouring countries. One such study, Hydraulic model studies for desilting basin for 500 MW Teesta-VI Hydro Electric Project, Sikkim is presented in this paper. Various design parameters such as length of desilting basin, length and bed slope of inlet transition, outlet transition, size of silt flushing tunnel, size and spacing of openings connecting main basin with silt flushing tunnel etc. were tested during the model studies. The results in terms of settling efficiency for 0.3 mm size particles were obtained.*

Keyword: *Desilting basin, suspended sediment, settling efficiency, silt flushing tunnel.*

1. INTRODUCTION

In India majority of untapped hydro power potential is in Himalayan Region. Many run-of-river hydro power schemes are on perennial Himalayan Rivers. Due to steep mountainous slopes, these rivers possess large head for power generation but steep slopes and fragile geology of the catchment cause erosion. Rivers carry heavy sediment load especially in monsoon/ high flood season *(Qamar, 2010)*. The settled bed load in the reservoir is periodically flushed out using large sluiceways provided near the river bed. However, the suspended sediment portion enters the power intake and causes heavy damage to turbines and other under water parts in the power house. To deal with this suspended sediment, desilting chambers are provided in the water conductor system. With increase in cross sectional area and reduction in forward velocity of flow, these chambers induce the settlement of suspended sediment. This minimizes entry of sediment to the power house and ensures the safety and longevity of turbines and other under water parts. The settling sediment in desilting chamber needs to be flushed out to maintain its settling efficiency. Continuous hydraulic flushing of desilting chambers is possible by allowing the deposited sediment and some water to escape from the bottom of the chamber. Such an approach would need water in excess of the design discharge (about20% extra) being admitted into the canal / tunnel at its head. The discharge downstream of the chamber would be the design discharge of the turbine. This continuous flushing of the desilting chamber is achieved by providing silt flushing tunnel below the chamber and connecting it with the main chamber by provision of openings at the bottom slab of the desilting chamber. A typical layout showing different components of desilting chamber and its cross sections shown in figure 1.

Central Water and Power Research Station, Pune has conducted several physical model studies for desilting basins for various hydro power projects in Himalayan Region. Some of the prominent ones include hydro power projects in India such as Chamera Stage I, II and III, Parbati Stage II and III, Uri Stage I& II, Teesta Stage IV and V, Nathpa Jhakri, etc. The hydro power projects outside India include Trishuli (Nepal) and Tala, Mangdechhu, Punatsangchhu Stage I and II and Kholongchhu (Bhutan).

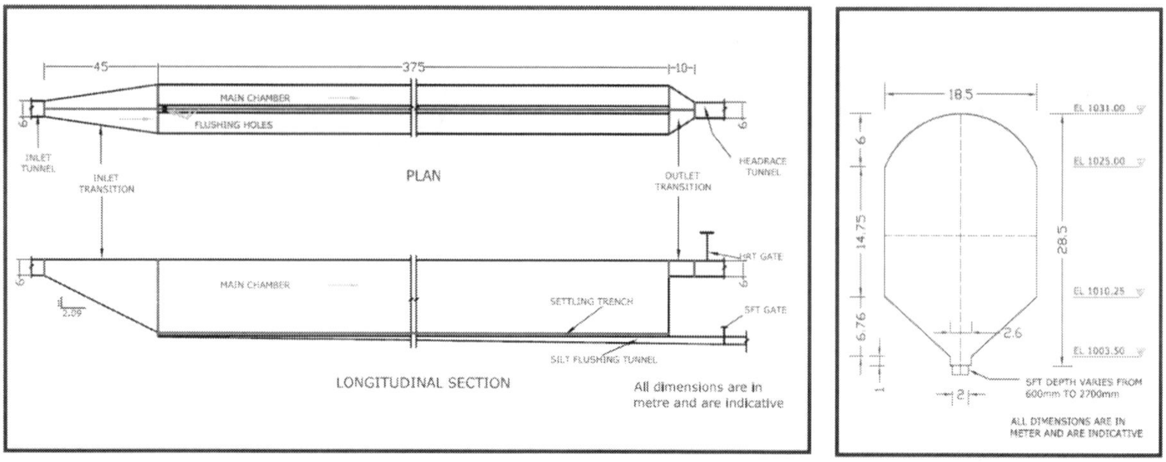

Figure 1.Typical layout and cross section of desilting chamber

These model studies are generally conducted on a geometric scale of 1:25 to 1:35. For simulation of suspended sediment low specific gravity crushed and sieved walnut shell powder is used. Desilting basins are generally designed for 90% removal of suspended sediment of size 0.2 mm and above for a maximum sediment concentration of 4000 to 5000 ppm. These model studies are conducted predominantly for following two purposes:

a. To find out the settling efficiency of desilting basin to the desired extent.
b. To find out the flushing efficacy of silt flushing tunnel below desilting basin,

Hydraulic model study for desilting basin for 500 MW Teesta - VI Hydro Electric Project, Sikkim, India is included in this paper.

2. PROJECT DETAILS

Teesta Stage-VI Hydroelectric Project located in Southern Sikkim is a run-of-the river scheme with a diversion barrage located about 100 m downstream of L. D. Kazi Bridge at Sirwani. The project has an installed capacity of 500 MW with four units of 125 MW each. The project is a part of the hydropower rich Teesta Cascade development scheme and is located downstream of Teesta–V project. The schematic diagram showing cascade development of Teesta river basin is shown in figure 2.

The project has been planned with a 26.5 m high barrage, two of desilting basins (surface) and silt flushing tunnel, two Intakes, two Head Race Tunnels and two surge shafts, four vertical steel lined pressure shafts, MIV cavern, Power house cavern, Transformer cavern and four tail race tunnels. The water conductor system and power house are located on the right bank of river Teesta. The regulation diversion structure is proposed to be 83 m wide, 26.5 m high and 130 m long barrage, comprising of five bays of size 15 m (W) x 17.5 m (H), across the river. Besides regulating the discharge, the barrage shall also raise the water level and provide a higher head for power generation. The top of the barrage is at El 369 m. FRL for the project has been kept at El 360 m and MDDL at El 354 m to provide storage for peaking during lean flow period and envisages utilization of a rated head of 105.4 m. The regulated releases from the Teesta Stage-V power station shall augment the peaking capacity of the project. The project shall be operated in tandem with Teesta-V Power station which is just upstream of barrage. The water in excess of the requirement of power generation shall be passed through spillway bays, regulated by radial gates. The crest level of the barrage has been kept at El 342.5 m near the river bed for reservoir flushing. A stilling basin has been provided for energy dissipation.

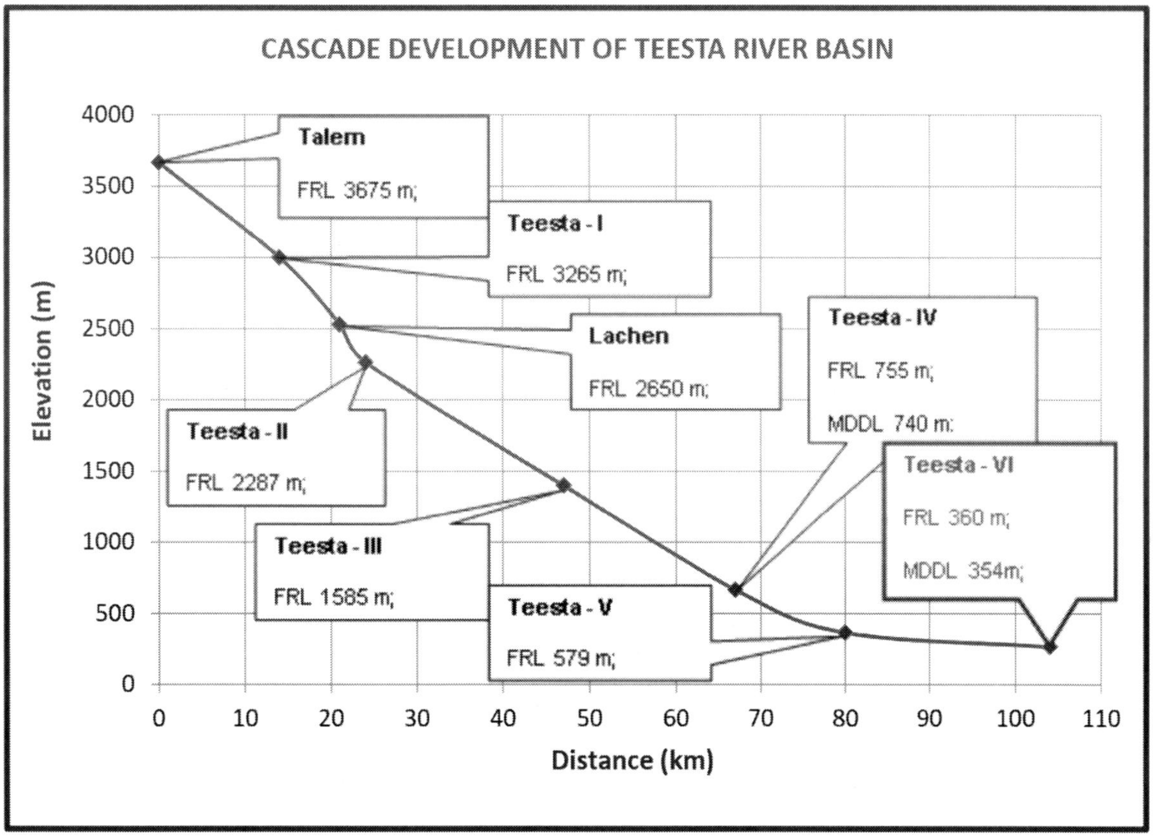

Figure 2: Cascade development of Teesta river basin

The head regulator consists of four bays with invert level at El 348.5 m and four gates of size 8 m x 11.5 m to pass an inlet discharge of 610.65 m^3/s (including 15% flushing discharge of 79.65 m3/s). Length of head regulator is 51 m on the barrage side and 130 m on the hill side. Two 225 m long desilting basins have been provided for exclusion of sediment size of 0.3 mm and above. Two intake structures are proposed in the downstream of desilting basin to divert the water in to HRT. Two modified horse shoe shaped Head Race Tunnel (HRT) of 9.8 m diameter and 13.76 km long are being excavated to cater a design discharge of 531 m3/s (265.5 m3/s through each tunnel). An underground power house with four generating units of 125 MW each is envisaged. The rated head is 105.4 m. A 247 m long, 8.5 m (W) x 7.5 m (H) D-Shaped Tail race tunnel conveys the turbine discharge back to the river. The project layout plan is shown in figure 4.

2.1. Desilting basins and sediment data

The proposed desilting arrangement comprises two 225 m long desilting basins including four cunettes for exclusion of suspended particles coarser than 0.3 mm size and maximum sediment concentration of 4000 ppm for efficient, trouble free and continuous operation of turbines with least possible wearing and erosion damages. The size of desilting basin (DB 1) is 25.25 m (W) X 25.57 m (H). The size of desilting basin (DB 2) is 25.25 m (W) X 23.65 m (H). The basins have been equipped with gates on downstream end for isolation and maintenance. It was expected that this basin would intercept the majority of the sediment bearing river flow at normal river discharges.

The inlet discharge for each desilting basin is 305.33 m^3/s including flushing discharge of 39.83 m^3/s. The average flow through velocity in the desilting basin works out to be 0.5 m/s. Average concentration of the suspended sediment computed on the basis of sediment data supplied by the project authority as percentage of coarse (> 0.2 mm), medium (0.075 mm - 0.2 mm) and fine (< 0.075 mm) sediment is 86%, 10.25% and 3.75% respectively. Analyzing this data based on Camp's criteria (Camp, 1946), the overall analytical settling efficiency of desilting chamber (prototype) worked out to be 72.96 % (RWL at FRL El 360 m). Based on this analysis, settling efficiency curve for various sizes of particles is shown in figure 3. From this curve, it is seen that the settling efficiency of 0.3 mm size particle would be 67%.

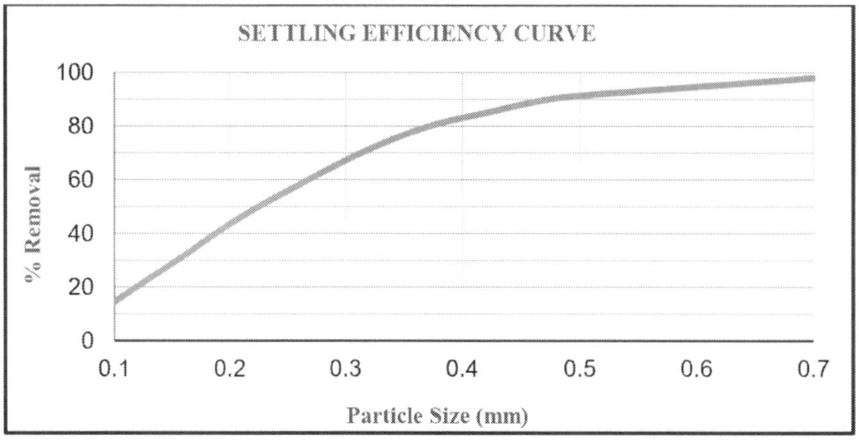

Figure 3: Analytical settling efficiency curve for prototype

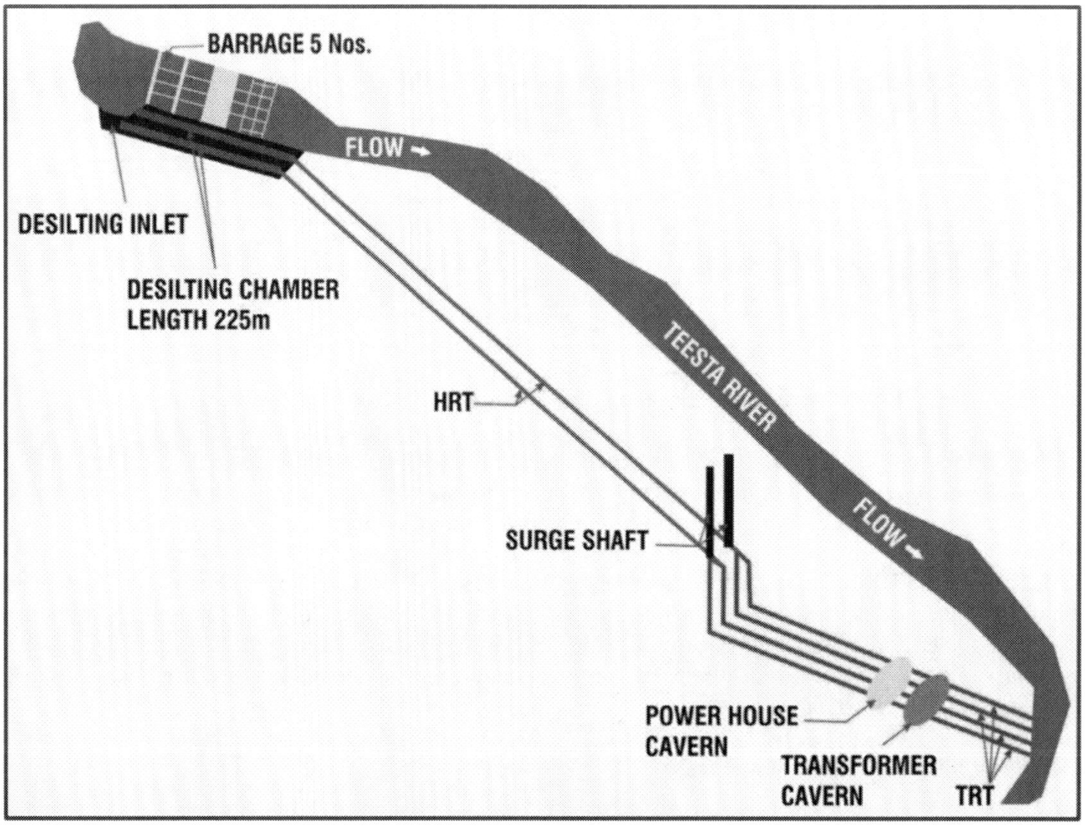

Figure 4: Project layout plan

3. MODEL STUDIES

One unit of desilting basin (DB1 with cunettes 1 & 2) was fabricated to a scale of 1:25 geometrically similar, in fiber glass with transparent perspex windows to observe the flow conditions and sediment movement / deposition pattern. Silt flushing tunnels below desilting basin were fabricated in full transparent sheets. The model installed at Sediment Management Division of CWPRS is shown in photo 1 to 4. Sediment traps of adequate sizes were provided on downstream of desilting basin at the outlet of head race tunnel (HRT) and silt flushing tunnel (SFT) for collecting the sediment flowing towards HRT and passing through flushing tunnel, separately (CWPRS 2021).

For measurement of inlet discharge, a standing wave flume was provided and triangular thin plate weirs were provided conforming to the Indian Standards on downstream end of traps for discharge measurement of HRT and flushing tunnel separately.

Photo 1: Downstream view of model

Photo 2: View of Inlet Transition

Photo 3: Upstream view of model

Photo 4: Model in running condition

3.1. Desilting basins and sediment data of Teesta VI HEP

The inlet transition with a bed slope of 1(V) : 2.114 (H) is provided before the start of main desilting basin to diffuse the flow and transport of sediment without deposition on its bed. It is seen from model experiments that entry of water is smooth from the head regulator and there was no return flow observed near the inlet transition in the initial reaches of desilting basin as shown in photo 5. Therefore, the flow conditions in inlet transition were satisfactory and it is adequate for desired flow diffusion and transport of sediment without deposition on its bed. To skim the sediment free top layers of water, a vertical wall at the end of the desilting basin and outlet transition have been provided. The performance of outlet transition was also found to be satisfactory.

Photo 5: Flow conditions near Inlet Transition

3.2. Velocity observation in the model

The velocities were measured in model in the entire length of 225 m long desilting basin at every 45 m interval, inlet transition, head regulator bays, at the upstream of head regulator and outlet transition and data acquisition was done with Acoustic Doppler Velocimeter (ADV). The velocities were taken at 0.2d, 0.6d and 0.8d on vertical depth. The velocities were also measured and confirmed in the model with Ott propeller model current meter. The velocities for reservoir water level at El 360 m (FRL) are shown in Table 1 and the average velocity in desilting basin is 0.523 m/s.

Table - 1

Velocity observation at El 360 m (FRL) in m/s					
Depth on vertical / Point of measurement	**0.2d**	**0.6d**	**0.8d**	**Average velocity**	
In front of head regulator (cunette 1)	0.595	0.600	0.690	0.628	
In front of head regulator (cunette 2)	1.250	0.940	0.690	0.960	
In bay of HR (cunette 1)	1.585	1.655	1.320	1.520	
In bay of HR (cunette 2)	2.000	2.015	1.910	1.975	
inlet transition	1.355	1.110	0.315	0.927	
At start of basin	1.334	0.662	0.221	0.739	
At 45 m from start of basin	0.960	0.662	0.456	0.693	
At 90 m from start of basin	0.610	0.533	0.610	0.584	0.523
At 135 m from start of basin	0.432	0.216	0.360	0.336	
At 180 m from start of basin	0.336	0.312	0.408	0.352	
At end of basin (225 m)	0.706	0.379	0.226	0.437	
At Outlet transition	0.465	1.325	0.880	0.890	

As per CWPRS Manual (CWPRS 2005) for design of desilting basins and past experience of conducting hydraulic model studies, if the limiting forward velocity in the desilting basin (for removal of 0.3 mm and above particle) is not more than 0.35 m/s, then the sediment particles once settled would not be thrown again in separation/suspension. It implies that cross sectional area provided for desilting basin should be sufficient to reduce the flow through velocity below 0.35 m/s.

If we consider the existing design of desilting basin system and site constraints in the present case as explained by project engineers, there is an unavoidable oblique/curved approach flow conditions at the inlet of head regulator.

Moreover, velocity observation shows that velocities are higher than the limiting velocity and with oblique approach flow conditions in this case, this observed velocity would not be favourable from sediment settlement point of view. Due to this, the settling efficiency would not be at par with conventional desilting basin.

3.3. Estimation of settling efficiency

For estimation of settling efficiency, the model experiments were conducted by simulating intake discharge equivalent to 305.3 m^3/s at the inlet of desilting basin, which comprises HRT discharge of 265.5 m^3/s and flushing discharge of 39.8 m^3/s. The water level was maintained at El 360 m (FRL). For conducting experiments, crushed and sieved walnut shell powder having low specific gravity of 1.32 was used in the model for simulation of prototype suspended sediment on the basis of fall velocity criteria. Studies were conducted with suspended sediment concentration of 4000 ppm.

Model experiments provided only the overall removal efficiency and not for different particle sizes. The inlet sediment quantity of 960 litres was injected at the rate of 4000 ppm. Sediment quantities of 768 litres and 192 litres were collected from SFT and HRT collection chambers respectively. The average overall settling efficiency works out to be 80% by physical model results. While conducting studies, it was observed that size of the flushing tunnel below desilting basin was adequate for efficient transport of settled sediment. No deposition was observed on side slopes of the hoppers of desilting basin. The dunes were accommodated within the settling trench and openings provided in model for flushing the settled sediment were working efficiently. The overall performance of desilting basin was found to be satisfactory during the conduct of experiments.

Analytical settling efficiency of the desilting basin for model parameters was calculated by Camp's criteria for intake discharge of 305.3 m^3/s including 39.8 m^3/s flushing discharge. The analysis was done for three samples of walnut shell powder used in model experimentation and average overall settling efficiency was calculated.

The average overall settling efficiency based on Camp's criteria worked out to be 77 % for model parameters. The settling efficiency by model experiments is 3% higher than that obtained by analytical estimation by Camp's criteria. On the basis of results for model parameters for three samples of walnut shell powder the settling efficiency curves of various sizes particles used in the model and its equivalent size of prototype material were prepared and are shown in figure 5. From curve for prototype, it would be seen that the settling efficiency of desilting basin would be of the order of 67.5% for particle having 0.3 mm size. If the difference of 3% is considered, the settling efficiency works out to be 70.5%.

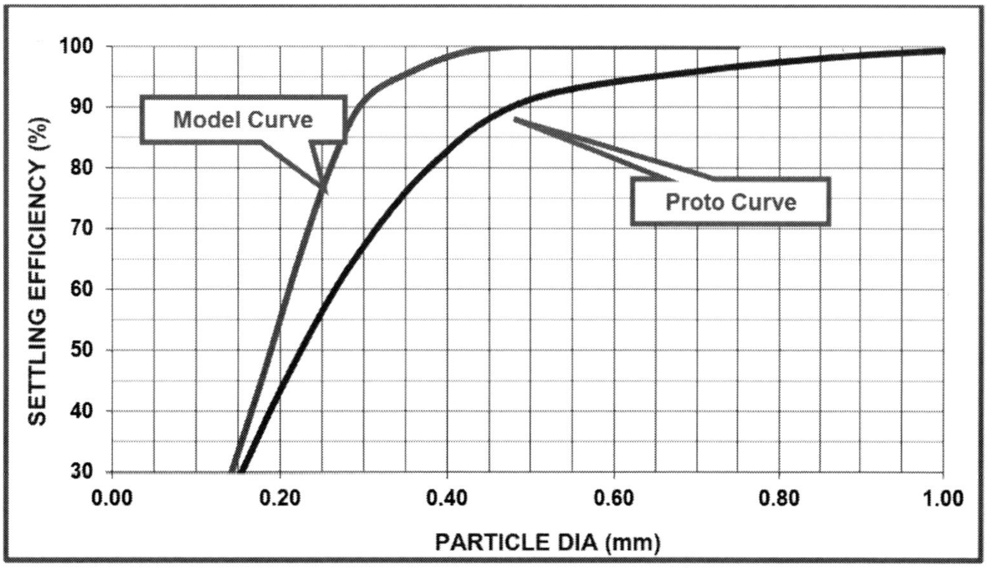

Figure 5: Analytical settling efficiency based on Camp's Method for model and its equivalent prototype

Model experiments were similarly conducted for inlet water level at El 357 m and 354 m (MDDL) and the settling efficiency for 0.3 mm particle size worked out to be 65.29 % and 63.45 % respectively.

4. CONCLUSIONS

The study for Teesta - VI H.E. Project, Sikkim included in this paper showed that the settling efficiency of desilting basin for particle size of 0.3 mm was estimated to be 70.5 % for reservoir water level maintained at El 360 m (FRL). Desilting basins are generally provided in hydro power projects to remove 90% of the designed size of sediment particle (0.3 mm in the present case) for efficient trouble-free operation of the project. However, the settling efficiency for 0.3 mm particle size is much lower than the desired value of 90%. Therefore, additional steps will have to be taken for protection of turbines and other under water parts from abrasion damage due to high quantity of incoming sediment of size larger than 0.3 mm. Also, the reservoir water level should be maintained close to FRL during the operation of the project because settling efficiency reduces with the reduction of water level.

Entry of water is smooth from the head regulator and there was no return flow observed near the inlet transition in the initial reaches of desilting basin. Various other components of desilting basin such as outlet transition, size of silt flushing tunnel, size and spacing of openings connecting main chamber with silt flushing tunnel etc. were also tested during the model studies. Their performance was found to be satisfactory.

5. ACKNOWLEDGMENTS

The authors sincerely thank Dr. R.S. Kankara, Director, CWPRS for constant encouragement, guidance and kind permission for publishing this paper. The authors are also thankful to project authorities for referring studies to CWPRS and making available the necessary data for conducting these studies.

6. REFERENCES

Camp, T.R. (1946)."*Sedimentation and the Design of settling tanks*", Trans ASCE, Vol 111, Paper No. 2285.
CWPRS (2005). CWPRS manual "*Guidelines for design of desilting basins (Pressure flow)*".
CWPRS (2021).Technical Report "*Hydraulic model studies for desilting basin of Teesta – VI H.E. Project, Sikkim*".
Qamar, M.Z. (2010). "*Silt Control in Reservoir using Diversion Tunnel and Blank panels*", M. Tech Dissertation at IIT, Roorkee.

DOI: https://doi.org/10.26077/24ae-7606

Non-intrusive measurements of air-water flow properties in supercritical flows down grass-lined spillways

H. Cui[1], S. Felder[2] & M. Kramer[1]
[1]*UNSW Canberra, School of Engineering and Information Technology (SEIT), Canberra, ACT 2610, Australia*
[2]*Water Research Laboratory, School of Civil and Environmental Engineering, UNSW Sydney, Manly Vale 2093, NSW, Australia*
E-mail: m.kramer@adfa.edu.au

Abstract: *Grass-lined spillways constitute an environmentally friendly alternative to conventional concrete structures with flat slope and therefore may contribute to water-sensitive urban design. Research into supercritical self-aerated flows down grass-lined spillways has commenced recently, but important knowledge gaps related to air entrainment, flow resistance and bubble-canopy interaction are yet to be filled. This study presents first results of an experimental investigation into flows over a rough grass-lined channel bed with embankment slope. The deployed measurement instruments included acoustic displacement meters (ADMs) and a conventional intrusive phase-detection probe. A transformation of measured free-surface elevations into multiple time series of instantaneous air concentrations allowed to non-intrusively compute air concentration distributions, dimensionless interface frequency distributions, and chord times. The results demonstrated that the amount of entrapped air was significantly larger than the amount of entrained air. It was further shown that air concentration distributions followed a recently developed semi-analytical solution with a characteristic length that was closely linked with the mean air concentration. The ADM measurements were limited to entrapped air and future research should explore other non-intrusive instrumentation to also explore measurements of entrained air.*

Keywords: *Grass-lined spillway, air entrainment, non-intrusive sensing technology, acoustic displacement meter, entrapped air, interface frequency*

1. Introduction

The deployment of grass-linings in supercritical flows and associated self-aeration have recently attracted increased research interest (Scheres et al. 2020, Bai et al. 2022, Cui et al. 2022a), because of ecological and environmental benefits, including embankment protection (Wilcock et al. 1999), natural integration into the landscape (Mekonnen et al. 2015), and increased ecological connectivity for benthic organisms (Yang and Li 2010), amongst others. These benefits may be counteracted by erosion risk due to shear stresses acting on the grass canopy (Samani and Kouwen 2002). Extensive investigations into thresholds for erosion stability in terms of mean velocities and critical shear stresses have been conducted assuming single-phase flow conditions (Temple 1987, Hewlett et al. 1987), while such guidelines are yet to be established for multi-phase flows, where the bubble-canopy interaction may lead to a modification of bed shear stresses. Scheres et al. (2020) used natural vegetation on soil which eroded during the supercritical air-water flow measurements. In previous physical models of supercritical spillway flows with artificial grass-linings, a flat channel bed was considered (Bai et al. 2022, Cui et al. 2022a), while the effects of underlying soil roughness on flow parameters are unknown.

Fundamental flow properties in two-phase air-water flows comprise time-averaged volumetric air concentrations ($\bar{c}$), interfacial velocities, and interface frequencies (F). The latter has typically been reported as bubble or particle count rate (equal to 0.5 F) but the term interface frequency seems more appropriate in flow situations where more air is transported through surface waves (entrapped air) than through air bubbles (entrained air), both transport modes defining the total volume of conveyed air (Wilhelms and Gulliver, 2005). The non-aerated upstream section of supercritical flows down a spillway is typically followed by an aerated region, both regions being separated by the so-called inception point of air entrainment (Fig. 1). There are several approaches for determination of this inception point, including the boundary layer intersection theory as well as analyses focusing on the development and growth of free-surface perturbations (Valero and Bung 2018).

Recent advances in the application of non-intrusive sensing technology for air-water flows in spillways have enabled researchers to remotely measure interfacial velocities, air concentrations and free-surface elevations using high-speed video cameras (Bung and Valero 2016, Zhang and Chanson 2018, Kramer and Chanson 2019, Kramer and Felder 2021), LiDAR technology (Kramer et al. 2020a), and acoustic displacement meters (ultrasonic sensors) (Bung 2013, Zhang et al. 2018). Some applications facilitated measurements of free-surface perturbations, including the studies of Valero and Bung (2016, 2018) and Wei et al. (2017).

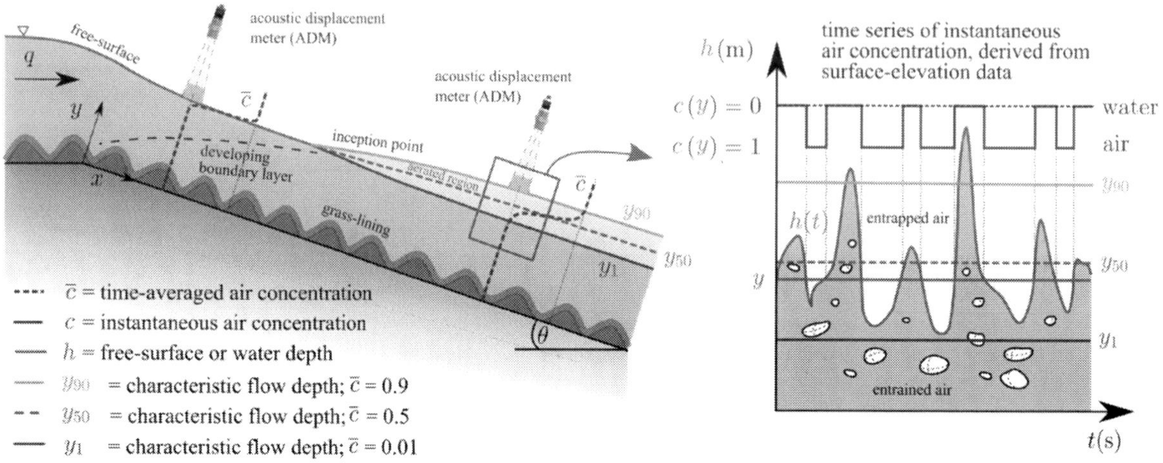

Figure 1. Schematic of a grass-lined spillway with rough channel bed; the figure on the right shows the concept of entrapped and entrained air including the transformation of the acoustic displacement meter (ADM) signals into a time series of instantaneous air concentrations as per Eq. (1), allowing to compute air concentration distributions, interface frequencies, and chord times; q = specific flow rate, θ = chute angle, c = instantaneous air concentration, x = streamwise coordinate, y = bed-normal coordinate, y_{90} = elevation where $\bar{c} = 0.9$ (notation applies for other characteristic elevations), h = free-surface or water depth, and t = time.

To improve our modelling capacity of such complicated air-water flows down grass-lined spillways, the present study deployed acoustic displacement meters and a phase-detection intrusive probe, focusing on the following two objectives:

- Investigate the effects of underlying roughness on air-water flow parameters of supercritical flows down grass-lined spillways.
- Determine the suitability of acoustic displacement meters for measurement of two-phase flow properties in flows with entrapped air, including air concentration distribution, interface frequency, and chord time distributions.

2. Methods

2.1. Experimental facility

Experiments were conducted in a spillway model at the UNSW Water Research Laboratory (WRL), which had dimensions of 0.8 m width, 8 m length, and 0.5 m sidewall height. The spillway chute was inclined at an angle θ = 10.8 degrees and a layer of corrugated iron sheets was installed on the smooth surface of the chute (Fig. 2*a*). The corrugated sheeting followed a sinusoidal wavy pattern with amplitude of 0.9 cm and wavenumber of 0.08 cm^{-1}, aiming to resemble small-scale microroughness of the underlying soil topography found in grass-lined waterways (Cui et al. 2022b). A dense layer of synthetic turf was attached to the top of the corrugated iron sheets (Fig. 2*b*). The artificial turf represented an idealized vegetation cover for grassed waterways with unit frontal area of 281 m^{-1} and undeflected grass height of 0.025 m, which corresponded to type E grass according to design guidelines (USDA, 1954). The same artificial turf was previously used in Cui et al. (2021, 2022a). For the present study, supercritical flows with q = 0.3125 m^2/s were provided by a recirculation system whose flow rate was monitored with an ABB electromagnetic flow meter to an accuracy of ± 0.4 % of the flow rate.

Figure 2. Experimental setup of the present study (*a*) upstream view of the physical model with corrugated iron sheets and artificial turf (*b*) sidewall view of the underlying roughness profile and the artificial turf (*c*) deployed dual-tip phase-detection conductivity probe (CP) measuring free-surface perturbations at q = 0.3125 m^2/s (*d*) deployed ADMs with measuring tubes recording free-surface data at q = 0.3125 m^2/s.

2.2. Instrumentation and signal processing

The current study used a WRL-manufactured dual-tip phase-detection conductivity (CP) probe with very thin leading and trailing tips (inner diameter = 0.125 mm; outer diameter = 0.6 mm) (Felder et al. 2019) (Fig. 2*c*). The tips were separated by 5.75 mm in the streamwise direction. Air-water interfacial properties were measured above the peaks of the corrugated profile at 7 locations along the chute, starting at x = 1.5 m with an interval of 1.0 m. At each location, vertical profiles of air concentrations ($\overline{c}$), interface frequencies (F) and chord times were measured with a sampling frequency of 40 kHz for 180 s, see Kramer et al. (2020b) for a detailed description on the evaluation of $\overline{c}$ and F using a phase-detection probe. Further, the adaptive window cross-correlation (AWCC) software (Kramer et al. 2019, Kramer et al. 2020b, Kramer et al. 2021) was applied to compute pseudo-instantaneous interfacial velocities from the raw data (results not shown herein).

At the same streamwise locations, an array of three acoustic displacement meters was deployed to record free-surface elevations. The ADMs (Microsonic Mic+35/IU/TC) were installed on a wooden platform, which was mounted on an aluminum rail along the chute sidewalls (Fig. 2*d*). Each ADM was calibrated with a digital gauge with 0.01 mm accuracy. Similar to Kramer and Chanson (2018), a circular extension tube (25 cm long) was attached to each sensor to narrow the measurement footprint of the ADMs, which is anticipated to increase accuracy in the estimation of free-surface elevations. Herein, ADMs were sampled for 20 minutes at 100 Hz, and a low-pass filter with a passband frequency of 25 Hz was applied to minimize aliasing effects, see discussion in Zhang et al. (2018). The ADMs provided time-varying free-surface profiles $h(t)$, which were converted for different bed-normal elevations (y) into instantaneous air concentrations using the transformation depicted in Fig. 1:

$$c(y,t) = \begin{cases} 0 \; if \; h(t) \geq y \\ 1 \; if \; h(t) < y \end{cases} \tag{1}$$

This novel approach (Eq. 1) allowed to convert ADM free-surface elevation data into time series of instantaneous air concentrations, which subsequently were analyzed using existing signal processing techniques developed for phase-detection probes, providing chord times, interface frequencies, and time-averaged air concentrations from ADM data, where the latter was calculated as:

$$\bar{c}(y) = \frac{1}{T}\int_{t=0}^{T} c\ dt \qquad (2)$$

with T being the sampling duration, which was significantly larger than the time scale of turbulent processes. Note that air concentrations derived from free-surface measurements with ADMs measured the amount of entrapped air, which is because entrained air bubbles are not detected by the acoustic beam (Fig. 1). A comparison with phase-detection intrusive measurements allows for a differentiation between entrapped and entrained air.

3. Results

3.1. Air-water flow properties

Measured air concentrations are presented in Figs. 3*a-c* for three selected streamwise locations of x = 2.5 m, 3.5 m, and 6.5 m. The air concentration distributions followed typical S-shaped profiles and there was an excellent agreement between intrusive phase-detection probe and top-view ADM measurements for all measurement locations along the flume, indicating that the volume of entrapped air was significantly larger than the volume of entrained air, and that the ADMs were able to accurately predict the air concentrations in flows characterized by free-surface perturbations. Time-averaged air concentration profiles were well represented through an analytical solution developed for entrapped air (Valero and Bung, 2016):

$$\bar{c}(y) = \frac{1}{2}\left(1 + erf\left(\frac{y/y_{50}-1}{D/y_{50}\sqrt{2}}\right)\right) \qquad (3)$$

where erf is the error function, and y_{50} is the elevation where $\bar{c} = 0.5$. For entrapped air, the parameter D represents the standard deviation of the water surface level distribution, which is assumed to follow a Gaussian distribution (Valero and Bung, 2016). Herein, Eq. (2) was applied to air concentration profiles measured with ADMs and CP. For profiles that also encompassed a small amount of entrained air, D was interpreted as characteristic length that describes the thickness/height of the aerated region. The solutions of Eq. (1) were added to Figs. 3*a*-c, and the characteristic length D was well represented by an empirical relationship from Felder et al. (2022), developed for depth-averaged air concentrations $\langle\bar{c}\rangle < 0.25$:

$$\frac{D}{y_{50}} = 1.19\,\langle\bar{c}\rangle^{1.17} \qquad (4)$$

Figures 3*d-f* show dimensionless distributions of the interface frequency (F/F_{max}). The results of the phase-detection intrusive probe and the ADMs were almost identical, although the maximum cross-sectional interface frequency F_{max} was between 55 (1/s) $< F_{max,PD} <$ 130 (1/s) for the phase-detection probe and between 2 (1/s) $< F_{max,ADM} <$ 30 (1/s) for the ADMs (Fig. 3g), due to the lower ADM sampling rate, as mentioned by Zhang et al. (2018). The self-similar dimensionless interface frequency distributions were well approximated by a parabolic relationship $F/F_{max} = 4\bar{c}(1-\bar{c})$ proposed by Toombes and Chanson (2007), which was combined with Eq. (1) to yield (see Figs. 2*d-f*):

$$\frac{F}{F_{max}} = 4\,\bar{c}\,(1-\bar{c}) = 2\left(1 + erf\left(\frac{y/y_{50}-1}{D/y_{50}\sqrt{2}}\right)\right)\left[1 - \frac{1}{2}\left(1 + erf\left(\frac{y/y_{50}-1}{D/y_{50}\sqrt{2}}\right)\right)\right] \qquad (5)$$

To demonstrate the effect of the sampling limitations of the ADMs, $F_{max,PD}$ was recalculated for downsampled phase-detection probe signals at a sampling rate of 25 Hz, representing the sampling capabilities of the acoustic displacement meters. These $F_{max,PD}$ results were added to Fig. 3g showing a much closer agreement in the maximum detected interface frequencies for the two measurement instruments. This also confirmed the small contribution of entrained air. Despite the restrictions in sampling rate, it is remarkable that ADM measurements were able to reproduce the self-similar dimensionless interface frequency distributions shown in Figs. 2*d-f*.

An analytical expression for the depth-averaged air concentration in flows dominated by entrapped air was derived through integration of Eq. (1) between the channel bed ($y = 0$) and y_{90}, i.e., the elevation where $\bar{c} = 0.9$, which led to the following solution:

$$\langle\bar{c}\rangle = \frac{1}{y_{90}} \int_{y=0}^{y_{90}} \bar{c}\; dy$$

$$= \frac{1}{2\, y_{90}} \left(\sqrt{\frac{2}{\pi}}\, D \left(e^{-(y_{50}-y_{90})^2/(2\,D)^2} - e^{-y_{50}^{\,2}/(2\,D^2)} \right) + (y_{90} - y_{50})\, erf\left(\frac{y_{90}-y_{50}}{\sqrt{2}\,D}\right) - y_{50}\, erf\left(\frac{y_{50}}{\sqrt{2}\,D}\right) + y_{90} \right) \qquad (6)$$

Equation (6) was compared with measured depth-averaged air concentrations (phase-detection probe), showing that the analytical solution performed well (Fig. 3*h*). In terms of applicability, Eq. (6) can be solved iteratively or numerically for the length scale D, which would provide a more exact relationship between D and $\langle\bar{c}\rangle$ when compared to the empirical Eq. (4). To perform this iteration, a relationship between y_{50} and y_{90} is required, which can be established using Eq. (3). For practical purposes, it is believed that Eq. (4) is sufficiently accurate.

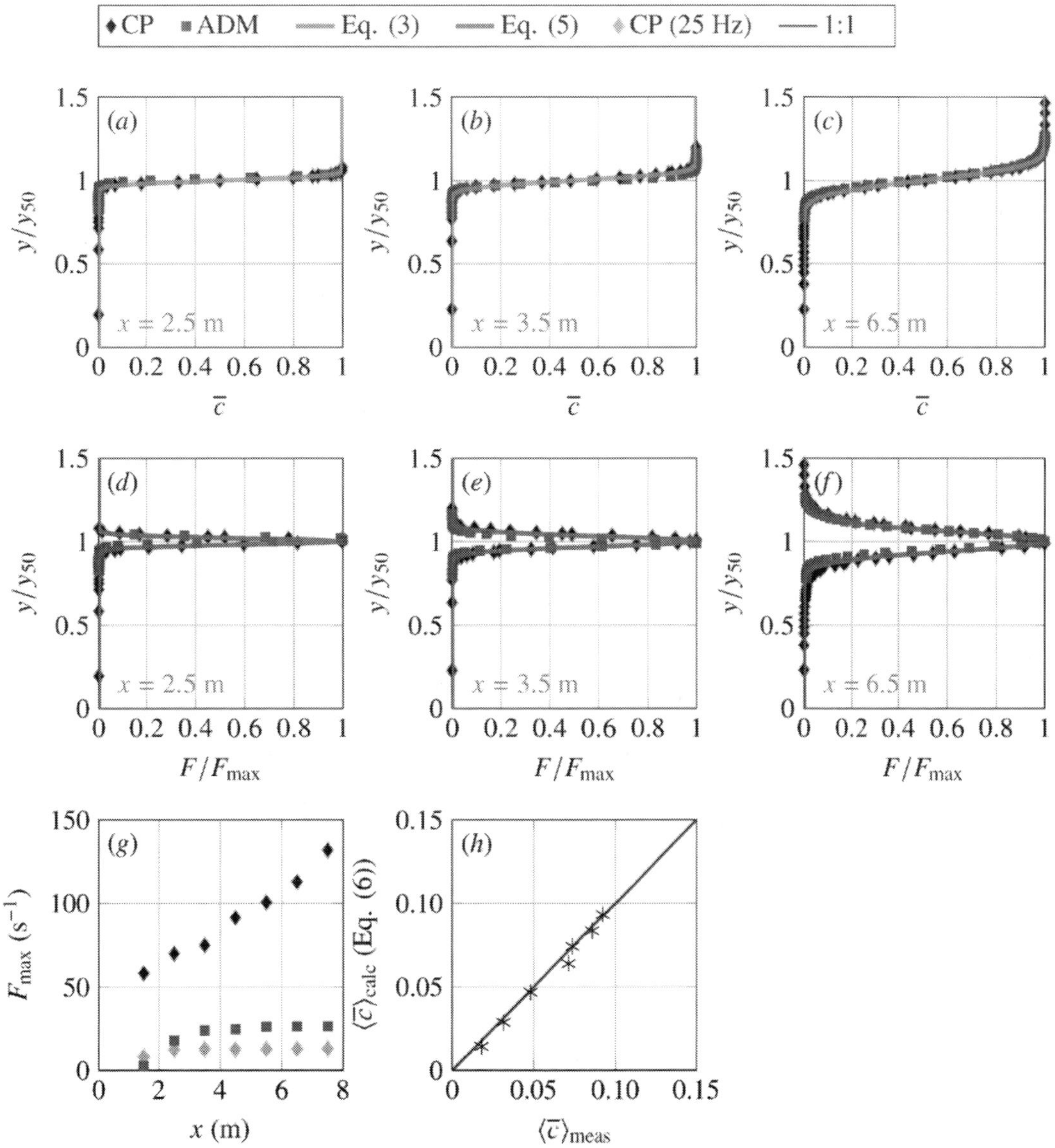

Figure 3. Air concentration and interface frequency distributions, measured with acoustic displacement meter (ADM) and phase-detection intrusive probe (CP) for q = 0.3125 m^2/s at different streamwise locations (*a-c*) air concentration distributions (*d-f*) dimensionless interface frequency distributions (*g*) measured F_{max} along the spillway chute (*h*) computed versus measured depth-averaged air concentration.

3.2. Chord time distributions

Figure 4 shows a comparison of chord time distributions measured with phase-detection intrusive probe and ADM for a streamwise location $x = 6.5$ m at three different vertical elevations, corresponding to $\overline{c} = 0.1$, 0.5, and 0.9. These distributions were representative examples for all present data sets. It was evident that the phase-detection probe was able to detect smaller air chords than the ADM, resulting from a higher sampling rate in combination with a small sensor tip. For air concentrations $\overline{c} = 0.1$ and 0.5, the estimated shape of the probability density distribution was similar for both instruments (Figs. 4*a,b*). For $\overline{c} = 0.9$, the chord time distributions tended to diverge (Fig. 4*c*). Despite these differences, the time-averaged air concentrations were identical, which was because small-sized air bubbles did not contribute much to the volumetric air concentration and the relative contribution of larger bubbles to the air concentration was similar for phase-detection probe and ADM. Note that the chord time distributions of the downsampled phase-detection probe data matched the ADM measurements well at all three locations.

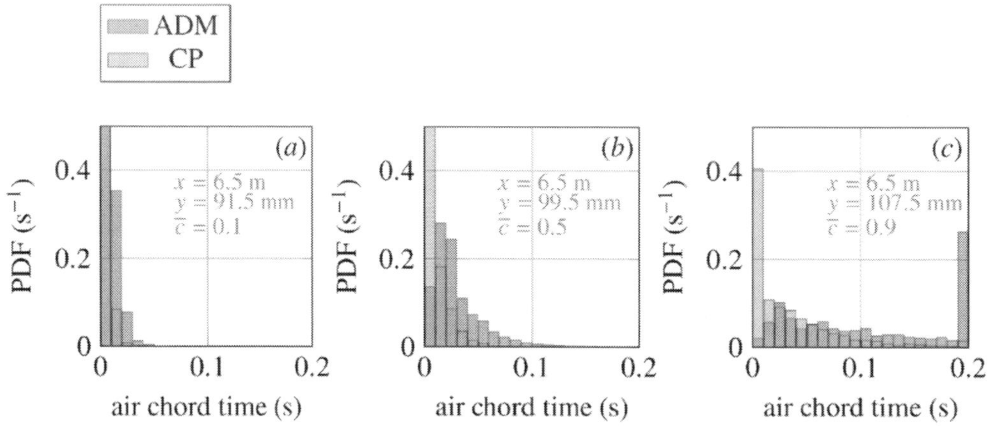

Figure 4. Representative air chord time distributions, measured with acoustic displacement meter (ADM) and phase-detection conductivity probe (CP) for $q = 0.3125$ m²/s at $x = 6.5$ m (*a*) $\overline{c} = 0.1$ at $y = 91.5$ mm (*b*) $\overline{c} = 0.5$ at $y = 99.5$ mm (*c*) $\overline{c} = 0.9$ at $y = 107.5$ mm.

4. Conclusion

This study proposed a novel method to transform signals of water surface elevations, such as acoustic displacement meter results, into time series of instantaneous air concentrations, thereby enabling the evaluation of common air-water flow properties of entrapped air, including air concentration distributions, interface frequencies, and chord times. This simple concept is anticipated to be useful as it facilitates direct comparison of air-water flow quantities derived from non-intrusive sensing techniques with phase-detection intrusive measurements, which in turn allows for a differentiation between entrapped and entrained air. It is further anticipated that the deployment of two non-intrusive sensors placed next to each other will enable the computation of interfacial velocity profiles through cross-correlation analysis of instantaneous air concentration time series at different bed-normal elevations.

The phase-detection probe and ADM measurements of flows over a rough grass-lined channel bed with embankment dam slope showed that air concentration and interface frequency distributions followed previously developed solutions for entrapped air involving the error function. The fact that both measurement techniques delivered similar results demonstrated that the volume of entrained air was minimal. Entrained air bubbles did not penetrate deep into the water column, and aeration properties were therefore not affected by the interaction with the macro-rough channel bed, which is however anticipated to change for steeper slopes. A meaningful extension of this work would comprise (i) an investigation of grass-lined spillways with different bed configurations on steeper slopes, (ii) the deployment of ADMs or LiDAR to test the ability of measuring interfacial velocity profiles remotely, and (iii) an extension of the conducted data analysis to include sensitivity of sampling parameters.

5. Acknowledgments

The authors thank Rob Jenkins and Larry Paice (WRL, UNSW Sydney) for their technical assistance. Dr Daniel Valero is thanked for visualization support.

6. Notation

$\bar{c}$	air concentration (time-averaged)	(-)
c	air concentration (instantaneous)	(-)
$\langle\bar{c}\rangle$	mean air concentration (depth-averaged)	(-)
D	characteristic length scale	(L)
F	interface frequency	(T^{-1})
F_{max}	maximum cross-sectional interface frequency	(T^{-1})
$F_{max,ADM}$	maximum cross-sectional interface frequency of ADM	(T^{-1})
$F_{max,PD}$	maximum cross-sectional interface frequency of phase detection probe	(T^{-1})
h	free-surface or water depth	(L)
q	specific flow rate	(L^2T^{-1})
t	time	(T)
T	sampling duration	(T)
x	streamwise coordinate	(L)
y	bed-normal coordinate	(L)
y_1	elevation where $\bar{c} = 0.01$	(L)
y_{50}	elevation where $\bar{c} = 0.5$	(L)
y_{90}	elevation where $\bar{c} = 0.9$	(L)
θ	chute angle	(°)

7. REFERENCES

Bai, R., Bai, Z., Wang, H., & Liu, S. (2022). "Air-water mixing in vegetated supercritical flow: effects of vegetation roughness and water temperature on flow self-aeration." *Water Resources Research*, 58(2), e2021WR031692.

Bung, D. B. and Valero, D. (2016). "Optical flow estimation in aerated flows." *Journal of Hydraulic Research*, 54(5), 575–580.

Cui, H., Felder, S., Kramer, M., (2021). "Modelling velocity profiles of aerated flows down grassed spillways." *8th International Junior Researcher and Engineer Workshop on Hydraulic Structures*, 1.

Cui, H., Kramer, M., and Felder, S. (2022a). "Multilayer velocity model predicting flow resistance of aerated flows down grass-lined spillway." *Journal of Hydraulic Engineering*, 148(10), 06022014.

Cui, H., Kramer, M., and Felder, S. (2022b). "Effects of underlying soil roughness on flow patterns in grassed waterways." *39th IAHR World Congress, Granada, Spain*, 19, 24.

Felder, S., Hohermuth, B., and Boes, R. (2019). "High-velocity air-water flows downstream of sluice gates including selection of optimum phase-detection probe." *International Journal of Multiphase Flow*, 116, 203-220.

Felder, S., Severi, A., and Kramer, M. (2022). "Self-aeration and flow resistance in high-velocity flows down spillways with micro-rough beds." *Journal of Hydraulic Engineering*, under review.

Hewlett, H., Boorman, L., and Bramley, M. (1987). *Design of reinforced grass waterways*. Construction Industry Research and Information Association (CIRIA), London, the United Kingdom.

Kramer, M., and Chanson, H. (2018). "Free-surface instabilities in high-velocity air-water flows down stepped chutes." Daniel Bung, Blake Tullis, 7th IAHR International Symposium on Hydraulic Structures, Aachen, Germany, 15-18 May. doi: 10.15142/T3XS8P (978-0-692-13277-7).

Kramer, M., and Chanson, H. (2019). "Optical flow estimations in aerated spillway flows: Filtering and discussion on sampling parameters." *Experimental Thermal and Fluid Science*, 103, 318-328.

Kramer, M., Valero, D., Chanson, H., and Bung, D. B. (2019). "Towards reliable turbulence estimations with phase-detection probes: an adaptive window cross-correlation technique." *Experiments in Fluids*, 60(1), 1-6.

Kramer, M., Chanson, H., and Felder, S. (2020a). "Can we improve the non-intrusive characterization of high-velocity air–water flows? application of lidar technology to stepped spillways." *Journal of Hydraulic Research*, 58(2), 350–362.

Kramer, M., Hohermuth, B., Valero, B., and Felder, S. (2020b). "Best practices for velocity estimations in highly aerated flows with dual-tip phase-detection probes." *International Journal of Multiphase Flow*, 126, 103228.

Kramer, M., Hohermuth, B., Valero, B., and Felder, S. (2021). "On velocity estimations in highly aerated flows with dual-tip phase-detection probes-closure." International Journal of Multiphase Flow, 134, 103475.

Kramer, M., and Felder, S. (2021). "Remote sensing of aerated flows at large dams: proof of concept." *Remote Sensing* 13(14), 2836.

Mekonnen, M., Keesstra, S. D., Stroosnijder, L., Baartman, J. E., and Maroulis, J. (2015). "Soil conservation through sediment trapping: a review." *Land Degradation & Development*, 26(6), 544–556.

Samani, J. and Kouwen, N. (2002). "Stability and erosion in grassed channels." *Journal of Hydraulic Engineering*, 128, 40–45.

Scheres, B., Schüttrumpf, H., and Felder, S. (2020). "Flow resistance and energy dissipation in supercritical air-water flows down vegetated chutes." *Water Resources Research*, 56(2), 1–18.

Severi, A. (2018). "Aeration performance and flow resistance in high-velocity flows over moderately sloped spillways with micro-rough bed." Ph.D. thesis, The University of New South Wales.

Temple, D. (1987). *Stability design of grass-lined open channels*. Number 667. US Department of Agriculture, Agricultural Research Service.

Toombes, L. and Chanson, H. (2007). "Surface waves and roughness in self-aerated supercritical flow." *Environmental Fluid Mechanics*, 7, 259–270.

USDA (1954). *Handbook Design Water of Channel for Soil and Water Conservation*. United States Department of Agriculture, Soil Condervation Service.

Wei, W., Deng, J., and Zhang, F. (2016). "Development of self-aeration process for supercritical chute flows." *International Journal of Multiphase Flow*, 79, 172–180.

Wilcock, R., Champion, P., Nagels, J., and Croker, G. (1999). "The influence of aquatic macrophytes on the hydraulic and physico-chemical properties of a New Zealand lowland stream." *Hydrobiologia*, 416, 203–214.

Valero, D. and Bung, D. B. (2016). "Development of the interfacial air layer in the non-aerated region of high-velocity spillway flows. Instabilities growth, entrapped air and influence on the self-aeration onset." *International Journal of Multiphase Flow*, 84, 66–74.

Valero, D. and Bung, D. B. (2018). "Reformulating self-aeration in hydraulic structures: Turbulent growth of free surface perturbations leading to air entrainment." *International Journal of Multiphase Flow*, 100, 127–142.

Wei, W., Deng, J., and Zhang, F. (2016). "Development of self-aeration process for supercritical chute flows." *International Journal of Multiphase Flow*, 79, 172–180.

Wilcock, R., Champion, P., Nagels, J., and Croker, G. (1999). "The influence of aquatic macrophytes on the hydraulic and physico-chemical properties of a New Zealand lowland stream." *Hydrobiologia*, 416, 203–214.

Wilhelms, S. C. and Gulliver, J. S. (2005). "Bubbles and waves description of self-aerated spillway flow." *Journal of Hydraulic Research*, 43(5), 522–531.

Yang, B. and Li, M. (2010). "Ecological engineering in a new town development: Drainage design in the woodlands, Texas." *Ecological Engineering*, 36(12), 1639–1650.

Zhang, G. and Chanson, H. (2018). "Application of local optical flow methods to high-velocity free-surface flows: Validation and application to stepped chutes." *Experimental Thermal and Fluid Science*, 90, 186–199.

Zhang, G., Valero, D., Bung, D. B., and Chanson, H. (2018). "On the estimation of free-surface turbulence using ultrasonic sensors." *Flow Measurement and Instrumentation*, 60, 171–184.

DOI: https://doi.org/10.26077/cb4c-fdc3

Numerical and Experimental Study on Energy Dissipation in Hydraulic Jump: A Comparison between Horizontal and Sloping Rough Channel Bed

Koroungamba Laishram[1], Thiyam Tamphasana Devi[1*], Ngangbam Romeji Singh[1] & Potsangbam Albino Kumar[1]
[1]*Department of Civil Engineering, National Institute of Technology, Manipur, Langol-795004, India*
E-mail: thiyam85@gmail.com (Corresponding address)

Abstract: *Hydraulic jumps are used as an energy dissipater, allowing excess energy from downstream flow to be dissipated through hydraulic structures, preventing scouring. Channel roughness is also necessary for good flow control, as it allows more excess energy to be dissipated downstream. To better understand the impact of channel roughness on dissipating energy in an open channel, this study investigates and compares the energy dissipation of hydraulic jumps based on roughened (gravel size 20mm) horizontal and sloping channel bed condition. The experiment was carried out in the Hydraulics Laboratory, Department of Civil Engineering, NIT Manipur (India) in a rectangular channel flume (16 m length, 0.6 m wide, and 0.8 m deep). The characteristics of hydraulic jumps (length of jump, upstream and downstream head, velocities, and subsequent depth ratio) were measured and using these characteristics, the amounts of energy dissipation are calculated. Results show that the sloping channel bed is much safer for the structure, dissipates more energy, and the control of the flow is much easier as compared to the horizontal condition. Thus, it can be suggested that sloping channel bed is an effective channel bed condition when energy dissipation is the main purpose. The experimental results were numerically simulated using CFD (Computational Fluid Dynamics) techniques and found to be within an acceptable range.*

Keywords: *CFD, Open channel flow, Hydraulic jumps, Bed roughness, Sloping channel, Energy dissipation.*

1. Introduction

A hydraulic jump is the sudden change of flow in the direction of flow under certain conditions, when the flow passes from supercritical flow to subcritical flow (Chow, 1959). Hydraulic jump is an energy dissipater for dissipating the excess energy of downstream flowing water. Types of channel roughness also plays an important role for effective control of flow in the channel, to guarantee safety of the structure, easy to operate, economy, easy to maintain and to dissipate more excess energy at the downstream. For avoiding the damage to structure at downstream due to high pressure and velocity, the Energy dissipator must be present.

Normally, the dimension of the stilling basin referred the dimension of hydraulic jump as an indicator (Ali and Mohamed 2010, Attashi et al. 2020, Perez et al. 2020) or as a design parameter. Behaviors of hydraulic jump can be determined by several factors such as rough (Parsamhr et al. 2017, Saghebian and Stanciu 2020) or smooth (Mahtabi et al. 2020); bed slope (Abbaspour et al. 2019); shape of channel (Omid et al. 2010, El-Seddik 2016, Nandi et al. 2020) and flow velocity/discharge, etc. Also, hydraulic jump forming on sloping channels were concerned while designing stilling. In view of the related limitations of Bélanger's theory (Belanger 1841), many experiments were performed by researchers such as Abdel-Azim 2003, Beirami 2006, Jan and Cheng 2009, Smith and Chen 2010, Abdel-Mageed 2015 in order to examine the hydraulic jump properties on smooth sloping beds and to apprehend the complication of the occurrence in a wider range of hydraulic and geometric structures (Bakhmeteff and Matzke 1938, Chow 1959, Rajaratnam 1967, Hager et al. 1990, Ohtsu and Yasuda 1991 and Husain et al. 1995).

In the study of flow characteristics of open channel, besides the experimental investigations, the employ of computational fluid dynamics (CFD) techniques have gained importance in recent years. Ma et al. (2001) conducted a numerical research on submerged hydraulic jump to find its turbulence characteristics. In the study he used the standard *k-ε* (SKE) turbulence closure model and Volume of Fluid (VOF) scheme to find out the free-surface profile. The data of submerged hydraulic jump found by simulation is then compared with the data from the experiment and result gives an understanding into the turbulence structure. Gumus (2015) used a laser Doppler anemometer (LDA) in a submerged hydraulic jump below a vertical sluice gate to determine the free-surface flow velocity field. The VOF-based two-dimensional CFD modeling was conducted for numerical simulation of turbulent flow. Finite-volume method by means of standard *k-ε*, realizable *k-ε*, renormalization group *k-ε*, Reynolds stress turbulence models, and shear stress transport *k-ω* were used for solving the governing equations of the flow. The experimental data are compared with the numerically computed data for the velocity field. The discretization error was expected to remain below 2%. Comparison results show that the horizontal velocities are well predicted by Reynolds stress model. Kim (2015) performed an experiment to study the precise hydraulic phenomena of flow with different weir type. The hydraulic jumps and parameters of flow in downstream based on a static weir and adjustable weir like sluice gate was compared, and also the characteristics of hydraulic jump are examined. It was found that flow with Energy dissipators for energy reduction dissipates more energy by 50% per unit length when compare to the flow without energy dissipator if fitted at a height of 10% for average water depth of river. Valero et al. (2018) research inside a Type III stilling basin of United States Bureau of Reclamation (USBR) to investigate the flow field. For numerical simulation the RNG $k-\varepsilon$ model and the VOF method were employed. The numerical results, anticipating 8 Froude numbers ranging from 3.12–9.52 were compared to the results of recent studies by Frizell and Svoboda (2012). Comparison results for sweep-off point and sequent depth relations were in good agreement. Al-Hashimi (2019) used ANSYS FLUENT, standard k–ε model, VOF method to investigate the submerged flow condition over a Broad crested weir for water surface profile. ANSYS FLUENT and a standard k–ε turbulent model were used to develop the numerical model. To evaluate the different head of water on the weir, the VOF method was used. The numerical results were compared with the experimental data collected. Result shows that the FLUENT model and the investigational results for head of water were in acceptable agreement. Nikmehr (2020) used the Flow-3D model to numerically examine the characteristics of hydraulic jump. Result shows that the hydraulic jump characteristics is properly able to be simulated by the numerical model found out by Flow-3D model. Also the roughness element distance and height considerably decreased the ratios of sequent depths of every Froude numbers with significance of MARE (Mean Absolute Relative Error) among the numerical and experimental sequent depth ratios are 4.8 % and 10.85 % for hydraulic jump length.

Many recent researchers have focused on hydraulic jump characteristics over smooth horizontal and sloping channel beds but few studies have conducted for comparing the energy dissipation of hydraulic jump between rough horizontal and rough sloping beds experimentally and numerically. The core objective of the investigation is to examine and compare the energy dissipation of hydraulic jumps on rough horizontal and sloping beds. The hydraulic jump on rough horizontal and sloping bed was numerically studied using ANSYSFLUENT 18.1 for different flow condition using the VOF method of multi-phase flow theory.The numerical results were compared with the experimental data for validation.

1.1. Basic Equation of Hydraulic Jump

Hydraulic jump is the sudden change in the direction of flow when moves from supercritical to subcritical. The toe of the jump is the head of water before the jumps while the head of water at downstream beyond the roller is the end of jump and the distance between them is called hydraulic jump length. The depth y_1 is the height of water before the jump and y_2 is the height after the jump, they are also called as alternate depth or sequent depths. Figure 1 shows the typical presentation of a hydraulic jump.

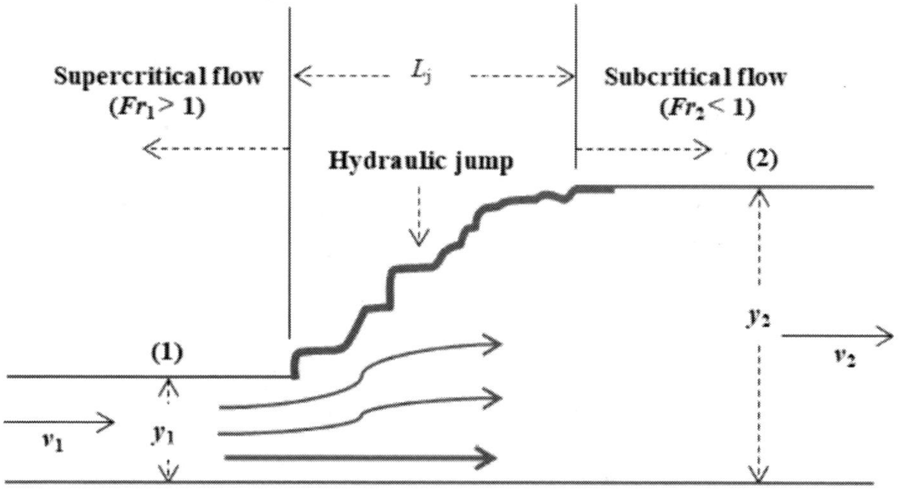

Figure 1. Typical representation of hydraulic jump

The loss of energy (ΔE) by the hydraulic jump is calculated as:

$$\Delta E = E_1 - E_2 = \left(y_1 + \frac{v_1^2}{2g} \right) - \left(y_2 + \frac{v_2^2}{2g} \right) \tag{1}$$

Where, v_1 and v_2 are the velocity of flow before and after the hydraulic jump respectively and g is the acceleration due to gravity which has constant value of 9.81m/s^2. For a rectangular open channel, energy loss in the hydraulic jump is based on upstream flow conditions. The energy loss (ΔE) is given as:

$$\Delta E = \frac{\left(y_2 - y_1 \right)^3}{4 y_2 y_1} \tag{2}$$

Where, E_1 is specific energy before the hydraulic and E_2 is specific energy after the hydraulic Length of jump (L_j) is given as:

$$L_j = 6\left(y_2 - y_1 \right) \tag{3}$$

Height of jump (H_j) is given as:

$$H_j = y_2 - y_1 \tag{4}$$

2. MATERIALS AND METHODS

2.1. Experimental set up

The experiment was conducted in the Rectangular channel flume in the Fluid mechanics and Hydraulics laboratory of Civil Engineering Department, National Institute of Technology, Manipur (India). A sluice gate (0.42m long, 0.82m wide and 0.7m height) shown in Figure 2, made by GUNT, Hamburg Germany is used in the channel to create a hydraulic jump. Velocitymeter were used to measure the velocity at downstream of the flow, Digital level gauge was used to find out the different head in the channel.

Figure 2. Sluice Gate

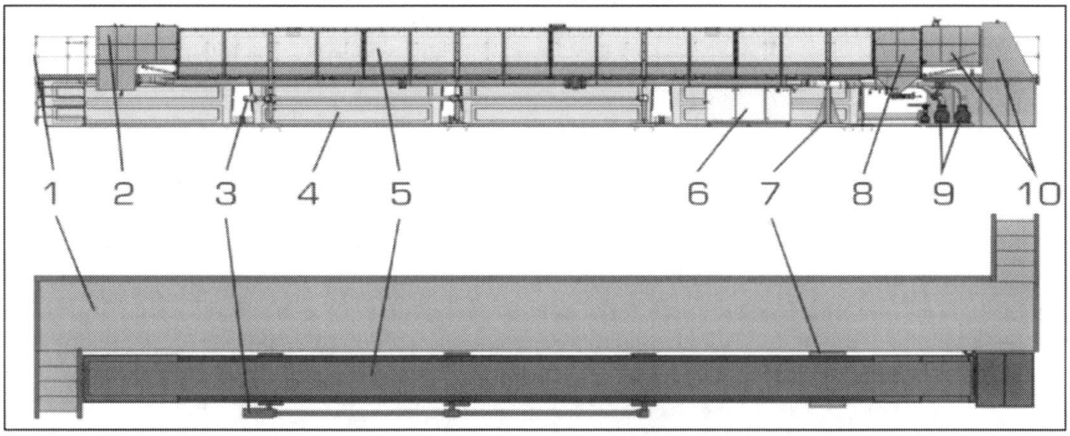

Figure 3. Channel Flume, Gunt Hm161, Cross Sectional view with (1) Detail sGallery (2) Inlet element (3) Jacking support with motorised inclination adjustment (4) Water tank (5) Experimental section (6) Switch cabinet (7) Fixed support (8) Sediment trap HM 161.72, (9) Pump (10) Outlet element.

Figure 4. Experimental Channel Flume

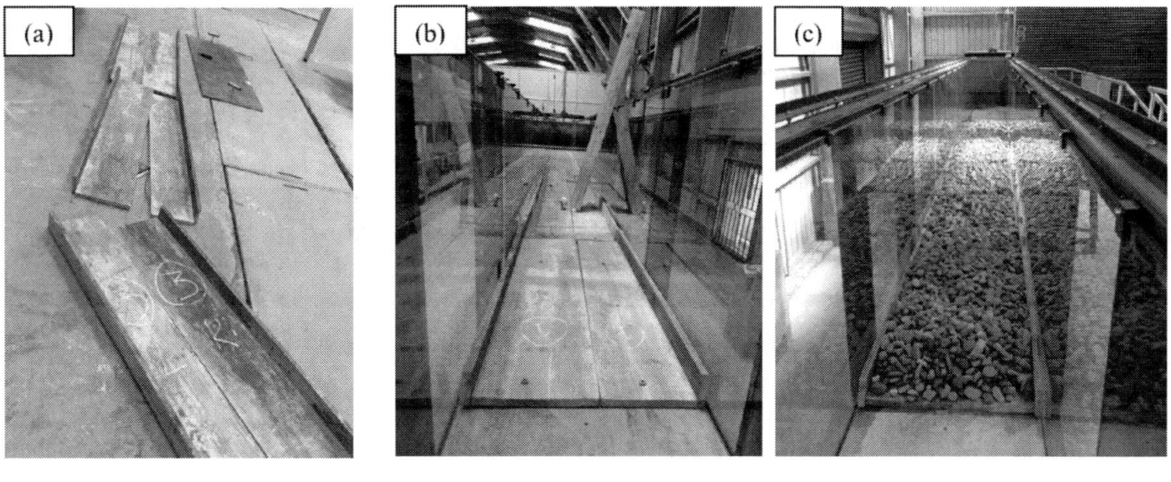

Figure 5. (a) Wooden tray for placing aggregate (b) After fixing tray on channel (c) Gravel bed arrangement for gravel size, k_s = 20mm

For preparation of the Roughen channel bed, a wooden rectangular tray of dimension 6m x 0.60m x 0.1m is used to protect the steel bed from corrosion as shown in Figure 5(a). The bed is placed at 4m from the inlet section of the flume. The tray is to be filled with gravel of size 30mm and is glued to the wooden tray Figure 5 (b) and 5 (c). The gravel is sieved and weighed before placing it in the tray so as to calculate the roughness co- efficient.

The flume was run for some pump discharge and characteristics of hydraulic jump such as sequent depths (y_2/y_1), critical depth, flow rate, amount of energy dissipated were measured for horizontal roughen bed at constant height of gate opening. The channel is then made sloping for 1.5% and same data were recorded at the same section of the channel with similar height of gate opening and same upstream head with that of horizontal bed.

2.2. Data collection

For horizontal roughen bed, the channel was made rough by pasting a gravels of size 20mm on wooden tray fixed on the channel. For creating the hydraulic jump, a sluice gate was fixed on the upstream section of channel. The gate is opened for some depth for formation of hydraulic jump. Different characteristics of hydraulic jump were measured at different section of the channel. Later, the channel is then tilted by 1.5% for making it sloping. The same procedure was used for collecting the data on the channel. Figure 6 shows the formation of hydraulic jump on sloping channel.

Figure 6. Hydraulic jump formation on rough sloping bed

The hydraulic jump characteristics such as velocity, sequent depth ratio, length of jump, height of jump and amount of energy dissipated over horizontal and sloping rough bed arrangement are summarized in the following Table 1.

Table 1. Characteristics of Hydraulic jump

Runs	Q	v_1	Fr_1	v_2	Fr_2	L_j	L_r	y_o	y_1	y_2	y_2/y_1	Energy loss (ΔE)
	m^3/sec	m/sec		m/sec		m	m	m	m	m	m	m
Horizontal Rough												
1	0.107	2.258	2.54	0.886	0.62	2.24	0.52	0.389	0.08	0.21	2.625	0.033
2	0.086	2.077	2.5	0.594	0.41	2.36	0.47	0.403	0.07	0.215	3.071	0.051
3	0.092	2.385	2.98	0.7	0.48	2.44	0.53	0.427	0.065	0.218	3.354	0.063
4	0.083	2.694	3.77	0.594	0.38	2.52	0.84	0.478	0.052	0.24	4.615	0.133
5	0.082	3.004	4.47	0.577	0.38	2.84	0.7	0.52	0.046	0.228	4.956	0.144
6	0.079	3.037	5.18	0.396	0.29	3.06	0.63	0.536	0.035	0.185	5.285	0.13
7	0.074	3.45	6.15	0.377	0.29	3.1	0.56	0.54	0.032	0.172	5.375	0.125
8	0.07	3.8	7	0.38	0.3	3.24	0.5	0.564	0.03	0.158	5.267	0.111
9	0.068	3.112	4.7	0.388	0.25	2.78	0.65	0.524	0.044	0.225	5.113	0.15
10	0.062	2.83	3.8	0.417	0.26	2.66	0.87	0.477	0.054	0.245	4.537	0.132
Sloping Rough												
1	0.078	3.28	5.23	0.313	0.21	2.725	0.855	0.562	0.04	0.225	5.625	0.176
2	0.079	2.953	4.44	0.28	0.19	2.82	0.831	0.465	0.045	0.22	4.89	0.135
3	0.082	2.877	4.19	0.37	0.255	2.425	0.732	0.425	0.048	0.215	4.48	0.185
4	0.089	2.365	3.008	0.313	0.219	2.52	0.42	0.345	0.063	0.208	3.301	0.058
5	0.075	2.17	2.877	0.443	0.316	2.225	0.525	0.284	0.058	0.2	3.448	0.062
6	0.069	1.943	2.532	0.313	0.229	1.96	0.345	0.277	0.06	0.191	3.183	0.049
7	0.074	1.895	2.355	0.495	0.363	2.125	0.29	0.248	0.066	0.19	2.879	0.038
8	0.075	2.996	4.667	0.384	0.258	2.78	0.92	0.485	0.042	0.225	5.357	0.162
9	0.07	3.177	5.273	0.542	0.369	2.75	1.125	0.558	0.037	0.22	5.946	0.188
10	0.069	1.657	2	0.542	0.395	1.944	0.272	0.234	0.07	0.192	2.743	0.034

2.3. Numerical Modelling

In the present research, hydraulic jump on rough horizontal and sloping bed was numerically studied using ANSYS-FLUENT 18.1 for different flow condition using the VOF method of multi-phase flow theory. Geometry was created using ANSYS SpaceClaim software as shown in the Figure 7 and Meshing was done using ANSYS Meshing. The VOF method was used to find out the water surface location and for simulating the turbulence modeling of hydraulic jump, the Standard k-ε model was used. The equations for Standard k-ε model are given as

$$\frac{\nabla}{\nabla t}(\rho k)+\frac{\nabla}{\nabla x_i}(\rho k u_i)=\frac{\nabla}{\nabla x_j}\left[\left(\mu+\frac{\mu_t}{\sigma_k}\right)\frac{\nabla k}{\nabla x_j}\right]+G_k+G_b-\rho\varepsilon-Y_M+S_k \tag{5}$$

And,

$$\frac{\nabla}{\nabla t}(\rho\varepsilon)+\frac{\nabla}{\nabla x_i}(\rho\varepsilon u_i)=\frac{\nabla}{\nabla x_j}\left[\left(\mu+\frac{\mu_t}{\sigma_\varepsilon}\right)\frac{\nabla\varepsilon}{\nabla x_j}\right]+C_{1\varepsilon}\frac{\varepsilon}{k}\left(G_k+C_{3\varepsilon}G_b\right)-C_{2\varepsilon}\rho\frac{\varepsilon^2}{k}+S_\varepsilon \tag{6}$$

Where, G_k is the development of TKE because of the average velocity gradients, G_b is the production of TKE because of buoyancy, Y_M is the contribution of changing dilation of compressible turbulence to the overall dissipation rate, S_k and S_ε are user-defined source terms. $C_{1\varepsilon}$, $C_{2\varepsilon}$, and$C_{3\varepsilon}$are constants, σ_k and σ_ε are the TPNs (Turbulent Prandtl Number) for k and ε, respectively.

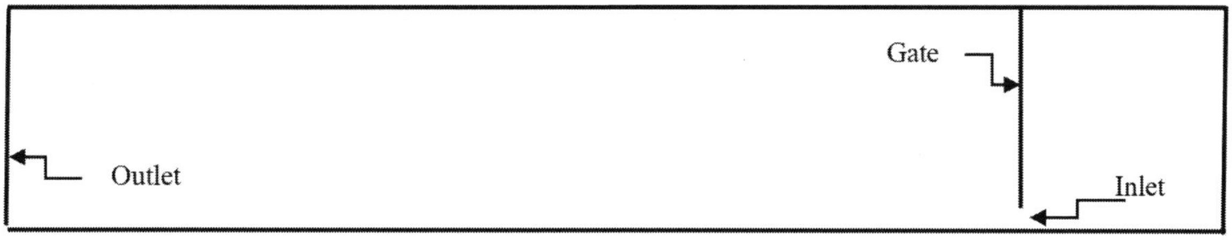

Figure 7. Geometry of Horizontal rough bed

For boundary conditions, at the inlet boundary the flow velocity is taken as u=0.46 m/s and v= 0 for horizontal rough bed were employed and for outlet boundary it was made as free overfall condition with $p = 0$. At the upper boundary of the domain, the air-filled pressure is taken as 0 and the no-slip wall condition was used for wall boundaries. For the free surface on top the 'symmetry' boundary condition is used. After assigning the flow conditions to the numerical model, the model calculation was done to find out the result. Figure 8 shows the numerical simulation for hydraulic jump in rough channel bed.

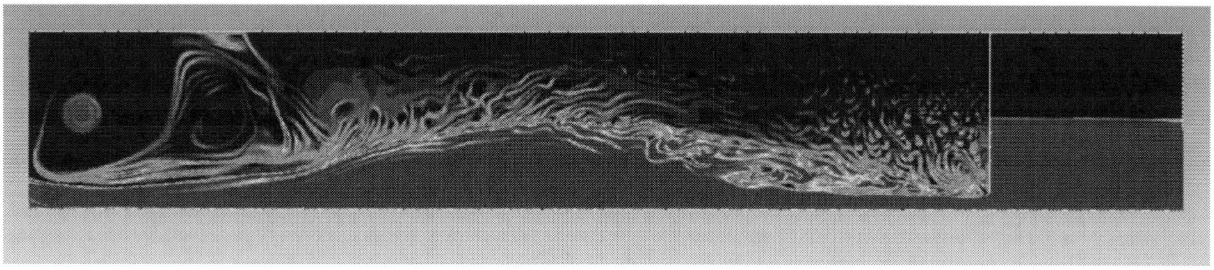

Figure 8. Numerical simulation of hydraulic jump in horizontal rough bed

For the case of sloping rough channel, inlet velocity was taken as 0.40 m/sec with reference to the experimental data. Other conditions are employed the same as that of horizontal rough bed. Simulation was done and figure 9 shows the simulated result for hydraulic jump in sloping channel bed.

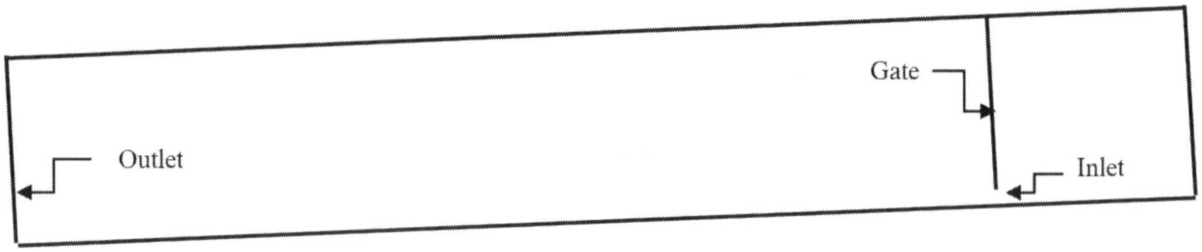

Figure 9. Geometry of sloping rough bed

2.4. Verification using experimental data

The numerical results obtained from simulation are then compared with the experimental data for both horizontal and sloping rough bed. In the present study the velocities for 1 trial from experimental data were used for validation of the numerical model. Table 2 shows the velocity difference of hydraulic jump for experimental and numerical model for different flow condition.

Table 2. Numerical data comparing with experimental data

Channel bed	Data	v_1 (m/s)	v_2 (m/s)
Horizontal bed	Experimental	2.26	0.88
	Numerical	2.8	0.6
Sloping bed	Experimental	2.36	0.313
	Numerical	2.2	0.8

3. RESULTS AND DISCUSSION

Jump characteristics such as Sequent depth ratio y_2/y_1, Length of jump L_j, flow rates Q were evaluated to examine the energy dissipation ability of hydraulic jump formed on Rough horizontal and sloping bed. Also, the experimental data are compared with the numerically simulated data for validation. The results are discussed in the following section.

3.1. Energy dissipation

Figure10 shows the values of Energy loss against velocity for Rough horizontal and rough sloping bed found experimentally. The figure shows that the amount of energy dissipated by rough sloping bed is much higher than that of rough horizontal at constant velocity.

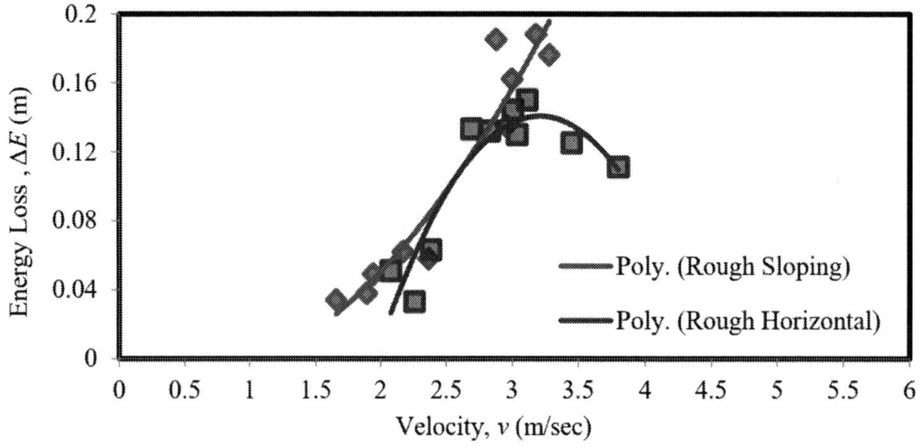

Figure 10. Comparison of energy loss between horizontal rough and sloping bed

3.2. Length of jump

The length of jump with respect to the upstream Froude number (Fr_1) for horizontal rough and sloping bed were shown in the Figure 11. It can be seen from the figure that length of jump from horizontal bed is much higher than that of sloping rough bed.

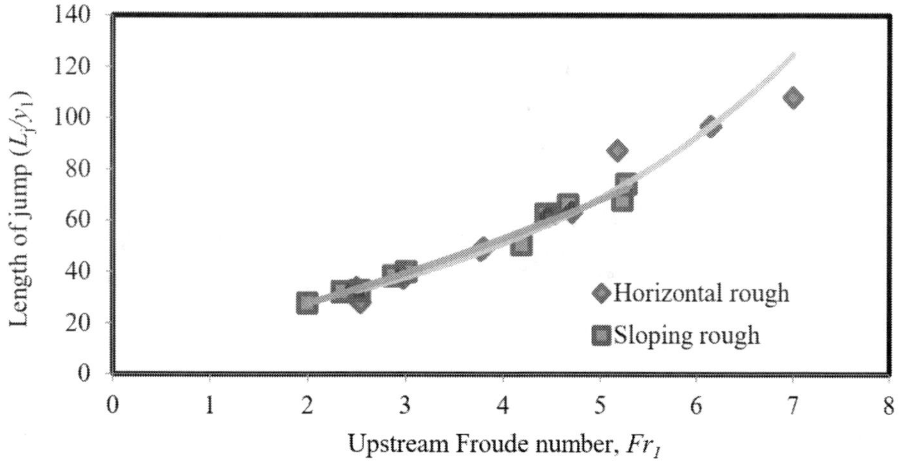

Figure 11. Comparison of Length of jump and upstream Froude number for horizontal and sloping rough bed

3.3. Velocity difference

A comparison between velocity of numerical and experimental data for horizontal rough and sloping bed were made as shown in the Figure 12 and 13. From these figures, it can be seen that the numerical and experimental results are in good agreement. It should be noted that the numerical results have underestimated the velocity of hydraulic jump compared to the experimental conditions.

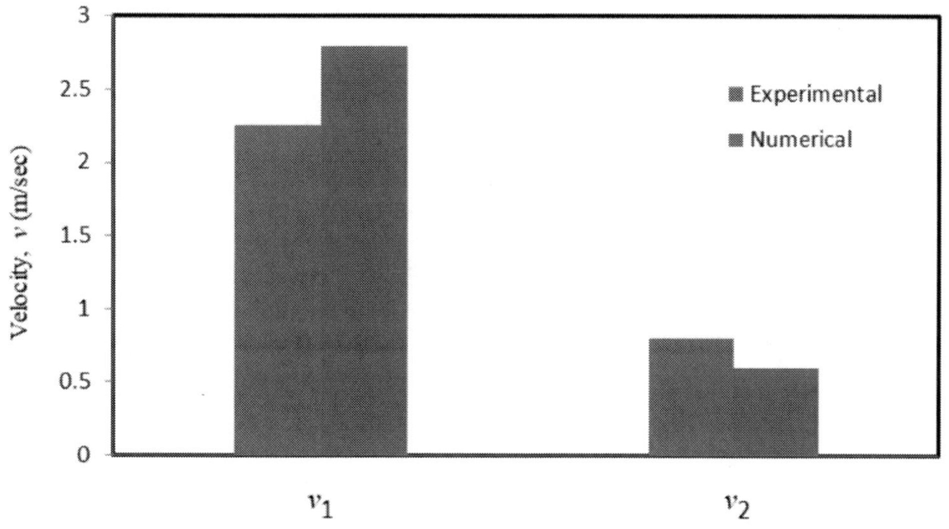

Figure 12. Comparison of experimental and numerical velocity for horizontal rough bed

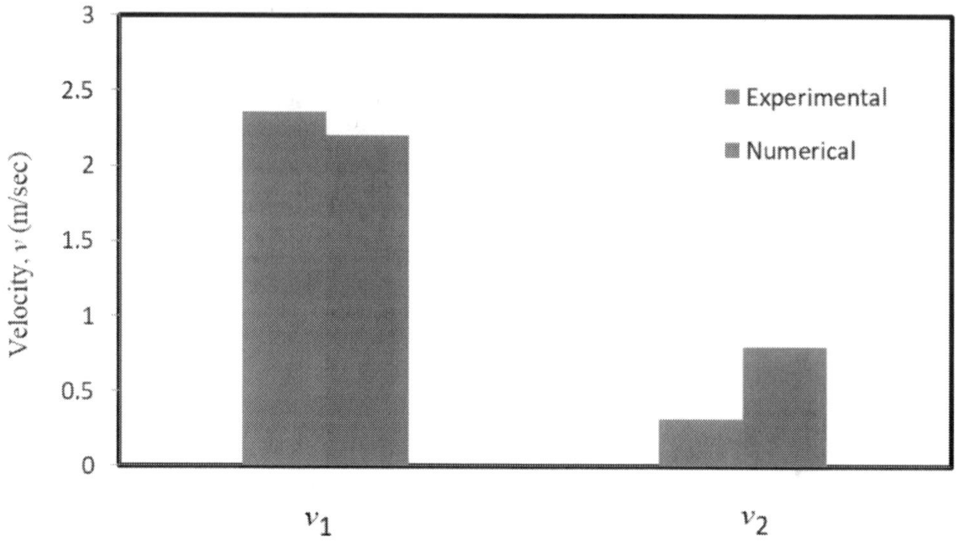

Figure 13. Comparison of experimental and numerical velocity for sloping rough bed

4. CONCLUSION

For the experimental study of energy dissipation of hydraulic jump between the horizontal and sloping roughen channel bed, the result shows that sloping channel bed dissipates more energy as compare to horizontal bed. Length of jump from horizontal rough bed is found to be much higher than that of sloping channel. The sloping rough bed is much safer for the structures and control of the flow is much easier as compared to the horizontal condition. Thus, it can be suggested that sloping channel bed is an effective channel bed condition when energy dissipation is the main purpose.Also, comparison of numerical and experimental results are found to be in good agreement.

REFERENCES

Abbaspour, A., Taghavianpour, T. and Arvanaghi, H. (2019). "Experimental study of the hydraulic jump on reverse bed with porous screens", *Appl.Water Sci.*, 91(55), 1-7.

Abdel-Azim M.N. (2003). "Analysis and formulation of hydraulic in sloping rectangular closed conduits". *Proc., Seventh International Water Technology Conference, Egypt, 1 – 12.*

Abdel-Mageed, N.B. (2015). "Effect of Channel Slope on Hydraulic Jump Characteristics". *Phys. Sci. Int. J.*, 7(4), 223–233.

Ali, M.A. and Mohamed, Y.A. (2010). "Effect of stilling basin shape on the hydraulic characteristics of the flow downstream radial gates", *Alexandria Eng. J.*, 49(4), 393-400.

Attashi, V., Lim, Y.H., Khajavi, M. and Bajestan, S.M. (2020). "Characteristics of Hydraulic Jump in Stilling Basins with Permeable Six-Legged Elements", *World Env. Water Res. Cong.*,65 – 75.

Bakhmeteff, B.A. and Matzke, A.E. (1946). "The hydraulic jump in terms of dynamic similarity", *Trans. Am. Soc. Civ. Eng*, 101, 630-680.

Beirami, M. and Chamani, M. (2006). "Hydraulic jumps in sloping channels: Sequent depth ratio". *J. Hydraul. Eng.*, 132(10), 1061–1068.

Belanger, J.B. (1841). Notes on Hydraulic Engineering, session 1841–1842.

Chow, V.T. (1956). Open channel hydraulics, McGraw-Hill, United States.

Jan, C.D., Chang, C.J. (2009). "Hydraulic jump in an inclined rectangular chute contraction". *J. Hydraul. Eng.*, 135(11), 949 – 958.

El-Seddik,M.M. (2017). "Hydraulic jump experiment in a rectangular open channel flume". *Ann. Civ. Environ. Eng.*, 1, 042-048.

Frizell, K., Svoboda, C and Matos, J. (2016). "Performance of Type III Stilling Basins for Stepped Spillways". *Proc., 2nd International Seminar on Dam Protection Against Overtopping.* Colorado, USA

Hager, W.H., Bremen, R. and Kawagoshi, N. (1990) "Classical Hydraulic Jump: Length of Roller". *J. Hydraul. Res.*, 28, 591-608.

Hussain, D., Alhamid, A.A. and Negm, Abdel-Azim M. (1995). "Length and depth of hydraulic jump in sloping channels", *J. Hydraulic. Res.* 33(3), 305-316.

Ma, F., Hou, Y., and Prinos, P. (2001). "Numerical calculation of submerged hydraulic jumps". *J. Hydraul. Res.*, 39(5), 493–503.

Mahtabi, G., Chaplot, B., Azamathulla, H.M. and Pal, M. (2020). "Classification of Hydraulic Jump in Rough Beds". *Water* 2020, 12, 2249.

Nandi, B., Das, S. and Mazumdar, A. (2020). "Experimental analysis and numerical simulation of hydraulic jump", *IOP Conf. Ser. Earth Environ. Sci.,* 505 (012024), 1-9.

Nikmehr, S., and Aminpour, Y. (2020). "Numerical Simulation of Hydraulic Jump over Rough Beds". *Period. Polytech. Civil Eng.,* 64 (2), 396 -467.

Omid, M.H., Varaki, M.E. and Narayanan, R. (2010). "Gradually expanding hydraulic jump in a trapezoidal channel". *J. Hydraul. Res.,* 45(4), 512-518.

Ohtsu, I. and Yasuda, Y. (1991). "Hydraulic Jump in Sloping Channels". *J. Hydraul. Eng.,* 117(7), 905–921.

Parsamehr, P., Farsadizadeh, D., Dalira, A.H., Abbaspour, A. and Esfahani, M.J. N. (2017). "Characteristics of hydraulic jump on rough bed with adverse slope". *J. Hydraul. Eng.*, 23(3), 301-307.

Rajaratnam, N. (1967). "Hydraulic jumps," Adv. Hydrosci.,V.T. Chow, ed.,4, Academic Press, New York,197-280.

Saghebian, A.M. and Stanciu, D.D. (2020). "Hydraulic jump energy dissipation evaluation in rough bed channels via kernel based approach". *Proc., 2nd Cilicia International Symposium in Engineering and Technology*, Mersin, Turkey, 110-114.

Smith, C.D. and Chen, W. (2010). "The hydraulic jump in a steeply sloping square conduit". *J. Hydraul. Res.,* 1989, 385 – 399.

Valero, D., Viti, N. and Gualtieri, C. (2018). "Numerical Simulation of Hydraulic Jumps. Part 1: Experimental Data for Modelling Performance Assessment". *Water*, 11(1), 36.

Gumus, V., Simsek, O., Soydan, N.G., Akoz, M.S. and Kirkgoz, M.S. (2016). "Numerical Modeling of Submerged Hydraulic Jump from a Sluice Gate". *J. Irrig. Drain Eng.*, 142(1): 04015037.

Kim, Y., Choi, G., Park, H. and Byeon, S. (2015). "Hydraulic Jump and Energy Dissipation with Sluice Gate". *Water,* 7, 5115-5133.

DOI: https://doi.org/10.26077/4900-5b11

Physical and Numerical model studies for Hydraulic design of Stilling basin as an Energy dissipator of a Spillway - A Case study

B.S. Sundarlal[1], Mrs. Vaishali Gadhe[1], Mrs. S.R. Patnaik[1], Amol H. Gaikwad[1]

[1]*Central Water and Power Research Station, Pune -411024, India*

E-mail: srinivasaslal@gmail.com

***Abstract** : The design of the stilling basin for a dam spillway is site specific and hydraulic model studies are required to be carried out for evolving the best possible design. A properly designed hydraulic jump stilling basin can ensure 60-70% dissipation of energy in the basin itself. Supported by hydraulic model studies and prototype experience, a suitable stilling basin caters to a variety of conditions covering discharge intensity, head, tail water variation, and topographical conditions. The most serious problem with hydraulic jump type dissipator is more structural strength rather than hydraulic efficiency. Experience in recent years gives many examples of stilling basins having damages due to uplift, vibration, cavitation and abrasion, all having their origin in the internal structure of hydraulic jump. Model studies play an important role in optimizing the length of the stilling basin, elevation of the stilling basin floor level for a high head spillway and height of the end sill and hydraulic jump characteristics. This paper discusses hydraulic model studies conducted at Central water and Power Research Station (CWPRS), Pune, India, for the hydraulic jump stilling basin as energy dissipator for Kotlibhel dam spillway stage 1-B, Uttarakhand, India, at the conceptual stage, which played an important role in enhancing the overall performance of spillway and energy dissipator by incorporating modifications such as lowering of stilling basin floor level and modifying the height of the end sill. Through recent advances in computing power and modeling software capabilities, it is now feasible to undertake complex three-dimensional analysis using Numerical/Computational Fluid Dynamics (CFD) modeling techniques. Numerical model studies were carried out to assess the performance of the stilling basin using Computational fluid dynamics software, FLOW-3D. The major benefit of the CFD modeling is that it allows early identification of problematic flow features and modifications to the design/layout could be tried rapidly and cost-effectively. The results obtained from the numerical model were compared with the results of the physical model studies. Comparison of the results from the two modeling approaches has provided a very good insight into the design process to optimize the stilling basin design. This use of physical and Numerical modeling (CFD modeling) techniques has provided invaluable insight and greater confidence for future use of standalone CFD analysis in spillway/energy dissipator design as complementary tool for physical model studies. After studying the performance of the stilling basin for Kotlibhel stage 1-B Hydro electric project, both through the physical and numerical techniques, it is found that CFD modeling can be used as a valuable tool to test the preliminary empirical design and in the optimization phase of the stilling basin design prior to conducting physical model studies. The details of physical and numerical model studies conducted for improving the overall performance of stilling basin are described in the paper.*

***Keywords:** Hydraulic jump, Numerical modeling physical modelling, Spillway, Stilling basin, Tailwater level.*

1. Introduction

The design of energy dissipators below a spillway of a dam occupies a vital role in the design construction and safe operation of the dam. The problem while designing energy dissipator is one essentially reducing the high velocity flow to a velocity low enough to minimize erosion of natural river bed. Otherwise, it would damage not just the downstream river reach but also cause local scour of the spillway or dam, which if significant and able to progress could present a dam safety issue. This reduction in velocity is accomplished by providing suitable energy dissipator at the base/downstream of the spillway depending upon the head, discharge intensity, tailwater conditions and type of bed rock or bed material downstream of the spillway. For spillways on weak bed rock, hydraulic jump type stilling basins are recommended. When the tailwater rating curve approximately follows the jump height curve or is slightly above or below it, then hydraulic jump type stilling basin with horizontal apron provides the best solution for energy dissipation. A well-formed stable and strong hydraulic jump contained within the stilling basin is an indication of proper energy dissipation downstream of the spillway.

Criteria recommended by Bureau of Indian Standards (BIS) code IS 4997-1968 for stilling basins of rectangular cross-section with horizontal apron hold good provided that the jet entering the basin is reasonably uniform in regard to both velocity and depth. Though the criteria are applicable for all cases, yet for heads greater than 15 m, discharge intensities greater than 30 $m^3/s/m$ and possible asymmetry of flow, the specific design should be tested on a hydraulic model. Extensive studies of stilling basin have been carried out by Peterka (1984), Higgs, (1996), (Johnson et al.2006), (Amorim et al. 2015) through physical as well as numerical models.

The present paper describes the physical model studies conducted for optimizing the performance of the stilling basin for Kotlibhel Stage-1B Hydroelectric Project, Uttarakhand, India on a 1:80 scale 3-D comprehensive model and the results were compared with the results obtained from the numerical model studies. Numerical model studies were carried out using Computational Fluid Dynamics, FLOW-3D Software.

1.1 Case Study / Project Details

Kotlibhel H.E. Project, Stage -1B, Uttarakhand, India, is located on river Alaknanda in the Pauri and TehriGarhwal district of Uttarakhand. The project envisages construction of a 90 m high and 348.2 m long concrete gravity dam 2.2 km upstream of Devprayag at the confluence of river Alakananda with Bhagirathi. The spillway consists of six spans each with an opening size of 15 m (W) x 21.2 m (H) with breast walls and has been designed to pass the design discharge of 26, 615 m^3/s at Maximum Water Level (MWL) at EL 522 m. The Full Reservoir Level (FRL)/ Minimum Draw Down level (MDDL) is at EL 521 m. The right training wall in the form of cellular wall has been provided between spillway and power house. Figures 1 and 2 show the general layout plan and section of the spillway, respectively. The energy dissipator is provided in the form of a stilling basin with floor level at EL 460.8 m. A surface powerhouse on the right bank has been provided with an installed capacity of 320 MW (4 units of 80 MW each) with a design discharge of 587.4 m^3/s. A 230 m long tail race channel has been provided downstream of the power house.

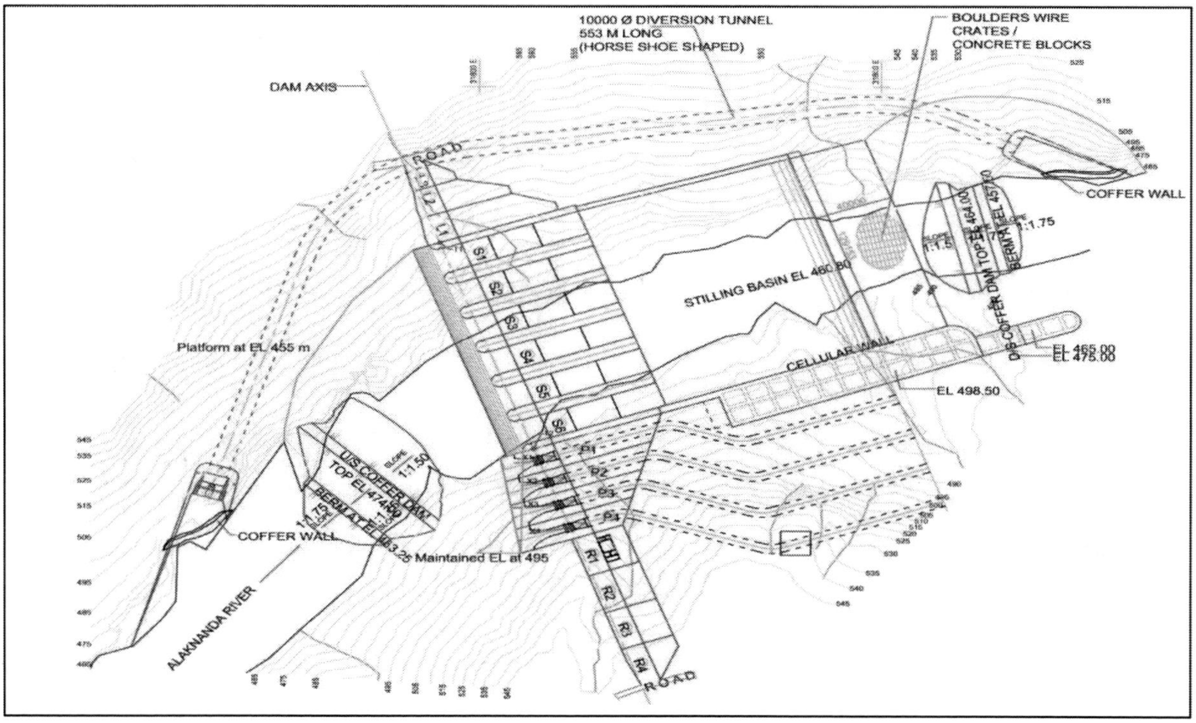

Figure 1. General layout plan

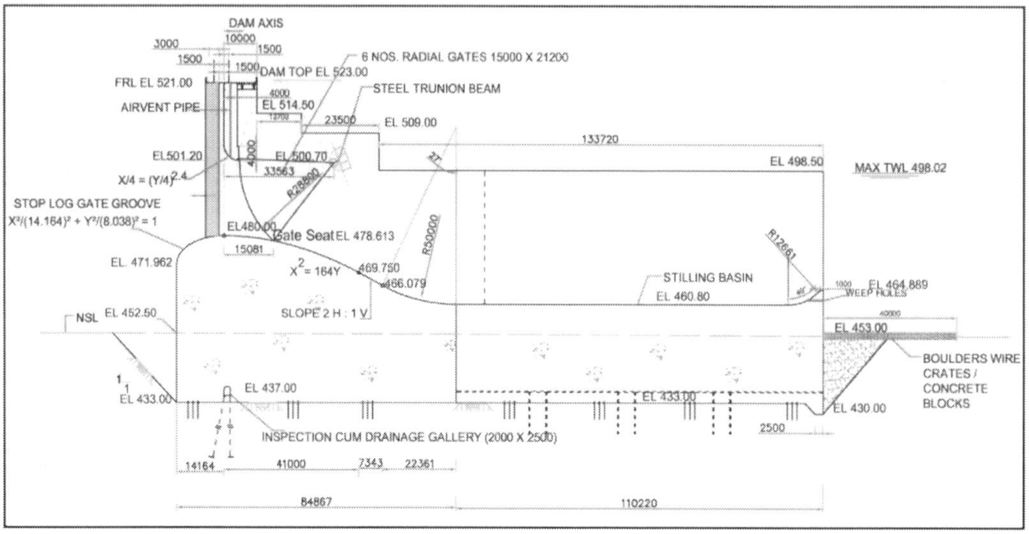

Figure 2. Section of the spillway

2. Physical Model Studies

A 1:80 geometrically similar scale 3-D comprehensive model was constructed after duly incorporating all the design features such as upstream & downstream river reach, spillway, stilling basin, piers, training walls and radial gates. Necessary arrangements were made for measurement of discharge, reservoir water levels and pressures. The accepted equations for similitude, based on Froudian similitude were used to express the mathematical relationship between the dimensions and hydraulic parameters of the model and the prototype. Figures 3 and 4 show the stilling basin with original and modified design of stilling basin constructed at CWPRS respectively.

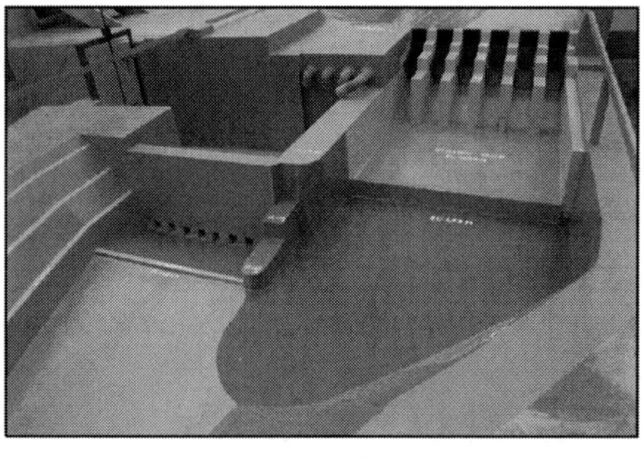

Figure 3. Downstream view of the model with stilling basin with floor level at El 460.8 m (Original Design)

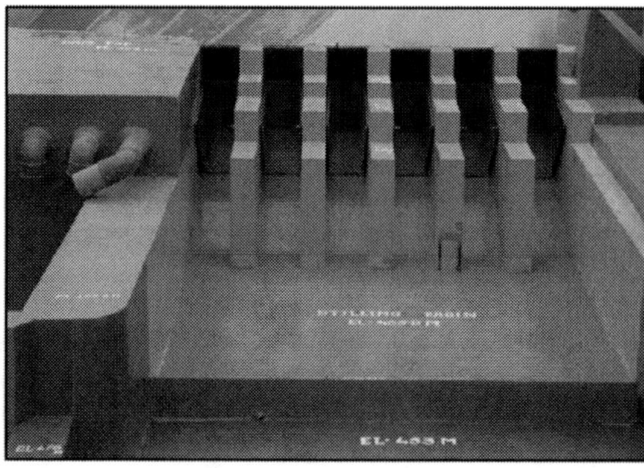

Figure 4. Downstream view of the model with stilling basin with floor level at El 455.8 m (Modified Design)

Hydraulic model studies were conducted to assess the performance of stilling basin in respect of pressure distribution along the bottom profile of the spillway for various operating conditions, water surface profiles and to observe the flow conditions in the stilling basin for entire range of discharges to assess its efficacy.

2.1 Performance of Stilling Basin with Original Design (Stilling Basin Floor Level at El. 460.8 m)

Hydraulic model studies were conducted for the original design of the stilling basin with floor level at El. 460.8 m for both the gated and ungated operation of the spillway for the entire range of discharges. Studies revealed that the performance of stilling basin was unsatisfactory as the hydraulic jump was sweeping out of the basin for the discharges of 12,000 m^3/s and below for gated operation of the spillway. It was seen that the tail water level was deficient for the formation of hydraulic jump for the discharges of 12,000 m^3/s and below. Figure 5 shows the sweep out condition for the discharge of 6653.75 m^3/s at FRL for gated operation of spillway.

Figure 5. Sweep out condition for the discharge of 6653.75 m³/s for gated operation of the spillway

2.2 Tailwater and Jump height analysis

The tailwater rating curve supplied by the project authorities as shown in Figure 6 was maintained at 200 m downstream of dam axis while conducting the model studies. Based on the analytical calculations for Froude number and sequent depth, it was found that by lowering the stilling basin by 5 m may a yield a sequent depth (D_2) which would match with the tail water requirements to form a stable and strong hydraulic jump. Jump height analysis was carried out by computing sequent depth with stilling basin floor levels at EL 460.8 m and El. 455.8 m for entire range of discharges with gated and ungated operation of spillway and the same have been superimposed for comparison as shown in Figure 6. For the stilling basin with floor level El. 460.8 m, it was noticed that the tail water rating curve was less than the sequent depth for lower discharges for gated and ungated operation of the spillway due to which sweeping out condition was taking place in the stilling basin. But with the lowering of the stilling basin floor level by 5m i.e. at El. 455.8 m, the tail water level was higher for entire range of discharges. Therefore, it was suggested to lower the stilling basin floor level by 5 m i.e. at El. 455.8 m so that the performance of the stilling basin could be improved.

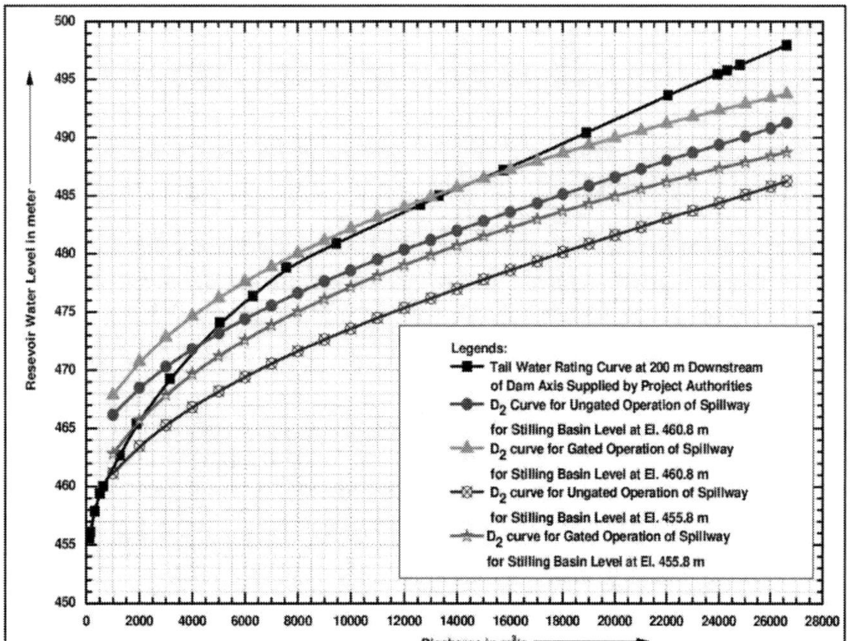

Figure 6. Tail water rating curve and sequent depth curves with stilling basin level at El. 460.8 m & El 455.8 m for ungated and gated operation of the spillway

2.3 Performance of the Stilling basin with Modified design (Stilling Basin Floor Level at El. 455.8 m)

Hydraulic model studies were carried out for the modified design of the stilling basin with floor level at El. 455.8 m to assess the performance of the stilling basin. It was observed that a stable and strong hydraulic jump was forming in the stilling basin for the entire range of discharges with gated and ungated operation of spillway as the tail water level was sufficient to form the hydraulic jump. Thus, the performance of the stilling basin was found to be satisfactory. Figure 7 shows the flow conditions in the stilling basin with stable and strong hydraulic jump formation for the discharge of 6653.75 m^3/s for gated operation of the spillway.

Figure 7. Formation of hydraulic jump in the stilling basin for discharge of 6653.75 m^3/s for gated operation of the spillway

3. Numerical Model Studies

In recent years, the computational fluid dynamics (CFD) method, through turbulence models, has become a useful tool to study complex environmental fluid-mechanics problems. The numerical modelling of a hydraulic jump, which involves fluctuating boundaries as well as a multiphase flow, is still challenging, considering its complexity. Furthermore, whereas most of the experiments provide measurements at a point or on a plane, the complete three-dimensional (3D) flow field supplied by a CFD simulation would enable us to have a deeper understanding of the dynamics of coherent structures that are responsible for free-surface fluctuations and aeration in hydraulic jumps. Several turbulence models have been proposed for solving the turbulent flow. In this regard, the K-epsilon, RNG K-epsilon, and large eddy simulation (LES) are the advanced turbulence models that recently have been widely used in hydraulic engineering (H.K. Versteeg, 2007). Several commercial software and open-source codes have been proposed for the comfortable use of numerical modelling in hydraulic engineering studies. In this regard, the commercial software Flow-3D is used for numerical model studies.

3.1 Numerical Model set up

The commercial software Flow-3D, developed by Flow Science, uses the finite-volume method to solve the Reynolds-averaged Navier –Stokes (RANS) equations over the computational domain (Amorim et al, 2004). The equations resulting from the time-averaging of Navier-Stokes equations are known as Reynolds-Averaged Navier-Stokes (RANS) equations and the numerical modeling based on these equations is known as RANS-based turbulence modeling. Tracking of the free surface is performed using the Volume-of-Fluid method. To represent the flow characteristics of the flow over the stilling basin, the Re-normalized Group (RNG) turbulence model was used before starting the solution, initial guesses like geometrical information of all grid points, the meshing of the geometry, the initial upstream boundary condition, and the number of time steps to achieve the steady-state solution have to be provided for the solution flow field. The numerical model studies were carried out to observe the performance of stilling basin for the original and modified design of the stilling basin. Therefore two geometries were generated with the help of AutoCAD software in the geometrically similar physical model scale of 1:80. The numerical simulations were carried out for the prototype discharge of 6653.7 m^3/s (25 percent of the design discharge 26615 m^3/s) which in the model scale comes out as 0.116 m^3/sec for gated condition by maintaining the tailwater level at El. 476.74 m. The geometry of the spillway profile including stilling basin is embedded in the computational grid by the pre-processor using the FAVOR concept (Hirt, C.W., 1985). In Flow 3D, grid generation is the most important issue for an accurate solution. Flow 3D uses a grid system of orthogonal meshes in the Cartesian or cylindrical coordinates. The Cartesian coordinate system is selected for this investigation. The flow region is subdivided into a mesh of fixed rectangular cells. The grid generation does not depend on geometry. The geometry in the domain is represented simply and properly without requiring a body-fitted grid system so that the geometry and grid generation are independent of each other. A uniform mesh is used for the first simulation trial, as it took more computational time non-uniform mesh system was generated for other trials to provide more grid refinement close to the block of the spillway. For non-uniform mesh systems, Flow 3D users manual advices to generate an appropriate grid. The most significant issue is to avoid large differences in sizes between adjacent cells. The size ratio between adjacent cells should be as close to unity as possible, and not exceed 1.25 for efficient results. Hence, in the upstream domain, cell size of 0.02 (total cell count of 2 million) was provided whereas in the spillway domain, fine mesh with a cell size of 0.01 (total cell count of 4 million) was provided and in the downstream, coarse mesh with a cell size of 0.02 (total cell count 1 million) were provided to optimize the use of computational resources and time. In this paper, input data to set the boundary conditions were used from thephysical model data to study the performance and location of hydraulic jump in the stilling basin. Figures 8 and 9 show the numerical model showing boundary conditions and grid distribution for the computational domain of the model respectively.

3.2 Boundary Conditions

Boundary conditions were set for the upstream and downstream of the domain and the interface between the fluid and solid boundary was considered a closed boundary. The upstream face of the geometry was set as the inlet discharge of 6653.7 m^3/s (0.116 m^3/s in the model scale) with gated condition which is recognized as Volume Flow Rate in the Flow 3D. In the downstream, the exit of the domain was set as a pressure outlet. The extent of the mesh in the upstream X-direction was adjusted until any further increases had a negligible effect on the discharge. Numbers of probes were provided at different locations on the surface of the spillway and stilling basin floor level for observing the performance of stilling basin and free surface elevation. Baffles were provided at the inlet and outlet boundary of the domain to observe the discharge passed through the spillway. For iterative convergence, it is generally stated that the normalized residuals for each equation in the numerical solutions must drop under at least 0.001. If the computed values within a time step exhibit variations over a distance comparable to cell width or time comparable to the time increment, the accuracy of the computed results is not reliable. It is said that the solution has not converged. Hence for accuracy mesh increments were chosen small enough to resolve the spatial variation in all dependent variables and finer grids were provided near the wall boundaries. The duration of the simulation was decided by running a typical simulation for a longer time of about 80 seconds. It was found that the inlet and outflow discharge was reaching steady-state condition after about 50 seconds. Therefore, a simulation time of 50 seconds was adopted for the remaining runs.

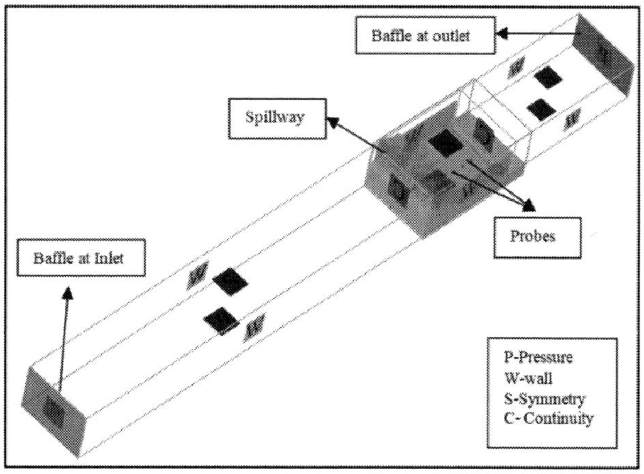

Figure 8. Numerical model showing boundary conditions

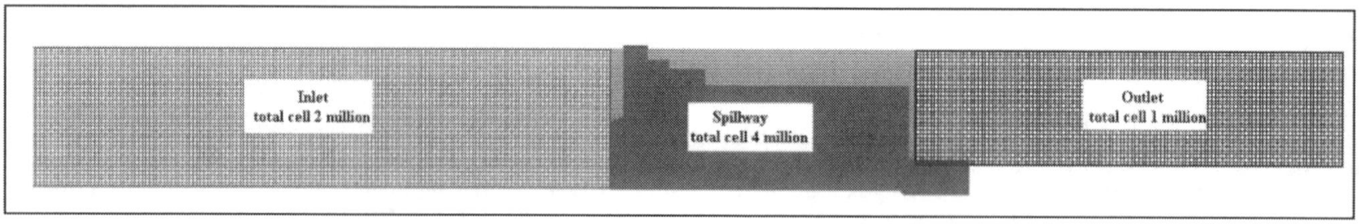

Figure 9. Grid distribution for computational domain of model

4. Comparison of results of Physical and Numerical Models

The prototype size of the spillway of the Kotlibhel H.E. Project is 138 m wide, 195 m long and 90 m high and the numerical model study for this large size prototype spillway is not possible with the available computational facility. Hence for the numerical model study physical model scale i.e., 1:80 was chosen. The numerical model study was carried out for the gated operation of the spillway for 6653.75 m^3/s (0.116 m^3/s model scale) for the original and revised design of stilling basin. Numerical Model results were validated with physical model results of the water surface profile and the pressures on the spillway. To correlate the model study results with the prototype, the results obtained from both the physical and numerical model study were converted into the prototype dimensions.

4.1 Water Surface Profiles

In the numerical model studies water surface profiles were obtained for a discharge of 6653.75 m^3/sec (0.116 m^3/s model scale) at FRL for the gated operation of the spillway for the revised design of the stilling basin with floor level at El. 455.8 m. Water surface profiles observed along the centre line of spillway, for a discharge of 6653.75 m^3/s at FRL for gated operation of the spillway from both the model studies were converted to prototype dimensions to compare the results as shown in figure 10. It could be observed from the figure that the hydraulic jump formed is contained well within the stilling basin. Comparison of the numerical model profile with the physical model, it was observed the numerical model profile almost follow the trend of physical model profile. However, there was around 1.8 % to 13% error in observed water surface elevations of the profiles. As the numerical model study was done on the 1:80 scale, in the simulation 0.1 m error stands for 8 m in the prototype result. Hence, a slight difference could be seen in the water profiles. The reasons for the error generation in the numerical model studies are primarily caused by the grid density, especially in the regions close to the interface. The errors can be reduced with a denser grid, but the computing time will be increased accordingly.

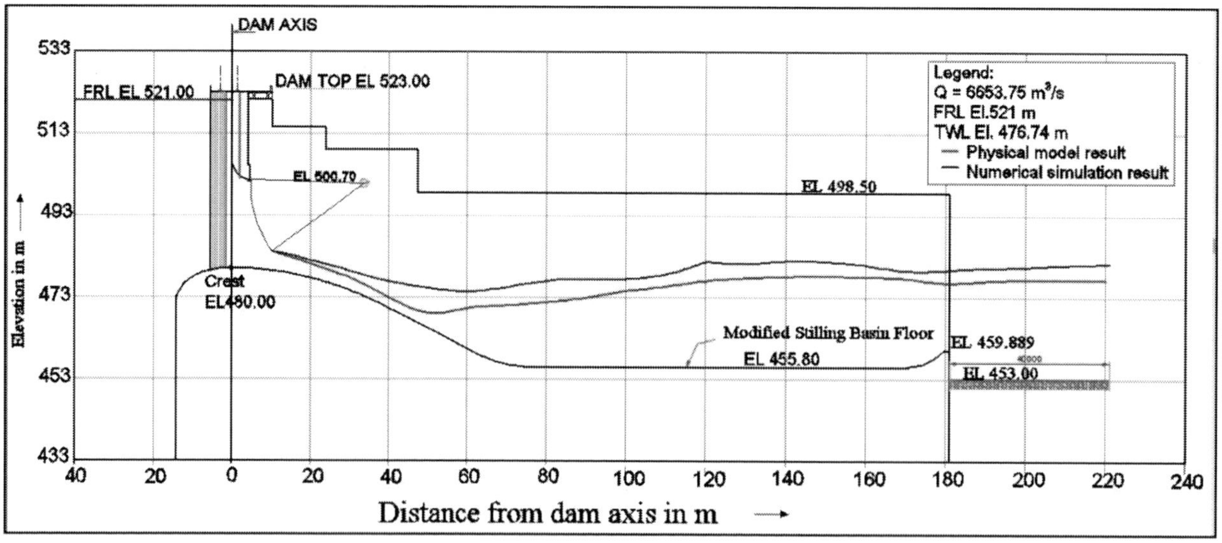

Figure 10. Comparison of water surface profile for the discharge of 6653.75 m^3/s for gated operation of spillway (Modified Design)

4.2 Pressure Profiles

Pressure observations on the physical model were made along the center line of the spillway up to the end of stilling basin with revised design for different discharges viz. 26615, 19961.25, 13307.5, 6653.75 m^3/s for ungated as well as for the gated operation of the spillway. The pressure profiles on the surface of the spillway and over the stilling basin indicated that the pressures were positive for the entire range of discharges except at very few locations for the discharge of 6653.75 m^3/sec passing under the gated operation of the spillway. However, the corresponding cavitation index works out to be above the critical cavitation index of 0.2, therefore, the pressures were found to be acceptable. As such the performance of the spillway profile along with the stilling basin in respect of pressures was considered to be satisfactory. The pressure profiles obtained from the modified design of stilling basin with floor level at El. 455.8 m, for a discharge of 6653.75 m^3/s for the gated operation of the spillway from the physical and numerical studies are depicted in figure 11. It could be seen from the figure that the pressure profiles are matching very well.

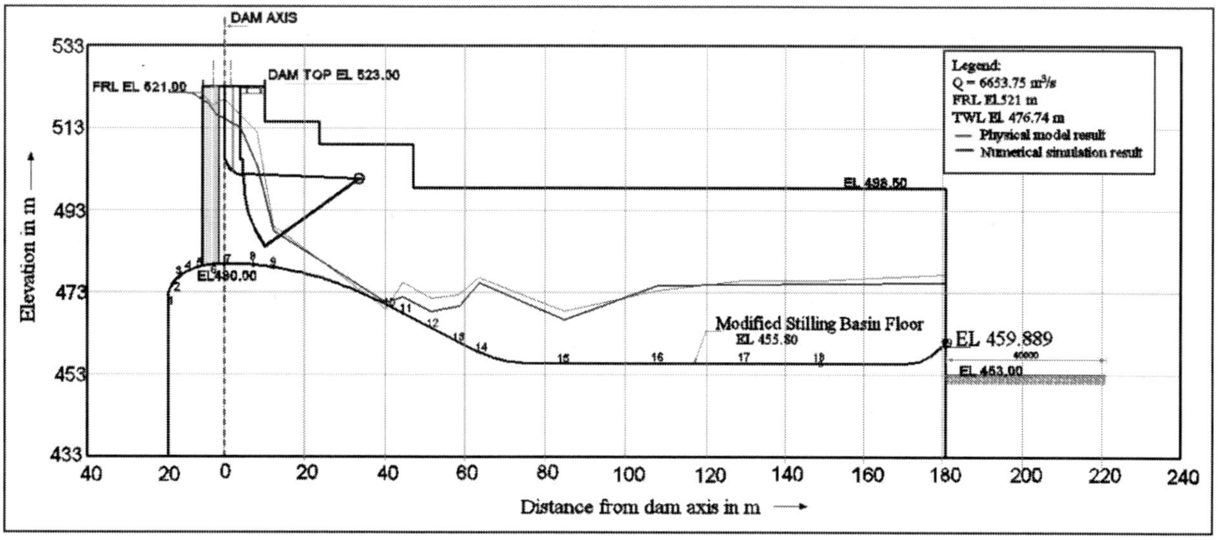

Figure 11. Comparison of pressure profiles for the discharge of 6653.75 m^3/s for gated operation of spillway (modified design)

4.3 Performance of the Stilling Basin with Original and Modified Design

Figure12 shows the phase diagram of sweep out of hydraulic jump in the stilling basin for a discharge of 6653.75 m^3/s for gated operation of the spillway in the Numerical simulation for the original design of the stilling basin. Figure 13 shows the phase diagram showing the formation of strong and stable hydraulic jump in the stilling basin for discharge of 6653.75 m^3/s for the gated operation of the spillway in the Numerical simulation for the revised design of the stilling basin. It shows that the performance of the modified design of stilling basin is satisfactory as a stable and strong hydraulic jump is forming in the stilling basin.

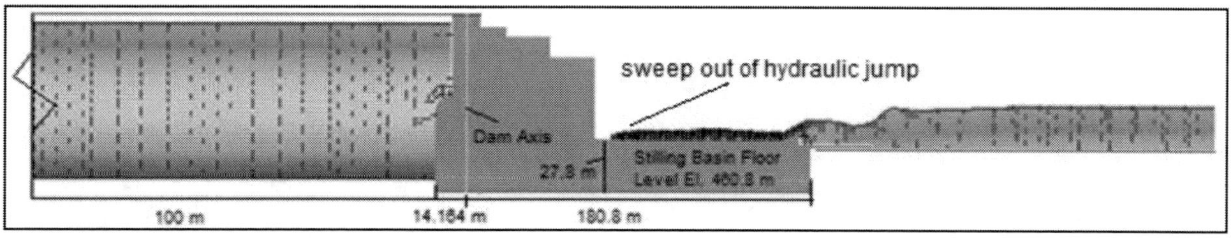

Figure 12. Phase diagram showing sweep out of hydraulic jump in the stilling basin for discharge of 6653.75 m^3/s for gated operation of the spillway by Numerical simulation (original design)

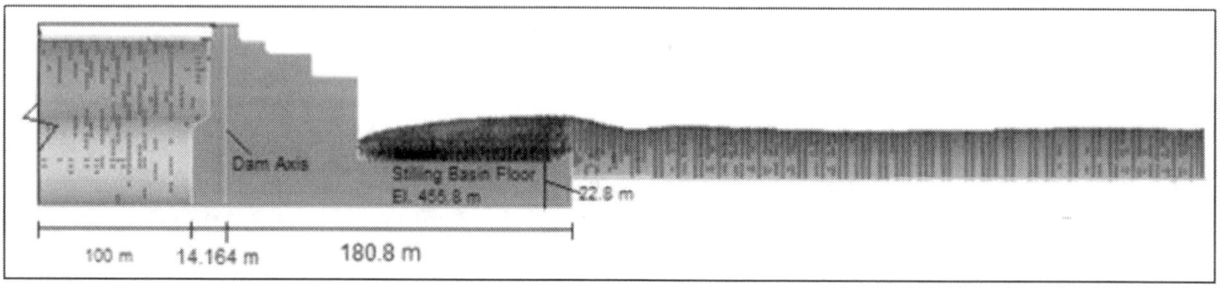

Figure 13. Phase diagram showing formation of hydraulic jump in the stilling basin for a discharge of 6653.75 m^3/s for gated operation of the spillway by numerical simulation (modified design)

5. Results and Discussions

Performance of the stilling basin of the spillway of proposed Kotlibhel H.E. Project stage 1-B was assessed by conducting hydraulic model studies for flow conditions in the stilling basin for the original and revised design of the stilling basin. The results of the water surface profiles and pressures over the spillway profile obtained by numerical studies are compared with the physical model studies. Flow conditions observed in the stilling basin from numerical model studies were compared with that of physical model studies to assess its performance for the original and modified design from both techniques. The following observations are made.

- Results showed that water surface profiles of the numerical model and physical model are in reasonably good agreement with one another. However, for the gated operation of the spillway, water surface profiles obtained from numerical modelling are slightly lower than that obtained by physical modelling. The difference in water surface profiles may be attributed to the formation of the Rooster tail in the 3D physical model, and lesser grid density in the numerical model.

- Pressures observed over the spillway profile and stilling basin were positive for the entire range of discharges with the gated and un-gated operation of the spillway in both physical as well as numerical models except at very few locations for the discharge of 6653.75 m^3/sec passing under the gated operation of the spillway. However, the corresponding cavitation index works out to be above the critical cavitation index of 0.2, therefore, the pressures were found to be acceptable. As such the performance of the spillway profile in respect of pressure was considered to be satisfactory. The pressures obtained from the numerical simulation are in good agreement with the physical model studies.

- Phase diagrams show that flow conditions/features observed on the physical model have been captured exactly in the numerical simulation. Sweep out of hydraulic jump due to deficient tail water level as observed in the physical model (shown in the photo, Figure 5), and formation of hydraulic jump with the lowering of the stilling basin floor with sufficient tail water level as observed in the physical model (shown in the photo, Figure 7), have been exactly simulated through the numerical model studies (Phase diagrams in figures 12 and 13). This highlights that the flow conditions observed in the physical model are very well simulated in the numerical model.

- The formation of the stable and strong hydraulic jump was seen in the stilling basin with the revised stilling basin floor level at El. 455.8 m, both in the physical and numerical model study results and hence it was concluded that its performance was satisfactory.

6. Conclusions

In the present study, the performance of stilling basin as an energy dissipator for the proposed Kotlibhel 1-B Dam spillway, Uttarakhand, India, was evaluated by physical and numerical model studies. The results of both the model studies are compared. It is concluded that the performance of the stilling basin with revised floor elevation at El. 455.8 m is satisfactory. It is to highlight that the flow conditions in the stilling basin are very well simulated in the numerical model studies when compared with the flow conditions in the physical model for both the original and modified design of the stilling basin. The results for pressure measurements and water level elevations obtained from numerical model studies are well corroborated with the physical model results. It is found that Numerical / CFD modelling software Flow-3D could effectively simulate the flow conditions over the spillway and in the stilling basin. A comparison of the results from the two modelling approaches has provided a very good insight into the design process to optimize the design of stilling basin. The numerical modelling technique has provided invaluable insight and greater confidence for future use of standalone CFD analysis as a complementary tool for physical model studies in optimizing the design of stilling basin as an energy dissipator for the spillway. After studying the performance of the stilling basin both through physical and numerical techniques, it is found that numerical / CFD modelling can generally be used as a valuable tool to test the preliminary design and in the optimization phase of the spillway/stilling basin design prior to taking up the physical model study. This will save time and cost involved in the physical model construction and studies at the conceptual stage of the project.

7. Acknowledgements

The authors are thankful to Dr. Roop Singh Kankara, Director, and Shri. Y. N. Srivastava, Additional Director, CW&PRS for the constant encouragement and guidance during the preparation of the technical paper. The authors are grateful to the project and design engineers from NHPC, for their constant interaction during the course of studies.

8. References

Bhate, R.R., Bhajantri, M. R., Bhosekar, V. V., (2015),"Numerical Simulation of Flow through Breastwall Spillway", Proceedings *of HYDRO International conference, Dec 17-19, IIT, Roorkee.*

Bureau of Indian Standards Code 4997 (1968), "Criteria for design of hydraulic jump type stilling basin with Horizontal and sloping apron". *Bureau of Indian Standards, New Delhi, India*

CWPRS Technical Report No. 4612 (March 2009) "Hydraulic model studies for spillway and power intake of Kotlibhel H. E. Project, Stage 1-B, Uttarakhand, 1:80 scale 3-D Comprehensive model".

CWPRS Technical Report No. 4786 (October 2010) "Hydraulic model studies for revised design of spillway and power intake of Kotlibhel H. E. Project, Stage 1-B, Uttarakhand, 1:80 scale 3-D Comprehensive model".

Computational Fluid Dynamics Software – Flow 3D from Flow Science, <http:// www.flow3d.com> (2013)

Versteeg, H.K. , Malalasekera , W. (2007), "An Introduction to Computational Fluid Dynamics: The Finite Volume Method", *Pearson Education Limited, Harlow, England.*

Hirt, C.W. and Sicilian, J.M. (1985), "A porosity technique for the definition of obstacles in rectangular cell meshes". *Proc. 4th Int. Conf. Ship Hydro, National Academy of Science, Washington, DC, September, pp. 1-19.*

Amorim, J.C.C, Amante, R. C. R., Barbosa V. D (2015), "Experimental and Numerical modelling of flow in a stilling basin", *published in the proceedings of 36th IAHR World Congress, The Hague, The Netherlands.*

Michael C. Johnson, and Bruce M. Savage, (2006). "Physical and Numerical Comparison of Flow over Ogee Spillway in the Presence of Tailwater", Journal *of Hydraulic Engineering, vol 132, pp. 1353-1357.*

Peterka, A.J., (1984), "Hydraulic design of stilling basin and energy dissipators", *8th edition, Government Printing Office, Denver, Colorado: United States.*

DOI: https://doi.org/10.26077/1d18-ff7d

Prediction of Discharge Coefficient of Circular Side Orifice through Machine Learning Technique

Syed Ahmed Noman[1], Dr. Ajmal Hussain[2], Mr. Mohd. Danish[3] & Prof. Rashid Ali[4]
[1]*MTech (Artificial Intelligence), Aligarh Muslim University, India*
[2]*Assistant Professor, Department of Civil Engineering, Aligarh Muslim University, India*
[3]*Assistant Professor, Civil Engineering Section (University Polytechnic) Aligarh Muslim University, India*
[4]*Professor, Department of Computer Engineering, Aligarh Muslim University, India*
E-mail: syedahmednoman@zhcet.ac.in

Abstract: *A sharp-crested circular side orifice is a crucial element when it comes to diverting flow from primary source to its subordinate source. Such a flow measurement instrument technique is of immense value in conservation and evaluation of drainage and irrigation networks. Usually, it is placed towards the side of a channel in order to regulate the flow of the fluid. Traditionally, coefficient of discharge was predicted through regression methods which are time-consuming and lack accuracy. Artificial Intelligence (AI) and its applications in this domain have bridged this gap by providing novel alternative methods which prove much more efficient. Repeated studies have pointed out that AI techniques generally give better results when it comes to a myriad of water variables such as rainfall-runoff, evaporation and evapotranspiration, streamflow, and dam water level changes. Total 261 dataset has been collected from the literature review comprising of the fully submerged orifice and for partially-submerged orifice with varying orifice diameter (D) of 5 cm, 10 cm and 15 cm. This study aims to provide a better estimate of prediction of discharge through circular sharp-crested orifice using Artificial Neural Network (ANN). The ANN model has been deployed to randomly select 80% of the data for training, 15% for validation and remaining 5% for testing. In the ANN model, Lavenberg-Marquardt algorithm was used as back-propagation step to assign weights in order to predict the output. The correlation coefficient (R), mean absolute error (MAE) and root mean squared error (RMSE) for complete data of fully and partially submerged circular side orifice are observed to be 0.9765, 0.0228 and 0.0172 respectively.*

Keywords: *Artificial neural network, Circular side orifice, Coefficient of discharge, Flow measurement.*

1. Introduction

Side orifices are utilised in irrigation, drainage, wastewater treatment plants to divert the flow from the main channel (Azimi et al. 2017). Side orifices are meant to check and adjust the flow in a channel. Different shapes of orifices are employed to control flow in an irrigation system. Various experimental and numerical studies have been conducted over recent years to determine the flow behaviour around orifice and its discharge coefficients. (Asadi et al 2020). Ramamurthy et al. (1986, 1987) conducted on rectangular side orifices where open channels were inspected to ascertain flow. (Gregg et al. 2003) and (Werth et al. 2005) conducted experiments to study the determination of discharge coefficients for hydraulic sparger design. Several researchers have proposed relationships for the coefficient of discharge of sharp crested side weir subjected to subcritical flow conditions (Zahiri et al. 2013; Fu et al. 2018. Emiroglu et al. in 2010, 2011 conducted experiments on straight and curved channels to measure the lateral outflow over triangular labyrinth and rectangular side weirs. Hussain et al. (2016) proposed the equations of discharge for large and small circular orifices.

When it comes to the amalgamation of computational techniques and discharge analysis, Artificial Neural Networks (ANN) proves to be an effective alternative to the costly yet labour intensive experimental process. This method of using Artificial Intelligence has been extensively used in hydraulics applications from estimating the factor of friction (Yuhong and Wenxin, 2009) to alluvial channel geometry (Riahi-Madvar et al. 2011). Various machine learning techniques have been used recently to solve the complicated and non-linear problems in water resources engineering. ANN and neuro-fuzzy inference system (Ebtehaj and Bonakdari 2014; Khoshbin et al. 2016) were used for evaluating the performance for sediment transport in sewers. (Sharifipour et al. 2016) compared genetic programming and radial basis function neural network for open channel junction velocity feild protection, Ebtehaj et al. (2018) developed discharge coefficient prediction equations for rectangular side weirs using adaptive neuro-fuzzy inference system and generalized group method of data handling. However, it has been observed from the literature review that ANN has not been used to predict the discharge coefficient of circular side orifice under fully and partially submerged flow conditions, therefore this study aims at predicting the coefficient of discharge for circular side orifice.

2. Artificial Neural Networks

The expression "artificial neural network" bases itself on the biologically induced subfield of artificial intelligence centering on the brain. Thus, the structure of the human brain is replicated through a computational network based on its biological counterpart and comes to be known as an artificial neural network. Just as neurons in the human brain are connected to each other, so are artificial neural networks that contain neurons are interconnected in a dedicated mesh of networks. The fundamental unit of the ANN is a neuron. all neurons in a layer are connected with other neurons of adjacent layers through weighted connections. Information is fed in the input layer through neurons and then after being multiplied by weight function, the information is carried forward to then hidden layer where, various nonlinear functions (e.g. sigmoid function, tanh, ReLU) change the inputs signals and push them forward to the next layer. The procedure is repeated till the output layer is reached. This network is an example of feed-forward neural networks (FFNN) (Schalkoff (1997)). The number of hidden layers and the number of neurons within this layer depend upon the intricacy of the model. Greater the number of hidden layers, the better will be the performance of the model in FFNN (Norgaard et al. (2000). The output achieved the end of the network is then compared with the actual output and the difference is termed as error.

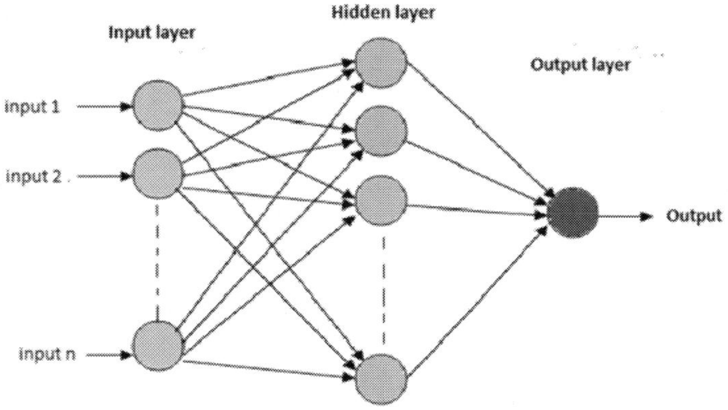

Figure 1. Artificial neural network model

Another important step for the ANN model development is training where the determination of appropriate weights and biases for the connections is estimated. This is further broken down into feed forward step and back-propagation step.

Feed-forward step is the one where the information moves forward from the input layer to the hidden layer, which can be mathematically written as:

$$Net_j = \sum_{i=1}^{n} W_{ij} X_i + b_0 \qquad (1)$$

where Net_j = input received by j^{th} neuron of hidden layer;
W_{ij}= Associated weight from i^{th} neuron to j^{th} neuron of the hidden layer
n= number of neurons
b_o= bias weight and X_i= value of ith neuron in input layer
The jth neuron receiving the input is changed to output through nonlinear sigmoid activation function.

Back-propagation step is the one where the total error function is curtailed through a back-propagation algorithm and the training is done after the weights are initialised. The error computed in this method is back propagated and the weights are readjusted in order to minimize the error. In this study, we would use the widely acknowledged Levenberg-Marquardt back-propagation algorithm as it is the most fruitful technique in ANN models.

2.1. Levenberg-Marquardt Algorithm

Developed in the 1960s, Lavenberg-Marquardt algorithm is a union of gauss newton algorithm and gradient descent algorithm and it is because the gauss newton algorithm is efficient in global convergence while the gradient descent works well in quick local convergence near the minimum. In the gradient descent method, the parameters are readjusted in the direction of the steepest descent, which reduces the sum of the squared errors. In the Gauss-Newton method, the process involves assumption that the least squared function is quadratic and evaluating the minimum of this quadratic in order to reduce squared errors. The iterative updating of LM can be written as:

$$X_{(b+1)} = x_b - (j^T j + \mu I)^{-1} g \qquad (2)$$

Where, I = identity matrix; μ = scalar; x_b and x_{b+1} = vectors of independent variables at b^{th} and $(b+1)^{th}$ iterations respectively.

The Gauss-Newton algorithm and gradient descent algorithm interpose with one another depending on the value of μ. Other training algorithms such as Bayesian regularization and Scaled conjugate gradient algorithms have not been conventionally used to train the ANN model due to their less accuracy and slower computational speed and therefore, Levenberg-Marquardt Algorithm is preferred over other training algorithms (Ayaz and Mansoor 2021).

3. Experimental Setup and Dataset

While performing the experiment, the diameter of the circular orifices were taken to be 5 cm, 10 cm, and 15 cm and the crest height of the orifice was set at W= 0, 0.05m, 0.10m, 0.15m, 0.20m and 0.25 m. More than two discharges Q_m in the main channel were there for every set of D and W. The downstream sluice gate was regulated through maintaining varied depths of flow in the main channel for every Q_m. Digital pointer gauge of ± 0.1 mm accuracy was used to measure flow depths in the main channel in surrounding areas of the orifice as well as head over the crests of weir-A and weir-B. The calibration of the weirs in order to estimate discharge through orifice in the main channel was done by ultrasonic flow meter. Since our experimental setup is in regard to fully and partially submerged orifice, a sluice gate was put downstream of the diversion channel in order to effectively achieve the level of submergence required. For every Q_m, the variation in the depth of submergence in the downstream direction was changed with the help of the aforementioned sluice gate. The measurements were taken for the flow depth in the main channel and head over the crests of weir A and weir B for each run. The experiment was conducted for subcritical flow condition in the main channel.

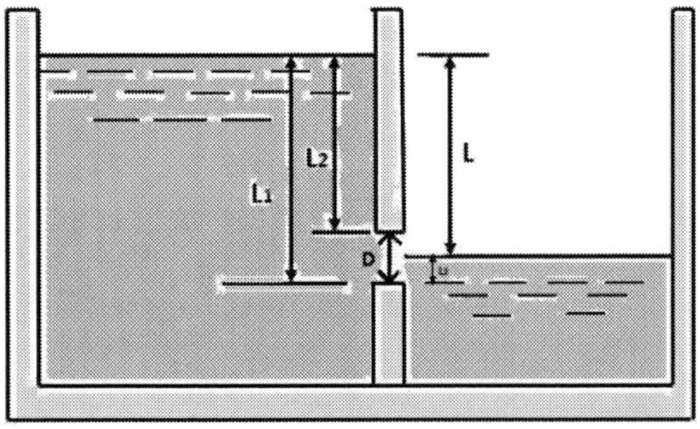

Figure 2. Fully-submerged orifice

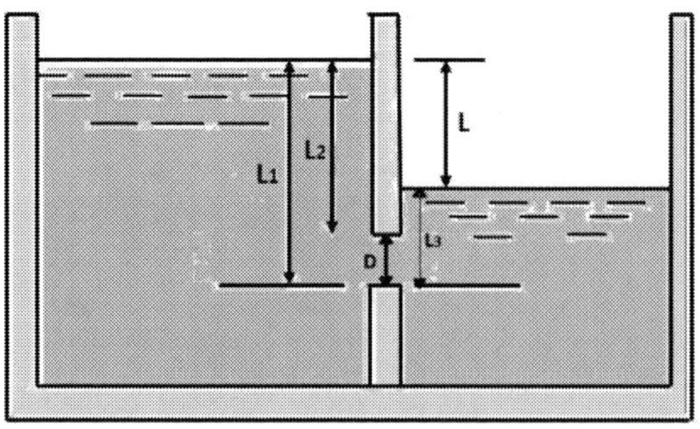

Figure 3. Partially-submerged orifice

directly. Floating frames are useful for convenient positioning of graphics. Original line drawings, photos, etc. should be scanned before with reference to the above figures (Fig 2 and Fig 3), where L_1 and L_2 are the height of the liquid on the upstream side of the bottom and top edge of the orifice respectively; D is the diameter of the orifice. Thus, when L is calculated as:

$L = L_1 - L_3$ and on comparing, when $L \leq (L_1 - D)$ then the orifice is fully submerged and when $L > (L_1 - D)$ then the orifice is partially submerged.

Thereby, the dataset has been classified into fully and partially submerged orifice. This is done on the relative length of L (refer Fig 2 and Fig 3). When L is smaller than L_2, then the orifice is fully submerged. When L is greater than or equal to L_2, then the orifice is partially submerged. For this study, these calculations were used to divide the datasets into fully (Dataset 1), partially submerged orifice (Dataset 2) and combined dataset of fully and partially submerged orifice (Dataset 3).

Dataset 1 contains 170 datapoints, dataset 2 comprises of 91 datapoints which combines to give the total no. of datapoints in dataset 3, i.e. 291 datapoints.

The summary of Dataset 3 has been presented below in Table no. 1:

Table 1- Summary of dataset 3

Parameters	Range of Data			
	Minimum	Maximum	Mean	Std. Deviation
B/D	3.3	9.9	5.1775	2.5459
H/D	0.1720	5.95	1.573851	1.300183
F_r	0.116326	0.995964	0.405612	0.178886
W/D	0	2.446	0.671019	0.656611
C_d	0.223203	0.640177	0.549501	0.079988

4. Model Development and Performance Measures

In any machine learning model, input selection of the variables is important for appropriate model development and application. As the literature puts forward by Hussain (2010) that the target output (coefficient of discharge (C_d)) is a function of non-dimensional parameters. i.e.

$$C_d = f\left[F_r, \frac{B}{D}, \frac{W}{D}, \frac{H}{D}\right] \quad (3)$$

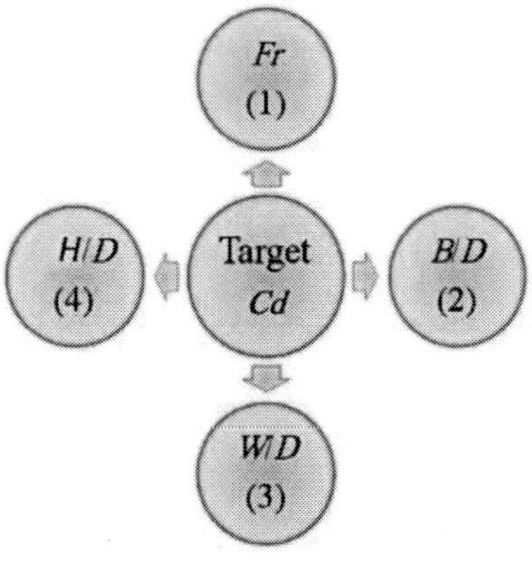

Figure 4. Input and target parameters

Where F r= Froud's number
B = Width of the main channel
H = Head of water above centreline of orifice
D = Diameter of the orifice

ANN models are considered powerful when it comes for predictions. Hence, we have developed the ANN model using the MATLAB software. After selecting the number of neurons for training from 1 to 15, number of neurons chosen in the hidden layer was 10 as it gave the best results and after which the error was stabilized. The activation function used for the model was the sigmoid function, and the training algorithm used for the backpropagation step is the Lavenberg-Marquardt algorithm. For a optimized model, ample amount of training data is needed. Hence, training, testing and validation data split was chosen as 80%, 5% and 15% respectively.

Therefore, for dataset 1, 135 datapoints were used for training, 9 for testing and 26 for validation.
For dataset 2, 72 datapoints were used for training,5 for testing and 14 for validation.
For dataset 3, 209 datapoints were used for training,13 for testing and 39 for validation.

The optimal ANN model architecture was ascertained to be 4-10-1.

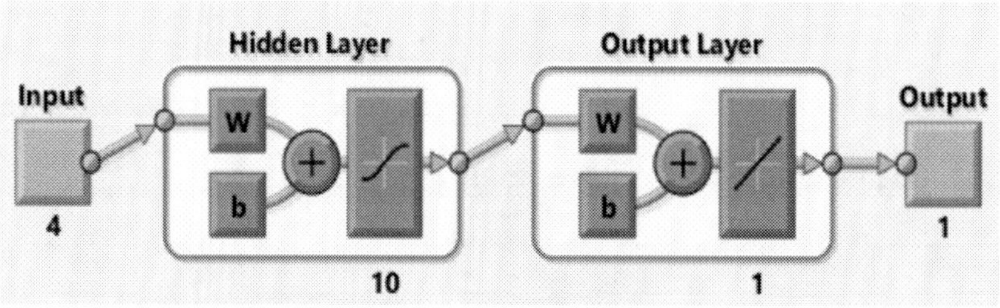

Figure 5. Developed ANN model illustration model in MATLAB

The performance measures for evaluating the developed model are-
R (Correlation Coefficient) = Correlation coefficient is used to assess the strength of the relationship between two variables. It illustrates how the output parameter changes with the inputs (-1 to +1)

$$R = \frac{Cov(a,b)}{\sqrt{p_a^2 p_b^2}} \tag{4}$$

Where, Cov(a,b) is the covariance, i.e., how afar is each observation (a,b) pair is from the mean of a and the mean of b concurrently, and p_a^2 and p_b^2 are the sample variances for a and b.

MAE (Mean Absolute Error) = It is the average difference between the observations (true values) and the model output (predictions). It gives an idea about the mean of residuals in the data.

$$\text{MAE} = \frac{1}{N}\sum_{i=1}^{N} |y_i - \hat{y}| \tag{5}$$

where, y_i are the predicted values, $\hat{y}$ are the true values, and N is the number of samples

RMSE (root mean squared error) = It is the square root of Mean Squared error. It calculates the standard deviation of residuals.

$$\text{RMSE}= \sqrt{\frac{1}{N}\sum_{i=1}^{N}(y_i - \hat{y})^2} \tag{6}$$

For the best assessment of the model, R value should be close 1 whereas the RMSE and MAE values should be closer to 0.

5. Results

For the current study, ANN model with 10 neurons in the hidden layer with activation function as sigmoid was adopted and applied to all the 3 datasets that we had. The total datapoints were 261 in dataset 3, out of which 170 were of fully submerged orifice, which we named dataset-1 and 91 datapoints for partially submerged orifice, which is dataset-2.
For the dataset 1, of fully submerged circular side orifice, R = 0.9883, MAE = 0.0112 and RMSE = 0.0091for the training data, R = 0.9816, MAE = 0.0254 and RMSE = 0.0210 for validation data, R = 0.9614, MAE = 0.0218 and RMSE = 0.0148 for testing data and R = 0.9849, MAE = 0.0138 and RMSE = 0.0120 for all data.
For the dataset 2, of partially submerged circular side orifice, R = 0. 0.9782, MAE = 0.0239 and RMSE = 0.0180 for the training data, R = 0.9830, MAE = 0.0344 and RMSE = 0.0226 for validation data, R = 0.9862, MAE = 0.0566 and RMSE = 0.0336 for testing data and R = 0.9750, MAE = 0.0272 and RMSE = 0.0199 for all data.
For the dataset 3, all data combined, R = 0. 0.9804, MAE = 0.0213 and RMSE = 0.0158 for the training data, R = 0.9587, MAE = 0.0279 and RMSE = 0.0220 for validation data, R = 0.9731, MAE = 0.0320 and RMSE = 0.0225 for testing data and R = 0.9765, MAE = 0.0228 and RMSE = 0.0172 for all data.
The regression plots show that most predicted data points lie on or near the line of agreement and the error histograms display that most points lie on or near the zero-error line

For Dataset 1-

Table 2. performance measures of dataset 1

	R	MAE	RMSE
Training data	0.9883	0.0112	0.0091
Validation data	0.9816	0.0254	0.0210
Testing data	0.9614	0.0218	0.0148
All data	0.9849	0.0138	0.0120

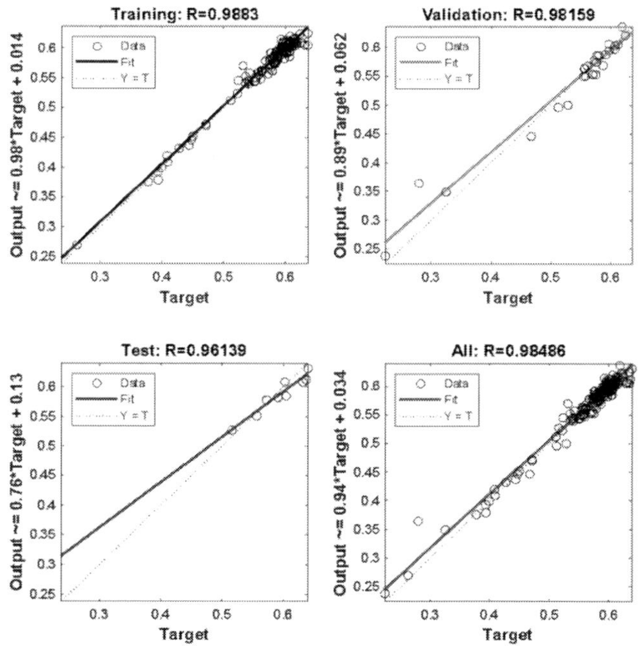

Figure 6. Regression plots for training, validation, testing and all data

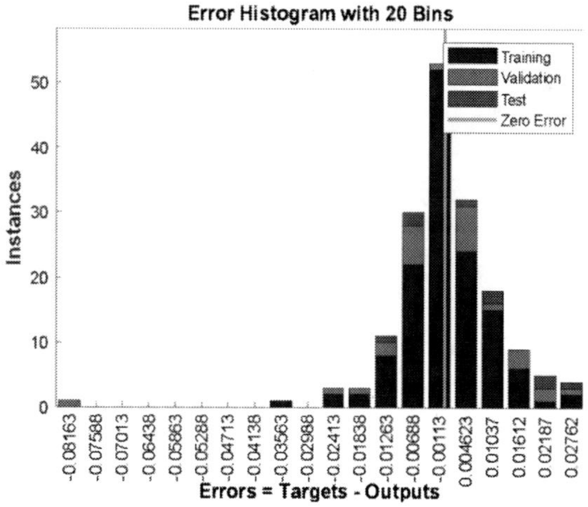

Figure 7. Error histogram with 20 bins

For Dataset 2-

Table 3. performance measures of dataset 2

	R	MAE	RMSE
Training data	0.9782	0.0239	0.0180
Validation data	0.9830	0.0344	0.0226
Testing data	0.9862	0.0566	0.0336
All data	0.9750	0.0272	0.0199

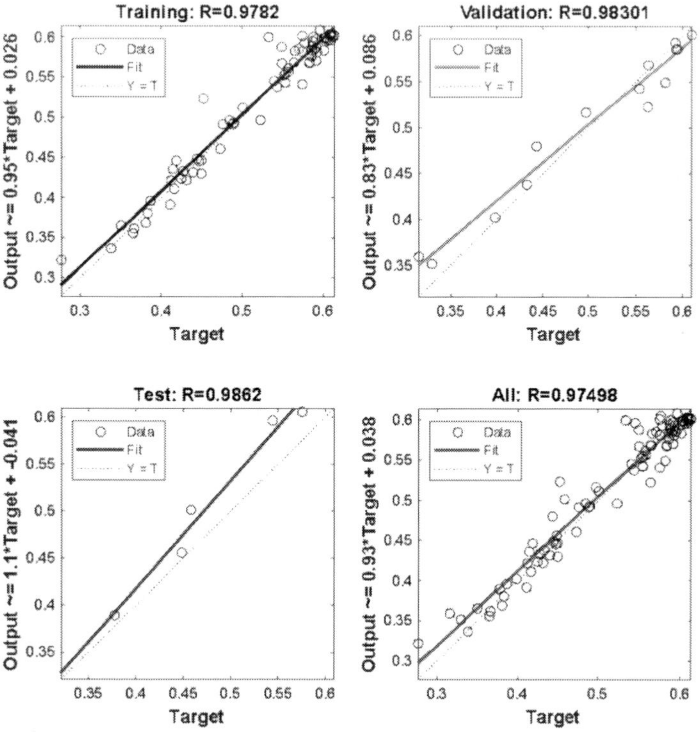

Figure 8. Regression plots for training, validation, testing and all data

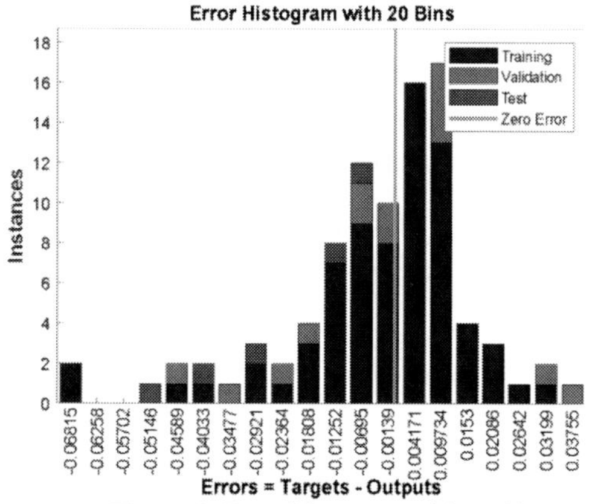

Figure 9. Error histogram with 20 bins

For Dataset 3-

Table 4. performance measures of dataset 3

	R	MAE	RMSE
Training data	0.9804	0.0213	0.0158
Validation data	0.9587	0.0279	0.0220
Testing data	0.9731	0.0320	0.0225
All data	0.9765	0.0228	0.0172

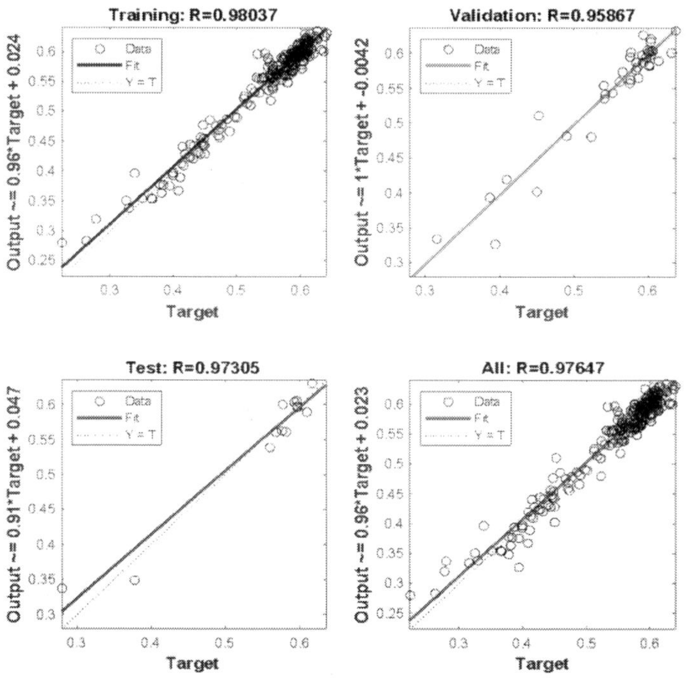

Figure 10. Regression plots for training, validation, testing and all data

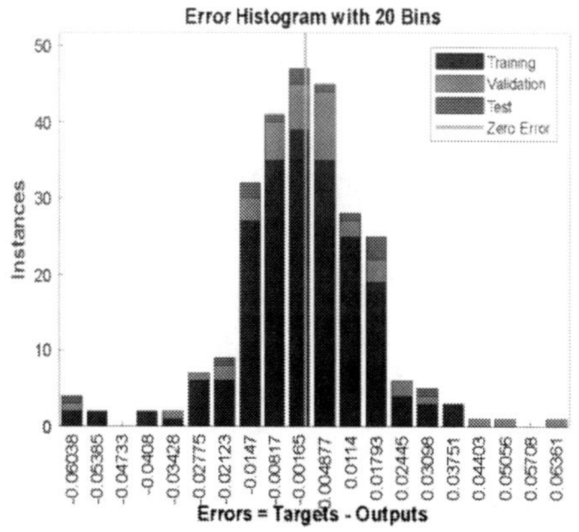

Figure 11. Error histogram with 20 bins

6. Conclusion

A circular side-orifice is a crucial aspect of discharge and indispensable to a hydraulic structure. The fundamental issue is to measure the discharge coefficient in order to better control the flow. Although multiple experiments have been conducted, there haven't been any studies on fully and partially submerged side orifice. This study employs ML techniques to rectify this research gap. The ANN model in this study was developed by employing Levenberg Marquart algorithm for back propagation step. The performance measures were evaluated by correlation coefficient (R), Mean Absolute Error (MAE) and Root Mean Square Error (RMSE). The developed model successfully provides satisfactory high level of accuracy in measuring the discharge coefficient as the R value is close to 1 in our models and MAE and RMSE value is close to 0. While empirical research can come up with results, it requires tremendous cost and labour to conduct. Reaching results empirically are definitely important; however, in order to achieve a functional formula, it requires numerous trial and error experiments.

Thus, this Machine Learning approach acts as faster and accurate alternative. This difficulty is efficiently resolved by Machine Learning methodology. These algorithms are faster and more accurate in predicting discharge coefficient. Along with experimental pursuits, this research can be utilised in predicting discharge coefficients in hydraulic structures that use fully and partially submerged circular side orifices.

7. REFERENCES

Asadi, M. J., Shabanlou, S., Najarchi, M., & Najafizadeh, M. m. (2020). A Hybrid Intelligent Model and Computational Fluid Dynamics to simulate Discharge Coefficient of Circular Side Orifices. *Iranian Journal of Science and Technology*.

Ayaz, M., & Mansoor, T. (2021). Development of ANN model for dischage prediction and optimal design of sharp crested triangular plan form weir for maximum discharge using linked ANN-optimization model. *Water supply journal*, 3027-3041.

Azimi, H., Shabanlou, S., Ebtehaj, I., Bonakdari, H., & Kardar, S. (2017). Combination of Computational Fluid Dynamics, adaptive nuero-fuzzy inference system and genetic algorithm for predicting discharge coefficient of rectangular side orifice. *Journal of Irrigation and Drainage Engineering*, 143.

Ebtehaj, I., & Bonakdari, H. (2014). Performance Evaluation of adaptive neuro fuzzy inference system for sediment transport in sewers. *Water resource management, 28*(13), 4765-4779.

Ebtehaj, I., Hossein, B., & Bahram, G. (2018). Development of more accurate discharge coefficient prediction equations for rectangular side weirs using adaptive neuro-fuzzy inference system and generalized group method of data handling. *Measurement, 116*, 473-482.

Emiroglu, Bilhan, M. E., & Kisi. (2011). Neural networks for estimation of discharge capacity of triangular labyrinth side- weir located on a straight channel. *Expert Systems with Applications, 38*, 867-874.

Emiroglu, Kaya, M., N., & Agaccioglu. (2010). Discharge capacity of labyrinth side weir located on a straight channel. *Journal of Irrigation Drainage Engineering, 136*, 37-46.

Fu, Z.-F., Cui, Z., Dai, W.-H., & Chen, Y.-J. (2018). Discharge Coefficient of Combined Orifice-Weir Flow. *Water journal, 10*, 1-17.

Gregg, W. B., Werth, D. E., & Frizzell, C. (2003). Determination of discharge coefficients for hydraulic sparger design. *J. Press. Vessel Technol., 126*(3), 354-359.

Hussain, A., Ahmad, Z., & Ojha, C. (2016). Flow through lateral circular orifice under free and submerged flow conditions. *Flow measurement and instrumentation, 52*, 57-66.

Hussain, A., Z.Ahmad, & G.L.Asawa. (2010). Discharge Characteristics of sharp-crested circlar side orifices in open channel. *Flow measurement and instrumentation, 21*, 418-424.

Khoshbin, F., Bonakdari, H., Ashraf Talesh, S., Ebtehaj, I., Zaji, A., & Azimi, H. (2016). Adaptive neuro fuzzy inference system multi-objective optimization using the genetic algorithm/singular value decomposition method for modelling the discharge coefficient in rectangular sharp-crested side weirs. . *Engg. optim., 48*(6), 933-948.

Nørgaard, Ravn, M. &., Poulsen, O. &., LK, N. &., & Hansen. (2000). *Neural Networks for Modelling and Control of Dynamic Systems: A Practitioner's Handbook.* Springer-Verlag.

Ramamurthy, A., Tim, U., & Rao, M. V. (1987). Weir Orifice Units for Uniform Flow Distribution. *Journal of Environmental Engineering, 113*, 155-166.

Ramamurthy, A., Tim, U., & Sarraf, S. (1986). Rectangular Lateral orifices in Open Channel. *Journal of Environmental Engineering, 112*, 292-300.

Riahi-Madvar, H., Ayyoubzadeh, S. A., & Atani, M. G. (2011). Developing an expert system for predicting alluvial channel geometry using ANN. *Expert Systems with Applications , 38*(1), 215-222.

Schalkoff, R. J. (1997). *Artificial neural networks.* New York: McGraw-Hill.

Sharifipour, M., Bonakdari, H., & Zaji, A. (2016). Comparison of genetic programming and radial basis function neural network for open channel junction velocity feild protection. *Neural comp. Appl., 30*(3), 855-864.

Werth, D. E., Khan, A. A., & Gregg, W. B. (2005). Experimental Study of wall curvature and bypass flow effects on orifice discharge coefficients. *Exp. Fluids, 39*, 45-491.

Yu-hong, Z., & Wen-xin, H. (2009). Application of artificial neural network to predict the friction factor of open channel flow. *Communications in Nonlinear Science and Numerical Simulation, 14*(5), 2373-2378.

Zahiri, A., Azamathullah, H. M., & Bagheri, S. (2013). Discharge coefficient for compound sharp crested side weirs in subcritical flow conditions. *Journal of hydraulics, 480*, 162-166.

DOI: https://doi.org/10.26077/9fa5-ed46

Preliminary Evaluation and Design of a New Energy Dissipation Stilling Basin via Numerical and Experimental Modeling

F. Zabaleta[1], F.A. Bombardelli[1], S.A. Valbuena[1]

[1] *Department of Civil and Environmental Engineering, University of California, Davis, USA*

E-mail: fede.zabaleta@gmail.com

Abstract: *Design of stilling basins for the dissipation of energy has been established by the seminal tests developed by the U.S. Bureau of Reclamation and the Saint Anthony Falls Laboratory, by mid XX century. Such designs basically include blocks and end sills, dentated or complete. However, those designs need variations in desertic zones, where the end sill is not allowed due to self-cleaning purposes, and no "drop" is possible for the same reason. In the case of the outlet structure for Perris Dam, located in the Riverside County, California, the design of the stilling basin was optimized using a Computational-Fluid-Dynamics (CFD) approach, and a physical model. Potential overspill due to unsteady oscillations of the free surface and energy dissipation were studied using a Volume of Fluid (VoF) method combined with a RANS turbulence closure, and then validated with a physical model. The combination of powerful numerical simulations with experimental models allows for an efficient analysis and design of complex hydraulic structures. The cheaper numerical simulations provide an accurate description of the flow that can be used to test several geometries and evaluate the most suited one to be represented on the physical model. From this first stage, an optimized geometry is implemented in a physical model which is used to calibrate and validate the numerical simulations as well as to optimize the geometries obtained from the numerical model. The new proposed structure promoted a high level of dissipation and a design that reduce the maintenance issues associated with desertic regions.*

Keywords: *Energy dissipator, hydraulic jump, numerical modeling, physical modeling.*

1. Introduction

Energy dissipation structures are needed downstream of spillways or outlet works to dissipate the excess of kinetic energy, preventing channel erosion. Energy dissipation structures can vary from very simple to highly complex designs. In general, the design must account for: The kinetic energy of the flow entering the dissipator, the duration and frequency of the flow, the tailwater conditions, economic factors, among others (FEMA 2010). Several dissipation structures in California located in desertic regions (in particular stilling basins) are currently experiencing maintenance issues due to the accumulation of sand. In these arid regions, soil particles are carried along by wind and are deposited in the lower parts of the structures. Because the structures are unused for several years, the accumulation of sediment becomes important, affecting the flow behavior for when they are needed. Traditional typologies for dissipation structures such as hydraulic jump stilling basins, impact basins or plunge basins cannot be implemented under these conditions due to these types of maintenance issues, or due to limitations of the maximum discharge. A new typology that could manage high flow rates without incurring in high cost of maintenance after several years without use is desirable for this region. The new design should not contain any depressed area or any section with a positive slope, reducing this way the accumulation of sediment in it.

This is the case of the energy dissipation structure in Perris Dam. Perris Dam is a 39.5 m (130 ft) tall earth-filled dam located in Riverside County in Southern California (Figure 1). Perris lake is located in the Bernasconi Hills, and it connects to the Perris Valley Drainage Channel, one mile to the west of the lake. The city of Perris, with a population of approximately 80,000 people, is located very close to the dam. The reservoir has a small drainage area of only 25.9 square kilometers, with thirty-six percent covered by the lake surface. It can be dewatered either over the spillway or through the outlet works. The spillway consists of a 4.8 m (16 ft) wide ogee crest with a concrete baffled chute and a concrete lined channel downstream. It has a maximum discharged of 9.06 m^3/s (320 cfs). The outlet works is located on the southern part of the reservoir. The tower has 10 valves to release water from five selected levels within the reservoir. The concrete tunnel takes the water to a steel lined delivery manifold to deliver water to the Municipal Water District. Downstream of the manifold there is a 1.8 m (6 ft) wide by 3.65 m (12 ft) tall operable slide gate followed by a non-operable bulkhead gate that would be opened during an emergency release event. The outlet works is designed for an emergency release capacity of 96.2 m^3/s (3,400 cfs), and a minimum discharge of 42.5 m^3/s (1,500 cfs).

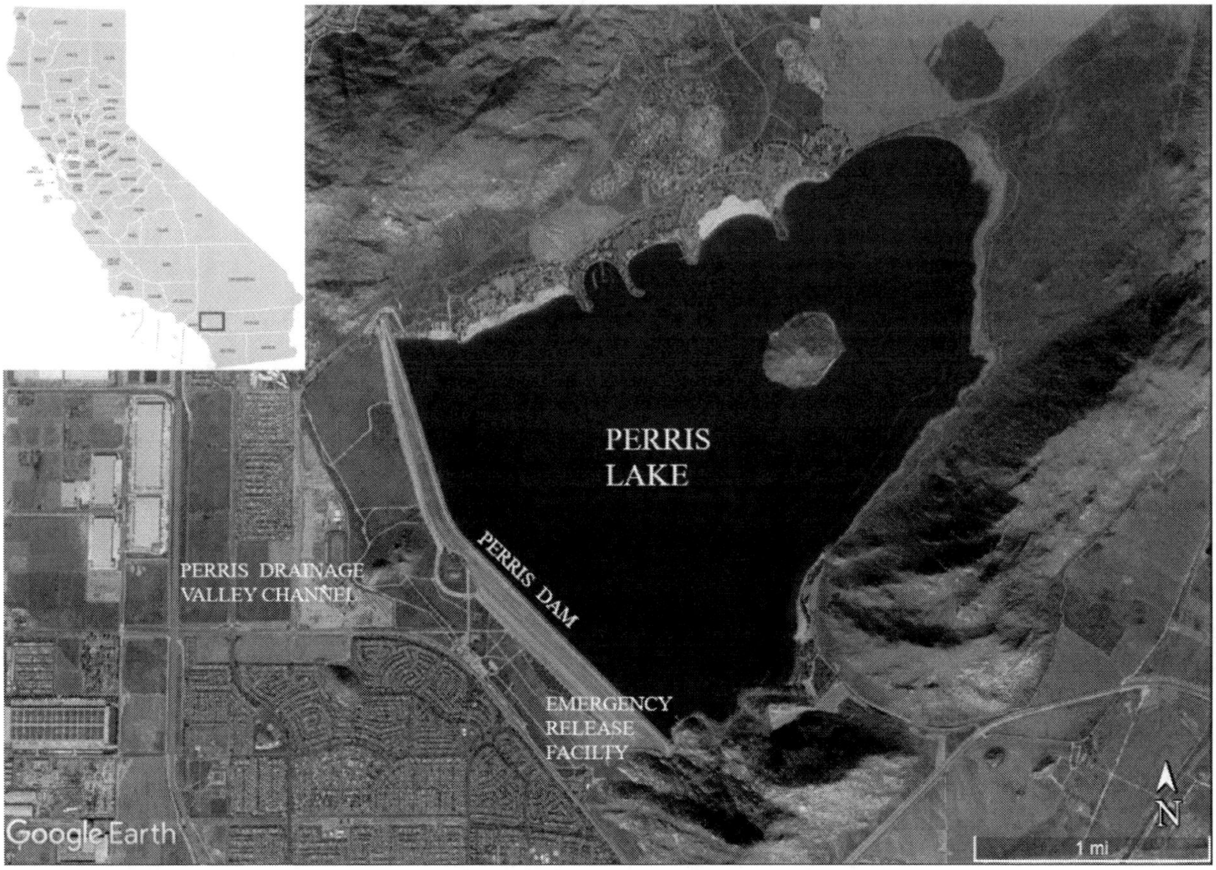

Figure 1. Aerial view of Perris Lake. Source of background image: Google Earth.

The main objective of this study is to develop a dissipation structure for the outlet works that provides a good flow behavior and an efficient dissipation of energy for the range of flow rates operated in Perris Dam. In addition, the design should not contain any depressed region or any section with a positive slope. The development of the new typology consisted of three steps:

1. Conceptual design: Proposal for a new typology and definition of the main features of the structure and of the flow.
2. Numerical modeling: Definition and optimization of the structure's dimension using numerical modeling.
3. Physical modeling: Validation of numerical simulations using a physical model, and analysis of flow features not captured in the simulations.

The paper is structured as follows. The governing equations and the numerical techniques used in the numerical model are described in Section 2 as well as the experimental setup and instrumentation. In Section 3, a description of the conceptual design and the main characteristics of the flow are discussed. A description of the flow using the numerical model is introduced in Section 4 as well as the optimized geometry. The results of the physical model and the comparisons with the numerical simulations are also presented in Section 4. Finally, conclusions and final remarks are found in Section 5.

2. Conceptual design

Based on the review of traditional energy dissipators, a novel design which combines a baffle drop and a hydraulic jump stilling basin was adopted. A schematic representation of such structure is presented in Figure 2. The main idea of this typology is to create a stable hydraulic jump by raising the water level through the use of baffles instead of reducing the elevation of the basin invert (as is done in traditional hydraulic jump stilling basins). The baffles raise the water level such that the hydraulic jump is formed upstream of the baffles. Most of the energy is dissipated in the jump, and then the remaining through changes in direction in between baffles. Baffles height decrease in the downstream direction to avoid the acceleration of the flow after the last row. The dissipation in the hydraulic jump should be smaller than the one that would occur in an equivalent hydraulic jump stilling basin due to the lack of homogeneity of the flow in the transverse direction, but this difference can be compensated by the action of the blocks (U.S. Army Corps of Engineers 1977). This structure allows for an efficient energy dissipation while following all the design restrictions mentioned above. In the side view profile of the structure (Figure 2, top), the location of the water surface in the structure can be seen; the rationale for the design is also apparent. Very importantly, because this is a non-traditional structure, there are no guidelines to define the dimensions of this structure. Some guidance can be obtained from manuals for stilling basins hosting hydraulic jumps with baffle drops, but many of these recommendations cannot be applied to this case. The design was developed by using the results from similar structures (U.S. Army Corps of Engineers 1980) as well as the experience of the authors.

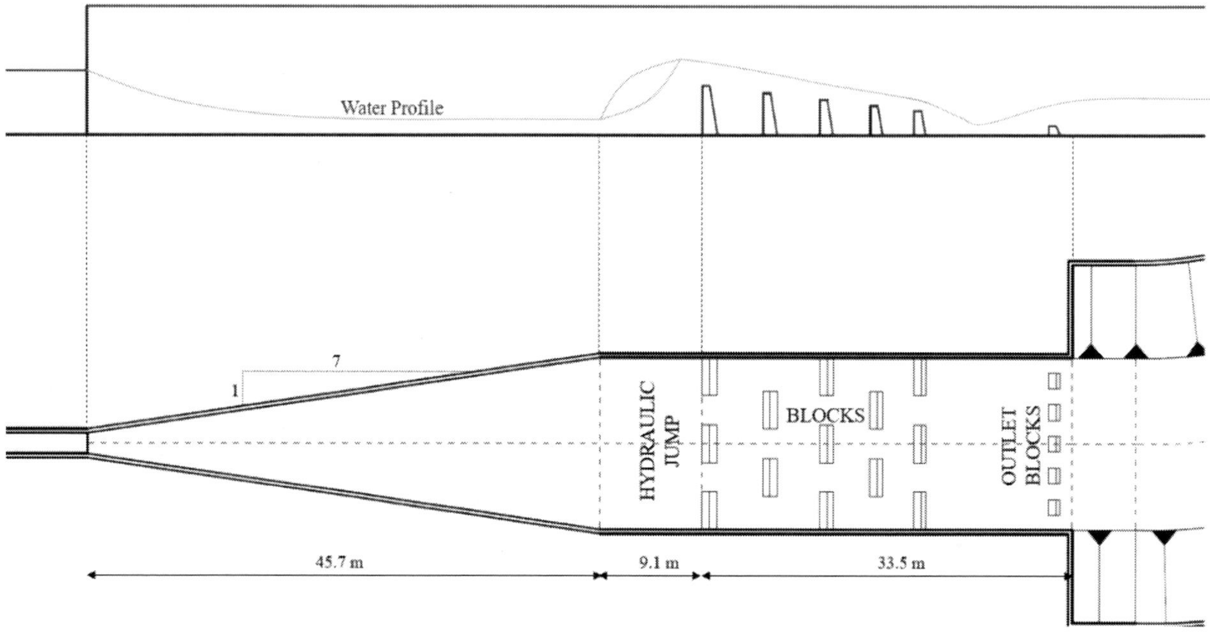

Figure 2. Proposed design for energy dissipation structure. Blocks' locations and dimensions are representative and not to scale.

3. Methodology

Advances in computational power, turbulent closures, and numerical methods, have allowed for the computation of three-dimensional flows under reasonable computational times. Solving the three-dimensional Reynolds-Averaged Navier-Stokes Equations (RANS), combined with an interface-capturing method, it is possible to obtain a detailed description of the flow behavior of many hydraulic structures such as those in spillways, stilling basins, open channels, etc. The implementation of a new typology requires significant time and geometry iterations that could only be undertaken via numerical simulations. A conceptual design is proposed based on traditional structures, and then its different features calculated and optimized using a three-dimensional numerical model.

RANS simulations only provide mean distributions of flow parameters, and the accuracy of the results may be limited by the turbulent closure. To validate the results of the numerical simulations, as well as to study the unsteady nature of the flow, the proposed design will also be studied using physical modeling with Froude similarity. Physical models allow for a detailed analysis of the characteristics of the flow. During the optimization procedure with the numerical model, the following features of the structure were analyzed and optimized: height, width, separation, and number of blocks; expansion rate of the rapid; and transition between structure and trapezoidal channel. The structure was optimized for the maximum flow rate (96.2 m^3/s), and verified for smaller flow rates (79.2, 59.4 and 42.4 m^3/s). Nearly 25 different geometries were analyzed to arrive to the final design. For reasons of brevity, only the analysis of the blocks height and the transition from the structure to the trapezoidal channel are presented on this article.

3.1. Numerical study

The numerical simulation was performed with the open-source CFD toolbox, OpenFOAM, using the solver interFoam. InterFoam solves the RANS equations (Kundu et al. 2015), as well as the Volume-of-Fluid (VOF) method (Hirt and Nichols 1981) to capture the location of the free surface. The RANS mass and momentum equations are, respectively:

$$\nabla \cdot \boldsymbol{u} = 0 \tag{1}$$

$$\frac{\partial \boldsymbol{u}}{\partial t} + \boldsymbol{u} \cdot \nabla \boldsymbol{u} = -\frac{1}{\rho}\nabla p + \frac{1}{\rho}\nabla \cdot (\boldsymbol{T} + \boldsymbol{T}^r) + \boldsymbol{g} \tag{2}$$

where ∇ represents the Nabla operator; $\boldsymbol{u}$ is the turbulence-averaged velocity; ρ denotes the density of the fluid; p indicates the turbulence-averaged pressure; $\boldsymbol{T}$ describes the tensor encompassing the turbulence-averaged viscous stresses; $\mathbf{T}^r$ denotes the Reynolds stresses; and $\boldsymbol{g}$ represents the gravity acceleration vector. The filtered viscous stresses can be expressed as a function of the turbulence-averaged velocity as

$$\boldsymbol{T} = \mu(\nabla \boldsymbol{u} + \nabla \boldsymbol{u}^T) \tag{3}$$

where μ represents the dynamic viscosity of the fluid. The residual stresses can be modeled using an eddy-viscosity model (μ_t), as follows:

$$\boldsymbol{T}^r = \mu_t(\nabla \boldsymbol{u} + \nabla \boldsymbol{u}^T) - \frac{2}{3}\rho\, k\boldsymbol{I} \tag{4}$$

where k is the turbulence kinetic energy and $\boldsymbol{I}$ is the identity tensor. The turbulence closure is achieved using the $k - \omega\ SST$ (Shear Stress Transport) model, whipch has been extensively used for free surface flows and better performance for separated flows than more traditional models like $k - \varepsilon$ (Pope 2000). To capture the location of the free surface, the modified VOF algorithm solves an equation for the water phase fraction (α_w). This parameter represents the volume fraction of water. The modified equation is:

$$\frac{\partial \alpha_w}{\partial t} + \nabla \cdot (\boldsymbol{u}\alpha_w) + \nabla \cdot \left(\boldsymbol{u}_c \alpha_w (1 - \alpha_w)\right) = 0, \tag{5}$$

where the last term is an artificial compression term to keep the interface sharp, and $\boldsymbol{u}_c$ is the compressive velocity that acts perpendicularly to the interface. Then, the density and kinematic viscosity of the momentum equation are calculated as:

$$\rho = \alpha_w \rho_w + (1 - \alpha_w)\rho_a \tag{6}$$

$$\mu = \alpha_w \mu_w + (1 - \alpha_w)\mu_a \tag{7}$$

where the subscript w indicates water and the subscript a indicates air. Given the presence of the time derivative and the flow features, these simulations can be interpreted as URANS, i.e., Unsteady RANS.

The meshes used in this project were generated using the *snappyHexMesh* tool, provided in the OpenFOAM framework. *SnappyHexMesh* generates polyhedral meshes based on a base mesh composed of hexahedral cells. The meshes were optimized to represent the regions of interest with more detail while maintaining the computational cost affordable. The number of cells varied from 2 million volumes in the coarsest simulations up to more than 23 million volumes in the finest ones. The inlet boundary condition (BC) was set with a specific flow rate, while the downstream BC was set with the uniform water level of the trapezoidal channel.

The initial simulation domain includes 18.3 m (60 ft) of inlet pipe, the totality of the structure, and 61 m (200 ft) of the downstream rectangular channel after the curve. The final design was tested in a finer mesh and in an extended domain to reduce the effects of the boundary conditions. This domain includes 35 m (115 ft) of pipe and 125 m (410 ft) of trapezoidal channel after the curve. The mesh used for the final design consisted of 23 million volumes, with a horizontal dimension of approximately 10 cm and vertical dimension of 3 cm. A portion of the mesh corresponding to the blocks is presented in Figure 3. Flow rates were specified as inlet boundary condition while water level was specified for the outlet. No slip boundary conditions were employed on the walls.

Figure 3. Portion of the mesh utilized in the final design.

3.2. Physical modeling

The physical modeling of the numerically optimized structure was performed in the Jaime Amorocho Hydraulics Laboratory (JHAL). The model was constructed using Froude similarity in a scale 1:12 to minimize the scale effects (Tropea et al. 2007). The inlet pipe, the expansion channel, and the walls on the region where blocks are located were constructed from acrylic. The blocks and restitution channel were constructed from wood and waterproofed with an epoxy paint. The discharge was controlled using a variable speed pump and controlled using a magnetic flow meter installed in the inlet pipe. The water was pumped from a water sump and transported into a head tank where the model pipe starts. The pipe was 2.4 m (8 ft) long to let the flow fully develop. The water level at the end of the model was controlled by a variable height weir. The water coming out of the model is collected by a tank and transported back into the sump creating a closed circuit.

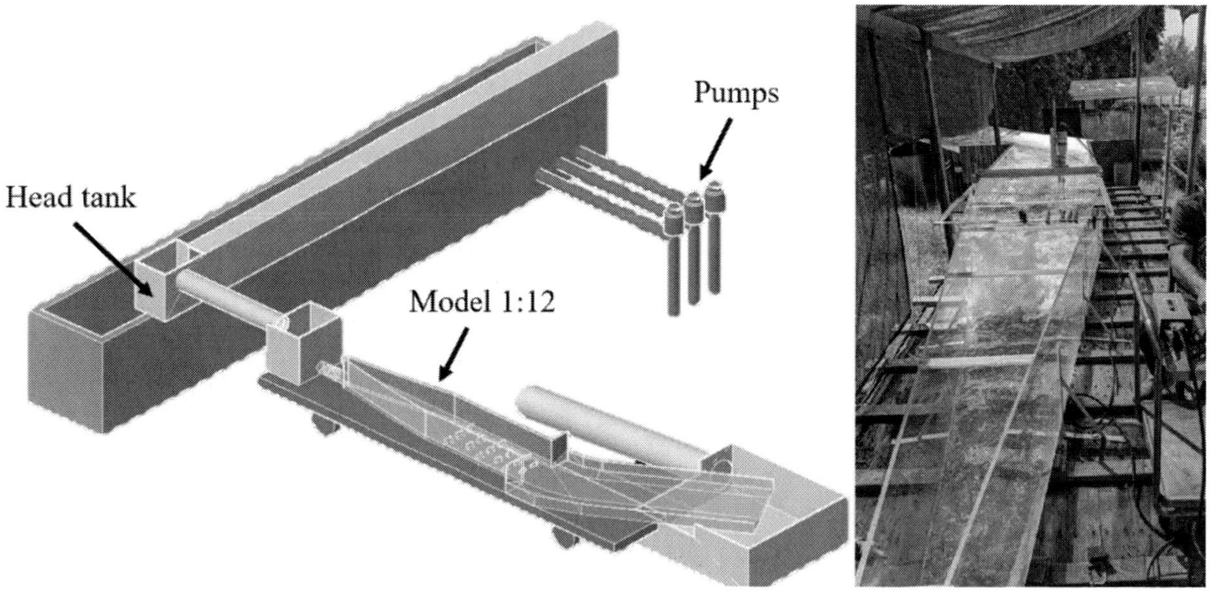

Figure 4. Scheme of the physical model (left). Picture of physical model (right).

Water level measurements near the walls were performed by taking videos of the acrylic walls next to a measuring ruler. An in-house post-processing software developed specifically for this study detects the free surface location and evaluates its position over time. This method allows to track water level fluctuations with a high frequency (equal to the video frame rate), not only in a point location but in the whole area captured by the frame. Water level measurements in the interior points were performed in a similar manner but using a ruler with a black background inserted into the flow, allowing the software to detect the free surface location with respect to the ruler. Velocity measurements were taken with an ADV Sontek at a 60 Hz sampling rate.

4. Results

4.1. Numerical modeling

The first parameter to optimize was the height of the blocks, and in particular the initial row of blocks. The role of the blocks is to increase the water level and to force the formation of a hydraulic jump upstream. The height of the initial row of blocks (*H*) as well as their separation severely affect the location and characteristics of the hydraulic jump. If the blocks are not high enough, the hydraulic jump may be too close to the blocks or even the supercritical flow may impact the blocks, leading to vibration, erosion, and other issues. If the blocks are too high, the hydraulic jump will form close to the outlet, reducing the amount of energy dissipation. The lateral separation of the blocks also affects the hydraulic jump location. If the blocks in the first rows are closer to each other, more flow will have to go over them, raising the water level. To show the effect of this design variables, we present the results of three simulations with different initial block heights: 3 m (10 ft), 2.4 m (8 ft), and 1.8 m (6 ft). The three structures have a block separation of 0.35*H*, 0.5*H* and 1.4*H*, respectively.

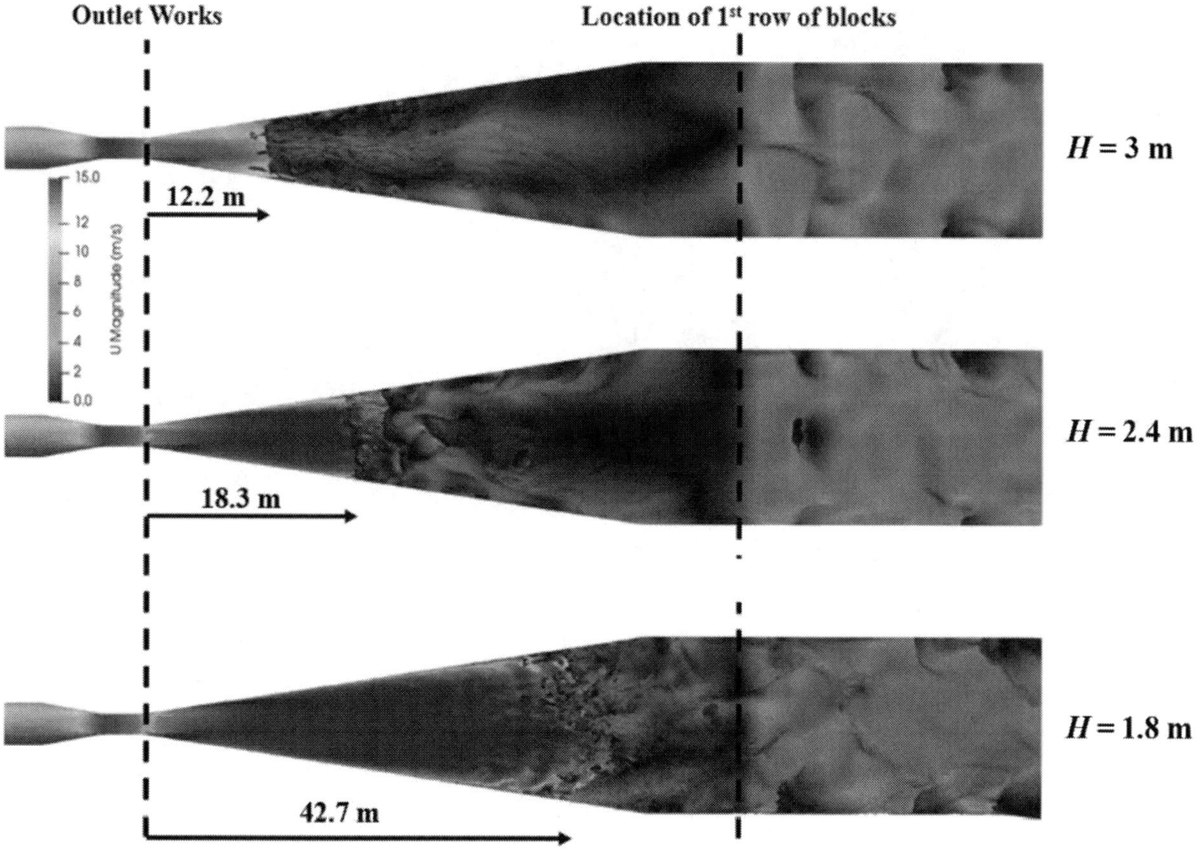

Figure 5. Velocity distribution at the free surface for three different height of blocks and a flow rate of 96.2 m^3/s.

In Figure 5, the velocity distribution at the free surface for the three cases are presented (for a flow rate of 96.2 m^3/s). It can be clearly seen that the structures with smaller blocks have the hydraulic jump closer to them. From a design standpoint of view, the toe of the hydraulic jump should be as close as possible to the region where the channel stops widening. For smaller flow rates, the hydraulic jump positions further upstream. Because it is beneficial to avoid a submerged hydraulic jump next to the outlet, the design forces the jump for the maximum flow rate to take place as close as possible to the blocks, without compromising their structural integrity. It was found that the best results were obtained with a first-row height of $H = 1.8$ m (6 ft) and a separation of $1.4H$.

After the last line of blocks, there is a transition between the flow in the structure and the quasi-uniform flow. The flow accelerates, the water level diminishes, and the flow may transition to supercritical, creating a region of high velocity immediately after the structure. The height of the last line of blocks (H_L) has a big influence in how this transition develops. Higher blocks create a higher water level, leading to higher acceleration and velocities. High velocities at the end of the structure will require more floor protection to avoid any type of erosion immediately after it. To evaluate the effect of this design parameter, we analyzed the results of three different configurations of the structure. In the first one, the height of the last line of blocks is 1.5 m (5 ft), while in the second and third structure is 1 m (3.3 ft). The last design includes a final set of blocks, separated from the main blocks, hereafter called "outlet blocks" (see Figure 6, right). The velocity fields for the three structures are presented in Figure 6. In the first two geometries, the supercritical flow extends up to the end of the structure, and then transitions to a subcritical flow with a low-Froude number hydraulic jump. The 1 m (3.3 ft) blocks generate smaller velocities than the 1.5 m (5 ft) blocks but, in both cases, they are still over 5 m/s, which may would require special protection. In the third design with the outlet blocks, the acceleration region is restricted within a smaller region inside the structure. The velocities on the free surface as well as close to the bottom are smaller than for the other two designs. After the outlet blocks, there is still some acceleration, but this time it is much smaller than in the other two cases, especially near the bottom. This design promotes the containment of the high velocity regions inside the structure, providing a much safer behavior for the structure. For all these reasons, the last design is adopted.

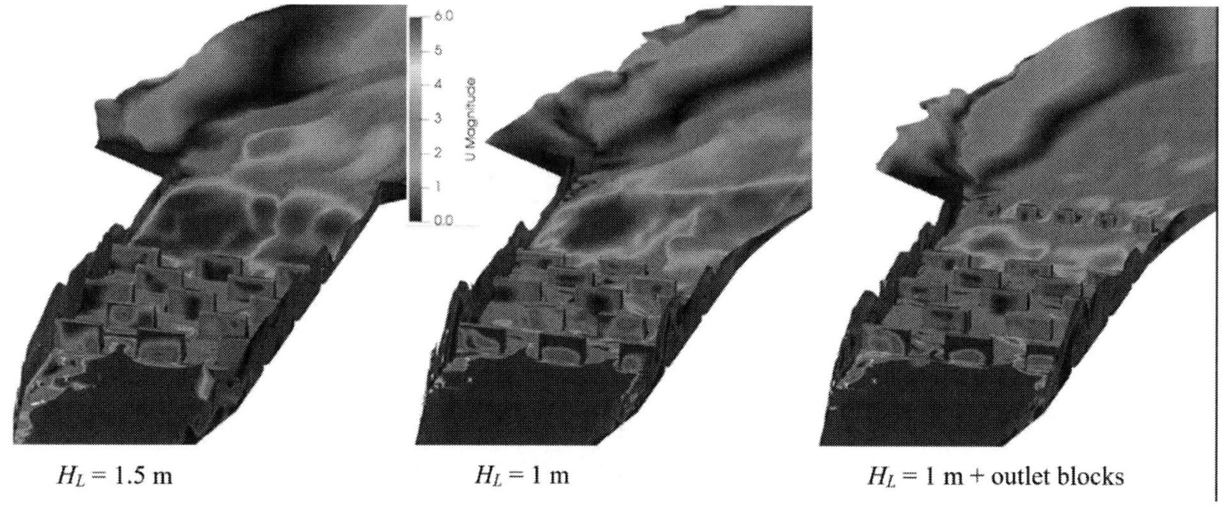

Figure 6. Velocity distribution near the walls for three designs with different block heights in the last row.

Other parameters optimized in the numerical model were the number of rows, the separation between rows and the transition shape between the dissipation structure and the trapezoidal channel. The analysis indicated that 5 rows of blocks offer an acceptable level of dissipation while reducing the length of the structure. A smooth transition between the structure and the channel provided more uniformly distributed flow along the channel, reducing regions of higher velocities after the structure. Using the separation between blocks recommended in the guidelines for baffled drops showed the best performance. Overall, this design featured a high level of dissipation (77% for the design flow rate) which results in adequate velocities in the channel downstream the structure. Test under lower flow rates showed an acceptable level of dissipation and good overall behavior of the flow. The results of the numerical model may be influenced by (small) numerical errors and the turbulence closure. Furthermore, RANS simulations do not capture fluctuations caused by turbulence, and therefore oscillations in the velocity and pressure field, as well as in the free surface are not fully captured. Additionally, the model, as currently implemented, does not include air entrainment. For all these reasons, physical modeling is required to fully verify and optimize the design.

4.2. Physical modeling

The physical model was run for four different flow rates varying form the minimum discharge (42.4 m^3/s) up the maximum flow rate at which the release facility can be operated (96.2 m^3/s). Two intermediate flow rates (59.5 m^3/s and 78.3 m^3/s) were also analyzed. Data was collected in 15 points evenly distributed in the domain, presented in Figure 7 (left). Comparison between physical and numerical model showed very satisfactory agreement. The comparison of water depth for 59.5 m^3/s (2,100 cfs) is presented in Figure 7 (right). The error bar in the numerical model data represents the numerical model uncertainty given by the VOF methodology that tends to diffuse the free surface between two or three cells (in this case represented by the height of two cells). The difference between measured and modelled depth lies within the uncertainty of the numerical model for most of the points in the domain. The representation of the free surface location is extremely accurate in the points located near the blocks as well as on the walls. The points located at the center of the hydraulic jump (2) and after the final blocks (15) have more scatter, suggesting that a finer mesh may be required in these regions.

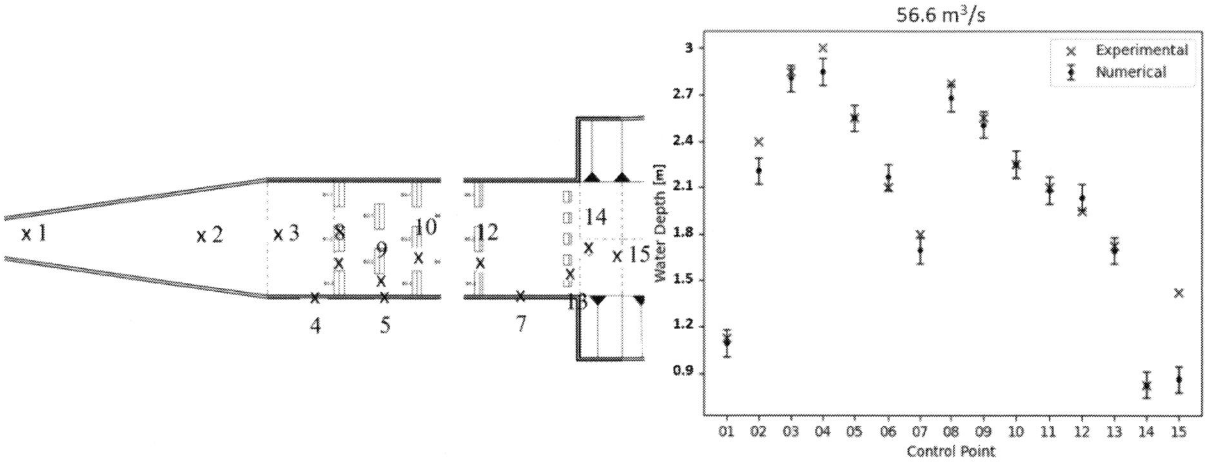

Figure 7. Measuring points location (left) and comparison of modeled and measured water level (right).

The physical model was not only used to validate the numerical model, but also to study the turbulent nature of the flow. A great example of this is the evaluation of the freeboard. The freeboard for dissipation structures is usually defined using empirical formulas based on mean flow parameters. Guidelines for similar structures like hydraulic jump stilling basins and baffle drops, suggest freeboards of 2.1 m (7 ft) and 1.5 m (5 ft) respectively. Extrapolating these values to a new design presents a significant number of questions and concerns that may be better addressed using the physical model.

Water levels reach their maximum values in measuring locations 4 and 5. Maximum water depth during the measuring interval were 2 m (6.5 feet) higher than the mean value. A probability distribution analysis suggest that the water level follows a log-normal probability distribution function (presented in Figure 8). This analysis shows that only 5% of the time the water level exceeds the mean value by 0.76 m (2.5 ft) and only 1% of the time exceeds the mean value by 1.22 m (4 ft). Rare events of splashing caused by jets of water coming from the hydraulic jump and impacting the blocks, cause a momentary but significant increase in water level that has to be taken into account when determining the wall heights.

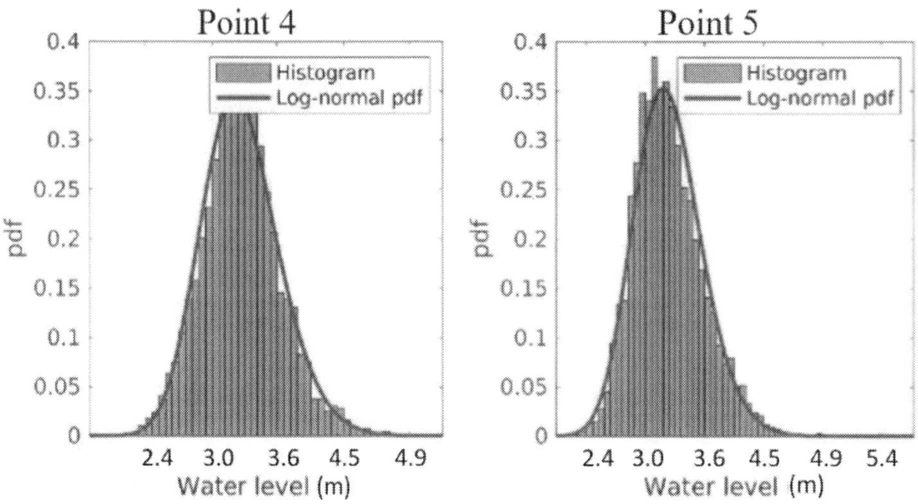

. Water level histograms superimposed with adjusted log-normal distribution. Inches in model corresponds to feet in prototype.

5. Conclusions

In this study, a preliminary design for the energy dissipation structure of Perris Dam has been proposed and tested using numerical and physical modeling. Given the restrictions imposed by the design, none of the traditional structures could be used; therefore, a novel typology had to be developed. The new design arises from a combination between a hydraulic jump stilling basin and a baffle drop. The main conclusions from this study are:

1. The novel design achieved an adequate level of dissipation while allowing self-cleaning capabilities.
2. The design reaches a high level of dissipation (77% for the design flow rate) which results in adequate velocities in the channel downstream of the structure.
3. When done properly, numerical modeling has proven to be a great tool that allows testing of several geometries and optimization of the dimensions of the structure before performing the more time and cost demanding physical modeling. Predictions obtained even without air entrainment can be trusted for design purposes.
4. Water levels obtained with numerical and physical model showed very satisfactory agreement.
5. Physical modeling provided great insight on the turbulent features of the flow such as the fluctuations of the free surface.

Although the design was undertaken for the conditions present in Perris Dam, we believe that it can be generalized for different flow rates and restitution values.

6. ACKNOWLEDGMENTS

We enormously thank the support by the Department of Water Resources of California and by Mike Driller as the Program Manager.

7. REFERENCES

Federal Emergency Management Agency (FEMA). (2010). Outlet Works Energy Dissipators. Best Practices for Design, Construction, Problem Identification and Evaluation, Inspection, Maintenance, Renovation, and Repair.

Hirt, C. W., Nichols, B. D. (1981). "Volume of fluid (VOF) method for the dynamics of free boundaries". *J. Comput. Phys*. 39(1). 201-225.

Kundu, P., Cohen, I., Dowling, D. (2015). Fluid Mechanics (Sixth Edition), Academic Press, Boston.

Peterka, J. (1984). Hydraulic Design of Stilling Basins and Energy Dissipators, United States Department of The Interior, Bureau of Reclamation, Denver, Colorado, USA.

Pope, S. (2000). Turbulent Flows, Cambridge University Press, Cambridge.

Tropea, C., Yarin, A. L., Foss, J. F. (2007). Handbook of Experimental Fluid Mechanics, Springer.

U.S. Army Corps of Engineers. (1977). Hydraulic Design Criteria, Washington, D. C.

U.S. Army Corps of Engineers. (1980). Hydraulic Design of Reservoirs Outlet Works, Washington, D. C.

DOI: https://doi.org/10.26077/dffd-1cfb

Recent challenges in design of spillway – An Indian scenario

R.R. Bhate[1]
[1]*Scientist 'C', Central Water and Power Research Station, Pune, India*
E-mail: ravindra.bhate@gmail.com

***Abstract**: There are about 5264 completed large dams in India. These dams are invariably equipped with some or other kind of spillway. As this dam building activity has been going on for many years, all easy sites have already been tapped and are left with sites where designers have to face challenges from nature. Recently, while harnessing the hydropower potential of Himalayan rivers, designers are facing issues like high sedimentation concentration flows, narrow valleys with a high magnitude of the flood, leading to the provision of an additional spillway or multi-tier spillways. Melting of glaciers resulting in increased rates of Glacial Lake Outburst Flow (GLOF) is challenging tasks to design a spillway for debris flow. On the other hand, for existing dams in India, there is a large upward revision in the design flood due to improvements in flood assessment techniques and the availability of huge data for flood assessment. This demands an increase in discharging capacity of the spillways by increasing crest length or providing breaching sections. India has got a large seashore and to utilize the freshwater from the rivers joining the sea, multipurpose dams across the sea in the vicinity of the Gulf can be considered. Special design considerations are to be employed for spillway and energy dissipation arrangements as their operational performance is influenced by cyclic tidal variations and storm surges. This paper makes an attempt to discuss the relevant hydraulic aspects for the design of spillways and energy dissipators for efficient and safe operation.*

***Keywords:** Orifice Spillway, Sediments, Flood, GLOF, Coastal reservoir, Hydraulic model*

1. Introduction

Since its independence, India has invested a huge amount in creating big infrastructures to store the surface runoff in the reservoirs for irrigation, drinking water supply and hydropower in the form of large, medium and small dams. The dam building activities over the years have been concentrated on the ideal and easy sites. This leaves the geologically and topographically complex and difficult sites for future activities. These sites have been identified mainly in the Himalayan and North Eastern regions. High mountains, deep narrow valleys, complex geological strata with occasional problems of stability of hill slopes, and generally high levels of seismicity characterize the Himalayan terrain. Majority of these rivers have a very steep bed gradient in the range of 1 in 30 to 1 in 100. These give rise to flash floods and sediment load. This sediment would ultimately settle in the reservoir resulting in the reduction of live storage capacity. To utilize the high heads and discharge available in this region for power generation this sediment should be flushed regularly. The overflow spillways using conventional methods i.e. high head storage dams for river valley projects in Himalayan regions are not suitable. The spillways considered in such topography required to perform dual roles i.e. passing the flood as well as flushing the sediments. Many spillways are also designed considering the flash floods caused by Glacial Lake Outburst Flood (GLOF). Due to the potential impacts of climate change on glacier dynamics, catastrophic GLOF may increase future flood risks for infrastructure and population. This is particularly relevant given the current development of hydropower projects in the Himalaya. These flash floods also contribute to the major part of the sediment in glacierized catchments. In such a complex hydrological scenario, assessing the hydraulic performance of spillways with a sediment flushing facility would be of vital importance.

In terms of large dams, India is now ranked third in the world after the United States of America and China (ICOLD, 2011). There are currently about 5260 completed large dams in India and 437 under construction (CWC, 2018). Many of these dams are decades old and for these dams, spillways were designed for original design floods based on empirical formulae with applied discretions by experienced designers. Checking and upgrading the dam design flood estimates incorporating the available extended data sets and latest flood estimation methods has been given technical priority in the national dam safety programme (Pandya et.al., 2014). Based on outcomes of the substantial sets of design reviews under Dam Rehabilitation and Improvement Programme (DRIP), it is found that in more than 50% of cases the original flood needs an upward revision of 50% to 100%. As per data collected by Indian Commission on Large Dams (ICOLD) on the failure of large dams, nearly 35% of dams have failed globally due to causes related to inadequate spillway capacity and hence this becomes a major concern in a Dam safety programme (CWC, 2018). This necessitates the need of redesigning the spillways and energy dissipators to cater to this revised flood as a structural measure.

India has got about 7515 km of coastline and the majority of surface runoff is directly delivered into the sea. This highlights the fact that there is enough water available but the deficiency is in storage. Hence, the solution lies in utilizing (or) conserving the abundant monsoon water which runs off into the sea (Yang, 2015). The coastal reservoir is a paradigm shift in water resources development from storing water in inland dams to storing freshwater by the coast. This converts flood water into valuable water resources. A coastal reservoir is a reservoir in the sea to store a portion of the river flood waters that joins the ocean during the monsoon period and use it during the drought periods. It is a unique hydraulic structure constructed at an estuary, gulf, or bay or in the sea (at the point where a river meets a sea) to store the portion of excess water at flood times. This seawall or dike structure may run for several kilometres joining the curved coastline in the form of chord length. The requirement for the design of spillways and energy dissipators for coastal reservoirs are unique. Tidal levels and storm surge downstream (seaside) plays important role in the design and operation of the spillway.

This paper makes an attempt to discuss the relevant hydraulic aspects under such challenging scenarios for design of spillway and energy dissipators for efficient and safe operation.

2. Challenges in design of spillways in Himalayan region

The major causes that are responsible for sediment-laden floods and flash-floods in Himalayan Rivers are:

1. Cloudburst in the catchment of the river
2. Heavy rainfall in the upper reaches of the river
3. Sudden breach or bursts of glacial lakes
4. Landslides and GLOF
5. Earthquake prone zone leading to slope failure and landslides

These rivers carry heavy sediment load during monsoon and snowmelt in summer, which has to be efficiently managed to ensure the long term sustainability of hydropower projects. Conventional designs viz. high ogee spillways in storage dams are not suitable to deal with floods carrying heavy sediment. Considering the experience of silting of dams and damages to power plants, the trend has been changed to designing run-of-the-river schemes which utilize the stream flow as it comes without any large storage. Innovative designs of spillways have evolved based on the concept of flushing. Recent trends in designing the spillways are by modifying the low level sluices with due consideration for flushing (Deolalikar et.al., 2008). Orifice spillways (Breastwall / sluice spillways) are thus evolved over the last few years to cater to both flood disposal and flushing of sediments. The important parameters required to be determined while designing orifice spillways are; Bottom profile of spillway crest conforming to the flow through orifice, the roof profile of the orifice, size and dimensions of the orifice, protection of spillway to resist abrasion and choice of energy dissipator. The performance of the orifice spillway is affected by the head over the opening, size of the sluice barrel, and entrance shape of the sluice bellmouth. The hydraulics of the orifice spillway change with the varying reservoir levels. The flow is free flow for reservoir water levels below the top of the roof of the sluice. For higher water levels the flow is orifice flow. The crest profile is required to be designed for orifice (pressurized) flow. Figure 1 shows the definition sketch of the orifice spillway. It has been experienced from the physical model studies that the flow through sluice barrel do not get fully developed and that flow separation takes place on the sluice roof profile resulting in negative pressure for high head spillways. Such high head sluice spillways should be checked against the presence of sub-atmospheric pressures on the bell-mouth roof to avoid cavitation damage. The roof profile has not been standardized with respect to the upstream head because of many hydraulic and structural parameters such as thickness of the breast wall and piers, the bottom profile of spillway, width of the opening and sediment flushing requirements etc. There is a need for evolving hydraulically efficient crest and roof profiles for high head orifice spillways. To address these short comings, systematic studies were carried out at Central Water and Power Research Station using physical and numerical modelling for evolving the design of orifice spillway (Bhosekar and Gadge, 2018).

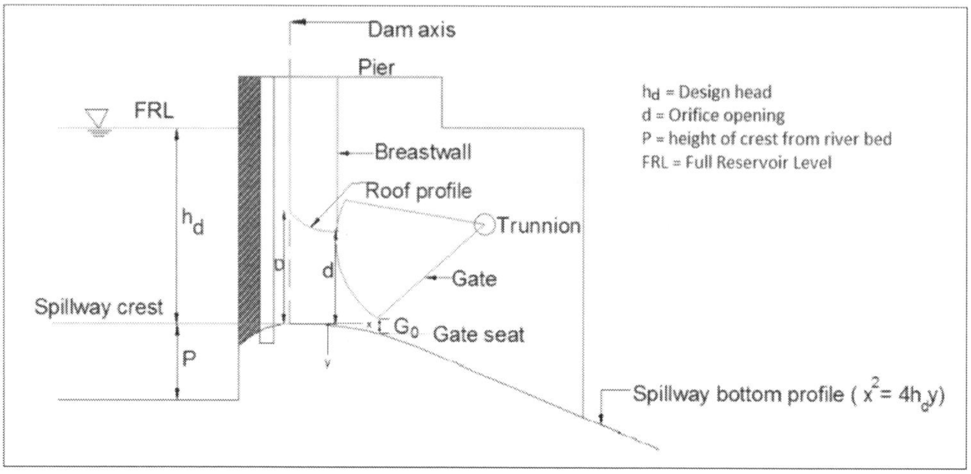

Figure 1. Definition Sketch of Orifice Spillway

Orifice spillways necessitate some special design considerations. The larger velocities associated with the high heads may increase the potential for cavitation and erosion damage to the structure. Adequate protection measures should be taken during the construction of the sluice barrel and breast wall spillway, to withstand the erosive power of the silt-laden water while flushing the reservoir and flood routing, by way of a special type of concreting (epoxy-bonded epoxy mortar, polymer concrete, or epoxy-bonded replacement concrete) or providing steel lining along the discharge channel of the spillway.

Orifice spillways have effectively been used on many run of the river hydro power projects in Himalaya since last two decades. Pandoh, Dulhasti, Nathpa Jhakari, Chamera-III, Sewa, Parbati-II, Uri-II, Teesta cascade projects etc are some of the projects where the orifice spillways have effectively been used for passing flood and flushing sediments. The experience from these projects has lead to improvement in refinement of design of orifice spillway and energy dissipator.

Based on site-specific constraints of geological and topographical considerations, innovative spillways are evolved. For example, rockfill or concrete faced rockfill dams are designed due to weak geology for the diversion dam. In such cases, the spillway is provided along the banks of the reservoir. Long chutes instead of the short spillway are provided for discharging the flow in the river downstream as in the case of the Kishanganga and Dhauliganga H. E. projects. Tunnel spillways are often provided when the river gorge is narrow and the discharge to be passed is not large enough for designing a two tier spillway. A tunnel spillway can also be provided for diverting silt-laden flows, e.g. Parbati-III project. Many a time diversion tunnels used during the construction stage of the dam are converted into spillway tunnels, e.g. Dhauliganga project, the Tehri project.

2.1. Two Tier Spillway

In the upper reaches of mighty rivers like the Brahmaputra, the river gorges are narrow and the discharges to be passed through spillway are high. Therefore, designing an orifice spillway with large opening becomes difficult due to construction and operational aspects. Multi-tier spillways have evolved to overcome this difficulty. Two/ three orifice spillways and an overflow spillway on the top can form a multi-tier spillway. The bottom orifice spillway is designed to take care of normal floods as well flushing of sediment and the upper spillways will cater to excess flood to be passed through reservoir. The energy dissipator for these spillways may be common or independent.

One such spillway has been designed for PMF of 60,315 m^3/s and studied at Central Water and Power Research Station, Pune for the Lower Siang Dam spillway in Arunachal Pradesh. A common energy dissipator in the form of stilling basin has been provided for both the orifice and overflow spillway. The design has evolved into a sectional model to solve the complex flow problems at the junction of the spillways. Figure 2 shows a view of the sectional model in operation.

Figure 2. View of the sectional model of two tier spillway

Sometimes, based on techno-economic feasibility considerations, power intake and powerhouse structure and tailrace channel have to be provided in the body of the dam, if the fragile geology of the Himalaya does not permit an independent layout for power intake, water conductor system, and huge underground powerhouse cavern. This type of two-tier spillway and dam foot powerhouse has been considered for the Siang Lower run-of-the-river H. E. Project in Arunachal Pradesh. The project envisages the construction of an 86 m high concrete gravity dam across river Siang to generate power of 2700 MW utilizing a net head of 55 m. The spillway arrangement and construction schedule envisaged are unique. A two-tier spillway comprising of overflow ogee spillway on the top tier and a sluice spillway below is proposed. Sluice spillway performs a dual function of disposing of the flood as well as flushing of sediment out of a reservoir of diversion dam of a run- of-the river scheme. The design of these structures cannot be standardized and depend on site-specific conditions. Extensive physical model studies were conducted on the 1:55 scale 2-D sectional model and 1:100 scale 3-D comprehensive model for finalizing/ optimizing the design. These studies played an important role in enhancing the overall performance of the spillway and energy dissipator by incorporating various modifications to the orifice roof & bottom profile, overflow spillway profile, length, and elevation of the stilling basin, finalizing the height of the cofferdam required during construction phases, etc.

2.2. Special design considerations for energy dissipator

The factors that govern the choice of the type of energy dissipator are hydraulic considerations, topography, geology, type of dam, layout and other associated structures, economic comparisons, frequency of usage etc. In Himalayan regions the topography is normally steep, fragile geology, high intensity of rain, high level of seismicity, etc. cause a high load of sediments. Since the spillway has to surplus both the flood and sediment, special considerations are required for the design of suitable energy dissipators. A ski-jump bucket is found to be the most suitable form of energy dissipator because of its obvious advantage during the flushing operation. The sediment passes down the spillway with the supercritical flow without deposition and churning in the bucket. Fortunately, steep bed slopes of the rivers result in low tail-water depth permitting two choices of energy dissipator. The water and scour profiles for the entire range of discharges and reservoir water levels observed on the model provide a comprehensive picture of ski jump jet and scour profiles. These studies play important role in deciding the location and size of a plunge pool. In many projects like Nathpa Jhakri, Tala, Chamera- I, Dhauliganga, and Ranganadi, a ski-jump bucket has been provided as an energy dissipator. If, however, geological conditions are not favorable, a hydraulic jump stilling basin may have to be adopted. The high unit discharge passing down low head results in a low Froude number conditions.

The stilling basins for the Froude number in the range of 2.5 – 4.5 are rather difficult to design to ensure satisfactory performance for the entire range of discharge. Because of the requirement of passing silt-laden flows, the use of energy dissipating appurtenances like chute and baffle blocks is not advisable. The resulting basin is excessively long and often deep-seated below the general river bed, making it vulnerable to deposition by silt during the flushing operation. Experience with stilling basin of the Chamera- II project shows that a trade-off is desirable between the hydraulic efficiency of energy dissipation and the self-cleansing potential of the stilling basin during the flushing operation. Cylindrical end-sills are generally preferred for easy movement of sediment out of the basin. The provision of a roller bucket is generally avoided as an energy dissipator due to the likelihood of abrasion damage to the bucket due to the churning of sediment.

3. Impact of Himalayan Glaciers and GLOF on spillway design flood

The Himalayas have the largest concentration of glaciers outside the polar region. These glaciers are a freshwater reserve for nine major river systems in Asia – a lifeline for almost one-third of humanity. The climatic change/variability in recent decades has made considerable impacts on the glacier life cycle in the Himalayan region. As a result, many big glaciers melted, forming a large number of glacial lakes e.g. Parechu Lake (Tibet), Dig Tsho Lake (Nepal) and Charobari Lake (Uttarakhand). Due to an increase in the rate at which ice and snow melted, the accumulation of water in these lakes started increasing. Sudden discharge of large volumes of water with debris from these lakes potentially causes glacial lake outburst floods (GLOFs) in valleys downstream. These in turn give rise to an increase in the potential threat of glacial lake outburst floods occurring. A number of hydroelectric projects in India, Bhutan and Nepal are being planned in the Himalayan regions. It has become necessary for the project planners and designers to account for the GLOF also along with the Probable Maximum Flood (PMF) due to rainfall for deciding the spillway capacity of projects. Table 1 shows the magnitude of GLOF considered for various hydropower projects in Himalaya in addition to Probable Maximum Flood (PMF) for which hydraulic model studies for spillways were carried out or being referred for studies at the CWPRS, Pune.

Table 1. Magnitude of GLOF and PMF for various projects in Himalayan region

Sl. No.	Name of the Project	PMF in m^3/s	GLOF in m^3/s	PMF + GLOF m^3/s	Percentage of GLOF wrt PMF
1	Punatsangchhu-I H.E. Project, Bhutan	11,500	4,300	15,800	37.4
2	Punatsangchhu-II H.E. Project, Bhutan	11,723	4,300	16,023	36.7
3	Etalin H.E. Project, Dri Limb, Ar. Pradesh	11,811	1,170	12,981	9.9
4	Etalin H.E. Project, Tangon Limb, Ar.Pradesh	10,218	2,143	12,361	20.97
5	Arun-III H.E. Project, Nepal	8,880	6,830	15,710	76.91
6	Chamakharchhu H.E. Project, Bhutan	9,406	5,112	14,518	54.35
7	Kwar H.E. Project, J & K	10534	620	11,154	5.88
8	Mangdechhu H.E. Project, Bhutan	6,900	3,715	10,615	53.84

Approximately 15,000 glaciers (covering an area of 33,340 sq.km.), and 9000 glacial lakes throughout Bhutan, Nepal and Pakistan, as well as selected river basins in China and India were documented in a baseline study conducted earlier by International Center for Integrated Mountain Development (ICIMOD), United Nations Development Programme (UNEP), and the Asia Pacific Network for Global Change Research (APN) (Bajaracharya et.al, 2006). There are about 2028 glacial lakes and water bodies which contributes the rivers flowing into India. GLOF events have adversely affected Nepal and Bhutan in the recent past and to date over 200 potentially dangerous glacial lakes have been documented across the Himalayan region (Bajracharya et.al., 2006). A GLOF event of 1985 originating from Dig Tsho, a glacial lake in the Khumbu Himal, Nepal, destroyed the Namche Small Hydel Project and caused extensive damage downstream. GLOFs have caused damage across national borders; outbursts originating in China have impacted areas in Nepal, India, and Bhutan (Ives et.al., 2010). June 2013 floods in Uttarakhand also have a big GLOF component. The Nathpa Jhakri Hydroelectric Project situated in Himachal Pradesh has dealt with success since its commissioning in 2000 the frequent flash floods of large magnitudes due to large sluice spillway openings create riverine flows and accurate warning system so that sufficient time is available for proper assessment of flood and depleting the Nathpa reservoir in anticipation of heavy inflows and timely operation of orifice spillway gates. The GLOF of the order of 4500 m^3/s was observed on 26th June 2005 due to a breach of an artificial lake formed on the Parechu river (in the Tibet region). Since necessary measures had already been taken, no major damages occurred to the project. During the floods, the river carried a lot of silt, boulders, gravel, & trash, etc. and the intake gates of the water conductor system of the powerhouse were closed to avert damages to the Power House in Jhakri. Urgent action by the international community to develop an even better scientific understanding of the consequences of global climate change on GLOF and to take corrective and precautionary measures is required. Monitoring and tracking of the lakes in West and North Sikkim have revealed that quite a few of them are expanding due to accelerated glacial retreat and melting due to climate change impacts. In addition, new lakes have also developed due to glacier retreat and melting. During the retreat, the glaciers leave behind moraines (accumulation of boulders, stones, or other debris) in the valley. Hydraulic modeling of the spillway is very important in the case of a Himalayan project to predict the discharging capacity of the spillway for PMF + GLOF condition and fix the important levels like the crest of the spillway, FRL, MDDL, invert of power intake, and top of the dam.

4. Upward revision of flood for existing dams

Previous design flood selection criteria considered factors such as dam height, storage volume, and downstream development and related design floods to downstream hazard classifications. The current practice is to classify dams based on the consequences of dam failure. This requires the identification of potential failure modes and a quantitative evaluation of the consequences of dam failure. The spillway design flood is based on these considerations. All existing dams whose failure would result in loss of life, significant economic or environmental damage or other unacceptable consequences for any condition will require corrective actions. Quantitative risk assessment includes the probability of adverse loading, the type of response, the probability of an adverse response and the consequences in terms of expected losses. It is seen that when such an analysis is applied to existing structures, in most of the cases, the spillway comes out to be of inadequate capacity. It is estimated that in India at least 50% of the existing spillways are inadequate in respect of the safety of the dams. Another consideration is that a dam constructed in the olden days was in remote areas whereas these days, the projects are very close to urban development and the potential damage due to the failure of such a dam is now many folds more severe than anticipated at the time of design or construction. The existing spillways are therefore required to be remodeled to handle larger floods. Increased spillway capacity to handle larger floods can be achieved by increasing spillway crest length, the addition of spillway bays, an increase in the operating head, or improving the discharge co-efficient. However, these measures are often governed by the topography and geology at the dam site. A modest increase in the discharge coefficient can generally be realized by refinement of the spillway crest shape and by channel improvement but at a greater cost. A more common modification to the existing dam to accommodate larger floods is an enlargement of the existing spillway or the construction of a new spillway. All these conventional methods have many limitations and call for design innovations.

Verifying and revising the dam design flood estimates- incorporating additional available data and modern methods of flood estimation-is the key technical priority in the national dam safety programme; and, this activity is also the prime requirement under the ongoing Dam Rehabilitation and Improvement Programme (DRIP). A review of design flood estimates has been completed for 217 dams. The review indicates that there is an upward revision of 50% for 58% of dams and an upward revision of 100% for 36% of dams. For a few dams, the revised flood values have exceeded their adopted design value by substantial orders. To name a few, the Kharadi dam of Madhya Pradesh exceeded by 929%, Sher tank (M.P.) exceeded by 503%, Manimukhanadhi dam in Tamilnadu exceeded 384% (Pillai et.al., 2013). Once the reviewed design flood is found to be significantly higher than the original flood, the adequacy of spillway capacity needs to be thoroughly reviewed. If this structural method is found to be inadequate, non-structural methods are required to be adopted. The solution adopted may vary from case to case. The following are the few alternatives that can be adopted to mitigate the increased design flood:

- Augmenting the existing spillway capacity by providing PK weirs, Labyrinth spillways
- Provision of breaching section or fuse plugs
- Increasing the freeboard above FRL by provision of Parapet wall etc.
- Implementation of early warning system
- Lowering the water levels in the reservoir before monsoon such that flood moderation can be enhanced

For an increase in the existing spillway capacity, structural options like the provision of an additional spillway, where the site is available, or an increase in spillway crest length by the provision of a P K Weir, labyrinth spillway or duckbill spillway, breaching section in the form of fuse-gates, etc can be thought of. Use of breach sections in the form of fuse gates etc can also be provided in the saddle on the reservoir rim from where flood discharge can be safely passed. The rate at which the fuse plug washes out is of primary importance and depends upon the fuse-plug geometry, the depth of flow, and the gradation and compaction of embankment material.

In the case of the dams having gated spillways, the provision of devices like P K Weir, labyrinth spillways etc may not be suitable. Although special arrangements for providing necessary waterway for such devices could be made through abutments or saddles in the reservoir, these may prove to be costly. For the gravity dams, especially concrete dams, allowing overtopping for floods of rare occurrences may prove to be more economical and effective provided necessary safety measures are taken care of. In the case of narrow valleys, options like tunnel spillways can augment the existing spillway capacity.

Under the DRIP programme, the design flood for Hirakud dam spillway, Odisha has been revised to 69,632 m^3/s from original design discharge 42,450 m^3/s i.e.upward revision by 27,182 m^3/s. To pass this increased flood, two additional spillways, one on right side and another on left is proposed to be constructed in the phased manner. CWPRS has been entrusted with hydraulic model studies for optimizing the design of the additional spillway and energy dissipator on the left dyke during phase-I. Two models, 1:40 scale 2-D sectional model and 1:100 scale 3-D comprehensive model have been constructed to assess the discharging capacity of spillway, overall performance of energy dissipator and spill channel and flow conditions in the river for various scenarios during combined operation of existing spillway on main Dam. Hydraulic model studies are a very important tool to understand the complex hydraulics of flow over spillways and appurtenant structures and the safety of mega projects like Hirakud Dam.

5. Spillway design considerations for coastal reservoirs

Coastal reservoirs have the potential to become a major source of fresh water in coastal areas. By definition, a coastal reservoir is freshwater storage located in coastal waters (river mouth, lagoon, gulf or protected bay) being fed by a sustainable freshwater flow. Coastal reservoirs have an impermeable barrier isolating the freshwater storage from intrusion by the surrounding brackish seawater. Depending on its location, water captured in a coastal reservoir will be of varying quality for domestic, agricultural or industrial water use. India has got about 7515 km of coastline and about 78% of surface runoff is directly delivered into the sea. For the benefit of the people living along the coastline, coastal reservoirs can be a better solution. Now, the world is using only 1/6th of total surface runoff and the other 5/6th part are directly discharging into the sea (Yang, 2015). This highlights the fact that there is enough water available but the deficiency is in storage. Hence, the solution lies in utilizing (or) conserving the abundant monsoon water which runs off into the sea. In the next 100 years, population and water demands may continue to increase significantly and the majority of the existing inland reservoirs will be deficient to fulfill the demand of increased population due to their structural life span and sedimentation. Hence, in the future, there will be more demand for coastal reservoirs. Coastal Reservoirs are already being used in China, South Korea, Hong Kong, and Singapore (Table 2) and have been largely successful, especially in places where there are no more opportunities to build inland reservoirs.

Table 2. Existing coastal reservoirs in the World (Yang, 2015)

Name	Catchment (km^2)	Dam length (m)	Capacity (10^6 m^3)	Year completed	Country/river
Qing Chaosha/Shanghai	1.8million	43,000	553	2011	China/Yangtze
Saemanguem	332	33000	530	2011	South Korea
Sihwa	---	12400	323	1994	South Korea
Marina Barrage	113	350	---	2008	Singapore
Chen Hang/Shanghai	1.8million	4700	8.3	1992	China/Yangtze
Yu Huan	166	1080	64.1	1998	China/ Zhejiang
Baogang/Shanghai	1.8million		12	1985	China/Yangtze
Plover Cove	45.9	2000	230	1968	Hong Kong

5.1 Coastal reservoirs in India

Thanneermukkom bund: Thanneermukkom bund was constructed as part of the Kuttanad lowland development scheme and the creation of a freshwater reservoir in the coastal area of Kerala, India (Figure 3). It divides Vembanad Lake into a freshwater lake fed by the rivers draining into the lake and a brackish water lake fed by ocean currents into the lowlands of Kuttanad. The four major rivers of Kerala, the Pamba, Meenachil, Achankovil, and Manimala, flow into the region before their confluence with the Arabian Sea. The regions receive a good amount of annual rainfall, which is above 3,000 mm, and these four rivers bring a large quantity of water into the lake before joining the sea. By constructing the saltwater barrier, a coastal reservoir having freshwater has been created for increasing agricultural activities in the area in addition to land development. The Thanneermukkom Bund has a total span of 1,410 metres and has a total of 90 shutters including 31 shutters each on the western and eastern sides and 28 shutters have replaced the old earthen bund.

Figure 3. View of Thanneermukkom Barrage

Kalpasar Project: The proposed gigantic Kalpasar project, also known as the Gulf of Khambhat development project is mainly a water resources project involving the creation of a freshwater reservoir in the Gulf of Khambhat for meeting the demands of irrigation, domestic and industrial water supply (Figure 4). Associated components related to the freshwater reservoir are the use of the top of the dam across the Gulf as a surface transport link, potential development of fisheries, and reclamation of saline land around the freshwater reservoir. The Kalpasar Project envisages building a 30 km long dam across the Gulf of Khambhat for establishing a huge freshwater reservoir for irrigation, drinking, and industrial purposes. Once constructed it will be one of the largest freshwater reservoirs in the sea with the highest priority for irrigation and drinking water in the region of Saurashtra and the Central Gujarat regions of India. (https://kalpasar.gujarat.gov.in/)

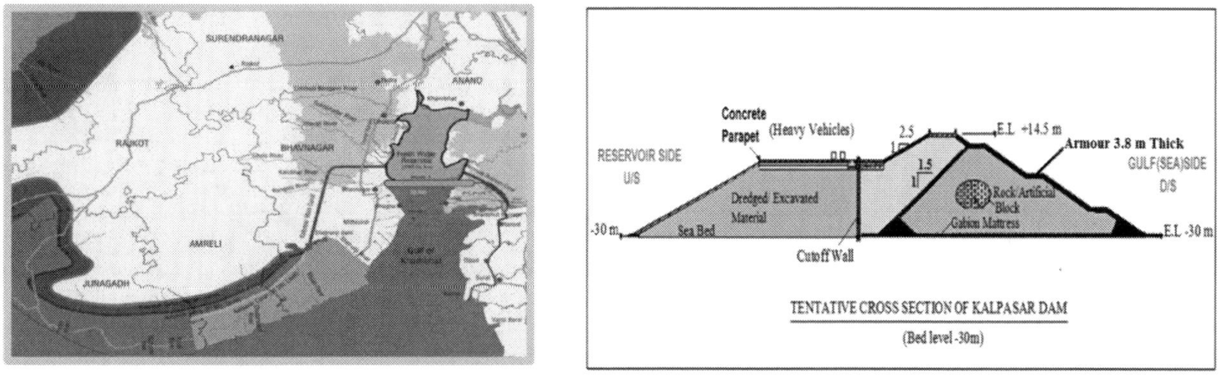

Figure 4. Tentative layout and cross section of Kalpasar dam ((https://kalpasar.gujarat.gov.in/)

5.2 Design considerations for coastal spillway and energy dissipation arrangement

Constructing a dam in the sea is a very challenging task as it involves understanding various aspects of natural phenomena like tides and waves, challenging foundation issues, complex hydraulics and ecology. Special design considerations are to be employed for spillway and energy dissipation arrangements as their operational performance is influenced by the cyclic tidal variations and storm surges. Stringent reservoir regulations are necessary to avoid the mixing of fresh water and saline water. For the design of a spillway, it is necessary to evolve the following parameters:

- Estimation of design flood
- Sizing of spillway capacity such that to avoid the flooding of the reclaimed areas without the water level in the reservoir exceeding certain maximum level
- Tidal variation, storm surge and wave data
- Dam safety aspect with tidal effects on the dam and proposed spillway
- Reservoir operation simulations according to inflow, reservoir water levels and tidal variations
- Prevention of saline water in to fresh water lake
- Energy dissipation arrangement to avoid entry of high velocity jets in the sea through spillway spans

From the schematic view shown in figure 5, it can be seen that it is possible to operate the spillway only during low tide when the sea water level is below or equal to the reservoir water level. This point is required to be kept in mind while designing the reservoir capacity considering design outflow limitation against the inflow flood. It is necessary to find out the discharging capacity of the spillway for free-flowing condition and partial downstream submergence conditions. Hydraulic model studies play an important role to optimize the spillway design and location and determine a various aspects of the operation of the coastal spillway as no readily available guidelines exist.

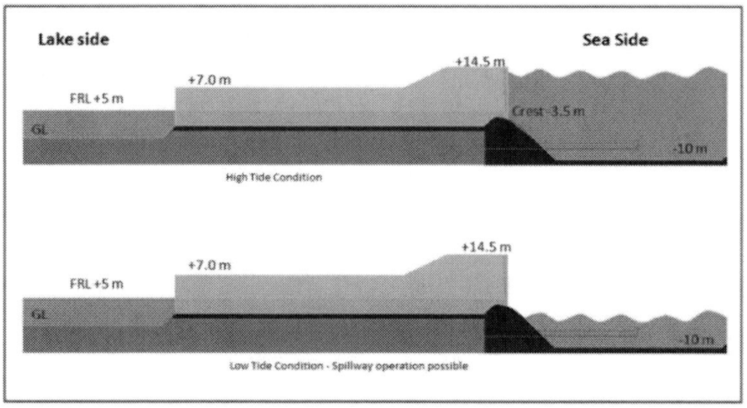

Figure 5. The schematic view of the coastal reservoir spillway

6. Conclusions

Spillways and energy dissipators are of paramount importance and the most integral part of dams. As the dam building activity is going on for years together, almost all easy sites have already been tapped. In this paper, an attempt has been made to identify the situations where there are challenges in front of the designer for future projects. Designing the spillways considering flash floods, GLOF and Himalayan fragile geology leading to high silt load are some of the challenges for the designers. The spillway will play a major role in designing sustainable projects in this region. Under DRIP, many old dams in India are assessed for design flood and there is an upward revision in the design flood for many of the dams. To minimize the risk of overtopping the dam, designers have to think of enhancing the existing spillway capacity and/or alternative methods to pass the revised flood. To conserve the excess runoff flowing into the sea, and to avoid submergence of landmass and rehabilitation issues in inland dam building, coastal reservoirs will be an attractive option in the future. As India got about 7500 km coastline, there is a future for building big coastal reservoirs for various purposes.Hydraulic model studies play a very important role in optimizing the design and layout of the spillways and associated appurtenant structures as scanty guidelines are available for designing spillways and energy dissipators under such situations.

7. ACKNOWLEDGEMENTS

The authors are grateful to Dr. R. S. Kankara, Director, CWPRS and Shri Y. N. Srivastava, Additional Director for encouragement and permission to publish this paper. The help of all the staff of SED division in conducting the studies and compiling the data is gratefully acknowledged.

8. REFERENCES

Bhopal Singh, (2013). "Hydrologic safety of dams in India." *J.INCOLD, volume 2, No. 1.*

Bajracharya, S. R., Mool, P. K., Shreshtha, B.R. (2006). "The impact of global warming on the glaciers of the Himalaya". *Proc. international symposium on geo-disasters, infrastructure management and protection of world heritage sites, Nepal*

Bhosekar, V. V. (Mrs.) and Gadge P.P. (Mrs.) (2018). Guidelines for hydraulic design of orifice spillway. CWPRS Technical Memorandum, Pune, India

Deolalikar, P. B., Bhosekar, V. V. (Mrs.) and Pethe, P. C. (2008). Research in to factors influencing the design of breastwall /sluice spillways. CWPRS Technical Memorandum, Pune, India

The gulf of Khambat development project (Kalpasar), <https://kalpasar.gujarat.gov.in/>

Ives, J D, Shrestha R B, Mool P. K. (2010). "Formation of glacial lakes in the Hindu Kush-Himalayas and GLOF risk assessment". *Proc. ICIMOD (international centre for integrated mountain development), Kathmandu*

Khatsuria, R. M. (2004). Hydraulics of Spillways and Energy Dissipators. Marcel Dekker Publication, N.Y.

Pandya, A.B., Goel, N.K., Pillai B.R.K. (2014). "Management of design flood issues in existing dams under climate change". *Proc. intern.symp Dams in global environmental challenges.,Bali, Indonasia,*

Pillai, B.R.K., Manoj Kumar, Nathan, L.A.V. (2013). "Design flood reviews of existing dams and related risk management issues under Dam Rehabilitation and Improvement Project". *Proc. Workshop Dam flood issues, CWC, New Delhi.*

Shu-Qing Yang . (2015). "Coastal reservoir-a technology that may dominate future water supply". *J. Water Resource and Hydraulic Engineering Oct. 2015, Vol. 4 Issue. 4, pp: 388-397.*

DOI: https://doi.org/10.26077/1d83-f83b

Stilling Basins Using Wedge-Shaped Baffle Blocks

N.N. Pillai[1] & M.L. Kansal[2]
[1]*Emeritus Professor, School of Engineering, Amrita University, Coimbatore, India*
[2]*Water Resource Development and Management, IIT Roorkee, Roorkee, 247667, India*
E-mail: mlk@wr.iitr.ac.in

Abstract. *The United States Bureau of Reclamation (USBR) extensively studied hydraulic jump-type stilling basins for various inflow and tailwater conditions. They recommend Basin III using chute blocks, rectangular baffle blocks, and sloping end sill for medium to high dams. However, a significant contribution to the knowledge regarding the production of turbulence in discontinuity layers, the dissipation of energy, the convection, and the decay of turbulence in hydraulic jump was made by subsequent researchers. However, while developing the stilling basin, the wake formed by the jets separating at the front corners of the rectangular baffle blocks in Basin III and reattaching the downstream of the blocks was not considered. The use of Basin III was found limited to lower dams since cavitation of the baffle blocks was observed in many dams. After that, the use of wedge-shaped baffle blocks below a model spillway was systematically investigated. It was noticed that baffle blocks with a vertex angle of 120^0 cut back at 90^0; the separating jets do not reattach downstream. Due to the wedge shape (super-cavitating), the low-pressure cavities do not collapse on the boundary. The new forms of baffle blocks are found to offer more drag to the flow when compared to rectangular blocks. Model tests were conducted to evolve stilling basins using the new baffle blocks and were highly successful. For example, in a stilling basin with an Inflow Froude Number of 7.07, with new blocks, a floor-length of about half the earlier suggested length is found to have better characteristics of scouring and sweep-out depth as compared to USBR Stilling Basin III.*

Keywords: *Wedge-shaped Baffle Blocks, Stilling Basins, hydraulic jump*

1. Introduction

Energy dissipaters of the hydraulic jump type for various inflow conditions have been evolved by the USBR and are being successfully used worldwide. Some of the rectangular baffle blocks have been affected by cavitation. In the light of new knowledge on the production process of turbulence and energy dissipation, improvement in the design of the baffle blocks and the stilling basins is sought in this paper.

1.1. The USBR stilling Basin III and USBR Stilling Basin II

The USBR, with its wide experience in many dams, have conducted extensive laboratory tests and evolved various designs of stilling basins for inflow Froude numbers of 5 to 9 (Bradley and Peterka, 1957). They recommended hydraulic jump type Stilling Basin Type III for medium head dams, using appurtenances, chute blocks, rectangular baffle blocks or baffle piers (with rectangular front face) and sloping end sill. For an inflow Froude number of nearly 7, the length of stilling basin will be 2.58 d_2, where d_2 is the conjugate depth of flow.

In dams like McNary, Bonneville, and Kanopolis, the pressures at the side of the rectangular Baffle piers were very low, leading to cavitation (Berryhill, 1957). It was recognized that the jets from the front side of the blocks got separated from the boundary but reattached to the side enclosing a fixed cavity due to Coanda Effect (Newman, 1961). The pressures in the cavity were very low in the case of very high inflow velocities. The USBR has recommended that Stilling Basin II without the baffle blocks may be adopted in such cases. The basin must be provided with chute blocks at the toe of the dam and a dentated end sill at the end. In case of an inflow Froude Number of 7, the length of the basin II should be about 4.2 d_2.

USBR conducted further experimental studies (Frizell and Svoboda, 2012) in connection with the Folsom dam near Sacramento, where the velocity at the toe of the dam was 25 - 37 m/s. To be safe against cavitation around the blocks, the velocity at the toe is reduced by providing steps in the chute, baffle piers, by providing a super-cavitating shape, and the provision of a floor ramp.

1.2. Studies on Process of Energy dissipation

Nemenyi (1940), in his discussion of the work of Kalinske (1940) on the statistical theory of turbulence, has made a classification of four types of discontinuity surfaces, as shown in Figure 1 (with sharp changes in velocity) and compared their efficiencies in the production of turbulence.

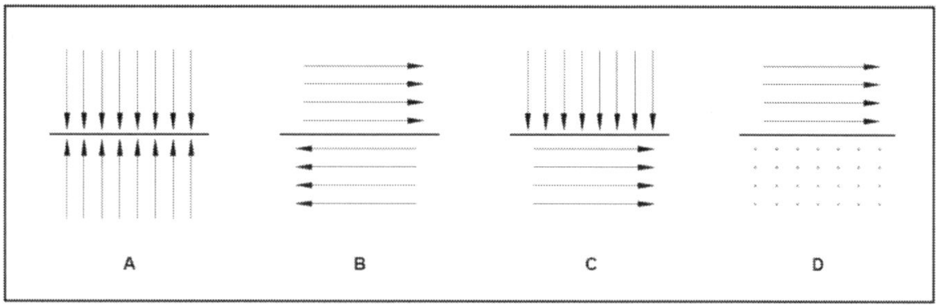

Figure 1. Types of discontinuities shown by Nemenyi (1940)

Kalinske (1946), in his studies on the energy transformation characteristics of pipe expansions, concluded that the principal loss of energy occurred outside the high-velocity stream as it passes into the expansion due to the development of very high shear. Also, he concluded that the total turbulence energy is a small part of the total energy change that is taking place.

In two common phenomena, the sudden expansion of pipe and the plain hydraulic jump, the energy dissipation is considered without the use of any appurtenances. The energy dissipation is caused solely by turbulence in the discontinuity layers. The production of turbulence and dissipation is far greater in pipe expansion due to type and areas of sharp discontinuity. In the case of the spatial hydraulic jump also, the energy loss is higher due to the additional discontinuity layers formed on the sides (Unny, 1961).

Major progress in understanding the energy dissipation process occurred when Rouse et al. (1959) published their work on the airflow model of the hydraulic jump. Using a hotwire anemometer, they could measure the various quantities of production, convection, and energy dissipation. It was shown that the turbulence produced is convected by the mean flow and diffused by the turbulent fluctuations. Since the kinetic energy of turbulence is small and the mixing in the mixing region is not high, the convection and diffusion terms remain very small. Hence, the dissipation of energy occurs near the production regions of turbulence. They found that the energy dissipation is slightly above the continuity region between the top eddies and the main flow, where the cross-velocity gradient is high. Mathematically, the work-energy equation for a hydraulic jump (secondary motion) was given as follows:

$$\int_0^d \frac{\overline{V'^2}}{2g}\overline{u^2}dy + \int_0^d \frac{\overline{V'^2}}{2g}u'dy + \int_0^d\int_0^x\left[\frac{\overline{u'v'}}{g}\left(\frac{\partial\bar{u}}{\partial y}+\frac{\partial\bar{v}}{\partial x}\right)+\frac{\overline{u'^2}-\overline{v'^2}}{g}\frac{\partial\bar{u}}{\partial x}\right]dy\,dx$$

$$= -\int_0^d \frac{P'}{\gamma}u'dy + \int_0^d \frac{\mu}{\gamma}\left[\frac{\partial}{\partial x}\left(\frac{\overline{V'^2}}{2}+\overline{u'^2}\right)+\frac{\partial\overline{u'v'}}{\partial y}\right]dy - \int_0^d\int_0^x K\frac{\mu}{\gamma}\overline{\left(\frac{\partial u'}{\partial x}\right)^2}\,dy\,dx \qquad (1)$$

Here, V is the total velocity V, V′ is the velocity fluctuation, μ is the dynamic viscosity, γ is the specific weight of water, u and v are the velocity components, and u′ and v′ are the velocity fluctuations. x is the distance of a point from the start of the jump, y is the height above the floor of the jump and d is the depth of flow at the cross section.

Further, in Eq. (1), the third component on the left-hand side is recognized as the turbulence production function. The third component on the right-hand side is the main dissipative function. It has a constant K which they found to be 12, 2, to 15 for the study they had conducted. Changing the spatial gradient in the dissipative function to a temporal one, the term can be written as $\int_0^d\int_0^x K\frac{\mu}{\gamma}\frac{1}{\bar{u}^2}\overline{\left(\frac{\partial u'}{\partial t}\right)^2}\,dy\,dx$, which indicates that the dissipation is maximum where high-frequency turbulent fluctuations occur.

2. Flow past rectangular baffle blocks

The USBR (Bradley and Peterka 1957), Kuttiammu and Rao (1951) and Ramamoorthy and Seetharamaiah (1959) tested the suitability of baffle piers with a rectangular face (Rectangular Baffle piers). The baffle piers offered considerable drag to the flow, and it was successfully used to reduce the length of stilling basin considerably.

Pillai and Unny (1964) studied the wake formed between two rectangular baffle blocks. It was found that the flow separated at the front corners of the blocks but got reattached downstream on the side, enclosing an attached eddy. It may be compared with the condition of flow between two rectangular plates.

Figure 2(a) schematically depicts two rectangular blocks of width w. The flow separates at the corners but reattaches on the side of the block. Hence, when they join with the block flow, the jet issues the gap between them. The wake formed is small.

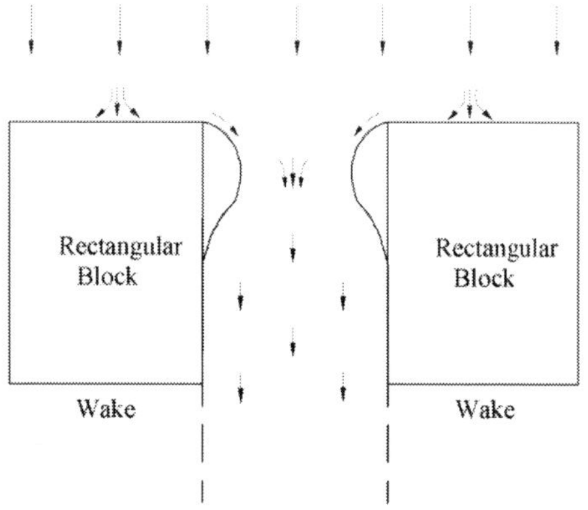

Figure 2(a). Flow between two rectangular baffle blocks

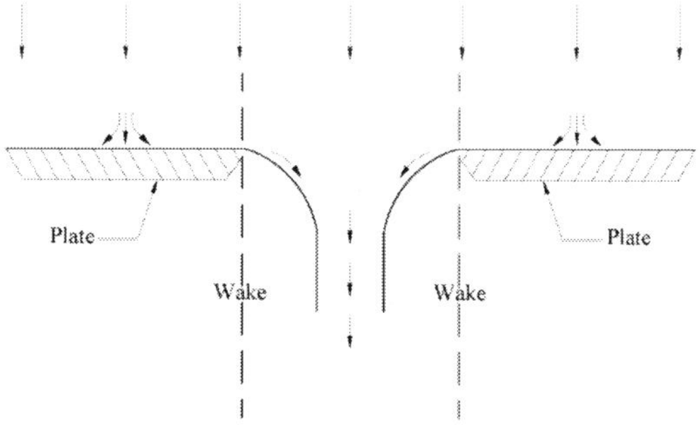

Figure 2(b). Flow between two parallel plates

Figure 2(b) schematically depicts the flow impinging on two rectangular flat plates of width w. The flow separates at the end of the plates and moves further to join the flow between the plates to form a sharp jet. The wake is big. In the case of rectangular blocks, the drag is much less than that in the case of plates of the same width.

3. Cavitation on rectangular baffle piers

It was found that in Bullstone, Bonneville, and McNary Dams (Pillai and Jayaraman (1967); Vennard (1947)), cavitation occurred on the separated regions of rectangular baffle blocks. Cavitation occurrence with thumping noise and vibration was observed in Glendo Dam. A detailed study of the pressures on the sides of the baffle blocks with partial streamlining was carried out and reported (Martin and Wagner (1961)) as shown in Figure 3(a-b).

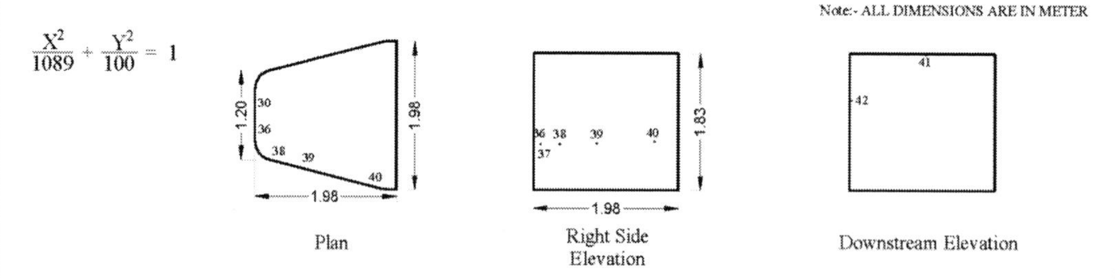

Piezometer	30,	36,	37,	38,	39,	40,	41,	42
Approx min. pressure (meter of water), (value read from oscillogram)	0;	< - 9.15,	- 6.1,	- 6.1,	+ 0.61,	+ 3.05,	+ 3.05,	+ 4.575

Figure 3(a). Baffle blocks pressure in the original design

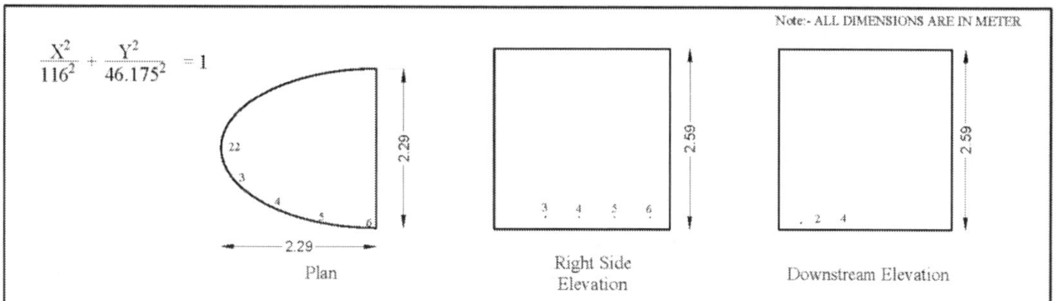

Piezometer	3,	4,	5,	6,	22,	24
Approx min. pressure (meter of water), (value read from oscillogram)	+ 1.525,	+ 1.525,	+ 2.135,	+ 3.05,	+ 4.575,	+ 0.915

Figure 3(b). Baffle pier pressure in a modified design

4. Tests on different shapes of baffle piers for stilling basins

Various baffle piers were tested to assess their suitability for use in the hydraulic jump. Many experimenters (Bradley and Peterka, 1957; Kuttiammu and Rao, 1951; Ramamoorthy and Seetharamaiah, 1959) found that the rectangular baffle piers offered the maximum drag to the flow and were suitable in stilling basins, among the many types of piers tested by them. In the case of rectangular blocks in operating conditions, pressure is reduced due to separation at the sides. In the Glendo dam, experimental studies have indicated the need for block streamlining to avoid cavitation, which can even lead to nearly a parabolic shape (Martin and Wagner, 1961). A series of tests were conducted by Pillai at the hydraulic laboratory of IIT, Kharagpur, to arrive at suitable baffle blocks for hydraulic jump stilling basins (Pillai and Unny, 1964).

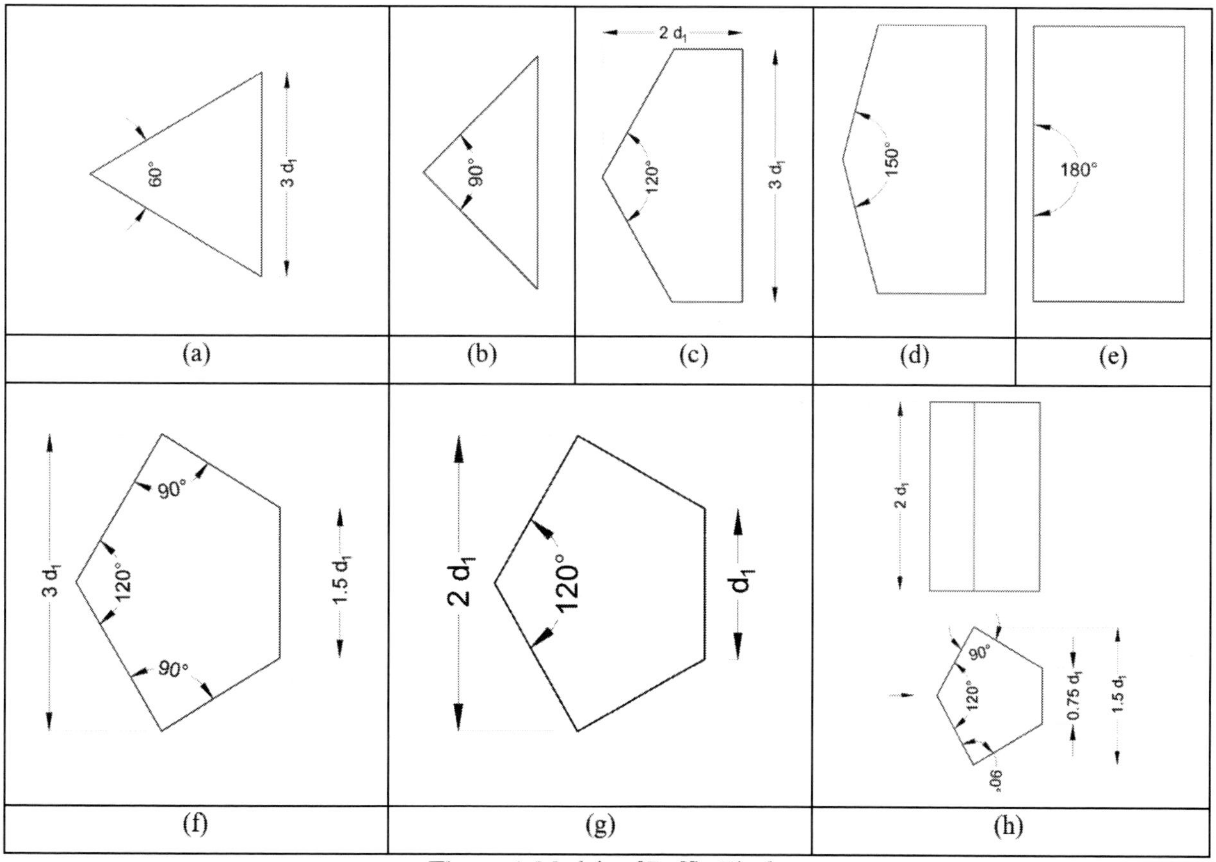

Figure 4. Models of Baffle Blocks

The form of the baffle blocks should be such that they would withstand the dynamic forces acting on them. The shape should give maximum high drag to the flow and be of practical dimension without acute angles, small projections or complicated curves. It was seen that a nearly parabolic shape was arrived at by streamlining the rectangular baffle blocks in Glendo Dam. A wedge shape of the same length and width can give greater drag than a parabolic shape and prevent cavitation problems (Tulin and Streeter 1961). The wedge-shaped blocks tested by Steele and Monroe (1929) and Smreek (1929) had small vertex angles and offered only low drag. A detailed test for blocks of higher vertex angles was taken up. As the vertex angle is increased to 120^0 or more, the material is added downstream to make it stable and strengthen the corners. Some of the model baffle piers tested are shown in Figure 4(a-e).

Later the block with vertex angle 120^0 (block-c) was tested along with blocks with the downstream face cut off (block-f) at 90^0; the drag increased considerably. The drag also increased when the block size was reduced (block-g), keeping the blocking ratio the same (Pillai and Unny, 1964). New baffle blocks of width 1.5 d_1 and height 2 d_1 were used in the model tests for the stilling basins as shown in Figure 4(f-h). The baffle blocks were placed at a distance of $0.5d_2$ away from the toe of the dam. The blocking ratio was 0.5 (Pillai,1969)

4.1. Criterion of comparing the scour in the model tests

The scour should not occur near the end of the floor, and the maximum scour should occur as far away from the sill as possible. The following criteria were used to compare the stilling basin's effectiveness and limit the scour at the end of the basin. The scour profile formed after half an hour of the test was assumed as a parabola, and the depth of maximum scour d_s, and the distance L_s away from the sill were measured. The quantity $tan\,\alpha = \frac{2\,d_s}{L_s}$ was used as an index of scouring for comparison as shown in Figure 5. For safety, the index should not be high.

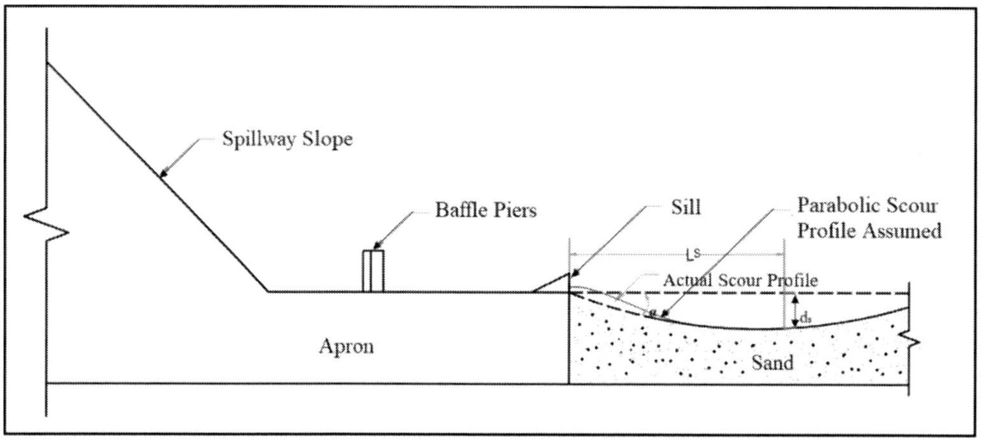

Figure 5. The criterion for comparison of scour $tan\,\alpha = \frac{2\,d_s}{L_s}$

4.2. Tests conducted for the development of Stilling basin for inflow Froude number nearly 7

Experiments were conducted to find the inflow depth at the toe of the test spillway to get a Froude Number of nearly 7.0. It is conveniently located when d_1 = 0.0174 m. The Froude Number was 7.07. The depth after the normal jump was found to be 0.165 m. A 2:1 sloping sill of height 0.1 d_2 was used in all the new basins tested. The basis for comparing the performance of the new stilling basin was the results obtained by tests on the USBR stilling basin III models. USBR stilling basin for F_1 = 7.07 made and tested. The length of the stilling basin was 2.58 d_2, and the rectangular baffle piers were placed at 0.8 d_2 from the toe. The chute blocks and the end sill were provided as per the recommendations. The performance of the stilling basin was good. Let us consider the condition of tailwater depth d'_2 less than the conjugate depth d_2. The jump left the chute blocks when the tailwater level was lower than d'_2= 0.8 d_2. The operation of the basin model when the tailwater level was 0.82 d_2 is shown in Figure 6. The scour index of the stilling basin operating with tailwater of 0.82 d_2 was 0. 596.

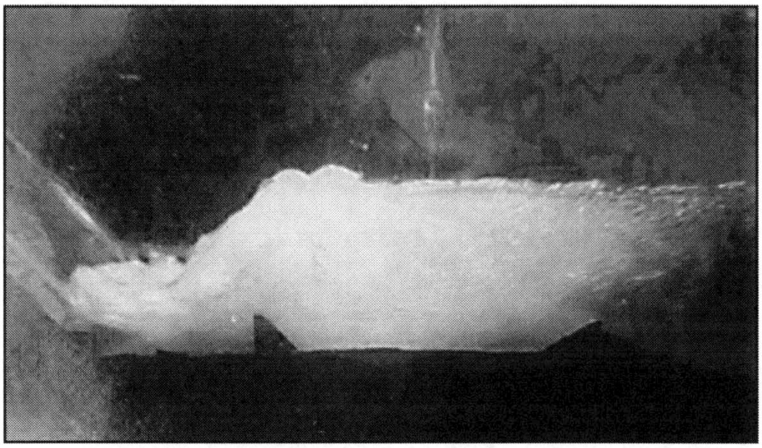

Figure 6. Working of USBR stilling Basin III model

$$(F_1 = 7.07;\ L = 2.58\ d_2\ ;\ \frac{d'_2}{d_2} = 0.82,\ \text{and}\ \tan\alpha\ =\ 0.596)$$

4.3. New Stilling Basin A

A stilling basin with no chute blocks but the same length as the USBR stilling basin with wedge-shaped blocks at 0.5 d_2 from the toe and a sloping end sill of 0.1 d_2 were made and tested. The operation was good for tailwater depths of d_2 and 0.82 d_2. The scour index when operating with tailwater of 0.82 d_2 was 0.423. The basin is termed Basin A. The cross-section is shown in Figure 7.

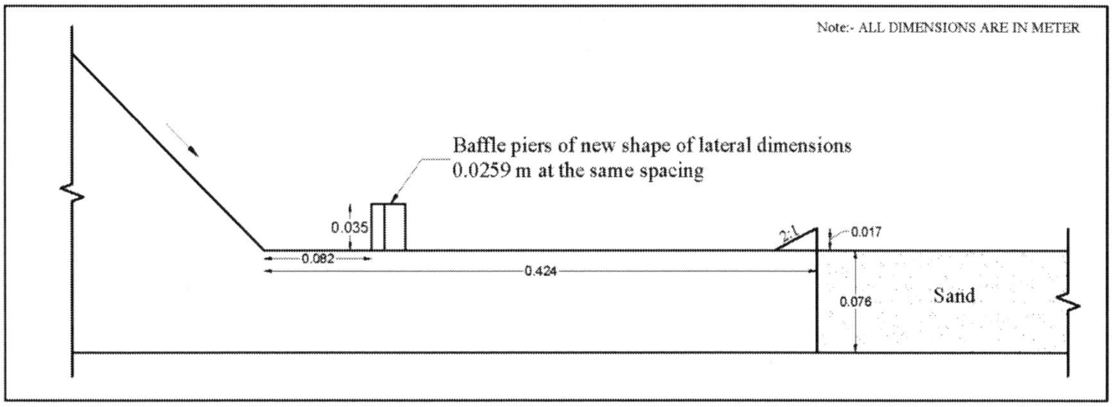

Figure 7. New Stilling basin A

$$(\text{For } F_1 = 7.07;\ L = 2.58\ d_2;\ \frac{d'_2}{d_2} = 0.82,\ \tan\alpha = 0.423)$$

4.4. New Stilling Basin B

It was noted while investigating the basin-A that a weak roller formed starting from the wedge-shaped blocks and dipped at about 2.4 d_2 from the toe. Hence, it was impossible to reduce the basin's length by a short length. But it was found that when a sloping sill of height 0.1 d_2 was placed 1.2 d_2 from the toe, the roller formed by the sill and the earlier roller got combined and dipped far away from the basin. The basin was terminated with the sloping sill at 1. 2 d_2. The stilling basin performed well when the tailwater level was d_2 and 0. 82 d_2. The scour index was 0.493. The basin is called Basin B. The cross-section is shown in Figure 8. The operation of the model stilling basin when the tailwater was 0.82 is shown in Figure 9.

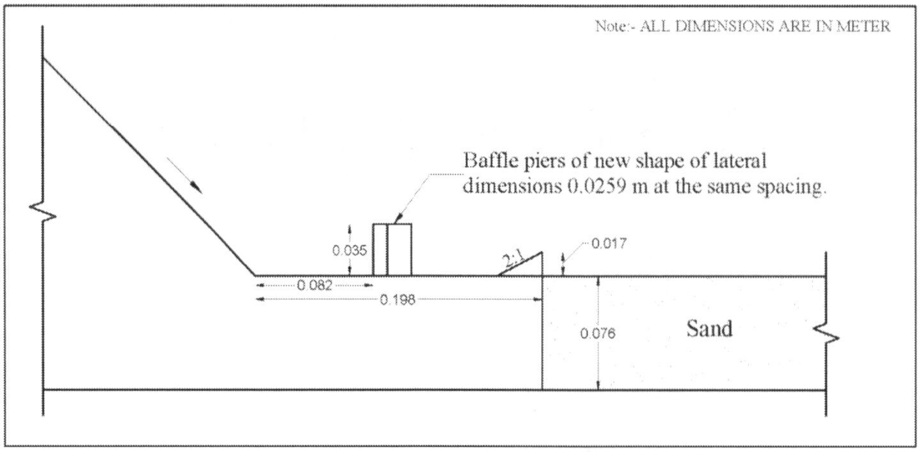

Figure 8. New stilling basin B

(For $F_1 = 7.07$; $L = 1.2\ d_2$; $\dfrac{d_2'}{d_2} = 0.82$; $\tan\alpha = 0.493$)

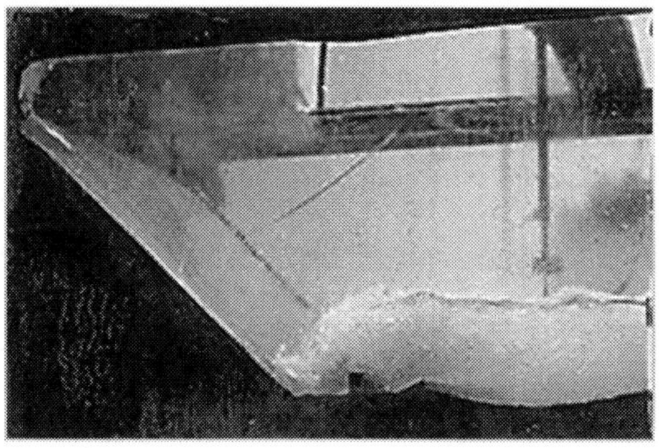

Figure 9. Working of stilling basin model B

4.5. New Stilling basin C

A stilling basin design was tested with a length of 2.32 d_2 (90% length of USBR Basin III design), and new wedge-shaped blocks and chute blocks arranged staggered with them. The basin ended with a sloping end sill. The basin operation was good when the tailwater level was d_2 and 0.82 d_2. The scour index while operating with tailwater depth was 0.82 d_2 was 0.319. The basin was termed New Basin C. The cross-section is shown in Figure 10, and the operation of the model basin is shown in Figure 11.

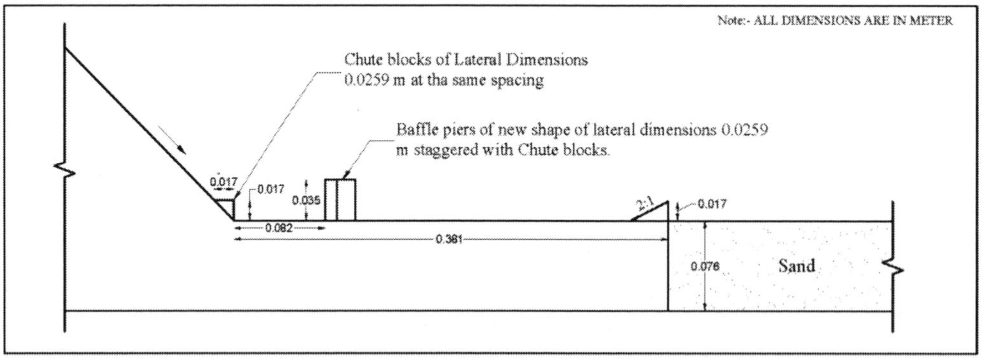

Figure 10. New Stilling basin C

(For $F_1 = 7.07$; $L = 2.32\ d_2$; $\dfrac{d_2'}{d_2} = 0.82$; $\tan\alpha = 0.319$)

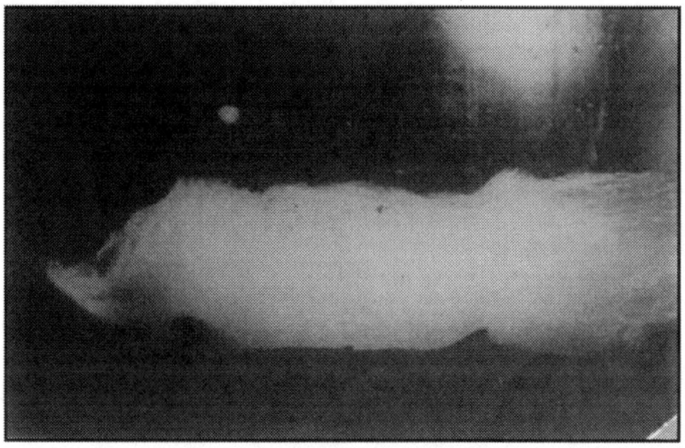

Figure 11. Working of new stilling basin C

(For F_1= 7.07, L = 2.32 d_2; $\frac{d'_2}{d_2}$ = 0.82 and $tan\,\alpha = 0.319$)

4.6. New stilling basins used wedge-shaped blocks when the Froude numbers were near 9 and 5.

A series of tests were conducted to evolve stilling basins when Froude numbers were 9.0 and 4.97. The blocks were placed at 0.5 d_2 from the toe. The blocks were of height 2 d_1 and width 1.5 d_1. Various basins and arrangements of blocks and sill were tried to arrive at a suitable design. The comparison was made with the study of model results on USBR designs, and the criterion of scouring index was followed. In the case of F_1 = 9, comparative results (L = 1.41 d_2 for basin B; and L = 2.32 d_2 for basin C) were obtained. For F_1 = 4.97, two rows of staggered baffle blocks at a blocking ratio of 31.5 % were effective for Basin B with a length of 1.41 d_2. In all cases, it was found that the new wedge-shaped blocks were very effective.

4.7. Comparison of drag offered by the rectangular baffle blocks and the new blocks

The average drag coefficient for the rectangular blocks has been nearly 0.7, whereas the wedge-shaped blocks' drag coefficient is nearly 0.82 (Pillai, 1967).

4.8. Stilling basins were developed for other cases.

It was found that the vertex angle of the wedge-shaped blocks in the stilling basin could be increased to 150^0 with a small increase in drag (Pillai et al. 1989). It has been possible to develop economical stilling basins for low Froude numbers (Pillai et al. 1989; Pillai and Goel, 1994). Stilling basins using the new shapes for baffle blocks could affect the economy in pipe outlets (Goel and Verma, 1999, Verma and Goel, 2003). Verma et al. (2004) found that economical basins could be developed in deep rectangular outlets.

5. Discussions

Studies on satisfactory baffle block shapes for stilling basins led USBR (Bradley and Peterka, 1957) to blocks with a rectangular surface facing the high-velocity flow. The drag on the rectangular block was expected to be very high due to the direct impact on the rectangular face. But model tests in the Glendo dam have shown that the jets at the side of blocks got separated and then reattached on the sides, enclosing a cavity as in Figure 5a. The total drag on the block does not remain high. Moreover, cavitation can affect the blocks at high velocities of inflow.

Pillai and Unny, (1964) had evolved a Super-cavitating shape for wedge-shaped baffle blocks of vertex angle 120° cut back at 90° at the back. Pillai et al. (1989) had also found that the vertex angle could be increased to 150° causing a small increase in drag. It may be observed that the USBR published its report on stilling basins (Bradley and Peterka, 1957) and recommended rectangular baffle blocks. In 2012 USBR published the result on Supercavitating baffle blocks with a rectangular face with a ramp (Frizell and Svoboda, 2012). It is of scientific interest to make a comparative study of the characteristics of the new rectangular blocks like the drag, safety against cavitation and cost with characteristics of the wedge-shaped blocks of vertex angle of 150°.

The appurtenances used in hydraulic jump-type stilling basins can cause a reduction in the tailwater depth required by 10% to 20%. When the tailwater depth is reduced by 15%, the outflow velocity increases, and the resulting increase in energy dissipation is only 5%. The increase in energy dissipation, even though not high, the appurtenances served well the purpose of imparting stability of the jump and ensured safety against sweep-out from the basin. In the USBR stilling basin III, the chute blocks and the baffle blocks created discontinuity layers in the flow, causing the production of turbulence and the big eddies, creating greater convection and diffusion, resulting in reducing the length of the basin to about 50% of the length of the hydraulic jump.

The rectangular blocks could not be kept closer to the toe because they cause a cascade and then dip the flow in the basin floor. The new wedge-shaped blocks can be placed at a 0.5 d_2 distance from the toe. The jet issuing at the front side of the blocks joined that from the adjoining block and created very sharp discontinuity surfaces with the wake. The jet issuing between the wedge-shaped blocks can be considered to form miniature expansions. The result is very fine-grained eddies of turbulence for energy dissipation and bigger eddies to effect high convective action. That further helps to develop new stilling basins using the new wedge-shaped blocks.

The model for USBR stilling basin for the selected Froude number was first tested. When the jump was about to leave the chute blocks, the tailwater depth and the scour index were noted after 30 minutes of working. Then the new wedge-shaped blocks blocking 50% of the flow width were placed 0.5 d_2 away from the toe. The end sill was of height 0.1 d_2 and upstream slope 2:1. A stilling basin like USBR stilling Basin III (without chute blocks) was developed for F_1 = 7.07 (Basin A). An attempt was made to reduce the length of the basin. It was successful. (Basin B) needed only a length of 1.2 d_2. This is possible only because of the convective action of the eddies formed between the wedge-shaped blocks. If a higher energy dissipation is desired, a row of chute blocks staggered with the baffle blocks could be used. The length of the basin could be 2.32d_2 (Basin C). The stilling Basins for F_1 = 9.0 and F_1 = 4.97 gave encouraging results. Further tests using the wedge-shaped blocks with a vertex angle of 150° were conducted for use in the case of low Froude numbers. The wedge-shaped blocks have been established in the case of flows with low Froude Numbers and pipe outlets (Pillai et al. 1989).

In a plane hydraulics jump with Froude number 7. 07 (depth ratio d_2/d_1=9.51), the energy loss in terms of the kinetic energy of the inflow stream will be 0.65. With a 15% reduction in tailwater depth in a stilling basin, the energy loss can rise only to 0.70 times the inflow kinetic energy. It was because when the tailwater depth decreased, the outflow velocity increased. In a sudden pipe expansion (with an expansion ratio of 9.51), the energy loss is about 0.80 times the inflow kinetic energy. The high variation of energy loss in the cases above can be attributed only to the type of discontinuity layers formed and the intensity and scale of turbulence developed therein.

6. Conclusions

Wedge-shaped baffle blocks with a vertex angle of 120° have been found to dissipate more energy of high-velocity flow in a hydraulic jump stilling basin than rectangular baffle blocks. The wedge-shaped blocks could be placed 0.5 d_2 away from the toe of the dam, increasing its effectiveness because it could avoid a cascade formation, as in the case of rectangular blocks. It is also found that a small amount could enhance the drag of the blocks by increasing the vertex angle of blocks to 150°. Due to the super-cavitating shape of blocks, they could be safe against cavitation. Experiments were conducted to evolve stilling basins with wedge-shaped baffle piers. The performance of the model of new basins for any Froude number, like the tailwater required for jump formation and the scour index, were compared with the performance results obtained for the USBR design. Three types of basins, A, B and C, could be evolved. They performed well and were more economical than the USBR stilling basins. The basins are recommended to be used with a tailwater depth of d_2 to be safe against usual fluctuations in the tailwater conditions.

Considering the natural energy dissipation in a hydraulic jump, the additional energy loss by using the baffle blocks is small. Still, it greatly helps in imparting stability to the jump. The jets from the wedge-shaped blocks create very sharp discontinuities in the flow and generate more intense turbulence and energy dissipation. Their convective action of bigger eddies helps shorten the length of stilling basins. More laboratory tests and prototype studies are required to explore the possibility of using the wedge-shaped baffle piers and new stilling basins.

REFERENCES

1. Berryhill, R. H. (1957). "Stilling basin experiences of the Corps of Engineers". Journal of the Hydraulics Division, 83(3), 1264-1.
2. Bradley, J. N., & Peterka, A. J. (1957). "Hydraulic design of stilling basins: Hydraulic jumps on a horizontal apron (Basin I)". Journal of the Hydraulics Division, 83(5), 1401-I-1401-24.
3. Frizell, K. W., & Svoboda, C. D. (2012)."Performance of type III stilling basins–stepped spillway studies". Hydraulic Laboratory Report HL-2012-02, US Bureau of Reclamation, Denver, USA.
4. Goel, A., & Verma, D. V. S. (1999). "Improved design of energy dissipaters for pipe outlets". *Irrigation and Drainage Systems,* 13(4), 313-320.
5. Kalinske, A. A. (1940). "Relation of the statistical theory of turbulence to hydraulics". *Transactions of the ASCE,* 105(1), 1547-1566.
6. Kalinske, A. A. (1946). "Conversion of kinetic to potential energy in flow expansions". *Transactions of the ASCE,* 111(1), 355-374.
7. Kuttiammu, T. P., & Rao, J. V. (1951). "Bhavani type stilling basin for spillways of large dams". In *IV ICOLD Congress,* New Delhi. V (44), 397-414
8. Martin, H. M., & Wagner, W. E. (1961). "Experience in turbulence in hydraulic structures". In *IX IAHR Congress,* Dubrovnik, 153-172
9. Nemenyi, P. (1940). Discussion on "Relation of the Statistical theory of turbulence in Hydraulics" by Kalinske, A.A., Paper No. 2091, *Trans. ASCE,* 105(1), 1547-1566.
10. Newman, B.G. (1961). " The Deflection of Plane Jets by Adjacent Boundaries - Coanda Effect, " Boundary Layer Control (Edited) by C.V. Lachmann, Pergamon Press, London.
11. Pillai, N. N., & Unny, T. E. (1964). "Shapes for appurtenances in stilling basins". *Journal of the Hydraulics Division,* 90(3), 1-21.
12. Pillai, N.N. (1967)." The drag of Baffle piers in a hydraulic jump". *Symposium on High-velocity flow,* IISc, Bangalore, India 1967.
13. Pillai, N.N. (1969)."Stilling basis with wedge-shaped baffle blocks". *Water and Water Engineering,* December, 506-509.
14. Pillai, N. N., Goel, A., & Dubey, A. K. (1989). "Hydraulic jump type stilling basin for low Froude numbers". *Journal of Hydraulic Engineering,* 115(7), 989-994.
15. Pillai, N. N., & Goel, A. (1994). "Stilling basins for low Froude number using wedge shaped blocks". In *Ninth Congress of Asian & Pacific Division of IAHR,* Singapore.
16. Pillai, N. N., & Jayaraman, V. (1967). "Cavitation on baffle piers in stilling basins". In *Symposium on High Velocity Flow G,* IISc. Bangalore.
17. Ramamoorthy, and A. S., Seetharamaiah, K. (1959). " Effect of Shapes on Energy Dissipation for high Velocity Flow". *Annual report, Dept. of Power Eng',* Indian Institute of Science, Bangalore, India.
18. Rouse, H., Siao, T. T., & Nagaratnam, S. (1959). Turbulence characteristics of the hydraulic jump". *Transactions of the ASCE,* 124(1), 926-950.
19. Smreek, A. (1929). "The Hydraulic Laboratory of Bohanian Technical University of Bruenn". Hydraulic Laboratory Practice (Edited by John R Freeman), ASME.
20. Steele, I. C., & Monroe, R. A. (1929)." Baffle-pier experiments on models of Pit River Dams". *Transactions of the American Society of Civil Engineers,* 93(1), 451-480.
21. Tulin, M. P., & Streeter, V. (1961). "Super-cavitating flows. Handbook of Fluid Dynamics", McGraw-Book Co., Inc., (edited by V. L Streeter), New York, N.Y 1224-1246.
22. Unny T.E. (1961). "The Spatial Hydraulic Jump", *Proceedings – IAHR, Ninth Convention,* Belgrade, Yugoslavia, I/10-1-10.
23. Vennard, J. K. (1947). "Cavitation in Hydraulic Structures: A Symposium". *Transactions of the ASCE,* 112(1), 1-15.
24. Verma, D. V. S., & Goel, A. (2003). "Development of efficient stilling basins for pipe outlets". *Journal of Irrigation and Drainage Engineering,* 129(3), 194-200.
25. Verma, D.V.S., Goel, A, and Rai, V (2004), "New Stilling Basin Designs for Deep Rectangular Outlets" *IJE Transactions A: Basics,* 17(1), 1-10.

DOI: https://doi.org/10.26077/010f-f6ea

Submerged flow regimes downstream of a weir at multiple slopes

K.S. Vikas[1] & A. Sarkar[2]
[1]*Interdisciplinary Centre for Water Research (ICWaR), IISc Bangalore, India.*
[2]*School of Infrastructure (SIF), IIT Bhubaneswar, Khorda, India.*
E-mail: asarkar@iitbbs.ac.in

Abstract: *The higher water levels unanticipated at the time of floods are not generally considered in the design of hydraulic structures. The present study emphasizes the experimental investigation for the characterization of submerged flow regimes downstream of a rectangular sharp-crested weir at three different slopes (0, 0.004, and 0.008) that are possible with a laboratory flume. The classical flow patterns of submerged regimes, namely impinging jet, surface wave, surface jump, and surface jet regime, were analyzed. The experiments were conducted to examine the water profiles and the variations of different parameters, representing surface jumps and surface waves, with a gradual rise in submergence. The range of t/h values (where t and h are the downstream and upstream depths above the crest of the weir) at which these regimes shift from one to another were observed to vary with the slope. The photographs of the patterns of the submerged flow regimes at three different slopes and six variants of discharge were captured, and image analysis was carried out. The profiles were extracted for each submergence level to develop the correlations for various characteristics of surface jumps and waves. With an increase in the submergence, the non-dimensional lengths of the surface jumps decreased at all the slopes and were within a range of 10 to 55, whereas the non-dimensional heights increased and were within a range of 0 to 0.7. The surface wave with a maximum normalized amplitude of 0.6 and a maximum normalized wavelength of 4 faded out gradually as the regime ends.*

Keywords: *Sharp-crested weir, Submergence, Flow regimes, Free Surface.*

1. Introduction

Water is one of the major means through which any change in the climate can be experienced. Many regions of the world are being adversely affected by climate change. Floods are one of those climate extremities and make a significant impact on hydraulic structures. Floods may cause higher water depths at the upstream of hydraulic structures leading to a sudden release of huge quantities of water to the downstream, which may significantly impact life and property. Flood waves, being turbulent in nature, exhibit complex flow physics at various hydraulic structures. Although the classical flow patterns and other interesting phenomena at various hydraulic structures were analyzed for better design and management, it is very much essential to consider the higher water levels and the fully submerged conditions of the hydraulic structures that occur at the time of floods. Weirs obstruct the flow and raise the water level upstream of the crest, which functionally gets related to the discharge. According to the mode of obstruction based on the shape of the crest, the classification of James (2020) distinguishes between broad-crested and sharp-crested (also known as thin-plate) weirs. Sharp-crested weirs are usually made up of thin metal sheets resulting in slim crests. Water flows over the crest and forms a sheet called nappe without clinging to the downstream portion of the weir. In general, these sharp-crested weirs are used at small irrigation channels, hydraulic laboratories, and industries where accurate flow measurement is required.

With the increase in the intensity of floods, the level of tail-water also increases. This abrupt rise in tail-water depth above the height of the weir (or any other hydraulic structure) is known as 'submergence'. A weir that submerges or drowns with a high downstream water depth is known as a 'submerged weir'. Various flow patterns such as impinging jet, surface wave, surface jump, and surface jet that are formed with gradual changes in the tail-water level above the weir height are identified as the 'submerged flow regimes'. When the tail water level is close to the weir height, the impinging jet regime occurs in which the flow over the weir dives towards the bed and gets diffused. As the tail water increases, the jet remains at the surface and results in the formation of a surface jump regime with strong turbulent mixing. As the tail water depth further increases, the jet changes into a wavy surface with a sudden reduction of air bubbles, causing the surface wave regime. At higher submergence, the wavy nature diminishes, and a flat surface jet moves to cause the surface jet regime. The effect of submergence on sharp-crested weirs was initially summarized by Vennard et al. (1943). Later, Skogerboe et al. (1967) analyzed the submerged flows downstream of a weir and discussed various discharge measurement methods. Rajaratnam (1967), who has contributed to a great extent in this field amongst researchers, explained the applicability and characteristics of a submerged hydraulic jump, such as the length of the jump, height of the jump, etc. Rajaratnam and Muralidhar (1969) examined the geometrics of the flow development, fully developed, and eddying regions in a surface jet regime. Their work concluded that the length of the flow development region is twice the tail-water depth above the crest and proposed a more simplified equation for the estimation of the discharge for higher submergence. Wu and Rajaratnam (1995) compared the flow structure of the free jump, submerged jump, and wall jets to make a distinction between the impinging jet regime and other surface flow regimes. They found that the submerged jump has a significant reverse flow with unique velocity patterns. Wu and Rajaratnam (1996) conducted experimental investigations by considering a wide range (in terms of the height of the weir) of the weirs and concluded that the transition of these regimes occurs in a particular fashion with an alteration in the direction of the tail-water depth. They have vividly demarcated the occurrences of the impinging jet and surface flow (i.e., surface jet, surface jump, and surface wave) regimes. Similar correlations between the tail-water depth and flow regimes were made by Wu and Rajaratnam (1998) for a vertical drop. For a finite crest length, discharge characteristics and water surface profiles were analyzed by Azimi et al. (2014) under submerged conditions. Azimi et al. (2016) also studied flow regimes over rectangular sharp-crested weirs and quantitatively defined the regime boundaries of occurrences and represented a better estimate for discharge estimation.

Ahmed et al. (2019) started an introductory analysis on surface hydraulic jumps showing the analogous nature to the classical hydraulic jumps by expressing the variations of different characteristics of the jumps with the relative submergence. Later, Bijankhan and Ferro (2020) supported these associations and determined the submergence zone at which surface jump occurs for different combinations of upstream and downstream water levels. However, the aforementioned works have only considered a horizontal slope. In contrast, the present work focuses on the effect of the longitudinal bed slope on the characteristics of submerged regimes such as surface jump and surface wave. A laboratory model was fabricated to form various drowned regimes downstream of a rectangular sharp-crested weir. Experiments were conducted with different combinations of discharge and slope to study the characteristics of surface hydraulic jumps and surface waves formed downstream of a sharp-crested weir in a rectangular flume at multiple slopes. The data observed was digitized to analyze the characteristics of the obtained flow regimes.

2. Experimental Setup

The experiments were performed in a recirculating flume of 10 m length, 0.4 m width, and 0.8 m depth with a stainless-steel bottom and glassed sidewalls through which the observations were made. The flume also has a provision for varying the slope (S) within the range of 0 to 0.01. To control the tail-water depth, tailgates were provided at the outlet of the flume. Fabrications were made at a section far away from the inlet inside the flume such that the weir fits into the section without any leakages. A photograph of the flume is presented in Figure 1a.

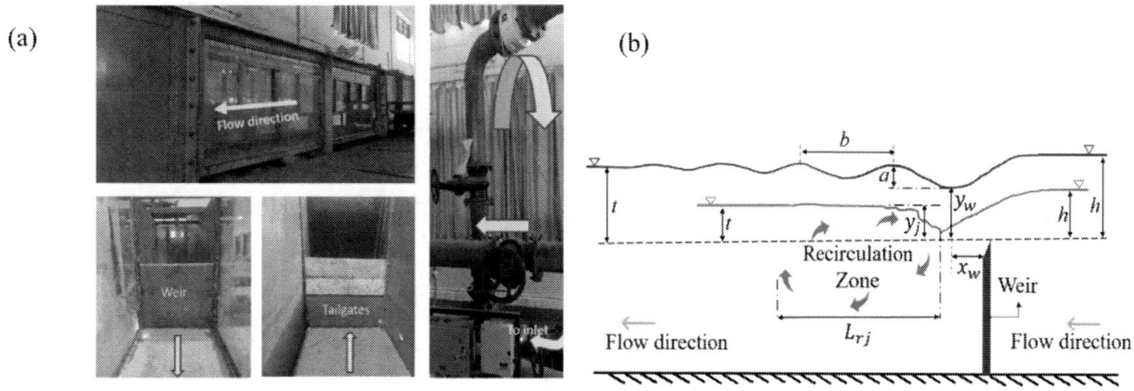

Figure 1. Experimental Setup Details. (a) Flume photographs; (b) Definition sketch of submerged flow

A rectangular sharp-crested weir of width 0.4 m and height of 0.4 m was used. The weir plate was fabricated from a mild steel sheet of a thickness of 5 mm. The crest of the weir was restricted to a thickness of 1 mm with a crest angle of 45°. Proper care was taken to ensure no leakage happened through the sides and bottom of the weir section. The circulating water supply scheme consisted of a pump, an inlet reservoir, an outlet reservoir, and a storage tank (provided below the flume). The pump was used to lift water from the storage tank to deliver at the desired discharge that can be visualized with the help of a digital flow meter connected to the inlet. A gate valve fitted at the junction of the inlet reservoir and storage tank was used to regulate the water supply at the inlet. As the flow setup is re-circulatory, water comes out of the outlet and moves into the storage tank. The pump takes the water from the storage tank at the bottom and delivers it to the flume inlet at the top. A bypass valve is also provided to regulate the inflow accordingly. Water levels upstream and downstream of the weir were measured with the help of two point-gauges which were placed on the top of the flume. Point gauges with an accuracy of 0.1 mm are used to measure the upstream and downstream water levels. The point gauges were moved to the desired location (section) within the length of the flume to measure the water levels in all the experimental runs. Considering the minimum and maximum slopes that are possible with the laboratory flume, the experiments were aimed to observe the change in characteristics of the submerged flow at three different slopes, i.e., 0, 0.004, and 0.008, respectively.

Once the slope of the channel was kept constant, a particular discharge was set by regulating the gate valve to ensure that the flow stabilizes and maintains the desired discharge value. Later, the tailgates were added at the downstream end of the flume to create submerged conditions by raising the tail-water depth beyond the weir height. Then, the value of '*t*', the tail-water depth above the crest of the weir, was measured with the help of point-gauge at a location where the free surface is stable without any turbulence. Similarly, the upstream head on the weir (which changes from the free flow head due to submergence) was also noted as '*h*'. All such symbols adopted for the study are represented in Figure 1b. The ratio (*t/h*) of tail-water depth and upstream head, both above the crest of the weir, was calculated. The tailgates were increased slowly such that ten (approximately) different *t/h* values were observed. Thus, for a particular discharge value, flow characteristics were captured such that all four submerged flow regimes were covered within the range of *t/h* values. In each slope, a variation of six discharges (i.e., 20, 25, 30, 35, 40 and 50 L/s) was carried out. For each discharge, *t/h* values between 0 and 1 were made by changing the tail-water depth gradually such that all the submerged flow regimes were observed to the possible extent. The experimental framework and runs made during the analysis are illustrated in Table 1.

Table 1. Experimental framework showing different runs.

S. No	Slope	Discharge (L/s)	*t/h* ratio
1.	0	20	0.11, 0.24, 0.30, 0.39, 0.49, 0.55, 0.60, 0.72, 0.81
		25	0.13, 0.22, 0.31, 0.36, 0.45, 0.57, 0.65, 0.72, 0.83, 0.85
		30	0.05, 0.14, 0.24, 0.34, 0.42, 0.50, 0.60, 0.67, 0.75, 0.81, 0.85
		35	0.13, 0.21, 0.30, 0.40, 0.45, 0.50, 0.52, 0.60, 0.66, 0.733, 0.78, 0.83, 0.85
		40	0.12, 0.24, 0.33. 0.41, 0.45, 0.53, 0.60, 0.69, 0.74, 0.81, 0.85, 0.89
		50	0.12, 0.21, 0.32, 0.40, 0.46, 0.49, 0.55, 0.62, 0.68, 0.71, 0.79, 0.83
2.	0.004	20	0.14, 0.24, 0.37, 0.45, 0.52, 0.57, 0.65, 0.74, 0.84, 0.91, 0.95
		25	0.14, 0.25, 0.36, 0.47, 0.55, 0.58, 0.67, 0.73, 0.85, 0.93, 0.97
		30	0.12, 0.22, 0.34, 0.42, 0.59, 0.65, 0.71, 0.80, 0.87, 0.96
		35	0.16, 0.27, 0.33, 0.43, 0.52, 0.61, 0.63, 0.71, 0.79, 0.86, 0.93
		40	0.14, 0.23, 0.33, 0.41, 0.48, 0.54, 0.62, 0.70, 0.78, 0.84, 0.89, 0.92, 0.97
		50	0.11, 0.21, 0.30, 0.43, 0.50, 0.59, 0.66, 0.74, 0.84, 0.92, 0.95
3.	0.008	20	0.18, 0.21, 0.42, 0.52, 0.70, 0.79, 0.84, 0.98
		25	0.11, 0.26, 0.36, 0.46, 0.51, 0.60, 0.76, 0.82, 0.90, 0.97
		30	0.10, 0.23, 0.35, 0.43, 0.55, 0.66, 0.75, 0.83, 0.90, 0.96

		35	0.19, 0.28, 0.40, 0.46, 0.60, 0.71, 0.82, 0.84, 0.90, 0.97
		40	0.13, 0.22, 0.39, 0.4, 0.571, 0.61, 0.73, 0.80, 0.86, 0.94
		50	0.12, 0.21, 0.32, 0.40, 0.55, 0.65, 0.76, 0.81, 0.86, 0.93

For a flow pertaining to each *t/h* value, photographs and videos were recorded with the help of a tripod stand and mobile camera setup. A set square was attached to the flume, as shown in Figure 2, such that two real coordinate points (X1, Y1) and (X2, Y2) on it were manually measured (considering the bottom of the weir as origin), and the entire setup was scaled. The photographs and videos captured for each run corresponding to a *t/h* value were individually plot-digitized and analyzed using the web plot digitizer (automeris.io/WebPlotDigitizer). The bottom of the weir, when considered as the origin, resulted in the following coordinate values (shown in Table 2) at the positions marked in Figure 2. Once the uploaded image was calibrated with an axis as per the absolute coordinates, data points were marked and extracted to obtain the flow characteristics such as free surface profile, surface jump, surface wave properties, etc.

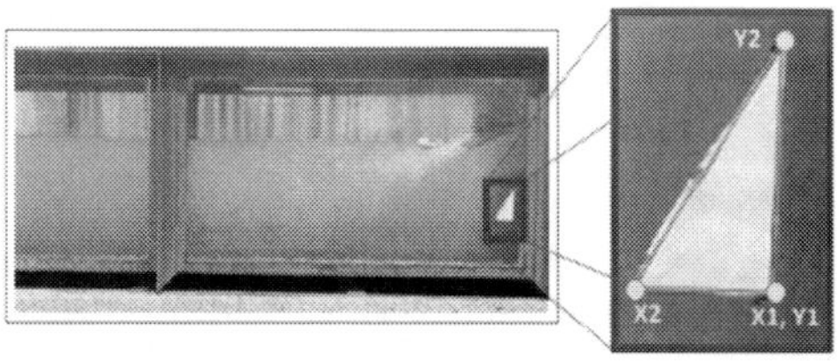

Figure 2. Setsquare attached to the flume for scaling the setup.

Table 2. Coordinates of the points used for scaling.

S. No	Coordinate	Value (in cm)
1.	X1	10
2.	X2	17
3.	Y1	20
4.	Y2	32

All the images for each *t/h* value were stored and uploaded for data extraction in the Automeris application. As the photographs include a predefined scale as per the actual distances, the axis was calibrated accordingly, as mentioned earlier. Datapoints were marked exactly on the free surface, as shown in Figure 3, and then the data was extracted to excel to plot the free surface profile separately.

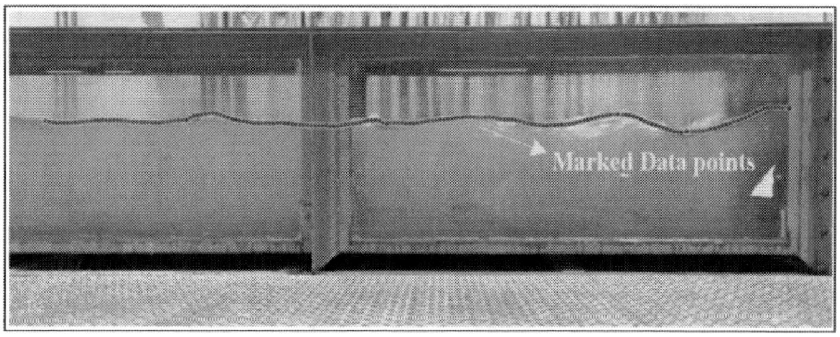

Figure 3. Data points marked on the water surface at Slope = 0, Q = 40 L/s, and t/h = 0.53.

3. Results and Discussion

Free surface profiles were extracted from the photographs for all the experimental runs. The photographs were taken once the flow conditions reached a steady state. Hence, the time-varying fluctuations of the free surface profiles are not considered for the analysis. Correlations between the length of the surface jump and the height of the surface jump were plotted against the *t/h* ratio. The wave properties, such as the amplitude of the wave, wavelength, and position of the first trough, were also extracted for the runs corresponding to a surface wave regime, and their variations with *t/h* ratio were observed. All the above-mentioned correlations are discussed below.

3.1. Free Surface Profiles

The interfaces between the air and water in an open channel flow are known as the free surfaces. They were examined to observe the changes in different regimes based on the existing flow conditions. Most of the characteristics, such as length and height of the surface jump, amplitude, and wavelength of the surface wave, can be directly extracted by analyzing a free surface profile. Although the past works (Azimi et al. 2016; Ahmed et al. 2019) have observed the free surface profiles of submerged flow regimes at the horizontal slope, the present study reports the free surface profiles at multiple slopes and discharge values. All the images of the experimental runs were digitized as per the procedure mentioned above. The obtained results are categorized slope-wise for all the different discharges and then represented below in Figure 4.

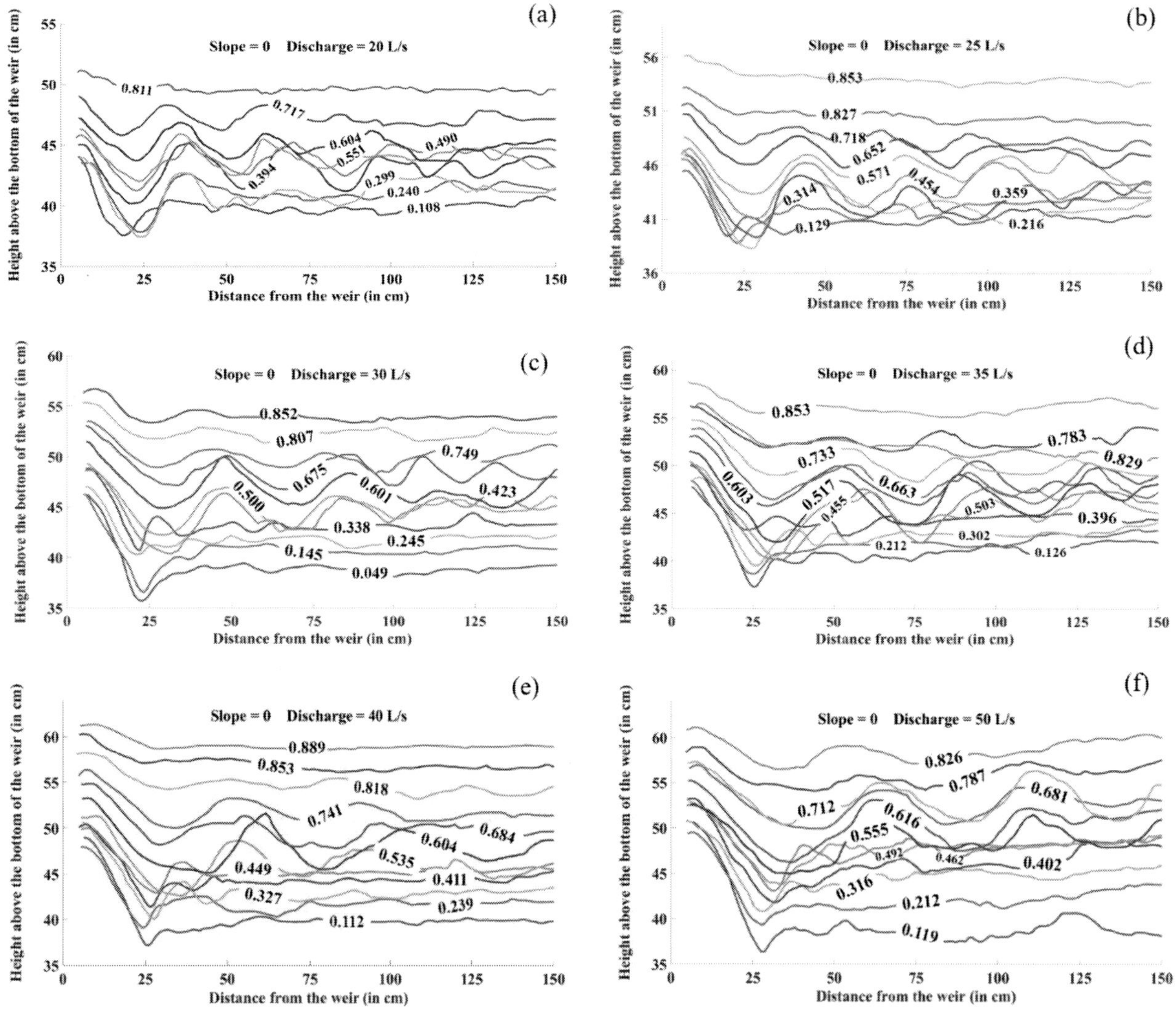

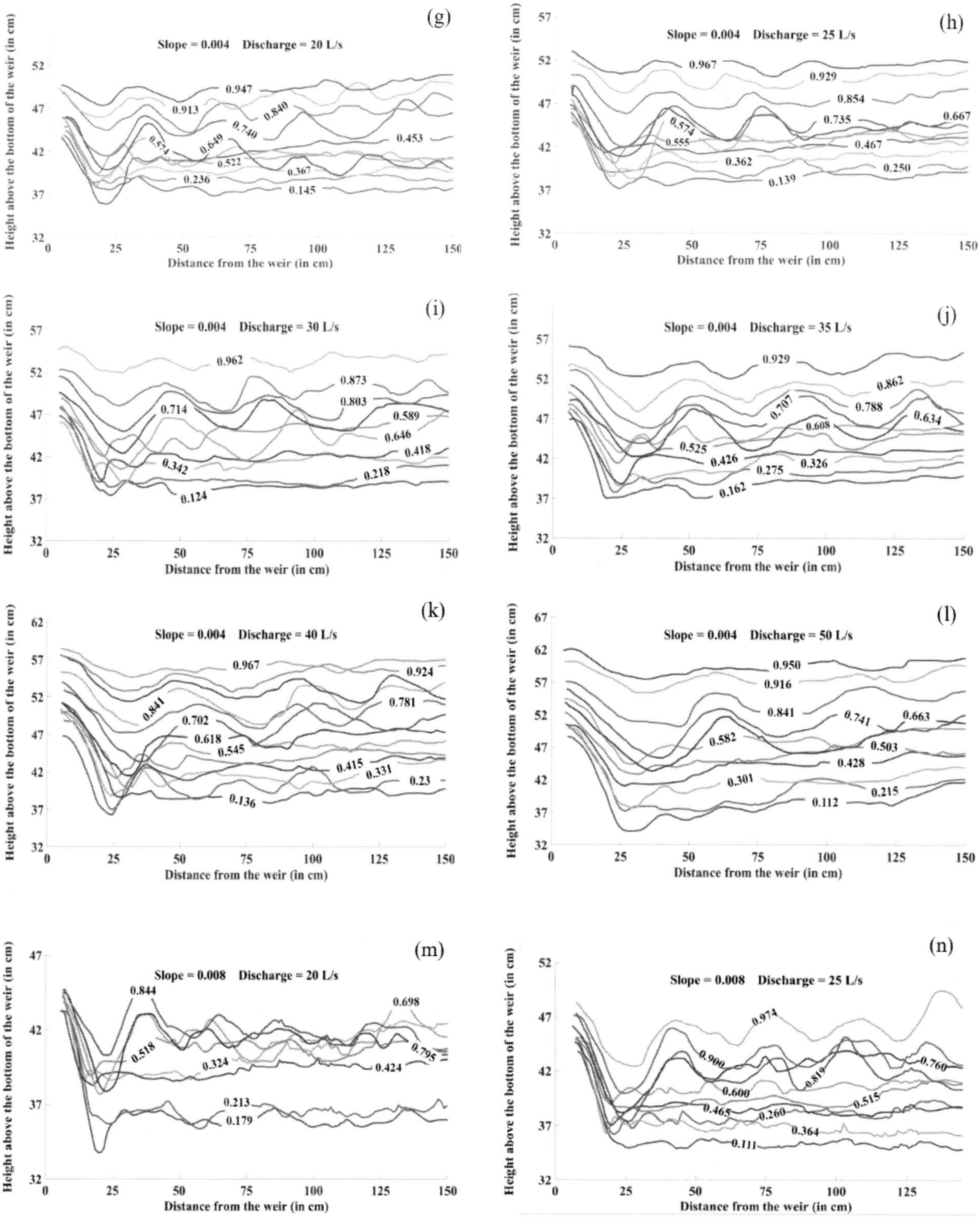
(g)
Slope = 0.004 Discharge = 20 L/s
0.947
0.913
0.840
0.740
0.574
0.649
0.453
0.522
0.367
0.236
0.145
Height above the bottom of the weir (in cm)
Distance from the weir (in cm)
(h)
Slope = 0.004 Discharge = 25 L/s
0.967
0.929
0.854
0.574
0.735
0.667
0.555
0.467
0.362
0.250
0.139
(i)
Slope = 0.004 Discharge = 30 L/s
0.962
0.873
0.803
0.714
0.589
0.646
0.418
0.342
0.218
0.124
(j)
Slope = 0.004 Discharge = 35 L/s
0.929
0.862
0.707
0.788
0.634
0.608
0.525
0.426
0.326
0.275
0.162
(k)
Slope = 0.004 Discharge = 40 L/s
0.967
0.924
0.841
0.781
0.702
0.618
0.545
0.415
0.331
0.23
0.136
(l)
Slope = 0.004 Discharge = 50 L/s
0.950
0.916
0.841
0.741
0.663
0.582
0.503
0.428
0.301
0.215
0.112
(m)
Slope = 0.008 Discharge = 20 L/s
0.844
0.698
0.518
0.795
0.324
0.424
0.213
0.179
(n)
Slope = 0.008 Discharge = 25 L/s
0.974
0.900
0.760
0.819
0.600
0.515
0.465
0.260
0.364
0.111

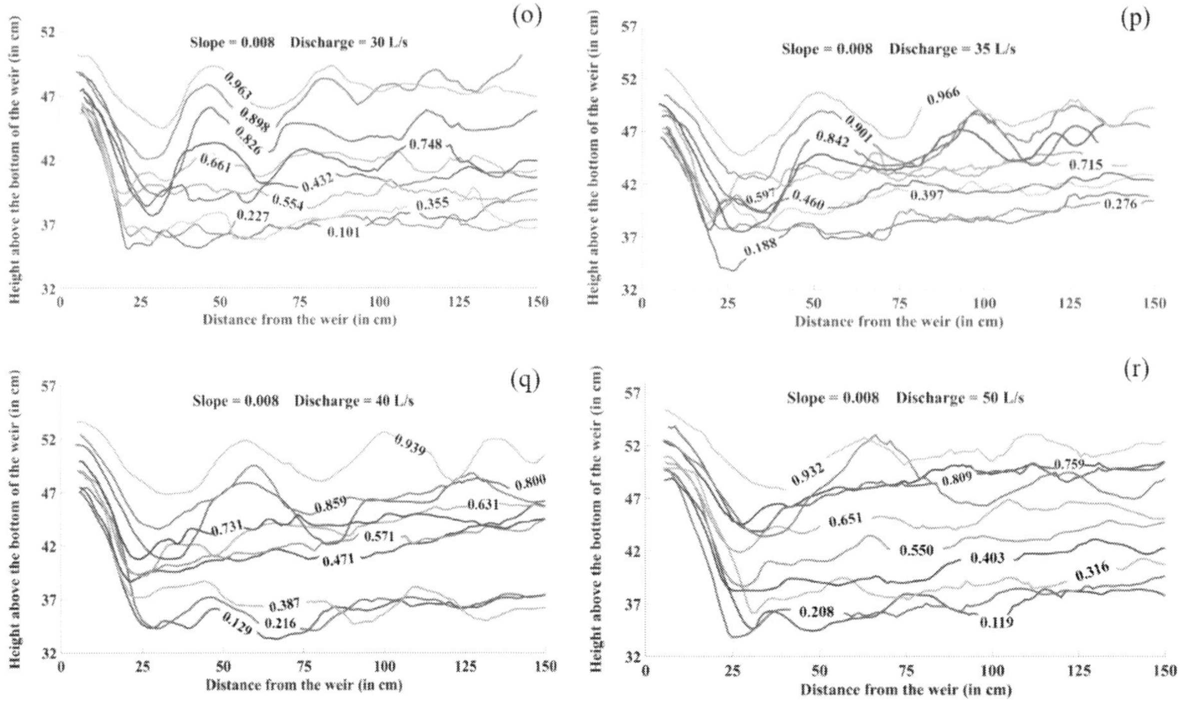

Figure 4. Free Surface Profiles (The *t/h* values are labelled on the corresponding profiles).

3.2. Occurrence of submerged flow regimes

The effect of variation in the slope has shown that the occurrence (in terms of *t/h*) of submerged flow regimes is not the same for any slope; instead changes from one slope to another. Hence, the behavior of the submerged flow regimes at multiple slopes in their corresponding zones of occurrence changes. Transitional zones were observed in terms of *t/h* ratio where the regime changes from one to another. Considering these transitional zones, Table 3 illustrates the submergence ratio ranges at which a particular flow regime gets established. Thus, quantitative criteria are set to identify which flow regime gets established at a particular *t/h* ratio. At higher slopes and discharges, it was observed that surface waves exist even when the *t/h* ratio is closer to 1. The obtained results showed that all the non-dimensional values of submerged flow for higher slopes lie within the range of values obtained corresponding to the horizontal slope. All the above-mentioned correlations are discussed below.

Table 3. Occurrence of submerged flow regimes.

Regime/Slope value	$S = 0$	$S = 0.004$	$S = 0.008$
Impinging Jet Regime	$t/h < 0.1$	$t/h < 0.3$	$t/h < 0.4$
Surface Jump Regime	$0.1 < t/h < 0.5$	$0.3 < t/h < 0.6$	$0.4 < t/h < 0.8$
Surface Wave Regime	$0.3 < t/h < 0.85$	$0.5 < t/h < 0.97$	$0.75 < t/h < 0.98$
Surface Jet Regime	$t/h > 0.85$	$t/h > 0.97$	$t/h > 0.98$

3.3. Surface Jump

As the tail-water level increases, the subcritical flow develops on the downstream side and meets a supercritical incoming jet on the upstream. This resulted in the formation of a surface jump with a strong recirculation zone of bubbles. The length (L_{rj}) and height (y_j) of a surface jump at all three slopes were observed and plotted to distinguish them. The length of the jump is defined as the distance between the point where the trough of the jump (taken as the section where the froth begins) to the section where the recirculation ends. The starting and end points of the surface jump were marked during the plot digitization process, and the process was repeated for all the experimental runs of the surface jump. The following excel graph was plotted between the non-dimensional length of the jump, i.e., L_{rj}/t and t/h, where t and h are the downstream depth and upstream elevation above the crest of the weir, respectively. The equations were fitted at best for each of the slopes and represented in polynomial equations. The variations in the non-dimensional length of the jump (L_{rj}/t) with t/h values at multiple slopes are shown in Figure 5a. The graph shows the non-dimensional length of the jump (L_{rj}/t) that varies from about 55 for t/h = 0.1 to 10 for t/h = 0.5 when the slope is horizontal. When the slope is 0.004, it varies from about 25 for t/h = 0.35 to 10 for t/h = 0.6. When the slope further increases to 0.008, it changes from about 25 for t/h = 0.4 to 10 for t/h = 0.8. The height of the jump is defined as the vertical distance between the highest point of the jump (considered the crest of the jump) and the crest of the weir. The crest positions of the surface jump were noted for all the experimental runs of the surface jump. The following graph was plotted between the non-dimensional height of the jump, i.e., y_j/h_0 and t/h ratio. Here, h_0 is the upstream head of the corresponding discharge at free flow conditions (without any submergence). The obtained polynomial equations for each of the three slopes are shown in Figure 5b. The graph shows the height of the jump (y_j/h_0) that increases from 0.1 for t/h = 0.1 to 0.55 for t/h = 0.5 when the slope is horizontal. When the slope is 0.004, it increases from 0.1 for t/h = 0.35 to 0.5 for t/h = 0.6. When the slope further increases to 0.008, it changes from a value closer to 0 for t/h = 0.4 to 0.65 for t/h = 0.8. As the tail water depth increases, the weakening of the recirculation zone is observed, due to which the jump length consistently drops, and the transition happens towards a surface wave. This weakening of the recirculation zone happens due to the modification in the velocity distribution with the increase in the tail water depth. This increasing trend in the height of the jump is caused mainly due to the reduction in velocity as the water level on the downstream side increases. The observations also reveal that, at higher slopes, surface jumps have increased lengths and decreased heights at particular upstream and downstream water depths (i.e., same t/h).

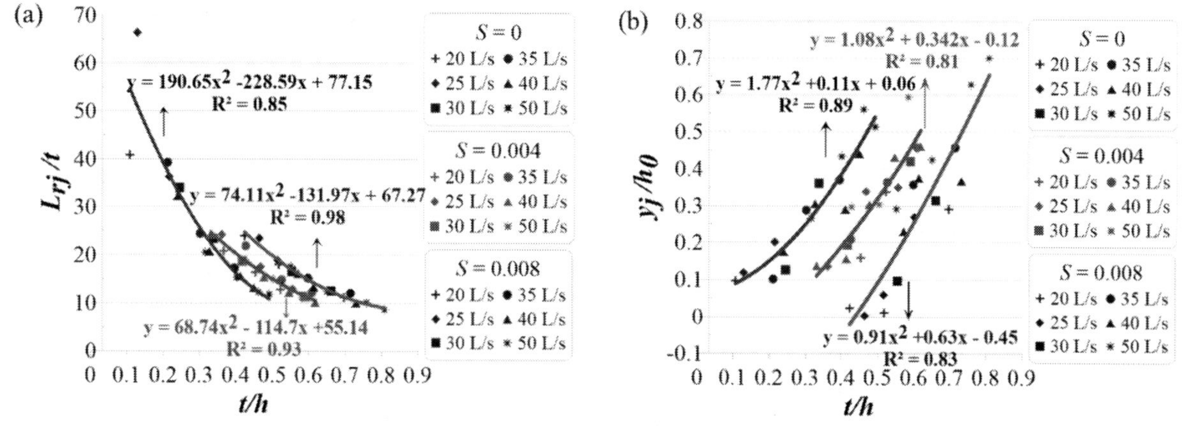

Figure 5. Variation of the surface jump characteristics at multiple slopes with t/h ratio. (a) non-dimensional length of the jump (L_{rj}/t); (b) non-dimensional height of the jump (y_j/h_0).

3.4. Surface Wave

When the tail-water rises beyond the margin of the surface jump, the entrainment gets reduced, and the free surface takes the form of continuous waves with a decreasing amplitude in the flow direction. The positions of the first troughs, amplitudes, and wavelengths of all the surface waves were examined, and the obtained results were used to develop the correlations at all three slopes. The horizontal and vertical distances of the minimum location of the surface wave from the crest of the weir were located by observing the position of the first trough. Both X and Y coordinates of the first trough were extracted during the plot digitization. The variation of these locations, i.e., non-dimensional horizontal position (x_w/h) and vertical position (y_w/h) from the top point of the weir were correlated against the t/h ratio. The obtained relations and graphs (in Figure 6) are shown below. The horizontal x_w/h and vertical distances y_w/h (non-dimensional) of the first trough show opposite trends to each other. x_w/h drops from 2.4 to 1.2 when S = 0, 2.2 to 1.4 when S = 0.004 and 2.25 to 1.75 when S = 0.008. Whereas y_w/h increases from 0.3 to 0.8 when S = 0, 0 to 0.7 when S = 0.004 and 0.1 to 0.6 when S = 0.008. The amplitude (a) is defined as the vertical distance between the first trough and the first crest of a surface wave. As the elevations of the first crest and trough are marked and noted, the differences in the depths between them provide the amplitude (a). The variation of the normalized amplitude (a/h) with the submergence ratio t/h was plotted and represented below in Figure 6c. The obtained results showed that the values of a/h decreased from 0.6 at all the slopes and dropped to values closer to 0.1. The wavelength (b) is defined as the horizontal distance between the first two successive crests of the surface wave. The longitudinal distance (or the difference) between the peak positions of the first and second crest was taken as wavelength (b). The wavelengths of all the experimental runs were measured in this manner. The normalized wavelengths, i.e., b/h, were plotted against the submergence ratio t/h, and correlations were developed, as shown below in Figure 6d. The obtained results showed that the values of b/h decrease from 3.5 to 1.5 at multiple slopes. With higher submergence, the trough position has changed up towards the weir due to the increase in the sub-critical nature that raises the upstream head slightly. The decreasing trends of wavelength and amplitude are due to the fading out of waves progressing towards a transition to a surface jet. At particular upstream and downstream water depths (same t/h), surface waves have increased amplitudes and wavelengths at higher slopes. For higher slopes, as the surface wave regimes occur even at submergence ratios closer to 1, the amplitude changes may happen beyond the currently obtained values.

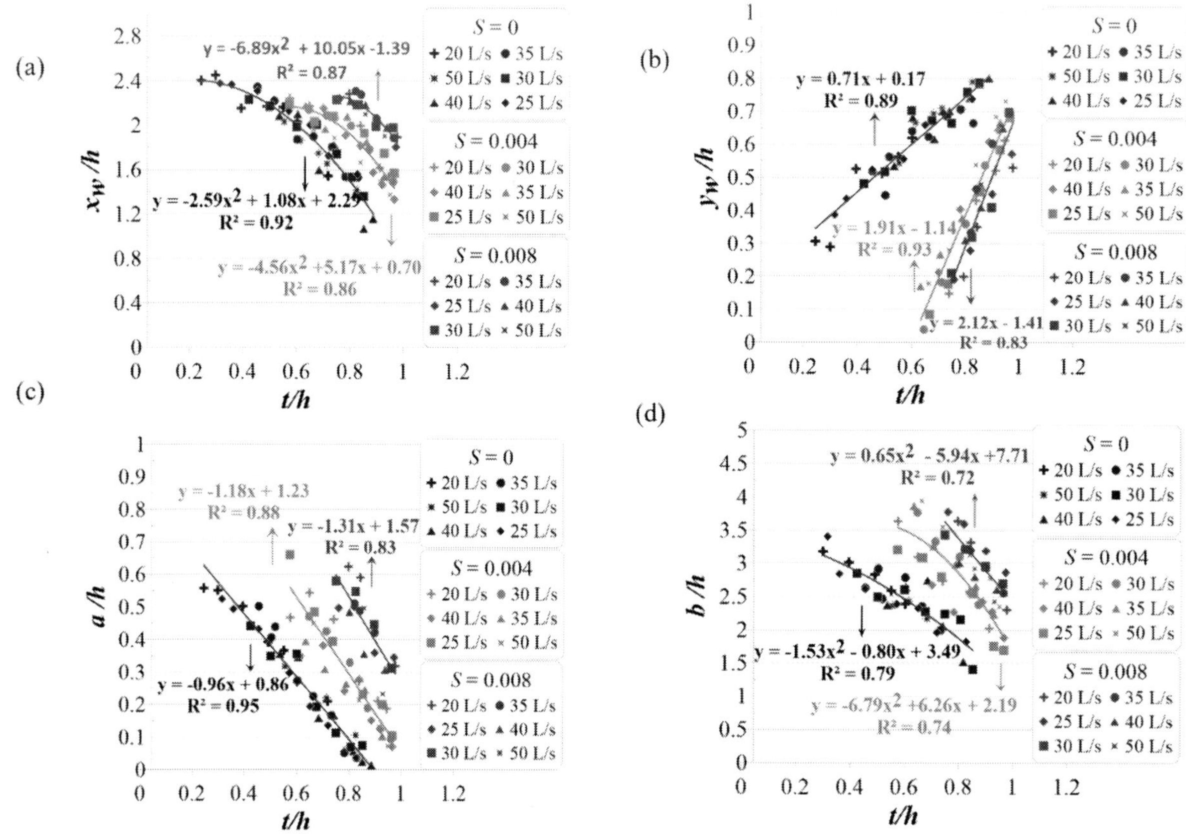

Figure 6. Variation of the surface wave characteristics at multiple slopes with t/h ratio. (a) non-dimensional horizontal position (x_w/h) of first trough; (b) non-dimensional vertical position (y_w/h) of first trough; (c) non-dimensional amplitude (a/h); (d) non-dimensional wavelength (b/h).

4. Conclusions

The experimental observations of the submerged flow regimes at multiple slopes have exhibited occurrence patterns in terms of the submergence ratio. Various significant characteristics of the surface jumps and surface waves help in the modifications of the design of hydraulic structures, particularly downstream of a sharp-crested weir. The occurrence of submerged flow regimes changes with a change in the slope. The transitional zones (in terms of t/h) have increased about 0.1 to 0.2 for each successive slope increment. The non-dimensional lengths of the surface jumps (decreased with an increase in the submergence) at all three slopes together lie within a range of 55 to 10. Similarly, the non-dimensional heights of all the surface jumps (increased with an increase in the submergence) lie within a range of 0 to 0.7. The surface wave with a maximum normalized amplitude of about 0.6 and a maximum normalized wavelength of about 4 started fading out gradually as the regime ended. The experimental observations at higher slopes have shown that surface waves exist even when the submergence ratio is closer to 1.

5. Limitations and Future Scope

A study of submerged flow regimes beyond a submergence ratio value of 1 (i.e., $t/h > 1$) would be interesting to investigate further. Also, the work can be extended by developing standard equations for the submerged flow characteristics using independent variable parameters (such as channel geometry). The three-dimensional aspects of the submerged phenomenon can be explored along with the simulations using CFD (Computational Fluid Dynamics) tools. Accurate simulations that represent the complex flow physics of submergence at multiple slopes can reduce the time and cost of the analysis. Observations can be further generalized by considering different channel geometry, moving a step closer to the real existing problems.

6. ACKNOWLEDGMENTS

All the experiments were performed in the Water Resources Engineering Laboratory of the Indian Institute of Technology Bhubaneswar, India. The authors are thankful for the technical assistance provided by the laboratory staff and security.

7. REFERENCES

Ahmed, S., Ye, Y., Liu, H., and Rajaratnam, N. (2019). "Preliminary Study of Surface Hydraulic Jumps." *Journal of Irrigation and Drainage Engineering*, 145 (3): 06019001. American Society of Civil Engineers (ASCE). https://doi.org/10.1061/(ASCE)IR.1943-4774.0001379.

Azimi, A.H., Rajaratnam, N., Zhu, D.Z. (2016). "Water Surface Characteristics of Submerged Rectangular Sharp-Crested Weirs." *Journal of Hydraulic Engineering*, 142 (5): 06016001. American Society of Civil Engineers. https://doi.org/10.1061/(ASCE)HY.1943-7900.0001110.

Azimi, A. H., Rajaratnam, N. and Zhu. D. Z. (2014). "Submerged Flows over Rectangular Weirs of Finite Crest Length." *Journal of Irrigation and Drainage Engineering*, 140 (5): 06014001. American Society of Civil Engineers. https://doi.org/10.1061/(ASCE)IR.1943-4774.0000728.

Bijankhan, M., and Ferro, V. (2020). "Discussion of 'Preliminary Study of Surface Hydraulic Jumps' by S. Ahmed, Y. Ye, H. Liu, and N. Rajaratnam." *Journal of Irrigation and Drainage Engineering*, 146 (4): 07020001. American Society of Civil Engineers. https://doi.org/10.1061/(ASCE)IR.1943-4774.0001452.

James, C.S. (2020). "Flow-Measuring Structures. In: Hydraulic Structures". Springer, Cham. https://doi.org/10.1007/978-3-030-34086-5_7.

Rajaratnam, N. (1967). Hydraulic Jumps, Advances in Hydroscience, 197–280. doi:10.1016/b978-1-4831-9935-1.50011-2

Rajaratnam, N., and Muralidhar, D. (1969). "Flow below deeply submerged rectangular weirs." *Journal of Hydraulic Research*, 7 (3): 355–374. https://doi.org/10.1080/00221686909500273.

Skogerboe, G., M. Hyatt, and L. Austin. (1967). "Design and Calibration of Submerged Open Channel Flow Measurement Structures: Part 4 - Weirs." Reports.

Venard, J. K. and Weston, R.F. (1943). "Submergence effect on sharp-crested weir." Engineering News Record. pp. 118-120.

Wu, S., and Rajaratnam, N. (1995). "Free jumps, submerged jumps and wall jets." *Journal of Hydraulic Research*, 33 (2): 197–212. https://doi.org/10.1080/00221689509498670.

Wu, S., and Rajaratnam, N. (1996). "Submerged Flow Regimes of Rectangular Sharp-Crested Weirs." *Journal of Hydraulic Engineering*, 122 (7): 412–414. American Society of Civil Engineers. https://doi.org/10.1061/(ASCE)0733-9429(1996)122:7(412).

Wu, S., and Rajaratnam, N. (1998). "Impinging Jet and Surface Flow Regimes at Drop." *Journal of Hydraulic Research*, 36(1): 69–74. https://doi.org/10.1080/00221689809498378.

DOI: https://doi.org/10.26077/11eb-458d

Surface Velocities and Free-surface Aeration in a Converging Smooth Chute during a Major Flood Event

H. Chanson[1] & C.J. Apelt[1]
[1]*The University of Queensland, School of Civil Engineering, Brisbane, Australia*
E-mail: h.chanson@uq.edu.au

Abstract: *In a free-surface spillway, the upstream flow is non-aerated and the flow becomes a strong air-water mix downstream of the onset location of air entrapment. Field observations were conducted over a large dam's spillway during a major flood event. The wastewaterway system was a smooth converging chute with a longitudinal slope of 11.3°: the Chinchilla Minimum Energy Loss weir on the Condamine River (Australia). Detailed quantitative measurements were undertaken in the high-speed chute flows with strong turbulence and high Reynolds numbers at the early part and later part of a major flood event. During this exceptional flood, the spillway passed nearly three times the design discharge capacity at the peak for the event. Down the smooth chute, the observations indicated that the onset of free-surface aeration was a complicated transient three-dimensional process. A robust optical flow (OF) technique was applied and delivered physically-meaningful surface velocities in the air-water flow region. The streamwise surface velocities were reasonably close to backwater calculations. The data presented large streamwise surface velocity fluctuations in the aerated flow region, with* Tu_s *~ 150-200%, on the centreline consistent with self-aerated flow measurements using dual-tip phase detection probe in laboratory. Overall, the study demonstrated the application of optical techniques to prototype smooth spillway flows, as well as some intrinsic difficulties with field investigations.*

Keywords: *Prototype velocity measurements; Smooth converging chute; Self-aeration; Optical techniques; Minimum Energy Loss weir.*

1. Introduction

In most overflow spillways, the upstream flow is non-aerated and a strong air-water mix develops downstream of the onset region of air entrapment (Halbronn et al. 1953, Wood et al. 1983). At the upstream end of an un-controlled spillway, the flow is rapidly accelerated as it changes from a critical motion at the weir crest to a supercritical motion on the chute in a relatively smooth manner (Halbronn 1952, Keller and Rastogi 1975). A turbulent boundary layer develops along the invert from the upstream end (Ippen et al. 1955, Chanson 1997a). On a steep spillway chute, the onset of free-surface aeration is typically observed when the outer edge of the boundary layer starts to interact with the free-surface (Lane 1936, Rao and Rajaratnam 1961). These interactions can be explosive with strong air-water ejections (Chanson 2013, Zabaleta and Bombardelli 2020). On flat chutes, in contrast, the inception region presents a rippled surface with small free-surface instabilities tending to become a choppy water surface once aeration takes place (Michels and Lovely 1953, Chanson 1997b). It was argued that the onset of aeration on flat spillways results from a combination of longitudinal vortices' breakdown and of gravity waves produced by roughness and irregularities at the boundaries (Levi 1965,1967, Anwar 1994). Yet, the two theories are not exclusive as discussed by Toro et al. (2017) and Chanson (2021), based upon three-dimensional CFD computations and detailed prototype observations respectively. The onset processes of free-surface aeration on spillway chutes are illustrated in Figure 1 by high-shutter speed photographs of the inception region in steep (Fig. 1A) and flat (Fig. 1B) prototype chutes.

During the last five decades, a number of overflow embankment designs were developed including earth dam spillway with precast concrete blocks (Pravdivets and Bramley 1989, Chanson 2001), concrete protection of the downstream embankment slope (McLean and Hansen 1993, FEMA 2014) and Minimum Energy Loss weir (McKay 1971). The Minimum Energy Loss (MEL) weir design was developed specifically for the river catchments affected by heavy tropical and sub-tropical rainfalls with very flat gradients, i.e. S_o ~ 0.1% (McKay 1971, Turnbull and McKay 1974).

In the current contribution, a series of field observations were conducted in a large protype MEL weir during a major flood event in late 2021. The data were analysed, and the results are discussed herein. Detailed quantitative measurements were performed in terms of the surface velocity field in the high-speed, high-Reynolds-number overflow. Some key features of the inception region of free-surface aeration were described, showing the onset of air entrapment being a complicated process, while the air-water flow was characterised by large streamwise surface velocity fluctuations.

Figure 1. High-shutter speed photographs of the inception region of free-surface aeration on prototype chute spillways. (A) Explosive burst at and air-water ejections above the water surface at Hinze dam (Australia) on 27 March 2021 ($\theta = 37.5°$ [at inception], $Re = 2.34 \times 10^7$, Shutter speed: 1/8,000 s). (B) Rippled/choked water surface at the water surface of the inception region at Chinchilla MEL weir (Australia) on 27 November 2021 ($\theta = 11.3°$, $Re = 2.3 \times 10^6$ [at inception], Shutter speed: 1/8,000 s).

2. Study site and Methodology

2.1. Study site

The Chinchilla MEL weir (QLD, Australia) is located on the Condamine River, on the northern part of the Murray-Darling basin. The weir was completed in 1973 to provide irrigation water, and it is listed as a "large dam" by ICOLD (1984). The catchment area is 19,192 km^2. The structure is a 14 m high earthfill embankment with a 410 m long dam crest (incl. abutments) and equipped with an overflow spillway (Figs. 2A & 2B) (Turnbull and McKay 1974). The design spillway capacity is 850 m^3/s corresponding to bank full. The spillway consists of a broad crest, 3.05 m long, and 213.4 m wide, followed by a 60.6 m long smooth converging chute with a 11.3° slope (1V:5H) and no dissipation structure was built. The downstream width of the chute is 81.2 m and the chute convergence is $\partial B/\partial x = -2.18$. The overflow section is concrete-lined, with a series of drains to reduce seepage pressure. The Chinchilla weir was designed to pass the design discharge ($Q_{des} = 850$ m^3/s, bank full) with an afflux of less than 1.83 m. Between 1973 and 2022, the weir was overtopped by a number of large flood events, including a number of overflows larger than the design flow. The weir operated safely and properly, and inspections after the flood showed no damage.

Figure 2. Chinchilla Minimum Energy Loss (MEL) weir (QLD, Australia). (A) Overflow on 27 November 2021 (Q = 121 m^3/s); (B) Hydraulic jump at the chute toe on 27 November 2021 (Q = 121 m^3/s, Shutter speed: 1/2,500 s); (C) Camera tripod setup on the lower right bank; (D) High-water marks on the right bank on 15 December 2021

2.2. Instrumentation

The water discharge was deduced from the measured reservoir elevation, using the predicted discharge coefficient, based upon the drawings as constructed (Irrigation and Water Supply Commission 1972).

Visual, photographic and cinematographic records were undertaken from the right bank of the spillway and from the right river bank downstream of the spillway toe. Figure 2 present hand-held photographs taken from several locations. The observations were conducted using three dSLR Pentax™ cameras with sensor resolutions between 12 Mpx and 24 Mpx, a digital camera Sony™ RC100VA and an iPhone XI. The dSLR cameras were equipped with full-frame prime lenses producing photographs and movies with negligible barrel distortion. The dSLR camera movies were recorded in high definition (1920×1080 px) at 30 fps and 60 fps.

The analyses of surface features were conducted using both high-shutter speed photographs and video movies. For all the cases, the tracking and measurements of the water surface features were conducted manually to guarantee the best quality control, owing to the complexity of the prototype flow's turbulent motion, characterised by very rapid and unpredictable changes with time and space. Further, a number of movies were taken from downstream and analysed using an optical flow (OF) technique, with the camera fixed on a sturdy professional-grade tripod (Fig. 2C). The Optical Flow (OF) is a set of tools, detecting the flow motion between consecutive frames based upon brightness intensity gradients (Bung and Valero 2017, Zhang and Chanson 2018). In the current study, Farnebäck's (2003) OF technique was applied to calculate the surface velocity field. The Farneback OF technique was applied with OF parameters previously validated in laboratory for surface velocity field (Arosquipa Nina et al. 2022) and used for the Hinze Dam prototype spillway data sets (Chanson 2022). It is acknowledged that the OF calculations were not truly 'calibrated' owing to the significant number of intrinsic difficulties with field observations during flood events, including un-controlled optical conditions (Chanson 2021). Since the camera field of view was perpendicular to the spillway chute, the raw data included the vertical and horizontal transverse surface velocity components recorded at various vertical elevations. The streamwise surface velocity was deduced from the vertical surface velocity component and invert slope based upon geometric considerations, assuming a two-dimensional flow.

2.3. Flow conditions

From mid-November 2021, some solid rain fell in the Condamine River catchment area upstream of Chinchilla. Some widespread major flooding was reported in the upstream catchment. The Chinchilla weir overflowed for more than one month, with the peak headwater observed on 5 December 2021 morning. Observations of overflow discharges are presented in Figure 3. For seven days, the spillway discharged a flow rate greater than the design discharge, with a peak flow about 1,930 m^3/s corresponding to 2.3 times the design discharge. Note that the inspection on 15 December 2021 confirmed that the abutments were overtopped by nearly 3 m of water at the peak of the flood, without damage (Fig. 2D)

The Chinchilla weir overflow was documented on 27 November 2021 and 15 December 2021 (Table 1). The first data set was obtained at the start of the flood event, while the second took place after the peak of the flood (Fig. 3). The flow conditions at the time of observations are detailed in Table 1. Note that the weir overflow could not be witnessed at the peak of the flood, because the access road become submerged when the tailwater elevation exceeds about 290-291 m AHD. The afflux, difference between recorded HWRL and TWRL, was 3.25 m at the peak of the flood event.

Table 1. Overflow observations at Chinchilla Minimum Energy Loss weir (QLD, Australia) during the major flood with its peak on 5 December 2021

Date	Time	Headwater (m AHD)	Tailwater (m AHD)	Q (m^3/s)	Re (at spillway crest)
27 November 2021	12:45-13:20	296.23	287.20	121	2.3×10^6
15 December 2021	12:30-13:30	296.29	287.52	144	2.7×10^6

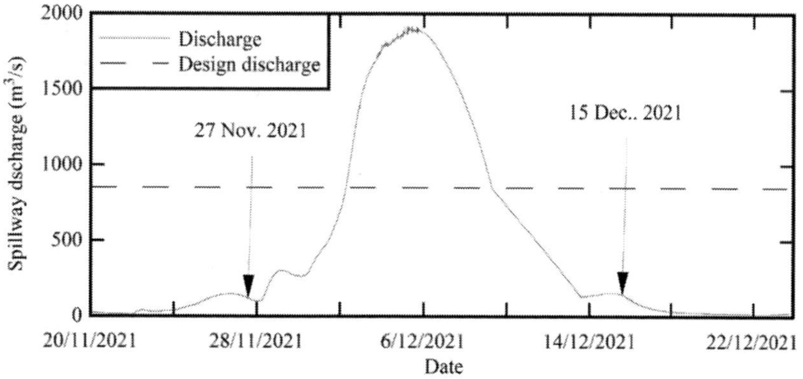

Figure 3. Spillway discharge at Chinchilla Minimum Energy Loss weir (QLD, Australia) during the major flood with its peak on 5 December 2021. Overflow discharge over the spillway chute, excluding the abutment overflows.

3. Basic observations

On both occasions (Table 1), the approach flow was very smooth, and this was confirmed by aerial footages on 21 December 2018 and 3 December 2021. The reservoir flow converged smoothly towards the weir crest and the water surface was waveless and almost still. On both 27 November and 15 December 2021, the flow was critical at the spillway crest which acted as a broad-crested weir, with a ratio of crest length to upstream head over crest of 6.2 and 5.54 respectively. The smooth change on water surface elevation between the upstream reservoir and spillway crest was nicely documented with photographs (Fig. 4A). Immediately downstream of the broad-crest, the water surface was smooth and glassy (Figs. 2A, 2B & 4A). On 27 November 2021, a few logs were trapped at the crest, and their wake was clearly seen at the surface of the non-aerated flow. Further downstream, the wake progressively disappeared in the self-aerated air-water flow region (Fig. 4B).

As the flow accelerated down the smooth converging chute, the glassy free-surface become rough and choppy, before becoming self-aerated, i.e. the inception region. The inception region showed a progressive transition of the surface roughness, from a smooth glassy surface, to a rough, coarse-sand-paper-like appearance and later to a very-rough choppy surface (Fig. 4). The self-aerated region presented a beige colour, evidence of a three-phase air-water sediment motion. Such an observation was well-documented visually on both 27 November and 15 December 2021, as well as during an earlier overflow on 8 November 1997 (Toombes and Chanson 2007).

The location x_I of the inception region of self-aeration was recorded during both overflows, and the data are reported in Table 2 (column 5). The corresponding water depth d_I was deduced for the backwater calculations (Table 2, column 6). The observations are compared to the correlations of Cain and Wood (1981) for a smooth prismatic channel, assuming k_s = 1 mm as for the backwater calculations (see below). The comparative results (Table 2) showed a close agreement in terms of the water depth at inception, while the observed location data presented more scatter, possibly caused by the wide spread of the inception region, while the flow convergence was expected to induce a slower growth in bottom boundary layer compared to prismatic chutes.

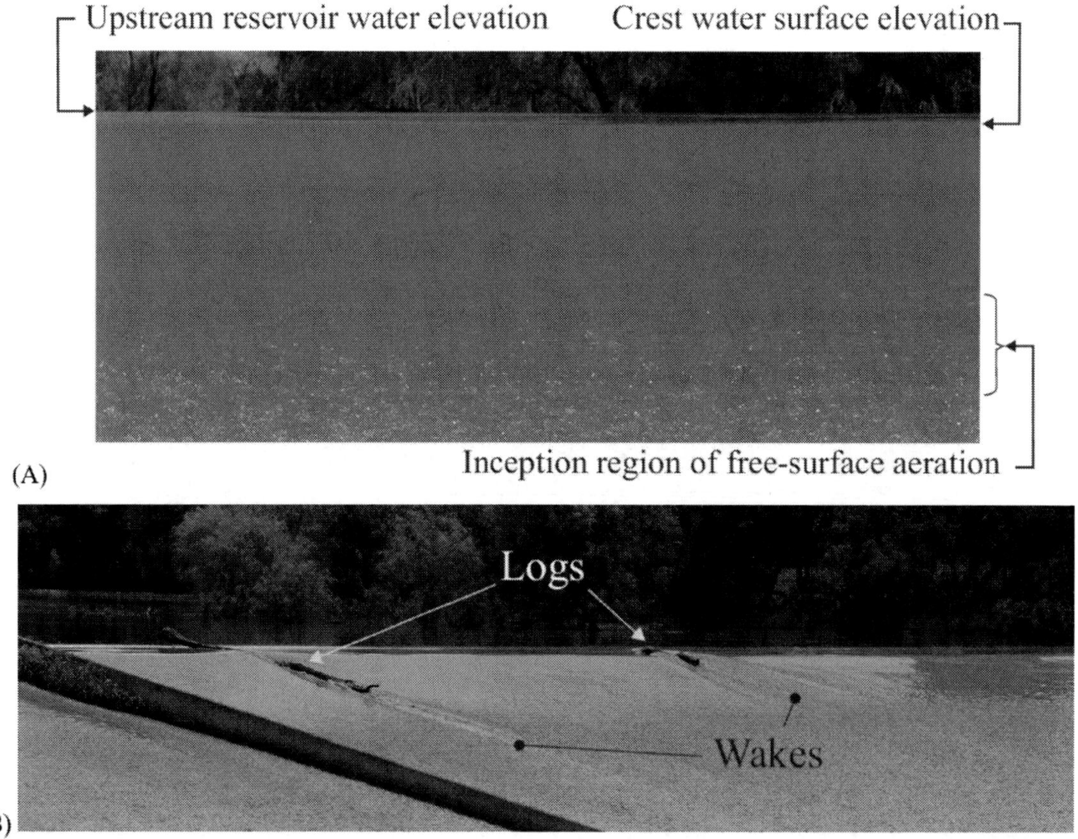

Figure 4. Photographic observations of the spillway operation at Chinchilla Minimum Energy Loss weir (QLD, Australia) during the major flood with its peak on 5 December 2021. (A) Upstream flow viewed from downstream on 15 December 2021; (B) Timber logs blocked at the spillway crest on 27 November 2021.

Table 2. Location and water depth at the inception of free-surface aeration at Chinchilla Minimum Energy Loss weir (QLD, Australia) during the major flood with its peak on 5 December 2021

Date	Time	Q (m^3/s)	Head above crest (m)	x_I ([1]) (m)	d_I ([2]) (m)	x_I predicted ([3]) (m)	d_I predicted ([3]) (m)
27 November 2021	12:45-13:20	121	0.49	7.5	0.11	10.2	0.10
15 December 2021	12:30-13:30	144	0.55	13.4	0.12	11.6	0.12

Notes: ([1]) statistical median of photographic and movie data; ([2]) water depth estimate deduced from present backwater calculation; ([3]) Correlations of Cain and Wood (1981).

At the toe of the spillway chute, the high-velocity self-aerated flow impinged into the tailwater and a hydraulic jump formed (Fig. 2C). The jump was stationary and very stable, located on the underwater sloping chute on both days. The position of the jump was controlled by the tailwater conditions and the very-flat Condamine River channel bed slope downstream of Chinchilla. The jump roller was highly turbulent and fluctuating about its mean position. Both visual and video movie observations indicated strong three-dimensional large-scale vortices in the roller. The interactions of these large turbulent structures with the roller free-surface created large scars and surface features, with length-scales comparable to the roller height.

4. Surface velocity measurements

Based upon the video movies taken from downstream, the Optical Flow (OF) data encompassed the surface velocity, measured in the plane parallel to the invert. Both the longitudinal velocity component V_s and the transverse surface velocity component V_t were extracted on the left chute centreline. In the current study, the data provided meaningless results in the non-aerated flow region. This was most likely caused by the shiny surface glare, seen in Figures 2 and 4, upstream of the inception region. Meaningful OF results were achieved in the self-aerated flow region (see below). The finding differed from the OF observations at the Hinze Dam spillway (Chanson 2021,2022).

For the left chute, the centreline OF surface velocities were compared to the ideal fluid flow theory and to the backwater equation. On an uncontrolled chute spillway, the flow is accelerated by the gravity force component in the flow direction, and the ideal velocity V_{max} at a sampling location on the chute is derived from the Bernoulli principle:

$$V_{max} = \sqrt{2 \times g \times (H_1 - d \times \cos\theta)} \qquad (1)$$

where H_1 is the upstream total head above the sampling point, θ is the channel slope and d is the water depth (Chanson 2004). For a non-uniform and steady real fluid flow, the backwater calculations are based upon the differential form of the energy equation and performed assuming that the flow is gradually varied. For a rectangular channel of non-constant width and slope, the backwater equation yields (Chanson 2004, pp. 377-381):

$$\frac{\partial d}{\partial x} \times (\cos\theta - Fr^2) = -\frac{\partial z_o}{\partial x} - S_f + d \times \sin\theta \times \frac{\partial \theta}{\partial x} + Fr^2 \times \frac{d}{B} \times \frac{\partial B}{\partial x} \qquad (2)$$

$$Fr = \frac{Q}{\sqrt{g \times d^3} \times B} \qquad (3)$$

With d the water depth, S_f the friction slope, θ the invert slope and B the free-surface width. At Chinchilla weir, the channel is rectangular, with constant slope and a reducing width. Thus, the backwater equation becomes:

$$\frac{\partial d}{\partial x} \times (\cos\theta - Fr^2) = -\frac{\partial z_o}{\partial x} - S_f + Fr^2 \times \frac{d}{B} \times \frac{\partial B}{\partial x} \qquad (4)$$

with $\cos\theta = 0.9806$, $\sin\theta = 0.196$, and $\partial B/\partial x = -2.1805$ at the Chinchilla MEL weir.

The results showed a relatively close agreement between the streamwise OF surface velocity and the backwater-calculated velocity downstream of the onset region, both qualitatively and quantitatively (Fig. 5). Figure 5 presents a typical comparison in terms of streamwise surface velocity component V_s, with data for one movie. The ideal fluid flow estimates are shown with a solid green curve and the backwater calculations are the dashed red curve. In Figure 5B, the streamwise surface velocity standard deviation v_s' and transverse surface velocity standard deviation v_t' are added for completeness. In Figure 5, the data were collected on the centreline of the field of view and the vertical axis is the vertical elevation in m AHD.

First, the OF streamwise surface velocity increased with decreasing vertical elevation (Fig. 5). Second the OF data showed some poor outcome in the non-aerated developing flow region. For $z_o > 294$ m AHD, the poor OF results were caused by the surface glare seen in Figures 2A and 4. Third, the surface velocity data agreed relatively well with the backwater calculations developed for the converging chute (see above). For $z_o < 288$ m AHD, a lack of match with backwater estimate might be caused by optical artifacts above the hydraulic jump roller, caused by spray and splashing seen in Figure 2B. Fourth, the OF data output quality was closely linked to the quality of the original movies. This encompassed the camera position ideally placed perpendicularly to the spillway chute, and the camera body and lens equipment, with all the present data set acquired with professional-grade full-frame prime lenses, following earlier studies (Arosquipa Nina et al. 2022, Chanson 2021,2022). Furthermore, the movie definition delivered improved outputs with higher spatial resolutions, i.e. less millimetres per pixel. It is worth to mention that both field data sets were filmed under good weather conditions, hence decent lighting conditions.

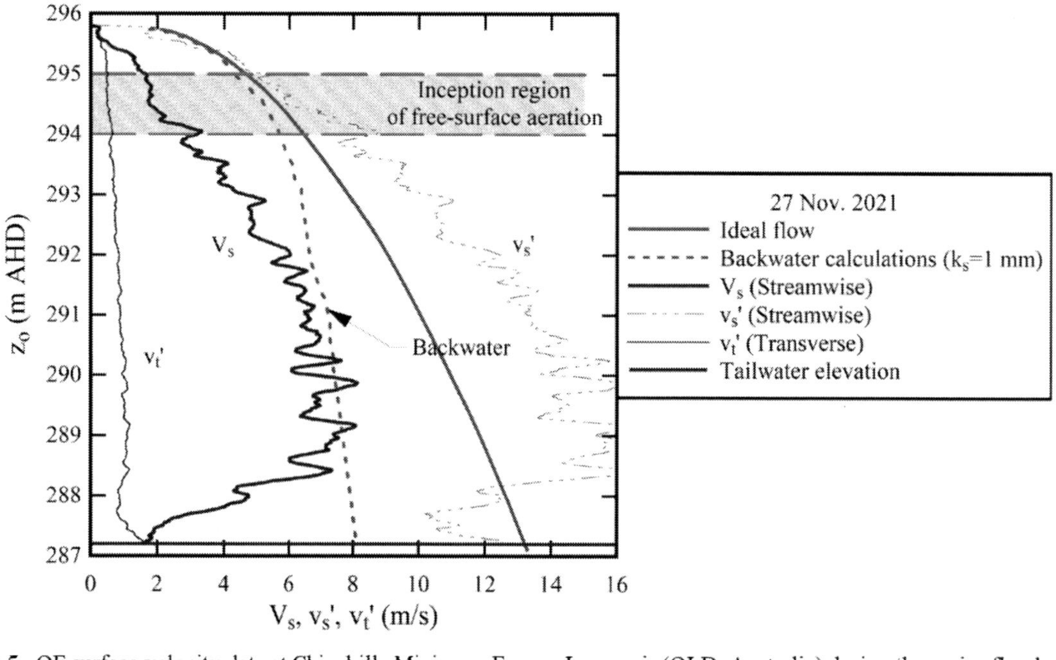

Figure 5. OF surface velocity data at Chinchilla Minimum Energy Loss weir (QLD, Australia) during the major flood with its peak on 5 December 2021. Comparison with ideal fluid flow velocity and backwater calculations. Analyses of image centreline. Streamwise surface velocity and standard deviations of streamwise and transverse velocity components on 27 November 2021, Camera: Pentax K-3 with Pentax FA*300mm f4.5 lens & HD TC 1.4x, HD movie (1,920×1080 px), 30 fps, Number of analysed frames:840, Spatial resolution: 5.15mm/px, Reference 1136.

The OF surface velocity characteristics were analysed in the self-aerated flow region, in terms of the standard deviations of the streamwise and transverse surface velocity. The weighted averages are presented in Table 3, together with the number of analysed frames, the frame rate and the spatial resolutions. The large number of analysed frames must be noted in the current study (Table 3, 4th column). The streamwise velocity data presented large surface turbulent intensity $Tu_s = v_s'/V_s \sim 1.7\text{-}2$ (Table 3). Such large values were likely caused by a combination of relatively large velocity fluctuations initiated by surface wave motion and of true turbulence. The data were quantitatively of the same magnitude as the field measurements in the Hinze Dam spillway (Chanson 2022) and as laboratory self-aerated flow data (Chanson and Toombes 2002, Arosquipa Nina et al. 2022). In the other hand, the transverse velocity data exhibited much lower transverse surface turbulent intensity $Tu_t = v_t'/V_s \sim 0.11\text{-}0.17$, likely induced by the smooth convergence of the flow. The ratio v_t'/ v_s' was between 0.06 and 0.1 (Table 3), implying a strong anisotropy of the surface turbulence. The observations were close to reported values of v_t'/v_s' in rough turbulent boundary layers (Schlichting and Gersten 2000) and on the Hinze Dam prototype spillway (Chanson 2022).

Table 3. OF surface turbulence characteristics of self-aerated flow at Chinchilla Minimum Energy Loss weir (QLD, Australia) during the major flood with its peak on 5 December 2021

Date	Q (m³/s)	Frame rate (fps)	Total Nb of analysed frames	Spatial resolution (mm/px)	v_s'/V_s (¹) (²) (³)	v_t'/V_s (¹) (²) (³)	v_t'/v_s' (¹) (²) (³)
27 November 2021	121	30 & 60	6,660	5.15 & 7.2	1.71	0.111	0.064
15 December 2021	144	30 & 60	92,310	5.8 & 8.3	2.09	0.165	0.096

Notes: V_s: time-averaged streamwise surface velocity; v_s': standard deviation of streamwise surface velocity; v_t': standard deviation of transverse surface velocity; (¹) weighted average for all analysed movies; (²) analyses of image centreline; (3) observations for 288 < z_o < 293 m AHD.

5. Conclusion

The Chinchilla weir in Queensland, Australia is a large dam equipped with an unusual overflow spillway system, i.e. a minimum energy loss (MEL) design. The approach flow is very smooth and streamlined before it reaches the broad-crest, and then continues down a converging smooth chute. In the current study, the spillway chute was observed and documented during two days, at the start and towards the end of a major flood event peaking on 5 December 2021.

The field observations showed the smooth streamlining of the reservoir waters into the spillway crest and chute. Downstream of the crest, the chute flow presented a very smooth and glassy free-surface. On 17 November 2021, the presence of large logs, blocked on the crest, had no major impact on the chute flow. Further downstream, the water surface became rough and choppy, immediately before the flow became self-aerated. The inception region presented a gradual variation of the surface roughness, from a smooth glassy surface, to a coarse-sand-paper-like appearance and then a very-rough choppy surface. Both the upstream non-aerated and self-aerated flow regions exhibited a brown colour indicating a large amount of suspended sediments.

An optical flow (OF) technique was applied to relatively long-duration movies recorded from a sturdy tripod facing the left section of the spillway chute. The time-averaged streamwise surface velocities were successfully compared to the application of the backwater calculations for a smooth converging chute, in the self-aerated flow region. The streamwise velocities were about 8-9 m/s at the chute toe. In contrast, the OF data were physically-meaningless in the developing flow region because of optical artifacts. The OF data presented large streamwise surface velocity fluctuations in the aerated flow region, with $Tu_s \sim 150\text{-}200\%$, consistent with the broad literature on self-aerated flow measurements using dual-tip phase detection probe in laboratory. The ratio of transverse to streamwise surface turbulence intensity v_t'/ v_s' was between 0.06 and 0.1, indicating a strong anisotropy of the free-surface turbulence.

6. Acknowledmennts

Hubert Chanson thanks Mr André Chanson and Ms Ya-Hui Chou for their assistance with the field works. He acknowledges Sunwater for providing technical details of the Chinchilla weir spillway, and Leigh Cook and Tobias Burwood (Western Downs Regional Council) for providing photographic information on the flood event. He also thanks Mr Barton Maher (QLD Department of Regional Development, Manufacturing and Water), Jeff Dann (Sunwater), Mr Rui (Ray) Shi (The University of Queensland), Dr Carlos Gonzalez (QLD Department of Transport and Main Road) for their assistance and advice.

In line with recommendations of the International Committee on Publication Ethics (COPE) and the Office of the Commonwealth Ombudsman (Australia), Hubert Chanson declares a major conflict of interest with Matthias Kramer (UNSW, Canberra).

7. References

Anwar, H.O. (1994). "Self-Aerated Flows on Chutes and Spillways - Discussion." *Journal of Hydraulic Engineering*, ASCE, Vol. 120, No. 6, pp. 778-779.

Arosquipa Nina, Y, Shi, R., Wüthrich, D., and Chanson, H. (2022). "Intrusive and Non-Intrusive Two-Phase Air-Water Measurements on Stepped Spillways: a Physical Study." *Experimental Thermal and Fluid Science*, Vol. 131, Paper 110545, 22 pages, 4 movies and a digital appendix (6 pages) (DOI: 10.1016/j.expthermflusci.2021.110545).

Bung, D.B., and Valero, D. (2016). "Optical Flow Estimation in Aerated Flows." *Journal of Hydraulic Research*, IAHR, Vol. 54, No. 5, pp. 575-580 (DOI: 10.1080/00221686.2016.1173600).

Cain, P., and Wood, I.R. (1981). "Measurements of Self-aerated Flow on a Spillway." *Journal Hydraulic Division*, ASCE, 107, HY11, pp. 1425-1444.

Chanson, H. (1997). "Air Bubble Entrainment in Open Channels. Flow Structure and Bubble Size Distributions." *International Journal of Multiphase Flow*, Vol. 23, No. 1, pp. 193-203 (DOI: 10.1016/S0301-9322(96)00063-8).

Chanson, H. (1997b). "Air Bubble Entrainment in Free-Surface Turbulent Shear Flows." *Academic Press*, London, UK, 401 pages.

Chanson, H. (2001). "The Hydraulics of Stepped Chutes and Spillways." *Balkema*, Lisse, The Netherlands.

Chanson, H. (2004). "The Hydraulics of Open Channel Flow: An Introduction." *Butterworth-Heinemann*, 2nd edition, Oxford, UK, 630 pages.

Chanson, H. (2013). "Interactions between a Developing Boundary Layer and the Free-Surface on a Stepped Spillway: Hinze Dam Spillway Operation in January 2013." *Proc. 8th International Conference on Multiphase Flow ICMF 2013*, Jeju, Korea, 26-31 May, Gallery Session ICMF2013-005 (Video duration: 2:15). {https://www.youtube.com/watch?v=cKHDrkNFfC0&t=3s}

Chanson, H. (2021). "Stepped Spillway Prototype Operation, Spillway Flow and Air Entrainment: the Hinze Dam, Australia." *Hydraulic Model Report No. CH123/21*, School of Civil Engineering, The University of Queensland, Brisbane, Australia, 183 pages (DOI: 10.14264/c8d5280).

Chanson, H. (2022). "On Air Entrapment Onset and Surface Velocity in High-Speed Turbulent Prototype Flows." *Flow Measurement and Instrumentation*, Vol. 83, Paper 102122, 9 pages (DOI: 10.1016/j.flowmeasinst.2022.102122).

Chanson, H., and Toombes, L. (2002). "Air-Water Flows down Stepped chutes: Turbulence and Flow Structure Observations." *International Journal of Multiphase Flow*, Vol. 28, No. 11, pp. 1737-1761 (DOI: 10.1016/S0301-9322(02)00089-7).

Farnebäck, G. (2003). "Two-frame motion estimation based on polynomial expansion." *Proc. Scandinavian conference on Image analysis*, Springer, Berlin, Heidelberg, pp. 363-370.

FEMA (2014). "Overtopping protection for dams." *Technical Manual P-1015*, US Bureau of Reclamation, 462 pages.

Halbronn, G. (1952). "Etude de la Mise en Régime des Ecoulements sur les Ouvrages à Forte Pente. Applications au Problème de l'Entraînement d'Air." *Journal La Houille Blanche*, No. 1, pp. 21-40; No. 3, pp. 347-371; No. 5, pp. 702-722 (in French).

Halbronn, G., Durand, R., and Cohen de Lara, G. (1953). "Air Entrainment in Steeply Sloping Flumes." *Proc. 5th IAHR Congress*, IAHR-ASCE, Minneapolis, USA, pp. 455-466.

ICOLD (1984). "World Register of Dams - Registre Mondial des barrages - ICOLD." *International Commission on Large Dams*, Paris, France, 753 pages.

Ippen, A.T., Tankin, R.S., and Raichekn, F. (1955). "Turbulence measurements in free- surface flow with an impact tube-pressure transducer combination." Technical Report No. 20, MIT, Department of Civil and Sanitary Engineering, 112 pages.

Irrigation and Water Supply Commission (1972). "Chinchilla weir layout. As constructed." *Drawing L44375 75313*, Version B, 16/5/1972.

Keller, R.J., and Rastogi, A.K. (1975). "Prediction of Flow Development on Spillways." *Journal of Hydraulic Division*, ASCE, Vol. 101, No. HY9, pp. 1171-1184.

Lane, E.W. (1936). "Recent Studies on Flow Conditions in Steep Chutes." *Engineering News-Record*, Jan. 2, pp. 5-7.

Levi, E. (1965). "Longitudinal Streaking in Liquid Currents." *Journal of Hydraulic Research*, IAHR, Vol. 3, No. 2, pp. 25-39.

Levi, E. (1967). "Macroturbulence Produced by Vortex Breakdown in High Velocity Flows." *Proc. 12th IAHR Congress, Colorado*, USA, Vol. 2, Paper B7, pp. 54-60.

McKay, G.R. (1971). "Design of Minimum Energy Culverts." *Research Report*, Dept of Civil Eng., University of Queensland, Brisbane, Australia, 29 pages & 7 plates.

McLean, F.G., and Hansen, K.D. (1993). "Roller Compacted Concrete for Embankment Overtopping Protection." *Proc. Spec. Conf. on Geotechnical Practice in Dam Rehabilitation*, ASCE, Raleigh NC, USA, L.R. Anderson editor, pp. 188-209.

Michels, V., and Lovely, M. (1953). "Some Prototype Observations of Air Entrained Flow." *Proc. 5th IAHR Congress*, IAHR-ASCE, Minneapolis, USA, pp. 403-414.

Pravdivets, Y.P., and Bramley, M.E. (1989). "Stepped Protection Blocks for Dam Spillways." *International Water Power and Dam Construction*, Vol. 41, No. 7, July, pp. 49-56.

Rao, N.S. Govinda, and Rajaratnam, N. (1961). "On the Inception of Air Entrainment in Open Channel Flows." *Proc. 9th IAHR Congress*, Dubrovnick, Yugoslavia, pp. 9-12.

Schlichting, H., and Gersten, K. (2000). "Boundary Layer Theory." *Springer Verlag*, Berlin, Germany, 8th edition., 707 pages.

Toombes, L., and Chanson, H. (2007). "Surface Waves and Roughness in Self-Aerated Supercritical Flow." *Environmental Fluid Mechanics*, Vol. 7, No. 3, pp. 259-270 (DOI 10.1007/s10652-007-9022-y).

Toro, J.P., Bombardelli, F.A., and Paik, J. (2017). "Detached Eddy Simulation of the Nonaerated Skimming Flow over a Stepped Spillway." *Journal of Hydraulic Engineering*, ASCE, Vol. 143, No. 9, Paper 04017032, 14 pages.

Turnbull, J.D., and McKay, G.R. (1974). "The Design and Construction of Chinchilla Weir - Condamine River Queensland." *Proc. 5th Australasian Conf. on Hydraulics and Fluid Mechanics*, Christchurch, New Zealand, Vol. II, pp. 1-8.

Wood, I.R., Ackers, P., and Loveless, J. (1983). "General Method for Critical Point on Spillways." *Journal of Hydraulic Engineering*, ASCE, Vol. 109, No. 2, pp. 308-312.

Zabaleta, F., and Bombardelli, F.A. (2020). "Eddy-resolving Simulation of Flows over Macroroughness." *Proceedings of RiverFlow2020*, CRC Press, Uijttewaal et al. Editors, pp. 1293-1299.

Zhang, G., and Chanson, H. (2018). "Application of Local Optical Flow Methods to High-Velocity Free-surface Flows: Validation and Application to Stepped Chutes." *Experimental Thermal and Fluid Science*, Vol. 90, pp. 186-199 (DOI: 10.1016/j.expthermflusci.2017.09.010)

DOI: https://doi.org/10.26077/f04b-f7cd

2-D simulation of flow structures over dunes for flow characteristics estimation

Mohammad Sharifi[1], Mohammad Reza Majdzadeh Tabatabai[2], Seyed Hossein Ghoreishi Najafabadi[2]
[1]*PhD Candidate, Department of Civil, Water and Environmental Engineering, Shahid Beheshti University, Tehran, Iran*
[2]*Assistant Professor, Department of Civil, Water and Environmental Engineering, Shahid Beheshti University, Tehran, Iran*
E-mail: m_majdzadeh@sbu.ac.ir

Abstract: *Rivers have undulated beds, which are called bedforms. Depending on the hydraulic conditions, the bedforms have different types that cause resistance in the flow. Despite various research and experiments on bedforms, the topic is still debatable and requires further research. The present study is mainly focused on simulating the flow motion numerically on dunes in open channels to evaluate the effect of dune geometry on flow structure. Twenty-nine simulations were conducted to study the effect of the geometry of five types of dunes with different angles and heights under different hydraulic conditions and bed roughnesses. RANS and DES turbulence models were used to simulate small and large-scale dunes, respectively. The results of the numerical model were compared with the experimental results of previous researchers to validate the work, indicating the appropriate accuracy of the numerical model. Then, an empirical equation was adopted to evaluate the effect of dune geometry on flow hydraulics by considering the equations of previous researchers. Finally, sensitivity analyses were carried out to determine the dependence of each parameter in this equation.*

Keywords: *Bedform, Dune, Numerical simulation, Turbulence models.*

1. Introduction

Generally, rivers play a crucial impact on their surrounding areas and are considered a crucial water supply source for humans. Therefore, any changes in the river flow regime could disturb the social-economic of the riparian zone and endanger their lives. In open channel flow with a fixed bed, it is possible to estimate a constant value to determine the roughness coefficient (Ahmad *et al.* 2017, Chow 1981) and use one of the flow resistance formulas to calculate the flow rate (Fenton 2010). In natural rivers with alluvial beds, bed deformation may occur with the entrainment of particles, which could be followed by bedforms, depending upon the flow conditions. The bed deformation could increase the hydraulic flow resistance, which may result in the particles (Strickler 1923, Aberle et al. 2008, Ji-Sung, et al. 2010) and bedforms roughness (Engelund and Hansen 1967, Hafez *et al.* 2001, Venditi 2007). Although various researchers have used different approaches to estimate the bedforms (Heydari *et al.* 2014, Attar and Li 2012, Karamisheva *et al.* 2005 and Julien and Klaassen 1995), their results are significantly different. This could be owing to the differences in laboratory conditions, lack of uniformity to predict bedform dimensions and the shortage of knowledge about the effect of bedform structure on the turbulence of sediment transport phenomena. This, in turn, necessitates further research on the complexity of the phenomena.

Many researchers have noticed the significant effect of the bedforms on the flow structure, by decomposing the total roughness into grain and form roughnesses (Cai *et al.* 2020). The term "dune" was initially introduced by Gilbert (1914). Soulsby (1989) suggested that a triangular, asymmetrical cross-sectional area generally forms sand waves with a sharp lee-side angle of 28 to 35° and a gentle stoss-side angle of about 5°. Dunes cause a flow separation zone at the lee-side, in which eddy flows are formed and dissipate much energy. By measuring the velocity in this zone, he observed that the velocity is approximately one-third to half of the average flow velocity in the stoss-side direction.

Giri and Shimizu (2006) presented a two-dimensional morphodynamic model for free surface flows on bedforms. The proposed model has a high ability to simulate average flow and turbulence structure. They validated the model with laboratory data in a mobile bed channel and compared it with previous works. They also found that the nonlinear $k-\varepsilon$ model showed better results than the linear model, particularly, in turbulent zones. Nabi et al. (2012) presented a three-dimensional high-resolution hydrodynamic model for unsteady incompressible flow over an evolving bed topography. The model used LES to solve the turbulence; their results were compared with the laboratory and experimental models of previous research works and showed reasonable agreement. Reisenbüchler *et al.* (2019) cited that Van Rijn (1984c) developed a diagram, based on a large number of laboratory and field data, to determine the bedform in open channels. Niazkar *et al.* (2019) developed a direct flow-dependent approach calculating roughness caused by the grain and bedform. They compared the accuracy of their method with the previous works.

In general, experimental studies to evaluate the effect of bedforms on flow structure is costly and time-consuming. At the same time, numerical studies provide additional information and reduce time and cost. Motamedi *et al.* (2012) investigated the length of the separation zone and the interaction of dune geometry on the flow structure. They concluded that the length of the flow separation zone could increase directly with the flow rate, dune height and bed material size, and decrease inversely with the flow depth. Lefebvre (2019) used Delft3D software to simulate 3D flow above natural bedforms. He initially verified the model against Maddux *et al.* (2003) laboratory data. Then, Parsons *et al.* (2005) data were used to model 3D flow velocities and turbulence above a dunes bedform field for Río Paraná (Argentina).

Although many research works have been undertaken on the resistance of dunes to the flow, the topic is still debatable as this signifies the necessity of further research on the dune geometry and its effect on flow structure. Furthermore, assessment of an appropriate turbulence model is of great importance to simulate the geometry of the bedforms under different hydraulic conditions.

In this regard, adapting empirical equations to estimate dunes dimensions enables us to gain a better understanding of the interaction between bedforms and flow structure. Generally, most previous research works used the Reynolds-averaged Navier–Stokes (RANS) turbulence model to simulate the flow structure because of its simplicity and low cost (Berrouk 2019). However, on account of its deficiency in computing velocity fluctuations, the model fails to detect accurately flow eddies (Spalart and Allmaras 1994 and Spalart *et al.* 1997) and leads to an inappropriate demonstration of flow separation at the dunes lee side. Hence, this study used the detached eddy simulation (DES) turbulence model in STAR-CCM+ to simulate flow on large-scale dunes (i. e. dune height to water depth ratio of greater than 0.25) with a high lee-side angle (38°), as the effect of the bedforms on the flow structure is significant in a shallow-water river. In contrast, the significance of bedform roughness decreases by increasing the ratio of water depth to dune height, which may reflect the flow separation formation at the dune lee-side (Mustaffa *et al.* 2016). Since riverbed conditions are one of the factors affecting floods and aquatic life cycles, this study attempts to identify the effective parameters on the bedforms to obtain an in-depth understanding of the interaction between them and the flow structure to be applied to the river by the engineers and practitioners. Besides, further efforts have been made to develop an empirical equation to predict the interaction of riverbed geometry and flow characteristics.

2. Methods

2.1. Governing equations

The flow of a viscous incompressible fluid with constant properties is governed by the Navier–Stokes equations (Alfonsi 2009):

$$\frac{\partial u_i}{\partial t} + \frac{\partial}{\partial x_j}\left(u_i u_j\right) = -\frac{\partial p}{\partial x_i} + v\frac{\partial^2 u_i}{\partial x_j \partial x_j}, \frac{\partial u_i}{\partial x_i} = 0 \qquad (1)$$

The Reynolds equation, which is called RANS, is one of the Navier-Stokes equations, where u = velocity, x = location, t = time, i and j = Cartesian indices, P = pressure, and v = kinematic viscosity of water. According to the concept of Reynolds decomposition, the dependent variables of equation (1) decomposed into mean and fluctuating parts:

$$u_i = \bar{u}_i + u_i', \quad p = \bar{p} + p' \qquad (2)$$

Where the sign (′) = the fluctuating component while the sign ($\overline{\ }$) = the time average component (Alfonsi 2009). The left-hand side of equation (1) represents the changes in the mean momentum caused by the unsteady flow and convective accelerations. It is notable that the term $\frac{\partial}{\partial t}$ is eliminated when the flow is steady.

2.2. Numerical setup

Sharifi et al. (2020) suggested that DES and RANS turbulence models give the most appropriate results for large and small-scale dunes, respectively. Star-CCM+ was applied as the tool to carry out the simulation. It is a state-of-the-art commercial code developed by CD-Adapco (Melville, New York, USA) which solves the fully 3-D hydrodynamic incompressible Navier–Stokes equations using a finite volume method on unstructured meshes (Wu *et al.* 2019). The realizable two-layer k-ε model was used to perform the RANS simulations. Based on previous studies (Shur *et al.* 1999, Bunge *et al.* 2007, Spalart *et al.* 1997), the k-ε model was applied as a principle for the RANS model in DES, which the constant is shown is Table 1. The no-slip condition was considered for bed and side walls while the layer of air with height equals twice the water depth was added to the computational domain to model free surface dynamics using the VOF. Furthermore, the volume of fraction method was applied to calculate the free surface. The numerical solution method in this model is implicit unsteady. The VOF method has been used for problems involving two or more unmixable fluids and the interface between the two fluids (Park *et al.* 2020). A simple but powerful method is described based on the concept of a fractional volume of fluid (VOF). This method is more flexible and efficient than other methods for treating complicated free boundary configurations (Hirt *et al.* 1981). By solving these equations, the number α is obtained to indicate the volumetric ratio of the two fluid phases within each of the computational cells. For example, $\alpha = 1$ indicates that the cell volume is full of water, $\alpha = 0$ shows that the cell volume is full of air, and $0 < \alpha < 1$ is the transition zone between the two water and air fluids. For an incompressible fluid, the kinematic volume fraction; is:

$$\frac{\partial F}{\partial t} + \frac{1}{V_F}\left[\frac{\partial(F A_i u_i)}{\partial x_i}\right] = \frac{1}{V_F}\left[\frac{\partial\left(\upsilon_F A_i \partial F / \partial x_i\right)}{\partial x_i}\right] \tag{3}$$

Where u_i = horizontal component of velocity; x_i = horizontal vector in the Cartesian coordinate system; V_F = volume fraction of fluid in each cell; A_i = fractional areas open to flow in the i-coordinate of the Cartesian system; υ_F = diffusion coefficient and F = the volume of fluid per unit volume (User Guide, STAR-CCM+ 2018).

Table 1. Constants of k-ε model

C_μ	$C_{\varepsilon 1}$	$C_{\varepsilon 2}$	σ_k	σ_ε
0.09	1.44	1.9	1	1.2

2.3. Experimental data

The laboratory model developed by Motamedi *et al.* (2012) was used to simulate the flow on dune bedforms. They conducted their experiments in a rectangular channel 12 m long, 0.75 m wide and 0.9 m high on fixed dunes with different dimensions and lee and stoss-side angles (Nelson *et al.* 1993, Allen 1985, Nasiri 2010 and Davarpanah 2011). The material of dunes was of two types of non-uniform sand with an average diameter of 5.8 and 13.2 mm, which were referred to as fine and coarse grains, respectively, in this research. The dunes had a wavelength of 1 m and heights of 4, 6, and 8 cm, which were divided into sharp and flat-crested classes. The former were of type 2 (Figure 1, Table 2) while the latter were of type 1, 3 and 4 (Figure 1, Table 2). Dune type 3 was a large-scale dune with a height to water depth ratio greater than 0.25 and a lee-side angle of 38°. However, other dunes were of the small-scale type with a height to water depth ratio of less than 0.25 and the lee-side angle less than 38°.

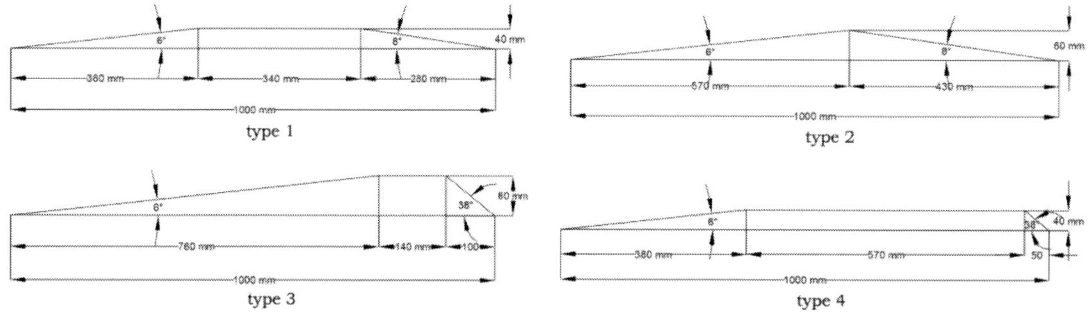

Figure 1. Used sharp-crested dune (type 2) and Used broad-crested dune (type 1, 3 and 4) (Motamedi et al. 2012)

The experiments were carried out with discharge values of 30 and 60 litres per second under depths of 32 and 20 cm (Motamedi *et al.* 2012). In addition, Balachandar and Patel (2008) data were also used to simulate the flow on the dune bedform; they applied Van Mierlo and de Ruiter (1988) dune geometry (Figure 2, Table 2) and created a train of 22 dunes in a laboratory channel with dimensions of 10 m long, 0.61 m wide and 0.61 m high. Experiments were conducted under the constant velocity of 0.4 m/s and the flow depth of 0.12 m (Balachandar and Patel 2008). They consider three modes of flat bed, the bed covered with a network of 0.72 mm stainless steel wires with a distance of 6.35 mm and bed glued with 18 mm sand grains, which was simulated in the present study.

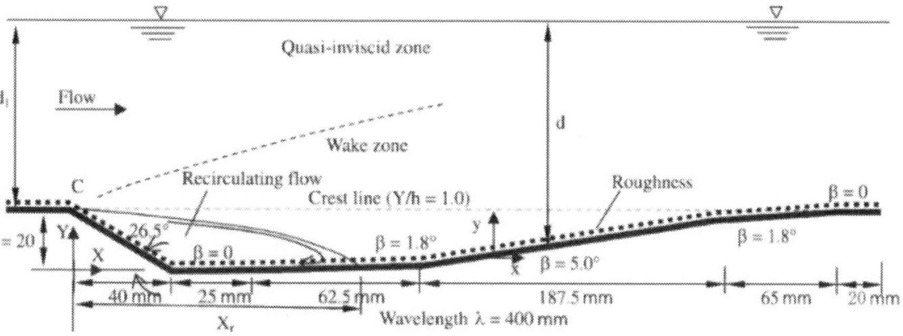

Figure 2. Experimental model of dune geometry (Balachandar and Patel 2008)

Table 2. Characteristics of geometry dimensions of simulated dunes

Dune length	Stoss side angle (Degree)	Crest length (cm)	Dune height (cm)	Lee side angle (Degree)	Crest type	Dune type
100	6	0	6	8	Sharp	2
100	6	34	4	8	Broad	1
100	6	14	8	38	Broad	3
100	6	57	4	38	Broad	4
40	5	2	2	26.5	Broad	Balachandar

2.4. Boundary conditions

The numerical model was designed with input and output boundary conditions in which "Velocity Inlet" option with constant velocity was selected in Star-CCM+ for the input boundary condition. For the output boundary condition, "Pressure Outlet" option was selected equal to nought, likewise, for the free water surface to address the effect of the atmosphere. The walls and bed of the channel were also considered with the "Wall" option; this was determined by equivalent roughness k_s to be $2.5D_{50}$and $3.5D_{84}$(Yang *et al.* 2005 and Motamedi *et al.* 2012) for the fine and coarse materials, respectively (Figure 3). The flow of water was assumed incompressible with a density of 1000 kg/m^3 and dynamic viscosity of $0.001\frac{kg}{m.s}$.

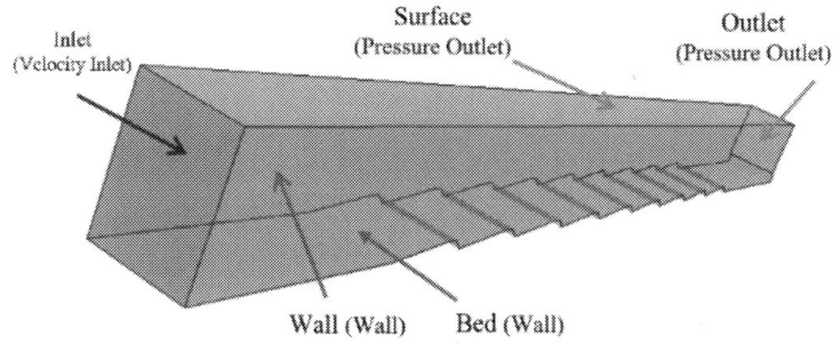

Figure 3. Numerical model boundary conditions in STAR-CCM + software

2.5. Mesh Generation

In general, selecting the most optimal mesh size is regarded as one of the most critical steps in numerical simulation. As coarse meshing increases, the gradient of variables variations as a result of which inappropriate outputs may be gained. Hence, fine meshing should be implemented so that all cases can be examined by condensing the mesh points in the zones where the flow variables are under sudden changes. Therefore, it enables us to achieve outputs more accurately by studying all the model variables in a computational domain. Nevertheless, should fining mesh sizes be performed with caution. This is because it would increase the number of computational cells and the extension of computational time, which may not be economical. By this means, mesh optimization should initially be implemented in numerical simulation.

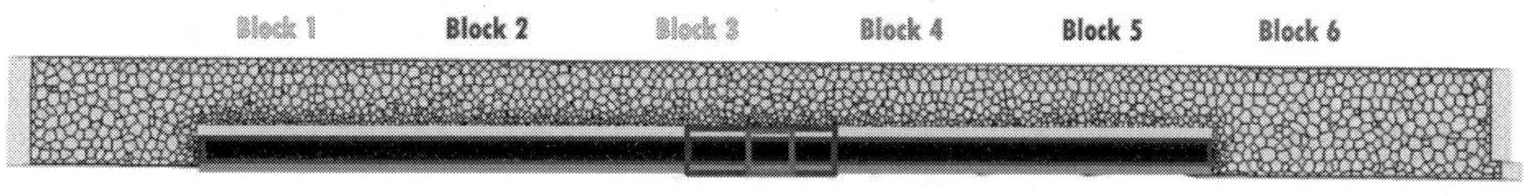

Figure 4. Meshing blocks

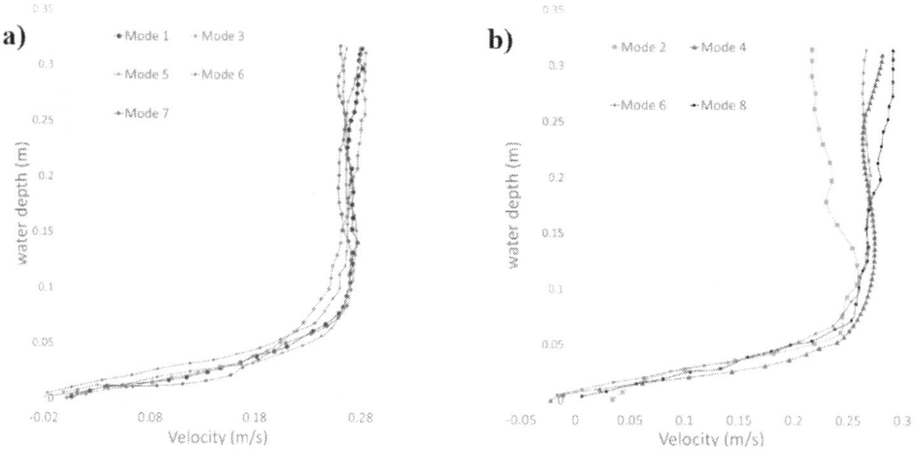

Figure 5. Profile changes velocity in the vertical direction. a) Couple states and b) odd states relative to the ideal state (6)

For example, simulation No. 25 (Dune type 4, 58 mm grain size, 30 l/s flow rate and 317 mm flow depth) was chosen arbitrarily for the optimal cell size selection tests. To measure the sensitivity of the numerical model to the computational cell size, as well as to optimize the run time and to promote the accuracy of the outputs, changes in the meshing size were considered in the form of multiple blocks in the sampling data locations on the dunes (Figure 4). Eight simulations were carried out to optimize the time and accuracy of the output results. Based on this, the model with the smallest cell size (i. e. mesh generation mode (6) was selected as the most accurate (100%) under a long run time of 98 hours. Given, the run time was not cost-effective; however, accuracy was desirable. Thus, other modes were validated by this in terms of velocity profile deviation in the vertical direction (Figure 5 and Table 3). Under these circumstances, mesh generation mode (1) worked out having optimized simulation with appropriate accuracy of 85% under a run time of 50 hours. It is worth pointing out that no significant change is observed in the accuracy of the model by reducing the mesh sizes. however, the run time increases significantly to reach stable time (Section 2-7). Equally, enlarging the mesh sizes could significantly increase the deviation of the vertical velocity profile (Figures 5a and b) and, subsequently, the divergence of the streamlines and further mixing of the water and air (Figures 6a and b). Hence, the mesh dimensions mode (1) is selected as the appropriate option (Figure 7). The average value of GCI is 1.025 approximately.

Table 3. Mesh generation sensitivity analyses

Mesh Generation Mode	Dimensions of grid in block 1 (mm)	Dimensions of grid in block 2 (mm)	Dimensions of grid in block 3 (mm)	Dimensions of grid in block 4 (mm)	Dimensions of grid in block 5 (mm)	Dimensions of grid in block 6 (mm)	Time required to sec in reach 60 simulation time (hour)	Percentage of flow deviation
1	100	20	10	20	10	5	50	15
2	100	40	20	20	10	5	40	27
3	100	20	10	5	10	5	66	13
4	100	20	10	10	10	1	82	9
5	100	20	10	10	10	2	74	11
6	100	20	10	10	5	1	98	8
7	100	20	10	10	5	5	61	14
8	100	20	10	10	10	5	55	15

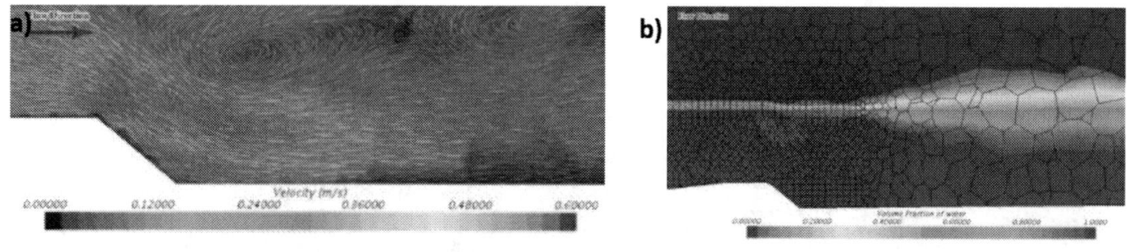

Figure 6. Distribution a) Streamlines and b) Mixing the weather in the vicinity of the last telescope with a large meshing of 10 cm

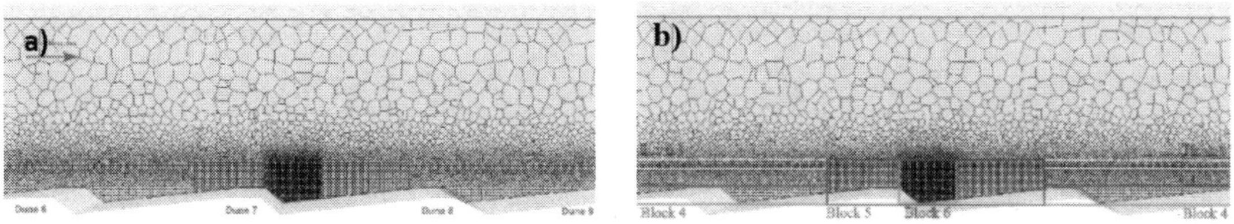

Figure 7. The meshing of numerical model sampling location a) Overview b) Location of blocks

2.6. Time of equilibrium

Time of equilibrium or stability time of the numerical model may vary depending upon the dimensions of the mesh sizes and the applied turbulence model. The accuracy of the simulation results is directly related to the run time of the model which may not be economical for a long run time. When the simulation accuracy reaches about 80 to 90% of its final value after a certain time, it could be accepted as the final result to save time and money.

In this way, experiments were carried out finding the optimal run time for numerical simulation. On account of the different simulation times for each turbulence model, two experiments were made with two different geometries under various hydraulic conditions, one of which with the RANS turbulence model (simulation No. 24 (Dune type 4, 132 mm grain size, 60 l/s flow rate and 195 mm flow depth)) and the other with the DES turbulence model (simulation No. 17 (Dune type 3, 132 mm grain size, 60 l/s flow rate and 317 mm flow depth)). Since RANS equations are based on time-averaged approach, these equations have fewer computational processes and their run time is shorter than that of the DES model. The model was run for 40, 50, 60, 70, 80, and 90 seconds to simulate the RANS model, and the velocity profiles were compared at the boundary between the fifth and sixth dunes (Figures 8a and b). The model has achieved acceptable accuracy from the very first seconds, enabling us to select the run time of 50 seconds as the stability time to achieve sufficient accuracy in the simulations of the RANS.

The run times of 40, 50, 60, 70, 80, and 90 seconds were made in the numerical model to find the optimal simulation time for the DES, and then, the velocity profiles were compared the same as before for the same location. Figure (8b) indicates that the velocity profiles differ by less than 7% in 60 and 70 seconds, therefore, the DES requires more time to achieve sufficient accuracy, which takes the run time of about 60 seconds. However, for more certainty, the run time of 70 seconds was selected for simulation to ensure that the process was appropriately in progress. The times obtained in the process can be considered as the time for the flow to reach stability because no significant differences have been noticed in the velocity profiles.

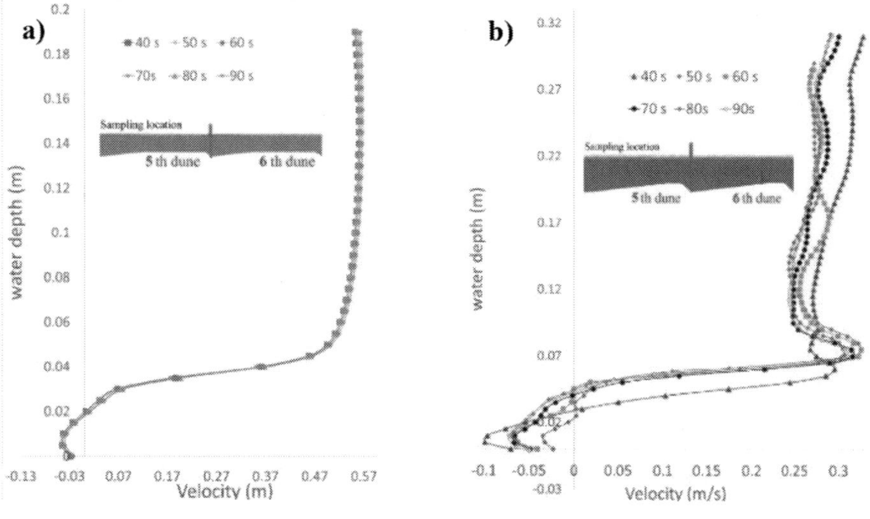

Figure 8. Flow velocity profiles at a distance of 7 meters of upstream at different times

2.7. Numerical model calibration

Twenty-nine experiments were simulated to evaluate the model by the measured velocity profiles in the vertical direction. The data of 11 experiments from Motamedi *et al.* (2012) (simulations No. 1, 2, 6, 7, 15, 21, 22, 23, 25, 26, 27) and one experiment from Balachandar and Patel (2008) (simulation No. 29 (Dune type 5, 18 mm grain size, 30 l/s flow rate and 120 mm flow depth)) were available. In all experiments, equation (4) was used to work out the percentage error of the velocity profile of the numerical model (Num) relative to the measured laboratory data (Obs).

$$PercentofError(\%) = \frac{\sum_{i=1}^{n}(Obs-Num)}{\sum_{i=1}^{n}(Obs)} \times 100 \quad (4)$$

Where Obs = measured laboratory data and Num = numerical model output.

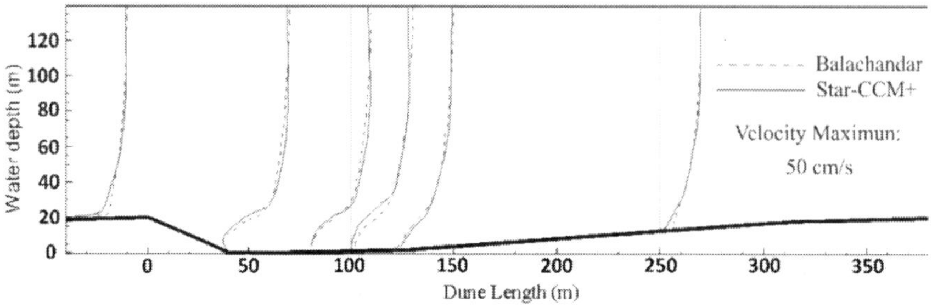

Figure 9. Comparative graph of flow velocity profiles in Experiment 29 (RANS turbulence model, Balachandar 2008 model geometry with a discharge of 36.7 litres per second and a 14 cm depth)

The results of the numerical model state that the difference between the predicted and measured values of the velocity profiles is on average 12.3% with Motamedi *et al.* (2012) and about 7.4% with Balachandar and Patel (2008) data. In total, the minimum and maximum differences are 7.4 and 16.8%, respectively, while the average is 11.9%, which highlights the appropriate agreement of the results of the present study with the previous research works (Figures 9 and Table 4).

Table 4. The average difference of vertical velocity profiles (%)

Simulation Number	1	2	6	7	15	21	22	23	25	26	27	29	Average
Observed and STAR-CCM+	11.4	9.5	13.2	15.5	14.5	9.4	10.2	16.8	12.2	9.6	12.8	7.4	11.9
STAR-CCM+ and SSIIM	21.6	18.9	23.2	26.5	17.3	15.2	14.1	24.5	19.3	13.2	27.7	-	20.1

The results of the STAR-CCM+ were also compared with the SSIM used by Motamedi *et al.* (2012) study for simulation No. 15 (Dune type 3, 132 mm grain size, 30 l/s flow rate and 317 mm flow depth) (Figure 10). The minimum and maximum differences between the results of the predictions by the two models are 13.2 and 27.7%, respectively, with an average of about 20.1%. Bearing in mind that there is no information available for Balachandar *and* Patel (2008) data in SSIIM.

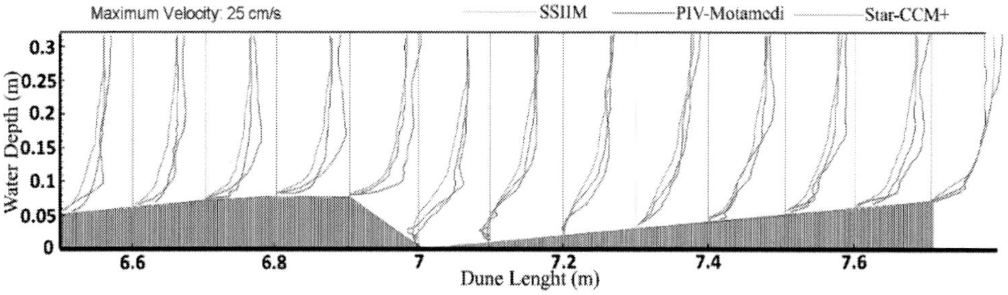

Figure 10. Comparative graph of flow velocity profiles in Experiment 15 (DES turbulence model, dune type 3 and 30 litres per second discharge and 32 cm depth)

3. Dimensional analysis

Given the significance of various variables in the formation of the bedform, it is possible to apply dimensional analysis to make these variables dimensionless. Generally, numerous variables are involved in developing the bedform, which are as follows.

$$f(V_*, H, \overline{V}, g, \mu, \rho, \rho_s, k_s, \lambda, \theta_{US}, \theta_{DS}, y, S, b, V_c, y_c) = 0 \quad (5)$$

Where V_* = shear velocity, H= dune height, D_{50} = median particle size, $\overline{V}$= mean flow velocity, g = gravity acceleration, μ= dynamic viscosity, ρ= water density, ρ_s= bed particle density, k_s= equivalent roughness height, λ = dune length, θ_{US} = dune stoss-side angle, θ_{DS}= dune lee-side angle, y = water depth, S = bed slope, τ_0= shear stress, b = channel width, V_c= critical velocity for particle movement, and y_c= critical depth for particle movement.

Totally, 18 variables are defined to form the dune, among which three repetitive variables are designated (i. e. H for the length, $\overline{V}$for the time and ρ for the mass).

Among the 29 available simulations, about 60% of them (i. e. 17 simulations) were used to obtain the coefficients of equation (6) while the remainders were for equation validation. The coefficients of the equation were calculated regression analyses with the coefficient of determination ofR^2 = 0.847.

$$\frac{\lambda}{H} = 1.47\theta_{us}^{-0.023}\theta_{ds}^{-0.089}(Fr^2)^{-0.072}\left(\frac{k_s}{y}\right)^{0.032}\left(\frac{V_*}{\overline{V}}\right)^{-1.188} \quad (6)$$

Furthermore, the rest of the data (i. e. 12 simulations) were used to validate equation (6). The$\frac{\lambda}{H}$ values obtained by Equation (6) have an average percentage error of about 11.25% compared to the observed values (Figure 11).

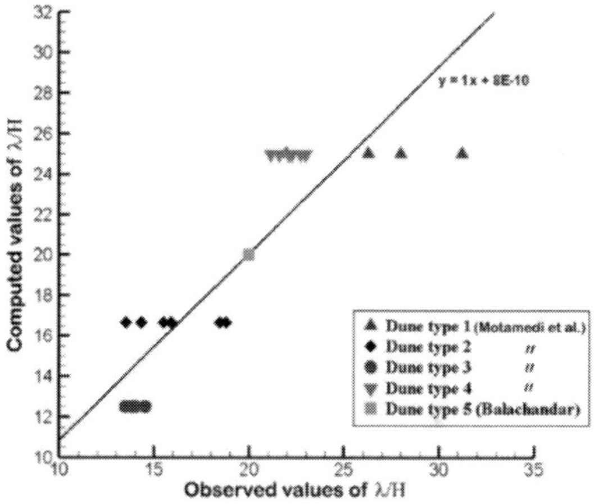

Figure 11. Comparison of the ratio of length to height observed and calculated according to equation (6)

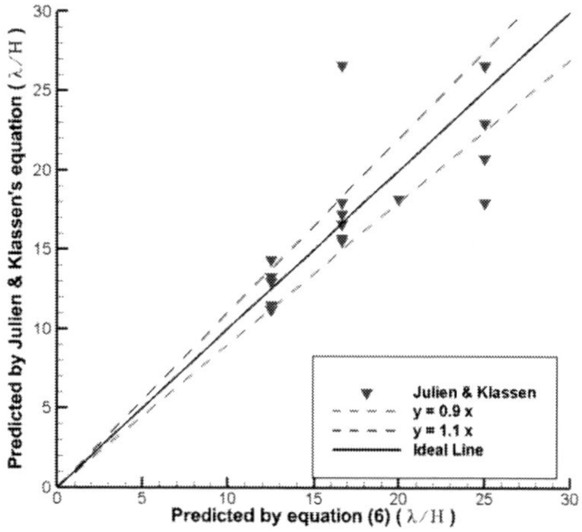

Figure 12. Comparison of λ/H values predicted by this study and Julien and Klassen 1995

Figure (12) provides the diagram of $\frac{\lambda}{H}$ values predicted by equation (6) versus those predicted by Julien and Klassen (1995). It is observed that the graphs are relatively symmetric for $\frac{\lambda}{H}$ values smaller than 20. In contrast, Equation (6) underestimates $\frac{\lambda}{H}$ values significantly for $\frac{\lambda}{H}$> 20 compared to Julien and Klassen (1995) equation with the average MND equals to 11.2.

3.1. Sensitivity analysis of the dimensionless numbers

Some researchers have used different variables to obtain the$\frac{\lambda}{H}$ ratio (Heydari *et al.* 2014; Attar and Li 2012; Karamisheva *et al.* 2005 and Julien and Klaassen 1995). This indicates that all dimensionless numbers in equation (6) may not necessarily appear in the equations developed by different researchers, as most studies have been carried out under controlled experimental conditions (Kwoll *et al.* 2016; Heydari *et al.* 2014 and Karamisheva *et al.* 2005). By conducting sensitivity analysis on each dimensionless number in equation (6), it is possible to conceive the significance of the numbers and their effects on estimating the dimensions of bedform and flow structure, which enables us to predict and control floods or probably to improve river ecosystem conditions (Naqshband and Hoitink 2020).

The sensitivity analysis was undertaken on dimensionless numbers of equation (6) for five different cases, such that, in each case, one dimensionless number was removed from the equation and the regression analysis was done to the data based on other dimensionless numbers (Table 5).

Table 5. Dependence of equation (13) dimensionless numbers

Deleted parameter	Formula	R^2	Percentage of deviation (MNE)
$\frac{k_s}{y}$	$\frac{\lambda}{H} = 2.54\theta_{us}^{0.326}\theta_{ds}^{-0.087}(Fr^2)^{-0.06}\left(\frac{V_*}{V}\right)^{-1.168}$	0.837	9.6
$\frac{V_*}{V}$	$\frac{\lambda}{H} = 19.22\theta_{us}^{0.98}\theta_{ds}^{-0.036}(Fr^2)^{-0.016}\left(\frac{k_s}{y}\right)^{0.036}$	0.03	26.2
Fr^2	$\frac{\lambda}{H} = 3.29\theta_{us}^{0.287}\theta_{ds}^{-0.095}\left(\frac{k_s}{y}\right)^{0.001}\left(\frac{V_*}{V}\right)^{-1.118}$	0.78	11
θ_{DS}	$\frac{\lambda}{H} = 2.49\theta_{us}^{0.24}(Fr^2)^{0.0236}\left(\frac{k_s}{y}\right)^{-0.076}\left(\frac{V_*}{V}\right)^{-1.137}$	0.786	10.8
θ_{US}	$\frac{\lambda}{H} = 1.52\theta_{ds}^{0.32}(Fr^2)^{-0.089}\left(\frac{k_s}{y}\right)^{-0.072}\left(\frac{V_*}{V}\right)^{-1.188}$	0.84	9.4

Table (5) displays that the ratio $\frac{V_*}{V}$ has the most significant effect on equation (6) as a result of which the shear velocity and hence shear stress has the highest effect on riverbed formation which is in conformance with Willett (2006).

4. Conclusion

Generally, the use of computer simulations to model and study flows can reduce cost and time. In this research, the numerical model was used to simulate the dune geometry. Then, the experimental results of Motamedi *et al.* (2012) and Balachandar and Patel (2008) were used to validate the model. The comparison of the two vertical velocity profiles showed an average difference of 11.9%, which indicates the appropriate accuracy of the numerical model. Flow structure was also simulated numerically on five different types of dunes, the results of which are as follows:

-There is a high possibility of flow separation occurrence in sharp-crested dunes. In two dunes with the same dimensions, the sharp-crested dunes create more changes in the flow structure. However, shear stress variations in flat-crested dunes are more gradual than those of the sharp-crested dunes; therefore, flow separation is less likely to occur.

-The results of flow simulation with the RANS model state that the model works only for small-scale dunes to simulate flow separation zone. In contrast, LES and DES models detect flow separation with adequate accuracy near large-scale dunes, whereas they cause relatively intensive and irregular turbulences in small-scale dunes. The difference between LES and DES models could be noticeable in the generation of irregular vortices by LES in the flow layers away from the dunes and the lower DES run time. In other words, RANS and DES are the most appropriate turbulence models in simulating small and large-scale dunes, respectively.

-The results have revealed that the proposed equation (Equation 6) has a reasonable accuracy due to the random nature of the bedforms formation, in which the $\frac{V_*}{V}$ ratio has the most significant effect. Therefore, changing that may allow making changes in the characteristics related to the river bedforms.

5. REFERENCES

Aberle, J., Dittrich, A., Nestmann, F., Novak, P., Rennie, C. D., & Millar, R. G. (1999). Estimation of gravel-bed river flow resistance. *Journal of Hydraulic Engineering, 125*(12), 1315–1319.

Ahmad, N., Saiful Bahry, S., Md Ali, Z., Mat Daud, A. and Musa, S. (2017). "Effect of Flow Resistance In Open Rectangular Channel". MATEC Web of Conferences, 97, p.01107.

Alfonsi, G. (2009). "Reynolds-averaged Navier-Stokes equations for turbulence modelling", Appl. Mech. Rev.,vol.62,pp. 040802.

Allen, J. R. L. (1985). "Principles of Physical Sedimentology". Chapman and Hall, PP. 272

Allen, J.R.L. (1978). "Computational methods for dune time-lag: Calculations using Stein's rule for dune height". Sedimentary Geology 20(3): 165-216.

Attar, S. and Li, S.S. (2012). "Numerical investigation of flow structure and shear stress over fixed dunes". International Conference on Fluvial Hydraulics River Flow 2012, San José, Costa Rica, 5–7 September 2012.

Balachandar, R., and Patel, V.(2008). "Flow over a fixed rough dune". Canadian Journal of Civil Engineering, 35(5), pp.511-520.

Berrouk, A. (2019). "Stochastic Lagrangian modeling for large eddy simulation of dispersed turbulent two-phase flows". Bentham Science Publishers, p 23.

Bunge, U., Mockett C., and Thiele F. (2007). "Guidelines for implementing Detached-Eddy Simulation using different models". Aerospace Science and Technology, Vol. 11, pp. 376-385.

Cai, R., Zhang, H., Zhang, Y., Zhang, L. and Huang, H. (2020). "Flow Resistance Equation in Sand-bed Rivers and Its Practical Application in the Yellow River". Water, 12(3), p.727.

Chow, V.T. (1981). "Open channel hydraulics". Mc Graw – Hill Limted, London, 680 p.

DavarPanah, Sh. (2011). "Investigation of Interaction of Straight Crested Gravel Bedforms and Vegetated Banks on Turbulent Flow Components", Master's Thesis, Faculty of Agriculture, Isfahan University of Technology, Isfahan, Iran. (in Persian)

Engelund, F., and Hansen, E.(1967). "A Monograph on Sediment Transport in Alluvial Streams". Teknisk Vorlag, Copenhagen, Denmark.

Fenton, J.D. (2010). "Calculating resistance to flow in open channels". Alternative Hydraulics Paper, 2, 1-7.

Gilbert, G. K.(1914). "The transition of debris by running water". U.S. Geol. Surv. Prof. Pap. 86, pp 263.

Giri, S. and Shimizu, Y., (2006). Numerical computation of sand dune migration with free surface flow. Water Resources Research, 42(10).

Hafez, Y.I., El-Gamal, F.S., and Soliman, W.R. (2001). "River Nile bed forms, resistance to flow and discharge prediction". Water Science, the 28th – 29th issue, October 2000-April 2001.

Heydari, H., Zarrati, A.R., and Karimaee Tabarestani, M. (2014). "Bedform characteristics in alive bed alluvial channel". Scientia Iranica,TransactionsA:CivilEngi- neering, 21(6), 1773–1780.

Hirt, C.W. and Nichols, B.D. (1981). "Volume of Fluid (VOF) Method for the Dynamics of Free Boundaries". Journal of Computational Physics, 39, 201-225.

Ji-Sung, K., Chan-Joo, L., Won, K. and Yong-Jeon, K., (2010). Roughness coefficient and its uncertainty in gravel-bed river. *Water Science and Engineering*, 3(2).

Julien, P.Y., and Klaassen, G.J. (1995). "Sand-dune geometry of large rivers during floods". Journal of Hydraulic Engineering, 121, 657-663.

Karamisheva, R., Lyness, J. F., Myers W. R. C., O'Sullivan J. and Cassells J. (2005). "Prediction of bed form height in straight and meandering compound channels". Proceedings of the 3rd International Conference on Water Resources Management, Algarve, Portugal, 11-13 April, 311-320.

Kwoll, E., Venditti, J.G., Bradley RW, Winter C. (2016). "Flow structure and resistance over subaqueous high- and low-angle dunes". Journal of Geophysical Research: Earth Surface 121(3): 545–564.

Lefebvre, Alice. (2019). "Three-Dimensional Flow Above River Bedforms: Insights From Numerical Modeling of a Natural Dune Field (Río Paraná, Argentina)". Journal Of Geophysical Research: Earth Surface, 124(8) 2241-2264.

Maddux, T. B., Nelson, J. M., & McLean, S. R. (2003). "Turbulent flow over three-dimensional dunes: 1. Free surface and flow response". Journal of Geophysical Research: Earth Surface, 108(F1).

Motamedi, A., Afzalimehr, H. Singh, V., and Dufresne, L.(2012). "Experimental Study on the Influence of Dune Dimensions on Flow Separation". Journal of Hydrologic Engineering, 19(1), pp.78-86.

Mustaffa, N., Ahmad, N., and Razi, M. (2016). "Variations of Roughness Coefficients with Flow Depth of Grassed Swale". IOP Conference Series: Materials Science And Engineering, 136, 012082.

Naqshband, S. and Hoitink, A. (2020). "Scale-Dependent Evanescence of River Dunes During Discharge Extremes". Geophysical Research Letters, 47(6).

Nasiri DehSorkhi, A. (2010). "Interaction of vegetation cover, bed forms and flow structure on distributions of velocity and turbulent intensities". Master's Thesis, Faculty of Agriculture, Isfahan University of Technology, Isfahan, Iran. (In Persian)

Nabi, M., de Vriend, H., Mosselman, E., Sloff, C. and Shimizu, Y., (2012). Detailed simulation of morphodynamics: 1. Hydrodynamic model. *Water Resources Research*, 48(12).

Nelson, J. M., McLean, R. and Wolfe,S. (1993). "Mean flow and turbulence over fixed, two-dimensional bed form. Water Resour. Res. 29: 3935-3953.

Niazkar, M., Talebbeydokhti, N., and Afzali, S. (2019). "Development of a New Flow-dependent Scheme for Calculating Grain and Form Roughness Coefficients". KSCE Journal Of Civil Engineering, 23(5), pp 2108-2116.

Park, J., Kim, B., Sohn, D., Choi, Y. and Lee, Y. (2020). "A Study on Flow Characteristics and Flow Uniformity for the Efficient Design of a Flow Frame in a Redox Flow Battery". Applied Sciences, 10(3), p.929.

Parsons, D. R., Best, J. L., Orfeo, O., Hardy, R. J., Kostaschuk, R., and Lane, S. N. (2005). "Morphology and flow fields of three-dimensional dunes, Rio Paraná, Argentina: results from simultaneous multibeam echo sounding and acoustic Doppler current profiling". Journal of Geophysical Research, 110(F4), F04S03.

Reisenbüchler, M., Bui, M., Skublics, D., and Rutschmann, P. (2019). "An integrated approach for investigating the correlation between floods and river morphology: A case study of the Saalach River, Germany". Science Of The Total Environment, Vol. 647, pp 814-826.

Shur, M., Spalart P.R., Strelets M., and Travin A. (1999). "Detached-eddy simulation of an airfoil at highangle of attack". Engineering Turbulence Modeling and Experiments, Vol. 4, pp. 669-678.

Soulsby, R. L. (1989). "Bedform migration in sandy estuaries". H.R Wallingford Res. Rep. No. SR 208, pp 16.

Spalart, P. R., Jou, W.-H., Strelets, M., and Allmaras, S. R. (1997). "Comments on the Feasibility of LES for Wings, and on a Hybrid RANS/LES Approach". Proceedings of the First ASOSR Conference on DNS/LES, Greyden Press, Columbus, OH.

Strickler, A. (1923). "Some contributions to the problem of the velocity formula and roughnessfactors for rivers, canals, and closed conduits". Bern, Switzerland, Mitt.Eidgeno assischen Amtes Wasserwirtschaft, no. 16.

Van Mierlo, M. C. L. M. and de Ruiter, J. C. C. (1988). "Turbulence measurements above artificial dunes, Report on measurements". Report Q789, WL | Delf Hydraulic, Delf, The Nederlands.

Van Rijn, L.C. (1984c). "Sediment Transport, Part III: Bed Forms and Alluvial Roughness". Journal of Hydraulic Engineering, ASCE, Vol. 110, No. 12.

Venditi, J. G. (2007). "Turbulent flow and drag over fixed two- and three-dimensional dunes". J. Geophys. Res. 112 F04008.

Willett, S. (2006). "Tectonics, Climate, And Landscape Evolution. Boulder, Colo". Geological Soc. of America, pp.83-84.

DOI: https://doi.org/10.26077/fdab-c7cd

A Numerical Investigation on Residual Energy of Labyrinth Weirs

P. Langohr[1], B.M. Crookston[2] & D.B. Bung[1]
[1]FH Aachen University of Applied Sciences, Aachen, Germany
[2]Utah Water Research Laboratory, Utah State University, Logan (UT), USA
E-mail: langohr@fh-aachen.de

Abstract: *The replacement of existing classical weirs with labyrinth weirs is a proven techno-economical solution and a means to increase the discharge capacity when rehabilitating existing structures. However, flows exiting a labyrinth weir are complex, three-dimensional, and aerated; additional information is needed regarding energy dissipated by such weirs. In this study, three labyrinth weirs with different crest lengths were simulated with FLOW-3D HYDRO. Reynolds-averaged Navier-Stokes (RANS) modeling with the use of finite-volume method and Re-Normalisation Group (RNG) k-ε turbulence closure were employed. An attempt was made to more precisely account for the flow field in the downstream region. Consequently, the velocity head was determined with both depth-averaged and section-averaged velocities. Additionally, the kinetic energy correction coefficient was considered. A comparison of the computational fluid dynamics (CFD) results with prior experimental data showed that the residual energy was influenced by factors such as probe position and geometric parameters. A minor influence was observed for the kinetic energy correction coefficient. Overall, the high amount of energy dissipation was underlined and an acceptable agreement between simulated and literature data was documented.*

Keywords: *CFD, energy dissipation, labyrinth weirs*

1. Introduction

Due to the aging of existing hydraulic structures and increased societal demands, rehabilitation techniques have gained importance. Deficiencies to be addressed by rehabilitation include not only hydraulic and structural issues, but also factors such as climate change, longer hydrologic records, and urbanization. Replacing classical, linear weirs with nonlinear weirs (e.g., labyrinth weir, piano key weir) is one possibility to increase the discharge capacity of an existing spillway while construction costs and space requirements may impede the widening of existing linear weirs. Regarding the number of published articles in scientific journals and realized prototypes, a continuous increase of interest in this type of hydraulic structure is eminent (Crookston et al. 2019; Hager et al. 2015; Schleiss 2011). Another positive aspect of nonlinear weirs is the wide range of application from river structures to flood protection spillways.

In general, nonlinear weirs are characterized by repetitive cycles in plan view (Fig. 1). Since their introduction, numerous design approaches (Erpicum et al. 2017; Erpicum et al. 2013; Erpicum et al. 2011; Crookston 2010; Falvey 2003; Hinchliff and Houston 1984; Hay and Taylor 1970; Taylor 1968) as well as innovative design ideas (Ben Said and Ouamane 2022; Akbari Kheir-Abadi et al. 2020; Ghaderi et al. 2020; Bilhan et al. 2018; Dabling and Tullis 2017; Crookston and Tullis 2012; Dabling and Crookston 2012) have been proposed. Possible cycle shapes are rectangular (A), trapezoidal (B) or triangular (C) (see Fig. 1) with trapezoidal most common. Generally, these weirs cause a three-dimensional flow field immediately downstream and such turbulence provides energy dissipation. Particularly downstream of the weir, significant free-surface fluctuations are encountered. Therefore, collection of relevant hydraulic parameters in the laboratory is challenging. To date, there is still limited information about the energy dissipation processes and their dependence on selected geometric parameters (Carrillo et al. 2020; Crookston 2020; Merkel et al. 2018; Lopes et al. 2011, 2008, 2006; Magalhaes and Lorena 1994). Specifically, how does a labyrinth weir influence the design of a stilling basin in a rehabilitation project? Therefore, in this study numerical simulations were conducted of quarter-round trapezoidal labyrinth weirs with varying cycle lengths to estimate the relative residual energy and study the influence of parameters such as probe position and different velocity head determination methods.

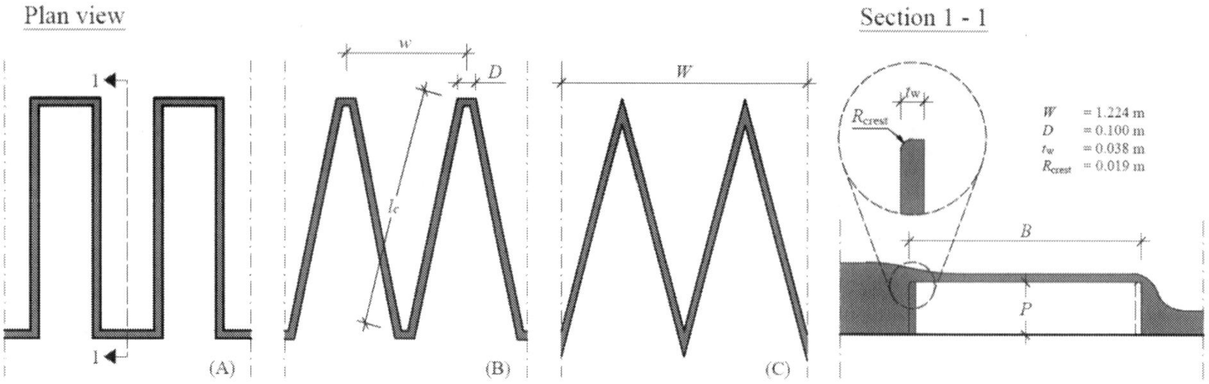

Figure 1. Plan view of typical nonlinear weir cycle shapes (rectangular (A), trapezoidal (B) or triangular (C))

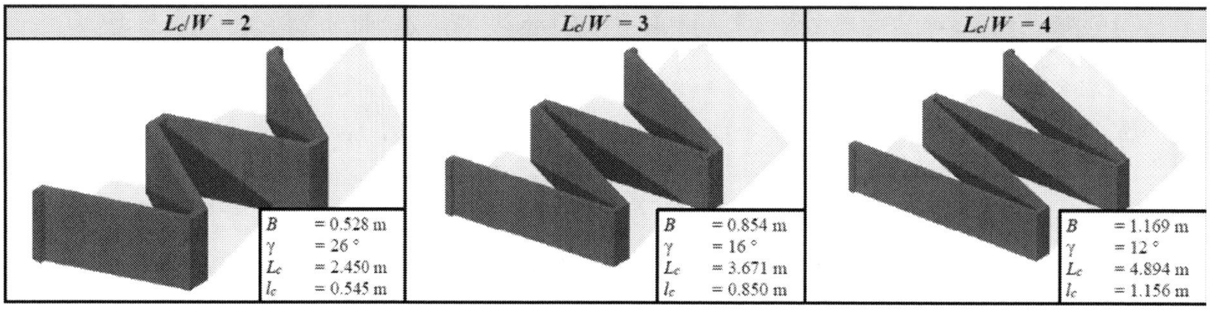

Figure 2. Geometries of the studied labyrinth weirs.

2. Numerical Model

2.1. Overview

Three different quarter-round trapezoidal labyrinth weirs were investigated. With a constant weir height of $P = 30.5$ cm and a total flume width of $W = 1.22$ m, only the ratio of total crest length to total width (L_c/W) is varied from 2 to 4. According to this increase of L_c/W, the corresponding sidewall angles γ were 26°, 16° and 12° (see Fig. 2). For each geometry a total of 5 discharges with $0.05 < Q \leq 0.25$ m³/s were investigated. Along the computational domain, several numerical probes for data extraction were considered placed in three transects each centered on an upstream labyrinth apex. The selected labyrinth weir dimensions were of laboratory scale model and were similar to published geometries by Crookston (2020, 2010) and Tullis et al. (2020; 2005).

2.2. Numerical Settings

For this study, the three-dimensional Reynolds-averaged Navier-Stokes (RANS) equations were solved by FLOW-3D HYDRO using a finite-volume method (Hirt and Nichols 1981) and Re-Normalisation Group (RNG) *k-ε* (Yakhot et al. 1992) turbulence closure with dynamically computed characteristic turbulence length. Further details of the numerical methods are described i.e. by Versteeg and Malalasekera (2007) or Wilcox (2006). The single fluid approach of FLOW-3D was chosen for this study (Crookston et al. 2018; Mansoori et al. 2017; Bayon et al. 2016). According to Flow Science, Inc. (2020) all values, except fractional areas and velocities, are computed at the center of each cell. The remaining values are calculated in the center of the hexahedral cell faces.

To minimize computation time, a nested multiblock mesh with varying cell sizes was used in this study (see. Fig. 3). The finest mesh resolution was considered for the vicinity of the structure. Referring to the upstream labyrinth weir face, the inlet and outlet boundaries were located 3 m upstream and 5.5 m downstream. A volume flow rate was defined as an upstream boundary condition (BC). The downstream BC was defined as a water surface elevation equal to zero. Both vertical sides as well as the horizontal floor were defined as smooth walls.

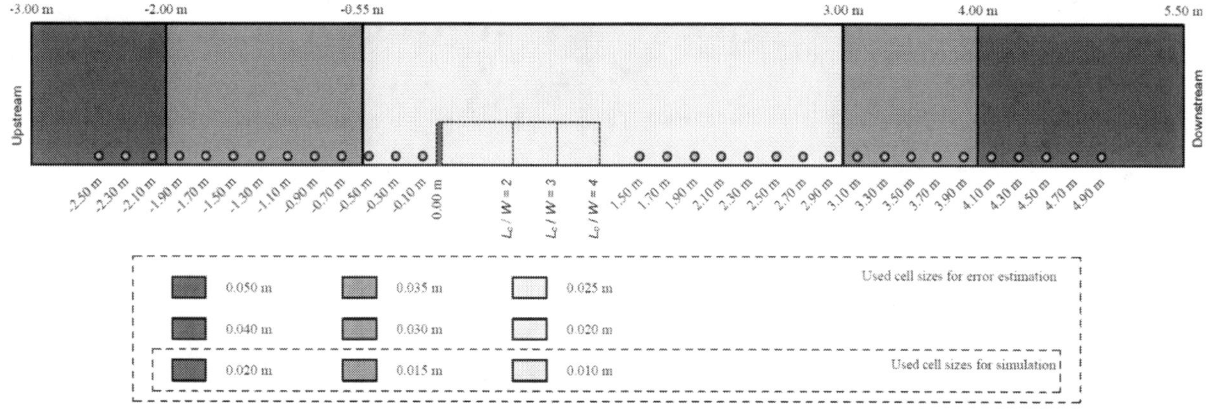

Figure 3. Sideview of the numerical domain, considered cell sizes for the nested multi-block mesh and location of numerical probes for error estimation and data analysis.

A mesh sensitivity analysis was conducted considering the grid convergence index (GCI; Roache 1994) and the methodology as proposed by Celik et al. (2008). Accordingly, three different representative cell sizes were assessed for each mesh (see Fig. 3). For verification, two geometries and discharges ($L_c/W = 2$ with $Q = 0.05$ m^3/s; $L_c/W = 4$ with $Q = 0.25$ m^3/s) were used. For both analyses, six GCI values (3 upstream, 3 downstream) for flow depth h, depth-averaged velocity U_{DA} and Froude number F_r were calculated. This results in a total of 36 GCI values with a mean GCI of 9.2 %. As expected, the GCI values are higher for low discharges and downstream of the weir. This may be caused by an increased inaccuracy of the volume of fluid method at low water levels downstream of the weir. An opposite trend is documented for higher discharges at the upstream area. The used mesh sizes were in a comparable size to several published studies (Ben Said and Ouamane 2022; Carrillo et al. 2020; Ghaderi et al. 2020; Carrillo et al. 2019; Crookston et al. 2018). As a result, the sensitivity analysis shows adequate values. With the used mesh sizes, each simulation required between three to seven days using a 8-core Intel(R) Xeon(R) CPU at 2.10GHz.

3. Results

3.1. Flow Depth

Resulting water levels from all configurations are illustrated in Figure 4. Downstream of the weir, no classic hydraulic jump was observed. The results show more of a plunging jet into the downstream water body. Air entrainment is disregarded in the employed single fluid approach; a related subgrid model was not activated. However, Valero et al. (2018) showed that energy dissipation processes at hydraulic structures can be modeled with reasonable accuracy neglecting air entrainment. Due to high-turbulence, an area with significant free-surface waves was found below the weir, with its length depending on the discharge as well as the weir geometry. With increasing distance from the weir, i.e. at ~15P, this free-surface pattern became smoother for all discharges and geometries. Overall, a similar trend of the free-surface along the domain was observed.

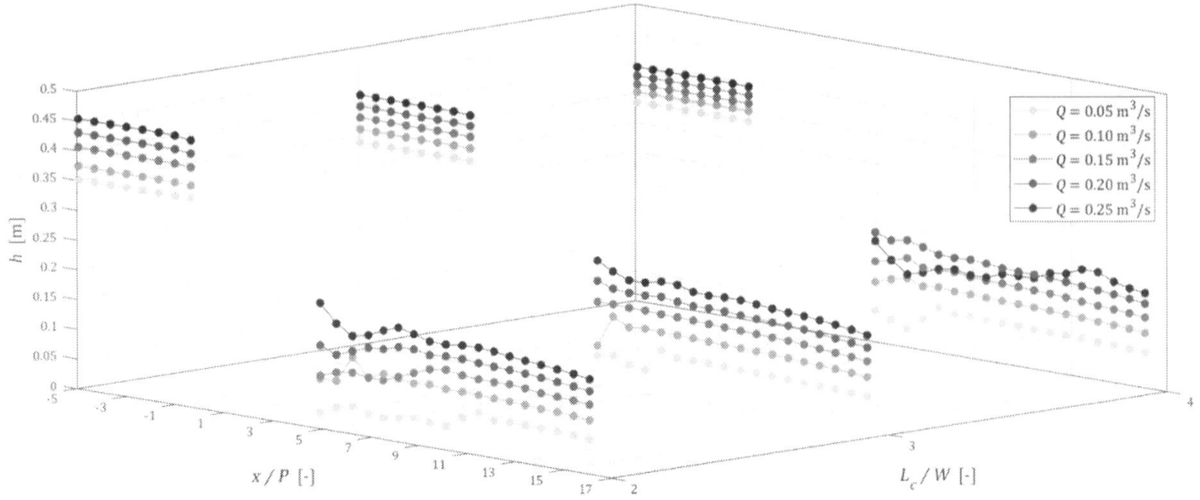

Figure 4. Flow depth (h) along the domain, expressed as a ratio of probe position to weir height (x/P) over all conducted discharges (Q) and geometries (L_c/W).

3.2. Residual Energy

Due to their folded geometry, nonlinear weirs cause a three-dimensional flow field in the outlet cycles and immediately downstream of the weir from nappe collision and thus, more complex energy dissipation processes than linear weirs. Based on laboratory experiments by Magalhaes and Lorena (1994), Lopes et al. (2006) proposed a design equation for estimation of relative residual energy downstream of trapezoidal labyrinth weirs (see. Eq. 1):

$$\frac{H_1}{H_0} = 0.571 + 0.254 \ln\left(\frac{H}{P}\right) + 0.199 \ln\left(\frac{L_c}{W}\right) \tag{1}$$

with H_1 = downstream total head, H_0 = upstream total head and H = upstream head over the weir crest. Eq. (1) was validated by physical model tests for quarter-round crested weirs with $P = 0.25$ m and $1.8 \leq L_c/W \leq 4$ (Lopes et al., 2011, 2008, 2006). The results regarding the residual energy from Lopes et al. (2011) are within the range expected. It is noteworthy that residual energy has been mostly estimated in literature by simple point measurements of flow depth with estimation of kinetic energy by consideration of the continuity equation. However, Carrillo et al. (2020) found notable differences in the velocity profiles depending on probe location downstream of the weir.

In this study, the influence of the three-dimensional downstream flow field on the relative residual energy was investigated. All presented data was extracted after a simulation time of $t = 75$ s, when steady conditions in relation to boundary conditions were obtained. Energy heads were determined as $H = h + U^2/2g$ (with flow depth h, velocity U and gravity acceleration g), accounting for both, depth-averaged (U_{DA}) and section-averaged (U_{SA}) velocities. Furthermore, the kinetic energy correction coefficient α was estimated according to Eq. (2) (with flow area A, sectional mean velocity U_{SA} and local streamwise velocity u). This coefficient was used to describe non-uniform velocity distributions more accurately (Singh et al. (2019). As expected, the coefficient was found to increase close to the weir (see Fig. 5) and to converge to unity further downstream. In detail, the velocity head correction factor was in the range of $1.0003 < \alpha \leq 1.2051$ and thus, was reasonable (Singh et al. 2019; Keshavarzi and Hamidifar 2018; Hulsing et al. 1966).

$$\alpha = \frac{1}{A\, U_{SA}^3} \int_A u^3 \, dA \tag{2}$$

Figure 5 illustrates the evolution of the relative residual energy expressed as a ratio of downstream total head H_1 to the upstream total head H_0. H_0 was measured at a fixed distance of $-4.92P$ (-1.5 m). As expected, the positioning of the probes affects the relative residual energy. Near the weir, significant differences between velocity heads from section-averaged as well as depth-averaged velocity profiles and from local flow depth measurements at the centerline and the continuity equation were observed. With a maximum α of 1.205, the highest influence of α was observed close to the weir as well (see Fig. 5 bottom left). With increasing distance from the weir towards the downstream boundary, a convergence of H_1/H_0 and α was noticed. If the measuring location is far downstream of the structure, a sufficient accuracy can be expected (see Fig. 5). A minimum distance of 15P should be respected for further investigations. At this point, the influence of α was small and can be neglected. Furthermore, a homogeneous progress of H_1/H_0 can be found beyond this minimum distance.

According to this evolution of the flow field, the probe position at 15.42P (4.7 m) was chosen for further comparison. Figure 6 compares selected results with published data from Crookston (2020), Merkel et al. (2018), Lopes et al. (2011, 2008; 2006) and Magalhaes and Lorena (1994). Comparing both velocities U_{DA} and U_{SA}, it was found that the depth-averaged velocity on the flume centerline is generally higher than the section-averaged velocity. Consequently, this fact yields an overestimation of the relative residual energy with depth-averaged data.

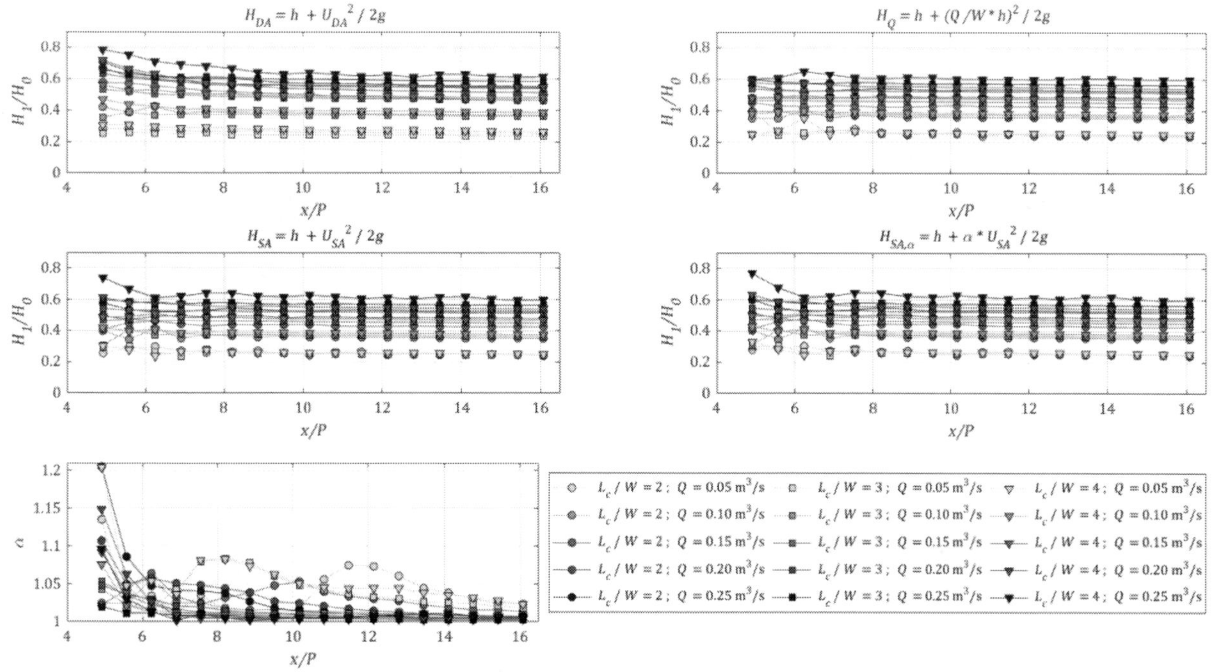

Figure 5. Evolution of relative residual energy along the downstream with different velocity head calculations (H_{DA}, H_Q, H_{SA} and $H_{SA,\alpha}$); evolution of the kinetic energy correction coefficient α along the entire domain.

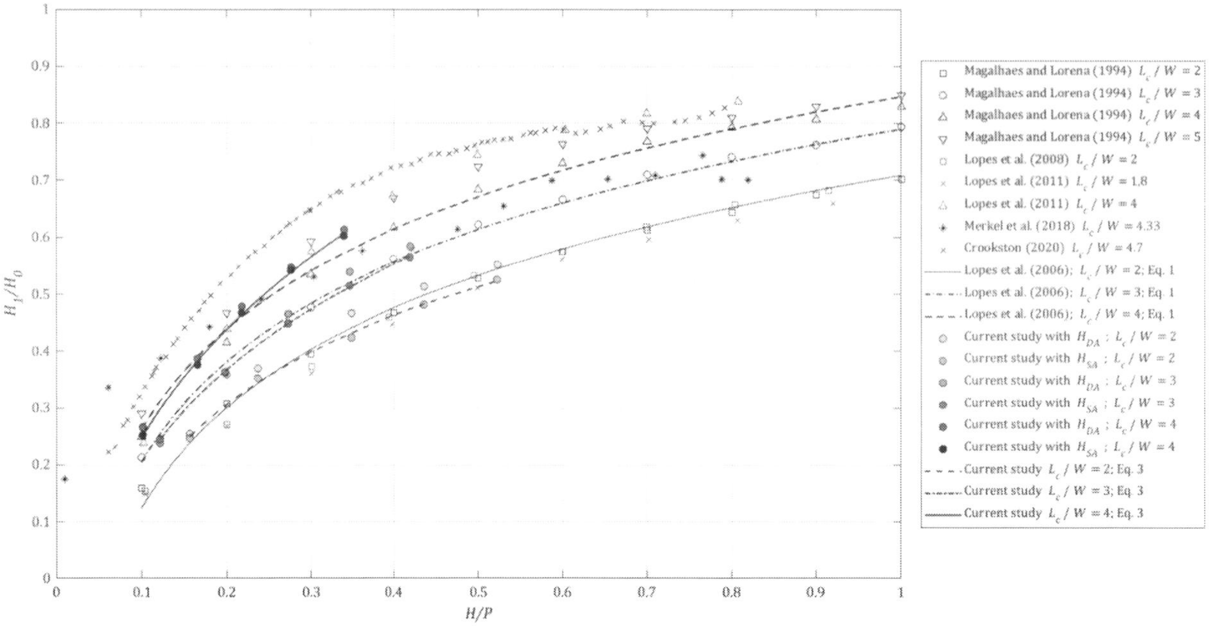

Figure 6. Residual energy of labyrinth weirs from the current study, downstream head estimation with depth-averaged velocities and section-averaged velocities, the kinetic energy correction coefficient α is neglected; rectangular labyrinth weir data from Merkel et al. (2018), trapezoidal labyrinth weir data from Lopes et al. (2011, 2008) and Magalhaes and Lorena (1994).

Besides this overestimation, some differences between simulated and published data are noticeable. They may be caused by the influence of different measuring positions (see Fig. 5), scale effects, different weir heights, sidewall angles, number of cycles or surface tension effects for small headwater ratios. The chosen one fluid approach and the resulting negligence of the air entrainment may have an influence as well. Equation (3) in combination with Table 1 summarizes the presented data of the relative residual energy (see also Fig. 6).

$$\frac{H_1}{H_0} = a\left(\frac{H}{P}\right)^b + c \tag{3}$$

Table 1. Curve-fit coefficients for quarter round trapezoidal labyrinth weirs, presented L_c/W ratios and section averaged velocity heads H_{SA} in Figure 6.

L_c/W	γ [°]	a	b	c	R^2	Validated range
2	26	-6.297	-0.0345	6.926	0.9972	$0.156 < H/P \leq 0.523$
3	16	2.345	0.135	-1.519	0.9998	$0.121 < H/P \leq 0.420$
3	12	1.817	0.244	0.7892	0.9989	$0.102 < H/P \leq 0.339$

4. Discussion and Conclusions

A numerical study of three trapezoidal labyrinth weirs with a constant height of 0.305 m and different ratios of total crest length to total width was conducted. The flume width was 1.22 m. For each geometry, five discharges were studied ($0.05 < Q \leq 0.25$ m³/s). For the employed mesh resolution, a mean GCI of 9.2% for different parameters was obtained. Some GCI values are higher than the GCI of 10 %, which was recommended by Celik et al. (2008), at small discharges. For the selected cell size, some mesh dependence thus may be present for small discharges and resulting low water levels. Nevertheless, results were found to be generally consistent to literature data for all discharges.

With the conducted simulations, an attempt was made to consider the 3D flow field in the downstream region more precisely. Accordingly, the velocity head was determined with depth-averaged and section-averaged velocities.

Additionally, the kinetic energy correction coefficient α was calculated. A minor influence of this parameter is observed when data is extracted in sufficient distance to the weir where turbulence is reduced and a redirection of the flow is obtained. For further investigations, it is recommended that a minimum distance of $15P$ is maintained to ensure convergence of H_1/H_0 and α. An empirical design equation is proposed for trapezoidal quarter-round labyrinth weirs. Even though the influence of the crest shape seems to be small, it should not be neglected. Beside the distinction of L_c/W ratios, a further distinction according to crest shape should be made. Furthermore, the presented equation takes the effective velocity profile into account and does not estimate the energy height by using the continuity equation or a depth-averaged velocity. The structure of the presented equation is similar to the equation of the discharge coefficient published by Crookston (2010). In general, high energy dissipation downstream of nonlinear weirs is found in agreement to literature data.

Currently, there is a lack of information regarding laboratory measurements of the described 3D velocity fields. In addition, there are only a few numerical studies on this problem. Future research should focus on extensive 3D laboratory measurements to expand and clarify the existing data basis for more geometries.

5. REFERENCES

Akbari Kheir-Abadi, M., Karami Moghadam, M., Sabzevari, T. and Ghadampour, Z. (2020). "An experimental study of the effects of the parapet walls geometry on the discharge coefficient of trapezoidal piano key weirs." *Flow. Meas. Instrum.,* 73(0955-5986), 101742.

Bayon, A., Valero, D., García-Bartual, R. and Valles-Moran, F. (2016). "Performance assessment of OpenFOAM and FLOW-3D in the numerical modeling of a low reynolds number hydraulic jump." *Environ. Model. Softw.* 80(1364-8152), 322–335.

Ben Said, M. and Ouamane, A. (2022). "Performance of rectangular labyrinth weir – an experimental and numerical study." *Water Supply*, 22(1606-9749), 3628-3644.

Bilhan, O., Cihan Aydin, M., Emin Emiroglu, M. and Miller, C. (2018). "Experimental and CFD analysis of circular labyrinth weirs." *J. Irrig. Drain. Eng.* 144(6), 4018007.

Carrillo, J., Matos, J., and Lopes, R. (2019). "Numerical modeling of free and submerged labyrinth weir flow for a large sidewall angle." *Environ. Fluid Mech.* 20(2020), 357–374.

Carrillo, J., Matos, J. and Lopes, R. (2020). "Numerical modelling of subcritical flow downstream of a labyrinth weir." *Proc., 8th IAHR ISHS 2020.* The University of Queensland, QLD., Santiago, Cl., 10.14264/uql.2020.616.

Celik, I., Ghia, U., Roache, P. and Freitas, C. (2008). "Procedure for estimation and reporting of uncertainty due to discretization in CFD applications." *J. Fluids Eng.* 130(7), 078001.

Crookston, B. (2010). Labyrinth weirs. All Graduate Theses and Dissertations. Paper 802. USU Library, Logan, Ga.

Crookston, B. (2020). "A laboratory investigation on residual energy of nonlinear weirs." *Proc., 8th IAHR ISHS 2020.* The University of Queensland, QLD., Santiago, Cl., 10.14264/uql.2020.614.

Crookston, B., Erpicum, S., Tullis, B. and Laugier, F. (2019). "Hydraulics of labyrinth and piano key weirs: 100 years of prototype structures, advancements, and future research needs." *J. Hydraul. Eng.* 145(12), 02519004.

Crookston, B. and Tullis, B. (2012). "Arced labyrinth weirs". *J. Hydraul. Eng.*138(6), 555-562.

Crookston, B., Anderson, R. and Tullis, B. (2018). "Free-flow discharge estimation method for piano key weir geometries." *J. Hydro-Environ. Res.* 19(1570-6443), 160-167.

Dabling, M. and Crookston, B. (2012). "Staged and notched labyrinth weir hydraulics." *Proc., 4th IJREWHS*, USU Library, Logan, Ga. 1-9.

Dabling, M. and Tullis, B. (2017). "Modifying the downstream hydrograph with staged labyrinth weirs." *J. Appl. Water Eng. Res.* 6(3), 183-190.

Erpicum, S., Laugier, F., Boillat, J., Pirotton, M., Reverchon, B. and Schleiss, A. (2011). Labyrinth and Piano Key Weirs – PKW 2011, CRC Press Taylor & Francis Group, London

Erpicum, S., Laugier, F., Pfister, M., Pirotton, M., Guy-Michel, C. and Schleiss, A. (2013). Labyrinth and Piano Key Weirs II – PKW 2013, CRC Press Taylor & Francis Group, London

Erpicum, S., Laugier, F., Ho Ta Khanh, M. and Pfister, M. (2017). Labyrinth and Piano Key Weirs III – PKW 2017, CRC Press Taylor & Francis Group, London.

Falvey, H. (2003). Hydraulic design of labyrinth weirs, ASCE, Reston

Flow Science, Inc. (2020). FLOW-3D HYDRO User Manual, Release 12.0. Santa Fe, Ca.

Ghaderi, A., Daneshfaraz, R., Abbasi, S. and Abraham, J. (2020). "Numerical analysis of the hydraulic characteristics of modified labyrinth weirs." *Int. J. Energ. Water. Res.* 4(2020), 425-436.

Hager, W., Pfister, M. and Tullis, B. (2015). "Labyrinth weirs: developments until 1985." *Proc., 36th IAHR World Congress*, IAHR, The Hague, NL.
Hay, N. and Taylor, G. (1970). "Performance and design of labyrinth weirs." *J. Hydraul. Div.* 96(11), 2337–2357.
Hinchliff, D. and Houston, K. (1984). Hydraulic design and application of labyrinth spillways, U. S. Bureau of Reclamation Denver, Co.
Hirt, C. and Nichols, B. (1981). "Volume of fluid (VOF) method for the dynamics of free boundaries." J. Comput. Phys. 39(1), 201–225.
Hulsing, H., Winchell, S. and Cobb, E. (1966). Velocity-head coefficients in open channels, U.S. G.P.O. Washington, US.
Keshavarzi, A. and Hamidifar, H. (2018). "Kinetic energy and momentum correction coefficients in compound open channels." *Nat. Hazards* 92(3), 1859–1869.
Lopes, R., Matos, J. and Melo, J. (2008). "Characteristic depths and energy dissipation downstream of a labyrinth weir." Proc., 2nd IJREWHS, Pisa University Press, Pisa, IT.
Lopes, R., Matos, J. and Melo, J. (2011). "Flow properties and residual energy downstream of labyrinth weirs." *Proc., Labyrinth and Piano Key Weirs – PKW 2011,* CRC Press Taylor & Francis Group, London, 97–104.
Lopes, R., Matos, J. and Falcao de Melo, J. (2006). "Discharge capacity and residual energy of labyrinth weirs." *Proc., 1st IJREWHS*, The University of Queensland, Montemor-o-Novo, PT, 47–55.
Magalhães, A. and Lorena, M. (1994). "Perdas de energia do escoamento sobre soleiras em labirinto." *Proc. 6 SILUSB/1 SILUSBA*, Lisboa, PT, 203-211. (in Portuguese)
Mansoori, A., Erfanian, S. and Khamchin Moghadam, F. (2017). "A study of the conditions of energy dissipation in stepped spillways with Λ-shaped step using FLOW-3D." *J. Civ. Eng.* 3(10), 856-867.
Merkel, J., Belzner, F., Gebhardt, M. and Thorenz, C. (2018). "Energy dissipation downstream of labyrinth weirs." *Proc. 7th IAHR ISHS 2018.* USU Library, Logan, Ga., 508-517.
Roache, P. (1994). "Perspective: A method for uniform reporting of grid refinement studies." *J. Fluids Eng.* 116 (3), 405-413.
Schleiss, A. (2011). "From labyrinth to piano key weirs - A historical review." *Proc., Labyrinth and Piano Key Weirs – PKW 2011*, CRC Press Taylor & Francis Group, London, 3-15.
Singh, P., Naik, B., Tang, X., Khatua, K., Kumar, A. and Banerjee, S. (2019). "Models for kinetic energy and momentum correction coefficients for non-prismatic compound channels using regression and gene expression programming." *SN Appl. Sci.* 1(2019), 1229.
Taylor, G. (1968). The performance of labyrinth weirs. PhD thesis, University of Nottingham, Nottingham.
Tullis, B., Willmore, C. and Wolfhope, J. (2005). "Improving performance of low-head labyrinth weirs." *Proc., Impacts of global climate change. World Water and Environmental Resources Congress 2005*, ASCE, Reston, Va., 1-9.
Tullis, B., Crookston, B., Brislin, J. and Seamons, T. (2020). "Geometric effects on discharge relationships for labyrinth weirs." *J. Hydraul. Eng.* 146(10), 4020066.
Valero, D., Bung, D. and Crookston, B. (2018). "Energy dissipation of a type III basin under design and adverse conditions for stepped and smooth spillways." *J. Hydraul. Eng.* 144 (7), 04018036.
Versteeg, H. and Malalasekera, W. (2007). An introduction to computational fluid dynamics. The finite volume method. 2nd ed. Pearson/Prentice Hall, Harlow.
Wilcox, D. (2006). Turbulence modeling for CFD. 3rd Ed., CW Industries, Los Angeles.
Yakhot, V., Orszag, S., Thangam, S., Gatski, T. and Speziale, C. (1992). "Development of turbulence models for shear flows by a double expansion technique." *Phys. Fluids A, Fluid. Dyn.* 4(7), 1510–1520.

DOI: https://doi.org/10.26077/3e3e-fcc4

An Experimental Study on Scour at Zero Degree Confluent Channels

Mohd Faisal Ansari[1], Mohammad Aamir[2] & Zulfequar Ahmad[1]
[1]*Indian Institute of Technology Roorkee, Roorkee, India*
[2]*Chaitanya Bharathi Institute of Technology, Hyderabad, India*
Email: fai.ansari1991@gmail.com

***Abstract**: This paper presents an experimental study on scour depth at the confluence of a mobile and a rigid bed open channel. Substantial momentum difference between the confluent channels controls flow hydrodynamics and mixing in the confluence hydrodynamic zone. Scour at the confluence of the channels is a threat to the stability of the structure separating the two channels. Results for bed morphology and maximum scour depth at the confluence for different hydraulic and geometric conditions in the channels are presented. Data analysis shows that the discharge ratio (ratio of discharge in mobile bed and total discharge downstream) and tailwater conditions are key factors affecting the bed morphology and maximum scour depth. The maximum scour depth decreases by increasing the discharge ratio and tailwater depth.*

***Keywords:** bed morphology, confluence, hydrodynamics, open channel, scour*

1. Introduction

The amalgamation of rivers at channel nodes or confluences produces a complex hydrodynamic and morphodynamic environment within the fluvial framework. An abrupt variation in flow structures and associated fluvial hydraulics are the key features of the confluence. A better understanding of sediment transport, hydrodynamics, and bed morphology are vital in flood control, local sedimentation, and river management systems. The confluence hydrodynamic zone, a zone in close proximity to the confluence where hydrodynamic conditions, redirects the effects of flow convergence (Kenworthy and Rhoads 1995), are strongly influenced by confluence symmetry, junction angle, and the momentum flux ratio of the confluent flows (Best 1986; 1988).

Plentiful research covering the fluvial morphology of the channel confluence has been carried out for a range of junction angles (Mosely 1976; Ashmore and Parker 1983; Borghei and Sahebari 2010; Giglou et al. 2016). Rhoads et al. (2015) found a secondary circulation induced by lateral flow while merging with the main channel, resulting in a scour hole along the central portion of the confluence. Yuan et al. (2018) delineated detailing of the morphology of the right angle confluence channel for different discharge ratios. Their findings include a corner scour hole, bank attached bar, mid-channel bar, and mid-channel scour hole. Liu et al. (2015) observed that in the case of sediment-free supply from inlets, there was formation of a central scour hole alongside the shear plane progressing towards the tributary channel. Khosravinia et al. (2019) considered the effectiveness of the main channel side slope in connection with the scour hole position. An increment of the main channel side slope angle results in supplementary advancement in scour hole through the main and lateral channels. However, the penetration length of scour hole through the main and lateral channels was shortened with the rise of the side slope angle of the main channel. Bryan and Kuhn (2002) observed more influence of junction planform than junction angle on bed scour. Symmetrical confluence scour was detected in symmetrical (Y-shaped) confluences. However, asymmetrical junctions stirred to take more complex scour.

Laboratory studies have been performed to describe the spatial development of shallow mixing layers developing between two parallel channels with different inflow velocities (Uijttewaal and Booij 2000; Chu et al. 1991). The direct relevance of these studies with parallel (zero degree) open channel confluences limits the momentum ratio close to unity. The momentum ratio between two joining channels is a key parameter governing the flow hydrodynamics and mixing in the confluence hydrodynamic zone. Miyawaki et al. (2009) numerically compared the asymmetric confluence for momentum ratio close to unity, and high momentum ratio (~5) to understand the consequences of momentum ratio on flow structures and sediment entrainment capacity in the confluence hydrodynamic zone. For the momentum ratio close to unity, the streamwise-oriented vortices developing on both sides of the mixing layer are the primary vortices responsible for developing the scour hole. As the momentum ratio between the two streams increases, the scour is mainly motivated by the primary streamwise-oriented vortices forming on the high momentum side of the confluence and by the more coherent eddies shed in the mixing layer.

The entering flows at confluences are characterized by unequal flow, temperatures, and suspended sediment. In the field, two channels with considerable differences in their discharges meet at some angles, like downstream of the dam where spillway flow meets with the power channel. The available literature contains a sparse study on the confluence where there has been difference in incoming flow and meeting at zero degree.

The present study deals with the open channel confluences where two incoming parallel channels separated by a wall meet with considerable differences in their discharges. The bed of the main Channel-1 was protected up to the confluence point, while Channel-2 has mobile bed throughout. This study investigates the bed features such as scour profile, bed morphology, and maximum scour depth adjacent to the confluence region for different discharge ratios (discharge in low discharge carrying channel i.e., Channel-2/ total discharge downstream) and tailwater depth conditions.

2. Experimental Setup

All the experiments were conducted in the hydraulics laboratory of the Civil Engineering Department, Indian Institute of Technology Roorkee, in a flume of 10 *m* length, 0.60 *m* width and 0.54 *m* depth. A photographic view of the setup is shown in Figure 1. Figure 2 shows the plan and sections of the experimental setup. The discharge in the flume was regulated with the help of valves provided in the inlet pipes coming from an overhead tank. Two parallel channels of equal width were separated by an acrylic sheet having thickness of 0.005 *m* extending up to the confluence point. An ultrasonic flow meter was used for measuring the discharge in the pipes. A point gauge of least counts of 1 *mm* attached to the mobile carriage was used to measure the bed levels in the longitudinal and transverse directions of the channel. A honeycomb masonry grid wall was provided at the upstream end of the flume to exterminate large size eddies in the flow entering into the flume. Flow straighteners were provided downstream of the honeycomb masonry wall for sustaining the flow parallel to flume walls. A sediment recess of 0.25 *m* depth and 1.0 *m* length was provided downstream of the confluence point. A power-driven tailgate arrangement was provided at the downstream end of the flume to maintain tailwater level d_t (0.07 *m*, 0.1 *m*, and 0.13 *m*).

For each experimental run, the mean size of grain (d_{50}) was taken as 1.92 *mm*, having a geometric standard deviation $\left(\sigma_g = \sqrt{\frac{d_{84}}{d_{16}}}\right)$ less than 1.4, confirming the uniformity of the grain (Marsh et al. 2004). A total of nine experiments were run for three different tailwater levels and discharge ratios. The experiment was performed for three discharges in the main Channel-1 i.e., $Q_1 = 0.019\ m^3/s$, $0.02\ m^3/s$, $0.022\ m^3/s$ and for two discharges in the channel-2 i.e., $Q_2 = 0.002\ m^3/s$ and $0.003\ m^3/s$. Total downstream discharge (Q_3) was maintained at $0.022\ m^3/s$ for each experimental run. Three discharge ratios [$Q_r\ (=Q_2/Q_3) = 0, 0.09, 0.14$] and three tailwater depths (d_t = 0.07 *m*, 0.1 *m* and 0.13 *m*) were taken into investigation to study the development of scour depth adjacent to the confluence region. Clear water condition prevailed in the flume. The bed material was sheltered with plywood till the steady state condition in the flume was reached, in order to avoid erosion of bed material in the initial stage of flume operation. Upon reaching the equilibrium scour condition, the flow was stopped in the flume, and scour profile was measured with the help of a point gauge.

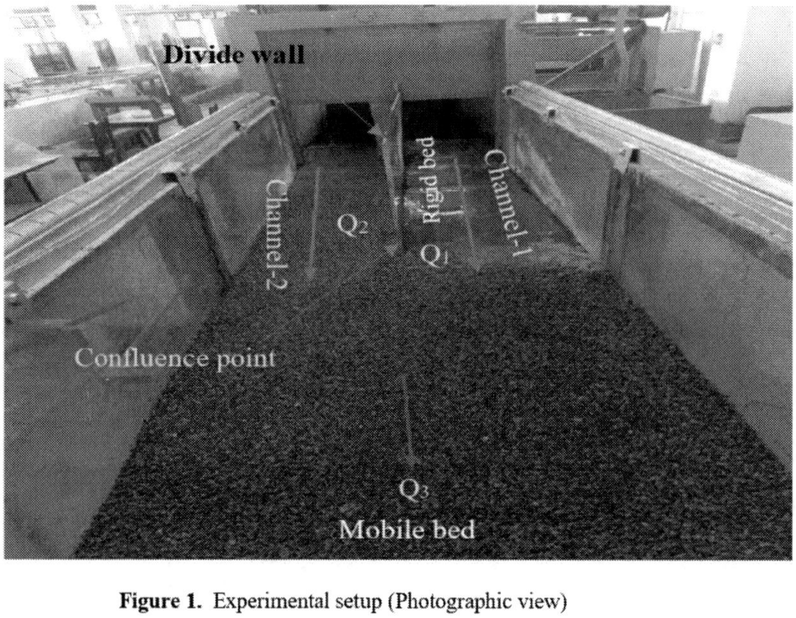

Figure 1. Experimental setup (Photographic view)

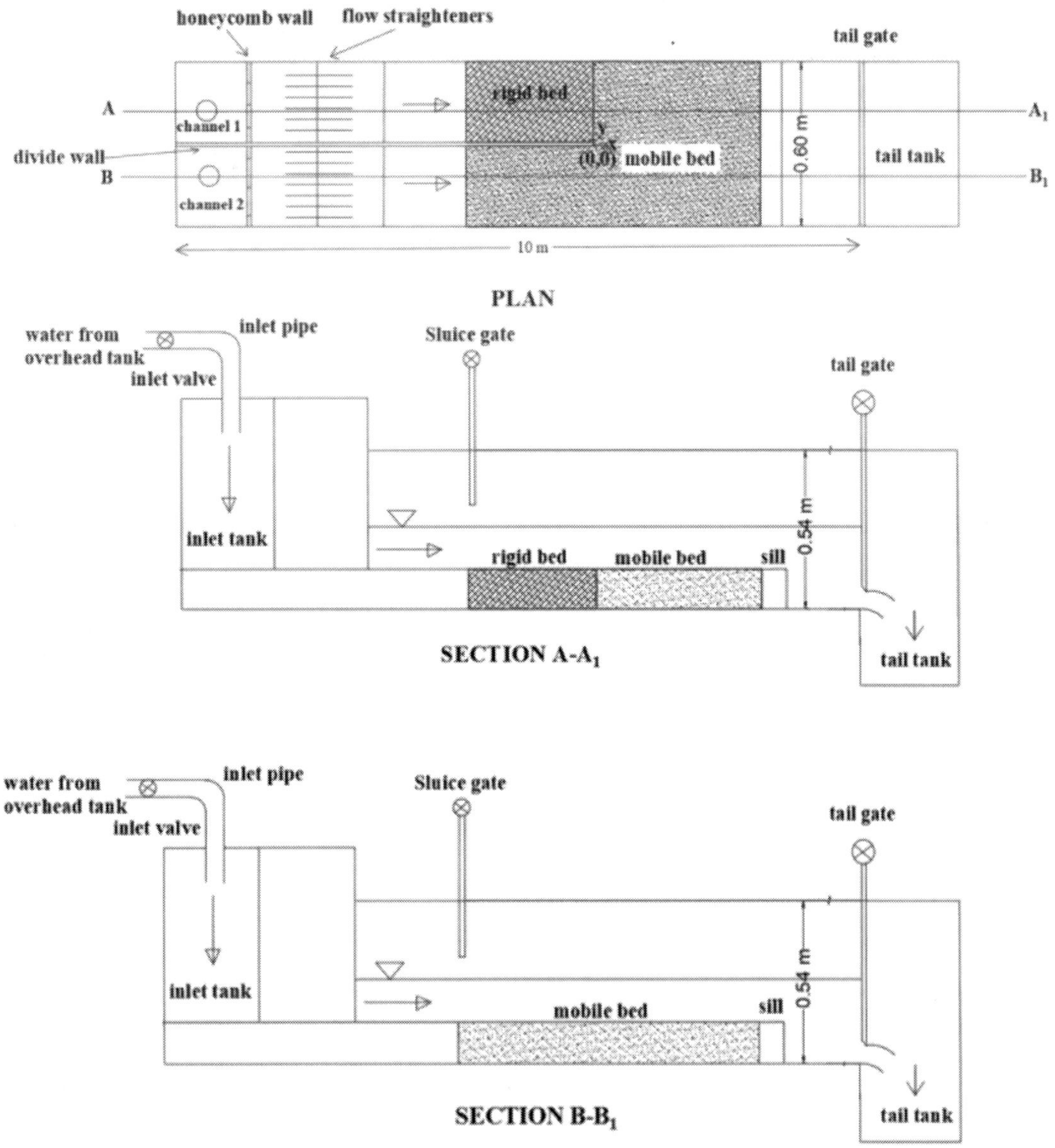

Figure 2. Schematic diagram of the experimental set up

3. Results and Discussion

3.1. Scour Profiles

In the experimental flume, the coordinate system was selected such that x-axis represents the streamwise distance along the length of the flume ($x = 0$ being the point of confluence), y-axis represents the lateral distance across the width of the flume ($y = 0$ being the point of confluence), and z-axis represents the depth (or height) in the vertical direction. Scour profiles were plotted at two longitudinal sections of the flume, one at $y = 0$ and the other at that section where the maximum scour depth occurred for a particular run. Figures 3 (a-c) show the scour profiles for constant d_t (= 0.1 m) and different values of Q_r. It is observed that for $Q_r = 0$, maximum scour depth occurs at $y = 0.11$ m; for $Q_r = 0.09$, maximum scour depth occurs at $y = 0.13$ m; and for $Q_r = 0.14$, maximum scour depth occurs at $y = 0.14$ m. Hence, it can be concluded that with increasing Q_r, the maximum scour depth shifts away from the confluence towards channel-1. It is also seen from the figures that at $y = 0$, a distinct scour hole geometry is not observed. Rather, the scour hole is in the form of ripples with two depressions. Other conditions remaining the same, recirculation length tends to increase with decreasing channel depth (Hen et.al 2020). Hence, more recirculation causes more scour in channel-2. Scour in channel-2 is primarily caused by the recirculating flow originating downstream of the confluence. Low discharge ratios and low tailwater depths are favorable conditions for the recirculation of flow in channel-2, consequently causing more scour which threaten the stability of the divide wall. Maximum extent of scour in channel-2 reached up to a section 0.25 *m* upstream from the confluence point, and scour depth was observed to increase on advancing towards the confluence.

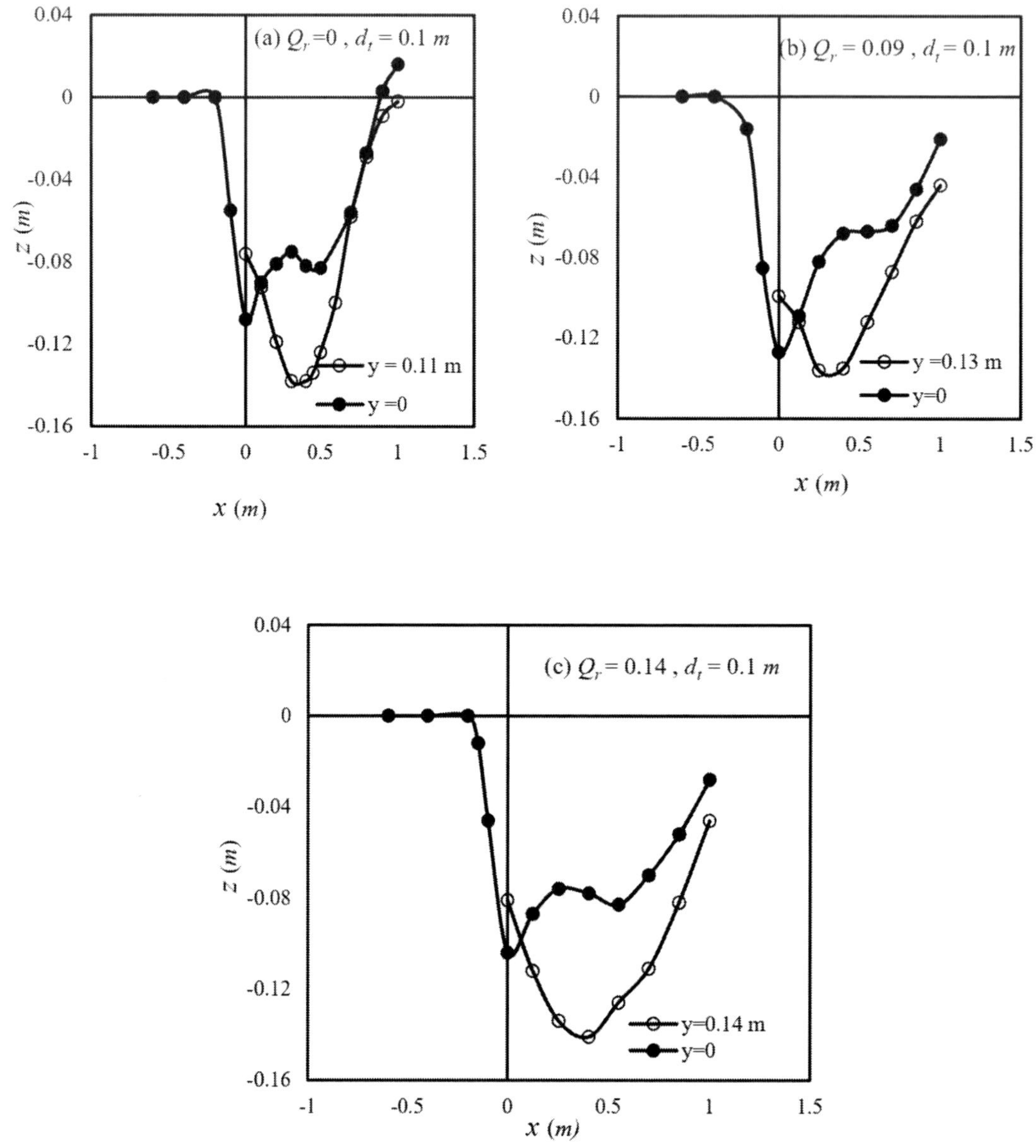

Figure 3. Streamwise scour profiles for $d_t = 0.1$ *m* and (a) $Q_r = 0$; (b) $Q_r = 0.09$; and (c) $Q_r = 0.14$

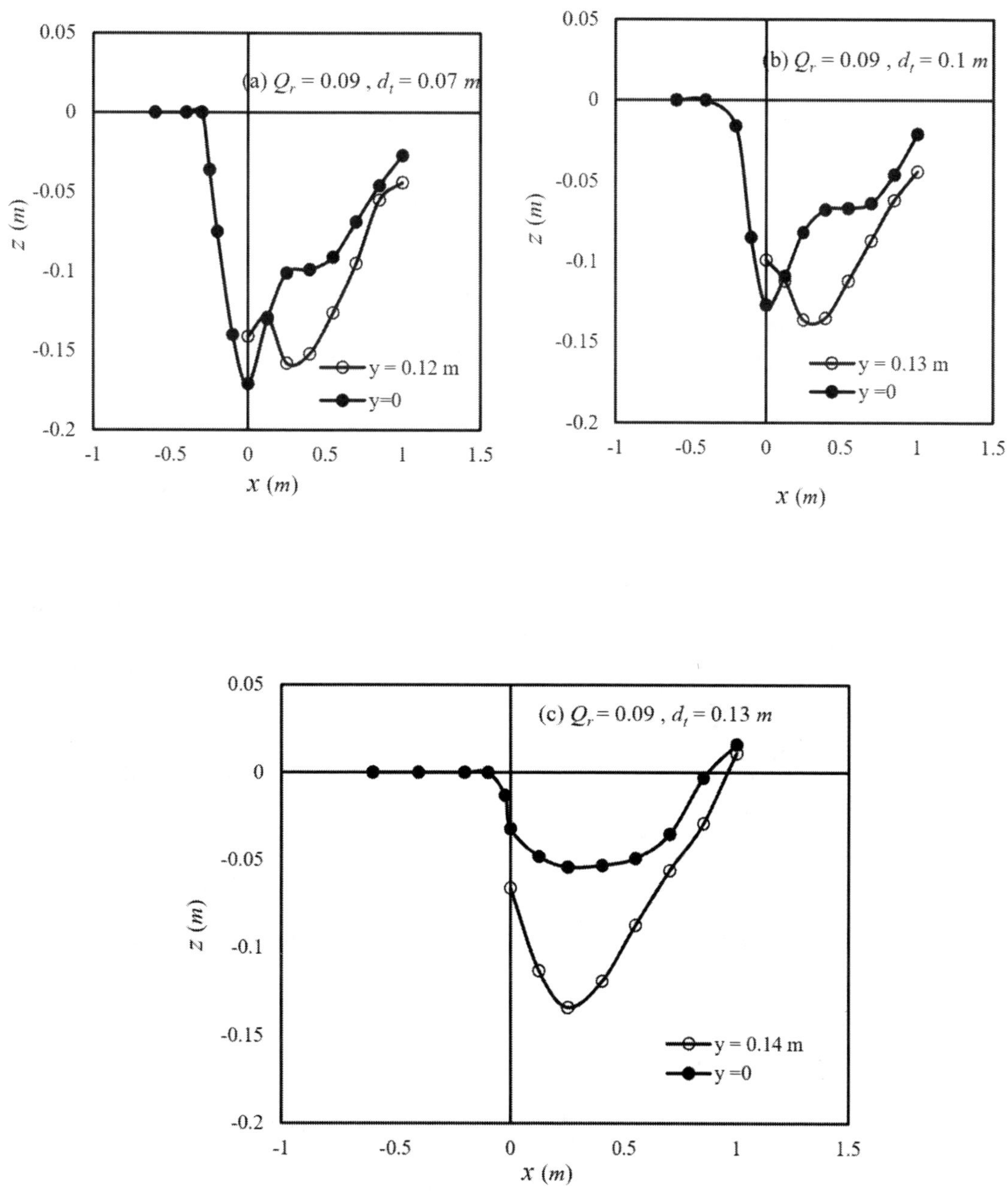

Figure 4. Streamwise scour profiles for *Qr* = 0.09 and (a) *dt* = 0.07 m; (b) *dt* = 0.1 m; and (c) *dt* = 0.13 m

Figures 4 (a-c) show the scour profiles for constant Q_r (= 0.09) and different values of d_t. It is observed that for d_t = 0.07 *m*, maximum scour depth occurs at y = 0.12 *m*; for d_t = 0.1 *m*, maximum scour depth occurs at y = 0.13 *m*; and for d_t = 0.13 *m*, maximum scour depth occurs at y = 0.14 *m*. Hence, it can be concluded that with increasing d_t, the maximum scour depth shifts away from the confluence towards channel-1. A distinct scour hole geometry is observed in all the three cases at y = 0 also. The streamwise distance for maximum scour depth was observed in the range of $0.3\ m \leq x \leq 0.5\ m$ for all the experimental runs. Maximum scouring capacity of the flow is exhausted in this range, as most of the energy is dissipated. Hence, the scouring capacity gets reduced further downstream. Some part of scoured materials gets deposited in the downstream of the channel. However, some amount of sediment gets washed away with the flow. For the condition of high tailwater depth, the scouring capacity of the flow gets reduced, while downstream deposition is increased. Maximum scour depth occurred for the condition of discharge ratio Q_r = 0, and tailwater depth d_t = 0.07 *m* in all the experimentation runs.

3.2. Bed Morphology

3.2.1. Effect of Discharge Ratio on Bed Morphology

Figures 5 (a,b,c) show the effect of discharge ratio on bed morphology. Where ratio of scour depth to channel width (z/b) represents the intensity of the scour and deposition in the channels. Maximum scour depth (z_{max}) for Q_r = 0, when no flow was supplied in channel-2, was observed at the confluence. As the flow in channel-2 increases, maximum scour depth shifts downstream of channel-1. After traversing the rigid bed, discharge (Q_1) flowing in channel-1 encounters the mobile bed, it results in the formation of scour hole downstream of the confluence. Momentum difference between two channels, carrying discharges Q_1 and Q_2, gives rise to the formation of coherent structure responsible for scour and deposition in the channel. An abrupt alteration in channel boundary gives rise to flow separation, which includes a region of strong transverse shear just downstream of the separating wall, giving rise to large turbulence intensities and a region of flow recirculation. Recirculating flow leads to deposition and scouring in the channel. The intensity of recirculating flow depends upon the discharge ratio and tailwater depth. In case of high discharge ratio and high tail water depth, a weak recirculating flow was formed in the channel. However, low discharge ratio and low tailwater depth enhance the intensity of recirculation of the flow, causing an increase in scour hole. As the discharge ratio was increased, an increase in deposition of sediment occurred near the right bank of the channel, while a reduction was observed in the maximum scour depth in the vicinity of the confluence. Moreover, a reduction in the upstream advancement of the scour hole was observed in channel-1 on increasing the discharge ratio. It was also observed that scour in channel-2 occurred only by the effect of flow in channel-1, either by direct impingement or recirculation of flow. Scour process begins from the point of confluence and proceeds laterally and longitudinally. Further downstream of the confluence (x=0.8 m), scour hole dimension decrease as transverse shear flow weakens.

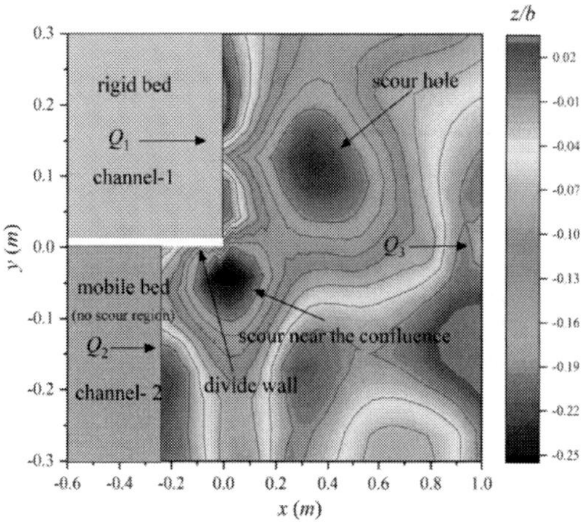

Figure 5 (a). Bed morphology for Q_r = 0

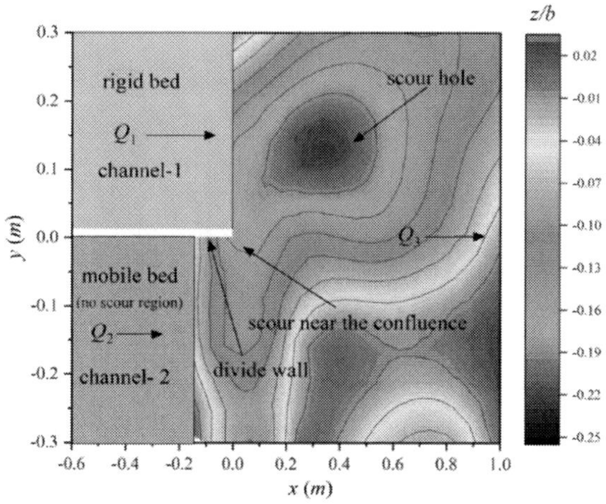

Figure 5 (b). Bed morphology for Q_r = 0.09

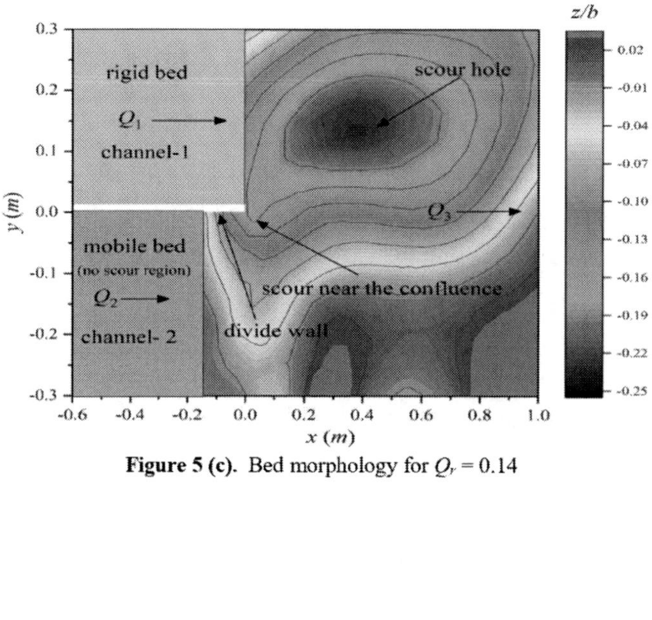

Figure 5 (c). Bed morphology for Q_r = 0.14

3.2.2. Effect of Tailwater Depth on Bed Morphology

Figures 6 (a,b,c) show the effect of tailwater depth on the bed morphology. Scour depth at the confluence and downstream of channel-1 increases on decreasing the tailwater depth. On increasing tailwater depth, scour depth decreases, and deposition of sediment on the right bank of the channels increases. Figure 6 (a) shows the bed morphology for tailwater depth $d_t = 0.07$ *m*, where two scour holes, one at the confluence and the other downstream of channel-1 were observed. Further, there was minor sediment deposition in the channels. Figure 6 (b) depicts two scour holes and deposition of sediment on the right bank. As the intensity of recirculation gets reduced on increasing the tailwater depth, deposition of sediment occurs downstream of the confluence. In Figure 6 (c), only one scour hole downstream of channel-1 was observed. Extent of scour upstream of the confluence point in channel-2 also increases by lowering the tailwater depth. Maximum scour extension upstream of confluence point in channel-2 was observed as 0.25 *m* at 0.07 *m* tailwater depth.

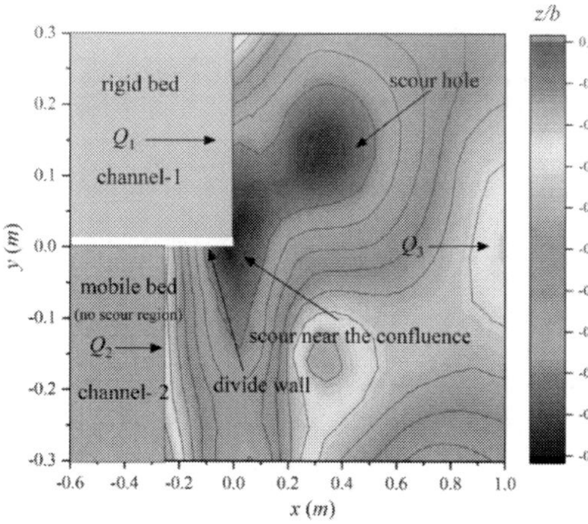

Figure 6 (a). Bed morphology for tailwater depth, $d_t = 0.07$ *m*

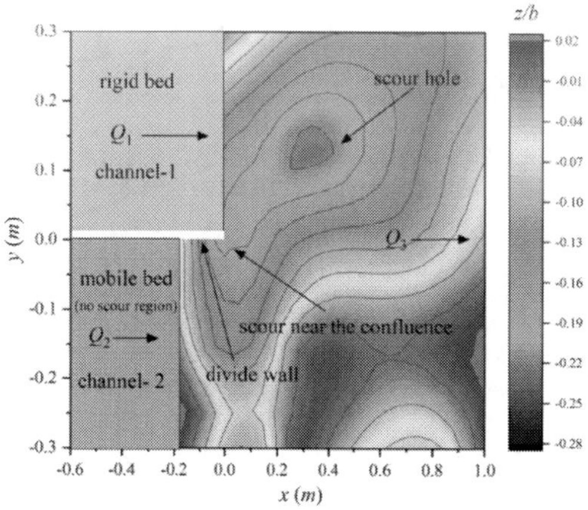

Figure 6 (b). Bed morphology for tailwater depth, $d_t = 0.1$ *m*

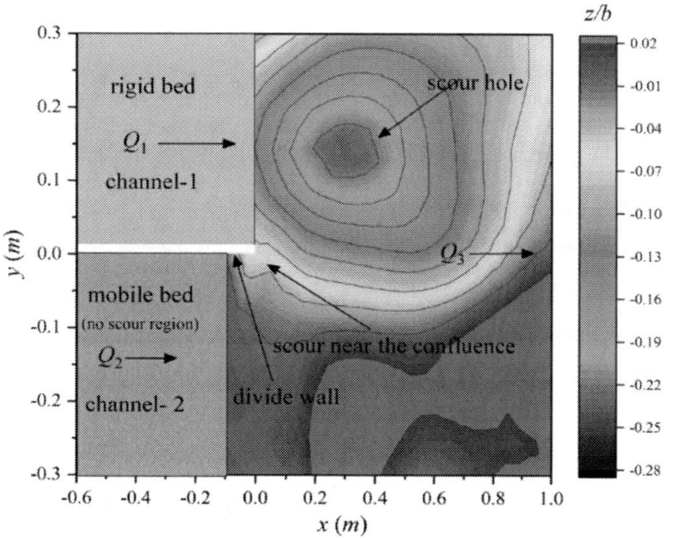

Figure 6 (c). Bed morphology for tailwater depth, $d_t = 0.13$ *m*

3.3. Effect of Discharge Ratio and Tailwater Depth on Maximum Scour Depth

Experiments were performed for three tailwater conditions. For each tailwater depth, three different values of discharge ratio were taken to study the maximum scour depth in the vicinity of the confluence. Figure 7 shows the variation of maximum scour depth with the discharge ratio for three different tail water conditions. Increasing the discharge ratio helps reduce the maximum scour depth because higher Q_2 prevents the advancement of recirculating flow towards the upstream of channel-2. Maximum scour depth(z_{max}) was observed increasing with decreasing tailwater depth, as shown in Figure 7.

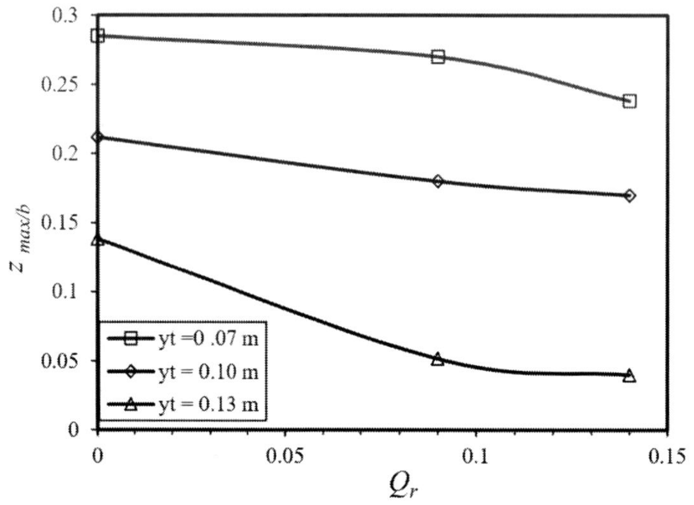

Figure 7. Showing the effect of discharge ratio on maximum scour depth

4. Conclusion

Development of scour hole in the vicinity of the confluence was investigated experimentally. Streamwise scour profiles passing through the confluence point and downstream of channel-1 were plotted for the different discharge ratios and tailwater depths. Scour profiles revealed that for high tailwater depth, maximum scour depth occurs at downstream of channel-1, However, no significant difference was observed in the maximum scour depth for lower tailwater conditions. Bed topography for three discharge ratios and three tailwater depths depicted a clear understanding of scouring patterns in the confluence and post confluence regions. Two scour holes were observed, one originating from the confluence point and advancing towards the upstream of channel-2, and the other occurring downstream of the solid apron of channel-1. More deposition of the sediment was also observed for high discharge ratio near the right bank of the channel. The maximum scour depth was found to decrease with increasing discharge ratio and decreasing tailwater depth.

5. REFERENCES

Ashmore, P.E., Parker, G. (1983). "Confluence scour in coarse braided streams." *Water Resour. Res.*, 19(2), 392-402.

Best, J.L. (1986). "The Morphology of river channel confluences." *Prog Phys Geogr.*, 10(2), 157-174.

Best, J.L. (1988). "Sediment transport and bed morphology at river channel confluences." *Sedimentology.*, 35(3), 481-498.

Borghei, S.M., and Sahebari, A.J. (2010). "Local scour at open-channel junctions." *J. Hydraul. Res.*, 48(4), 538-542.

Bryan, R. B., and Kuhn, N. J. (2002). "Hydraulic conditions in experimental rill confluences and scour in erodible soils." *Water Resour. Res.*, 38(5), 21-1.

Chu, V.H., and Babarutsi, S. (1988). "Confinement and Bed-Friction Effects in Shallow Turbulent Mixing Layers." *J. Hydraul. Eng.*, 114(10), 1257-1274.

Han, L., Riviere, N., Chatelain, M., Mignot, E. (2020). "Recirculation zone downstream lateral expansions of open channel flow." *Phys. Fluids.*, 32(11), 115119.

Giglou, A.N., Sahebari, A.J., Shakibaeinia, A., Borghei, S.M. (2016). "An experimental study of sediment transport in channel confluences." *Int. J. Sediment Res.*, 31(1), 87–96.

Khosravinia, P., Nikpour, M.R., Malekpour, A., Hosseinzadeh Dalir, A. (2019). "Effect of side slope of main channels on formation and penetration of scour hole in confluences." *River Res. Appl.*, 35(2), 159-168.

Liu, T., Chen, L., Fan, B. (2012). "Experimental study on flow pattern and sediment transportation at a 90° open-channel confluence." *Int. J. Sediment Res.*, 27(2), 178-187.

Liu, T., Fan, B., Lu, J. (2015). "Sediment–flow interactions at channel confluences: a flume study." *Adv. Mech. Eng.*, 7(6), 1-9.

Marsh, N.A., Western, A.W., Grayson, R.B. (2004). "Comparison of methods for predicting incipient motion for sand beds." *J. Hydraul. Eng.*, 130(7), 616-621.

Miyawaki, S., Constantinescu, G., Rhoads, B., Sukhodolov, A. (2010). "Changes in three-dimensional flow structure at a river confluence with changes in momentum ratio." *River flow* (2010)., S. 225-232.

Mosley, M.P. (1976). "An Experimental Study of Channel Confluences." *J. Geol.*, 84(5), 535-562.

Rhoads, B.L., and Kenworthy, S.T. (1995). "Flow structure at an asymmetrical stream confluence." *Geomorphology.*, 11(4), 273-293.

Riley, J.D., Rhoads, B.L., Parsons, D.R., Johnson, K.K. (2015). "Influence of junction angle on three-dimensional flow structure and bed morphology at confluent meander bends during different hydrological conditions." *Earth Surf. Process. Landf.*, 40(2), 252-271.

Uijttewaal, W.S.J., and Booij, R. (2000). "Effects of shallowness on the development of free-surface mixing layers." *Phys. Fluids.*, 12(2), 392-402.

Yuan, S., Tang, H., Xiao, Y., Qiu, X., Xia, Y. (2018). "Water flow and sediment transport at open-channel confluences: An experimental study." *J. Hydraul. Res.*, 56(3), 333–350.

Guillen-Ludena,S.,Franca,M.J., Cardoso, A. H., Schleiss, A. J. (2016). "Evolution of the hydromorphodynamics of mountain river confluences for varying discharge ratios and junction angles." *Geomorphology.*, 255(2016),1-15.

Leite Ribeiro, M., Blanckaert, K., Roy, A. G., Schleiss, A. J. (2012). "Flow and sediment dynamics in channel confluences." *J. Geophys. Res.,*117 (2012), 1-19.

Leite Ribeiro, M. L., Blanckaert, K., Roy, A. G., Schleiss, A.J. (2012). "Hydromorphological implications of local tributary widening for river rehabilitation." *Water Resour. Res.,*48(10), 1-19.

Xie, Q., Yang, J., Lundstrom, S., Dai, W. (2018). "Understanding Morphodynamic Changes of a Tidal River Confluence through Field Measurements and Numerical Modeling". *Water.,*10(10), 1424.

Xie, Q., Yang, J., Lundstrom, T.S. (2020). "Flow and sediment behaviours and morpho-dynamics of a diffluence-Confluence unit." *River Res. Appl.,* 36(8),1515-1528.

DOI: https://doi.org/10.26077/60b8-4263

Bank scour protection using spur dyke in a meandering channel under low flow velocity

Nishank Agrawal[1] & Ellora Padhi[1]
[1]*Shiv Nadar University, Greater Noida, Uttar Pradesh, India*
E-mail: na886@snu.edu.in

Abstract: *In geophysical flows, the channel often starts to meander when it encounters a plan terrain. Owing to higher flow velocity, it tends to erode the channel boundary (bed and banks) that comes across its path. Scouring in a meandering channel primarily depends on channel boundary, flow condition, and sediment properties. To date, numerous works have been conducted to understand the velocity distribution and scouring pattern in a meandering channel under varying discharge conditions. However, only a handful of work has been reported on the scour pattern under low discharge conditions in a meandering channel. Therefore, in this paper, an experimental study was conducted under a very low discharge in an 80° (central angle θ) bend to analyse the scour pattern. Investigation of the scour profile reveals that maximum scour occurs at the downstream side of the bend apex rather than the bend apex. Hence, an attempt has been made to counter the scour using the spur dyke. In doing so, a non-submerged spur dyke was installed at three different locations θ = 0°, 40°, and 60° of bends with three orientations α = 60°, 90°, and 120° for approaching and developed flow conditions to identify the best location and orientation of the spur dyke. Observation suggests that spur dyke at θ = 60° with α = 90° and spur dyke at 'θ' = 40° with 'α' = 60° provide the best results for approaching and developed flow conditions, respectively, in terms of preventing scour under low flow velocity.*

Keywords: *Scouring, Meandering channel, Spur dyke, Dyke orientation, Low flow velocity.*

1. Introduction

In the present age, government agencies are emphasizing more for making inland waterways for navigation. For making waterways, hydraulic engineers have to face several problems such as uneven bed topography of the rivers, low discharge or low water volume in the river, and sometimes meandering river patterns. Among these problems, the meandering river pattern makes the task more challenging for the engineers, as the river continuously changes its boundary (bed and banks) over the years. Several theories and arguments are directly associated with river meandering, as mentioned by Rhoads and Welford (1991). These arguments include Coriolis force, helical and secondary flows, turbulent flow, energy arguments, bank erosion, sediment effect, the interaction between flow and mobile channel, bar theory, and bend theory. According to Seminara (2006), instability mechanisms due to the flow and erodible boundary interaction are the main cause of meandering, and he summarized that "*According to bend instability theory, any minor random perturbation of channel alignment gradually increases, resulting in a meandering pattern*". In a meandering river, flow erodes the soil in the outer bank and deposits it to the inner bank, and this process goes on uninterruptedly. To overcome this issue, the installation of spur dykes is come out as an effective solution since the placement of spur dyke at the concave side of the river bank deviates the flow and saves the banks from further erosion. Spur dyke, also called groin, is a very important hydraulic structure with one end projecting across stream current and another embedded in the river bank. Depending upon the spur dyke projecting face, the flow gets modified; for example, a spur dyke with an upstream projected face repels the incoming flow, whereas a spur dyke with a downstream projected face attracts the flow; however, a spur dyke that remains perpendicular to the streamwise direction always diverts the flow towards the center of the channel. At the toe of the spur dyke, intense vortex movement occurs due to the continuous interaction between dyke projected face and the incoming flow. Owing to these intermittent vortices, the bed materials remain in suspension, making it easier for the river to take them towards the downstream side. The scour hole that forms around the spur dike, and the angle of repose of the bed material, are the key determinants of extent of bank erosion in the dyke's vicinity. Therefore, an estimate of expected scour at the spur dyke's nose is required to provide spur dyke depth greater than scour hole near dyke (Copeland (1983)). Scouring around spur dyke is a very complex phenomenon to understand. To understand this phenomenon, many researches have been carried out to ascertain the flow pattern, spatial and temporal variation of local scour around dykes in straight and curved channels (Garde et al. 1961, Gill 1972, Rajaratnam and Nwachukwu 1982, Kothyari and Ranga Raju 2001, Haltigin et al. 2007, Karami et al. 2012, Fang et al. 2014, Pandey et al. 2021). Garde et al. (1961) proposed that the ratio of scour depth near dyke to the approach depth is a function of Froude number, length to width ratio (*L*/*B*) of the dyke, and densimetric Froude number. Gill (1972) did experiments on scour of sand bed around spur dyke and suggested that maximum scour depth depends on geometric contraction ratio and the ratio of sediment size to approach depth. Rajaratnam and Nwachukwu (1982) stated that shear stress at the upstream corner of dyke is five times of approaching bed shear stress. Zaghloul (1982) saw the effect of upstream flow condition, sediment characteristics, and spur dyke geometry on the maximum scour depth and scour pattern around a dyke and concluded that maximum scour depth is affected by Froude number and the opening ratio at spur dyke location. Also, in his study, it was found that the depth of scouring progresses linearly with the logarithm of time during the development of the hole. Mayerle et al. (1995) verified the three-dimensional numerical modal simulation to see the effect of the eddy viscosity field on velocities, flow depths, and reattachment lengths. Elawady et al. (2001) conducted an experimental study on movable bed scour around submerged spur dyke and observed that the combination of longer dyke length and low flow depth results in a larger scour area than the combination of shorter dyke length and higher flow depth. Kothyari and Ranga Raju (2001) proposed that scour mechanism around dyke was similar to scouring around the pier. Kuhnle et al. (2002) did experiments to see the effect of spur dyke orientation and contraction ratio on the volume of scour and potential aquatic habitat. Haltigin et al. (2007), Karami et al. (2012), and Fang et al. (2014) performed numerical studies in the straight channel to know the temporal and spatial variation of scour around spur dyke by changing flow properties (velocity, discharge, Froude number, densimetric Froude number), sediment properties (Median diameter d_{50}, geometric standard deviation $(d_{84}/d_{16})^{0.5}$, etc.), and dyke's dimension (length, thickness, orientation, spacing, shape, contraction ratio, etc.). Pandey et al. (2021) used the stacked generalization ensemble tree-based frameworks machine learning technique to predict the scour around dyke.

When spur dyke is placed in channel bend, then flow behaviour, local scour around dyke become difficult to understand. The complexity arises due to circulatory flow, transverse rotational flow, and secondary flow due to centrifugal acceleration (Pandey et al., 2018). Quite a few investigations were carried out for experimental work to know the flow behaviour and local scour around the dyke in the curved channel (Fazli et al., 2008, Vaghefi et al., 2009, Masjedi et al., 2011, Radan and Vaghefi, 2016, Karki et al., 2018, Sharma and Mohapatra, 2020, Tripathi and Pandey, 2021). Fazli et al. (2008), Vaghefi et al. (2009), Vaghefi et al. (2015), and Radan and Vaghefi (2016) studied the scour phenomenon in 90° channel bend around T shape spur dyke. Masjedi et al. (2010, 2011) conducted experiments in 180° channel bend to study scour around T shape spur dyke. Sharma and Mohapatra (2020), studied the effect of bed roughness on flow field around spur dyke in meandering channel and inferred that turbulence intensities decrease around the dyke with an increase in roughness. Tripathi and Pandey (2021) worked on local scour around T shape in a meandering channel and proposed a non-dimensional empirical equation for temporal and maximum scour depth as a function of Froude number and location of the dyke (Measured from entry to meander).

To date, many studies exist on spur dyke installed in straight, curved, and meandering channels, but they are mainly confined to local scour around the dyke and the flow behaviour around the dyke in straight and curved channels only. In fact, very minimal work has been reported for the design of spur dyke in the meandering channel. Although the knowledge of scouring and flow behaviour around dyke help in designing of spur dyke, the effect of spur dyke on bed morphology will give a better understanding on the installation of spur dyke, such as dyke location (where to install), orientation (what should be the angle of the dyke with respect to the flow) and spacing between the dykes (minimum and maximum distance between two dyke locations). Karki et al. (2018) did experiments in sine-generated erodible meandering channels with and without groynes by changing the sinuosity and curvature of the channel to know the evolution of flow and morphological variation. They observed that low sinuosity and low central radius to channel width ratio (R_c/W) are more susceptible to erosion and morphological change. However, the authors did not show how the orientation and location of spur dykes affect the channel bed morphology.

Moreover, all the research studies related to spur dyke have been carried out so far under high discharge conditions, and such consideration does not match the reality. In a practical scenario, rivers do not always flow under high discharge conditions. In fact, except monsoon season, most of the meandering rivers flow under low flow velocity. Hence, it is also important to see the behaviour of the spur dyke under such circumstances. Therefore, in the present study, an attempt has been made to show the impact of dyke orientation and location on the bed morphology in the meandering channel for a low discharge condition. The specific objectives of the research have been to study the bed variation with and without installation of spur dyke; to minimize the erosion at the outer bank with the help of spur dyke; to determine the best location and orientation of spur dyke to reduce the scouring near the bank.

2. Experiments

Experimental runs were conducted in the Advanced Post Graduate Hydraulics Laboratory, Department of Civil Engineering, Zakir Hussain College of Engineering and Technology, Aligarh Muslim University, Aligarh. The data were collected in an open horizontal rectangular sinuous (meandering/curved) channel (0.35 m wide and 0.43 m deep) made up of 0.5 mm thick tin sheet, carefully installed in an open horizontal rectangular flume (0.76 m wide and 0.60 m deep and 10.5 m long) prismatic glass walled channel with cement plastered bottom. In the channel, the bed is prepared with 0.22 m height of sand throughout the channel. Sand size d_{50} as 0.27 mm was used for preparing the channel bed. A schematic diagram and photograph of the experimental setup are shown in Fig. 1a and Fig. 1b, respectively. The model had a straight upstream reach of 1.68 m and a straight downstream reach of 0.6 m. Both upstream and the downstream reach were connected by two sinuous bends having the same dimensions. Further, the two sinuous bends comprised four 80° curved channel bends and were provided in a series. Each bend had a rectangular cross-section with 0.35 m width (W) and 0.43 m height. The central angle of each bend is 80°, and the central radius of the channel (R_c) is 0.705 m. The Ratio of central radius to width of the channel $R_c/W = 2.014$ (< 3), and a straight transition of 0.05 m was maintained between each bend.

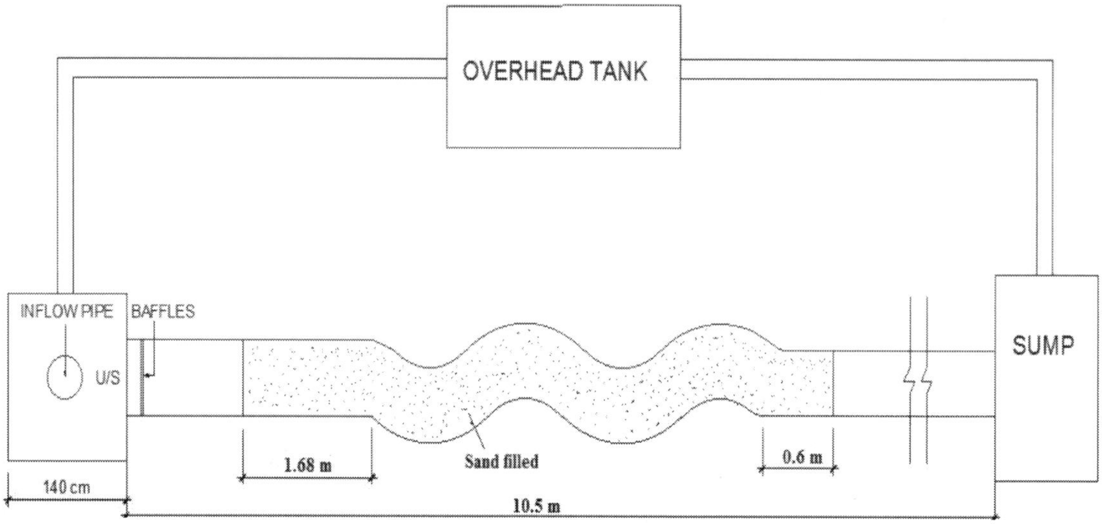

Figure 1a. Schematic diagram of the experimental channel.

Figure 1b. Photograph of the experimental flume.

In the present study, a spur dyke was made by a wooden sheet, having width b and thickness t equal 0.075 m and 0.015 m, respectively. The width of the spur dyke was decided on the basis of the contraction ratio (Ratio of dyke's width to channel width $b/W = 0.214$). Breusers and Raudkivi (1991) demonstrated that if the contraction ratio is less than 0.4, it has no influence on local scour. Özyaman et al. (2022) concluded that the effect of the opposite bank was observed for the contraction ratios of 0.29 for uniform sediment and 0.36 for non-uniform sediment. Spur dyke of non-submerged condition was installed one by one at three different locations based on the angular displacement by varying orientation of dyke at an angle α (measured from the tangent of the right-hand side of the channel in the clockwise direction) equals 60°, 90°and120° in Bend 1 and Bend 3 simultaneously. The details of bend showing angular displacement and orientation of spur dyke are given in Fig 2.

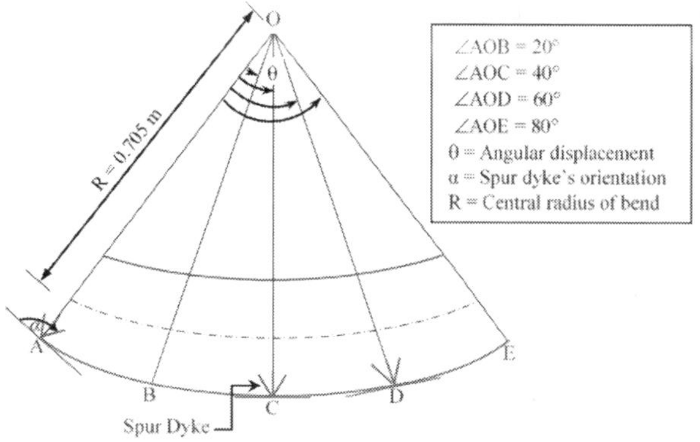

Figure 2. Details of the channel bend.

For the discharge measurement, a sharp crested calibrated rectangular weir was provided at the end of the downstream drain channel. A point gauge (with 0.10 mm accuracy) was used to record the head on the weir. In all the experimental runs, discharge Q was recorded as 4.5×10^{-3} m^3 s^{-1}, flow depth h was kept as 0.045 m, cross-sectional averaged velocity V and the flow Froude number was found to be 0.286 m s^{-1}, and 0.43, respectively, indicating the flow condition to be subcritical.

At first, the test bed was prepared by filling the well graded sand having d_{50} as 0.27 mm with a uniform compaction method. It is pertinent to mention that in the present study, both b/d_{50} (= 277.78) and t/d_{50} (= 55.56) are greater than 50. Hence scale effect of sediments can be neglected. According to Raudkivi (1986), scouring around flow obstruction becomes independent when b/d_{50} is greater than 50. These results had also been verified by Lee and Sturm (2009) by considering both field and laboratory data. It was ensured that the test bed was properly compacted and smoothened. Then, water is allowed to flow continuously at a rate of 4.5×10^{-3} m^3 s^{-1} over the bed for 3-4 hours. Thereafter, the flow was stopped, and the changes in the bed surface elevations were measured at different locations along the longitudinal and transverse directions. In the second phase, a spur dyke was installed in Bend 1 and Bend 3 simultaneously, keeping the dyke location and orientation the same. Then, another experimental run carried out under the same flow condition until the equilibrium state was achieved, and the duration of the run was around 12 hours. After that the bed surface elevation was measured. To understand the impact of spur dyke location and orientation on the sediment transport process, additional experimental runs were carried out for different angles of displacements and orientations for the same discharge $Q = 4.5 \times 10^{-3}$ m^3 s^{-1} and bed elevation data under each run were collected for analysis. The details of the experimental runs are presented in Table 1.

Table 1. Experimental Scheme

Run no.	Location of dyke	Orientation of Dyke
1	Without Spur Dyke	-
2		$\alpha = 60°$
3	Location 1 ($\theta = 0°$)	$\alpha = 90°$
4		$\alpha = 120°$
5		$\alpha = 60°$
6	Location 2 ($\theta = 40°$)	$\alpha = 90°$
7		$\alpha = 120°$
8		$\alpha = 60°$
9	Location 3 ($\theta = 60°$)	$\alpha = 90°$
10		$\alpha = 120°$

3. Results and Analysis

Figure. 3 depicts the contours of bed surface elevation of Bend 1 and Bend 3 in the absence of spur dyke. Here, the abscissa represents the streamwise/longitudinal direction of the flow, while the ordinate shows the transverse direction of flow. Further, the colour scale shows sediment erosion and deposition in centimeters. The blue colour indicates the bed surface degradation/ erosion, whereas the brown colour shows sediment deposition. It is evident from Fig.3 that the bed morphology is different under approaching flow conditions (flow comes from straight to curved channel, that is, Bend 1) than that under fully developed flow conditions (flow through curved boundaries, that is, Bend 3). Under approaching flow conditions, the bed pattern changes significantly downstream of the bend apex, that is, $\theta > 40°$ of the bend (Fig. 3a), while for fully developed flow conditions, the bed pattern changes throughout the bend (Fig. 3b). The reason for such occurrence in Bend 1 is attributed to the channel transition (straight channel to curved channel), because when flow comes from straight to curved channel, it continues to behave like a straight channel. As a result, instead of hitting at the channel bend apex ($\theta = 40°$), it strikes at the locations lying beyond the bend apex, that is, $\theta = 60°$ to 80° of the concave side. Therefore, maximum scouring in Bend 1 occurs at $\theta = 60°$ to 80° of the concave side, while deposition occurs at 60° to 80° of the convex side.

On the other hand, scouring is comparatively less at the bend apex, as the flow doesn't get notable acceleration. Unlike Bend 1, in Bend 3, the degradation of bed surface elevation starts close to the bend apex and increases towards the downstream of the bend apex. Further, it is observed that the scouring, as well as deposition process, happened over a larger area ($\theta = 50°$ to 80°) as compared to that of Bend 1. This infers that for the same rate of flow, there is no scouring in the straight channel, but significant scouring is observed on the concave side of the meandering channel. The reason for such occurrence is accredited to the spiral flow, transverse rotational flow, and secondary flow that prevailed in the channel resulting from the centrifugal acceleration.

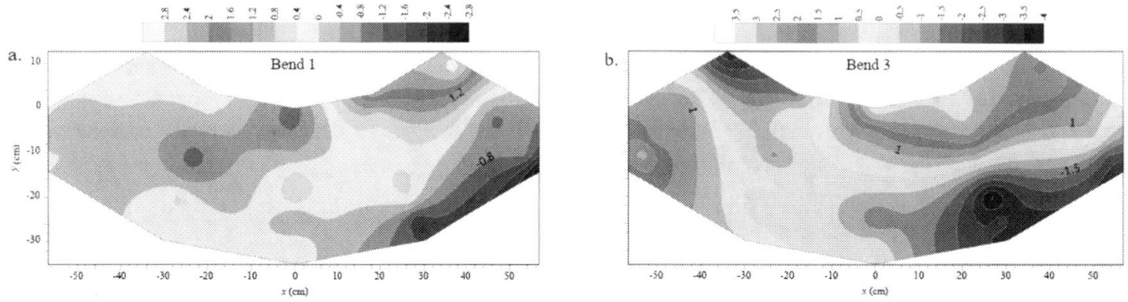

Figure 3. Bed contour lines at bend 1 and 3 in the absence of spur dikes, and colour scale shows sediment erosion and deposition in centimeters.

To counter the scouring process, a spur dyke was installed at location 1 (that is, at $\theta = 0°$) of both Bend 1 and Bend 3. In addition, to understand the impact of spur dyke orientation on the scoring process, in both bends, a spur dyke was installed for $\alpha = 60°$, 90°, and 120° (as discussed in section 2). Contours of bed topography in Bend 1 and Bend 3 for various spur dyke orientations are plotted and shown in Fig.4. When spur dyke was installed at location 1 oriented at $\alpha = 60°$ in Bend 1 (Fig. 4a), it was observed that scouring occurs near the dyke and gets stretched in the upstream direction (that is, along with the orientation of the dyke). However, the flow is not able to propagate along with the degraded bed material due to low energy resulting from flow velocity. As a result, the scoured bed materials get deposited just at the downstream side of the dyke (at near $\theta = 20°$ of the concave side). Nevertheless, the presence of 60° oriented spur dyke could not restrict the scouring at the far downstream end (at $\theta = 60°$ and 80°). Besides, when dyke is oriented at $\alpha = 90°$ at location 1 (Fig. 4b), localized scouring occurs in front of the dyke and gradually gets stretched up to the other side of the bank in the direction of dyke orientation. Though, it is able to restrict the scour in the immediate downstream ($\theta = 20°$), it does not have a significant impact on the reduction of scour at $\theta = 60°$ to 80°. Then, in Fig. 4c, it is seen that when dyke is oriented at $\alpha = 120°$ at location 1, it attracts the flow and increases the extent of scouring from $\theta = 40°$ to 80° and also reduces the deposition amount at the convex side. A comparison study reveals that with an increase in orientation angle, deposition of bed materials at the immediate downstream of location 1 is reduced. In contrast, at the farther downstream, the erosion of bed material is enhanced. Local scour around the spur dyke is more for $\alpha = 90°$ orientation, as compared to $\alpha = 60°$ and 120° orientations. Unlike Bend 1, in Bend 3, scouring around spur dyke does not happen at location 1; rather, deposition occurs, because the concave side of Bend 3 is in continuation with the convex side of Bend 2, resulting in a deposition at location 1. However, in case of Bend 3, one can notice local scour for $\alpha = 90°$ is similar to that in Bend 1. This infers that for $\alpha = 90°$ orientation, the spur dyke is acting as an obstruction to the flow under approaching as well as developed flow conditions, resulting in local scour around it. A comparison study shows that in Bend 3, the amount of erosion and deposition increases as compared to that in Bend 1. The probable reason for such occurrence is attributed to the flow curvilinearity in Bend 3.

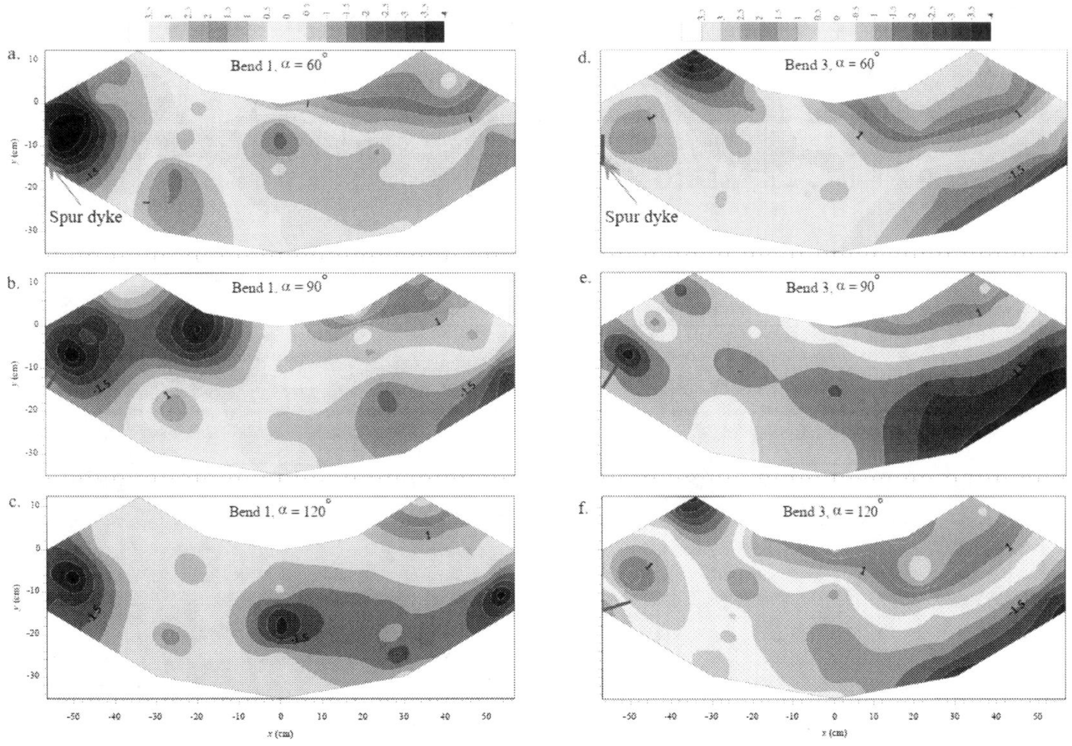

Figure 4. Bed contour lines at bend 1 and 3 when spur dyke at location 1($\theta = 0°$), and colour scale shows sediment erosion and deposition in centimeters.

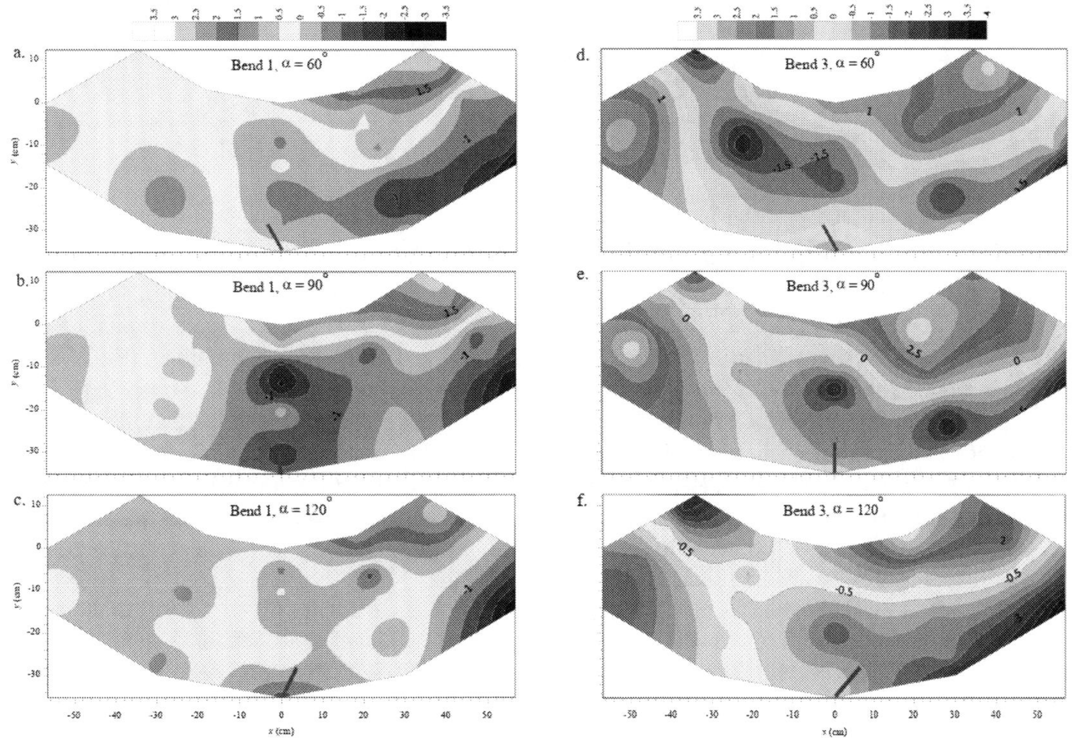

Figure 5. Bed contour lines at bend 1 and 3 when spur dyke at location 2 ($\theta = 40°$), and colour scale shows sediment erosion and deposition in centimeters.

Figure 5. shows the contours of bed surface elevation of Bend 1 and Bend 3, when dyke was installed at location 2 ($\theta = 40°$) of both bends. It is apparent that the local scour around spur dyke at location 2 for all three spur dyke orientations $\alpha = 60°$, 90°, and 120° (Fig. 5a, 5b, and 5c) are similar to that at location 1 for Bend 1, but with less intensity of scouring. Like Location 1, spur dyke at location 2 oriented at $\alpha = 90°$ shows maximum scouring around dyke (Fig. 5b), resulting in deposition of sediments in the immediate downstream. However, it is not able to reduce the scouring process at far downstream (near $\theta = 80°$). Comparing Fig. 5a, 5b, and 5c, it can be found that dyke oriented at $\alpha = 120°$ prevents scouring more as compared to dyke oriented at $\alpha = 60°$ and 90°. On the contrary, for the same flow condition in Bend 3, when spur dyke oriented at 60° at location 2 (Fig. 5d), it reduces the downstream scouring more in comparison to spur dyke oriented at $\alpha = 90°$ and 120°. Although dyke oriented at 90° reduces the significant downstream scouring of the concave side (Fig. 5e), spur dyke oriented at $\alpha = 60°$ is proven to be better. Besides, dyke oriented at $\alpha = 120°$ (Fig. 5f) attracts the flow and increases the scouring, and thus, this orientation ($\alpha = 120°$) is not suitable to counter the downstream scouring of the concave side.

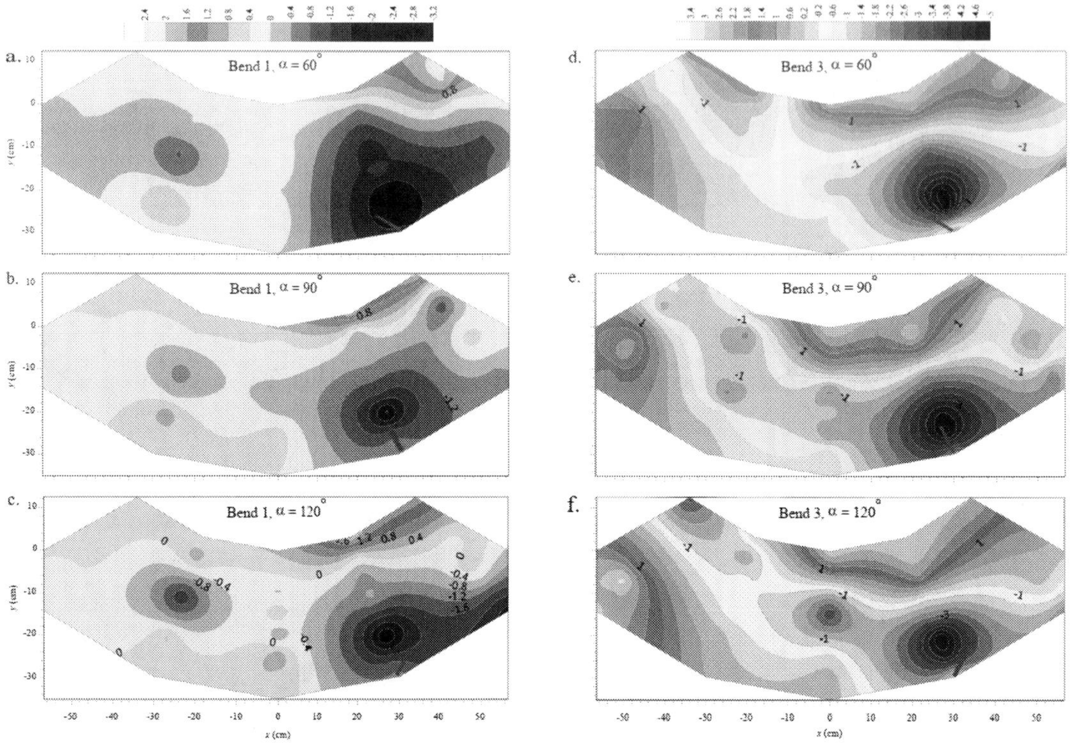

Figure 6. Bed contour lines at bend 1 and 3 when spur dyke at location 3($\theta = 60°$), and colour scale shows sediment erosion and deposition in centimeters.

The contours of bed surface elevation resulted from the installation of the spur dyke at location 3 ($\theta = 60°$) in Bend 1 and Bend 3, are depicted in Figure 6. Unlike locations 1 and 2, at location 3, local scour around dyke at $\alpha = 60°$ orientation in Bend 1 shows more scouring than that at $\alpha = 90°$ and 120° dyke orientations (Fig. 6a, 6b, and 6c). Though dyke oriented at $\alpha = 60°$ in Bend 1 (Fig. 6a) reduces the scouring near $\theta = 80°$, at $\alpha = 90°$ (Fig. 6b) reduces the scouring in bend 1 in comparison to other dyke's orientation. Hence, it can be inferred that the dyke orientation $\alpha = 90°$ at $\theta = 60°$ is the best orientation and location under approaching flow conditions for reducing the scouring in the channel bend. On the other hand, in Bend 3, when dyke is located at $\theta = 60°$ and oriented at $\alpha = 60°$ and $\alpha = 90°$ (Fig. 6d and 6e), it repels or deflects the flow to more distance as compared to Bend 1, because flow gets developed in Bend 3 and has sufficient kinetic energy to scour the nearby location. Hence, spur dyke at location 3 would not be beneficial in reducing the downstream scouring. Similar to locations 1 and 2, in location 3, dyke oriented at $\alpha = 120°$ (Fig. 6e) attracts the flow and increases the downstream scouring. Hence, comparing the results of all three spur dyke installation locations, it can be inferred that the spur dyke should be installed at location 3 (near zone of scouring) oriented at $\alpha = 90°$ for Bend 1, whereas for Bend 3, location 2 ($\theta = 40°$) with dyke oriented at $\alpha = 60°$ is sufficient for reducing the scour.

For better visualization of deposition and scouring pattern, three-dimensional topographical surface elevation of Bend 1 and Bend 3 without spur dyke are depicted in Fig. 7. In both Bend 1 and 3, an amount of degradation and deposition can be seen which occurred at far downstream. However, in Bend 3 (Fig. 7b), the area of deposition and scouring is large as compared to that in Bend 1.

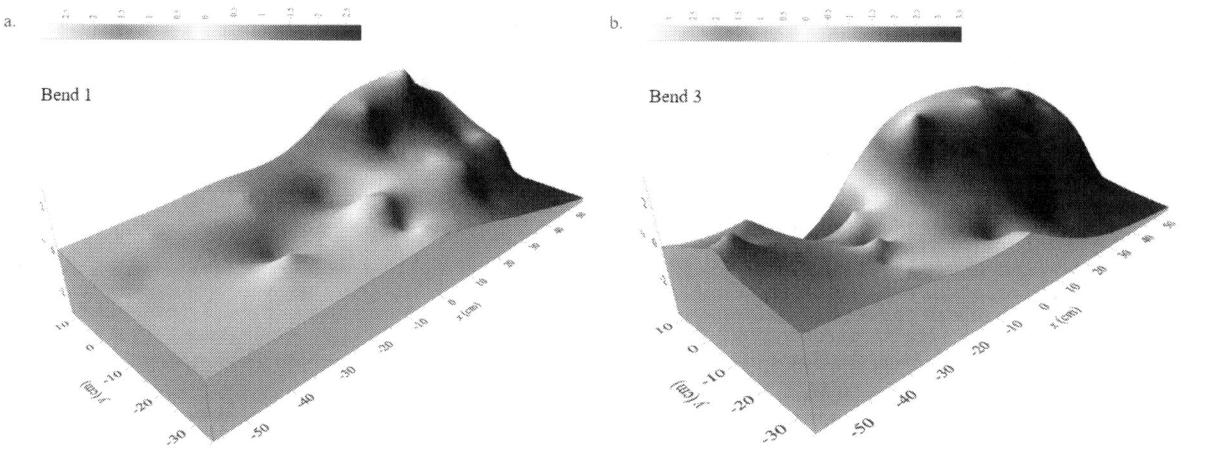

Figure 7. Three-dimensional topographical bed profile of (a) Bend 1 and (b) Bend 3 without spur dyke, and colour scale shows sediment erosion and deposition in centimeters.

Figure 8a and 8b show the three-dimensional topographical surface elevations of Bend 1 (at location 3 dyke oriented at $\alpha = 90°$) and Bend 3 (at location 2 dyke oriented at $\alpha = 60°$), respectively. It may be noted that spur dyke at location 3 oriented at $\alpha = 90°$ for Bend 1 and spur dyke at location 2 oriented at $\alpha = 60°$ for Bend 3, reduces the scouring as compared to other installation locations and orientations of spur dyke. Comparing Fig. 7 and 8, one can notice how the erosion and deposition patterns get modified when the dykes are installed at their best location and orientation in both the bends.

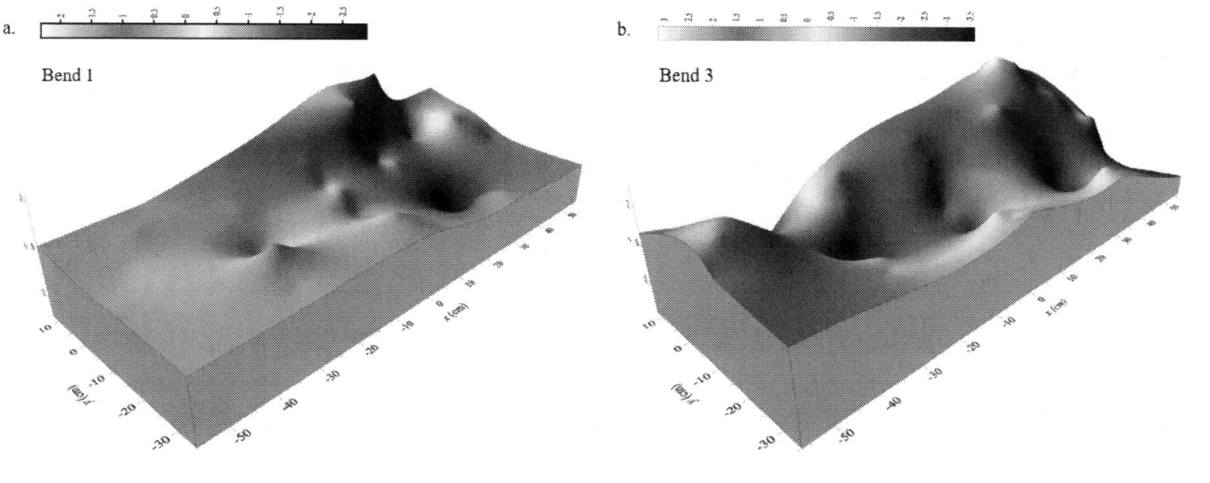

Figure 8. Three-dimensional topographical bed profiles of (a) Bend 1 and (b) Bend 3 based on the best location and orientation of the spur dyke, and colour scale shows sediment erosion and deposition in centimeters.

4. Conclusion

Scouring in open channels depends on channel geometry (sinuosity, curvature), flow properties, and sediment characteristics. To restrict the scour and to protect the bank in the meandering channel, a spur dyke is considered to be one of the effective tools. In the present study, experimental runs were carried out to understand the impact of the location and orientation of spur dyke on the channel bed morphology. Bed surface elevation was measured under low discharge conditions in Bend 1 (flow comes from straight to curved channel) and Bend 3 (flow through curved boundaries). It was found that when the flow was moving through the straight channel, no scouring was noticed, whereas, in the curved channel, significant scouring was seen for the same flow condition. The probable reason is attributed to the formation of centrifugal acceleration in the channel bend. Generally, it is seen that scouring occurs in the mid portion of the concave side of the bend, but in the present case, scouring occurs at the last portion of the concave side of the bend because of the low discharge and sharpness of the bend. Moreover, the scouring phenomenon is found to be different in bend 1 and bend 3 due to approaching and developed flow conditions, respectively. In Bend 1, scouring occur in $\theta = 60°$ to $80°$, while in bend 3, scouring occur at $\theta = 50°$ to $80°$. To reduce this scour, a spur dyke was installed at Bend 1 and Bend 3. Three different locations with varying dyke orientations were chosen for the assessment of the impact of spur dyke location and orientation on bed morphology. The effect of spur dyke for reducing scouring was seen more in Bend 3 than that in Bend 1, because dyke diverts the flow more in Bend 3 due to sufficient kinetic energy resulting from developed flow conditions. A comparison study reveals that in Bend 3, a spur dyke at location 2 oriented at $\alpha = 60°$ is sufficient for reducing scouring, whereas, in Bend 1, dyke has to be installed at location 3 ($\theta = 60°$) oriented at $\alpha = 90°$ to reduce the erosion.

Hence, it can be concluded that for low discharge conditions, a spur dyke should be installed at the downstream of the bend apex, keeping the orientation angle as 90° for approaching flow, while dyke can be installed at the bend apex, keeping the orientation angle as 60° for developed flow condition. Generally, upstream face dyke ($\alpha = 60°$) repels the flow more, but for approaching flow, dyke perpendicular to the flow direction ($\alpha = 90°$) is proven to be better for scouring reduction.

In this study, very interesting observations were found, and one of them is: In approaching flow condition, scouring around spur dyke at Bend 1was high with respect to the scouring that occurred in Bend 1 without installation of spur dyke. Now the real question arises, that is, will spur dyke be useful for approaching flow conditions (Bend 1) to reduce the scour?

5. REFERENCES

Athar, M., Mansoor, T., and Aggarwal, N. (2021). "Effects of Spur Dyke's Orientation on Bed Variation in Channel Bend." *Hydro Sci. Mar. Eng.*, 3(2), 1-9.

Breusers, H.N.C., and Raudkivi, A.J. (1991). "Scouring." IAHR hydraulic structures design manual 2, AA Balkema, Rotterdam, The Netherlands.

Copeland, R.R. (1983). "Bank protection techniques using spur dikes." *Army engineer waterways experiment station vicksburg ms hydraulics lab.*

Elawady, E., Michiue, M., and Hinokidani, O. (2001). "Movable bed scour around submerged spur-dikes." *Annu. J. Hydraul. Eng.*, 45, 373-378.

Fang, H., Bai, J., He, G., and Zhao, H. (2014). "Calculations of nonsubmerged groin flow in a shallow open channel by large-eddy simulation." *J. Eng. Mech.*, 140(5), 04014016.

Fazli, M., Ghodsian, M., and Neyshabouri, S.A.A.S. (2008). "Scour and flow field around a spur dike in a 90° bend." *Int. J. Sediment Res.*, 23(1), 56-68.

Garde, R.J., Subramanya, K., and Nambudripad, K.D. (1961). "Study of scour around spur-dikes." *J. Hydraul. Div.*, 87(6), 23-37.

Gill, M.A. (1972). "Erosion of sand beds around spur dikes." *J. Hydraul. Div.*, 98(9), 1587-1602.

Haltigin, T.W., Biron, P.M., and Lapointe, M.F. (2007). "Predicting equilibrium scour-hole geometry near angled stream deflectors using a three-dimensional numerical flow model." *J. Hydraul. Eng.*, 133(8), 983-988.

Karami, H., Ardeshir, A., Saneie, M., and Salamatian, S.A. (2012). "Prediction of time variation of scour depth around spur dikes using neural networks." *J. Hydroinformatics,* 14(1), 180-191.

Karki, S., Nakagawa, H., Kawaike, K., and Hashimoto, M. (2018). "Experimental study on the performance of slope-crested groynes in erodible meandering channels of different sinuosity." *J. Jpn. Soc. Nat. Disaster Sci.*, 37(S05), 93-105.

Kothyari, U.C., and RangaRaju, K.G. (2001). "Scour around spur dikes and bridge abutments." *J. Hydraul. Res.*, 39(4), 367-374.

Kuhnle, R.A., Alonso, C.V., and Shields Jr, F.D. (2002). "Local scour associated with angled spur dikes." *J. Hydraul. Eng.*, 128(12), 1087-1093.

Lee, S.O., and Sturm, T.W. (2009). "Effect of sediment size scaling on physical modeling of bridge pier scour." *J. Hydraul. Eng.*,135(10), 793-802.

Masjedi, A., Dehkordi, V., Alinejadi, M., and Taeedi, A. (2010). "Experimental study on scour depth in around a T-shape spur dike in a 180-degree bend." *World Appl. Sci. J.*, 10(10), 1146-1152.

Mayerle, R., Wang, S.S.Y., and Toro, F.M. (1995). "Verification of a three-dimensional numerical model simulation of the flow in the vicinity of spur dikes." *J. Hydraul. Res.*, 33(2), 243-256.

Özyaman, C., Yerdelen, C., Eris, E., and Daneshfaraz, R. (2022). "Experimental investigation of scouring around a single spur under clear water conditions." *Water Supply*, 22(3), 3484-3497.

Pandey, M., Ahmad, Z., and Sharma, P.K. (2018). "Scour around impermeable spur dikes: a review." *ISH J. Hydraul. Eng.*, 24(1), 25-44.

Pandey, M., Jamei, M., Karbasi, M., Ahmadianfar, I., and Chu, X. (2021). "Prediction of maximum scour depth near spur dikes in uniform bed sediment using stacked generalization ensemble tree-based frameworks." *J. Irrig. Drain. Eng.*, 147(11), 04021050.

Radan, P., and Vaghefi, M. (2016). "Flow and scour pattern around submerged and non-submerged T-shaped spur dikes in a 90° bend using the SSIIM model." *Int. J. River Basin Manag.*, 14(2), 219-232.

Rajaratnam, N., and Nwachukwu, B.A. (1983). "Flow near groin-like structures." *J. Hydraul. Eng.*, 109(3), 463-480.
Raudkivi, A.J. (1986). "Functional trends of scour at bridge piers." *J. Hydraul. Eng.*, 112(1), 1-13.
Rhoads, B.L., and Welford, M.R. (1991). "Initiation of river meandering." *Prog. Phys. Geogr.*, 15(2), 127-156.
Seminara, G. (2006). "Meanders." *J. Fluid Mech.*, 554, 271-297.
Sharma, K., and Mohapatra, P.K. (2020). "Effect of bed roughness on flow field around spur dyke in a rigid bed meandering channel." *ISH J. Hydraul. Eng.*, issue and vol. 1-8.
Tripathi, R.P., and Pandey, K.K. (2021). "Experimental study of local scour around T-shaped spur dike in a meandering channel." *Water Supply*, 21(2), 542-552.
Vaghefi, M., Ghodsian, M., and Neyshaboori, S.A.A.S. (2009). "Experimental study on the effect of a T-shaped spur dike length on scour in a 90° channel bend." *Arab. J. Sci. Eng.*, 34(2B), 337-348.
Zaghloul, N.A. (1983). "Local scour around spur-dikes." *J. Hydrol.*, 60, 123-140.

DOI: https://doi.org/10.26077/5174-4426

Bed Load Computation For Gravel Particles In Gravel-Silt Mixture

U.K. Singh[1], S. Kumar[1] & Z. Ahmad[2]
[1] *Koneru Lakshmaiah Education Foundation, Vaddeswaram, India*
[2] *Indian Institute of Technology Roorkee, Roorkee, India*
E-mail: umesh.ais@gmail.com

Abstract: *Fluvial hydraulics deals with the process of erosion, transportation, and deposition of sediment in channels by the action of flowing water. This paper presents experimental results on bed load transport rate of gravel particles transported from mobile channel bed made of gravel-silt mixture. Gravel and sand particles were observed transported as bed load due to their coarser size while silt transported as suspended load. Bed load was collected at regular time interval of like 15 minutes, 30 minutes, 60 minutes, etc. Bed surface profile and water surface profile were measured spatially at an interval of 50 cm at the center line of the test section along the flow. It was observed that collected bed load decreases with the time passes and the run was continued till the equilibrium condition i.e., low amount of bed load collected for a long duration and simultaneously bed surface profile and water surface profile come into nearly stable state. This condition is here treated as equilibrium condition as there is no feeding of sediment in the channel and test section also tends to in stable condition with nearly no outflow of sediment. At the end of run, layer of gravel particles was found on the top surface of the channel bed in the mixture of gravel-silt; however, high degradation was observed towards the upstream of test section. A relationship has been developed to compute the total bed load which found in good agreement with the observed data.*

Keywords: *Bed load, Cohesionless sediment, Equilibrium condition, Gravel particles, Hydraulics, Sediment transport.*

1. Introduction

Fluvial hydraulics deals with the process of erosion, transportation, and deposition of sediment in channels by the action of flowing water. The understanding of transport phenomena is important as transport of sediment may lead to harmful consequences. The sediment transport in alluvial channel significantly affects the life of hydraulic structures made under alluvial channels. Deposition of sediments upstream of dams, degradation of channel bed below dams, local scour around hydraulic structures, etc. are some examples of changes in river morphology due to human activity. The quantity of sediment coming with flow in alluvial channel plays a significant role in the erosion of channel bed in alluvial rivers. Little or negligible sediment in the incoming flow in channel may result in hungry water or sediment-starved water that leads to channel incision and may cause the undermining of bridges and other structures. Garde and Ranga Raju (2000) mentioned the failure of Islam weir on the Sutlej River due to degradation.

When the shear stress due to stream flow exceeds the submergence weight of particles in case of cohesionless sediment then the detachment of sediment starts from the cohesionless channel bed. Further increase in shear stress due to flow causes the transport of sediment along with the flow. The degradation of channel bed is continued till a stable bed condition reached (in case of sufficient bed depth) further which there is no transport of sediment, and this stable condition is treated as equilibrium condition as there is no incoming sediment and no outgoing sediment. A stream in equilibrium is defined as the one in which channel dimensions and slope are so adjusted over a period of time that it carries incoming sediment load and water without appreciable erosion or deposition (Mackin 1948). Garde and Ranga Raju (2000) also defined the equilibrium condition as "a certain length of an alluvial stream is said to be in equilibrium if the amount of sediment coming into this reach is equal to the sediment going out from the same". Stream equilibrium is significantly important in the study of sediment transport as it responses to degradation and aggradation in the channel bed. When equilibrium condition in stream deviates then a change occurs in alluvial channel geometry which may cause the lowering of channel bed till the equilibrium condition of flow reaches so that it balances between inflowing and outflowing water and sediment discharges. Julian (2002) reported a formation of armor layer as a representative of stable bed conditions in cohesionless channel bed when the incoming flow in the channel is clear water (i.e., no feeding of sediment in the channel). The degradation occurred in response to carrying capacity of stream till a stable bed condition reached. Further there is no transport of sediment due to development of an armor layer and this corresponds to equilibrium condition. The rate of degradation of channel bed decreases with time (Vogel et al. 1992) i.e., the total load increases with time in a slower manner. Formulation for the computation of total bed load may serve as one of the important tools in the modeling of sediment transport study.

Misri et al. (1984) proposed methods for the computation of the bed load transport rates of different fractions in a mixture. They carried the experimental work in a tilting flume having length 16.0 m, width 0.75 m, and depth 0.48 m with four coarse uniform materials (sand and gravel) and nine sediment mixtures as the bed material. The eroded sediment form the channel bed were collected in a trap at the downstream end of the flume and then fed back manually at the upstream end of the flume at a constant rate to maintain sediment equilibrium. They also assessed the accuracy of existing methods for the computation of bed load transport rate and concluded that the bed load transport of uniform sediment satisfactorily determined, however, Einstein's method doesn't predict satisfactorily the bed load transport rates of individual fractions in a mixture. Wu et al. (2003) proposed the methods for the computation of transport rates for non-uniform sediment in sand-bed channels based on the Transport Capacity Fraction concept. They reported that this method has the advantage of avoiding discrepancies due to the distribution and number of class intervals in computing bed-material load, and it limits the errors in computing concentrations for individual size fractions.

A number of formulations have been derived for the computation of bed load transport rate of cohesionless sediment by various researchers (Meyer-Peter and Müller 1948, Einstein 1950, Ashida and Michiue 1972, Engelund and Fredsøe 1976, Parker 1979, Misri et al. 1984, Parchure and Mehta 1985, Samaga et al. 1986, Mehta et al. 1989, Jepsen et al. 1997, Wong and Parker 2006, Xu et al. 2014, etc.). However, cumulative bed load transported from the channel bed has not been reported. This paper presents an experimental study which leads to the formulation of the cumulative bed load eroded from the channel bed till the stable bed condition reached for gravel-silt mixture. This paper also discusses the bed profiles at an equilibrium condition for gravel-silt mixture. Formulation for the computation of total bed load from the channel bed may use in the modeling of sediment transport study.

1.1. Experimental Set-up

The experiments were conducted using sediments of gravel and silt on a tilting flume of 16 m length, 0.75 m width and 0.50 m depth in Hydraulic Engineering Laboratory, Civil Engineering Department, Indian Institute of Technology Roorkee, Roorkee, India. The channel had a test section of 6.0 m length, 0.75 m width and 0.18 m depth starting at a distance of 7.0 m from the channel entrance. The depth of the test section in the channel for sediment filling is 0.18 m which is considered as general level of the flume bed. The water flow in the flume was regulated with the help of a valve provided in the inlet pipe coming from the overhead tank and the discharge measurement in the flume was done volumetrically through a tank provided at the end of the flume. For the collection of bed load a rectangular trap was used which covered with a net-clothing having fine pores so that bed load was retained on the trap cover and this rectangular trap was placed at the end of the flume just after the tail gate. The test section profile of channel bed was measured with the help of two-dimensional bed level profiler. The channel bed profile was also measured by flat gauge of least count 0.10 mm and the water surface profile was measured with a pointer gauge having a least count of 0.10 mm. The slope of the channel bed is computed with the help of two containers which are connected at their bottom by a long plastic tube. The sediment used in the present study was gravel and sand. The size distribution of these sediment shown in Fig. 1. The median size (d_{50}) of the gravel and silt are 5.50 mm and 0.062 mm respectively. The geometric standard deviation ($\sigma_g [= \frac{1}{2}\sqrt{(d_{84}/d_{50}) + (d_{50}/d_{16})}]$) for all the sediment was computed as per Garde and Ranga Raju (2000) which was 1.31 and 1.18 for gravel and silt respectively; where d_{84}, d_{50} and d_{16} are the sediment size such that 84%, 50% and 16% of material is finer than that size by dry weight respectively.

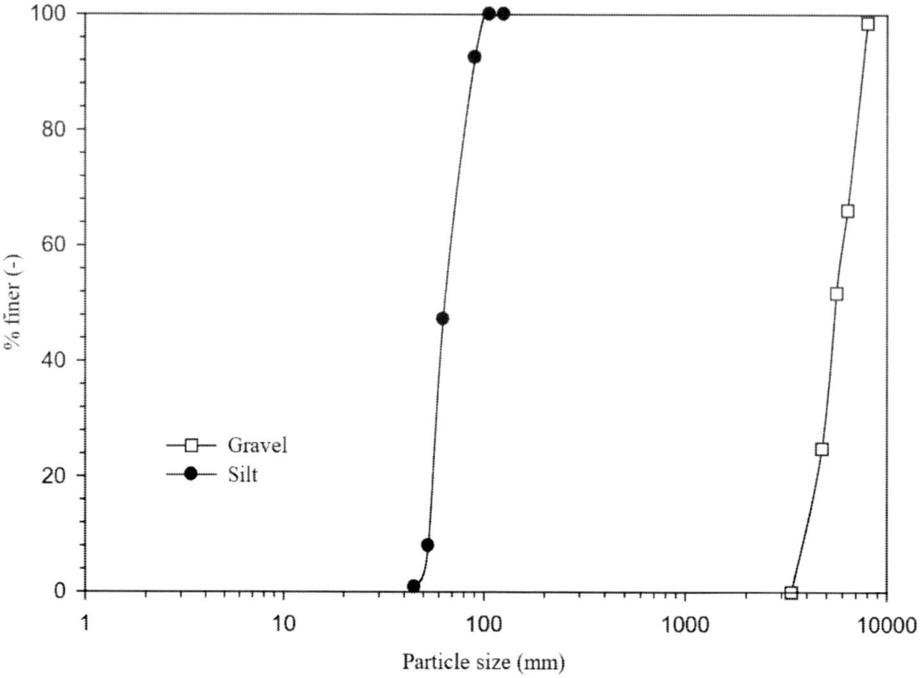

Fig. 1. Size distribution for sediments used in the present study

1.2. Channel Bed Preparation

The dried sediments were weighted as per proportion and mixed together manually and then water was added to sediment mixture and mixed them thoroughly well. The mixture was covered with polythene and left for around 24 hours so that the moisture was distributed uniformly throughout the mixture. This prepared sediment mixture was filled in the test section and compacted in three layers by dynamic compaction method. Each layer was compacted with a cylindrical roller having weight equal to 400 N and sides of the channel was compacted by hand rammer having rectangular bottom. To ensure bonding among different layers, the top surface was roughened by trowel before laying the next layer over it. After compacting all the three layers, extra sediments were chiseled off using sharp edge large knife. Before the beginning of experimental run, bed was saturated for 24 hours in order to achieve the field's condition. For all the experimental runs, the sediment beds were prepared afresh. A prepared channel bed has been shown in Fig. 2 for gravel-silt mixture of the present study.

Fig. 2. Prepared channel bed for gravel-silt mixture

1.3. Experimental Procedure

The developed shear stress on the channel bed due to incoming flow was large enough to detach the sediment particles from the test section. The detached gravel and silt particles transported as bed load and suspended load respectively. Gravel and sand particles were observed transported as bed load due to their coarser size while silt transported as suspended load. Bed load was collected at regular time interval. Initially, it was collected at an interval of around 15 minutes as the more transportation of sediment was observed which may be in response of higher erosion rate than later on the time interval was increased as the erosion rate comes down with the time passes. Measurements of bed surface profile, water surface profile, and suspended load was taken simultaneously with the bed load. Bed surface profile and water surface profile were measured spatially at an interval of 50cm at the center line of the test section along the flow. It was observed that collected bed load decreases with the time passes and the run was continued till the low amount of bed load collected for a long duration and simultaneously bed surface profile and water surface profile come into nearly stable state. This condition is here treated as equilibrium condition as there is no feeding of sediment in the channel and test section also tends to in stable condition with nearly no outflow of sediment.

Gravel particles were dominating on the top surface of the test section of the channel bed for the mixture of gravel-silt. The gravel particles were come out on the top surface and finally coarser gravel particles were developed on the top surface of test section as an armor layer which was large enough to resist the movement against the flow and bed comes in stable condition and hence after the end of run layer of gravel particles were found on the top surface of the channel bed in the mixture of gravel-silt. Figure 3 shows the appearance of gravel layer on the top surface of gravel-silt mixture after the end of run. More degradation was observed towards the upstream of test section.

Fig. 3. Top view of channel bed after end of run

1.4. Results and Discussions

There are several parameters which affect the cumulative bed load collected (till the equilibrium condition reached) for the sediment mixture of gravel-silt. However, only the parameters which show good results for the formulation were considered and represented in the analysis as below:

$$M^e = f(\tau, \tau_{cc}, q, \rho, \rho_s, d_a, b, l, g) \quad (1)$$

Here, M^e is the cumulative bed load (Kg) collected till the equilibrium condition reached; τ is the shear stress (N/m^2) developed on the test section of the channel bed due to the incoming flow; τ_{cc} is the critical shear stress (N/m^2) for sediment mixture; q is the unit discharge (m^2/s) i.e. discharge per unit width of the channel; ρ_s and ρ are particle and fluid densities (Kg/m^3) respectively; d_a is arithmetic mean size (m) of the cohesive sediment mixture; b is the width (m) of the test section; l is the length (m) of the test section; and g is gravitational acceleration (m/s^2).

Equation (1) represented in the dimensionless form using dimensional analysis as below:

$$M^{e*} = f(\tau^{e*}, h^{e*}) \quad (2)$$

Here, M^{e*} is the dimensionless cumulative bed load till equilibrium condition of flow reached which can be computed as

$$M^{e*} = \frac{(M^e g/bl)}{(\rho_s - \rho) g d_a} \quad (3)$$

τ^{e*} is the dimensionless excess shear stress on the test section of channel bed and computed as

$$\tau^{e*} = \frac{\tau - \tau_{cc}}{(\rho_s - \rho) g d_a} \quad (4)$$

h^{e*} is the dimensionless form of flow depth in terms of unit discharge and computed as

$$h^{e*} = \frac{(q^2/g)^{1/3}}{d_a} \quad (5)$$

Here, $d_a = \dfrac{\sum(d_{50} P)}{P}$ (6)

P is the percentage of the individual sediment in the sediment mixture.

Observed data of τ_{cc} from the experimentation is used for the formulation development and τ is computed as

$$\tau = (\rho_s - \rho) g R S_0 \quad (7)$$

Here, S_0 is the bed slope of the channel; and R is the hydraulic radius (m) which is computed as below:

$$R = \frac{A}{P} \quad (8)$$

$$A = bh \quad (9)$$

$$P = b + 2h \quad (10)$$

Here, A is flow cross-sectional area, P is wetted parameter, and h is the flow depth corresponding to the incoming flow in the channel at solid face before the start of test section.

$$AR^{2/3} = \frac{nQ}{S_0^{1/2}} \quad (11)$$

n is manning's roughness coefficient which computed as

$$n = \frac{(d_a)^{1/6}}{24} \quad (12)$$

With the help of Eqs. (8), (9), (10), (11), and (12) the value of h can be computed for a known Q, S_0, and d_a and hence, with computed h the value of R is also computable and so τ can be computed as per Eq. (7) with computed value of R.

Equation (2) represents the functional relationship for the dimensionless cumulative bed load. i.e., M^{e^*} with the dimensionless parameter τ_e^* and h^*. Variation of parameters τ_e^* and h^* has been plotted against M^{e^*} which shows the increasing trend as illustrated in Fig. 4 and Fig. 5 respectively.

Large number of trials led into the following relationships as formulation for the computation of M^{e^*} for the sediment mixture of gravel-silt in the present study:

$$M^{e^*} = 0.097(\tau^{e^*})^{0.455}(h^{e^*})^{1.467} \tag{13}$$

The computed value of M^{e^*} were plotted against the observed ones for the computation of regression coefficient that shows a very good regression coefficient, and it indicates a very good match between observed and computed data as shown in Fig. 6.

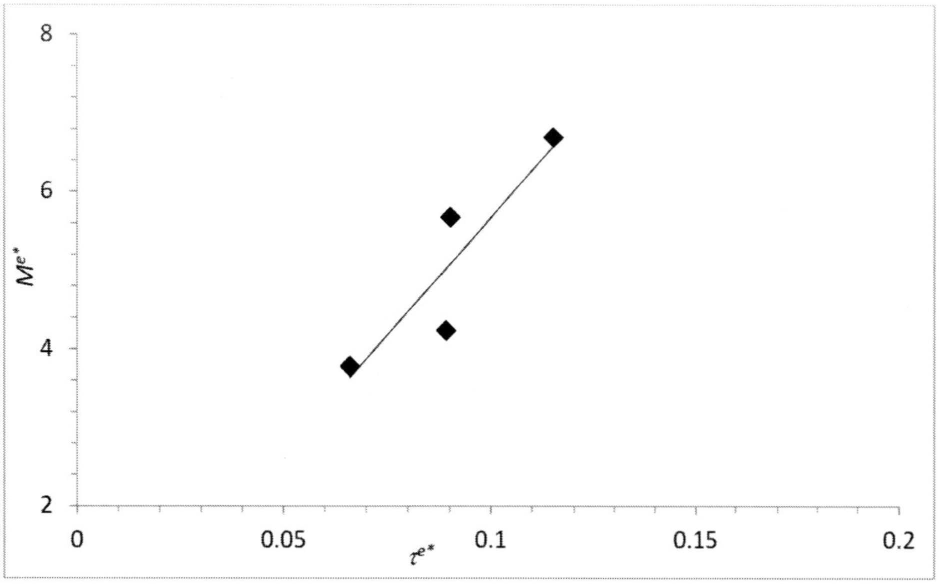

Fig. 4. Variation of τ_e^* with M^{e^*}

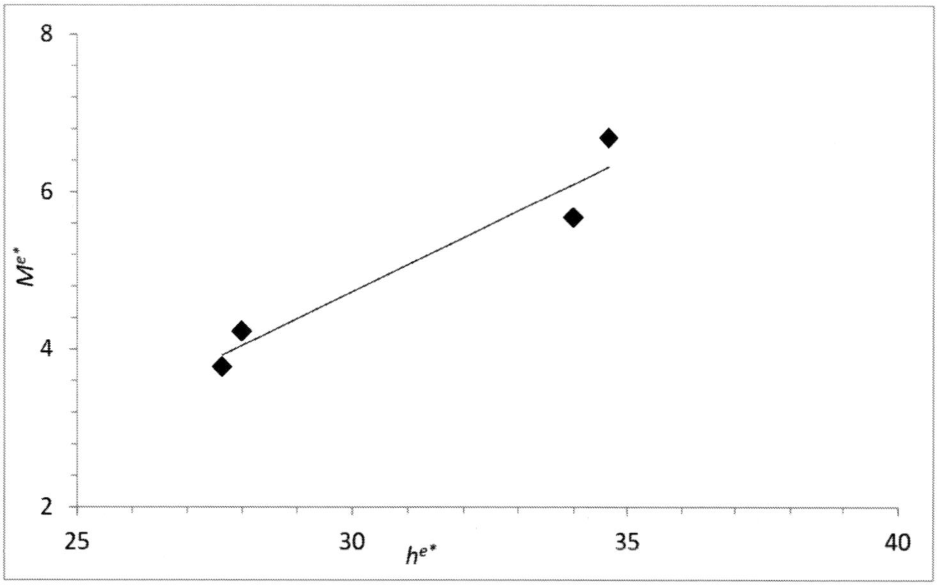

Fig. 5. Variation of h^* with M^{e^*}

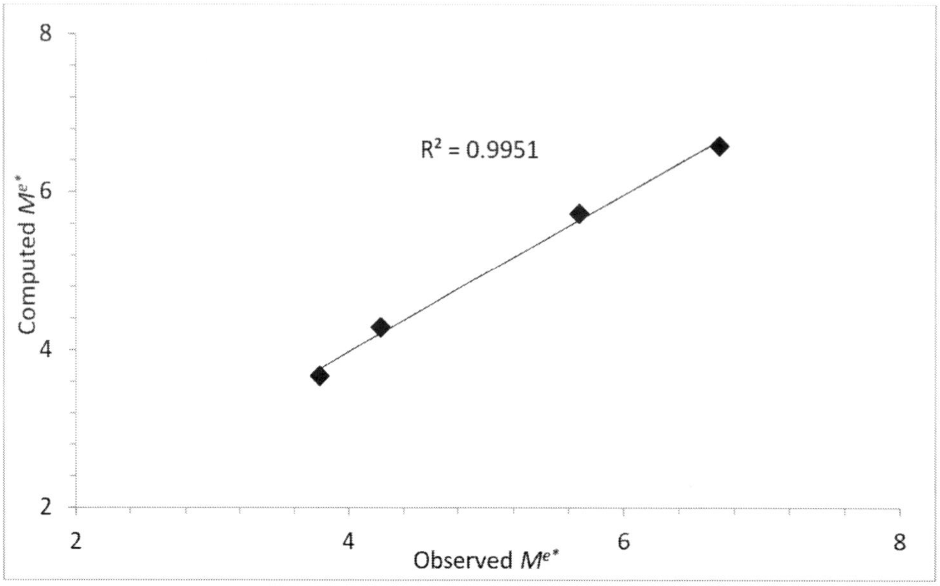

Fig. 6. Comparison between computed and observed value of M^{e^*}

1.5. Conclusions

The experimental study has been conducted to compute cumulative bed load that transported after the erosion from the channel bed. The cumulative bed load is the total bed load collected till the flow attained the equilibrium condition i.e. no transport of sediment with the flow as there is no feeding of sediment in the upstream of the channel bed. The channel degradation occurs due to the excess shear stress developed on the test section of the channel bed by the incoming clear water flow. The visual observations were taken after the end of run which shows high degradation in the upstream section than that of downstream for grave-silt mixture. A relationship has been developed for the computation of cumulative bed load in the form of Eqs. (13), for gravel-silt mixture. The factors governing the proposed equations were found to be dimensionless excess shear stress and dimensionless flow depth in terms of unit discharge. High value of regression coefficient shows the good match between observed and computed value for relationship proposed for the computation of cumulative bed load.

2. REFERENCES

Ashida, K., and Michiue, M. (1972). "Study on hydraulic resistance and bed-load transport rate in alluvial streams." *Proceedings of the Japan Society of Civil Engineers*, 59–69.

Einstein, H. A. (1950). *The bed-load function for sediment transportation in open channel flows. Technical Bulletin No. 1026, USDA, Soil Conservation Service, Washington.*

Engelund, F., and Fredsøe, J. (1976). "A sediment transport model for straight alluvial channels." *Hydrology Research*, 7(5), 293–306.

Garde, R. J., and Ranga Raju, K. G. (2000). *Mechanics of sediment transportation and alluvial stream problems*. New Age International, New Delhi.

Jepsen, R., Roberts, J., and Lick, W. (1997). "Effects of bulk density on sediment erosion rates." *Water, Air and Soil Pollution*, 99(1), 21–31.

Julien, P. Y. (2002). *River Mechanics. Water*, Cambridge University Press, New York, 434 pp.

Mackin, J. H. (1948). "Concept of the graded river." *Bulletin of the Geological Society of America*, 59, 463–511.

Mehta, A. J., Hayter, E. J., Parker, W. R., Krone, R. B., and Teeter, A. M. (1989). "Cohesive sediment transport. I: process description." *Journal of Hydraulic Engineering*, 115(8), 1076–1093.

Meyer-Peter, E., and Müller, R. (1948). "Formulas for bed-load transport." *Proceedings of the 2nd Meeting of the International Association of Hydraulic Research*, IAHR, Stockholm, 39–64.

Misri, R. L., Garde, R. J., and Ranga Raju, K. G. (1984). "Bed load transport of coarse nonuniform sediment." *Journal of Hydraulic Engineering*, 110(3), 312–328.

Parchure, T. M., and Mehta, A. J. (1985). "Erosion of soft cohesive sediment deposits." *Journal of Hydraulic Engineering*, 111(10), 1308–1326.

Parker, G. (1979). "Hydraulic geometry of active gravel rivers." *Journal of Hydraulic Division*, 105(9), 1185–1201.

Samaga, B. R., Ranga Raju, K. G., and Garde, R. J. (1986). "Bed load transport of sediment mixtures." *Journal of Hydraulic Engineering*, 112(11), 1003–1018.

Vogel, K. R., Niekerk, A. van, Slingerland, R. L., and Bridge, J. S. (1992). "Routing of heterogeneous sediments over movable bed: Model verification." *Journal of Hydraulic Engineering*, 118(2), 263–279.

Wong, M., and Parker, G. (2006). "Reanalysis and correction of bed-load relation of Meyer-Peter and Müller using their own database." *Journal*

of Hydraulic Engineering, 132(11), 1159–1168.
Wu, B., Molinas, A., and Shu, A. (2003). "Fractional transport of sediment mixtures." *International Journal of Sediment Research*, 18(3), 232–247.
Xu, Y., Jiand, H., Chu, F., and Liu, C. (2014). "Fractal model for surface erosion of cohesive sediments." *Fractals*, 22(3), 1440006.

DOI: https://doi.org/10.26077/6945-08ed

Calculating Local Pier Scour in Coarse-bed Streams

D.C. Froehlich[1]
[1]*Consulting Engineer, Cary, North Carolina, USA*
E-mail: dcfroehlich@aol.com

Abstract: *Streams can be divided into two broad classes based on the predominant composition of the alluvial material in which they are formed: sand-bed and coarse-bed. Coarse-bed streams are formed primarily in gravel and larger sediments. They are common in mountain and foothill regions, mainly where igneous and metamorphic bedrock exists, and in areas of glacial-fluvial deposits. A feature that characterizes nearly all coarse-bed streams is a surface covered by large sediment particles compared to those forming the substratum –an armor layer. The presence of an armor layer has a considerable influence on the physical processes that scour material from around a bridge pier. Several methods exist for estimating local scour depth at bridge piers. Data collected in small-scale laboratory experiments in sand-filled flumes, which hardly resemble conditions in coarse-bed streams, are the basis for most. Fortunately, recent on-site measurements of local pier scour in coarse-bed channels supply a collective data set used in this analysis to develop a scour prediction relation for application in streams where armor layers are present. Nonlinear quantile regression provides prediction models for quantile levels of 0.5 (which gives the median value), 0.95, and 0.99. Estimates for the larger quantiles (corresponding to the 95 and 99 percentiles) offer statistically sound safety margins needed for a reliable bridge design that is not overly excessive. The predictive ability of this study's local pier-scour relations for coarse-bed streams provides significant improvements over other commonly used empirical equations.*

Keywords: *Bridge, pier, scour, coarse-bed, empirical model, quantile regression, safety margin*

1. Introduction

After a bridge spanning the River Tyne at Hexham, England collapsed during a large flood in 1782, John Smeaton, designer of the bridge, and widely regarded as the founder of the civil engineering profession in Great Britain, wrote: "...and all the experience I have gained is, not to attempt to hold a bridge upon a gravel bottom in a river subject to such violent rapidity" (Smiles 1861, page 97). Scour of the coarse sediment surrounding the central pier footing was the cause of the abrupt failure. Nevertheless, despite Smeaton's reluctance to span such a river again, a rebuilt bridge is still used today. Although much has been learned about scour since Smeaton's time, constructing bridge foundations to resist every vagary of a river's current continues to present challenges.

Scour develops around a bridge pier when the capacity of the approaching streamflow to wash away the surrounding bed sediment exceeds the rate at which material refills the affected area. As a scour hole deepens, it becomes more difficult for the flow to move deposits out of the excavated region. If the approach flow is not transporting sediment, clear-water conditions exist in the stream, and sediment loss continues until the hydrodynamic forces at the bottom of the scoured area can no longer remove the bed material. In contrast, live-bed conditions exist if the stream actively conveys sediment and the bed surrounding a pier erodes until the rates at which material is carried into and through the hollow balance.

The maximum scoured depth may be significantly influenced by the type of streambed in which a pier is founded. Based on the predominant composition of the alluvial material in which streams are formed, they can be divided into two broad classes: sand-bed and coarse-bed. Sand-bed streams consist almost entirely of cohesionless shifting sand with highly erodible substrates. They are usually located on wide floodplains and have small longitudinal gradients. In contrast, coarse-bed streams, such as the River Tyne at Hexham, are formed chiefly in gravel and larger-sized sediments and are common in mountain and foothill regions, mainly where igneous and metamorphic bedrock exists, and in areas of glacial-fluvial deposits. A feature that characterizes nearly all coarse-bed streams is a surface cover of particles (called an armor layer) that are large compared to the bulk mixture in the substratum (Leopold 1994, page 26).

Armor layers are typically about as thick as the diameter of the largest grains supplied to a channel (Froehlich 1995). However, the coarse surface cover may vary significantly in composition within a short distance and may not be continuous across the stream's entire width. The substrates of coarse-bed streams usually contain a high proportion of sand and granules – grain sizes less than 4 mm – and, as a result, they have a considerably different makeup than the surface cover. The small-sized substrate material would be entrained and transported quite easily if not for the rate-limiting effects of armor layers. Coarse particles within an armor layer hinder the movement of its smaller grains, thereby restricting their mobility.

Many empirical formulas exist for estimating local scour depth at bridge piers – see, for example, Vonkeman and Basson (2019) compare 30 such equations. Data collected in small-scale laboratory experiments in sand-filled flumes, conditions that scarcely resemble natural coarse-bed steams, are the basis for most. However, to make practical use of small-scale experimental data, Laursen (1955) long-ago pointed out that on-site measurements of local scour are needed to establish model-prototype conformity. Moreover, from a comprehensive review of practices for assessing scouring at bridge waterways, Jones (1970, page 22) concluded that "the first priority for research on scour problems should be given to field measurements. No relationship for predicting scour can be used with confidence until proof has been shown that it does predict what happens with reasonable accuracy and reliability." Two decades later, Melville et al. (1989) again emphasized collecting and analyzing on-site measurements to identify the deficiencies of scour-depth predictions developed from scale-model experiments alone. This awareness is vital when applying those relations to predict local scour at bridge piers founded in course-bed streams.

Figure 1. A remnant scour hole around a cylindrical column of circular cross-section in a coarse-bed stream. (Photograph courtesy of Stephen R. Holnbeck, U.S. Geological Survey)

Fortunately, over the past half-century, several data collection programs have provided a sizeable, combined set of on-site measurements of local pier scour in streams in which armor layers have been present. These data are assembled and analyzed in this study using quantile regression to develop formulas for estimating local pier-scour that significantly reduce, but do not eliminate, dependence on scale model experiments. Prediction equations are prepared for three quantile levels (the proportion of the population) of the local scour depth to provide a means of evaluating model uncertainty. As a result, one can assign statistically well-reasoned safety margins that ensure a reliable bridge design.

2. Characteristics of Coarse-bed Streams that Affect Local Pier-Scour

Average shear stresses acting on the beds of coarse-bed streams at their bankfull flows are at most two to three times the values needed to move the armor layers' median size particles (Mueller et al. 2005). While the distribution of grain sizes in surface armor layers of coarse-bed streams varies spatially and temporally, the subsurface distribution is more consistent. For example, Ryan and Emmett (2002, pages 7-8) found that for Little Granite Creek, a well-armored stream in Wyoming, the median particle diameter D_{50} of the bed surface ranged from 55 mm to 100 mm over six years, with an average of about 90 mm, while the subsurface material distribution was nearly constant.

From experimental studies carried out in small-scale flumes, Raudkivi and Ettema (1985) and Dey and Barbhuiya (2004) show that armor layers increase the critical shear velocity for the threshold motion of bed particles, thereby maintaining an extended clear-water scour condition up to the limiting stability of surface particles. Consequently, local scour at piers is greater in clear-water flows when armor layers are present if scattered secondary armoring forms within scour holes. Melville (1975), Neill and Morris (1980), and Raudkivi and Ettema (1985) also note this phenomenon.

Another influence of armor layers is their effects on scour hole persistence. Because of an armor layer's impact on bed-material movement, scour holes may remain nearly unchanged from their maximum depths and sizes as flood flows decrease and sediment transport rates fall quickly to negligible values (see Fig. 1). In contrast, in sand-bed streams, scour holes refill quickly as flow rates drop, with depths reflecting the ongoing hydrodynamic conditions, which usually produce non-negligible sediment transport rates even at low flows. For this reason, one needs to carefully evaluate on-site measurements of local pier scour made in coarse-bed streams to ensure that the measured scour depths and flow parameters are reasonably harmonious. However, even in coarse-bed streams, scour holes usually refill when sediment transported during a flood accumulates as the discharge subsides. This replenishment process, if it occurs, complicates evaluating the maximum scour depth from streambed sections measured after the flood. Recent investigations have successfully used various geophysical methods to detect the presence of infilling.

3. Coarse-bed Stream Pier Scour Data

Mueller and Wagner (2005) assembled 493 measurements of local pier scour at 79 bridges in 17 States collected by the U.S. Geological Survey (USGS) in cooperation with the U.S. Federal Highway Administration (FHWA) and many State highway agencies. Benedict and Caldwell (2014) extended this database with measurements made outside the United States and those from USGS studies not included in Mueller and Wagner (2005). They also supplemented some data with personal communications that provide essential information not found in the published sources.

In this study, 228 on-site measurements of local scour around the piers of bridges that span coarse-bed stream channels have been drawn mainly from the following sources: Norman (1975), Zhuravlyov (1978), Davoren (1985), Jarrett and Boyle (1986), Butch (1991), Williamson (1993), Becker (1994), Gao et al. (1994), Wilson (1995), Hayes (1996), Jackson (1997), Chase and Holnbeck (2004), and Holnbeck (2011). Relevant data not found in the original publications were obtained from the comprehensive database prepared by Benedict and Caldwell (2014).

Local scour measurements extracted from the cited references and analyzed in this study are summarized in Table 1. The assembled data consist of the pier width b, the median diameter of the surface bed-surface material D_{50}, the water depth upstream of the pier y, the depth-averaged velocity of the flow upstream of the pier V, and the scour depth d_s below the ambient bed level (see the schematic in Fig. 2).

The following analysis is limited to natural coarse-bed streams having reported median particle diameters of 8 mm (medium gravel) or larger. Only measurements made at cylindrical columns having a constant circular cross-section and at other piers aligned with the approaching streamflow that had upstream ends (noses) that were either round or pointed (that is, sharp-nosed) are included in the data set (none of the included piers had square noses). Additionally, measured depths of less than 0.3 m were omitted from the collection to minimize the inclusion of remnant scour-hole data.

Table 1. Ranges of data from on-site measurements of local scour at bridge piers founded in coarse-bed streams used in this study

Source	Location	Count	b (m)	D_{50} (mm)	y (m)	V (m/s)	d_s (m)
Becker (1994)	USA (Missouri)	6	1.04-1.52	9-13	2.41-9.88	0.85-2.59	0.21-1.25
Butch (1991)[a]	USA (New York)	20	0.91-1.68	22-46	3.66-9.72	1.98-4.27	0.52-1.55
Chase and Holnbeck (2004)	USA (Montana)	4	0.91- 1.04	8-73	0.46-2.99	1.34-2.96	0.32-1.22
Davoren (1985)	New Zealand	10	1.50	20	0.70-3.10	2.24-2.69	0.40-1.31
Gao et al. (1994)	China	38	1.00-3.50	21-69	0.22-6.10	1.26-4.70	0.35-2.24
Hayes (1996)	USA (Virginia and Maryland)	30	0.61-1.52	22-108	0.61-3.54	0.69-2.63	0.30-0.82
Holnbeck (2011)	USA (Montana)	53	0.51-3.05	10-108	0.49-4.85	0.92-3.20	0.30-2.01
Jackson (1997)	USA (Ohio)	12	0.76-1.65	10-54	1.95-3.38	1.16-2.13	0.76-1.65
Jarret and Boyle (1986)	USA (Colorado)	1	0.94	8	1.40	1.54	0.37
Norman (1975)	USA (Alaska)	10	0.98-4.57	8-90	1.70 to 6.71	1.60-3.51	0.61-1.83
Williamson (1993)	Canada (Alberta)	3	1.19-4.91	48-53	2.19-8.29	1.80-1.80	1.49-2.10
Wilson (1995)	USA (Mississippi)	8	2.44	8	2.59-3.93	1.75-2.26	0.88-1.95
Zhuravlyov (1978)	Uzbekistan	33	2.00-2.50	90-150	0.80-5.90	1.02-3.84	0.66-2.93
All Data		228	0.51-4.91	8-150	0.23-9.88	0.69-4.70	0.21-2.93

[a] Benedict and Caldwell (2014) supplemented the data with information obtained by written communication.

4. Local Pier Scour Model

The regression model of local scour depth $\hat{d}_s$ for quantile level $\alpha\,(0 \le \alpha \le 1)$ devised in this study is

$$\hat{d}_s(\alpha) = \theta_0 \times K_s(\theta_1) \times K_I(\theta_2,\theta_3) \times K_y(\theta_4,\theta_5,\theta_6) \times K_D(\theta_7,\theta_8,\theta_9) \times b \tag{1}$$

where K_s = pier shape factor (normalized about a coefficient of 1 for round-nosed piers, K_I = approach-flow intensity factor, K_y = approach-flow depth factor, K_D = bed sediment factor based on surface-layer measurements, and $\theta_j\,(j = 1..9)$ = parameters to be determined, all of which are considered functions of the quantile level α of the response. Based on other evaluations of local pier scour, such as those developed by Melville (1997), expressions for the scour model factors are formulated as follows:

$$K_s = \begin{cases} 1; \text{ for round-nosed piers} \\ \theta_1(\alpha); \text{ for sharp-nosed piers} \end{cases} \tag{2}$$

$$K_I = \begin{cases} 0; \text{ for } V \le V_{ic} \text{ where } V_{ic} = \theta_2(\alpha) \times V_c \\ \left(\dfrac{V - V_{ic}}{V_c - V_{ic}}\right)^{\theta_3(\alpha)}; \text{ for } V > V_{ic} \end{cases} \tag{3}$$

$$K_y = \begin{cases} \left(\dfrac{1}{\theta_4(\alpha)}\dfrac{y}{b}\right)^{\theta_5(\alpha)}; \text{ for } \dfrac{y}{b} < \theta_4(\alpha) \text{ (shallow flow)} \\ \left(\dfrac{1}{\theta_4(\alpha)}\dfrac{y}{b}\right)^{\theta_6(\alpha)}; \text{ for } \dfrac{y}{b} \ge \theta_4(\alpha) \text{ (deep flow)} \end{cases} \tag{4}$$

$$K_D = \begin{cases} \left(\dfrac{1}{\theta_7(\alpha)}\dfrac{b}{D_{50}}\right)^{\theta_8(\alpha)}; \text{ for } \dfrac{b}{D_{50}} < \theta_7(\alpha) \text{ (narrow pier)} \\ \left(\dfrac{1}{\theta_7(\alpha)}\dfrac{b}{D_{50}}\right)^{\theta_9(\alpha)}; \text{ for } \dfrac{b}{D_{50}} \ge \theta_7(\alpha) \text{ (wide pier)} \end{cases} \tag{5}$$

where V = the approach flow depth-averaged velocity, V_c = the critical approach flow velocity at which general sediment motion along the streambed begins (a formula for V_c is derived in the Appendix), and $V_{ic} = \theta_2(\alpha) \times V_c$ = the depth-averaged approach flow velocity at which scour initiates around a pier, y = the approach flow depth, b = the pier width, and D_{50} = the median bed-surface particle diameter, b = the pier width, and ϕ = the angle at which streamflow approaches a pier ($\phi = 0°$ for flow aligned with a pier). Characteristics of both armor layer sediment and shallow subsurface bed material should be considered when the surface layer influences local pier scour. However, because only information about bed-surface particles is known for most coarse-bed channel measurements, variables describing the substratum are not included in the following analysis.

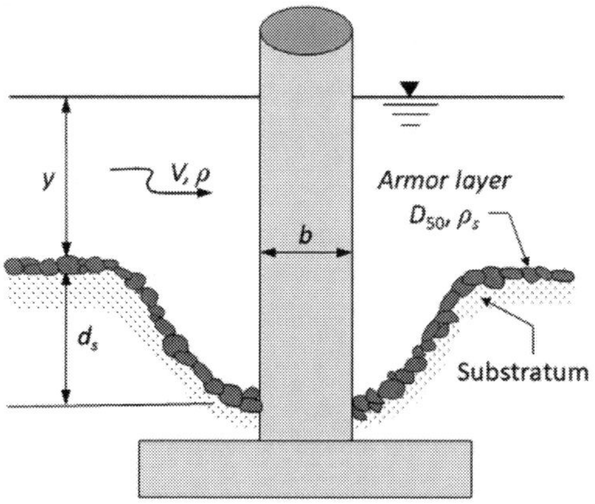

Figure. 2. Schematic of local pier scour around a cylindrical column of circular cross-section founded in a coarse-bed stream showing salient parameters.

Because the flow accelerates as it passes around a pier $0 < \theta_2 < 1$. Therefore, for $V \le V_{ic}$, $K_I = 0$ and no scouring occurs. The factor K_y shows the difference between comparatively shallow or deep flows in relation to the pier width. The factor K_D differentiates between narrow and wide piers in relation to the median surface-layer particle diameter.

4.1. Quantile Regression

The complexity and uncertainty of coarse-bed stream pier scour data requires a versatile and robust method to build explanatory and predictive statistical models. Quantile regression (Koenker 2005) meets these requirements by fitting conditional quantiles of the response with a general nonlinear model that assumes no parametric form for the conditional distribution of the response. The method provides information not obtained directly from standard least-squares regression methods. Quantile regression yields valuable insights where answers to essential questions lie in modeling the tails of the conditional distribution. In this study, regression models were developed for quantile levels $\alpha = 0.5, 0.95$, and 0.99. Equations for $\alpha = 0.95$ and 0.99 give immediate solutions for upper-level predictions of local pier scour, which provide adequate error margins that ensure a safe design estimate.

The $\theta_j(\alpha)$ parameters were established by solving the nonlinear minimization problem based on the $m = 228$ on-site measurements summarized in Table 1:

$$\min_{\theta_0(\alpha),\ldots,\theta_9(\alpha)} \sum_{i=1}^{m} \omega_\alpha \left(d_{s,i} - \hat{d}_{s,i}\right) \tag{6}$$

where

$$\omega_\alpha(r_i) = \alpha \times \max(r_i, 0) + (1-\alpha) \times \max(-r_i, 0) \tag{7}$$

and $r_i = d_{s,i} - \hat{d}_{s,i}$ is the prediction residual. Hence, for each quantile level α, the solution to the minimization problem yields a distinct set of regression coefficients. The median regression is produced with $\alpha = 0.5$ and $2 \times \omega_{0.5}(r_i) =$ the absolute value of the residual.

The minimization problem was solved numerically for the sought-after parameters using a variant of the generalized reduced gradient (GRG) algorithm of Smith and Lasdon (1992). A multi-start procedure was applied in which solutions were generated repeatedly using randomly selected starting values between specified bounds to increase the chance that the answers provided were globally optimal. The best of the locally optimal solutions was chosen as the proposed globally optimal solution.

4.2. Scour Equation Factors

Most of the nine model parameters were rounded slightly, some to the nearest simple fraction, to make the values easier to understand and reduce the complexity of the resulting formulas, which are marginally less accurate but are simpler to apply. For example, θ_1 for sharp-nosed piers aligned with the approach flow was close to 0.7 for each quantile level. Similarly, $\theta_2 \approx 1/3$, $\theta_6 \approx 0$, $\theta_7 \approx 13$, and $\theta_9 \approx 0$ for all three quantiles evaluated. Only θ_0, θ_3, θ_4, θ_5, and θ_9 were found to vary to a noticeable degree between the different levels.

With $\theta_2 = 1/3$, the flow intensity factor K_I is

$$K_I = \left(\frac{V - 1/3 \times V_c}{V_c - 1/3 \times V_c}\right)^{\theta_3(\alpha)} = \left(\frac{V/V_c - 1/3}{2/3}\right)^{\theta_3(\alpha)} = \left[\frac{1}{2}\left(3\frac{V}{V_c} - 1\right)\right]^{\theta_3(\alpha)} \tag{8}$$

when $V > V_{ic}$. Making use of Eq. *(23)* found in the Appendix, the ratio V/V_c is

$$\frac{V}{V_c} = \frac{k_n}{\sqrt{\tau_{co*}(S_s - 1)}} \times \frac{V}{\sqrt{gy}} \times \left(\frac{y}{D_{50}}\right)^{1/3} \tag{9}$$

which, letting $k_n = 0.148$, $\tau_{co*} = 0.03$, and $S_s = \rho_s/\rho = 2.65$ becomes

$$\frac{V}{V_c} = \frac{2}{3} \times Fr \times \left(\frac{y}{D_{50}}\right)^{1/3} \tag{10}$$

where $Fr = V/\sqrt{gy} =$ the approach flow Froude number. Substituting for V/V_c in Eq. *(8)* gives

$$K_I = \left[Fr\left(\frac{y}{D_{50}}\right)^{\frac{1}{3}} - 0.5\right]^{\theta_3} \tag{11}$$

The adjustment factor is like those proposed by Hancu (1971), Melville and Sutherland (1988), Breusers and Raudkivi (1991, page 77), Froehlich and Arneson (1995), and the Academy of Railway Sciences of China (Gao et al. 1994). The ratio $V_{ic}/V_c = \theta_2 = 1/3$, which is slightly less than similar values found by others (Ettema et al. 2011), reveals that scour initiates around a pier well before the general movement of the particles forming the armor layer begins.

Because $\theta_6 = 0$ for the three quantile levels, the relative depth factor K_y simplifies to

$$K_y = \begin{cases} \left(\dfrac{1}{\theta_4(\alpha)} \dfrac{y}{b} \right)^{\theta_5(\alpha)}; \text{ for } \dfrac{y}{b} < \theta_4(\alpha) \text{ (shallow flow)} \\ 1; \text{ for } \dfrac{y}{b} \geq \theta_4(\alpha) \text{ (deep flow)} \end{cases} \tag{12}$$

The factor adjusts only for shallow flows based on the coefficients θ_4 and θ_5, both functions of α. It shows the same tendency as those given by Melville and Sutherland (1988), Breusers and Raudkivi (1991, pages 70-71), and Richardson and Davis (2001, page 6.4).

With $\theta_7 = 13$ and $\theta_8 = 0$ for the three quantile levels, the factor accounting for armor-layer particle size becomes

$$K_D = \begin{cases} 1; \text{ for } \dfrac{b}{D_{50}} < 13 \text{ (narrow pier)} \\ \left(\dfrac{1}{13} \dfrac{b}{D_{50}} \right)^{\theta_9(\alpha)}; \text{ for } \dfrac{b}{D_{50}} \geq 13 \text{ (wide pier)} \end{cases} \tag{13}$$

For a specified pier width and $b/D_{50} < 13$, scour depth is not affected by bed particle size. However, for $b/D_{50} > 13$, as D_{50} increases, so does K_D. The influence of the ratio b/D_{50} agrees with Hancu's (1971) findings, who carried out live-bed experiments in gravel-mixture channels. As D_{50} increases, so does V_c, thereby boosting the probability of clear-water flow conditions during which bed sediment is not swept into a developing scour hole. Moreover, once $V > V_c$ and the armor layer is mobilized, the sediment transport rate is lessened with a larger D_{50}. Both effects contribute to enhanced scouring as D_{50} increases.

4.3. Equation Summary

Inserting Eqs. *(11)*, *(12)*, and *(13)* into Eq. *(1)* for K_I, K_y, and K_D gives, after some rearrangement, the following equations for quantile estimates of the local scour depth $\hat{d}_s(\alpha)$ for $\alpha = 0.5, 0.95$, and 0.99 when $V > V_{ic}$:

$$\hat{d}_s(\alpha = 0.5) = \begin{cases} 0.85 \times K_s \times \left[Fr \times \left(\dfrac{y}{D_{50}} \right)^{1/3} - 0.5 \right]^{1/3} \times \left(\dfrac{y}{b} \right)^{1/6} \times b; \text{ for } \dfrac{y}{b} < 2 \text{ and } \dfrac{b}{D_{50}} \leq 13 \\ 2.00 \times K_s \times \left[Fr \times \left(\dfrac{y}{D_{50}} \right)^{1/3} - 0.5 \right]^{1/3} \times \left(\dfrac{y}{b} \right)^{1/6} \times \left(\dfrac{b}{D_{50}} \right)^{-1/3} \times b; \text{ for } \dfrac{y}{b} < 2 \text{ and } \dfrac{b}{D_{50}} > 13 \\ 0.95 \times K_s \times \left[Fr \times \left(\dfrac{y}{D_{50}} \right)^{1/3} - 0.5 \right]^{1/3} \times b; \text{ for } \dfrac{y}{b} \geq 2 \text{ and } \dfrac{b}{D_{50}} \leq 13 \\ 2.24 \times K_s \times \left[Fr \times \left(\dfrac{y}{D_{50}} \right)^{1/3} - 0.5 \right]^{1/3} \left(\dfrac{b}{D_{50}} \right)^{-1/3} \times b; \text{ for } \dfrac{y}{b} \geq 2 \text{ and } \dfrac{b}{D_{50}} > 13 \end{cases} \tag{14}$$

$$\hat{d}_s(\alpha = 0.95) = \begin{cases} 1.13 \times K_s \times \left[Fr \times \left(\dfrac{y}{D_{50}} \right)^{1/3} - 0.5 \right]^{1/5} \times \left(\dfrac{y}{b} \right)^{1/3} \times b; \text{ for } \dfrac{y}{b} < 2.5 \text{ and } \dfrac{b}{D_{50}} \leq 13 \\ 1.74 \times K_s \times \left[Fr \times \left(\dfrac{y}{D_{50}} \right)^{1/3} - 0.5 \right]^{1/5} \times \left(\dfrac{y}{b} \right)^{1/3} \times \left(\dfrac{b}{D_{50}} \right)^{-1/6} \times b; \text{ for } \dfrac{y}{b} < 2.5 \text{ and } \dfrac{b}{D_{50}} > 13 \\ 1.54 \times K_s \times \left[Fr \times \left(\dfrac{y}{D_{50}} \right)^{1/3} - 0.5 \right]^{1/5} \times b; \text{ for } \dfrac{y}{b} \geq 2.5 \text{ and } \dfrac{b}{D_{50}} \leq 13 \\ 2.36 \times K_s \times \left[Fr \times \left(\dfrac{y}{D_{50}} \right)^{1/3} - 0.5 \right]^{1/5} \times \left(\dfrac{b}{D_{50}} \right)^{-1/6} \times b; \text{ for } \dfrac{y}{b} \geq 2.5 \text{ and } \dfrac{b}{D_{50}} > 13 \end{cases} \tag{15}$$

and

$$\hat{d}_s(\alpha=0.99)=\begin{cases} 1.26\times K_s\times\left[Fr\times\left(\dfrac{y}{D_{50}}\right)^{1/3}-0.5\right]^{1/6}\times\left(\dfrac{y}{b}\right)^{1/3}\times b;\ \text{for } \dfrac{y}{b}<3 \text{ and } \dfrac{b}{D_{50}}\le 13 \\ 2.10\times K_s\times\left[Fr\times\left(\dfrac{y}{D_{50}}\right)^{1/3}-0.5\right]^{1/6}\times\left(\dfrac{y}{b}\right)^{1/3}\times\left(\dfrac{b}{D_{50}}\right)^{-1/5}\times b;\ \text{for } \dfrac{y}{b}<3 \text{ and } \dfrac{b}{D_{50}}>13 \\ 1.82\times K_s\times\left[Fr\times\left(\dfrac{y}{D_{50}}\right)^{1/3}-0.5\right]^{1/6}\times b;\ \text{for } \dfrac{y}{b}\ge 3 \text{ and } \dfrac{b}{D_{50}}\le 13 \\ 3.04\times K_s\times\left[Fr\times\left(\dfrac{y}{D_{50}}\right)^{1/3}-0.5\right]^{1/6}\times\left(\dfrac{b}{D_{50}}\right)^{-1/5}\times b;\ \text{for } \dfrac{y}{b}\ge 3 \text{ and } \dfrac{b}{D_{50}}>13 \end{cases} \quad (16)$$

Four expressions are presented for each quantile level because the relations for K_I (a function of y/b) and K_D (a function of b/D_{50}) are interrupted at specified values of y/b and b/D_{50} to identify significant changes in the response.

Measured scour depths are graphed against the expected values given by Eqs. *(14)*, *(15)*, and *(16)* in Fig. 3. Just 11 of the measured scour depths are underpredicted at the 0.95 quantile level and only by insignificant amounts. The 0.99 quantile estimates underpredict just five measured depths. Consequently, the 0.95 quantile prediction formulas provide reliable safety margins that are not excessive.

4.4. Pier Shape and Approach Flow Alignment

With the pier shape factor standardized based on K_s = 1.0 for cylindrical columns of circular cross-section and other round-nosed piers aligned with the approach flow, the slightly rounded optimized K_s for sharp-nosed piers is 0.7. Because the data assembly lacked square-nodes pier data, a shape factor K_s = 1.3 is taken from a previous study (Froehlich 1988) that evaluated a collection of on-site measurements made in sand-bed and coarse-bed streams for square-nosed piers, thereby giving

$$K_s=\begin{cases}1.0;\ \text{for round-nosed piers}\\0.7;\ \text{for sharp-nosed piers}\\1.3;\ \text{for square-nosed piers}\end{cases} \quad (17)$$

Others, such as Arneson et al. (2012, page 7.4) and Hoffmans and Verheij (2021, page 193), suggest slightly different coefficients for sharp-nosed and square-nosed piers aligned with the approaching streamflow.

For pier shapes other than circular cylinders, local scour depths are significantly influenced by a pier's alignment with the approaching streamflow. Scour depths increase as the pier's width projected normal to the flow direction expands. The projected width and resulting scour intensification can be substantial even for small approach-flow angles and typical pier length-to-width ratios. However, separating the effects of pier shape and approach flow alignment is challenging because the nose shape character changes as the flow angle increases. One must consider the combined influence of a pier's shape, its length-to-width ratio ℓ/b, and the approach flow direction.

The standard tactic – see, for example, Arneson et al. (2012, page 7.4) or Hoffmans and Verheij (2021, pages 192-193) – is to maintain separate influences of pier shape and flow alignment by adding to Eq. *(1)* an approach-flow alignment factor

$$K_\phi(\phi,\ell/b)=\left(\cos\phi+\frac{\ell}{b}\sin\phi\right)^{0.65} \quad (18)$$

that considers the expanded pier width presented to the approaching streamflow. This term is multiplied by the scour depth calculated for $\phi=0^o$ to adjust for flow alignment. However, once ϕ exceeds 5°, Arneson et al. (2012, page 7.5) recommend treating the pier as round-nosed no matter the shape by setting $K_s=1.0$. The Florida Department of Transportation (FDOT 2021, pages 23-24) takes a different approach by replacing b in Eqs. (14), (15), and (16) with the projected pier width $b'=b\cos\phi+\ell\sin\phi$.

5. Summary and Conclusions

Understanding the relevant physics and predicting scour at bridges remains challenging because of complex interactions between hydrodynamic forces, channel beds' dynamics, and uncertainties of natural streams. Although many empirical formulas have been developed for estimating local scour depth at bridge piers, small-scale laboratory experiments in sand-filled flumes are the basis for most. However, these conditions scarcely resemble those in natural coarse-bed steams, characterized by a surface cover of particles (an armor layer) that are large compared to the bulk mixture in the substratum.

An obvious way of dealing with this problem is to base scour depth predictions on measurements made at full-scale bridge piers founded in natural coarse-bed streams. Making on-site measurements of local pier scour during floods is hazardous and logistically demanding. Fortunately, recent efforts to measure local pier scour have provided a collective data set consisting of 228 on-site samples collected worldwide in streams with median bed surface particle diameters of 8 mm or larger. The assembled data used in this study are summarized in Table 1. Only measurements made at cylindrical piers with a circular cross-section and other piers aligned with the approaching streamflow were included.

Quantile regression, a versatile and robust method, was used to build predictive statistical models of local pier scour based on the collected samples. Coefficients of nonlinear equations for scour depth were fit to the data for three quantile levels $\alpha = 0.5$, 0.95, and 0.99. The median estimate (the expected scour depth) is given by $\alpha = 0.5$. Equations for quantile levels 0.95 and 0.99 predict scour depths that are not likely to be exceeded, thus providing the safety margins needed to design reliable bridges. The predictive models show that for a specified pier width b, scour does not depend on water depth for deep flows ($y/b > 2$ to 3). Also, scour was found to deepen as the size of particles forming the bed armor layer increased with other variables (b, y, and V) constant.

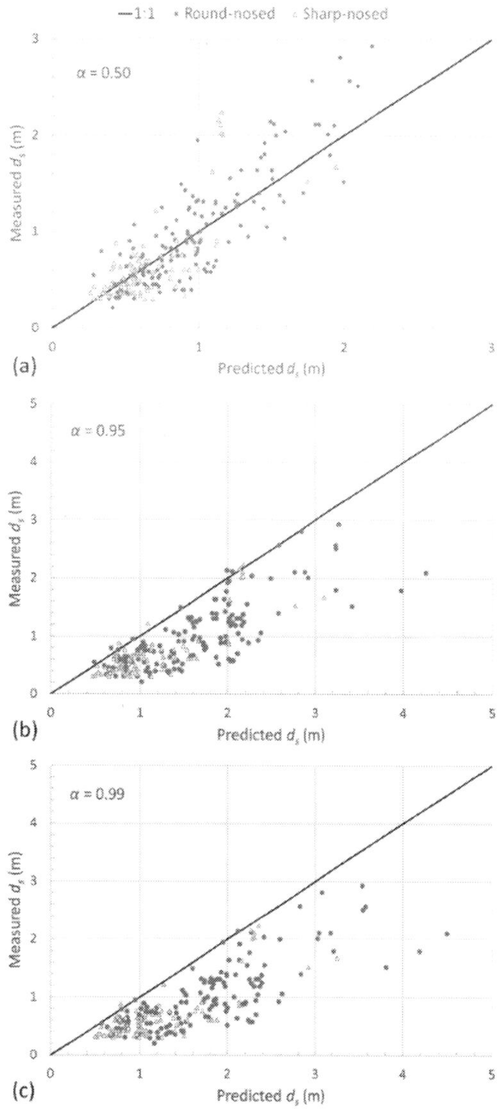

Figure. 3. Graphs of measured ds versus predicted ds for quantile levels $\alpha = 0.5$, 0.95, and 0.99 for the 228 on-site measurements of local pier scour evaluated in this study.

6. Appendix -- Critical Approach Flow Velocity

A relation for the velocity of approaching streamflow V_c that creates general sediment motion is needed to assess the difference between clear-water and live-bed scour conditions. Manning's formula (Chow 1959, page 98), which is uncomplicated and accepted widely, gives bed shear stress τ for uniform flow in a wide channel as

$$\tau = \rho \frac{gn^2}{k_u^2 y^{1/3}} V^2, \tag{19}$$

where ρ = the fluid (water) mass density, g = gravitational acceleration (approximately 9.81 m/s^2), y = the approach flow depth, V = the approach flow depth-averaged velocity, n =Manning's roughness coefficient and k_u = unit conversion factor (1.0 for SI units, 1.486 for US Customary units). An expression for n as a function of the surface bed material median diameter D_{50} given by Strickler (1923), which has been tested and found suitable for gravel-bed streams by Maynord (1991), is

$$n = \frac{k_u k_n}{\sqrt{g}} D_{50}^{1/6}, \tag{20}$$

where k_n = 0.148 = a dimensionless coefficient found by Strickler for natural gravel and cobble bed channels. Combining Eqs. *(19)* and *(20)* provides the following relation for bed shear stress as a function of the size of the bed-surface particles and the approach flow velocity and depth:

$$\tau = \rho k_n^2 \left(\frac{D_{50}}{y} \right)^{1/3} V^2. \tag{21}$$

The shear stress on a nearly horizontal streambed at which widespread sediment motion begins, that is, the critical bed shear stress, is

$$\tau_{co} = \tau_{co*} \left(S_s - 1 \right) \rho g D_{50}, \tag{22}$$

where τ_{co*} = the dimensionless critical bed shear stress coefficient or the critical Shields parameter, $S_s = \rho_s/\rho$ = the specific gravity of the bed sediment, and ρ_s = the bed sediment mass density. Most coarse-bed deposits consist of quartz or feldspar particles, for which $S_s \approx 2.65$, a value applied in this study. Combining Eqs. *(21)* and *(22)* gives the critical approach flow velocity at which general sediment transport begins:

$$V_c = \frac{\sqrt{\tau_{co*}}}{k_n} \sqrt{\left(S_s - 1 \right) g D_{50}} \left(\frac{y}{D_{50}} \right)^{1/6} \tag{23}$$

A systematic analysis of eight decades of incipient motion studies by Buffington and Montgomery (1997) found that Shield's parameter varied from 0.02 to 0.086, with an average value of 0.046 for gravel. After reviewing estimates of the dimensionless critical shear stress, $\tau_{co*} = 0.030$ was adopted for use in this analysis.

7. REFERENCES

Arneson, L.A., L.W. Zevenbergen, P.F. Lagasse, and P.E. Clopper (2012). "Evaluating scour at bridges (5th ed)." *Publication No. FHWA-HIF-12-003, Hydraulic Engineering Circular No. 18*, Office of Bridge Technology, Federal Highway Administration, Washington, D.C.

Becker, L.D. (1994). "Investigation of bridge scour at selected sites on Missouri streams." *Water-Resources Investigations Report 94-4200*, U.S. Geological Survey, Rolla, Missouri.

Benedict, S.T., and A.W. Caldwell (2014). "A pier-scour database: 2,427 field and laboratory measurements of pier scour." *Data Series 845*, U.S. Geological Survey, Reston, Virginia.

Breusers, H.N.C., and A.J. Raudkivi (1991). *Scouring*. A. A. Balkema, Rotterdam, Netherlands.

Buffington, J.M., and D.R. Montgomery (1997). "A systematic analysis of eight decades of incipient motion studies, with special reference to gravel-bedded rivers." *Water Resources Research*, 33(8), 1993-2029.

Butch, G.K. (1991). "Measurement of bridge scour at selected sites in New York, excluding Long Island." *Water-Resources Investigations Report 91–4083*, U.S. Geological Survey, Albany, New York.

Chase, K.J., and S.R. Holnbeck (2004). "Evaluation of pier-scour equations for coarse-bed streams." *Scientific Investigations Report 2004-5111*, U.S. Geological Survey, Reston, Virginia.

Chow, V.T. (1959). *Open-channel hydraulics*. McGraw-Hill, New York, New York.

Davoren, A. (1985). "Local scour around a cylindrical bridge pier" *Publication No. 3*, Hydrology Centre, Christchurch, New Zealand.

Dey, S., and A.K. Barbhuiya (2004). "Clear-water scour at abutments in thinly armored beds." *Journal of Hydraulic Engineering*, 130(7), 622–634.

Ettema, R., G. Constantinescu, and B. Melville (2011). "Evaluation of bridge pier scour research: scour processes and prediction," *Web-Only Document 175 (Contractor's Final Report for NCHRP Project 24-27)*, National Cooperative Highway Research Program, Transportation Research Board, National Academy of Sciences, Washington, D.C.

FDOT (2021). *Bridge scour manual*. Florida Department of Transportation, Central Office, Tallahassee, Florida.

Froehlich, D.C. (1988). "Analysis of on-site measurements of scour at piers." *Proceedings of the 1988 National Conference on Hydraulic Engineering*, Colorado Springs, Colorado, August 15-19, 1988, American Society of Civil Engineers, New York, 826-831.

Froehlich, D.C. (1995). "Armor-limited clear-water contraction scour at bridges." *Journal of Hydraulic Engineering*, 121(6), 490-493.

Froehlich, D.C., and L.A. Arneson (1995). "Local scour at bridge piers in coarse-bed channels." *Proceedings of the First International Conference on Water Resources Engineering, ASCE*, San Antonio, Texas, August 14-18, 1995, American Society of Civil Engineers, New York, 966-970.

Gao, D.G., L. Posada, and C.F. Nordin (1994). "Pier scour equations used in the People's Republic of China."

Publication No. FHWA SA-93-076, Federal Highway Administration, U.S. Department of Transportation, Washington, D.C.
Hancu, S. (1971). "Sur le calcul des affouillements locaux dans la zone des piles du pont." *Proceedings of the 14th Congress of the International Association for Hydraulic Research*, Aug 29-Sep 3, 1971, Vol. 3, 299-305.
Hayes, D.C. (1996). "Scour at bridge sites in Delaware, Maryland, and Virginia." *Water-Resources Investigations Report 96–4089*, U.S. Geological Survey, Richmond, Virginia.
Hoffmans, G.J.C.M., and H.J. Verheij (2021). *Scour manual – current-related erosion (2nd edition)*. CRC Press, Boca Raton, Florida.
Holnbeck, S.R. (2011). "Investigation of pier scour in coarse-bed streams in Montana, 2001 through 2007." Scientific Investigations Report 2011–5107, U.S. Geological Survey, Reston, Virginia.
Jackson, K.S. (1997). "Evaluation of bridge-scour data at selected sites in Ohio." *Water-Resources Investigations Report 97-4182*, U.S. Geological Survey Columbus, Ohio.
Jarrett, R.D., and J.M. Boyle (1986). "Pilot study for collection of bridge-scour data." *Water-Resources Investigations Report 86-4030*, U.S. Geological Survey, Denver, Colorado.
Jones, G.D. (1970). "Scour at bridge waterways." *Synthesis of Highway Practice 5*, National Cooperative Highway Research Program, Highway Research Board, National Academy of Sciences, Washington, D.C.
Koenker, R. (2005). *Quantile regression*. Cambridge University Press, New York.
Laursen, E.M. (1955). "Model-prototype comparison of bridge scour." in F. Burggraf, (ed.), *Proceedings of the 34th Annual Meeting*, Highway Research Board; National Academy of Sciences, Washington, D.C., 188-193.
Leopold, L.B. (1994). *A view of the river*. Harvard University Press, Cambridge, Massachusetts.
Maynord, S.T. (1991). "Flow resistance of riprap." *Journal of Hydraulic Engineering*, 117(6), 687-696.
Melville, B.W. (1975). "Local scour at bridge sites: New Zealand." *Project Report No.* 117, University of Auckland, School of Engineering, Auckland, New Zealand.
Melville, B.W., R. Ettema, and S.C. Jain (1989). "Measurement of bridge scour." In *Proceedings of the Bridge Scour Symposium*, Dah-Cheng Woo and Shou-Shan Fan (eds.), Turner-Fairbank Research Center, Mclean, Virginia, October 17-19, 1989, Published as *Report No. FHWA-RD-90-035*, Federal Highway Administration, U.S. Department of Transportation, Washington, D.C., 183-194.
Melville, B.W., and A.J. Sutherland (1988). "Design method for local scour at bridge piers." *Journal of Hydraulic Engineering*, 114(10), 1210-1226.
Mueller, D.S., and C.R. Wagner (2005). "Field observations and evaluations of streambed scour at bridges." *Publication No. FHWA-RD-03-052*, U.S. Department of Transportation, Federal Highway Administration, Office of Infrastructure Research and Development, Turner-Fairbank Highway Research Center, McLean, Virginia.
Neill, C.R., and L.R. Morris (1980). "Scour problems at railway bridges on the Thompson River, B.C." *Canadian Journal of Civil Engineering*, 7(2), 357-372.
Norman, V.W. (1975). "Scour at selected bridge sites in Alaska." *U.S. Geological Survey Water-Resources Investigations Report 32-75*, Anchorage, Alaska.
Raudkivi, A.J., and R. Ettema (1985). "Scour at cylindrical bridge piers in armored beds." *Journal of Hydraulic Engineering*, 111(4), 713-731.
Richardson, E.V., and S.R. Davis (2001). "Evaluating scour at bridges (4th edition)." *Publication No. FHWA NHI 01*-001, *Hydraulic Engineering Circular No. 18*, Federal Highway Administration, U.S. Department of Transportation, Washington, D.C.
Ryan, S.E., and W.W. Emmett (2002). "The nature of flow and sediment movement in Little Granite Creek near Bondurant, Wyoming." *General Technical Report RMRS-GTR-90*, U.S. Department of Agriculture, Forest Service, Rocky Mountain Research Station, Fort Collins, Colorado.
Smiles, S. (1861). *Lives of the engineers, vol II*. John Murray, London, England.
Smith, S. and L. Lasdon (1992). "Solving large sparse nonlinear programs using GRG." *ORSA Journal on Computing*, 4(1), 2-15.
Strickler, A. (1923). "Beitrage zur Frage der Geshwindigkeitsformel und der Rauhigkeitszahlen fur Strome, Kanale und geschlossene Leitungen." *Mitteilungen des eidgenossischen Amtes fur Wasserwirtschaft*, Bern, Switzerland, 16, 12-13.
Vonkeman, J.K., and G.R. Basson (2019). "Evaluation of empirical equations to predict bridge pier scour in a non-cohesive bed under clear-water conditions." *Journal of the South African Institution of Civil Engineering*, 61(2), 2-20.
Williamson, D. (1993). "Local scour measurements at piers in Alberta." In *Hydraulic Engineering '93*, American Society of Civil Engineers, New York, New York, 2057-2062.
Wilson, K.V., Jr. (1995). "Scour at selected bridge sites in Mississippi." *Water-Resources Investigations Report 94–4241*, U.S. Geological Survey, Jackson, Mississippi.
Zhuravlyov, M.M. (1978). "New method for estimation of local scour due to bridge piers and its substantiation." *Transactions*, Ministry of Transport and Construction, Moscow, Russia.

DOI: https://doi.org/10.26077/d3a8-d233

CFD-Based Modeling for Computing Discharge Coefficient of an Ogee Spillway

Hariom Gautam[1], Zulfequar Ahmad[1], & Pramod Kumar Sharma[1]
[1]*Indian Institute of Technology, Roorkee, India*
E-mail: h_gautam@ce.iitr.ac.in

Abstract: *The basic functionality of the spillway is to control the flow and safely convey the excess flow while keeping the structure stable. An ogee spillway is commonly used due to its high discharging capacity. The discharge coefficient of an ogee spillway is greatly affected by the ratio of an operating head and design head. Since physical models are expensive and have limitations in terms of scale effect, a comprehensive numerical simulation was performed in this study to estimate the discharge coefficient for an ogee spillway under various head ratios using CFD-based numerical modeling. The numerical simulations were done using FLOW-3D. In the present study, renormalized group (RNG) $k-\varepsilon$ turbulence closure model and the Volume of Fluid (VOF) algorithm were used for the simulation of flow over an ogee spillway. The discharge coefficient of the ogee spillway computed using CFD simulation for various head ratios shows a reasonable agreement with the values reported in the literature. Also, the water surface profiles, pressure distribution, discharge rating curve, and velocity distribution over the ogee spillway for the different heads were plotted and analyzed.*

Keywords: *Ogee spillway, Numerical modeling, CFD, Discharge coefficient, FLOW-3D.*

1. Introduction

The basic functionality of the spillway is to control the flow in a reservoir and to provide smooth conveyance of the discharge from upstream to downstream. Out of various spillways, the ogee spillway is commonly used in reservoirs and dams to control flow depths and discharge because of its ability to pass the flood safely. Muller was probably the first who suggested the idea of the ogee spillway in 1908 (Micheal and Walker 1948). Since spillways are the most crucial structure of the dam, its safety must be ensured by evaluating various parameters related to its hydraulic and structural stability. The discharge capacity is usually associated empirically with the discharge coefficient (C), such that a spillway with higher C is preferred. The profile of the ogee spillway corresponds to the path of the bottom nappe of free-falling water from the sharp-crested spillway (Hager 1987). Ogee spillway is designed for a single upstream head known as the designed head (H_d). The C is greatly affected by the head ratio defined as the operating head (H_e) to the design head (H_d). As the operating head differs from the design head, it changes C and pressure over the ogee-spillway. It may lead to cavitation when the local pressure falls below the water vapor pressure and may cause detachment of the lower nappe from the spillway surface (Erpicum et al. 2018). So far, several investigations have been performed to examine the hydrodynamics of the ogee spillway (Kline 1953, Roushangar et al. 2019, Saneie et al. 2016, Imanian and Mohammadian 2019). USBR (1987) performed detailed experiments to study the flow profiles over spillways for a wide range of head ratios and upstream face slopes at the Water Experimental Station (WES) and produced manuals for the spillway design. Their spillway design is based on an upstream slope and design head to get zero pressure over the spillway. Vermeyen (1992), and Imanian and Mohammadian (2019) concluded that under ideal conditions, C increases up to the head ratio (H_e/H_d) of five due to a reduction in pressure beyond the separation zone.

Apart from various experimental studies, many authors also carried out numerical simulations and compared the results (Johnsen and Savage 2006, Kocaer and Yarar 2020). Physical models are expensive and have limitations in scale effect analysis. Albeit, numerical simulations can be cost-effective with acceptable accuracy. The appropriate optimization and turbulence closure scheme must be applied to get reasonably accurate results from numerical simulations (Imanian and Mohammadian 2019). Savage and Johnson (2001) carried out a physical study over a standard ogee spillway and compared the results with the numerical models using finite element CFD solver FLOW-3D. Dargahi (2006) compared the water surface profiles and discharge coefficient (C) obtained by numerical simulation and physical experiments. Chanel and Doering (2008) investigated three case studies with different head ratios using the FLOW-3D CFD solver. Rad (2016) stated that the RNG (Renormalized group) turbulence model and VOF method combination are most suitable for analyzing the flow through the spillways. Yarar and Kocaer (2020) compared the experimental flow results over the ogee spillway with the numerical results. The physical models were numerically simulated using ANSYS-Fluent and OpenFOAM with the two different turbulence closure models ($K-\varepsilon$, and $K-\omega\ SST$ turbulence model).

The literature survey shows a research gap in the context of a numerical study of the ogee crest spillway. First, in the context of numerical estimation of discharge coefficient (C) for the design discharge and discharge different than the design discharge in depth using a FLOW-3D. Second, very few numerical studies have been done to assess the vertical pressure distribution over the ogee spillway surface under different head ratios in detail. In this research, flow over a standard ogee-shaped spillway (USBR 1987) was simulated numerically using the CFD solver FLOW-3D. The C of the ogee spillway under various head ratios (H_e/H_d) were investigated numerically. The computed results were compared to the USBR (1987), and USACE-WES published data. Apart from this, water surface profiles, vertical pressure distribution over the ogee spillway, discharge rating curve, and velocity profiles were analyzed using the CFD technique.

2. Methodology

2.1. Ogee Spillway

The profile of the ogee spillway is based on a trajectory of the lower nappe in a free-falling flow over a fully aerated sharp-crested spillway. It can be estimated by the law of projectile and having a shape similar to 'ogee' (S – shape curve). Since lower nappe profiles vary with head over the sharp-crested spillway, the design of the ogee crest is taken for a maximum head over the spillway. To achieve the objective of no negative pressure under the design head (H_d), several standardized ogee spillway profiles have been suggested by the Corps of Engineers (1970) (USBR 1987). The profile of the downstream section of the ogee spillway can be obtained by using Eq. (1) (USBR 1987).

$$x^n = K.H_d^{n-1}.z \tag{1}$$

In this x and z are the stream-wise and upward direction coordinates of the downstream profile from the crest of the spillway. H_d is the designed head consisting of the approach velocity head. K and n are constant and depend on the upstream slope and approach velocity. As in the present study, the upstream face is considered vertical, and the approach velocity head is negligible, so the corresponding K and n values are 2.0 and 1.85, respectively (USBR 1987). Thus, the profile for the downstream section is given by Eq. (2).

$$x^{1.85} = 2.H_d^{0.85}.z \tag{2}$$

In the present research, an ogee spillway with the design head (H_d) of 0.06 m, and height (P) of 0.12 m, is considered for the numerical modeling. The upstream section of the ogee spillway is generally constructed with two circular arcs with a radius of $0.53 \times H_d$ and $0.235 \times H_d$, Respectively (USBR 1987). The general flow profile over the ogee spillway is shown in Fig. 1.

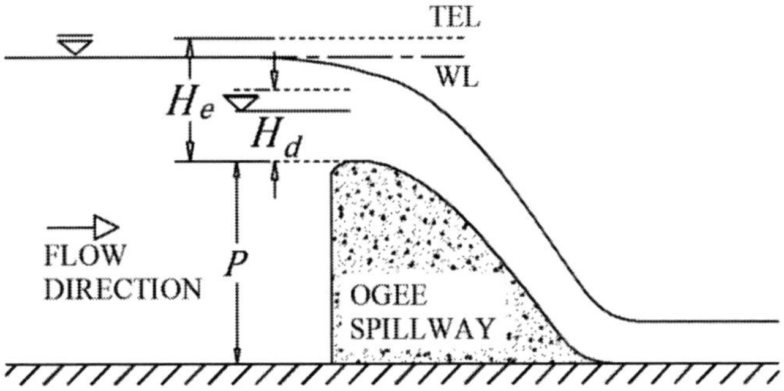

Figure 1. Definition of flow over ogee spillway.

The geometrical profile of the ogee spillway used in this study is as per USBR (1987), as shown in Fig. 2. The geometrical parameters of the ogee spillway are given in table 1. The discharge equation can be written as (USBR 1987)

$$Q = C.L_e.H_e^{3/2} \tag{3}$$

Where Q is the discharge in m^3/s, L_e is the effective crest length (m), C is the discharge coefficient, and H_e is energy head above the crest (m). In the present study, the discharge coefficient C is represented as C_0 for head equal to the design head (H_d) and taken 2.173 (USBR 1987). The value of C depends on various parameters like crest height, operating head, the upstream slope of the spillway, downstream apron, and submergence and is thus obtained experimentally. In this paper, the C is computed numerically using CFD solver FLOW-3D, and the results are compared with the USBR (1987) and USACE-WES results. The various head ratios (H_e/H_d) considered in this study are 0.1, 0.2, 0.4, 0.5, 0.6, 0.8, 1.0, 1.2, 1.33, 1.4, 1.6, and 2.

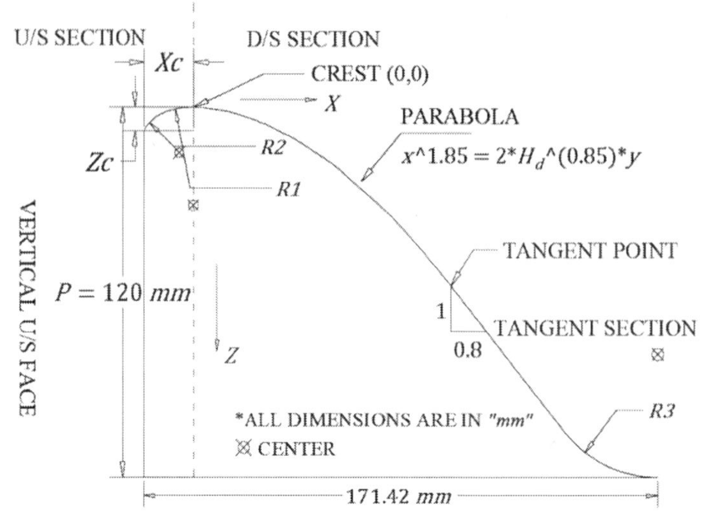

Figure 2. Design sketch of an ogee-crested spillway used in numerical modeling.

Table 1. Geometrical parameters of the ogee spillway.

Spillway Height (P) (mm)	Crest Length (L_e) (mm)	Xc (mm)	Zc (mm)	$R1$ (mm)	$R2$ (mm)	$R3$ (mm)	Slope $(V{:}H)$	Tangent Point (x, z) (mm)
120	390	16.92	7.62	31.8	14.1	30	1:0.8	85.51, 57.78

2.2. Numerical Methodology

FLOW-3D is a commercial computational fluid dynamics (CFD) solver that can simulate complex fluid flow problems by implementing various numerical schemes and turbulence models. FLOW-3D uses a volume of fluid (VOF) method to track the free surface interface (Hirt and Nichols 1981). It is based on the assumption that two or more fluids should not be mixed. This two-phase technique represents both the water and air in the grid. The approach is based on the idea that each cell contains a fraction of water (F), which is unity when the cell is filled with water and zero when it is empty. Cells may be half full at the free surface and assigned a value equal to the fluid volume divided by cell volume (between zero and unity). The program uses a cell porosity technique called the fractional area/volume obstacle representation (FAVOR) method to define the flow geometry or to incorporate the model shape into the governing equations (Hirt and Sicilian 1985). In the computational cells, flow obstacles are represented as a fractional value between zero and unity. Therefore, the fractional value of the area is equal to unity if the whole cell is filled with obstacles. In FLOW-3D, the finite volume approach is used in conjunction with standard rectangular grids to numerically solve the three-dimensional Reynolds'-averaged Navier-Stokes (RANS) equations of fluid motion. For an incompressible flow comprising the FAVOR and VOF variables, the general governing continuity and RANS equations can be written, respectively as

$$\frac{\partial}{\partial x_i}(u_i A_i) = 0 \tag{4}$$

$$\frac{\partial u_i}{\partial t} + \frac{1}{V_F}\left(u_j A_j \frac{\partial u_i}{\partial x_j}\right) = -\frac{1}{\rho}\frac{\partial P_i}{\partial x_j} + g_i + f_i \tag{5}$$

Where g_i is the local fluid acceleration, f_i denotes the Reynolds' stresses (frictional/viscous forces) in the respective directions for which a turbulence closure model is required, P is the hydrostatic pressure, V_F is fluid volume fraction in each cell open to flow in the FAVOR method, u_i denotes velocities components (in x, y, and z directions), t represents time, A_i is the fractional (cross-sectional) area open to flow along the flow direction, and ρ is the fluid density. Numerical simulations based on the RANS equations are more reasonable in hydraulics engineering. A turbulence closure model is necessary to approximate turbulent Reynolds' stress tensor in RANS. The standard $k-\epsilon$ model is a commonly used RANS turbulence closure model. This model is based on a linear constitutive relationship to solve the shear stress in RANS. In addition to the $k-\epsilon$ model, the RNG $k-\epsilon$ model solves the same differential equations as $k-\epsilon$. Moreover, turbulence equation constants are explicitly determined in the RNG model, whereas the traditional $k-\epsilon$ model has empirical formation. RNG $k-\epsilon$ model does not disintegrate turbulent structures in space and time. Previous studies have shown that an RNG $k-\epsilon$ turbulence closure model is appropriate for the numerical simulation of flow through hydraulic structures. Therefore, in this study, the RNG $k-\epsilon$ model is used. The RNG $k-\epsilon$ model is a two-equation model with the following expressions:

$$\frac{\partial K}{\partial t}+\frac{1}{V_F}\left(uA_x\frac{\partial K}{\partial x}+vA_y\frac{\partial K}{\partial y}+wA_z\frac{\partial K}{\partial z}\right)=P_T+B_T+D_K-\varepsilon \quad (6)$$

$$\frac{\partial \varepsilon}{\partial t}+\frac{1}{V_F}\left(uA_x\frac{\partial \varepsilon}{\partial x}+vA_y\frac{\partial \varepsilon}{\partial y}+wA_z\frac{\partial \varepsilon}{\partial z}\right)=\frac{\eta.\varepsilon}{K}(P_T+\alpha.B_T)+D_\varepsilon-\beta\frac{\varepsilon^2}{K} \quad (7)$$

Where K is the turbulent kinetic energy, P_T is the turbulent kinetic energy production due to mean velocity, while V_F, A_x, A_y, and A_z are the FLOW-3D FAVOR functions, u, v, and w are the velocities in x, y, and z directions, respectively. B_T is the turbulent kinetic energy production due to buoyancy, ε is the rate of turbulent kinetic energy dissipation, D_K and D_ε are the diffusion term for K and ε, respectively. η, α, and β are the dimensionless user-adjustable parameters, and for the RNG $k-\epsilon$ model, default values of η, and β are 1.42 and 1.6, respectively. Whereas α computed from the K and P_T terms. Thus, the turbulent viscosity (ν_T) is modeled as:

$$\nu_T=C_\mu\frac{K^2}{\varepsilon} \quad (8)$$

Where, C_μ is the constant and equal to 0.085 in RNG $k-\epsilon$ model.

2.3. Numerical Setup

In order to get a steady and stable solution, the numerical model's finish time was set to 20 s. The length unit was chosen to be in the SI system, while the temperature unit was in degrees Celsius ($°C$). From the fluid database, water with a temperature of 20 $°C$ was chosen in the model. Incompressible fluid flow with sharp-interface mode was activated. The gravity option was enabled, and z direction gravitational acceleration was set at $-9.81\ m/s^2$. The viscosity and turbulence options were also activated with the RNG $k-\epsilon$ turbulence closure model. The ogee model was created in AutoCAD-3D and then exported as a Stereo Lithography (.stl) file for use in FLOW-3D, as shown in Fig. 3. The FAVOR or VOF approach defines the interface between geometry and water. A no-slip condition over the ogee surface was taken in the numerical model. A roughness height of 0.003 m was assigned to the ogee surface.

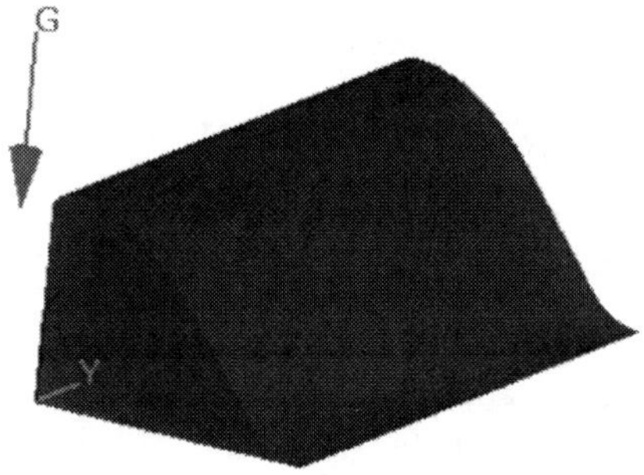

Figure 3. Ogee spillway 3D model used in numerical simulation.

A structured and orthogonal cartesian mesh with hexahedral cells was used in FLOW-3D. The uniform mesh is ideal for complex shape models and free-surface flows, including hydraulic jumps (Jacobsen and Olsen 2010). It is also precise and stable over time (Uçar and Kumcu 2018). A single uniform mesh block was used to discretize the domain in the present study. The mesh and cell size were compared in a sensitivity study, and the 4 mm cell size outperformed the others in terms of outcomes and time spent on solving equations. A three-dimensional grid with cell sizes of 4 mm in the x, y, and z dimensions was constructed, as shown in Fig. 4. A further reduction in cell size has no significant effect on the accuracy of Q test results. As given in table 2, the computed discharge and the discharge obtained from the USBR were compared under various cell sizes for the $P=0.12\ m$ and $H_e=H_d=0.06\ m$. Fig. 5 illustrates the mean relative error as a function of cell size for the discharge calculation over the ogee spillway. The main setup was identical for all the models.

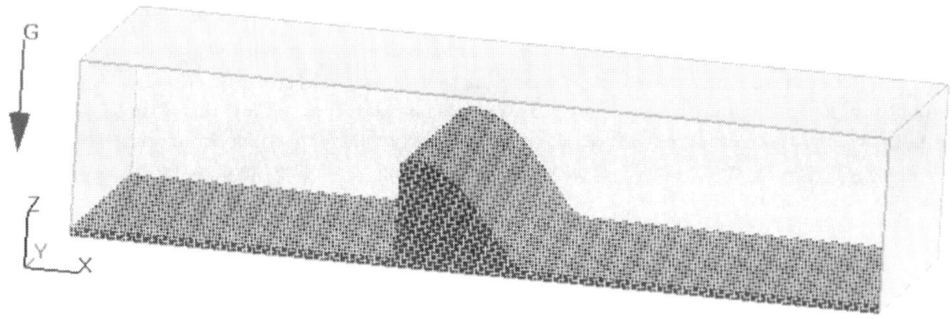

Figure 4. Mesh geometry.

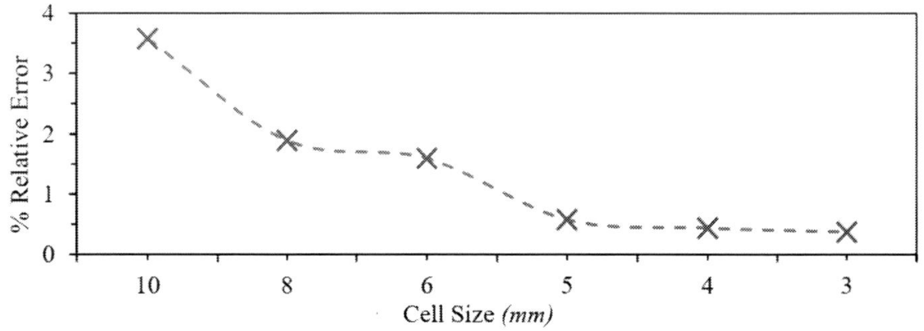

Figure 5. Variations of % relative error of discharge values versus cell size.

Table 2. Mesh-sensitivity analysis for the present study.

Cell size (mm)	Total cell number	Elapsed time (HH:MM)	Relative error (%) $\left(\frac{Q(USBR) - Q(computed)}{Q(USBR)}\right) \times 100$
10	101354	00:12	3.575
8	189314	00:42	1.892
6	432822	01:20	1.594
5	730802	03:12	0.575
4	1416802	06:14	0.438
3	3237698	24:42	0.373

To correctly replicate the physical model in a numerical simulation, boundary and initial conditions must be carefully specified. The different types of boundary conditions available in the FLOW-3D can be used to establish the individual state of the models. As indicated in Fig. 6, the x component indicates the longitudinal direction along with the flow, the z component represents the vertical direction, and the y component represents the lateral direction with respect to the origin. Because the flow domain is defined as a hexahedron in the Cartesian system, the primary grid system comprises six separate $(Xmin,\ Xmax,\ Ymin,\ Ymax,\ Zmin,\ Zmax)$ boundaries. The distance of the $Xmin$ from the ogee crest was taken greater than $4 \times H_e$ to avoid waves and get smooth reservoir conditions. The same for the $Xmax$ was set to $10 \times H_e$. The distance between $Ymax$ and $Ymin$ was taken as equal to L_e. $Zmin$ was taken at the bottom of the bed, and $Zmax$ was set to $(P + H_e + 0.02)\ m$ from the model origin.

Water is supplied to the channel at the upstream boundary $(Xmin)$, and it is defined as specified stagnation pressure (P) depending on the head over an ogee spillway with zero approach velocity. This specified stagnation pressure entered as a fluid elevation in the computational domain, and it creates a head over the ogee spillway. Therefore, with this approach, different flow depths can be modeled. The downstream boundary condition $(Xmax)$ is defined as continuative (c) at the end of the computational domain, indicating that the flow will continue to flow smoothly through the boundary. At the top boundary $(Zmax)$, specified pressure with the fluid fraction zero $(F = 0)$ was set to depict the free surface flow condition. The bottom boundary $(Zmin)$ is defined as the wall (W). In the wall boundary condition, a no-slip condition was employed. Side boundaries $(Ymax, Ymin)$ are defined as symmetric (S), indicating there is no drag because the flow is similar on both sides of the boundary. Fig. 6 depicts the computational domain used in this research and associated boundary conditions on the $x, y,$ and z planes.

Figure 6. Boundary conditions used in the numerical model.

3. Results and Discussions

This study is intended to check the capabilities of the FLOW-3D, CFD model for the computation of discharge coefficient (C) for the design discharge and discharge different than the design discharge in depth. In addition to the C other aspects like water surface profile, flow rates over the spillway crest, velocity distributions, and pressure distributions over the spillway are studied in detail. Since the developed model on FLOW-3D had performed well thus, such a model can be extended for the other geometry and flow conditions of the spillways. The results of numerical simulations (FLOW-3D) generated for various head ratios (H_e/H_d) for the flow over an ogee spillway is compared and presented in this section. Turbulence quantities like turbulence intensity and turbulent kinetic energy can be simulated numerically through the RNG $k - \epsilon$ model. However, due to the limitation of the number of pages, the same are not included.

3.1. Discharge coefficient (C)

The discharge coefficient (C) for $P = 0.12\ m$, and $H_d = 0.06\ m$ under various head ratios (H_e/H_d) are computed using FLOW-3D. The C is normalized with the C_0 (equal to 2.173), which is a discharge coefficient under the design head (H_d). Fig. 7 shows a comparison of discharge coefficient (C) for the various head ratios (H_e/H_d) with the USBR (1987) published data. It is observed that the discharge coefficient (C) obtained numerically shows a good agreement with the USBR values.

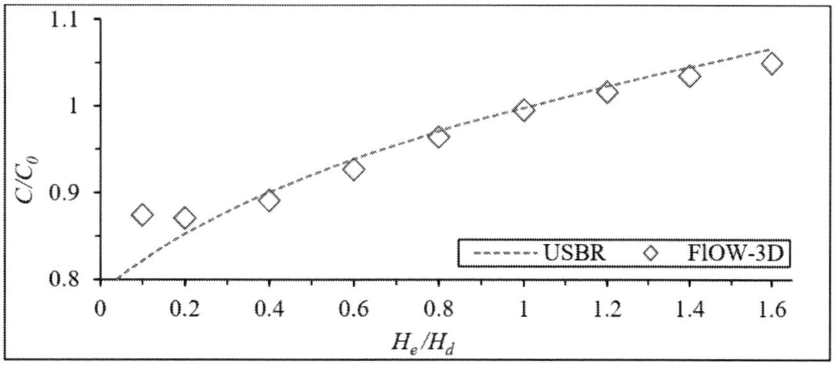

Figure 7. Comparison of discharge coefficient under various head ratios.

The values of C (computed) and C (USBR) with percentage Relative Error ($\%\ RE$) for $0.1 \leq H_e/H_d \leq 1.6$ are tabulated in table 3. The percentage relative error in C at a given H_e/H_d, is defined as below:

$$\%\ RE = \frac{C(computed) - C(USBR)}{C(USBR)} \tag{9}$$

With increase in H_e/H_d over unity, C also increases as pressure reduces over the spillway surface, whereas a decrease in the head ratio below unity results in a reduced C value due to positive pressure buildup over the spillway surface. For the $0.4 \leq H_e/H_d \leq 1.4$, both values are very close to each other, with the minimum value of $\% RE = 0.442$ at $H_e/H_d = 1.0$. It is also observed that for $0.4 \leq H_e/H_d \leq 1.4$, the C (computed) is underestimated as compared to the C (USBR). For the $H_e/H_d \geq 1.0$, the $\% RE$ increases gradually and reaches 1.516 at $H_e/H_d = 1.6$. When the $H_e/H_d < 0.4$, C (computed) are overestimated as compared to C (USBR) and show a maximum $\% RE$ of -6.4 for the $H_e/H_d = 0.10$. This overestimation of C (computed) for lower head ratios may be due to the scale effect or due to the $RNG\ k - \varepsilon$ turbulence model used in the present study. The use of the LES model can give better results for the lower head ratios as it captures eddies or scales in complete detail, whereas they are modeled in the RANS model (Rajaa and Kamel 2020).

Table 3. Percentage Relative Error.

H_e/H_d	C (computed)	C (USBR)	% Relative Error (% RE)
0.100	1.899	1.785	-6.402
0.200	1.892	1.852	-2.138
0.400	1.935	1.956	1.054
0.600	2.014	2.039	1.245
0.800	2.095	2.110	0.687
1.000	2.163	2.173	0.442
1.200	2.208	2.222	0.644
1.400	2.249	2.270	0.937
1.600	2.281	2.316	1.516

3.2. Discharge (Q)

The computed discharge from FLOW-3D is compared with the published discharge data of USACE-WES as shown in Fig. 8 for the head ratios ($0.1 \leq H_e/H_d \leq 1.6$). As shown in Fig. 8, discharge increases with an increase in head over the crest. This comparison shows good agreement, and the maximum difference between the normalized USACE-WES discharge and the computed discharge is 6.40%. The results show that FLOW-3D can accurately compute discharge for the flow over an ogee spillway.

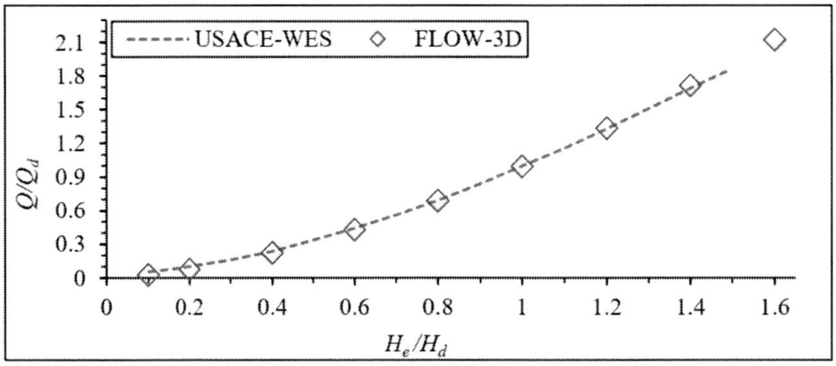

Figure 8. Comparison of discharge computed from FLOW-3D and USACE-WES.

3.3. Water Surface Profiles

The water surface profiles are obtained numerically from FLOW-3D for the various head ratios ($0.1 \leq H_e/H_d \leq 2.0$). These profiles are computed at the mid-way of crest length ($y = 0.195\ m$) and then plotted for the computational domain of $-0.3\ m \leq x \leq 0.5\ m$, as shown in Fig. 9. It was observed that for the higher head ratios ($H_e/H_d > 0.6$), the incoming flow to downstream of the spillway propels the flow in downstream, and due to that, a hydraulic jump was not formed within the computational domain ($-0.3\ m \leq x \leq 0.5\ m$). As shown in Fig. 9, for $H_e/H_d > 0.6$ both upstream and downstream free surface elevation increases with the increase in head ratio (H_e/H_d). However, for low head ratios ($H_e/H_d \leq 0.6$) due to low specific force, a hydraulic jump forms near the toe of the spillway, i.e., within the computational domain ($-0.3\ m \leq x \leq 0.5\ m$). Therefore, the downstream water surface profiles for the $H_e/H_d \leq 0.6$ are different than of $H_e/H_d > 0.6$. It was observed that for $H_e/H_d \leq 0.6$ the upstream free surface elevation shows a similar trend as of $H_e/H_d > 0.6$. The results show that FLOW-3D can capture hydraulics jump phenomena and free surface elevation for the flow over an ogee spillway.

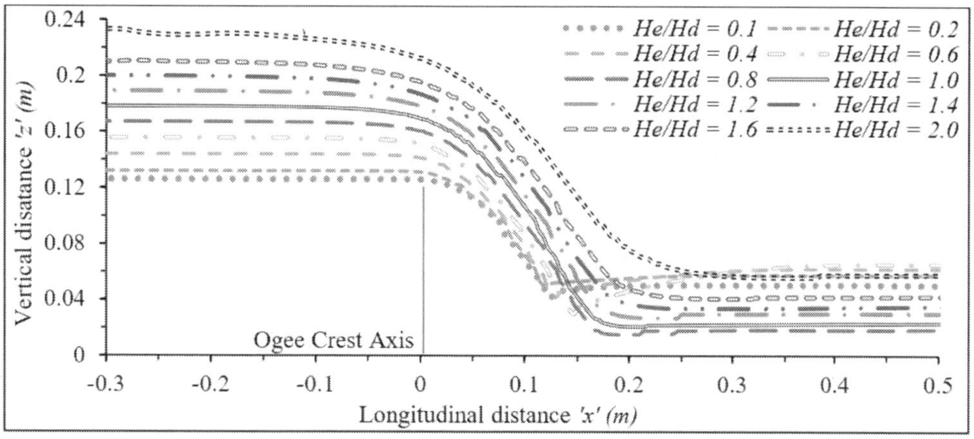

Figure 9. Water surface profiles for various head ratio.

3.4. Pressure distributions

The computed pressure head (H_p) at the ogee spillway surface is compared with the USACE-WES pressure distribution in the computational domain of $x/H_d = -0.3$ (Upstream face) to $x/H_d = 1.2$, as shown in Fig. 10.

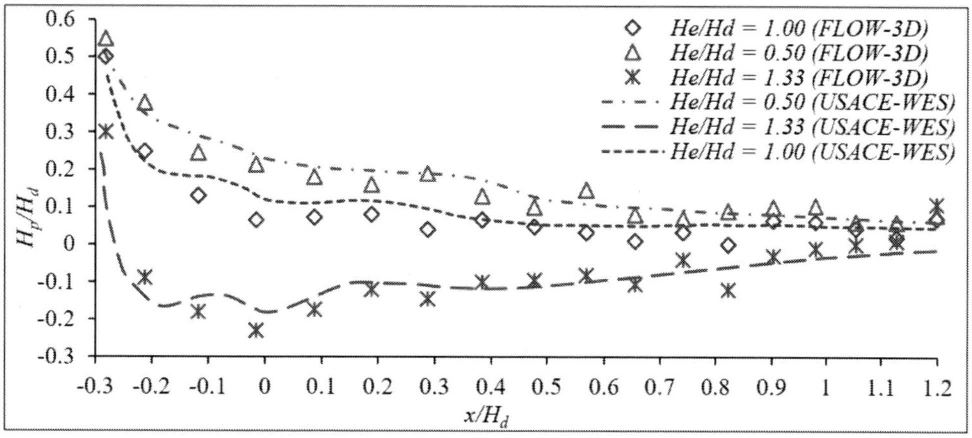

Figure 10. Pressure distributions for $H_e/H_d = 0.5, 1.0$, and 1.33.

The H_p over the ogee spillway is obtained numerically from FLOW-3D for the head ratios $(H_e/H_d) = 0.5, 1.0$, and 1.33. This comparison shows a good agreement. As shown in Fig. 10, H_p at the crest $(x/H_d = 0)$ decreases with an increase in head ratio. It is observed that pressure at the upstream section of the spillway is higher than in the downstream segment. For $H_e/H_d = 0.5$, a positive pressure buildup over the entire ogee spillway surface. When the $H_e/H_d = 1.0$, almost zero pressure is observed at the downstream segment of the spillway. Whereas a slight positive pressure buildup at the upstream segment indicates no negative pressure for the design head. For $H_e/H_d = 1.33$, a small negative pressure zone is developing at the top of the ogee crest profile. This zone extends from the upstream segment to the downstream segment of the ogee crest. While the head rises above the design head, the pressure may fall below the water vapor pressure, causing cavitation and nappe instabilities (Erpicum et al. 2018). The results show that FLOW-3D can accurately compute the pressure head (H_p) over the ogee spillway surface.

3.5 Velocity distributions

Figs. 11 (a) and (b) depict velocity vectors over the computational domain $(-0.2\ m \leq x \leq 0.3\ m)$ for the head ratios, $H_e/H_d = 1$ and 2, respectively. The velocity magnitude is proportional to the vector length. The maximum velocity for $H_e/H_d = 1$ was 1.62 m/s, whereas it was 1.87 m/s for $H_e/H_d = 2$. These maximum velocities were observed at the downstream portion of the spillway. The velocity at the heel of the upstream section and the toe of the downstream section is less as compared to other regions due to vortex formation and high pressure in these regions. However, this velocity reduction is more for $H_e/H_d = 2$ than $H_e/H_d = 1$. It was observed that velocity is more at the crest of the spillway for $H_e/H_d = 2$, because the depth is more and relative pressure is less over the crest for the $H_e/H_d = 2$ than $H_e/H_d = 1$.

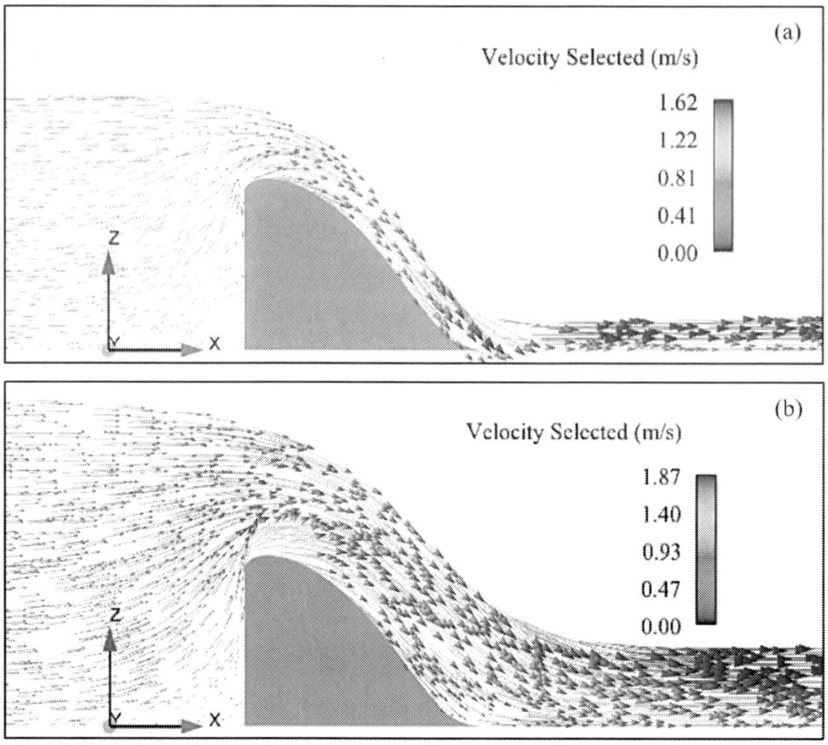

Figure 11. Velocity vector distribution for (a) $H_e/H_d = 1$, and (b) $H_e/H_d = 2$.

4. Conclusions

In the present study, a numerical simulation of flow over an ogee spillway has been carried out for various head ratios using a CFD solver FLOW-3D. The parameters, including flow rates over the spillway crest, discharge coefficient, free surface elevations, velocity distributions, and pressure distributions over the spillway, were computed and compared with USBR (1987) and USACE-WES published data. It shows a good agreement with the published data. The discharge coefficient obtained from numerical simulation and USBR (1987) show good relation with maximum and minimum % relative error of 6.40 and 0.44, respectively. It is observed that for $H_e/H_d \leq 0.6$ the hydraulic jump was formed within the computational domain. The discharge rating curve computed from numerical results shows a similar trend as USBR (1987) results. The velocity over the crest is higher for the higher head ratio. When the hydraulic head equals the design head ($H_e/H_d = 1$), no negative pressure has occurred. Whereas a minor negative pressure zone was observed developing at the top of the ogee crest profile for $H_e/H_d = 1.33$. In the present study, the numerical simulation has been carried out for the scaled model due to limitations in computing facilities. However, this can be extended to prototypes or models of different scales to study the scale effect.

5. Acknowledgment

We acknowledge with thanks Flow Science, India, for providing the academic research license of the FLOW-3D.

6. References

Bureau of Reclamation. (1987). Design of Small Dams, United States Department of the Interior, United States Government Printing Office, Denver, 3rd Ed., 365–387.

Chanel, P. G., and Doering, J. C. (2008). "Assessment of spillway modeling using computational fluid dynamics." *Can. J. Civ. Eng.*, 35(12), 1481-1485.

Dargahi, B. (2006). "Experimental study and 3D numerical simulations for a free-overflow spillway." *J. Hydraul. Eng.*, 132(9), 899-907.

Erpicum, S., Blancher, B., Peltier, Y., Vermeulen, J., Archambeau, P., Dewals, B., and Pirotton, M. (2018). "Experimental study of ogee crested spillway operation above the design head and influence of the upstream quadrant geometry." *Proc., 7th International Symposium on Hydraulic Structures,* IAHR, Aachen.

Flow Science, Inc. (2021). "FLOW-3D® HYDRO" Santa Fe, NM, USA.

Hager, W. H. (1987). "Continuous crest profile for standard spillway." *J. Hydraul. Eng.*, 113(11), 1453-1457.

Hirt, C. W., and Nichols, B. D. (1981). "Volume of fluid (VOF) method for the dynamics of free boundaries." *J. Comput. Phys.*, 39(1), 201-225.

Hirt, C. W., and Sicilian, J. M. (1985). "A porosity technique for the definition of obstacles in rectangular cell meshes." *Proc., In International Conference on Numerical Ship Hydrodynamics,* Washington, DC, IVa.

Imanian, H., and Mohammadian, A. (2019). "Numerical simulation of flow over ogee crested spillways under high hydraulic head ratio." *Eng. Appl. Comput. Fluid Mech.*, 13(1), 983-1000.

Jacobsen, J., and Olsen, N. R. B. (2010). "Numerical modelling of the capacity for a complex spillway." *Proc., the Institution of Civil Engineers,* Water Management, 163(6), 283-288.

Johnson, M. C., and Savage, B. M. (2006). "Physical and numerical comparison of flow over ogee spillway in the presence of Tailwater." *J. Hydraul. Eng.*, 132(12), 1353-1357.

Kline, S.J. (1953). "Describing uncertainty in single sample experiments." *Mech. Eng.,* 75, 3-8.

Kocaer, Ö., and Yarar, A. (2020). "Experimental and numerical investigation of flow over Ogee Spillway." *Water Resour. Manag.*, 34(13), 3949-3965.

Maynord, S.T. (1985). "General Spillway Investigation: Hydraulic model investigation." *Technical Report, US Army Engineering Waterways Experiment Station,* Vicksburg, Mississippi, 85(1).

Michael, W. S., and Walker, R. Y. (1948). "Studies of crests for over fall dams." *Boulder Canon project final reports,* USBR, Denver, Colorado.

Rad, I. N. (2016). "Application of numerical methods in design of hydraulic structures." *Commun. Appl. Math. Comput. Sci.*, 1-15.

Rajaa, A. I., and Kamel, A. H. (2020). "Comparison among the Turbulence Models to Simulate Flow Pattern over ogee Spillway Case Study (Mandali dam in Iraq)." *Iraqi J. Civ.*, 14(2).

Roushangar, K., Foroudi Khowr, A., and Saneie, M. (2019). "Experimental study and artificial intelligence-based modeling of discharge coefficient of converging ogee spillways." *J. Hydraul. Eng.*, 27(sup1), 97-104.

Saneie, M., Sheikh, Kazemi, J., and Azhdary, Moghaddam, M. (2016). "Scale effects on the discharge coefficient of ogee spillway with an arc in plan and converging training walls." *Civ. eng. infrastruct. j.,* 49(2), 361-374.

Savage, B. M., and Johnson, M. C. (2001). "Flow over ogee spillway: Physical and numerical model case study." *J. Hydraul. Eng.*, 127(8), 640-649.

U. S. Army Corps of Engineers, Waterways Experiment Station, (USACE-WES). (1952). Corps of Engineers Hydraulic Design Criteria, Vicksburg, Mississippi.

Ucar, M., and Kumcu, S. Y. (2018). "Comparison of physical modelling and CFD simulation of flow over spillway in the Arkun Dam." *Proc., International Symp. on Dam Safety.*, Va.

Vermeyen, T. B. (1992). "Uncontrolled ogee crest research." *US Bureau of Reclamation.,* PAP-0609.

DOI: https://doi.org/10.26077/3051-fc31

Designing Smooth Mixed-Geometry Canal Transition

D.C. Froehlich[1]

[1]*Consulting Engineer, Cary, North Carolina, USA*

E-mail: dcfroehlich@aol.com

Abstract: *When a canal's size or shape changes, usually over a short distance, a section of the channel, known as a transition structure, is needed to connect the waterway's two stretches. Fifth-degree parametric equations are developed to calculate the cross-section dimensions and bed centerline elevations (thus, the geometric surface coordinates) between the two ends of a warped transition structure in a water-supply canal. The parametric modeling approach provides a smooth representation of the mixed geometry that results from terminal sections having vastly different shapes. A generalized cross-section defined by four parameters enables a straightforward model of various forms ranging from trapezoids to semi-circles. This approach significantly simplifies the interpolation of surface coordinates between the terminal points of a transition structure. It also maintains a smoothness that helps avoid undesirable consequences of channel contractions and expansions. An example is presented that applies the parametric modeling approach to design a significant canal transition where the cross-section changes from a standard trapezoidal shape with rounded bottom vertices to a rectangular section in a steeper aqueduct that carries the flow across a broad valley.*

Keywords: *Canal, water supply, transition, parametrically smooth, mixed-geometry*

1. Introduction

An artificial canal supplying water for irrigation or public consumption may wind hundreds of kilometers from its source to its destination. Along the way, the flow rate delivered by the waterway usually decreases because of diversions, evaporation, and seepage losses, which reduces the size of the channel cross-section needed to convey the flow. The longitudinal slope might also vary, increasing or decreasing, accompanied by a change in cross-section shape. The channel cross-section also changes at flumes, siphons, and aqueducts, where contractions and expansions of the flow occur (Aisenbrey et al. 1978).

When the size or shape of a canal changes, usually over a short distance, a section of the channel, known as a transition structure, is needed to connect the waterway's two stretches. The sidewalls' geometrical shape (for example, conical frustum) and the physical appearance of the sidewalls (flared, splayed, straight, or streamlined) characterize the transition type. The most common types of transitions connecting channels having trapezoidal and rectangular cross-sections are the cylindrical quadrant, wedge, and semi-elliptical (Laycock 2007, page 154), which is like the so-called "warped" or twisted structure proposed by Hinds (1928) nearly a century ago. In Hinds' warped transition, a surface is generated between the end (terminal) sections (which usually have flat walls either sloped or vertical) by a straight line moving so that no two consecutive positions are in the same plane.

In comparatively small concrete-lined canals where energy losses at transitions are tolerable, the special formwork and skilled labor required to build elaborate structures are usually not justified economically. The most straightforward construction is suitable (Simmons 1964, page 1). However, for large canals on gentle slopes where energy losses at transitions must be kept small, the design typically tries to provide a gradual changeover that avoids turbulence caused by flow contractions and expansions and keeps the generation of surface waves to an acceptable level. Scobey (1933) notes that a cylindrical quadrant transition works well for contractions with 2.4 m/s or less velocities. However, when the velocity-head changes by 0.6 m or more, he finds that a warped structure is best. Ippen (1949) found that warped transitions produce a minor energy loss in expansions.

The procedure developed by Hinds (1928) for designing a warped transition relies on the hypothesis that the longitudinal water-surface profile consists of two reverse parabolas tangent to each other at the transition mid-length and merge tangentially with the profiles at the entry and exit sections. It also assumes a constant value of the energy head-loss coefficient for the entire length of the transition and a dimensioning of the intermediate cross-section that is subjective to a large extent. Detailed explanations of the strategy are given by Chow (1959, page 319), Vittal and Chiranjeevi (1983), and Mazumder (2020, pages 82-85 and 157-167). However, contemporary methods for evaluating canal transition hydraulic performance use computational techniques more advanced than those devised by Hinds (1928). These approaches involve the numerical solution of differential equations that describe the canal flow in one, two, or three dimensions. For this reason, the technique developed by Hinds (1928) for computing water-surface profiles through warped transitions now has little practical relevance.

In this investigation, the focus is not on the hydraulic calculations that go into designing a canal transition but on creating a layout that mirrors Hinds' (1928) idea of connecting the terminal sections with a warped structure. Without abrupt changes in geometry, such a shape keeps energy losses small for contracting and expanding flows. A parametric modeling procedure produces a smooth, gradually-varying warped transition of mixed geometry. The coordinates of the intermediate cross-sections are calculated by mathematical expressions that employ specified coefficients to create a three-dimensional representation or parameterization of the structure surfaces.

A fifth-degree polynomial equation is applied that provides a high measure of smoothness of the transition surfaces between the end sections. A similar fifth-degree polynomial equation interpolates the longitudinal variation of the channel centerline bed elevation. This approach dramatically simplifies the cross-section layout for warped transitions connecting channels with a wide range of conventional shapes. Although this investigation focuses on open channel transitions in water-supply canals, the same principle exists for waterways used for other purposes, such as diversion structures at dams (Hager et al. 2020, Chapter 7) and hydroelectric plant intake structures (Gemperline and Crane 1995).

2. Generalized Canal Cross-Section

Continuous excavating, trimming, and lining equipment is the most economical method for building large canals lined with erosion-resistant material. Widespread acceptance and use of canal cross-sections that conform to standard trapezoidal shapes enable manufacturers to build excavating and lining equipment more efficiently and lower replacement parts and service costs. For these reasons, the Bureau of Indian Standards (BIS 2004) recommends model cross-sections that are trapezoidal with rounded bottom vertices, emphasizing those lined with concrete. The BIS standard cross-sections set the bottom vertices' radii to the full-supply-depth (that is, the typical channel depth at the design discharge) for comparatively shallow excavation and one-half of the supply depth where deep cuts are needed.

In this analysis, the development of smooth, mixed geometry transitions is based on a generalization of the BIS standard cross-section shapes introduced by the writer (Froehlich 2008). The cross-section is defined by four parameters (see Fig. 1): b = the horizontal bottom width, h = the full-supply depth, m = the side-slope ratio (horizontal to vertical), and $\kappa = r \div h$, where r = the radius-of-curvature of the bottom vertices and $0 \le \kappa \le 1$. Cross-sections of various shapes ranging from trapezoids to semi-circles can be formed with different combinations of the four parameters, as illustrated in Table 1. With $b > 0$, setting κ to 1 and 0.5 gives the two BIS standard canal cross-sections. With $\kappa = 0$, a trapezoid with sharp bottom vertices is produced.

The flow area A for the generalized cross-section is calculated as

$$A = (b + \alpha h)h, \tag{1}$$

the wetted perimeter P (that is, the boundary length of the section that is in contact with the flow) is given by

$$P = b + \beta h, \tag{2}$$

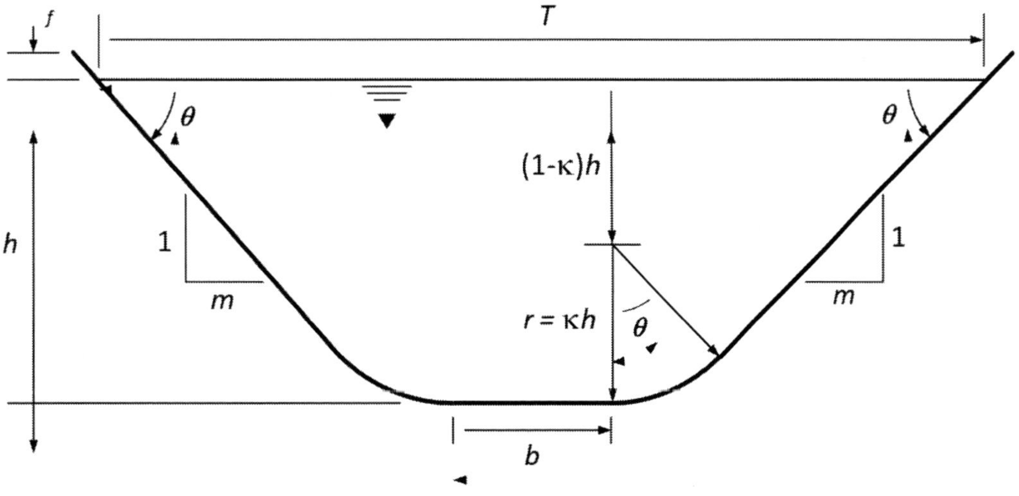

Figure 1. Generalized trapezoidal cross-section.

Table 1. Generalized cross-section shapes for various values of b, m, and κ.

Bottom width	Corner curvature	Side Slope (Horizontal/Vertical) $m = 0$	Side Slope (Horizontal/Vertical) $m = 1/\sqrt{3}$
$b > 0$	$\kappa = 0$		
	$\kappa = 1/2$		
	$\kappa = 1$		
$b = 0$	$\kappa = 0$	--[a]	
	$\kappa = 1/2$		
	$\kappa = 1$		

[a]Rectangular cross-section with $b = 0$ is not possible.

and the section topwidth T is found as

$$T = b + \delta h, \tag{3}$$

where the coefficients α, β, and δ are calculated as follows:

$$\alpha(\kappa, m) = \left[1 - 2\kappa(1-\kappa)\right]m + 2\kappa(1-\kappa)\sqrt{1+m^2} + \kappa^2 \arctan\frac{1}{m}, \tag{4}$$

$$\beta(\kappa, m) = 2\left[\kappa m + (1-\kappa)\sqrt{1+m^2} + \kappa \arctan\frac{1}{m}\right], \tag{5}$$

$$\delta(\kappa, m) = 2\left[(1-\kappa)m + \kappa\sqrt{1+m^2}\right]. \tag{6}$$

When $\kappa = 0$, the expressions for A, P, and T reduce to those for a trapezoidal section with sharp bottom vertices if $b > 0$, or a triangular section if $b = 0$.

Freeboard is an extra height (f shown in Fig. 1) added to the full-supply depth to set the top elevation of the channel's lining. The freeboard aims to accommodate waves, a flood surcharge, or a flow surge caused by a faulty operation. It is also a safety margin that can absorb the effects of low construction tolerances or inaccurate estimates of the channel lining surface roughness or longitudinal bed slope, factors affecting the calculated full-supply depth (Liria Montañés 2006, pages 164-170).

The best hydraulic section provides the maximum flow-carrying capacity for a fixed area or the minimum cross-sectional area and perimeter to pass a given discharge. As it is a minimum area and minimum perimeter section, it provides the maximum hydraulic radius and, therefore, requires the least excavation and lining costs. The most efficient cross-section shape for carrying open-channel flows is a semi-circle and a regular half-hexagon when a sharp-cornered trapezoid (Froehlich 1994). A shape's hydraulic efficiency can be judged by comparing its non-dimensional hydraulic radius $R_* = R/\sqrt{A} = (A/P)/\sqrt{A} = \sqrt{A}/P$ to the value for by a semi-circle ($1/\sqrt{2\pi}$). For the generalized cross-section given by Eqs. *(1)* and *(2)*, $R_* = \sqrt{(b+\alpha h)h}/(b+\beta h)$.

3. Curve and Surface Continuity and Smoothness

Transition structures must efficiently connect the terminal cross-sections with continuous wall and bed surfaces to achieve the goals of small energy loss production and minimal generation of surface waves. Continuity is associated with the smoothness of the geometry and describes how two items come together. These items may be two curves that meet in some way or two portions of the same curve. The function derivatives define the concept of the smoothness of curves. A straight curve (a line) has a zero second derivative, and a nonlinear curve with a sharp kink has a high second derivative at the abrupt bend.

This analysis evaluates the smoothness of transition surfaces using a parametric equation that interpolates between the generalized terminal cross-sections, whose shapes are defined by the parameters b, h, m, and κ in Eqs. *(1)* through (6). Parametric continuity is determined by the n-th derivative of the defining expression, so-called C^n continuity. If a curve or surface is continuous at the n-th derivative, it is said to have an n-th degree of continuity (or degree of continuity n). A function n times continuously differentiable is n-th order parametric continuous, with smoothness increasing with n (Veltkamp 1992). A curve is said to possess a particular degree of continuity when the continuity is at least that degree for all points on the curve's interior. The same holds for surfaces.

C^1 continuity means the parametric equation associated with the curve or surface exhibits first derivative continuity. It follows that C^1 continuity implies C^0 continuity; if a curve or surface is C^1, it is also C^0 continuous. Similarly, C^2 continuity implies C^1 continuity. More precisely, C^0 continuity means that a curve or surface is continuous (that is, it has no gaps) but may exhibit kinks or sharp bends. C^1 continuity means that the parametric equation associated with the curve possesses first derivative continuity – in addition to C^0 continuity – so that the tangents are identical. Likewise, C^2 continuity implies parametric second derivative continuity and being C^1 continuous. The curvature is said to be continuous if its entities are C^2.

While it may be evident that a curve would require C^1 continuity to appear smooth, higher geometric continuity levels are required for pleasing aesthetics (Barnhill 1985, Boehm 1988). For example, a reflective surface, such as an automobile's body, does not appear smooth unless it has C^2 continuity. C^2 means that the curvature is continuous across two connected curves. Again, all previous conditions must be fulfilled for C^2 to be possible. Although the reflective property of light from the surface of a concrete-lined transition is not a concern, the reflection of surface waves from its sidewalls is worrying. To avoid the generation of cross-waves and runup along the sidewalls, enforcing C^2 continuity is a prudent design condition.

4. Transition Parameterization

Parametric modeling, also known as constraint modeling, automatically merges two structures into a single unit governed by the union of all the constraints in a coordinated way. A parametric equation defines a group of entities as functions of one or more independent variables called parameters. Parametric equations are commonly used to express the coordinates of the points that make up a geometric object, such as a curve or surface. The equations are collectively called a parametric representation or parameterization of the object. In this analysis, the item analyzed is a canal transition surface formed from the cross-section properties and the longitudinal bed profile as they vary between the ends of the structure.

4.1. Cross Sections

Considering a transition of length L, constraints are imposed at the terminal sections on the values of the generalized cross-section factors b, h, m, and κ along with their first and second derivatives with respect to channel distance x (as measured along the channel centerline.) A general profile shape function $F(\xi)$, where $\xi = x/L$ is defined so that

$$F(0) = F'(0) = F''(0) = F'(1) = F''(1) = 0 \text{ and } F(1) = 1, \tag{7}$$

where $\xi = 0$ is at the upstream end of the transition and $\xi = 1$ at the downstream end. The following fifth-degree parametric polynomial equation meets these requirements, as can easily be verified:

$$F(\xi) = 10\xi^3 - 15\xi^4 + 6\xi^5;\ 0 \le \xi \le 1. \tag{8}$$

The function is fifth-order parametric continuous. Moreover, because the function is C^2 continuous, if $F''(\xi_0) = 0$ and $F'''(\xi_0) \ne 0$ for some ξ_0 – that is, if $F''(\xi) < 0$ on one side of ξ_0 and $F''(\xi) > 0$ on the other side), F has an inflection point at ξ_0. At an inflection point, the curvature is continuous, and the surface smoothness is assured. Because these conditions are satisfied at $\xi = 0$ and $\xi = 1$, there exist inflection points at the terminal cross-sections. Consequently, the transition surface curvature is in line with the upstream and downstream channels, which is a critical property because any discontinuities in curvature at the end sections, or if $F''(0) \ne 0$ or $F''(1) \ne 0$, can lead to flow separation and sudden steep increases in the local fluid pressure field.

On the plane $\xi = 0$ (that is, the upstream end of the transition), the four cross-section parameters are b_0, h_0, κ_0, and m_0, and on the plane $\xi = 1$ (the downstream end), they are b_L, h_L, κ_L, and m_L. The cross-section parameters between the terminal sections $(0 \le \xi \le 1)$ interpolated with the same properties attributed to $F(\xi)$ are found as follows:

$$\begin{aligned} b(\xi) &= (b_L - b_0)F(\xi) + b_0 \\ h(\xi) &= (h_L - h_0)F(\xi) + h_0 \\ \kappa(\xi) &= (\kappa_L - \kappa_0)F(\xi) + \kappa_0 \\ m(\xi) &= (m_L - m_0)F(\xi) + m_0 \end{aligned} \tag{9}$$

The parametric equations give points on the warped transition surface between the two end sections of the structure. Transition surface schematics in Table 2 illustrate three combinations of entrance and exit sections.

4.2. Bed Centerline Profile

The centerline bed profile of a lined canal is usually composed of stretches with constant slopes connected by vertical curves. Consequently, the transition centerline bed elevation (z_b) is parameterized slightly differently than cross-sections to allow non-zero values of the longitudinal bed slope to be specified at the terminal sections. The following fifth-degree polynomial is used to interpolate z_b between the entrance and exit sections at $\xi = 0$ and 1:

$$z_b(\xi) = z_{b0} + z'_{b0}\xi + (10k_1 - 4k_2)\xi^3 - (15k_1 - 7k_2)\xi^4 + (6k_1 - 3k_2)\xi^5 \tag{10}$$

where $z'_b = dz_b/d\xi = L \times dz_b/dx$, and the coefficients k_1 and k_2 are

$$k_1 = z_{bL} - z_{b0} - z'_{b0} \text{ and } k_2 = z'_{bL} - z'_{b0} \tag{11}$$

The conditions $z_b(0) = z_{b0}$, $z_b(1) = z_{bL}$, $z'_b(0) = z'_{b0}$, $z'_b(1) = z'_{bf}$, and $z''_b(0) = z''_b(1) = 0$ as a function of ξ are satisfied and are easily verified.

Sharp concave (that is, the tangent line to the longitudinal profile lies below the bed in the vicinity of a point) and convex (the tangent line lies above the bed) vertical curves create undesirable conditions usually avoided in a water-supply canal. Concave curves should have an amply long radius-of-curvature to control the dynamic pressures exerted on the floor by the centrifugal force resulting from the flow direction change. Convex curves should be flat enough to maintain positive pressures and prevent the flow's tendency to separate or spring free from the channel's floor, as described by the U.S. Bureau of Reclamation (USBR 1987, page 384). The radius-of-curvature of the bed centerline profile is

$$r_c = \frac{\left[1 + (dz_b/dx)^2\right]^{3/2}}{d^2z_b/dx^2} \tag{12}$$

where from Eq. *(10)*,

$$\frac{dz_b}{dx} = \frac{dz_b}{d\xi}\frac{d\xi}{dx} = \left[z'_{b0} + 3(10k_1 - 4k_2)\xi^2 - 4(15k_1 - 7k_2)\xi^3 + 5(6k_1 - 3k_2)\xi^4\right]\frac{1}{L} \tag{13}$$

and

Table 2. Transition surface schematics for three combinations of entrance and exit sections.

Rotation angle about the vertical axis (deg)	Terminal Section Parameters		
	$b_0 = 20\text{ m}, m_0 = 1, \kappa_0 = 0$ $b_L = 15\text{ m}, m_L = 0, \kappa_L = 0$	$b_0 = 20\text{ m}, m_0 = 0, \kappa_0 = 1$ $b_L = 0\text{ m}, m_L = 0, \kappa_L = 1$	$b_0 = 15\text{ m}, m_0 = 0, \kappa_0 = 0$ $b_L = 20\text{ m}, m_L = 1, \kappa_L = 1/2$
45	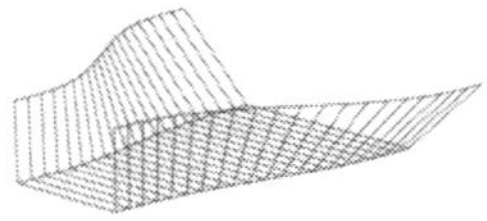	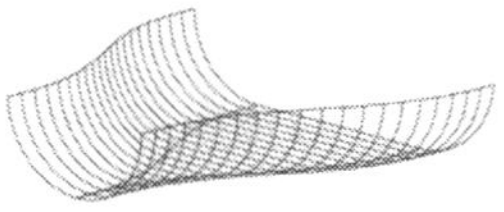	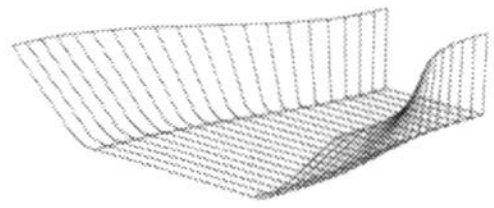
30	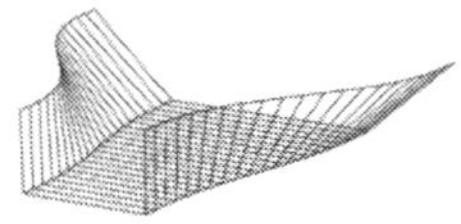	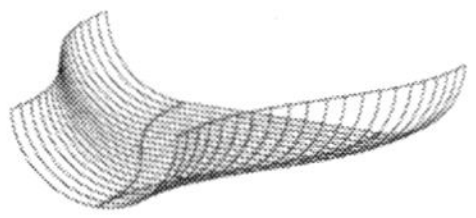	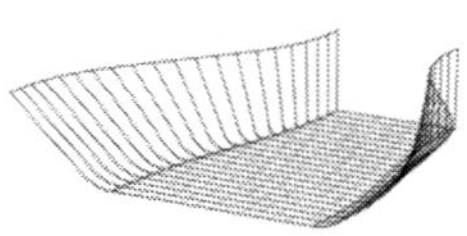
15	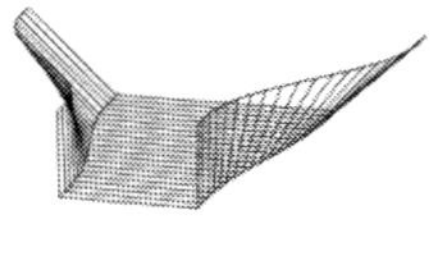	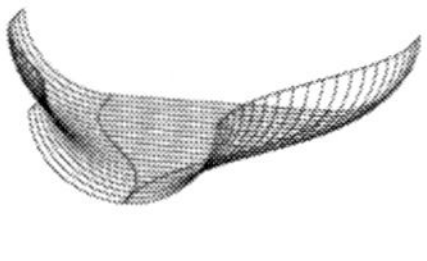	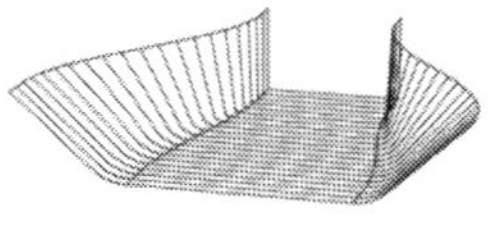
0	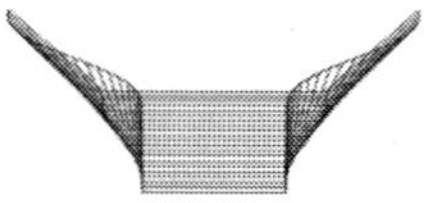		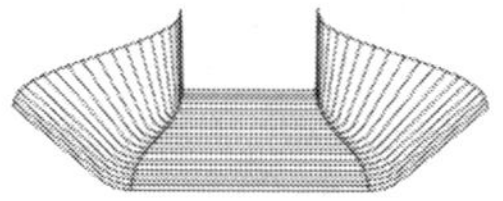

[a]All views have an elevation angle (that is, the angle made with a plane normal to the vertical axis) of 30 degrees. The azimuth angle defines the horizontal rotation of the view about the vertical axis.

$$\frac{d^2 z_b}{dx^2} = \frac{d^2 z_b}{d\xi^2}\frac{d^2 \xi}{dx^2} = \left[6(10k_1 - 4k_2)\xi - 12(15k_1 - 7k_2)\xi^2 + 20(6k_1 - 3k_2)\xi^3\right]\frac{1}{L^2} \quad (14)$$

A positive value of r_c indicates a concave vertical curve, and a negative value one that is convex. Concave profiles present no difficulty provided the curvature radius is not smaller than $10 \times y_1$, where y_1 = the flow depth at the start of the curve. However, a larger radius is better.

5. Example Application

The design of a smooth transition connecting a concrete-lined water-supply canal on a gentle slope to an aqueduct with a steeper incline that crosses a broad valley illustrates the geometric surfaces' parametric representation. Only development of the geometric layout is presented. Procedures for computing water-surface profiles in contracting and expanding transitions can be found elsewhere (see, for example, Swamee and Chahar 2015). Standard software that numerically solves differential equations that describe flow in open channels (in one, two, or three dimensions) can be used for such calculations.

The canal cross-section is typical of large waterways in India (BIS 2004), with rounded bottom vertices having a curvature radius equal to the full-supply water depth h_0 = 4.5 m with a design discharge Q = 100 m^3/s. The canal has a gentle longitudinal slope $dz_{b0}/dx = -1/15500$ m/m, a horizontal bottom width b_0 = 14.6 m, a side-slope ratio m_0 = 2 (horizontal to vertical), and a freeboard f = 0.75 m. The entrance section wall height is then h_{w0} = 4.5 + 0.75 = 5.25 m.

Table 3. Example transition structure terminal section parameters.

Section Parameters	Terminal Section Parameter Value	
	Entrance (canal)	Exit (aqueduct)
Bottom width b (m)	14.6	10.0
Lined bank height h (m)	5.25	6.0
Side slope m (horizontal:vertical)	2	0
Bottom vertex curvature ration κ	1	0
Channel bottom elevation z_b (m)	7.5	6.6
Full-supply water depth h_n (m)	4.5	5.25
Longitudinal channel slope dz_b/dx (m/m)	1/15500	1/5000

The longitudinal slope of the channel increases to $dz_{bL}/dx = -1/5000$ m/m in the aqueduct, which has a sharp-cornered rectangular cross-section with $b_L = 10.0$ m, $h_L = 5.25$ m, $h_{wL} = 5.25 + 0.75 = 6.0$ m, $m_L = 0$, and $\kappa_L = 0$. The bed elevation drops 0.9 m through the transition ($z_{b0} = 7.5$ m and $z_{bL} = 6.6$ m), which is 50 meters in length ($L = 50$ m), giving $z'_{b0} = L \times dz_{b0}/dx = 50 \times (-1/15500) = -0.003223$ m/m, and $z'_{bL} = L \times dz_{bL}/dx = 50 \times (-1/5000) = -0.01$ m/m. Dimensions of the canal and the aqueduct are summarized in Table 3.

The cross-section measures b, h, m, and κ given by Eq. *(9)*, the centerline bed elevation z_b obtained from Eq. *(10)*, the first and second derivatives of z_b with respect to the channel distance x, and the radius-of-curvature of the bed centerline r_c from Eq. (12), are listed at five-meter spacings in Table 4. For the calculation of z_b from Eq. *(10)*, the coefficients $k_1 = 6.6 - 7.5 - (-0.00323) = -0.897$ and $k_2 = -0.01 - (-0.00323) = -0.00677$. The minimum radius-of-curvature of the bed $r_{c\_min} = 481.2$ m at $\xi = 0.789$ ($x = 39.5$ m), which was found by a binary search, and is concave. A three-dimensional plot of the smooth transition surface is presented in Fig. 2.

6. Summary and Conclusions

Fifth-degree parametric equations are developed to calculate the cross-section dimensions and bed centerline elevations (thus, the geometric surface coordinates) between the two ends of a warped transition in a water-supply canal. The parametric modeling approach provides a smooth representation of the mixed geometry that results from terminal sections having vastly different shapes. Additionally, modifications to the transition structure parameters during the design process can be easily implemented and evaluated using this technique.

Table 4. Example transition structure cross-section and bed centerline parameters

Transition distance x (m)	$\xi = x/L$ (--)	$F(\xi)$ (--)	Cross-section Shape Parameters: b (m)	h (m)	m (m/m)	κ (--)	Bed Centerline Profile: z_b (m)	dz_b/dx (m/m)	d^2z_b/dx^2 (m/m²)	r_c[a,b] (m)
0	0.0	0.000	14.600	4.500	2.000	1.000	7.500	-6.667×10^{-5}	0	∞
5	0.1	0.009	14.561	4.506	1.983	0.991	7.492	-0.0044	-0.00155	-647
10	0.2	0.058	14.334	4.543	1.884	0.942	7.448	-0.0138	-0.00206	-485
15	0.3	0.163	13.850	4.622	1.674	0.837	7.353	-0.0237	-0.00180	-555
20	0.4	0.317	13.140	4.738	1.365	0.683	7.215	-0.0310	-0.00103	-969
25	0.5	0.500	12.300	4.875	1.000	0.500	7.051	-0.0336	0.00000	-2.50×10^{5}
30	0.6	0.683	11.460	5.012	0.635	0.317	6.887	-0.0310	0.00103	977
35	0.7	0.837	10.750	5.128	0.3226	0.163	6.749	-0.0238	0.00180	557
40	0.8	0.942	10.266	5.207	0.116	0.058	6.654	-0.0139	0.00206	487
45	0.9	0.991	10.039	5.244	0.017	0.009	6.609	-0.0045	0.00154	648
50	1.0	1.000	10.000	5.250	0.000	0.000	6.600	-0.0002	0	∞

[a]Negative values of r_c show convex curvature; positive values are concave.
[b]The minimum $r_c = 481.2$ m at $\xi = 0.789$.

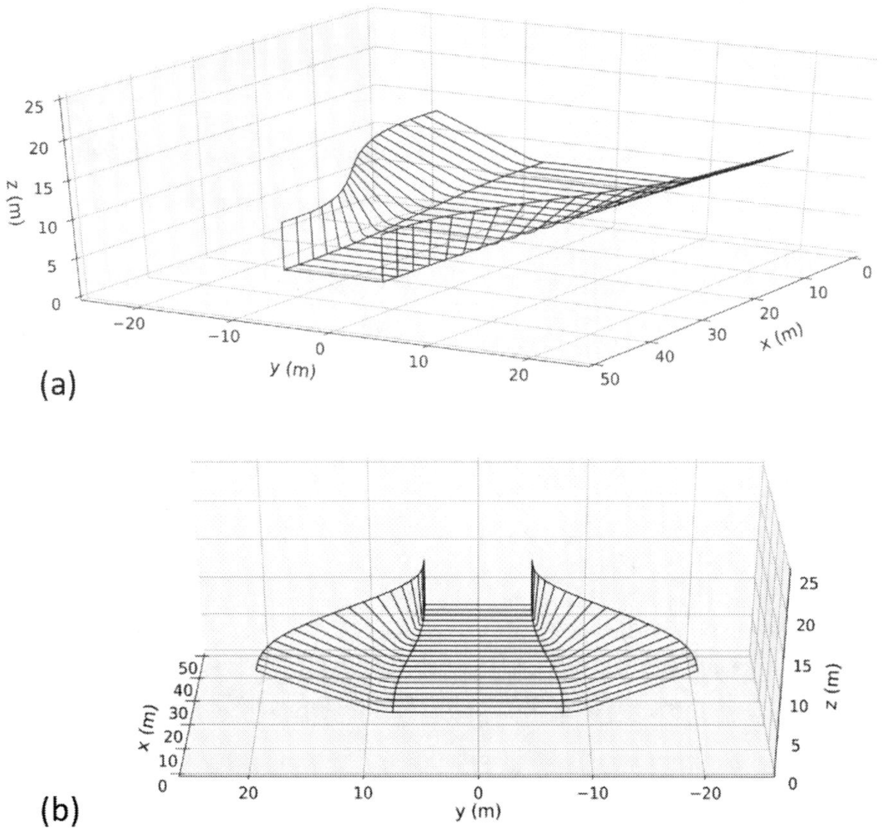

Figure 2. A three-dimensional representation of the smooth transition surface: (a) The view from downstream (aqueduct) end with a 45° rotation about the vertical axis, and (b) the view from the upstream (canal) end with no rotation about the vertical axis. Both views have an elevation angle (that is, the angle made with a plane normal to the vertical axis) of 30°.

The generalized cross-section shape introduced by the writer (Froehlich 2008) for canal design significantly simplifies the interpolation of sections between the terminal points of a transition structure and its geometric surface representation. The standard shape (see Fig. 1) is defined by four parameters (b, h, m, and κ), which enables a straightforward representation of cross-sections having various forms ranging from trapezoids to semi-circles (see Table 1). With $b > 0$, setting κ to 1 and 0.5 gives the two standard cross-sections used for major canals in India (BIS 2004).

Higher-degree polynomial formulas could be used for the parameterization. For example, Wilson (2005) uses a seventh-degree expression to develop smooth profiles of nozzles and transition ducts. However, in this analysis, the fifth-degree equation is sufficient to deliver a high level of surface smoothness that is esthetically pleasing (see Fig. 2). Additionally, the parametrization guarantees that inflection points exist along the warped sides at the terminal cross-sections. Consequently, the transition surface curvature is in line with the upstream and downstream channels, which avoids discontinuities in curvature at the end sections that can lead to flow separation, sudden steep increases in the local fluid pressure field, and generation of cross-waves and runup along the sidewalls.

An example is presented that applies the parametric modeling approach to design a significant canal transition where the cross-section changes from a standard trapezoidal shape with rounded bottom vertices to a rectangular section in a steeper aqueduct that carries the flow across a broad valley. The bed centerline elevations are also calculated to ensure a smooth changeover and that the vertical curvature is not excessive. The example shows that the parametric approach dramatically simplifies the geometric design of warped transitions connecting channels having a wide range of conventional shapes.

Because the focus of this analysis is creating a smooth mixed-geometry layout that embodies Hinds' (1928) idea of connecting the terminal sections with a warped transition structure, it is a designer's task to evaluate energy losses using (most likely) numerical models of flow in the open channel (possibly a three-dimensional numerical hydrodynamic model). A small-scale physical model could also be constructed and used to evaluate the design. The ability of warped transition structures to minimize energy losses is documented by Hinds (1928), Scobey (1933), and Ippen (1949).

7. REFERENCES

Aisenbrey, A.J. Jr., Hayes, R.B., Warren, H.J., Winsett, D.L., and Young, R.B. (1978). *Design of Small Canal Structures*. U.S. Bureau of Reclamation, Denver, Colorado.

Barnhill, R. (1985). "Surface in computer-aided geometric design: a survey with new results." *Computer-Aided Geometric Design*, 2(1), 1-17.

Boehm, W. (1988). "Visual continuity." *Computer-Aided Design*, 20(6), 307-311.

BIS (2004). "Criteria for design of lined canals and guidelines for selection of type of lining (1st revision)." *IS:10430:2000*, New Delhi, India.

Chow, V.T. (1959). *Open-channel Hydraulics*, McGraw-Hill, New York, New York.

Froehlich, D.C. (1994). "Width-and-depth constrained best trapezoidal cross-section." Journal of Irrigation and Drainage Engineering, 120(4), 828-835.

Froehlich, D.C. (2008). "Most hydraulically efficient standard lined canal sections." *Journal of Irrigation and Drainage Engineering,* 134(4), 462-470.

Gemperline, G. and Crane, C. (1995). "Hydraulic design." Chapter 3 in *Guidelines for Design of Intakes for Hydroelectric Plants*, by the Committee on Hydropower Intakes of the Energy Division, American Society of *Civil* Engineers, New York, 16-111.

Hager, W.H., Schleiss, A.J., Boes, R.M., and Pfister, M. (2020). *Hydraulic Engineering of Dams*. CRC Press, London, United Kingdom.

Hinds, J. (1928). "The hydraulic design of flume and siphon transitions." *Transactions of the American Society of Civil Engineers*, 92(1), 1433-1459.

Ippen, A.T. (1949). "Channel transitions and controls." *Engineering Hydraulics*, Proceedings of the Fourth Hydraulics Conference Iowa Institute of Hydraulic Research, 12-15 June 1949, 496-588.

Laycock, A. (2007). *Irrigation Systems: Design, Planning, and Construction*. CAB International, Wallingford, Oxfordshire, United Kingdom.

Liria Montañés, J. (2006). *Hydraulic Canals: Design, Construction, Regulation, and Maintenance*. Taylor & Francis, London, United Kingdom.

Mazumder, S.K. (2020). *Flow Transition Design in Hydraulic Structures*. CRC Press, Boca Raton, Florida.

Scobey, F.C. (1933). "The flow of waters in flumes." *Technical Bulletin 393*, USDA, Washington, DC.

Simmons, W.P. Jr. (1964). "Hydraulic design of transitions for small canals." *Engineering Monograph No. 33*, U.S. Bureau of Reclamation, Denver, Colorado.

Swamee, P. K., and Chahar, B. R. (2015). *Design of Canals*. Springer India, New Delhi, India.

USBR (1987). *Design of Small Dams (3rd Edition)*. Denver, Colorado.

Veltkamp, R.C. (1992). "Survey of continuities of curves and surfaces." *Report CS-R9202*, Computer Science/Department of Interactive Systems, Centrum voor Wiskunde en Informatica, Amsterdam, Netherlands.

Vittal, N., and Chiranjeevi, V.V. (1983). "Open channel transitions: Rational method of design." *Journal of Hydraulic Engineering*, 109(1), 99-115.

Wilson, J.F. (2005). "Smooth mixed geometry profiles of nozzles and transition ducts." *Journal of Mechanical Design*, 127(6), 1211-1213

DOI: https://doi.org/10.26077/aa4f-76f0

Effect of Diameter and Inlet-depth on Hydro-suction performance of a Suction pipe

A. Jaiswal[1], Z. Ahmad[1] & S.K. Mishra[1]
[1]*IIT Roorkee, Roorkee, India*
E-mail: ajaiswal@wr.iitr.ac.in

Abstract: *Sedimentation in rivers and reservoirs leads to inundation of surrounding areas, topsoil degradation, low depth for navigation, loss of reservoir capacity, etc. Hydro-suction is a process of sediment removal in which sediment is sucked along with water using a suction pipe placed vertically on/above/below the sediment bed. This paper deals with the effect of diameter and inlet-depth of suction pipe on performance of hydro-suction. A series of experiments are performed using five suction pipes of diameter of* 5.08×10^{-2} *m,* 7.62×10^{-2} *m,* 10.16×10^{-2} *m,* 12.70×10^{-2} *m, and* 15.24×10^{-2} *m, placed at inlet depth of 0.015 m and 0.03 m, under discharge ranging from* 0.5×10^{-3} m^3/s *to* 3×10^{-3} m^3/s *and median sediment size of 0.33 mm. Hydro-suction performance is evaluated by the sediment volume removed, which is calculated from the experimental data of scour profile. The investigation inferred that for a constant diameter, hydro-suction performance decreases with an increase in suction inlet depth. An increase in hydro-suction performance is seen up to suction pipe diameter of* 10.16×10^{-2} *m, and any further increase in suction pipe diameter decreases the hydro-suction performance.*

Keywords: *Hydro-suction, Suction pipe diameter, Suction inlet-depth, Scour profile, Scour volume.*

1. Introduction

Deposition of sediment over time decreases the reservoir capacity, thus creating an acute problem for its functioning. Sedimentation induces several adverse impacts within reservoirs and upstream and downstream areas of river systems with dams (CWC Handbook, 2019). May it be a passage of ships through a port, water storage capacity of dams, electricity production, or flow in a river. It ultimately leads to a loss in the economy and the purpose of the built structure. The removal of deposited bed materials is still a frantic problem. Removing bed material using manual dredging machines is worthwhile only after emptying the water bodies. Instead, hydro-suction removal can be a more effective and economical solution.

A suction pipe is a dredging tool used to deepen the ports and harbors, reclamation of dams and reservoirs, and sediment removal from a riverbed. In hydro-suction, a pipe is placed vertically above/below/on deposited material, and through suction, the bed material is removed. In hydro-suction, fluid flow is arranged in a pattern that covers the shortest path with minimal total energy loss (Mousa et al., 2020). Flow patterns near the suction inlet and its periphery play a vital role in sediment removal efficiency. The flow field, created by velocity variation near the suction inlet, produces a lift force that creates the bed material in resuspension, which is then removed by suction (Brahme et al., 1986). The experimental studies have been reported by Brahme et al. (1896), Hotchkiss et al. (1995), Ullah et al. (2005), Su-Chin et al. (2010), Shrestha (2012), Zhou et al. (2013), Sadatomi (2015), Wun-Tao Ke et al. (2016), Asiaban et al. (2017), Pishgar et al. (2018), and Pu Yang et al. (2020) on scour pattern formed during hydro-suction removal of sediment.

This study investigates the effect of the diameter and inlet depth of the suction pipe on its hydro-suction performance. A suction pipe of five different diameters was placed vertically above the sediment bed at different suction inlet depth. A series of experiments were performed under a discharge range of 0.5×10^{-3} m^3/s to 3×10^{-3} m^3/s, and the scour profile for each run was noted. The hydro-suction performance of the suction pipe was evaluated by the sediment volume removed during hydro-suction. Sediment volume removed was calculated by considering maximum scour depth and scour radius.

A definition sketch of a hydro-suction is shown in Fig. 1, where V_i is inlet velocity, D is suction pipe diameter, C is suction inlet depth (the distance between the suction inlet and sediment bed level), d_{50} is sediment median size, R_s is scour radius, D_s is scour diameter ($2R_s$), d_s is scour depth just below the suction pipe, R_{sm} is scour radius at maximum scour depth (d_{sm}), and d_r is scour depth at any scour radius R_r.

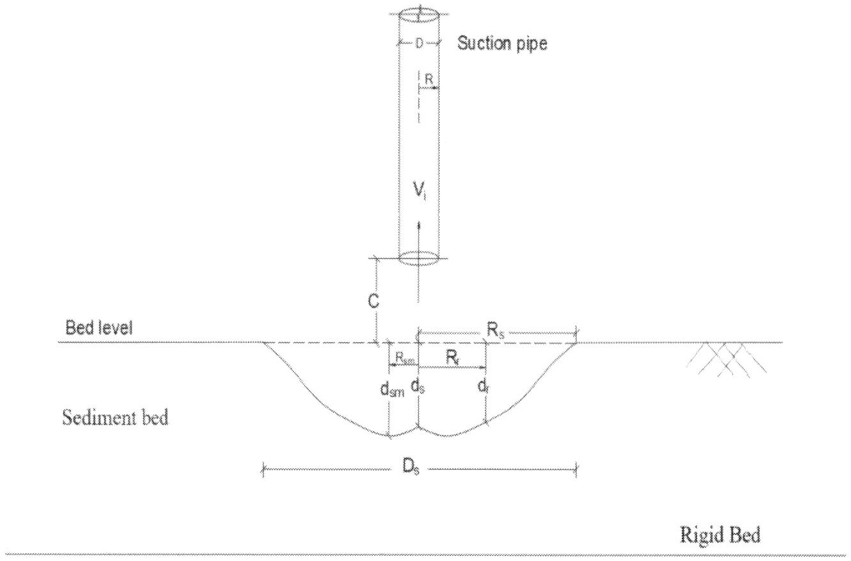

Figure 1. Definition sketch of hydro-suction.

2. Experimental Setup and Methodology

The experiments were performed in the Hydraulic engineering laboratory of the Department of Civil Engineering, IIT Roorkee. A tank of 2.25 m length, 1.25 m depth, and 1.25 m width was used for the experiment. The tank was divided into two parts by a perforated baffle wall. The test portion was 1.25 m long, 1.25 m wide and 1.15 m depth and filled with sediment in a thickness of 0.45 m. The experimental set up was re-circulating as shown in the Figs. 2a and 2b. Water was supplied to the second portion of the tank and the same was re-circulated. Depth of water above the sediment bed was kept 0.55 m for all the experimental runs. The suction pipe was placed vertically above the center of the test portion of the tank. Water was sucked using a centrifugal pump connected to the suction pipe at one end and recirculated the water in the same tank through a pipe connected to the other end.

The discharge was measured using an orifice meter fitted in the suction pipe. An ultrasonic flow meter (UFM) was used to calibrate the orifice meter.

In this study, suction pipes of diameter (D) of 5.08×10^{-2} m, 7.62×10^{-2} m, 10.16×10^{-2} m, 12.70×10^{-2} m, and 15.24×10^{-2} m, placed vertically above the sediment bed at two inlet depths (C) of 0.015 m, and 0.03 m were used for investigation. Sediment has a median diameter, d_{50} of 0.33 mm. As per Raudkivi (1988), the grain size distribution of sediment is considered uniform if, $d_{95}/d_5 < 4$ and $s_g = d_{84.1}/d_{15.9} < 1.35$. In this study, the geometric standard deviation of bed material used was 1.33, thus representing uniform gradation.

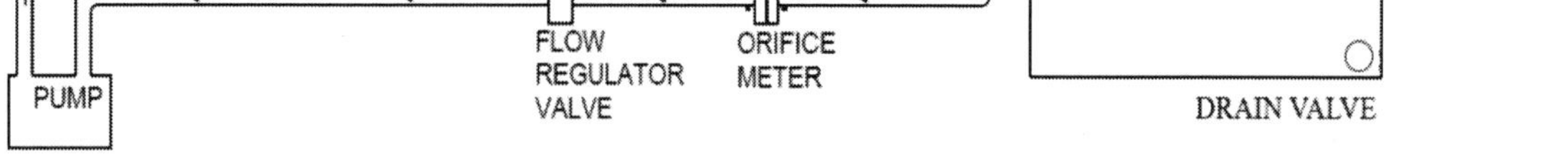

Figure 2 (a). Sketch of a side view of the experimental setup.

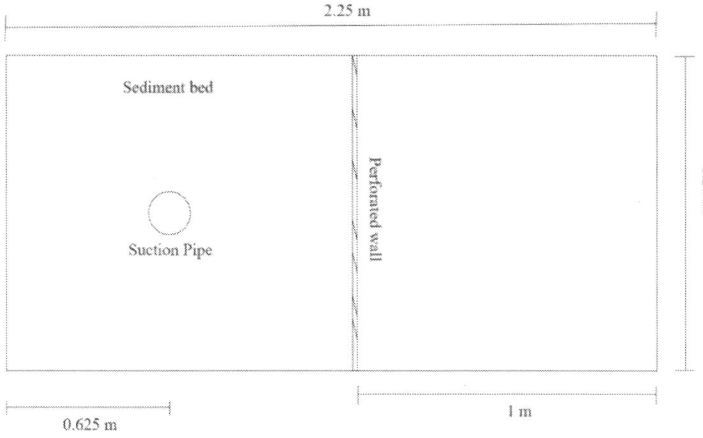

Figure 2 (b). Sketch of a top view of the experimental setup.

A series of experiments were performed for a discharge range of 0.5×10^{-3} m^3/s to 3×10^{-3} m^3/s, and the scour profile for each run was measured. Scour depth was measured throughout the scour diameter at spacing of 0.01 m to get the scour profile. Fig. 3 below shows the matrix of the total observations taken during this study.

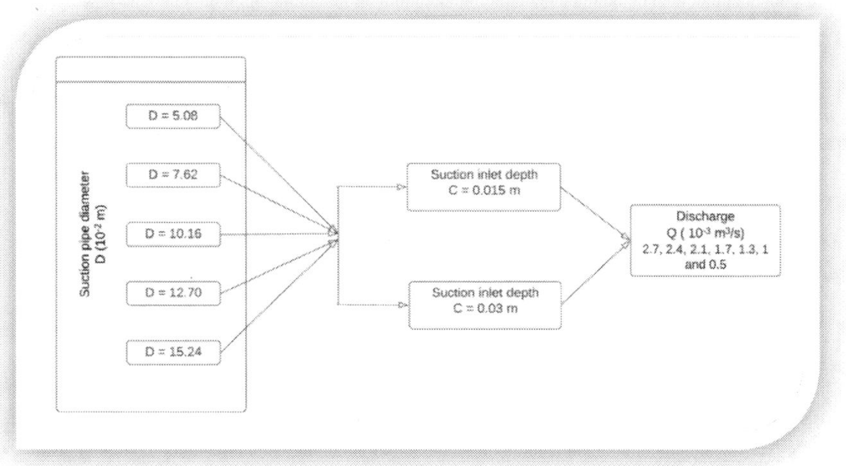

Figure 3. Matrix for total observation taken during experimentation.

A total of 70 observations were taken during the investigation. Tables 1 and 2 provide the experimental values of maximum scour depth and scour radius for different D and Q, at C = 0.015 m and 0.030 m, respectively.

Table 1. Collected experimental data for maximum scour depth and scour radius for C = 0.015 m.

D (10^{-2} m) → / Q (10^{-3} m^3/s) ↓	5.08		7.62		10.16		12.7		15.24	
	d_{sm} (10^{-2} m)	R_r (10^{-2} m)	d_{sm} (10^{-2} m)	R_r (10^{-2} m)	d_{sm} (10^{-2} m)	R_r (10^{-2} m)	d_{sm} (10^{-2} m)	R_r (10^{-2} m)	d_{sm} (10^{-2} m)	R_r (10^{-2} m)
2.7	-6.9	12	-8.5	18	-11.5	20	-5	14	-4.5	14
2.4	-6	12	-8	14	9.5	18	-4.1	12	-2	12
2.1	-5.1	10	-6.9	12	-8.4	18	-2.5	12	0	0
1.7	-4.3	10	-5.9	12	-7.6	15	0	0	0	0
1.3	-3.8	10	-4.5	9	-6	12	0	0	0	0
1	-3.2	9	-3.3	9	-5.1	10	0	0	0	0
0.5	-2.4	8	-2.9	8	-4.2	10	0	0	0	0

Table 2. Collected experimental data for maximum scour depth and scour radius for C = 0.03 m.

D (10^{-2} m) → / Q (10^{-3} m^3/s) ↓	5.08		7.62		10.16		12.7		15.24	
	d_{sm} (10^{-2} m)	R_r (10^{-2} m)	d_{sm} (10^{-2} m)	R_r (10^{-2} m)	d_{sm} (10^{-2} m)	R_r (10^{-2} m)	d_{sm} (10^{-2} m)	R_r (10^{-2} m)	d_{sm} (10^{-2} m)	R_r (10^{-2} m)
2.7	-5.5	10	-6.9	16	-11	20	-4	14	-3.3	11
2.4	-4.6	9	-6	12	-9	18	-2.6	12	0	0
2.1	-4.3	9	-5.5	10	-7.8	16	-1.9	12	0	0
1.7	-4	9	-5	10	-5.6	14	0	0	0	0
1.3	-3.5	9	-3.9	10	-4.6	12	0	0	0	0
1	-2.9	8	-2.6	9	-3.6	10	0	0	0	0
0.5	-2.1	8	-2.3	8	-3.3	10	0	0	0	0

3. Results and Discussion

Effect of diameter and inlet depth of suction pipe on the scour profile and scour volume during hydro-suction were analysed and discussed below.

3.1. Scour Profile

Scour profile is resembled by scour depths taken at 1cm intervals along the scour diameter during experimentation. Since the scour hole is symmetric, a 2D plot of scour profile at any section along its diameter will be sufficient to understand its variation for different suction pipes. Scour depth values at 1cm intervals are taken throughout the scour diameter, and these scour depths are plotted along with the scour diameter to get the scour profile.

3.1.1. Effect of diameter of suction pipe on scour profile

Scour profile graphs are plotted for all the five suction pipes taking constant values of Q and C. Scour profile approximately resembles a semi-circle, having a hump at the midsection in cases having the lowest discharge. Fig. 4 a, b, c, d, e, and f show the scour profile variation with suction pipes diameter at varying discharges and inlet depths. Fig. 4 a, c, and e are plots of scour profiles varying with suction pipe diameters, placed at a suction inlet depth of 0.015 m for discharges Q = 2.7×10^{-3} m^3/s, 1.3×10^{-3} m^3/s, and 0.3×10^{-3} m^3/s, respectively. And Fig. 4 b, d, and f are plots of scour profiles varying with suction pipe diameters, placed at a suction inlet depth of 0.03 m for discharges Q = 2.7×10^{-3} m^3/s, 1.3×10^{-3} m^3/s, and 0.3×10^{-3} m^3/s, respectively. These plots show the change in scour depth along the scour diameter and its variation for different suction pipe diameters. From Fig. 4 a, it is observed that scour depth is maximum at the center of the scour hole, and it decreases along the scour radius while going away from the center. The scour depth is maximum for D = 10.16×10^{-2} m, followed by D = 7.62×10^{-2} m, 5.08×10^{-2} m, and 12.7×10^{-2} m, and it is minimum for D = 15.24×10^{-2} m. A small hump at the mid of scour profile is seen for the discharge = 0.5×10^{-3} m^3/s. Similar observations of scour profile are seen in all the figures in Fig. 4. Fig. 4 c, d, e, and f, show no scour for suction pipe diameters of 12.7×10^{-2} m and 15.24×10^{-2} m.

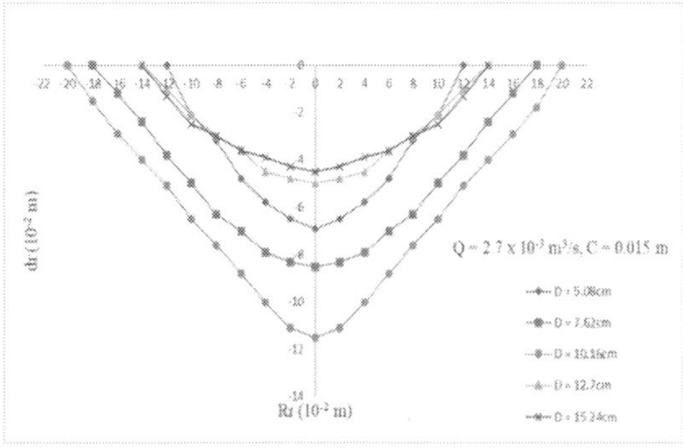

(a)

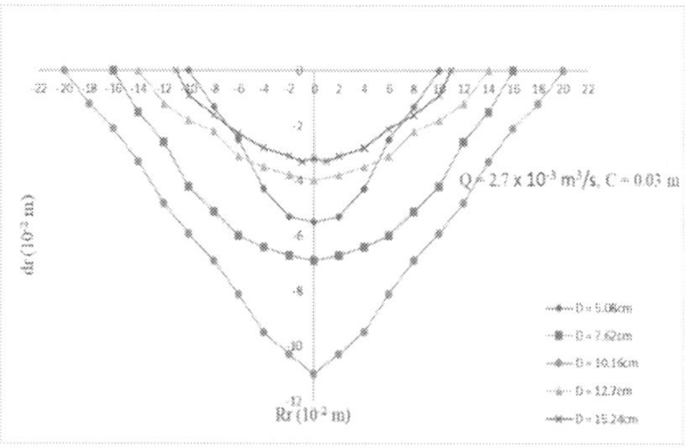

(b)

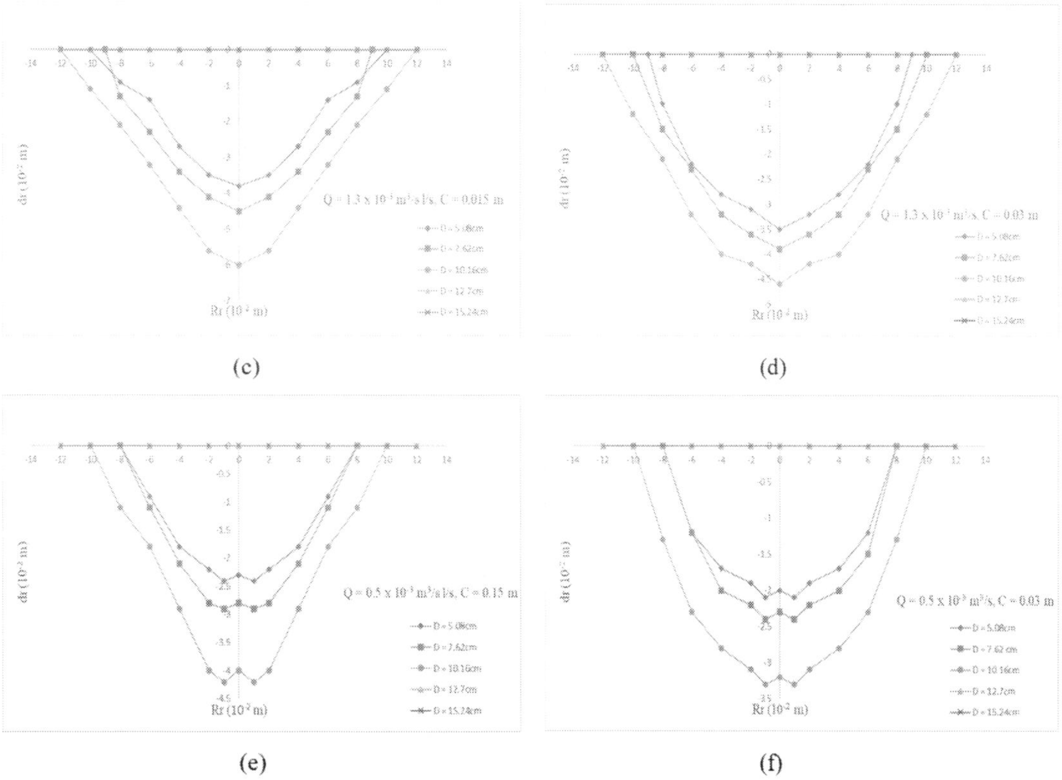

Figure 4. Scour profile at C = 1.5cm, (a) Q = 2.7×10^{-3} m^3/s, (c) Q = 1.3×10^{-3} m^3/s, and (e) Q = 0.5×10^{-3} m^3/s, and at C = 3 cm, (b) Q = 2.7×10^{-3} m^3/s, (d) Q = 1.3×10^{-3} m^3/s, and (f) Q = 0.5×10^{-3} m^3/s.

It can be inferred from the above figures that maximum scour depth increases with suction pipe diameter up to D = 10.16 ×10^{-2} m, and any further increase in pipe diameter decreases the maximum scour depth. It is also seen that scour depth decreases with discharge for a constant diameter and suction inlet depth.

3.1.2. Effect of suction inlet depth on scour profile

The effect of suction inlet depth on scour profile is observed by comparing Fig. 4 a and b, c and d, e and f of Fig. 4. A decrease in scour depth throughout the scour diameter is seen in Fig. 4 a and b, as suction inlet depth increases from 0.015 m to 0.03 m. Maximum scour depth, d_{sm} is decreased from 0.069 m to 0.055 m when suction inlet depth increases from 0.015 m to 0.030 m at a discharge of 2.7×10^{-3} m^3/s as shown in Fig. 4a and 4b. Similarly, for the discharge of 1.5×10^{-3} m^3/s, d_{sm} is decreased from 0.038 m to 0.035 m when suction inlet depth increases from 0.015 m to 0.030 m as shown in Fig 4 c and 4 d. The same observations are by comparing Fig. 4e and 4f.

Comparing the plots in Fig. 4, it can be inferred that maximum scour depth decreases with an increase in suction inlet depth at constant D and Q.

3.2. Scour volume

Maximum the scour volume is indicator of high hydro-suction efficiency. Graphs of the scour volume with other parameters are plotted and analysed below:

3.2.1. Effect of diameter of suction pipe on scour volume

Variation of scour volume with discharge and suction pipe diameter at C = 1.5 cm is shown in Fig. 5. At a discharge of 2.7×10^{-3} m^3/s and C = 0.015 m, scour volume is maximum for D = 10.16×10^{-2} m, followed by D = 7.62×10^{-2} m, 5.08×10^{-2} m,12.7×10^{-2} m, and 15.24×10^{-2} m, with scour volume of ∀ (10^{-6} m^3) = 9630, 2080, 5765, 2051 and 1846, respectively. A similar variation in scour volume with suction pipe diameter is observed for other discharges also.

It can be concluded that at a constant value of Q and C, scour volume increases with suction pipe diameter up to D = 10.16×10^{-2} m, after which it decreases with any further increase in D.

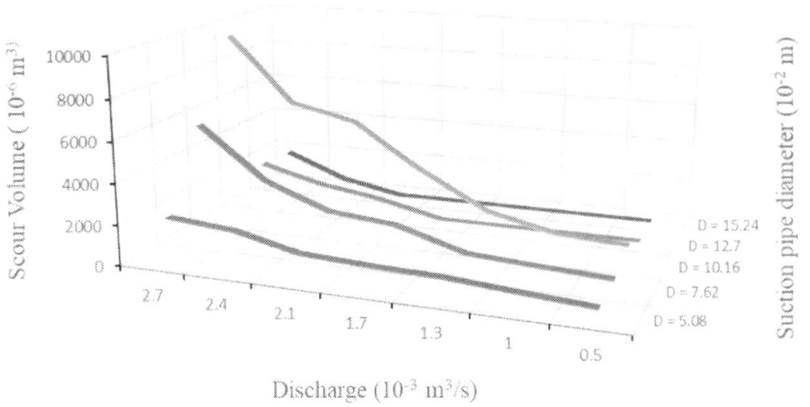

Figure 5. Effect of suction pipe diameter on scour volume.

3.2.2. Effect of suction inlet depth on scour volume

The effect of suction inlet depth on scour volume is analyzed herein. Figs. 6 a, b, c, d, and e show variation in scour volume with suction inlet depth at a constant suction pipe diameter, $D = 5.08 \times 10^{-2}$ m, 7.62×10^{-2} m, 10.16×10^{-2} m, 12.7×10^{-2} m, and 15.24×10^{-2} m. Fig 6 a shows that scour volume decreases from 2080×10^{-6} m³ at C = 0.015 m to 1151×10^{-6} m³ at C = 0.03 m for $D = 5.08 \times 10^{-2}$ m and $Q = 2.7 \times 10^{-3}$ m³/s. A similar trend of the decrease in scour volume with an increase in suction inlet depth is seen in other plots of Fig. 6.

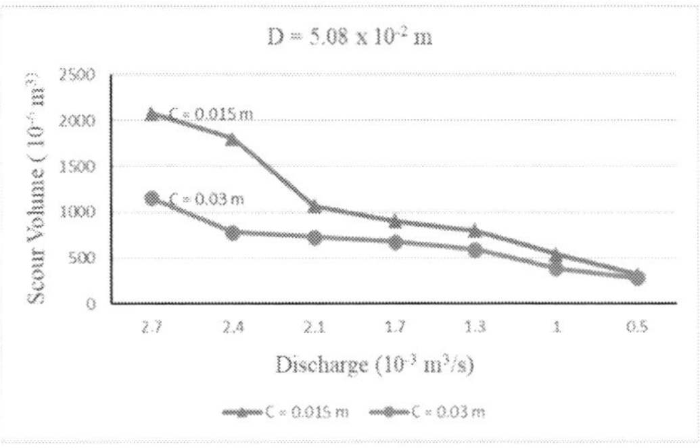

(a)

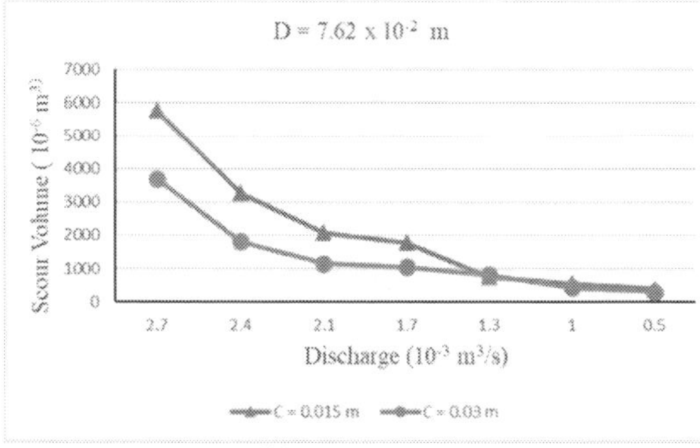

(b)

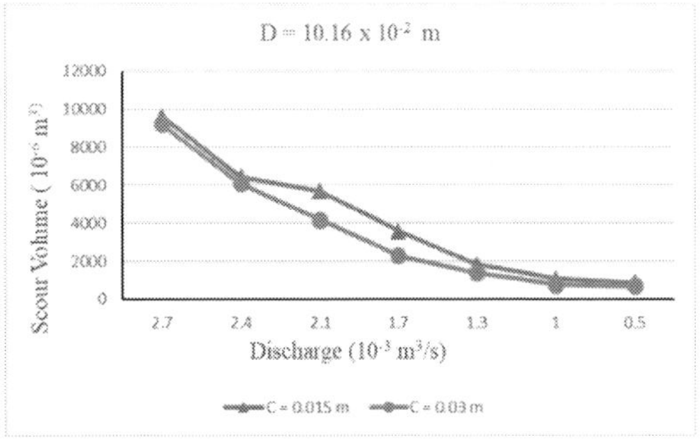

(c)

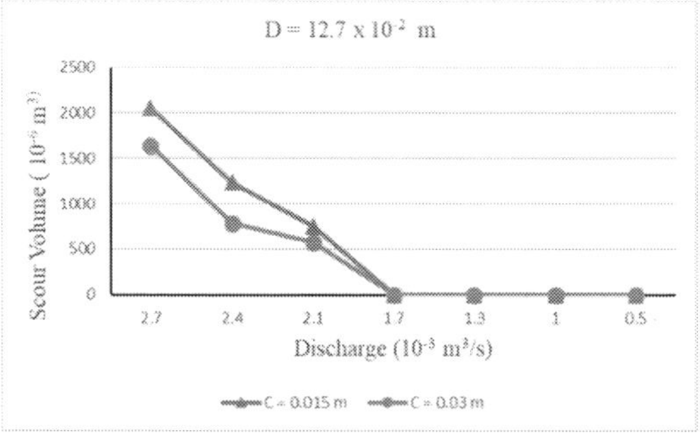

(d)

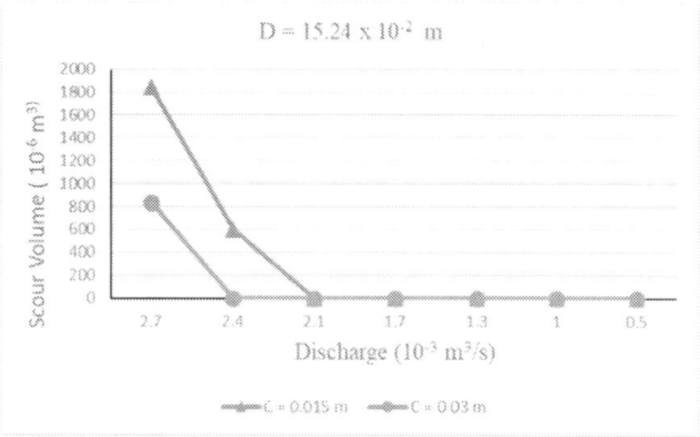

(e)

Figure 6. Effect of suction inlet depth on scour volume at a constant suction pipe diameter, a) D = 7.62×10^{-2} m, b) D =5.08×10^{-2} m, c) D = 10.16×10^{-2} m, d) D =12.7×10^{-2} m, and e) D = 15.24×10^{-2} m.

4. Conclusions

This study concludes that both scour profile and scour volume are a function of suction pipe diameter and suction inlet depth. The following conclusions are made on the effect of diameter and inlet depth on the hydro-suction performance of a suction pipe

a. For a given discharge and suction inlet depth, with an increase in suction pipe diameter up to D = 0.1016 m, hydro-suction performance increases, and any further increase in D leads to low hydro-suction performance.
b. For a given discharge and diameter of pipe, hydro-suction performance decreases with an increase in suction inlet depth.

This investigation is limited to cohesionless sediments, and also the suction inlet depth should not be less than zero. The hydro suction method can be positively used to remove sediment from the river for ship passage, storage reclamation, allowing the stream to flow through some hydraulic structure, unclogging of irrigational inlets, etc. The Hydro-suction method can be effectively used for sediment removal without stopping the functioning of the structures connected, such as inlets to electrical power plants and irrigational channels.

5. References

Asiaban, P., Kouchakzadeh, S. and Asiaban, S., (2017). "Enhanced Hydro-suction Performance for Cohesive Sediment Removal in Low-head Reservoirs." *Ain Shams Engineering Journal*, 8(4), 491-497.

Patra, B., Gir, S., Narayan, P. (2019), "Reservoir Sedimentation in Indian Dams: Trends and challenges." *International Dam Safety Conference*., Bhubaneswar, India, Sr. 6.

Brahme, S.B. and Herbich, J.B.(1986). "Hydraulic Model Studies for Suction Cutterheads." *Journal of waterway, port, coastal, and ocean engineering*, 112(5), 591-606.

Dey, S., Ali, S.Z. and Padhi, E., (2020). "Hydrodynamic Lift on Sediment Particles at Entrainment: Present status and its prospect." *Journal of Hydraulic Engineering*, 146(6), 03120001.

Elgamal, M. and Fouli, H., (2020). "Sediment Removal From Dam Reservoirs Using Syphon Suction Action." *Arabian Journal of Geosciences*, 13(18), 1-10.

Handbook for Assessing and Managing Reservoir Sedimentation. February (2019). CWC. Ministry of Water Resources River Development & Ganga Rejuvenation, Government of India. Doc. No. CDSO_GUD_DS_04_v1.0.

Hotchkiss, Rollin H., and Xi Huang (1995). "Hydrosuction Sediment-removal Systems (HSRS): principles and field test." *Journal of Hydraulic Engineering*, 121(6), 479-489.

Ke, W.T., Chen, Y.W., Hsu, H.C., Toigo, K., Weng, W.C. and Capart, H., (2016). "Influence of Sediment Consolidation on Hydrosuction Performance.", *Journal of Hydraulic Engineering*, 142(10), 04016037.

Tajima, N., (2010). "Dredging of Sediment in Dam Utilizing Siphonage with Sliding Outer Tube. *Japanese Journal of Multiphase Flow*, 24(1), 70-76.

Obialor, C.A., Okeke, O.C., Onunkwo, A.A., Fagorite, V.I. and Ehujuo, N.N., (2019), "Reservoir Sedimentation: Causes, Effects, And Mitigation.", *International Journal of Advanced Academic Research, 5(10), 92-109.*

Pishgar, R., Ayyoubzadeh, S.A., Ghodsian, M. and Saneie, M., (2018). "The Influence of Burrowing-Type Suction Pipe Geometrical and Mechanical Specifications an the Hydro-Suction Method Performance". *ISH Journal of Hydraulic Engineering*, 27(6),1-10.

Raudkivi, A.J., (1988). "The Roughness Height Under Waves." *Journal of Hydraulic Research*, *26*(5), 569-584.

Rehbinder, G., (1994). "Sediment Removal with a Siphon at Critical Flux." *Journal of Hydraulic Research*, 32(6), 845-860.

Rehbinder, G., (1976). "Some aspects on the mechanism of erosion of rock with a high speed water jet." *Proc. of the 3rd International Symp. on Jet Cutting Technique,* Chicago, E1 1-20.

Sadatomi, M., Nagano, T. and Kawahara, A., (2015). "Siphonic Removal of Sediments in Water Reservoirs--Additional Experiment for Model Revision." *International Journal of Environmental Science and Development*, 6(6), 409.

Shrestha, H.S., (2012). "Application of Hydrosuction Sediment Removal System (HSRS) on Peaking Ponds."*Hydro Nepal: Journal of Water, Energy and Environment*, 11, 43-48.

Su-Chin, C.H.E.N., Shun-Chang, W.A.N.G. and Chun-Hung, W.U., (2010). "Sediment removal efficiency of siphon dredging with wedge-type suction head and float tank." *International Journal of Sediment Research*, 25(2), 149-160.

Ullah, S.M., Mazurek, K.A., Rajaratnam, N. and Reitsma, S., (2005). "Siphon removal of cohesionless materials."*Journal of waterway, port, coastal, and ocean engineering*, 131(3), 115-122.

Yang, P., Wang, G. and Zhong, L., (2020). "Suction Removal of Cohesionless Sediment." *Energies*, 13(20), 5436.

Zhou, Y., Zhang, Y., Tang, P., Chen, Y. and Zhu, D.Z., (2013). "Experimental study of the performance of a siphon sediment cleansing set in a CSO chamber." *Water science and technology*, 68(1), 184-191.

DOI: https://doi.org/10.26077/4735-1995

Effect of sinuosity variation on flow characteristics of sand mined sinuous channel using numerical modeling

O.P. Maurya[1], *K.K. Nandi*[1], *S. Modalavalasa*[1] & *S. Dutta*[1]
[1]*Indian Institute of Technology, Guwahati, India*
E-mail: om1996@iitg.ac.in

Abstract: *Anthropogenic activities within the river corridor can have a long-term influence on river health if not managed effectively. Uncontrolled practices of sand mining, a common but aggressive activity can cause severe river bed degradation and, as a result, river incisions, lateral channel instability, and bed armoring. This can further affect the downstream structure such as the bridge pier as well as the morphology of the river. The purpose of this study is to examine, with the help of a numerical model, the impact of a mining pit on the downstream pier and the main channel hydrodynamics if it is present on the flood plain of a sinuous channel. To carry out the study we have considered a meandering channel with two different sinuousness (1.1, 1.25). Additionally, a rectangular mining pit and a downstream circular bridge pier were considered on the flood plains of the channel. The numerical model used for the study was Flow 3D HYDRO with the renormalized group turbulence solver (RNG). As sinuosity increases, the highest zone of streamwise velocity narrows near the mining pit and widens out just downstream of the bridge pier, according to the findings. Near the sand mining pit, the shear layer zone shifted towards the main channel from the near bank. The secondary current at the outer bank near the mining pit is more concentrated in both cases. The channel having more sinuosity has shown a significant change in turbulence kinetic energy (TKE) as compared to the other channel. Additionally, this study can provide insight into the morphodynamics of meandering rivers under different in-channel disturbance conditions and thus be helpful in the proper planning and management of river health.*

Keywords: *(Sinuous channel, Floodplain, Sand mining pit, Bridge pier, Flow-3D, Flow characteristics).*

1. Introduction

In nature, a sinuous river is a typical type of river that arises naturally and occasionally by human intervention. These human interventions or any type of anthropogenic activities such as indiscriminate sand mining, channel dredging, and bridge construction are the principal drivers of river planform alteration (Pradhan et al. 2021; Nandi et al. 2022). If it is close to the bridge pier, it may harm the bridge pier due to upstream nick point migration (Kondolf et al. 1997) and downstream mining pit migration (Zolghadr et al. 2021). Many experiments and theoretical analyses have been carried out by researchers to investigate flow characteristics around bridge piers and agree that downflow, wakes, and horseshoe vortices comprise the basic flow structure around a bridge pier, which causes failure of the bridge pier (Laird 1971; Dargahi 1989; Ahmed and Rajaratnam 1998; Ansari et al. 2002; Graf and Istiarto 2002). The Talar and Balaroud Bridges in Iran are two examples of bridges that were destroyed due to scouring caused by river material harvesting (Daneshfaraz et al. 2021). Haghnazar et al. (2020) suggested that the mining pit depth should be 70% of the flow depth to ensure the entire pit is refilled.

The cross-sectional variation of flow velocity at the apex bends tends to increase as the sinuousness of the channel increases (D. Xu and Y. Bia 2013). Local impediments such as bridge piers, abutments, and roadway embankments, according to (Johnson et al. 2010) cause the flow to be extremely three-dimensional. Using a one-dimensional numerical model to predict such flow characteristics is insufficient because the key preconditions have been violated. Two-dimensional free surface models have limited utility in the horizontal directions because they do not account for diffusion or turbulent mixing (Erduran et al. 2002). Therefore, Flow 3D was selected because it is an effective hydraulic engineering tool for modeling free surface flow (Johnson and Savage 2006).

A lot of research has been done to evaluate the influence of flow characteristics around a bridge pier caused by sand mining in open channels (Modalavalasa et al. 2022a). However, they are limited to straight channels. The majority of natural channels are meandering, but there has been little research into the effects of sand mining on bridge piers in sinuous channels. Therefore, it is necessary to conduct a thorough investigation of the influence of flow characteristics around a bridge pier caused by upstream floodplain sand mining. This study investigated the effect of sinuousness (sinuosity index 1.1 and 1.25) on flow characteristics around a circular bridge pier in a sand-mined channel using CFD-based numerical modeling software (Flow-3D).

2. Material and method

2.1 Flow-3D

FLOW-3D is a general-purpose computational fluid dynamics (CFD) software. It employs specially developed numerical techniques (Volume of Fluid method) to solve the equations of motion for fluids to obtain transient, three-dimensional solutions to multi-scale, multi-physics flow problems. Typically, a numerical model starts with a computational mesh or grid. It consists of several interconnected elements or cells. These cells subdivide the physical space into small volumes with several nodes associated with each such volume. The FLOW-3D approach is to subdivide the flow domain into a grid of rectangular cells. Researchers find it more usable because of the advantages such as reduced time, costs, and no limitation in scale constraints.

2.2 Governing equations

The governing equations of the flow include continuity and conservation of momentum. The continuity equation is expressed as follows for the Cartesian coordinate and incompressible flow:

$$\frac{\partial}{\partial x}(uA_x) + \frac{\partial}{\partial y}(vA_y) + \frac{\partial}{\partial z}(wA_z) = \frac{R_{SOR}}{\rho} \tag{1}$$

where, ρ is the fluid density, and R_{SOR} is a mass source. The velocity components (u, v, w) are in the coordinate directions (x, y, z). A_x is the fractional area open to flow in the x-direction, A_y and A_z are similar area fractions for flow in the y and z directions, respectively.

The equations of motion for the fluid velocity components (u, v, w) in the three coordinate directions are the Navier-Stokes equations with some additional terms

$$\frac{\partial u_i}{\partial t} + \frac{1}{V_F}\left(u_j A_j \frac{\partial u_i}{\partial x_j}\right) = -\frac{\partial P}{\partial x_i} + g_i + f_i \tag{2}$$

where,

- V_F is the fractional volume open to flow,
- g_i is body accelerations,
- f_i is viscous accelerations,

2.3 Numerical setup

The channel geometry, built-in Solid works CAD software, is based on the dimension of the flume setup of the IIT Guwahati Fluvial hydro-ecological laboratory with a dimension of 18 m long, 1 m wide, and 0.3 m deep.

Table 1. Detail of Numerical setup with all the dimension of model setup and flow conditions

Sinuosity index (SI)	Flood plain height (m)	Mining pit size (m^3)	Bridge pier diameter (m)	Flow discharge(m^3/sec)	Depth of water (m)	Bed roughness(m)
1.1	0.08	0.4*0.2*0.06	0.04	0.01536	0.14	0.0016
1.25	0.08	0.4*0.2*0.06	0.04	0.01536	0.14	0.0016

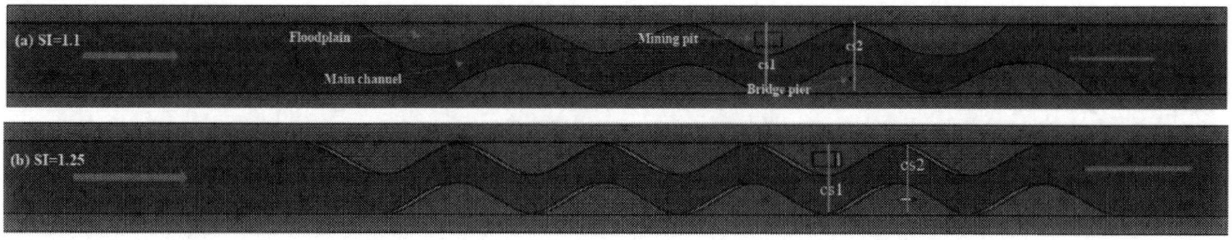

Figure 1. CAD Modelling of numerical setup for two different sinuousness of 1.1 and 1.25 with a sand mining pit on the upstream flood plain and a bridge pier on its downstream flood plain

3. Numerical modeling

The user can simulate multi-physical scenarios using Flow 3D software. In this study, the required physical models are: (1) Gravity and Non-inertial, (2) Turbulence and viscosity

3.1 Gravity and non-inertial

This model takes into account both constant body forces (such as gravity) and non-constant, non-uniform body forces caused by an accelerating reference frame. There is no accelerating reference frame in this investigation. The numerical setup's slope (1:1000) can be used by introducing gravity components in the direction of the slope (x,y,z).

3.2 Turbulence and viscosity model

By including the viscosity and turbulence model, the computation of viscous stresses and turbulence quantities is also included in the solution. Turbulence occurs whenever there are insufficient stabilizing viscous forces that prevent fluids from flowing in an organized, predictable manner. The natural instabilities that exist within a flow at high Reynolds numbers do not get dampened, and this leads to the formation of eddies of various sizes. The striations visible on the free surface of a faucet or a fast-moving stream are a good example of this behaviour.

In an ideal world, we'd be able to replicate the entire spectrum of turbulent fluctuations using mass and momentum conservation equations. For solving the nonlinear term of the Navier-Stokes equation and the Reynolds Averaged Navier Stokes equation turbulence model is required. In Flow-3D there are six turbulence models are given (Prandtl mixing length, turbulence energy model, $K-\omega$, $K-\varepsilon$, RNG $k-\varepsilon$, and LES). In this study RNG $k-\varepsilon$ model is used because of its best agreement with the experiment (Karami et al. 2014).

3.3 Boundary conditions and initial conditions

There are four different boundary conditions are defined at mesh faces in this study:

- Specified pressure: Specify the pressure at the boundary. If the fluid elevation is specified, the pressure at the boundary will follow a hydrostatic distribution (applied at outlet and free surface of flow along with fluid elevation and fluid fraction).
- Symmetry: Applies a zero-gradient condition at the boundary as well as a zero-velocity condition normal to the boundary (applied at smooth side walls)
- Volume flow rate: Applies the specified flow rate at the boundary (Discharge at inlet = 0.01536 m^3/s).
- Wall: Applies the no-slip condition at the boundary as well as a zero-velocity condition normal to the boundary (applied at bed of the setup).

4. Results and discussions

4.1 Flow characteristics near the sand mining pit (cross-section 1)

Figure 2 represents the comparison of the flow characteristics (streamwise velocity, secondary velocity, secondary current, and turbulence kinetic energy) at cross-section (1) of two different sinuous (SI = 1.1 and 1.25) channels. The flow characteristics is measured after the simulation equilibrium (at 180 s). The streamwise velocity distribution along the inner bank shows a high velocity (at cross-section1). After passing by the apex, the main flow velocity moves to near the opposite side (at cross-section 2), see figure 2 and 3. This is a common aspect of a sinuous or meandering river's flow (D. Xu and Y. Bia 2013; Blanckaert 2011; Abad and Garcia 2009; Whiting and Dietrich 1993). Cross-sectional velocity redistribution, which forms a point bar in the low velocity area and a pool on the opposite side, has been widely proven and considered vital to the formation and maintenance of river meanders in fluvial environments (Federici and Seminara 2003; Braudrick et al. 2009). As sinuosity increases, the greatest streamwise velocity zone at cross-section 1 narrows see figure 2. As a result, the shear layer zone in the main stream is slightly more sifted, which may result in aquatic life being shifted in the main stream as well as sedimentation in the main stream. As sinuosity rises, secondary velocity and secondary current become highly concentrated near the outer bank. Turbulence kinetic energy is more concentrated in the sand mining pit in the sinuosity 1.1 case (it indicates the high velocity fluctuation into the mining pit), but it is more concentrated in the inner bank in the sinuosity 1.25 case (as a result, there may be more chance of erosion).

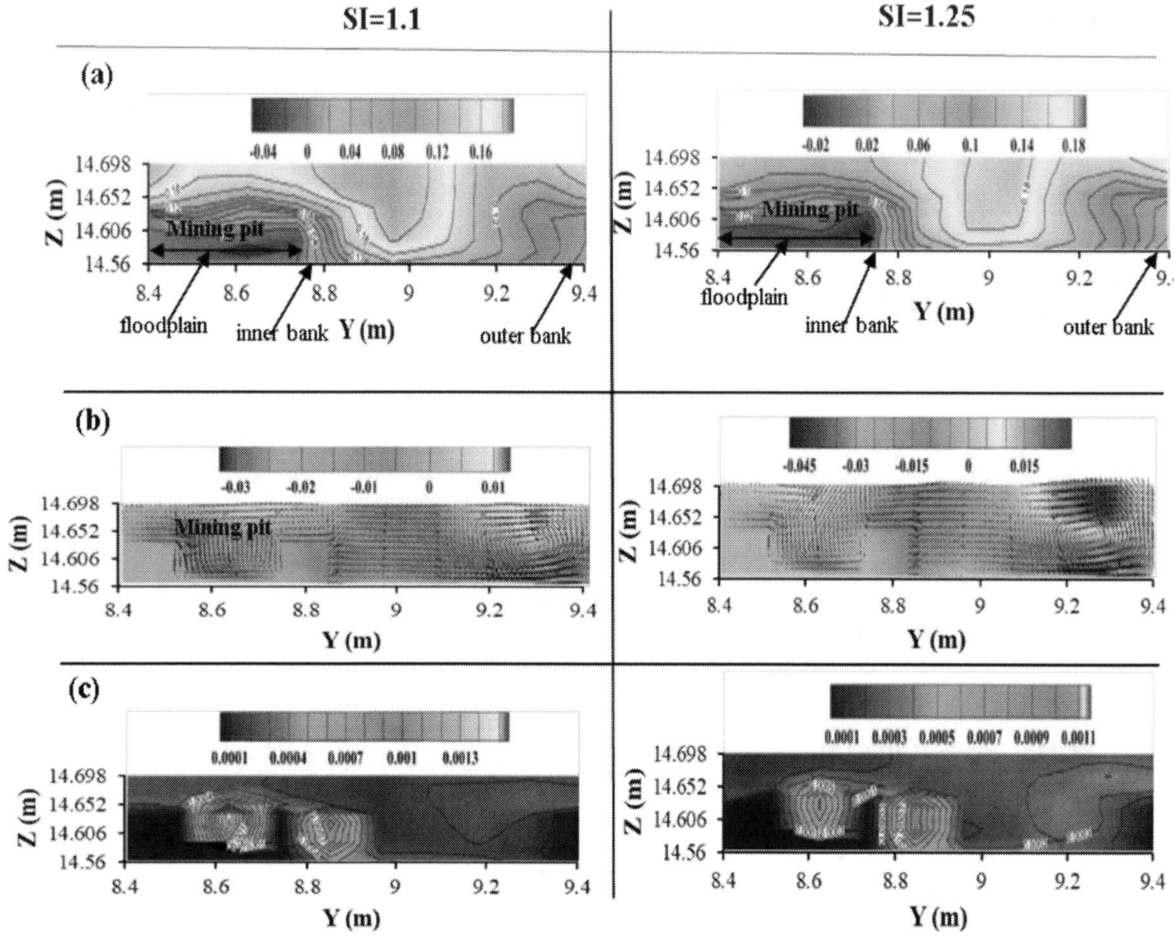

Figure 2. Contour plots at CS2, (a) Streamwise velocity (m/s), (b) Secondary velocity (m/s) and secondary current, (c) Turbulence kinetic energy (J/kg)

4.2 Flow structure just downstream of bridge pier (cross-section 2)

Figure 3. represents the comparison of the flow characteristics (streamwise velocity, secondary velocity, secondary current, and turbulence kinetic energy) at cross-section (1) of two different sinuous (SI = 1.1 and 1.25) channels. The flow characteristics is measured after the simulation equilibrium (at 180 s. The plots clearly show that just downstream of the bridge pier, as increasing the sinuosity, the maximum streamwise velocity zone sifts towards the flood plain, and in the vicinity of the bridge pier, a low streamwise velocity zone and vortex are created (Modalavalasa et al. 2022b). And at the outer bank, a highly low velocity zone is created in the case of sinuosity 1.25, which indicates the increase in the siltation phenomenon at the outer bank. At the location of bridge pier flow is diverted and secondary current in the main stream almost happens in the same pattern, but near the inner zone of sinuosity 1.25, a negative secondary current (anticlockwise) zone is generated (which may cause a highly low-pressure zone). Turbulence kinetic energy is more concentrated near the inner bank in the case of sinuosity 1.1 and on the flood plain in the case of sinuosity 1.25 (which may cause high turbulence generated by increasing the sinuosity near the downstream of the bridge pier). It is little bit increases in the main channel as sinuosity increased see figure 3.

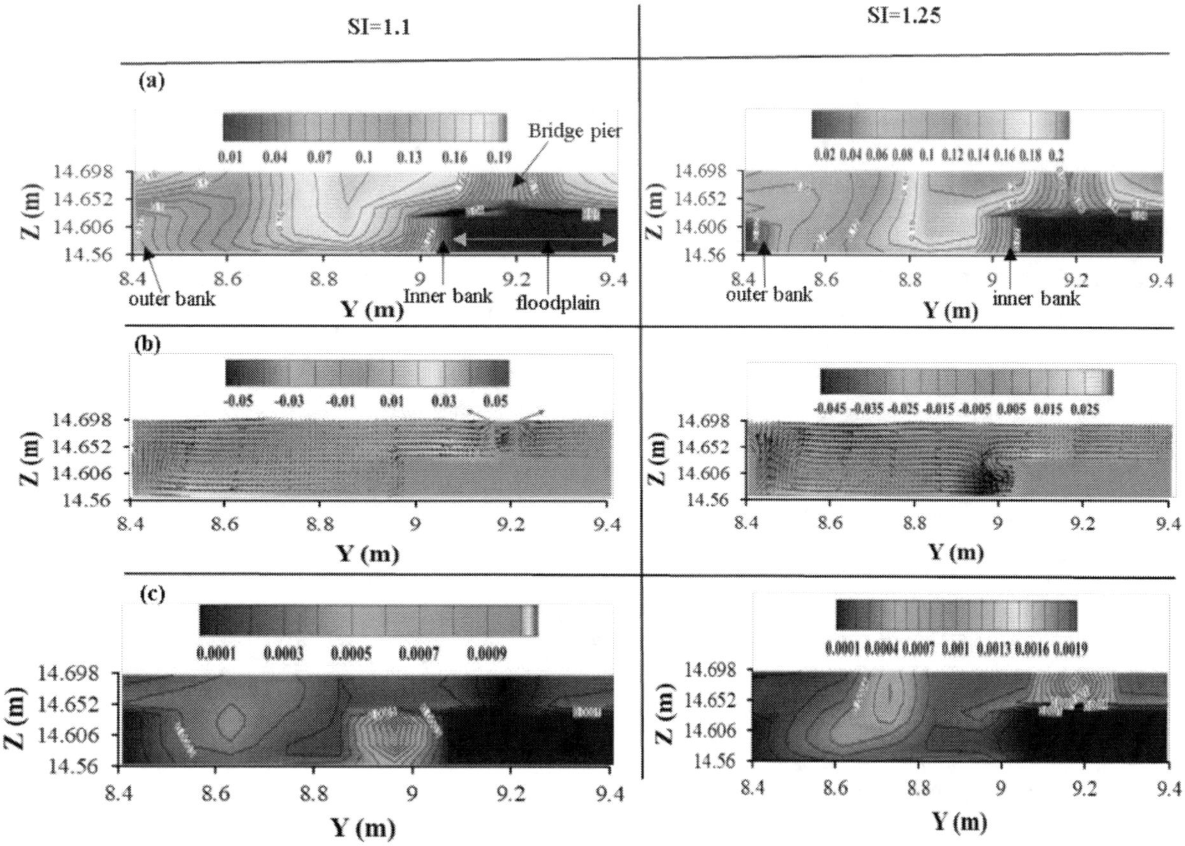

Figure 3. Contour plots at CS2, (a) Streamwise velocity (m/s), (b) Secondary velocity (m/s) and secondary current, (c) Turbulence kinetic energy (J/kg)

5. Conclusion

The impact of sinuosity variation (SI = 1.1 and 1.25) on the flow parameters of a sinuous channel with a flood plain sand mining pit and bridge pier was explored in this study. Flow-3D software was used to conduct a numerical study for this aim.

- As sinuosity increased, the shear layer zone shifted from the inner bank to the main stream, and the maximum zone of streamwise velocity narrowed near the mining pit (at cross-section 1), but the maximum zone of streamwise velocity shifted a little bit towards the flood plain just downstream of the bridge pier.
- The secondary current pattern is almost the same near the mining pit, but downstream of the bridge pier, an anticlockwise secondary current is generated near the inner bank of the main stream in the case of sinuosity 1.25.
- Turbulence kinetic energy is concentrated near the inner bank and into the mining pit in both cases at cross-section 1, but just downstream of the bridge pier, turbulence kinetic energy is concentrated on the flood plain in the case of sinuosity 1.25.

References

Journals:
Ahmed, Ferdous, and Nallamuthu Rajaratnam. "Flow around bridge piers." *Journal of Hydraulic Engineering* 124.3 (1998): 288-300.
Ansari, S. A., U. C. Kothyari, and K. G. Ranga Raju. "Influence of cohesion on scour around bridge piers." *Journal of Hydraulic Research* 40.6 (2002): 717-729.
Abad, Jorge D., and Marcelo H. Garcia. "Experiments in a high-amplitude Kinoshita meandering channel: 1. Implications of bend orientation on mean and turbulent flow structure." *Water Resources Research* 45.2 (2009).
Braudrick, Christian A., et al. "Experimental evidence for the conditions necessary to sustain meandering in coarse-bedded rivers." *Proceedings of the National Academy of Sciences* 106.40 (2009): 16936-16941.
Blanckaert, Koen. "Hydrodynamic processes in sharp meander bends and their morphological implications." *Journal of Geophysical Research: Earth Surface* 116.F1 (2011).
Pradhan, Chandan, et al. "Regulated rivers in India: research progress and future directions." *ISH Journal of Hydraulic Engineering* (2021): 1-13.
Dargahi, Bijan. "The turbulent flow field around a circular cylinder." *Experiments in Fluids* 8.1 (1989): 1-12.
Daneshfaraz, Rasoul, et al. "Investigation of local scouring around hydrodynamic and circular pile groups under the influence of river material harvesting pits." *Water* 13.16 (2021): 2192.
Erduran, K. S., V. Kutija, and C. J. M. Hewett. "Performance of finite volume solutions to the shallow water equations with shock-capturing schemes." *International journal for numerical methods in fluids* 40.10 (2002): 1237-1273.
Federici, B., Seminara, R., 2003. On the convective nature of bar instability. Journal of Fluid Mechanics 487, 125e145
Graf, W. H., and I. Istiarto. "Flow pattern in the scour hole around a cylinder." *Journal of Hydraulic Research* 40.1 (2002): 13-20.
Haghnazar, Hamed, et al. "Evaluation of infilling and replenishment of river sand mining pits." *Environmental Earth Sciences* 79.14 (2020): 1-18.
Johnson, Michael C., and Bruce M. Savage. "Physical and numerical comparison of flow over ogee spillway in the presence of tailwater." *Journal of hydraulic engineering* 132.12 (2006): 1353-1357.
Johnson, Peggy A., Scott A. Sheeder, and Jessica T. Newlin. "Waterway transitions at US bridges." *Water and Environment Journal* 24.4 (2010): 274-281.
Kondolf, G. Mathias. "PROFILE: hungry water: effects of dams and gravel mining on river channels." *Environmental management* 21.4 (1997): 533-551.
Karami, Hojat, et al. "Verification of numerical study of scour around spur dikes using experimental data." *Water and environment journal* 28.1 (2014): 124-134.
Laird, A. D. (1971). "Eddy formation behind circular cylinders." J. Hydraul. Div., 97(6), 763–775.
Modalavalasa, Suresh, et al. "Combined effect of bridge piers and floodplain vegetation on main channel hydraulics." *Experimental Thermal and Fluid Science* (2022b): 110669.
Nandi, Ketan Kumar, et al. "How dynamic is the Brahmaputra? Understanding the process–form–vegetation interactions for hierarchies of energy dissipation." *Ecohydrology* 15.3 (2022): e2416.
Whiting, Peter J., and William E. Dietrich. "Experimental studies of bed topography and flow patterns in large-amplitude meanders: 1. Observations." *Water Resources Research* 29.11 (1993): 3605-3614.
Xu, Dong, and Yuchuan Bai. "Experimental study on the bed topography evolution in alluvial meandering rivers with various sinuousnesses." *Journal of Hydro-environment Research* 7.2 (2013): 92-102.
Zolghadr, Masih, et al. "Migration of sand mining pit in rivers: An experimental, numerical and case study." *Measurement* 172 (2021): 108944.
Book Reference:
Modalavalasa, S., et al. "Numerical and Experimental Investigation of Effect of Green River Corridor on Main Channel Hydraulics." *Recent Trends in River Corridor Management.* Springer, Singapore, (2022a). 165-176.

DOI: https://doi.org/10.26077/442f-a552

Estimation of Discharging Capacity and assessing the performance of Overflow Spillway with Full and Partial Operation of Gates using Numerical Modelling

Dr. Prajakta P. Gadge[1]& Kunal Kapur[2]
[1]*Scientist C, Central Water and Power Research Station, Pune, India*
[2]*Research Assistant, Central Water and Power Research Station, Pune, India*
E-mail: prajaktagadge@gmail.com

Abstract: *The spillway is among the most important structures of a dam project. The operation of spillway gates during the flood and fulfilling the downstream requirement is one of the main problems in reservoir management. Generally, the equal opening of all the gates during spillway operation is mostly preferred for maintaining an equal distribution of discharge in the energy dissipator and at the downstream side. A systematic study is required to determine the ability of the spillway to pass the discharge at various reservoir water levels through a specific opening of gates and assess the performance for the safety of the structure. The design of each project is unique and site-specific. A physical model is an indispensable tool to optimise the hydraulically efficient and economical design of spillways. However, the evaluation of spillway performance especially for the partial operation of gates is a time-consuming task on the physical model. Nowadays, the computational fluid dynamics technique is becoming popular in modelling spillway flows. This paper discusses the studies carried out for overflow spillway for gated and ungated operations. The Computational fluid dynamics software FLOW-3D was used for numerical simulation. The numerical model was validated by comparing the results with the physical model for ungated operations. The results in terms of design discharge passed over the spillway and corresponding pressures over the spillway surface computed using a numerical model were found closer to the results obtained from the physical model. Numerical model studies were extended further for the gated operation. Large data was generated which was used to create discharging capacity plots and evaluate the performance of the spillway for operating the spillway with different openings of gates at various reservoir water levels.*

***Keywords**: Overflow spillway, spillway gates, physical model, numerical model, discharging capacity.*

1. Introduction

The safety of dams plays an important role in water resources management in many countries because their failure can cause disasters at the downstream. The design of a dam with gates provides greater flexibility in the operation and management. Due to the importance of dam safety, the proper operation of gated spillways remains an obvious challenge in reservoir management and operation during a flood event. The gated operation of a spillway is the art of releasing water through gates during high inflow/flood. Generally, operating all the gates equally and simultaneously for all the discharges is the safest mode of operation since it ensures equitable distribution of discharge in the energy dissipator and the river downstream. Reservoir control is achieved by the operation of spillway gates.

For many years physical hydraulic modelling has been a standard design tool for hydraulic engineering of dams and their associated flow control structures. Recent development in numerical simulation models includes the advancement of computational techniques and understanding of physical processes within the models, and improved modelling approaches through the coupling of models with graphic interface, visualization, and parameter estimation methods. However, a comparison of the numerical results with experimental or prototype data is still required for calibration and validation. There are various CFD codes available for modelling the spillway flows. Jai Hong Lee et al (2019) studied the interference of Dual Spillways (service and emergency spillway) operation by investigating the various parameters using a physical model and CFD programme FLOW-3D. Gadge et al. (2018) investigated the flow through orifice spillway using physical model and CFD code FLUENT and developed the equation for the design of the roof profile. Serife Yurdagul Kumcu (2016) discusses the ability of the FLOW-3D model in the simulation of flow over spillways and stilling basins by comparing the results with a physical model. Jothiprakash et al. (2015) studied the flow characteristics of orifice spillway aerators using a Computational Fluid Dynamic Software FLUENT. The air concentration, pressure and velocity profiles upstream and downstream of the aerator with and without a ramp were analyzed across the depth and in the flow direction. Chanel and Doering (2007) discussed the capability of the three-dimensional numerical modelling for several spillway configurations using CFD software Flow-3D. Savage and Johnson (2001) studied and compared the results of flow characteristics of an ogee spillway from the physical and numerical models with the existing literature from USBR and USACE.

The results were analysed for 10 different flow conditions. The commercially available CFD package Flow-3D was used to numerically simulate the flow conditions.

The main aim of the present study is to determine the discharging capacity of the surface spillway and assess the performance in terms of determining the cavitation susceptibility for the partial operation of spillway gates at different reservoir water levels using CFD code FLOW-3D.

2. Numerical Modelling

With increasing computer processing capacity, numerical simulations for hydrodynamic processes become attractive, including flow over spillways. For hydraulics applications, the CFD modelling is implemented by solving the governing equations describing the incompressible flow behaviour based on the conservation of mass and momentum, e.g., continuity equations and momentum equations known as Reynolds-averaged Navier-Stokes equations. For the present study, the CFD software FLOW-3D was used for simulating the flow over the overflow type of spillway. For tracking the free surface of air and water interface concerning time and space, the volume of fluid (VOF) technique was used. The turbulence models such as RNG k-ε, k-ω and large eddy are available in the FLOW 3D software. The Renormalized Group (RNG) k-ε model (Yakhot& Orszag 1986) is a more robust version of the two-equation k-ε model, and is recommended for most industrial problems. It extends the capabilities of the standard k-ε model to provide better coverage of transitionally-turbulent flows, curving flows, wall heat transfer, and mass transfer (Flow 3D). In the present study, the renormalized group (RNG) model was selected as the turbulence closure model because this model has been widely used and found to be suitable for modelling the spillway flows.

When conducting the simulation of spillway flow, it is most important that the model of spillway structure should be replicated accurately. Accordingly, the geometry of the model consisting of two spans with intermediate pier, radial gates, and training walls was generated in AUTOCAD software, taking into account their influence on the spillway overflow characteristics. The said geometry was prepared in stereo lithographic (STL) format and imported into FLOW-3D software for simulation studies. The geometry of the spillway is shown in Figure 1a. The extent of the domain was considered 100 m upstream of the spillway crest and 140 m downstream of the spillway bucket to form fully developed flow conditions. The insufficient length at the upstream and downstream of the spillway greatly affects the flow conditions over the spillway surface resulting in inaccuracy in the measurement of various hydraulic parameters. The next step in numerical simulation is to generate the grid, define the boundary condition and apply the physics to get the final solution. Consequently, the grid was generated with a uniform size of 0.01 m in the x, y and z-direction in the main area of interest i.e the region of high-velocity flow over the ogee-crest spillway between the crest and lip of the bucket. The fine grid was adopted in this region taking into consideration the complexity of flow behaviour i.e. transition of flow from subcritical to supercritical state and back to subcritical. The total mesh count in this region was about 3638320. The coarser grid of size 0.015 was generated in the upstream and downstream channels, which was the area of low interest. The total mesh count in these regions was about 1434720. Figure 1b shows the grid generation near the spillway gates. Boundary conditions were established for each side of the computational domain. At domain inlet pressure inlet boundary condition defining the free surface elevation ranging from maximum water level up to the level at which free flow occurs was defined. The outflow boundary condition was defined at the end of the domain. The extent of the outflow boundary condition was selected in such a way that the complete ski-jump should be formed in the domain and water is allowed to flow free at the downstream side without causing any backwater effect. Pressure inlet boundary with gauge pressure equal to zero (atmospheric) was defined at the top of the domain. No-slip condition for all the sidewalls including the gate, pier walls and spillway surface of the domain were applied for the simulation of flow. The CFD modelling uses unsteady state simulation to reach a steady state. For running the simulations, computation modules of viscosity, turbulence and gravity were activated for all cases studied. The data from the physical model was taken for validation of the numerical model.

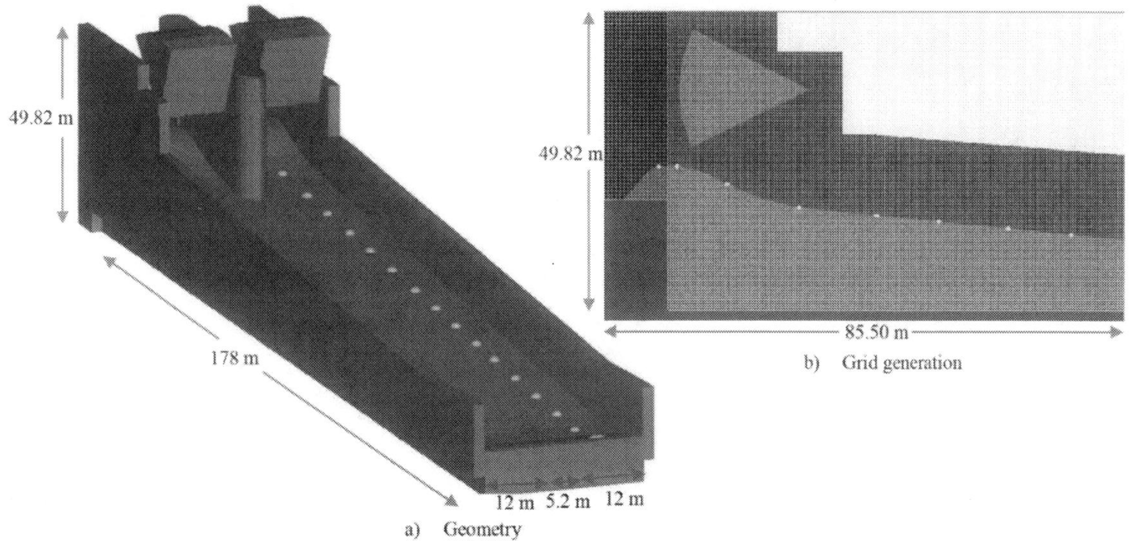

Figure 1 Geometry and grid generation for spillway structure

Numerical model studies were carried out for overflow type of spillway structure. This spillway was designed to pass the discharge of 4000 m^3/s at MWL El. 1703 m through all two spans equipped with radial gates of size 12 m wide x 16 m high. The crest of the spillway and maximum water level (MWL) is at EL. 1684 m and 1703 m respectively. A flip bucket is provided for energy dissipation. Physical model studies were conducted on a 1:40 scale 2-D sectional model for assessing the performance of the spillway in respect of determining the discharging capacity of the spillway, pressures over spillway surface, water surface profiles and performance of energy dissipator for 25%, 50%, 75% and 100 % of design discharge for the ungated and gated operation. Based on the findings, the design of the spillway and energy dissipator was finalised. Evaluating the performance of spillway for the different operations of spillway gates was a time-consuming task on the physical model. Hence, the numerical model studies were carried out in assessing the performance of spillway for gated operations. Studies were carried out for 2m, 4m, 6m, 8m, 10m, 12m and 14m openings of the gates at different reservoir water levels starting from maximum water level to the level at which free flow occur. The results computed using the numerical model studies have been discussed in the following section.

3. Results

The discharges passed over the spillway were computed for the full and partial operation of gates. At the start, a numerical model was validated by comparing the results with the physical model for operating the spillway for 4000 m^3/s (100 % design discharge) and 1000 m^3/s (25% design discharge). Simulations were run till the steady state is achieved. Figure 2 shows the convergence plot of the discharge value for 100% and 25% of design discharge. Figure indicates that the solution was converged after 7 sec as there was no change in the discharge values in both the cases.

Physical model studies indicated that the discharges of 4000 m^3/s and 1000 m^3/s could be passed at RWL El. 1703.6 m and RWL El. 1691.6 m respectively through all the two spans. In the numerical model, the discharges of 4050 m^3/s and 993 m^3/s could be passed at RWL El. 1703.6 m and RWL El. 1691.6 m respectively through all two spans. The comparison shows that there is a good agreement in the results computed in the numerical and physical models in terms of discharging capacity of the spillway. The relative difference between the CFD results and the experimental data was found to be 1.25 % and 0.76% for 100% and 25% design discharges respectively.

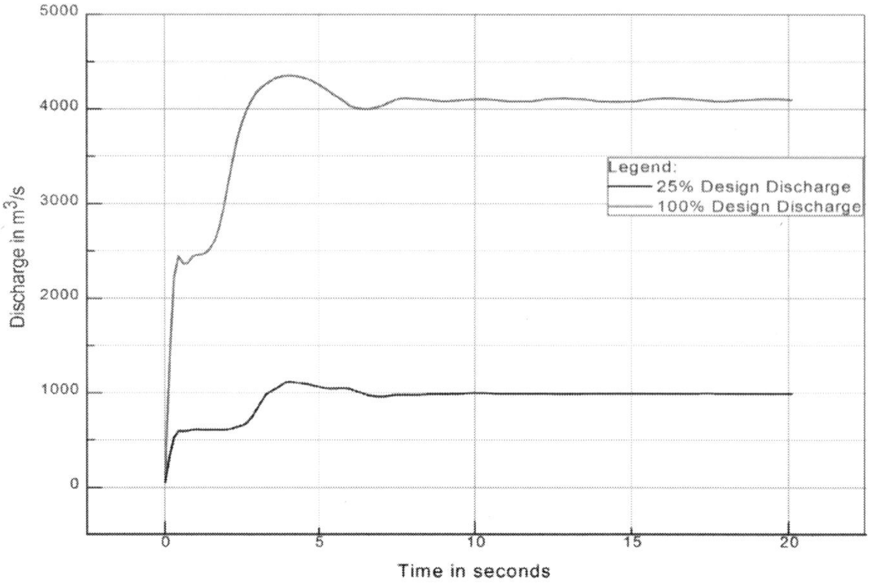

Figure 2 Convergence plot for 100% and 25% of design discharge

Flow condition over spillway surface was also compared between numerical and physical model. A minor rooster tail was seen along the centreline of the pier forming due to the interaction of flow emerging from adjoining spans in the physical and numerical model for both the discharges. Figure 3 shows the comparison of flow condition over the spillway surface between numerical and physical model for 100% design discharge. Figure 4 shows the comparison of pressure over the spillway surface between numerical and physical model for 100% design discharge. The maximum relative difference between the CFD results and the experimental data was found to be 6%. It was observed that the pressures drop down gradually from crest to the beginning of ski jump bucket and then start rising due to the centrifugal action of flow in the bucket. The trend and the values of pressures over the spillway surface computed in the numerical model were found closer to the pressures measured in the physical model.

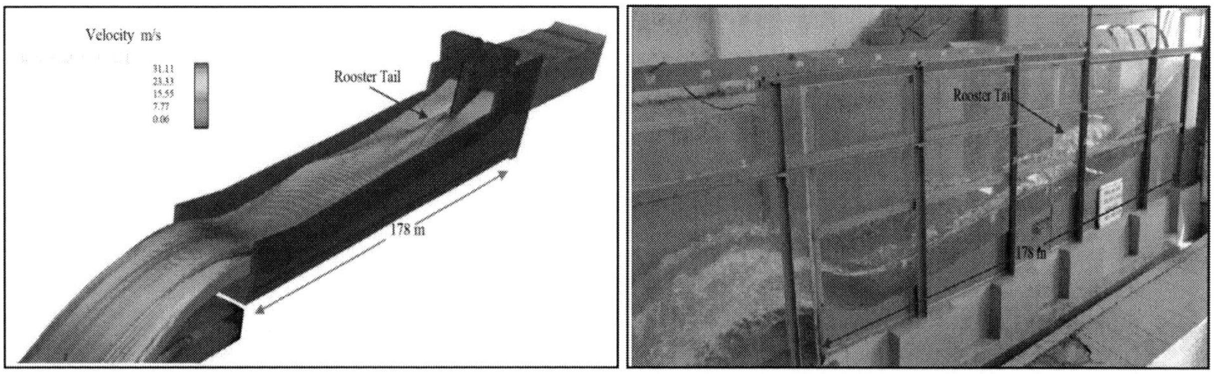

a) Numerical model

b) Physical model

Figure 3 Flow conditions in numerical and physical model for 100% design discharge

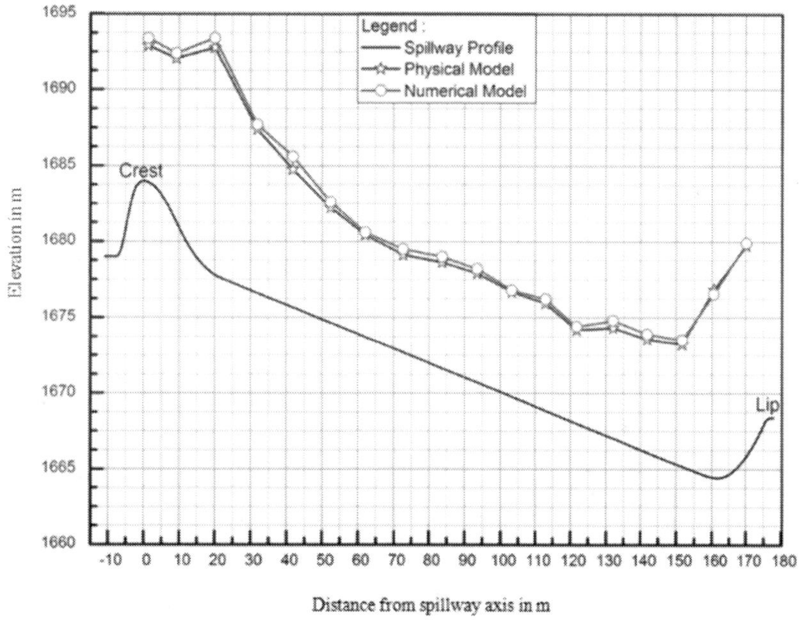

Figure 4 Comparison of pressure profile in numerical and physical model for 100% design discharge

Based on the validation of the model, the numerical model data was extracted further for computing the discharges passed over the spillway for the partial operation of spillway gates. Spillway gates were operated equally for the openings ranging from 2m to 14m for various reservoir water levels up to MWL El. 1703 m. Figure 5 shows the discharging capacity curve for the full and partial operation of spillway gates. The curve showing the discharges for full gate openings was generated using experimental data, whereas the plots for different gate opening shown in red colour are generated using numerical model studies.

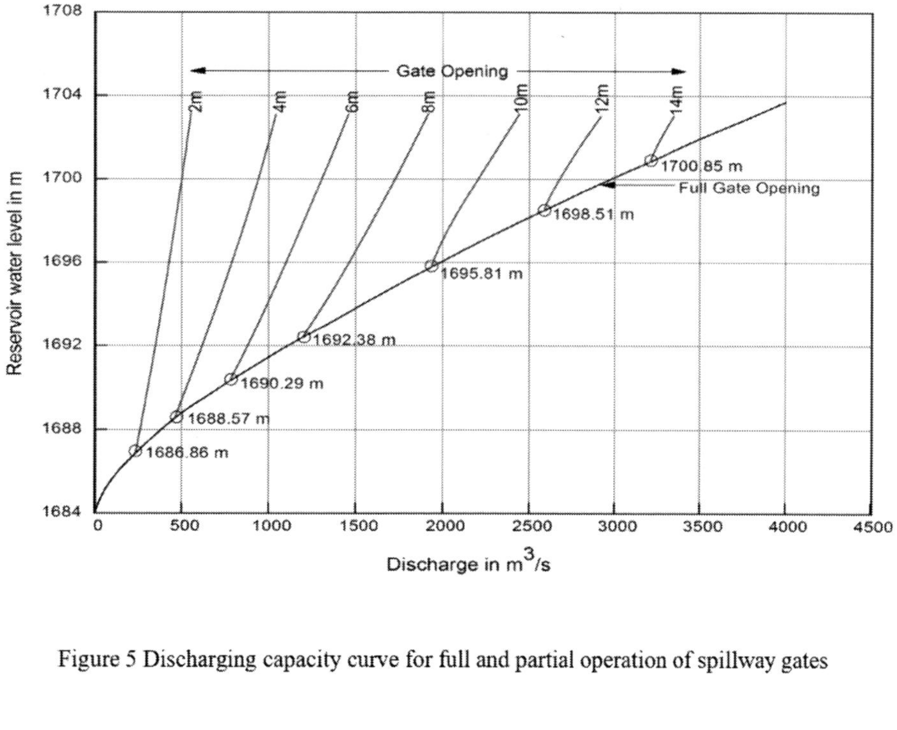

Figure 5 Discharging capacity curve for full and partial operation of spillway gates

Large data was generated giving the discharges passed over the spillway for various reservoir levels and gate openings as shown in Figure 5. These results are useful in planning the operation of the spillway after the completion of the project. The numerical model also helped in assessing the performance of the spillway in terms of determining the cavitation susceptibility of the spillway surface for the gated operation which was a time-consuming task in the physical model. For different gate operations of the spillway, the pressures over the spillway surface and corresponding velocity were computed and the cavitation Index was calculated. Cavitation is a complex hydrodynamic phenomenon and can cause serious damage to the spillway surface and is governed by pressures, velocities and duration of spillway operation. The inception of cavitation damage can be assessed by the cavitation index which can be calculated using Eq. (1).

$$\sigma = \frac{(P_0 - P_V)}{\left(\frac{\rho V^2}{2}\right)} \tag{1}$$

Where, P_0 = Reference pressure, V= Flow velocity, P_v = vapour pressure of water, ρ = density of water

Figure 6 shows the pressure profile along the spillway surface and corresponding cavitation Index for MWL El. 1703 m for various gate openings. The pressures were found to be positive and corresponding cavitation indices were above the value of the critical cavitation index of 0.2 for various spillway operations (Falvey, 1990). Hence, the design of the spillway was found to be satisfactory and there was no possibility of occurrence of cavitation damage to the spillway surface. The design was also checked for different reservoir water levels below 1703 m with the different opening of gates and found to be satisfactory. Figure 6 indicates that as the flow leaves the gate, the pressure over the spillway surface decreases rapidly and increases further in the ski-jump bucket due to centrifugal action. The same observations are also represented in graphical format in terms of pressure and velocity contours in the vicinity of the spillway gate as shown in Figure 7. The colour scale represents the variation in the parameters between the blue colour showing the minimum value and the red colour showing the maximum value. The figure indicates that as the flow leaves the gate, the pressure over the spillway surface decreases with a corresponding increase in the velocity due to the transition of flow from subcritical to supercritical state. Hence, the physics of the flow is well captured in the numerical model. The flow regime over the spillway was also determined by computing the Froude numbers at different locations over the surface for all the gate openings as shown in Figure 8. Figure indicates the transition of flow from subcritical (Fr <1) to supercritical (Fr >1) state. At the downstream of spillway crest, the flow was subcritical upto the distance of about 4m. Afterwards, the flow was found to be supercritical till the end of bucket lip with Froude number varying from 1 to 7 for different gate openings.

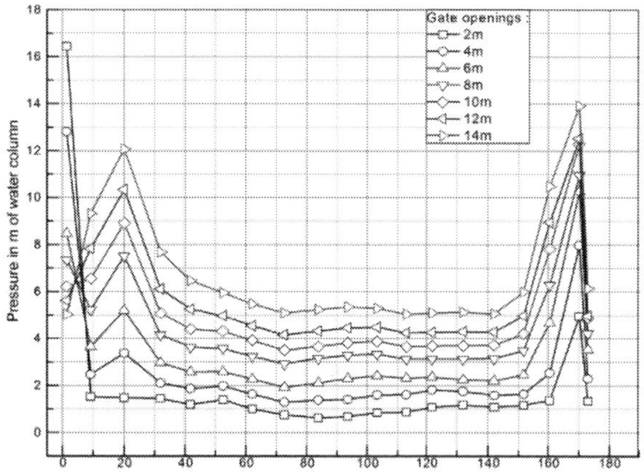

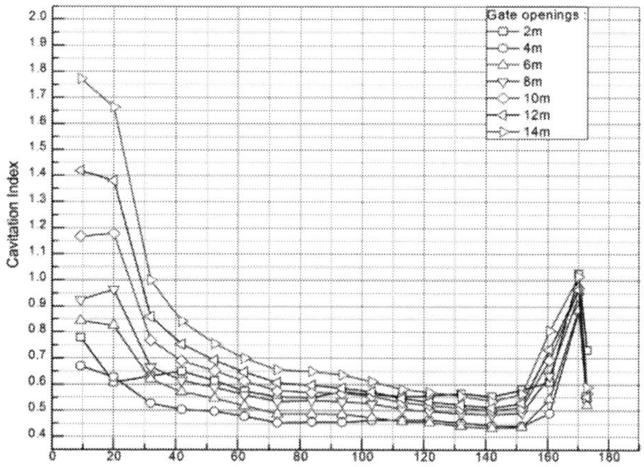

Figure 6 Pressure profile along the spillway surface and corresponding cavitation Index for MWL El. 1703 m

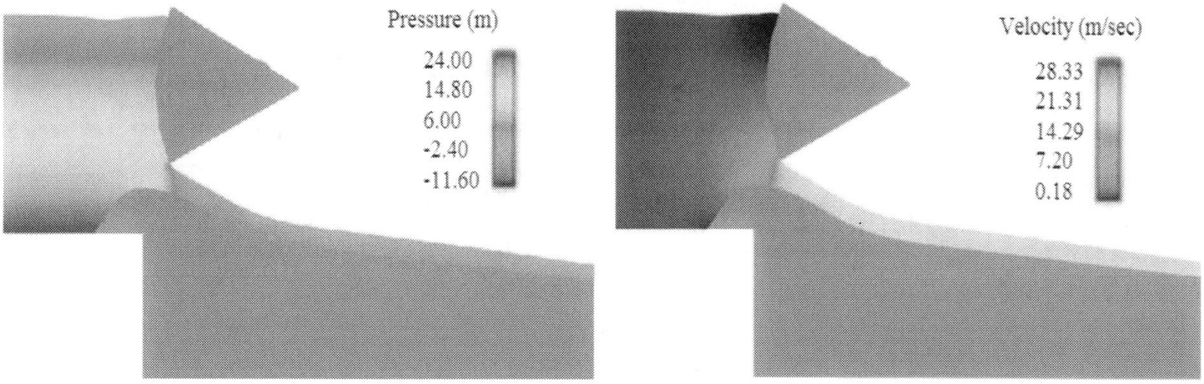

a) Pressure Contour

b) Velocity Contour

Figure 7 Pressure and velocity contours in the vicinity of spillway gate

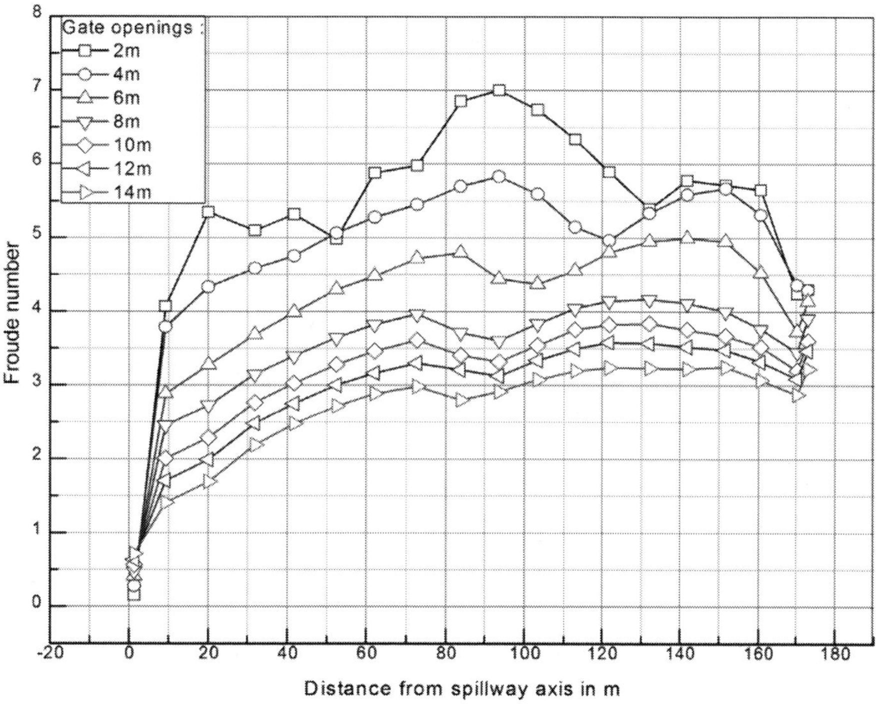

Figure 8 Froude number at different locations along spillway surface

4. Conclusion

Taking into consideration the safety of dams, the spillway should be designed in such a way that it should have adequate discharging capacity during the flood and can be used systematically as per the requirement of the downstream population during the lean period. Physical modelling has been widely used in the design and investigation of hydraulic structures. However, evaluating the performance of a spillway, especially for the operation of the spillway with different gate openings and reservoir water levels is a time-consuming task in a physical model. Due to advancements in computers, the computational fluid dynamic technique has been used increasingly for modelling the spillway flows. The CFD software FLOW-3D used in the study became useful in simulating the spillway flows and assessing its performance in terms of evaluating the discharging capacity and determining the cavitation susceptibility of the spillway surface for partial operation of the spillway at different reservoir water levels.

5. ACKNOWLEDGEMENTS

The authors are thankful to Shri Y. N. Srivastava, Additional Director, CW&PRS for his constant encouragement and valuable guidance in the preparation of this paper. The authors are grateful to the Dr R. S. Kankara, Director, CWPRS for the consent to publish this paper and provide the facility to conduct physical model studies at CWPRS.

6. REFERENCES

Chanel, P. G. and Doering, J. C. (2007). "An evaluation of computational fluid dynamics for spillway modelling." *16th Australian Fluid Mechanics Congress*, 1201-1206.

Falvey, H. T. (1990). Cavitation in chutes and spillways, Engineering Monograph No. 42, USBR publication, Denver, CO.

Flow-3D, User Manual, Version 11.03. Flow Science Inc.

Gadge, P. P., Jothiprakash, V. and Bhosekar, V. V. (2018). "Hydraulic investigation and design of roof profile of an orifice spillway using experimental and numerical models."*J. of Appl. Wat. Engine. and Resear., IAHR*, 6(2), 85–94.

Jai Hong Lee, Pierre Y. Julien, and Christopher I. Thornton (2019). "Interference of dual spillways operations." *J. of Hydrau. Engine., ASCE,* 145(5), DOI:10.1061/(ASCE)HY.1943-7900.0001593.

Jothiprakash, V., Bhosekar, V. V. and Deolalikar, P. B. (2015). "Flow characteristics of orifice spillway aerator: numerical model studies." *J. of Hydrau. Engine., ISH,* 21(2), 216-230.

Savage, B. M. and Johnson, M. C. (2001). "Flow over ogee spillway: physical and numerical model case study." *J. of Hydra. Engine., ASCE*, 127 (8), 640-649.

Serife Yurdagul Kumcu (2016). "Investigation of flow over spillway modelling and comparison between experimental data and CFD analysis." *J. of Civil Engine, KSCE* , DOI:10.1007/s12205-016-1257-z.

Yakhot, V. and Orszag, S. A. (1986). "Renormalization group analysis of turbulence I. Basic theory." *J. of Scient. Compu.,* 1(1), 3-51.

DOI: https://doi.org/10.26077/2802-25c6

Estimation of Maximum Scour Depth and Scour Pattern Around Submerged Spur-dike

Manish Pandey[1] and Manali Pal[2]
[1,2]*Department of Civil Engineering, National Institute of Technology Warangal, Warangal, India*
E-mail: *manishpandey3aug@gmail.com, manalipal@nitw.ac.in*

Abstract: *Scouring due to turbulent flows is a process that naturally happens around river structures, including river training or riverbank protection works such as spur dikes, submerged vanes, etc. present experimental study aims to estimate the maximum scour depth around submerged spur dikes, which have not been studied as extensively. In this study, we checked the influence of different flow conditions, spur dike dimensions and submergence conations on maximum scour depth at equilibrium scour conditions. The dependence of the criteria for scour initiation and scour patterns on the dike dimensions, such as the dike length and height, has been assessed. We also checked the previously proposed equations for maximum scour depth at submerged spur dike using present and previous (collected from literature) experimental data. The dependency of maximum scour depth on some hydraulic and geometric control parameters is also discussed. In addition, a new equation is also proposed to calculate the maximum scour depth around submerged spur dike at equilibrium condition, which shows better agreements with present and previous experimental datasets than other equations.*

Keywords: *Clear-water condition, equilibrium scour condition, maximum scour depth, submerged spur dike.*

1. Introduction

A spur dike is also known as groin, a man-made elongated hydraulic structure with one end on the bank of a stream and another projecting into the current (Elawady et al. 2001) with fixed perpendicular or at an angle to the flow direction. Rivers and streams are often referred to as dynamic systems, which means they are constantly changing with complex and interrelated factors controlling river and stream formation (Jain et al. 2021). These factors include the amount and rate of water supply and sediments with the extent of vegetation in the catchment area. They are objected to changing over time in response to river systems altering their shape, form, and location. But the rate of these changes in stable streams is generally slow and imperceptible. In order to limit the movement of the bank of a meandering river, specific structures are constructed on the riverbank known as riverbank protection works. Spur dike is also one of the river protection works. Several types of spur dike can be used for river bed protection, i.e., submerged and non-submerged, permeable and non-permeable, attracting deflecting and repelling. Submerged spur dikes are entirely water sink, and water flows over them, but non-submerged spur dikes are not entirely drowned, but some of the height is above water level. For the case of submerged spur dike, submergence ratio (s) plays an important role, which can be defined as submergence ratio, $s = \left(\frac{y-d}{y}\right)$. where, y is the flow depth and d is the height of the spur dike. Spur dikes built of stones, gravels, local soils, rocks are preamble and impermeable spur dikes made of steel timber, bamboo, or RCC. Keeping the view of shape spur dike may be classified as T-shape, L-shape and hockey shape. As per economic point of view, spur dikes are often constructed riprap with submerged design to operate during high flow. Spur dike Protects banks from erosion caused by flow concentration and trains the river along the desired course to reduce the flow concentration at the point of attack. Restoring fish habitat and mass transport processes throughout the channel (Elawady et al. 2001 and Pandey at el. 2022). The use of groyne is also done to limit coastal erosion due to wave action and long-shore currents (Raudkivi 1996).

The past studies show that the scour for spur dike has similarities with the abutment scour mechanism. The formation of the horse-shoe vortex or primary current formation at upstream of spur dike, while wake vortex and secondary current formation at downstream corner of the spur dike creates an interaction between the bed materials and flowing water. When the flow of water approaches to spur dike, one part accelerates and separates the main channel and the other part goes towards the side of the flume and forms an anti-clockwise eddy at the corner. Due to velocity difference, the downward dynamic pressure gradient forms and a force on spur dike also acts. Flow around spur dike is also categorized into three zones viz. main flow zone, wake zone and mixing zone (as can be seen in Fig. 1).

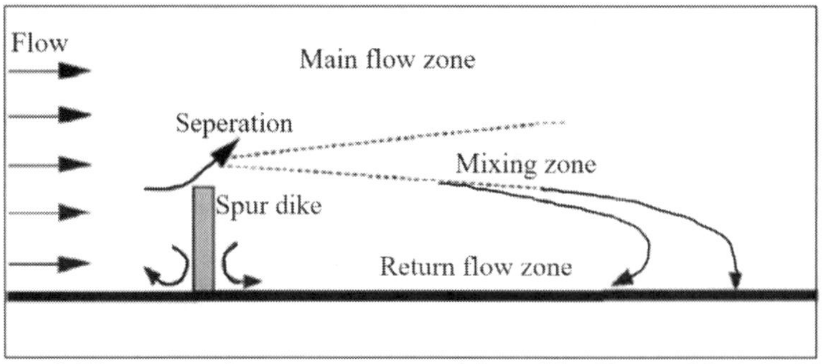

Figure 1. Flow separation processes around Spur dike (Source Masjedi; 2012)

For submerged spur dike case, the development of scour pit can be divided into three stages viz. initial stage, main scour stage and balance stage (Pandey et al. 2018). It has been observed by researchers that the most of scour is completed in first two stages (Kothyari et al. 2001; Lodhi et al. 2016 and Pandey et al. 2018, 2021). In the main scour stage, the intensity enhancement of submerged flow and horseshoe vortex increase the scour pit depth rapidly (Lodhi et al. 2016).

Various studies have done on the flow vicinity of non-submerged spur dikes (Pandey et al., 2018; Rajaratnam and Nwachukwu 1983 and Grade et al., 1961). Researchers examined the velocity field around the groyne-like structure in non-submerged conditions and found that the formation of horseshoe vortex surrounding the upstream and lateral side of piers (Melville and Raudkivi 1977 and Graf and Yulistiyanto 1998). Lodhi et al. (2016) stated that submerged spur dike has more applications than non-submerged like it reduces more erosion by reducing the adverse effects during floods. He also found that the submerged spur dike reduces more scour than the non-submerged spur dike. Elawady et al. (2001) checked the effect of spur dike's dimension on maximum scour depth and found that the maximum scour depth increases with the transverse length of spur dike. Fang et al. (2006) proposed a relationship for calculating the maximum scour depth around submerged spur dike. Kunhle et al. (2002) analyzed the scour hole dependency with time and used this approach to restore the pool for aquatic habitats. Researchers also found that the use of spur dike as an effective pool habitat as well as bed protection (Kunhle et al. 2002 and Fang et al. 2006). Since most of the studies have been carried out for non-submerged spur dike. Hence, this study aims to perform experiments around submerged spur dikes under submerged conditions and provide an approach to predict the maximum equilibrium scour depth around submerged spur dikes.

2. Experimental Set-Up and Methodology

All experiments were carried out in a 20 m long, 1 m wide and 0.30 m deep fixed bed rectangular channel, which was located in the Hydraulic Laboratory of Indian Institute of Technology Roorkee. The longitudinal slope of the flume was taken equal to 0.005. The flow rate in the flume was measured with the help of a sharp-crested weir, which was provided at the end of the flume. A 4 m long 1 m wide and 0.25 m deep working section of channel was selected for the experimental work. This working section of the flume was started from 12 m downstream to the flume entrance. Two impermeable spur dikes under submerged condition having transverse lengths, $l = 8$ cm and 13 cm were used as spur dike models. The effective height of submerged spur dikes, $d = 2$cm, 4cm, and 6cm were taken. All the spur dike models were fixed at an angle of 90° to the flume wall. Figures 2 (a-b) illustrate the sketch and photometric view of the experimental setup. Spur dike model was fixed at the centre of test section i.e. 14 m from the flume entrance. Total 24 experimental runs were conducted in this study. Maximum scour depth at equilibrium scour condition measured at the upstream nose of the dike. For all experiments, fine uniform fine sand having median diameter, $d_{50} = 0.27$ mm and geometric standard deviation $\sigma = (d_{84}/d_{16})^{0.5} = 1.23$ was used as bed materials. d_{84} and d_{16} are particle size at 84% finer and 16% finer, respectively

After attaining the preferable flow condition, scour depth at different time intervals was measured using a Vernier point gauge at the upstream and downstream nose of the spur dike. It was assumed that the equilibrium condition for all the experiments were reached approximately in 12 hours. It was observed that the maximum scour depths always occurred at upstream nose of spur dike. In Table 1, R1-R12 represents spur dike length l = 8 cm and R13-R24 represent spur dike length l = 13 cm. All experimental runs were performed under clear water scour condition i.e. $U/U_c<1$. U is the approach flow velocity and U_c is the threshold velocity of sand particles and was calculated using Lauchlan and Melville (2001)

$$\frac{U_c}{U_{c*}} = 5.75\log\left(\frac{y}{k_s}\right)+6 \qquad (1)$$

where, U_{*c} is critical threshold shear velocity, y is approach flow depth, k_s is equivalent roughness height, d is depth or height of spur dike, α is opening ratio and calculated as $\left(1-\frac{l\times b}{B\times y}\right)$, Q is flow rate, B is channel width and d_{se} is maximum scour depth at spur dike's nose.

Table 1 Maximum equilibrium scour depth data

Run no	l (cm)	d (cm)	Submergence ratio, $s=\left(\frac{y-d}{y}\right)$	y (cm)	α	Q(m³/s)	d_{se}
R1	8	2	0	2	0.92	0.0052	4.1
R2	8	2	0.5	4	0.96	0.0095	3.8
R3	8	2	0.667	6	0.97	0.0136	2.7
R4	8	2	0.75	8	0.98	0.0159	2.3
R5	8	4	0	4	0.92	0.0095	5
R6	8	4	0.333	6	0.95	0.0136	3.3
R7	8	4	0.5	8	0.96	0.0159	3.1
R8	8	4	0.6	10	0.97	0.0182	3.4
R9	8	6	0	6	0.92	0.0136	4.9
R10	8	6	0.25	8	0.94	0.0159	4.2
R11	8	6	0.4	10	0.95	0.0182	4.6
R12	8	6	0.5	12	0.96	0.02	5.2
R13	13	2	0	2	0.87	0.0052	4.7
R14	13	2	0.5	4	0.94	0.0095	4
R15	13	2	0.667	6	0.96	0.0136	3.2
R16	13	2	0.75	8	0.97	0.0159	2.6
R17	13	4	0	4	0.87	0.0095	5.2
R18	13	4	0.333	6	0.91	0.0136	3.8
R19	13	4	0.5	8	0.94	0.0159	3.4
R20	13	4	0.6	10	0.95	0.0182	3.7
R21	13	6	0	6	0.87	0.0136	5
R22	13	6	2	8	0.90	0.0159	4.4
R23	13	6	4	10	0.92	0.0182	5.1
R24	13	6	6	12	0.94	0.02	5.3

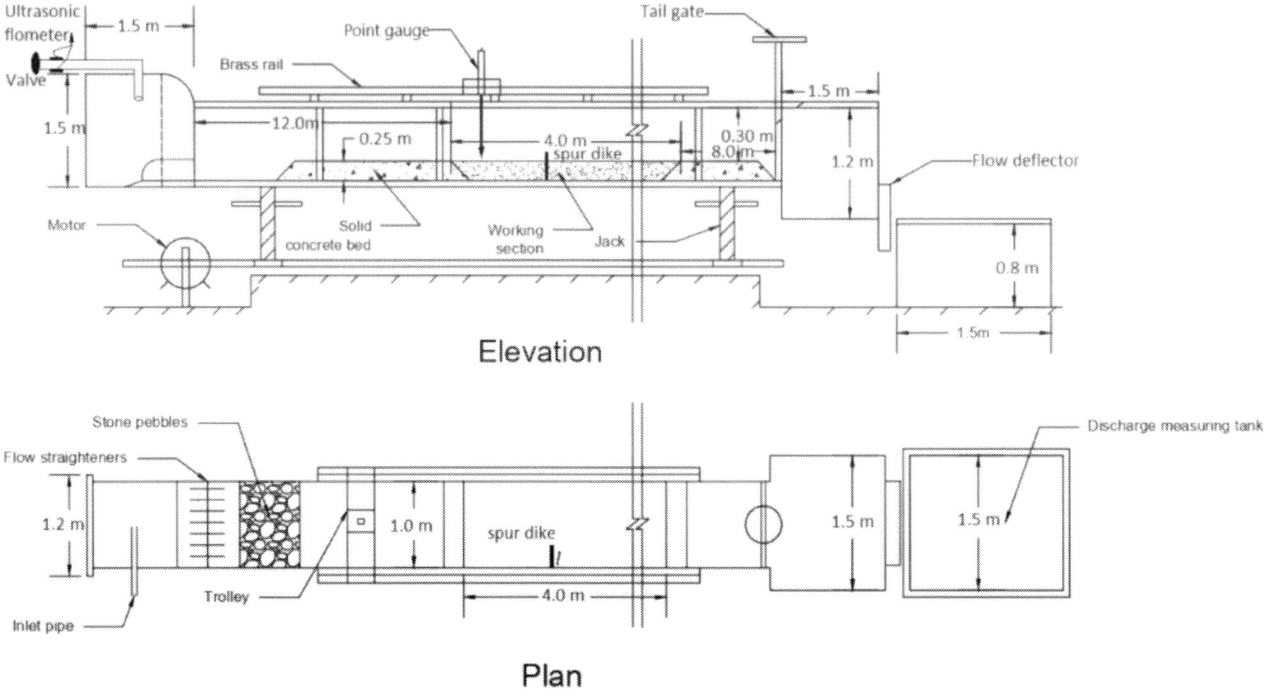

Figure 2 (a) Sketch diagram of experimental setup

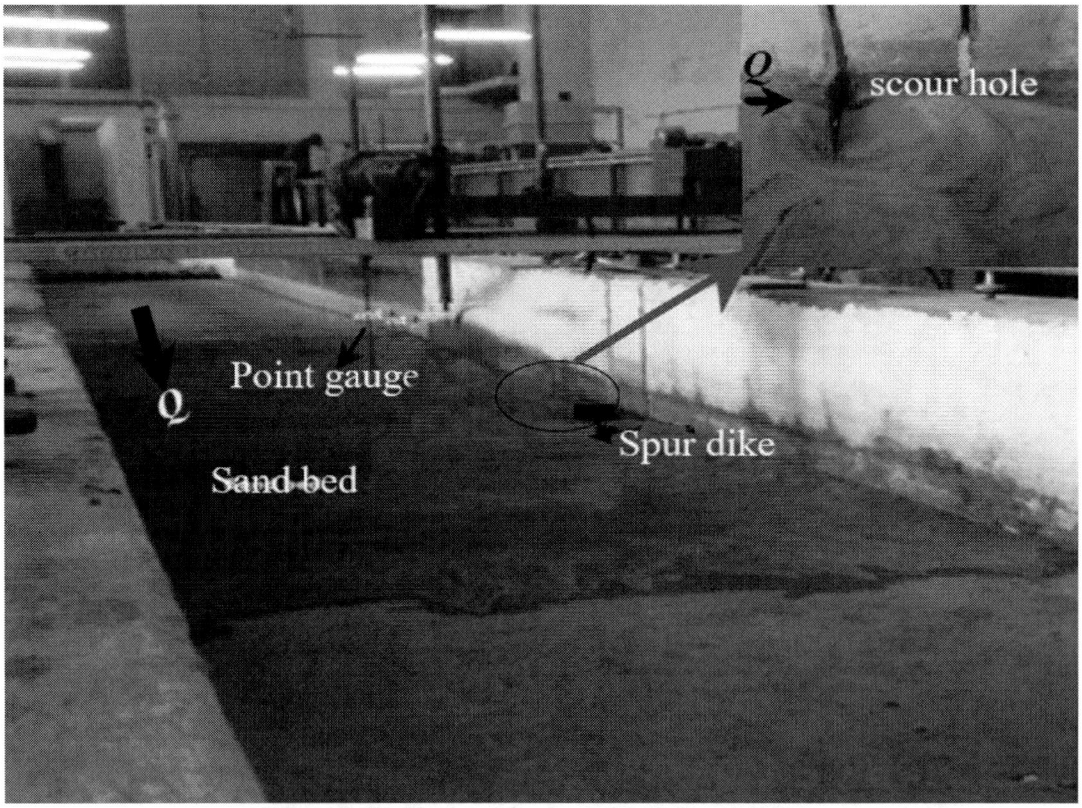

Figure 2. (b) Photometric view of experimental setup

3. Data Analysis, Results and Discussion

3.1 Scour Pattern around Submerged Spur Dikes

To find the location of maximum equilibrium scour depth, the scour hole must be observed using equilibrium scour hole pattern. Scour pattern is drawn by measuring the scour depth by taking X and Y grid points with the help of a point gauge having ±0.1 cm accuracy. The surfer software is used to analyze all the experimental data. A 3-D view of scour hole is drawn with the help of surfer and the scour depth is measured in the vicinity of flow. 3-D view of scour holes show that maximum scour depth always occurs at the upstream nose of the spur dike and eroded sediment deposits at downstream of the spur dike as can be seen in Figs. (3-4). Also, it can be clearly identified that the maximum scour depth increases with the increase in transverse length of the spur dike, and scour hole is much deeper for larger transverse length. It can also state that the maximum scour depth and volume of scour hole increase with the increase in discharge and flow depth.

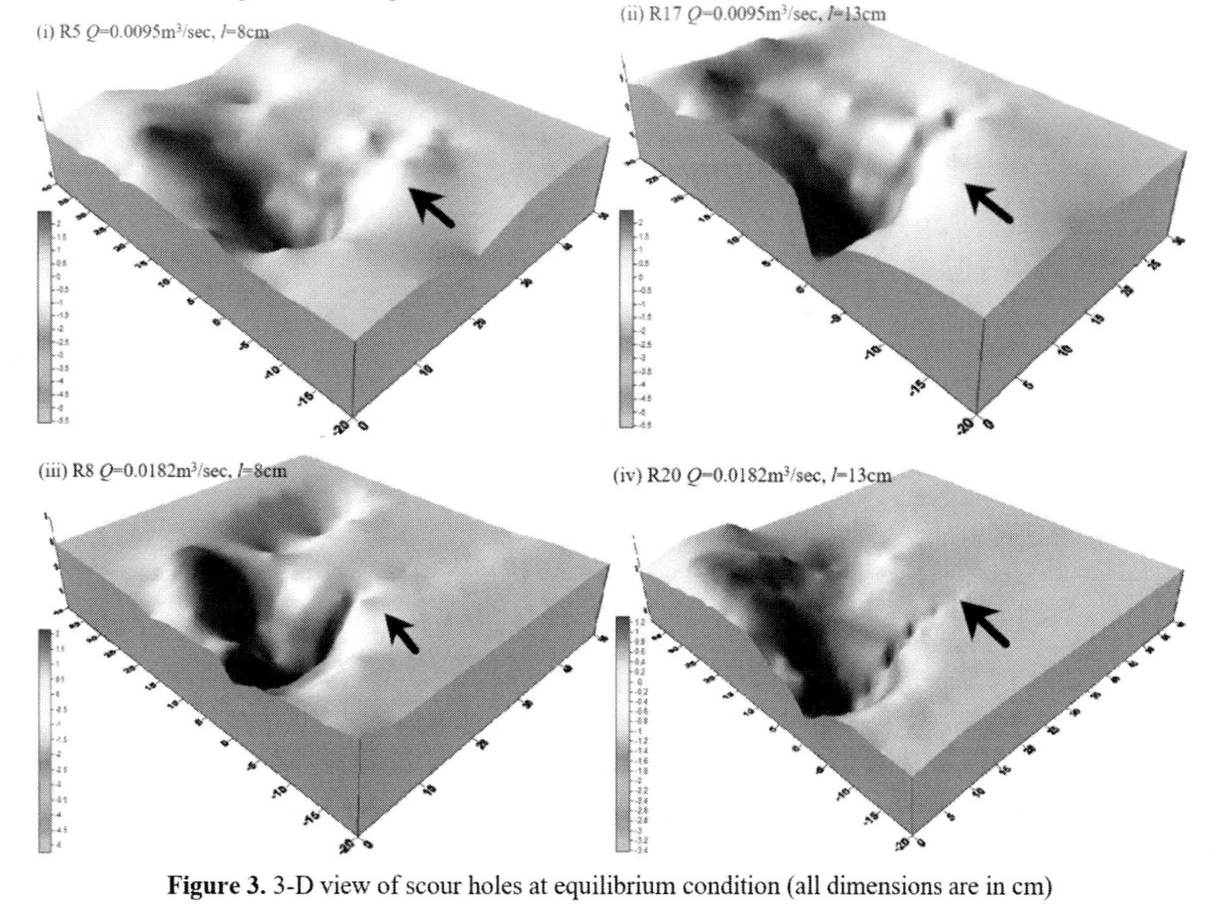

Figure 3. 3-D view of scour holes at equilibrium condition (all dimensions are in cm)

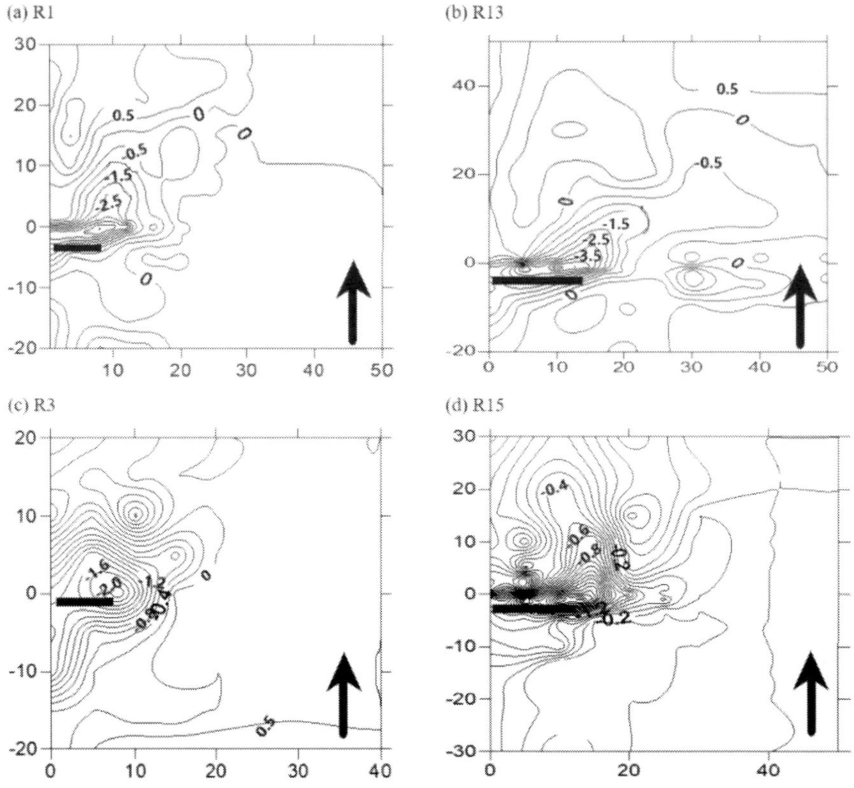

Figure 4. Scour depth contours at equilibrium scour condition

3.2 Effect of Froude Number on Maximum Scour Depth

A normalized scour depth (d_{se}/y) is used to check the dependency of scour depth on approach Froude number, as can be seen in Fig. (5). Using a constant transverse length of spur dike, we varied the depth of spur dike viz. 2 cm, 4 cm and 6 cm. A plot between normalized scour depth and (d_{se}/y) and Froude number is drawn, which shows maximum scour depth increases with an increase in Froude number. For a particular Froude number and discharge, maximum scour depth increases with spur dike height.

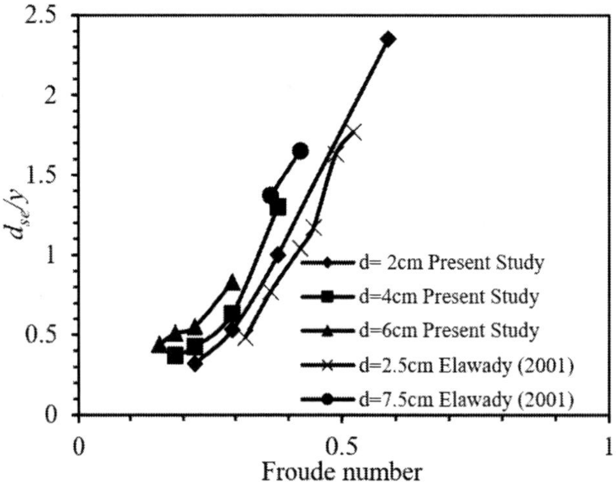

Figure 5. Effect of Froude number on maximum equilibrium scour depth using present and Eawady's et al. (2001) experimental data.

3.3 Impact of Opening Ratio on Maximum Scour Depth

Opening ratio α represents the unblocked area and is defined as $\left(1-\frac{l \times b}{B \times y}\right)$. Figure 6 shows scour depth increase with decrease in opening ratio or can say that scour depth increases with increase in spur dike area, as can be seen in Fig. (6). Larger the area of spur dike volume of scouring is also increasing and making it appropriate for aquatic habitats.

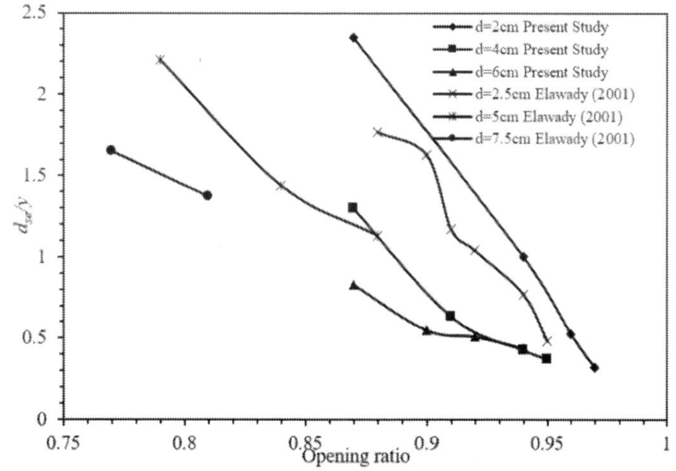

Figure 6. Effect of opening ratio on maximum equilibrium scour depth using present and Eawady's et al. (2001) experimental data

3.4 Effect of Submergence Ratio on Maximum Scour Depth

Submergence ratio (s) defines as $\left(\frac{y-d}{y}\right)$, which is mainly function of flow depth and spur dike height. Graphical representation shows the effect of the submergence ratio on maximum scour depth. Figure (7) shows maximum scour depth increases with decrease in submergence ratio. For same submergence ratio, scour depth is more for larger transverse length of spur dike. Similar results were obtained by Elawady et al. (2001)

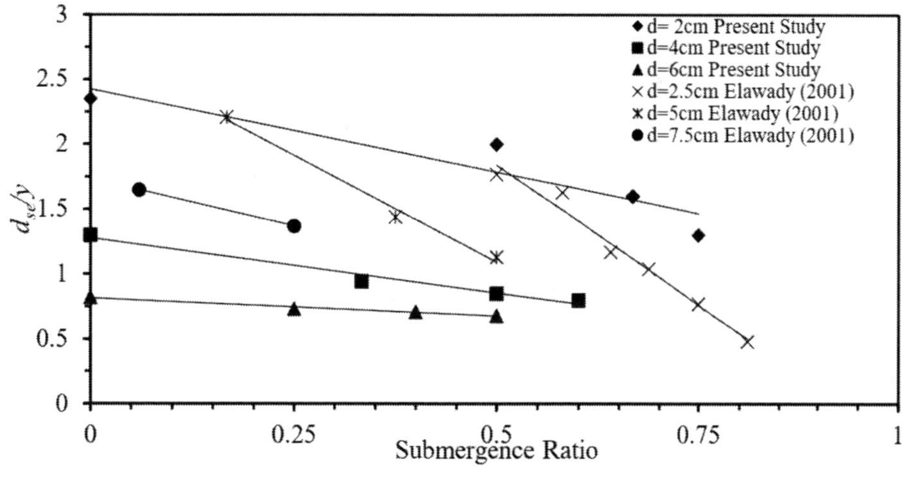

Figure 7. Effect of submergence ratio on maximum equilibrium scour depth using present and Eawady's et al. (2001) experimental data.

3.5 Dimensionless Scour Depth Approach

Hydraulic parameters and intrinsic properties are the prime factor for scouring. Scouring and deposition of bed material depend on channel's dimensional parameters such as channel width, bed material properties like median diameter of bed material and type of particles are used as bed material. Hydraulic parameters like discharge and water depth impart a significant role in scouring. Spur dike dimensions like spur dike height, spur dike width are essential to estimate the scour around the spur dike under submerged conditions. The condition of spur dike changes the whole approach, whether submerged or non-submerged. This study revolves around the maximum scour depth under submerged spur dike. Hence, as an important parameter, maximum scour depth's dependency on other-dimensional, hydraulic, and intrinsic parameter have been established here:

$$d_{se} = f\left(Q, B, l, y, d, d_{50}, s\right) \tag{2}$$

Eq. (2) can be re-written in dimensionless form as

$$\frac{d_{se}}{y} = f\left(F_r, \frac{l}{B}, \frac{d}{h}, \frac{d_{50}}{B}, s\right) \tag{3}$$

Using above functional relationship, Fang et al. (2006) proposed a maximum spur dike scour depth equation under submerged flow case, as given by Eq. (4).

$$\frac{d_{se}}{y} = 0.52F_r^{0.61}\left(\frac{d_{50}}{B}\right)^{-0.22}\left(\frac{A_d}{A}\right)^{0.65} + 0.394 \tag{4}$$

where, F_r is the Froude number $\left(F_r = \frac{U}{\sqrt{gy}}\right)$, A_d is the area of submerged spur dike and A is cross sectional area.

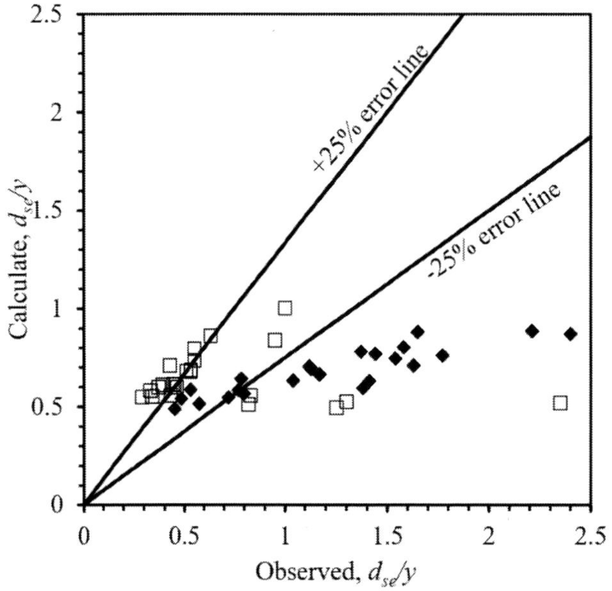

Figure 8. Comparison between computed and observed dimensionless scour depth for present data and data collected from Elawady et al. (2001) using Fang's et al. (2006) equation.

3.6 New Relationship for Dimensionless Scour Depth

Fang et al. (2006) derived a relationship for calculating the dimensionless scour depth at spur dike's nose, given by Eq. (2). First, the accuracy of Fang's et al. (2006) equation was checked using present experimental data and data collected from Elawady's et al. (2001) study. Graphically, it was found that Fang's et al. (2006) equation shows only 50% data within ±25% error. Most of their experimental data lied under ±25% errors. For present experimental data, one can easily identify that most of the datasets to be found outside the error band. Hence, a new relationship is proposed for calculating the dimensionless scour depth using present and previous experimental datasets. Multiple linear regression analysis is done to propose maximum scour depth equation using above influencing parameters viz. approach Froude number (F_r) and opening ration (α).

$$\frac{d_{se}}{y} = 3.2F_r^{1.42}\alpha^{-3.45} \tag{5}$$

The value of the coefficient of determination (R^2) for the presently derived equation is 0.96, much larger than Fang et al. (2006). The present equation is derived using present experimental data and data collected from Elawady's et al. (2001) experimental data. The coefficient of determination (R^2) for Fang's et al (2006) equation is 0.80, which we calculated using both present and Elawady's et al. (2001) experimental data. Figure (9) shows the comparison between the computed and observed maximum dimensionless scour depth and plot shows good agreements with observed dimensionless spur dike scour depths. More than 85% of datasets are to be found under fold lines, and only a few datasets under predict the scour depth, as can be seen in Fig. (9).

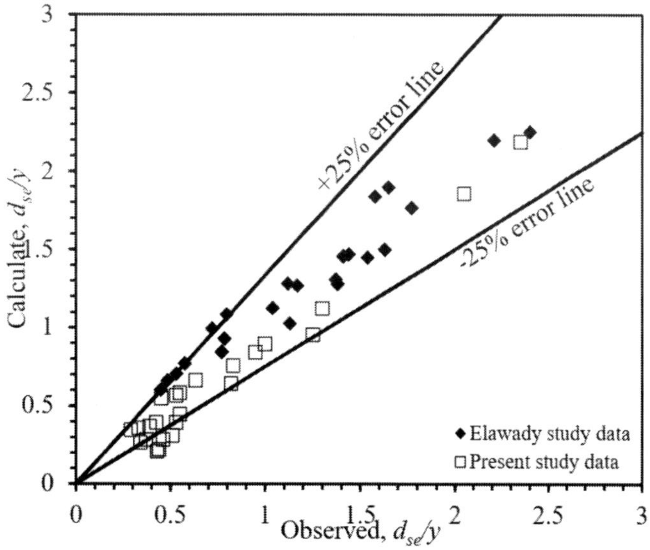

Figure 9. Comparison between computed and observed dimensionless scour depth for present data and data collected from Elawady et al. (2001) using presently proposed equation.

4. Conclusion

This experimental study presents new spur dike scour data obtained using a range of proper instrumentation (for measuring water and bed surface elevation and flow rate). Further, the use of these data for deriving a new equation for spur dike maximum scour depth calculation and compared to other existing equations. The dependency of maximum scour depth on some hydraulic and geometric control parameters is also discussed. Nonlinear regression analysis is done to derive a new equation using present and Elawady's et al. (2001) experimental datasets. Also, the present experimental data is compared with the previously proposed equation i.e., Fang's et al. (2006) equation. Fang's et al. (2006) equation is used for checking the present and previous experimental; datasets and a comparison is made using present and previous datasets. After completing all analyses, the following conclusions are made

- Maximum scour depth always occurs at the upstream nose of the submerged spur dike for all submergence conditions.
- Scour pattern around spur dike under submerged conditions is analyzed using scour depth contours and 3D view of scour holes.
- Maximum scour depth increases with an increase in Froude number, transverse length of spur dike, submergence ratio, and channel discharge.
- For the same submergence, maximum scour depth increases with spur dike height.
- A comparison between observed and calculated maximum scour depth using Fang's et al. (2006) equation was done and found that the 50% data are within ± 25%.
- A new relationship is proposed for calculating the dimensionless scour depth, which is a function of Froude number and opening ratio.

5 REFERENCES

Elawady, E., Michiue, M., and Hinokidani, O. (2001). "Movable bed scour around submerged spur-dikes." *Proceedings of Hyd.c Engg.* 45, 373-378.

Fang, D., Sui, J., Thring, R.W., and Zhang, H. (2006). "Impacts of dimension and slope of submerged spur dikes on local scour processes- an experimental study." *Int. J. of Sed. Res.* 21(2), 89-100.

Garde, R. J., Subramanya, K., and Nambudripad, K.D. (1961). "Study of scour around spur-dikes." *J. of the Hyd. Div.* 87(6), 23-37.

Graf, W.H., and Yulistiyanto, B. (1998). "Experiments on flow around a cylinder; the velocity and vorticity fields." *J. of Hyd. Res.* 36(4), 637-654.

Jain, R., Lodhi, A.S., Oliveto, G., and Pandey, M. (2021). "Influence of cohesion on scour at piers founded in clay–sand–gravel mixtures." *J.l of Irr. and Dra.e Engg.* 147(10), 04021046.

Kuhnle, R.A., Alonso, C.V., and Shields Jr, F.D. (2002). "Local scour associated with angled spur dikes." *J. of Hyd. Engg.* 128(12), 1087-1093.

Lodhi, A.S., Jain, R K., and Sharma, P.K. (2016). "Influence of cohesion on scour around submerged dike founded in clay–sand–gravel mixtures." *ISH J. of Hyd. Engg.* 22(1), 70-87.

Kothyari, U.C., and Ranga Raju, K.G. (2001). "Scour around spur dikes and bridge abutments." *J.l of Hyd. Res.* 39(4). 367-374.

Melville, B.W., and Raudkivi, A.J. (1977). "Flow characteristics in local scour at bridge piers." *J. of Hyd. Res.* 15(4). 373-380.

Pandey, M., Ahmad, Z., and Sharma, P.K. (2018). "Scour around impermeable spur dikes: a review." *ISH J. of Hyd. Engg.* 24(1). 25-44.

Pandey, M., Jamei, M., Karbasi, M., Ahmadianfar, I., and Chu, X. (2021). "Prediction of maximum scour depth near spur dikes in uniform bed sediment using stacked generalization ensemble tree-based frameworks." *Journal of Irr. and D. Engg.* 147(11), 04021050.

Pandey, M., Jamei, M., Ahmadianfar, I., Karbasi, M., Lodhi, A.S., and Chu, X. (2022). "Assessment of scouring around spur dike in cohesive sediment mixtures: A comparative study on three rigorous machine learning models." *J. of Hydrology*. 606, 127330.

Rajaratnam, N., and Nwachukwu, B.A. (1983). "Erosion near groyne-like structures." *J. of Hyd. Res.* 21(4), 277-287.

Raudkivi, A. J. (1996). "Permeable pile groins." *J. of Waterway, Port, Coastal, and Ocean Engg.* 122(6), 267-272.

DOI: https://doi.org/10.26077/5fff-2753

Experimental Evaluation of Drag Force on Different Shapes of Pontoons at Different Water Stream Velocities at Specified Loads and Submergence

Ms. R. B. Deogade[1], Smt. H. R. Khandagale[2], Shri. M. Someshwara[3]
[1, 2, 3]*Central Water and Power Research Station, Pune, India*
[1]*deogade_rb@cwprs.gov.in,* [2]*harshali.rk@cwprs.gov.in,* [3]*milan.s@cwprs.gov.in*

Abstract: *The ability to cross both natural and man-made obstacles have always been crucial for good mobility and is vitally important for various operations. The series of Pontoons are used as floating bridge to allow the transport vehicles and personnel over water and marshy grounds. Floating bridges are cost-effective solutions for crossing large bodies of water with unusual depth. Friction drag is observed when relative motion exists between pontoon and fluid. In view of this drag force is required to be evaluated correctly so that arrangement to fix the pontoon can be accurately designed considering the load and safety factors. For this required to carry out experiment to measure drag forces acting on the pontoons due to water resistance at various weights and water velocities. To study and collect numerous data, three pontoons of various shapes and materials were chosen for experiments. The drag force of pontoons is measured at 0, 5 and 10 degrees by add on various loads and submergence levels. During the experiments, a pontoon freely suspended to Rating Trolley when towed at a predetermine speed in straight open water tank, the drag force induced is measured by precise load cell. Experimental outcomes of all three pontoons for drag forces at 0, 5 and 10 degrees at various submergences and velocities will limit its application in specific velocity range and weight applied on it. Experimental analysis indicates the drag force is directly proportional to the frontal resistive area, water velocity and depth of submergence.*

Keywords: *Drag force, Pontoon, Submergence, Rating Trolley.*

1. Introduction

Transport links between rivers, flooded areas and offshore areas are dependent on bridges. Large water bodies with unusual depths can be crossed with floating bridges. A floating bridge is a more economical and conspicuous alternative, as the self-weight of bridge and vehicle loads are supported by the buoyant forces from pontoons. Furthermore, a floating bridge made for vehicles, bicycles and pedestrians to travel can be easily removed when relocation is necessary. The design of pontoons should be lightweight in order to reduce buoyancy requirements but to sustain for weight of transportation over it.

When a pontoon travels through fluid, some resistances are exerted by the water body namely frictional resistance, eddy resistance and wave resistance. The drag force is the resistance offered by the flowing fluid to the floating body. Considering this drag force evaluation is required to find out the hydrodynamic forces acting on the pontoon, which in turn determines the power required to drive the floating object or resistive force required to hold the object under induced drag force. Various methods have been proposed to calculate frictional drag. The physical experimental tests under different conditions applied on the pontoons are proved to be the most reliable test for desired design consideration. This study aims to determine the drag force acting on the different shape, size and self-weight pontoons at various velocities with predefined loads and mounting angle. Based on the experiment results the design of pontoons can be finalize as per suitable application.

2. Experimental Setup, Methodology and Test Procedure

2.1. Experimental Setup

Central Water and Power Research Station (CWPRS), Pune established facility in 1955 for testing of ship models as well as for calibration of current meters. A rating tank having dimensions of 228 m long, 3.66 m wide and 2.13 m deep in which the Current Meter Rating Trolley (CMRT) is placed which is self-propelled type and runs on two parallel straight steel rails 4.267 m apart, accurately aligned over the length of tank. The Rating Trolley is equipped with servo motors and electronic drive and state of the art technology to achieve precise speed in the range of 0.01 m/s to 7.5 m/s. A customized dedicated software is provided to operate the Rating Trolley for predefined speed profile and acquiring output of test device. Figure 1 shows the testing of ship model and current meter calibration facility at CWPRS. This infrastructure is used to conduct several experiments to evaluate the drag force on pontoons. [https:/www.cwprs.gov.in]

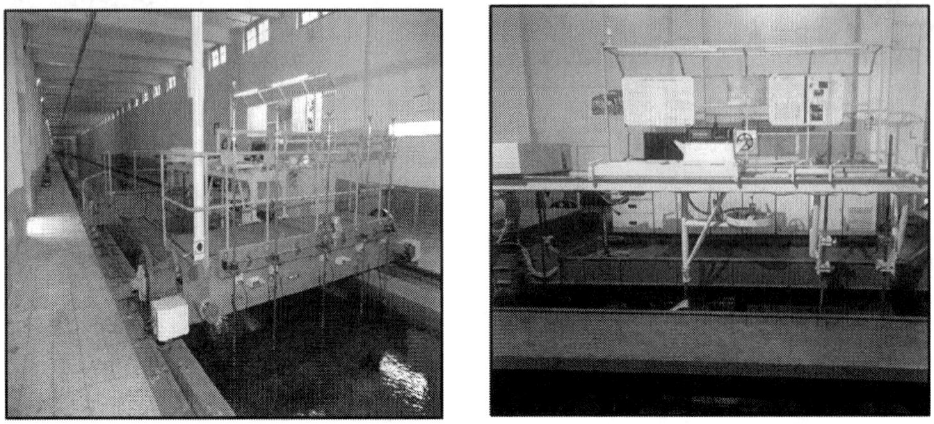

Figure 1. Ship model testing and current meter calibration facility at CWPRS

The drag force evaluation study was carried out on three types of pontoons. Table 1 describes the details of pontoons like overall dimensions of pontoons, self-weight of pontoons and velocity under which the drag of pontoons observed during this experiment. [CWPRS Technical Report No. 5999, February 2022]

Table 1. Details of Pontoons

Parameter	Pontoon-I	Pontoon-II	Pontoon-III
Length (m)	2.25	2.0	2.0
Width (m)	1.0	1.0	1.0
Depth (m)	0.245	0.3	0.28
Self-Weight (kg)	49.8	36.18	41.76
Velocity Range (m/s)	0.5 - 1.5	0.5 - 3.0	0.5 - 3.0

To conduct various experiments, the pontoons freely suspended on Rating Trolley with the help of compatible brackets for different angle as desired for 0°, 5° and 10°. Due to supporting brackets for mounting, the overall weight of pontoons increased. On pontoon-I, five different weights were selected for experiments and on each weight condition, six velocity profile were selected for only 0° angle condition to measure the drag force. On pontoon-II and pontoon-III, five and six different weights were selected respectively for experiments under 0°, 5° and 10° angle condition and on each weight and angle condition six velocity profiles were selected to measure the drag force. This is described briefly in Table 2.

Table 2. Detail Experiments profile for drag force measurement on Pontoons

Pontoon-I: Mounting Angle at 0°		
Sr. No.	weights of pontoon (kg)	Submergence of pontoon in water (m)
1	60.1 (Self-weight + bracket)	0.027
2	185.1 (60.1 + 125)	0.083
3	263.5 (60.1 + 203.4)	0.105
4	280.1 (60.1 + 220)	0.130
5	396.4 (60.1 + 336.3)	0.175

Pontoon-II:				
	Mounting Angle at 0°	Mounting Angle at 5°	Mounting Angle at 10°	
Sr. No.	weights of pontoon (kg)	weights of pontoon (kg)	weights of pontoon (kg)	Submergence of pontoon in water (m)
1	46.78 (Self-weight + bracket weight)	46.78 (Self-weight + bracket weight)	56.6 (Self-weight + bracket weight)	0.026, 0.026, 0.035 for 0°, 5° &10°respectively
2	51 (46.78 + 4.22)	51 (46.78 + 4.22)	--	0.041(for 0° & 5°)
3	117.45 (46.78 + 70.67)	117.45 (46.78 + 70.67)	117.45 (46.78 + 70.67)	0.080
4	182.14(46.78 + 135.36)	182.14(46.78 + 135.36)	182.14(46.78 + 135.36)	0.110
5	250.18 (46.78 + 203.4)	250.18 (46.78 + 203.4)	250.18 (46.78 + 203.4)	0.160
6	333.3 (46.78 + 286.52)	333.3 (46.78 + 286.52)	333.3 (46.78 + 286.52)	0.200

Pontoon-III:				
	Mounting Angle at 0°	Mounting Angle at 5°	Mounting Angle at 10°	
Sr. No.	weights of pontoon (kg)	weights of pontoon (kg)	weights of pontoon (kg)	Submergence of pontoon in water (m)
1	53.88 (Self-weight + bracket weight)	60. (Self-weight + bracket weight)	62.09 (Self-weight + bracket weight)	0.032, 0.036, 0.040 for 0°, 5° &10°respectively
2	76.2 (53.88 + 22.32)	76.2 (60.82 + 15.38)	76.2 (62.09 + 14.11)	0.045
3	117.45(53.88 + 63.57)	117.45 (60.82 + 56.63)	117.45 (62.09 + 55.36)	0.080
4	182.14(53.88 + 128.26)	182.14(60.82 + 121.32)	182.11(62.09 + 120.02)	0.110
5	250.18 (53.88 + 196.3)	250.18(60.82 + 189.36)	250.23(62.09 + 188.14)	0.160
6	339.36(53.88 + 285.48)	339.3(60.82 + 278.48)	339.41(62.09 + 277.32)	0.200

During the experiment concentrated weights were used as huge load over a small flat surface area of the pontoons. Hence flat steel plates and blocks with a higher density were chosen. After weighing all the pontoons, the Centre of Gravity (C.G.) of each pontoon was marked on the respective pontoon. Figure 2(a) shows dimensional details of three pontoons. All dimensions shown are in mm. The frontal resistive area in case of pontoon- I being rectangular in shape, longer side as 2250 mm was selected and for pontoon -II and pontoon-III nose end was selected for towing pontoons while conducting experiments.

Figure 2(b) shows a schematic diagram of experimental setup of drag force evaluation of pontoons. The mounting brackets were fixed such that, they should not be displaced under the induced water force during the experiment. The pontoon is freely suspended between two rods of Rating Trolley and the string is tied between C.G of the pontoon position and the load cell. The load cell is equipped with a spring to absorb the shocks developed while change in movement or velocity of water body transmitted to load cell through string.

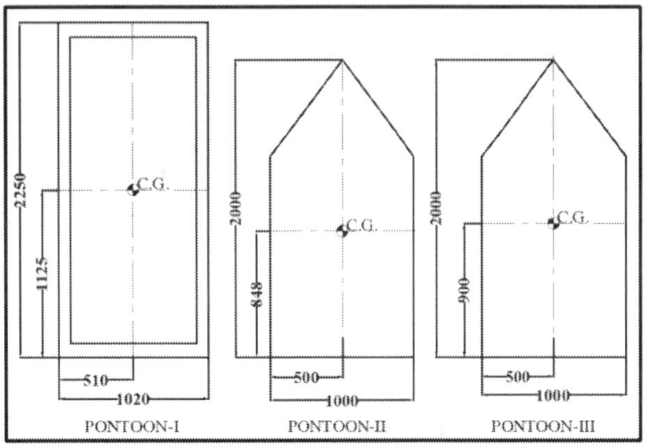

Figure 2. (a) Location of Centre of Gravity (C.G.) of Pontoons (All dimensions are in mm.)

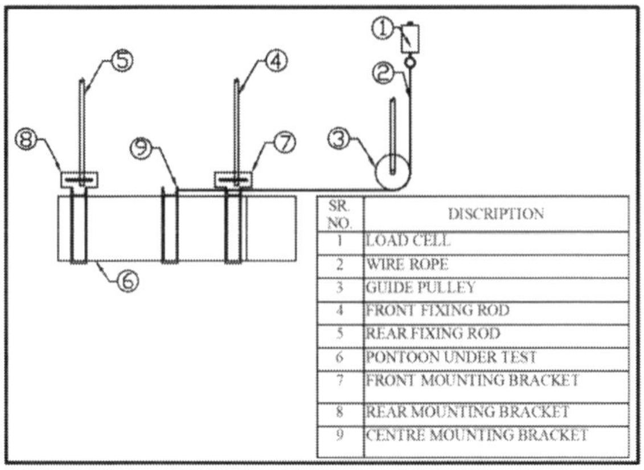

Figure 2. (b) Experimental Setup of Drag Force Evaluation

In Figure 3(a) the pontoon-III is mounted on Rating Trolley for 5° with self-weight condition and Figure 3(b) the pontoon-II under test for 5° with 250 kg total load condition.

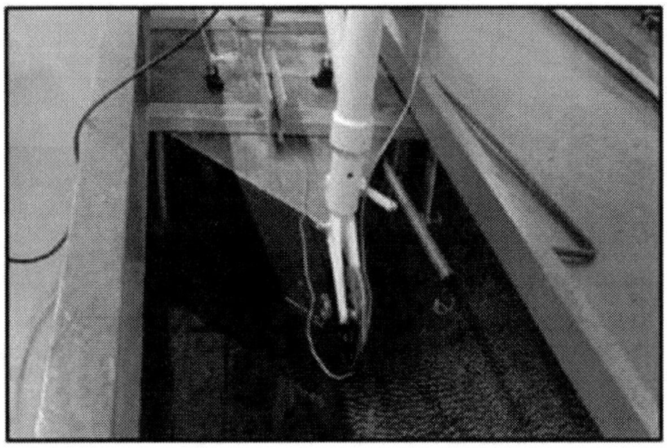

Figure 3. (a) Pontoon-III on CMRT at 5° with Self-weight

Figure 3. (b) Pontoon-II on CMRT at 5° with 250 kg

2.2. Experimental Methods and Test Procedure

During the experiment the following procedure was adopted for all pontoons at different depth of submergences, velocities and various angles as specified in the Table 2

1. The pontoon under test was freely suspended on the Rating Trolley with the help two rods and the string is connected between pontoon and load cell. The output cable of the load cell was connected to the Real Time Data Acquisition System (RTDAS) for acquiring the data. Figure 4 shows experimental set up along with load cell and data acquisition system.
2. Pontoon was immersed in the rating tank and the corresponding submergence of the pontoon was measured.
3. As the Rating Trolley was towed in the forward direction, the pontoon experiences backward drag force due to the relative motion of the water and the pontoon. The exerted drag can be directly measured by the load cell whenever stable movement of the pontoon is achieved. The Rating Trolley is move at different pre-defined velocities and corresponding drag is acquired.

4. During experiments, stilling of water was observed so that water in the rating tank will be tranquil before starting each run of a Rating Trolley.

Figure 4. Experimental setup along with load cell and data acquisition system

5. For addition of weights, the suitable pre-decided weights were placed on the pontoon. Considering C.G. of the pontoons, weights were added on the pontoon to keep it in balance condition.
6. The above procedure as mentioned in 1 to 5 was followed for all the specified weights when the pontoon axis was at 0^0, 5^0 and 10^0 with the towing direction of the Rating Trolley and corresponding drag forces were measured for specified velocities.

3. Results and Discussion

Table 3 shows drag forces induced on pontoon-I at various submergence levels, velocities and its graphical representation is shown in Figure 5.

Table 3. Drag force observed at various conditions on the pontoon-I

Total Weight on Pontoon Including Self-weight in kg	Submergence in m	Rating Trolley Velocity in m/s					
		0.5	0.6	0.8	1	1.2	1.5
		Drag Force in kg					
60.1 (Self-weight + bracket)	0.027	0.75	1.28	3.07	4.04	4.96	10.13
185.1 (60.1 + 125)	0.083	2.31	3.78	4.87	8.08	11.89	25.79
263.5 (60.1 + 203.4)	0.105	3.87	4.80	5.93	9.19	12.80	27.83
280.1 (60.1 + 220)	0.130	5.90	6.58	7.84	11.48	17.19	41.37
396.4 (60.1 + 336.3)	0.175	6.47	8.11	11.79	20.08	33.51	*
Maximum Drag Force		6.47	8.11	11.79	20.08	33.51	**41.37**
Minimum Drag Force		**0.75**	1.28	3.07	4.04	4.96	10.13

*It was observed during the experiment that for the speed of Rating Trolley 1.5 m/s and add-on weight on pontoon was approximately 396 kg, instability was noticed resulting in the tilting of the pontoon.

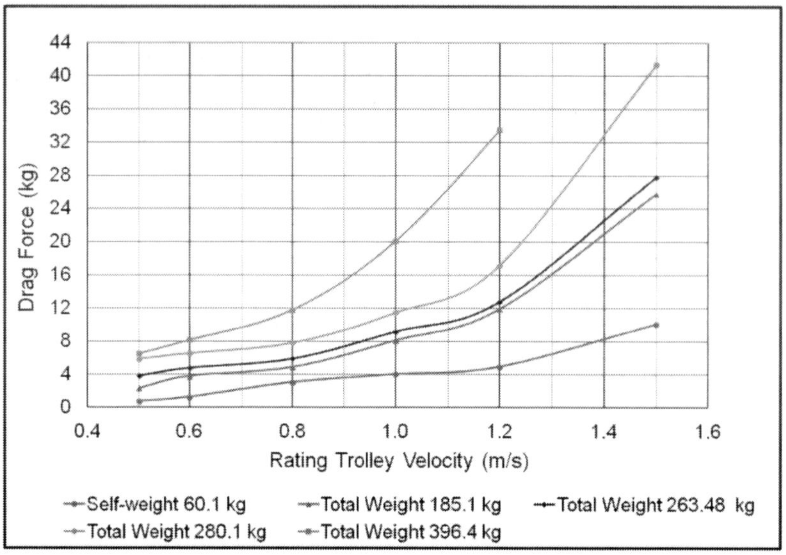

Figure 5. Drag force vs. Rating trolley velocity at various loads on pontoon I

From this experiments it was observed that the maximum drag force is induced 41.47 kg at 1.5 m/s velocity of the Rating Trolley for 280.1 kg load and 0.130 m submergence level, and the minimum drag force is induced 0.75 kg at 0.5 m/s velocity of the Rating Trolley for 60.1 kg load and 0.027 m submergence level on the pontoon-I at 0°.

Table 4, Table 5 and Table 6 show drag forces induced on pontoon-II when its axis was 0°, 5° and 10° with towing direction respectively. The graphical representation of the same are shown in Figure 6(a), Figure 6(b) and Figure 6(c) respectively.

Table 4. Drag force observed under various load on pontoon II when pontoon axis is at 0° with towing direction

Total Weight on Pontoon Including Self-weight in kg	Submergence in m	Rating Trolley Velocity in m/s					
		0.5	1	1.5	2	2.5	3
		Drag Force in kg					
46.78 (Self-weight + bracket)	0.026	0.500	1.703	4.280	9.542	19.270	34.098
51 (46.78 + 4.22)	0.041	0.710	1.907	4.311	9.640	20.146	35.379
117.45 (46.78 + 70.67)	0.080	0.839	1.949	6.539	13.427	29.399	52.579
182.14 (46.78 + 135.36)	0.110	0.957	3.860	11.759	23.388	46.161	63.829
250.18 (46.78 + 203.4)	0.160	1.387	5.589	12.925	32.064	63.559	81.392
333.3 (46.78 + 286.52)	0.200	1.877	6.498	20.204	47.055	82.981	108.696
Maximum Drag Force		1.877	6.498	20.204	47.055	82.981	**108.696**
Minimum Drag Force		**0.500**	1.703	4.280	9.542	19.270	34.098

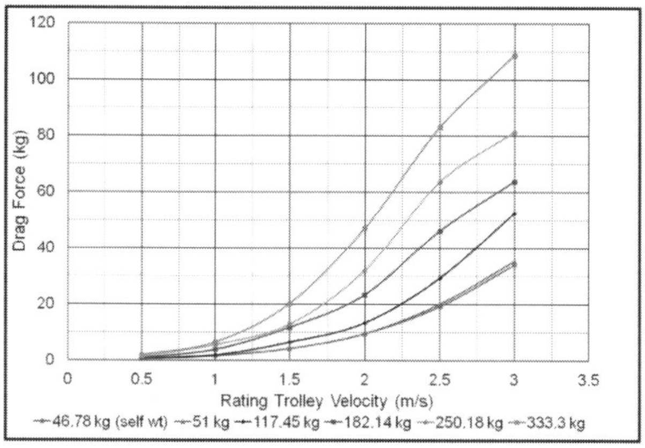

Figure 6. (a) Drag force vs. Rating trolley velocity at various loads on pontoon-II at 0°

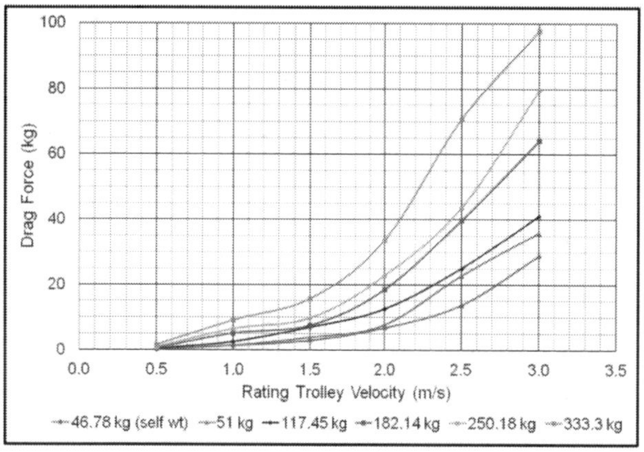

Figure 6. (b) Drag force vs. Rating trolley velocity at various loads on pontoon-II at 5°

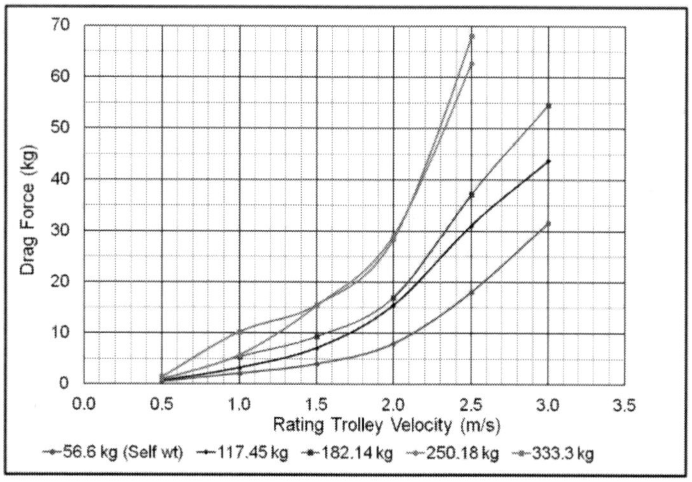

Figure 6. (c) Drag force vs. Rating trolley velocity at various loads on pontoon-II at 10°

Table 5. Drag force observed under various loads on pontoon-II when pontoon axis at 5° with towing direction

Total Weight on Pontoon Including Self-weight in kg	Submergence in m	Rating Trolley Velocity in m/s					
		0.5	1	1.5	2	2.5	3
		Drag Force in kg					
46.78 (Self-weight + bracket)	0.026	0.258	1.374	2.937	6.784	13.707	28.773
51 (46.78 + 4.22)	0.041	0.284	1.457	3.910	7.758	22.736	35.965
117.45 (46.78 + 70.67)	0.080	0.373	2.519	6.954	12.567	25.012	41.156
182.14 (46.78 + 135.36)	0.110	0.670	5.239	7.628	18.560	39.748	64.110
250.18 (46.78 + 203.4)	0.160	0.875	6.463	9.784	22.743	43.595	79.637
333.3 (46.78 + 286.52)	0.200	1.512	9.078	15.641	33.585	70.602	97.826
Maximum Drag Force		1.512	9.078	15.641	33.585	70.602	**97.826**
Minimum Drag Force		**0.258**	1.374	2.937	6.784	13.707	28.773

Table 6. Drag force observed under various loads on pontoon-II when pontoon axis at 10° with towing direction

Total Weight on Pontoon Including Self-weight in kg	Submergence in m	Rating Trolley Velocity in m/s					
		0.5	1	1.5	2	2.5	3
		Drag Force in kg					
56.6 (Self-weight + bracket)	0.035	0.548	2.190	4.067	8.054	18.110	31.616
117.45 (56.6 + 60.85)	0.080	0.593	3.243	7.126	15.490	31.142	43.734
182.14 (56.6 + 125.54)	0.110	0.798	5.403	9.365	17.008	37.175	54.561
250.18 (56.6 + 193.58)	0.160	0.938	5.723	15.323	29.093	62.870	*
333.3 (56.6 + 276.7)	0.200	1.434	10.287	15.548	28.464	68.054	*
Maximum Drag Force		1.434	10.287	15.548	29.093	**68.054**	54.561
Minimum Drag Force		**0.548**	2.190	4.067	8.054	18.110	31.616

*In case of pontoon- II, for the condition of experiment at higher loads and Rating Trolley velocity of 3 m/s, the water overtopped on the pontoon and the pontoon get tilted drastically which results in falling of weights even when tied properly. In view of this, beyond 250.18 kg at 3 m/s speed, it is not feasible to conduct experiments hence no data is obtained.

From the experiments it was observed that (i) the maximum drag force is induced 108.696 kg at 3 m/s velocity of the Rating Trolley for 333.3 kg load and 0.200 m submergence level and the minimum drag force is induced 0.500 kg at 0.5 m/s velocity of the Rating Trolley for 46.78 kg load 0.026 m submergence level on a pontoon-II at 0° (ii) The maximum drag force is induced 97.826 kg at 3 m/s velocity of the Rating Trolley for 333.3 kg load and 0.200 m submergence level and the minimum drag force is induced 0.258 kg at 0.5 m/s velocity of the Rating Trolley for 47.78 kg load and 0.026m submergence level on a pontoon-II at 5°. (iii) The maximum drag force is induced 68.054 kg at 2.5 m/s velocity of the Rating Trolley for 333.3 kg load and 0.2 m submergence level and the minimum drag force is induced 0.548 kg at 0.5 m/s velocity of the Rating Trolley for 56.6 kg load and 0.035 m submergence level on a pontoon-II at 10°. Table 7, Table 8 and Table 9 show drag forces induced on pontoon-III when its axis is 0°, 5° and 10° with towing direction respectively. The graphical representation of the same are shown in Figure 7(a), Figure 7(b) and Figure 7(c) respectively.

Table 7. Drag force observed under various loads on pontoon-III when pontoon axis at 0° with towing direction

Total Weight on Pontoon Including Self-weight in kg	Submergence in m	Rating Trolley Velocity in m/s					
		0.5	1	1.5	2	2.5	3
		Drag Force in kg					
53.88 (Self-weight + bracket)	0.032	0.300	1.582	4.817	9.289	19.861	23.364
76.2 (53.88 + 22.32)	0.045	0.450	2.100	5.544	14.338	21.867	37.902
117.45 (53.88 + 63.57)	0.080	0.618	3.200	10.122	23.719	33.018	53.432
182.14 (53.88 + 128.26)	0.110	0.678	5.836	11.152	22.879	45.368	65.556
250.18 (53.88 + 196.3)	0.160	1.287	6.339	16.621	28.747	53.977	81.150
339.36 (53.88 + 285.48)	0.200	1.632	6.665	18.669	49.535	121.231	128.479
Maximum Drag Force		1.632	6.665	18.669	49.535	121.231	**128.479**
Minimum Drag Force		**0.300**	1.582	4.817	9.289	19.861	23.364

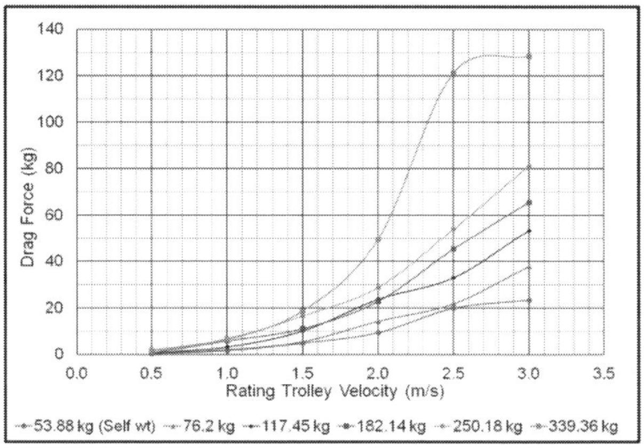

Figure 7. (a) Drag force vs. Rating trolley velocity at various loads on pontoon-III at 0°

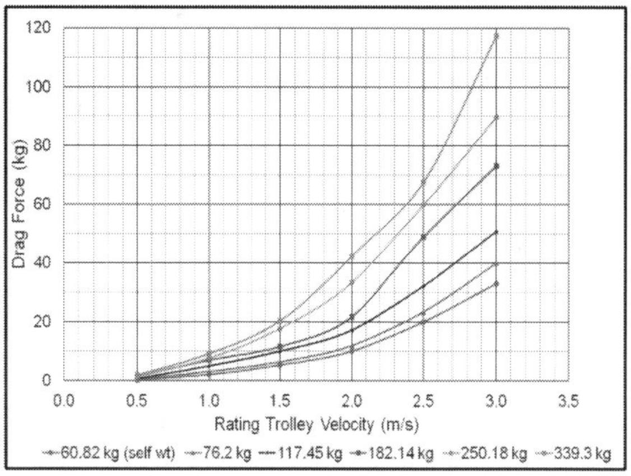

Figure 7. (b) Drag force vs. Rating trolley velocity at various loads on pontoon-III at 5°

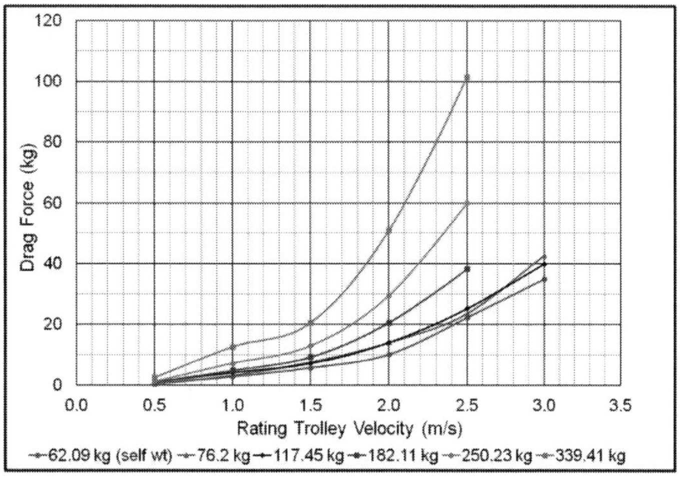

Figure 7. (c) Drag force vs. Rating trolley velocity at various loads on pontoon-III at 10°

Table 8. Drag force observed under various cases on pontoon-III when pontoon axis at 5° with towing direction

Total Weight on Pontoon Including Self-weight in kg	Submergence in m	Rating Trolley Velocity in m/s					
		0.5	1	1.5	2	2.5	3
		Drag Force in kg					
60.82 (Self-weight + bracket)	0.036	0.215	2.011	5.242	10.054	20.074	33.161
76.2 (60.82 + 15.38)	0.045	0.412	3.034	6.300	11.935	23.451	40.002
117.45 (60.82 + 56.63)	0.080	0.646	4.961	9.974	17.223	32.238	50.870
182.14 (60.82 + 121.32)	0.110	1.482	6.976	11.617	21.846	49.023	73.206
250.18 (60.82 + 189.36)	0.160	1.561	7.585	17.672	33.469	59.936	89.799
339.3 (60.82 + 278.48)	0.200	1.841	9.169	20.293	42.434	67.665	117.482
Maximum Drag Force		1.841	9.169	20.293	42.434	67.665	**117.482**
Minimum Drag Force		**0.215**	2.011	5.242	10.054	20.074	33.161

Table 9. Drag force observed under various loads on pontoon-III when pontoon axis at 10° with towing direction

Total Weight on Pontoon Including Self-weight in kg	Submergence in m	Rating Trolley Velocity in m/s					
		0.5	1	1.5	2	2.5	3
		Drag Force in kg					
62.09 (Self-weight + bracket)	0.040	0.381	2.730	5.636	10.086	22.266	34.913
76.2 (62.09 + 14.11)	0.045	0.479	3.217	7.607	14.014	23.590	42.616
117.45 (62.09 + 55.36)	0.080	1.026	4.169	7.198	14.081	25.265	39.937
182.11 (62.09 + 120.02)	0.110	1.030	4.832	9.298	20.626	38.315	*
250.23 (62.09 + 188.14)	0.160	1.068	7.131	12.916	29.420	60.001	*
339.41 (62.09 + 277.32)	0.200	2.352	12.560	20.444	50.781	101.339	*
Maximum Drag Force		2.352	12.560	20.444	50.781	**101.339**	42.616
Minimum Drag Force		**0.381**	2.730	5.636	10.086	22.266	34.913

*- In case of pontoon-III, at higher loads and 3 m/s speed, the water overtopped on the pontoon and it tilted drastically which resulted in falling of weights even when tied properly. Hence under 182.11 kg, 250.23 kg, and 339.41 kg at 3 m/s speed no data was obtained.

From the experiments it was observed that (i) the maximum drag force is induced 128.479 kg at 3 m/s velocity of the Rating Trolley velocity for 339.36 kg load and 0.200 m submergence level and the minimum drag force is induced 0.3 kg at 0.5 m/s velocity of the Rating Trolley for 53.88 kg load and 0.032 m submergence level on a pontoon-III at 0°.(ii) The maximum drag force is induced 117.482 kg at 3 m/s velocity of the Rating Trolley for at 339.3 kg load and 0.200 m submergence level and the minimum drag force is induced 0.215 kg at 0.5 m/s velocity of the Rating Trolley for 60.82 kg load and 0.036 m submergence level on a pontoon-III at 5° (iii) The maximum drag force is induced 101.339 kg at 2.5 m/s velocity of the Rating Trolley for the 339.41 kg load and 0.02 m submergence level and the minimum drag force is induced 0.381 kg at 0.5 m/s velocity of the Rating Trolley for 62.09 kg and 0.040 m submergence level on a pontoon-III at 10°

4. Conclusion

- The drag force measurement is varied in direct relation with the water stream velocity, load carrying and submergence level at 0 degree angle of pontoons.
- Recommendation for the design of the Pontoon I for the maximum permissible weight is approximately 396 kg for the velocity 1.2 m/s and 280kg for 1.5 m/s the velocity.
- Considering the results of experiments while designing of Pontoon II, at 0 degree the maximum permissible weight is approximately 333 kg for the velocity 3.0 m/s, at 5 degree the maximum permissible weight is approximately 333 kg for the velocity 3.0 m/s, and at 10 degree the maximum permissible weight is approximately 182 kg for velocity 3.0 m/s or 333 kg for velocity 2.5 m/s. These values are more important while designing pontoons and selecting during its site application.
- Considering the results of experiments while designing of Pontoon III, at 0 degree the maximum permissible weight is approximately 339 kg for the velocity 3.0 m/s, pontoon II at 5 degree the maximum permissible weight is approximately 339 kg for the velocity 3.0 m/s, and Pontoon II at 10 degree the maximum permissible weight is approximately 182 kg for velocity 3.0 m/s or 101 kg for velocity 2.5 m/s. These values are more important while designing pontoons and selecting during its site application.

5. ACKNOWLEDGEMENT

We wish to express our deep sense of gratitude to Dr. R. S. Kankara, Director, CW&PRS, for constant encouragement and valuable suggestions.

6. REFERENCES

https:/www.cwprs.gov.in and CWPRS Technical Report No. 5999, February 2022

DOI: https://doi.org/10.26077/7527-742f

Experimental optimization of a gravel trap in a gallery

S. Erpicum[1], N. Claude[2], B. Dewals[1], P. Archambeau[1] & M. Pirotton[1]
[1]*Liege University – Hydraulics in Environmental and Civil Engineering Research Group, Liege, Belgium*
[2]*Electricité de France – Hydro Engineering Center, La Motte Servolex, France*
E-mail: s.erpicum@uliege.be

Abstract: *A partly unlined pressurized gallery supplies water to a hydropower plant. Since rocks and gravel may detach from the gallery walls, two gravel traps have been built on the gallery invert, downstream of the unlined sections. During gallery and powerplant inspections, it appears that the gravel traps are empty while gravel is found in the tail channel downstream of the turbines. Clear evidence of turbine damaging by solids impact is also observed. To understand why the gravel traps are ineffective and to define solutions to be implemented on site to increase their trapping efficiency, a 1:7.7 scale physical model of the gravel traps has been built and operated. The tests clearly demonstrated that the short-term trapping efficiency of the gravel traps is good since 98% of the gravel arriving upstream falls into the traps. However, the gravel traps are ineffective in keeping the gravel over long periods. This is due to two effects. In pressurized conditions, helicoidal vertical currents developing at the downstream extremities of the trap and turbulent velocity fluctuations extract the gravel from the trap. In addition, during gallery dewatering (for maintenance and inspection), temporary free surface flow can transport gravel downstream of the trap. Non-structural solutions to the problem have been identified, optimized and validated on the scale model. They consist in creating above the traps a pseudo-bottom aligned with the gallery invert, using horizontal bars placed transversally to the flow, and in compartmenting the traps with vertical walls. The paper presents in detail the physical model and all the tests and measurements performed to quantify the problem and assess the efficiency of the modification solutions.*

Keywords: *Hydropower, Physical modeling, Rock trap, Scale modeling, Trapping efficiency, Unlined tunnel.*

1. Introduction

Removing solid particles entrained by water flowing through hydropower plants enables to prevent damages to turbines and then decreases maintenance and reparation requirements. Besides floating debris that are stopped at inlet using trash racks, sediments transported in suspension or on the bed are removed using a sedimentation process favored in so-called desilting basins, sand traps, gravel traps or rock traps, depending on the type and size of the solid particles.

This paper focuses on two gravel traps located in an around 6-km long pressurized gallery supplying water to a hydropower plant. Since the gallery is partly unlined, rocks and gravel may detach from the gallery walls. Traps were then built downstream of these sections, on the gallery bottom. They simply consist in prismatic volumes created in the gallery invert, 0.3 or 0.45 m deep and with varied horizontal dimensions ranging from 1.2 to 3 m in width and 2.5 to 7.65 m in length.

The circular gallery diameter is 3 m in lined sections and around 3.5 m in unlined sections. The equipment discharge is 15 m^3/s (average flow velocity of 2.1 m/s). The gallery slope in the sections where the gravel traps are located varies from 0.64 to 4.77 %. The gallery is linear except two sections curved with a 500 and 800 m radius. While the gallery operates in pressurized flow conditions, it experiences free surface flow during dewatering process before inspection. Dewatering process is very slow, with a pressure gradient limited to 1.05 m/h, but leakages are prone to create continuous discharge with free surface flow in the gallery, estimated from 0.05 to 0.2 m^3/s.

During gallery and powerplant inspections, it appears that the gravel traps are almost empty, while 20 to 50 mm diameter gravel was found in the tail channel downstream of the turbines. Decametric blocks were also found in the gallery, upstream of the gravel traps. The volume of gravel found downstream of the turbines is small (a few liters per year) but clear evidence of turbine damaging by solids impact is observed. Gravel traps seem thus unable to block gravel transported by the flow.

The objective of the study depicted in this paper was to understand why the gravel traps are ineffective and to define technically and economically feasible solutions to be implemented on site to increase their trapping efficiency for gravel from 10 to 50 mm (density equals 2.65).

While pressurized rock and gravel traps have been constructed in many hydropower plants, scientific literature on the topic is limited. This is attributed to the narrowness of the topic and to the lack of research within the last 30 years (Havrevoll et al., 2021). Literature focuses on scale modeling research carried out in the framework of specific sites studies in the US (Mattimoe et al., 1964) and Australia (Griffiths and Brett, 1974). Recently, a similar study was depicted by Hachem (2009) for a site in France. These studies provide tips for design of pressurized rock traps that have proved to be efficient on site. These tips can be summarized as follow, leading to the CELSEP trap concept (Mattimoe et al., 1964; Havrevoll et al., 2021): rock trap should consist in trapping cells on the gallery invert, free of any current, and with a sharp geometry (gradual flow expansion must be avoided). These cells should be covered with a horizontal staging at the level of the gallery invert, made of a succession of open sections, horizontal widely spaced round bars and solid plates. The cover aims at maintaining a uniform gallery cross section over the cells and at separating the flow in the gallery from the water in the cells. It should create as few turbulence as possible, since turbulent velocity fluctuations in the traps were shown to cause solid particles ejection.

The existing design of the rock traps depicted in this paper does not respect the above-mentioned guidelines. In particular, the traps are not divided in cells and there is no cover. This may explain why they are ineffective. To deepen the understanding of the rock traps behavior and to test the ability of the design recommended in the literature to improve their trapping efficiency, a scale physical model study of the gravel traps has been carried out. The model is depicted in section 2 while the results of the tests are detailed and discussed in section 3.

2. Scale model characteristics

A scale physical model of both gravel traps has been implemented at the Engineering Hydraulics Laboratory of Liege University. The model included two parallel Perspex pipes featuring linear sections of the gallery, connected upstream and downstream to a rectangular reservoir (Fig. 1a). One of the gravel traps was included in each pipe (Fig. 1b). Water was pumped in the upstream reservoir and discharged using a perforated pipe. A repartition screen separated the perforated pipe from the gallery inlets (Fig. 1c). A plate was used to obstruct separately each gallery inlets depending on the trap to be tested. In the downstream reservoir, 4 morning glories controlled the water level in the system and evacuated the pumped discharge. Morning glories level has been adjusted to operate both gallery in pressurized conditions and to avoid vortex formation in the upstream reservoir. A low-level pipes and valves system in the downstream reservoir enabled to lower the water level and to operate the galleries in free surface flow conditions to model dewatering operation. A fixed 40 mm diameter PVC pipe was used to inject solid particles directly into the galleries' upstream extremity (Fig. 1c).

Figure 1. Main components of the 1:7.7 scale model of the gravel traps. a) General view of the model; b) downstream gravel trap and c) upstream reservoir.

Scale models for pressurized water flows must respect Reynolds similitude for full dynamic similarity. To model free surface flows such as during the dewatering process, Froude similitude must be respected. Both dimensionless numbers can only be conserved at full scale (Ettema et al., 2000). The model has been built at the largest geometric scale factor (1:7.7) that was technically and economically reasonable. The scale factor has been tuned to get a gallery diameter on the scale model (390 mm) in agreement with standard commercial Perspex pipes inner diameter and then drastically ease construction. Considering Froude similitude to define the velocity scale, Reynolds number on the scale model in pressurized configuration was on the order of 3 10^5, while it is 6.4 10^6 on the prototype. In both cases, the flow is in the turbulence transition zone since walls are relatively smooth (relative roughness is in the order of 3 10^{-5} on the model and 3 10^{-4} on the prototype). Then, it has been decided to scale the discharge considering Froude similitude.

Regarding gravel scaling, conservation of Shields number for grain Reynolds numbers higher than 200 means that sediment size can be scaled according to model geometric scale if sediment density is the same in model and prototype. The same relation is found when considering settling velocity of solid particles larger than about 2 mm diameter, since it is proportional to the square root of the grain size. Sizing 10 to 50 mm solid particles according to the selected scale factor gives model grains ranging from 1.3 to 6.5 mm. Except for the lowest diameters, such a range respects the above-mentioned criteria on grain Reynolds numbers and grain size.

The water discharge in the model (91.2 l/s for plant equipment discharge – 1.2 to 0.3 l/s for leakage discharge during dewatering) was measured using either an electromagnetic dischargemeter Siemens MagFlow (accuracy of 0.5% FS) or volumetric gauging. 1 MHz UVP probes (accuracy of 1%) were used to measure flow velocity in the gravel traps. Sieves were used to grade solid particles before weighting them with a precision balance. All the solid particles were carefully dried before weighting, before and after the tests.

3. Results and discussion

Only the results gained on the downstream trap are detailed in the following. Similar results have been gained with the upstream trap, despite a more complex geometry.

All the tests with gravel used a homogenous mix of particles covering the whole range of diameter from 1.3 to 6 mm at model scale (10 to 46 mm at prototype scale).

The first tests focused on the existing geometry of the traps. They are depicted in section 3.1. Once the reasons for poor performance of the traps have been understood, remediation solutions have been tested, with the objective to avoid too much important modifications of the traps, i.e. to avoid increasing the traps volume (Section 3.2).

3.1. Existing geometry and problem characterization

As a first step, 1 kg of gravel (equivalent to 0.3 m^3 at prototype scale, i.e. 28% of the downstream trap volume) was injected in the model with a water discharge equivalent to 15 m^3/s (design discharge) (test A). The test was stopped once all the gravel has been injected in the model, i.e. it lasted only a few minutes. It has been repeated twice. 98.2% (standard deviation SD = 1.4%) of the injected gravel mass was found in the trap at the end of the test. It was located in the upstream part of the trap (Fig. 2). No gravel was found in the gallery; the missing gravel was in the downstream reservoir. Only small size gravel was not trapped. Visual observation during the tests showed that this gravel did not pass over the trap but travelled along the cat wall on each side of the trap or on gallery banks. These first results showed that the existing traps can stop most of the gravel transported in the gallery.

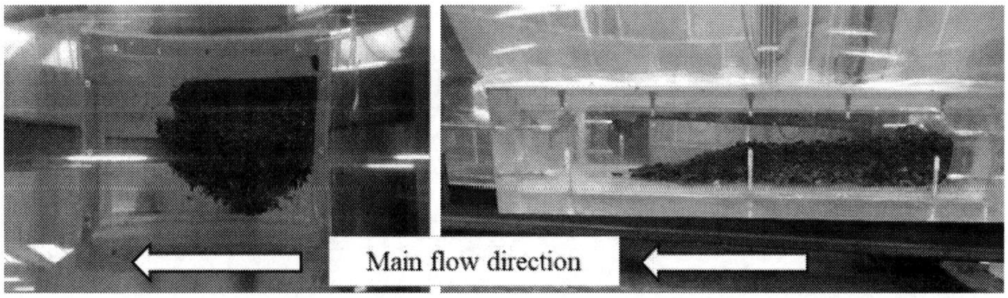

Figure 2. Gravel deposit in the trap during tests A.

In a second step, longer tests (tests B) have been carried out considering the same quantity of gravel and injection time but operating the model at constant water discharge during 1 or 4 hours (at model scale) after gravel injection. While the gravel was trapped shortly after injection, it moved slowly downstream once in the trap (Fig. 3). When grains reached the downstream extremity of the trap, vertical helicoidal currents in the edges of the cavity (Fig. 4) uplifted them out of the trap. They were then transported downstream. After 1 h, 96.9% of the gravel was still in the trap. After 4 h, it was 80.7%. These results showed that, if the existing traps can capture the gravel transported in the gallery, this trapping is not efficient over long periods of time. However, some residual deposits were stable in the upstream part of the trap, while such deposits have not been observed in the prototype.

These results are fully consistent with the findings of Mattimoe et al. (1964) related to a 1/12 scale model study of a horseshoe gallery of similar diameter (4.2 m), with the same mean water flow velocity (2.1 m/s) and same range of solid particles diameter as the present test case.

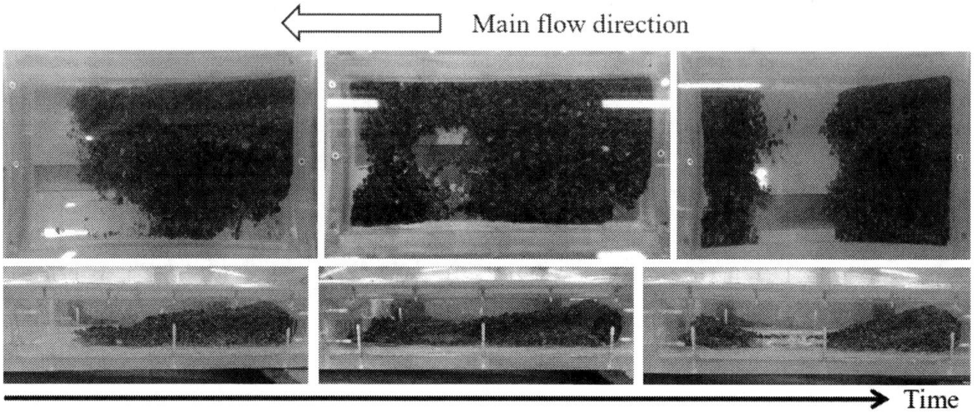

Figure 3. Slow motion of gravel deposit in the trap during tests B. Time increases from left to right.

Figure 4. Helicoidal currents with gravel ejection on both downstream edges of the trap during tests B.

To document flow conditions during the gallery dewatering process, a third type of tests has been done (tests C). It started similarly to the tests A and B, but once the gravel has been injected, the water discharge was reduced and the gallery was emptied slowly to reproduce a free surface flow with a water discharge varying from 0.05 to 0.2 m^3/s, i.e. in the range of the residual discharges suspected to take place along the gallery because of leakages. When free surface flow took place in the gallery, a hydraulic jump or jet impact (depending on discharge) occurred in the trap, pushing gravel in the downstream section, and then eventually transporting them in the downstream gallery (Fig. 5). These free surface flows were able to transport most of the gravel laying at the trap entrance. This may explain why stable deposits showed on figure 3 were not observed in the prototype during gallery inspections. Since the residual gravel at the end of the dewatering test was in the downstream section of the trap, it was prone to be ejected out of the trap by the helicoidal currents once the gallery was operated again in pressurized conditions.

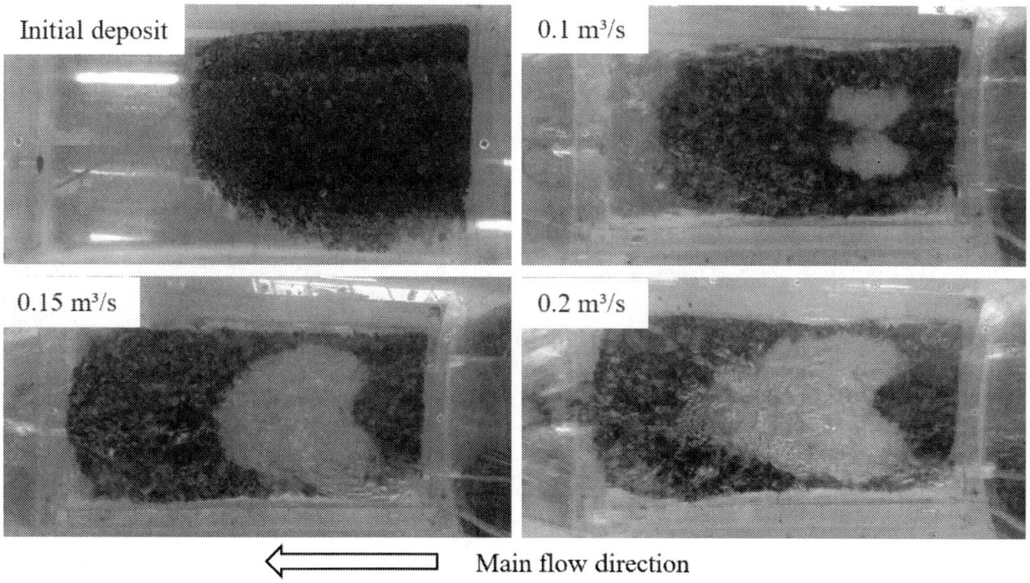

Figure 5. Evolution of gravel deposit in the trap during tests C, depending on leakages discharge.

The tests performed with the existing configuration showed that the problem didn't lie in the capacity of the traps to stop gravel transported in the gallery but was in the capacity of the trapped gravel to stay inside the traps over long periods, either because of turbulence effects during normal pressurized operation conditions or when free surface flow took place in the gallery due to leakages after dewatering. While the problem of free surface flow has not been documented yet in the literature, the inability of simple prismatic traps without cover created in the gallery invert to keep gravel along time was shown 60 years ago by Mattimoe et al. (1964) on a project very similar to the one considered in this study.

3.2. Modified configurations and solution

The main difference between the geometry of the traps considered in this study and the one recommended by Mattimoe et al. (1964) is the absence of trap cover aiming at maintaining a uniform gallery cross section over the traps and at separating the flow in the gallery from the water in the traps. Since the traps have a short length, full cover with rebars (Fig. 6) has been considered to not decrease the length over which gravel can fall into the traps. 30 mm diameter linear rebars (prototype scale) have been used. They have been placed horizontally, at the elevation of the gallery invert. Such a scheme does not keep perfectly constant the gallery section over the trap and creates a sharp section transition in the gallery invert on both sides of the trap. However, it has been found easier to build since rebars don't need to be curved. As gravel was observed to move downstream along the cat walls on each side of the trap or along the gallery banks, guiding elements 0.1 m high and 1.07 m long, inclined by 27° on the gallery axis, have been placed upstream to redirect gravel onto the trap (Fig. 6).

Before conducting tests with gravel, flow velocity measurements inside the trap have been performed using 3 UVP ultrasonic sensors placed in the upstream and downstream vertical walls of the trap, at mid depth of the trap. The two sensors placed in the upstream wall were looking downstream and were placed close to the trap banks, one on each side of the trap, while the sensor placed in the downstream wall was looking upstream and was located on the trap longitudinal axis. The sensors enabled to measure the distribution along the trap of the mean longitudinal flow velocity and its standard deviation in time. Measures have been done during 10 s per test with a frequency of 10Hz.

Velocity measurements results have been used to define the rebars spacing. The configuration without rebars has been considered as reference. In this case, backwater flow is observed on half of the trap length (Fig. 7a). It occupies the whole trap width with velocity of 0.3 m/s on the sides (prototype scale). Flow velocity fluctuations, characterized by the standard deviation of the measurements, are high with values close to or above 0.1 m/s on each section (Table 1).

With 0.15 m rebars spacing (bars spacing recommended in Mattimoe et al., 1964), the backwater flow amplitude and intensity decreased. The backwater flow almost completely disappeared with 0.1 m rebars spacing (Fig. 7b), but the maximum flow velocity in the downstream direction increased compared to the configuration with 0.15 m rebars spacing (Table 1). In every case, the smaller the rebars spacing, the smaller the flow velocity fluctuations.

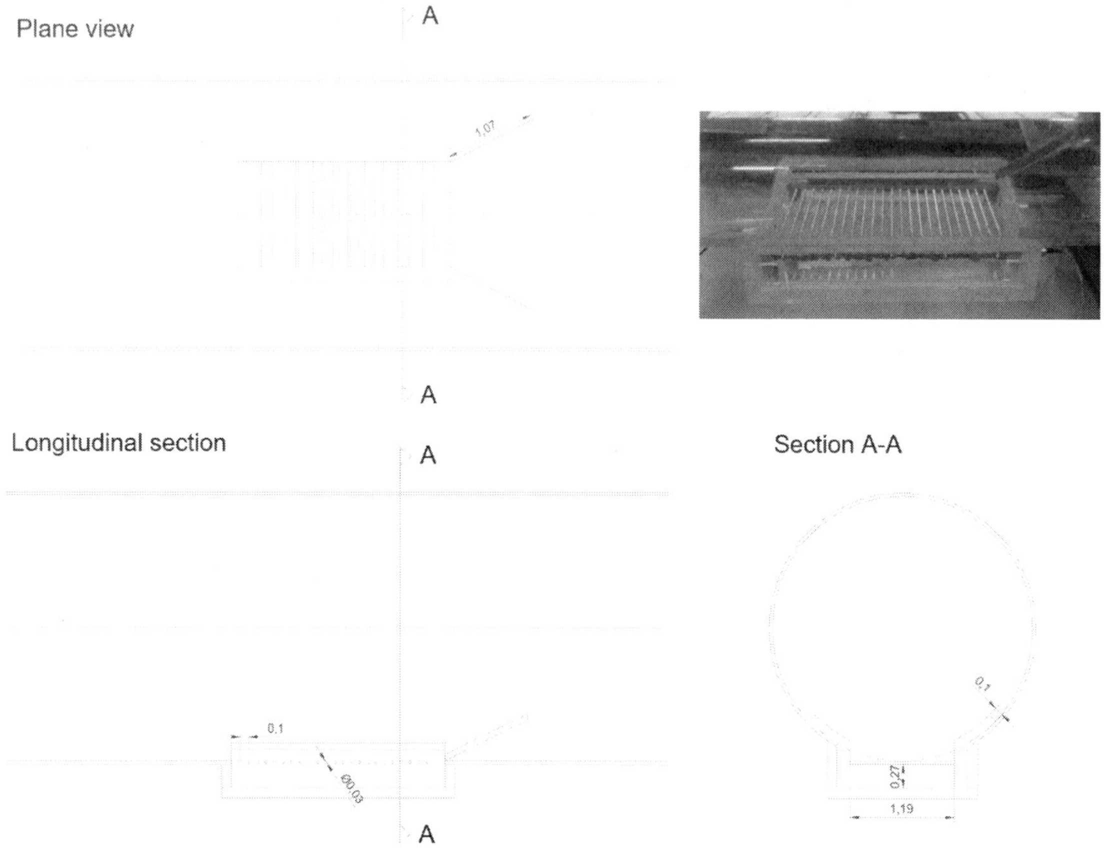

Figure 6. Principle of gravel trap covering with linear rebars with 0.1 m spacing. Dimensions in m at prototype scale. Flow from right to left.

With 0.1 m rebars spacing, it was observed that some large solid particles (corresponding to 40 to 50 mm particles) did not fall into the trap since they rebound on the rebars and were then transported downstream of the trap. To overcome this problem, some rebars have been removed on the downstream part of the trap to double the spacing. Removing 3 or 4 rebars, i.e. creating 3 or 4 areas with 0.2 m spacing, has been tested. In both cases, the flow velocity field was not degraded. On the contrary, maximum flow velocity in the downstream direction was decreased when some rebars were removed (Table 1).

Table 1. Maximum and minimum longitudinal flow velocity values at mid-depth in the trap and mean standard deviation depending on trap geometry. Dimensions at prototype scale.

	V_{max} [m/s]			V_{min} [m/s]			SD [m/s]		
	Left side	**Center**	**Right side**	**Left side**	**Center**	**Right side**	**Left side**	**Center**	**Right side**
Existing geometry	0.21	0.23	0.45	-0.30	-0.16	-0.29	0.086	0.237	0.120
0.15 m spacing rebars	0.04	0.03	0.10	-0.18	-0.16	-0.16	0.049	0.065	0.045
0.1 m spacing rebars	0.22	0.18	0.32	-0.04	0.00	-0.09	0.033	0.058	0.034
0.1 m spacing rebars - 3 bars removed	0.17	0.23	0.19	-0.06	0.00	-0.09	0.024	0.060	0.024
0.1 m spacing rebars - 4 bars removed	0.18	0.26	0.23	-0.05	0.00	-0.03	0.025	0.062	0.025

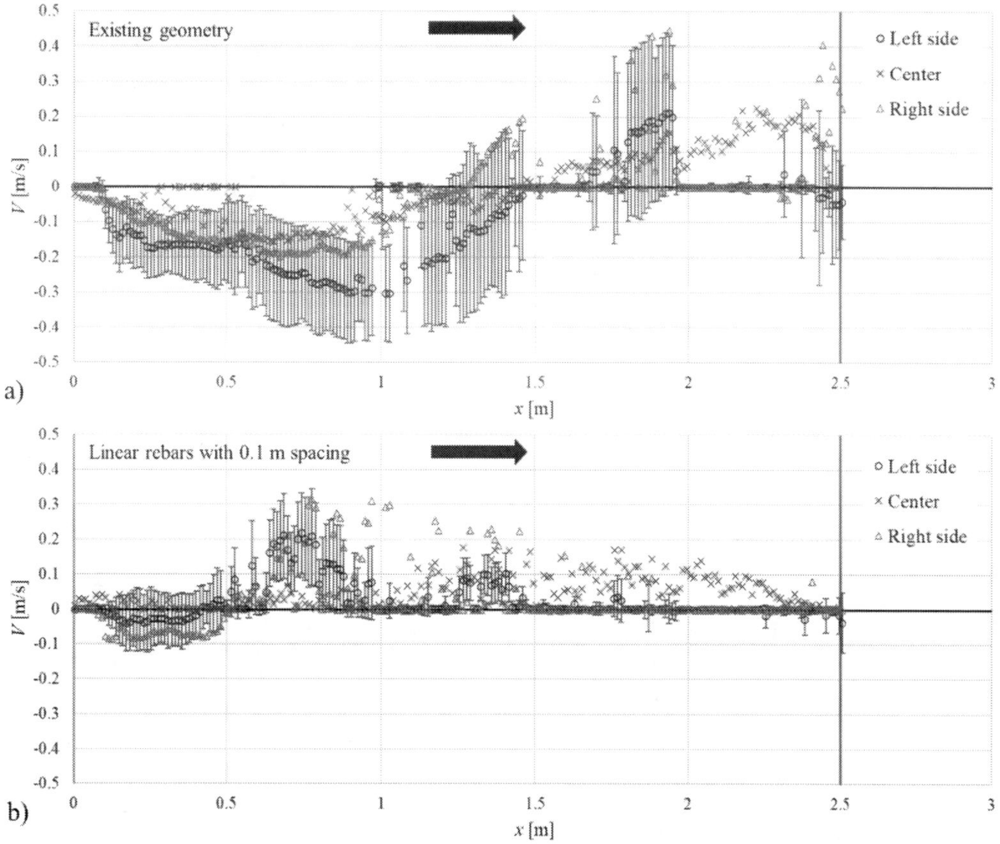

Figure 7. Longitudinal flow velocity distribution along the trap at mid-depth with standard deviation of measures on the left side of the trap. a) without rebars and b) with rebars spaced by 0.1 m. Dimensions at prototype scale.

The configuration with 0.1 m rebars spacing and 4 rebars removed has been considered for the tests with gravel. Two tests A with 1 kg of gravel injected in the model with a water discharge corresponding to 15 m³/s showed that 99.1 % (SD= 0.05%) of the injected gravel mass was stopped in the trap just after injection. Longer tests (tests B – 15 and 24h duration at model scale) proved that gravel deposits were stable over time in the modified trap configuration (Fig. 8 – Left). 99.6% of the injected gravel mass (SD= 0.05%) was found in the trap at the end of the tests. Moreover, it was observed that gravel did not move into the trap once it falls in. Finally, tests with free surface flow in the gallery with a discharge varying in the estimated leakage discharge range showed that rebars prevent the formation of a hydraulic jump in the trap and that only a small jet, broken by the rebars, may persist. It only had very limited impact on the gravel deposits' location and shape (Fig. 9 – Left). In this case, trapping efficiency was found to be 99.7%.

To satisfy a request by the plant operator to have possible rocks more than 0.2 m stopped by the trap, the latter has been divided in two cells by a vertical wall 0.24 m high. The upstream cell was 1.5 m long and was fully covered by rebars with 0.1 m spacing except on its downstream extremity were spacing was double. The downstream cell was 1 m long and was only covered with a few rebars to create a 0.7 m long opening and then enable the trapping of larger particles. With this configuration, 99.7% of the injected gravel (SD=0.3%) was found in the trap, and mainly in the first cell, after 24 and 88h operation (at model scale) of the scale model with a water discharge corresponding to 15 m³/s (Fig. 8 – Right). For test with leakage discharges, trapping efficiency was found to be 99.5% and gravel deposits were only a bit modified, with some gravel existing the upstream cell to fall into the downstream one, where it stays stable just downstream of the separating wall (Fig. 9 – Right).

Figure 8. Deposit in the trap after tests B with 0.1 m spacing rebars. Configuration with 1 cell and 4 rebars removed (left) and configuration with 2 cells and sparse downstream cover (right).

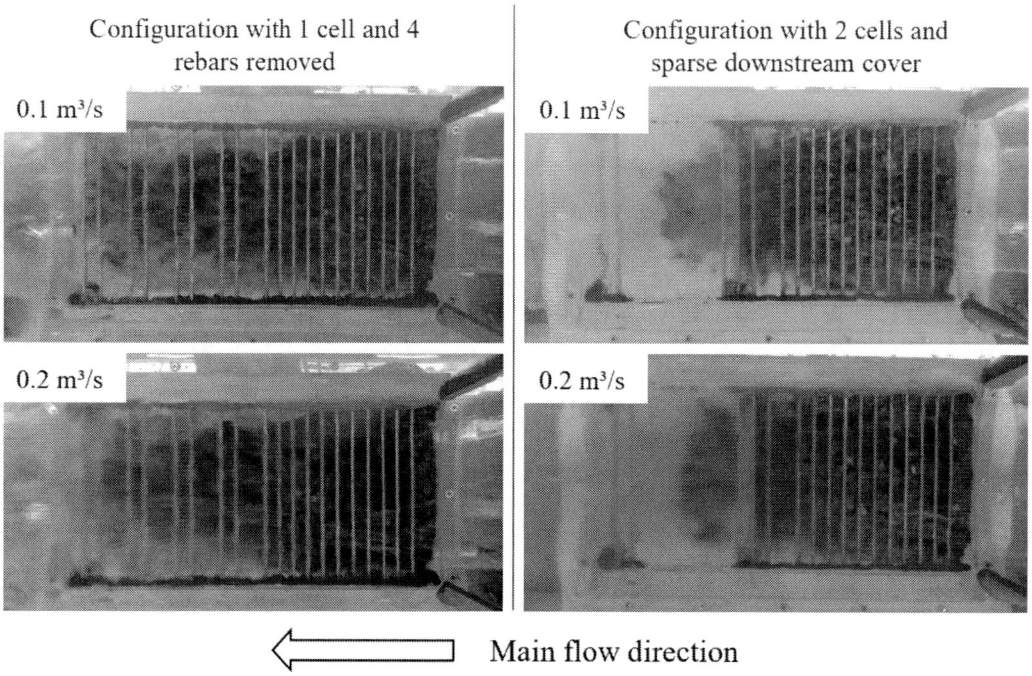

Figure 9. Evolution of gravel deposit in the trap covered with 0.1 m spacing rebars, depending on leakage discharge (tests C) – Configuration with 1 cell and 4 rebars removed (left) and configuration with 2 cells and sparse downstream cover (right).

4. Conclusion

Tests on a 1:7.7 scale physical model showed that gravel traps designed as prismatic volumes in the invert of a pressurized circular gallery have a trapping efficiency as high as 98%. However, as already mentioned in the literature, such traps are ineffective in keeping the gravel over long periods since intermittent motion of the particles in the trap associated to turbulent fluctuations, and helicoidal vertical currents developing at the downstream extremities of the trap, can slowly extract the gravel from the traps. In addition, for the specific case considered in this study, temporary free surface flows taking place in the gallery because of leakages during gallery dewatering can flush the gravel out of the traps and then decrease again the overall trapping efficiency when considering the whole of the flow conditions existing in the system.

Following recommendations from the literature, a covering of the traps with rebars, spaced by 0.1 m and aligned with the gallery invert, proved to be effective in suppressing undesirable flow conditions in the traps, and then proved to be a good solution to drastically increase long term trapping efficiency. This rebars cover was also effective in case of free surface flow conditions.

The cover doesn't need to be continuous since larger spacings on the downstream half of the trap did not affect the trapping efficiency. Dividing the trap in successive trapping cells improved the overall gravel deposits stability in the trap.

This study showed that additional research on gravel traps design in pressurized and non-pressurized systems is required to fully understand flow conditions causing trapped gravel ejection and to provide complete detailed design rules and recommendations.

5. REFERENCES

Ettema, R., Arndt, R., Roberts, P., & Wahl, T. (2000). *Hydraulic modeling: Concepts and practice*. Reston, VI: American Society of Civil Engineers ASCE.

Griffiths P. T. A., Brett T. M. (1974). "A scouring rock trap in Lemonthyme power tunnel." *Proc., 5th Australasian conference on hydraulics and fluid mechanics*, University of Canterbury, Christchurch, New Zealand, 49-55.

Hachem, F. (2009). "Aménagement de Castirla-Galerie d'amenée - Etude expérimentale sur modèle physique du piège à cailloux aval de la galerie d'amenée. » (In French). <https://www.epfl.ch/labs/lch/wp-content/uploads/2018/11/Etude-de-Castirla-Galerie.-Etude-exprimentale-sur-modle-physique-du-pige-cailloux-aval-de-la-galerie-damene.pdf> (May 4, 2022).

Havrevoll, O. H., Vereide, K., Lia, L. (2021). "Efficiency of Pressurized Rock Traps for Unlined Hydropower Tunnels." *Energies*, 14(14): 4344. https://doi.org/10.3390/en14144344

Mattimoe, J. J., Tinney, E. R., Wolcott W. W. (1964). "Rock trap experience in unlined tunnels." *J. Power Div.*, 90:29-45.

DOI: https://doi.org/10.26077/f13f-9ef3

Experimental Study on Three-Dimensional Velocity near Piano Key Weirs

Binit Kumar[1] & Zulfequar Ahmad[2]
[1]*Former PhD candidate, Department of Civil Engineering, Indian Institute of Technology Roorkee, Roorkee 247667, India*
[1]*Postdoctoral fellow, Department of Civil Engineering, National Institute of Technology Warangal, Warangal-506004, India*
[2]*Professor, Department of Civil Engineering, Indian Institute of Technology Roorkee, Roorkee 247667, India*
E-mail: bkumar@ce.iitr.ac.in (Corresponding Author)

Abstract: *A Piano key weir (PK weir) is a non-rectilinear spillway commonly used to augment the discharge capacity at the dams and diversion structures. Installation of such hydraulic structures may alter the flow characteristics, sediments transport and sedimentation in the river channels. The present study deals with an experimental investigation of three dimensional velocities near the two different laboratory models of PK weir having noses with an outlet overhangs. The purpose of this study is to observe the flow field in the vicinity of a one-cycle and two-cycle PK weir to understand the flow behavior of the weir that may be useful for sediment movement over weir. The velocity profile was measured at various grid points for both models considering unit discharges of* 48.5×10^{-3} *m²/s and* 49×10^{-3} *m²/s. The velocity fields near both inlet and outlet keys were examined and it was found that there is a considerable increase in the vertical (y-component) and lateral (z-component) velocities. It was observed that the rise in the mean velocity along longitudinal, vertical and lateral direction was more for two cycle PK weir due to increased crest length. For both discharge, the maximum velocity was observed near the inlet key. Computational Fluid Dynamics (CFD) simulations were also used to understand the flow configuration over the PK weir. It was seen that the magnitude of velocity increased due to vertical contraction near the weir and the flow velocity is very high immediately downstream of PK weir.*

Keywords: *CFD, lateral-direction, outlet key, Piano Key Weir, velocity.*

1. Introduction

The Piano key weir (PK weir) is an upgraded labyrinth weir basically installed to increase the discharging capacity of the existing dams and diversion structures. Construction of any transverse hydraulic structures can modify the interaction involving flow and sediment transport (Mazumdar and Sarkar 2014). Application of transverse hydraulic structures like rectangular or triangular weirs have essential roles in preserving depth of flow in the upstream which is useful in conserving the water resources for best possible use (Kabiri-Samani and Javaheri 2012). But, a few hydraulic structures disturb sediment movement because the flow velocity is reduced due to the enhanced depth of flow (Bai and Duan 2014; Fan and Morris 1992). Due to the disturbance in the sediment continuity, sediment gets deposited upstream of a weir affecting hydro-ecological habitats such as aquatic life, geographical physiognomy and drastic alteration to the riparian ecosystem (Kim et al. 2014).

Safety of dams is ensured by the spillways for the high flood condition. Thirty three percent of the dam failures take place due to the less number of spillways (Schleiss 2012). A PK weir was suggested in the early 21st century by Hydrocoop in cooperation with Biskra University (Algeria), the Hydraulic Laboratory of Electricité de France (France) and IIT Roorkee (India) which is a modified traditional weir to enrich the discharging capacity of the weir (Erpicum et al. 2017; Lemperiere and Ouamane 2007). The flow pattern of incoming water is almost different in the PK weir than other weirs because of its unique shape. The approaching flow divides into two segments due to the presence of inlet key and outlet key. The first part of approaching flow is passed by the inlet key and the second section is carried by outlet key which is falling as a jet towards the base of the key (Leite Ribeiro et al. 2007; Ouamane and Lemperiere 2006). The schematic diagram of PK weir is shown in Fig.1 where '*a*' and '*b*' are inlet and outlet key measurements; the height of weir denoted by '*P*'; '*P'*' is the approach flow height beneath the inlet key; channel width is represented by '*W*'; Sidewall crest length is given by '*B*' and the overhang lengths in the downstream and upstream denoted by B_i and B_o respectively (Pralong et al. 2011).

Velocity is very significant factor that governs the flow pattern, mechanics of sediments transport, suspension of sediments and sedimentation. The estimation of velocity is a very robust task as it depends on different parameters such as discharge, bed characteristics, depth of the water and the channel slope. Previous investigations were carried out on the discharging capacity of the weir based on the different types of PK weirs having various geometries but limited literatures are available for the flow characteristics and sediment passing capacity near PK weirs. Leite Ribeiro et al. (2012) performed laboratory experiments on scaled physical models and compared the discharging capacity between PK weir and normal sharp crested weir. Based on their evaluation, a design procedure of PK weir was proposed. The effect of the sloping key floors was studied by Anderson and Tullis (2012) through a series of experimentations on laboratory-based PK weir models and also compared the discharging capacity of this weir with a rectangular labyrinth weir. A PK weir having noses below the outlet apex passes higher discharge than a regular PK weir type (Anderson and Tullis 2012). Anderson and Tullis (2013) used rounded noses and observed 2-3% augmentation in the discharging capacity of PK weir. Belaabed and Ouamane (2013) conducted experiments using triangular shaped noses and also observed a rise in the discharging capacity. Crookston et al. (2018) evaluated five empirical discharge equations with their CFD simulations based on various turbulence models like LES and RNG k-ε. Kumar et al. (2019) evaluated the four given empirical formulae of different investigators for the discharge equations of PK weirs. Sharma and Tiwari (2013) found that vertical component (z-component) of the approaching flow velocity near the PK weir enhances gradually which increases the movement of sediment particles and vertical velocity component reduced to 14% on moving from higher flow to intermediate flow. Increase in discharge from lower to medium flow, z-component velocity is getting increased by about 40 percent. It indicates that at lower discharge, upward velocity component (z-component) is appreciably more which may be helpful for significant amount of sediment suspension at lower discharge also. Recently, Kumar et al. (2021a) studied the effect of upstream siltation on the discharging capacity of PK weirs having different cycles and proposed empirical equations for the same. The mechanics of sediment movement over the inlet key of PK weirs were also investigated by Kumar et al. (2021b). Kumar and Ahmad (2022) observed that scour profiles are dissimilar at the inlet and outlet keys due to the intricacy in shape of the weir. Moreover, it was seen that the jets approaching from the outlet and inlet keys are unlike and these jets are liable for the progress of dips and ridges. The dips developed in front of the outlet keys due to impinging jets whereas ridges occurred in front of the inlet keys by falling jets. This is due to the fact that impinging jets might be stronger than falling jets. Ghanbari and Heidarnejad (2020) simulated three-dimensional flow fields with the help of FLOW-3D and the hydraulics of flow was understood and the rates of discharge were also compared for different models. Densys and Bossons (2018) found that there is a formation of vortex sheet in an inlet key due to the development of a re-circulating zone occurring just behind the shear layer at the entrance edge. They also noticed that the zone of negative pressure occurs at outlet key because of the intersection of the longitudinal outlet key flow and the transverse sidewall nappe. The present study deals with an experimental investigation on three dimensional velocities near two different laboratory based physical models of PK weirs having noses below the outlet overhangs to understand the flow field in the vicinity of PK weir. The velocity profile over the inlet key could not be measured precisely, probably due to flow complexity so computational fluid dynamics simulations (CFD) were performed to understand the flow pattern along the inlet key of PK weir.

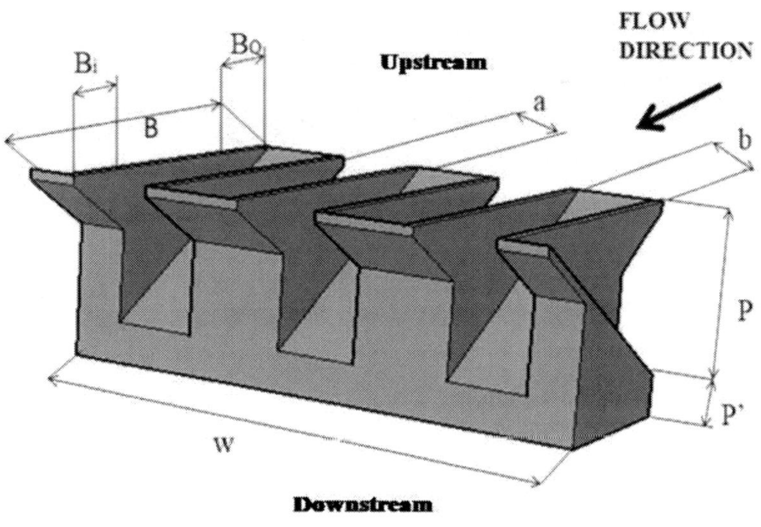

Figure 1. A Schematic diagram of PK weir (Pralong et al. 2011)

2. Experimental Setup

All the experimental runs were carried out in the Hydraulics Laboratory of the Department Civil Engineering, IIT Roorkee in a 15.0 m long tilted flume with a cross-section 0.39 m × 0.5 m having a glass-walled in its main segment for better visualisation of the flow. A schematic plan (three-dimensional view) is shown in Fig. 2(a). The upstream part of the flume was provided by a honeycomb grid wall, flow straighteners and wooden floater to minimise flow disturbance and cross-currents. A storage tank was available for regular supply of clear water. An inlet pipe of diameter 0.1 m was mounted with a valve for discharge regulation. Ultrasonic flowmeter up to accuracy ±1% and point gauge (least count = 0.0001m) were used to record discharge and upstream water depth, respectively. The flowmeter was connected to the inlet pipe and the point gauge was placed at 0.2 m upstream from the outlet key. Acoustic- Doppler velocimeter (ADV) of Nortek-AS having 10 megahertz (MHz) was used to study the flow pattern in vicinity of PK weir with nose as shown in Fig. 2(b). Two physical models of PK weir having different cycles were used in the present study. Both the models of PK weir having nose below the outlet apexes (PKN) were fabricated by 0.006 m thick acrylic sheet as shown in Fig. 2(c) and the details of these models are listed in Table 1.

(a)

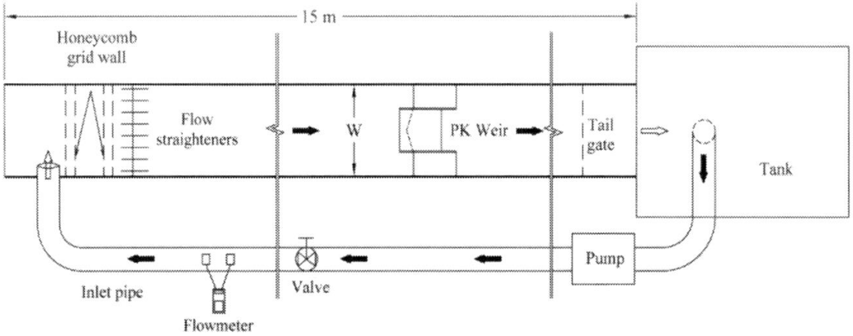

(b)

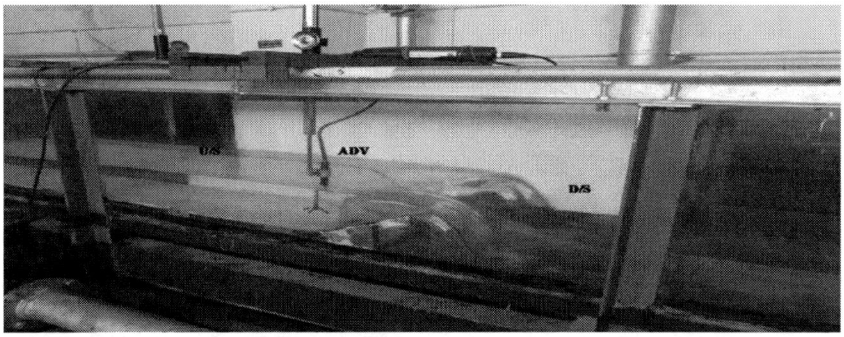

(c)

Figure 2. Experimental activities: (a) Schematic plan of the setup, (b) Photometric view of setup with ADV, (c) PK weir models with nose.

Table 1. Detailed dimensions of a PK weir.

Model	No. of cycles	L(m)	W(m)	P(m)	a(m)	b(m)	B_b(m)	$B_i = B_o$(m)	$S_i = S_o$
PKN_1	One	0.898	0.39	0.105	0.2	0.178	0.125	0.064	5(V):9(H)
PKN_2	Two	1.402			0.1	0.083			

3. Computational Fluid Dynamics (CFD) Simulations

CFD simulations were carried out using CFX solver which requires less computational cost and space than Fluent solver due to a lesser number of degrees of freedom (Berggren et al. 2009; Kadia et al. 2020). The broadly used standard k-ε turbulence model (an eddy-viscosity model) which was introduced by Launder and Spalding (1972 1974) was used in the present simulation. Turbulence models are essential because it is one of the best elements in CFD related numerical modelling.

The present CFD study is on PK weir models considering Ansys CFX solver within Ansys 19.1 academic research version (ANSYS 2018). The Ansys CFX solver applies the finite volume method and vertex-centered method (Acharya 2016; Berggren et al. 2009). In CFD simulations the well-known Reynolds-Averaged Navier-Stokes equations (continuity and momentum) are considered as the basic equations.

4. Methodology

Two experimental tests were conducted with unit discharges of 48.5 $\times 10^{-3}$ m^2/s and 49.0 $\times 10^{-3}$ m^2/s respectively for each PK weir model in the study. The velocity data were obtained for different grid points (X_0Z_1, $X_1'Z_1$, $X_1'Z_2$ $X_1'Z_3$, $X_2'Z_1$, $X_2'Z_2$, $X_3'Z_1$, $X_3'Z_2$, $X_3'Z_3$) as shown in Fig. 3, for PK weir having one cycle. The data were collected at the sample rate of 50 Hz for one minute. For each local point, the vertical grid points along the flow depth were varied from downwards to upwards i.e. 0.5 mm close to the bed and 10-20 mm near the free surface. After the collection of data, the recorded raw data were filtered on the basis of minimum SNR=15 and minimum COR=70 (Voulgaris and Trowbridge 1998). These filtered data were further analysed to calculate the mean velocities along XY plane for the development of vector diagrams using MATLAB software. The velocity profile over the inlet key could not be measured precisely, probably due to flow complexity, high turbulence, unsteadiness of flow and vortices within inlet key which was also observed by Denys and Basson (2020). So, CFD simulations were performed to understand the flow pattern over the key of PK weir.

For the CFD analysis, the whole simulation for conducting the velocity on both the physical models consists of geometry formation, development of meshing, boundary condition of all the three setups, simulation runs and the post processing. The simulation was performed at a unit discharge of 49.0 $\times$ 10^{-3} m^2/s for both the models. The geometry of the physical models were basically constructed and designed in AutoCAD 3D with the scale of 1:1 to the experimental setup and later it was imported to the Ansys workbench, as shown in Fig. 4(a). The dimension of flume used in this present simulation study is 1.1 m×0.39 m×0.3 m respectively. Structured multi-block tetrahedron mesh for 3D domains has been adopted for the present simulation studies that can be seen in Fig. 4(b). All the meshes were created using tetrahedral meshing in ANSYS CFX solver. The meshing components of all the models were mainly imported to CFX-pre-processing unit of workbench. Figure 4(c) shows the domain with boundary conditions. The total simulation time and time steps were selected as 20 s and 0.05 s, respectively. ANSYS CFD-POST was applied to generate the images of flow velocity, streamlines, velocity contours along the inlet key of all the models.

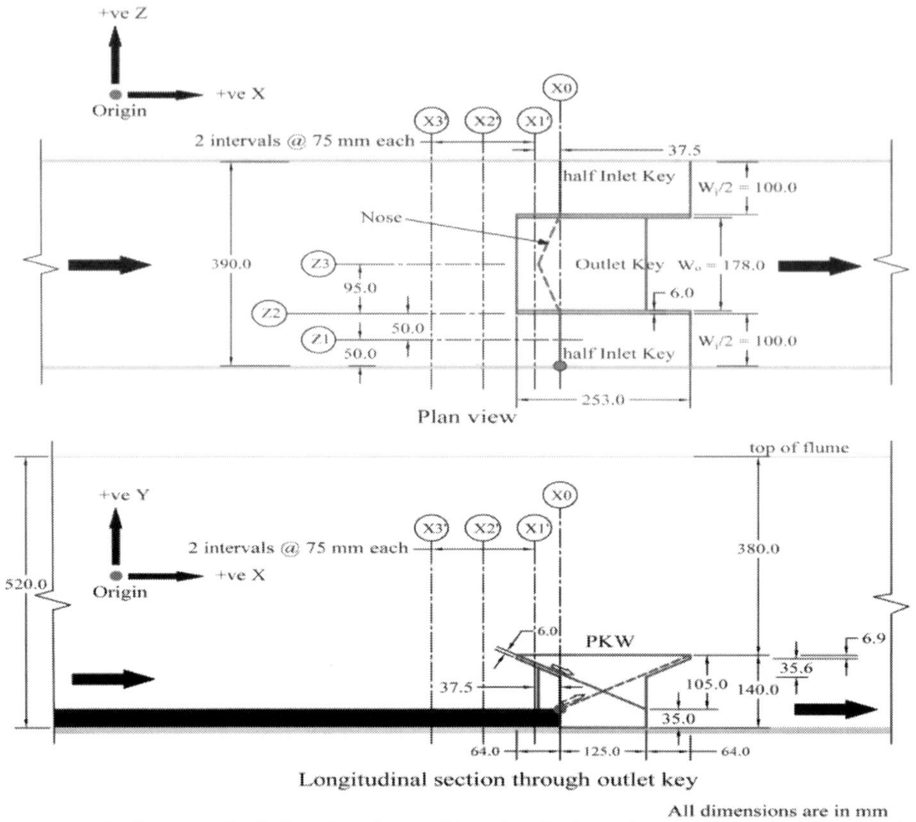

Figure 3. Index map depicting the plan and longitudinal section of PK weir with a cycle

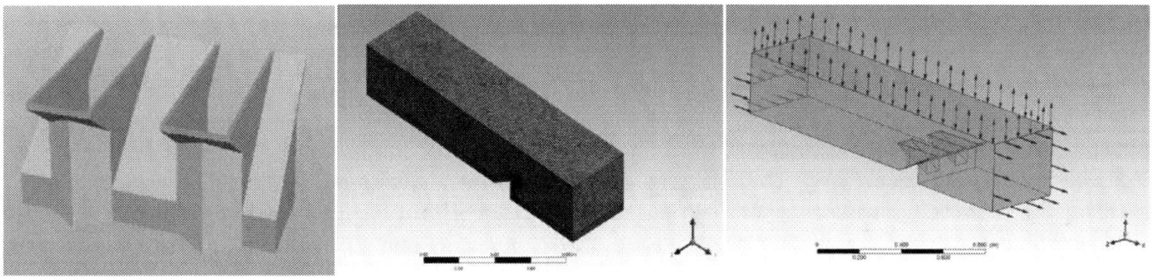

Figure 4. Simulation activities: (a) PK weir imported model with two cycles, (b) Meshing domain, (c) Boundary conditions of setup

5. Results and Discussions

The velocity profiles were measured for the eight grid points as shown in Fig. 3 for unit discharges of 48.5×10^{-3} m^2/s for PK weir having one cycle whereas ten grid points were observed for discharge of 49.0×10^{-3} m^2/s in the case of PK weir having two cycles. However, for the grid point X_0Z_1, the results were not accurate as the data obtained after filtrations and analysis were not up to the mark. Similarly, the inaccurate data obtained from the grid point for two cycles PK weir were ignored. The velocity fields near inlet key and outlet key were found and it was observed that there are considerable enhancement in the vertical (y-component) and lateral (z-component). For the case of unit discharge of 48.5×10^{-3} m^2/s, the enhancement in the mean velocity of the longitudinal, vertical and lateral direction were found to be 4.8 cm/s, 6.9 cm/s and 5.27 cm/s respectively when moving towards the PK weir (i.e. X_3' to X_1'). Furthermore, it was seen that the mean vertical velocity was more near the outlet key than the mean vertical velocity near inlet key whereas mean lateral and longitudinal velocities were low in front of the outlet key with respect to the velocities near inlet key. The maximum velocity was observed near the inlet key (i.e. at the grid point X_1'Z_1). The velocity components in all the three direction goes on increasing as the flow approaches the inlet key i.e. from X_3'Z_1

to X_1'Z_1 as depicted in Fig. 5(a-c). In this case of PK weir with two cycles and unit discharge of 49.0×10^{-3} m^2/s, the rise in the mean velocity along longitudinal, vertical and lateral direction was observed to be 3.22 cm/s, 8.96 cm/s and 1.00 cm/s respectively when moving towards the weir (i.e. X_3' to X_1'). Added to this, it was observed that the mean vertical, lateral and longitudinal velocities were high in front of the inlet key as compared to in front of outlet key. Therefore, both the discharges velocity vector diagrams are similar in nature.

(a)

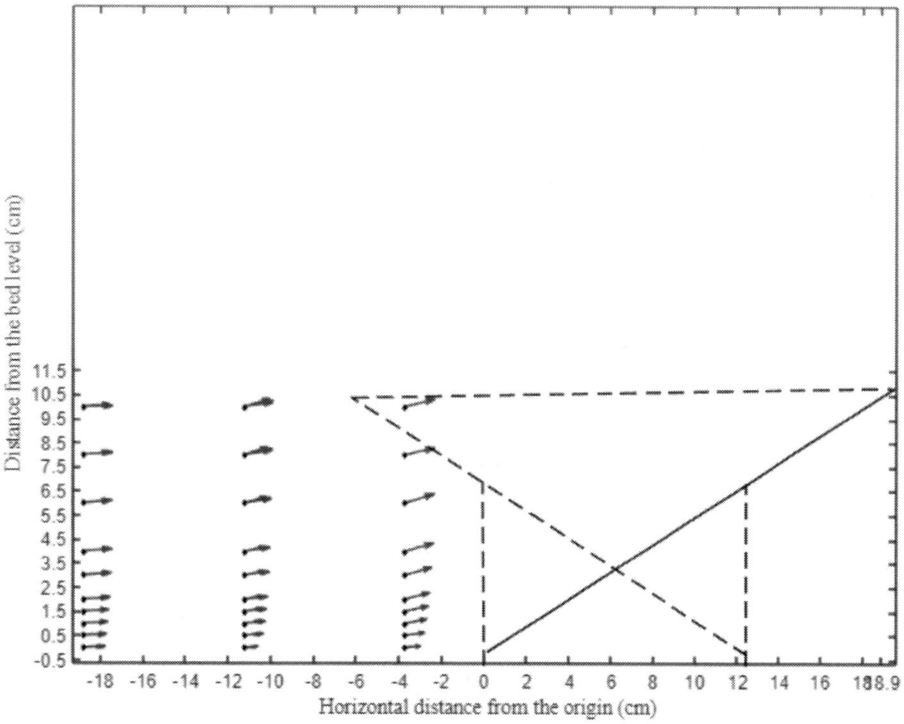

(b)

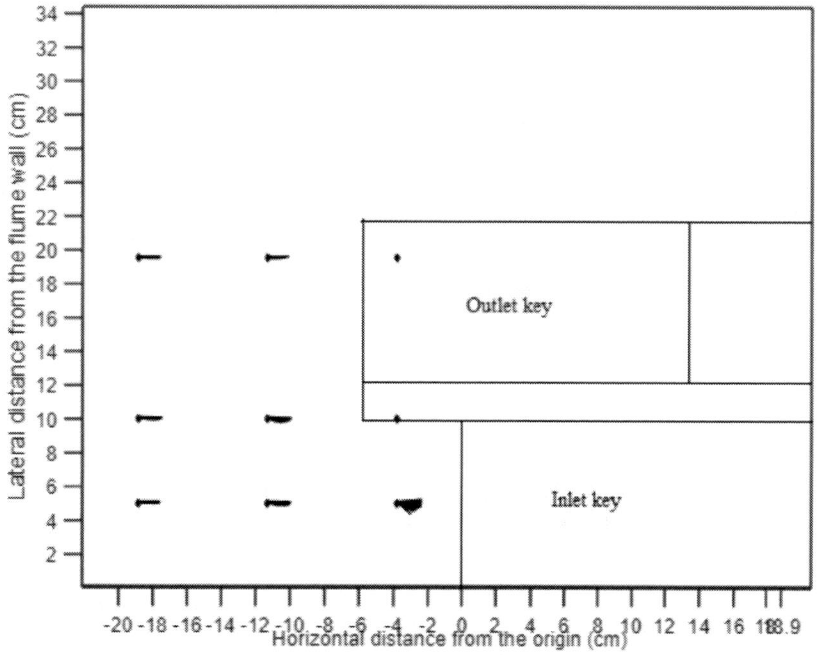

(c)

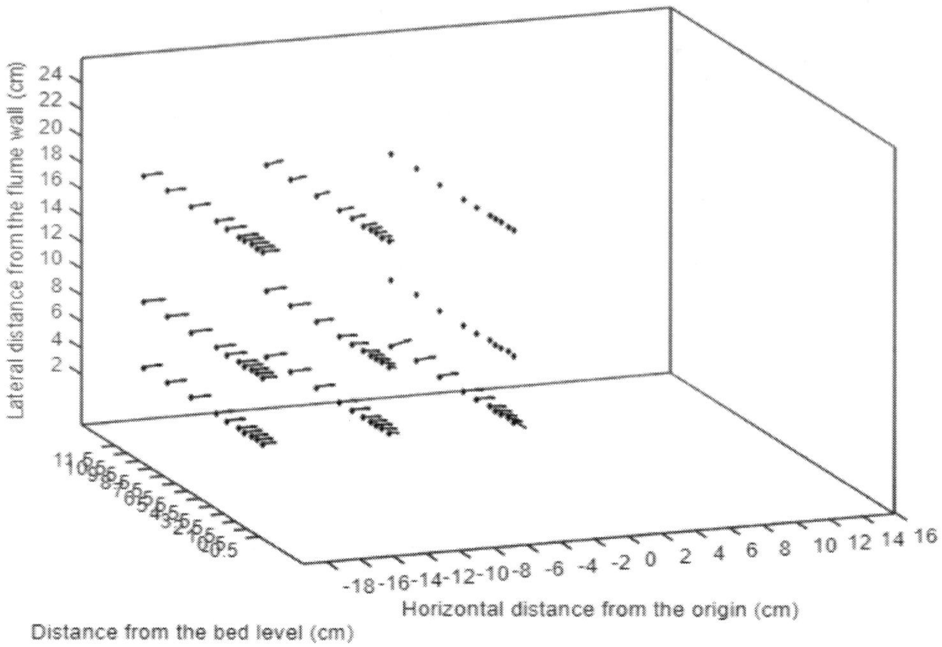

Figure 5. Resultant Velocity vector diagram for discharge 48.5×10^{-3} m^3/s (a) longitudinal plane (b) horizontal plane (c) Three-dimensional plane

It was noticed that the flow field along the inlet and outlet key of the weir couldn't be obtained by ADV due to disturbed flows. So CFD simulations were performed to observe the flow field over keys of PK weir having one cycle and two cycles. Figure 6 shows the velocity vector along the inlet key of PK weir. It was noticed that the low velocity magnitude was present at the bed level and upstream of PK weir. It was also observed that magnitude of velocity got increased due to vertical contraction near the PK weir.

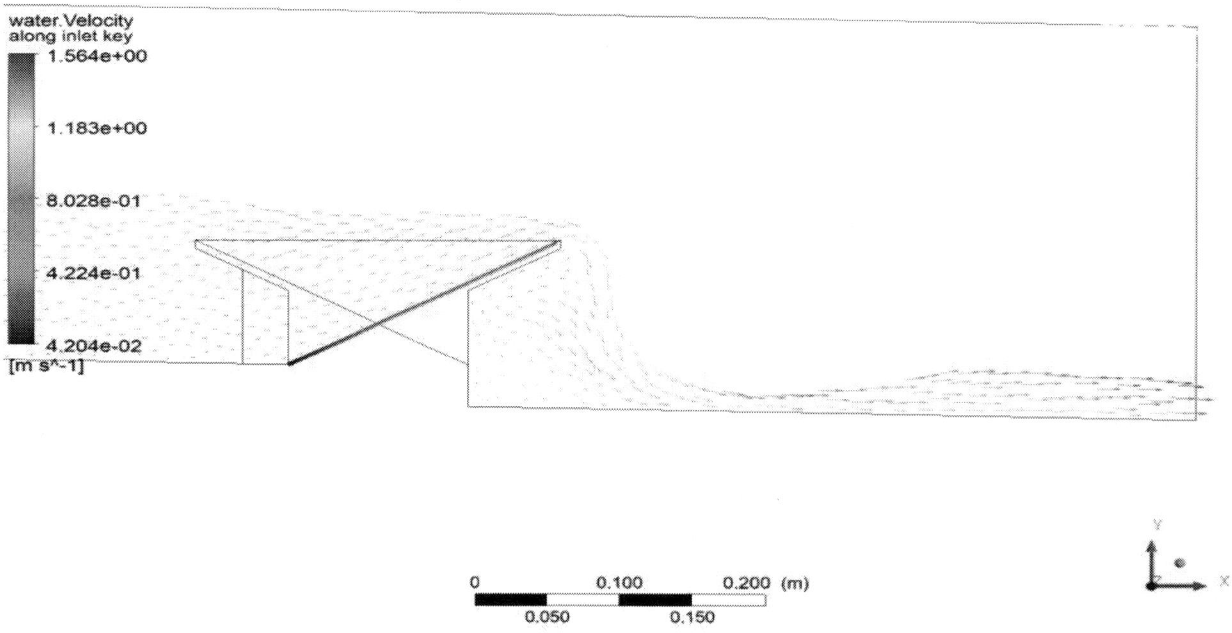

Figure 6. Velocity vector along the inlet key of PK weir having one cycle

6. Conclusions

Experimental and simulation studies were carried out to observe the flow pattern in vicinity of PK weirs with noses below the outlet apexes. Two physical models of PK weirs were used for the experimentation. The velocity components in all the three direction goes on increasing as the flow approaches the inlet key i.e. from $X_3'Z_1$ to $X_1'Z_1$. It was observed that there is significant rise in the vertical (y-component) and lateral (z-component) velocities near PK weir which may be responsible for sediment suspension and movement over the weir. Moreover, it was found that the mean vertical velocity was more in front of the outlet key as compared to in front of inlet key while mean lateral and longitudinal velocities were low. The maximum velocity was observed in front of the inlet key of PK weir. The findings of the two PK weir models with different cycle were presented in this study. Further additional research can be done considering more physical models. Authors are also carrying out their study for other values of discharges and sediment passing capacity of PK weirs.

7. Acknowledgements

The first author is highly obliged to the Ministry of Human Resource Development, Govt. of India for granting research scholarship.

8. References

- Acharya, R. (2016). *Investigation of Differences in Ansys Solvers CFX and Fluent.* Master Thesis, KTH Royal Institute of Technology, Stockholm.
- ANSYS. (2018). *ANSYS Academic Research Mechanical and CFD, Release 19.1, Canonsburg, PA.* Canonsburg, PA.
- Anderson, R.M., and Tullis, B.P. (2012). "Comparison of Piano Key and Rectangular Labyrinth Weir Hydraulics". *Journal of Hydraulic Engineering,* 138 (2012), 358–361, doi:10.1061/(asce)hy.1943-7900.0000509.
- Anderson, R.M., and Tullis, B.P. (2013). "Piano Key Weir Hydraulics and Labyrinth Weir Comparison." *Journal of Irrigation and Drainage Engineerin,* 139 (2013) 246–253, doi:10.1061/(ASCE)IR.
- Berggren, M., Ekström, S.E., and Nordström, J. (2009). "A discontinuous Galerkin extension of the vertex-centered edge-based finite volume method." *Communications in Computational Physics, 5(2-4)*(2–4), 456–468. https://doi.org/10.1093/dnares/dsu046
- Bai, Y., and Duan, J.G. (2014). "Simulating unsteady flow and sediment transport in vegetated channel network." *Journal of Hydrology,* 515, 90–102.
- Belaabed, F., and Ouamane, A. (2013). "Submerged flow regimes of Piano Key weir" *Proceedings of the Second International Workshop on Labyrinth and Piano Key Weirs -PKW 2013,* Paris, France, CRC Press, London, 2013: pp. 85–92.
- Crookston, B.M., Anderson, R.M., and Tullis, B.P. (2018). "Free-flow discharge estimation method for Piano Key weir geometries" *Journal of Hydro-Environment Research,* 19 (2018), 160–167. doi:10.1016/j.jher.2017.10.003.
- Denys, F.J.M., and Basson, G.R. (2020). "Unsteady Hydrodynamic Behavior at Piano Key Weirs." *Journal of Hydraulic Engineering, 146*(5), 04020028, https://doi.org/10.1061/(ASCE)HY.1943-7900.0001729.

- Denys, F. (2018) "Transient Hydrodynamics of Piano Key Weirs." Daniel Bung, Blake Tullis, 7th IAHR International Symposium on Hydraulic Structures, Aachen, Germany, 15-18 May. doi: 10.15142/T30M0J (978-0-692-13277-7).
- Erpicum, S., Archambeau, P., Dewals, B., and Pirotton, M. (2017). "Hydraulics of Piano Key Weirs: A Review" *Proceedings of the 3rd International Workshop on Labyrinth and Piano Key Weirs.* Qui Nhon, Vietnam, Taylor & Francis Group, London, 2017: pp. 27–36.
- Fan, J., and Morris, G. (1992) "Reservoir sedimentation I: delta and density current deposits."*Journal of Hydraulic Engineering,* 118 (1992) 354–3369.
- Ghanbari, R., and Heidarnejad, M. (2020). Experimental and numerical analysis of flow hydraulics in triangular and rectangular piano key weirs,. Water Science. Retrieved from doi: 10.1080/11104929.2020.1724649
- Kabiri-Samani, A., and Javaheri, A. (2012). "Discharge coefficients for free and submerged flow over Piano Key weirs". *Journal of Hydraulic Research.* 50 114–120. doi:10.1080/00221686.2011.647888.
- Kumar, B., Kadia, S., and Ahmad, Z. (2021a). "Discharge characteristics of PK weirs with and without upstream siltation." *International Journal of civil engineering. Springer,* 19, 1043-1054. https://doi.org/10.1007/s40999-021-00607-x.
- Kumar, B., Kadia, S., and Ahmad, Z. (2021b). "Study on mechanics of sediment movement over Type-A Piano key weirs." *Journal of Irrigation and Drainage Engineering,* ASCE, 147(6). https://doi.org/10.1061/(ASCE)IR.1943- 4774.0001561.
- Kadia, S., Kumar, B., and Ahmad, Z. (2020). "Discharge Characteristics of Triangular Weir with Upstream Ramp and Its CFD Modelling Using Ansys CFX Module." In M. B. Kalinowska, M. M. Mrokowska, & P. M. Rowiński (Eds.), *Recent Trends in Environmental Hydraulics. GeoPlanet: Earth and Planetary Sciences* (1st ed., pp. 77–90). https://doi.org/10.1007/978-3-030-37105-0_7.
- Kumar, B., Kadia, S., and Ahmad, Z. (2019). "Evaluation of discharge equations of the Piano Key Weirs." *Flow Measurement and Instrumentation,* Elsevier, 68(2019), 101577. doi:10.1016/J.FLOWMEASINST.2019.101577.
- Kim, S., Im, J., and Lee, S. (2014). "Assessment of sediment exclusion efficiency for several modified Labyrinth weirs." *Paddy Water Environment,* 1–8.
- Lempérière, F., and Ouamane, A. (2003). "The Piano Keys weir: A new cost-effective solution for spillways." *International Journal on Hydropower and Dams.* 10(5), 144–149.
- Launder, B. E., and Spalding, D. B. (1972). "Lectures in mathematical models of turbulence." London, New York: Academic Press.
- Launder, B. E., and Spalding, D. B. (1974). "The Numerical Computation of Turbulent Flows." *Computer Methods in Applied Mechanics and Engineering, 3*(1974), 269–289. https://doi.org/10.1016/0045-7825(74)90029-2
- Leite Ribeiro, M., Boillat, J.-L., Schleiss, A., Laugier, F., and Albalat, C. (2007). "Rehabilitation of St-Marc Dam Experimental Optimization of a Piano Key Weir." *Proceedings of the 32nd Congress of IAHR.*

(2007). https://infoscience.epfl.ch/record/112265 (accessed August 17, 2019).

- Leite Ribeiro, M., Pfister, M., Schleiss, A.J., and Boillat, J.L. (2012). "Hydraulic design of a-type piano key weirs". *Journal of Hydraulic Research,* 50, 400–408. doi:10.1080/00221686.2012.695041.
- Mazumdar, B., and Sarkar, K. (2014) "Turbulence flow characteristics and drag over 2D forward -facing dune shaped structures with two diffrent stross-side slopes." *Environmental Fluid Mechanics.* 14 , 617–645.
- Ouamane, A., and Lempériѐre, F. (2006) "Design of a new economic shape of weir." *International Symposium on Dams in the Societies of the 21st Century; Dams and Reservoirs, Societies and Environment in the 21st Century ICOLD-SPANCOLD*; 2006; Barcelona, Spain, Taylor & Francis Group, spain, 2006: pp. 463–470.
- Pralong, J., Vermeulen, J., Blancher, B., and Schleiss, A. J. (2011). "A naming convention for the Piano key weirs geometrical prameters." *Labyrinth and Piano Key Weirs*, 271–278.
- Sharma, N., and Tiwari, H. (2013). "Experimental study on vertical velocity and submergence depth near Piano key weir" *Labyrinth and Piano Key Weirs II-PKW.*
- Schleiss, A. (2012). "From Labyrinth to Piano Key Weirs – A historical review, Labyrinth and Piano Key Weirs.", 3–15, doi:10.1201/b12349-3.

DOI: https://doi.org/10.26077/66b9-64d9

Flow Over Orifice Spillway: Physical and Numerical Study for Spillway Profile Design of Hydroelectric Project

Sushma Vyas[1] & Y. N. Srivastava[2]
[1]*Scientist C, Central water and power research station, Pune, India*
[2]*Additional Director, Central water and power research station, Pune, India*
E-mail: vyassush13@gmail.com

Abstract: *Many high dams on the upper Himalayan rivers which have very steep bed slopes are constructed with orifice type of Spillways with low level sluices for flood disposal and flushing of sediments. To assess the performance of spillway physical model studies were conducted on a 1:50 scale 2-D sectional model. Physical model studies can have a wide range of outcomes depending on the purposes of the model study, although many physical models frequently analyse properties such as velocity patterns, discharge rating curves, water surface profiles, and pressures. The physical measurements highlighted that the spillway is subjected to negative pressures. To overcome this, different spillway profile options were verified using numerical methods. For this, a commercially available computational fluid dynamics (CFD) program FLOW 3D was used which solves the Reynolds-averaged Navier-Stokes equations along with RNG turbulence closure model. Numerical modelling is quite beneficial when used by the designer in combination with the physical modelling. Numerical modelling enables designers to evaluate several possibilities until they find the hydraulically optimal solution, which can then be evaluated on a physical model, reducing the expense of physical model studies. In order to validate the numerical modelling results, the measured data from physical model studies for the original profile of spillway is used. Discharge passing through spillways, water surface profiles and pressures were used to compare the results of the physical model and the numerical model. When it was seen that there is reasonably good agreement between the physical and numerical model results, then all other alternatives were analyzed numerically and an alternative having a hydraulically optimal solution was selected to study on the physical model.*

Keywords: *CFD, Flow 3D, Numerical modelling, Physical modelling, Spillway*

1. Introduction

Dams and reservoirs effectively used to regulate river levels and flooding downstream of the dam by temporarily storing the flood volume and releasing it later. A spillway may be located either within the body of the dam or at one end of the dam or entirely away from the dam as an independent structure. Spillways must be operated in such a way that the water which was released to the downstream will not erode or damage the downstream toes of the dam. The surface of the spillway should also be designed to withstand erosion or scouring due to the very high velocities generated during the passage of a flow through the spillway.

Physical modelling is widely used to verify the spillway design and to study various operational issues due to complex flows. In physical modelling, scaling laws were used to convert model flow information into full scale prototype values. Due to latest advancements in computational fluid dynamics and hardware technology the use of numerical techniques for the analysis of flow over a spillway is now possible.

In the present study, a physical model was built to the geometrically similar scale of 1:50 based on Froude number criteria for assessing the hydraulic performance of a spillway in respect of discharging capacity of spillway, flow over spillway profiles and the vertical pressure distribution on the spillway profile. From the preliminary studies, it was observed that the spillway profile was subjected to very high negative pressures at 25 % and below opening of the spillway gate. To eliminate negative pressures, three different spillway profiles were numerically analyzed, and the best solution was chosen for physical modelling. For numerical simulations, authors selected gated conditions with 25% of PMF discharge. The results of numerical simulation were compared to those of physical modelling.

The following were the objectives of the present study:

a) To validate the physical model results with numerical model results for original design of spillway profile for 25% opening of radial gate.
b) Comparison of numerical model results for three alternative spillway profile in terms of discharge passing through spillway, water surface profile and pressure acting on spillway profile.
c) Selected alternative was fabricated and tested in physical model for 25% opening of radial gate of spillway.

d) Validation of the physical model results with numerical model results for selected alternative spillway profile for 25% opening of radial gate.

2. Experimental facility

The studies were conducted on a 1:50 scale two-dimensional geometrically similar physical model of a spillway based on Froude number similarity located in the Union Territory of Jammu and Kashmir. The project envisages construction of four orifice spillways and two overflow crest spillways. The orifice spillway consists of 4 spans with an elliptical bottom profile of breast walls. The energy dissipator is provided in the form of a ski jump bucket with a preformed plunge pool. Figure 1 (a) shows the side view of the dry model of spillway provided with a preformed plunge pool. The model includes a part of the upstream river, orifice spillway, ski jump bucket profile and downstream plunge pool. The model was fabricated using transparent Acrylic and PVC foam sheets. Piezometers were provided along the centre line of the centre span of the spillway for piezometric pressure measurement. Sharp crested Rehbock weir was used for measurement of discharge. Figure 1(b) shows a longitudinal section of the spillway for the original and three alternatives of spillway profile.

Figure 1 (a). Side view of the dry physical model (1:50 scale) of spillway and plunge pool for original design of spillway.

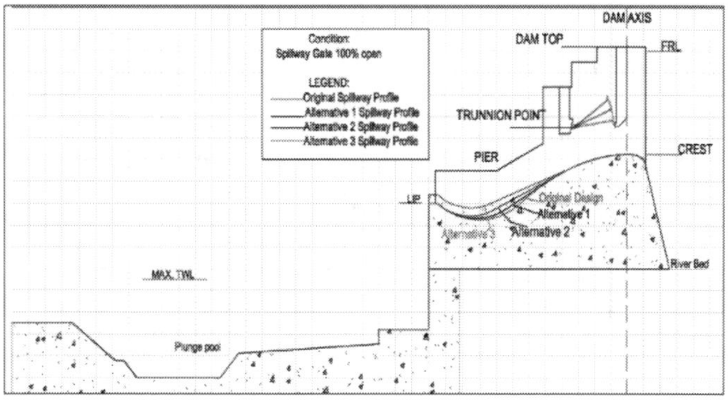

Figure 1 (b). A longitudinal section of the Prototype spillway for the original and three alternate spillway profiles.

3. Numerical methodology

Computational fluid dynamics (CFD) program Flow 3D is widely used to analyze spillway flows. The program solves the Reynolds Averaged Navier-Stokes (RANS) equations along with RNG turbulence closure model and computes free surface through volume of fluid method to track free surface motion given by Hirt and Nichols (1981). The fractional area/volume obstacle representation (FAVOR) technique given by Hirt and Sicilian (1985) is used to determine the void or flow region within each cell. One means of tracking the free surface is the VOF method. High resolution in non-flowing areas can be achieved using finer grid spacing and nested mesh technique. Complex flow physics such as short lengths of spillway, large variation in reservoir levels, high flow depths and wide range of Froude numbers varying from 3 to 9 are covered by FLOW 3D.

Savage and Johson (2001) simulated the flow over an ogee spillway using Flow-3D and found a reasonably good agreement between physical and numerical model results in terms of both pressures and discharges. However, the Flow-3D slightly overestimated the discharges. Chanel and Doering (2008) used FLOW-3D and concluded that CFD should not be considered a complete replacement for physical modelling. Ho et al. (2006), Ho et al. (2003a) and (2003b) concluded that the computed results from CFD over-estimated the velocity and hence underestimated the pressure distribution along the spillway. Chatila and Tabbara (2004), Dan Gessler (2005), Ho and Riddete (2010), Zohaib et al. (2015), Nguyen and Wang (2015) and Erpicum et al. (2017) validated the numerical predictions with experimental data with minor errors.

3.1. Numerical model setup

To simulate the given flow, a three dimensional geometry was created using AutoCad software of the similar scale as prototype and exported as a stereo lithographic (Stl) file and then this file was imported into CFD (Flow-3D) software for numerical analysis. The computational domain for numerical modeling is shown in Fig. 2. Two porous baffles were also included in the mesh as a means of computing the flow rate in Flow-3D software by defining them as flux surfaces and are arbitrarily placed at x=-33 m and x =75 m from dam axis. A manual multi grid method was adopted to speed up convergence to a steady-state solution. A rectangular Catersian grid type mesh with a uniform cell size in all directions was used for all mesh blocks. The mesh is extended upstream and downstream of the structure to view fluid movement. Coarse mesh (2 cm) selected for upstream and downstream sections and finer (1 cm) for spillway portion to reduce the computation time. Computational multi grid meshing with uniform cell size for numerical modelling is shown in Fig. 3.

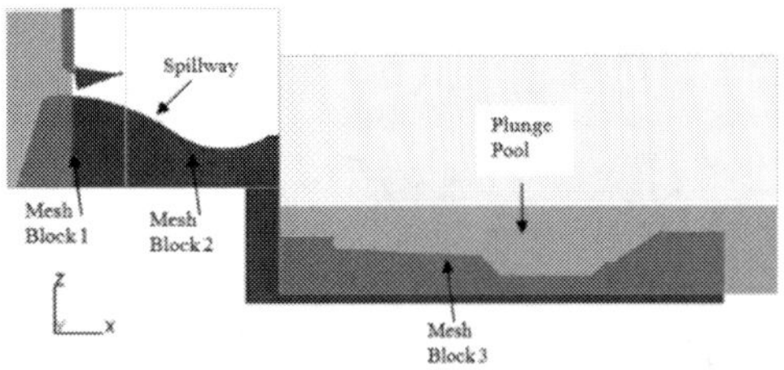

Figure 2. Computational domain for numerical modeling

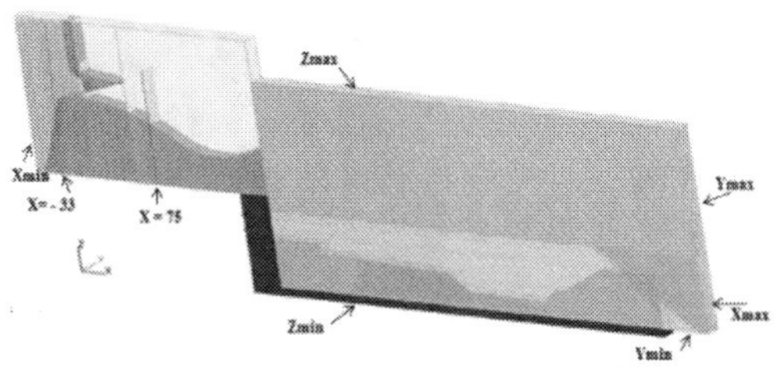

Figure 3. Computational multi grid meshing with uniform cell size for numerical modelling

To simulate given flow, it is important that the boundary conditions accurately represent what is physically occurring. Because the flow is defined in Catersian coordinates, there are six different boundaries on the computational mesh domain. To handle mesh boundary conditions, the program adds fictitious boundary cells to the grid with their values either fixed at a given value or updated as the calculations continue. The boundaries on the mesh and their coordinate directions were set as follows: Xmin – Specified pressure and fluid elevation with a hydrostatic pressure distribution; Xmax – outflow; Ymin and Ymax – wall with no slip condition; Zmin – wall with no slip condition, Zmax – pressure boundary with a gauge pressure equal to zero (atmospheric) (see figure 3). In running the Flow-3D CFD software, computation modules of viscosity, turbulence and gravity are activated for all case studies. After pre-processing, fluid flow solver option was selected. The post processing was carried out to extract the results, using the "Flow Sight" post processing software.

3.2. Validation of numerical model

In this study, prototype data was not available to compare the results of the numerical model, so we performed a validation of numerical model using the results of the physical model. Since we had already done experiments on the physical model of the original design of the spillway profile, we first did numerical modeling and validation for the same. Numerical model studies were conducted for only one span of spillway. Numerical and physical model results were compared in terms of discharge, water surface profile, and pressures along the center span of the spillway profile.

It was observed that water profile and discharge are in reasonably good agreement with each other. However, for the pressure values at few locations, a significant difference in low and high pressure values was observed, but it follows similar trend and magnitudes as observed in the physical model results (refer figure no. 4 (a) and (b)).

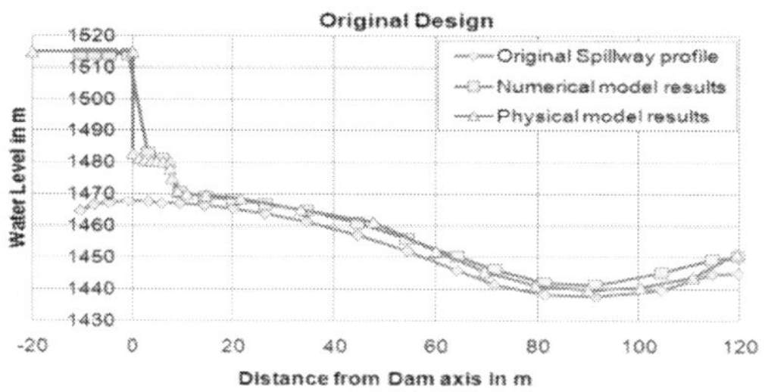

Figure 4(a). Comparison of water surface profile results from physical and numerical models for the original spillway profile

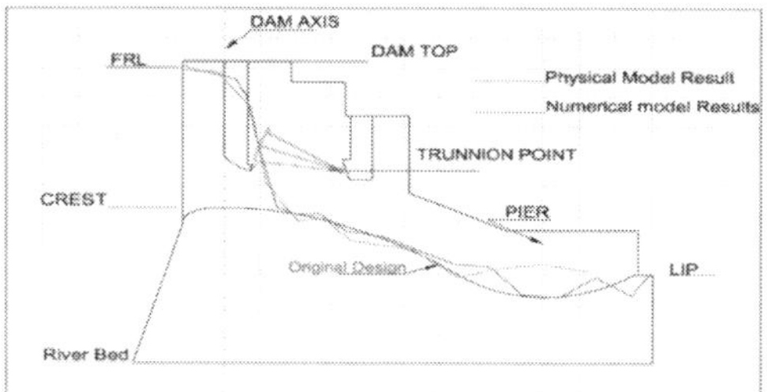

Figure 4(b). Comparison of pressures acting on spillway profile results from physical and numerical models for original spillway profile

4. Results and discussions

According to preliminary studies for the original spillway design, the spillway profile was subjected to extremely high negative pressures (ranging from - 4 m to - 9 m) at 25% and below opening of the spillway gate. Therefore, for numerical simulations, authors selected gated conditions with 25% of PMF discharge. To eliminate negative pressures and to prevent wear and tear of spillway profile due to cavitations, three alternative profiles of spillway were analysed numerically and the most effective and optimal solution was selected for physical modelling. Table 1 show the various model operating conditions considered for comparison of results.

The longitudinal section of spillway with original and all three alternative spillway profiles were shown in figure no. 5. After validating the original design, all three alternatives were numerically simulated using CFD Flow 3D software. The numerical analysis results were compared in terms of discharge, water surface profile and pressures on spillway profile. Table 1 show the various model operating conditions considered for comparison of results.

Table 1. Model operating conditions.

Spillway Profile	Operating condition in %	Required Spillway Discharge through one span in m^3/sec	Head above spillway crest in m	Froude Number (F_o)	Physical Modelling	Numerical Modelling
Original Design	25	2704	47.5	2.218	Yes	Yes
Alternative I	25	2704	47.5	2.081	No.	Yes
Alternative II	25	2704	47.5	2.191	No.	Yes
Alternative III	25	2704	47.5	2.823	Yes	Yes

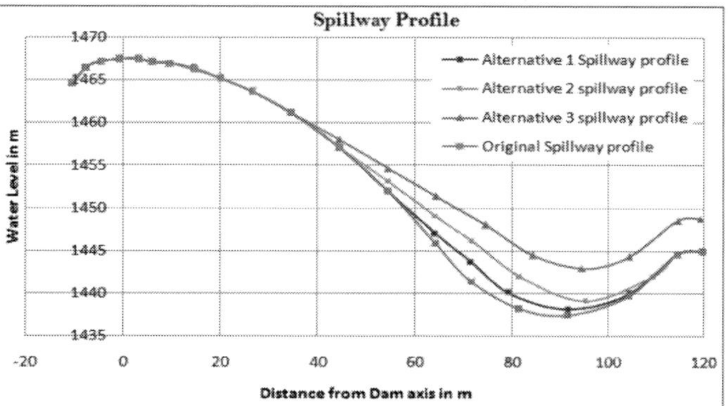

Figure 5. Longitudinal section of spillway with original and all three alternative spillway profiles

4.1. Water surface profile

Water surface profiles are important to ensure that flood water is not interfering with other structures such as gate trunnion, bridges at crest or raised gates or overtopping the spillway divide walls. From numerical analysis it is observed that for all alternatives, the water surface profiles are following spillway profile and it is not interfering with the gate trunnion or raised gates or is not overtopping the divide walls throughout the length of spillway chute (refer figure no. 6 to 9)

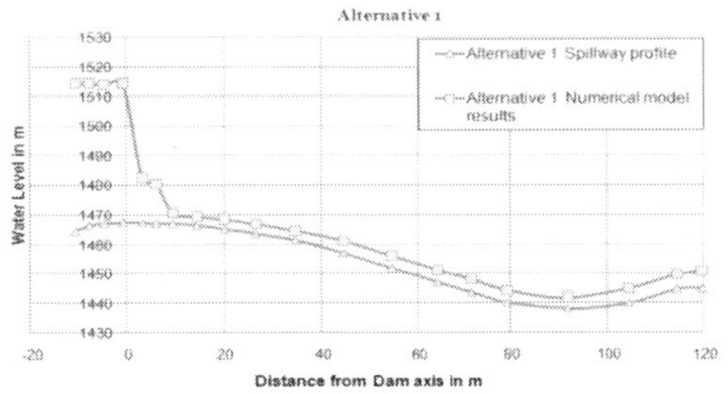

Figure 6. Plot for water surface profile for Alternative I obtained from numerical analysis.

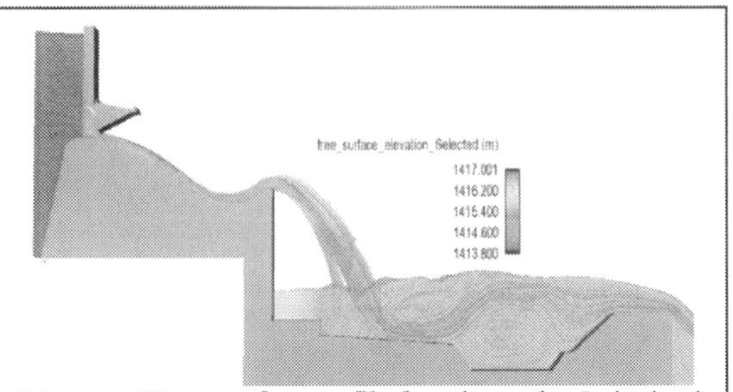

Figure 7. Water surface profile for Alternative I obtained from numerical analysis.

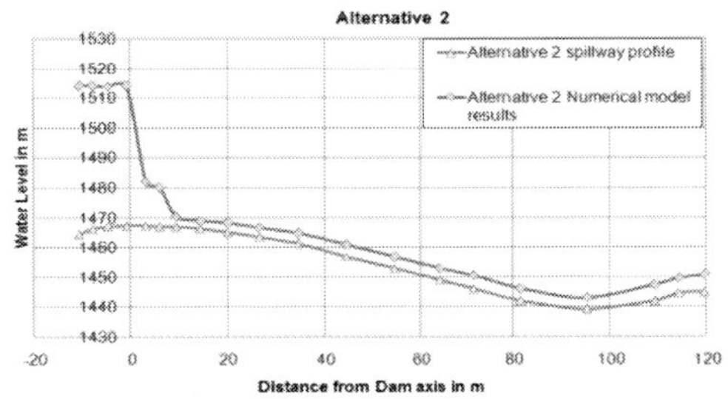

Figure 8. Water surface profile for Alternative II obtained from numerical analysis

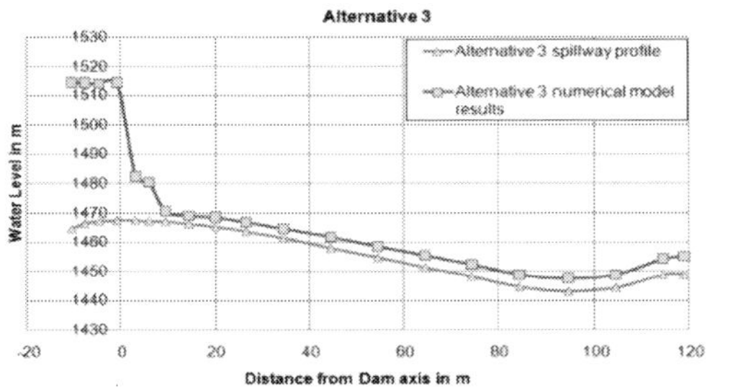

Figure 9. Water surface profile for Alternative III obtained from numerical analysis

4.2. Discharge passing through spillway opening

Table no. 2 shows a comparison between numerically obtained discharge and required discharge at 25% gate opening at reservoir level at FRL for all three alternatives. When the physical and numerical model results for the discharge passing through the spillway for a 25% gate opening were analyzed, it was found that the discharge percentage difference for the original profile was 5.31 percent, while it was 6.72 percent for Alternative 3. It indicates that the more discharge is passing through Alternative 3 profile as compared to original profile. The discharge through the Alternative III spillway profile is nearly 1.38 % more than that passes through the Alternative I spillway profile.

Table 2. Comparison of discharge passing through each alternative from numerical analysis.

Spillway profile	Operating condition	Required Model Discharge passing through spillway in m^3/s	Numerical model Discharge in m^3/s	Percentage Difference between numerical and physical model discharge
Original	Reservoir upstream water level at FRL and 25% opening of spillway gate	2,704.00	2,847.55	5.31%
Alternative 1			2846.16	5.26%
Alternative 2			2848.97	5.36%
Alternative 3			2885.72	6.72%

4.3. Pressure distribution on spillway profile

The possibility for cavitation damage caused from excessive sub atmospheric pressure was analyzed using pressure simulations along the spillway chute. Figure 10 to 13 shows the pressure distributions over spillway profile for all three alternatives for 25% opening of gate. The spillway profile was raised at a tangent point at the junction of the spillway downstream profile and bucket, to achieve positive pressure values on the spillway profile.

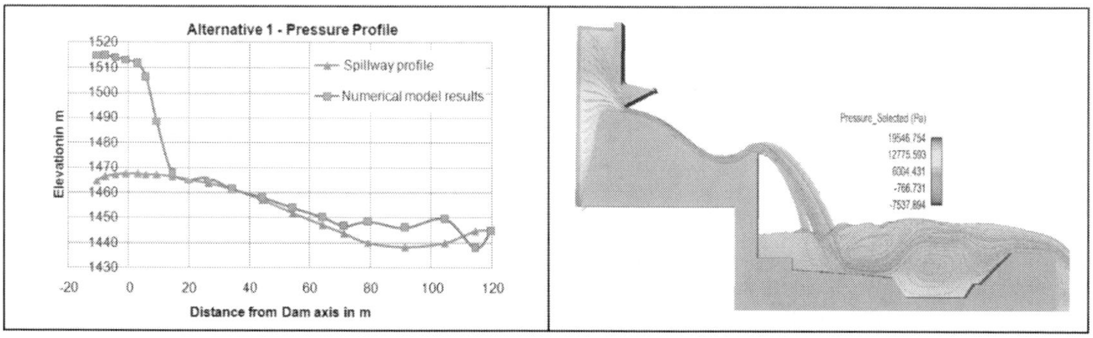

Figure 10. Pressure distribution over spillway profile for Alternative 1

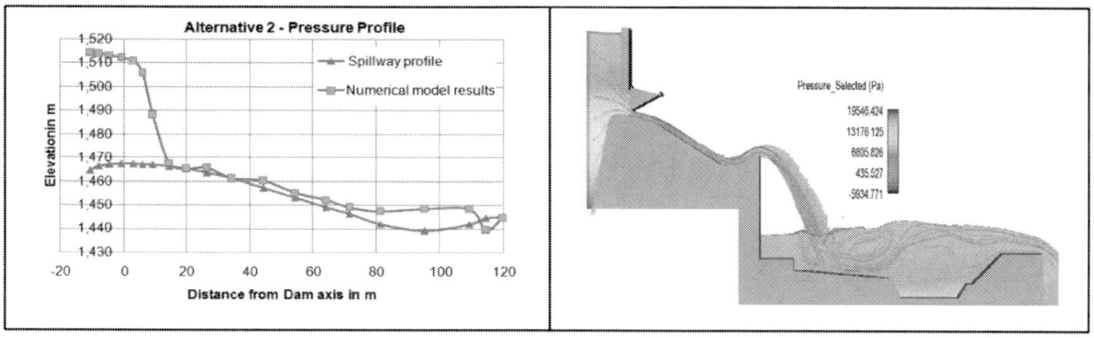

Figure 11. Pressure distribution over spillway profile for Alternative 2

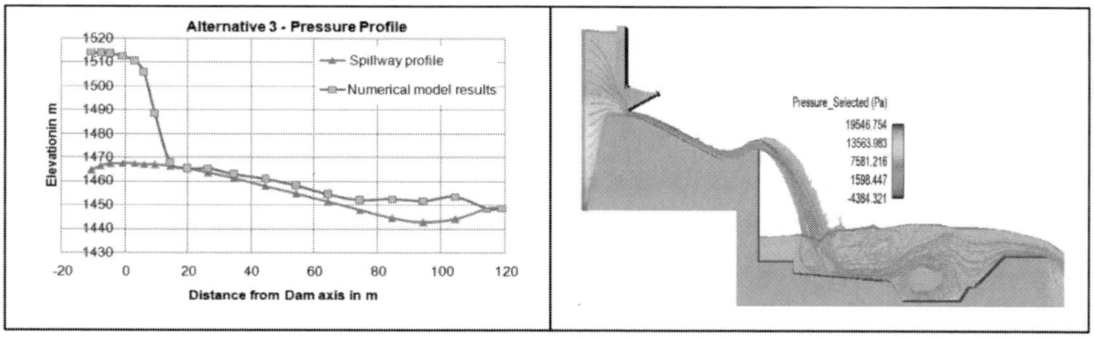

Figure 12. Pressure distribution over spillway profile for Alternative 3

In the bucket part, there is more difference in the results of the physical model and the results of the numerical model, this may be due to shorter simulation time because of which steady state is not achieved or due to coarser mesh. From Figure 10, we can see that for alternative 1, the pressure values improved due to lifting of the bucket at the tangent point. For alternative 2, the bucket is lifted further (more than alternative 1) at the tangent point, due to which the pressure values are further improved (see Figure 11). For alternative 3, the spillway profile and bucket lip both are raised. Figure 12, shows the pressure distribution over spillway profile for alternative 3.

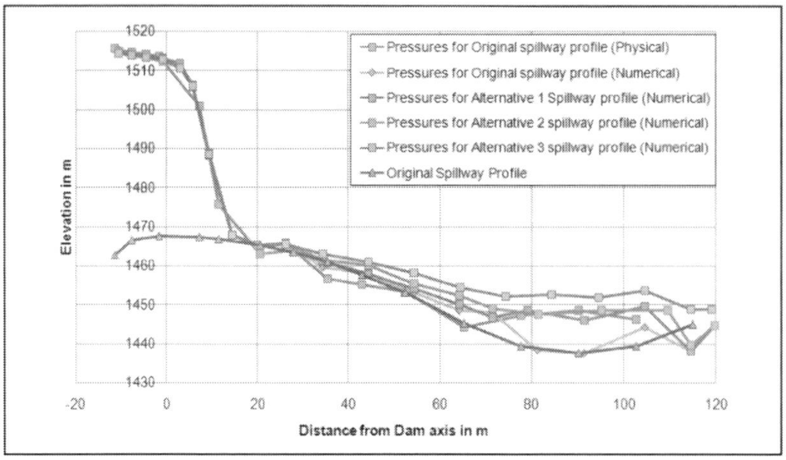

Figure 13. Comparison of pressures acting on spillway profile for all three alternatives

From figure 12, it can be seen that the pressure values are fairly positive as compared to alternative 1 and 2. Therefore, alternative 3 is selected for further study on physical model. Figure 13 shows the comparison of numerical and physical model results for original and all three alternatives.

4.4. Selection of Alternative and comparison with physical model results

After comparing numerical model results for all three alternatives, Alternative 3 was selected for physical model fabrication. After fabrication experiments for various operating conditions were repeated on a physical model, and the findings were compared once again with the numerical model results. Figure 14 shows a comparison of pressures acting on the spillway profile in a physical and numerical model. Figure 14 shows that the model predictions are in good agreement with the experimental data. There is a difference in pressure spikes at a few locations between the physical model and the numerical model results. This is due to the use of a relatively coarse mesh, which was chosen to reduce simulation time.

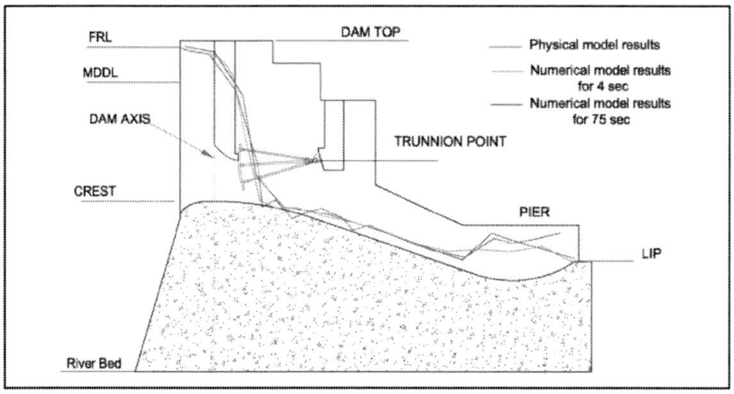

Figure 14. Comparison of results for pressures acting on the spillway profile obtained from physical and numerical model studies for Alternative 3

4.5. Comparison of physical model results for Original and selected design

The physical model results obtained for original design of spillway profile and selected Alternative 3 spillway profile were compared and it is observed that more discharge is passing through alternative 3 than through original design. The discharge passing through the Alternative 3 spillway profile is nearly 1.34 % more than that passes through the original spillway profile. From figure 15 and 16 it can be seen that the water surface profile and pressure profile are improved significantly.

From above study, it can be concluded that, the use of numerical modelling as a complement to physical modelling is extremely advantageous to obtain hydraulically optimal solutions and lowering the costs of physical model studies.

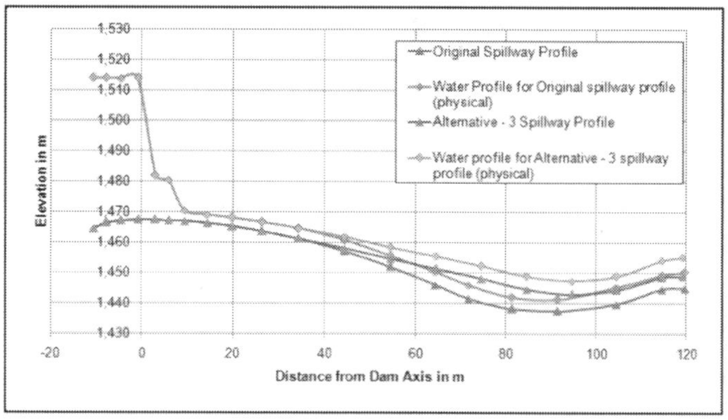

Figure 15. Comparison of water surface profile obtained from physical model results for Original and Alternative 3 spillway profile.

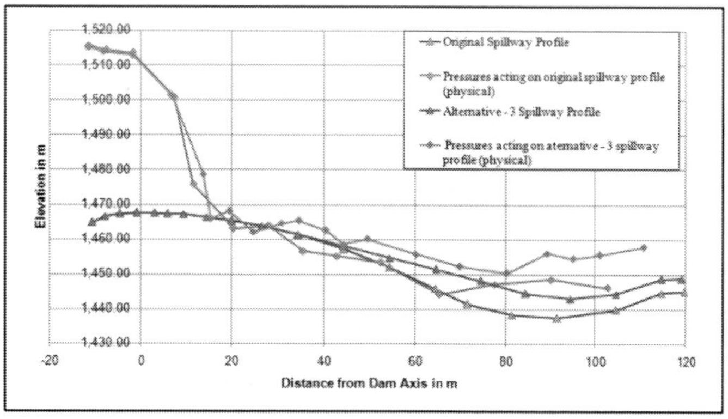

Figure 16. Comparison of pressures acting on spillway profile obtained from physical model results for Original and Alternative 3 spillway profile

5. Conclusion

This paper investigated the use of CFD Flow 3D modelling in combination with physical modelling. In the physical model study, it was observed that at 25% and lower opening of the spillway gate, the spillway profile was subjected to extremely high negative pressures (rages between (-) 4 m to (-) 9 m) for the original spillway profile. As a result, the authors chose a gated condition with a PMF discharge of 25% for numerical simulations. In order to eliminate negative pressures and prevent cavitation damage to the spillway profile, three alternative spillway profiles were numerically analysed, and the most effective and optimal solution was chosen for physical modelling.

According to the numerical analysis, the water surface profiles for all three alternatives follow the spillway profile and are not interfering with the gate trunnion or raised gates, nor are they overtopping the divide walls along the length of the spillway chute.

It is also observed that more discharge passes through alternative 3 than through alternatives 1 and 2. The discharge through the Alternative 3 spillway profile is nearly 1.34 % more than that passes through the original design of spillway profile.

The numerical simulation shows that the alternative 3 spillway profile has marginally improved negative pressures. As a result, alternative 3 was chosen and fabricated in the physical model for further studies on 2-D sectional models. The physical model results were compared to the numerical model results once more, and it was observed that the physical and numerical model results are in reasonably good agreement. Further for more refinement of results and reduce errors, we can also introduce more fine mesh, other relevant turbulence models and air entrainment models. Therefore, it can be concluded that the use of numerical modelling as a complement to the physical modelling is extremely advantageous for obtaining hydraulically optimal solutions and lowering the costs of physical model studies.

6. ACKNOWLEDGMENT

The authors are thankful to Dr. R. S. Kankara, Director, Central Water and Power Research Station, Pune for his constant encouragement and kind permission to publish the paper. The authors are also thankful to the project and design engineers for constant interaction during the course of studies.

7. REFERENCES

Chanel, P.G and Doering, J. C. (2008). "Assessment of spillway modelling using computational fluid dynamics." *Canadian Journal of Civil Engineering*, pp: 1481-1485.

Chatila, J., and Tabbara, M. (2004). "Computational Modeling of Flow over an Ogee Spillway." *Computers and Structures* 82. 1805-1812.

Gessler, D. (2005). "CFD Modelling of Spillway performance." *Proceedings of world Water and Environmental Resources Congress 2005, American Society of Civil Engineer, Anchorage, Alaska.*

Hirt, C. W and Nichols, B. D. (1981). "Volume of fluid method for dynamic of free boundaries." *Journal of Computational Physics*, Vol. 39, pp. 201-225.

Hirt, C. W. and Sicilian, J. M. (1985). "A Porosity technique for the definition of obstacles in rectangular cell mesh." *Proceedings of 4th international conference of ship dynamic, National Academy of Science, Washington D.C*, 1985.

Ho, D. K. H., Boyes, K. M. And Donohoo S. M. (2001). "Investigation of Spillway Behaviour under Increased Maximum Flood by Computational Fluid Dynamics Technique." *14th Australasian Fluid Mechanics Conference Adelaide University, Adelaide, Australia* 10-14 December 2001.

Ho, D. K. H. and Riddette, K. M. (2010). "Application of computational fluid dynamics to evaluate hydraulic performance of spillway in Australia." *Australia Journal of Civil Engineering*, Vol. 6, No.1, pp: 81-104.

Ho, D., Boyes, K., Donohoo, S., and Cooper, B. (2003b). "Numerical Flow Analysis for Spillways." *Proc. 43rd ANCOLD Conference, Hobart, Tasmania*, Oct 24-29, 2003: 227-237.

Ho, D. K. H., Cooper, B. W., Riddette, K. M., and Donohoo, S. M. (2006). "Application of Numerical Modelling to Spillways in Australia." *Dams and Reservoirs, Societies and Environment in the 21st Century: Proc. of the International Symposium on Dams in the Societies of the 21st Century, 22nd International Congress on Large Dams* (ICOLD), Barcelona, Spain, June 18, 2006, L. Berga et al., eds., Taylor & Francis Group, London, UK.

Ho, D. K. H., Donohoo, S. M., Boyes, K. M., and Lock, C. C. (2003a). "Numerical Analysis and the Real World: It Looks Pretty But Is It Right?" *Proc. of the NAFEMS World Congress, Orlando, FL*, May 27-31, 2003.

Thanh, N.C., Ling-Ling, W. (2015) "Physical and numerical model of flow through the spillways with a breast wall.". *KSCE J Civ Eng 19*, 2317–2324 (2015). https://doi.org/10.1007/s12205-015-0742-0

Savage, B. M. and Johnson, M. C. (2001). "Flow over ogee spillway: Physical and Numerical Model Case Study." *Journal of Hydraulic Engineering*, Vol. 127, No. 8, pp: 640-649.

Peltier, Yann & Dewals, Benjamin & Archambeau, Pierre & Pirotton, M. & Erpicum, Sébastien. (2017). "Pressure and velocity on an ogee spillway crest operating at high head ratio: Experimental measurements and validation.". *Journal of Hydro-environment Research*. 19. 10.1016/j.jher.2017.03.002.

Zohaib, N. , Sarwar, M. K. , Nabi Ghulam, (2015). "Hydraulic performance assessment of an orifice spillways using CFD modeling." *Sci. Int.(Lahore)*,27(2),1303-1308.

DOI: https://doi.org/10.26077/d163-390d

Hydraulic Performance of Piano Key Weir with Linear and Curvilinear Profiles of Outlet Keys

Prakash Kumar Sinha[1], Binit Kumar[2] & Zulfequar Ahmad[3]
[1]*Research Scholar, Indian Institute of Technology Roorkee, Roorkee 247667, India*
[2]*Former PhD, Indian Institute of Technology Roorkee, Roorkee 247667, India*
[3]*Professor, Indian Institute of Technology Roorkee, Roorkee 247667, India*
E-mail:pk_sinha@ce.iitr.ac.in (Corresponding Author)

Abstract *With rising demand for storing water in reservoirs, many spillways require modification by optimizing their shape for enhancing their discharge capacity. A Piano Key Weir (PK weir) is a modified weir and it is installed at the dams to augment the discharge capacity of the weir. Due to different geometrical configurations of PK weirs, the hydraulic performance of the weir is still in evolving research. In this study, hydraulic performance of two different models of PK weir is studied experimentally. All the configurations of both the PK weir are same except the shape of the outlet keys. The outlet key of the first model is linear whereas for the second model a curvilinear ogee shape is adopted. Ten discharges between 0.0075-0.03 m³/s and ratio of head over crest to weir height between 0.17-0.58 m were considered for each model. Water surface profiles were plotted along longitudinal and transverse directions at various sections of the PK weir. The effect of the weir geometry on discharging capacity was studied and it is found that the PK weir with curvilinear shaped outlet key is hydraulically more efficient than the linear shaped key. The coefficient of discharge of PK weir with curvilinear shaped outlet key is 12.41-18.15% more than the linear shaped key.*

Keywords: *Curvilinear, Hydraulic-performance, Piano Key Weir, Water flow profile*

1. Introduction

A Piano Key Weir (PK weir) is currently designed to increase the overflow discharge capacity of dams. It is one of the types of a modified version of the labyrinth weir structures. It is preferred over labyrinth weirs in rehabilitation purpose for existing dams. It increases the discharge capacity of the dams. An universal design technique of PK weirs is not available now due to the complexity of its shape (Ribeiro et al. 2012). It is essential to estimate the coefficient of discharge of the PK weir with high clarity in order to ensure a safe and cost-effective design of the PK weir. The precision of the coefficient of discharge estimation is very important since it is required to deal with the dams' overflow capacity. A PK weir is considered to be a novel and realistic alternative for dam rehabilitation and construction of new projects with limited space.

Hydrocoop (France), in partnership with three institutions, Biskra University (Algeria), the Hydraulic Laboratory in Electricite de France (France) and IIT Roorkee (India) proposed the PK weir at the turn of the 21st century to replace the long-established weir so that it can augment the discharging capacity of the weir (Erpicum et al. 2017; Lemperiere and Ouamane 2003). The PK weir was originally developed by Blanc and Lemperiere (2001) to improve the performance of labyrinth-type weirs. At equal head, initial model tests showed that a PK Weir enhances discharge capacity by four times that of traditional frontal weirs (Lemperiere and Ouamane 2003). The 1st prototype PK weir was constructed in France at the Goulours Dam (Laugier 2007). The 2nd PK weir was built on the St. Marc Dam, whereas the 3rd was installed on Gloriettes Dam (Bieri et al.2009). The design procedure was first reported by Machiels et al. (2011b). Historical reviews on the evolution from the labyrinth to the PK weir are summarized by Schleiss (2011) and Lemperiere et al. (2011).

PK Weir's geometry is complicated since it is regulated by a vast number of parameters. In certain laboratories, a naming standard was created particularly to combine the notations (Pralong et al. 2011). The major geometric parameters are W_i and W_o, the width of the inlet and outlet keys, respectively, P is the height of weir, P' is the depth below inlet key, W_u is the width of one cycle or PKW unit, N_u is the total number of PK weir units, B is the total stream wise crest length, the lengths B_o and B_i of the upstream and downstream overhangs, B_b is the base length and T is the thickness of wall. The width of a PK weir unit W_u is equal to $W_i + W_o + 2T$ and the total width W of the weir is equal to $N_u \times W_u$. The developed crest length L_u of a PKW unit is equal to $W_u + 2B$ and the total developed crest length L of the weir is calculated as $N_u \times L_u$ (Erpicum et al. 2017). All the geometrical parameters are shown in the Figure 1.

To determine the discharge capacity of the PK weir, many researchers have performed a series of both experimental and numerical research (Erpicum et al. 2017). Ribeiro et al. (2012) carried out a series of experiments on numerous PKWs with upstream and downstream overhangs (Type-A) and identified the characteristics that had the greatest impact on the flow. They came up with a head–discharge relationship for PK weir by expressing the enhanced discharge ratio of PKW in relation to a rectangular sharp-crested weir. Nowadays extensive and systematic model studies have yielded more elaborated design equations with reasonable accuracy within the parameters given by the researchers (Cicero & Delisle 2013; Crookston et al. 2018 and Kumar et al. 2019). PK weir are classified according to the presence or absence of overhangs in the upstream and downstream of the weir (Lemperiere et al. 2011), if upstream and downstream, both part of weir contains overhang, and the overhangs are identical then that is called type A PK weir. When only upstream of the PK weir Includes overhang then that is called the type B PK weir. Whereas PK weir only downstream overhang is called as Type C PK weir. When piano key weir does not contain any overhangs, it has only an inclined floor then that is called type D PK weir. Kabiri-Samani and Javaheri (2012) conducted a series of experiments and derived equation for coefficient of discharge for all types of PK weir (Type A, B, C & D) for both free and submerged condition.

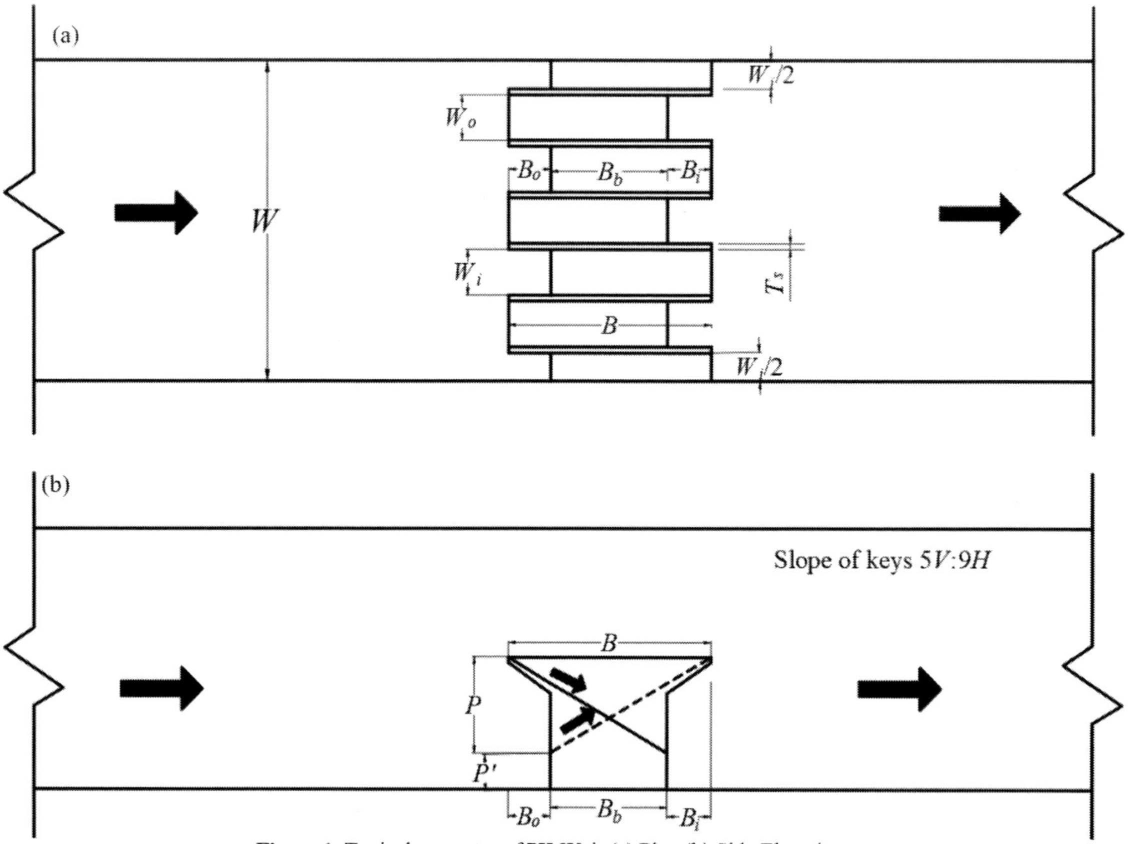

Figure 1. Typical geometry of PK Weir (a) Plan (b) Side Elevation

Previous investigations indicate that almost all the studies till date considered the linear inlet outlet keys of PK weir. Limited literatures are available on the discharge coefficient for the linear and curvilinear outlet keys of PK weir. The present study deals with the experimental result of discharge coefficient for the linear and curvilinear outlet keys of PK weir.

2. Methodology

2.1 Geometry of models

In this paper, two models of Type-A PK weir with different shape of outlet keys have been studied. The first model of piano key weir with three cycle consists of linear shape of both inlet and outlet keys (indicated as L-PKW). All the parameters of the second model are similar as L-PKW except the shape of the outlet key which is curvilinear and indicated as CL-PKW. The curvilinear shape of the outlet key was selected based on waterway experimental station (WES) standard spillway shapes given by U.S. Army corps of engineers USBR (1987). The curvilinear profile of the outlet key slope of CL-PKW was selected as ogee downstream profile based on WES standard spillway shapes. The equation given by the WES for the ogee shape is as follows.

$$X^n = kH_d^{n-1}Y \tag{1}$$

Where (X, Y) are the coordinates of points on the outlet key with the origin at highest point on key, H_d is the design head including the velocity head, k and n are the constants depending upon the slope of the upstream face. The values of k and n given by WES for vertical upstream face is 2 and 1.85 respectively. The reverse curve radius r has been taken as half of the design head and the internal angle made is 60^0. Figure 2 shows a schematic diagram of the curvilinear profile. This form is developed from the lower surface of a free-falling nappe and is recognized for its efficiency and safety in passing flows (Savage & Johnson 2001). At the bottom of the main slope, a reverse curve was added to smoothly direct the falling jet downstream. To obtain the slope profile for the model with curvilinear outlet keys, the design head required to be fixed above the crest. Most of these weirs constructed or built till date have a total design head to weir height ratio of less than one, because PKWs are very efficient at lower heads (Crookston et al. 2019). Therefore, for the present study, the profile of the outlet slope of the curvilinear model were designed with $\frac{H_d}{P}$ equal to 0.25.

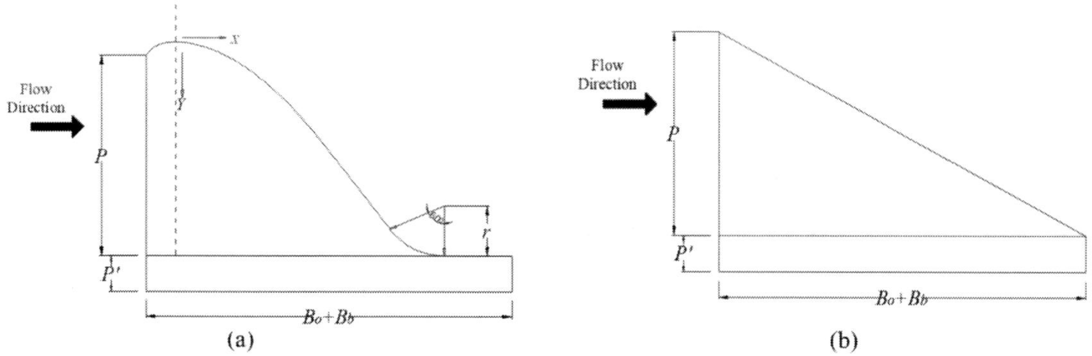

Figure 2. Profile of outlet key for (a) CL-PKW (b) L-PKW

For the calculation of discharge coefficient, the general formula for a rectangular sharp crested weir were used (Kabiri-Samani and Javaheri 2012, Kumar et al. 2019) i.e.,

$$Q_{PKW} = \frac{2}{3} C_{PKW} W \sqrt{2g} H^{\frac{3}{2}} \tag{2}$$

Where Q_{PKW} is discharge, H is upstream head above the crest of the weir, W is width of weir (in transverse direction) and C_{PKW} as discharge coefficient of PK weir. Ribeiro et al., (2012) calculated the discharge enhancement ratio 'r' between PK weir discharge Q_{PKW} and a sharp crested weir discharge Q_S and has been defined as:

$$r = \frac{Q_{PKW}}{Q_s} = \frac{C_{PK} L \sqrt{2g} H^{\frac{3}{2}}}{C_d W \sqrt{2g} H^{\frac{3}{2}}} = f\left(\frac{L}{W}\right) \tag{3}$$

Where W is total width of the PK weir. The discharge coefficient C_d of the sharp crested standard weir can be assumed as almost constant with C_d =0.42 (Hager and Schleiss 2009). Anderson (2011), Anderson and Tullis (2011,2012,2013) conducted experimental studies on various PK weir shapes, and Crookston et al., (2018) later developed an empirical formula for the coefficient of discharge. While comparing their study with Eq. (2), C_{PKW} comes out to be

$$C_{PKW} = \left[1\Big/\left\{a_1 + b_1\left(\frac{H}{P}\right) + c_1\Big/\left(\frac{H}{P}\right)\right\} + d_1\right]\left(\frac{L}{W}\right) \tag{4}$$

Cicero and Delisle (2013) conducted experimental studies on the various piano key weir and proposed the equation for the coefficient of discharge.

$$C_{PKW} = \frac{3}{2}[a_2 + a_3\left(\frac{H}{P}\right) + a_4\left(\frac{H}{P}\right)^2 + a_5\left(\frac{H}{P}\right)^3 + a_6\left(\frac{H}{P}\right)^4] \tag{5}$$

Where, $a_1, b_1, c_1, d_1, a_2, a_3, a_4, a_5, a_6$ are the empirical coefficients and can be found in the literature. Eq. (5) and (6) has been used to plot the graph in Figure 6.

2.2 Experimental Setup

All the experiments were conducted in the Hydraulics laboratory of Civil Engineering Department at Indian Institute of Technology, Roorkee. Two PK weir models were installed in a flume of length of 15 m, width of 0.39 m and height of 0.5 m, respectively. The flume has a 13 m long glass side wall for observing the flow over the weir and visual observation over the weir. A rectangular tank of length 2.5 m, width 1.15 m, depth 1.0 m was provided at the end of the flume for the storage of water and recirculation of flow. A pump was installed from the tank to the flume inlet for supplying the water. A control valve and an electro-magnetic flow meter were installed on the pump to control the flow and measure the discharge, respectively. At the entrance of flume honeycomb grid walls were installed to minimize the disturbances in the flow. A flow straightener and wave suppressor were also provided at the entrance of the flume to suppress cross current and surface disturbances respectively. A point gauge of ±0.0001 m least count was installed to measure the water level in the flume. The experimental set up is shown in the Figure 3.

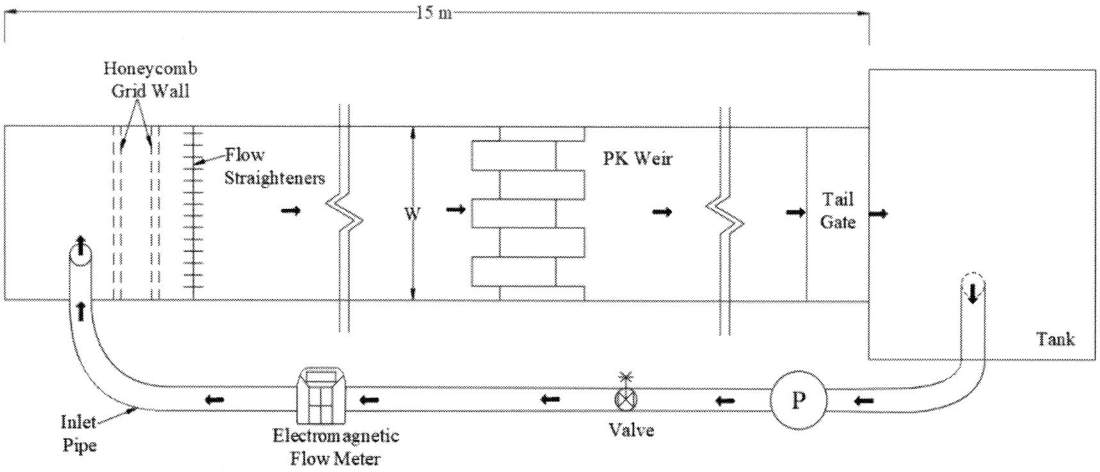

Figure 3. Schematic diagram of the experimental setup

The PK weir model made of 0.006 m thick acrylic sheet and installed at the midway of the flume. To measure the water level, the point gauge system was placed at $2P$, i.e., 0.210 m, upstream from the outflow crest. Ten discharges between 0.0075 to 0.03 m^3/s were used at each model of PK weir. The range of H/P ratio for linear and curvilinear models were 0.19-0.63 and 0.17-0.58 respectively. After the flow is uniform and constant for at least 20 minutes, the point gauge was utilized to measure the water level for each discharge. At least three water level readings were measured across a section to ensure that a steady state flow had been reached. The minimum value of H is kept higher than 0.03 m to eliminate the scale effects as mentioned by Erpicum et al., (2016).

Table1. Geometrical parameter of PK weir models

S.N.	Parameters	Linear model (m)	Curvilinear model (m)
1	Channel width (W)	0.39	0.39
2	Side wall overflowing length (B)	0.254	0.254
3	Inlet key width (W_i)	0.06	0.06
4	Outlet key width (W_o)	0.06	0.06
5	Inlet key overhang length (B_i)	0.064	0.064
6	Outlet key overhang length (B_o)	0.064	0.064
7	Crest thickness (T)	0.006	0.006
8	Weir cycle number (N)	3	3

9	Total crest length (*L*)	1.884	1.884
10	Height of weir (*P*)	0.105	0.105
11	Approach flow depth (*P'*)	0.035	0.035
12	Inlet key slope (S_i)	29.06^0	-
13	Outlet key slope (S_o)	29.06^0	-
14	*H*/*P* ratio	0.19-0.63	0.17-0.58

3. Results and Discussion

3.1 Water surface profile over PK weir

For both the models i.e., L-PKW and CL-PKW water surface profiles were analyzed. In the PK weir, there are two types of keys that is outlet key and inlet key, the water surface profile was analyzed on both the keys separately. At every section of PK weir, the water surface profile is not same. In this paper, the water depth at Centre of each outlet and inlet keys were taken and analyzed. Measurement points for water flow profile plot were taken from 0.75 m upstream to 1.05 m downstream from the upstream crest of the piano key weir at 0.05 m of interval and at each section, a minimum of three readings were noted and average value were plotted for the greater accuracy.

3.1.1 Water surface profile over outlet key

The Water flow profile at the center of the outlet key for the different discharges are shown in the figure 4. It shows constant head in the upstream to the start point of the piano key weir and then small drop of head at the start of the weir crest. Figure 4 shows that the head for the L-PKW is more as compared to the CL-PKW for the same discharge. In the far downstream, the flow is more gradual and steadier in the CL-PKW as compared to the other.

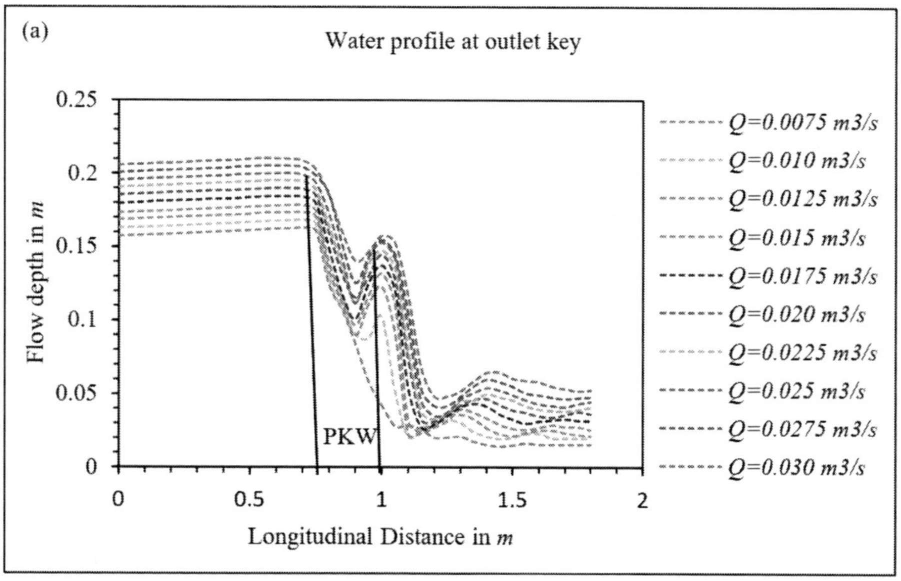

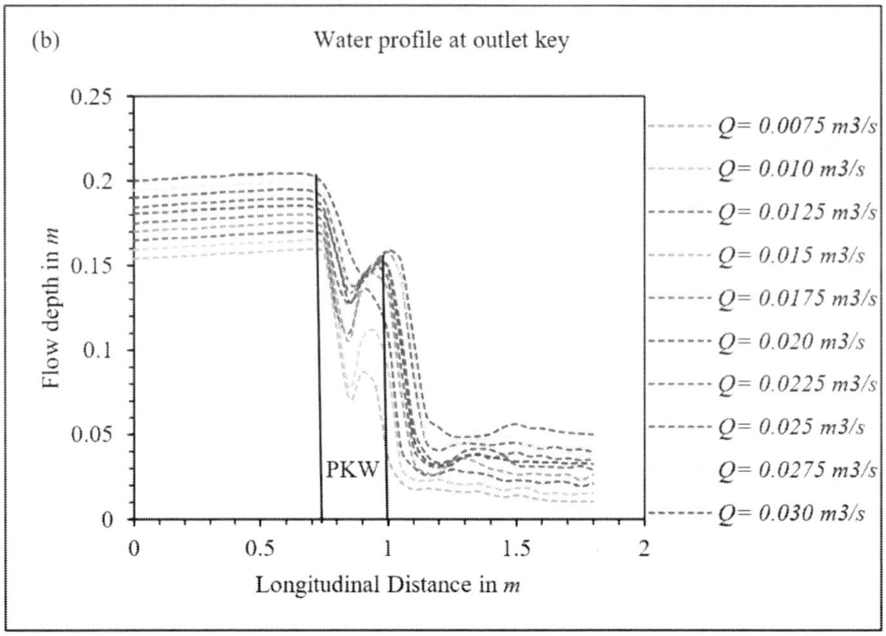

Figure 4. Water surface profile at outlet key for (a) L-PKW (b) CL-PKW

3.1.2 Water surface profile over Inlet key

The water surface profile at the center of inlet key for different discharges are shown in the Figure 5. It shows constant head in the upstream to the start point of the piano key weir and then small drop of head at start of the weir crest. Figure 5 shows that the head for the L-PKW is more as compared to the CL-PKW at the same discharge. In the far downstream, the flow is more gradual and steadier in the CL-PKW as compared to the other. In the downstream when water is falling from the inlet key the air entrainment takes place between piano key weir and flow nappe, which is more in L-PKW compared to CL-PKW.

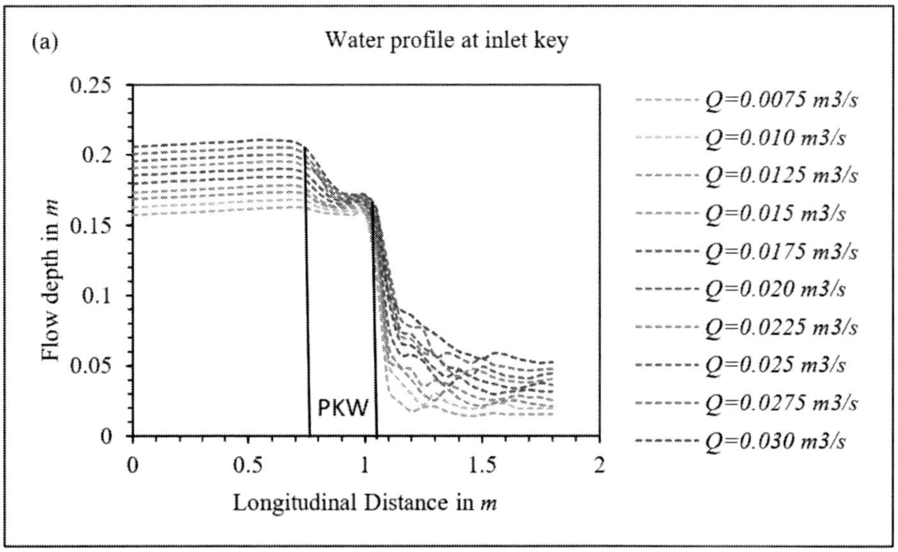

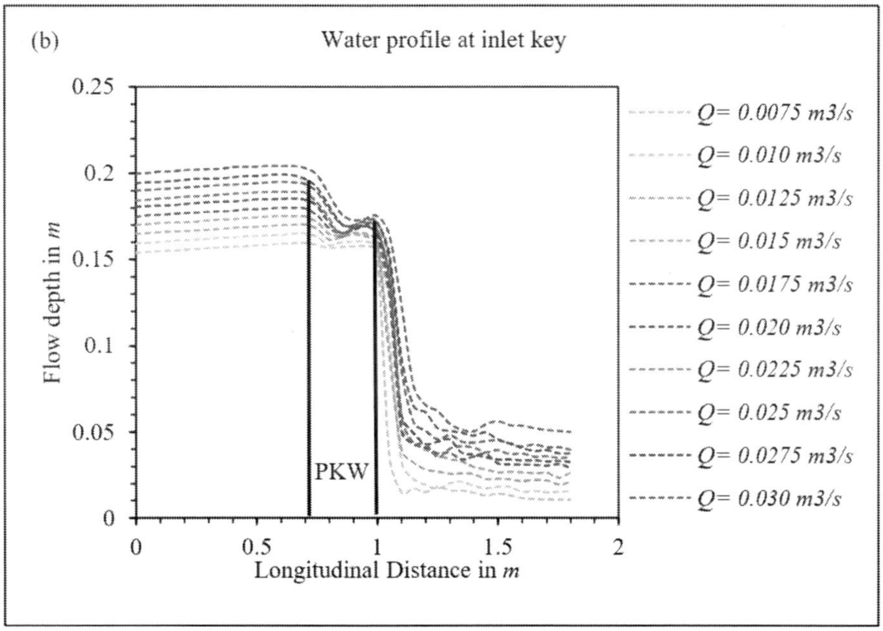

Figure 5. Water flow profile at inlet key for (a) L-PKW (b) CL-PKW

3.2 Discharge Characteristics

The discharge capacity of piano key weir can be described using the conventional formula for the weir which is mentioned in Eq. (2). The coefficient of discharge for the different discharges were calculated using equation (2) and plotted with head to weir height ratio, as shown in the Figure 6. From the figure it is clear that the discharge coefficient for the L-PKW is less as compared to the CL-PKW. For low head C_{PKW} is very high for CL-PKW model as compared to the L-PKW but as head increases the value of C_{PKW} for both the models nearing the same value for the given range of the data in the Table 1. As H/P tends to 0.6, the C_{PKW} value for both the models are almost equal.

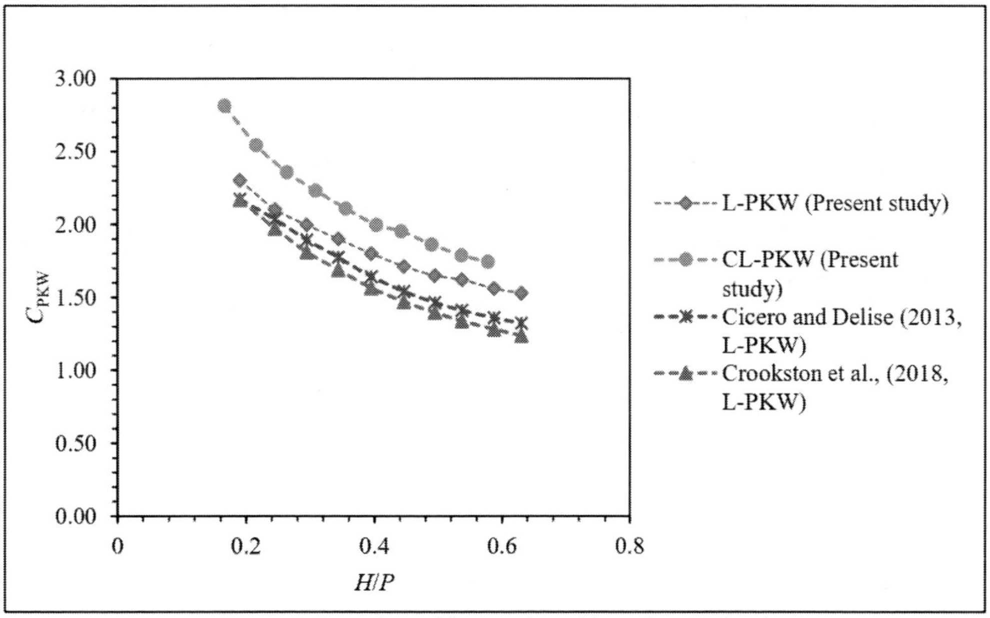

Figure 6. Comparison of linear and curvilinear shaped PK weir

For the given range of data, the relative increase in the coefficient of discharge for CL-PKW is 12.41-18.15% which is shown in the Table 2. Present study shows the similar trend as in literature by Cicero and Delise (2013) and Crookston et al. (2018) [Eq. (4) and (5)] that is plotted in Figure 6.

Table 2. Relative increase in C_{PKW} of CL-PKW

S.N.	Discharge (m^3/s)	C_{PKW} of L-PKW [1]	C_{PKW} of CL-PKW [2]	Relative increase in Percentage ([2]-[1])/ [2]×100
1	0.0075	2.30	2.81	18.15
2	0.01	2.10	2.54	17.33
3	0.0125	2.00	2.36	15.41
4	0.015	1.90	2.23	14.97
5	0.0175	1.80	2.11	14.69
6	0.02	1.71	2.00	14.25
7	0.0225	1.65	1.95	15.54
8	0.025	1.62	1.86	13.00
9	0.0275	1.56	1.79	12.70
10	0.03	1.53	1.74	12.41

From the Figure 6. it is also clear that both the models show the similar trend for the coefficient of discharge. From Table 2, it is apparent that as head increases the CL-PKW for both the models decreases that's why the PK weir is most suitable for the low head flow. Also, from Table 2, CL-PKW is more efficient than the L-PKW model.

The flow passing through a PK weir is three-dimensional, with a longitudinal jet flow over the upstream and downstream crests and a lateral jet flow over the sidewalls. The longitudinal and lateral jets pouring into the outlet key are near to the walls at low upstream heads. These jets spring merge with each other in the downstream area of the outlet key as the head increases.

This interaction of jets downstream of the outlet crest decreases the discharging capacity of outlet key. As a result, the nappe interference may be blamed for the considerable decrease in efficiency of both PKW models as head increases. It has been also shown in the literature that lateral crests submergence because of outlet flow is a key parameter to explain decrease of efficiency with increasing head.

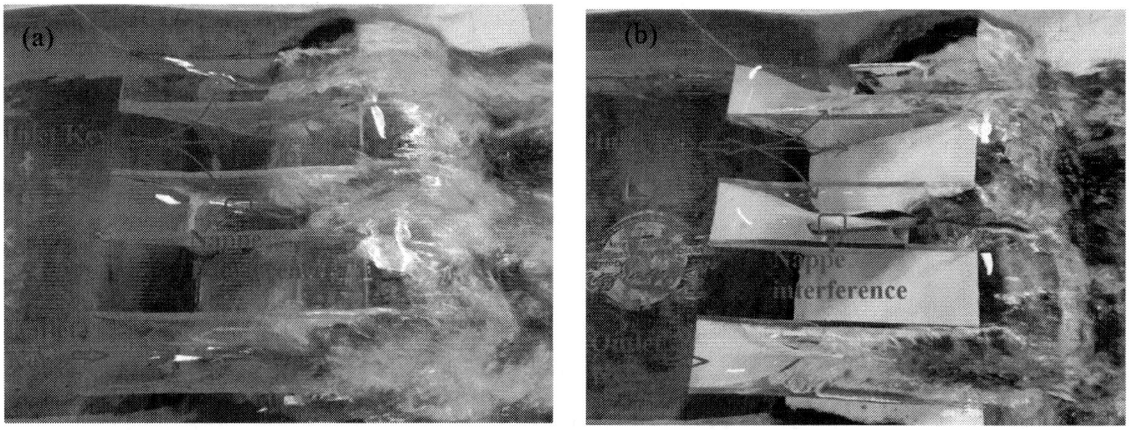

Figure 7. Flow over PK weir and Nappe Interference at Q=15×10^{-3} m^3/s (a) L-PKW (H/P=0.34) (b) CL-PKW (H/P=0.31)

The impact of curvilinear outlet key profiles on discharge efficiency of PK weir may be described as follows: The lower surface of an aerated nappe flowing over a sharp crested weir for a single upstream head determines the form of the ogee profile. The ogee spillway having a S-shaped crest profile which provides a high discharge coefficient without causing cavitation. Furthermore, across most of the outlet key, the curvilinear profiles have higher slopes than the linear profiles, allowing for quick stream escape. As a result, at identical high flow circumstances, the limits of nappe interference for CL-PKW were seen to have shifted less towards the downstream than for L-PKW, as shown in Figure 7.

4. Conclusion

In this paper two experimental models with linear slope (L-PKW) and curvilinear slope (CL-PKW) of outlet key were examined to see how the form of the outlet key slope affects the discharge capacity of piano key weir. The water surface profile for both the models were also analysed at the outlet and inlet keys of the piano key weir separately. The conclusions based on the results of this experimental study is as follows:

- The coefficient of discharge for the CL-PKW is higher than the L-PKW in the given range of data. For low head the coefficient of discharge is quite higher for CL-PKW as compared to L-PKW. The maximum relative increase in coefficient of discharge was observed to be 18.15% for $0.17 < H/P < 0.58$.
- Both the model shows similar trend with increase in head, the coefficient of discharge decreases because of the nappe interference at the outlet key and the lateral crest submergence.
- Nappe interference were observed more downstream side on L-PKW in comparison to CL-PKW.
- Water surface profile shows that for the same discharge, the hydraulic head (H) is less in the case of CL-PKW, that's why the coefficient of discharge is more for CL-PKW.

The findings of a single PK weir model with ogee shape outlet keys are presented in this study. To enhance and reinforce the findings of this paper, more tests and modifications on the curvilinear shape of PK weir can be performed. Further additional research can be done on the curvilinear shape PK weir with different design head to weir height ratio to enhance the discharge capacity of the weir. More experimental study will be helpful to design an optimal shape of these weirs.

5. REFERENCES

Anderson, R.M., Tullis, B.P. (2012). "Comparison of piano key and rectangular labyrinth weir hydraulics." *J. Hydraul. Eng.*, 138 (4), 358–361.

Anderson, R.M., Tullis, B.P. (2011). "Influence of Piano Key Weir geometry on discharge." *Proceedings of the International Conference on Labyrinth and Piano Key Weirs -PKW 2011*, Taylor & Francis Group, London, 75–80.

Anderson, R.M., Tullis, B.P. (2013). "Piano key weir hydraulics and labyrinth weir comparison." *J. Irrig. Drain. Eng.*, 139 (3), 246–253.

Anderson, R.M. (2011). "Piano Key Weir Head Discharge Relationships (Master's Thesis)." Utah State University, Logan, Utah.

Crookston, B.M., Anderson, R.M., and Tullis, B.P. (2018). "Free-flow discharge estimation method for Piano Key weir geometries." *J. Hydro-Environ. Res.*, *19*, 160-167.

Crookston, B.M., Erpicum, S., Tullis, B.P., and Laugier, F. (2019). "Hydraulics of labyrinth and piano key weirs: 100 years of prototype structures, advancements, and future research needs." *J. Hydraul. Eng.*, *145*(12), 02519004.

Cicero, G.M., Delisle, J.R., Lefebvre, V., and Vermeulen, J. (2013). "Experimental and numerical study of the hydraulic performance of a trapezoidal Piano Key weir." *Labyrinth and Piano Key Weirs II*, CRC press, London, UK, 265-272.

Cicero, G. and Delisle, J. (2013). "Effects of the crest shape on the discharge efficiency of a type A Piano Key Weir." *Labyrinth and Piano Key Weirs II*, CRC press, London, UK, 41-48.

Erpicum, S., Archambeau, P., Dewals, B., and Pirotton, M. (2017). "Hydraulics of piano key weirs: a review." *Labyrinth and Piano Key Weirs III*, CRC press, London, UK, 27-36.

Kumar, B., Kadia, S., and Ahmad, Z. (2019). "Evaluation of discharge equations of the Piano Key Weirs." *Flow Meas. Instrum.*, *68*, 101577.

Kumar, B., Kadia, S., and Ahmad, Z. (2021). "Sediment movement over type A piano key weirs." *J. Irrig. Drain. Eng.*, *147*(6), 04021018.

Kumar, B., Kadia, S., and Ahmad, Z. (2021). "Discharge characteristics of piano key weirs with and without upstream siltation." *Int. J. Civ. Eng.*, *19*(9), 1043-1054.

Kabiri-Samani, A., and Javaheri, A. (2012). "Discharge coefficients for free and submerged flow over Piano Key weirs." *J. Hydraul. Res.*, *50*(1), 114-120.

Lemperiere, F., and Ouamane, A. (2003). "The Piano Keys weir: a new cost-effective solution for spillways." *Int. J. Hydropower Dams*, *10*(5), 144-149.

Lemperiere, F., Royet, P., and Blanc, P. (2001). "Low-cost methods for safety improvement of medium and small dams." In *Dams in a european content ICOLD European symposium, Geiranger, NOR,* 443-450.

Laugier, F. (2007). "Design and construction of the first Piano Key Weir spillway at Goulours dam." *Int. J. Hydropower Dams*, *14*(5), 94.

Leite Ribeiro, M., Bieri, M., Boillat, J.L., Schleiss, A., Delorme, F., and Laugier, F. (2009). "Hydraulic capacity improvement of existing spillways–design of a piano key weirs." In *Proceedings (on CD) of the 23rd Congress of the Int. Commission on Large Dams CIGB-ICOLD* (Vol. 2, No. CONF), 100-118.

Lemperiere, F., Vigny, J.P., and Ouamane, A. (2011). "General comments on Labyrinth and Piano Key Weirs: The past and present." *Labyrinth and Piano Key Weirs I*, CRC press, London, UK, 17-24.

Leite Ribeiro, M., Bieri, M., Boillat, J.L., Schleiss, A.J., Singhal, G., and Sharma, N. (2012). "Discharge capacity of piano key weirs." *J. Hydraul. Eng.*, *138*(2), 199-203.

Machiels, O., Erpicum, S., Archambeau, P., Dewals, B. J., and Pirotton, M. (2011). "Piano key weir preliminary design method: Application to a new dam project." *Labyrinth and Piano Key Weirs I*, CRC press, London, UK, 199-206.

Mishra, R.K., and Ahmad, Z. (2021). "Discharge Capacity of a Piano Key Weir with Curvilinear Keys." 8th IJREWHS, Galway, Ireland.

Pralong, J., Vermeulen, J., Blancher, B., Laugier, F., Erpicum, S., Machiels, O., Pirotton, M., Boillat, J.L., Leite Ribeiro, M., and Schleiss, A.J. (2011). "A naming convention for the Piano Key Weirs geometrical parameters." *Labyrinth and Piano Key Weirs I*, CRC press, London, UK, 271-278.

Ribeiro, M.L., Pfister, M., and Schleiss, A.J., (2013). "Overview of Piano Key weir prototypes and scientific model investigations." *Labyrinth and Piano Key Weirs II*, CRC press, London, UK, 273-281.

Ribeiro, M.L., Pfister, M., Schleiss, A.J., and Boillat, J.L. (2012). "Hydraulic design of A-type piano key weirs." *J. Hydraul. Res.*, *50*(4), 400-408.

Schleiss, A.J. (2011). "From labyrinth to piano key weirs: A historical review." In *Proc. Int. Conf. Labyrinth and Piano Key Weirs I, CRC press, London, UK*, 3-15.

Savage, B.M., and Johnson, M.C. (2001). "Flow over ogee spillway: Physical and numerical model case study." *J. Hydraul. Eng.*, *127*(8), 640-649.

USBR (United States Bureau of Reclamation). (1987). "Design of small dams." *Water resources technical publication*, 860.

DOI: https://doi.org/10.26077/d9ec-010f

Hydrodynamics of Permeable Horseshoe Obstacle in Fluvial Environment: a Physical Modelling

H. Chanson[1], H.K. Lee[1] & W. Johnson[1]
[1]*The University of Queensland, School of Civil Engineering, Brisbane, Australia*
E-mail: h.chanson@uq.edu.au

Abstract: *Along Australian inland waterways, a number of man-made horseshoe obstacles were built for centuries. The permanent structures interacted with the streamflow across a wide range of water discharges including when they are fully-submerged. Their use ranged from water holes and fish trap at low to moderate flows, to large bed roughness and turbulent manipulation at large water discharges. The aim of the study was to gain a sound physical understanding of the hydraulic operation of horseshoe obstacle across a broad range of submergence ratio, ranging from low flows and emergent obstacle to major floods with fully-submerged structures. In this laboratory study, the physical modelling of permeable horseshoe obstacle was undertaken under controlled flow conditions, based upon a Froude similitude. Two physical models were built corresponding to a 5:1 and 20:1 geometric scaling ratio for typical riverine structures. The scale models were 3D-printed with random pattern and a porosity of 0.27, close to the porosity of rockfill material. The physical observations included visual observations, three-dimensional free-surface profiles and detailed velocity measurements. The physical observations showed a broad range of flow patterns, depending upon the submergence ratio. The porosity of the obstacle facilitated some interactions between the seepage and recirculation region, leading to changes in the wake region and its turbulence, compared to a impervious obstacle with the same shape.*

Keywords: *Physical modelling; Horseshoe obstacle; Wake region; Riverine interactions.*

1. Introduction

Along inland Australian waterways, indigenous Australians built hydraulic structures over a long period time, and there are a wide variety of waterworks (Bandler 1995, 2003). One type of structure was the rock fish trap (Mathews 1903, Flecker 1951, Dargin 1976, Bandler 2007). Although a wide range of shapes were in usage, a common shape was a horseshoe element opened downstream. Most structures were built of stones set and key without mortar or clay (Dargin 1976). The porous construction facilitated the resistance to the current and prevented siltation during floods. The permanent structures have interacted with the fluvial flow for a broad range of flow rates over log period of times, including when the elements were fully-submerged. Plates and photographs suggested that the rock fish traps were used pro-actively for fishing when the water levels were less than waist height (Fig. 1) (Dargin 1976). The constraint derived directly from the stability and safety of the fishermen/women in the river waters, since higher water depths would prevent their ability to stand, move and fish in the riverine environment (Takahashi et al. 1992, Xia et al. 2014, Chanson and Brown 2018). For other flow conditions, the horseshoe elements could retain water in the trap and rock holes during dry periods, contributing to groundwater recharge and creating drinking holes for fauna. During large flows, the structures interacted with the flood flow, generating strong secondary motion.

At low flows, the operation of horseshoe structures presents some similarity with the overflow of horseshoe waterfalls (Pasternack et al. 2006,2007, Lapotre and Lamb 2015) and man-made horseshoe weirs (Stamataki et al. 2014). At high flows, the fluid flow around the horseshoe fish trap exhibits some analogy with atmospheric boundary layer flows past canopy patches (Taddeil et al. 2016) and past shallow islands (Branson et al. 2019). The obstacle porosity plays a key role, affecting the wake flow, reducing the total drag on the structure and the risks of downstream scour (Wüthrich et al. 2020, Gillies et al. 2021). Yet, the literature on the hydrodynamic of porous horseshoe structure is nil.

The aim of the study was to gain a sound physical understanding of the hydrodynamics of porous horseshoe obstacle in rivers, across a broad range of submergence ratio. The physical investigation covered a broad range of conditions, from low flows and emergent obstacle to major floods with fully-submerged structures. The physical modelling of permeable horseshoe obstacle was undertaken under controlled flow conditions, based upon a Froude similitude with a 5:1 and 20:1 geometric scaling ratio for a typical riverine structure.

Figure 1. Historic (probably posed) photo of a man fishing in the Brewarrina Aboriginal rock fish traps / Baiame's Ngunnhu, published in The Dawn 1893 - Main flow direction from bottom right to top left.

2. Physical modelling, facility and instrumentation

New experiments were performed in two 3.2 m long 0.4 m wide channels and a 15 m long 0.5 m wide channel located in the AEB Hydraulics Laboratory of The University of Queensland (Australia). In each flume, the test section consisted of smooth PVC bed and glass walls. The bed slope was horizontal and the waters were supplied by a constant head tank. In one 0.4 m wide channel, the flow rates were recorded with a volume per time technique, well-suited for small to medium flows. In the other 0.4 m wide flume, the discharges were measured with an orifice meter installed along the supply line connected the reticulation system of the laboratory. In the 0.5 m wide flume, the flow rates were recorded with a Venturi meter. The water depths were recorded with pointer gauge. Velocity measurements were undertaken with a Dwyer® 166 Series Prandtl-Pitot tube with a 3.18 mm diameter tube made of corrosion resistant stainless steel, featuring a hemispherical total pressure tapping (Ø = 1.19 mm) at the tip and four equally spaced static pressure tappings (Ø = 0.51 mm) located 25.4 mm behind the tip. The tip design met AMCA and ASHRAE specifications and the tube did not require calibration. Additional measurements were conducted with a total head tube (Ø = 0.915 mm). The translation of the velocity probes in the vertical direction was controlled by a fine adjustment travelling mechanism connected to a HAFCO™ digital scale unit. Flow visualisations using dye injection were conducted with digital cameras, operating in movie mode at 120 fps, 100 fps and 30 fps.

Table 1. Horseshoe structure model characteristics

Reference	Height h (m)	Outer diameter D (m)	Annular thickness t (m)	Porosity	Comment
L4	0.089	0.174	0.034	0.27	Random pattern
M4	0.051	0.100	0.020	0.27	Random pattern
M3				0.29	8mm hole lattice
M2				0.64	4mm hole lattice
M1				0	Plain

Table 2. Experimental test conditions

Channel	S_o	Width B (m)	Discharge measurement technique	Discharge Q (m^3/s)	x_o (m)	Horseshoe structure model	Instrumentation
B	0	0.5	Venturi meter	0.01-0.10	8.16	L4	Pointer gauge, Prandtl-Pitot tube, total head tube
2		0.4	Orifice meter	0.010-0.05	1.185	M1 to M4	Pointer gauge
1			Volume per time	0-0.012	1.185	M1 to M4	Pointer gauge

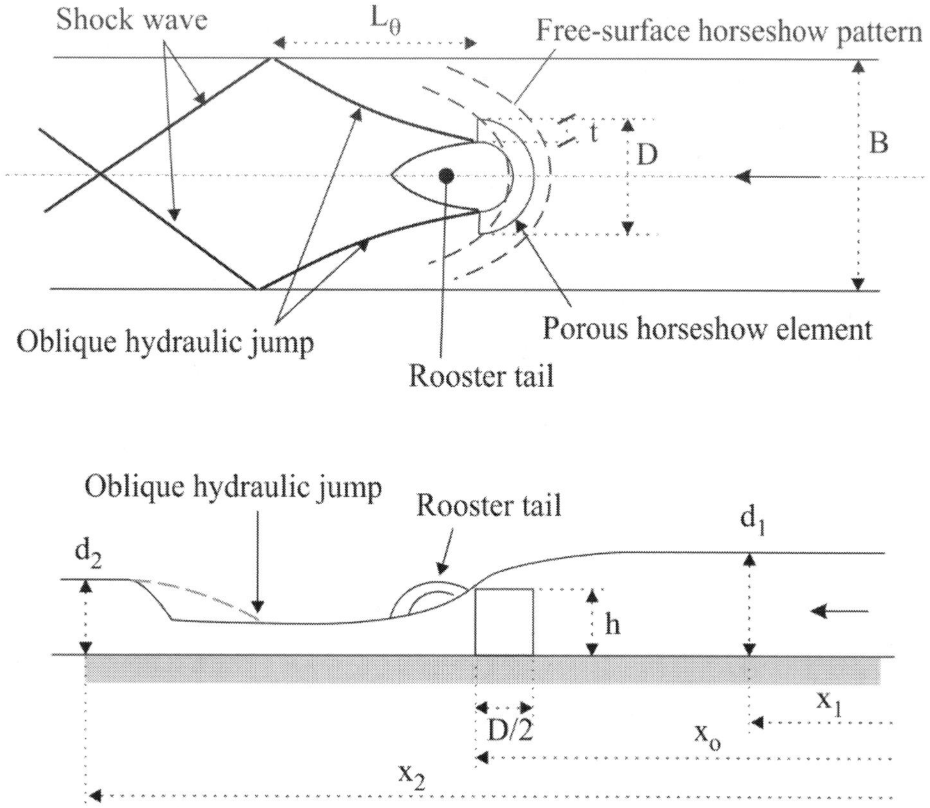

Figure 2. Definition sketch of the flow past a submerged porous horseshoe element ($d_1/H \sim 1{,}2\text{-}1{,}3$) in channels 1 & 2.

The rock fish traps were modelled with simple semi-circular horseshoe elements (Table 1). The selection of the two model sizes was mostly related to the measurement techniques. Initial tests in the small-size facility provided detailed information on the free-surface patterns. The large-size facility was used to undertake velocity measurements, with minimum blockage and impact of the Pitot tube. Since most Australian indigenous rock fish traps were between 0.5 m and 1 m high (Fig. 1), the two models corresponded to 5:1 down to 20:1 geometric scale models. The models were 3D printed out of a resin with density 1164.2 kg/m^3. A metallic plate was attached to the footing of each model, and attached to the flume invert with a magnetic strip. The elements were placed on the channel centreline. All the experiments will be conducted with zero bed slope, and the downstream gate was fully-opened, i.e. no tailwater effect. The flow conditions are summarised in Table 2 and a definition sketch is presented in Figure 2.

3. Free-surface flow patterns

The free-surface flow patterns were investigated in the three flumes for all flow conditions (Table 2), although a greater focus was given for inflow conditions such that $0.4 < d_1/h < 2$ with d_1 the inflow depth and h the vertical height of the element.

With the horseshoe element being partially-submerged, the upstream free-surface developed a horseshoe surface pattern when critical flow conditions took place around the element. This three-dimensional surface pattern is illustrated in Figure 3. A strong recirculation took place in the near-wake of the element. With the porous elements (Fig. 3A & 3B), the recirculation was fed by the seepage through the element, leading to a more stable and relatively less intense recirculation region, than for the impervious model (Fig. 3C). Interestingly, Wŭthrich et al. (2020) observed a similar surface pattern around an un-submerged rectangular building.

When the upstream flow depth was about the height of the element, some difference in flow pattern was seen between the impervious model (M1) and the porous models. With the impervious element, the overflow above the element top plunged into the recirculation and modified the wake region. This was observed for $1 < d_1/h < 1.2$-1.3. Such an overflow-recirculation was not observed with the porous elements, with a more gentle flow pattern transition from partially-submerged to fully-submerged element.

When the horseshoe element was fully-submerged, a marked horseshoe free-surface pattern was observed and shock waves separated from the element and propagated downstream (Fig. 4). At very large flow rates ($d_1/h > 2.5$), the element no longer interacted with the free-surface and acted as an isolated roughness patch

The impact of the horseshoe element on the flow was tested by systematically recording the water depths along the channel, with and without element. The inflow depth d_1 was consistently measured at $(x_1-x_o)/h = -5$ and the downstream depth d_2 at $(x_2-x_o)/h = +25$, with x_o the longitudinal location of the element (Fig. 2). Both depths were recorded on the channel centreline. Although the accuracy on the inflow depth was within ±0.5 mm, the error on the downstream depth data was larger because of the three-dimensional flow patterns, e.g. shock waves, hydraulic jumps (Figs. 2, 3 & 4). Typical results are presented in Figure 5, in terms of the dimensionless inflow depth d_1/h and dimensionless differences in water depths $(d_1-d_2)/h$.

Overall, the data showed a number of key observations. First, the presence of the horseshoe element increased the upstream depth, compared to the channel without element, for the same flow rate and channel (Fig. 5A). Second, the inflow depth d_1 was with the plain (M1), impervious element than with the porous elements (M2, M3, M4) for the same flow rate (Data not shown). The finding implied that the porous elements generated lesser flow resistance and associated energy dissipation. Third, some differences in terms of the dimensionless inflow and downstream depths, d_1/h and d_2/h, were observed between the porous elements L4 and M4, designed with the same porosity (Fig. 5B). While the difference was small in terms of the upstream depth, the data in terms of $(d_1-d_2)/h$ was much larger, This is seen in Figure 5 B for $d_c/h > 0.7$ and $d_1/h > 1$. Such differences were likely linked to (a) the different channel breadths B/h: i.e., B/h = 5.6 and 8 for the small and large channels respectively, ad (b) some difference in tailwater conditions, with a marked hydraulic jump seen in the larger channel downstream of model L4.

Figure 3. Flow patterns around a partially-submerged horseshoe element with $d_1/H \sim 0.75$. Dye injection highlights the near-wake recirculation region. (A) Model L4, $d_1/H = 0.725$, $Q = 0.012$ m^3/s; (B) Model M4, $d_1/H = 0.77$, $Q = 0.0042$ m^3/s; (C) Model M1, $d_1/H = 0.74$, $Q = 0.0040$ m^3/s.

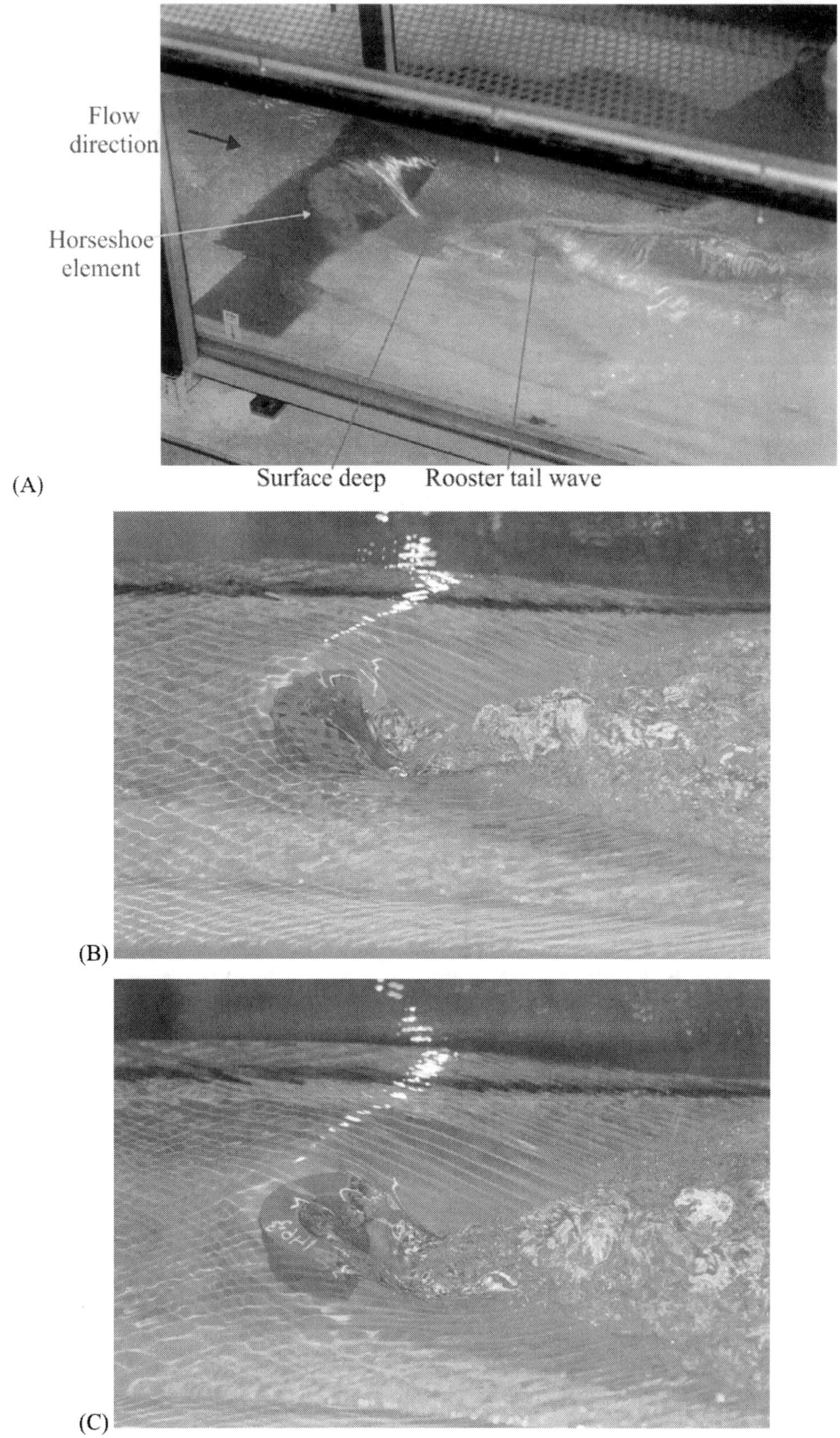

Figure 4. Flow patterns around a fully-submerged horseshoe element for $d_1/H \sim 1,4$. (A) Model L4, $d_1/H = 1.445$, $Q = 0.0447$ m^3/s; (B) Model M4, $d_1/H = 1.38$, $Q = 0.0121$ m^3/s; (C) Model M1, $d_1/H = 1.4$, $Q = 0.0128$ m^3/s.

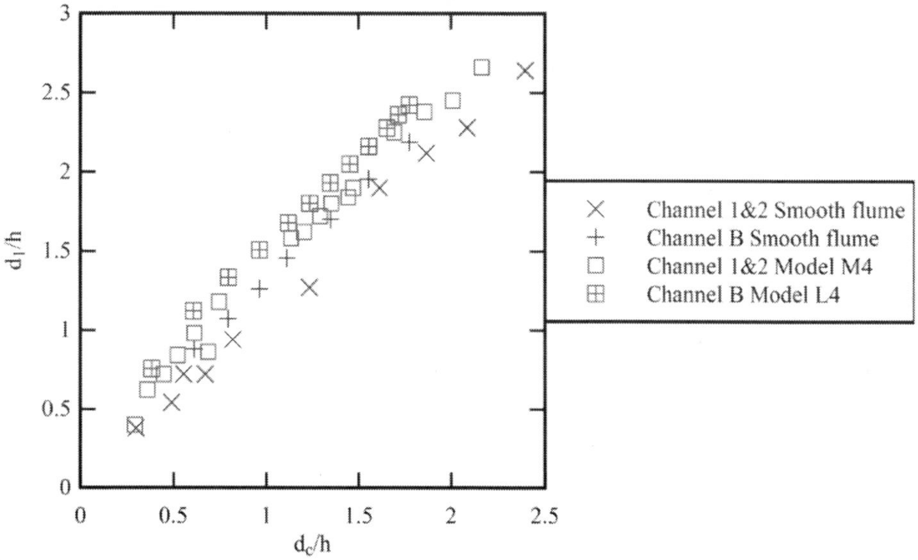

(A)

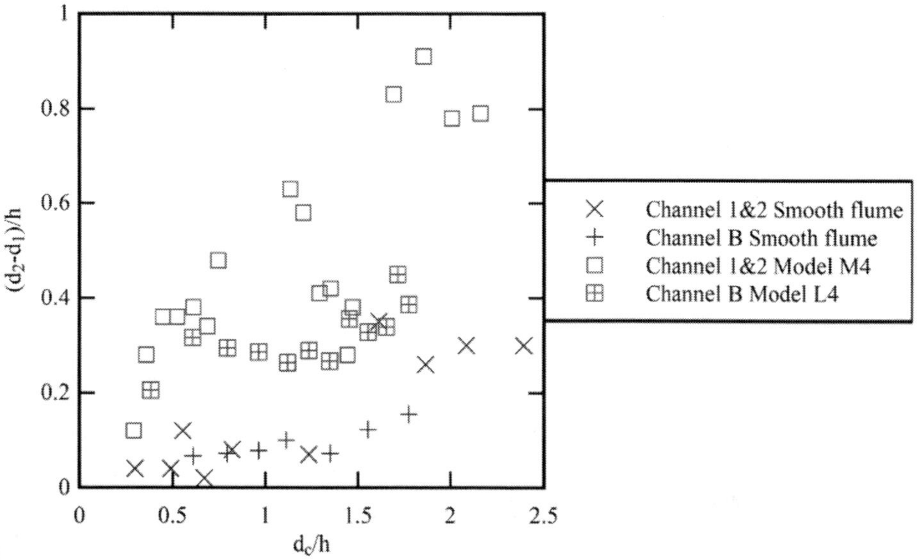

(B)

Figure 5. Impact of the porous horseshoe elements L4 and M4 (Porosity: 27%) on the inflow depth and difference in flow depths. (A) Dimensionless inflow depth d_1/h as a function of the dimensionless discharge d_c/h; (B) Dimensionless decrease in flow depth $(d_1-d_2)/h$ as a function of the dimensionless discharge d_c/h.

4. Velocity measurements

Pressure and velocity measurements were conducted in the larger channel B with the large porous element model L4, seen in Figures 3A and 4A. While the large majority of the data were collected with the Prandtl-Pitot tube, a smaller total head tube was used in regions where the streamline direction deviated from the channel centreline, as well as in the near-wake of the element. Typical velocity data are presented in Figures 6 and 7.

First, the pressure measurements showed that the pressure gradient was hydrostatic at all measurement locations. Overall, the streamline curvature was small in the entire flow field, except possibly at the downstream edges of the semi-circular element, where the pressure could not be measured locally with the Prandtl-Pitot tube.

Second, the velocity measurements showed that the inflow conditions were partially developed. The approach flow around the horseshoe element presented a streamline pattern in agreement with irrotational flow theory of ideal fluid, e.g. similar to the streamline pattern around a circular cylinder for $(x-x_o) < 0$ (Rouse 1959, Vallentine 1969, Chanson 2014). Downstream of the horseshoe element, the supercritical flow interacted with the recirculation region, with two marked shear zone, seen in Figure 6. Figure 6A presents a typical transverse distribution of the longitudinal velocity. A marked velocity gradient $\partial V_x/\partial y$ is observed at $y \pm D/2$. Figure 7 illustrates the major change in velocity filed around the porous horseshoe element. Figure 7 shows velocity contour maps in the accelerating flow region around the element (Fig. 7A) and immediately downstream of the horseshoe element (Fig. 7B).

Third, some negative velocity was observed in the recirculation zone. While the Prandtl-Pitot tube was not designed to measure negative velocities, a lower reading in the dynamic tapping was observed, compared with the static tube, at these locations. In addition, dye injection confirmed qualitatively the observations. Past studies derived a correlation between the negative Pitot reading and negative velocity (Cabonce et al. 2019). Although the quantitative magnitude was questionable in the current study, the findings could not be ignored, as reported in Figure 7B.

Overall, the velocity measurements highlighted a complicated three-dimensional velocity field around the porous horseshoe element.

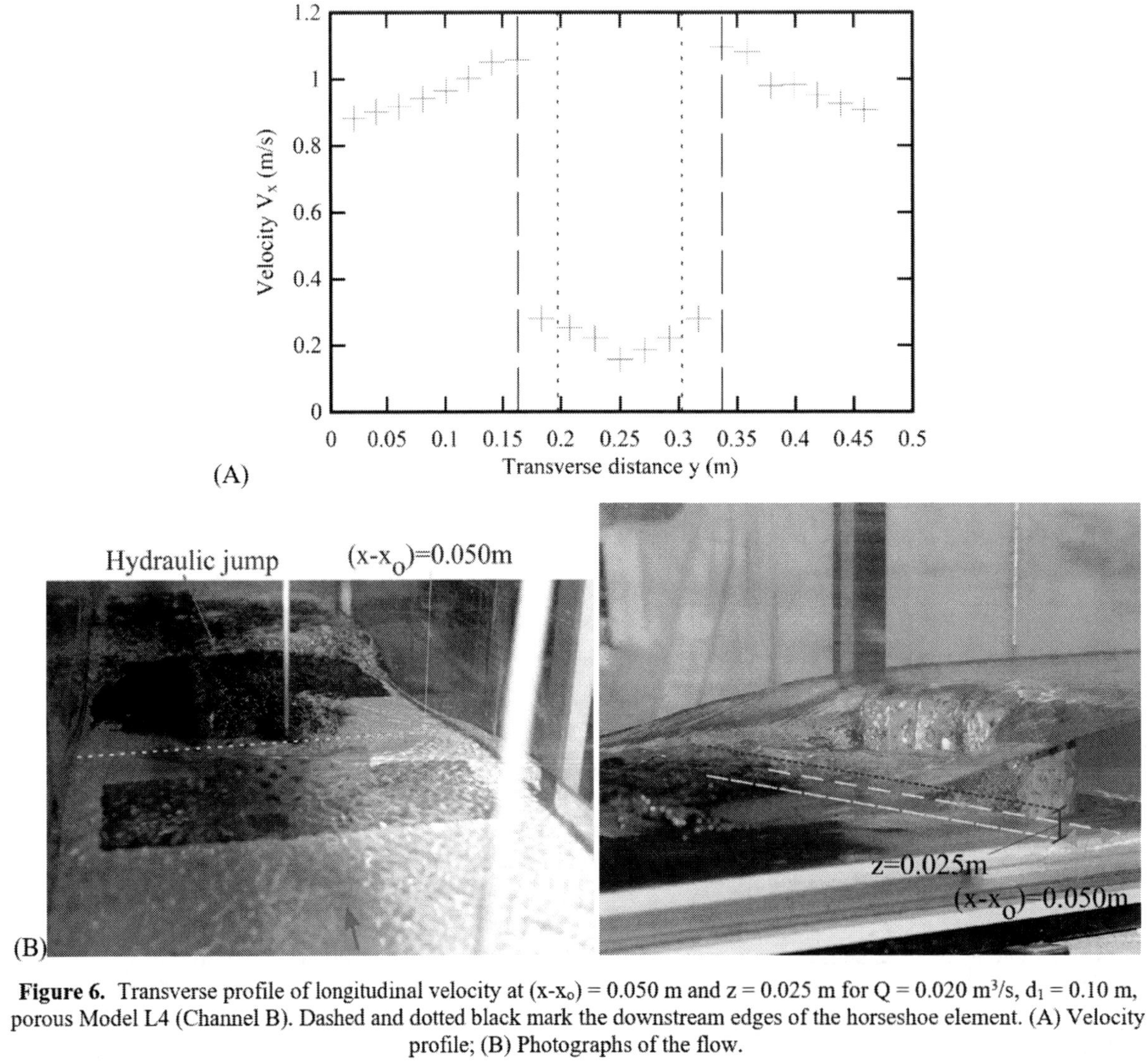

Figure 6. Transverse profile of longitudinal velocity at $(x-x_o) = 0.050$ m and $z = 0.025$ m for $Q = 0.020$ m^3/s, $d_1 = 0.10$ m, porous Model L4 (Channel B). Dashed and dotted black mark the downstream edges of the horseshoe element. (A) Velocity profile; (B) Photographs of the flow.

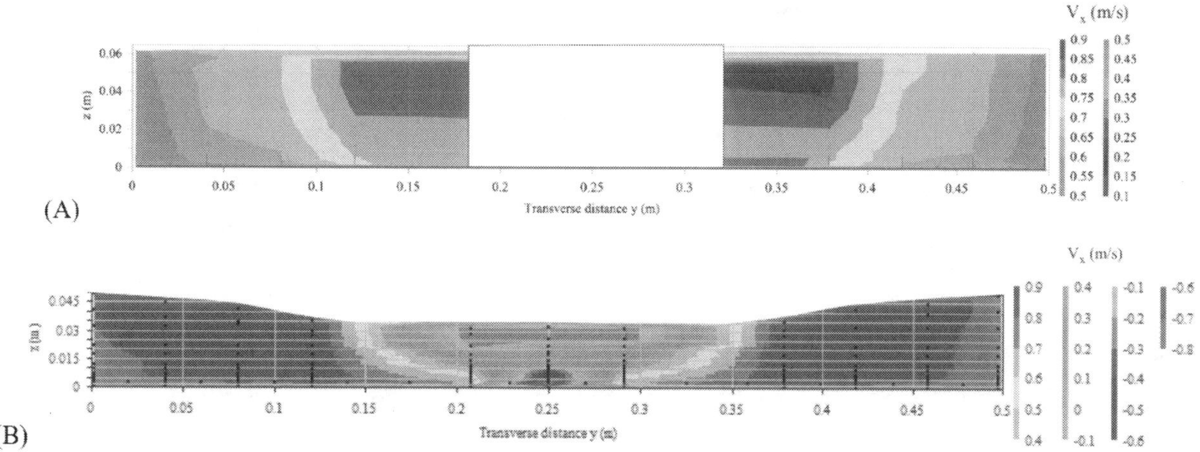

Figure 7. Un-distorted velocity contour maps for Q = 0.012 m^3/s, porous Model L4 (Channel B). (A) $(x-x_o) = -0.050$ m; (B) $(x-x_o) = +0.050$ m.

5. Conclusion

A physical study of horseshoe obstacles was undertaken, to gain a greater understanding of man-made structures built along Australian inland waterways for centuries. The physical modelling of permeable horseshoe obstacle was undertaken under controlled flow conditions, based upon a Froude similitude at two scales, corresponding to between 5:1 and 20:1 for typical riverine structures. The scale models were 3D. The physical observations included visual observations, three-dimensional free-surface profiles and detailed pressure and velocity measurements.

The physical observations showed a broad range of flow patterns, depending upon the submergence ratio. The porosity of the obstacle facilitated some interactions between the seepage and recirculation region, leading to changes in the wake region and its turbulence, compared to a impervious obstacle with the same shape. The velocity measurements presented a complicated three-dimensional velocity field around the porous horseshoe element, with an accelerating flow motion as the flow divides around the obstacle, followed by complicated three-dimensional flow patterns downstream of the elements

The present work paves the way for more detailed studies of real-life prototype structures Further investigations should test the impact of adjacent horseshoe structures (e.g. Fig. 1), different tailwater conditions, the interactions with aquatic life, as well as sediment processes.

6. Acknowledgments

The first author acknowledges some helpful exchanges with Dr Duncan Keenan-Jones (UQ School of Historical and Philosophical Inquiry), Sarah Martin, and Dr Michael Westaway (UQ School of Social Sciences) in alphabetical listing. The financial support of the University of Queensland, School of Civil Engineering is acknowledged.

7. References

Bandler, H. (1995). "Water resources exploitation in Australian prehistory environment." *The Environmentalist*, Vol. 15, pp. 97-107.

Bandler, H. (2003). "Expertise in Finding Water and Exploiting Water Resources in Australian Prehistory." in "Wasserhistorische Forschungen. Schwerpunkt Antike", *Schriften der Deutschen Wasserhistorischen Gesellschaft (DWhG) e.V. Siegburg*, C. P. J. Ohlig Editor, pp. 81-96.

Bandler, H. (2007). "Expertise in finding water and exploiting water resources in Australian prehistory." *Australian Journal of Water Resources*, Vol. 11, No. 1, pp. 1-11.

Branson, P.M., Ghisalberti, M., and Yvey, G.N. (2019). "Three-dimensionality of shallow island wakes." *Environmental Fluid Mechanics*, Vol. 19, pp. 1393-1416 (DOI: 10.1007/s10652-019-09661-5).

Cabonce, J., Fernando, R., Wang, H., and Chanson, H. (2019). "Using Small Triangular Baffles to Facilitate Upstream Fish Passage in Standard Box Culverts." *Environmental Fluid Mechanics*, Vol. 19, No. 1, pp. 157–179 (DOI: 10.1007/s10652-018-9604-x)

Chanson, H. (2014). "Applied Hydrodynamics: An Introduction." *CRC Press*, Taylor & Francis Group, Leiden, The Netherlands, 448 pages & 21 video movies.

Chanson, H., and Brown, R. (2018). "Stability of Individuals during Urban Inundations: What Should We Learn from Field Observations?" *Geosciences*, Vol. 8, No. 9, Paper 341, 9 pages (DOI: 10.3390/geosciences8090341).

Dargin, P. (1976). "Aboriginal Fisheries of the Darling-Barwon Rivers." *Brewarrina Historical Society*, Brewarrina NSW, Australia, 77 pages.

Flecker, P.O. (1951). "Remains of Aboriginal Habitation on the Great Barrier Wall." *The North Queensland Naturalist*, Vol. 19, No. 97, pp. 1-3.

Gillies, J.A., McKenna Neumann, C., and O'Brien, P. (2021). "Flow around surface-mounted permeable cubes on solid and deformable surfaces." *Environmental Fluid Mechanics*, Vol. 21, pp. 619-641 (DOI: 10.1007/s10652-021-09789-3).

Lapotre, M.G.A., and Lamb, M.P. (2015). "Hydraulics of floods upstream of horseshoe canyons and waterfalls." *Journal of Geophysical Research Earth Surfaces*, Vol. 120, pp. 1227-1250 (DOI: 10.1002/2014JF003412).

Mathews, R.H. (1903). "The Aboriginal Fisheries at Brewarrina." *Journal and Proceedings of the Royal Society of New South Wales*, Vol. 37, pp. 146-156.

Pasternack, G.B., Ellis, C.R., Leier, K.A., Valle, B.L., and Marr, J.D. (2006). "Convergent hydraulics at horseshoe steps in bedrock rivers." *Geomorphology*, Vol. 82, pp. 126-145.

Pasternack, G.B., Ellis, C.R., and Marr, J.D. (2007). " Jet and hydraulic jump near-bed stresses below a horseshoe waterfall." *Water Resources Research*, Vol. 43, Paper W07449, 14 pages.

Rouse, H. (1959). "Advanced Mechanics of Fluids." *John Wiley*, New York, USA, 444 pages.

Stamataki, I., Zang, J., Bazeley, W.D., and Morgan, G.C.J. (2014). "Study of Flow Over Weirs such as Pulteney Weir." *Proc. 11th International Conference on Hydroscience & Engineering ICHE 2014*, Hamburg, Germany, Lehfeldt & Kopmann Editors, pp. 295-302.

Taddei, S., Manes, C., and Ganapathisubramani, D. (2016). "Characterisation of drag and wake properties of canopy patches immersed in turbulent boundary layers." *Journal of Fluid Mechanics*, Vol. 798, pp. 27-49 (DOI: 10.1017/jfm.2016.312).

Takahashi, S., Endoh, K., and Muro, Z.I., (1992). "Experimental Study on People's Safety against Overtopping Waves on Breakwaters." *Report on the Port and Harbour Institute*, Vol. 34, No. 4, pp. 4-31 (in Japanese).

Vallentine, H.R. (1969). "Applied Hydrodynamics." *Butterworths*, London, UK, SI edition.

Wüthrich, D., Pfister, M., and Schleiss, A.J. (2020). "Forces on buildings with openings and orientation in a steady post-tsunami free-surface flow." *Coastal Engineering*, Vol. 161, Paper 103753, 11 pages.

Xia, J., Falconer, R.A., Wang, Y., and Xiao, X. (2014). "New Criterion for the Stability of a Human Body in Floodwaters." *Journal of Hydraulic Research*, IAHR, Vol. 52, No. 1, pp. 93-104.

DOI: https://doi.org/10.26077/c41f-39ff

Importance of Risk and Hazard Assessment of River Projects in India

Jitesh N. Vyas, Supriya Nath and R. B. Deogade
Central Water and Power Research Station, Pune-411024
E-mail: jiteshvyas33@gmail.com

Abstract: *Rivers are the lifeline for every country as they provide water for proliferation, sustenance and propagation of life since time immemorial. With increasing population pressure, water demand increased for drinking, irrigation, and industrial consumption. This water deficit required utilization of available water resources in the best possible manner through various hydrological projects like construction of dams, weirs, barrage, etc. There is a disparity in water availability between north and south India as most of the perennial rivers are situated in the north while seasonal rivers in south India. To resolve this disproportionate distribution and to maintain the balance of water between flood and drought-prone rivers, interlinking of rivers is proposed. This proposal has garnered a lot of attention from scientists and environmentalists around the world. These human interventions affect the ecological flow of rivers. There are various risks associated with these projects which require prior assessment. In this study, risks and environmental hazards associated with river projects in India have been discussed with a special focus on the ongoing river projects like the interlinking of rivers. Various factors which determine the feasibility of the proposed projects were studied. There are many risks associated with the change in gradient and environmental flow of rivers with the construction of such projects. We have reviewed risks on humans, like displacement of population and impact on their livelihood as reported for various projects. There are various social implications of such projects which need to be assessed and their mitigation measures need to be undertaken.*

Keywords: *environmental flow, gradient, hydrological projects, interlinking of rivers, social implications, water quality*

1. Introduction

Human civilizations have thrived on the water banks of rivers which have played a significant role in every aspect of the expansion of civilisation-social, political, economic, and ecological. There may be the difference in opinion about the level of influence but every scholar agrees that the early civilization was dependent on water supply for all their basic needs- irrigation for cultivation, navigation for trade and commerce and water for consumption of humans and cattles. Rapid industrialization and advanced technology created need for production of electricity and for this purpose water of rivers was harnessed through hydroelectric power projects. Humans devised various methods to overcome the challenges posed by natural barriers. Construction of canals, dams, rainwater harvesting structures, etc are testimony to the fact that human have found ways to meet their growing needs and diversion of river water to the required areas are not new. Even there is evidence of elaborate canal and drainage system during Harappan civilization. There is huge water scarcity in our country as the per capita water availability (1486 cubic meters in 2021) making it water stressed country. According to The Water Gap – The State of the World's Water, 2018, India ranks amongst the top 10 countries where there is lowest access to the clean source of water nearby home. This Interlinking of rivers is however in nascent stage of development as this requires elaborate changes in course of a river to meet another river where there is water deficit. This manmade intervention with natural course is associated with many risks which require proper assessment. It also requires proper assessment of feasibility of such projects since they involve huge economic, social and ecological costs.

2. Interlinking of Rivers in India

A British Irrigation Engineer, Sir A. Cotton in 1839 first proposed to link rivers of Southern parts of India to promote inland navigation. However it was only partially implemented due to growth of railways network (Mirza et al. 2008). The interlinking of rivers in independent India has been first proposed by Dr. K. L. Rao in 1972 to link the Ganga River with the southern Cauvery River (Ganga-Cauvery Link)but it was not implemented due to economic constraints (Bandyopadhyay and Perveen 2003; Mirza et al. 2008). Captain D.J. Dastur proposed the 'Garland Canal scheme' to connect Himalayan and peninsular rivers at two points (Delhi and Patna) but it was found to be technically and economically infeasible (Bandyopadhyay and Perveen 2003). National Water Development Agency (NWDA) was set up in 1982 for study and execution of interlinking of river projects. Its major function is to prepare Pre–Feasibility Reports (PFRs)/Feasibility Reports (FRs) and Detailed Project Report (DPR) of River Link proposals under the National Perspective Plan (NPP). Since 2002, this project has garnered lot of media attention as Supreme Court asked the government to complete the

ISBN: 978-1-958416-07-5 DOI: https://doi.org/10.26077/c41f-39ff

project within 12 years (Bandyopadhyay and Perveen 2003). In the present budget, it was announced that Rs 43billion were allocated in Revised Estimates 2021-22 and Rs 14 billion in 2022-23 have been made for the project.

National Perspective Plan has two components

1. Himalayan Rivers Development
2. Peninsular Rivers Development

A map of the proposed scheme of links on both the components of ILR is shown in Fig. 1 and Fig. 2.

2.1. Himalayan Rivers Development

There are 14 links in the Himalayan component of which, NWDA has completed feasibility studies of 4 links (http://www.nwda.gov.in/content/index.php):

1. Sarda-Yamuna Link (Indian portion),
2. Ghaghara-Yamuna Link (Indian portion),
3. Manas-Sankosh-Tista-Ganga (MSTG) Link
4. Farakka-Sunderban (FS) Link.

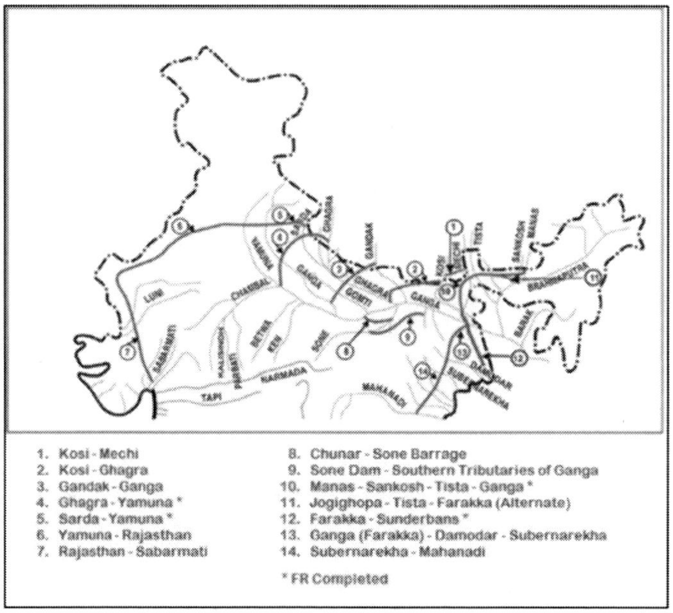

Figure 1. Proposed Himalayan component of ILR (Source: NWDA).

2.2. Peninsular Rivers Development

NWDA had proposed 16 links of peninsular component of which PFR for 14 are complete except for two links viz.Netravati - Hemavati and Bedti-Varda links (http://www.nwda.gov.in/content/index.php).

ISBN: 978-1-958416-07-5 DOI: https://doi.org/10.26077/c41f-39ff

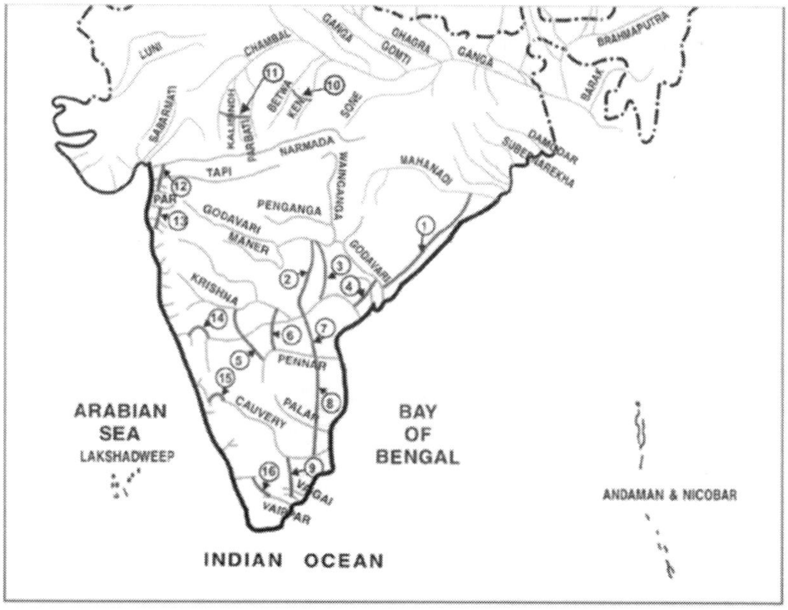

1. Mahanadi (Manibhadra) – Godavari (Dowlaiswaram) *
2. Godavari (Inchampalli) – Krishna (Nagarjunasagar) *
3. Godavari (Inchampalli) – Krishna (Pulichintala) *
4. Godavari (Polavaram) – Krishna (Vijayawada) *
5. Krishna (Almatti) – Pennar *
6. Krishna (Srisailam) – Pennar *
7. Krishna (Nagarjunasagar) – Pennar (Somasila) *
8. Pennar (Somasila)–Palar- Cauvery (Grand Anicut) *
9. Cauvery (Kattalai) – Vaigai – Gundar
10. Ken – Betwa *
11. Parbati – Kalisindh – Chambal *
12. Par – Tapi – Narmada *
13. Damanganga – Pinjal *
14. Bedti – Varda
15. Netravati – Hemavati
16. Pamba – Achankovil – Vaippar *

* FR Completed

Figure 2. Proposed peninsular component of ILR (Source: http://nwda.gov.in/upload/uploadfiles/files/1.JPG).

3. Advantages of interlinking of Rivers

The interlinking of rivers will resolve the long standing issue of twin problems of flood and drought. The water scarcity faced by southern part of the country due to dependency on monsoon. The inter-basin transfer will ensure that the water deficient region gets water from water surplus regions (Jain et al. 2008), fulfilling the water demands of that region and simultaneously ensuring that the regions which were getting flooded are not affected by floods. It will also help in the generation of electricity and provide means for transportation which will in turn promote economic development (Mukhopadhyay 2008). There will be employment generation during this project as it will require lots of human resources during infrastructure development and it will also promote industrialization near water which will further boost the job prospects of many.

4. Probable risks associated with Interlinking of Rivers

There are various environmental and social costs of the interlinking of rivers. There should be proper impact assessment and redressal of the environmental and social issues before implementing these projects. Environmental impacts are deforestation, soil erosion, disturbance to the biodiversity, loss of habitat, degradation of water quality in downstream, etc. Massive deforestation for building canals may result in negative impact on rainfall and in turn affect biodiversity (Babu and Padmavathi 2016).The interlinking of rivers may lead to an increased frequency of earthquake (Mehta and Mehta 2018) as a result of the construction of large dams for interlinking. There are also concerns over pollution load which may be shifted to a less polluted river due to interlinking of the heavily polluted river to relatively less polluted one. In case the river pollution has some toxic element, this may prove to be disastrous for the biodiversity as well as humans (Mehta and Mehta 2018). It has also been reported that the reduction in volume of fresh water in the sea will negatively affect the biodiversity of the estuaries. Biodiversity of fishes will also be affected as there may arise dominance of invasive species from northern rivers by displacing the endemic species of peninsular rivers after these rivers are interlinked (Babu and Padmavathi 2016). There will be influence on turbidity, flow and depth of rivers which will further causes changes in breeding grounds of fishes causing their migration (Babu and Padmavathi 2016). There are various social impacts such as displacement of local people who needs proper rehabilitation

ISBN: 978-1-958416-07-5 DOI: https://doi.org/10.26077/c41f-39ff

and proper compensation for the loss of livelihood and land. This project will also strain our relation with our neighbouring countries like Bangladesh, Bhutan, and Nepal as they are apprehensive of loss of water availability to them which might occur due to ILR projects. This will lead to submergence of land, forest, destruction of rivers biodiversity, concentration of pollution, destruction of groundwater, etc.

Rivers change their course naturally mainly due to flood (Mahmood et al. 2015, Maurya and Yadav 2016)., so the entire exercise done for interlinking will prove futile and the whole process will have to be repeated to link the rivers again which will again involve large economic, social and environmental costs. NRLP has determined annual flow volume according to which they have classified the surplus and deficit area. But there is seasonal flow variability that needs to be considered as most of the peninsular rivers are rain-fed rivers (Smakhtin et al. 2008). There are concern about the coastal erosion and salt water intrusion (Smakhtin et al. 2008) which reduces land productivity. Bharati et al. (2008) warned that water deficits areas may be interchanged and suggested introduction of low water intensive crops during dry season as an alternative to ILR project.

5. Present projects and their status

There are 5 priority project links identified by NWDA:

1. Ken-Betwa Link Project
2. Damanganga-Pinjal Link Project
3. Godavari (Inchampalli)-Cauvery (Grand Anicut) Link Project
4. Par-Tapi-Narmada Link Project
5. Kosi-Mechi Intra State Link Project

5.1. Ken–Betwa basin

Ken-Betwa River interlink, which will link Ken River with Betwa river in Bundelkhand region (as shown in Fig. 3), is one of the first interlink which was initiated. It will provide irrigation around 10.62 billion square meter area of which 8.11 billion square meter is in Madhya Pradesh and 2.51 billion square meter lies in Uttar Pradesh. It will also provide domestic water supply for 6.2 million (MP-4.1 million, UP-2.1 million) population and will also generate 103 MW of Hydro and 27 MW of Solar power. However there are several environmental and social issues which are associated with the project. Construction of Daudhan dam will lead to 90 million square meter area getting submerged, including 58.03 million square meter of Panna Tiger Reserve. This constitutes 7.6% of the total Panna Tiger Reserve area of which 41.41 million square meter is of forest area (NWDA). This will lead to great loss of habitat for various species of plants and animals. The loss of biodiversity cannot be compensated.

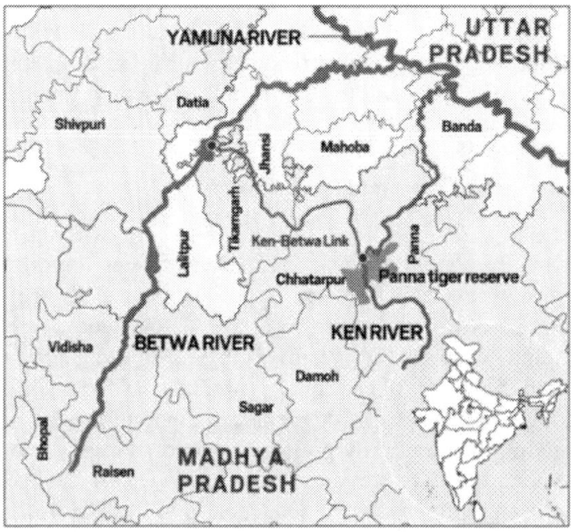

Figure 3. Proposed Ken Betwa River Link canal.

ISBN: 978-1-958416-07-5 DOI: https://doi.org/10.26077/c41f-39ff

Various studies have assessed the risks and hazards due to the interlinking of Ken-Betwa. Study by Avtar et al. (2011) found that the interlinking of Ken-Betwa will negatively impact the downstream of the Ken River and the highest impact will be found near the dam region and the Panna tiger reserve forest. However, the Betwa basin will be the least impacted negatively. The rest of the area comes under the medium impact zone. The study also highlighted the fact that waterlogging and construction of dam will cause disturbance in the natural ecological balance (Avtar et al. 2011) which will further cause land degradation which in turn affect cropping pattern. In this project, the reservoir (Greater Gangau Weir) will negatively impact the land in the vicinity of the reservoir edge (Avtar et al. 2011).

Similarly Goparaju et al. (2017) analysed the Landsat 8 OLI data of the proposed Daudhan/Greater Gangau Dam and found lacunae in the results of EIA– EMP report of the Ken-Betwa link project, Phase 1when they compared it with their study. They found there is a risk of submergence of the Panna Tiger Reserve which will endanger the lives of various animals and plants (Goparaju et al. 2017). They suggested that Geospatial technology(including remote sensing and GIS) will help in minimizing the loss as they will help in better prediction of spatial view of the proposed submergence area which will help in taking mitigation measures (Avtar et al. 2011; Goparaju et al. 2017). This will resolve not only environmental issues related to the project but will also find out solutions to socio- economic issues related to the project (Avtar et al. 2011).

5.2. Damanganga-Pinjal Link Project

It will link Damanganga River at Bhugad and Khargihill reservoir to Pinjal reservoir in Vaitarna basin as shown in Fig. 4. The collective water of Damanganga and Pinjal rivers is diverted to Mumbai to meet the domestic requirement of Mumbai. The total water supply from this link and Pinjal project shall be 2451 MLD (millions of liter per day) along with the generation of 5MW of hydropower. NWDA has completed the Detailed Project Report (DPR) but there are issues between Govt. of Maharashtra and Gujarat which needs to be resolved for building consensus. There are problems regarding the construction of link tunnel as the presence of plateau in the Sahyadri Hills is interspersed with many weak zones (Thanvellu et al.). For example, Thanvellu et al. reported seven weak zones in the Bhugad-Khargihill and Khargihill - Pinjal section has 12 weak zones where link tunnel will be constructed (Thanvellu et al.). According to their study, 0.4% of the total length of the tunnel was found to be hazardous due to very poor tunneling media (Thanvellu et al.).

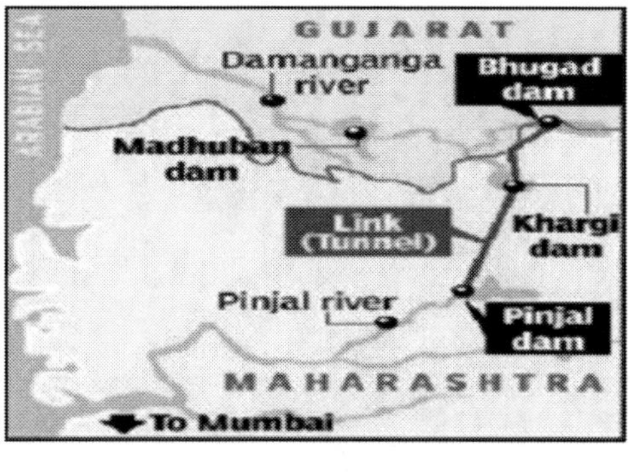

Figure 4. Proposed Damanganga-Pinjal Link canal.

5.3. Godavari (Inchampalli)-Cauvery (Grand Anicut) Link Project

It comprises of three links (namely Inchampalli, Nagarjunasagar and Somasila) to connect Godavari – Cauvery as shown in Fig. 5. This link project connects Godavari, Krishna, Pennar, Palar and Cauvery basins and will cover the States of Telangana, Andhra Pradesh and Tamil Nadu. According to NWDA, it will divert 7000 million cubic meters (MCM) of water annually from Godavari basin to Krishna, Pennar and Cauvery basins. It will provide Annual irrigation of 9.44 billion square meter. However, there are disagreements between the party states which need to be resolved for the implementation of this link. Smakhtin et al. (2008) found that this project will have adverse effect on the Singaram Sanctuary and cause submergence of 650,000 square meter land of the Indravati National Park but it does not have any endangered species. Coastal erosion and sea water intrusion may also be the possible result of these projects in Godavari and Krishna basin (Smakhtin et al. 2008). Inchampali Dam faces the problem of loss of water due to evaporation and sedimentation causing reduction in flow of river as well as shrinking of deltas (Smakhtin et al. 2008).

ISBN: 978-1-958416-07-5 DOI: https://doi.org/10.26077/c41f-39ff

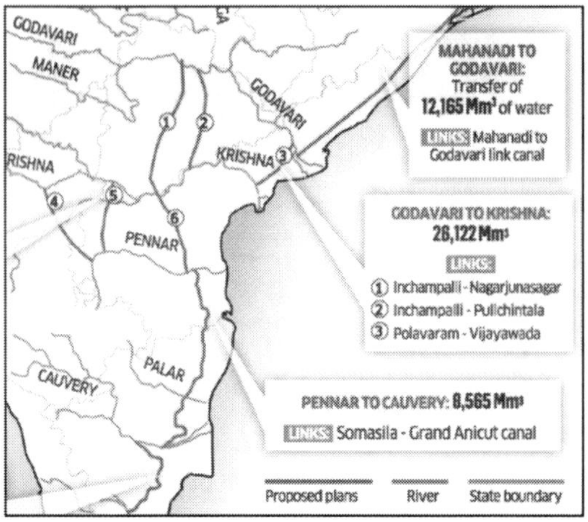

Figure 5. Proposed Godavari (Inchampalli)-Cauvery (Grand Anicut) Link.

5.4. Par-Tapi-Narmada Link Project

Par-Tapi-Narmada and Damanganga-Pinjal Link Projects are both associated with Maharashtra and Gujarat (as shown in Fig. 6). Par-Tapi Narmada Link Project will benefit Gujarat while Damanganga-Pinjal Link Project will benefit Maharashtra. According to NWDA, this link proposes to provide 1330 MCM of water annually and provide Annual Irrigation of 2.32 billion square meter (2.28 billion square meter in Gujarat State and 0.04 billion square meter in Maharashtra State), 76 MCM of drinking water supply to about 2.75 million human population and 60 MCM of water for local domestic use from each reservoir and 21 MW of hydropower.

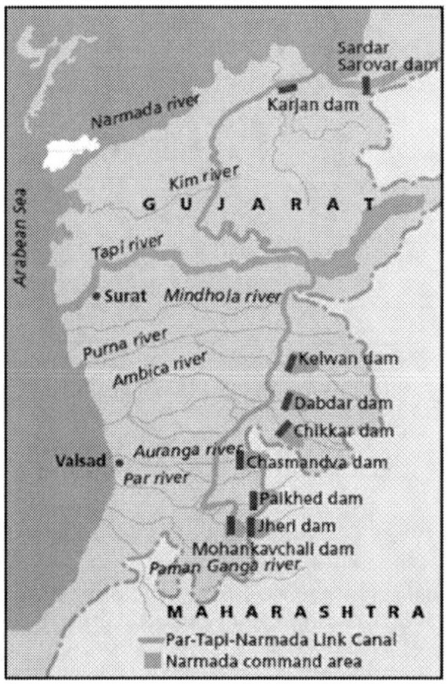

Figure 6. Proposed Par-Tapi-Narmada Link.

The DPR of Par-Tapi-Narmada link project was completed in 2015 which was modified according to suggestions of Govt. of Gujarat to include more tribal areas in the command area (NWDA).

There are conflicts regarding water sharing between the Government of Maharashtra and Gujarat which is causing delay in implementation of the project. This delay due to interstate political issues causes huge loss economically since there is increase in the cost of construction of reservoirs and dams due to the inflation in the cost of raw material, labour rate. Mehmood et al. (2014) has also reiterated the fact that there will be severe impact on land use pattern due to construction of seven reservoirs. Kelwan dam will cause submergence of villages Bhondya and Engnipada (Kolbari) villages completely at proposed FRL 144 m (Mehmood et al. 2014). The flood in Par basin has been found to occur more when there is above average rainfall (Patil and Hire 2020) while during below average rainfall it experiences very few floods. In case of deficiency in rainfall the river Par will not have surplus water to provide to the link canal.

5.5. Kosi-Mechi Intra State Link Project

It proposes diversion of surplus 148 cumecs of water from Eastern Kosi Head Regulator for irrigating 2.15 billion square metere area in Araria, Purnia, Kishanganj and Katihar districts of Bihar (depicted in Fig. 7). The DPR of Kosi – Mechi intra-State link project was completed in March 2014. However Kosi is infamous for changing it course frequently which will make the link worthless and new link will be required to be constructed every time the river changes its course. Another major concern is that submergence of fertile land of Nepal and displacement of people over 300 km of very large link canals (Kosi-Mechi and Kosi-Ghagra Link) (Pun 2009).

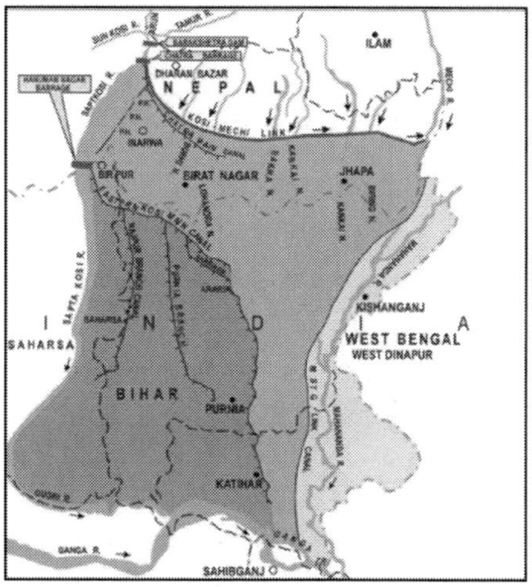

Figure 7. Proposed Kosi-Mechi Intra State Link Project.

6. Risk associated with various river projects

Various river projects like hydropower projects are associated with numerous risks and environmental hazards which are unforeseen. They are subjected to sudden natural disasters like flood, earthquake, landslides which may cause irreparable damage to the project causing loss of life and property. There are serious concerns like problem of sedimentation, reservoir induced seismicity, loss of agricultural land which causes social distress due to loss of livelihood, resettlement issues, loss of wildlife and aquatic biodiversity (Chaurasiya et al. 2013). Hydropower projects causes change in flow and water quality of river that may lead to restriction in movement of aquatic fauna (Opperman et al. 2022). Climate change induced water scarcity and drought may lead to failure of many hydropower projects (Opperman et al. 2022). Huber et al. (2017) has emphasized on the social divide induced by various hydropower projects that causes vulnerable and poor section to face the dam safety risks.

ISBN: 978-1-958416-07-5 DOI: https://doi.org/10.26077/c41f-39ff

7. Conclusion

There are numerous risks and hazards associated with the interlinking of rivers which need to be properly assessed and taken care of while implementing the project. The Government of India is serious about implementation of the interlinking project and has prioritized 5 projects. These projects have their own issues and challenges which are hindering the timely completion of the project. The common issue with interstate projects is disagreement between the states regarding sharing of water as surplus states do not want to share water with deficit state fearing that they will loose their precious resource. Some of the projects are facing environmental challenge as this may lead to submergence of forest area, like Panna tiger Reserve in case of Ken-Betwa link, which will be huge loss to environment and biodiversity. The social risks associated with these projects are displacement of people due to submergence of numerous villages needs to be rehabilitated properly by providing them compensation and employment so as to reduce the distress faced by them due to sudden relocation and loss of livelihood. Even after completion of the project there will be serious risks associated with flooding caused by deposition of sediments in reservoir. NWDA while implementing the projects need to seriously consider all these factors and take measures to address them. It also needs to address the concern of various states project wise which will help them bring consensus and cooperation for timely completion of these projects.

8. ACKNOWLEDGEMENTS

The authors are thankful and wish to express deep sense of gratitude to Dr. R. S. Kankara, Director, Central Water and Power Research Station (CW&PRS) for constant encouragement and valuable suggestions.

9. REFERENCES

Avtar, R., Singh, C. K., Shashtri, S., and Mukherjee, S. (2011). "Identification of erosional and inundation hazard zones in Ken–Betwa river linking area, India, using remote sensing and GIS." Environ. Monit. Assess., 182(1), 341-360.

Babu, R., and Padmavathi, P. (2016). "Interlinking of Krishna and Godavari rivers: An ecological study".

Bandyopadhyay, J., and Perveen, S. (2003). "The interlinking of Indian rivers: some questions on the scientific, economic and environmental dimensions of the proposal. In Seminar on Interlinking Indian Rivers: Bane or Boon".

Bharati, L., Anand, B. K., and Smakhtin, V. (2008). "Analysis of the Inter-basin Water Transfer Scheme in India: a case study of the Godavari-Krishna link" (No. 614-2016-40883).

Chaurasiya, P. K., Prasad, V., and Khare, R. (2013). "Scenario and risk of hydro power projects in India." Int. J. Chemtech Res., 5(2), 1068-1075.

Goparaju, L., Ahmad, F., and Thakkar, H. (2017). "Submergence analysis of the proposed Ken Betwa Dam (Madhya Pradesh) India, using geospatial technology in Environmental Impact Assessments." Environ. Socio-Econ. Stud., 5(4), 18-28.

Huber, A., Gorostiza, S., Kotsila, P., Beltrán, M. J., and Armiero, M. (2017). "Beyond "socially constructed" disasters: Re-politicizing the debate on large dams through a political ecology of risk." Capital. Nat. Social., 28(3), 48-68.

Jain, S. K., Kumar, V., and Panigrahy, N. (2008). "Some issues on interlinking of rivers in India." Curr. Sci., 728-735.

Mahmood, R., Babel, M.S., and JIA, S. (2015). "Assessment of temporal and spatial changes of future climate in the Jhelum river basin, Pakistan and India." Weather and Climate Extremes, 10, 40–55.

Maurya, S. P., and Yadav, A. K. (2016). "Evaluation of course change detection of Ramganga river using remote sensing and GIS, India." Weather and climate extremes, 13, 68-72.

Mehta, D., and Mehta, N. K. (2018). "Interlinking of rivers in India: Issues & challenges." Geo-Eco-Marina, 19, 137-144.

Mehmood, K., Patel, A. , Singh, V. , Prajapati, S. , Kalubarme, M. , Prakash, I. and Gupta, K. (2014). "Submergence Analysis Using Geo-Informatics Technology for Proposed Dam Reservoirs of Par-Tapi-Narmada River Link Project, Gujarat State, India." Int. J. Geol., 5, 622-633. doi: 10.4236/ijg.2014.56057.

ISBN: 978-1-958416-07-5 DOI: https://doi.org/10.26077/c41f-39ff

Mirza, M. M. Q., Ahmed, A. U., and Ahmad, Q. K. (2008). "Interlinking of rivers in India: Issues and concerns." CRC Press.

Mukhopadhyay, R. (2008). "Inter-Linking of Rivers: Are we on the right side?".

NWDA <http://www.nwda.gov.in/content/index.php> (April 7, 2022)

Opperman, J. J., Camargo, R. R., Laporte-Bisquit, A., Zarfl, C., and Morgan, A. J. (2022). "Using the WWF Water Risk Filter to Screen Existing and Projected Hydropower Projects for Climate and Biodiversity Risks." Water, 14(5), 721.

Patil, A. D., and Hire, P. S. (2020). "Flood hydrometeorological situations associated with monsoon floods on the Par River in Western India." Mausam, 71(4), 687-698.

Pun, S. B. (2009). "The Kosi Pralaya: could the catastrophe have been averted? and what next." Hydro Nepal, 4(7).

Smakhtin, V., Gamage, N., and Bharati, L. (2008). "Hydrological and environmental issues of inter-basin water transfers in India: a case study of the Krishna River Basin. Strategic Analyses of the National River Linking Project (NRLP) of India." Series 2, 79.

Thanvellu, C., Dhakate, M. V., Chavan, R. H., and Sharma, V. K. "Rock Mass Prognostication at the Tunnel Grade: Case Study of Damanganga-Pinjal River Link Project, Western Ghats". <http://isegindia.org/pdfs/Abstracts/thanavellu%20damanganga%20edited.pdf> (April 8, 2022).

The Water Gap – The State of the World's Water (2018). WaterAid <https://www.wateraid.org/in/publications/the-water-gap-the-state-of-the-worlds-water-2018> (April 7, 2022).

DOI: https://doi.org/10.26077/6d47-2e35

Introduction and investigation of a large-scale piano key weir fabricated via rapid prototyping

M. Oertel[1], B. Belay[2], X. Shen[3] & H. Willems[4]
[1,2,3,4]*Hydraulic Engineering Section, Helmut-Schmidt-University, Hamburg, Germany*
[1]*E-mail: mario.oertel@hsu-hh.de*

Abstract: *Piano key weirs (PKW) are novel hydraulic structures, characterized by an increased discharge capacity, compared to regular weirs. Through the last decades, experimental studies have mainly focused on efficiency aspects and design regulations. Therefore, often small scaled experiments were conducted in hydraulics laboratories in which model geometries were produced with materials like acrylic glass. But also in recent practices, 3D printing techniques are being used to plot small-scaled piano key weir configurations up to 30 or 40 cm width. The present investigation focuses on a large-scale 3D printed PKW with a total width of 1 m featuring quarter round crest shape in upstream direction. The weight of the printed structure is approximately 30 kg and 8 days continuous plotting process were necessary to produce the weir – consequently, it is the largest 3D printed PKW up-to-date. The plotted PKW was installed at Helmut-Schmidt-University's new hydraulics laboratory within a 20 m long and 1 m wide flume. Discharges ranging from 20 to 330 L/s were used to investigate the hydraulic performance of this printed PKW with a height of 40 cm. Results depict that the scale difference between models of 30 cm and 40 cm in height is not significant if the objective is to study discharge coefficients. 30 cm models can reproduce similar flow properties at a relatively smaller cost and less preparation time. Moreover, patterns of nape detachment were observed at the downstream crest lip, indicating the influence of the printing layer height in inducing undesired flow separation for relative upstream heads above 0.33 that could likely affect the nape trajectory.*

Keywords: *Discharge coefficients, Large-scaled model, Piano key weir, 3D printing*

1. Introduction

For their advantages in terms of achieving better hydraulic performance, piano key weir (PKW) research became a topic of interest in contemporary studies of hydraulic structures. These hydraulic structures allow highly effective passing of a range of discharges with smaller upstream heads compared with regular weirs (Eslinger and Crookston 2020). Recent laboratory scale studies are carried out to assess hydraulic performance and scale effects among piano key weirs of different sizes ranging from 1:7 to 1:25 relative to a prototype (Erpicum et al. 2016). Several investigations were carried out in the past with the aim to optimize the geometry of PKW while maintaining favorable hydraulic performance (Shen and Oertel 2021; Cicero and Delisle 2014; Machiels et al. 2014; Ouamane and Lempérière 2006), mainly discharge coefficients and energy dissipation processes. The implementation of PKW in flood release structures at dams or in-channel applications has been encouraged through intensive experimental and numerical studies, while some research (Ribeiro et al. 2012) even proposed a preliminary design guideline for practical applications. The use of large-scale laboratory models is ideally preferred so that adverse scale effects can be minimized, and different flow fields or geometric variables can be thoroughly investigated. However, this is often limited by available flume sizes and especially discharge capacities of a particular hydraulics laboratory. Some flow field properties, such as air entrainment, are better understood in laboratories when large-scale models are used. In that case, the effect of surface tension force near the free surface gets less significant compared with the inertial force and turbulence stress, especially in prototypes (Oertel 2019; Erpicum et al. 2016; Heller 2011; Chanson 2009). Hence, while maintaining the minimum criteria, for instance $H_T \geq 0.03$ m (total head above the weir) or Weber number ≥ 54 (Erpicum et al. 2016), such large-scale models are more convenient to visualize and study air entrainment as well as nape trajectory. In the absence of large-scale modeling capabilities, to investigate discharge coefficients and nape behavior at low heads for small scaled PKW models, Oertel (2019) has also proposed a surface tension correction solution so that underestimation of discharge coefficients for small heads can be minimized for such experimental campaigns.

The current study aims on introducing a large 3D printed PKW model that has been developed at Helmut-Schmidt University's (HSU) hydraulics laboratory (Figure 1). The 900 m^2 laboratory is currently under construction with a final discharge capacity of up to 1500 L/s. A first horizontal flume was installed with a length of L_f = 20 m, a width of W_f = 1.0 m, and a height of H_f = 1.0 m. A 500 mm pipe system and two 175 kW pumps connected to a 300 m^3

storage basin provide the discharge. Moreover, this investigation was motivated by the advantages of time- and cost-effective rapid prototyping practices in hydraulics laboratories. The practice of large-scale model preparation of complex geometries has been currently standardized in the newly established HSU's hydraulics laboratory.

Figure 1. HSU's new hydraulics laboratory, left: two 175 kW pumps and DN500 pipe system, right: BigRep One 3D printer with 1 m^3 plot volume (1×1×1 m) and partially finished 40 cm high and 100 cm wide PKW model in printing process.

2. Model preparation

The current experimental measurement campaign was commenced by 3D printing a trapezoidal PKW to fabricate the desired experimental model (Figure 1, right). A BigRep One 3D printer was used to plot the large-scaled PKW with a width of 1000 mm and a height of 400 mm (8 days 24/7 printing time, ~30 kg filament weight), representing the largest 3D printed PKW model up-to-date (see Figures 1 and 2). Polylactic acid (PLA) was used with a layer height of 0.6 mm in horizontal orientation with an installed nozzle diameter of 1 mm. The model was designed to have a quarter round upstream crest lip shape with a radius of 20 mm. A developed innovative system was used to fix the model within the flume without any flow disturbing influences. Table 1 shows all model dimensions for the new plotted PKW model and for all compared previous models. The notation P_{40} is used to indicate the weir height (40 cm), likewise for the other three models.

The current PKW model was plotted as an extension of previous models in which all the geometries and test discharges were scaled and parametrized through Froude similarity relative to the other models' main parameters, presented in Table 1. It is imperative to mention that the printing layer heights are independent of the scaling process.

Table 1. Geometric features of studied and compared PKW models

		Model dimension			
Geometric parameter	**Unit**	**P_{40}**	**P_{30*}**	**P_{20*}**	**P_{10*}**
P (weir height)	[mm]	400	300	200	100
B (weir length)	[mm]	640	480	320	160
L (total crest length)	[mm]	3677	1103	1103	1103
W (weir width)	[mm]	1000	300	300	300
W_i/W_o	[-]	1	1	1	1
T (wall thickness)	[mm]	40	30	20	10
R (crest radius)	[mm]	20	15	10	5
printing layer height	[mm]	0.6	0.4	0.4	0.2

* *Previous models, printed on a Creality CR-10 max.*

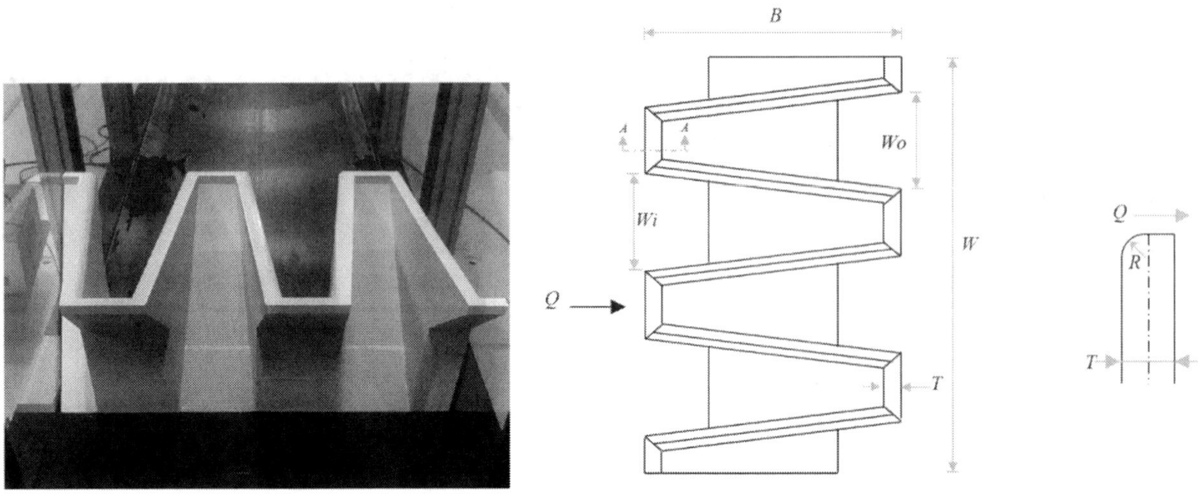

Figure 2. Investigated PKW model, left: flume installation (looking in upstream direction), right: geometric features

3. Experimental setup

Model runs were conducted in the previously mentioned rectangular flume (L_f = 20 m, W_f = 1.0 m, H_f = 1.0 m) at the hydraulics laboratory of Helmut-Schmidt-University, Hamburg, Germany. The maximum discharge capacity for the current experimental setup was approximately 350 L/s. Flow depth measurements were performed using an ultrasonic sensor (Microsonic; mic+130/IU/TC, signal range: 200 to 1300 mm, accuracy: ± 1%) installed 2 m (5*P*, where *P* is the weir height) upstream of the PKW. Discharges were measured using a magnetic inductive flow meter (MID, Krohne Optiflux, accuracy ± 0.3%). The flume entrance included perforated vertical steel plates within the inlet box as flow straighteners to guarantee a smooth approaching flow. No control gates were used at the downstream end of the flume and a free outflow was achieved. Measurements were conducted for discharges from Q = 20 to 330 L/s. Each flow depth for varying discharges was measured for a duration of 30 seconds with a sampling frequency of 75 Hz. Appropriate data filtering (removing outliers by standard deviation criteria) was undertaken to obtain an average flow depth value from recorded time series. Additionally, image acquisition was carried out using a GoPro Hero 10 camera and pictures are then used for qualitative assessment of flow behavior at the PKW crest under the proposed test discharges.

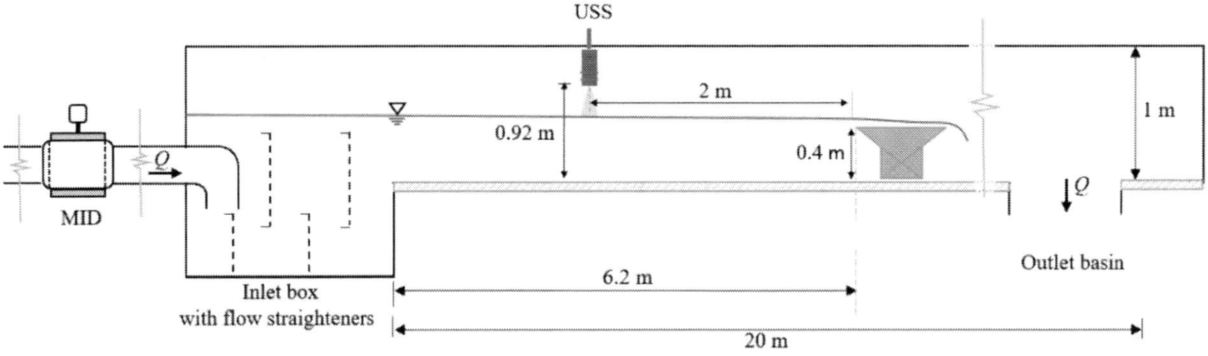

Figure 3. Experimental model setup sketch, including PKW and USS positions (non-scaled drawing)

4. Results and discussion

4.1. Scale effects

A comparison of discharge coefficients between the current 40 cm model and previous experiments (for 10, 20, and 30 cm models) was carried out based on H_tP^{-1} values ranging from 0.054 to 0.5. Resulting discharge coefficients indicated the expected phenomenon at low heads in which the current large-scaled 40 cm model prevails with a relatively higher C_{dL} value, while no significant differences were observed for $0.15 < H_tP^{-1} < 0.35$, as presented in Figure 4. In some areas for $H_tP^{-1} > 0.35$, small deviations can be found that may be caused by small nape detachments (flow separation) at the tip of the outlet keys which will be discussed within Chapter 4.2. Discharge coefficients were calculated using the Du Buat equation:

$$C_{dL} = \frac{3Q}{2L\sqrt{2g}H_t^{1.5}} \tag{1}$$

Where: C_{dL} = discharge coefficient (related to total crest length), Q = discharge, L = total crest length, g = acceleration due to gravity, H_t = total upstream energy head = $v^2(2g)^{-1} + h_t$, v = depth-averaged upstream velocity, h_t = overfall height.

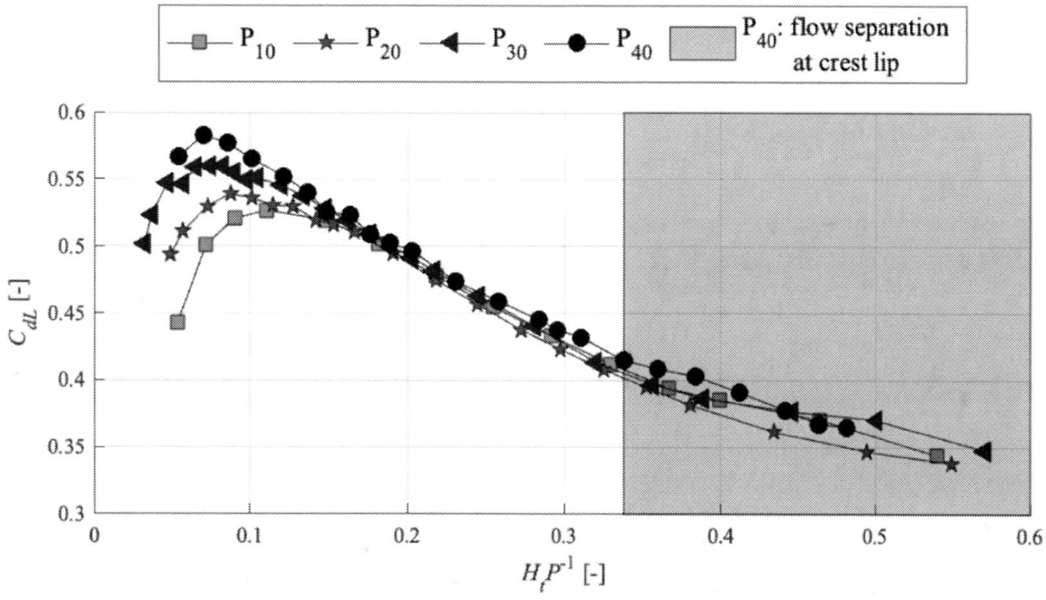

Figure 4. Discharge coefficients of investigated PKW models

Often, large scale modeling in hydraulic laboratories is taken as a better option but comes with financial and technical challenges. Ideally, near prototype physical models in hydraulic laboratories reproduce the flow parameters efficiently as the scale effect appears negligible compared with smaller ones. However, limitations still exist on the aspects of preparation cost and time when it comes to the largest possible model size that can be fabricated. In case of PKWs as investigated in the current study, unlike the 10 cm and 20 cm models, the results illustrate that an increment in model scale or size from 30 cm to 40 cm did not depict adverse deviations in terms of discharge coefficients at low heads when compared at same discharges and H_tP^{-1}. This indicates the use of a 30 cm high PKW for laboratory scales could be taken as sufficient size to study discharge coefficients instead of 40 cm models, which saves preparation cost and time. However, because this investigation didn't include measurements of other complex parameters such as air-water flows downstream, a remark could not be put whether these 30 cm and 40 cm high PKW models reproduce relatively similar results (like discharge coefficients).

Furthermore, the influence of surface tension at low heads was evident (supporting previous studies) while comparing the two scale extremes of this study: 10 cm model vs. current 40 cm model (Figure 5). In this case, a one-to-one comparison shows a slight departure of C_{dL} values (beyond a 5% error bound) against the Froude similarity at low heads ($0.054 < H_tP^{-1} < 0.07$) as the model size gets smaller relative to the current P_{40} model (Figure 5).

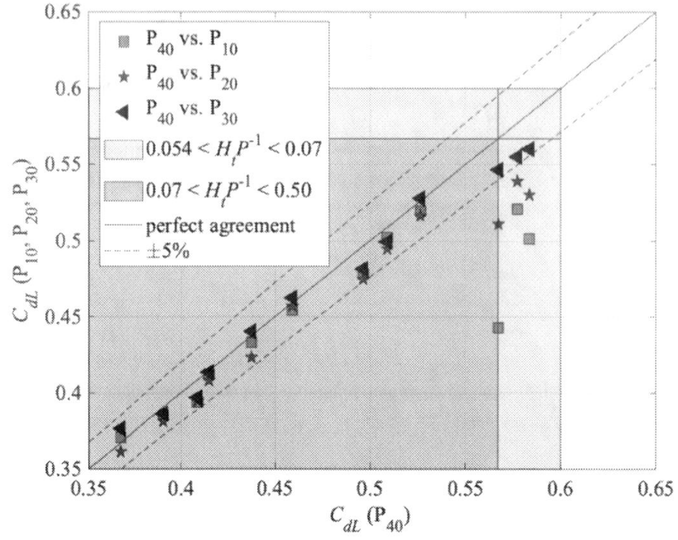

Figure 5. Relative comparison of discharge coefficients indicating scale effects (at relatively similar values of H_tP^{-1})

Apparent differences in nape trajectories were also noticed while comparing the flow behavior of the 20 cm with the current 40 cm model (see Figure 6), depicting the adversity of scale effects prevailing at small PKW models. This effect was reflected by the outcome of corresponding discharge coefficients at low heads ($0.054 < H_tP^{-1} < 0.07$) for these two models as shown in Figure 5.

Figure 6. Scale effects and its influence on nape trajectory for comparable discharges ($H_tP^{-1} = 0.05$), left: P_{20} model, right: P_{40} model

4.2. Effect of printing layer height on flow separation

Under the current setup, as discharges increased above 220 L/s ($H_tP^{-1} \geq 0.338$ m), significantly identifiable flow separation was observed at the tip of the crest (Figure 7). With close qualitative examination, the printing layer height (0.6 mm) of the 40 cm PKW depicted a direct influence on the nape trajectory at the upstream side of the crest lip. This behavior indicated the presence of undesired flow fields at higher discharges and the effect tends to slightly elevate the nape trajectory with unevenly distributed cavities in the spanwise direction over the weir crest (Figure 7).

For 3D printed models it must be noted that these represent a time- and cost-effective fabrication process for hydraulic laboratories. But the optimum printing layer height should be carefully chosen so that the crest lip roughness influence is maintained as minimum as possible. The printing layer height was seen to only affect the nape at the crest lip of the outlet keys where the flow accelerates compared to the side crest and the inlet keys. This result probes a hint for future model preparation methods using 3D printers that, it is essential to avoid or minimize forced roughness that could arise from the choice of printing layer heights so that undesired interferences on the nape behavior could be reduced. The observed flow separation beyond the specified discharge did not induce a major change in resulting discharge coefficients (see Figure 4), however this should be taken into account when investigating the nape trajectory. Smaller printing layer heights or surface finish mechanisms might reduce layer influences. Nonetheless, it could be expected that such flow separation might also occur on side walls if the upstream head further increases. This could lead to a slightly rise of the nape, reducing the discharge coefficient compared with model runs without flow separation.

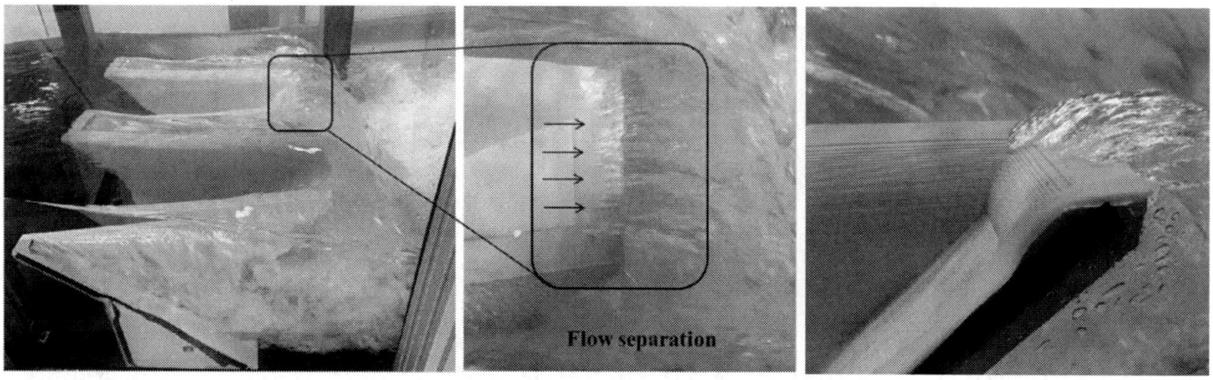

Figure 7. Observed flow separation at the upstream side of the crest lip (P_{40} model), left: overview, middle: top view detail, right: side view detail

5. Conclusions

With the aim of evaluating the advantages from developing and implementing experimentations using large scale models in hydraulic laboratories, this study undertook an assessment of a 40 cm high piano key weir model prepared via 3D printing. The test discharges covered a range in which the hydraulic performance can be clearly compared with other smaller models of 30 cm, 20 cm, 10 cm height. The following remarks are learnt from this experimental study of 3D printed PKWs:

- To investigate common hydraulic performance parameters of PKWs (in this case, discharge coefficients and rating curves), 30 cm high PKW models can be considered adequate to provide comparable results as the 40 cm models with minimum scale effects between them. Hence, 30 cm high PKW models can simply be used to study discharge coefficients while maintaining minimum cost and preparation time. Although this can be seen as an advantage, the size adequacy of these models (30 cm or 40 cm high PKW) may not be taken strictly sufficient to study every hydraulic parameter (such as air entrainment) because such complex flow filed properties are not covered in the current investigation. Hence, further studies are also encouraged to fill this gap in which optimum model sizes can be specified for air entrainment studies at minimum preparation cost and time of a physical model via 3D printing.
- The adverse deviations of discharge coefficient values for the same H_tP^{-1} among 40 cm and 10 cm models were evident at low heads, indicating the effect of surface tension will be important for small model scales. Considering the 40 cm PKW is a large-scale model, the comparison enhanced visualization of such scale effects.
- Furthermore, printing layer height (0.6 mm) of the 40 cm weir was observed to intervene with the flow pattern right at the upstream side of the crest lip at the outlet keys for larger discharges ($Q > 220$ L/s, $H_tP^{-1} > 0.338$). Although, this did not significantly alter the trend of the discharge coefficient for high heads, the behavior gives an indication that the nape trajectory could be directly affected depending on 3D plot parameters. It should be avoided by using relatively smaller layer heights or by an implementation of a smooth surface finish during model preparation. Hence, for model heights above 30 cm, the use of printing layers height less than 0.6 mm can be considered as a reasonable measure to minimize the presence of undesired flow pattern at the crest lip.

6. REFERENCES

Chanson, H. (2009). "Turbulent air-water flows in hydraulic structures: Dynamic similarity and scale effects". *Environmental Fluid Mechanics*, 9(2), 125–142. https://doi.org/10.1007/s10652-008-9078-3

Cicero, G. M., and Delisle, J. R. (2013). "Experimental and numerical study of the hydraulic performance of a trapezoidal piano key weir". *Labyrinth and Piano Key Weirs II*, 265–272. https://doi.org/10.1201/b15985

Erpicum, S., Tullis, B. P., Lodomez, M., Archambeau, P., Dewals, B. J., and Pirotton, M. (2016). "Scale effects in physical piano key weirs models". *Journal of Hydraulic Research*, 54(6), 692–698. https://doi.org/10.1080/00221686.2016.1211562

Eslinger, K. R., and Crookston, B. M. (2020). "Energy dissipation of type a piano key weirs". *Water* (Switzerland), 12(5). https://doi.org/10.3390/W12051253

Heller, V. (2011). "Scale effects in physical hydraulic engineering models". *Journal of Hydraulic Research*, 49(3), 293–306. https://doi.org/10.1080/00221686.2011.578914

Leite Ribeiro, M., Bieri, M., Boillat, J.-L., Schleiss, A. J., Singhal, G., and Sharma, N. (2012). "Discharge Capacity of Piano Key Weirs". *Journal of Hydraulic Engineering*, 138(2), 199–203. https://doi.org/10.1061/(asce)hy.1943-7900.0000490

Machiels, O., Pirotton, M., Pierre, A., Dewals, B., and Erpicum, S. (2014). "Experimental parametric study and design of Piano Key Weirs". *Journal of Hydraulic Research*, 52(3), 326–335. https://doi.org/10.1080/00221686.2013.875070

Oertel, M. (2019). "Size-Scale Effects of an a-Type Piano Key Weir". *Proc. 38th IAHR World Congress*, 38(1), 1930–1939. https://doi.org/10.3850/38wc092019-0954

Ouamane, A., and Lempérière, F. (2006). "Design of a new economic shape of weir". *Proc. of the International Symposium on Dams in the Societies of the 21st Century*, ICOLD-SPANCOLD – Dams and Reservoirs, Societies and Environment in the 21st Century, 1, 463–470. https://doi.org/10.1201/b16818-74

Shen, X., and Oertel, M. (2021). "Comparative Study of Nonsymmetrical Trapezoidal and Rectangular Piano Key Weirs with Varying Key Width Ratios". *Journal of Hydraulic Engineering*, 147(11), 04021045. https://doi.org/10.1061/(asce)hy.1943-7900.0001942

DOI: https://doi.org/10.26077/3268-7a17

Investigation on near bed flow features over a water-worked gravel bed

Ellora Padhi[1] & Nishank Agrawal[1]
[1]*Shiv Nadar University, Greater Noida, Uttar Pradesh, India*
E-mail: ellora.padhi@snu.edu.in

Abstract: *One of the key problems in the research on fluvial hydraulics is to correctly reproduce the flow dynamics over a riverbed in laboratory experiments. Generally, while performing experiments, attention is paid to specific hydraulic conditions related to the flow regime, sediment transport, and bed erosion. In this context, the sediment size of the riverbed to be analysed plays an important role. It is a common practice to create the bed manually using a screeded mixture of sediments with random orientation that could be very different from that observed in natural riverbeds. To overcome this issue, a manually created gravel bed is required to be water worked before performing the experiments to get a bed surface, which resembles to that of a natural gravel bed river. Such a bed is termed as water-worked gravel bed (WGB). In recent past, even though the bed topography of WGB has been intensively studied, its effects on the flow features, which govern the near-bed flow field, are still to be explored under varying flow rates. In this study, the double averaged (DA) turbulence characteristics such as velocity, Reynolds shear, and normal stresses, and form induced shear and normal stresses, are analysed for three different flow rates. A comparative study infers that with an increase in flow rate the magnitude of all flow features increases, especially in the near-bed flow zone. The reason for such occurrence is attributed to a higher flow rate that enhances the near-bed turbulence, and in turn the velocity fluctuations.*

Keywords: *Double-averaging methodology, Turbulence characteristics, Water-worked gravel bed.*

1. Introduction

The geophysical flows are often found to be rough owing to the presence of the roughness elements, such as, gravels, boulders etc. The existence of gravels in a stream puts significant impacts on the velocity flow fields, especially in the near-bed flow zone, owing to the generation of coherent turbulence structures. Coherent structures are irregular, quasi-deterministic, and repetitive structures which help to characterize the flow and govern the mean flow velocity, stress, and other statistical properties of the flow. Owing to the action of the coherent structures, the flow gets decelerated at the upstream of the gravel, whereas recirculation of the flow occurs downstream of the gravel (Roy et al. 2004; Adrian and Marusic 2012). It implies that coherent structures comprise of sequence of flow driving events, such as, ejections and sweeps. Ejections cause the flow to decelerate by pushing the low momentum fluid parcels away from the bed, whereas sweeps cause the flow to accelerate by moving high momentum fluid parcels toward the channel boundary. Hence, it can be inferred that coherent structures are responsible for the incipient motion of the gravels (in turn, transport of sediment particles).

By understanding the importance of the coherent structures in the flow, several researchers put forward their interest to study the behaviour of these structures (Kelin et al. 1967; Falco 1977; Cantwell 1981; Hussain 1983; Adrian and Moin 1988). Kelin et al. (1967) reported to visualize the eruption of near-wall low speed coherent structures and also claimed that these structures are responsible for vertical outward transport of momentum. Further, they also argued that determination of the flow properties such as, Reynolds shear stress (RSS) and turbulent kinetic energy (TKE) production are governed by these logical structures. Later, Kim et al. (1971) extended the work of Kelin et al. (1967) and confirmed that the TKE production in the near-bed flow zone occurs only during the intermittency bursting events. Inspired by these findings, various studies were carried out on coherent structures for smooth wall flows (Kim et al. 1987; Zhou et al. 1996; Adrian et al. 2000a and Adrian et al. 2000b). To understand the effects of surface roughness on the turbulence structures, Grass et al. (1971) performed experimental analysis for smooth as well as rough wall open channel flows. They found that behaviour of the coherent structures is influenced by the channel surface roughness. Further, Drake et al. (1988) captured the bedload transport process using high speed camera and visualized the sediment motions that were affected by the downward moving high-speed fluid parcels (sweep events). Nezu and Nakagawa (1993) studied the coherent structures and distinguished the near bed flow structures (bursting-structures) from the far bed large-scale vortical structures. Later on, by experimentation or by numerical simulation many researchers tried to analyse the behaviour of the coherent structures considering several open channel flow conditions, such as, permeable wall flow (Manes et al. 2012; Lamb et al. 2017; Voermans et al. 2017), flow within vegetation canopy (Raupach et al.1996; Ghisalberti et al. 2002; Nezu and Sonjou 2008; Bailey and Stoll 2016) mobile bed flow (Nelson et al. 1995, Santos et al. 2014, Keylock et al. 2014), gravel bed flow (Roy et al. 2004; Hurther, et al. 2007; Hardy et al. 2009; Detert et al. 2010; Guala et al. 2012; Voermans et al. 2018), etc.

Moreover, in case of gravel bed flow, although significant progress has so far been made to study the effects of coherent structures on turbulence parameters, only a few important turbulence parameters were taken into consideration. To be specific, Buffin-Bélanger et al. (2000) performed space time correlation analysis considering the velocity fluctuations to show the presence of large-scale flow structure in the gravel bed. Roy et al. (2004) conducted field experiments to quantify the vertical, lateral and longitudinal extent of the flow structures and to estimate their advecting velocities. Hurther et al. (2007) tried to show the presence of large-scale coherent structures in the gravel bed flow using the instantaneous velocity. In addition, they showed the effects of coherent structures on the turbulent kinetic energy (TKE) production and TKE fluxes. Detert et al. (2010) tried to show the presence of coherent structures in the flow by the wedge like structures which are formed by the faster and slower moving fluid streaks. They also showed the sequence of coherent structures by considering the vertical and horizontal planes. However, in their study they have not analyzed the effects of coherent structures on any turbulence parameters. Further, Guala et al. (2012) confirmed the existence of vortex packets in the outer layer of flow over rough walls by performing experimental study on sparse roughness elements, which also causes a shortening of the streamwise length scale in the near-wall region. Recently, Voermas et al. (2018) used conditional sampling method to separate the coherent structure motions based on the different bursting events, However, the study was limited to instantaneous flow velocity and Reynolds shear stress (RSS) and hence, the impact these conditionally sampled coherent structures on the double averaged flow parameters were not taken into consideration. Moreover, in a gravel bed flow, near-bed flow is locally three-dimensional and heterogeneous in space owing to continuous fluid-gravel interaction, making the near-bed turbulence parameters analysis complicated. Understanding the complexity, several researchers investigated the gravel bed flow fields and analysed important turbulence parameters (double averaged streamwise velocity, spatially averaged Reynolds shear and normal stresses, TKE fluxes, TKE budget, etc.) (Nikora et al. 2004, Nikora et al. 2007; Pokrajac et al. 2008; Mignot et al. 2009; Dey et al. 2011; Dey and Das 2012; Ferraro et al. 2016, etc).

Though these literatures provide an insight into the near-bed flow field of gravel bed flows, visualization of coherent structures motions and the variation of the time-averaged flow parameters over a space domain, level of turbulence, degree of anisotropy, and second-order correlation coefficients are yet to be explored. To date, the gravel bed experimental studies that are reported in the past were created by dispersing the gravels mixture in the flume and then screeded them manually. The bed topography of such screeded gravel beds is dissimilar from the beds that are observed in natural gravel-bed rivers. In the natural rivers, due to repetitive flood, continuous erosion and deposition of gravels occur. As a result, a sedimentary bed roughness structure is created by the flowing water and such a bed is commonly termed water-worked gravel bed (henceforth WGB). The bed surface in the WGB possesses spatially multifaceted, three-dimensional, well organized roughness structure. Hence, such bed surface (WGB) is completely deferent than the bed surface that is created manually in the laboratory flumes (Cooper and Tait 2009; Padhi et al. 2018). Recently, Padhi et al. (2018, 2019) reported that the surface topography is one of the significant parameters that influences the near-bed flow field greatly. Hence, to visualize the motions of coherent structures and to understand its impact on the near-bed turbulence parameters more accurately, it is important to conduct an experimental study over a WGB.

Therefore, the aim of this study was to capture the fluid parcel motion by analysing the time averaged velocity and vorticity field in space domain and also to understand their impact on the turbulence parameters. For doing so, three experimental runs were carried out in a WGB by varying the flow Froude number, in turn shear Reynolds number, using a particle image velocimetry (PIV) system. Besides, double averaged streamwise velocity, spatially averaged Reynolds Shear Stresses (RSS), form induced shear stress (FISS), spatially averaged TKE, form induced kinetic energy (FIKE), and quadrant analysis of spatial velocity fluctuations for all three Runs over WGB were analysed and the results were compared. Thus, this study helps us to understand the near-bed flow physics governed by the coherent structures and also the impact of shear Reynolds number on the turbulence parameters of the experimental runs over WGB.

2. Experimental Setup and Procedure

An experimental campaign was carried out in the *Grandi Modelli Idraulici* laboratory at the Università della Calabria, Italy. A rectangular tilting flume of length 9.6 m, width 0.485 m, and depth 0.5 m was used for the experimentation. The flume inlet consisted of a stilling tank, an uphill slipway, and a honeycomb designed to dampen the disturbance generated by the flow at the entrance of the flume. A tank was attached downstream of the tailgate to collect the outflow. A calibrated Thomson weir fitted at the outlet of the tank in order to measure the flow discharge. The coarse gravels with a median diameter d_{50} = 4.81 mm were dispersed to create the experimental bed. The flume walls were made of glass to enable the flow to be assessed visually. All measurements were taken within a 1 m long zone 6.3 m to 7.3 m from the entrance of the flume. The uniformity of flow in the test section was ascertained by measuring the flow depth with a point gauge. A detailed description of the setup has been given in Padhi et al. (2018).

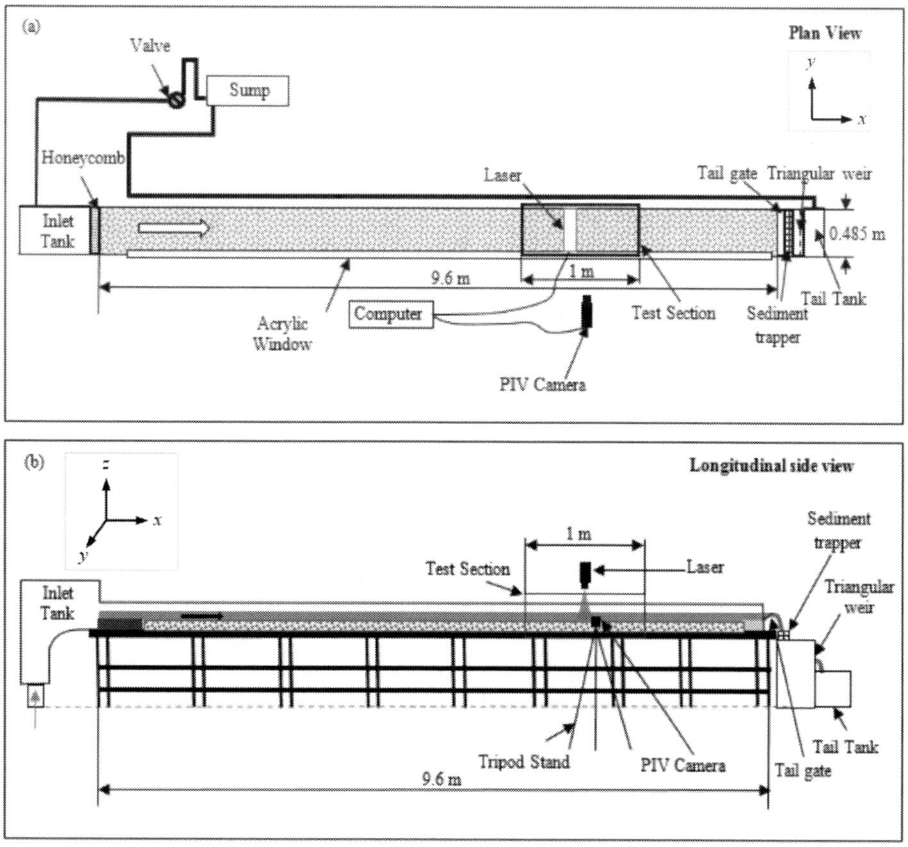

Figure 1. Schematic diagram of the experimental setup a) plan view and (b) longitudinal side view

The experiments were conducted in two phases. In the first phase, sediments were placed in the flume and screeded manually to form a screeded gravel bed with an average bed slope of 0.7%. Next, a WGB was created by initiating a flow over the screeded gravel bed in a manner that could transport surface sediment particles. Please refer Padhi et al. (2018), for detailed description on preparation of the WGB. Once the WGB was created, the bed surface fluctuations were measured using the laser scanner and the average bed slope was found to be 0.4%. Thereafter, three experimental runs (that is, WGB-1, WGB-2 and WGB-3) were carried out over the WGB for three different flow conditions in order to show the effects of flow Froude number on the turbulent parameters.

Table 1. Hydraulics characteristics of the experimental runs over the WGB.

Runs	h (m)	U (m s^{-1})	u_c (m s^{-1})	u_* (m s^{-1})	ν (m^2 s^{-1})	Fr	R	R_*
WGB-1	0.1	0.43	0.6	0.068	1.10×10^6	0.432	1.72×10^5	78
WGB-2	0.1	0.31	0.6	0.067	1.10×10^6	0.312	1.13×10^5	76
WGB-3	0.1	0.31	0.6	0.065	1.10×10^6	0.312	1.13×10^5	75

Table 1 reports the hydraulic parameters of the three experimental runs (WGB–1, WGB–2, WGB–3) that had been carried out over WGB: the flow depth h, the average flow velocity U, the critical flow velocity u_c, the shear velocities u_*, the flow Froude number Fr (=$U/(gh)^{0.5}$, where g is the gravitational acceleration), the flow Reynolds number R (= $4Uh/\nu$; where ν is the kinematic viscosity at 20°C), the shear particle Reynolds numbers R_* (=$u_*\Delta_k/\nu$, where Δ_k is the average roughness height). Importantly, even though the bed topography remains the same in three cases, due to a change in the rate of flow, the shear velocity changes. Therefore, to visualize the impact of change in the rate of flow, u_* is used as a scaling parameter. In all the experimental runs, the critical flow velocity was found to be higher than the average flow velocity, this indicates a no motion condition. Further, in all the cases, the values of R_* were greater than 70, which confirmed that the rough-turbulent flow prevailed.

The flow field over the gravel bed was measured using a two-dimensional particle image velocimetry manufactured by TSI. It consisted of a Nikon 12-bit CCD camera with 2048 × 2048 square pixels and a frame rate of 15 Hz, as well as a double-pulse Nd: YAG laser with a pulse energy of 200 mJ. Titanium dioxide particles with a mean diameter 3 μm and mass density of 4.26×10^3 kg m^{-3} was used as seeding particles. The laser sheet illuminated the test section in order to visualize the particle motion captured by the camera placed parallel to the laser sheet. An interrogation area (IA) of 32 × 32 pixels, each with a size of about 2.7 × 2.7 mm^2 was considered for the flow measurements. Thus, the overall area imaged was 170 × 170 mm^2. Overlapping of IAs was not considered. Altogether, 3000 pairs of images were captured along the flume centerline in order to measure the flow field over a period of 414 s at a streamwise distance of 0.94 m from the starting point of the test section. The image analysis provided 46 velocity profiles in the streamwise direction, covering an area of 120 × 100 mm^2 with a spatial resolution of 2.7 mm in both vertical and streamwise directions at the measuring location. No data cleaning process was used in analyzing the data.

3. Results and Discussion

3.1. Double Averaged Velocity and Time Averaged Velocity Field

Variations of nondimensional double averaged streamwise velocity profiles with the nondimensional vertical distance z/h in Runs WGB–1, WGB–2, and WGB–3 are shown in Fig. 2. Note that all of the flow quantities with an angle bracket represent the double averaged quantities. Here, u_* is considered as a scaling parameter in order to show the impact of the rate of flow as well as the roughness structure on the flow parameters. Near the bed, the streamwise velocity $\langle \bar{u} \rangle$ is less, but with the increase in depth, it also increases and attains the maximum velocity just below the free surface. An inflectional shape seen in the double averaged streamwise velocity profiles is an effect of the roughness element rather than the effect of mixing layer type nature of the near-bed flow. The log layer does not start immediately above the crest level, rather it starts from the form-induced sublayer. The reason is attributed to the presence of gravels that influences the flow in the near bed region. Further, comparison study reveals that the streamwise velocity of WGB–1 is higher as compared to WGB–2 and WGB–3. This happens due to the higher shear Reynolds number in the open channel, and hence, it accelerates and gains higher velocity for WGB–1.

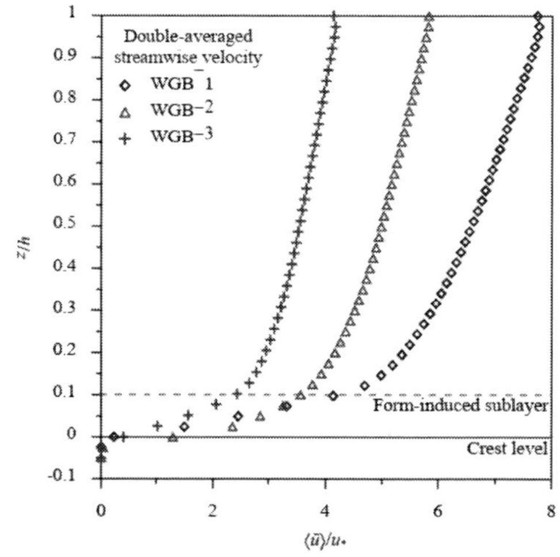

Figure 2. Variations of nondimensional double averaged streamwise velocity profiles with the nondimensional vertical distance z/h in Runs WGB-1, WGB-2, and WGB-3.

To visualize the time-averaged flow velocity field in a spatial domain, the contours of time-averaged accelerated and decelerated flow in a vertical plane along the flume centerline in the streamwise direction in Runs WGB–1, WGB–2, and WGB–3 are plotted in Fig. 3. In this study, the time-averaged flow is considered to be accelerated if $\langle \bar{u}_{dif} \rangle > 0$ and vice versa. Here, $\langle \bar{u}_{dif} \rangle = (\bar{u}_{res} - \langle \bar{u} \rangle_{res})/u_*$, where $\bar{u}_{res}$ is the time averaged resultant velocity and can be calculated as $\sqrt{\bar{u}^2 + \bar{w}^2}$, and $\langle \bar{u} \rangle_{res}$ is the double averaged resultant velocity, computed as $\sqrt{\langle \bar{u} \rangle^2 + \langle \bar{w} \rangle^2}$. The red colored contours define the accelerated flow, while the green colored contours show the decelerated flow. It is apparent from Fig. 3a) that in case of WGB–1, the time-averaged flow is observed to be highly accelerated in the initial part of the space domain, then slows down up to some extent, and thereafter, it again gets accelerated. However, owing to sudden change in bed surface topography at downstream of the space domain, a recirculation zone is created, which causes to decelerate the flow. Similar trend is also observed in case of WGB–2, but with less magnitude. However, in case of WGB–3, the initial stretch is dominated by decelerated flow, and then, followed by accelerated flow. In all three cases, the accelerated and decelerated flow pattern exists in alternate manner. This implies that when the flow rate is low, the flow structure changes its patterns smoothly, but when the flow rates get increased, fluid mixing process gets enhanced. As a result, the flow structures get stretched and expanded over the space domain for higher shear Reynolds number flow.

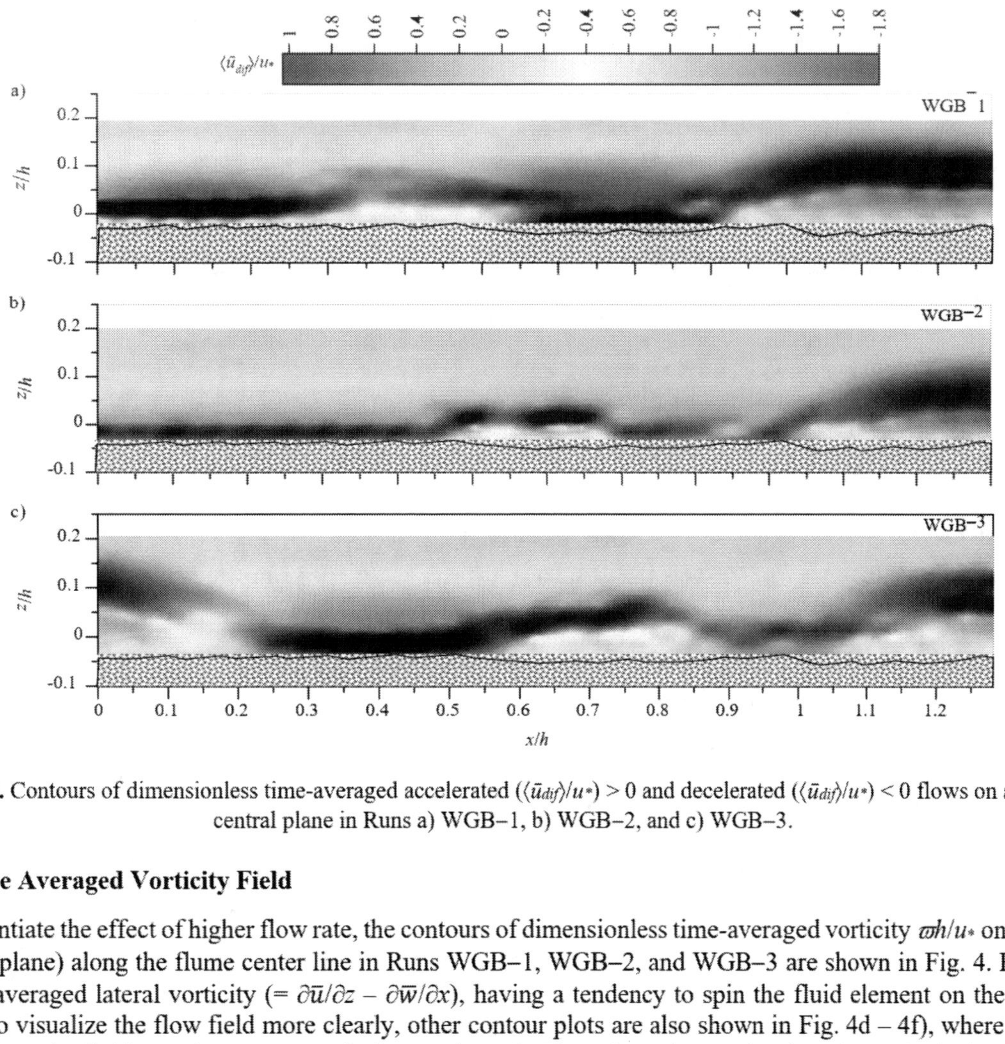

Figure 3. Contours of dimensionless time-averaged accelerated ($\langle \tilde{u}_{dif} \rangle / u_*$) > 0 and decelerated ($\langle \tilde{u}_{dif} \rangle / u_*$) < 0 flows on a vertical central plane in Runs a) WGB–1, b) WGB–2, and c) WGB–3.

3.2. Time Averaged Vorticity Field

To substantiate the effect of higher flow rate, the contours of dimensionless time-averaged vorticity $\varpi h/u_*$ on a vertical plane (xz plane) along the flume center line in Runs WGB–1, WGB–2, and WGB–3 are shown in Fig. 4. Here, ϖ is the time-averaged lateral vorticity (= $\partial \bar{u}/\partial z - \partial \bar{w}/\partial x$), having a tendency to spin the fluid element on the xz plane. Further, to visualize the flow field more clearly, other contour plots are also shown in Fig. 4d – 4f), where the black patches show the fluid parcel movement. Similar to Fig. 3, in Fig. 4a), ϖ changes its sign alternatively from negative to positive throughout the measurement section in all three cases. The cyan colour contour shows the positive vorticity, indicating the fluid parcel entrainment in to the main flow layer, whereas, the dark blue colour contour shows the negative vorticity that accelerates the flow, indicating the inrush of fluid parcel towards the bed. This implies that the presence of gravels causes the flow to be heterogeneous. As a result, flow perturbation occurs, causing the fluid parcels to move in a clockwise (positive motion) and anticlockwise (negative motion) manner.

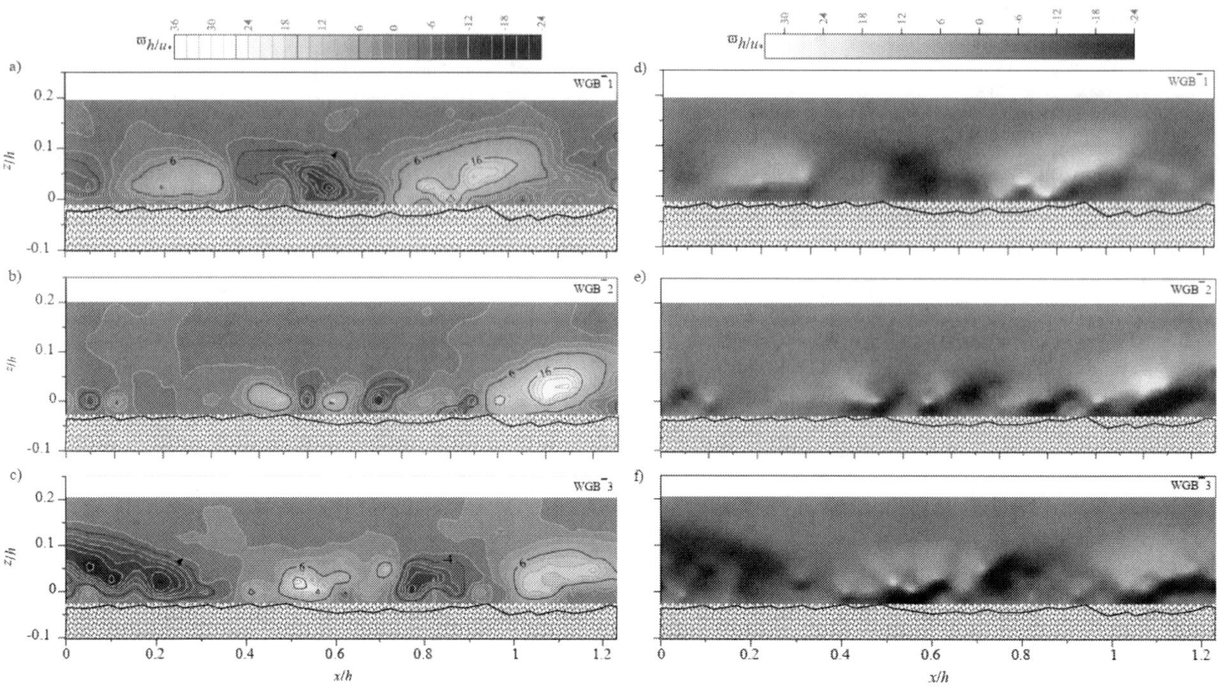

Figure 4. Contours of dimensionless time-averaged vorticity $\bar{\omega}h/u_*$ on a vertical central plane in Runs a) WGB-1, b) WGB-2, and c) WGB-3. Visualization of fluid parcel movement (black patch) based on dimensionless time-averaged vorticity $\bar{\omega}h/u_*$ on a vertical central plane in Runs d) WGB-1, e) WGB-2, and f) WGB-3.

3.3. Spatially Averaged Reynolds and Form Induced Shear Stresses

The variations of spatially averaged Reynolds Shear Stresses (RSS) $\langle -\overline{u'w'}\rangle/u_*^2$ with z/h for Runs WGB–1, WGB–2, and WGB–3 is shown in the Fig. 5a). The spatially-averaged RSS is the governing shear stress across the flow depth. Figure 4(a) reveals that for WGB-1, $\langle -\overline{u'w'}\rangle/u_*^2$ attains its peak at z/h = 0.18 m while, for WGB–2 and WGB–3, it attains its peak at z/h = 0.13 and 0.17 m, respectively, and then, it decreases with an increase in z. In all three cases the $\langle -\overline{u'w'}\rangle/u_*^2$ follows a linear law of RSS distribution, because the contribution from other flow governing shear stresses such as Form-induced shear stress (FISS) and the viscous shear stress are negligible above the log layer. Near the bed region, the $\langle -\overline{u'w'}\rangle/u_*^2$ decreases with decrease in z owing to a reduction in temporal velocity fluctuations u' and w'. Further in this region the damping in $\langle -\overline{u'w'}\rangle/u_*^2$ is partially compensated by the appearance of FISS ($\langle -\tilde{u}\tilde{w}\rangle/u_*^2$). The detailed analysis related to FISS is discussed in the next paragraph. Comparing the $\langle -\overline{u'w'}\rangle/u_*^2$ profiles of three runs, it is found that in the near bed region, $\langle -\overline{u'w'}\rangle/u_*^2$ is higher for Run WGB–1 as compared to that for Runs WGB-2, and WGB–3. The possible reason behind this is attributed to the effect of rate of flow on temporal variation of streamwise fluctuation.

As the time-averaged flow close to the gravel crest is spatially heterogeneous, the FISS starts developing within the form induced sublayers and grows sharply with a decrease in z. Variations of FISS $\langle -\tilde{u}\tilde{w}\rangle/u_*^2$ with z for all three Runs WGB–1, WGB–2, and WGB–3 is shown in the Fig. 5b). The $\langle -\tilde{u}\tilde{w}\rangle/u_*^2$ initiates developing just above the roughness crest at z = 0.17 in WGB–1, while in WGB–2, and WGB–3, it starts developing at z = 0.1 and 0.15 m, respectively. It is noticeable that for all three Runs, the FISS attains its peak just above the roughness crest (z = 0.0025 m), but with different magnitude. Similar kind of results was obtained for Manes et al. (2007), Cooper and Tait (2009). Within the form induced sublayer and a zone below the crest level, the spatial velocity fluctuations ($\tilde{u}$ and $\tilde{w}$) are high, giving rise to higher magnitude of $\langle -\tilde{u}\tilde{w}\rangle/u_*^2$. In the interfacial sublayer ($z \leq 0$), the damping in the $\langle -\tilde{u}\tilde{w}\rangle/u_*^2$ distribution is associated with a decrease in spatial velocity fluctuations ($\tilde{u}$ and $\tilde{w}$). This indicates that the flow becomes spatially less heterogeneous in the zone for $z \leq 0$ than that in form induced sublayer. Above the log layer, the spatial fluctuations ($\tilde{u}$ and $\tilde{w}$) are negligible, making the $\langle -\tilde{u}\tilde{w}\rangle/u_*^2 = 0$. The comparison of FISS profile for all three infers that the $\langle -\tilde{u}\tilde{w}\rangle/u_*^2$ in WGB–1 is much higher than that for the other two cases. The possible reason behind this is attributed to the reduction in spatial fluctuations in both streamwise and vertical directions resulting from slower flow rates in WGB–2, and WGB–3 as compared to WGB–1.

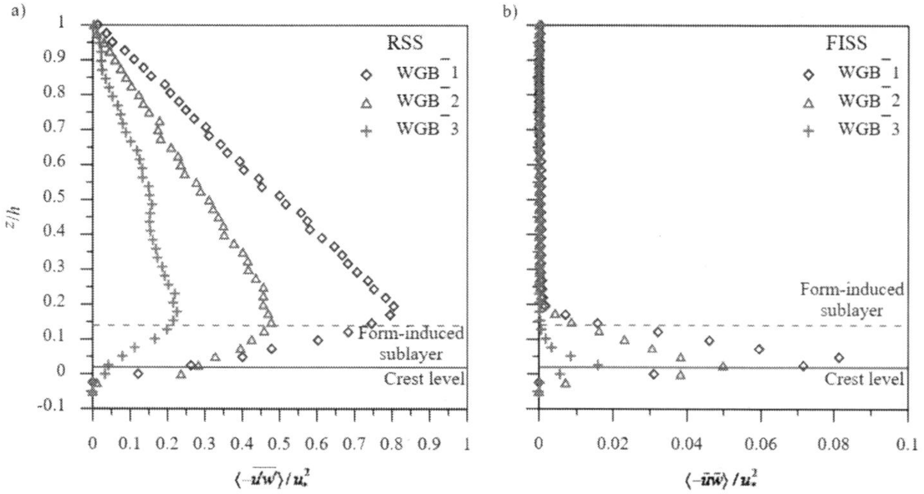

Figure 5. Variations of nondimensional a) spatially averaged Reynolds shear stress profiles and b) Form-induced shear stress profiles with the nondimensional vertical distance z/h in Runs WGB–1, WGB–2, and WGB–3.

3.4. Spatially Averaged and Form Induced Turbulent Kinetic Energy

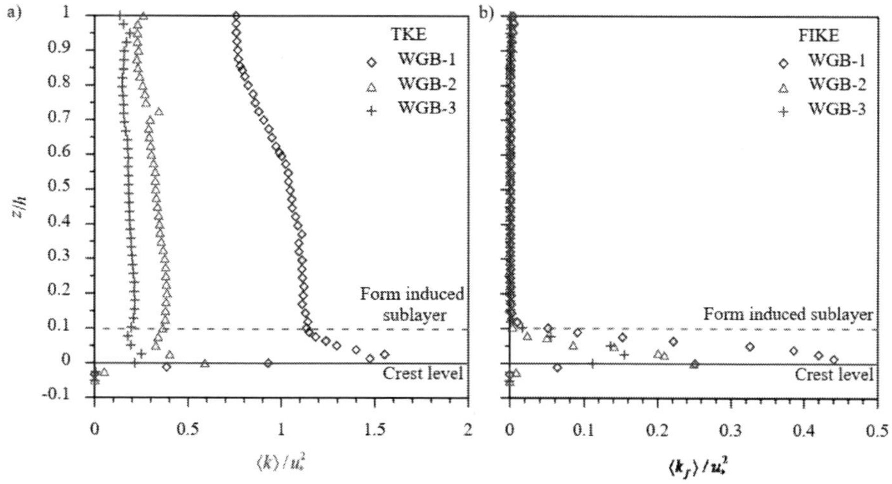

Figure 6. Variations of nondimensional a) spatially averaged Turbulent kinetic energy profiles and b) Form-induced Turbulent kinetic energy profiles with the nondimensional vertical distance z/h in Runs WGB–1, WGB–2, and WGB–3.

The spatially averaged Turbulent Kinetic Energy (TKE) is expressed as $\langle k\rangle = 0.5(\langle\overline{u'u'}\rangle + \langle\overline{v'v'}\rangle + \langle\overline{w'w'}\rangle)$, where v' is the fluctuations of the spanwise velocity component with respect to its time averaged value. However, due to limitation of two-dimensional PIV, the velocity in the spanwise direction was not possible. Moreover, from the examination of the double averaged streamwise velocity and spatially averaged RSS profiles it is confirmed that at least in the central portion of the flume (± 0.075 off the centerline) the two-dimensionality of the flow was satisfactorily preserved (Padhi et al. 2018). This implies that the impact of spanwise velocity v on the flow is insignificant as compared to the streamwise velocity u and the vertical velocity w. Therefore, in this study, spanwise velocity v and in turn the $\langle\overline{v'v'}\rangle$, can be neglected and the spatially averaged TKE can be calculated as $\langle k\rangle = 0.5(\langle\overline{u'u'}\rangle + \langle\overline{w'w'}\rangle)$. The variations of nondimensional spatially averaged TKE $\langle k\rangle/u_*^2$ with z/h for all three Runs WGB–1, WGB–2, and WGB–3 is shown in the Fig. 6a). For all three cases, it is noticeable that the $\langle k\rangle/u_*^2$ decreases with increase in z/h and attains its peak at or above the gravel crest and after that it decreases with increase in z. It suggests that in the near-bed flow zone, the fluid mixing process induced by the gravel roughness is to increase the temporal velocity fluctuations u' and w'. Hence, the magnitudes of $\langle k\rangle/u_*^2$ are greater in the near-bed flow zone than in the rest of the flow zone. It is apparent from the comparative study that the magnitudes of $\langle k\rangle/u_*^2$ follow a sequence of Run WGB–1 > Run WGB–2 > Run WGB–3. Owing to the larger shear Reynolds number R_* in Run WGB–1 as compared to those in Runs WGB–2 and

WGB–3, the u' and w' for former are greater than those for two latter runs. In essence, the SA TKE increases with an increase in the shear Reynolds number in the WGBs.

To understand the influence of spatial fluctuations on kinetic energy, Form-induced kinetic energy (FIKE) for all three cases is analyzed. Similar to spatially averaged TKE, FIKE can be computed as $\langle k_f \rangle = 0.5(\langle \tilde{u}\tilde{u} \rangle + \langle \tilde{w}\tilde{w} \rangle)$ and made nondimensional by multiplying u_*^{-2}. The results for all three cases are illustrated in Fig. 6b). It is evident from the figure that owing to higher spatial fluctuations, FIKE remains higher in the near the bed region. Further, FIKE for WGB–1 is higher than those for WGB–2 and WGB–3. Near the bed, the reason behind this may be due to the higher shear Reynolds number in WGB–1.

3.5. Quadrant Plots of Spatial Velocity Fluctuations

A quadrant analysis of the spatial velocity fluctuations, $\tilde{u}$ and $\tilde{w}$, in the $\tilde{u}\tilde{w}$ domain was performed for the WGB–1, WGB–2 and WGB–3 for different vertical distances z/h (= 0.15, 0.07, 0.025, 0, –0.025). The results for all three runs are shown in Figs. 7a), 7b) and 7c), respectively. It is apparent that for $z/h < 0.05$, the data form a scatter plot, indicating that the flow is highly heterogenous in the near bed zone due to the presence of gravel. This indicates that the form-induced shear stress is at its maximum when $z/h = 0.025$, decreasing when z/h either increases or decreases. However, above $z/h = 0.05$, the data create cluster plot, showing that the flow regains its homogeneity. Upon comparing Figs. 7a), 7b) and 7c), one can see that the $\tilde{u}$ and $\tilde{w}$ in the WGB–1 appear to be more scattered and have higher magnitudes than those in WGB–2 and WGB–3. This implies that higher rate of flow causes significant disturbances in the near bed flow region in presence of gravels.

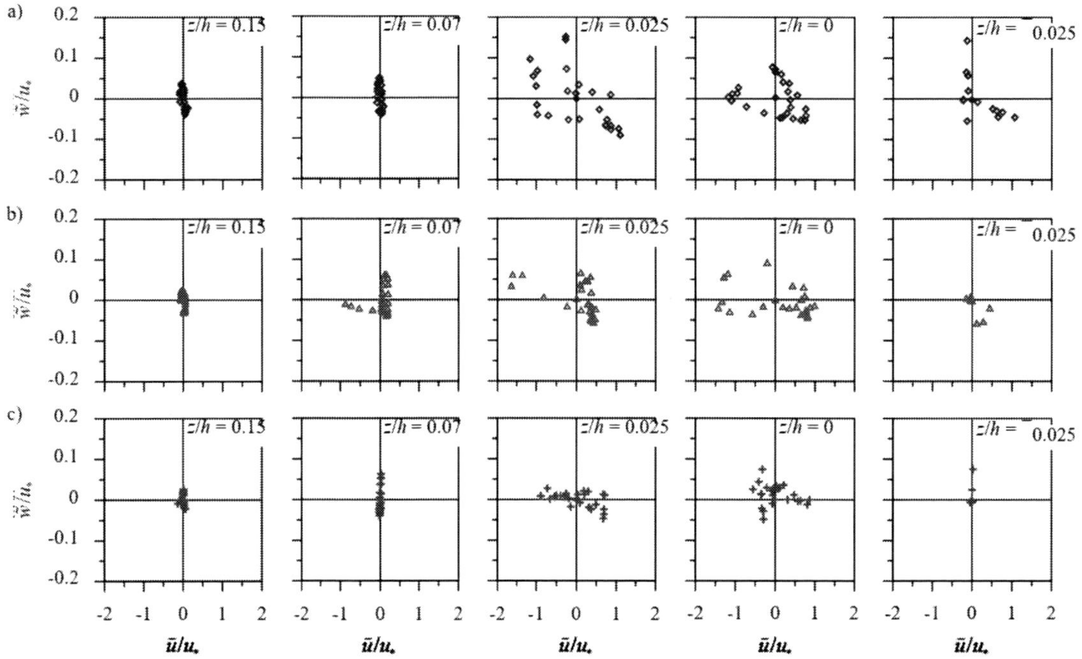

Figure 7. Quadrant plots of spatial velocity fluctuations at different vertical distances z/h in Runs a) WGB–1, b) WGB–2, and c) WGB–3

4. Conclusions

Experimental campaign was conducted to analyze the near-bed flow velocity and vorticity fields, and other turbulence characteristics using a two-dimensional PIV system. In order to understand effects of water work on the flow field and other turbulence characteristics, flow measurements were carried out in the WGB for three different flow Froude number. The main findings of this study are follows:

For all three cases, above the form induced sublayer, the vertical distribution of double averaged streamwise velocity follows the log law, while within the form induced and interfacial sublayer, it follows the linear law and third order polynomial function, respectively. The analysis of contours plots infer that the motions of the fluid parcels are the repetitive in nature. In near-bed flow zone, vorticity field shows the clockwise and anticlockwise motions of the lateral vortices which help to decelerate and accelerate the flow, respectively. With increase in flow rate, the size of the vorticity also increases. The spatially averaged RSS attains its peak above the form induced sublayer for all three cases. On the other hand, the FISS attains its peak just above the crest for all three cases. For both the shear stresses profiles WGB–1 is having higher magnitude as compared to that of WGB–2 and WGB–3. From the spatially averaged TKE and FIKE profiles, it can be concluded that with an increase in shear Reynolds number, magnitude of spatially averaged TKE and FIKE also increases. The spatial velocity fluctuations for all three cases were plotted. In the near-bed flow zone, the plotted clusters formed pseudo-elliptical shapes. The clusters became small beyond the form-induced sublayer, indicating the disappearance of spatial velocity fluctuations. The quadrant plots of spatial velocity fluctuations in the WGB–1 was more scattered than those in the WGB–2 and WGB–3, indicating the influence of difference in shear Reynolds number between the experimental runs.

The experimental results for the WGB allow us to understand the interaction between flow and natural gravel bed. This study reveals that with increase in flow rate, the fluid parcels movement get enhanced and frequency of alternate lateral vortices also gets increased. Moreover, it is suggested that the future experiments are to be performed for the WGB, which is more representative of the actual riverbed condition, in order to get an accurate estimation of near-bed turbulence characteristics. On the other hand, for the SGB, the results of the near-bed turbulent parameters obtained from an experiment are to be handled with care.

5. REFERENCES

Adrian, R., Christensen, K., and Liu, Z. (2000a). "Analysis and interpretation of instantaneous turbulent velocity fields," *Exp. Fluids*, 29(3), 275-290.

Adrian, R.J. (2013). "Structure of turbulent boundary layers," in Coherent Flow Structures at Earth's Surface, edited by J. G. Venditti, J. L. Best, M. Church, and R. J. Hardy, JohnWiley & Sons, Chichester, UK.

Adrian, R.J. and Marusic, I. (2012). "Coherent structures in flow over hydraulic engineering surfaces." *J. Hydraul. Res.*, 50(5), 451-464.

Adrian, R.J. and Moin, P. (1988) "Stochastic estimation of organized turbulent structure: homogeneous shear flow," *J. Fluid Mech.*, 190(1), 531-559.

Adrian, R.J., Meinhart, C.D., and Tomkins, C.D. (2000b). "Vortex organization in the outer region of the turbulent boundary layer," *J. Fluid Mech.*, 422, 1-54.

Bailey, B.N., and Stoll, R. (2016). "The creation and evolution of coherent structures in plant canopy flows and their role in turbulent transport," *J. Fluid Mech.*, 789, 425-460.

Buffin-Bélanger, T., Roy, A.G., Lamarre, H., and Kirkbride, A.D. (2000). "On large-scale flow structures in a gravel-bed river," Geomorphology 32(3), 417-435.

Cantwell B.J. (1981). "Organized motion in turbulent flow," *Annu. Rev. Fluid Mech.*, 13, 457-515.

Cooper, J.R., and Tait, S.J. (2009). "Water-worked gravel beds in laboratory flumes–a natural analogue?," Earth Surf. Processes Landforms, 34(3), 384-397.

Detert, M., Nikora, V., and Jirka, G.H. (2010). "Synoptic velocity and pressure fields at the water-sediment interface of streambeds," *J. Fluid Mech.*, 660, 55-86.

Dey, S., and Das, R. (2012). "Gravel-bed hydrodynamics: Double-averaging approach," *J. Hydraul. Eng.*, 138(8), 707-725.

Dey, S., Sarkar, S., and Ballio, F. (2011). "Double-averaging turbulence characteristics in seeping rough-bed streams," *J. Geophys. Res.*, 116(F3), F03020.

Drake, T.G., Shreve, R.L., Dietrich, W.E., and Whiting, P. J. (1988). "Bedload transport of fine gravel observed by motion-picture photography," *J. Fluid Mech.*, 192, 193-217.

Falco, R.E. (1977). "Coherent motions in the outer region of turbulent boundary layers," *Phys. Fluids*, 20(10), S124-S132.

Ferraro, D., Servidio, S., Carbone, V., Dey, S., and Gaudio, R. (2016). "Turbulence laws in natural bed flows," *J. Fluid Mech.*, 798, 540-571.

Ghisalberti, M., and Nepf, H.M. (2002). "Mixing layers and coherent structures in vegetated aquatic flows," *J. Geophys. Res.*, 107(C2), 3011.

Grass, A.J. (1971). "Structural features of turbulent flow over smooth and rough boundaries," *J. Fluid Mech.*, 50, 233-255.

Guala, M., Tomkins, C.D., Christensen, K.T., and Adrian R. J. (2012). "Vortex organization in a turbulent boundary layer overlying sparse roughness elements," *J. Hydraul. Res.*, 50(5), 465-481.

Hardy, R.J., Best, J.L., Lane, S.N., and Carbonneau, P.E. (2009). "Coherent flow structures in a depth-limited flow over a gravel surface: the role of near-bed turbulence and influence of Reynolds number," *J. Geophys. Res.*, 115(F3), 1-18.
Hurther, D., Lemmin, U., and Terray E.A. (2007). "Turbulent transport in the outer region of rough-wall open-channel flows," *J. Fluid Mech.*, 574, 465-493.
Hussain A.K.M.F. (1983). "Coherent structures–reality and myth," *Phys. Fluids*, 26(10), 2816-2850.
Keylock, C.J., Lane, S.N., and Richards, K.S. (2014). "Quadrant/octant sequencing and the role of coherent structures in bed load sediment entrainment," *J. Geophys. Res.*, 119(2), 264-286.
Kim, J., Kline, S.J., and Reynolds, W.C. (1971). "The production of turbulence near a smooth wall in a turbulent boundary layer," *J. Fluid Mech.*, 50, 133-160.
Kim, J., Moin, P., and Moser, R. (1987). "Turbulence statistics in fully developed channel flow at low Reynolds number," *J. Fluid Mech.*, 177, 133-166.
Kline, S.J., Reynolds, W.C., Schraub, R.A., and Runstadler, P.W. (1967). "The structure of turbulent boundary layers," *J. Fluid Mech.*, 30, 741-773.
Lamb, M.P., Brian, F.B., and Fuller, M. (2017). "Hydrodynamics of steep streams with planar coarse-grained beds: Turbulence, flow resistance, and implications for sediment transport," *Water Resour. Res.*, 53(3), 2240-2263.
Manes C., Pokrajac, D., and McEwan, I. (2007). "Double-averaged open-channel flows with small relative submergence," *J. Hydraul. Eng.*, 133(8), 896–904
Manes, C., Ridolfi, L., and Katul, G. (2012). "A phenomenological model to describe turbulent friction in permeable-wall flows," *Geophys. Res. Lett.*, 39(14), L14403.
Mignot, E., Barthelemy, E., and Hurther, D. (2009). "Double-averaging analysis and local flow characterization of near-bed turbulence in gravel-bed channel flows," *J. Fluid Mech.*, 618, 279-303.
Neill, C.R. (1967). "Mean-velocity criterion for scour of coarse uniform bed material," in *Proc. International Association of Hydraulic Research 12th Congress*, Fort Collins, Colorado, Vol. 1, 46-54.
Nelson, J.M., Shreve, R.L., and McLean, S.R. (1995). "Role of near-bed turbulence structure in bed load transport and bed form mechanics," *Water Resour. Res.*, 31(8), 2071-2086.
Nezu, I., and Nakagawa, H. (1993). Turbulence in Open-Channel Flows, International Association of Hydraulic Engineering and Research Monograph, Rotterdam, Netherlands.
Nezu, I., and Sanjou, M. (2008). "Turbulence structure and coherent motion in vegetated canopy open-channel flows," J. *Hydro-environ. Res.*, 2(2), 62-90.
Nikora, V., Koll K., I. McEwan, McLean, S., and Dittrich A. (2004). "Velocity distribution in the form induced layer of rough-bed flows," *J. Hydraul. Eng.*, 130(10), 1036-1042.
Nikora, V., McLean, S., S., Coleman, D., Pokrajac, I., McEwan, L., Campbell, Aberle, J., Clunie, D., and Kantika, K. (2007). "Double-averaging concept for rough-bed open-channel and overland flows: Applications," *J. Hydraul.* Eng., 133(8), 884-895.
Padhi, E., Penna, N., Dey, S., and Gaudio, R. (2018). "Hydrodynamics of water-worked and screeded gravel beds: A comparative study," *Phys. Fluids*, 30(8), 085105.
Padhi, E., Penna, N., Dey, S., and Gaudio, R. (2019). "Near-bed turbulence structures in water-worked and screeded gravel-bed flows," *Phys. Fluids*, 31(4), 045107.
Pokrajac, D., McEwan, I., and Nikora, V. (2008). "Spatially averaged turbulent stress and its partitioning," Exp. Fluids 45(7), 73-83.
Raupach, M.R., Finnigan, J.J., and Brunet, Y. (1996). "Coherent eddies and turbulence in vegetation canopies: the mixing layer analogy," *Boundary-Layer Met.*, 78(3), 351-382.
Roy, A.G., Buffin-Bélanger, T., Lamarre, H., and Kirkbride, A.D. (2004). "Size, shape and dynamics of large-scale turbulent flow structures in a gravel-bed river." *J. Fluid Mech.*, 500, 67-99.
Santos, B.O., Franca M.J., and Ferreira, R.M.L. (2014) "Coherent structures in open channel flows with bedload transport over a hydraulically rough bed," in *Proc. International Conference on Fluvial Hydraulics, River Flow* Lausanne, Switzerland, 883-890.
Voermans, J.J., Ghisalberti, M., and Ivey, G.N. (2017). "The variation of flow and turbulence across the sediment-water interface," *J. Fluid Mech.*, 824, 413-437.
Voermans, J.J., Ghisalberti, M., and Ivey, G.N. (2018). "The hydrodynamic response of the sediment-water interface to coherent turbulent motions," *Geophys. Res. Lett.* 45(19), 10520-10527.
Zhou, J., Adrian, R.J., and Balachandar, S. (1996). "Auto generation of near-wall vortical structures in channel flow," *Phys. Fluids*, 8(1), 288-290.

DOI: https://doi.org/10.26077/e166-a040

Jet Scour in Cohesive and Non-Cohesive Soil

J. Di Nardi[1], M. Palermo[1], S. Pagliara[1] & F.A. Bombardelli[2]
[1]*Department of Energy, Systems, Territory and Construction Engineering, University of Pisa, Italy*
[2]*Department of Civil and Environmental Engineering, University of California, Davis (CA), USA*
E-mail: s.pagliara@ing.unipi.it

Abstract: *A safe design of hydraulic structures depends on reliable estimations of the scour generated by the interaction between flow, structure, and bed sediments. Non-cohesive and cohesive soils are characterized by very different scour processes. Consequently, the behavior of the two soil types under scour has usually been studied separately. More specifically, cohesive soils are subjected to significant interparticle forces, which affect the properties of the material, resulting in different erosion mechanisms. Because of the complexity of the problem, erosion processes have been mainly investigated experimentally, resulting in a variety of case-dependent predicting equations. Recent theoretical advancements contributed to clarify the physics of scour in granular materials, paving the way to the development of general tools for its assessment. Nevertheless, a systematic characterization of local erosion in different types of soil is still lacking. The aim of this paper is to present an updated summary of current knowledge on scour caused by circular jets for cohesive and non-cohesive soils, highlighting differences and similarities of the scour phenomenon. In so doing, we also point out the still-challenging aspects and the future lines of research.*

Keywords: *Cohesive soil, granular soil, hydraulic structures, scour, turbulent jet.*

1. Introduction

A reliable assessment of the erosive capability of turbulent jets is crucial to ensure a safe design of hydraulic structures such as dam spillways, grade-control structures, and culverts. Thus, the erosional behaviour of both cohesive and cohesionless soils must be carefully assessed to avoid structural failures. Most studies on erosion downstream of hydraulic structures focused on cohesionless soils (e.g., sand and gravel), which typically originate from rock fracturing and mechanical weathering. Conversely, the erosion of *cohesive* sediment has received much less attention, despite their pervasive presence in riverine environments. Cohesive soils result from deposition and consolidation of suspended sediment. Thus, they accumulate in river deltas and in navigation channels, and constitute valuable fertile floodplains (Winterwerp & Van Kesteren 2004). In addition, their erosion might release large quantities of pollutants (Grabowski et al., 2011). The resistance of cohesionless soils is due to the friction between the grains and the submerged weight of particles (Partheniades 2009, Walder 2015). On the other hand, that of cohesive soils depends on interparticle and interaggregate forces. More specifically, internal forces depend on a wide number of physical, chemical, and biological properties of both soil and water, and contribute to determine the structure and properties of the material (Partheniades 2009, Shan et al. 2015). Only a few works addressed erosion of both cohesive and cohesionless soils (e.g., Mirtskhulava et al. 1967, Ansari et al. 2003, Van Ledden et al. 2004, Chen et al. 2018). Among these, very less focused on erosion by jets (Stein et al. 1993, Mazurek & Hossain 2007). This paper aims at presenting a critical review of current knowledge on scour caused by circular jets for cohesive and non-cohesive soils. To this end, we first describe the general features of circular jet-induced scour processes. Then, we summarize present knowledge on scour mechanisms, highlighting differences and similarities and pointing out challenging aspects and future lines of research.

2. Characteristics of jet-induced scour

Vertical or subvertical plunging jets may originate from different hydraulic structures. After jet impacts onto the downstream water pool, a dissipative process for the jet takes place (Hoffmans & Verheij 1997). Figure 1 shows a sketch of a jet with average velocity U_0, impinging angle α, and falling height h. D represents the water depth of the pool. A scour hole with maximum scour depth Δ and length L originates and enlarges up to the equilibrium configuration. The maximum scour depth at equilibrium is termed Δ_{eq}. For a circular jet, q indicates the total discharge and d_{jet} the jet diameter whereas, for a two-dimensional jet, q indicates the discharge per unit width of the jet and d_{jet} represents the jet thickness.

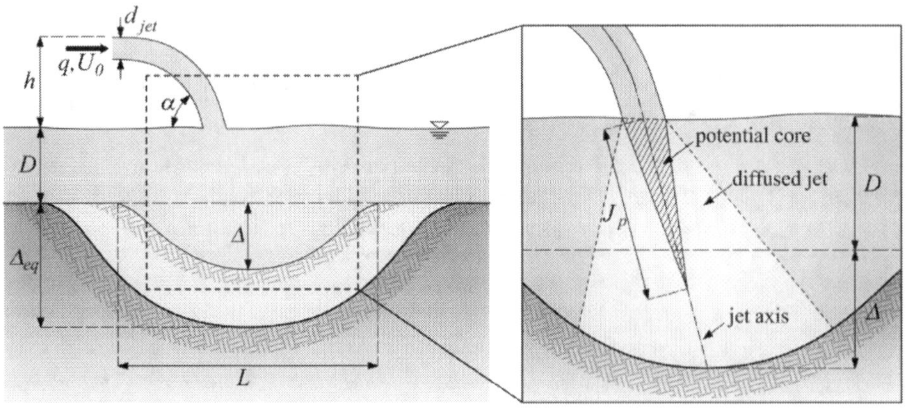

Figure 1. Diagram sketch of a scour hole generated by a plunging jet.

When a jet plunges into a pool, it diffuses radially in the water as it moves away from the impinging point. Experiments have shown that for a length J_p, the centerline velocity remains constant and equal to U_0. For circular jets, a cone of flow with unchanged velocity is formed, commonly referred to as "potential core". Beyond the length J_p, the centerline velocity starts decreasing along the jet trajectory. The centerline velocity U for a circular jet can be expressed as (Rajaratnam 1976, Stein et al. 1993):

$$\frac{U}{U_0} = C_d\sqrt{\frac{d_{jet}}{J}} \tag{1}$$

where J indicates the distance from the impinging point measured along the jet centerline, and C_d is a diffusion constant (for a free jet, Rajaratnam 1976 proposed $C_d = 2.47$). Note that the length of the potential core J_p can be derived from Eq. (1) with $U = U_0$, leading to $J_p = C_d^2 d_{jet}$. When the jet impacts the bottom surface, three main flow regions can be distinguished (Beltaos & Rajaratnam 1973): (a) the "free jet region", in which the jet is not affected by the horizontal boundary, (b) the "impingement region", which includes the stagnation point, and (c) the "wall jet region", where the fluid flows parallel to the boundary. Furthermore, if the potential core does not impact the boundary, the jet is termed "fully developed". Experimental evidence showed that such case occurs for values $D/d_{jet} > 8.3$ (i.e., large impingement heights, Beltaos & Rajaratnam 1977). In the "impingement region", experiments show that Eq. (1) is still valid, but different values of C_d must be used to account for the jet deviation due to the horizontal boundary ($C_d \approx 2.6$ according to Stein et al. 1993). It is worth noticing that this description, although useful to clarify the behavior of an impinging jet, is only valid for a flat bed. Consequently, it successfully describes the flow dynamics at the beginning of the scour process, when the undisturbed bed can be assumed to be horizontal. Nevertheless, it does not reflect the complex hydrodynamics inside the scour hole, where the jet is subjected to a significant confinement (Ghaneeizad et al. 2015, Karamigolbaghi et al. 2017).

The shear stress τ exerted on the soil by water is the main responsible parameter for erosion. Erosion occurs when forces are unbalanced, i.e., when the shear stress acting on the bed material exceeds the critical bed shear stress τ_c. Consequently, scour hole enlarges for $\tau > \tau_c$ and the equilibrium condition (i.e., scour depth equal to Δ_{eq}) is reached when $\tau = \tau_c$ (Stein et al. 1993, Hoffmans 2009). The rate of erosion can be expressed as a function of the excess shear stress (τ - τ_c). It is worth mentioning that scour processes are also influenced by shear stress fluctuations. Nevertheless, to date, scour prediction approaches based on the excess of shear stresses are still the most used. Sediment continuity implies that the rate of sediment mass eroded per unit area E at maximum scour depth location should be equal to sediment bulk density B (mass of solid per unit volume of soil) times the rate of change in scour depth. By equating sediment continuity and excess shear stress equation, Stein et al. (1993) obtained:

$$E = B\frac{d\Delta}{dt} = x(\tau_m - \tau_c)^{\xi} \tag{2}$$

where x and ξ are constants, and τ_m indicates the maximum bed shear stress at the bottom of the scour hole. Equation (2) is generally adopted to model scour evolution problems both in cohesive and non-cohesive soils. According to Stein et al. (1993), τ_m can be expressed as $C_f \rho U^2$, where ρ is the density of the fluid and C_f is a friction coefficient, depending on the flow Reynolds number Re. Therefore, from Eq. (1) it is evident that within the potential core (i.e., for $J \leq J_p$) $\tau_m = C_f \rho U_0^2$, whereas in the impingement region (i.e., for $J > J_p$) $\tau_m = C_d^2 C_f \rho U_0^2 d_{jet}/J$. Overall, both τ_m and τ_c depend on soil type, as we will discuss in the following sections.

3. Jet scour in non-cohesive soils

3.1. Characteristics of the scour

When soil particles have small surface area and a relatively large mass, their behaviour is governed by the particle weight, buoyancy, drag, and lift forces (Hoffmans & Verheij 1997, Chen et al. 2018). This is the case of non-cohesive soils (e.g., sand and gravel, characterized by particle diameter $d > 63$ μm). Such forces depend on the size, shape, and density of particles ρ_s, and on the hydrodynamic characteristics of the jet. Consequently, scour process is caused by the displacement of single grains, via sliding and/or rolling mechanisms. Soil particles removed from the bed are either kept in suspension/rotation within the scour hole or transported downstream, where they may accumulate forming a dune. The presence of a dune might interfere with the scour hole development, limiting its expansion. Likewise, suspended sediment kept in rotation by the jet in the scour hole also influences scour evolution, given that a certain amount of flow energy is needed to keep the sediment in rotation (Palermo et al. 2020). Aderibigbe & Rajaratnam (1996) distinguished a "weakly deflected" and a "strongly deflected" scour hole shape for vertical jets depending on the value of $Fr_0/(D/d_{jet})$, where Fr_0 indicates the densimetric Froude number $Fr_0 = U_0[gd(\rho_s-\rho)/\rho)]^{0.5}$, with g indicating the gravitational acceleration, and d the characteristic sediment size. For $Fr_0/(D/d_{jet}) > 2.0$, the jet is strongly deflected, and the scour hole is narrow and deep. Conversely, for $Fr_0/(D/d_{jet}) < 0.35$, the jet is weakly deflected, and the scour hole is wide and shallow. Aderibigbe & Rajaratnam (1996) also found that geometric similarity applies to all scour profiles, regardless of the two regimes, and the maximum scour depth occurs near the jet impact point.

Experimental tests conducted with different structures corroborate that scour features evolve logarithmically. Namely, the evolution is characterized by an initial rapid phase followed by a slow increase. A practical equilibrium condition is reached when the rate of scour tends to zero. Furthermore, for non-cohesive materials, two scour equilibrium configurations may occur, i.e., *dynamic* and *static*. The former refers to the maximum scour reached when the jet is on, whereas the latter indicates the configuration achieved when the jet flow ceases, and the material kept in suspension/rotation deposits into the scour hole. The two configurations may substantially differ, depending on the hydraulic and geometric characteristics of the process (Pagliara et al. 2008). For example, in a strongly deflected regime, the flow strongly re-circulates within the scour hole, and a considerable amount of sediment remains suspended. Therefore, in this case the dynamic scour depth might be much larger than the static counterpart (Mazurek & Hossain 2007). Nevertheless, measuring scour depth under dynamic conditions is very difficult, and the scour geometry in prototypes can only be assessed at the end of a flood event.

3.2. Experimental studies

Because of the complexity of erosion processes, the analysis of the scour caused by plunging jets has been traditionally conducted at laboratory scale, i.e., using physical models. A large number of empirical studies investigated scour processes in granular and rocky beds for different types of hydraulic structures, providing designers with useful tools to estimate the main characteristics of the scour hole. Among others, Mason & Arumugam (1985) and Hoffmans & Verheij (1997) have provided compilations of the numerous existing methods to predict jet-induced scour depth at equilibrium Δ_{eq}. Most formulas found in literature are special cases of the following general equation:

$$D + \Delta_{eq} = K q^{e_q} h^{e_h} g^{e_g} d^{e_d} \left(\frac{\rho}{\rho_s - \rho}\right)^{e_\rho} \tag{3}$$

where q indicates the total discharge for 3D jets and the discharge per unit width for 2D jets, K is a coefficient, and e_q, e_h, e_g, e_d, and e_ρ are exponents variously determined or set to zero by different researchers. In Table 1, the values of exponents proposed by different authors are reported. Listed works encompass a variety of different hydraulic structures, including flip buckets, grade control structures, and tunnel outlets. Experimental formulas (indicated with

Exp in Table 1) were obtained by mean of dimensional analysis and the values of the exponents were determined by fitting empirical results. Given the diversity of exponents in Table 1, very different results may be obtained. In fact, such formulas are not based on first principles, but are derived by fitting experimental data that are case-dependent. Moreover, experiments used to derive those formulas where not always conducted up to equilibrium. Overall, the validity of each formula is limited to tested geometries, hydraulic conditions, and model configurations.

Table 1. Values of exponents of Eq. (3) proposed by different authors pertaining to 2D and 3D cases. Abbreviations *Exp*, *S-Th*, and *Th* refer to experimental, semi-theoretical, and theoretical approaches, respectively (adapted from Bombardelli et al. 2018).

	Author(s)	Approach	e_q	e_h	e_g	e_d	e_p
2D	Kotoulas (1967)	*Exp*	0.7	0.35	-0.35	-0.4	0
	Chee & Padiyar (1969)	*Exp*	0.67	0.18	0	-0.063	0
	Ivanisevich Machado (1980)	*Exp*	0.5	0.3145	0	-0.0645	0
	Bormann & Julien (1991)	*S-Th*	0.6	0.5	-0.3	-0.4	0.8
	Hoffmans (2009) (d_{90} > 0.0125m)	*S-Th*	0.5	0.25	-0.25	0	0
	Bombardelli & Gioia (2005) Gioia & Bombardelli (2005)	*Th*	2/3	2/3	-1/3	-2/3	1
3D	Abt et al. (1984)	*Exp*	0.345	0.1425	-0.17	0	0
	Aderibigbe & Rajaratnam (1996)	*Exp*	0.5	0.25	-0.25	-0.5	0.5
	Bombardelli & Gioia (2006)	*Th*	0.4	0.4	-0.2	-0.4	0.6

3.3. Semi-theoretical and theoretical approaches

The lack of generality of empirical methods has been partially overcome by more recent, semi-theoretical approaches. Semi-theoretical methods (*S-Th* in Table 1) are generally based on the jet diffusion theory and the Shields criterion to evaluate the critical shear stress $\tau_c = \theta_c(\rho_s-\rho)gd$, where θ_c is the critical value of the Shields number (Shields 1936). As reported in Table 1, two semi-theoretical works are herein considered. Bormann & Julien (1991) examined grade control structures with a downstream sloping face. They expressed τ_m as a function of θ_c and of relative roughness (see Eq. (7) of Bormann & Julien 1991), and calculated τ_c using the Shields diagram. Then, they assumed some hypotheses based on experimental evidence to apply jet diffusion theory and derive the equilibrium scour depth. On the other hand, Hoffmans (2009) applied the linear momentum equation to a control volume and estimated the horizontal component of the drag force following the approach suggested by Forchheimer (see Eq. (9) of Hoffmans 2009). By balancing the forces acting on the control volume, he derived an equation to estimate $D+\Delta_{eq}$. Very importantly, both semi-theoretical formulas have the same algebraic structure [i.e., Eq. (3)], but different values of exponents. Such exponents are consistent with most of experimental formulas, but they also take into account the effect of α.

The development of fully theoretical methods for scour prediction, (i.e., based on first principles) represented a further step towards a physical understanding of the interaction between grains and water in jet-induced scour processes. The pioneering works of Bombardelli & Gioia (2005, 2006) and Gioia & Bombardelli (2002, 2005) paved the way to the analysis of scour problems from the perspective of the Phenomenological Theory of Turbulence (PTT), revealing the scaling relationship for the shear stress acting on a granular bed:

$$\tau \sim \frac{1}{(D+\Delta)^z}\rho(qgh)^{2/3}d^{1/3} \quad (4)$$

with $z = 1$ and $z = 5/3$ for 2D and 3D geometries, respectively. Unlike the semi-theoretical approaches discussed above, in this case the expression of τ does not require empirical assumptions. By imposing $\tau = \tau_c$, Bombardelli & Gioia (2005, 2006) and Gioia & Bombardelli (2005) proposed equations to estimate $D+\Delta_{eq}$, for 2D and 3D cases, respectively. The analytical structure of theoretical formulas is essentially the same of Eq. (3). The values of the exponents are consistent with those of the most-known empirical and semi-theoretical equations. In particular, $D+\Delta_{eq}$ decreases with d, decreases with $(\rho_s-\rho)/\rho$, and increases with both q and h. Bombardelli et al. (2018) furnished the values of the multiplicative constant K for 2D and 3D case, respectively. Furthermore, they distinguished two phases characterizing the scour evolution: a rapid "developing" phase, during which the scour hole and its dune are shaped, followed by a "developing phase", characterized by a homothetic expansion of the scour hole. The transition between the two phases mainly depends on the impinging angle. Based on the findings of Gioia & Bombardelli (2005) and Bombardelli & Gioia (2005, 2006), Bombardelli et al. (2018) also proposed two scour evolution models for 2D and 3D configurations. Their approach is consistent with that suggested by Stein et al. (1993) for non-cohesive material, i.e., the rate of scour is a function of the excess shear stress with ξ equal to 1.5 [Eq. (2)]. An example of application of this methodology can be found in Palermo et al. (2021).

4. Jet scour in cohesive soils

4.1. Characteristics of the scour

There is no universally accepted definition of cohesive sediment (Winterwerp & Van Kesteren 2004). Generally, in hydraulic engineering applications, cohesive sediment indicates a water mixture of sediment with diameter predominantly less than 63 μm (Mehta 2002). The mixture might also contain a variable percentage of sand and/or of organic material. Cohesion and plasticity are the main properties of such soils. Cohesion is the "capacity to withstand a finite shear stress without confinement"; whereas plasticity is the ability of a soil to bear plastic deformation and is quantified through the Atterberg limits (Partheniades 2009). Fluvial beds are typically characterized by a mixture of sand, clay, and silt. Experimental evidence showed that a soil exhibits cohesive properties when the presence of fines (i.e., silt and clays) in a mixture exceeds (approximately) 30%, although this percentage may vary with clay mineral composition and properties of the remaining components (Van Ledden et al. 2004, Van Rijn 2020). The clay fraction ($d < 2$ μm) is the main source of cohesiveness. In fact, clay particles are characterized by high specific surface area and small mass, so that surface forces dominate over inertial counterpart. In the presence of water, surface forces tend to keep particles together, resulting in the formation of flocs and higher order aggregates (Winterwerp & Van Kesteren 2004), that substantially contribute to the erosion resistance of the soil.

Cohesive soil exhibits different mechanisms of erosion, depending on the properties of the bed and the shear stress exerted by water. In literature, various classifications are available. For vertical jet scour in consolidated cohesive soils, Mazurek et al. (2001) described three types of erosion: mass erosion, flake erosion, and rapid surface erosion. Mass erosion is the dominant mechanism and was observed in many other works (e.g., Moore & Masch 1962, Mirtskhulava et al. 1967, Mehta 2002). It is a large-scale failure process consisting in the removal of lumps of bed material with angular shape and variable size. Such mechanism results in the largest increase of scour depth caused by jet and we are still far from a complete understanding of its physics (Winterwerp & Van Kesteren 2004). Flake erosion occurs when thin flakes of material are disrupted from the bed. Mazurek et al. (2001) observed circular flakes less than 0.5 mm thick and 1-3 mm diameter and attributed their occurrence to the preparation of the surface sample. Moreover, they argued that such mechanism occurs when the shear stress exceeds a critical value of about 15 Pa, much lower than the threshold value for mass erosion. This erosion mode likely corresponds to the "floc erosion" described by Winterwerp & Van Kesteren (2004) for flume erosion. Finally, rapid surface erosion was observed in correspondence with high applied stresses (> 300 Pa) and was depicted as a particle-by-particle erosion. However, this last erosion mode does not appear to be related to the "surface erosion" identified by Winterwerp & Van Kesteren (2004). In this regard, it is worth mentioning that the ratio of the time scale of bed response to the time scale of applied stress also affects the mode of erosion, because it allows to establish if the soil has a *drained* or *undrained* behavior. Namely, cohesive soils are drained when the deformation rate of the bed is small compared to the time scale of water flow through pores, and variations in water pressure result in pore water flow. Conversely, when the deformation rate of the bed is high, pressure gradients are not dissipated through pore water flow and soil exhibits an undrained behavior. Consequently, mass erosion can be classified as an undrained failure mechanism, i.e., pressure gradients cause local failures of the soil and the formation of cracks, thus leading to the removal of large chunks (Winterwerp & Van Kesteren 2004). The existence of *drained* and *undrained* behaviors, depending on the ratio of the time scales involved in the process, suggests that different modes of erosion might take place during the evolution of the scour.

Usually, a dune does not form in cohesive soil, and no material is kept in suspension in the hole. This is due to the fact that eroded soil is quickly transported downstream by the flow. Consequently, *dynamic* and *static* configurations are coincident. Moore & Masch (1962) and Hollick (1976) reported two types of scour hole shape, very similar to the regimes identified by Aderibigbe & Rajaratnam (1996) for non-cohesive soil. Moore & Masch (1962) suggested that the scour hole is deep and narrow for values of relative impingement height $D/d_{jet} < 7$, whereas, for $D/d_{jet} > 7$, it is wide and shallow. Conversely, Mazurek et al. (2001) argued that the relative impingement height is not the only parameter affecting the scour hole geometry. They suggested that the scour hole regime gradually varies from "weakly deflected" to "strongly deflected" with the value of an excess shear stress term $(X\text{-}X_c)/X_c$, with $X = \rho U_0^2(d_{jet}/D)^2$ and X_c indicating the value of X below which no mass erosion occurs. Dunn (1959) and Mazurek et al. (2001) observed that the maximum scour depth in cohesive soil does not always occur in correspondence with the jet centerline. Such behavior is consistent with the findings of Beltaos and Rajaratnam (1974), who showed that the maximum bed shear stress takes place at a certain distance from the jet axis. Most researchers agree that jet-scour in cohesive material has a logarithmic evolution, although the intermittent nature of mass erosion can produce a discontinuous behavior (Mazurek et al. 2001). In fact, when a large chunk of cohesive soil is eroded from the bed due to mass erosion, the scour hole size suddenly increases, and a deviation from the logarithmic trend might occur.

4.2. Experimental and semi-theoretical studies

Almost a century of research on cohesive sediment have highlighted that a wide number of factors influence the erodibility of cohesive soils, including clay content and mineralogy, structure of the soil, physical and chemical properties of pore and water (Winterwerp & Van Kesteren 2004, Van Rijn, 2020). In particular, most experimental evidence indicate that the erodibility of cohesive soil increases with soil and water temperature, Sodium Absorption Rate (SAR) of pore water, and soil void ratio, whereas it decreases with clay content, soil unit weight, density, shear strength, and plasticity index (Arneson et al. 2012). Moreover, the effects of other properties such as soil water content (W), saturation, mean grain size, and chemical characteristics of eroding water are still unclear and there is no agreement on the most significant factors to be considered when dealing with cohesive soil erosion (Grabowski et al. 2011, Arneson et al. 2012). The presence of stratifications, fissures, and other inhomogeneities further influences the scour mechanisms of cohesive soils, resulting in preferential erosion patterns.

Numerous studies have attempted to develop equations relating cohesive soil erodibility to its physical properties such as bulk density, void ratio, clay content, water content, plasticity index, and vane shear strength (see, among others, Dunn 1959, Kuti & Yen 1976, Hedges 1990, Zhu et al. 2008). Nevertheless, such "general" methods are still debated. Consequently, experimental approaches have been traditionally adopted to investigate scour of cohesive soils. Many researchers followed the approach proposed by Moore & Masch (1962), who tested cohesive sediment samples under the action of a fully turbulent submerged vertical circular jet at large impingement heights. At given time intervals, they weighed the soil sample of known density to determine its volume loss. Then, they calculated the mean scour depth Δ_{mean} as the cube root of the volume loss. Via dimensional analysis, they argued that Δ_{mean}/D can be expressed as a function of four terms:

$$\frac{\Delta_{mean}}{D} = f\left(\frac{t\mu}{\rho d^2}, \frac{\rho U_0 d_{jet}}{\mu}, \frac{D}{d_{jet}}, \frac{\sigma_s \rho d_{jet}^2}{\mu^2}\right) \tag{5}$$

with t indicating the time duration of scouring process, μ the dynamic viscosity of the fluid, and σ_s a sediment resistance not defined by the authors (and variously interpreted by other researchers, e.g., Hollick 1976, Hedges 1990, Hanson 1990). Conversely, for large impingement heights and fully turbulent jets, Mazurek et al. (2001) suggested that at equilibrium the maximum scour hole depth, the centerline scour depth, and the mean scour depth can be expressed as function of the excess shear stress term $(X\text{-}X_c)/X_c$, where X_c was experimentally determined. Following Beltaos & Rajaratnam (1974), Mazurek et al. (2001) estimated the maximum shear stress acting on the bed as $\tau = 0.16X$ and consequently $\tau_c = 0.16X_c$. They proposed the following Eq. (6) to predict the maximum scour depth at equilibrium:

$$\frac{\Delta_{eq}}{D} = 0.19\left(\frac{X-X_c}{X_c}\right)^{0.74} \tag{6}$$

Mazurek et al. (2006) complemented such results providing equations for the evolution of scour depth as a function of the same excess shear stress term. Ansari et al. (2003) analyzed vertical jet scour in non-cohesive and cohesive soils obtained by mixing sand and clays. They proposed an equation to estimate the equilibrium scour depth in cohesive soil as a function of soil properties (including initial water content W, clay content, mean sediment size, angle of internal friction of sand and clay, cohesion, and soil density) and of the *static* scour depth obtained in non-cohesive soil under the same hydraulic conditions (see Eqs (8) and (10) of that work). An important drawback of this method is that its predictive capability is also affected by the method used to determine the scour hole size in non-cohesive sediment. Moreover, for clay contents higher than about 20%, data are very spread, evidencing no (clear) correlation between scour depth and soil properties. Based on the results of Moore & Masch (1962), Mazurek et al. (2001), and Ansari et al. (2003), Dong et al. (2020) developed the following empirical equation:

$$\frac{\Delta}{D} = 0.27\left[ln\left(\frac{t\mu}{\rho d_{jet}^2}\right)\right]^{0.507}\left(\frac{W}{W_L}\right)^{1.993}\left(\frac{\tau-\tau_c}{\tau_c}\right)^{0.704} \quad (7)$$

where W_L indicates the liquid limit of the soil. Note that τ and τ_c were calculated according to Mazurek et al. (2001). Although the equation was calibrated on tests with short durations, validation using data of Mazurek (2001) for longer tests revealed a good predictive capability.

A simplified approach is based on the application of Eq. (2), calibrated using experimental results. To this end, Eq. (2) is re-written as follows, for cohesive sediment:

$$\frac{d\Delta}{dt} = k_d(\tau_m - \tau_c)^{\xi} \quad (8)$$

where k_d is an experimentally determined coefficient referred to as erodibility coefficient or erosion rate parameter, and ξ is usually assumed to be equal to 1. In particular, Stein et al. (1993) found that a good agreement with experimental data is achieved with $\xi = 1$. Nevertheless, some researchers have recently argued that a power law could improve predictions (Knapen et al. 2007, Walder 2015). Following Moore & Masch (1962), Hanson (1990) developed an experimental equation for Δ_{mean} valid for constant D/d_{jet}. In so doing, the last term of Eq. (5) was replaced by $\mu k_d/d_{jet}$, where k_d was obtained by fitting Eq. (8) to experimental results for open channels. Successively, Hanson & Cook (2004) proposed a portable device named Jet Erosion Test (JET) to test soil erodibility and provided a semi-theoretical method to estimate k_d and τ_c in Eq. (8). Since then, the JET has become one of the most widely used devices to evaluate erosion of cohesive soils, both in the field and in laboratory, because of its high versatility. The estimation of k_d and τ_c is obtained by applying jet diffusion principles and assuming $\xi = 1$ in Eq. (8), with scour rate equal to zero at equilibrium (i.e., when $\tau = \tau_c$). Because the testing time does not allow to reach equilibrium, Δ_{eq} is extrapolated from data, by assuming a hyperbolic logarithmic equation (Blaisdell et al. 1981). Hanson & Cook (2004) assumed that the bed is flat and smooth during the entire scour evolution and the environment is not confined. These assumptions appear to be not consistent with experimental evidence (Ghaneeizad et al. 2015, Karamigolbaghi et al. 2017). Moreover, several researchers highlighted that the erodibility coefficient and the critical shear stress obtained with JET test depend on the device and methodology adopted for the test and on the extrapolation technique used to estimate equilibrium. In particular, different test durations and extrapolation techniques lead to radically different results in terms of τ_c and Δ_{eq} (Cossette et al. 2012). In this regard, it is worth mentioning that improved JET techniques and alternative methods to interpret JET tests have recently been proposed to overcome some of the mentioned shortcomings. Among these, Marot et al. (2011) proposed an energy-based model to classify soil erodibility, whereas Rose et al. (2017) based their analysis on the estimation of the amount of energy needed to erode a unit mass of soil. Nevertheless, the traditional method of Hanson & Cook (2004) is still the most largely used in practice.

5. Differences and similarities

It is apparent that the main difference between granular and cohesive soils can be attributed to the behavior of particles in contact with water. The interaction between coarse sediments is limited to a mechanical contact; therefore the eroding unit consists of a single grain, and erosion thresholds can be established based on a representative grain size (Partheniades 2009). Conversely, cohesive sediments are characterized by surface forces responsible for flocculation. In this case, the erosion unit consists of flocs or higher order aggregates and the densimetric Froude number is no longer representative of the soil behavior because its resistance to erosion does not depend solely on friction and particle weight (Mazurek & Hossain 2007). Consequently, for cohesive sediment, τ_c also depends on the internal cohesion among particles. In this regard, some efforts to overcome such limitations are found in literature. Recently, Chen et al. (2018) proposed a unified theoretical formula for the critical shear stress of sand, mud, and sand-mud mixtures by introducing a modified critical Shields parameter that reflects the cohesive strength of the mud and exposed sediment bed features. On the other hand, Brunier-Coulin et al. (2020) proposed a generalized Shields criterion accounting for cohesion effect and valid for weakly cohesive granular materials by analyzing jet scour in artificial materials with controlled cohesion. It is also worth noticing that different modes of erosion can take place in cohesive soils, making more complicated the assessment of an erosion threshold. Different thresholds for different modes of erosion should be used, although for jet-induced scour the main mechanism observed is mass erosion.

In non-cohesive soils, scour evolves quite rapidly. Experimental and theoretical evidence suggests that for localized scour phenomena the rate of erosion increases with *d*, being the ratio Δ/Δ_{eq} constant (Dey 2014, Bombardelli et al. 2018). Often, for practical applications, the time evolution of scour is neglected, and equilibrium is assumed to occur immediately after the critical shear stress is exceeded, especially for coarse sediment (Arneson et al. 2012). Conversely, the rate of erosion above critical shear stress of cohesive soils is much lower (of the order of a few millimeters per hour), as suggested by the different values of ξ proposed by Stein et al. (1993) in Eq. (2) for cohesive and non-cohesive soil (ξ = 1 and 1.5, respectively). Therefore, the time-dependence of scour phenomena and their long-term behavior becomes relevant. Such occurrence is even more important if we consider that in cohesive soil, *dynamic* and *static* equilibrium configurations coincide. Consequently, scour process has no memory, and the concept of superposition applies. Assessing the scour behavior under multiple floods, rather than only for a single flood event, becomes crucial for designers (Arneson et al. 2012). Furthermore, in both soils, scour evolution follows a logarithmic law. Nevertheless, the existence of different modes of erosion in cohesive soils is responsible for sudden increases of the scour hole size observed in many experiments (Mazurek et al. 2001). The discontinuous nature of mass erosion should be further analyzed, to obtain more reliable predictions of scour evolution.

It is worth mentioning that experimental investigation of scour in cohesive soils is particularly challenging because of the limited control on soil properties during the preparation of samples. For instance, Mazurek et al. (2001) observed that tests conducted with the same hydraulic conditions exhibited different time evolutions, although their equilibrium scour depth and volume was comparable. The authors attributed this difference to slight variations of water content and saturation in samples. Recent investigations in material sciences, which led to the creation of artificial granular materials with controlled cohesion, could help in overcoming such problems and contribute to assess the influence of cohesion on erosion problems (Gans et al. 2020, Gong et al. 2021). Finally, unlike granular soil, the effect of impinging angle on the scour process of cohesive soils is still far from being understood. To the best of the authors' knowledge, only Hedges (1990), Stein et al. (1993), and Hou et al. (2016) analyzed inclined jets impinging on cohesive material, but their findings are still partial and not exhaustive.

6. Conclusions

This paper presented an updated summary of current knowledge on scour caused by circular jets for cohesive and non-cohesive soils. At first, we have introduced the main characteristics of scour induced by circular jets. Then, we have summarized the findings of studies dealing with jet-induced scour in non-cohesive and cohesive soils. In doing so, we have pointed out differences and similarities of the erosion mechanisms and highlighted still-challenging aspects worth of further investigations. Many questions remain open, particularly in the case of cohesive soils, for which a theoretical modelling is still lacking. Furthermore, it appears necessary to provide a better understanding of scour evolution and insights on the physics of scour process in cohesive soils.

7. ACKNOWLEDGMENTS

This work did not receive any specific grant from funding agencies in the public, commercial, or not-for-profit sectors.

8. REFERENCES

Abt, S.R., Kloberdanz, R.L., and Mendoza, C. (1984). "Unified culvert scour determination." *J. Hydraul. Eng.*, 110(10): 1475-1479.

Aderibigbe, O.O. and Rajaratnam, N. (1996). "Erosion of loose beds by submerged circular impinging vertical turbulent jets." *J. Hydraul. Res.*, 34(1): 19–34.

Ansari. S.A., Kothyari, U.C., and Ranga Raju, K.G. (2003). "Influence of Cohesion on Scour Under Submerged Circular Vertical Jets." *J. Hydraul. Eng.*, 129(12): 1014-1019.

Arneson, L.A., Zevenbergen, L.W., Lagasse, P.F., and Clopper, P.E. (2012). Evaluating Scour at Bridges, Fifth Edition. Hydraulic Engineering Circular No. 18 (HEC-18), Federal Highway Administration, Washington D.C.

Beltaos, S. and Rajaratnam, N. (1973). "Plane turbulent impinging jets." *J. Hydraul. Res.*, 100(10), 1313-1328.

Beltaos, S. and Rajaratnam, N. (1974). "Impinging circular turbulent jets." *J. Hydraul. Div. ASCE*, 11(1), 29-59.

Beltaos, S. and Rajaratnam, N. (1977). "Impingement of axisymmetric developing jets." *J. Hydraul. Res.*, 34(1), 19-33.

Blaisdell, F.W., Clayton, L.A., and Hebaus, C.G. (1981). "Ultimate dimension of local scour." *J. Hydraul. Div. ASCE*, 107(3): 327-337.

Bombardelli, F. A., and Gioia, G. (2005). "Towards a theoretical model for scour phenomena." *In Proc. RCEM 2005, 4th IAHR Symposium on River, Coastal, and Estuarine Morphodynamics*, Vol. 2, 931-936, Urbana, IL, USA.

Bombardelli, F.A. and Gioia, G. (2006). "Scouring of granular beds by jet-driven axisymmetric turbulent cauldrons." *Phys. Fluids*, 18, 088101.

Bombardelli, F.A., Palermo, M., and Pagliara, S. (2018). "Temporal evolution of jet induced scour depth in cohesionless granular beds and the phenomenological theory of turbulence." *Phys. Fluids*, 30(8), 085109.

Bormann, N.E. and Julien, P.Y. (1991). "Scour downstream of grade-control structures.". *J. Hydraul. Eng.*, 117(5), 579–594.

Brunier-Coulin, F., Cuéllar, P., and Philippe, P. (2020). "Generalized Shields criterion for weakly cohesive granular materials." *Phys. Rev. Fluids*, 5, 034308.

Chee, S.P. and Padiyar, P.V., (1969). "Erosion at the base of flip buckets." *Eng. J. Inst. Can.* 52(11), 22–24 .

Chen, D., Wang, Y., Melville, B., Huang, H., and Zhang, W. (2018). "Unified formula for critical shear stress for erosion of sand, mud, and sand-mud mixtures." *J. Hydraul. Eng.*, 144:04018046.

Cossette, D., Mazurek, K.A., and Rennie, C.D. (2012). "Critical shear stress from varied methods of analysis of a submerged circular turbulent impinging jet test for determining erosion resistance of cohesive soils." *In Proc. 6th Intl. Conf. on Scour and Erosion (ICSE6)*, Paris, France, SHF.

Dey, S. (2014). Fluvial Hydrodynamics. Springer, Berlin, Heidelberg.

Dong, C., Yu, G., Zhang, H., Zhang, M. (2020). "Scouring by submerged steady water jet vertically impinging on a cohesive bed." *Ocean Eng.*, 196, 106781.

Dunn, I.S. (1959). "Tractive Resistance of Cohesive Channels." *Jour. Soil Mech. and Found. Div.*, Proc. ASCE, Vol. 85, No. SM3, Proc. Paper 2062, pp. 1-24.

Gans, A., Pouliquen, O., and Nicolas, M. (2020). "Cohesion-controlled granular material.", *Phys. Rev. E*, 101, 032904.

Ghaneeizad, S.M., Atkinson, J.F., and Bennett, S.J. (2015). "Effect of flow confinement on the hydrodynamics of circular impinging jets: implications for erosion assessment". *Environ. Fluid. Mech.*, 15, 1–25.

Gioia, G. and Bombardelli, F.A. (2002). "Scaling and similarity in rough channel flows." *Phys. Rev. Lett.* 88, 014501.

Gioia, G. and Bombardelli, F.A. (2005). "Localized turbulent flows on scouring granular beds." *Phys. Rev. Lett.*, 95, 014501.

Gong, M., Azadi, S., Gans, A., Gondret, P. and Sauret, A. (2021). "Erosion of a cohesive granular material by an impinging turbulent jet." *EPJ Web Conf.*, 249, 08011.

Grabowski, R.C., Droppo, I.G., and Wharton G. (2011). "Erodibility of cohesive sediment: the importance of sediment properties." *Earth Sci. Rev.*, 105 (3–4).

Hanson, G.J. (1990). "Surface Erodibility of Earthern Channels at High Stresses: Part II Developing an In Situ Testing Device." *Trans. of ASAE*, 33(1): 132-137.

Hanson, G.J. and Cook, K.R. (2004). "Apparatus, test procedures, and analytical methods to measure soil erodibility in situ". *Appl. Eng. Agric.*, 20, 455.

Hedges, J.D. (1990). The Scour of Cohesive Soils by an Inclined Submerged Water Jet. Master's thesis, Texas A&M University.

Hoffmans, G.J.C.M. (2009). "Closure problem to jet scour." *J. Hydraul. Res.*, 47(1), 100-109.

Hoffmans, G.J.C.M. and Verheij, H.J. (1997). Scour manual. Balkema, The Netherlands.

Hollick, M. (1976). "Towards a routine assessment of the critical tractive forces of cohesive soils." *Trans. ASAE*, St. Joseph, Mich., 19, 1076–1081.

Hou, J., Zhang, L., Gong, Y., Ning, D., Zhang, Z. (2016). "Theoretical and experimental study of scour depth by submerged water jet." *Adv. Mech. Eng.*, 8, 1687814016682392.

Ivanisevich Machado, L. (1980). "Formulas to calculate the scour limit on granular or rock beds," *in Proc. XIII National Workshop on Large Dams*, Rio de Janeiro, pp. 35–52.

Karamigolbaghi, M., Ghaneeizad, S.M., Atkinson, J.F., Bennett, S.J., Wells, R.R. (2017). "Critical assessment of jet erosion test methodologies for cohesive soil and sediment". *Geomorphology*, 295, pp. 529-536.

Knapen, A., Poesen, J., Govers, G., Gyssels, G., and Nachtergaele, J. (2007). "Resistance of Soils to Concentrated Flow Erosion: A Review." *Earth-Sci. Rev.*, 80:75-109.

Kotoulas, D., (1967). "Das Kolkproblem unter besonderer berucksichtigung der faktoren zeit und geschiebemischung in rahmen der wildbachverbauung." Dissertation, Prom. 3983. Technischen Hochschule Zurich.

Kuti, E.O. and Yen, C.L. (1976). "Scouring of Cohesive Soils". *J. Hydraul. Res.*, 14(3), 195-206.

Marot, D., Regazzoni, P.-L., Wahl, T. (2011). "Energy-based method for providing soil surface erodibility rankings." *J. Geotech. Geoenviron. Eng.*, 137(12): 1290–1293.

Mason, P.J. and Arumugam, K. (1985). "Free jet scour below dams and flip buckets." *J. Hydraul. Eng.*, 111(2), 220-235.

Mazurek, K.A., Rajaratnam, N., and Sego, D.C. (2001). "Scour of a Cohesive Soil by Submerged Circular Turbulent Impinging Jets." *J. Hydraul. Eng.*, 127(7): 598-606.

Mazurek, K.A., Rajaratnam, N., and Sego, D.C. (2006). "Time Development of Scour in a Cohesive Material due to a Submerged Circular Turbulent Impinging Jet." *In Proc. 3rd Int. Conf. on Scour and Erosion (ICSE-3)*. November 1-3, 2006, Amsterdam, The Netherlands.

Mazurek, K.A. and Hossain, T. (2007). "Scour by Jets in Cohesionless and Cohesive Soil." *Can. J. Civ. Eng.*, 34(6): 744-751.

Mehta, A.J. (2002). "Mudshore dynamics and controls", in Muddy Coasts of the World, ed. Healy, T., Wang, Y., and Healy, J.-A. Elsevier, Proceedings in Marine Science, 4, 19-60.

Mirtskhulava, TS.E., Dolidze, I.V., and Magomedova, A.V. (1967). "Mechanism and computation of local and general scour in non-cohesive, cohesive soils and rock beds." *In Proc. 12th Congr. of IAHR*, Fort Collins, Colorado.

Moore, W.L. and Masch, F.D.J. (1962). "Experiments on the scour resistance of cohesive sediments." *J. Geophys. Res.*, 67(4), 1437-1449.

Pagliara, S., Amidei, M., and Hager, W.H. (2008). "Hydraulics of 3D plunge pool scour." *J. Hydraul. Eng.*, 134(9), 1275-1284.

Palermo, M., Pagliara, S., and Bombardelli, F.A. (2020). "Theoretical approach for shear-stress estimation at 2D equilibrium scour holes in granular material due to subvertical plunging jets." *J. Hydraul. Eng.*, 146(4): 04020009.

Palermo, M., Bombardelli, F.A., Pagliara, S., and Kuroiwa, J. (2021). "Time-dependent scour processes on granular beds at large scale." *Environ. Fluid Mech.*, 21(4), 791–816.

Partheniades, E. (2009). Cohesive Sediments in Open Channels. Butterworth-Heinemann, Oxford.

Rajaratnam, N. (1976). Turbulent Jets, Elsevier, Amsterdam.

Rose, C.W., Olley, J.M., Haddadchi, A., Brooks, A.P., and McMahon, J. (2018). "An alternative method for interpreting jet erosion test (JET) data: part 1. Theory." *Earth Surf. Process. Landf.*, 43:735– 742.

Shan, H., Shen, J., Kilgore, R., and Kerenyi, K. (2015). Scour in Cohesive Soils. Publication No. FHWA-HRT-15-033, Federal Highway Administration, U.S. Department of Transportation. McLean, VA, USA.

Shields, A. (1936). "Application of similarity principles and turbulence research to bed-load movement". transl. by Ott, W.P. and Van Uchelen, J.C., *Hydrodyn. Lab. Publ.* No. 167. Washington, DC: USDA.

Stein, O.R., Julien, P.Y., and Alonso, C.V. (1993). "Mechanics of Jet Scour downstream of a Headcut." *J. Hydraul. Res.*, 31(6), 723–738.

Van Rijn, L.C. (2020). "Erodibility of Mud-Sand Bed Mixtures". *J. Hydraul. Eng.*, 146(1), 04019050.

Van Ledden, M., Van Kesteren, W.G.M., and Winterwerp, J.C. (2004). "A conceptual framework for the erosion behaviour of sand-mud mixtures." *Cont. Shelf Res.*, 24(1): 1–11.

Walder, J.S. (2015). "Dimensionless Erosion Laws for Cohesive Sediment." *J. Hydraul. Eng.*, 142(2), 04015047.

Winterwerp, J.C. and van Kesteren, W.G.M. (2004). Introduction to the Physics of Cohesive Sediments in the Marine Environment., Dev. Sedimentol., vol. 56, Elsevier, New York.

Zhu, Y, Lu, J., Liao, H., Wang, J., Fan, B., and Yao, S. (2008) "Research on cohesive sediment erosion by flow: An overview." *Sci. China Technol. Sci.*, 51, 2001-2012.

DOI: https://doi.org/10.26077/bd87-97c0

Laws of turbulence and the estimation of turbulent kinetic energy budget for flow through a degraded channel-bed

Mosedul Sarkar[1] & Sankar Sarkar[2]
[1]Indian Statistical Institute, Kolkata, India 700108
[2] Indian Statistical Institute, Kolkata, India 700108
E-mail: mosedulsarkar2014@gmail.com

Abstract: *This present study focuses on the laws of turbulence and estimation of the turbulent kinetic energy (TKE) budget for flow over a bimodal degraded bed under equilibrium conditions. The analyses of turbulence were performed based on the three-dimensional velocity data captured by using a Vectrino velocimeter. The streamwise velocity and Reynolds shear stress profiles show an excellent agreement with the existing literature. In view of the advancement in the direction of bed degradation, the measured velocity data captured at different streamwise locations (before, on and after the maximum equilibrium depth of degradation) were further processed to explore the scaling laws of turbulence and TKE budget based on structure functions data. To this end, the second-, third-, and mixed third-order velocity structure functions were estimated. The structure functions data indicated the existence of inertial subrange within lower ranges of the time-lags and confirmed their agreement with the Kolmororov's 2/3-, and 4/5-, and Monin–Yaglom's 4/3-laws. However, the TKE dissipation rates estimated by Kolmororov's 4/5- and Monin–Yaglom's 4/3-laws of turbulence show significant differences in their values. The TKE budget results show that in the vicinity of the degraded bed, TKE dissipation rate is more than that of TKE production and turbulent diffusion rates. On the other hand, the pressure diffusion has larger negative values in the near-bed. However, all the turbulence parameters show no significant differences at as one moves from the initial bed-level (bed-level measured before the detachment started) to the free-surface.*

Keywords: *Degraded bed, Open channel flow, structure functions, Scaling laws, turbulent flow, TKE budget*

1. Introduction

Turbulent flow in open channels has been a topic of considerable attention to different scientific communities in physics, applied mathematics, environmental hydraulics, and so on. The sedimentary beds of open channels are in general comprised of uniform and nonuniform sizes of clay, sand, gravel, boulder, etc. in different proportions which make the beds bimodal or more specifically polymodal in nature. Based upon the flow and bed roughness conditions, the sediment particles get eroded from the bed and deposited on to the bed naturally that eventually form the degraded and aggraded beds, respectively which play important roles in the sedimentation process of a water-body. It is therefore quite interesting to study different parameters of turbulent flow through a degraded channel bed. However, a detailed review of literature suggests that the flow through degraded and mobile beds has not been explored so intensively in comparison to flow over rigid bed that possesses different flow characteristics [Nikora and Goring 2000]. In mobile bed hydrodynamics, Nikora and Goring (2000) observed a reduction in the values of von Kãrmãn constant in flow over a weakly mobile bed which is quite different from rigid bed. Later, Dey et al. (2012) observed a reduction in the near-bed velocity fluctuations and an excessive near-bed damping in Reynolds shear stress (RSS) distributions for flow over mobile bed and found significant differences in the flow characteristics when compared to rigid bed. They also found that the von Kármán coefficient decreases for flow over mobile bed. The turbulent kinetic energy budget results showed that in the vicinity of the mobile-bed, the pressure energy diffusion rate decreases sharply to a large negative value, signifying a gain in the turbulence production. In case of a degraded bed, the hydrodynamics is quite different depends on the bed roughness and the proportions of bed particles that are present in the sedimentary bed. For instance, in an experimental work, Shvidchenko and Pender (2001) observed the flow characteristics over mobile gravel beds made up of with various sizes of gravels and observed the most important roles for the movement of fluid particles are played by the lift and drag forces that are produced due to eddy motion of fluid. Experimental study of flow over a bed of immobile coarse sediments feeded with fine sand by Grams and Wilcock (2007) showed that the amount of sediment transportation is strongly affected by the proportion of a sand and gravel. In another study, Wren et al. (2011) observed that due to the addition of sand in a gravel-bed, the flow properties change quite remarkably. It was evident from the experimental findings that because of the addition of different sediment over a sedimentary bed, the bed roughness changes abruptly, which actively changes the individual velocity profiles significantly. Also, due to an increase in the elevation of sand layer over the gravel bed, the bed shear stress and the Reynolds stress decrease. However, none of those studies are on the degraded bed hydrodynamics. Interestingly, Jain et al. (2015) carried out an experimental work to understand the flow characteristics over a degraded bed comprised of sand-gravel mixture. The main findings of the study were the peak values of Reynolds stresses were at the initial undisturbed bed-level (bed-level that was measured before the commencement of bed erosion). However, the study was chiefly confined upto the streamwise velocity, Reynolds stresses, turbulent kinetic energy (TKE), and turbulent bursting etc., where no precise information of the velocity structure functions, scaling laws of turbulence, TKE budget, was found. Quite recently, Sarkar et al. (2021) has done similar kind of experimental study over degraded bed where scaling laws of turbulence were noticed. It is important to accept that the present paper is a replica of Sarkar et al. (2021). The scaling laws of turbulence were explored using the structure functions methodology that was pioneered by Kolmogorov (1941a,b) and later modified and extended by several researchers and used in hydraulics by Ferraro et al. (2016), Coscarella et al. (2017), Padhi et al. (2019), Penna et al. (2020) and so on.

2. Experimentation, data capturing and post-processing

There were six experiments carried out in the Fluvial Mechanics Laboratory of Indian Statistical Institute, Kolkata, India out of which data from two experimental runs namely, Set 1 and Set 2 were used in the present study. The experimental flume is 20 m long, 0.5 m wide and 0.5 m deep. The test section was located at a distance of 5 m downstream of the entrance of the flume where the visual observation of the experiments was enabled through the transparent side-walls. The test zone was 6 m × 0.5 m × 0.5 m in dimension. Two solid aprons each of 1 m length × 0.5 m width were placed at the up and downstream of the test zone. Two uniformly graded sediment samples of d_{50} = 0.25 mm and 3.5 mm were mixed thoroughly with equal proportions by weight and the mixture was placed on the flume covering the test zone to prepare a bimodal sedimentary bed. Each of the sedimentary beds was approximately 0.2 m thick. Once the bed preparation was finalized, a pump was in use to supply water into the flume for the experiments under required flow depth.

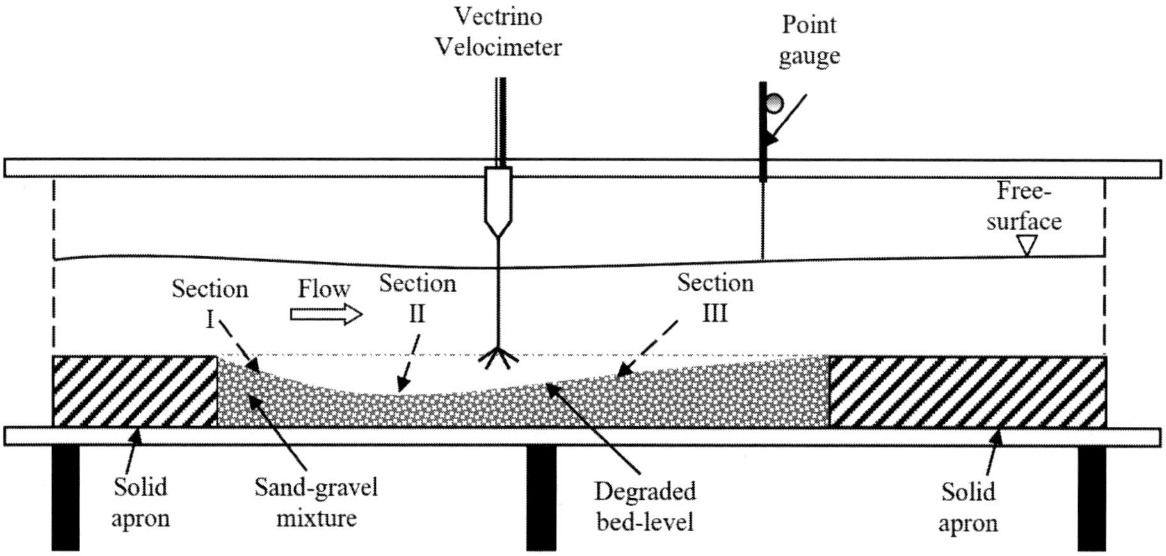

Figure 1. Experimental setup

Table 1: Experimental parameters

Set	d_{50} (mm)	D_{50} (mm)	d (m)	h(m)	S	U (m/s)	u_* (m/s)
1	0.25	3.5	0.098	0.105	0.0045	0.65	0.068
2	0.25	3.5	0.076	0.12	0.0045	0.61	0.073

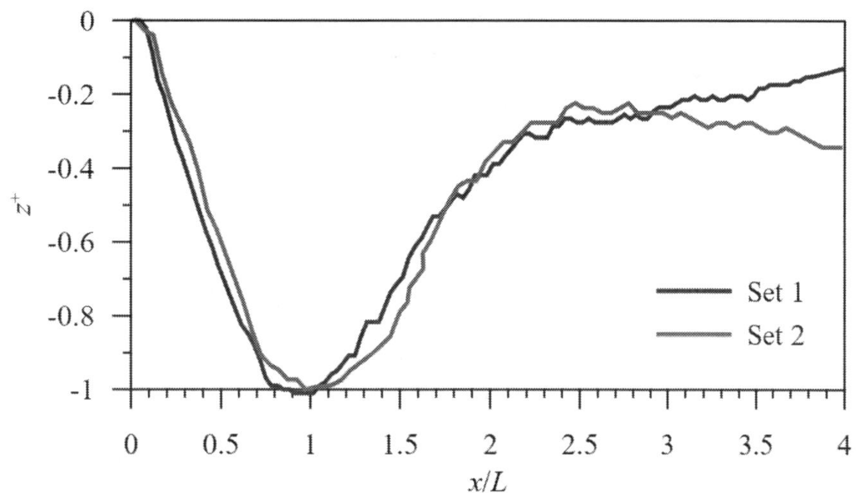

Figure 2. Profiles of the degraded bed under equilibrium condition.

The bimodal sedimentary bed started degrading and the height and shape of the sedimentary bed started changing until it attained the hydraulically equilibrium condition when no changes of the bed shape and elevations were further observed. Under such an equilibrium condition, the 3-D velocity components were measure using a Vectrino velocimter, a kind of acoustic Doppler velocimeter. Both the experimental runs were performed under $b/h < 2$ (b = width of the channel and h = flow depth measured before detachment of the sediments). Three test sections were chosen for each run in such a way that the first section, namely Section I was at the upstream of the maximum equilibrium depth of degradation, whereas Section II was chosen at the location where maximum depth of degradation took place under equilibrium condition and Section III was at the sufficient downstream of Section II where the effect of degradation was quite less and the flow profiles were started regaining the undisturbed features as in Section I or any rough plane bed.

The three dimensional velocity components in streamwise, lateral, and vertical directions (x, y, and z-directions, respectively) are written in this paper as u, v, and w, respectively. For all the locations velocity was measured for a duration of t = 180 s to achieve time-invariant velocity components. The measurement was taken carefully by maintaining of SNR (signal and noise ratio) value above 18 and correlation coefficient of more than 70. The velocity was measured along the centerline of the flume under shallow depth condition. Hence, it can be assumed that the data were not affected by the secondary currents. The temporal fluctuations of as u, v, and w were mentioned as u', v', and w' in x, y, and z-directions were calculated through subtracting u, v, and w from the corresponding time-averaged values $\overline{u}$, $\overline{v}$, and $\overline{w}$, respectively. Until unless required the raw data were used for the analysis. However, very close to the bed, data were infected by spikes which were detected and replaced using phase-space thresholding methodology following Goring and Nikora (2002). Table 1 shows the experimental parameters for experimental Set 1 and Set 2. Figures 1 and 2 show the experimental set up and the equilibrium degraded bed profiles, respectively. In Figure 2, $z^+ = z/d$ represents the nondimensional vertical distance, where d = maximum equilibrium depth of degradation, x/L represents the nondimensional streamwise distance where L = distance between the starting and commencement of maximum equilibrium depth of degradation i.e., d.

3. Result and discussions

3.1. Time-averaged streamwise velocity

Figure 3 presents the vertical profiles of nondimensional time-averaged streamwise velocity u^+ (= $\overline{u}/u_*$) at sections I, II, and III for experimental Set 1 and Set 2. As in the present experiments, beds were deformed during experiments, the shear velocity used here was calculated from the undisturbed initial bed-slope using the formula: $u_* = (ghS)^{0.5}$, where S is the bed slope measured initially and the vertical distance z was scaled by the maximum equilibrium depth of degradation d (i.e., $z^+ = z/d$) for a run.

It was observed from Figure 3 that in the vicinity of the bed, the u^+ starts with a small value which increases monotonically with increase in the vertical distance at all three sections for both the runs until it achieves an almost peak value at $z^+ \approx 0.6$. Beyond $z^+ \approx 0.6$, the u^+ becomes almost asymptotic. Interestingly, the changes of u^+ with respect to vertical distance happens more rapidly below $z^+ = 0$ indicating a higher velocity gradient within the equilibrium depth of degradation. Also, the values of u^+ are quite different when they were measured at different sections. On the other hand, above $z^+ = 0$, not so much differences in the profiles and values of u^+ were observed. However, above $z^+ = 0$, the variations in the u^+ were more pronounced for the data Set 1 than that of Set 2 which is due to the effect of the b/h ratio. The higher the values of b/h ratio, the more are the differences in the values in the present cases. Moreover, in the vicinity of the bed, negative streamwise velocity was observed signifying the flow reversal which may be a strong contribution to form vortex structure, although that was confined within very near zone of the bed. The findings are in conformation with the Jain et al. (2015). However, there values may be different due to the differences in the experimental parameters. The conclusion that can be drawn from the streamwise velocity is the bed degradation has a considerable influence over the streamwise velocity in an open channel flow.

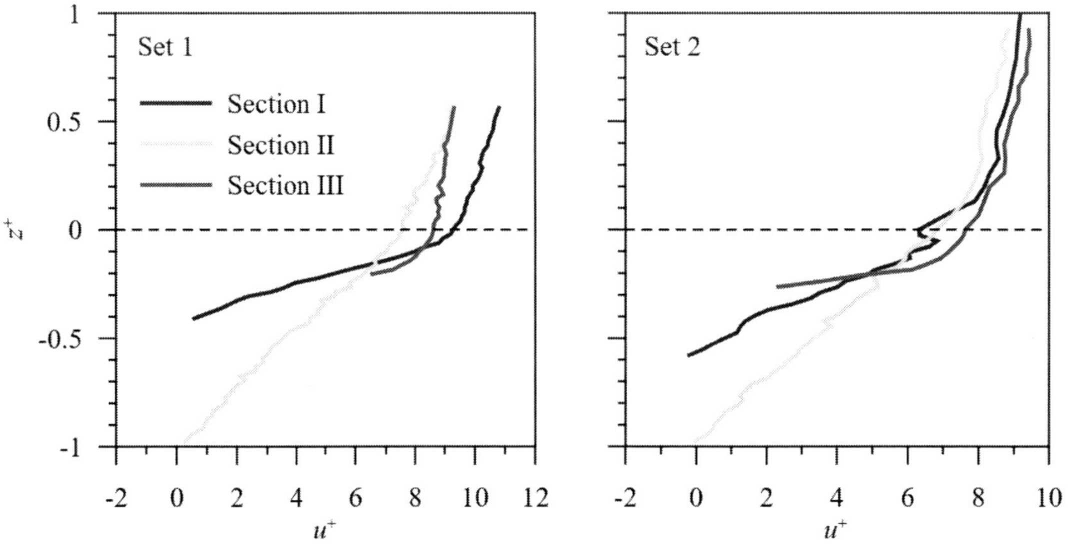

Figure 3. Vertical profiles of normalized time-averaged streamwise velocity u^+ at sections I, II, and III for Set 1 and Set 2 data.

3.2. Time-averaged Reynolds shear and normal stresses

The vertical profiles of nondimensional Reynolds shear stress (RSS) $\tau^+ (=\tau/u_*^2)$ at Section I, II, and III for Set 1 and Set 2 were illustrated in Figure 4. The Reynolds shear stress $\tau (= -\rho\overline{u'w'}$, where $-\rho\overline{u'w'}$ is the RSS and ρ is the mass density of fluid) is nondimensionalised by dividing it by u_*^2, where u_* is the shear velocity as explained earlier. In general, the normalized RSS τ^+-distribution for an open channel flow follows the gravity line (i.e., $\tau^+ = 1$ at $z^+ = 0$ and $\tau^+ = 0$ at $z^+ = 1$) with some scatter in values occurred due to the laboratory measurements. However, due to the changes in the bed and experimental conditions, the τ^+-distribution changes significantly. However, it was observed from Figure 4 that in the vicinity of the degraded bed, the τ^+ starts with a value that goes on increasing until it attains the peak value at $z^+ \approx 0$. Above $z^+ = 0$, the τ^+ starts decreasing and as one moves towards the free-surface, the τ^+ deceases more and follows the gravity line of τ^+-distributions. It is interesting to notice that for both the experimental sets the τ^+-profiles are similar. However, at $z^+ \approx 0$, the peak value of τ^+ is the maximum at Section I and the lowest is at Section III which signifies more prominent vortex structure at the upstream of the maximum depth of degraation which get reduced as one move towards the downstream signifying the weakening of the vortex structures.

The nondimensional Reynolds normal stress in streamwise and vertical directions σ_{uu}^+ and σ_{ww}^+ follow the similar profiles of that of τ^+ with different values. Here, $\sigma_{uu} = -\rho\overline{u'u'}$ and $\sigma_{ww} = -\rho\overline{w'w'}$, respectively. The nondimensional forms of σ_{uu}, σ_{vv}, and σ_{ww} are: σ^+_{uu}, $\sigma^+_{ww} = (\sigma_{uu}, \sigma_{ww})/(\rho u_*^2)$. They initiate with a lower value very close to degraded bed that increases with an increase in the vertical distance until attains the maximum value at the bed-level or slightly below the undegraded bed-levels. Interestingly, the $\sigma^+_{ww} \approx 0.5–0.6$ times of σ^+_{uu} irrespective of the locations and experimental runs. However, the present findings are in agreement with those of Jain et al. (2015).

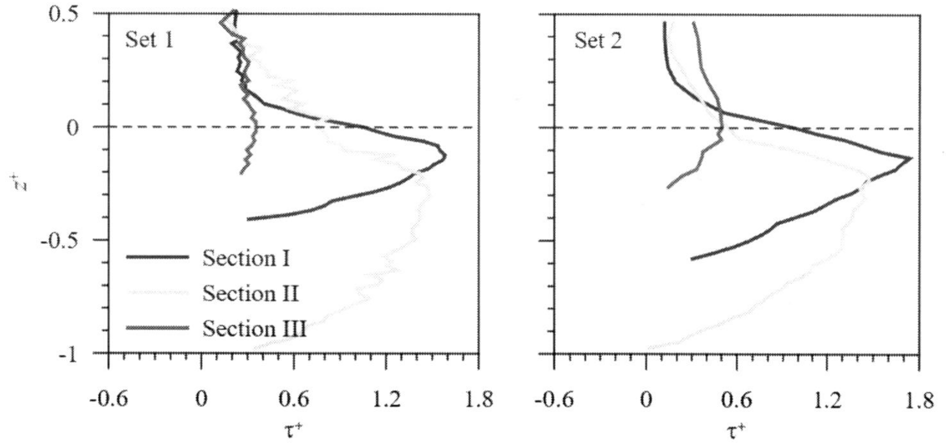

Figure 4. Vertical profiles of nondimensional RSS uw^+ at sections I, II, and III for Set 1 and 2.

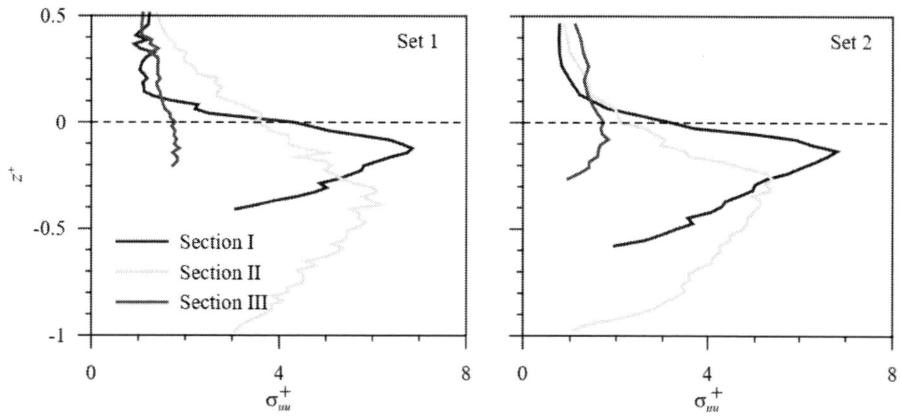

Figure 5. Vertical profiles of nondimensional streamwise Reynolds normal stress σ^+_{uu} at sections I, II, and III for Set 1 and 2.

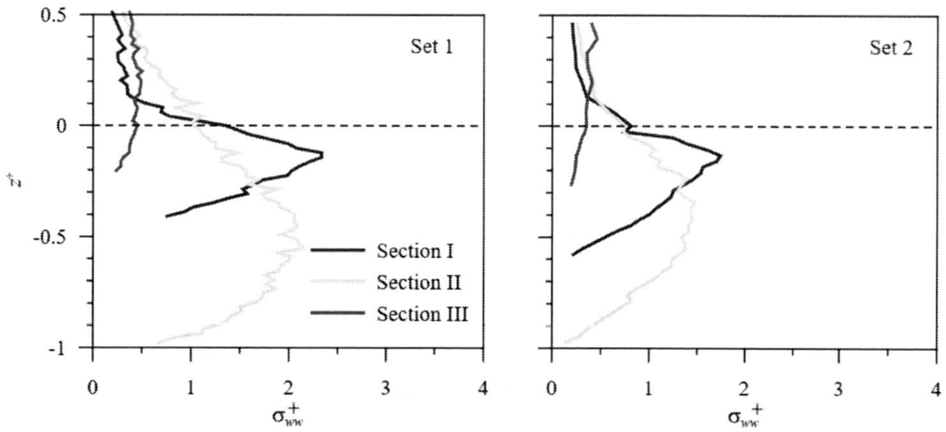

Figure 6. Vertical profiles of nondimensional vertical Reynolds normal stress σ^+_{ww} at sections I, II, and III for Set 1 and 2.

3.3. Scaling laws of turbulence

The scaling laws of turbulence is one of the most important aspects in understanding the flow in open channels which can more scientifically be observed with the help of structure functions methodology. The structure functions methodology is in principle based on the pioneering work of Kolmogorov (1941 a,b) also called as K41. Although the method was later modified and used by several researchers, still structure functions of second-, and third-order are most widely accepted tools in describing the scaling laws and estimating the turbulent dissipation rates.

The general equation of calculating structure functions of any order is written as:

$$S_p(\mathbf{r}) = \langle u(\mathbf{x} + \mathbf{r}) - u(\mathbf{x})\rangle^p, \quad (1)$$

for any $p > 0$, where $\mathbf{x}$, $\mathbf{x} + \mathbf{r}$ are points in a flow field, u is the velocity in the direction of $\mathbf{r}$, r is the length $|\mathbf{r}|$ of $\mathbf{r}$, and the angle brackets denote an average [Arenas et al. 2006]. However, in the present study the spatial increment r was calculated by adopting the Taylor frozen-in hypothesis i.e., $r(x, z) = |\bar{u}(x, z)|\Delta t$, where Δt is the time lag.

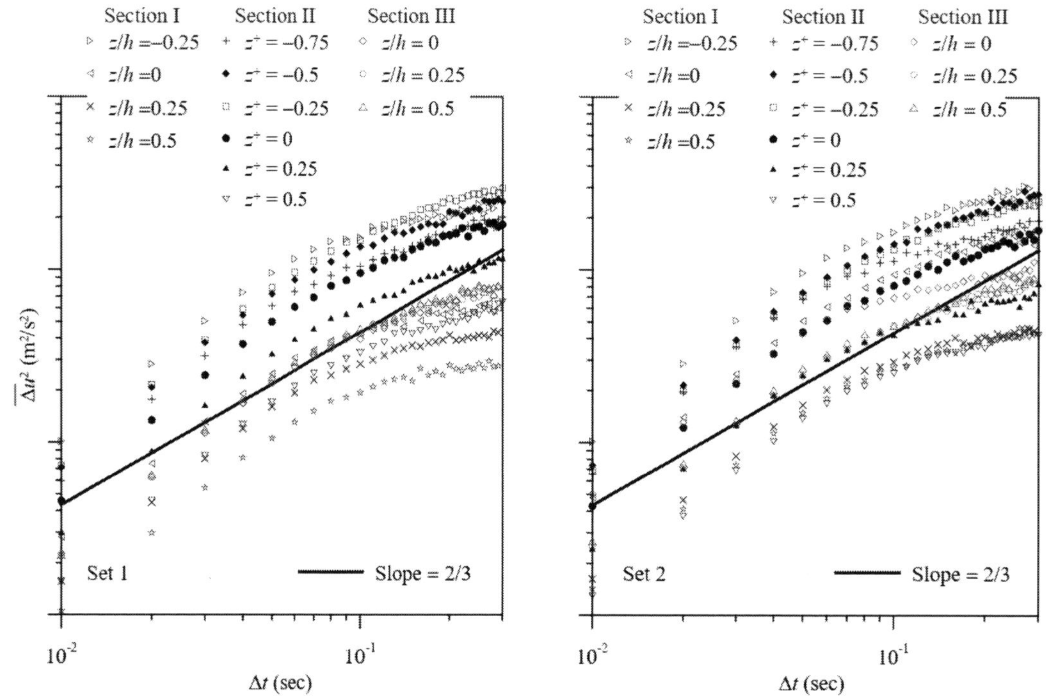

Figure 7. Variations of second-order velocity structure functions with time lag at sections I, II, and III at different vertical distances. Data of Set 1 and 2 are presented.

Figure 7 presents the variations of second-order structure functions with respect to time lag at different sections I, II, and III for the experimental data of Set 1 and Set 2. It is important to note that in Figure 7, the $\overline{\Delta u^2}$ is presented with respect to time lag which is different from the conventional way of presentation where the $\overline{\Delta u^2}$ is shown with respect to the spatial increment r. This was done due to the advantages of presenting them on a same horizontal axis. A watchful observation of $\overline{\Delta u^2}$ data reveals that the $\overline{\Delta u^2}$ follows the Kolmogorov's 2/3-law within a time lag span of approximately 0.5 s for all the data sets. For data at Section I in both the cases, the $\overline{\Delta u^2}$ obeying 2/3-slope lasts for a shorter span of time lag than that for Section II and III which signifies that due to the initiation of the bed degradation, the second-order structure functions loss its validity with 2/3-slope much earlier. However, with increase in the streamwise distance, second-order structure function follows the 2/3-slope for a larger span of time lags than that of Section I indicating the increase in the length of inertial subrange. However, at higher values of time lags, the significance of the second-order structure function is lost irrespective of the locations and $\overline{\Delta u^2}$ becomes almost asymptotic.

Figure 8 illustrates the third-order structure functions data as a function of time lag at different vertical distances of sections I, II, and III for Set1 and Set 2. In general, the plots indicate that within a shorter span of time-lag, the third-order structure functions follow Kolmogorov's 4/5-law. However, as the time-lag increase, the third-order structure functions losses validity with Kolmogorov's 4/5-law. Notably, close to the bed at Section I, the third-order structure functions data stop obeying 4/5-slope more rapidly than data of sections II and III which suggests that the third-order statistics are controlled by the degradation of a sedimentary bed. Interestingly, close to the bed, the third-order structure functions losses its validity more rapidly than that over the remaining depth of the flow. It is to mention that in case of the screeded gravel beds, Padhi et al. (2018) also observed the similar behavior at the near-bed region.

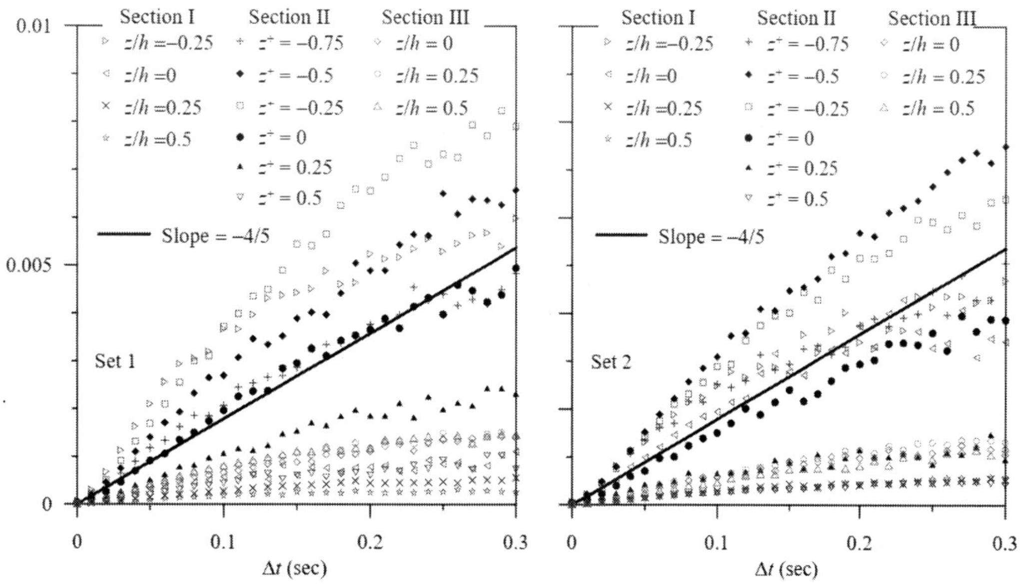

Figure 8. Variations of third-order velocity structure function with time lag at sections I, II, and III at different vertical distances. Data of Set 1 and 2 are presented.

Furthermore, the mixed third-order structure functions were calculated based on the velocity data. The mixed third-order structure functions are calculated as $\overline{\Delta u[(\Delta u)^2 + (\Delta v)^2 + (\Delta w)^2]}$. Figure 9 shows the mixed third-order structure functions at different vertical and streamwise locations at Sections I, II, and III for data sets 1 and 2. It was evident from Figure 9 that the mixed third-order structure functions follow Monin-Yaglom's 4/3-law for quite a shorter span of time-lag in the vicinity of the bed which increases with increase in the vertical distance. However, with increase in the value of time lag, the mixed third-order structure functions shows less validity with Monin-Yaglom's 4/3-law irrespective of the streamwise and vertical locations.

The third- and mixed third-order structure functions data were used to estimate the TKE dissipation rates ε using Kolmogorov's 4/5-law and Monin-Yaglom's 4/3-law. According to Kolmogorov's 4/5-law, the ε can be obtained from the equation given below:

$$\langle \overline{\Delta u^3} \rangle = (-5/4\varepsilon r), \tag{2}$$

The ε can also be estimated from the third-order mixed structure function using Monin-Yaglom's 4/3-law [Antonia et al. 1997, de Kármán 1938].

From mixed third-order structure function, ε can be calculated from the equation:

$$\varepsilon = (-4/3r) \langle \overline{\Delta u[(\Delta u)^2 + (\Delta v)^2 + (\Delta w)^2]} \rangle, \tag{3}$$

Figure 10 indicates significant disparity in the values of nondimensional ε [i.e., $E_D = \varepsilon d / u_*^3$] estimated by third-, and mixed-third order structure functions indicating anisotropy in the near-bed zone of the degraded bed.

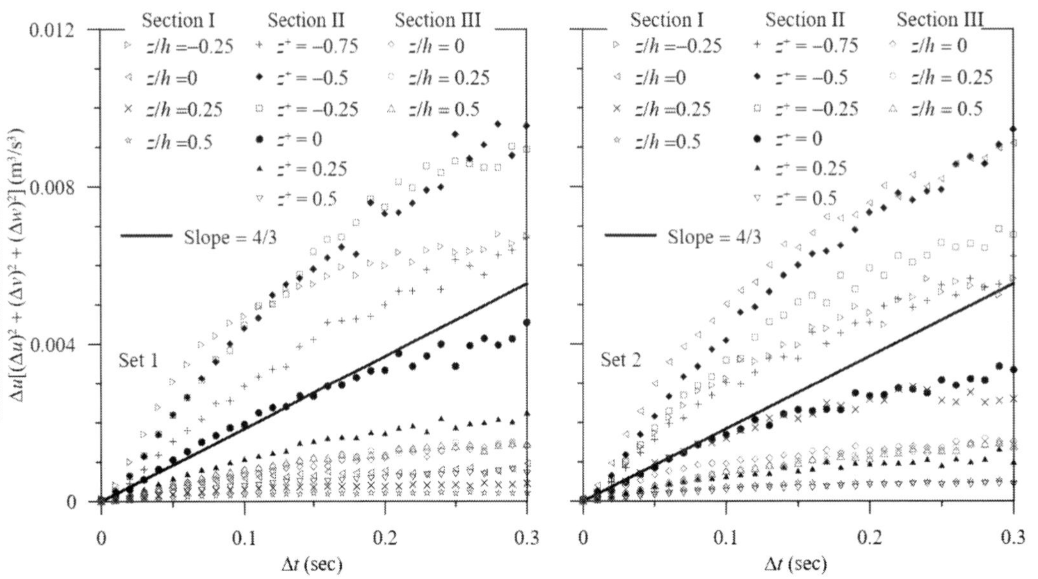

Figure 9. Variations of third-order mixed structure function with time lag at sections I, II, and III at different vertical distances for Set 1 and Set 2 data.

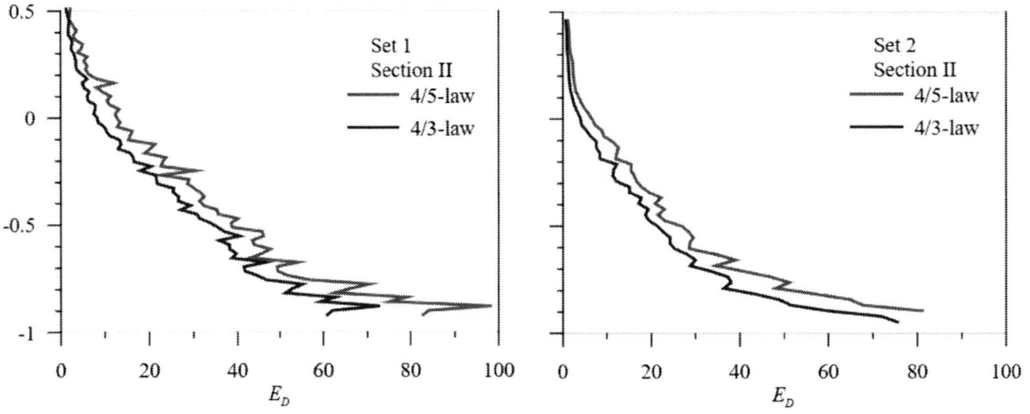

Figure 10. The values of E_D estimated from the third-order and mixed third-order structure functions for Set 1 and Set 2 data.

3.4. Time-averaged TKE budget

According to Nezu and Nakagawa (1993), the TKE budget for a uniform two-dimensional open-channel flow is calculated by:

$$\underbrace{-\overline{u'w'}\frac{\partial \overline{u}}{\partial z}}_{\substack{\text{TKE production}\\ t_P}} = \underbrace{\varepsilon}_{\substack{\text{TKE}\\ \text{dissipation}}} + \underbrace{\frac{\partial f_{kw}}{\partial z}}_{\substack{\text{TKE}\\ \text{diffusion}\\ t_D}} + \underbrace{\frac{1}{\rho}\frac{\partial \overline{p'w'}}{\partial z}}_{\substack{\text{Pressure energy}\\ \text{diffusion}\\ p_D}} - \underbrace{\upsilon\frac{\partial^2 k}{\partial z^2}}_{\substack{\text{Viscous}\\ \text{diffusion}\\ \upsilon_D}} \tag{4}$$

The υ_D is very negligible due to very low value of υ in comparison with the other parameters and be neglected. Other budget parameters can be calculated from the experimental data. The t_P was calculated from the Reynolds shear stress and velocity gradient, the ε used for this purpose was taken from their values estimated from third-order structure functions, the t_D was calculated from the vertical TKE flux gradient; whereas the p_D was estimated from Eq. (4).

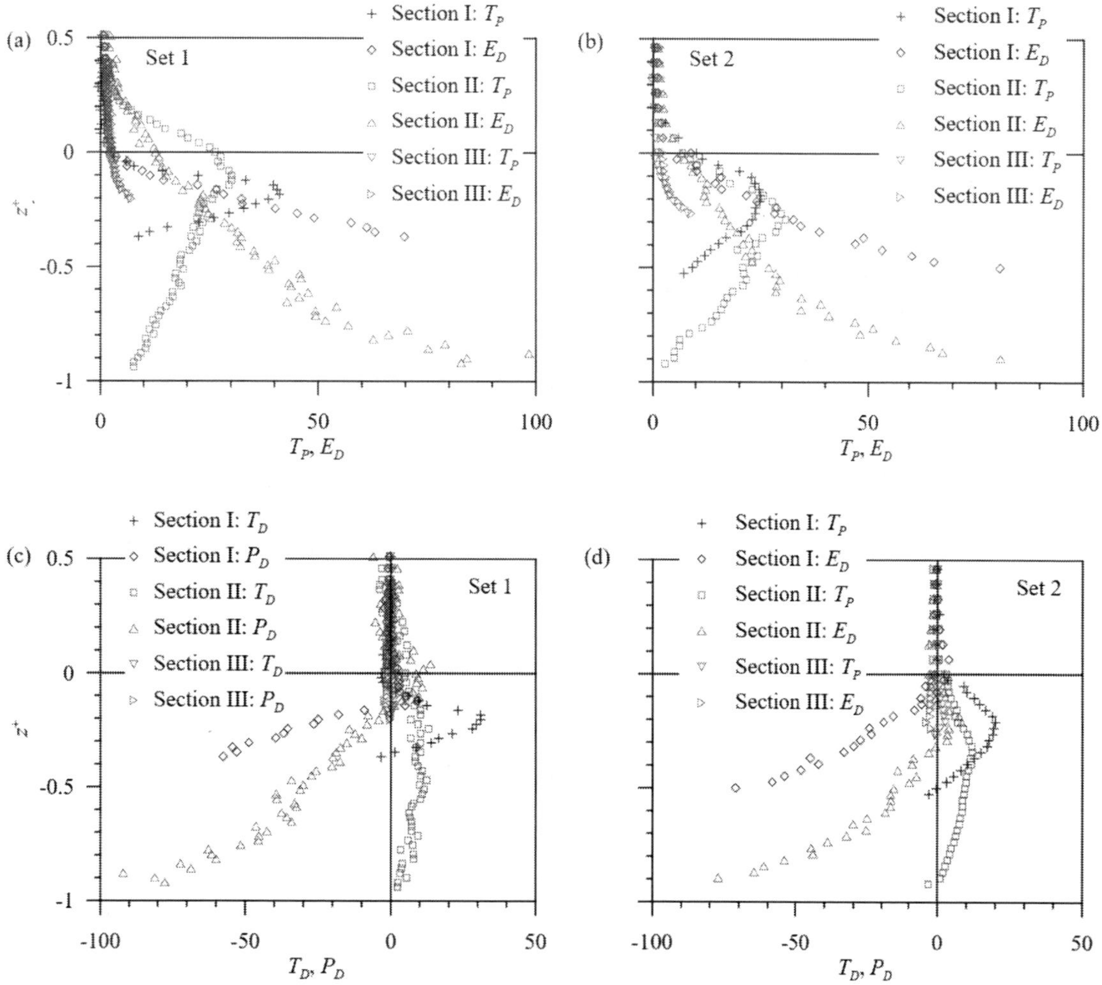

Figure 11. Vertical distribution of nondimensional TKE budget parameters T_P and E_D; and T_D and P_D at sections I, II, and III for Set 1 and Set 2.

The nondimensional form of TKE budget parameters calculated as T_P, E_D, T_D, $P_D = (t_P, \varepsilon, t_D, p_D) \times (d/u_*^3)$ were shown in Figure 11. A clear observation of Figure 11 suggests that below $z^+ = 0$, at Sections I and II, E_D is greater than T_P and T_D, which starts from the near-bed and continues until $z^+ \approx -0.1$. At $z^+ \approx -0.1$, T_P and E_D cross each other for both the data Set 1 and 2. Above $z^+ \approx 0$, T_P remains the maximum among all the budget parameters for the throughout the depth. On the other hand, T_D is the lowest throughout the depth and P_D possess large negative values near the bed that decrease continuously with increase in the vertical distance. Notably, above $z^+ \approx 0.2$, all the TKE budget parameters decrease significantly with increase in vertical distance. It can therefore be concluded that in the vicinity of the degraded bed, TKE dissipation rate is the maximum among all the budget parameters and the pressure diffusion has larger negative values.

4.Conclusions

Experimental observations of flow through a degraded bed were reported here. The results including streamwise velocity, Reynolds stresses implied that the degradation has a considerable impact on the flow characteristics. Importantly, a reversal flow takes place in the vicinity of the degraded bed; whereas beyond that level, the streamwise velocity does not show any significant differences. The peak of the Reynolds shear and normal stresses were observed at the undisturbed bed-level. The Second-, third-, and mixed third-order velocity structure functions data indicated the existence of inertial subrange within a short span of time lags. The values of TKE dissipation rates estimated from third-, and mixed third-order structure functions show differences in values in the near-bed. In the vicinity of the degraded bed, turbulent dissipation is more than turbulent kinetic energy production which is balanced by the turbulent diffusion and the pressure energy diffusion; whereas pressure energy diffusion remains the lowest throughout the depth.

5. REFERENCES

Antonia, R., Ould-Rouis, M., Anselmet, F., & Zhu, Y. (1997). "Analogy between predictions of Kolmogorov and Yaglom." *J. Fluid Mech.*, 332, 395-409.

Arenas, A., and Chorin, A. J. (2006). "On the existence and scaling of structure functions in turbulence according to the data." *Proc. Natl. Acad. Sci. USA.*, 103(12), 4352-4355.

Coscarella, F., Servidio, S., Ferraro, D., Carbone, V., and Gaudio, R. (2017). "Turbulent energy dissipation rate in a tilting flume with a highly rough bed." *Phys. Fluids.*, 29(8), 085101.

de Kármán T., and Howarth, L. (1938). "On the statistical theory of isotropic turbulence." *Proc. R. Soc. Lond.* A164192–215.

Dey, S., Das, R., Gaudio, R., and Bose, S. K. (2012). "Turbulence in mobile-bed streams." *Acta Geophys.*, 60, 1547–1588.

Ferraro, D., Servidio, S., Carbone, V., Dey, S., and Gaudio, R. (2016). "Turbulence laws in natural bed flows." *J. Fluid. Mech.*, 798, 540–571.

Goring, D. G., and Nikora, V. I. (2002). "Despiking acoustic Doppler velocimeter data." *J. Hydraul. Eng.*, 128(1), 117–126.

Grams, P. E., and Wilcock, P. R. (2007). "Equilibrium entrainment of fine sediment over a coarse immobile bed." *Water. Resour. Res.*, 43, W10420.

Jain, R. K., Kumar, A., and Kothyari, U. C. (2015). "Turbulence statistics of flow through degraded channel bed of sand-gravel mixture." *J. Hyd. Environ. Res.*, 9(4), 508-518.

Kolmogorov, A. N. (1941a). "Local structure of turbulence in an incompressible fluid for very large Reynolds numbers." *Dokl. Akad. Nauk, SSSR.*, 30, 299–303.

Kolmogorov, A. N. (1941b). "Dissipation of energy in the locally isotropic turbulence." *Dokl. Akad. Nauk SSSR.*, 32, 19–211.

Nezu, I., and Nakagawa, H. (1993). Turbulence in open-channel flows, Balkema.

Nikora, V., and Goring, D. (2000). " Flow turbulence over fixed and weakly mobile gravel beds." *J. Hydraul. Eng.*, 126 (9), 679e690.

Padhi, E., Penna, N., Dey, S., and Gaudio, R. (2019). "Near-bed turbulence structures in water-worked and screeded gravel-bed flows." *Phys. Fluids.*, 31(4), 045107.

Padhi, E., Penna, N., Dey, S., and Gaudio, R. (2018). "Spatially averaged dissipation rate in flows over water-worked and screeded gravel beds." *Phys. Fluids.*, 30(12), 125106.

Penna, N., Padhi, E., Dey, S., and Gaudio, R. (2020). "Structure functions and invariants of the anisotropic Reynolds stress tensor in turbulent flows on water-worked gravel beds." *Phys. Fluids.*, 32(5), 055106.

Sarkar, M, Maurya, S. K., Gopmandal, P. P., Sarkar, S. (2021). "Hydrodynamics of flow through a degraded channel bed." *J. Turbulence*, 22(10), 1-29.

Shvidchenko, A. B., and Pender, G. (2001). "Macroturbulent structure of open-channel flow over gravel beds." *Water Resour. Res.*, 37(3), 709-719.

Wren, D. G., Langendoen, E. J, and Kuhnle, R. A. (2011). "Effects of sand addition on turbulent flow over an immobile gravel bed." *J. Geophys. Res. (Earth Surface)*, 116(F1), 01018.

DOI: https://doi.org/10.26077/39db-f88b

Local Scour Downstream of Type-A Trapezoidal Piano Key Weir

Ch. A. Chooplou[1], A. Ghafouri[1], M. Ghodsian[1*] & M. Vaghefi[2]

[1]*Department of Civil and Environmental Engineering, Tarbiat Modares University, Tehran, Iran*
[2]*Department of Civil Engineering, Persian Gulf University, Bushehr, Iran*
**E-mail : ghods@modares.ac.ir*

***Abstract**: A major issue with weirs over erodible beds is the scour downstream, which can seriously challenge their stability. This paper presents experimental results on trapezoidal type-A piano key weirs having two values of bottom slopes keys (28° and 47°). Experiments were conducted in a recirculating channel 10 m long, 0.75 m wide and 0.8 m high. Approach flow depths in the range of 0.032 to 0.07 m, and tailwater depths in the range of 0.08 to 0.18 m were evaluated. It was observed that the bottom slope of weir keys significantly influences the scour depth, such that a 67.8% increase in the slope, increased the average maximum scour depth by about 65%. The maximum scour depth occurred downstream of the outlet keys. Moreover, the scour increased with an increase in the approach flow depth and decreased with the tailwater depth. Distance of location of the maximum scour depth from the weir toe, and length of the scour hole in the weir with key bottom slope of 47° were respectively 5 and 21% greater than the weir with key bottom slope of 28°. In general, the geometric dimensions of the scour hole which were influenced by tailwater depth were more obvious in the model with smaller key bottom slope, such that with a 125% increase in the tailwater depth, the average variations of the scour parameters (maximum scour depth, its distance from the weir toe, and the scour hole length) in the model with smaller key bottom slope were respectively 1.5, 2.5 and 2.7 times those of the other model.*

***Keywords:** Experimental Study, Scour, Piano Key Weir, Weir Geometry, Key Bed Slope*

1. Introduction

Piano key weirs (PKWs) represent a new form of nonlinear weirs that are economically viable in addition to having higher discharge capacity compared to linear weirs (Mehboudi et al., 2014). PKWs increase the discharge by at least 4 times compared to linear weirs, and incur lower maintenance costs (Machiels, 2012). This type of weir is mainly applied for increasing the discharge capacity of available weirs as well as being used as a control structure in newly built weirs (Laugier et al. 2013; Phillips and Lesleighter 2013). These weirs also serve several purposes when applied in rivers. PKWs, for instance, are utilized in rivers as control structures, a structure for adjusting the water level or as short diversion dams in river beds. PKWs are categorized into rectangular, triangular and trapezoidal types, in accordance to their angle of lateral walls. The trapezoidal PKWs are more efficient than the rectangular ones (Mehboudi et al., 2017; Mehboudi et al., 2016). Figure (1) shows the geometric characteristics of a trapezoidal PKW, where P is the vertical height, B represents the streamwise length, T_s is the lateral wall thickness, L denotes the weir crest length, W is the lateral width and w shows the width of a cycle. In addition, the subscript i refers to the inlet key and the subscript o refers to the outlet key. These subscripts are used for the streamwise length B and the weir width W.

Several researchers have conducted diverse experimental and numerical studies with the purpose of understanding the effects of PKWs parameters on flow behavior over the recent years. For example, Lempérière and Ouamane (2003), Barcouda et al. (2006), Hien et al. (2006), Ouamane and Lempérière (2006), Laugier et al. (2011), and Machiels et al. (2011) demonstrated that for a low hydraulic head, increasing the number of weir cycles increases the weir's efficiency. Ouamane and Lempérière (2006), Hien et al. (2006), and Barcouda et al. (2006) indicated that if the inlet key of weir is considered larger than the outlet key, it will increase the discharge capacity over the weir. Increasing weir length has a significant effect on the discharge capacity of weir; nevertheless, an excessive length lowers such an effect (Kabiri-Samani and Javaheri, 2012). Machiels et al. (2014) examined the flow parameters on PKWs and proposed relations to determine the discharge over these weirs. Khassaf and Baghdadi (2015) expressed the advantages of PKWs over linear weirs. Anderson (2011) tested different laboratory models of PKWs and sloped labyrinth weirs. In Anderson's research, the PKW slope had nearly the same effect as the labyrinth weir.

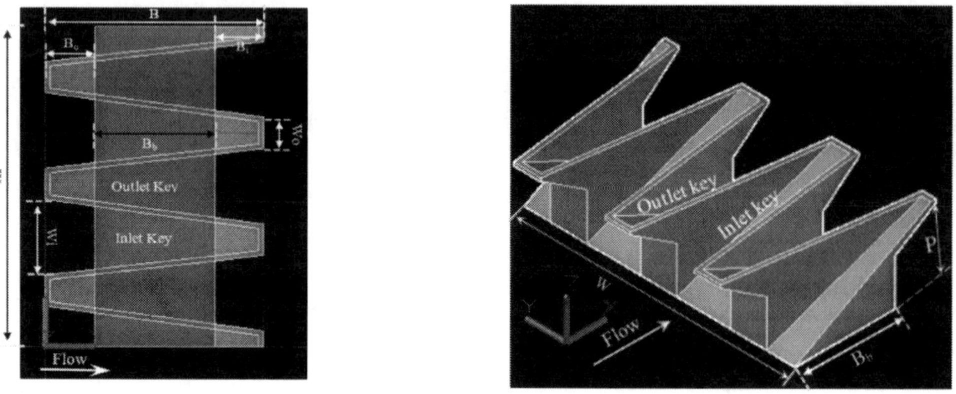

Figure 1. Geometric parameters of PKW

Review of available literature shows that most of the researchers have studied the hydraulics characteristics of PKWs, some of which were mentioned above. However, few studies and evaluations focused on scouring downstream of these weirs. Jüstrich et al. (2016) conducted a laboratory study on scouring downstream of a rectangular PKW and considered the effects of parameters such as discharge, drop height over the weir, sediments size downstream of the weir and tailwater depth on the dimensions of the scour hole and the sedimentary ridge. They found that increasing the discharge and decreasing the tailwater depth significantly increased the dimensions of the scour hole and proposed relations to estimate the dimensions of the scour hole. Ghafouri et al. (2020) conducted an experimental study on the effects of discharge and tailwater depth on the downstream bed topography of a trapezoidal PKW with the inlet and outlet key slopes of 28° and a height of 0.2 m. The results of their research revealed that the maximum scour depth and scour hole area increased by about 29.5 and 18.5% with a 66.7% increase in discharge at a constant tailwater depth, respectively. Kumar and Ahmad (2020) modeled the scour pattern by installing the noses below the outlet overhangs of rectangular PKWs with and without a fixed apron. The results of their research indicated that the length and depth of scouring are significant for higher discharge rates and lower tailwater depths. Yazdi et al. (2020) conducted a laboratory study of scouring downstream of rectangular and trapezoidal PKWs and stated that the maximum depth and length of the scour hole were greater in the rectangular weir. A comparison of scouring downstream of trapezoidal and triangular PKWs was made by Ghodsian et al. (2021). They indicated that the maximum scour depths downstream of the triangular PKW were higher than those of the trapezoidal PKW. Lantz et al. (2021) studied apron and cutoff wall scour protection for PKWs. It was observed that a horizontal apron deflects the plunging jets originating from the PKW, thus significantly reducing scour. Kumar et al. (2021) explored coarse sand and gravel quartz particles at the riverbed upstream of and over the inlet key of rectangular PKW. They observed that the particles slowed down as they further approached the inlet key, but the flow contraction and the rapid increase in the shear stress caused a sudden acceleration in the motion of the particles passing over the inlet key. However, the shear stress in their models had to be increased by another 17 to 43% to carry the sediment over the weir keys. Bodaghi et al. (2022) experimentally investigated the scouring downstream of a type A PKW under free and submerged flow conditions. They showed that with increasing 18% in the submersion ratio with the densimetric Froude number 0.29 and increasing 96% in the submersion ratio with the desnsimetric Froude number of 0.58, the maximum relative scour depth decreases by 92 and 56%, respectively. Lantz et al. (2022) conducted a laboratory study of scouring downstream of a rectangular PKW by varying discharge, tailwater depth and sediment size. Their results indicated that under certain hydraulic conditions, the tailwater level had a significant effect on the scouring mechanisms and the equilibrium morphology downstream of the PKW. They also proposed relations for estimating the maximum scour depth downstream of the weir.

A review of previous research shows that the majority of studies on PKWs focused on the flow hydraulics and their influencing parameters, while one of the most important parameters on safety and stability of dams is downstream scouring, which has received less attention so far. Considering the fact that the constructed PKWs are newer (with less than 15 years of construction history) than the linear weirs, there is limited knowledge on erosion and sediment transport on such weirs (Oertel 2018). This necessitates an evaluation and analysis of erosion processes downstream of these weirs. In this research, the effects of variations in discharge and the tailwater depth on scouring downstream of a trapezoidal PKW are studied. Two bottom slopes (28° and 47°) of the inlet and outlet keys were tested.

2. Materials and Methods

2.1. Dimensional Analysis

Considering the previous research and results on scouring of nonlinear weirs, the factors affecting the scour downstream of PKWs, including geometric and hydraulic parameters, could be demonstrated as follows (Gebhardt et al., 2018; Noseda et al., 2019).

$$f_1(\Psi;\ \Delta H; H_d; q; P;\ d_{50}; g;\ \rho_w; \rho_s; \mu) = 0 \tag{1}$$

In the above equation, ψ denotes the geometric characteristics of the scour hole, as shown in Figure (2), including the maximum scour depth Z_S, the distance of maximum depth from the weir toe X_S, and the length of the scour hole X_O. Independent variables in Equation (1) include the discharge per unit width of weir q, weir height P, the flow head downstream of the weir H_d, the hydraulic head difference upstream and downstream of the weir ΔH, the average particle diameter d_{50}, the acceleration of gravity g, sediment and water densities (respectively ρ_s and ρ_w), and the water dynamic viscosity μ.

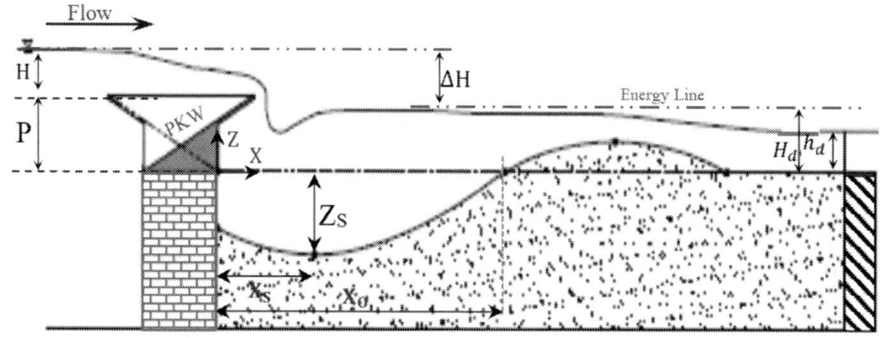

Figure 2. Flow and scour parameters along PKW

The dimensionless parameters affecting the scouring downstream of the weir are obtained using the Buckingham π theorem:

$$f_2(\frac{\Psi}{P}.\frac{\Delta H}{P}.\frac{H_d}{P}.\frac{q}{\sqrt{g\Delta H^3}}.\frac{\mu}{\rho_w q}.\frac{\rho_s-\rho_w}{\rho_w}.\frac{d_{50}}{\Delta H}) = 0 \tag{2}$$

where $\frac{q}{\sqrt{g\Delta H^3}}$ and $\frac{\mu}{\rho_w q}$ respectively represent the Froude and the Reynolds numbers. Since the Reynolds number was more than 40000 in all the present tests, the flow was fully turbulent, and the Reynolds number could be ignored (Henderson, 1986). Furthermore, the relative water head over the weir H/P was measured larger than 0.1 in light of the tested discharges to eliminate the effect of surface tension (Novak and Cabelka, 1981). Following the design procedure, the tests were done such that the surface tension and the dynamic viscosity were ineffective. Combining three parameters $\frac{q}{\sqrt{g\Delta H^3}}$, $\frac{d_{50}}{\Delta H}$ and $\frac{\rho_s-\rho_w}{\rho_w}$, the densimetric Froude number Fr_{d50} was obtained as $\frac{q}{\Delta H\sqrt{(\frac{\rho_s-\rho_w}{\rho_w})gd_{50}}}$. Hence, the dimensionless parameters affecting the downstream scour of the PKW with a specific bottom slope of keys are as given in Equation (3):

$$\frac{\Psi}{P} = F(\frac{\Delta H}{H_d}.Fd_{50}) \tag{3}$$

2.2. Laboratory Setup

All the tests were run in the hydraulic laboratory of Tarbiat Modares University, in a 10-m-long and 0.75-m-wide channel. The channel is 0.80 m high with bed slope of 0.001. The channel bed is made of galvanized sheets with security glass walls. A view of the laboratory channel is illustrated in Figure (3).

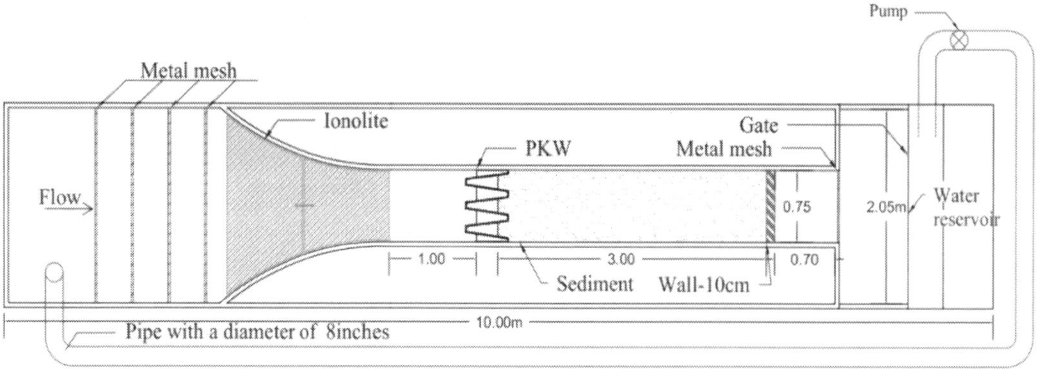

Figure 3. View of the laboratory channel

The water depth was adjusted by a gate installed at the end of the channel. The water flow was supplied using a pump with maximum discharge of 85 l/s.

Two geometries of a trapezoidal PKW, as shown in Figure (4), were used for the intended tests. The specifications of PKWs are presented in Table (1). The tests were carried out using two PKWs with bottom slope of inlet and out keys equal to 47° and 28°. These models are referred to as PKW1 and PKW2 further in the study.

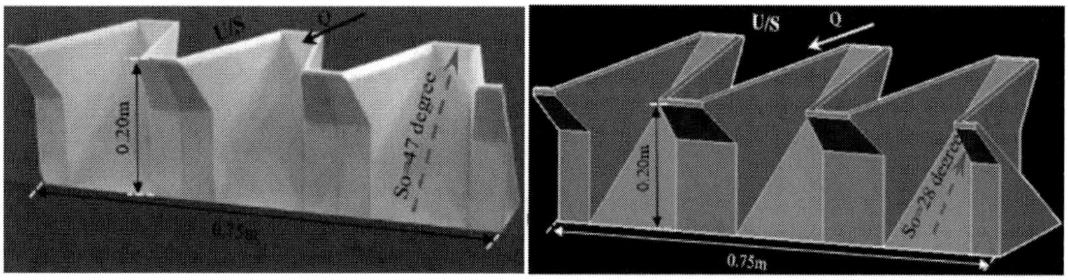

a) PKW1 b) PKW2

Figure 4. The PKWs used in this study

According to Pagliara et al. (2006), the scale effect due to sediment size will be negligible if the minimum particle size of the bed (d_{10}) is more than 1 mm. On the other hand, as reported by Jüstrich et al. (2016), the value of d_{50}/P for a number of constructed dams falls in the range of $0.01 \leq d_{50}/P \leq 0.04$. Hence, given the above criteria and use of a weir with a height of P = 0.2 m, the size selected for the bed materials is 2.2 mm, which was covered at a length of 3 m downstream of weir. The initial sediment bed surface was horizontal and approximately aligned with the weir toe.

Table 1. Specifications of the PKWs

Parameters / Weir models	P (m)	L (m)	W (m)	w (m)	$S_i = S_o$ (degree)	B (m)	B_b (m)	$B_i = B_o$ (m)	B/w	L/W
PKW1	0.20	1.85	0.75	0.25	47	0.25	0.125	0.0625	1	2.47
PKW2	0.20	3.33	0.75	0.25	28	0.50	0.25	0.125	2	4.44

Scouring phenomenon is time-dependent, expanding in time and finally approaching equilibrium state. According to Kumar et al. (1999), when the scour change is less than 1 mm after 3 hours, equilibrium state is achieved in the scour hole geometry. Temporal measurements of the scour depth were performed in a test with discharge of 0.05

m^3/s for duration of 15 hr, as shown in Figure (5). According to Figure 5, 96% of scouring occurs in the initial 300 minutes. Kumar et al. (1999) criteria for equilibrium state is achieved after about 300 minutes from the initiation of the experiment. Therefore, the duration of 300 minutes was selected further in the experiment.

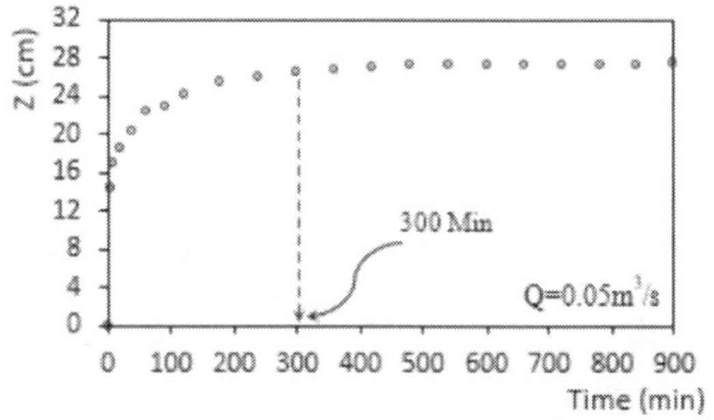

Figure 5. Temporal evolution of the scour depth

The tests were conducted by preparing the channel, then the weir model was installed inside the channel and was completely sealed using silicone sealant, then pouring sediment particles downstream of the weir, and eventually using a flattener device to make the sediments surface fully flat and horizontal. At the beginning of every test, a thin metal plate was used to cover the sediment bed downstream of the weir in order to prevent influence of the model start-up. The test began after stabilizing the specified discharge and the tailwater depth. When each test was completed, the water was evacuated from the channel and the bed topography was measured in rectangular meshes in longitudinal and lateral directions using a bed topography collection laser device (made by Bosch) with a precision of ±1 mm. The resulting scour was developed under clear water for flow conditions without the installed weir. The discharge varied between 0.030 and 0.050 m^3/s. An ultrasonic flowmeter, Model TFM3100-F1, was utilized which could measure the discharge with an accuracy of ±0.1 l/s. The flow depth was measured using a point gauge with an accuracy of ±1 mm. Table (2) presents a summary of the flow conditions and the number of tests performed.

Table 2. The initial hydraulic conditions of the tests

Test	Model	Q (m^3/s)	q (m^2/s)	H (m)	h_d (m)	Fr_{50} (-)	ΔH (m)	$\Delta H / H_d$
1	PKW1	0.03	0.04	0.046	0.08	1.33	0.17	0.27
2	PKW1	0.03	0.04	0.046	0.18	3.34	0.07	0.09
3	PKW1	0.04	0.053	0.057	0.08	1.67	0.18	0.28
4	PKW1	0.04	0.053	0.057	0.18	3.82	0.08	0.11
5	PKW1	0.05	0.066	0.070	0.08	1.94	0.19	0.30
6	PKW1	0.05	0.066	0.070	0.18	4.09	0.09	0.13
7	PKW2	0.03	0.04	0.032	0.08	1.46	0.15	0.25
8	PKW2	0.03	0.04	0.032	0.18	4.23	0.05	0.07
9	PKW2	0.04	0.053	0.040	0.08	1.85	0.16	0.26
10	PKW2	0.04	0.053	0.040	0.18	4.99	0.06	0.09
11	PKW2	0.05	0.066	0.050	0.08	2.17	0.17	0.27
12	PKW2	0.05	0.066	0.050	0.18	5.25	0.07	0.10

3. Results and Discussion

This study considered scouring downstream of PKWs while altering discharge and the tailwater depth. The bed topography was measured at the end of the tests, for a distance of 3 meters. Figure (6) demonstrates how the geometry and the bottom slope of the weir keys affects the bed topography downstream of PKWs. The difference in scouring at the downstream of inlet and outlet keys are evident in this figure. This is specifically a result of the outflow jets from the outlet keys, which lead to more scour by generating rotating flows beneath the outlet keys. In fact, the interaction between the outflows from the inlet and outlet keys and the created lateral flows augmented the flow turbulence in the vicinity of the outlet keys; as a result, local variations in scouring occurred.

a) PKW1

b) PKW2

Figure 6. Scouring downstream of PKWs

Effects of the relative drop height $\Delta H/H_d$ and the densimetric Froude number Fr_{d50} on the longitudinal bed profiles downstream of the outlet keys, for both the PKW1 and PKW2 models are compared in Figure (7). It can be seen that in the PKW1 (with bottom bed slope of 28°) model, the topographic changes are greater compared to the second model. It can be argued that in the PKW1 model (with bottom bed slope of 47°), inclined jets caused by the outlet keys, hit the tailwater surface with higher intensity and speed, resulting in greater bed scour. The values of relative maximum scour depth in the PKW2 model varied in the range of 0.6 to 1.67, while in the PKW1 model they varied in the range of 1.28 to 2.62. As shown in Figure (7), for each tailwater depth, by increasing $\Delta H/H_d$ and Fr_{d50}, which is associated with increased drop height, the scour depth as well as the location of maximum scour depth from the weir toe increases. The length of the scour hole also increases. According to Figure (7), by increasing $\Delta H/H_d$ and Fr_{d50}, the scour hole length in the PKW1 model for tailwater depth of 0.18 m increases by about 20% compared to that in the PKW2 model, while it remains almost the same in both models for tailwater depth of 0.08 m. Furthermore, it is observed that in the PKW1 model, by increasing Fr_{d50} from 1.33 to 1.94, the location of maximum scour depth from the weir toe increased by about 25%, while in the PKW2 model increasing Fr_{d50} from 1.46 to 2.17 led to about 37.5% increase in the location of maximum scour depth from the weir toe. It is also clear that the scour hole length is also significantly influenced by variations of the tailwater depth, such that a 125% increase in the tailwater depth, reduced the scour hole length by about 34.5% in the PKW2 model and by approximately 13% in the PKW1 model. Further, the water head in both the models show a marginal effect on the maximum height of sedimentary ridge height and length downstream of scour hole.

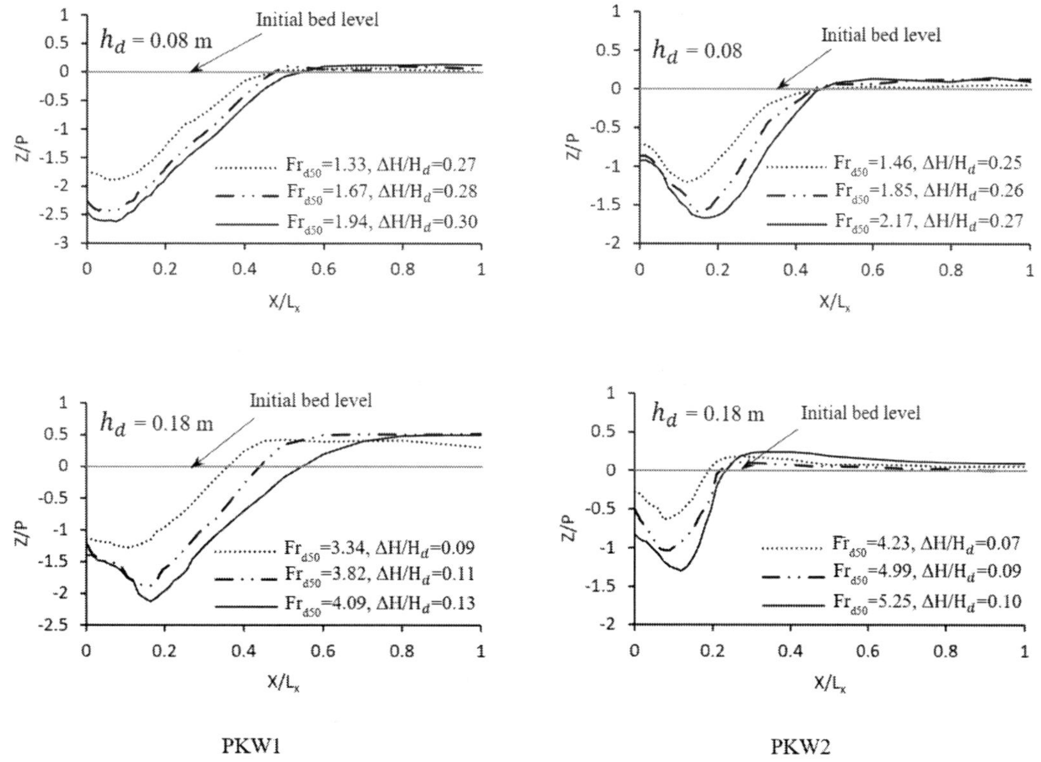

Figure 7. Longitudinal bed profiles for different $\Delta H/H_d$ and Fr_{d50}

Figure (8) compares the longitudinal bed profiles for both the PKW models. This Figure shows that the values of scour parameters of the PKW1 model are larger compared to the PKW2 model. In general, the maximum scour depth of the PKW1 model is on average 65% greater than that of the PKW2 model. Moreover, for a specific value of Fr_{d50} and tailwater depth, the distance of maximum scour depth from the weir toe and the length of the scour hole in PKW1 model were respectively 5 and 21% larger in the PKW1 model compared to the PKW2 model. In addition, the ridge height difference between the two models was reduced by increasing Fr_{d50}. According to Figure (8), for the tailwater depth of 0.18 m, the difference between the scour holes in the PKW models decreases by about 18% compared to the tailwater depth of 0.08. This difference is also reduced when Fr_{d50} is decreased. It was observed that, for a constant discharge, increasing the tailwater depth decreases the maximum scour depth and its distance from the weir toe, as well as the scour hole length. In fact, the tailwater depth plays a considerable role in dissipating the energy of the jets leaving the outlet keys. Hence, the scour depth is reduced when the tailwater depth is increased.

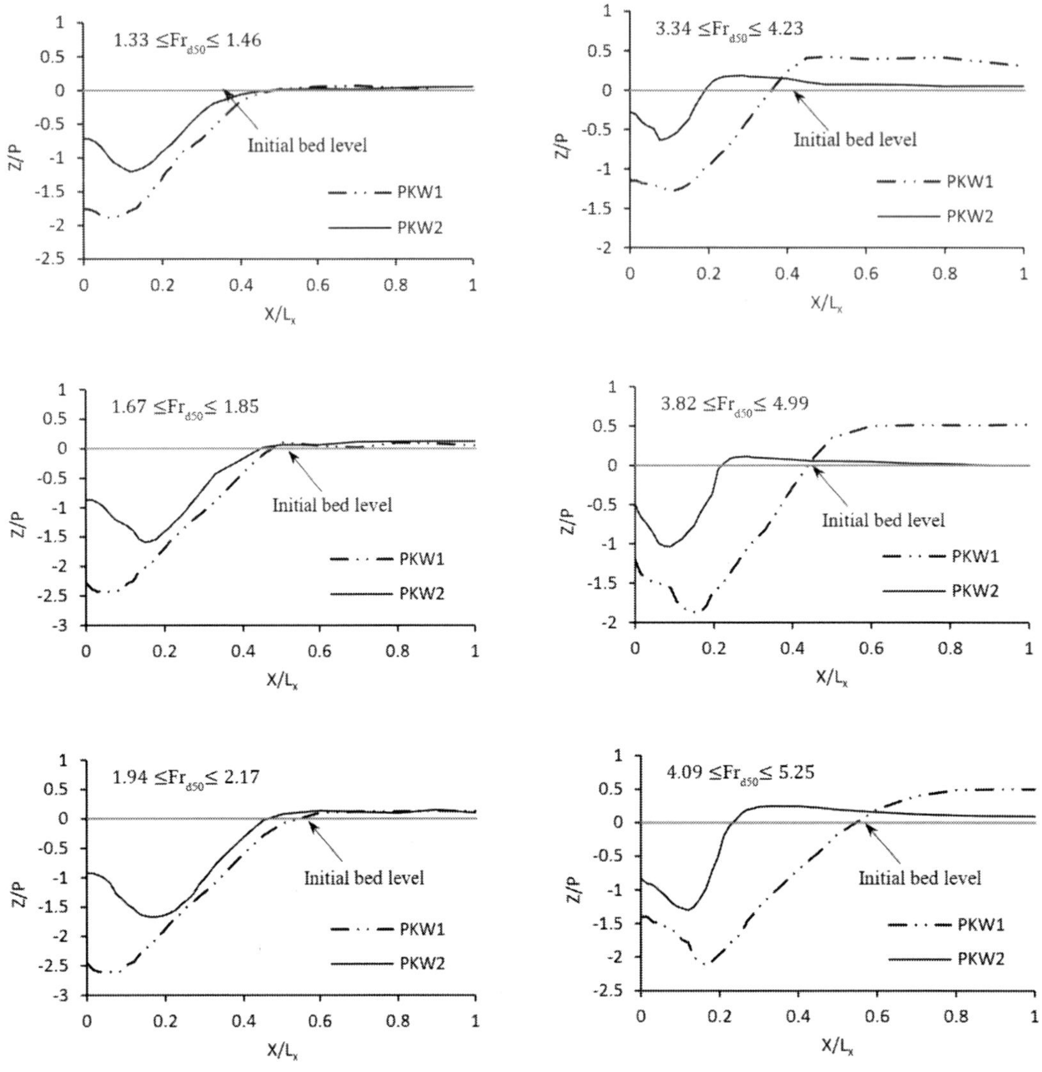

$h_d = 0.08$ m $h_d = 0.18$ m

Figure 8. Comparison of longitudinal bed profiles for models PKW1and PKW2

By scrutinizing Figures (7) and (8), an inference could be drawn that the longitudinal scour hole profiles have similar trend and follow a cubic equation. Figure (9) demonstrates the general trend of the dimensionless scour hole profiles for all the tests, plotted in terms of Z/Z_S and X/X_S, where Z and X are vertical and horizontal distances respectively (Fig. 2). The lower envelope line in Figure 9 represents Equation 4 with correlation coefficient $R^2 = 0.95$ and the symbols represent measured data points. This figure indicates the nearly similar trend of the longitudinal bed profiles in all the tests. Equation 4 is valid for $1.33 < Fr_{d50} < 5.25$ and $0.07 < \Delta H/H_d < 0.3$.

$$\frac{Z}{Z_S} = -2.95\left(\frac{X}{X_S}\right)^3 + 6.89\left(\frac{X}{X_S}\right)^2 - 3.62\left(\frac{X}{X_S}\right) - 0.45 \quad (4)$$

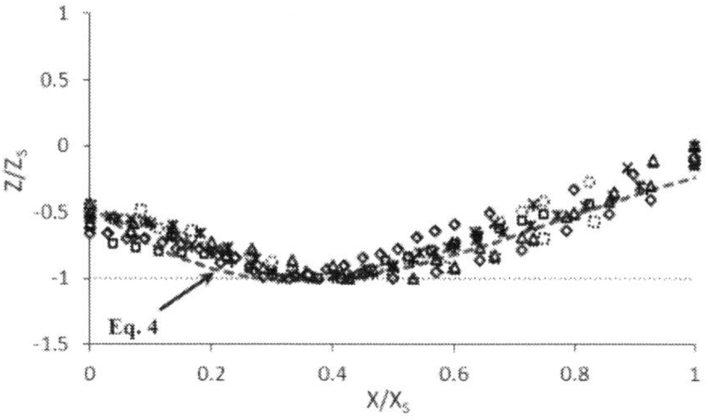

Figure 9. Dimensionless longitudinal profile of the scour hole

4. Conclusion

This study investigated the geometry of the scour hole downstream of two models of PKWs with different geometry. The conclusions from this experimental study are summarized as follows:

- The maximum scour depth occurred downstream of the outlet keys.
- The geometric parameters of the scour hole downstream of the PKWs depended on the geometry of the weir, discharge and the tailwater depth.
- The values of relative maximum scour depth in the PKW2 model were in the range of 0.6 to 1.67, while they were in the range of 1.28 to 2.62 for the PKW1 model.
- By increasing the densimetric Froude number and the relative drop height, the maximum scour depth, its distance from the weir toe, and the scour hole length increased.
- By increasing the tailwater depth by 125%, the maximum scour depth decreased by about 24% in the PKW1 model, and by approximately 34% in the PKW2 model.
- The variations of scour parameters due to tailwater depth was more significant in the PKW2 model compared to PKW1 model.
- By increasing the tailwater depth by 125%, the scour length decreased by 34.5% in the PKW1 model while it decreased by approximately 13% in the PKW2 model.
- The dimensions of scour parameters in the PKW2 model were on the average 65, 5, and 21% larger than those in the PKW1 model.

5. References

Anderson, R.M. (2011). "Piano key weir head discharge relationships." *A thesis submitted in partial fulfillment of the requirements for the degree of Master of science in civil and environmental Engineering*, Utah State university Logan, Utah, USA.

Barcouda, M., Cazaillet, O., Cochet, P., Jones, B.A., Lacroix, S., Laugier, F., Odeyer, C., and Vigny, J.P. (2006). "Cost-effective increase in storage and safety of most dams using fuse gates or P. K. weirs." *Proc. 22nd ICOLD Congress*. Barcelona, Spain.

Bodaghi, E., Ghodsian, M., and Abdi Chooplou, Ch. (2022). "The experimental study of downstream scouring of trapezoidal Piano key weir type A under free and submerged flow." *Journal of Hydraulics*, 10.30482/jhyd.2022.330389.1590.

Gebhardt, M., Herbst, J., Merkel, J., and Belzner, F. (2018). "Sedimentation at labyrinth weirs–an experimental study of the self-cleaning process." *Journal of Hydraulic Research*, 57(4), 1-12.

Ghafouri, A., Ghodsian, M., and Abdi Chooplou, Ch. (2020). "Experimental Study on the Effect of Discharge and Tailwater Depth on Bed Topography Downstream of a Piano Key Weir." *Journal of Hydraulics*, 15(3), 107-122.

Ghodsian, M., Abdi Chooplou, Ch., and Ghafouri, A. (2021). "Scouring downstream of triangular and trapezoidal pianos key weir." *Iranian Hydraulic Association Journal of Hydraulics*, 16(2), 43-58.

Henderson, F.M. (1986). Open channel flow, McMillan Publishing Company, New York.

Hien, T.C., Son, H.T., and Hotakhanh, M. (2006). "Results of some piano keys weir hydraulic model tests in Vietnam." *Proc. 22nd ICOLD Congress*, CIGB/ICOLD, Barcelona Q87(R39). 581–596.

Jüstrich, S., Pfister, M., and Schleiss, A.J. (2016). "Mobile riverbed scours downstream of a piano key weir." *Journal of Hydraulic Engineering*, 142(11), 04016043.

Kabiri-Samani, A., and Javaheri, A. (2012). "Discharge coefficients for free and submerged flow over Piano Key weirs." *Journal of Hydraulic Research*, 50(1), 114-120.

Khassaf, S.I., and Al-Baghdadi, M.B. (2015). "Experimental study of non-rectangular piano key weir discharge coefficient." *Journal homepage*: www. IJEE. IEE Foundation. Org., 6(5), 425-436.

Kumar, B., and Ahmad, Z. (2020). "Experimental study on scour downstream of a piano key weir with nose." *8th IAHR ISHS*, Santiago, Chile.

Kumar, B., Kadia, S., and Ahmad, Z. (2021). "Sediment movement over type A piano key weir." *Journal of Irrigation and Drainage Engineering*, 147(6), 04021018.

Kumar, V., Ranga Raju, K.G., and Vittal, N. (1999). "Reduction of local scour around bridge piers using slot and collar." *Journal of Hydraulic Engineering*, ASCE, 125(12), 1302–1305

Lantz, W., Crookston, B.M., and Palermo, M. (2021). "Apron and cut off wall scour protection for piano key weirs." *Water*, *13*(17), 2332.

Lantz, W.D., Crookston, B. M., and Palermo, M. (2022). "Evolution of local scour downstream of type A PK weir in non-cohesive sediments." *Journal of Hydrology and Hydromechanics*, 70(1), 103-113.

Laugier, F., Pralong, J., and Blancher, B. (2011). "Influence of structural thickness of sidewalls on PKW spillway discharge capacity" *Proc. Int. Conf. Labyrinth and piano key weirs*, CRC Press, Liège, Belgium. 159-165.

Laugier, F., Vermeulen, J., and Lefebvre, V. (2013). "Overview of piano key weirs experience developed at EDF during the past few years." *Labyrinth and piano key weirs II*, CRC, Boca Raton, FL, 213–226.

Lempérière, F., and Ouamane, A. (2003). "The piano keys weir: a new cost-effective solution for spillways." *International Journal on Hydropower & Dams*, 10(5), 144-149.

Machiels, O. (2012). "Experimental study of the hydraulic behavior of piano key weirs." *PhD Dissertation*, Faculty of Applied Science, University of Liège.

Machiels, O., Erpicum, S., Archambeau, P., Dewals, B.J., and Pirotton, M. (2011). "Piano Key weir preliminary design method - Application to a new dam project." *Proc. Int. conf. Labyrinth and Piano Key Weirs*, Liege, Belgium. 199-206.

Machiels, O., Erpicum, S., Dewals, B.J., Archambeau, P., and Pirotton, M. (2014). "Large scale experimental study of piano key weir." *Journal of Hydraulic Research*, 50(2), 101-108.

Mehboudi, A., Attari, J., and Hosseini, A. (2017). "Flow regimes over trapezoidal piano key weirs." *Labyrinth and piano key weirs III: Proceedings of the 3rd International Workshop on Labyrinth and Piano Key Weirs (PKW 2017)*, February 22-24, 2017, Qui Nhon, Vietnam, CRC Press.

Mehboudi, A., Attari, J., and Hosseini, S.A. (2016). "Experimental study of discharge coefficient for trapezoidal piano key weirs. Flow Measurement and Instrumentation, 50, 65-72.

Mehboudi, A., Attari, J., Seyed Abbas Hosseini, S.A., and Jalili Ghazizadeh, M.R. (2014). "Experimental study comparing piano key weir with artillery and linear Weirs." *13th Iranian Hydraulic Conference November 2014*, University of Tabriz, Iran.

Noseda, M., Stojnic, I., Pfister, M., and Schleiss, A. J. (2019). "Upstream Erosion and Sediment passage at piano key weirs." *Journal of Hydraulic Engineering*, 145(8), 04019029.

Novak, P., and Cabelka, J. (1981). "Models in hydraulic engineering: Physical principles and design applications." (Vol. 4). Pitman Publishing. *Monographs & surveys in water resources engineering*.

Oertel, M. (2018). "Piano key weir research: State-of-the-art and future challenges." *International Symposium on Hydraulic Structures -ISHS2018*, Aachen, Germany.

Ouamane, A., and Lempérière, F. (2006, June). "Design of a new economic shape of weir." *Proceedings of the International symposium on dams in the societies of the 21st Century*, 18, 463-470.

Pagliara, S., Hager, W.H., and Minor, H.E. (2006). "Hydraulics of plane plunge pool scour." *Journal of Hydraulic Engineering*, 10.1061/(ASCE)0733-9429 (2006)132:5(450), 450–461.

Phillips, M.A., and Lesleighter, E.J. (2013). "Piano key weir spillway, Upgrade option for a major dam." *Labyrinth and piano key weirs II* ,CRC/Balkema, Leiden, Netherlands, 159–168.

Yazdi, A.R., Hoseini, S.A., Nazari, S., and Amanian, N. (2020). "Comparison of downstream Scour of the rectangular and trapezoidal piano key weirs." *Iranian Hydraulic Association Journal of Hydraulics*, 15(2), 95-102.

DOI: https://doi.org/10.26077/9f7b-59e2

Near-full-scale physical modelling and open-channel flow velocity in a fish-friendly culvert with full-height sidewall baffles

J. Hu[1], Y. Li[1] & H. Chanson[1]
[1]*The University of Queensland, School of Civil Engineering, Brisbane, Australia*
E-mail: youkai.li@uq.edu.au

Abstract: *A near-full-scale physical modelling was performed on the oscillation and instability of open-channel flow in a fish-friendly culvert equipped with full-height sidewall baffles. High-resolution measurements of the instantaneous flow velocity were conducted using a Vectrino+ acoustic Doppler velocimeter (ADV). The physical results were marked by the existence of some low-frequency seiche phenomenon. A well-established triple decomposition technique was applied to the time series of free-surface and velocity time-series. The low-pass components confirmed a unique flow structure, consisting of a high-velocity zone in the main channel and a low-velocity flow reversal within the lateral cavities. The band-pass components corresponded to the low-frequency flow oscillations, highlighting the complicated transverse interactions between the lateral cavity and the main channel. The high-pass velocity components were related to the 'true' turbulence characteristics. The current study provides some further insights into the sustainable design of culverts to assist with upstream fish migration in man-made and natural fast waterways.*

Keywords: *Physical modelling; Sidewall baffle; Fish-friendly culvert; Hydrodynamics; Triple decomposition.*

1. Introduction

In eastern Australia, road crossings constitute a major hinderance for the upstream passage of small-body fish and juveniles of large fish, according to the Queensland and New South Wales Governments. These small individuals, less than 100 mm total length, have very limited swimming performance, typically with characteristic swimming speeds less than 0.3 m/s (Chanson and Leng 2021). Despite diverging opinions, a consensus suggests that the high-water velocities impact most adversely on the upstream passage of small-body weak-swimming fish (Pavlov et al. 1994, Chanson and Leng 2021). The installation of full-height sidewall baffles in standard box culverts is one of the proposed practical solutions to assist upstream passage for small-body fish species (Fig. 1). At low flow rates, the lateral baffles create a reduced flow velocity in the main flow as well as deliver rest areas in the lateral cavities, all of which assist upstream fish traversability (Duguay and Lacey 2014). Small-body fish and juveniles of large fish were observed to use the recirculation and stagnation zones, respectively behind and in front of sidewall baffles, to progress upstream and rest (Marsden 2015, Cabonce et al. 2018 and 2019). However, at larger flows, the full-height sidewall baffles were observed to drastically reduce the culvert conveyance and to cause hydrodynamic instabilities (Leng and Chanson 2020a).

The presence of lateral cavities in open channels leads to the formation of strong vortices in the separated shear zones, resulting in a range of water quality and ecological effects (Valentine and Wood 1979, Hill 2014). The hydrodynamic instabilities may be highly coherent when some coupling occurs between the separation layer and the natural wave modes in the channel and lateral cavity (Tuna et al. 2013, Perrot-Minot et al. 2020).Despite a wealth of research on subcritical flows past lateral cavities (Ohmoto et al. 2005, Yossef and De Vriend 2011), the study of flows past successive lateral cavities, including the interactions between cavities, has attracted a lesser interest, except a few studies (Meile et al. 2011, Tuna and Rockwell 2014, Leng and Chanson 2020b). Herein, new laboratory experiments were performed at the University of Queensland, in a relatively large and near-full-scale fish-friendly culvert model. The focus of the measurements was the free-surface instabilities and seiching, as well as the coupling between free-surface and velocity fluctuations.

2. Experimental setup and methods

2.1. Facility, instrumentation and flow conditions

The present laboratory experiments were conducted in a relatively-large-size facility acting at a near-full-scale box culvert barrel. The extrapolation of the results to larger prototype structures would be based upon a combined Froude and Morton similitude, with a geometric scale from 1:1 to 1:4 corresponding to typical precast concrete boxes between 0.5 m and 2 m in width for culvert construction. The physical testing was performed in a 15 m long 0.5 m wide (B = 0.50 m) horizontal flume (Fig. 2). The invert of the flume was made of smooth PVC and the sidewalls were made of transparent tempered glass. The discharge was fed to the flume by a 2.0 m long 1.25 m wide intake basin, equipped with baffles, flow straighteners and three-dimensional convergent leading to the 0.5 m wide channel. The water was supplied by a constant head tank. The flume ended with a free overfall at the downstream end.

Figure 1. Multicell box culvert equipped with sidewall baffles in Flagstone QLD (Australia) on 2 October 2021 - Details of culvert inlet and barrel start, with sidewall baffles (baffle width: h_b = 0.150 m, cavity length: L_b = 0.60 m) installed to assist with upstream fish passage

Figure 2. Experimental facility and instrumentation at the University of Queensland - Flow conditions: Q = 0.055 m^3/s, h_b =0.167 m, L_b = 0.67 m, flow direction from bottom right to top left as indicated by the blue arrow.

The channel was equipped with relatively large sidewall baffles (h_b = 0.167 m) placed on the right sidewall, with h_b the baffle size in the direction normal to the sidewall. The baffles were plain rectangular, and the longitudinal spacing interval was L_b = 0.67 m and 0.33 m. A constant baffle spacing was used for a given set of experiments. The baffles were fixed to the false floor and held at the top, and clay was used to seal the gaps between the baffles and sidewall.

The water discharge was measured with a Venturi meter placed on the pipeline of the water reticulation system. The meter was designed and built based upon British Standard (1943). The error of the discharge was less than 2%. The centreline water elevation was measured with a pointer gauge, with an accuracy of ±0.5 mm. Unsteady free-surface recordings were performed with two independent techniques: backlighting through the glass sidewalls and ultrasonic sensors mounted above the channel. In the former, two high-speed video cameras: Casio™ EX-10 at 120 fps and Sony™ RX100V at 100 fps were utilised. In the latter, a set of nine Microsonic™ Mic+25 ultrasonic acoustic displacement meters (ADMs) were sampled at 200 Hz.

The velocities were measured with a Nortek™ Vectrino+ acoustic Doppler velocimeter (ADV), equipped with a side-looking head. The ADV signal was sampled at 200 Hz for 180 s at each location. All ADV signal data were post-processed to remove erroneous data and spikes. Data with an average correlation of less than 60% and an average SNR less than 5 dB were removed, and the signal was "despiked" using a phase-space thresholding technique (Goring and Nikora 2002, Wahl 2003, Mori et al. 2007, Chanson et al. 2008).

2.2. Triple decomposition of free-surface and velocity data

Both water elevation and velocity data showed the presence of some seiching in the form of low-frequency oscillations with periods about 1 s to 3 s, depending upon the discharge and boundary conditions. A typical illustration is presented in Figure 3. A triple decomposition of the instantaneous water elevation and velocity component data was applied following earlier works in riverine systems (Hussian and Reynolds 1972, Fox et al. 2005, Brown and Chanson 2013). That is, the instantaneous time-series may be represented as a superposition of three components:

$$V = < V > + [V] + \{V\} \tag{1}$$

where V is the instantaneous measurement, <V> is the low-pass filtered contribution, [V] is the slow fluctuating component, or band-pass filtered, and {V} is the high-pass filtered component, corresponding to 'true' turbulence. A detailed sensitivity analysis was undertaken in terms of the selection of physically meaningful cutoff frequencies, f_1 and f_2. Herein, f_1 is the upper cutoff frequency of the low-pass filtered component, and f_2 is the lower cutoff frequency of the high-pass filtered signal. The results of the sensitivity analysis showed that the mean contribution <V> was little affected by a cut-off frequency $f_1 < 1/3 \times f_d$, with f_d the dominant frequency (Fig. 3), while the 'true' turbulent component {V} was nearly independent of an upper cut-off frequency $f_2 > 3 \times f_d$. Thus, the triple decomposition was applied to both water elevations and velocity components using: $f_1 = 1/3 \times f_d$ and $f_2 = 3 \times f_d$. Note that the ADV signal data were post-processed to remove erroneous data and spikes, before applying the triple decomposition analysis.

(A)

(B)

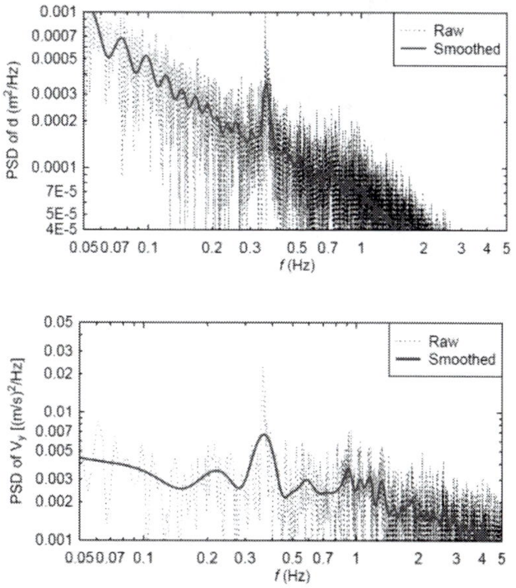

Figure 3. Fast Fourier transform (FFT) analyses of water elevation and transverse velocity signals: (A) Water elevation measured on the channel centreline at half-distance between two baffles $x-x_b = 0.335$ m, $y = 0.25$ m; (B) Transverse velocity component at $x-x_b = 0.335$ m, $y = 0.25$ m, $z = 0.0078$ m. – Herein, x is the longitudinal coordinate starting from the upstream end of the flume, y is the transverse coordinate starting from right sidewall, z is the vertical coordinate starting from the PVC false floor, $x_b = 8.20$ m indicates the longitudinal location of the upstream baffle of the test cavity.

3. Basic flow patterns

Visual, photographic and video observations were undertaken to record the basic open channel flow patterns for 0.029 $m^3/s < Q < 0.0925$ m^3/s with $L_b = 0.67$ m and for 0.029 $m^3/s < Q < 0.056$ m^3/s with $L_b = 0.33$ m. The observations were conducted about 6.2 m $< x < 9$ m, with x the longitudinal distance from the upstream end of the glass-walled channel.

The open channel flow passed beside the lateral cavities sequentially and presented a strong turbulent motion (Fig. 2). The main stream propagated downstream, while some recirculation was seen in the cavity regions between adjacent sidewall baffles. Some water stagnation occurred on the upstream side of the baffles, locally creating an increased free-surface elevation at the upstream corner. Separation took place at the outer edge of each baffle, inducing cross-waves propagating downstream. Overall, the free-surface was very rough and unstable (Fig. 2). Large-scale turbulent structures and surface scars were seen through the whole channel. Dye injection revealed some complicated recirculation motion behind the baffles for the full cavity length between baffles. These were investigated more specifically using neutral-buoyancy particles (Li and Chanson 2020).

Free-surface waves and instabilities were observed for the full length of the channel, both in the main stream and in the lateral cavities (Fig. 4). The water depth constantly fluctuated with time, and relatively large wave heights were observed in both the mainstream and the cavities between baffles. The unstable water surface created some considerable amount of individual air bubble entrainment, especially at larger discharges. Based upon the non-intrusive instantaneous water recordings, the dominant frequency f_d of water surface oscillations was found to be identical in the main stream and lateral cavity. Typical results are reported in Table 1. For three flow conditions, the dominant frequency results were found to be identical for both water elevation and instantaneous velocity component data.

Furthermore, the mean wave height and maximum wave height were also documented, based on the ADM sampling over the centreline at the middle longitudinal location of the test cavity for 0.029 $m^3/s < Q < 0.0925$ m^3/s with L_b = 0.67 m. For increasing flow rates during the current experiments, both the mean wave height and maximum wave height increased. For each flow rate, the maximum wave height was about twice the value of the mean wave height. The maximum wave heights were approximately 20% of the mean water depths. The results are reported in Figure 5.

Figure 4. Sidewall baffled channel in operation for Q = 0.0925 m^3/s, h_b = 0.167 m, L_b = 0.667 m, S_o = 0 - Sideview through the (smooth) left sidewall, with flow direction from right to left as indicated by the arrow (shutter speed: 1/80 s)

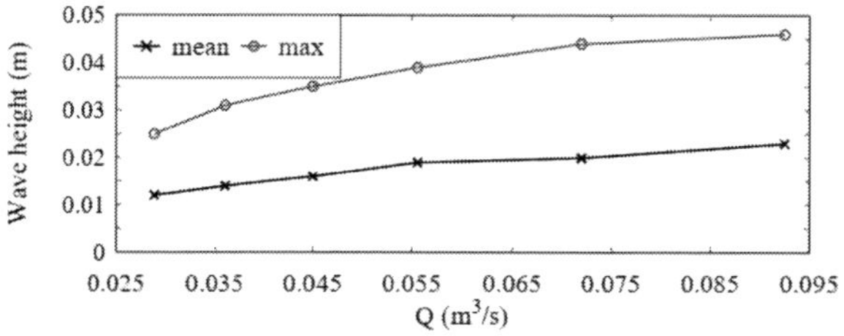

Figure 5. The average and maximum wave height from surface elevation measurements using ADM – Flow conditions: 0.029 $m^3/s < Q < 0.0925$ m^3/s, L_b = 0.67 m.

Table 1. Dominant frequencies f_d of instantaneous free-surface oscillations derived FFT analyses (for x_b = 8.20 m)

Q (m^3/s)	L_b (m)	h_b (m)	d_o (m)	V_o (m/s)	f_d (Hz) (¹)	Re	$f_d \times L_b/V_o$(¹)
0.0289	0.33	0.167	0.125	0.46	0.83	1.54×10^5	0.59
0.0289	0.67	0.167	0.132	0.44	0.36	1.51×10^5	0.55
0.0360	0.67	0.167	0.156	0.46	0.38	1.77×10^5	0.55
0.0450	0.67	0.167	0.180	0.50	0.42	2.09×10^5	0.56
0.0556	0.33	0.167	0.192	0.58	0.47	2.51×10^5	0.27
0.0556	0.67	0.167	0.204	0.54	0.44	2.44×10^5	0.54
0.0720	0.67	0.167	0.245	0.59	0.47	2.90×10^5	0.53
0.0925	0.67	0.167	0.286	0.65	0.50	3.45×10^5	0.52

Notes: d_o: characteristic water depth in the main stream; Re: Reynolds number defined in terms of the hydraulic diameter; V_o: cross-sectional-averaged bulk velocity; (¹) Dominant frequencies for instantaneous free-surface and/or velocity oscillations.

4. Velocity field

4.1. Mean velocity field

Velocity measurements were undertaken in the sidewall-baffled channel at three longitudinal locations and 7 or 10 transverse positions, with a minimum of 30 to 35 points per vertical profiles, leading to over 200 sampling points per cross-section at a specific longitudinal location, for a given flow condition. Herein, V_x is the longitudinal velocity component positive downstream, V_y is the transverse velocity component positive towards the left sidewall and V_z is the vertical velocity component positive upwards. Similarly, x is the longitudinal coordinate positive downstream with x = 0 at the upstream end of the glass-sidewalled channel, y is the transverse coordinate with y = 0 at the right sidewall and z is the vertical coordinate with z = 0 at the invert. The longitudinal location of upstream baffle of the test cavity is x_b.

The longitudinal velocity datasets presented a mean trend which was consistent with previous Prandtl-Pitot tube data obtained with a slightly different baffle configuration (h_b = 0.083 m, L_b = 0.333 m, Q = 0.054 m^3/s) (Leng and Chanson 2020b). The present data delivered a much greater level of details because of the high temporal resolution of the ADV system and its ability to measure instantaneous negative velocities. All the velocity data highlighted a major impact of the lateral baffles on the entire flow field, with some overall slower velocity and higher water depth, compared to a smooth (un-baffled) channel data. The presence of sidewall baffles modified the vertical profiles of longitudinal velocity component on the channel centreline (Fig. 6). While the centreline velocity data on a smooth (un-baffled) invert was characterised by a partially developed turbulent boundary layer flow with an ideal fluid flow region above (Leng and Chanson 2020b), the present centreline velocity distributions presented a velocity maximum at about $z/d \sim 0.45$, with a decreasing longitudinal velocity with increasing elevation for $0.45 < z/d < 1$, in the baffled channel, with d the water depth. That is, the data showed a marked velocity dip next to the free surface. Moreover, negative velocities were recorded in the wake of the lateral baffle, i.e. $0 < z << h_b$.

Typical velocity contour maps are shown in Figure 7. The longitudinal velocity component data showed three distinctive regions: (1) a well-defined high-velocity flow region for $h_b << y < B$, (2) a shear zone with large transverse velocity gradient $\partial Vx/\partial y$ downstream of the baffle's outer edge ($y \sim h_b$) and (3) a recirculation region with negative longitudinal velocities for $0 < y << h_b$ (Fig. 7A). The time-averaged negative velocities reached values as low as -0.15 m/s. The transverse velocities showed magnitudes smaller than the longitudinal velocity components, and their contour maps implied some strong secondary currents, especially in the shear zone (Fig. 7B).

The velocity data were checked for the conservation of mass. The water discharge Q was compared to the integration of the time-averaged longitudinal velocity data assuming a no-slip condition at the invert and on both sidewalls:

$$Q = \iint_A V_x \times dy \times dz \qquad (2)$$

with A the cross-section area of the flow. The results demonstrated the conservation of mass to be within less than 7% at all longitudinal locations. The time-averaged transverse velocity data were also integrated over the cross-section area to ensure the no-flow-through condition:

$$0 = \iint_A V_y \times dy \times dz \qquad (3)$$

The results showed a reasonable agreement. For completeness, a continuity check could not be conducted meaningfully in terms of the vertical velocity component, because of the large water elevation oscillations.

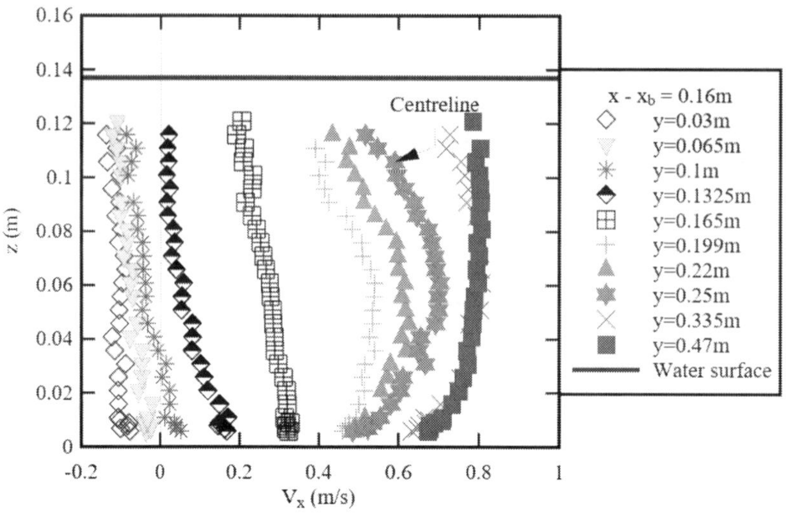

Figure 6. Vertical profiles of time-averaged longitudinal velocity for $Q = 0.029\ m^3/s$, $x_b = 8.2$ m, $h_b = 0.167$ m, $L_b = 0.667$ m, $(x-x_b)/L_b = 0.25$.

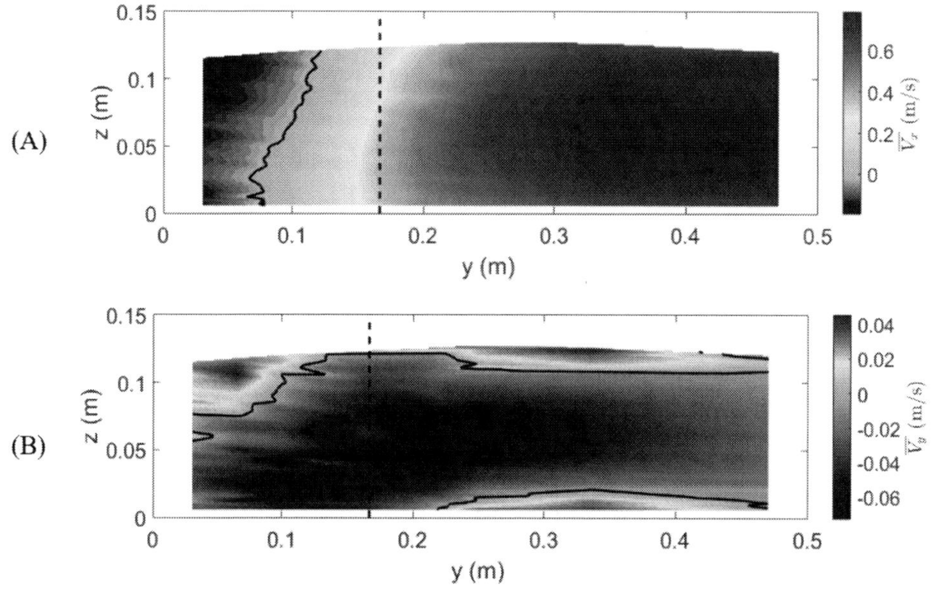

Figure 7. Contour maps of time-averaged longitudinal (A) and transverse (B) velocity components for $Q = 0.029\ m^3/s$, $x_b = 8.2$ m, $h_b = 0.167$ m, $L_b = 0.667$ m, $(x-x_b)/L_b = 0.5$ - The dashed black vertical line is the outer edge of the baffle; the solid black line corresponds to $\bar{V} = 0$.

4.2. Shear zone and recirculation

The flow between two baffles might be idealised as a two-dimensional supported jet, with a jet discharging from a nozzle of breadth $(B - h_b)$ and expanding into a confined space with a total width B. Neglecting energy and momentum loss, the equations of conservation of mass and momentum imply between two baffles:

$$Q = \int_{y=0}^{B} \int_{z=0}^{d} V_x \times dz \times dy = \text{cons tan t} \tag{4a}$$

$$M = \int_{y=0}^{B} \int_{z=0}^{d} \rho \times V_x^2 \times dz \times dy = \text{cons tan t} \quad (4b)$$

where Q is the water discharge and M the momentum flux. Some theoretical models of the flow in recirculation cavities were developed (Kimura and Hosoda 1997, Mizumura and Yamasaka 2020). Hill (2014) proposed a simplistic theoretical model of the recirculation velocity profile, associated with gyres and dead zones in river channels with lateral cavity. Assuming that the gyre diameter equals the baffle size h_b, and that the dimensionless interfacial shear along the cavity is approximately the dimensionless total drag, the predicted velocity distribution in the lateral cavity is presented in Fig. 8. Figure 8 compares further the theoretical model with the measured depth-averaged velocity. Although the gyre model was not developed for long lateral cavities, it showed a relatively close agreement with the data (Fig. 8).

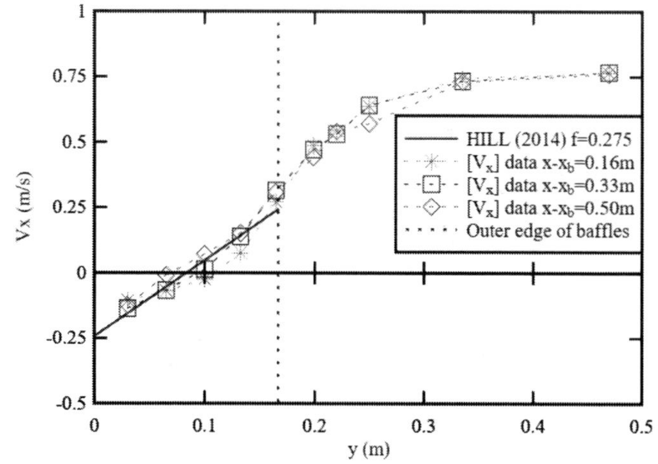

Figure 8. Transverse distribution of depth-averaged longitudinal velocity V_x in channel with sidewall baffles - Comparison with the theoretical model of cavity recirculation by Hill (2014) assuming f = 0.275 - The dashed black vertical line marks the outer edge of the sidewall baffle. – Flow conditions: Q = 0.029 m³/s, x_b = 8.2 m, h_b = 0.167 m, L_b = 0.667 m

4.3. Velocity fluctuations

The flow was very turbulent in the sidewall-baffled channel. Large velocity fluctuations were observed in all three directions, with the largest fluctuations typically in the shear zone located in the wake of the outer edge of the baffle. Large turbulent intensities were recorded with V_x'/V_o up to 70%, V_y'/V_o up to 30% and V_z'/V_o up to 100%.

For each velocity measurement experiment, the three velocity components were recorded at three longitudinal cross-sections, and the triple-decomposition of the signals was applied to determine the contribution of the slow water surface oscillations on the turbulent fluctuations. For all the experiments, [V]' presented higher values on the right (baffle-side) half of channel. This was directly caused by the impact of the baffles. In contrast, the high velocity zone presented smaller [V]' values. Basically, the slow oscillations contributed more to turbulent fluctuations in the cavity and shear regions than in the main flow region. Typical data of velocity fluctuations [V]' induced by slow oscillations are shown in Figure 9. A dominant feature was that all data {V}' presented a higher value region in the shear zone (Fig. 10). There, i.e. in the wake of the outer edge of baffle, the transverse velocity data $\{V_y\}'$ always presented some significantly increased values in the shear zone because of the secondary currents associated with transverse momentum transfer (Fig. 10B). The transverse velocity data $\{V_z\}'$ showed high values in the near vicinity of the-free surface, as a direct consequence of free-surface unsteadiness. Figure 10 shows some 'true' turbulence generated by high-frequency velocity fluctuations. Overall, the standard deviations of the high-pass velocity components were the largest for the longitudinal and vertical components, with $\{V_x\}'/V_x' \sim 62\%$ to 77% in average and $\{V_z\}'/V_z' \sim 83\%$ to 90% in average. For the transverse velocity component, the contribution from band-pass and high-pass components were comparable: i.e., $[V_y]'/V_y' \sim \{V_y\}'/V_y'$. For comparison, the band-pass filtered fluctuations of the water elevation were the largest component with $[d]'/d' \sim 85\%$ in average.

(A)

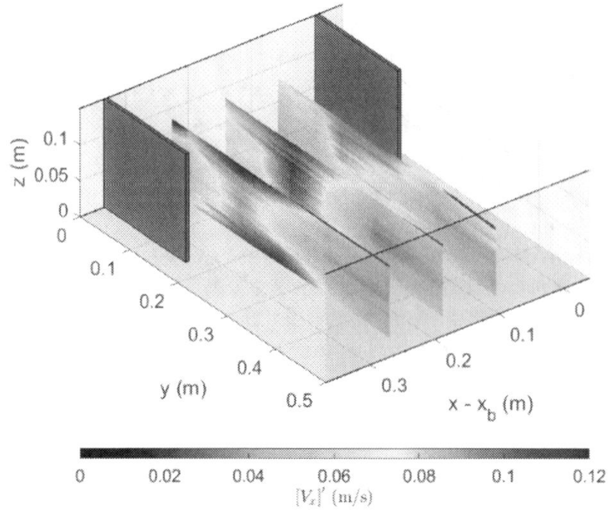

(B)

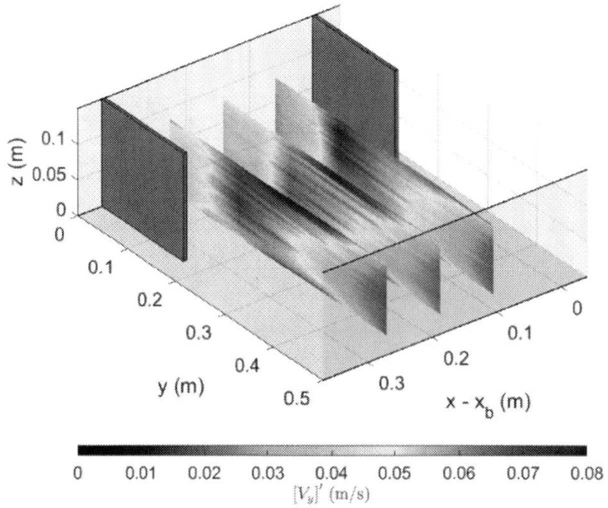

Figure 9. Standard deviation of band-pass filtered (0.3 -3.0 Hz) longitudinal (A)and transverse (B) velocity components for Q = 0.029 m^3/s, $x_b = 8.2$ m, $h_b = 0.167$ m, $L_b = 0.333$ m, $d_o = 0.125$ m - Flow direction from top right to bottom left

(A)

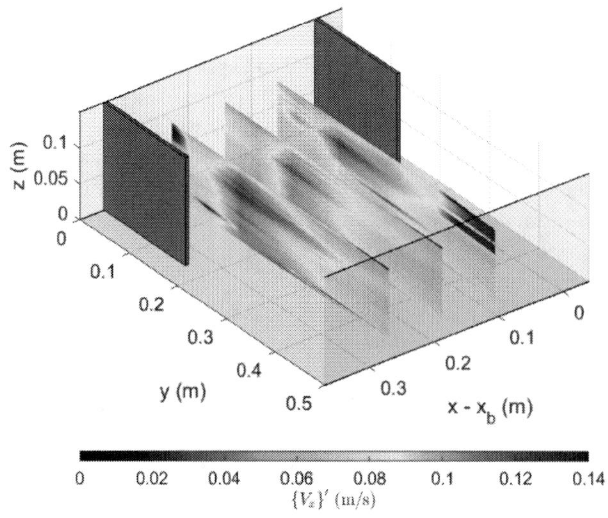

(B)

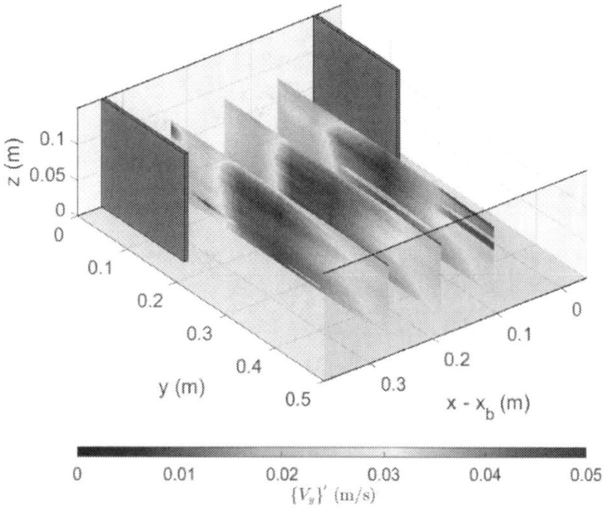

Figure 10. Standard deviation of high-pass filtered (3 -100 Hz) longitudinal (A)and transverse (B) velocity components for Q = 0.029 m³/s, x_b = 8.2 m, h_b = 0.167 m, L_b = 0.333 m, d_o =0.125 m - Flow direction from top right to bottom left

5. Conclusion

Detailed instantaneous measurements of water surface elevation and velocity were recorded in a near-full-scale box culvert barrel equipped with sidewall baffles on one side. While the design was developed to assist upstream passage of small fish at low flows, the sidewall baffles induced strong hydrodynamic instabilities at medium to large discharges. These instabilities include relatively slow oscillations of the water surface in the entire channel, with maximum amplitude up to 20% of the average water depth, as well as oscillations of all three velocity components.

The mean velocity field showed a high velocity zone in the main stream besides the baffles, a recirculation zone behind the baffles with marked negative velocities, and a shear region in between characterised by intense secondary motion. The water depth and velocity signals were analysed using a triple decomposition approach, to quantify the impact of the slow oscillation on the velocity fluctuations. The slow oscillations were the major source of free-surface turbulence and turbulent velocity in the transverse direction. For the longitudinal and vertical velocity component, the high-frequency fluctuations were dominant.

In practical design applications, the impact of baffles encompasses an increased fish traversability at low flows and a greater total cost linked to the installation costs of baffles, the increase in the number of barrel cells to achieve the same design discharge and afflux, and operational maintenance. In practice, sediment and debris trapped by the baffles must be mitigated since these may drastically reduce the conveyance of flow and impede upstream fish passage. Besides, based on the communication with Queensland Fire and Emergency Services, the baffles, typically manufactured in steel, may also create a potential hazard to swift water rescue and emergency personnel during rescue operations. All the above problems need to be considered for both further research and prototype designs.

6. Acknowledgements

The authors thank Jason Van Der Gevel and Stewart Matthews (The University of Queensland) for their technical assistance.

7. References

British Standard (1943). "Flow Measurement." *British Standard Code BS 1042:1943*, British Standard Institution, London.

Brown, R, and Chanson, H. (2013). "Turbulence and Suspended Sediment Measurements in an Urban Environment during the Brisbane River Flood of January 2011." *Journal of Hydraulic Engineering*, Vol. 139, No. 2, pp. 244-252.

Cabonce, J., Wang, H., and Chanson, H. (2018). "Ventilated Corner Baffles to Assist Upstream Passage of Small - Bodied Fish in Box Culverts." *Journal of Irrigation and Drainage Engineering, ASCE,* Vol. 144, No. 8, Paper 0418020, 8 pages.

Chanson, H., and Leng, X. (2021). "Fish Swimming in Turbulent Waters. Hydraulics Guidelines to assist Upstream Fish Passage in Box Culverts." CRC Press, Leiden, The Netherlands, 202 pages and 19 video movies.

Chanson, H., Trevethan, M., and Aoki, S. (2008). "Acoustic Doppler Velocimetry (ADV) in Small Estuary: Field Experience and Signal Post-Processing." *Flow Measurement and Instrumentation*, Vol. 19, No. 5, pp. 307-313.

Duguay, J., and Lacey, R.W.J. (2014). "Effect of Fish Baffles on the Hydraulic Roughness of Slip-Lined Culverts." *Journal of Hydraulic Engineering*, ASCE, Vol. 141, No. 1, Paper 04014065, 10 pages.

Fox, J.F., Papanicolaou, A. N., and Kios, L. (2005). "Eddy taxonomy methodology around submerged barb obstacle within a fixed rough bed." *Journal of Engineering Mechanics*, ASCE, Vol. 131, No. 10, pp. 1082-1101.

Goring, D.G., and Nikora, V.I. (2002). "Despiking Acoustic Doppler Velocimeter Data." *Journal of Hydraulic Engineering*, ASCE, Vol. 128, No. 1, pp. 117-126.

Hill, D.F. (2014). "Simple Model for the Recirculation Velocity of Open-Channel Embayments." *Journal of Hydraulic Engineering*, ASCE, Vol. 140, No. 4, Paper 06014004, 5 pages.

Hussian, A.K.M.F., and Reynolds, W.C. (1972). "The Mechanics of an Organized Wave in Turbulent Shear Flow. Part2: Experimental Results." *Journal of Fluid Mechanics*, Vol. 54, Part 2, pp. 241-261.

Kimura, I., and Hosoda, T. (1997). "Fundamental Properties of Flows in Open Channels with Dead Zone." *Journal of Hydraulic Engineering*, ASCE, Vol. 123, No. 2, pp. 98-107.

Leng, X., and Chanson, H. (2020a). "How Full-Height Sidewall Baffles affect Box Culvert Capacity: Balancing Fish Passage and Discharge Requirements." *Australasian Journal of Water Resources,* Vol. 24, No. 2, pp. 248-256.

Leng, X., and Chanson, H. (2020b). "Asymmetrical Wall Baffles to Assist Upstream Fish Passage in Box Culvert: Physical Modeling." *Journal of Irrigation and Drainage Engineering*, Vol. 146, No. 12, Paper 04020037, 10 pages.

Li, Y., and Chanson, H. (2020). "Hydrodynamic Instabilities in Open-Channel Flow past Lateral Cavities." *Proceedings of 22nd Australasian Fluid Mechanics Conference AFMC2020*, Brisbane, Australia, 7-10 December.

Marsden, T. (2015). "Common Rail Proof of Concept and Baffle Field Trial Assessment Report." *Report to OceanWatch Australia*, Australasian Fish Passage Services, 26 pages.

Meile, T., Boillat, J.L., and Schleiss, A.J. (2011). "Water-surface oscillations in channels with axi-symmetric cavities." *Journal of Hydraulic Research*, IAHR, Vol. 49, No. 1, pp. 73-81.

Mizumura, K., and Yamasaka, M. (2002). "Flow in Open-Channel Embayment." *Journal of Hydraulic Engineering*, ASCE, Vol. 128, No. 12, pp. 1098-1101.

Mori, N., Suzuki, T., and Kakuno, S. (2007). "Noise of acoustic Doppler velocimeter data in bubbly flows." *Journal of Engineering Mechanics*, ASCE, Vol. 133, No. 1, pp. 122-125.

Ohmoto, T., Hirakawa, R., and Watanabe, K. (2005). "Interaction between Water Surface Oscillations and Large Eddies in an Open Channel with Spur dikes." *Proc. 31th Biennial IAHR Congress*, Seoul, Korea.

Pavlov, D.S., LupandinN, A.I., and Skorobogatov, M.A. (1994). "Influence of Flow Turbulence on Critical Flow Velocity for Gudgeon (*Gobio gobio*)." *Doklady Biological Sciences*, Vol. 336, pp. 215-217.

Perrot-Minot, C., Mignot, E., Perkins, R., Lopez, D., and Riviere, N. (2020). "Vortex shedding frequency in open-channel lateral cavity." *Journal of Fluid Mechanics*, Vol. 892, A25, 11 pages.

Tuna, B.A., Tinar, E., and Rockwell, D. (2013). "Shallow flow past a cavity: globally coupled oscillations as a function of depth." *Experiments in Fluids*, Vol. 54, Paper 1586, 20 pages.

Tuna, B.A., and Rockwell, D. (2014). "Self-sustained oscillations of shallow flow past sequential cavities." *Journal of Fluid Mechanics*, Vol. 758, pp. 655-685.

Valentine, E.M., and Wood, I.R. (1979). "Experiments in Longitudinal Dispersion with Dead Zones." *Journal of Hydraulic Division*, ASCE, Vol. 105, No. HY9, pp. 999-1016.

Wahl, T.L. (2003). "Despiking Acoustic Doppler Velocimeter Data. Discussion." *Journal of Hydraulic Engineering*, ASCE, Vol. 129, No. 6, pp. 484-487.

Yossef, M.F.M., and De Vriend, H.J. (2011). "Flow Details near River Groynes: Experimental Investigation." *Journal of Hydraulic Engineering*, ASCE, Vol. 137, No. 5, pp. 504-517.

DOI: https://doi.org/10.26077/b079-8746

Numerical Modeling for Optimization of the Aspect Ratio of Submerged Vanes for the Purpose of Sediment Deflection in Rivers

V. Chauhan[1], R. Chavan[2]&G.D. Singhal[1]
[1]Shiv Nadar University, Greater Noida, India
[2]Maulana Azad National Institute of Technology, Bhopal, India
E-mail: vc651@snu.edu.in

Abstract: *In order to create a secondary circulation (spiral flow) in the downstream direction to their trailing edge, submerged vanes are tiny aspect ratio (H/L) devices that are positioned vertically on riverbeds. Since this secondary circulation is driven by the vane-induced tip vortex, submerged vanes are best recognized for changing the alluvial river-bed profile. Finding the right aspect ratio as a function of the angle of attack with the approach flow is one of the current issues with using the submerged vane technique. In this context, the current study aims to investigate the best vane length to be used, based on a specific angle of attack with the flow. A pair of rectangular vanes with a set height is among the studied vanes. Only 15^0 and 40^0 angles will be examined in the current study because these are the ideal minimum and maximum recommendations made by earlier studies. The vanes' length to height ratios that were taken into consideration for this investigation were 2, 3, and 4. The purpose of the study is to determine how these variables affect the strength of the secondary circulation and sediment deflection downstream of the vanes. The flow field around and past the submerged vane is investigated using the Flow-3D Hydro numerical model. The most ideal aspect ratios for a low Froude number of 0.168 are 0.25 and 0.33 at 40 and 15 degrees of attack, respectively. The most ideal aspect ratios with a high Froude number of 0.303 are 0.33 and 0.25 at 40 and 15 degrees of attack, respectively.*

Keywords: *inland navigation, morphology, scouring, secondary circulation, sediment deflection, vortex.*

1. Introduction

Small aspect ratio devices called submerged vanes are frequently utilized for river training (Odgaard and Wang 1991b). Sediment deflection is a crucial component of river training due to the increasing amount of sediment entering river systems. To deflect, redirect, and manage the river sediment load, submerged vanes can be used to impart transverse momentum to the approach flow. The vane-induced tip vortex is what causes the transverse momentum. Determining the strength of the tip vortex is therefore essential to assessing the effectiveness of submerged vanes for sediment deflection. The strength of the tip vortex produced by vanes has been measured in existing research using the "moment of momentum" metric (Marelius and Sinha 1998; Gupta et al. 2006; Gupta et al. 2007).

A higher thickness (more than 1mm) is frequently required for a cheaper material to manufacture vanes, but this result in considerably less lift developed per unit area (due to the weir effect), hence a greater length and consequently smaller aspect ratio *(H/L)* is needed for such materials (Odgaard 2009). Since longer vanes produce two or more secondary flow cells instead of just one, they are less efficient (Odgaard and Mosconi 1987). The height of the vane is a crucial design factor for sediment deflection. The length of the vane to be used should be determined by the vane's cost and anticipated effectiveness in managing sediment. Ouyang (2009) suggested a design table to find the ideal vane height at design length of submerged vanes for clear water conditions.

The stress field of the submerged vane extends stream-wise by almost three times the vane height. According to Ouyang (2009), who used their numerical approach, "the difference in performance between a vane height of 0.4 times flow depth and a vane height of 0.65 times flow depth is minimal when either the sediment Froude number or angle of attack is low." These are the desirable effects of vane design since the angle of attack is frequently unstable and subject to fluctuate depending on the design flow and because the vane-induced transverse bed-shear stress tends to carry more bed materials with a higher sediment Froude number. Navigation may also place restrictions on the height of vanes. The height of submerged vanes is therefore kept at 0.4 times the local water depth as an useful rule of thumb. Tan et al. (2005) discovered that for mobile bed conditions, an optimal vane height of two to three times the height of the bed-form can stop the bed-load from "escaping" over the bed-form crest. Additionally, a vane higher than this would severely restrict the flow and hinder sediment mobility; conversely, a vane lower than this would allow bed-load particles to climb over it, decreasing the effectiveness of the vane's sediment deflection. The highest vortex strength was reported by Gupta et al. (2007) to occur at an aspect ratio of 0.25. They carried out their studies using vanes made of superior material with a thickness of less than 1 cm (Chauhan et al. 2022). To determine the proper aspect ratios for thicker vanes, the current study will use a vane thickness of 5 cm.

The methodologies involved in analytical research of submerged vanes mainly differ with respect to the method used to quantify the vane induced transverse velocity. Odgaard and Spoljaric (1986) used "wing theory" (potential flow method), Marelius and Sinha (1998) used "Moment of Momentum", Sinha and Marelius (2000) used $k - \epsilon$ turbulence model, Flokstra (2006) used the mathematical model of Odgaard and Wang (1991a) to update morphological modelling system Delft3D-MOR (Struiksma et al. 1985) in order to simulate vane-induced bed level variations and later Ouyang (2009) used "Panel method of aerodynamics". Gravity causes pressure in an irrotational flow. Pressure develops velocity gradients which in presence of viscosity generates varying shear stress in fluids. Adding these viscous effects to the pressure is always a challenging problem in river mechanics, particularly for the flow around submerged streamlined objects such as submerged vanes. Therefore "Potential flow theory" used by Odgaard and Spoljaric (1986) seems to underperform in comparison to the "Panel Method" in aerodynamics (Chauhan et al. 2022).

Since normal stresses were not adequately represented, Sinha and Marelius' (2000) numerical investigation to solve the three-dimensional Reynold's-averaged Navier-Stokes equations with the turbulence closure as "standard $k - \epsilon$ model" was deemed to have produced unacceptable results. According to Patel and Yoon (1995), the influence of surface roughness can be taken into consideration using the *k-w* turbulence model. To simulate the flow past and around the submerged vanes, the viscous stress tensor in the Navier-Stokes equation can be derived using the k-w turbulence model. The range of possibilities for optimizing the performance of submerged vanes has increased as a result of recent developments in turbulence monitoring and simulation. Previous attempts (Odgaard 1991a; Sinha and Marelius 2000; Flokstra 2006) to create a model for submerged vane performance in sediment deflection did not perform accurately because the potential flow approach served as the theoretical underpinning. The problem has been how to apply the viscous contribution to the pressure in an irrotational flow. The model developed by (Ouyang et al. 2008; Ouyang 2009) is the only one that has been successful in creating a numerical model for submerged vane performance because it more effectively incorporates the effects of bound vortex and combines the viscous contributions to pressure using the Biot-Savart law (Chauhan et al. 2022). Moreover, it is usually advantageous to employ the "moment of momentum" method if the investigation is experimental and focuses on optimizing the vane parameters rather than characterizing them. It is because it is a precise gauge of vortex intensity and is readily quantifiable using turbulence measurements.

Higher angles of attack with the approach flow, such as 40 degrees, have been observed to increase the strength of the tip vortex in the literature (Marelius and Sinha 1998). On the other hand, for angles greater than 15 degrees, the local scouring near submerged vanes also amplifies. Increased local scouring around vanes necessitates deeper installation depths, which is more expensive. Therefore, with regard to approach flow, these two angles of attack—15^0 and 40^0—are the best options for installing vanes. There should be a specific advice to choose the right aspect ratio at these angles of attack because the flow fields obtained with the aforementioned vanes' angles of attack are very different. A recommendation of this kind is being sought to be established in the current investigation. A streamlined "moment of momentum" criterion is used to evaluate the vortex intensity for sediment deflection. The investigation will use two different downstream cross-sections of vanes to measure the *2D* velocity magnitude of the *y-z* plane. To shed light on the occurrence of local scouring around vanes, the flow field around the vanes will also be examined.

2. Numerical Method

2.1 Governing equations of numerical model

Turbulence in water flow is both unpredictable yet coherent at the same time (Nezu et al. 1993). The experimental measurement of turbulence quantities requires a lot of time and effort. As a result, turbulence models were created in order to associate the turbulence quantities with the flow properties. The Renormalized group (*RNG*) model was used in the current simulation investigation utilizing Flow-3*D* Hydro software. Based on kinetic energy k and its rate of dissipation ϵ, the eddy viscosity is calculated using the turbulence closure equations 1 and 2. Because it required less computing than other superior models like the k-ω and Large Eddy Simulation (*LES*) model, the adoption of the *RNG* model is preferred. Additionally, it has been noted that the *RNG* model is useful for incorporating the impacts of vortices and sediment transport. To mimic the changes in bed morphology, we use the Van Rijn equation (Van 1984) of bed load transfer rate.

$$\frac{\partial}{\partial t}(\rho k) + \frac{\partial}{\partial x_i}(\rho k u_i) = \frac{\partial}{\partial x_j}\left(\alpha_k \mu_{eff} \frac{\partial k}{\partial x_j}\right) + G_k + G_b - \rho\epsilon - Y_M + S_k \quad (1)$$

$$\frac{\partial}{\partial t}(\rho\epsilon) + \frac{\partial}{\partial x_i}(\rho\epsilon u_i) = \frac{\partial}{\partial x_j}\left(\alpha_\epsilon \mu_{eff} \frac{\partial \epsilon}{\partial x_j}\right) + C_{1\epsilon}\frac{\epsilon}{k}(G_k + C_{3\epsilon}G_b) - C_{2\epsilon}\rho\frac{\epsilon^2}{k} - R_\epsilon + S_\epsilon \quad (2)$$

Here, G_k = represents the generation of the turbulence kinetic energy due to the mean velocity gradients; G_b= represents the generation of the turbulence kinetic energy due to buoyancy; Y_M = represents the contribution of the fluctuating dilatation in compressible turbulence to the overall dissipation rate; $\alpha_k$and α_ϵ= inverse effective Prandtl numbers for k and ϵ respectively; S_k and S_ϵ = user-defined source terms.

2.2 Model setup

Two separate model setups have been created to simulate two different uniform flow depth settings at a fixed discharge of 0.021 m^3/s. These two model setups were chosen so that they might result in two cases of clear water research and mobile bed study, with Froude numbers of 0.168 and 0.303, respectively. Using Flow-3*D* Hydro's built-in geometry function, the study began by designing the geometries (Fig. 1) for a rectangular open channel bed and submerged vanes. The geometry of the channel bed, which was 0.5 m broad, 5 m long and 0.2 m deep, was classified as "packed sediment." Next, a pair of defined cubical volume submerged vanes was positioned at a set angle of attack with the approach flow. The two "General (Solid)" rectangular vanes in each model setup, with a lateral separation of 10 cm, were positioned in the centre of the created channel geometry. The cubical volume had dimensions of 5 cm in thickness, 0.4*d* in height (where d is the depth of the water), and lengths of 2 × 0.4*d*, 3 × 0.4*d* and 4 × 0.4*d*. Table 1 includes a list of the various model parameters employed in the current study. The submerged vanes model was also used in this work in a variety of geometrical configurations, which are listed in Table 2 along with the relevant designations, geometries, and flow parameters. The table displays 12 alternative submerged vanes configurations that required 12 numerical runs with various model geometries and flow conditions.

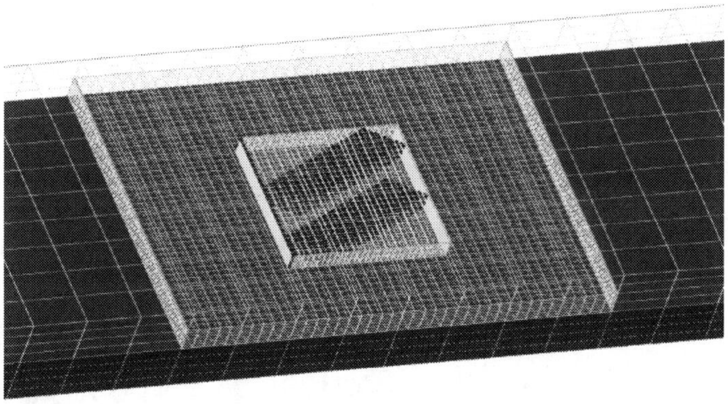

Figure 1. Model and meshing of submerged vanes generated by Flow-3*D* Hydro (*L=2H, α=40⁰ and d=18.5*)

The geometry was made, and then the meshing process was started. With each succeeding decrease in result variation, various mesh sizes were used. The appropriate mesh size is the one "at which the difference in acquired results was less than 5% from the previous coarser mesh size," according to the selection criteria. Based on this standard, the simulation study's mesh sizes were 0.05 m, 0.01 m, and 0.005 m, as shown in Fig.1. To capture the minute flow features at the sharp edges of the vanes, the smallest mesh size is used where the vanes are located. To preserve stability and make sure fluid fraction advection did not exceed computational cell volumes, the time-step size was automatically changed. For all of the taken into consideration geometrical configurations of the submerged vanes, it was discovered that with the 600 s simulation of the flow, the solution became fully converged and the steady state condition was achieved. Using the built-in "*FAVOUR*" feature of the Flow-3*D* Hydro software, the model is evaluated for usability after completion of the geometry and meshing. Then, based on their responsibilities in establishing the underlying physical flow phenomena, boundary conditions were set to each surface of the geometry. A uniform flow boundary condition is imposed for the inlet boundary condition. Initial flow depths at the inlet border are set to 12.5 m and 18.5 m at a discharge value of 0.021 m^3/s. A pressure boundary condition is imposed for the outflow boundary condition. As pressure boundary conditions, 12.5 m and 18.5 m of normal flow depth were used. The boundary condition "symmetry" is applied to the two x-z plane borders. While "atmospheric pressure" is the boundary condition for the highest elevation in the x-y plane, "symmetry" is the boundary condition for the lowest elevation. The volume of fluid (*VOF*) method is used to monitor the surface of the water.

Table 1. List of model parameters used in the simulation study

S. No.	Model parameters	Range
1	Width of the channel	0.5 m
2	Length of the channel	5 m
3	Depth of flow	12.5 cm and 18.5 cm
4	Distance between center of vane pair and inlet	2.5 m
5	Height of vane	5 cm and 7.4 cm
6	Thickness of vane	5 cm
7	Length of vane	For 5 cm vane height- 10 cm, 15 cm and 20 cm For 7.4 cm vane height- 14.8 cm, 22.2 cm and 29.6 cm.
8	Angle of attack of vane	15^0 and 40^0
9	Froude number	0.303 (for d = 12.5 cm) and 0.168 (for d = 18.5 cm)
10	Turbulence modeling used	*RNG* model
11	Sediment size (d_{50})	0.0003 m
12	Sediment model	Van Rijn model

Table 2. List of designations for different model configurations with their corresponding geometries and flow parameters used in the simulation study

Designation of (*d. α.L/H*)	Flow depth (d), in cm	Length of vanes (L), in cm	Height of vanes (H = $0.4d$), in cm	Aspect ratio (H/L)	Angle of attack (α), in degrees	Discharge, in m^3/s	Froude Number
A.1.2		10		0.5			
A.1.3		15		0.33	15		
A.1.4		20		0.25			
A.2.2	12.5	10	5	0.5			0.303
A.2.3		15		0.33	40		
A.2.4		20		0.25			
B.1.2		14.8		0.5		0.021	
B.1.3		22.2		0.33	15		

B.1.4		29.6		0.25			
B.2.2	18.5	14.8	7.4	0.5			0.168
B.2.3		22.2		0.33	40		
B.2.4		29.6		0.25			

2.3 Accuracy of numerical model

The tendencies of changes in transverse velocity in the z-direction are in good agreement with the study's experimental findings (Odgaard and Spoljaric 1986). It should be noted that the inclusion of the "weir effect" for the thicker vanes utilized in this study only caused the trends to be similar, although the transverse velocity magnitudes were found to be unequal. It is evident that the magnitude of the transverse velocity has significantly decreased. It is as a result of the model's geometrical features in the current simulation being different from those employed in the experimental study (Odgaard and Spoljaric 1986). Thus, the aspect ratio of submerged vanes was effectively optimized with the help of Flow-3*D* Hydro software.

3. Results and Discussion

The complex velocity patterns produced by the vortices surrounding the vanes help to define the flow field they cause. Without any characterization, the current study's sole goal is to estimate the vortex strength brought on by various vanes' model configurations. To measure and compare the vortex strength among the many configurations used in the current study, a streamlined criterion of "moment of momentum" is applied. Submerged vanes are thought to function better at deflecting sediments when their "moment of momentum" is higher. The magnitude of the velocity perpendicular to the centre of the vortex is precisely proportional to the physical attribute known as "moment of momentum." Therefore, it is hypothesized that the 2*D* velocity magnitudes in the y-z plane are proportional to the "moment of momentum" and are thus indicative of how well the vane performs in deflecting sediment. Table 3 provides a summary of the simulation findings for the various geometric configurations employed in this investigation. By contrasting the simulation results, vortex strength, local scour, and vane's sediment deflection performance are evaluated and assessed.

3.1 Local scouring around vanes

By contrasting the simulation results of 2-*D* velocity magnitudes in the y-z plane at the location of vanes, local scouring around vanes is evaluated and assessed. It is determined that the configuration *A.1.2* (Table 3) has excessive local scour while the configuration *B.1.4* has reduced local scour (Fig. 2). This means that at high Froude number values, the larger aspect ratio of 0.5 should be avoided, whereas at low Froude number values, the smaller aspect ratio of 0.25 should be favored. High Froude numbers typically result in more severe local scoring near vanes. Therefore, simply comparing the simulation results for configurations *A.1.4* (Fig. 3) and *A.2.4* (Fig. 4) is used to assess the impact of angle of attack on local scoring. According to Marelius and Sinha (1998), the comparison shows that there are multiple vortex cores immediately downstream of the vanes when the angle of attack is 40 degrees, but only one core is visible when the angle of attack is 15 degrees. Marelius and Sinha (1998) measured the intensity of the tip vortex but ignored its size. The simulation results for the submerged vanes positioned at the 40^0 angle of attack support their contention that there are two suction side vortices and one tip vortex. It appears that the transverse extent of the tip vortex for the vane pair positioned at the 40^0 angle of attack is two times greater than that of the vane pair at the 15^0 angle of attack, which is exhibiting roughly 15 cm of transverse extent of the vortex core. At 40^0, it is displaying a transverse extent of the vortex core of roughly 30 cm. Additionally, the comparison shows that the vane pair oriented at 15^0 has a more coherent vortex. Further analysis of the vortex patterns reveals that the suction side of the vanes installed at the 40^0 angle of attack had more severe local scouring than the vanes installed at the 15^0 angle of attack. Marelius and Sinha (1998) further demonstrated that the horseshoe vortex's extent for vanes positioned at a 40^0 angle of attack reaches toward the suction side in this situation.

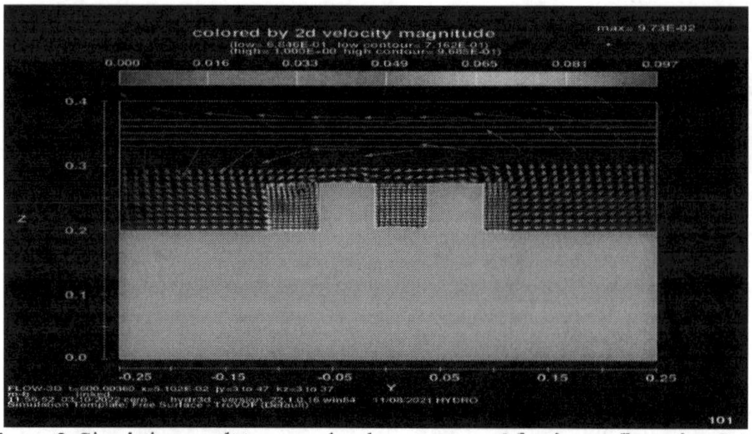

Figure 2. Simulation result to assess local scour at $x = 0$ for the configuration *B.1.4*

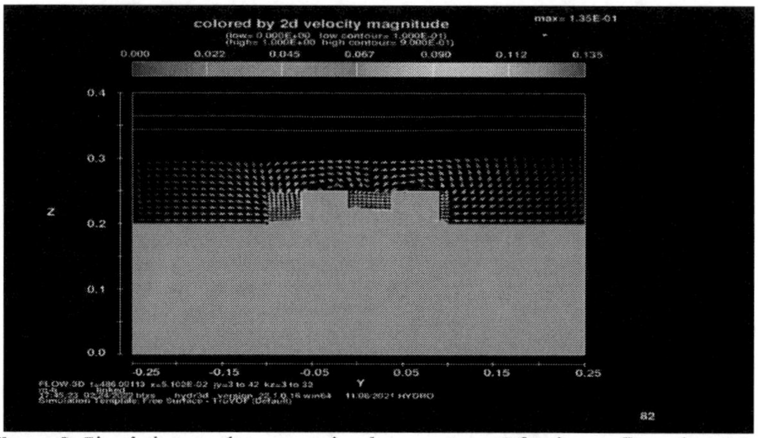

Figure 3. Simulation result to assess local scour at $x = 0$ for the configuration *A.1.4*

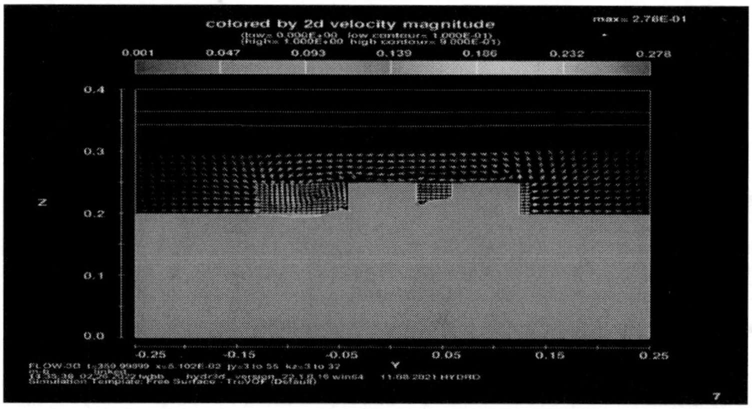

Figure 4. Simulation result to assess local scour at $x = 0$ for the configuration *A.2.4*

3.2 Optimization of aspect ratio based on quantification of local scour and vortex strength

By contrasting the simulation results of the 2-*D* velocity magnitudes in the *y-z* plane at two sites downstream from the vanes, the optimization with regard to the vortex strength is carried out. The centre of vane's pair is 5 cm upstream from one of the downstream sites, and the point with the highest *2D* velocity magnitude in the *y-z* plane is the other. In order to provide the comparisons specifically for the various Froude numbers with two extreme angles of attack, they are divided into the following sections. Comparing several numerical runs demonstrates that installing the vanes at a 40 degree angle of attack, which necessitates deeper installation, is necessary to optimize the performance of sediment deflection. By positioning the vanes at a greater angle of attack with the approach flow, the additional cost of deeper installation should be compensated for.

3.2.1 Lower Froude number value of 0.168

It is determined that an aspect ratio of 0.25 is best to employ at a greater angle of attack of 40 degrees since the best configuration *B.2.4* (Fig. 5) has produced the strongest vortex and the least amount of local scour. An aspect ratio of 0.25 is suitable for lowering scour for an angle of attack as low as 15 degrees, whereas an aspect ratio of 0.33 is preferable to improve the performance of vanes in deflecting sediment. For the 15 degree instance, *B.1.4* (Table 3) and B.1.3 (Fig. 6) are the best geometrical configurations because they reduce scour and improve deflection, respectively. With a tiny aspect ratio of 0.25 and a low Froude number, the vortex must travel further downstream to strengthen. Also, the maximum vortex strength shifts downstream at low Froude numbers when a greater aspect ratio of 0.5 is offered.

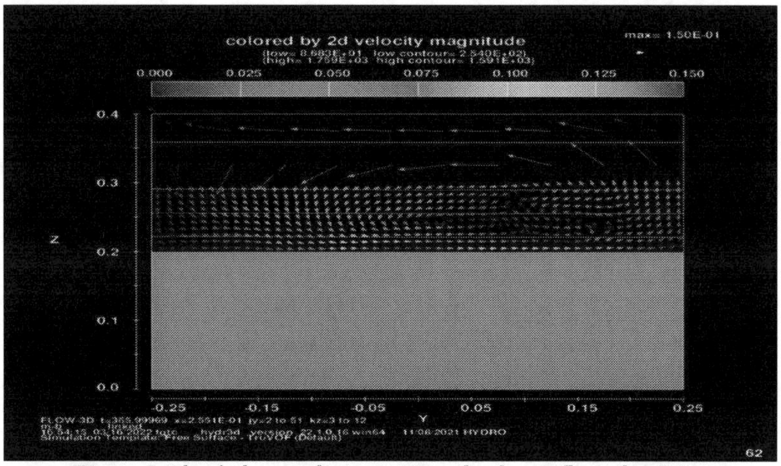

Figure 5. Simulation result at $x = 15$ cm for the configuration *B.2.4*

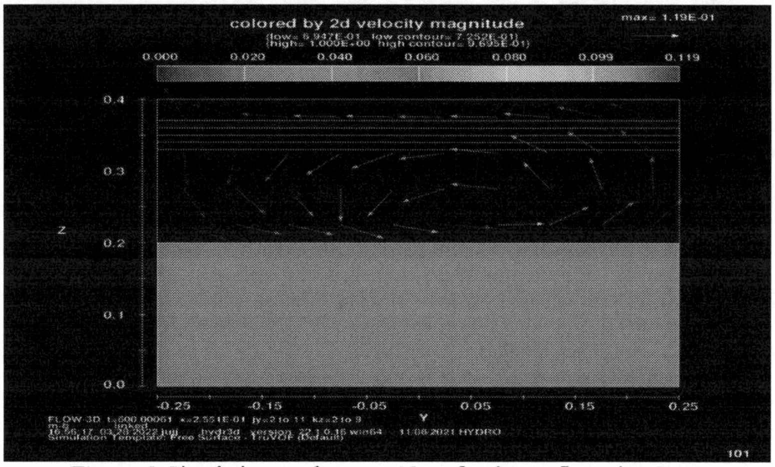

Figure 6. Simulation result at $x = 15\ cm$ for the configuration *B.1.3*

3.2.2 Higher Froude number value of 0.313

It is determined that an aspect ratio of 0.33 is the most appropriate to adopt at a greater angle of attack of 40 degrees. An aspect ratio of 0.25 is ideal at a lower angle of attack of 15 degrees. They have produced the strongest vortex and the least amount of local scour, therefore this explains why, *A.2.3* (Fig. 7) and *A.1.4* (Fig. 8) are the ideal arrangements for the 40 degree and 15 degree instances, respectively. At a high Froude number and a reduced aspect ratio of 0.25, it is seen that the maximum vortex intensity is shifting downstream. Different bed-forms come into play as the Froude number rises above the threshold required for the start of sediment motion, and the scouring patterns change quickly within the high Froude number value range. As a result, only when the Froude number is specific to around 0.3 should the results of the current study be considered accurate. Tan et al. (2005) provided a sizable dataset so that researchers could examine how well vanes performed at deflecting sediment over a wider range of Froude numbers in the context of mobile beds.

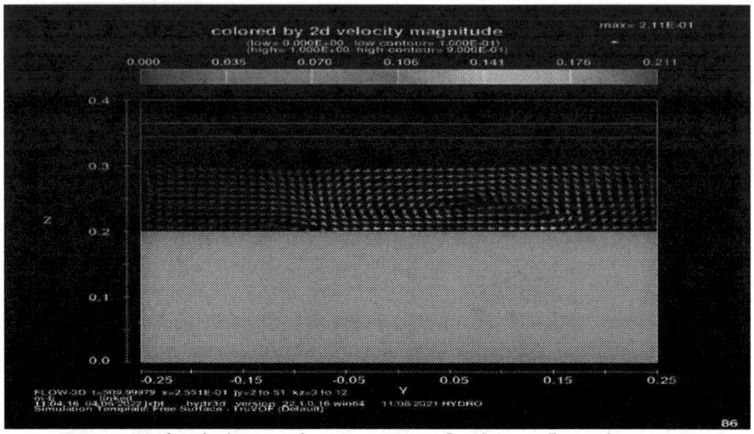

Figure 7. Simulation result at $x = 15$ cm for the configuration *A.2.3*

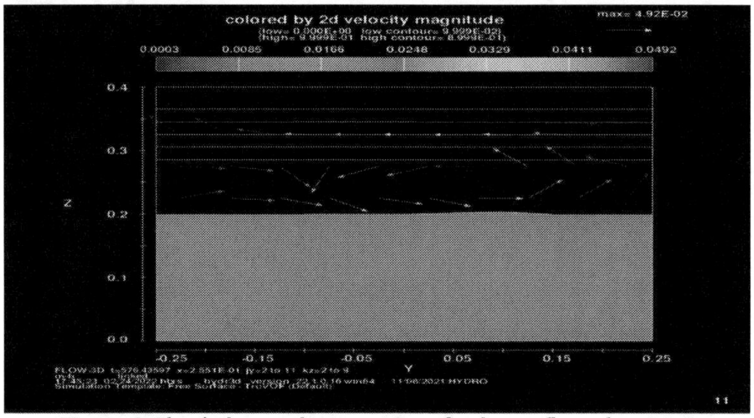

Figure 8. Simulation result at $x = 15$ cm for the configuration *A.1.4*

Table 3. Flow 3D Hydro simulation results for different model configurations used in this study

Designation	Maximum value of 2D velocity magnitude in y-z plane at x = 0 (center of vane's pair), in cm/sec	Maximum value of 2D velocity magnitude in y-z plane at x = 5 cm, in cm/sec	Maximum value of 2D velocity magnitude in y-z plane between x = 10 cm and x = 250 cm, in cm/sec
A.1.2	32.2	23.1	4.14
A.1.3	13.7	16.5	4.8
A.1.4	13.5	21	4.92
A.2.2	27.4	18.5	6.02
A.2.3	24.4	25.7	21.1
A.2.4	27.8	26.9	8.7
B.1.2	10.6	11	6.08

B.1.3	11	17.8	11.9
B.1.4	9.7	10.5	10.7
B.2.2	17.7	11.1	5.15
B.2.3	18.6	19.7	9.3
B.2.4	18.2	16.1	15

4. Conclusions

From this numerical study the following conclusions can be made:

1. Comparing the simulation results for the vane pair with an aspect ratio of 0.25 and under high Froude number, it seems that the transverse extent of the tip vortex for the 40^0 angle of attack becomes twice to that in case of the 15^0 angle of attack and also this comparison clearly indicate a more coherent vortex for the vane pair placed at 15 degree.
2. The comparison of vortex patterns also confirmed that the extent of the horseshoe vortex extends towards suction side for the vanes placed at the 40^0 angle of attack and therefore more severe local scouring is observed at the suction side of the vanes placed at the 40^0 angle of attack compared to the vanes at the 15^0 angle of attack.
3. At a low Froude number of 0.168, the most optimum aspect ratios for enhanced sediment deflection are 0.25 and 0.33 at 40 degree and 15 degree angles of attack respectively.
4. For a high Froude number of 0.303, the most optimum aspect ratios for enhanced sediment deflection are 0.33 and 0.25 at 40 degree and 15 degree angles of attack respectively.
5. The larger aspect ratio of 0.5 should be avoided at high Froude number value and smaller aspect ratio of 0.25 should be preferred at low Froude number value

ACKNOWLEDGMENTS

This work was conducted in collaboration with the departments of civil engineering at Shiv Nadar University and Maulana Azad National Institute of Technology.

REFERENCES

1. Chauhan, V., Singhal, G.D. and Chavan, R. (2022). "A review of sediment deflection in rivers using submerged vanes." *ISH J. Hydraul. Eng.*, 1-17.
2. Flokstra, C. (2006). "Modelling of submerged vanes." *J. Hydraul. Research*, *44*(5), 591-602.
3. Gupta, U.P., Ojha, C.S.P. and Sharma, N. (2006). "Vorticity with different shapes of submerged vanes." *ISH J. Hydraul. Eng.*, *12*(1), 13-26.
4. Gupta, U.P., Sharma, N. and Ojha, C.S.P. (2007). "Performance evaluation of aspect ratio of submerged vanes." *Water and Energy International*, *64*(2), 20-26.
5. Marelius, F. and Sinha, S.K. (1998). "Experimental investigation of flow past submerged vanes." *J. Hydraul. Eng.*, *124*(5), 542-545.
6. Nezu, I., Tominaga, A. and Nakagawa, H., 1993. Field measurements of secondary currents in straight rivers. *Journal of Hydraulic Engineering*, *119*(5), pp.598-614.
7. Odgaard, A.J. and Wang, Y. (1991a). "Sediment management with submerged vanes. I: Theory." *J. Hydraul. Eng.*, *117*(3), 267-267.
8. Odgaard, A.J. and Wang, Y. (1991b). "Sediment management with submerged vanes. II: Applications." *J. Hydraul. Eng.*, *117*(3), 284-302.
9. Odgaard, A.J. and Mosconi, C.E. (1987). "Streambank protection by submerged vanes." *J. Hydraul. Eng.*, *113*(4), 520-536

10. Odgaard, A.J., 2009, April. River training and sediment management with submerged vanes. American Society of Civil Engineers.
11. Ouyang, H.T., Lai, J.S., Yu, H. and Lu, C.H. (2008). "Interaction between submerged vanes for sediment management." *J. hydraul. reserach*, *46*(5), 620-627.
12. Ouyang, H.T. (2009). "Investigation on the dimensions and shape of a submerged vane for sediment management in alluvial channels." *J. Hydraul. Eng.*, *135*(3), 209-217.
13. Patel, V.C. and Yoon, J.Y. (1995). "Application of turbulence models to separated flow over rough surfaces." *ASME.J. Fluids Eng.*, 117(2), 234-241.
14. Sinha, S.K. and Marelius, F. (2000). "Analysis of flow past submerged vanes." *J. of Hydraul. Research*, *38*(1), 65-71.
15. Tan, S.K., Yu, G., Lim, S.Y. and Ong, M.C. (2005). "Flow structure and sediment motion around submerged vanes in open channel." *J. of waterway, port, coastal, and ocean eng.*, *131*(3), pp.132-136.
16. Van Rijn, L.C. (1984). "Sediment transport, part I: bed load transport." *Journal of hydraulic engineering*, *110*(10), 1431-1456.

DOI: https://doi.org/10.26077/6756-2055

Numerical modelling of scour and flow field for two different arrangements of piers using SSIIM model

Sushant Kumar Biswal[1] & Animesh Das[1]
[1]*National Institute of Technology Agartala, Tripura, India*
E-mail: sushantb69@gmail.com
E-mail: animesh.nit14@gmail.com

***Abstract**: In this work, a numerical model Sediment Simulation In Intakes with Multiblock option (SSIIM) is developed to analyse the effect of pier spacing on the scour process around side by side and staggered piers. Both $\kappa - \varepsilon$ and $\kappa - \omega$ turbulence models are employed to solve the eddy viscosity flow equations and their performances are evaluated. From the initial observations it is concluded that, when the spacing between the piers is negligible, then both the piers behave like a single pier and the scour depth is found to be more than 42% than that of a single pier. However, when the clear spacing between piers is more than 1.5 times pier diameter (D) for both arrangements the piers behaved independently. Furthermore, the numerical model is skill-assessed for two different piles spacing (S/D) in side by side piers and in staggered arrangements. While the model is effective in reproducing the scour depth in front of piers for side by side piers, it overestimated the bed scour depth between the piers, however, it underestimated the scour depth in front piers in the staggered arrangements. The results show that the SSIIM program can be used efficiently for simulating the scouring in natural rivers and good agreements are achieved between the piers in different arrangements.*

***Keywords** : Circular pier, Flow dynamics, Numerical method, Scour, SSIIM2.0*

1. Introduction

When a structure is placed in a hydraulic or marine environment, the presence of the structure significantly changes the flow pattern in its immediate neighbourhood, resulting in the contraction of flow; the formation of a horseshoe vortex, lee-wake vortices, the generation of turbulence and the occurrence of wave breaking. These changes commonly bring about an increase in the local sediment transport capacity and thus lead to scour. The scour around bridge pier is extremely difficult to track due to the complexity of the flow structure around hydraulic structure. Typically, three types of scouring around the piers can be differentiated: (1) general scour in which the quantity of sediment of a stream is carried downstream by changing in water flow velocity owing to erosion in the bed or banks of the stream; (2) contraction scour as a result of narrowing flow sections of the river primarily occurs in the vicinity of the pier; (3) Local scour.

Local scour is a common risk to structures such as bridge piers, abutments, and dikes that are constructed on natural rivers. To reduce the threat of foundation failure, the understanding of local scour phenomenon around hydraulic structures is significant. Scour around bridge piers in the flow is an unavoidable issue. Prediction of scour depth is an important feature of the design of hydraulic structures. Most of the researches are focused on the experimental study and field measurements to investigate the flow pattern and the scour hole around uniform bridge pier rather than compound piers. These types of pier not only can considerably reduce construction costs but are more practical and efficient (Ataie-Ashtiani and Beheshti 2006). However, the mechanisms of the scouring action around pier groups are considerably more complicated.

Many researchers like (Ataie-Ashtiani and Beheshti 2006; Ataie-Ashtiani et al. 2010; Ataie-Ashtiani and Aslani-Kordkandi 2012; Kumar and Kothyari 2012) have investigated the flow structure and local scour for different pier arrangements with different bridge pier spacing through the physical model. They presented that the scour depth is deeper between the two piers than that on the other side of the piers, and the horseshoe vortices are extended further to the downstream, while downflow is more intense than that in outer sides of the gap. The effect of pile spacing on the scour depth for a group of two piers under clear water condition has been comprehensively studied through laboratory experiments and develops empirical equations to calculate the utmost scour depth (Ettema et al. 2006; Sheppard et al. 2004). Khaple et al. (2017) conducted experiments on groups of pier in tandem and staggered arrangement to calculate the scour depth and they reported that in a staggered arrangement, as the distance between the piers increases, the maximum scour depth at the rear piers decreases.

One of the questions raised by location of a bridge pier is the blockage of the flow cross section. The blockage or wall interference refers to the influence of the side walls of a test section on the local scour depth. Chiew (1984) suggested that if the ratio of pier size to channel width is less than 10 percent ($D/B < 10\%$), the sidewalls will have no significant impact on flow characteristics and the scour profile. In the current study, the ratio of D/B is 7% (approx.). Further, as suggested by Yanmaz (2002) relative scour depth is not affected by the particle size when $D/D_{50} > 50$. In the present study the ratio of D/D_{50} is 168 in order to avoid the impact of D/D_{50}(the relative pier size) on scour depth.

However, owing to numerous laboratory conditions and limitations on data acquisition, especially in small-scale laboratory flumes, most of the empirical equations for estimating scour depth have a great deal of uncertainty and substantial limitations. In recent years with the development of computer system, several numerical models have been significantly improved for simulating the $3D$ flow field and bed variations around bridge piers and are now widely applied in river engineering. The numbers of scour model software programs have been constructed, such as Flow $3D$, CCHE3D, SSIIM and ANSYS Fluent, which are now being applied in many fields. In recent year, SSIIM is a free numerical CFD model that has been widely used in hydraulic structures and river engineering, which solves the Navier-Stokes equations using the various turbulence models (Olsen 1996) based on the finite volume method. One of the main improvements of SSIIM model is the capability to use adaptive grid which changes in each time step with changes in the bed and water level. SSIIM numerical model was adopted to predict the local scour around bridge pier for different pier shapes, and the study confirmed that the model simulated the temporal evolution of local scour at the narrowing location of a bridge effectively (Abouzeid et al. 2007; Aarabi et al. 2011; Ehteram and Mahdavi Meymand 2015). SSIIM model was used by (Nasr-Allah et al. 2017) to validate and comprehend the consequence of different widths and lengths of collar around bridge abutment on local scour depth pattern.

Realizing the fact that though large amount of literature is presently available on the topic of scour around uniform piers and its control. However, relatively few studies have been carried out so far on the topic of flow field and scour around compound piers. Therefore, a numerical study is planned and conducted to have further insight in this direction. In this study, the SSIIM numerical model is considered for simulating the flow pattern and local scour depth around side-by-side group piers and about staggered pier arrangements. The performance of two turbulence models ($k - \varepsilon$ and $k - \omega$) in solving Navier-Stokes equations is presented.

2. Numerical Model

2.1. Model Setup

The Numerical model in the present work is verified with the experimental results conducted in a rectangular flume with dimension of 15.0m long, 1.5m width and 0.9m depth. A channel section with 9.0m long × 1.34m width × 0.60m deep with two side by side piers and staggered pier located at the central channel which represents the test area is considered in the simulation. The working section of the flume was filled with sand to a uniform thickness of 0.20 m. The bed material (sand particles) size was $D_{50} = 0.54$ mm which are measured from the sieve analysis test using a vibrating shaker. The flow depth, z is normalized by constant flow depth (h) at the outlet. Being in staggered arrangement, the rear piers are placed at an angle of 60° with respect to the front pier. The length of the channel provides enough distance downstream of pier to ensure no interference between the turbulent flow and the downstream boundary.

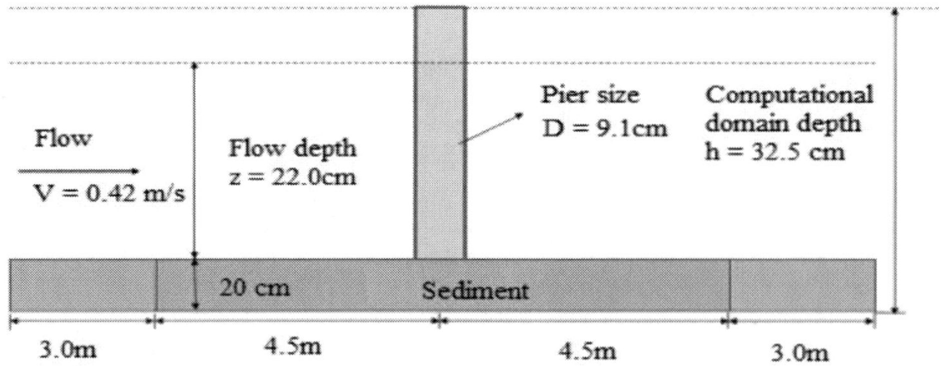

Figure1.The layout of the computational domain (side view)

Fine cells with mesh size of $1 \times 1cm$ are used nearly piers to provide precise results whereas $2.8 \times 1cm$ cells are employed away from piers to optimize the computational time. The sensitivity analysis revealed that the no significant changes are observed in bed scour with using smaller mesh. Fig. 2 shows the computational mesh used for simulation of side by side piers and staggered piers. The overall mesh size is $180 \times 134 \times 14$. The first order upstream scheme is adopted for solving the discretization of the convection terms. A time step of 10 second is being employed. The water level of $h = 32.5\ cm$ at the outlet is used for calculating water surface in the channel using one dimensional backwater. Where h is the constant flow depth at the outlet. Adaptive grid is used which means grid was updated according to new water level and bed level at each time iteration. Both turbulence models $\kappa - \varepsilon$ and $\kappa - \omega$ were used and their results were compared. In this study, simulations were repeated for several $(S/D)$for piers, and results of comparable to corresponding results of one pier.

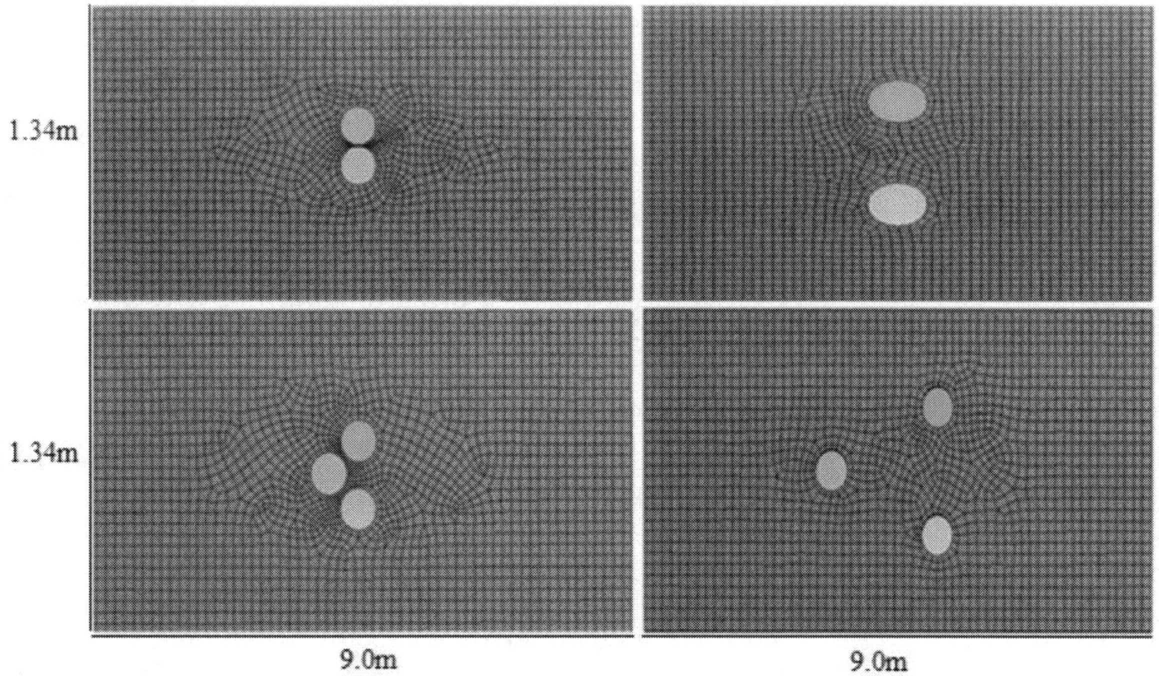

Figure 2. Computational mesh close to the side by side and staggered piers employed for numerical model.

2.2. Model Equations

The governing equation of the CFD model is the $3D$ Navier-Stokes equations for incompressible flow, which discretize using the finite volume method.

$$\frac{\partial \rho U_i}{\partial t} + \frac{\partial \rho U_i U_j}{\partial x_j} = -\frac{\partial p}{\partial x_i} + \frac{\partial}{\partial x_j}\left[(\mu + \mu_t)\left(\frac{\partial U_i}{\partial x_j} + \frac{\partial U_j}{\partial x_i}\right)\right] \quad (1)$$

In which $i = 1, 2,\ 3$ and U_i is the i th component of velocity; t is the time; x is spatial geometrical scale, x_j is the Cartesian coordinates; In this model, the Semi-Implicit Method for Pressure-Linked Equation (SIMPLE) is used for unknown pressure field, and an implicit finite volume scheme is employed to solve Eq. (1) to produce velocity field. In this study, both $\kappa - \varepsilon$ and $\kappa - \omega$ turbulence models are used to calculate the eddy viscosity (μ_t) and the equations are:

$$\mu_t = \rho C_\mu \frac{\kappa}{\varepsilon^2} \quad (2)$$

$$\mu_t = \rho \frac{\kappa}{\omega} \quad (3)$$

Here, κ is the turbulence kinetic energy, ε is the dissipation rate of turbulence and ω is the specific rate of dissipation, ρ is the fluid density; μ is the viscosity and C_μ is a constant equal to 0.09. SSIIM uses wall laws equation for rough boundaries to evaluate velocity in cells near the bed and boundary of channel. The values are considered as boundary condition for Navier-Stokes equation. The wall law equation is expressed as:

$$\frac{\bar{u}}{u_*} = \frac{1}{\kappa} ln\left(\frac{30y}{k_s}\right) \tag{4}$$

k_s is the roughness height, κ is the von Karman constant equals to 0.4, $\bar{u}$ is the mean velocity, u_* is the shear velocity and y is the height above the bed. In this method, the initially produced free surface is employed for the successive computations as a fixed lid for zero gradients for all variables. Sediment transport is divided as: suspended load and bed load. The suspended sediment transport is calculated by solving the transient convection – diffusion equation for sediment concentration (c):

$$\frac{\partial c}{\partial t} + u_j \frac{\partial c}{\partial x_j} + w_s \frac{\partial c}{\partial z} = \frac{\partial}{\partial x_j}\left(\Gamma \frac{\partial c}{\partial x_j}\right) \tag{5}$$

where c is sediment concentration, w_s is fall velocity of sediment particles, and Γ is diffusive coefficient. The diffusive coefficient is set to the eddy viscosity taken from turbulence model ($\Gamma = v_t/S_c$). S_c is the Schmidt number which is assumed to be unity in this study. Here, the effect of the turbulence on setting velocity of the sediment is considered. Sediment transport calculation is performed on all cells exclude the cell nearly the bed, where the concentration is specified by Van Rijn's equation [30]. The sediment concentration at bed load is

$$C_{bed} = 0.015\left(\frac{d}{a}\right)^{0.3} \frac{\left[\frac{\tau-\tau_c}{\tau_c}\right]^{1.5}}{\left[\frac{(\rho_s-\rho_w)g}{\rho_w v^2}\right]^{0.1}} \tag{6}$$

C_{bed} is the sediment concentration (*kg/kg*), d is the sediment particle diameter (*m*), a is the reference level set equal to the roughness height (m), τ is the bed shear stress (pa), τ_c is the critical bed shear stress for movement of sediment particles according to Shield's curve (Pa), ρ_w is the density of water (kg/m^3), ρ_s is the density of sediment (kg/m^3), v is the viscosity of the water (Pa.s), and g is the acceleration due to gravity (m/s^2). In addition to suspended load, the bed load discharge per unit width (q_b) can be calculated using Van Rijn's formula as follows:

$$\frac{q_b}{D_{50}^{1.5}\sqrt{\frac{(\rho_s-\rho_w)g}{\rho_w}}} = 0.053 \frac{\left[\frac{\tau-\tau_c}{\tau_c}\right]^{1.5}}{D_{50}^{03}\left[\frac{(\rho_s-\rho_w)g}{\rho_w v^2}\right]^{0.1}} \tag{7}$$

where D_{50} is the mean size of the particle (mm).

2.3. Boundary Condition

The boundary conditions that are specified are the water discharge, the initial water level, and the sediment size. The upstream boundary condition is given by the discharge amount from the experiments. At the downstream boundary condition, zero gradients are to be given to prevent instabilities, which mean that the water discharges at the downstream boundary was not specified. The size of the sediment $D_{50}$is taken as 0.54 mm. The side and bottom are specified as walls and the top and downstream boundaries as pressure outlets. At the farthest outlet a fixed water depth is taken into account, and other variables are extrapolated from interior solution domain. At the free surface, the influence of wind shear is ignored and the pressure is assumed to be the atmospheric value. The surface of the pier was set with a no-slippery condition, and remaining sides of the domain are set to be symmetrical.

3. Results and Discussion

3.1. Flow field

The velocity contour and vector fields around the piers in the side by side pier are illustrated in Fig. 3 (a, b) at $z/h = 0.25$ and $z/h = 0.5$ plane at the end of simulation time for both turbulence model $[k - \varepsilon, k - \omega]$. Here, z is the distance from the bed before scouring and h is the flow depth. Velocity vectors are evaluated to study the interference and downflow computed by the numerical model. The results show that the magnitude of horizontal velocity around the piers becomes increased at zero spacing ($S = 0D$), and attained the value of 0.46 m/s on the sides of the piers. In this case the velocity varied from 0.34 to 0.46 m/s. However, when the piers moved from centre to towards the boundary walls ($S = 1.5D$) the velocity of flow started to decrease.

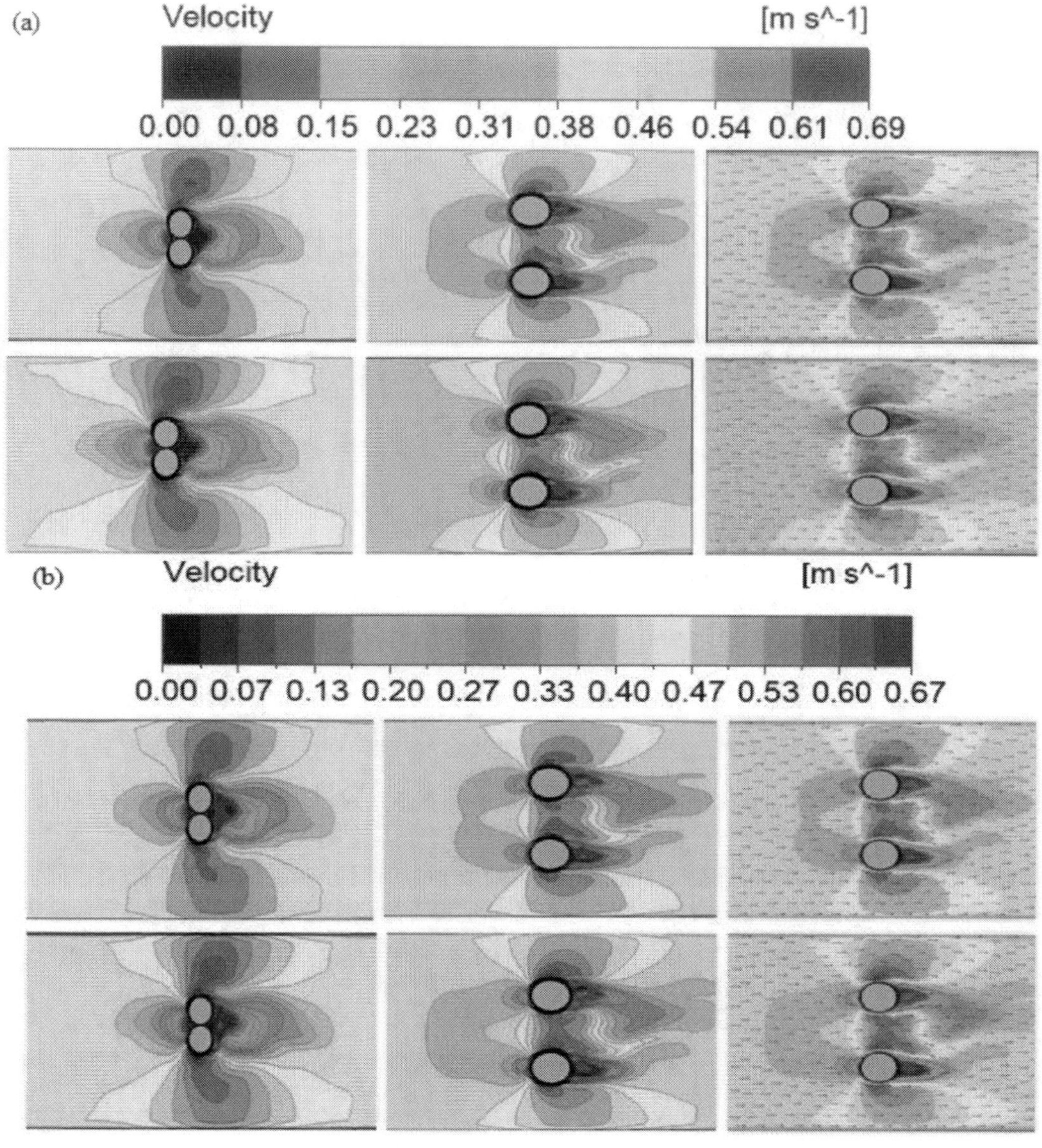

Figure 3. Horizontal velocity contours (m/s) and vector field in the plane $z/h = 0.25$ *and* $z/h = 0.5$ at the end of simulation time for (a) $k - \varepsilon$, and (b) $k - \omega$ turbulence model.

Figure 3(a, b) also shows the uniformity of flow fields around the bridge pier when two piers are present in the side by side arrangement at spacing of $1.5D$. It is clear from the figure that when the distance between the piers is ($S = 0D$) the piers behaves like a single pier of front projected size of double the diameter of the pier. Owing to this projected area the flow rate and magnitude becomes high. Because, the diameter of the bridge pier is the main factor to impedes the flow around bridge piers. The results show that the value of horizontal velocity nearly bed is increased between piers, most likely owing to contraction flow. It is worth mentioning that the downflow is more intense at areas between piers than outer sides of piers, and is stronger close to the bed than midwater depth. It is clear that high velocities at the area between two piers make it a critical area of erosion.

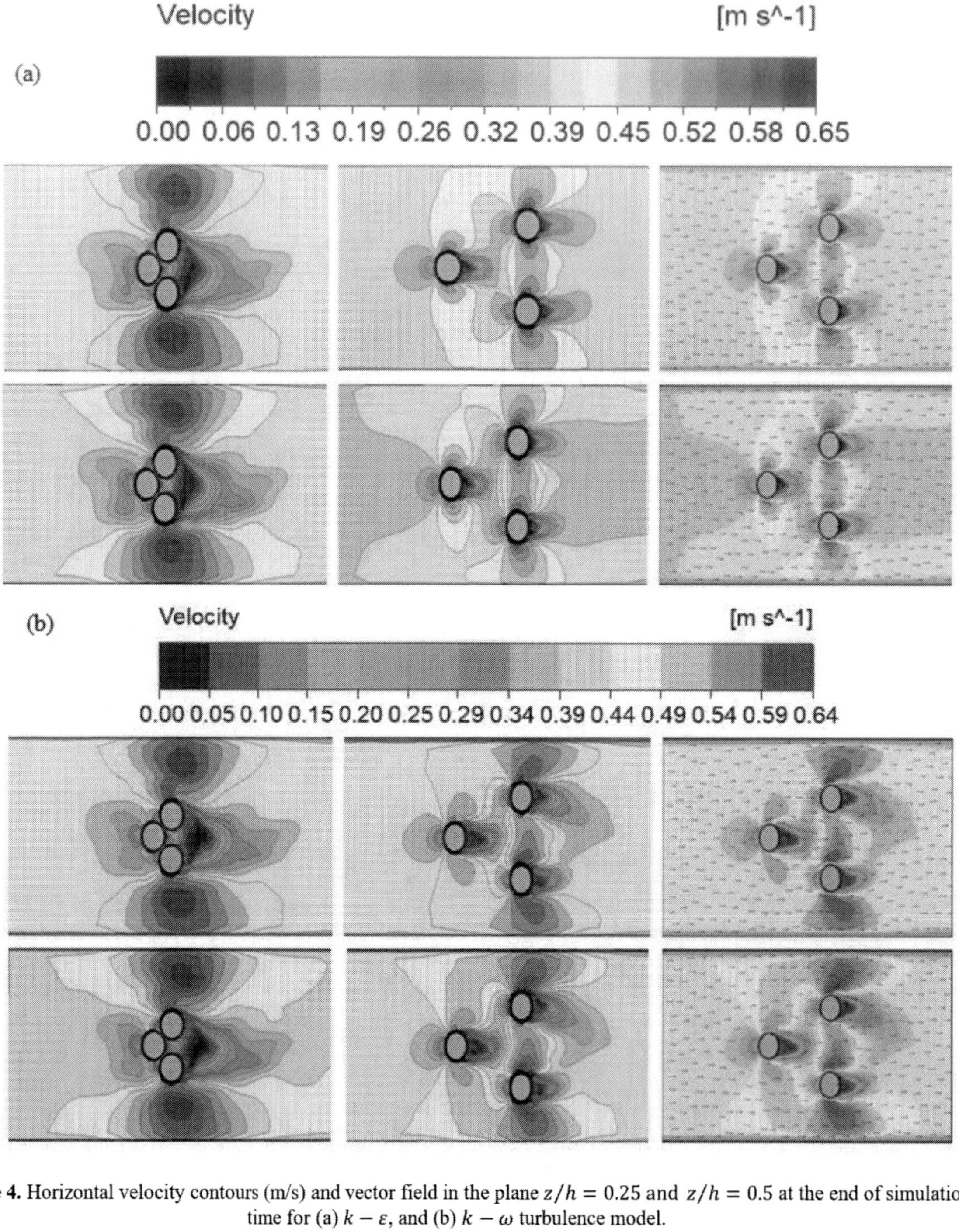

Figure 4. Horizontal velocity contours (m/s) and vector field in the plane $z/h = 0.25$ and $z/h = 0.5$ at the end of simulation time for (a) $k - \varepsilon$, and (b) $k - \omega$ turbulence model.

Figure 4(a, b) gives the velocity contour and flow field around the piers in the staggered arrangement at $z/h = 0.25$ and $z/h = 0.5$ plane at the end of simulation time for both turbulence model $[k - \varepsilon, k - \omega]$. Being in staggered arrangement, the rear piers are positioned at an angle of 60° with respect to the front pier. From this figure, it may be seen that the spacing between the flow fields is close to the piers however when the spacing between the piers is expanded, the separation between the flow fields also expanded.

3.2. Bed level changes

In the present study, simulations are repeated using two separate turbulence models. Contours of bed level change at the end of simulation computed using $k - \varepsilon$ turbulence model and $k - \omega$ turbulence model are illustrated in Fig. (5-7). It is clear from the figure that when the spacing between the piers is negligible i.e. $(S = 0D)$ the piers behave like a single pier of front projected size of double the diameter of the pier. Owing to this projected area the size of the obstruction is increased, consequently the scouring rate increases. Scour depth around the set of two piers with zero spacing in between is 42% more than that of a single pier. A partial sheltering influence happens in the gap between the piers. As the spacing between the piers increases the shielding effect of the front pier on the downstream pier reduces, subsequently the scour depth began to decrease overall. After the removal of shielding effect both the piers show independent behaviour. Although, the upstream pier registered higher scour depth in comparison to the downstream pier at any spacing. This is due to the fact that the upstream pier protects the downstream pier from excessive velocities and provides a shelter for the downstream pier, a phenomenon, identified by Hannah (1978).

The time series of the maximum scour depth calculated by the model are illustrated in Fig. 5c. The SSIIM numerical model is time-dependent model, and based on these results; ~54% of final scour depth (almost 148 mm) has been occurred during first hour of simulation. Some momentary reduction is detected in the figure which is initiated by variation in the scour pattern in which shifting of the location of the maximum scour occurred between the piers. The upstream pier shows more scour depth in comparison to the downstream pier. The sediment motion from the base of the front pier into the scour hole at the rear pier may lead to smaller scour depth in front of the rear pier. The results indicate that, the maximum scour depths always occurred at the upstream, front face of the bridge piers. The downflow generated by the flow hitting the upstream face of the pier, acts like a vertical jet which erodes a groove in front of the pier. The eroded sand particles are carried around the pier by the combined action of accelerating flow and the spiral motion of the horseshoe vortex (Hafez 2016). The deepest point of the scour hole is clearly at the front face of the bridge piers and the deposition mound is located downstream of the bridge piers. As it can be seen that when $S > 1.5D$, the bed level change at point nearly pier is almost independent of the arrangement of piers. In other words, the scour mechanism in this case is like to a single pier.

The results related to staggered arrangement have been observed that when the spacing between the piers is $S/D = 0$, the piers behave like a single pier and the arrangement showed high scour depth. But when spacing between the bridge piers is increased in staggered arrangement the rear pier showed a decreasing pattern of scour depth. The scour depth around rear pier with zero spacing in between is 48% more than that of a single pier.

Both the investigations show that at the front pier indicates more scour depth in comparison to the rear bridge pier. The difference in the magnitude of the scour depth in the two studies might be because of the arrangement of the bridge pier models and the median size of the sediment. Finally, it is inferred that as the spacing between the piers increases the scour depth decreases, thereby establishing the adverse interference effect of piers in close proximity.

In the current study, the Froude number ranges from 0.072 to 0.270 and S/D ranges from 0 to 1.5. According to Fig. 6-7, the ratio of the maximum scours depth to the depth of approaching flow (y_{max}/y_0termed as the relative MSD) decreases with the increase in the pier spacing ratio S/D. Hopkins et al. (1980) stated that the strength of the horseshoe vortex system is a function of the pier Reynolds number ($Re_P = UD/\nu$). Therefore, it can be concluded that the strength of horseshoe vortices, which is a function of (Re_P), decreases with the increase in the pier spacing ratio. Therefore, the smaller the pier size(D)and the larger the pier spacing (S), the weaker the horseshoe vortices around bridge piers, which will result in shallower scour depths around the bridge piers. Where U the average velocity of the approaching is flow; D is the diameter of the bridge pier, and v is the kinematic viscosity. It is clear that at early times of simulation the scour pattern was symmetrical. As time passes, erosion is more critical between piers than outside region.

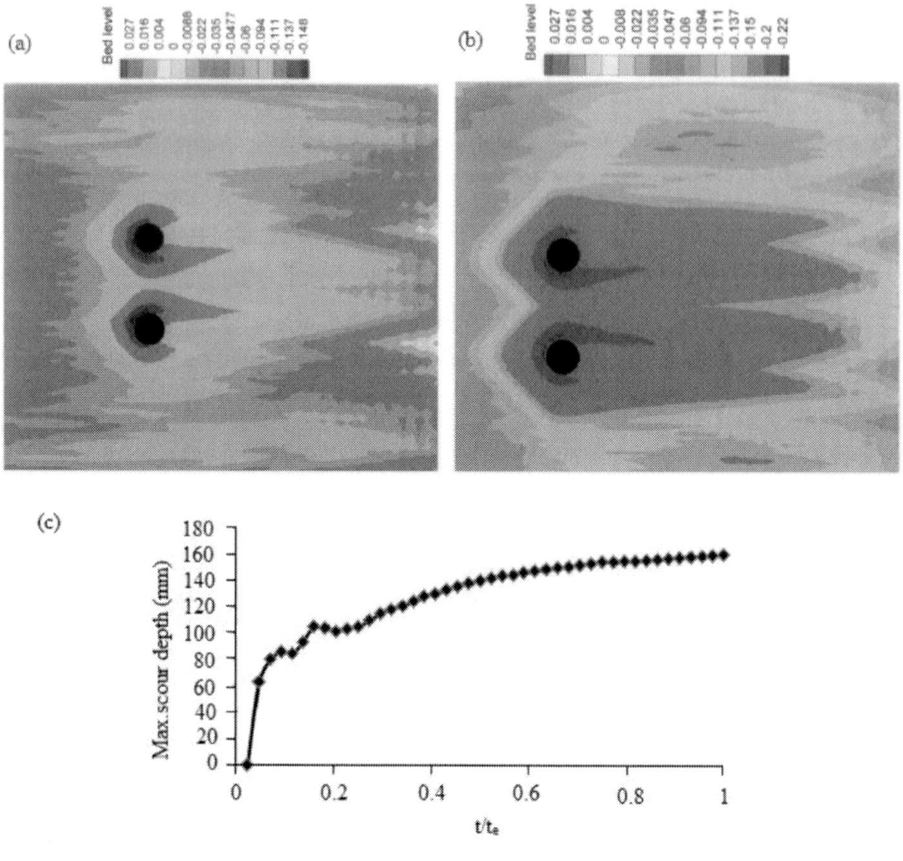

Figure 5. Contours of scour depths (m) close to piers at simulated using $k - \varepsilon$, and $k - \omega$ turbulence model at the end of simulation for $S = 1.35D$, and $1.5D$ and Time series of the maximum scour depth (mm) simulation

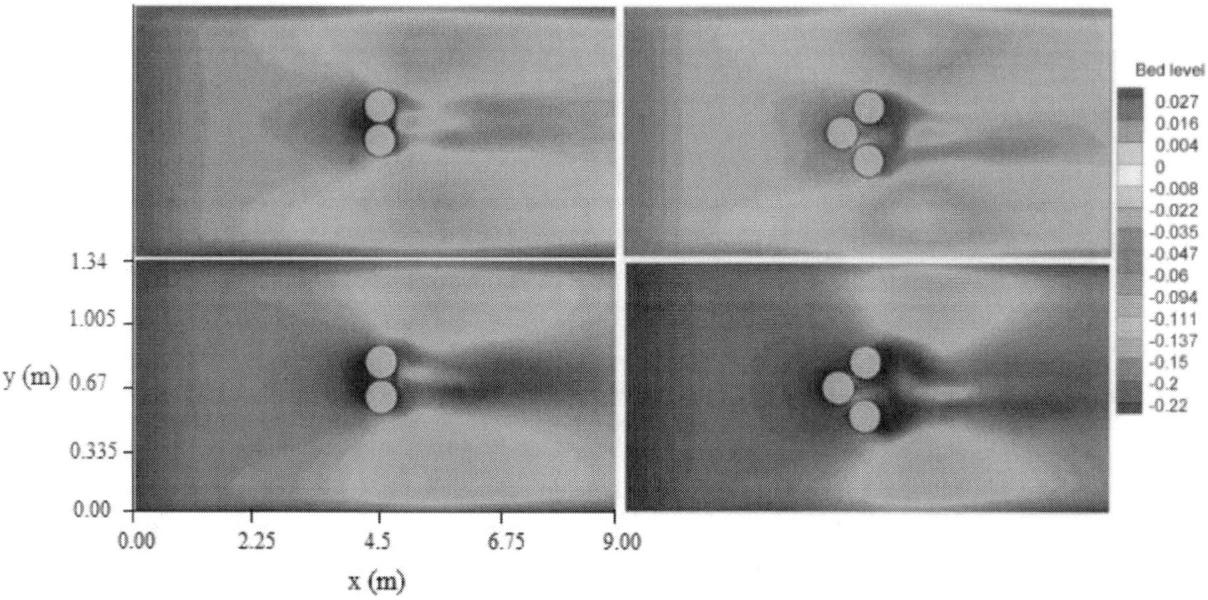

Figure 6. Contours of bed level (m) close to piers, simulated using $k-\varepsilon$ (1st row) and $k-\omega$ (2nd row) turbulence model at the end of simulation for $S=0D$.

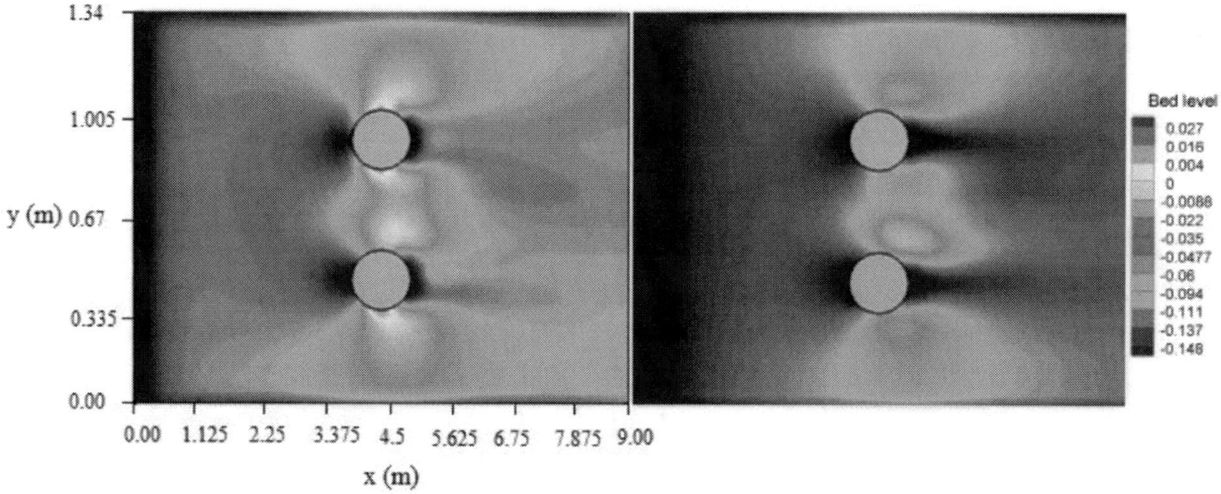

Figure 7. Contours of bed level (m) close to piers, simulated using $k-\varepsilon$ (left) and $k-\omega$ (right) turbulence model at the end of simulation for $S=1.5D$.

3.3. Turbulence model

As stated earlier, simulations were repeated using two distinct turbulence models. In both models, scour depths are deeper between two piers than other sides of piers. The root mean square (RMS) value is computed from bed levels at cells of simulation model. The RMS value of bed level change for $k-\varepsilon$ model is 0.0316 m, while this value for $k-\omega$ model is 0.0846 m.

As shown in Fig. 5, the erosion area is much more elongated in $k-\omega$ model results in comparison to $k-\varepsilon$ model. The reason is the model deficiency in replicating precise downflow and horseshoe vortices. No sign of sediment deposition is observed at downstream of piles in $k-\omega$ model. Though, little sedimentation has been detected at rearward region of pier in $k-\varepsilon$ model but it seems to be till less. Perez (2010) argued that the fall velocity of sediment particles did not vary with sediment concentration in SSIIM numerical model, and may be responsible for such discrepancies between model results. This evaluation proves that the $k-\varepsilon$ model is more favorable than $k-\omega$ model in reproducing flow field and bed dynamics.

4. Conclusion

In the present study the SSIIM numerical model is employed to solve Navier-Stokes equations for flow field and bed transport equations. Two turbulence models $k-\varepsilon$ and $k-\omega$ are employed for computing eddy viscosity of the flow field and it is shown that the $k-\varepsilon$ turbulence model perform better than $k-\omega$ model when bed level changes from the model. The time-dependent numerical analysis for bed level scour pattern illustrated that at early stages of erosion, the scour pattern is symmetric close to each pier; however, with passage of the time, the deepest bed level is moved to the region between two piers. Simulation results illustrated increase of horizontal velocities between the piers. The model results demonstrated that SSIIM is capable of calculating the approaching downflow and bed level scour at the front of pier; but, this model overestimated the bed scour depth at a position between two piers. Furthermore, contours map of bed level variations revealed that the numerical model underestimated the sediment deposition at downstream of the piers mainly due to a weakness of the entrainment-deposition model. In addition, in staggered arrangements SSIIM model underestimated the bed erosion at front pier while well agreement and good correlation are achieved between piers. When $S/D=0$ the both the piers behave like a single pier and the scour depth is observed to be the most extreme. When $S>1.5D$, both the piers exhibit independent behaviour. The maximum scours depths identified for two piers in side by side arrangement is 42% greater than that of single bridge pier. However, when $S>1.5D$ between the piers in staggered arrangement, the scour hole around the piers start reducing. When $S=0D$, all the three piers in staggered behave like a single pier and the maximum scour depth around the piers is 48% greater than that of single bridge pier. Additional studies of simulation of group piers may support to assess SSIIM model in flow field computation and sediment transport; especially in rear piers.

5. REFERENCES

Aarabi, M.J. (2011). "Numerical simulation of temporal evolution of local scour in bridge pier with nonuniform sediment." *In: Advanced materials research. Trans Tech Publ.*

Abouzeid, G., Mohamed, H.I., and Ali, S.M. (2007). "3-D numerical simulation of flow and clear water scour by interaction between bridge piers." *J. Eng. Sci.*, 35, 891–907.

Ataie-Ashtiani, B., and Beheshti, A. (2006). "Experimental investigation of clear-water local scour at pile groups." *J. Hydraul Eng.*, 132(10), 1100–4.

Ataie-Ashtiani, B., Baratian-Ghorghi, Z., and Beheshti, A. (2010). "Experimental investigation of clear-water local scour of compound piers." *J. Hydraul. Eng.*, 136 (6), 343–51.

Ataie-Ashtiani, B., and Aslani-Kordkandi, A. (2012). "Flow field around side-by-side piers with and without a scour hole." *Eur. J. Mech-B/Fluids*, 36, 152–66.

Chiew, Y. M. (1984). "Local scour at bridge piers (Ph.D. dissertation)." Department of Civil Engineering, University of Auckland, New Zealand.

Ehteram and Mahdavi Meymand. (2015). "Numerical modeling of scour depth at side piers of the bridge." *J. Comput. Appl. Math.*, 280, 68–79.

Ettema, R., Kirkil, G., and Muste, M. (2006). "Similitude of large scale turbulence in experiments on local scour at cylinders." *J. Hydraul. Eng.*, 132 (1), 33– 40.

Hafez, Y.I. (2016). "Mathematical Modelling of Local Scour at Slender and Wide Bridge Piers." *J. Fluids.*, 5, 1-19.

Hannah, C.R. (1978). Scour at pile groups, Christchurch: Civil Engineering Department, University of Canterbury, *Research Rep.* No. 28– 3, 92.

Khaple, S., Prashanth, R.H., Roberto, G., and Dey, S. (2017). "Interference of an upstream pier on local scour at downstream piers." *J. Acta Geophys.* 65(1), 29–46.

Kumar, A., and Kothyari, U.C. (2012). "Three-dimensional flow characteristics within the scour hole around circular uniform and compound piers." *J. Hydraul. Eng.*, 138(5), 420–429.

Nasr-Allah. T., Moussa, Y., Abdel-Aal, G., and Awad, A. (2017). “Experimental and numerical simulation of scour at bridge abutment provided with different arrangements of collars.” *Alexandria Eng. J.*, 55(2),40-52.
Olsen, N. R. B. (1996). “A three-dimensional numerical model for simulation of sediment movement in water intakes with multi-block option.” *SSIIM User Manual Version* 1.4.
Sheppard, D.M., Odeh, M., Glasser, T. (2004). “Large Scale Clear-Water Local Pier Scour Experiments.” *J. Hydraul. Eng.*, 130, 957–963.
Yanmaz, A. M. (2002). “Dynamic reliability in bridge pier scouring.” *Turkish Journal of Engineering and Environmental Sciences.*, 26(4), 367-376.

DOI: https://doi.org/10.26077/2c7e-69c6

Numerical Simulation of Air Entrainment in Stepped Spillways

F. Zabaleta[1], S. Márquez Damián[2], F.A. Bombardelli[1]
[1]*Department of Civil and Environmental Engineering, University of California, Davis, USA*
[2]*Research Center for Computational Methods (CIMEC), Santa Fe, Argentina*
E-mail: fede.zabaleta@gmail.com

Abstract: *Existing methods to simulate air entrainment in stepped spillways are based on a sub-grid air-entrainment function that accounts for the incorporation of air through the air-water interface; a transport equation for the bubbly phase is then solved inside the water domain. Unfortunately, these techniques are unable to represent bulking, i.e., the increase in water depths after the inception point. In this work, we introduce a novel method for the simulation of flow in stepped spillways (including the aerated region). The new air-entrainment model is implemented in the OpenFOAM platform and couples an air-concentration-transport equation with the traditional Volume-of-Fluid (VoF) method to account for bulking. The model was calibrated and validated using the experimental data by Amador (2005) and Felder and Chanson (2013). Air concentration and mean free surface location in the aerated region (defined as where the concentration of air is 90%) are analyzed on the top of the steps as well as inside the cavities. Very satisfactory agreement in the wall-normal distribution of air concentrations between model and experiments was obtained. The model showed to be capable of predicting bulking in stepped spillways, without the need for ad-hoc treatments*

Keywords: *Air entrainment, numerical modeling, stepped spillways.*

1. Introduction

Stepped spillways are widely employed around the world to dissipate energy downstream of dams. Compared to their smooth counterpart, they have various advantages such as a lower risk of cavitation due to the presence of bubbles in the flow, an easier process of construction, a higher concentration of dissolved air at the bottom, and a higher level of energy dissipation (Chanson 2002, Matos 1999, Ward 2002). Flow over stepped spillways are classified into three different regimes: nappe, transition, and skimming flow (Amador 2005). In the nappe regime the water falls from step to step, splashing on each of them dissipating energy. It usually occurs for the lower flow rates. In the skimming regime, water flows over the steps and recirculating fluid is entrapped between them. This regime occurs for the higher flow rates. The transition regime occurs for intermediate flow rates. On this regime flow from step to step and skimming flow coexist showing chaotic behavior with numerous droplet ejections downstream of the inception point. Stepped spillways are usually designed to operate in the skimming regime, and therefore our focus on this study will be solely on this regime. Two distinct zones can be identified in the skimming flow: the non-aerated region, characterized by a smooth profile with some small disturbances close to the end; and the aerated region, beginning at the inception point (region where air entrainment begins) and characterized by a large volume of air entrained into the flow.

Air entrainment is the process of incorporation of air into the flow due to the interaction of a highly-turbulent, free-surface flow structures with the attendant air flow. When turbulence near the free surface is strong, the air-water interface becomes very distorted, overcoming the effects of gravity and surface tension and incorporating air into the flow. The pockets of air experience processes of break up and coalescence, and become transported in the direction of the flow and normal to the free surface. It is widely accepted that the inception point occurs when the turbulent boundary layer reaches the free surface (Halbronn 1954, Bauer 1954) and eddies are sufficiently energetic to deform the free surface (Meireles et al. 2014). Zabaleta et al. (2020) showed that air entrainment occurs when turbulent eddies with sufficient energy reach the free surface with enough frequency and that this effect is highly correlated to the growth of the boundary layer.

Numerical modeling of self-aerated flows has proven to be very difficult to undertake. Ideally, a two-phase self-aerated flow should be addressed by solving mass, momentum (and energy if needed) conservation equations inside both the water and air domain, while tracking the position of the interface. This would require a mesh fine enough to model every scale of turbulence and every bubble in the flow (with sufficient mesh resolution) over the domain. Some authors were able to represent self-aeration using direct numerical simulation (DNS) for small domains (Deike et al. 2016, Mortazavi et al. 2016). This approach is very useful from a scientific perspective but unfeasible for field-scale Reynolds numbers and practical applications. Existing methods to represent self-aeration are based on an air-entrainment function (Souders and Hirt 2004, Ma et al. 2011) that incorporates bubbles into the flow which are represented as a passive scalar. They have been popular due to their relatively simple formulation and a lower computational cost when compared to other methods that track each bubble individually. Air entrainment functions are usually proportional to the turbulent kinetic energy and the velocity of the flow. One of their main disadvantages is that they need calibration for each individual case, and the calibration parameter tends to be very case specific. In addition, in channel flows such as flow over smooth and stepped spillways, these types of models cannot represent bulking (a process where the high volume of air entrained raises the water level).

In this study we present a new air-entrainment model that couples the bubble transport equation with the traditional Volume of Fluid (VOF) method to account for the bulking effect. The new model also proposes a method to calculate air entrainment that does not require calibration. In Section 2 the governing equations and the numerical framework used in this study are presented. In Section 3 the computational implementation of the model as well as a description of the experimental setup used for validation are described. In Section 4, the results from the computational model are presented, discussed, and compared with the experiments. Finally, in Section 5 the main conclusions and highlights of this study are discussed.

2. Governing equations and numerical framework

The mathematical model proposed in this study is based on a mixture formulation. Three different phases will be considered in this study: continuous water, continuous air, and dispersed air (or bubbles). Even though formally continuous air and dispersed air represent the same phase, the physics governing their dynamics from a macroscopical perspective are different, and for this reason they will be treated separately. Governing equations are derived from mass and momentum conservation for each phase (Drew 2014). The equations are manipulated to derive a momentum conservation for the mixture and mass conservation for each phase as a function of the mixture velocity. The water and bubble phases are combined into a two-phase mixture phase that will be used for the numerical treatment of the free surface. The final equations are:

$$\frac{\partial \alpha_m}{\partial t} + \nabla \cdot (\alpha_m \boldsymbol{u}) + \nabla \cdot \alpha_m (1-\alpha_m)\boldsymbol{u}_{m-a}] = S_b \tag{1}$$

$$\frac{\partial \alpha_b}{\partial t} + \nabla \cdot (\alpha_b \boldsymbol{u}) + \nabla \cdot \left[\alpha_b \left((1-\alpha_m)\boldsymbol{u}_{m-a} + \frac{\alpha_w}{\alpha_m}\boldsymbol{v}_{b-w}\right)\right] = S_b \tag{2}$$

$$\frac{\partial \rho v}{\partial t} + \nabla \cdot (\rho \boldsymbol{v}\boldsymbol{v}) = -\nabla p_{rgh} + \nabla \cdot (\mu_{eff}(\nabla \boldsymbol{u} + \nabla \boldsymbol{u}^T) + \boldsymbol{T}_{b-w} + \boldsymbol{T}_{m-a}) - \boldsymbol{g} \cdot x \nabla \rho + \boldsymbol{f}_\sigma \tag{3}$$

where Eq. (1) represents the mass conservation for the water-bubble mixture phase, Eq. (2) the mass conservation for the bubble phase and Eq. (3) the momentum conservation for the mixture. In this equations, α_m is the two-phase mixture phase fraction defined as $\alpha_m = \alpha_w + \alpha_b$ where α_w is the water phase fraction and α_b is the bubble phase fraction; $\boldsymbol{u}$ represents the volume averaged mixture velocity vector; $\boldsymbol{u}_{m-a}$ the relative velocity vector between the two-phase mixture phase and the continuous air phase; S_b the phase transfer term between continuous air and dispersed air; ρ the mixture density; $\boldsymbol{v}$ the mass averaged mixture velocity vector; p_{rgh} the reduced pressure; μ_{eff} the sum of viscous and turbulent viscosity of the three-phase mixture; and $\boldsymbol{T}_{b-w}$ and $\boldsymbol{T}_{m-a}$ are the momentum diffusion due to the relative motion between phases and are defined as:

$$\boldsymbol{T}_{b-w} = -\alpha_b \left(1 - \frac{\alpha_b}{\alpha_m}\right) \frac{\rho_b \rho_w}{\rho_m} \boldsymbol{v}_{b-w} \boldsymbol{v}_{b-w}, \tag{4}$$

$$\boldsymbol{T}_{m-a} = -\alpha_m \alpha_a \frac{\rho_a \rho_m}{\rho} \boldsymbol{v}_{m-a} \boldsymbol{v}_{m-a}, \tag{5}$$

$\boldsymbol{g}$ is the acceleration of gravity vector, and $\boldsymbol{f}_\sigma$ the surface tension force. The turbulent closure was performed using the $k-\varepsilon$ model. The $k-\varepsilon$ model uses the concept of eddy viscosity (ν_T) and assumes that the Reynolds stresses are proportional to the mean strain rate. Relative velocities ($\boldsymbol{u}_{m-a}$, $\boldsymbol{v}_{b-w}$) are based on algebraic models (Wüest et al. 1992).

The mass transfer term represents the transfer of air between the continuous air phase and the bubble phase. If it is positive, bubbles are incorporated into the flow in a process of air entrainment. If it is negative, bubbles are being incorporated into the atmosphere in a process of degassing. Traditional approaches in self aerated flows define the mean free-surface location where the bubble concentration is equal to 0.9 (Chanson 1996). For air entrainment, the source term is calculated following the same concept: entraining air into the flow until 0.9 concentration is reached. For degassing, the source term is calculated proportional to the flux of bubbles normal to the free surface. The mass transfer term is only applied in a small region surrounding the free surface. Several authors have proposed different methods to determine the occurrence of air entrainment based on mean parameters of the flow (Ervine et al. 1980, Souders and Hirt 2004, Meireles et al. 2014, Valero 2018). In this study, for the sake of simplicity, the regions where air is entrained will be assigned manually based on the experimental data.

3. Experimental setup and computational implementation

The model was validated using the experiments of Amador (2005) for the non-aerated region and of Felder and Chanson (2013) for the aerated region. The experiments by Amador (2005) were conducted at the Laboratory of Hydraulics at the University of A Coruña. The rectangular stepped spillway was composed of 37 steps, 5 cm high and 4 cm wide each. The width of the spillway was 0.5 m. The first three steps had variable dimensions, emulating the geometry of a WES crest profile. The crest of the spillway also followed a WES crest profile. The flow rate was 0.055 m^3/s and velocity measurements were performed using a PIV technique. The experiments conducted by Felder and Chanson (2013) were conducted at the McKay Hydraulics Laboratory at the University of Queensland. The rectangular stepped spillway was 1.0 m wide, 2.0 m long and 1.0 m high. It was composed of 10 steps, 10 cm high and 20 cm wide. The crest of the spillway consisted of a simple horizontal crest. The discharge was 0.143 m^3/s and air concentration measurements were performed using a double tip conductivity prove. None of the spillways were gated. A schematic representation of the spillways is presented in Figure 1.

The model was implemented in the OpenFOAM framework. OpenFOAM is an open-source computational fluid dynamics (CFD) software implemented in C++ which offers a large library of solvers that include single and multiphase flows, incompressible and compressible flows, heat transfer and combustion, between others. One of the major advantages of OpenFOAM is that their libraries can be reutilized by the user to implement new solvers doing modifications only on the higher levels of the code. The new code implementation is based on *interFoam* solver, which is the existing solver for two incompressible and immiscible fluids using the VOF method.

Simulations for both spillways were done in two-dimensional domains with similar mesh resolutions and boundary conditions (BC). The domains consist of the spillway and a small portion upstream of it. A schematic description of them is presented in Figure 1. The inlet BC specifies a uniform vertical velocity based on the flow rate of the experiment. The boundary condition at the upstream portion of the domain is set up with a slip BC to avoid creating resistance in the flow. The atmosphere BC is defined with a uniform pressure and Neumann boundary condition for the rest of the variables. The wall boundaries are set up with no-slip BC and wall functions for the turbulent parameters. For the outlet, a mixed BC which consist of a Neumann BC when the flow is leaving the domain and a no-slip BC when the flow is entering the domain was used. Both meshes were generated using the tools *blockMesh* and *snappyHexMesh* provided by OpenFOAM. The mesh corresponding to the spillway used by Amador (2005) had 63,000 cells and a mean value of $y^+ = u^* y/\nu$ equal to 71, while the spillway corresponding to Felder and Chanson's (2013) experiments has 110,000 cells and a mean value of y^+ equal to 200. The resolution in both meshes were of approximately 2 mm. A special refinement was done near the walls for the Felder and Chanson (2013) case to obtain a lower value of y^+. Mesh convergence tests were performed, achieving mesh independence in both cases. Schemes with first order accuracy were used for time discretization and second order accuracy for spatial discretizations.

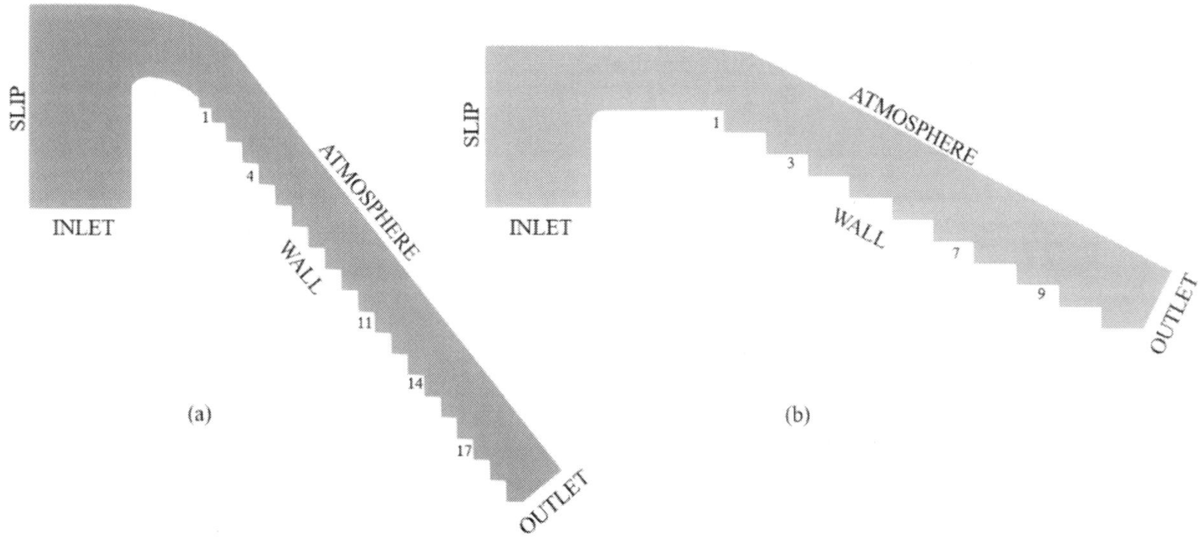

Figure 1. Schematic representation of the domains with boundary conditions and step numbering. (a) Spillway by Amador (2005). (b) Spillway by Felder and Chanson (2013).

4. Results

First, the results of the numerical simulation of the spillway corresponding to the experiments of Amador (2005) are compared with the experimental data. The comparison between water depths in the non-aerated region are presented for four different steps in Table 1. The results show very promising agreement with the experimental data. The numerical simulation tends to overpredict the water depths, with a mean error below 5%. The maximum difference between numerical and experimental results are 3 mm, which is of the order of the mesh size (2 mm).

Table 1. Water depth in the experimental test by Amador (2005) and results from numerical model.

Step	Experimental [cm]	Numerical [cm]
29	3.9	4.0
31	4.0	4.2
33	4.2	4.4
34	4.3	4.6

The dimensional velocity profile in step 10 (nondimensionalized by the mean velocity, U_{mean}, and the water depth, H) obtained with the numerical simulation as well as experimentally is presented in Figure 2. Velocity profiles in the non-aerated region present self-similarity, following a power law, agreeing with the numerous analyses done by other authors (Toro et al. 2017). Simulated velocity profiles show great agreement with the experimental data. The agreement is of the same nature for all cross-sections. A slight underprediction of the mean velocity is in agreement with the small overprediction of the water level. The difference between simulated and experimental values is in general below 5%. When no bubbles are present, the results converge to the *interFoam* solver. The comparison between these profiles and the ones obtained by Toro et al. (2017) show that the results are almost identical between the two solvers. Overall, the results of the model are very promising and show that it is capable of representing the non-aerated region of stepped spillways.

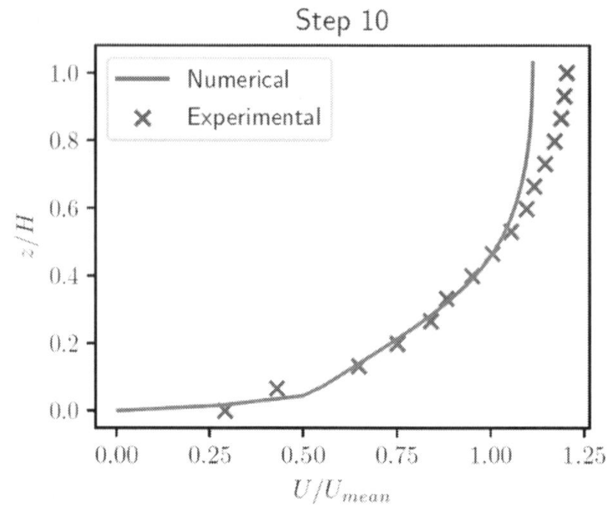

Figure 2. Measured mean velocity profile at step 10 in the experimental test by Amador (2005) superimposed with the results from the numerical simulation.

The experiments of Felder and Chanson (2013) were simulated considering air entrainment. The inception point was located between steps 5 and 6, based on the experimental reports. Free surface location as well as bubble phase fraction are presented in Figure 3. At the inception point, bubble concentration quickly reaches the value of 0.9, but is constrained to a small region near the free surface. As we move downstream, the bubble concentration at the free surface remain almost uniform (between 0.8 and 0.9), but the bubbles reach deeper locations in the flow. Near the pseudo-bottom, the bubble concentration remains close to zero over the domain, in agreement with the experimental data.

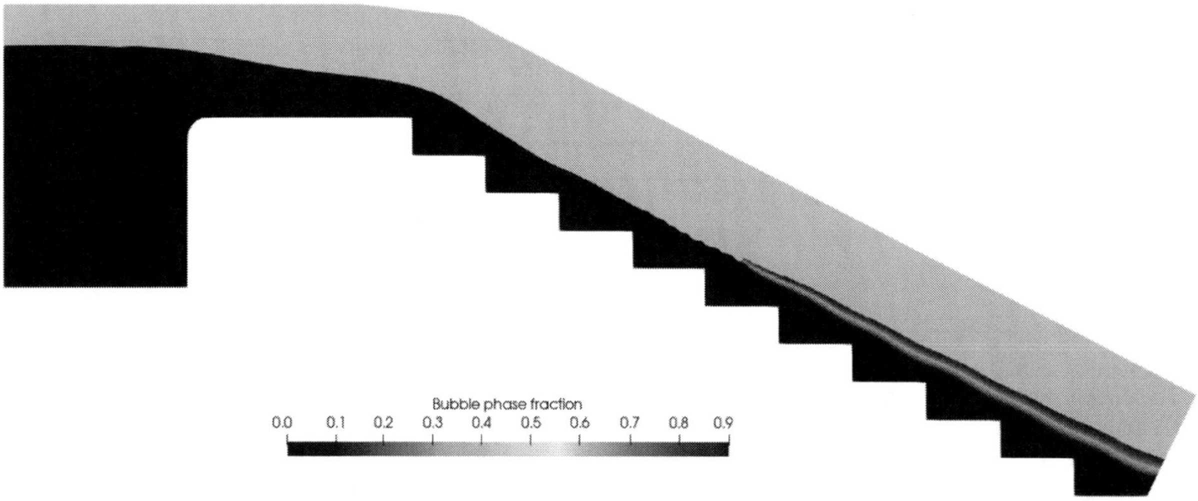

Figure 3. Free surface location superimposed with bubble phase fraction concentration for the simulation corresponding to the experiments of Felder and Chanson (2013).

Figure 3 also shows how the model is capable of representing bulking. Between steps 1 and 5, the flow accelerates and therefore the water depth decreases. At step 6, air starts to be incorporated into the flow and the water depth defined by the location where the air concentration is 0.9 increases even though the flow keeps accelerating. In Table 2, the distance from the pseudo-bottom to the free surface (z_{90}), the equivalent water depth (H_W) and the percentage of bulking are presented for the experiments and the simulations for steps 6 to 10. The equivalent water depth is defined as:

$$H_w = \int_0^\infty (1 - \alpha_w) dy, \tag{8}$$

and the percentage of bulking is defined as:

$$\text{Bulking}\,[\%] = (z_{90} - H_w)/z_{90} \times 100. \tag{9}$$

Numerical and physical model show similar levels of bulking. The numerical model tends to overpredict bulking at first steps, but after step 08, numerical and physical model provide similar values of bulking. z_{90} and H_w are slightly overpredicted by the model, a tendency that is similar to what was observed in the non-aerated region.

Table 2. Mean free surface location, equivalent water depth and bulking for the experiments of Felder and Chanson (2013) obtained from experiments and numerical simulations.

Step	$z_{90,exp}$ [cm]	$H_{w,exp}$ [cm]	Bulking exp.	$z_{90,num}$ [cm]	$H_{w,num}$ [cm]	Bulking num.
06	6.4	5.5	17%	7.5	6.1	23%
07	6.0	5.2	15%	7.5	5.8	29%
08	6.8	5.3	29%	7.5	5.8	30%
09	6.8	5.3	29%	8.0	5.9	37%
10	7.3	5.3	37%	8.0	6.1	31%

Air concentration distribution profiles in steps 6 and 8 are presented in Figure 4. Modeled and measured data show great agreement, with an error under 3% for most of the cases. The agreement is better on the upper portion of the flow than in the lower portions. In steps 8, the model predicts almost no mean concentration of bubbles below $0.3y/z_{90}$, but measured data shows that some bubbles reach positions close to the pseudo-bottom. The shape of the simulated air concentration matches the traditional air distribution in channel flows described by a hyperbolic tangent (Chanson 1996). By the air concentration profiles it can be seen how air penetrates further into the flow and closer to the pseudo-bottom as we move downstream, but the mean concentration stays the same at the free surface (by definition). The novel method to calculate the source term in the bubble transport equation automatically matches numerical and experimental results near the free surface and shows to be more accurate than traditional air entrainment functions to describe self-aerated flows.

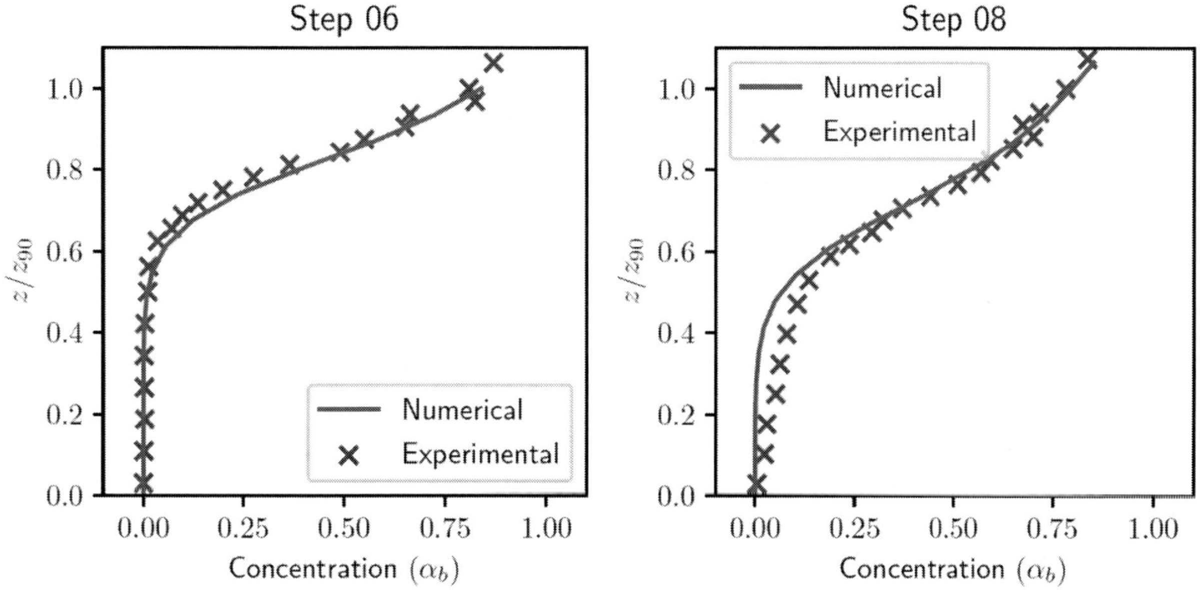

Figure 4. Measured mean air concentration at steps 6 and 8 in the experimental test by Felder and Chanson (2013) superimposed with the results from the numerical simulation.

5. Conclusions

In this study, a new model was proposed for the simulation of self-aerated flows in a RANS framework; it was validated for the flow over a stepped spillway. The new method to calculate the entrainment of air that requires no calibration and it couples the VOF method with a bubble transport equation to account for bulking. The results show that the model is capable of representing the aerated as well as the non-aerated regions of the flow. The modified VOF method converges to the traditional VOF when bubbles are not present. Velocity profiles, water level, air concentration distribution as well as total bulking are well represented by the model.

6. REFERENCES

Amador, A. (2005). "Comportamiento hidráulico de los aliviaderos escalonados en presas de hormigón compactado". Ph.D. thesis, Polytechnic Univ. of Catalonia (UPC), Barcelona, Spain.

Bauer, W.J. (1954). "Turbulent boundary layer on steep slopes". *Trans*, ASCE 119, 1212-1233.

Chanson, H. (1996). "Air Bubble Entrainment in Free-Surface Turbulent Shear Flows". Elsevier.

Chanson, H. (2002). "The hydraulics of stepped chutes and spillways". Taylor & Francis.

Deike, L., Melville, W.K. and Popinet, S. (2016). "Air entrainment and bubble statistics in breaking waves". *J. Fluid Mech.*. 801, 91–129.

Drew, D. and Passman, S. (2014). "Theory of Multicomponent Fluids". Springer, Verlag New York.

Ervine, D.A., McKeogh, E., Elsawy, E.M. (1980). "Effect of turbulence intensity on the rate of air entrainment by plunging water jets". *Proc. Inst. Civil Eng. Part 2*. 69, 425-445.

Felder, S., Chanson, H. (2013). "Aeration, flow instabilities, and residual energy on pooled stepped spillways of embankment dams". *J. Irrig. Drain. Eng.* 139, 880–887.

Halbronn, G. (1954). "Discussion to Turbulent boundary layer on steep slopes". *Trans*, ASCE 119, 1234-1240.

Ma, J., Oberai, A.A., Drew, D.A., Richard T. Lahey, J. and Hyman, M.C. (2011). "A Comprehensive Sub-Grid Air Entrainment Model for RaNS Modeling of Free-Surface Bubbly Flows". *J. Comput. Multiph. Flows*. 3, 41–56.

Matos, J. (1999). "Air entrainment and energy dissipation in flow over stepped spillways". PhD thesis, IST, Portugal.

Meireles I., Bombardelli, F.A., Matos, J. (2014). "Air entrainment onset in skimming flows on steep stepped spillways: an analysis". *J. Hydraul. Res.* 52, 375-385.

Mortazavi, M., Le Chenadec, V., Moin, P. and Mani, A. (2016). "Direct numerical simulation of a turbulent hydraulic jump: turbulence statistics and air entrainment". *J. Fluid Mech.* 797, 60–94.

Souders, D.T. and Hirt, C.W. (2004). "Modeling Entrainment of Air at Turbulent Free Surfaces in Critical Transitions". *Water Resour. Manag.*, ASCE. Reston, VA.

Toro, J.P., Bombardelli, F.A., Paik, J. (2017). "Detached eddy simulation of the nonaerated skimming flow over a stepped spillway". *J. Hydraul. Eng.* 143 (9).

Valero, D. (2018). "Reformulating self-aeration in hydraulic structures: turbulent growth of free surface perturbations leading to air entrainment". *Int. J. Multiph. Flow.* 16 (6).

Ward, J. (2002). "Hydraulic design of stepped spillways". PhD thesis, Colorado State University

Wüest, A., Brooks, N.H. and Imboden, D.M. (1992). "Bubble plume modeling for lake restoration". *Water Resour. Res.* 28, 3235–3250.

Zabaleta, F., Bombardelli, F.A. and Toro, J.P. (2020). "Towards an understanding of the mechanisms leading to air entrainment in the skimming flow over stepped spillways". *Environ. Fluid Mech.* 20, 375–392.

DOI: https://doi.org/10.26077/369c-74c5

Performance Testing of Acoustic Doppler Current Profiler Used For Stream Flow Measurement

Ms. R.B. Deogade[1], Smt. H. R. Khandagale[2] & Shri. M. Someshwara[3]
[1, 2, 3]*Central Water & Power Research Station, Pune, India*
[1]*deogade_rb@cwprs.gov.in*, [2]*harshali.rk@cwprs.gov.in*, [3]*milan.s@cwprs.gov.in*

Abstract: *As water demand is steadily increasing worldwide, it is very important to measure the discharge accurately for sustainable water resources management such as flood control, water distribution, hydraulic structures design, and hydro-environment management. Since decades the cup and propeller type current meters are used for the measurement of stream velocity, but due to emergent technology, Electromagnetic, Acoustic Doppler Current Profiler (ADCP), etc. are used for measuring the stream flows. These instruments have higher edge in accuracy, consistency, and reliability. Quantitative information on performance and validation in the field of discharge measurement is rare. Hence performance testing of such instrument is essential prior to conducting measurements in fields. The objective of the study is to ascertain the accuracy in velocity measurement of these instruments with the standards. The ADCP transmits a constant high frequency sound wave into the water that ricochets off particles suspended in the moving water and reflect back to the instrument with a slightly lowered frequency due to the Doppler Effect. The performance testing of these sophisticated instruments is done in the straight open channel by Current Meter Rating Trolley. The ADCP deployed on the trolley will be towed at a predefined speed of the Rating Trolley and the corresponding velocity indicated by the sensor will establish the relation. This indicates the performance of the ADCP and its appropriate working in terms of accuracy.*

Keywords: *Acoustic Doppler Current Profiler, Current Meter Rating Trolley, Doppler Effect, Water Velocity.*

1. Introduction

The importance of water conservation and environmental friendly technologies is growing today. Accurate measurement of water discharge is important for a country since water has become a limiting resource. For open channel flow measurement, mechanical type current meters are used since a long time but may not be suitable in certain circumstances to overcome this and adopting advance technology, new hydraulic instruments having very good accuracy are used to measure water flow. Acoustic Doppler Current Profiler (ADCP) is one of the water discharge measuring instrument giving very accurate results. Instruments like this were originally designed to study ocean currents - tracking them and calculating velocity profiles - and perform other oceanographic work. Since then, it has been adapted for use in estuaries and rivers. Furthermore, it can be employed to measure flows in tidal rivers and, to some extent, monitor sediment fluxes. As a preferred method for measuring current flow in an open channel, ADCPs have gained traction. Designed especially for shallow-water applications and improves performance for canals, estuaries, lakes, rivers, and coastal areas.

ADCP can be used during mobile vehicle applications and fixed position deployments. Mobile vehicle applications utilize ADCPs to profile currents and track the sea or river bottom. ADCP can be mounted to ships, vessels and boats to measure velocity profiles to determine flow or discharge in an open channel viz. oceans, rivers or canals and bathymetry studies. Vessel Mount systems have been used to measure currents for climate change and ocean mapping studies, as well as other more specialized research. Depending on the mounting type ADCP can be generalized side looking, up-looking and down-looking and accordingly deployment position can be fixed. In the case of down-looking systems, the ADCP can be installed at the surface or below the surface. For up-looking systems, the ADCP is installed below the surface and measures currents towards the surface. Bottom mounted systems measure a variety of parameters using an ADCP. Most oceanographic applications systems are up-looking and measure currents and waves (direction and wave height). For side-looking position, ADCPs are typically utilized in large channels and in harbors to monitor currents.

At CWPRS, the conventional rotating element current-meters are calibrated using the Current Meter Rating Trolley (CMRT), as per to ISO 3455: 2021. Therefore, it is not unusual to use the CMRT to carry out the performance testing of these sophisticated instruments. These instruments are rarely subjected to quantitative evaluation on their accuracy, validity and performance in the field. In this case, the instruments under test were installed on boat and placed at the central portion of the CMRT and side looking ADCP were attached to rigid rod. The rating trolley is towed with known stable velocity and log the velocity of sensor under test. The purpose of this study is to examine the accuracy of speed measurement of ADCP, ultimately accuracy of discharge estimation can be ensured.

2. Calibration Facility at CWPRS, Pune

Central Water and Power Research Station (CWPRS), Pune, has a rating tank with dimensions of 228 m long, 3.66 m wide, and 2.13 m deep. The Current Meter Rating Trolley (CMRT) is a self-propelled trolley that runs on two parallel straight rails across the length of the rating tank. The rating trolley is equipped with servo motors to run along the track at various speeds. The rating trolley facility has an electronic drive to achieve precise speed in the range from 0.01 m/s to 7.5 m/s and uncertainties in the measurement of calibration parameters within +/- 0.1% at 95% confidence level. Figure 1 shows the internationally adopted Current Meter Rating trolley (CMRT) facility at CWPRS.

Figure 1. Current Meter Rating Trolley (CMRT) facility at CWPRS

Acoustic Sensor type current profiler is suspended and drawn through stagnant water at numerous steady speeds of the Rating Trolley. The velocity indicated by its display unit is established relation with the corresponding Rating Trolley velocity to know the deviation in measurement.

3. Working Principle

3.1. Principle of Acoustic Doppler Current Profiler (ADCP)

The flow measurement technology of the ADCP is based on the Doppler shift phenomenon to measure velocity in flowing water. The sensor transmits a high frequency (hundreds of kHz) ultra- sound wave into the water. This sound wave is reflected back towards the sensor by suspended particles as shown in Figure 2. The reflection of sound waves from a particle, which is moving within a fluid, causes a change in the frequency of the reflected sound wave. The difference in frequency between the transmitted and reflected sound wave is known as the Doppler shift. The ADCPs can be found with 2, 3, 4, 5 or even 9 beams. Figure 3 shows schematic of nine beam, M9-ADCP. Further components of an ADCP are an electronic amplifier, a receiver, a clock to measure the travelling time, a temperature sensor, a compass to know the heading, and a pitch/roll sensor to know the orientation.

An analog-to-digital converter and a digital signal processor are required to sample the returning. [https://www.ysi.com/riversurveyor-s5-m9]

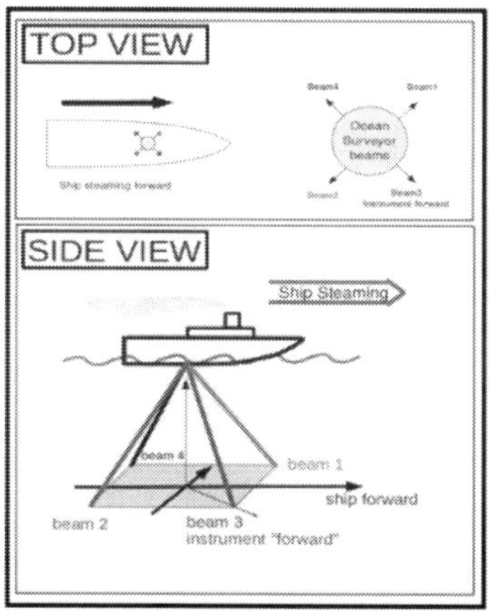

Figure 2. Working Principle of ADCP [Hummon, J. 2009]

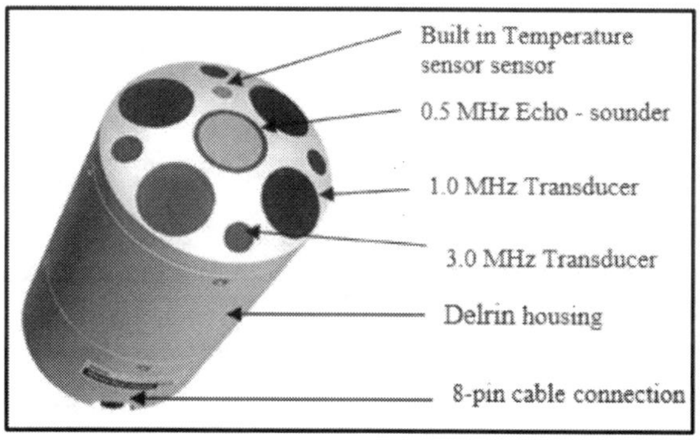

Figure 3. Schematic of M9- ADCP [ysi.com]

4. Experimental Setup

The performance testing of the said ADCP was mounted on the boat. The compass within the ADCP is calibrated by holding a moving boat at two points lengthwise and given clockwise and anticlockwise movement. An experimental set up for ADCP on CMRT in which the moving boat is held rigidly between two rods at the central portion of the rating trolley is shown in Figure 4. As the rating trolley is moved at a specific velocity, the moving boat will also be dragged in the same direction of motion of the rating trolley. The ADCP attached to the boat will measure the velocity of the moving boat and will be recorded on a stand-alone data acquisition system. The data acquisition system acquires real time speed of the boat, water depth, water temperature, and conductivity in engineering unit.

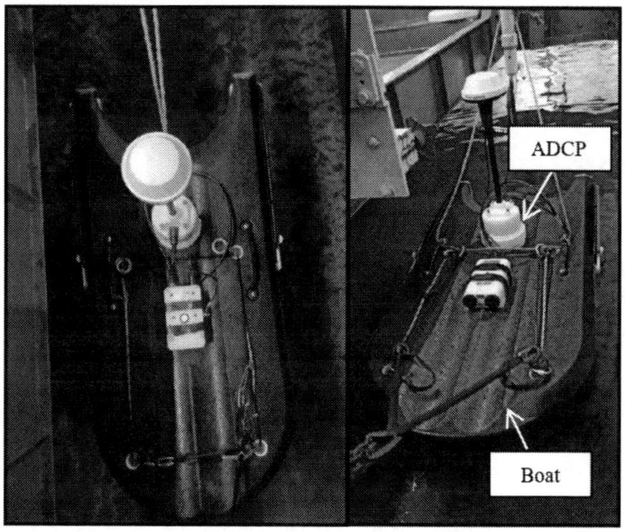

Figure 4. Mounting of moving Boat with ADCP

Figure 5. Mounting of side looking ADCP

5. Performance Testing Procedure

Performance testing of the ADCP is carried out at CMRT facility.

5.1. Procedure for performance testing of boat mounting ADCP is adopted

1. Attach moving boat with ADCP at the central portion of CMRT by means of a wire rope which should have sufficient tension to overcome drag.
2. Now run the rating trolley with pre-defined speed and maintains the speed steadily.
3. Note the velocity of rating trolley and the velocity of moving boat is acquired by real time data acquisition system.
4. Run the trolley to the pre-defined next speed.
5. Stilling of water should be observed during each rating trolley run.
6. Repeat the procedure for defined velocity range.
7. Prepare set of observations of the rating trolley velocity and the corresponding boat velocity measured by the ADCP and calculate the percentage deviation.

5.2. Procedure for performance testing of side looking ADCP is adopted

- Attach the side looking ADCP by rigid rod suspension on Rating Trolley as shown in Figure 5.
- Follows the procedure as mentioned in 5.1 sr. no (2) to (7).

6. Repeatability

To check the consistency of ADCP an experiment was carried for performance testing twice for same serial number of ADCP. It is found that first and second attempts of testing showing overlapping results. Table No. 1 & 2 show the result of repeatability of ADCP and its graphical representation in Figure 6.

Table 1. Performance testing of ADCP for Sr. No. M9 - 4123

Type of Current Meter	Acoustic Doppler Current Profiler (ADCP)		
Sensor Make & Model	Sontek, River Surveyor M9		
Calibration Range	0.01 – 3.0 m/s		
Type of Suspension	Boat Mounting		
Sr. No.	Rating Trolley velocity (m/s)	Boat Velocity indicated by Data Acquisition System (m/s)	Percentage Deviation (%)
	V_T	V_B	$[(V_T - V_B) / V_T]$ x 100
1	0.100	0.100	0.00
2	0.500	0.501	-0.20
3	0.700	0.700	0.00
4	0.900	0.900	0.00
5	1.000	1.000	0.00
6	1.200	1.202	-0.17
7	1.400	1.400	0.00
8	1.800	1.801	-0.06
9	2.000	2.000	0.00
10	2.300	2.300	0.00
11	2.700	2.700	0.00
12	3.000	3.003	-0.10

Table 2. Repeated performance testing of ADCP for Sr. No. M9 - 4123

Type of Current Meter	Acoustic Doppler Current Profiler (ADCP)		
Sensor Make & Model	Sontek, River Surveyor M9		
Calibration Range	0.1 – 3.0 m/s		
Type of Suspension	Boat Mounting		
Sr. No.	Rating Trolley velocity (m/s)	Boat Velocity indicated by Data Acquisition System (m/s)	Percentage Deviation (%)
	V_T	V_B	$[(V_T - V_B) / V_T]$ x 100
1	0.100	0.100	0.00
2	0.500	0.500	0.00
3	0.600	0.601	-0.17
4	0.900	0.900	0.00
5	1.000	1.000	0.00
6	1.200	1.200	0.00
7	1.500	1.500	0.00
8	1.800	1.802	-0.11
9	2.000	2.004	-0.20
10	2.300	2.300	0.00
11	2.500	2.508	-0.32
12	3.000	3.005	-0.17

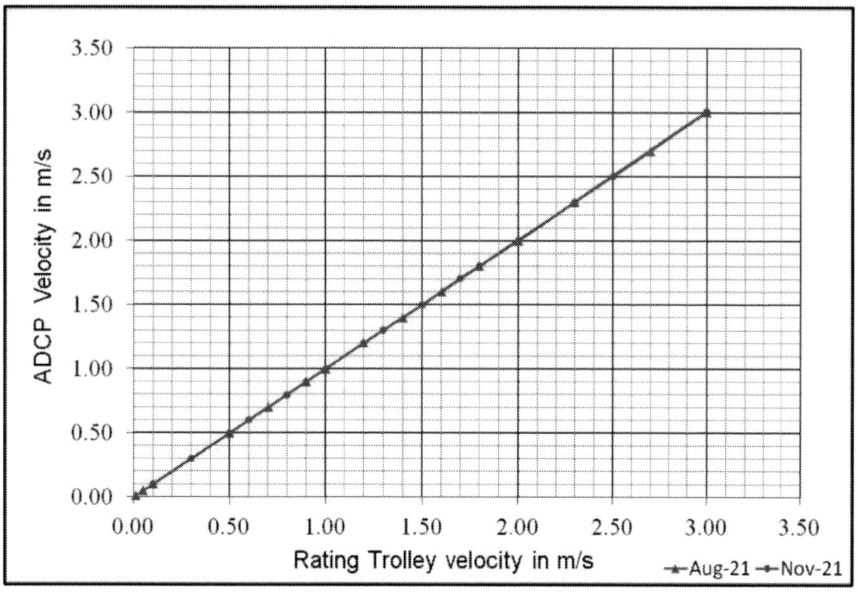

Figure 6. Performance testing of ADCP for Sr. No. 4123 – Repeatability

7. Results and Discussion

The performance testing of various types of ADCP were carried out in CWPRS whose results and their graphical representation is described in Table 3 to Table 8 and Figure 6 to Figure 12 respectively.

Table 3. Performance testing of ADCP for Sr. No. M9 - 6184

Type of Current Meter		Acoustic Doppler Current Profiler (ADCP)	
Sensor Make & Model		Sontek , River Surveyor M9	
Calibration Range		0.05 – 3.5 m/s	
Type of Suspension		Boat Mounting	
Sr. No.	Rating Trolley velocity (m/s)	Boat Velocity indicated by Data Acquisition System (m/s)	Percentage Deviation (%)
	V_T	V_B	$[(V_T - V_B) / V_T]$ x 100
1	0.050	0.050	0.00
2	0.100	0.100	0.00
3	0.500	0.503	-0.60
4	0.700	0.701	-0.14
5	0.900	0.899	0.11
6	1.000	1.000	0.00
7	1.200	1.201	-0.08
8	1.400	1.399	0.07
9	1.600	1.598	0.13
10	1.800	1.801	-0.06
11	2.000	2.001	-0.05
12	2.250	2.250	0.00
13	2.500	2.500	0.00
14	2.750	2.751	-0.04
15	3.000	2.998	0.07
16	3.200	3.203	-0.09
17	3.500	3.499	0.03

Table 4. Performance testing of ADCP for Sr. No. 1723 KDP 1141

Type of Current Meter		Acoustic Doppler Current Profiler (ADCP)	
Sensor Make & Model		MECH KDO, OE Model	
Calibration Range		0.1 – 3.0 m/s	
Type of Suspension		Rigid Rod	
Sr. No.	Rating Trolley velocity (m/s)	Boat Velocity indicated by Data Acquisition System (m/s)	Percentage Deviation (%)
	V_T	V_B	$[(V_T - V_B) / V_T]$ x 100
1	0.111	0.111	0.00
2	0.221	0.220	0.45
3	0.302	0.302	0.00
4	0.404	0.404	0.00
5	0.604	0.603	0.17
6	0.706	0.705	0.14
7	0.806	0.808	-0.25
8	0.908	0.909	-0.11
9	1.014	1.020	-0.59
10	1.202	1.213	-0.92
11	1.418	1.413	0.35
12	1.609	1.609	0.00
13	1.822	1.820	0.11
14	2.013	2.020	-0.35
15	2.224	2.220	0.18
16	2.505	2.520	-0.60
17	3.011	3.010	0.03

Figure 7. Performance Testing of ADCP of Sr. No. M9 – 6184:

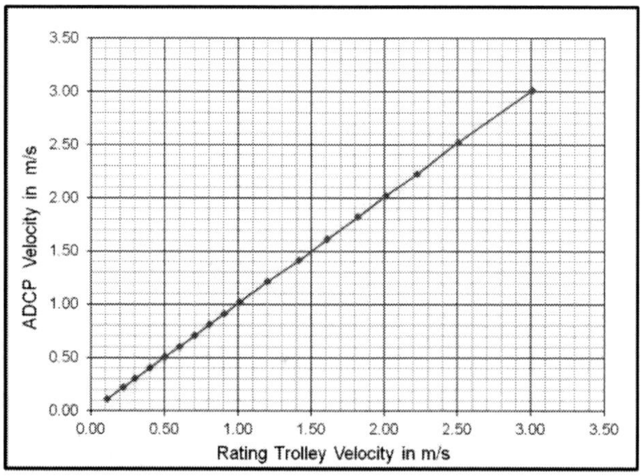

Figure 8. Performance Testing of ADCP for Sr. No. 1723KDO1141

Table 5. Performance testing of ADCP for Sr. No. M9 - 6921

Type of Current Meter		Acoustic Doppler Current Profiler (ADCP)	
Sensor Make & Model		Sontek , River Surveyor M9	
Calibration Range		0.05 – 5.0 m/s	
Type of Suspension		Boat Mounting	
Sr. No.	Rating Trolley velocity (m/s)	Boat Velocity indicated by Data Acquisition System (m/s)	Percentage Deviation (%)
	V_T	V_B	$[(V_T - V_B) / V_T]$ x 100
1	0.050	0.050	0.00
2	0.100	0.100	0.00
3	0.400	0.400	0.00
4	0.500	0.501	-0.20
5	1.000	0.999	0.10
6	1.200	1.201	-0.08
7	1.501	1.501	0.00
8	1.800	1.801	-0.06
9	2.000	2.001	-0.05
10	2.300	2.300	0.00
11	2.499	2.507	-0.32
12	2.700	2.701	-0.04
13	3.198	3.202	-0.13
14	3.500	3.501	-0.03
15	3.990	4.001	-0.28
16	4.500	4.501	-0.02
17	5.000	5.010	-0.20

Table 6. Performance testing of ADCP for Sr. No.M9-7043

Type of Current Meter		Acoustic Doppler Current Profiler (ADCP)	
Sensor Make & Model		Sontek , River Surveyor M9	
Calibration Range		0.1 – 5.0 m/s	
Type of Suspension		Boat Mounting	
Sr. No.	Rating Trolley velocity (m/s)	Boat Velocity indicated by Data Acquisition System (m/s)	Percentage Deviation (%)
	V_T	V_B	$[(V_T - V_B) / V_T]$ x 100
1	0.100	0.100	0.00
2	0.501	0.503	-0.40
3	0.700	0.700	0.00
4	1.000	1.000	0.00
5	1.200	1.205	-0.42
6	1.400	1.398	0.14
7	2.000	2.000	0.00
8	2.300	2.307	-0.30
9	2.500	2.496	0.16
10	2.700	2.701	-0.04
11	3.000	2.996	0.13
12	3.200	3.197	0.09
13	3.500	3.506	-0.17
14	4.000	4.009	-0.23
15	4.300	4.297	0.07
16	4.500	4.491	0.20
17	5.000	5.014	-0.28

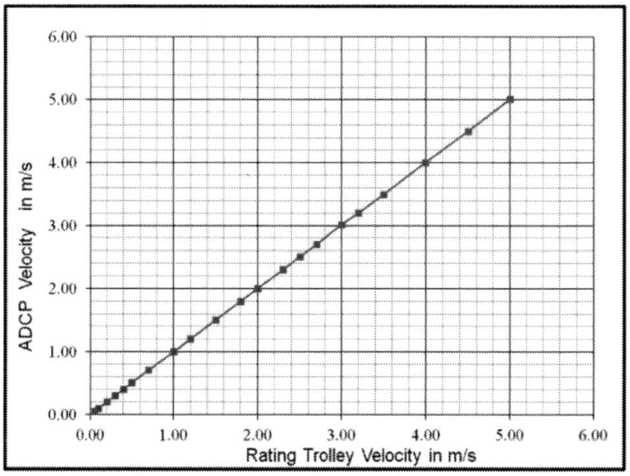

Figure 9. Performance testing of ADCP for Sr. No. M9 – 6921

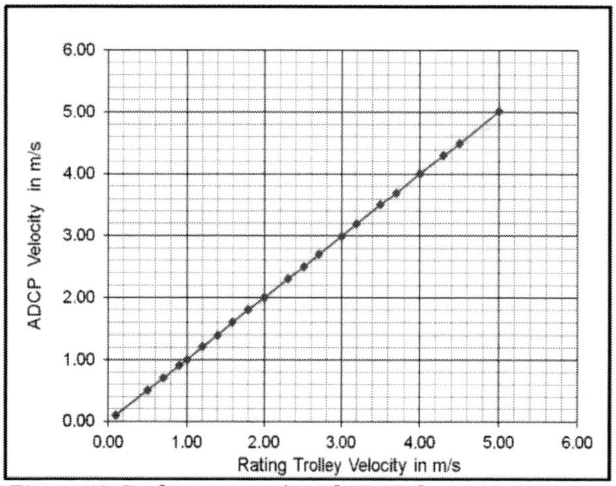

Figure 10. Performance testing of ADCP for Sr. No M9-7043

Table 7. Performance testing for Side Looking ADCP Sr. No. E - 2927

Type of Current Meter		Acoustic Doppler Current Profiler (ADCP)	
Sensor Make & Model		Sontek , SL - 500	
Calibration Range		0.2 – 3.5 m/s	
Type of Suspension		Rigid Rod	
Sr. No.	Rating Trolley velocity (m/s)	Velocity indicated by Data Acquisition System (m/s)	Percentage Deviation (%)
	V_T	V_B	$[(V_T - V_B) / V_T]$ x 100
1	0.217	0.219	-0.92
2	0.609	0.609	0.07
3	0.797	0.792	0.69
4	1.008	1.005	0.35
5	1.213	1.204	0.78
6	1.401	1.406	-0.36
7	1.591	1.610	-1.19
8	1.802	1.798	0.22
9	2.013	2.033	-0.99
10	2.201	2.213	-0.52
11	2.409	2.395	0.60
12	2.602	2.600	0.08
13	2.792	2.803	-0.39
14	3.015	3.023	-0.27
15	3.202	3.202	0.02
16	3.516	3.533	-0.48

Table 8. Performance testing for Side Looking ADCP Sr. No. 1510009

Type of Current Meter		Acoustic Doppler Current Profiler (ADCP)	
Sensor Make & Model		Sontek , SL-1500	
Calibration Range		0.2 – 3.5 m/s	
Type of Suspension		Rigid Rod	
Sr. No.	Rating Trolley velocity (m/s)	Velocity indicated by Data Acquisition System (m/s)	Percentage Deviation (%)
	V_T	V_B	$[(V_T - V_B) / V_T]$ x 100
1	0.196	0.196	0.00
2	0.407	0.406	0.25
3	0.612	0.61	0.33
4	0.800	0.795	0.63
5	1.002	0.992	1.00
6	1.198	1.1825	1.29
7	1.406	1.41	-0.28
8	1.598	1.598	0.00
9	1.790	1.793	-0.17
10	1.995	1.995	0.00
11	2.192	2.193	-0.05
12	2.391	2.395	-0.17
13	2.605	2.605	0.00
14	2.801	2.805	-0.14
15	3.030	3.015	0.50
16	3.499	3.486	0.37

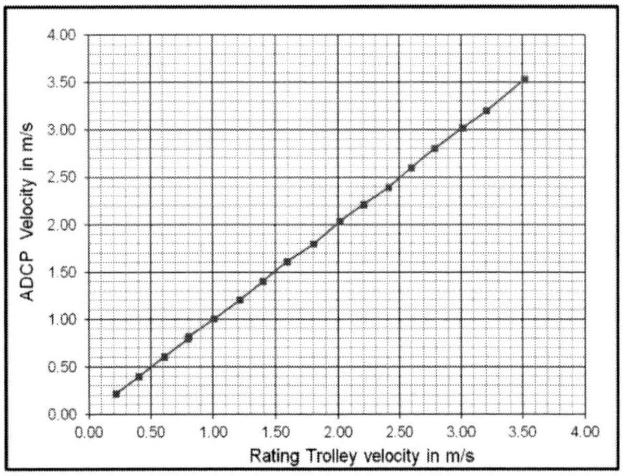

Figure 11. Performance Testing of Side Looking ADCP for SL 500

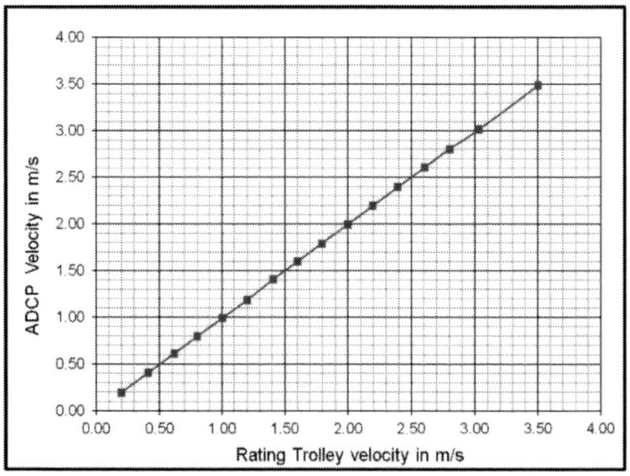

Figure 12. Performance Testing of Side Looking ADCP for SL 15100

8. Conclusion

The instruments like ADCP sensors are used extensively by water management bodies for a variety of hydraulic measurements. An acoustic instrument has an edge over a conventional current meter as it measures extreme low-flow conditions and has the ability to measure velocity in streams affected by backwater, tides etc. In this laboratory study, performance testing of acoustic sensors has been carried out for their velocity component. From this study, it is concluded that;

- Performance testing of the acoustic sensor is necessary to compare it with standards to estimate precision and consistency.
- The maximum deviation found by Acoustic Doppler Current Profiler (River Surveyor) is within the specified accuracy limits.
- The performance testing of ADCP shows adequate matching during repeatability.
- To ensure repeatability and consistency in precision, it is advisable to test the performance of these sensors periodically.
- The percentage deviation for the entire study was found between -1.19 to +1.29.

9. ACKNOWLEDGEMENT

We wish to express our deep sense of gratitude to Dr. R. S. Kankara, Director, CW&PRS, for constant encouragement and valuable suggestions.

10. REFERENCES

Hummon, J. (2009). "Calculating Ocean velocity from ADCP data" https://currents.soest.hawaii.edu/docs/adcp_doc_alpha (April, 25, 2020).

https://www.ysi.com/riversurveyor-s5-m9

https://www.cwprs.gov.in

DOI: https://doi.org/10.26077/1bf0-acd4

Physical and Numerical Model Studies of Hirakud Dam Additional Spillway –A case study

V. P. Gadhe[1] & S. R. Patnaik[1]

[1]*Central Water and Power Research Station, Pune -411024, India*
E-mail: vpgadhe@gmail.com, sangeeta_patnaik@rediffmail.com

Abstract: *Stilling basins are frequently used as energy dissipaters to protect the downstream hydraulic structures from scouring. The stilling basin uses the hydraulic jump as the hydraulic element for dissipating energy in a specific structure. A Stilling basin type of energy dissipator is mainly used for economic reasons; therefore, the Stilling Basin II type has been used on high and earth dam spillways and large canal structures. Basin II type design aims to reduce the hydraulic jump length by installing accessories, such as chute blocks and dentated sill in the stilling basin.*
Hirakud Dam is one of the multipurpose projects. In order to safely pass the reassessed additional flood of Hirakud dam, two additional spillways at suitable locations were proposed. One of these was proposed on the main spillway's left bank (Phase-I). Hydraulic model studies of this left bank additional spillway was carried out at CWPRS. This paper is focused on the physical and numerical model studies to assess the performance of the proposed BASIN II type energy dissipator and to finalize the design of the additional spillway on the left bank (Phase-I). Physical model study results revealed that the performance of the original spillway design with energy dissipators was unsatisfactory as the length of the stilling basin provided was insufficient to contain the hydraulic jump. Hence the modification in the length of stilling basin was suggested by the CWPRS. The physical model studies have also been carried out on the modified design and finalized the design of spillway and energy dissipator. The numerical model studies validate the simulation results with the physical model study results for the original and modified design. Numerical Model studies have been carried out on the licence copy of Flow 3D CFD software at CWPRS.

Keywords: *Additional Spillway, Chute block, Dentated sill, Flow 3D CFD, Type II Stilling Basin,*

1. Introduction

Although hundreds of stilling basins and energy dissipating devices have been designed in conjunction with spillways, outlet works, and canal structures, it is often necessary to make model studies of individual structures to be certain that these devices will operate as anticipated. The reason for these repetitive tests is that uncertainty exists regarding the overall performance characteristics of energy dissipators. In this paper, the authors have discussed the contribution of the hydraulic model study in finalizing the design of the spillway and energy dissipator of the additional spillway of Hirakud Dam. Hirakud Dam is more than 60 years old and is the longest earthen Dam in the Asia. The total length of the Dam, including Dyke is about 25.80 km. It is constructed across the River Mahanadi in the state of Odisha. In the existing Dam, there are two spillways: left and right. The overall length of these two spillways is 4.8 km, and the total discharging capacity of both the spillways is 42450 m^3/s. In 2014 the inflow design flood was reassessed by Central Water Commission (CWC) and it has been increased from 42450 m^3/s to 69632 m^3/s. Therefore, the Central Water Commission (CWC) advised constructing additional spillways at suitable locations. Based on the CWC recommendation, two additional spillways on the left and right sides of the existing spillway of discharge, 9122 m^3/s and 13571 m^3/s, respectively, have been proposed to pass the revised flood safely. Hydraulic model studies of this left bank additional spillway are carried out at CWPRS on a sectional model (1:40 geometric scale) to assess the performance of the original design of the spillway and energy dissipator and finalize the design of the additional spillway on the left bank (Phase-I).

1.1. Physical model

The proposed left bank additional spillway is consisting of five nos. of spillway bays of size 15 m x 15 m each to discharge 9122 m^3/s at FRL El. 192.024 m and MWL El. 192.454 m. A glass-sided one-meter-wide flume was used to incorporate the prototype design into a 1:40 geometrically similar 2-D sectional model. In the 2-D sectional model, one full and two half-span spillways with two full piers along with the stilling basin were incorporated. Necessary arrangements were made to measure the hydraulic parameters over the spillway and energy dissipator. The piezometers were used to measure pressures over the spillway surface; the piezometers were provided along the center of the spillway's span and the side of the pier (CWPRS, 2018). Figures 1, 2, 3a, and 3b show the general layout plan, the section of the spillway (original design), and the upstream and downstream view of the physical model (original design), respectively. The hydraulic parameters like discharging capacity of the spillway, water surface profiles, pressures over the spillway surface, performance of the spillway, and energy dissipator were studied in the model for various discharges for gated and ungated flow. Physical model study results revealed that the performance of the original spillway design with energy dissipators was unsatisfactory as the length of the stilling basin provided was insufficient to contain the hydraulic jump. Hence to improve its performance of the stilling basin the length of the stilling basin was suggested to increase from 71 m to 90 m. The physical model studies were also carried out on the modified design of stilling basin and finalized the design of the spillway and energy dissipators. In this paper, the physical and numerical model studies of spillway and energy dissipator were discussed for the prototype design discharge of 9122 m^3/sec; in the model scale, the design discharge is worked out as 0.36 m^3/sec for one full and two half bays of the spillway.

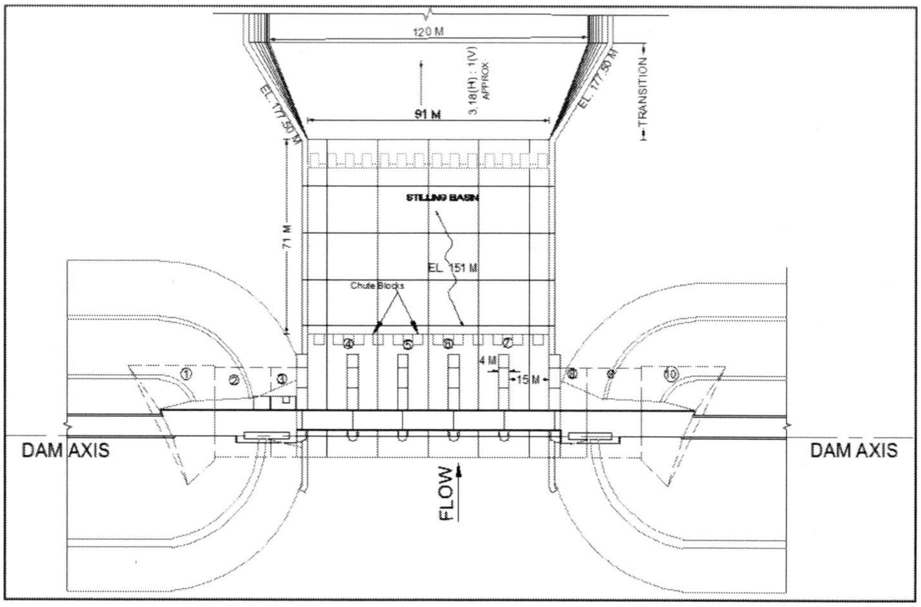

Figure 1. General layout plan

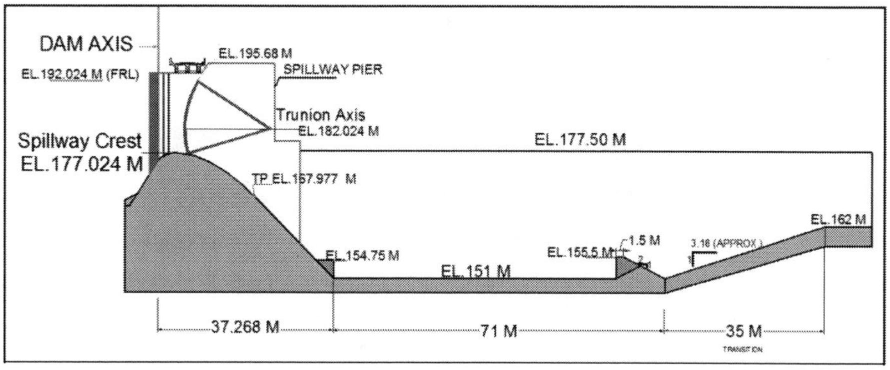

Figure 2. Original section of the additional spillway

Figure 3a. Upstream view of the model (original design)

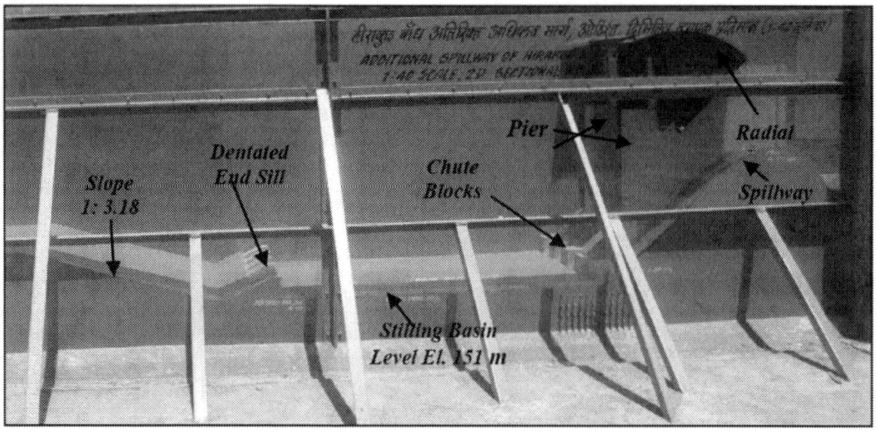

Figure 3b. Downstream view of the model (original design)

1.2. Numerical model

Hydraulic jump is the abrupt transition from supercritical to subcritical flow in open channels. This phenomenon involves strong velocity and pressure fluctuations, intense air entrainment, and significant energy dissipation. For decades, an experimental approach to studying the hydraulic jump has been used. In addition to the experimental study, computational fluid dynamics (CFD) techniques have emerged as a helpful tool over the last two decades. Numerical studies have addressed the hydraulic jump developed in stilling basins, analysing its characteristics and the influence of the energy dissipation devices (Carvalho et al. 2008; Macián-Pérez et al. 2020a; Valero et al. 2018). Nevertheless, CFD techniques still have some limitations when simulating complex hydraulic phenomena (Blocken and Gualtieri 2012; Bombardelli 2012). Therefore, the support of physical modeling to provide validated numerical models remains of paramount importance (Valero et al. 2018).

Numerical simulation is based on the governing equations of the flow, which are usually Navier–Stokes equations together with turbulence models. Finite volume and finite element methods are mainly used to solve turbulence problems. Several turbulence models have been proposed for solving the turbulent flow. In this regard, the K-epsilon, RNG K-epsilon, and large eddy simulation (LES) are the advanced turbulence models that recently have been widely used in hydraulic engineering (Versteeg 2007). In this regard, the commercial software Flow-3D, developed by Flow Science, has been used for the numerical modeling of the flow.

1.2.1. Numerical model setup

The numerical model studies were carried out to observe the performance of the spillway and energy dissipator for original and modified design of the stilling basin; therefore, two geometries were generated with the help of AutoCAD software in the 1:40 geometrically similar physical model scale. The numerical simulations were carried out for the prototype design discharge of 9122 m^3/sec (0.36 m^3/sec in the model scale) for the ungated operation of the spillway. In the Flow 3D software, geometry files are imported from Auto-Cad software. stl format. The geometry of the spillway profile, including stilling basin, is embedded in the computational grid by the pre-processor using the FAVOR concept (Hirt 1985). In a CFD numerical model, a mesh is a subdivision of the flow domain into relatively small regions called cells, in which numerical values such as velocity and pressure are computed. Determining the appropriate mesh domain and a suitable mesh cell size is critical to any numerical model simulation. Mesh and cell size can affect both the accuracy of the results and the simulation time, so it is essential to minimize the number of cells while including enough resolution to capture the critical features of the geometry as well as sufficient flow detail. An effective way to determine the critical mesh size is to start with a relatively large mesh and then progressively reduce the mesh size until the desired output no longer changes significantly with any further reductions in mesh size. The unsteady free surface calculations require fine grid spacing and a small initial time step. Hence in the upstream domain cell size of 0.02 (total cell count of 1 million) was provided whereas in the spillway domain fine mesh with a cell size of 0.01 (total cell count of 3.2 million) was provided and in the downstream coarse mesh with a cell size of 0.02 (total cell count 0.8 million) was provided to optimize the use of computational resources and time. Figures 4a and 4b show the numerical model setup in Flow-3D for the Hirakud

2-D sectional model for the original design of the energy dissipator, whereas figure 5 shows grid distribution for the computational domain of the model. Physical model study data was used for numerical simulation. Numerical simulations were carried out to study the performance and location of hydraulic jump in the stilling basin. Energy dissipation after the jump was also calculated from the residual velocity. The validation was carried out for the original and modified design of the energy dissipator for the design discharge of 9122 m^3/s (0.36 m^3/sec in the model scale) for the ungated operation of the spillway.

1.2.2. Boundary conditions

Setting up the appropriate boundary conditions can significantly impact whether the numerical model results reflect the actual situation one is trying to simulate. In this case, flow data from free surface flows is desired, so the outer edge of the upstream and downstream of the domain was specified as a wall. The bottom boundary was set at the bed of the model geometry in order to capture the channel bed, while the top boundary was set just above the highest water elevation. Since the goal of the simulation was to model flow rate over spillway with respective head water / tail water levels for comparison to physical model data, the upstream boundary was set as a volume flow rate, and the downstream boundary was set as a pressure outlet. Numbers of the probe were provided at different locations on the surface of the spillway for observing different parameters of the flow, such as pressures over the spillway profile and free surface elevation, velocity, etc. Baffles were provided at the inlet and outlet boundary of the domain to observe the discharge passing through the spillway. Each simulation was conducted for up to 50 seconds, which was enough for flow stabilisation.

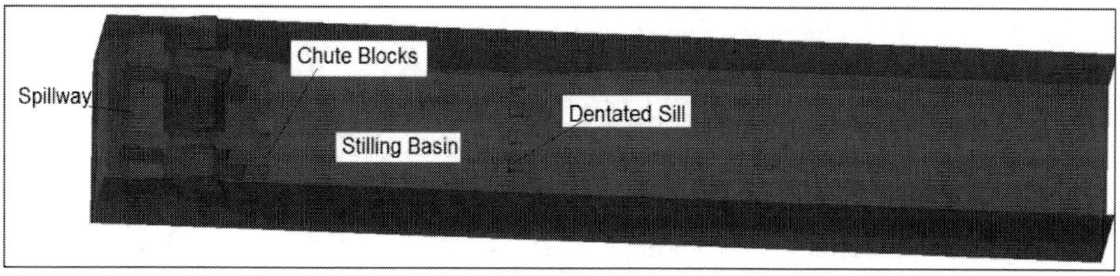

Figure 4a. Numerical model setup of Hirakud Dam additional spillway for the original design of energy dissipator

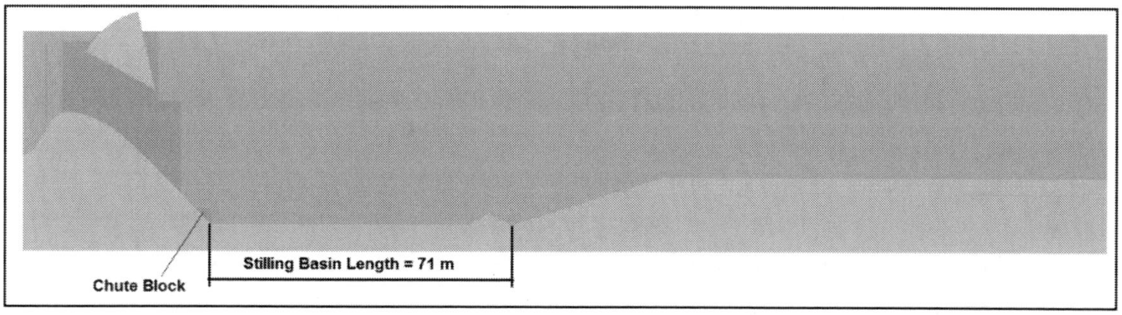

Figure 4b. Details of the Hirakud Dam additional spillway for the original design of energy dissipator

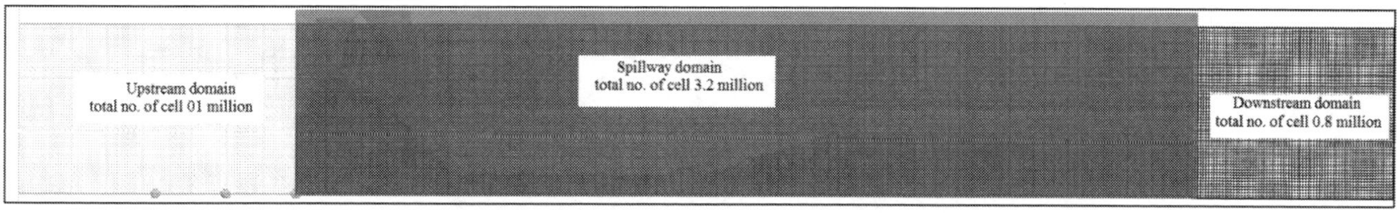

Figure 5. Grid distribution for the computational domain of the model

2. Comparison between numerical model results with physical model results for the original design of spillway

Numerical model results were validated with physical model results of the water surface profile, pressure, and surface velocity over the spillway. To observe the results into prototype design. The physical and numerical model study results were converted into the prototype scale as shown in the following figures. Therefore in the text, prototype and model scale both the dimensions have been mentioned.

2.1. Water surface profiles over the spillway for fully open gate operation

Water surface profiles were observed along the spillway's center line for ungated spillway operation. It was observed that the trunnion axis of the radial gates was well above the water surface for design discharge of 9122 m^3/s (0.36 m^3/sec in the model scale) for the ungated operation of the spillway. As such, the elevation of trunnion was found to be in order. The water surface profile was also not overtopping the training walls. Therefore, both the training walls' heights were sufficient to contain the flow. The heights of the training walls may be finalized based on the water surface profiles, bulking of flow due to air entrainment in the prototype, and freeboard. Numerical studies were carried out for the design discharge of 9122 m^3/s (0.36 m^3/sec in the model scale) under similar conditions. Water surface profile results of both the model study were converted to prototype scale for five no. of spillway bays to compare the results, as shown in figure 6. The numerical results also showed similar observations and close agreement with the physical model study.

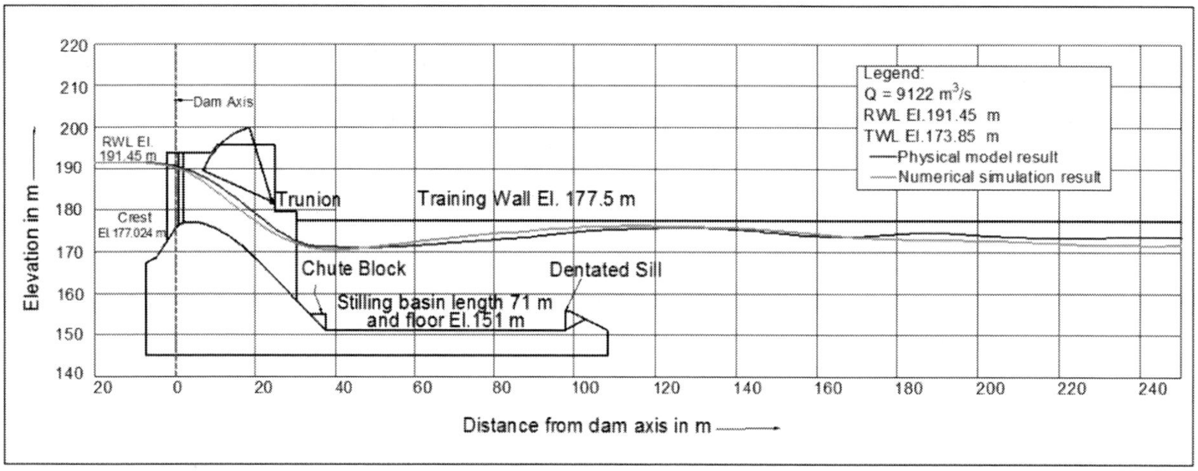

Figure 6. Comparison of water surface profile between physical and numerical model result converted to prototype scale for the discharge of 9122 m^3/s for the ungated condition of the spillway (original design)

2.2. Pressure profile

Studies were conducted to observe hydrostatic pressures along the center line of the spillway. It was found positive along the center line throughout the spillway profile for the discharge of 9122 m^3/s (0.36 m^3/sec in the model scale) under ungated condition. Therefore, the performance of the spillway profile in respect of pressures was considered satisfactory. The pressures observed on the chute blocks and dentated end sills were also found to be acceptable. In the numerical model study to find out the pressure profile over the spillway, probes were provided along the center line of the spillway up to 110 m (2.75 m in the model) total of 23 nos. of the probe were provided at the same of physical model locations over the spillway profile and stilling basin to calculate pressures. Pressure profile results of both the model study were converted into prototype size to compare, as shown in figure 7. The pressures obtained from the numerical model were validated and found to be in agreement with the results of the physical model.

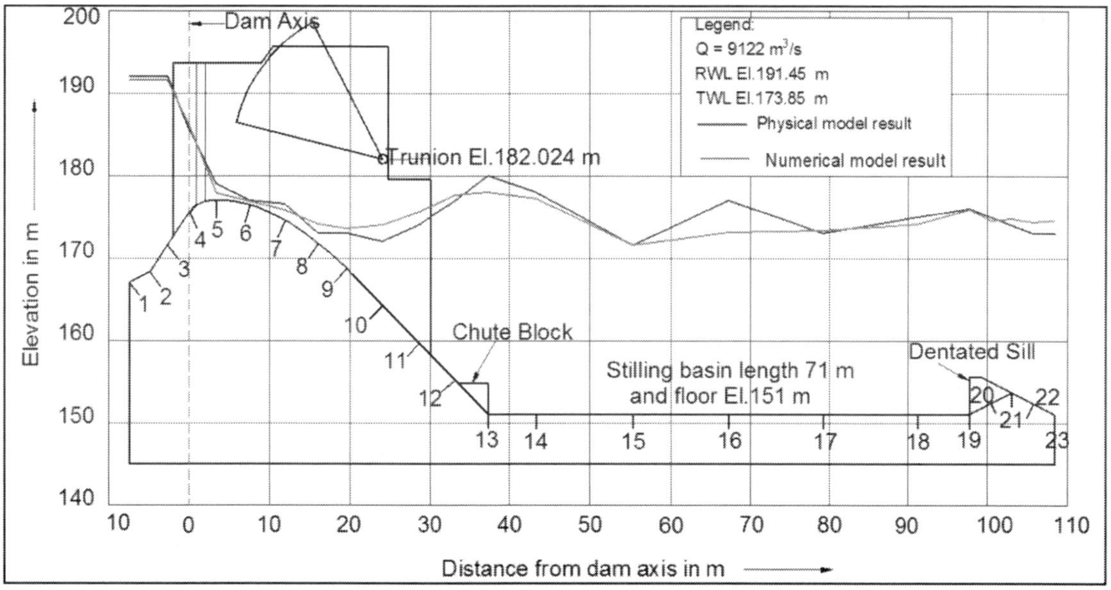

Figure 7. Comparison of pressure profile between physical and numerical model result converted to prototype scale for the design discharge of 9122 m³/s for the ungated condition of the spillway (original design)

2.3. Performance of stilling basin

The performance of stilling basin was observed for the ungated operation of the spillway for the discharge of 9122 m^3/s (0.36 m^3/sec in the model scale). It was observed that the performance of stilling basin was unsatisfactory as the length of the stilling basin provided was insufficient to contain the hydraulic jump. Therefore, the length of the stilling basin of 71 m (1.775 m in the model) was insufficient for energy dissipation. Hence it was recommended to increase the length of stilling basin from 71 m to 90 m (1.775 m to 2.25 m in the model) to improve its performance of the stilling basin. The performance of the stilling basin was also observed in the numerical simulation. Figures 8 and 9 show the flow conditions observed in the stilling basin for the discharge of 9122 m^3/s (0.36 m^3/sec in the model scale) for the ungated spillway operation in the physical model and numerical simulation, respectively. Velocities were measured up to chainage of 50 m (1.25 m in the model) in the upstream and chainage up to 110 m (2.75 m in the model) downstream from the dam axis with an interval of 20 m (0.5 m in the model) along the central span of the spillway. Table 1 compares prototype surface velocities in m/sec observed in the physical model and numerical simulation.

Figure 8. Flow conditions in the stilling basin for the ungated operation of the spillway for the discharge of 9122 m³/s (100%) (0.36 m³/sec in the model scale) in the physical model

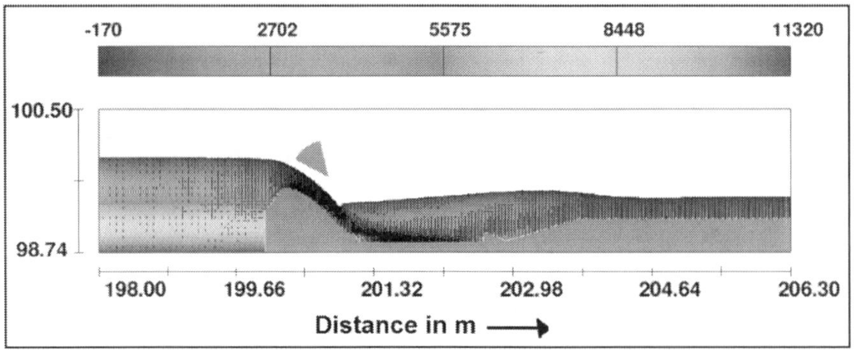

Figure 9. Flow condition in the stilling basin for the ungated operation of the spillway for the discharge of 9122 m^3/s (100%) (0.36 m^3/sec in the model scale) in numerical simulation

Table - 1 Surface velocities observed in physical model and numerical simulation (original design) for the design discharge of 9122 m^3/sec		
Chainage from Dam Axis (m)	**Physical model result velocity (m/s)**	**Numerical simulation result velocity (m/s)**
Upstream		
50	3.4	4.0
30	3.0	3.5
10	3.3	4.5
Dam Axis	6.8	7.1
Downstream		
10	10.7	12.65
30	17.5	18.08
50	7.8	8.22
70	8.4	9.61
90	4.7	5.06
110	5.1	5.26
130	4.2	4.7
150	6.5	6.7

3. Revised design of stilling basin

Taking due cognizance of the result of the model studies, a 1:40 scale geometrically similar 2-D sectional model was modified, incorporating a revised length of stilling basin of 90 m (2.25 m in the model) and stilling basin floor level El. 151 m for assessing the performance of stilling basin (CWPRS, 2019). Figure 10 shows the downstream view of the model.

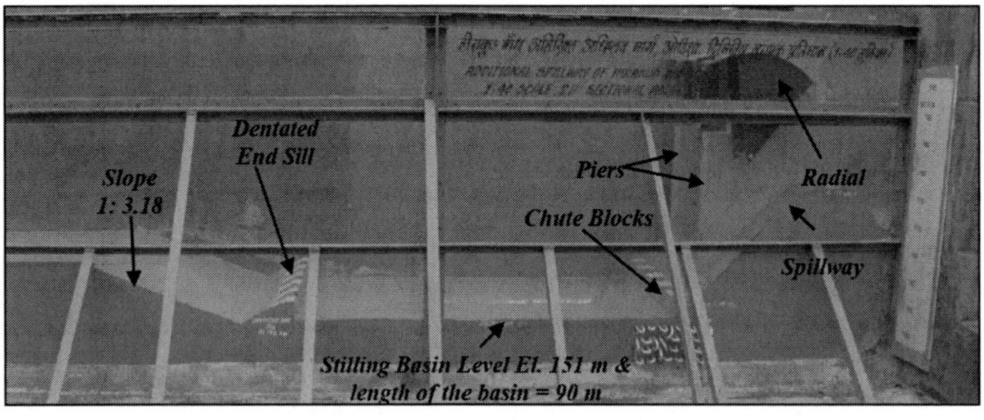

Figure 10. Downstream view of the revised design of the model (stilling basin length = 90 m)

3.1. Numerical model of the modified design of stilling basin

Modification of the stilling basin was also incorporated into the numerical model. Findings of the physical model studies of the revised design were also compared with the numerical simulation. Figure 11 shows the numerical model setup in Flow-3D for the Hirakud 2D sectional model for the modified design of the additional spillway. Grid distribution was done the same as the original numerical model set up.

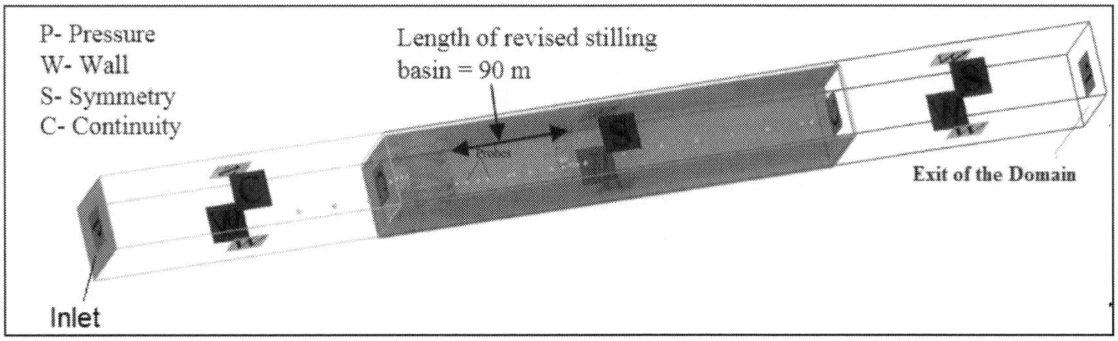

Figure 11. Numerical model setup of additional spillway of the Hirakud Dam for the modified design of energy dissipator

3.2. Performance of modified design of stilling basin

The model studies were conducted by incorporating the modified length of the basin in the existing model. The studies indicated that the performance of the stilling basin was improved as the hydraulic jump was contained within the stilling basin. Therefore, the length of the stilling basin of 90 m with stilling basin floor level El. 151 m was found to be satisfactory. The stilling basin's performance was also marginally compared to the original design. The numerical simulation also observed the hydraulic jump within the stilling basin. Figures 12 and 13 show the performance of the modified stilling basin for the discharge of 9122 m^3/s for the ungated spillway operation of the physical model and numerical simulation, respectively. Velocities were also measured upstream and downstream of the spillway along the central span at various chainages from the dam axis. It was observed that, compared to the residual velocity of the modified design, the original design showed marginal improvement. The numerical simulation also improved the residual velocity over the original design. Table 2 shows the surface velocities observed in the physical model and numerical simulation for the discharge of 9122 m^3/s ungated spillway operation for the modified stilling basin design.

Figure 12. Flow conditions in the modified stilling basin for the ungated operation of the spillway for the discharge of 9122 m^3/s (100%) in the physical model (stilling basin length = 90 m)

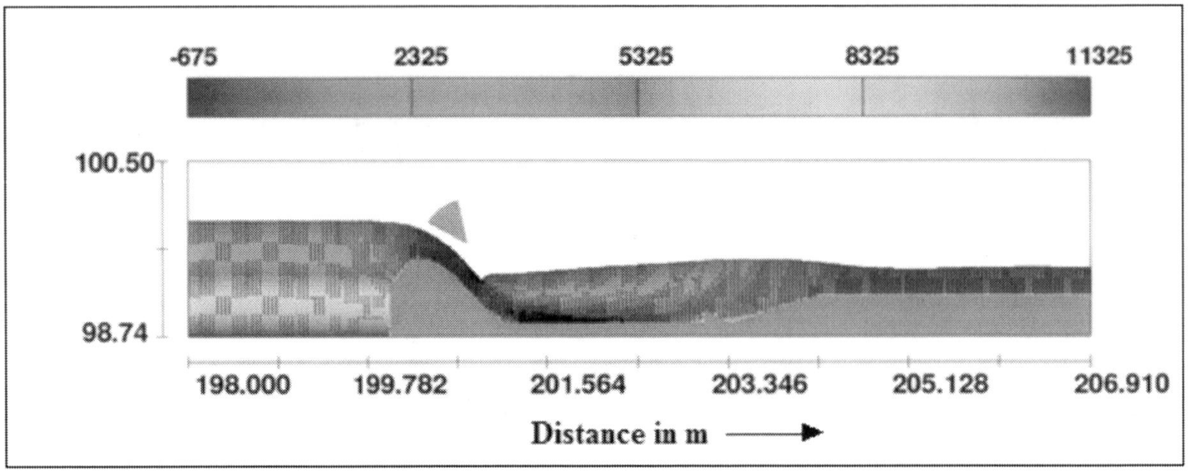

Figure 13. Flow conditions in the modified stilling basin for the ungated operation of the spillway for the discharge of 9122 m³/s (100%) in numerical simulation

Table - 2		
Surface velocities observed in physical model and numerical simulation (modified design) for the design discharge of 9122 m³/sec		
Chainage from Dam Axis (m)	**Gates Fully Open**	
	Physical model result velocity (m/s)	**Numerical simulation result velocity (m/s)**
Upstream		
50	3.4	3.9
30	3.0	3.5
10	3.3	4.1
Dam Axis	6.8	7.1
Downstream		
10	10.6	11.0
30	16.7	17.1
50	7.7	7.8
70	7.7	7.9
90	4.1	4.3
110	4.8	3.6
130	2.0	3.0
150	3.9	4.2

4. Conclusions

Model studies of the original spillway design indicated that the trunnion axis of the radial gates was well above the water surface; hence, the elevation of the trunnion was found to be in order. The water surface profile was also not overtopping the training walls. Therefore, both the training walls' heights were sufficient to contain the flow. The heights of the training walls may be finalized based on the water surface profiles, bulking of flow due to air entrainment in the prototype, and freeboard. The performance of the spillway profile in respect of pressures was considered satisfactory. The performance of stilling basin was found to be unsatisfactory as the length of the stilling basin provided was insufficient to contain the hydraulic jump. As the length of the stilling basin of 71 m was insufficient for energy dissipation, it was recommended to increase the length of the stilling basin from 71 m to 90 m to improve its performance of stilling basin. Modified design of stilling basin was also incorporated into the model. It was observed that the hydraulic jump was forming in the stilling basin. Hence, the stilling basin's performance was improved marginally compared to the original design. Therefore, the length of the stilling basin of 90 m with stilling basin floor level El. 151 m was found to be satisfactory. Thus, the model studies played an important role in enhancing the overall performance of the spillway and energy dissipator arrangement, which will ultimately help the project in achieving a techno-economic design of spillway. With the physical model studies, numerical simulation is also used as a complementary tool for assessing the hydraulic performance of structures. Utmost expertise is necessary for the analysis and interpretation of the results of numerical simulation.

5. ACKNOWLEDGEMENT

The authors are thankful to Dr. R. S. Kankara, Director, and Shri. Y. N. Srivastava, Additional Director, CW&PRS, for constant encouragement and guidance in preparing the technical paper. The authors are grateful to the project to the project engineers from Hirakud project authorities, Odisha, CWC, and Central Project Management Experts of DRIP for their constant interaction during studies. The help of all the staff in conducting the studies and compiling the data is gratefully acknowledged.

6. REFERENCES

- Blocken, B., and Gualtieri, C. (2012). "Ten iterative steps for model development and evaluation applied to Computational Fluid Dynamics for Environmental Fluid Mechanics." Environmental Modelling and Software, 33 (none), 1-22.

- Bombardelli, F. (2012). "Computational Multi-Phase Fluid Dynamics To Address Flows Past Hydraulic Structures." 4th IAHR International Symposium on Hydraulic Structures, Porto, Portugal.

- Bhajantri M. R., Patnaik S. R., Gadhe V. P., and Bhosekar V. V. (2019). "Hydraulic model studies for the optimum design of Additional spillway of Hirakud Dam, Odisha." International Dam Safety Conference, Bhubaneshwar, Odisha.

- CWPRS Technical Report No. 5627 (September 2018). "Hydraulic Model Studies for Additional Spillway of Hirakud Dam, Odisha. 1:40 Scale 2-D Sectional Model."

- CWPRS Technical Report No. 5743 (October 2019). "Hydraulic Model Studies for Revised Design of Additional Spillway of Hirakud Dam, Odisha; 1:40 Scale 2-D Sectional Model."

- Carvalho, R.F., Lemos, C.M., and Ramos, C.M. (2008). "Numerical Computation of The Flow in Hydraulic Jump Stilling Basins." Journal of Hydraulic Research, 46 (6), 739-752.

- Computational Fluid Dynamics Software – Flow 3D from Flow Science, <http:// www.flow3d.com> (2013)

- Macián-Pérez, J.F., Vallés-Morán, F.J., Sánchez-Gómez, S., De-Rossi-Estrada, M., and García-Bartual, R. (2020). "Experimental characterization of the hydraulic Jump profile and velocity distribution in A stilling basin physical model." Water, 12 (6), 1758.

- Parsaie A., Haghiabi A., and Moradinejad A. (2015). "CFD modeling of flow pattern in Spillway's approach channel." Sustain. Water Resour. Manag., 1 (3), 245–251.

- Versteeg H.K., Malalasekera W. (2007). An Introduction to Computational Fluid Dynamics: The Finite Volume Method, Pearson Education Limited, United Kingdome.

- Valero, D., Bung, D.B., and Crookston, B.M. (2018). "Energy dissipation of a Type III basin under design and adverse conditions for stepped and smooth spillways." Journal of Hydraulic Engineering, 144 (7), 04018036.

- Valero, D., Viti, N., Gualtieri, C. (2018). "Numerical simulation of hydraulic jumps. Part 1: Experimental data for modelling performance assessment." Water, 11 (1), 36.

DOI: https://doi.org/10.26077/1180-7bd6

Plunge Pool Rock Scour Assessment by Fluid-Solid Coupling

E.F.R. Bollaert[1] & M. Grünzner[2]
[1]*AquaVision Engineering Llc, Ecublens, Switzerland*
[2]*Flow Science Deutschland GmbH, Rottenburg, Germany*
E-mail: erik.bollaert@aquavision-eng.ch

Abstract: *This paper presents a novel approach for detailed 2D numerical assessment of rock scour in plunge pools or unlined channels downstream of hydraulic structures. The approach is based on an automated coupling between the 3D hydraulics and the turbulence of an impacting jet (or any other turbulent flow environment) and the detailed bathymetric evolution with time of the rocky bottom that is being scoured, for a given flood event. Detailed flow hydraulics are thereby computed by the FLOW-3D® CFD software, allowing sound implementation of air entrainment and defining the main flow parameters at the water-rock interface. Bathymetric evolution with time of the eroding rock mass downstream of the hydraulic structure is computed by the rocsc@r® software, a cloud-based digital tool implementing the latest and most widely used computational methods for scour prediction. This novel fluid-solid interactive approach has been tested and applied to large-scale laboratory experiments of scour hole formation in broken-up rock, generated by rectangular-shaped oblique jets impinging in a plunge pool made of cobblestones, performed at the laboratory of the Civil Engineering Department at Stellenbosch University, South Africa (Bosman and Basson (2020), Bosman (2021)). The scour formation observed during the experiments has been reproduced by the FLOW-3D® - rocsc@r® numerical coupling. Benefitting from a dedicated, user customizable and fully automated interface between both software programs, this coupling requests only very short computational times and constitutes a cutting-edge tool available to engineers for real-life applications and projects. The results obtained for different computational methods of scour formation and a set of CFD modelling parameters are pointed out.*

Keywords: *fluid-solid coupling, novel software, plunge pool, rock scour*

1. Introduction

Rock scour downstream of high-head dams and in unlined channels and stilling basins is more and more frequent. Climate change and related regulatory requirements generate more frequent functioning of dam spillways. Hence, unlined rock masses experience more action of hydrodynamic pressures. Sound prediction of rock scour potential becomes increasingly pertinent, especially for cases where the scour may potentially regress towards the dam.

This paper presents a novel digital software platform that has been developed with the aim to bundle and offer to its users the current state-of-the-art in the field of numerical predictions of scour of rock downstream of dams and in unlined channels and stilling basins. Users will be able to apply and compare a wide range of proven as well as new methods within one single numerical environment, based on user-defined tailored parametric settings of the dam, the turbulent flow and the rock mass.

Moreover, a novel digital database platform has been developed complementary to the software platform, allowing to freely consult any type of worldwide rock scour case studies that have been performed and published, either real-life or laboratory generated. As such, this platform aims at offering a practical and continuously updated database with the potential of informing specialists and engineers on the most relevant results, parameters and calibrations that have been used for each of the available scour computational methods.

The platform focuses on enhancing parametric pertinence and experience. It aims widespread distribution and exchange, so that shared benefits may be expected.

It is believed that this is the first attempt in this field to combine on one single digital platform both powerful and cutting-edge computational potential and directly related up-to-date database availability of published practical experiences and results.

2. The rocsc@r® cloud

2.1. General

The rocsc@r® cloud contains two digital platforms: X_pl@re is a licensed computational platform, while X_ch@nge is a free database platform (Bollaert, 2021). Both platforms solely offer web-based remote access and storage. Cloud users benefit from personalized data encryption. Data results can be stored on user-specific allocated space in the cloud, or on the user's local IT infrastructure.

X_pl@re comprises the rocsc@r® software, i.e. a suite of the latest computational methods for numerical prediction of rock scour downstream of dam spillways and in unlined channels and stilling basins. The range of methods covers both oblique and vertical impingement of turbulent jets, and is compatible with any other turbulent flow at the water-rock interface, such as free surface flows and hydraulic jumps. X_ch@nge offers a free-access database platform for any type of publicly available rock scour cases at hydraulic structures. For each case, specific geometric, flow and rock mass parameters are provided, in parallel to the observed or computed scour for the related flow event.

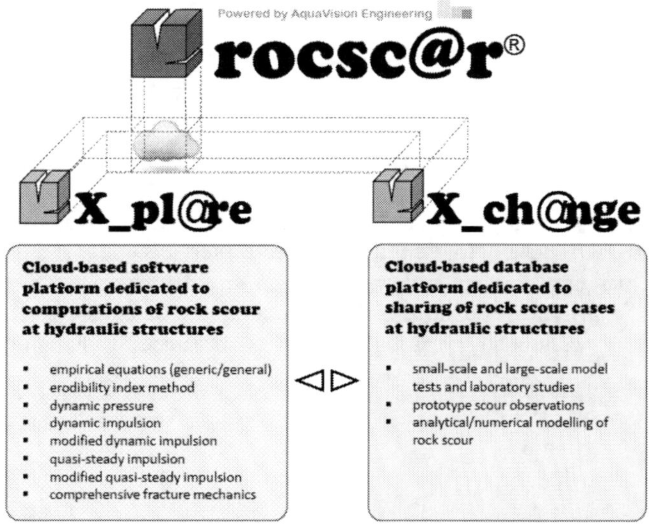

Figure 1. The rocsc@r® cloud and its main digital platforms implementing scour software (X_pl@re) and worldwide database of scour cases (X_ch@nge) (Bollaert, 2021)

2.2. Computational methods

The rocsc@r® software contains both novel and existing computational methods, expressing the different break-up mechanisms of fractured rock masses. Moreover, these methods have been implemented such that users may freely customize the related parameters and equations. The following methods are currently available:

- *Empirical equations*: A wide range of well-known empirical formulae are available. These formulae distinguish between a generic formula that can be user-defined, and more general and simplified formulas (Castillo and Carrillo 2017). More information can be found in Bollaert (2021).
- *Erodibility Index method (EIM)*: The Erodibility Index Method is based on Kirsten's ripability index and has been adapted by several researchers to scour in unlined plunge pool and stilling basins, based on a large dataset of case studies. This method relates the energy dissipation rate or stream power to the resistance of the rock as expressed by the erodibility index (EI), i.e. a geomechanical index incorporating the mass strength and different characteristics related to the size of the block, the joint roughness and alteration, and the structure of the discontinuities. information can be found in Bollaert (2021).
- *Dynamic Impulsion methods (DI/MDI):* The DI method developed by Bollaert and Schleiss (2005) considers the maximum dynamic impulsion underneath a rock block, obtained by time integration of the pressure forces under and over the block, of the immerged weight and of shear forces on the block. The net uplift impulsion considers a time duration of forces and uses a net uplift coefficient Cup [-] determined based on large-scale

laboratory experiments. Second, the MDI method developed by Bollaert (2021) uses similar physics of block uplift but computes the net uplift force as a function of the RMS values of the pressure fluctuations at the block upper surface and a multiplication coefficient based on large-scale 3D block uplift experiments. Both methods express block failure by comparing the vertical block displacement to the block height.

- *Dynamic Pressure method (DP):* Bollaert (2021) developed a 2D implementation of the block uplift method developed by Maleki and Fiorotto (2019). This method determines the net uplift force on a block based on analytical determination of jet characteristics, as well as by accounting for the RMS coefficient of pressure fluctuations at the block upper face multiplied by a surface reduction coefficient. This approach defines the necessary thickness for a block to be stable, accounting for its submerged weight.
- *Quasi-Steady Impulsion methods (QSI/MQSI):* The QSI method developed by Bollaert (2012) computes the scour potential by detachment or peeling off of protruding rock blocks, due to wall jet flows deviated by such blocks at the water-rock interface and generating quasi-steady lift forces on the blocks. The MQSI developed by Bollaert (2021) extends the previous method for both circular and rectangular jets and obliquely impinging jets, and allows to compute both quasi-steady and fluctuating detachment forces. The related lift forces are defined by an uplift coefficient Cup. The stability of the block is expressed by a force-balance equation.
- *Comprehensive Fracture Mechanics method (CFM):* This method developed by Bollaert and Schleiss (2005) computes progressive fracturing of existing joints in rock masses as a function of time, based on linear elastic fracture mechanics theory and the cyclic character of dynamic pressures generated by turbulent flows. The method expresses scour evolution as a function of time.

2.3. Plunging turbulent flows

The hydraulic parameters of 2D plunging turbulent flows (Fig. 1) are determined analytically along the initial water-rock interface. This process is repeated for similar fictitious interfaces that are progressively situated further downwards, increasing the total tailwater depth, vertically down into the rock mass. The obtained values with depth are stored as 2D matrices. The analytical equations make the assumption of a quasi-horizontal and quasi-flat interface between the rock and the water, without any perturbating elements or suddenly changing interface shapes.

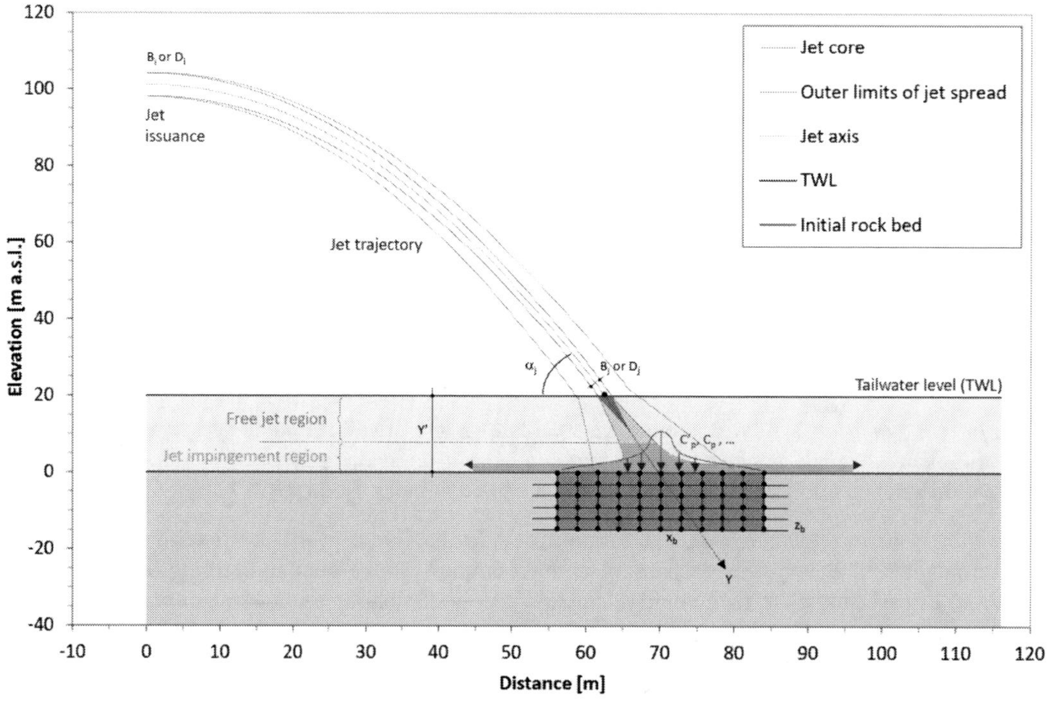

Figure 2. Sketch of obliquely plunging turbulent flow diffusing through plunge pool and impacting the water-rock interface (source: rocsc@r® Technical Manual, AquaVision Engineering 2022)

The hydraulic parameter values stored in the 2D matrices cannot be adapted during the computations as a function of the scour development, and are only applicable for simplified interfaces that respect the assumptions upon which the analytical equations are based. Use of predefined 2D matrices is strictly speaking only valid for cases where the change in interface shape has no significant influence on the hydraulic parameter values that were defined at start.

The analytical equations describing plunging turbulent flows allow predefining and storing the values of the main hydraulic parameters in 2D matrices. These matrices respect the network of computational nodes of the 2D vertical profiles, along which the scour computations are being performed (Fig. 2). As such, in every computational node, the hydraulic parameters that are needed for the scour computations are being predefined before start. The corresponding values account for the progressively increasing flow depth during scour formation. The following hydraulic parameters are predefined with depth and stored in 2D matrices, based on analytical solutions of the flow environment:

Table 1. List of hydraulic parameters pre-defined in 2D flow matrices

Parameter	Description
Cp	Mean dynamic pressure coefficient
C'p	Fluctuating (RMS) dynamic pressure coefficient
Cmax	Maximum dynamic pressure coefficient
CI	Dynamic impulsion coefficient
VEL	Average flow velocity quasi-parallel to interface
SP	Average stream power at interface

The pressure coefficients have to be multiplied by $\rho V^2/2g$ to obtain the pressure values in [N/m²], and the impulsion coefficient by $\rho A V^2 L/c$ to obtain impulsions in [Ns], with V the jet velocity at impact in the pool, c the wave celerity, L the joint length and A the horizontal rock block area. Fig. 3 illustrates an example of the average dynamic pressure coefficient Cp at the interface, due to a rectangular jet impacting the pool under an angle of 70° with the horizontal. The spatial distribution of the pressure at the rock surface is represented by the black curve (right-hand side axis), while the evolution with depth is represented by a graded heatmap (legend on the right).

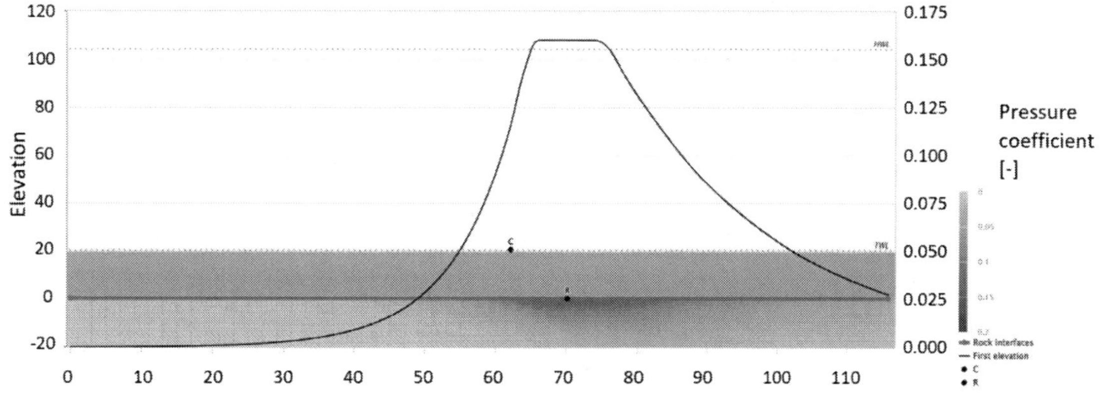

Figure 3. Example of 2D determination of average dynamic pressure at water-rock interface generated by an obliquely impacting jet (source: rocsc@r® Technical Manual, AquaVision Engineering 2022)

Similar curves are defined in the model for all hydraulic parameters listed at Table 1. This 2D matrix system covers both circular and rectangular-shaped jets, as well as vertically or obliquely impinging jets for angles between 20 and 90°. Moreover, both compact and broken-up jets are accounted for, as well as the air concentration and degree of break-up of the jets at impact, and the initial turbulence intensity at issuance from the dam.

3. Fluid-solid interactive scour computations

3.1. RemoteSc@r™

RemoteSc@r™ is an API (Application Programming Interface) that offers a dedicated interface between the FLOW-3D® commercial CFD software (Flow-3D, 2019) and the rocsc@r® software. The current version only allows 2D computations, but a 3D version is under development. RemoteSc@r™ allows to perform fluid-solid interactive scour computations by sequential coupling. Fig. 4 illustrates the main parameters to be defined for both software.

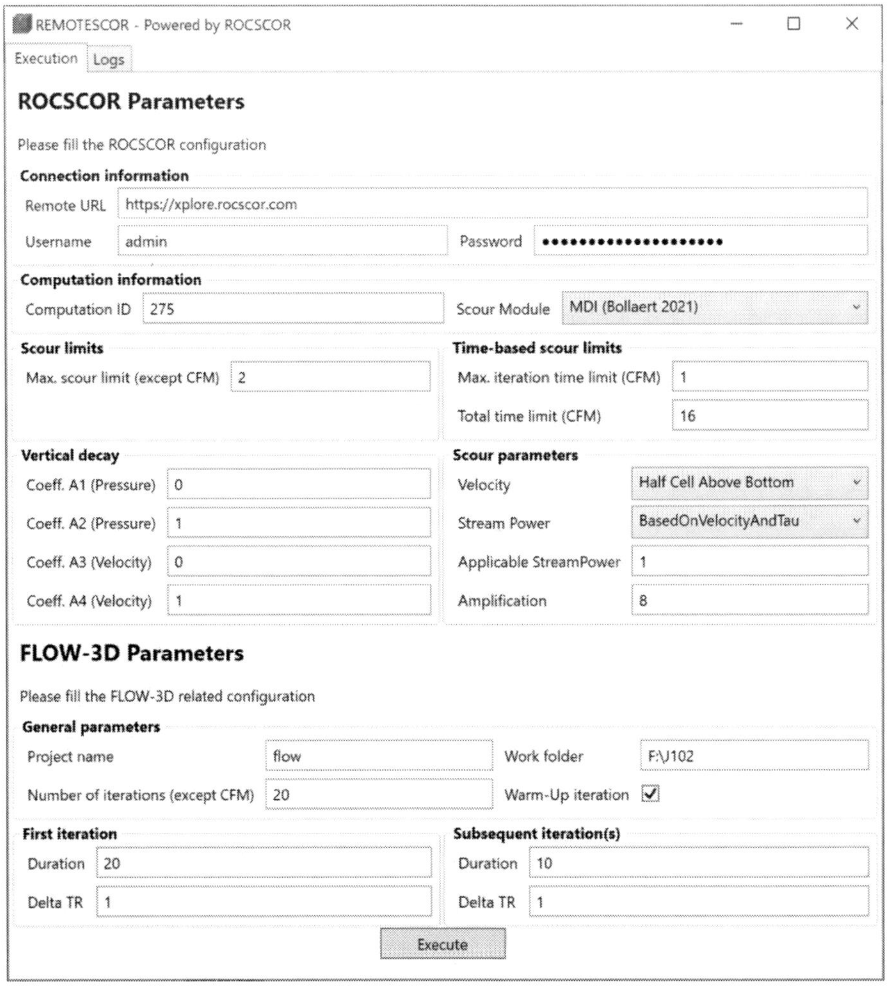

Figure 4. User Interface of the RemoteSc@r™ API (RemoteSc@r manual 2022)

3.2. Methodology

The fluid-solid interactive computations follow the following methodology:

1. FLOW-3D® transient simulation to reach a quasi steady-state of main hydraulic parameters at the water-rock interface
2. Transfer of hydraulic parameters to rocsc@r® by use of a generic exchange file (csv format)
3. Computation of scour in rocsc@r® based on a user-defined max. layer or a max. time duration (CFM method)
4. Transfer of scoured water-rock interface to FLOW-3D® by use of a STL-file
5. Repetition of steps 1 to 5 until no further scour occurs anymore

4. Case study

4.1. The Stellenbosch scour tests

Large-scale laboratory experiments of scour hole formation in broken-up rock, generated by rectangular-shaped oblique jets impinging in a plunge pool, have been performed in the laboratory of the Civil Engineering Department at Stellenbosch University, South Africa (Bosman and Basson (2020), Bosman (2021)). This scour formation has been reproduced by the rocscor® software.

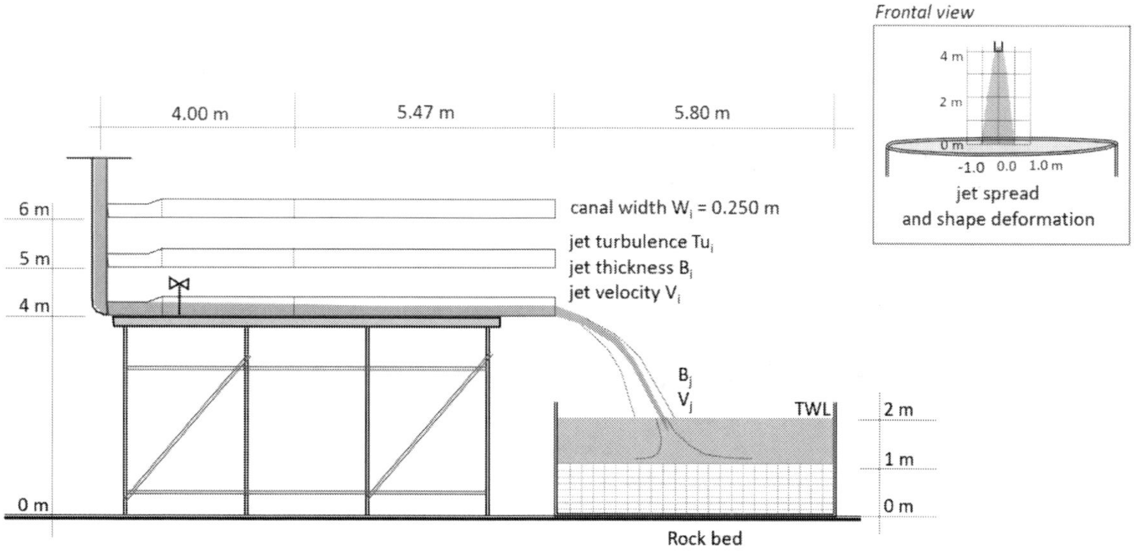

Figure 5. Experimental set-up of plunging jets impinging on broken-up rock bed (Bollaert 2021, adapted from Bosman, 2021)

The physical model is illustrated in Fig. 5 (adapted from Bosman, 2021) and consists of a zero-sloped 0.25 m wide rectangular canal replicating an uncontrolled dam spillway. The canal could be adjusted to three different fixed heights above the movable rock bed: 3, 4 and 5 m (prototype heights for 1:20 model scale: 60, 80 and 100 m). The tailwater depth at the plunge pool was adjustable between 0.5 m and 1 m (prototype: 10 m and 20 m). The plunging jets generated unit discharges between 0.100 and 0.224 $m^3/s/m$ (prototype: 35 to 80 $m^3/s/m$).

The broken-up rock bed was modelled by using tightly hand-packed concrete paver blocks (cobblestones), generating a uniform three-dimensional open-ended horizontal and vertical rock joint network. The two rock sizes tested were rectangular concrete pavers with x, y, z dimensions of 0.1 m × 0.1 m × 0.05 m and 0.1 m × 0.1 m × 0.075 m (prototype: 2 m × 2 m × 1 m and 2 m × 2m × 1.5 m respectively). The paver block densities used were 2'355.4 kg/m^3 (for 1m high blocks) and 2'388.1 kg/m^3 (for 1.5 m high blocks) respectively. A total of 31 experimental tests were carried out by the authors.

From these, one scour test has been chosen here to be reproduced by the different computational methods that are proposed by the rocscor® software, with and without coupling with the FLOW-3D® CFD software. The test involves a jet velocity at issuance of 12 m/s and a jet thickness at issuance of 6 m, for a tailwater level of 20 m.

4.2. Scour computations without FLOW-3D® coupling

Firstly, the scour computations have been performed without coupling with FLOW-3D®, i.e. by using the 2D flow matrices generated by rocsc@r® at start of the computations. Fig. 6 illustrates the computed scour for the DI, MDI, QSI and MQSI methods and shows good agreement with the scour observed during the laboratory experiments. Both the deepest scour and the shape of the scour hole are soundly reproduced by the rocsc@r® software. More details on calibration parameters can be found in Bollaert (2021).

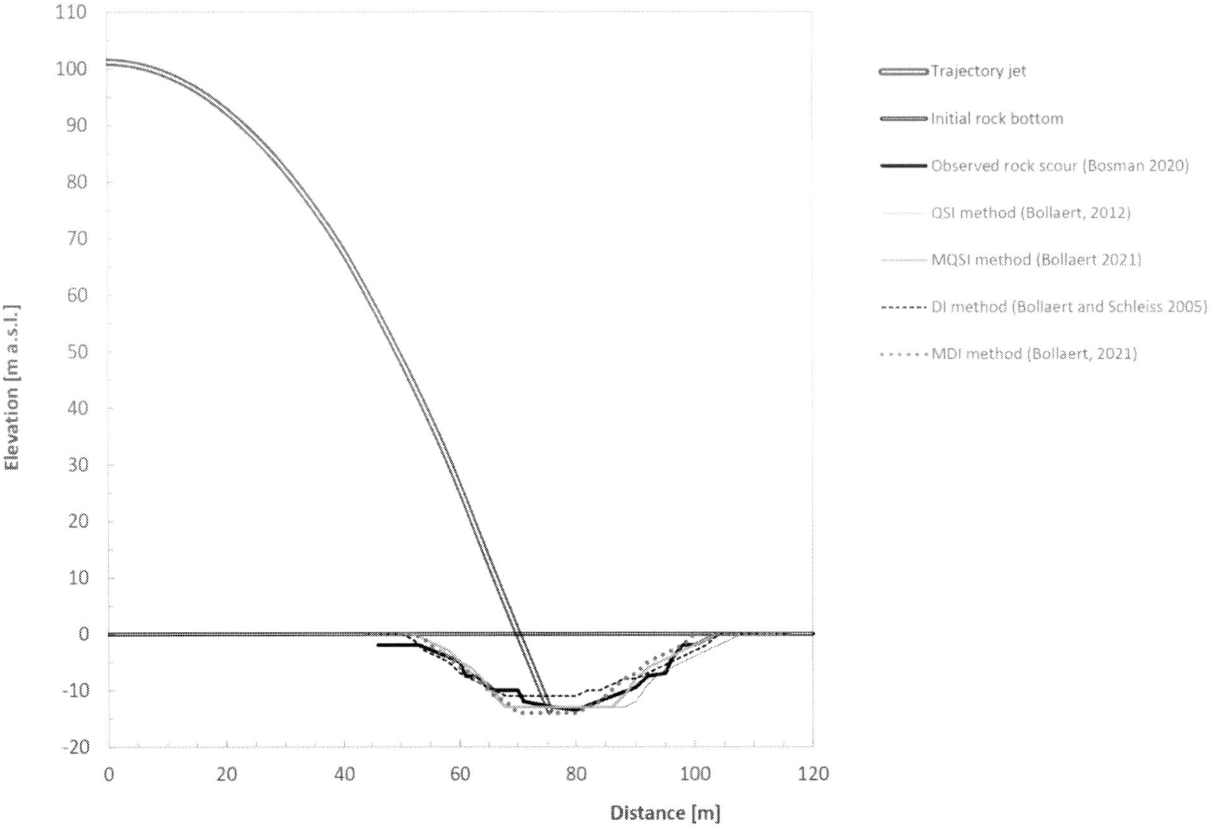

Figure 6. Results of rocsc@r® computations without coupling (DI, MDI, QSI, MQSI methods) (RemoteSc@r manual 2022)

4.3. Scour computations with FLOW-3D® coupling

Next, the scour computations have been repeated by coupling rocsc@r® to FLOW-3D®. The modelling parameters are summarized at Table 2. A Large Eddy Simulation (LES) turbulence model has been used, for a mesh size of 0.5 m. Furthermore, the model incorporates air entrainment of the jet and the plunge pool, as well as related flow bulking. The surface roughness height of the rock was defined at 1 m, which corresponds to plausible estimates given the physical height of the perturbing cobblestones. Values of rock roughness heights for practice would be typically between 0.10 and 1.0 m, depending on the joint sets and typical in-situ rock block heights. A no-slip boundary condition was applied at the water-rock interface.

Fig. 7 illustrates the scour results obtained during consecutive (numbered) iterations between rocsc@r® and FLOW-3D®, as well as the jet trajectory and plunge pool flow velocities computed by FLOW-3D® in a vertical 2D profile. The laboratory observed scour formation is presented by the thick dotted grey line.

For a max. scour value of 1 m per iteration (i.e. equal to the block height) and using the MDI method, good agreement is observed between measurements and computations. Not only the area of deepest scour formation is well modelled, but also the slopes of the scour hole towards upstream and downstream are adequately reproduced. The computation converged in 35 iterations and took 1h45 min. on a standard INTEL i5 processor. The critical net uplift height ratio n_b equaled 0.20, i.e. within the range of usual values based on experience from practice, but nevertheless slightly higher than the value of 0.10 that was proposed by using the internal 2D flow matrices instead of FLOW-3D®. This is merely because of the slightly different RMS pressure fluctuations and their different vertical decay computed by FLOW-3D®, as compared with the values as provided by internal 2D flow matrices in rocsc@r®.

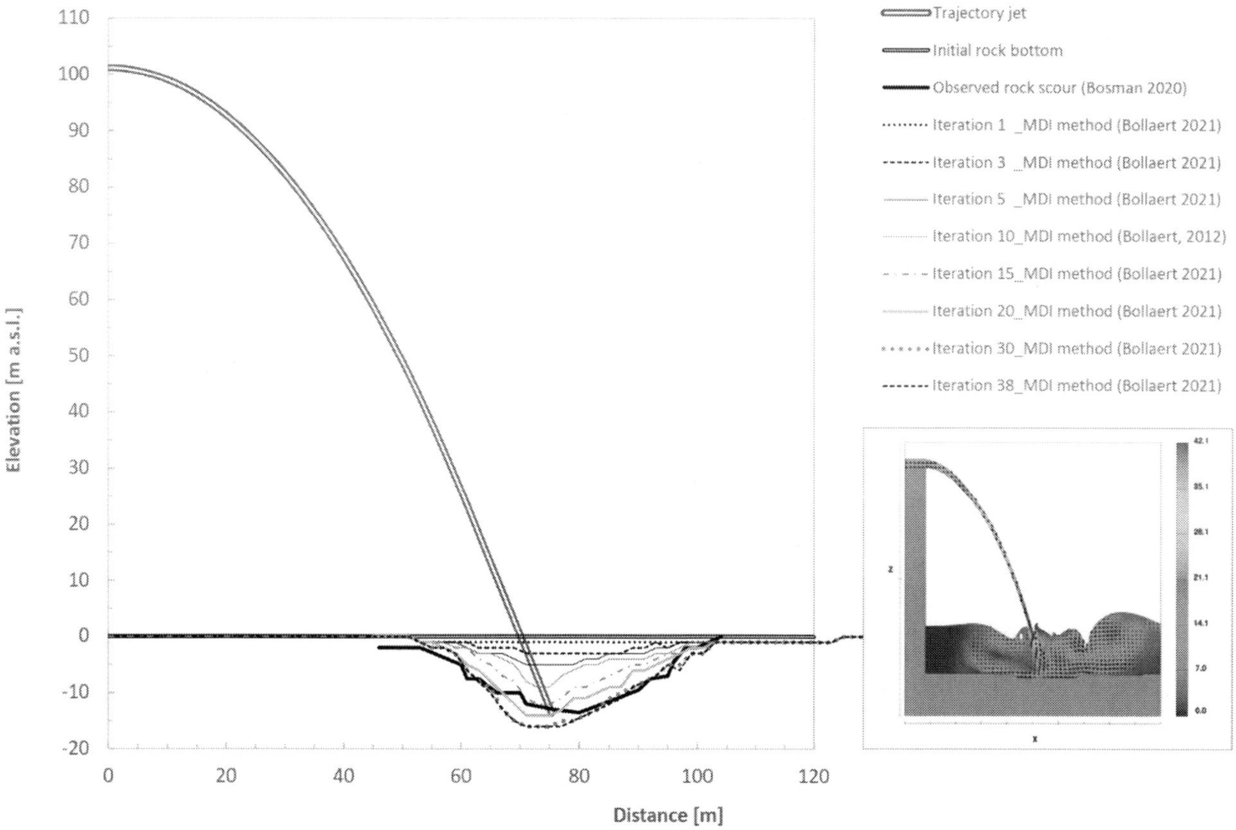

Figure 7. Results of rocsc@r® computations with FLOW-3D® coupling (MDI method) (RemoteSc@r manual 2022)

Table 2. FLOW-3D® modelling parameters

Turbulence model	LES
Volume of fluid adv.	Automatic
Momentum adv.	First order approx.
Time step control	Stability and convergence
Mesh size	0.5 m
Pressure solver	Implicit GMRES
Air entrainment	Yes, with flow bulking and buoyancy
Roughness dam	0.01 m
Roughness rock	0.10-1.0 m

4.4. Parametric comparison between FLOW-3D and rocsc@r

Finally, a direct comparison has been made of the main hydraulic parameters between FLOW-3D® and the internal 2D flow matrix system of rocsc@r®. The fluctuating part of the dynamic pressures generated by the impacting jet (root-mean-square or RMS values) at the water-rock interface is presented and compared in Fig. 8. The FLOW-3D® computed values are in good agreement with the rocsc@r® computed values at the point of jet impact, but slightly differ outside of this area. The lateral decay of the RMS values at first follows a very strong slope, in agreement with the rocsc@r® analytical values for a concentrated dynamic pressure curve. Next, this lateral decay strongly diminishes and the RMS values remain almost constant with increasing lateral distance from the impact point.

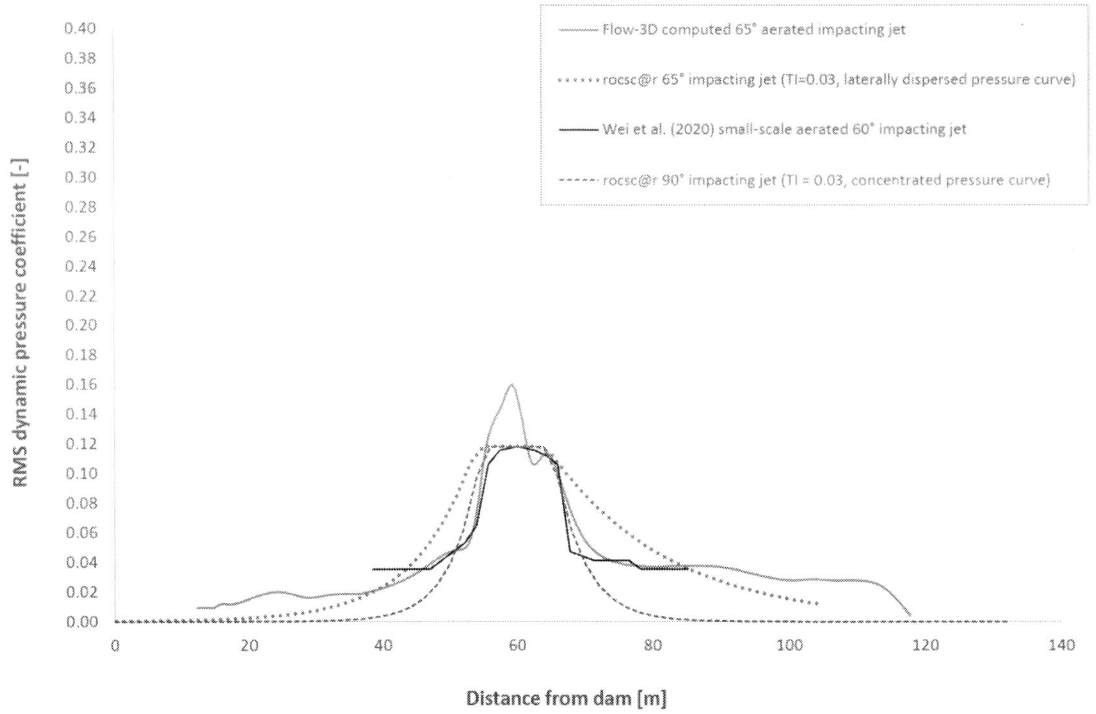

Figure 8. Comparison of RMS pressure fluctuations computed by rocsc@r® and by FLOW-3D®

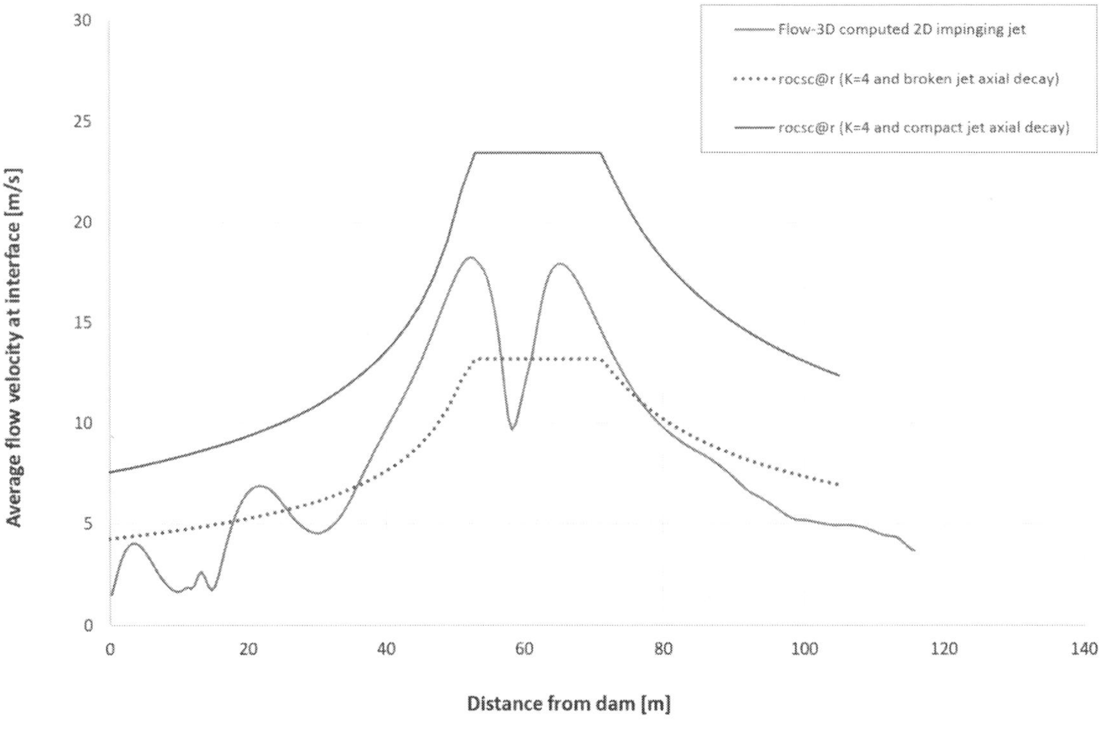

Figure 9. Comparison of averaged flow velocities at interface computed by rocsc@r® and by FLOW-3D®

The latter behavior is much more in agreement with rocsc@r® values for a strongly dispersed dynamic pressure curve, i.e. for which RMS values remain significant even at large distances. The latter curves are typically observed during laboratory observations involving jet impingement. Moreover, the FLOW-3D® computed RMS values are in excellent agreement with the measurements made by Wei et al. (2020) for a comparable case of an obliquely impinging aerated laboratory jet. The agreement can be noticed by a similar RMS increase at jet impact, and by the quite constant RMS values computed laterally outwards from the point of jet impact, which were also measured during in the laboratory.

Second, Fig. 9 compares the time-averaged wall jet velocities that develop along the water-rock interface following jet impingement. The FLOW-3D® computed velocity curve is of similar shape than the rocsc@r® computed flow velocity curve for broken-up jets and a jet core extension K of 4 times the jet diameter at impact. However, in the area of jet impingement, the FLOW-3D® computed flow velocities are somewhat higher and are situated in between values valid for broken-up and compact jets. Furthermore, the FLOW-3D® computed velocities show a dip at the point of jet impact where significant stagnation pressure builds up. This is logic because this area is not part of the wall jets, that generally start at a lateral distance of 1-2 jet diameters. The rocsc@r® computed velocities do not show this dip because it is automatically removed from the analytical equations for convenience of scour computations.

5. Conclusions

Rock scour computations downstream of hydraulic structures have been performed along vertical 2D profiles by a novel fluid-solid coupling between the rocsc@r® software and the FLOW-3D® CFD software. The former computes detailed 2D rock break-up and scour formation in rock masses with time, based on a large set of computational methods, while the latter computes the 2D or even 3D flow parameters and turbulence at the water-rock interface. Both programs are linked together by using the RemoteSc@r™ API connection benefitting from a dedicated, user customizable interface. This coupling requests relatively short computational times and constitutes a cutting-edge tool available to engineers for real-life applications and projects. Application of this fluid-solid coupling to the laboratory tests of scour performed at the University of Stellenbosch (Bosman & Basson 2020) shows first of all that the observed scour is well reproduced. Furthermore, the hydraulic parameters computed by FLOW-3D® at the water-rock interface are in reasonable to good agreement with the corresponding analytical equations based on 2D jet diffusion theory. This points out the adequacy of the LES turbulence model used by FLOW-3D® to generate these parameters.

6. REFERENCES

AquaVision Engineering (2022). "rocsc@r Technical Manual" and "RemoteSc@r Manual", unpublished.

Bollaert, E.F.R. (2012). "Wall Jet rock scour in plunge pools: a quasi-3D prediction model." *Intl. J. on Hydr. & Dams,* 19(4), 70-77.

Bollaert, E.F.R. (2021). "The rocsc@r™ cloud: An innovative digital platform to compute and record rock scour." *Intl. J. on Hydr. and Dams*, 28(5), 60-70.

Bollaert, E.F.R. and Schleiss, A.J. (2005). "A physically-based model for evaluation of rock scour due to high-velocity jet impact." *J. of Hydr. Engng*, https://doi.org/10.1061/(ASCE)0733-9429(2005)131:3(153).

Bosman, A. (2021). "High dam scour hole geometry prediction for fully developed jets plunging into shallow pools on bedrock." PhD Thesis, *University of Stellenbosch*, Cape Town, South Africa.

Bosman, A. and Basson, G.R. (2020). "Physical model study of bedrock scour downstream of dams due to spillway plunging jets." *J. of the S. African Inst. of Civil Engng*, Vol 62 No 3, Sept. 2020, 36–52, Paper 1000.

FLOW-3D® Version 12.0 [Computer software]. (2019). Santa Fe, NM: Flow Science, Inc. *https://www.flow3d.com.*

Maleki, S. and Fiorotto, V. (2019). "Blocks Stability in Plunge Pools under Turbulent Rectangular Jets." *J. of Hydr. Engng*. 145(4), https://doi.org/10.1061/(ASCE)HY.1943-7900.0001573.

Wei, W., Xu W., Deng J. and Liu B. (2020). "Experimental Study of Impact Pressures on Deep Plunge Pool Floors Generated by Submerged Inclined Jets with Controlled Aeration." *J. of Hydr. Engng.,* 2020, 146(4): 04020021, https://doi.org/10.1061/(ASCE)HY.1943-7900.0001704.

DOI: https://doi.org/10.26077/fc6d-a5c7

Porous Concrete Slabs Underlain by Polypropylene Fabric for Restoration of Beaches

N.N. Pillai[1], M.L. Kansal[2], & R. Ramkrishnan[3],
[1]*Emeritus Professor, School of Engineering, Amrita University, Coimbatore, India*
[2]*Water Resources Development and Management, I.I.T. Roorkee, Roorkee, 247667, India*
[3]*Scientist C, Directorate of Forest Education, Burnihat, 793101, Assam, India*

Corresponding E-mail: mlk@wr.iitr.ac.in

Abstract *Beaches in India and elsewhere are getting destroyed at an alarming rate as many small ports are being built and sea walls are coming up on the banks. There is an urgent need to stop it. In 1966, Posey used cylinders of net filled with rock (called rock sausages) underlain by a reverse gravel filter for safety against scour in hydraulic structures. Pillai used a bed of stones enclosed in nets underlain by a reverse filter to allow overflow over a sand embankment. Later it was understood that a Polypropylene Fabric (P.F.) could be effective as a filter, and a bed of polypropylene bag filled with sand (like geo-tubes) can be used to prevent scour. However, the P.F. get weathered when exposed to sunlight. A breakthrough came when porous concrete blocks with high porosity, high strength and seawater resistance were developed and were used underlined by a P.F. An attempt was made in an experimental study on the Cheriathura Beach at Trivandrum. Porous concrete blocks 0.5 m X 0.5m, 10 cm thick were cast and arranged at a time of low tide, covering an area of 10 m x 5 m (with the P.F. underneath). When the waves acted frontally during high tide, this rode smoothly on the pavement, had their run-up, and later receded without displacing the slabs. This showed that porous concrete slabs could be a good armour for protecting the beach. The side slabs got displaced with lateral waves because no sidewall could be provided in the narrow strip of protection. It is felt that the porous concrete slabs can be placed deep in the beach so that during the erosive phase, the slab could function as armour, and when there is accretion, there would be deposition of sand on the slabs maintaining the beach.*

Keywords: *Beach restoration, Porous Concrete, Polypropylene fabric.*

1. Introduction

India is gifted with many beautiful beaches on the seacoast. It supports the life of a large fishing community and attracts many domestic and foreign tourists. Many harbors, big and small, constructed along the coast along with the encroachment by the resort owners and businesspeople have adversely affected the regime of the sea and coastline, thereby destroying the coast and the beach. Many constructions have been carried out very near the coast. When the sea became rough during the monsoons, the waves attacked the coast, and many buildings got damaged or destroyed. To protect the infrastructure and the properties, the owners put up sandbags, stone walls or sea walls. Waves were reflected by the protection provided, and the reflected seaward currents eroded the beaches. The media has reported that about 50% of the beaches have been destroyed. Central water commission, government of India issued the guidelines for the protection and control of coastal erosion in India (CSIR-NIO, 2020).

A good part of the incoming wave energy is spent on the beach before reaching the banks when the beach is in position. But otherwise, the full energy of the waves acts on the banks, and the protective structures get damaged. Thus, it is necessary to take suitable measures to protect the valuable beaches.

Investigations for preventive methods of beach erosion were reported by Helen Robertson (2010) and concluded that "Present beach erosion prevention methods include dunes, vegetation, sea walls, sandbags and sand fences. Based on the research conducted, it is evident that new ways to prevent erosion must be obtained."

2. Experiments on protection against erosion

2.1. Use of rock sausage for scour protection

Basic studies on protection against erosion due to fast-flowing water were carried out at the Rocky Mountain Hydraulic Laboratory, Allens Park, Colorado, USA in 1960s. Researchers used the stones enclosed in cylindrical plastic nets, which was underlain by a reverse gravel filter conforming to the Terzaghi-Vicksburg specifications and was named as 'rock sausage'. It was used to protect scour against fast-flowing water through hydraulic structures in rivers and the sea (Posey, 1973, 1974). Even though the rock sausages with the stones enclosed in cylindrical nets are flexible, they cannot conform to the bed of the sand surface when it settles. Gaps can develop between the sand sausage and may lead to scouring. Further, laying the sausages over a large area is a long and tedious process. Laying a bed of stones enclosed in nets and providing a reverse filter could be easier and more effective.

2.2. Protection of overflowing sand bund.

From 1972 to 1994, series of experiments were conducted at National Institute of Technology (NIT) Kurukshetra, India (formerly, Regional Engineering College Kurukshetra) to study the use of stone beds in the nets. The nets used were of fine polypropylene strands. A sand spillway model with a bed of stones. covered in nets underlain by a reverse sand filter was tested. (Figure 1(a)).

The overflow was allowed like a spillway for 30 minutes (Figure 1(b)), and after the test, a bed of stones enclosing the nets on one side of the spillway was removed, and the underlain filter was found to be intact (Figure1(c)). It demonstrated that even a sand bund subjected to overflow could be protected by giving a cover of a stones bed enclosed in nets. The flowing current did not damage the underlying filter. The filter, the stones enclosing the nets and the nets act as part of the embankment without being damaged.

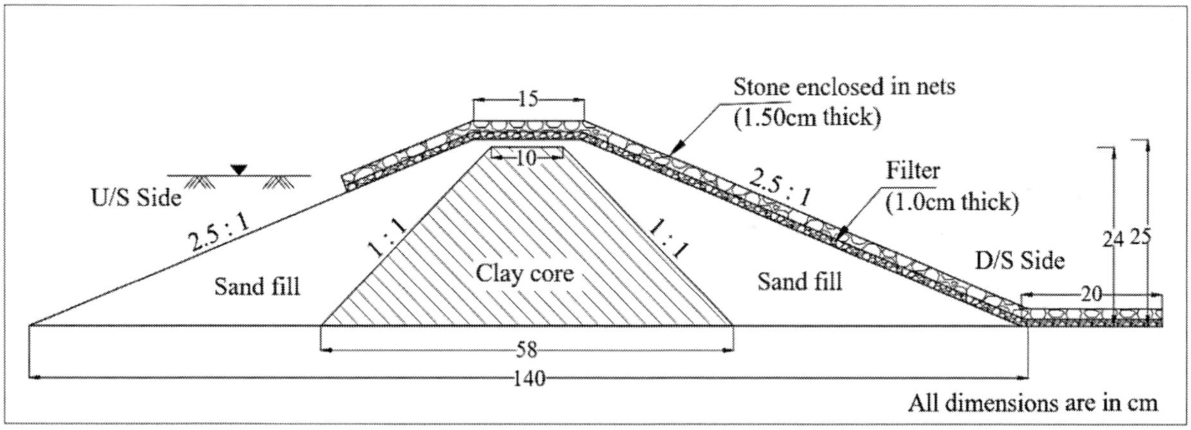

Figure 1 (a). Cross-section of sand bund tested

Figure 1 (b). Overflowing the bund protected by stones in nets; **1 (c)**. Stones enclosed in nets were partly removed

2.3. High-velocity flow over a sand slope

In another test, a sand slope of 3:1 was prepared in a flume and given a covering of stones enclosed in coir nets. (Pillai and Verma, 1977a). The sand-gravel reverse filter under-laid the nets (Figure 2). A hosepipe delivering a water jet was made to play on the nets at the top of the embankment. The sand embankment was subjected to overflow. It terminated in a hydraulic jump with accompanying high turbulence and eddies. It was safe, and the net material used was coir fiber. After the test was continued for 30 minutes, the jet was stopped, the coir nets were removed, and the filter layer was inspected. There was no damage to the filter layer.

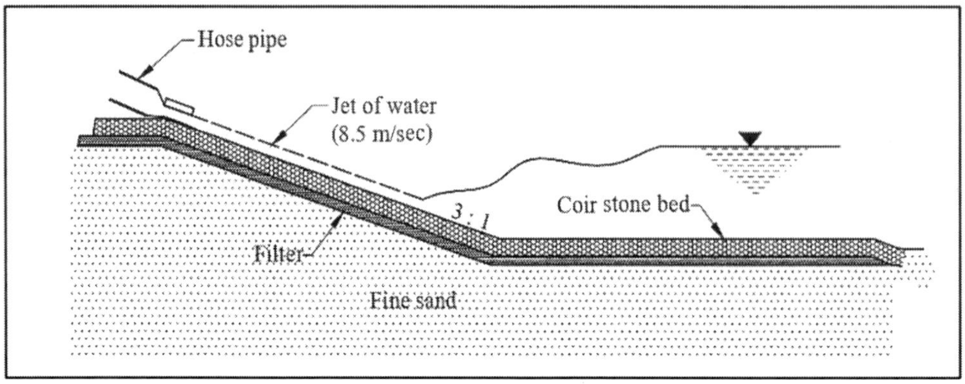

Figure 2. Test on a sand slope protected by stones enclosed coir nets.

2.4. Experiments conducted in a wave flume

A wave generator of flap-type was fitted in a glass-sided steel flume 12 m long, 40 cm wide, and 60 cm wide (Pillai and Verma, 1977b). A slope of sandbank 3:1was made. The flume was filled with water to a depth of 43.6 cm. The period of waves and wavelength etc., are given in Table 1.

Table 1. Details of Experiment on a wave flume

Flume size	40 cm wide, 60 cm deep and 12 m long
Wave Generator	Flap Type
Depth of water	43.6 cm
Wave Height	6.92 cm
Period	1.46 sec
Wavelength	4.046 m
The median diameter of sand used	0.2 mm
Sand slope	1 in 3

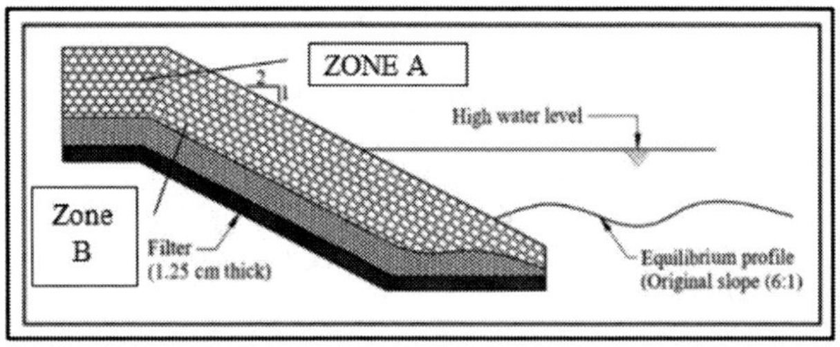

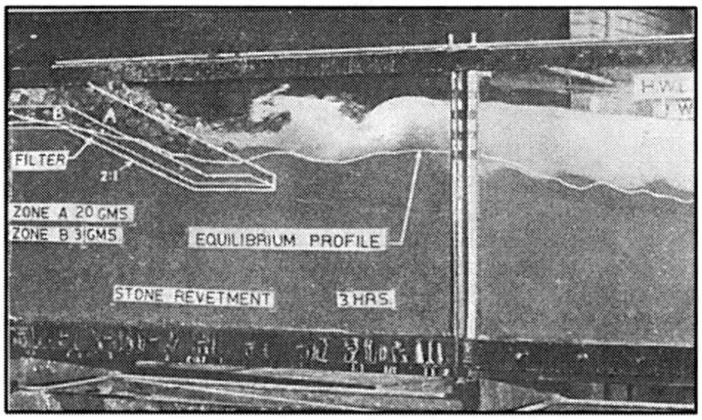

Figure 3 (a). Shore protection using a stone revetment **Figure 3 (b)**.Waves acting on the stone revetment

2.4.1. Waves acting on a revetment

A model of revetment for a sea wall was made of sones with a gravel filter placed underneath. The stone wall consisted of an outer zone A of 20-gram stones to a thickness of 6. 5 cm. and an inner zone B of 3-gram stones 2.5 cm. thick (Figure 3 (a) and (b)). The waves were allowed to act on the protection. It was observed that a few of the 20 gm stones were displaced after a run of 3 hours. The stones act independently and are moved by the waves.

2.4.2. Waves acting on stones enclosed in a net

The effectiveness of stones enclosed in nets for combating sea wave erosion on a beach was tested in a model study. The details of the polypropylene nets are given in Table 2.

Table 2. Properties of the net used

	Direction		**Remark**
	Warp	Weft	
Size of strand	264 deniers	264 deniers	Denier: weight (g) for 9000 m of strand
Number of strands per cm	9	9	
Tensile strength breaking load	1.30 kg	1.30 kg	Gauge length 5 cm. temp 230 C, crosshead speed 5 cm/min

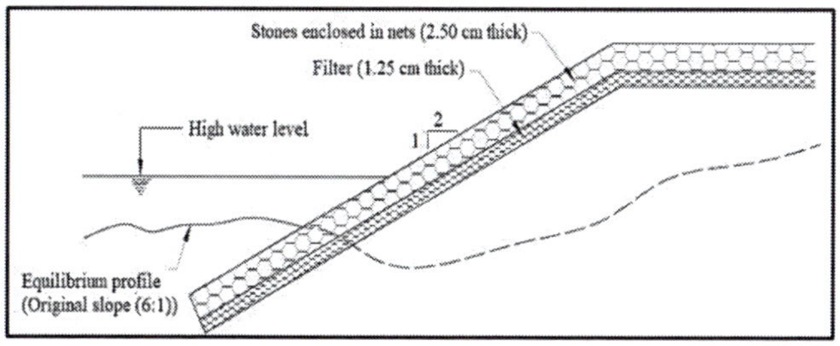

Figure 4 (a). Slope protected by stones enclosed in a net

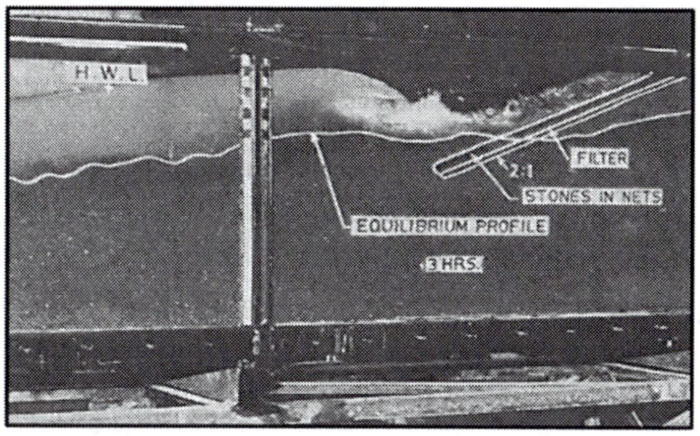

Figure 4 (b). Waves acting on a slope protected by stones enclosed in nets

The stones in the net were of an average weight of 0.45 g.; the mat was of the thickness of 2.5 cm. laid to a slope of 3:1, and the graded reverse gravel filter had a thickness of 1 cm. The waves were allowed to act on the slope for 3 hours, and protection had proved good. The filter effectively guards against the migration of sand upward, giving support to the stones and the net. The net was holding the stones intact, and the weight of the stones was holding the net from getting washed away by flowing water.

It is also inferred that if the waves entered tangentially over a suitable armour, it could move and ride on it safely to the natural run-up and recede without causing erosion.

2.5. Use of polypropylene sand-filled pillows

It has been shown that plastic filters can do the function of graded sand filters well and could do it cheaper (Barrett, 1966). Polypropylene fabric was being used commonly. Sand-filled Polypropylene bags or pillows were used to protect against the scour on the coast and river banks.

Before the monsoon in 1998, a three-meter long and one-meter high sand embankment with a clay core was made across a small stream on the Amrita University campus. It was covered and protected by polypropylene pillows to function as a spillway. Care was taken to see that no sand got exposed to the water current. During the monsoon, the stream passed a discharge of about 12 cm. of the flow depth over the sand spillway. Figure 5(a) shows the stream after a few showers. Figure 5(b) depicts the flowing stream.

Figure 5 (a). Sand embankment in the stream protected by polypropylene bags filled with sand

Figure 5 (b). Water flowing over the stream during the monsoon

2.6. Investigations for developing porous concrete for the armour of the beach.

The studies using polypropylene pillows are significant in many ways. (i) The lower surface of the pillows acted as a reverse filter, and it did not allow sand to escape upward and get washed out by the water current; ii) the sand filled in the pillows gave them the required weight so that it did not move away by the flowing water (at the same time highly pervious bed avoiding pore pressure being developed underneath) and (iii) the upper surface of the pillows functioning as armour so that the pillow surface is not eroded. Porous concrete slabs can also do the various functions that the fabric pillows perform under the Poly-fabric. Furthermore, porous concrete is not affected by exposure to sunlight because of its high permeability and capacity for drainage. The porous concrete may be called draincrete, and when underlain by polypropylene fabric, it may be termed DUPF. A detailed study on porous concrete was done by Ramkrishnan et al. (2017).

3. Selection of materials

In-depth studies on the effect of mineral admixtures and aggregate sizes on the strength, durability and permeability characteristics of pervious concrete were made by Ramkrishnan et al. (2017). Chettinad Cement O.P.C. 53 was selected for the experiment. The properties of the cement are shown in Table 3.

Table 3. Cement used for the porous concrete blocks

Physical properties	**Requirement (ISI2269:2013)**	**Chettinad OPC 53**
Fineness (m2/g)	225 (min)	285 (min)
Initial setting time (minutes)	30 (min)	161 (min)
Final setting time (minutes)	600 (max)	256 (max)
Soundness (Le chaterlier method) (mm)	10 (max)	0.9
Compressive strength (28 day) MPa	53 (min)	58.6

Studies have shown that exempting fine aggregates and coarse aggregates passing 20 mm and retaining on 12.5 mm sieves improved concrete strength and permeability (Ramkrishnan et al. 2017). Ultra-fine Ground Granulated Blast Furnace Slag (UFGGBS), commercially available as Alccofine 1203, was used as an admixture to increase the hydration process (Saurav, 2014). The manufacturer specifications/ properties are shown in Table 4.

Table 4. Properties of UFGGBS

Properties	**Values**
Particle size (microns)	4-6
Specific gravity	2.86
Fineness (cm2/g)	12000
Bulk density (kg/m3)	600 – 700

Glenium B233, which distinguishes from other super-plasticizers, greatly improves the cement dispersion (Ramkrishnan et al. 2017; Ramyalekha and Mallikarjuna Reddy, 2017), was used. Zhang et al. (2013) found that the water-cement ratio substantially impacts the sulphate attack on concrete. The higher the water-cement ratio greater the expansion and the more severe the deterioration of concrete samples. The water-cement ratio was reduced to 0.3 in the design mix to make the concrete safe against sulphate attack. Mix for Porous Concrete developed was used to form an armour in the tests conducted. The mix was done with a water-cement ratio of 0.3 and kept for curing for 28 days. 200 porous concrete blocks size 50 cm X 50 cm with 10 cm thick slabs were cast. Curing was done by immersing the cast slabs completely in a curing tank to avoid the development of shrinkage cracks. The properties of the resulting concrete are listed in Table 5.

Table 5. Properties of porous concrete.

Properties	**Values**
Compressive Strength	16.63 MPa
Porosity	24.84%
Density	1931.25 kg/m3
Coefficient of Permeability	1.05 cm/s

4. Proposed Scheme of protection of the beach

It is proposed to study the normal beach profile and also estimate. The profile after the scouring is based on previous experience. During the season when the sea is calm, the sand may be imported or brought by nourishment from the sea to fill the depth of scouring up to a level about 50 cm below the normal beach level and consolidated. The polypropylene fabric (P.F.) may be laid over the sand bed, as shown in Figure 6. A layer of porous concrete blocks may be arranged to form a bed. During the monsoon, the porous concrete layer acts as armour in the scouring season, and during the season of accretion, sand gets deposited on the porous concrete slabs. Thus the beach would be restored.

Wave action and its effects on a beach are schematically represented in Figure 6. Profile 1 shows the natural profile of the beach before a major erosion occurs. In Profile 2, the eroded shore's cross-section due to the waves' continuous action can be seen. As a solution to this, in the proposed model, the sand lost by scouring is made up by bringing sand from other areas or nourishment from the sea. The profile is made so that the waves enter tangentially in the beach over the porous concrete underlain by the P.F. layer and ride up to the natural run-up, as shown in profile 3.

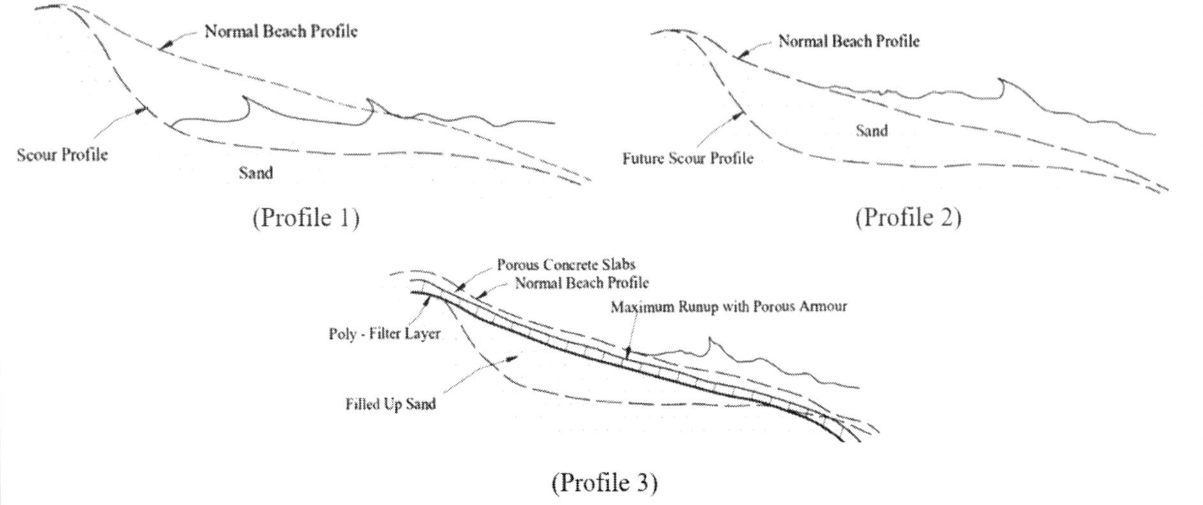

Figure 6. Schematic diagrams to show scouring and laying of porous armor

5. Experimental study on the use of porous concrete armour for beach protection.

The work was taken up as a student's project by the Amrita School of Engineering under the supervision of Ramkrishnan R. It was based on a seminal idea in the model study carried out by Pillai and Verma in 1977 to allow the waves to enter tangentially over a suitable armour and ride on it safely to the length of the run-up (Figure 4(a). and Figure 4(b)). (Ramkrishnan et al. 2018a, b)

The permission obtained for laying the experimental protective armour was at Cheriathura beach in Trivandrum. The height of waves was estimated at 1.2 m, and the period was 90 s. Only a small 5 m wide and 10 m long strip was covered. A diurnal variation of accretion during the night and erosion during the day was noticed. Three days of observation before placing the armour showed that the scour at the top point was about 0.26 m and at the point of entry at the bottom was about 0.48 m. It was planned to place the armour deeper than the erosion level noted. But an earthmover, J.C.B., could not reach the site because of a rubble mound at the bank. The work could not be postponed because a major storm was coming. Hence normal manual digging was resorted to. In addition, anchoring some of the blocks was done. The Laying of the Polypropylene Fabric (P.F.) and arrangement of the porous concrete blocks on the P.F. were done during the interval of low tide. The Laying of the fabric and the slabs is shown in Figure 7(a). and Figure 7(b).

Figure 7 (a). Covering the test area by P. F. and arranging the porous concrete slabs

Figure 7 (b). Covering the test area by P. F. and arranging the porous concrete slabs

At the entry point of the protection where the waves entered the pavement, care was taken to place the slabs at a lower level so that the water entered the slabs tangentially. The row of slabs was fastened laterally by stainless steel strips of 0.003 m thick and 0.05 m. wide. The final installation is shown in Figure 8. The accretion phase covered most of the test area with fine sand. Further, as the sea waves touched the pavement, they laterally brought much fine red sand and deposited it on the pavement, as shown in Figure 9(a). and 9(b).

Figure 8. The final installation of the porous concrete armour.

Figure 9(a). Fine sand covering the test area during the accretion

Figure 9(b). Fine sand covering the test area during the accretion

Afterwards, waves normal to the beach waves appeared as in Figure 10. The waves entered the pavement tangentially and were riding smoothly on the pavement. It washed away the fine sand deposited, moved up the slope and had a run-up almost to the end of the pavement. This riding and receding occurred many times (Figure 11).

Figure 10. Frontal Waves before entry to the armour

Figure 11. Frontal Waves riding on the armour and receding

Meanwhile, it was noticed that scour started occurring on the pavement side, and the blocks on the sides were getting displaced. Initially, they could not be placed deep to the scour depth and the anchoring was not sufficient to hold them. However, the centre portion of the pavement was not affected for a long time, and the waves continued to move up the pavement and receded. The pavement was very narrow and was not protected enough on the sides. It would not have been damaged if a wide revetment of porous concrete blocks were provided with a side wall or abutment.

6. Discussions

When Prof. Posey devised the rock sausages for scour prevention, the following facts were known. i) the stones used for protection should be bigger and heavier so that they do not move by the water flowing at high velocities near the bed; ii) if the stones are enclosed in nets, the size and weight of the stones could be greatly reduced; iii) there should be a filter between the bed and the stones to prevent migration of bed material through the stones and iv) if a net is used it should not get moved by the flowing current by the drag force. Experiment B in small sand spillway models illustrated the use of beds of stones in nets and how the protection would be effective (Figure 1(a), 1(b) and 1(c)). The experiment in C shows how the protection by stones in coir nets (Figure 2) worked even when the outflow had a hydraulic jump with high turbulence intensity.

Experiment D1 was conducted in a wave flume. A model of a sea wall to protect the slope was placed, and waves were made to act on it. The protection was adequate except that small displacement of the outer bigger stones has been noticed. (Figures 3(a) and 3(b).) In experiment D2. a thin layer of small stones enclosed in nets underlain by a thin layer of the reverse sand filter was given as protection. (Figure 4(a)). It may be noted that the wave height was 6.92 cm and the average stone size in the net was only 0.46 g. The experiment also showed that the entry of waves to the protection was tangential to the bank slope, and the waves ride on the slope up to the maximum run-up and recede. The protection given was quite safe (Figure 4(b)).

A major innovation in erosion control was utilizing plastic and polypropylene fabric as filter materials. The geo-fabric was started to be used in many geotechnical applications. The model studies on a spillway over a small sand embankment (Figure 5(a) and Figure 5(b)) illustrated how the polypropylene pillows could be used for protection against scour. Porous concrete slabs can also do the various functions the fabric pillows serve under the polypropylene fabric. Furthermore, the porous concrete is not affected by exposure to sunlight.

A detailed study on porous concrete was taken up by Ramkrishnan, 2017 could help design a mix with high strength, permeability, and durability.

Figure 6. shows a scheme which can be adopted to protect the beach by providing an armour of porous concrete. After scouring occurs, the profile of the beach may first be made up by importing sand from outside like sand nourishment from the sea to a level a few cm below the normal beach level. Then a layer of P.F. may be laid, and porous concrete slabs may be arranged over it. Turbulence in the breaking wave and higher friction offered to ride waves account for reducing the kinetic energy of the incoming waves. The armour is designed to resist the bottom shear due to the waves while it moves along the slope.

The arrangement of the polypropylene fabric (P.F) and the porous concrete slabs are shown in Figure 7a and Figure 7b. Figure 8 shows the final installation. Earlier, waves reaching the pavement brought an accretion of fine red sand on the pavement (Figure 9(a). and Figure 9(b).) As the intensity of waves increased, they became normal to the beach, and they moved on the pavement and removed the sand accumulated by accretion. (Figure 10)

The waves rode on the pavement and receded after going up to the run-up (Figure 11). This was repeated many times. After some time, it was noticed that some of the concrete slabs at the sides of the pavement were getting displaced. The scouring took place at the side of the pavement, but the wave runs continued for a few hours at the central portion of the pavement. Slowly the undermining of the slabs worked towards the central portion also.

It may be noted that there was a limitation that the blocks could not be placed deep down the beach below the scour depth. The damage could have been avoided if a side wall or groyne had been provided.

7. Conclusions

Using porous concrete slabs as an armour underlain by polypropylene fabric as a filter was a good solution to the problem of erosion control. The verification of the method aims to investigate the problem of beach erosion. A study by Ramkrishnan et al. (2017) has made it possible to arrive at a mix with admixtures to get high strength and permeability for porous concrete. It could be durable and of good density (Table 5).

Protection of beach with porous concrete underlain by Poly-filter was done on an experimental basis with a small width of 5 m. In a limited time of the study. The wave action on the central portion of the pavement was observed to assess whether the wave could ride on and later recede on the pavement without damage. The edge slabs at the seaside were affected, but the sand below the central region of the revetment was completely intact for a considerable period. Hence, if the pavement was wider and deeper with a side wall, the pavement underlain by a Poly-filter could be effectively used as a feasible solution to beach erosion problems. (Patent pending Application Number 201741031218). Further studies are required to find the thickness of porous concrete slabs, embedment depths and sidewall designs.

8. REFERENCES

1. Barrett, R. J. (1967). "Use of plastic filters in coastal structures". *In Coastal Engineering* 1966 (pp. 1048-1067).
2. Council of Scientific and Industrial Research - National Institute of Oceanography (CSIR-NIO) (2020). "Guideline for Protection and Control of Coastal Erosion in India". Goa, India.
3. Helen Robertson. (2010). "The prevention Methods of Beach erosion". *Summer Ventures in Science and Mathematics*.
4. Pillai, N. N., & Verma, D. V. S. (1977a). "Use of coir nets in hydraulic structures", *International Symposium on International Group on Hard Fibres*, New Delhi
5. Pillai, N. N., & Verma, D. V. S. (1977b). "Shore protection using stones enclosed in nets". *Coastal Engineering*, 1, 349-358.
6. Posey, C. J. (1973). "Erosion-proofing Drainage Channels". *Soil and Water Conservation*, 28, 2, 1973, pp. 93-95.
7. Posey, C. J. (1974). "Test of scour protection for bridge piers". *Journal of the Hydraulics Division*, 100(12), 1773-1783.
8. Ramkrishnan, R., Abilash, B., Trivedi, M., Varsha, P., Varun, P., & Vishanth, S. (2018a). "Effect of mineral admixtures on pervious concrete". *Materials Today: Proceedings*, 5(11), 24014-24023.
9. Ramkrishnan, R., Pillai, N.N. Aziz, A., & Gowri, R. (2018b). "Utilization of geotextile fabric and permeable concrete to prevent coastal erosion". *5th international Conference*, Hangzhou, Zhejiang University, (pp. 144-160), 2018.
10. Ramkrishnan, R., Pillai, N.N.: "A system and a method for preventing beach erosion". India Patent Pending (2017), Application Number 201741031218.
11. Ramyalekha, B., & Mallikarjuna Reddy, J. (2017). "Study on properties of fresh and hardened self-compacting concrete with varied percentages of metakaolin as mineral admixture (M40 Grade)". *IOSR J. Mech. Civ. Eng*, 14(1), 15-21.
12. Saurav, A.K.G.: "Experimental study of strength relationship of concrete cube and concrete cylinder using ultrafine slag Alccofine". *Int. J. Sci. Eng. Res.* 5(5), 102 (2014) Manufacturing Applications. Bengaluru, August 2017.
13. Zhang, M., Chen, J., Lv, Y., Wang, D., & Ye, J. (2013). "Study on the expansion of concrete under attack of sulfate and sulfate–chloride ions". *Construction and Building Materials*, 39, 26-32.

DOI: https://doi.org/10.26077/4026-ecaa

Predicting Density-driven Exchange Flows through the West Crack Breach of the Great Salt Lake Causeway with CFD and ANN

B.M. Crookston[1], M. Rasmussen[1], E. Larsen[2], H. Kartchner[2] & S. Dutta[2]
[1]*Department Civil and Environmental Engineering, Utah Water Research Laboratory, Utah State University, Logan, Utah, USA*
[2]*Department Mechanical and Aerospace Engineering, Utah State University, Logan, Utah, USA*
E-mail:brian.crookston@usu.edu

Abstract: *The Great Salt Lake is of environmental and economic value yet is threatened by various factors including drought and water diversion for irrigation purposes. Management efforts are required to preserve this saline lake; such efforts include accurately estimating the exchange flow through an opening in a railroad causeway that divides the lake. This study investigated two modeling approaches for predicting these discharges, a physics-based computational fluid dynamics model and a data-driven artificial neural network model. Good agreement was found between both models, and the advantages each provides to water management efforts are noted. Results indicate that, regardless of the modeling tool, accurate field data is invaluable when studying a hydraulic structure.*

Keywords: *Exchange Flow, CFD, ANN, Great Salt Lake.*

1. Introduction

Saline lakes are found on all seven continents and comprise approximately 44% of the volume of all lakes on earth (Messager et al. 2016; Williams 2002). Although historically perceived by some as unimportant, saline lakes and their associated wetlands are recognized for their biodiversity and valued habitat for many waterbirds (Pendleton et al. 2020; Zadereev et al. 2020). Many salt lakes also have thriving fishing, mineral extraction, and recreational industries (Demnati et al. 2017). Unfortunately, some saline lakes have been exploited for the abundant evaporitic minerals (e.g., sodium chloride, sodium carbonate, magnesium, potassium) that have a wide range of uses in medicine, manufacturing, chemical, agricultural, and construction industries (Waiser and Robarts 2009).

Of the various saline lakes scattered across the globe, the Great Salt Lake (GSL) is the largest saline lake in the western hemisphere and fourth largest in the world (Fig. 1a) (Arnow and Stephens 1990). As with other saline lakes, the GSL is currently under threat and requires additional conservation and management efforts. Threats to the GSL include climate shifts and irrigation demands that have reduced river inflow to the lake (Klotz and Miller 2010; Wine et al. 2019; Wurtsbaugh 2014; Wurtsbaugh et al. 2016, 2017). To highlight, in 2016 the GSL was below the historic low lake elevation of 1277.5 m and since then the lake elevation has further declined. Drought and flow diversion threaten an estimated economic output of over $1.3 billion via the mineral extraction, brine shrimp (*Artemia franciscana*) harvesting, and recreation industries (Bink 2021, Bedford 2005, Bioeconomics, Inc. 2012) and also the millions of migratory birds that use the GSL annually as a migratory staging area (Frank 2016).

When considering the past 100-years of the GSL, anthropogenic activities have permanently altered the lake in part from a rockfill railroad causeway (Fig. 1a,b,c) that was constructed (in 1956-1959) East-to-West that divided GSL into north and south sections (Great Salt Lake 1980). Construction of the causeway included two 4.6 m wide × 6.1 m high (15×20 ft) culverts with the purpose of providing flow exchange between the divided sections. The two sections quickly became physically and chemically different despite the culverts and the semi-porous causeway fill material (Madison 1970). Because 95% of freshwater flows into the GSL south of the causeway (Bear, Weber, and Jordan Rivers), the salinity of the northern section increased while the southern section became more dilute (Hahl and Handy 1969). In addition to salinity and the corresponding water density differences, a water surface gradient formed with a higher lake elevation in the southern section that has persisted even during periods of drought.

The railroad causeway has been modified several times since construction to manage the lake and allow passage of flows between the northern and southern sections. To reduce flooding of Salt Lake City in 1984, a 91.4 m wide breach was constructed in the causeway near GSL's west shoreline (Gwynn and Sturm 1987) (Fig. 1b). This breach does not experience flows below a lake elevation of about 1,278.5 m (Baskin 2005). Deterioration of the culverts eventually resulted in their closure and abandonment (west culvert 2012, east culvert in 2013, (Waddell et al. 2014)) with a new breach and bridge, known as the West Crack Breach (WC Breach) added and completed in December 2016 to replace the culverts (Fig. 1c,d).

Figure 1. Overview of the Great Salt Lake located in Northern Utah, USA (circa 1984) (a); GSL's Lake-Side Breach constructed in 1984 to alleviate flooding (b), GSL's West Crack (WC) Breach (8 km east of the 1984 breach) in operation since January 2017 (c), with a view of the bridge over the WC Breach looking south-west (d).

Due to the aforementioned differences in lake water surface elevation (ΔWSE) and density ($\Delta\rho$) between the northern and southern lake sections, a unique flow behavior occurs through the breach for the majority of the year. Waters flowed simultaneously in both directions (north and south); this density-stratified bi-direction flow is often referred to as a gravity-driven exchange flow. Such a hydraulic feature is known to occur in similar lake and ocean settings (Turner 1973). At the WC Breach, the bi-directional flow is characterized by a plunging current of northern water beneath a surface current of less-dense southern water. The less-dense south-to-north surface flow travels more than a kilometer north as it gradually expands laterally. The saltier north-to-south flow is mostly unmixed, passes beneath the surface flow, and connects to a deep brine layer in the southern section of the lake (Naftz 2017).

Predicting the exchange of flow and salt through the WC Breach is critical to current lake management and conservation efforts because salt concentrations in each section directly impact the environment and GSL industries/economies. Prior to the WC Breach, 1D computational hydraulic model of the original culverts were created to estimate discharge and salt exchange to support lake management efforts (Holley et al. 1976). Using this culvert model, the WC Breach was designed to duplicate (as closely as possible) the transfer of water previously provided by the culverts (HDR Engineering, Inc. 2018) but this model does not consider the WC Breach geometry and the resulting three-dimensional (3D) flow. Thus, a new model is needed to improve prediction accuracy. Therefore, what modeling tools might be suitable to improve discharge estimates through the Great Salt Lake Causeway WC Breach and how might viable tools be able to provide further insights into the bi-directional flow regimes as a function of lake conditions?

One common approach for simulating water interacting with a structure such as the West Crack Breach is a physics-based model, such as using computational fluid dynamics (CFD). Various open-source and proprietary CFD solvers have been used to study the interaction of water with control structures and bridges. For example, Crookston et al.

(2018) described the use of CFD to estimate discharge rating curves of piano key weirs. Similarly, Torres et al. (2021) presented the successful use of CFD models in free-surface flows over a weir and spillway to estimate a rating curve. Another study by Chanel and Doering (2008) compared CFD predictions with physical modeling of spillway flow with general success. CFD modeling has also been successfully applied to complex stratified and multiphase flows. Dutta et al. (2014) used CFD models to reveal particle trapping secondary currents in sewer conduits with single and multi-phase flows. Also, An et al. (2012) simulated particle-driven gravity currents that agreed well with similar laboratory experiments and Stancanelli et al. (2018) applied CFD models to analyze the classic lock exchange experiment.

Another approach for predicting discharge through a structure is to use a data-based model, such as machine learning models, which include artificial neural networks (ANN) and deep learning neural networks (DNN). This is a data-driven approach that, through regression and back propagation, formulates a predictive model. Such models are trained with specific data so that the key input parameters can be related to observed patterns in the dataset. Recently, ANN has been used for discharge prediction and flow behavior at a structure (Dutta et al. 2014; Khosravi et al. 2022a,b; Hassanpour et al. 2022; Salazar and Crookston 2019). ANN has also been used to study other fluid problems such as flood patterns (Tsakiri et al. 2018) and sediment loads (Gudino-Elizondo et al. 2019). Thus, in this study these two approaches were evaluated regarding the ability of each model to predict exchange flow rates through the West Crack Breach and provide insights into flow to support GSL management efforts.

In the current paper we present models developed using both the physics-based approach (CFD) and the data-driven approach (ANN), and discuss the scenarios under which a particular approach is advantageous.

2. METHODS

Field data for both modeling efforts relied on USGS data made available to Utah State University, which included monthly specific conductance σ, and velocity profiles, along with water surface elevations, wind speed, wind direction at (and around) the WC Breach. This data was supplemented by a Utah State University field campaign that included continual specific conductance measurements since November 2020 (Dutta et al., 2021), monthly TDS samples, and a temperature array located within the breach to identify the interface in the exchange flow. The USU field campaign also included an exploratory effort to help identify special variability of the flow field adjacent to the WC Breach to inform placement of CFD model boundaries. These field data were used for CFD model development, validation and for ANN model training and testing.

2.1. CFD Model

In this study, FLOW-3D™ by Flow Science, Inc.© was utilized to numerically simulate flow in the WC Breach (Flow Science Inc. 2022). The CFD software uses numerical methods to approximate the Reynold-averaged Navier-Stokes (RANS) equations that are solved with the Boussinesq assumption to simplify the compressible form of the RANS equations.

Bathymetry data for CFD model geometry were compiled from HDR Engineering Inc. (2021) who provided data (from time of construction completion) for the entire excavated section of the WC breach. Supplemental data for the remaining portion of the numerical domain was provided by the USGS (Baskin 2005) or derived from topographic maps. The specific geometry of the bridge was created using bridge structural drawings (HDR Engineering, Inc. 2016).

The CFD modeling in this study had 1) a sectional model phase and 2) a full-width model phase. Phase 1 was of a single-cell width through the center of WC Breach and allowed for an efficient initial study of mesh size, governing equations, boundary conditions, numerical methods for consideration in Phase 2. The Sectional mesh boundaries were assigned as an overlap boundary and pressure types for the E/W and N/S bounds, respectively. Pressure boundaries required input for both fluid elevation and ρ. Grid convergence with 8 different cell sizes was performed and focused on minimal changes in the velocity profile at the USGS measurement locations with numerical velocity probes assigned at locations identical to the USGS acoustic doppler velocimeter (ADV) measurements. In this study, four turbulence schemes provided by FLOW-3D were considered: k-ε, Renormalized Group k-ε (RNG- k-ε), k-ω, and Large Eddy Simulation (LES) (Flow Science, 2008). The four turbulence schemes were tested in Phase 1 considering different model coefficients, and the results indicated better performance by the LES and k-ω, which were then considered in Phase 2.

The following density evaluation options by the FLOW-3D solver considered herein include: 1) density as a function of hydraulic parameters (i.e., temperature, etc.), 2) a first-order approximation, and 3) a second-order monotonicity preserving approximation. Although a fourth option was available (a constant uniform density) it was not considered appropriate for this study.

For Phase 2, the computational mesh was expanded in the east and west directions. In order to continue the grid convergence study, a nested block approach was adopted with two primary or main mesh blocks to represent the north and south sections of the lake with an overlap boundary condition within the WC Breach. Similar to Phase 1, additional boundary conditions were assigned pressure conditions with *WSE* and ρ inputs (Fig. 2). Nested blocks with refined cell size were centered at the WC Breach due to the primary focus of discharge estimation and available velocity data. The total number of computational cells was limited to a maximum of around 25 million due to computing power limitations (four well-equipped desktop machines). To obtain mesh independence in the 3D domain, three mesh configurations were tested with a 2:1 cell size ratio between blocks: 1) 0.3 m cells without a nested mesh block, 2) with a nested 0.15 m mesh block around the breach, and 3) with both a 0.15 m mesh block around the breach and a smaller 0.08 m mesh block. Multiple simulations were performed to determine the appropriate grid size, turbulence closure, density modules, etc. More details can be obtained in Rasmussen (2022) and Dutta et al. (2021).

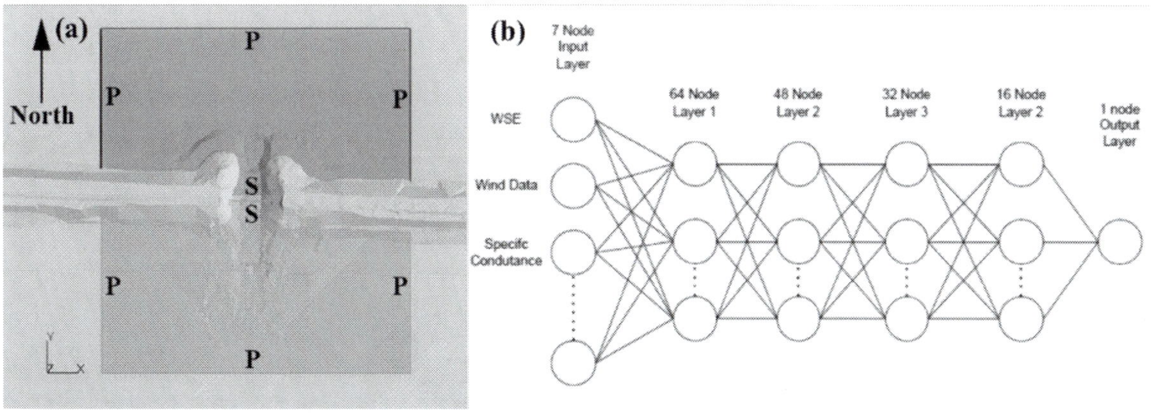

Figure 2. The computational domain for the CFD model with the boundary conditions, where P referring to a pressure boundary (water surface elevation) with assigned density and S referring to an internal overlap boundary (a) and structure of the most accurate ANN model with 4 hidden layers (b). The inputs and output layers have seven and one nodes respectively.

2.2. ANN Model

The ANN model developed herein (see Fig. 2b) used data measured by USGS. For the study, the feed forward network architecture was chosen, along with back-propagation (BP) for learning. These methods have been successful in the past for hydraulic structure (Dutta et al., 2014). Two separate networks were trained, both having the same inputs, but one predicting North-to-South (NS) flow and the other predicting the South-to-North (SN) flow. Three inputs for the model are related to water surface elevation, North of the breach, South of the breach and at the breach. One of the main drivers of the exchange flow is density difference between the North and South sides of the breach, which primarily depends on the salinity of the water. The salinity of N and S sides of the breach is accounted in the model using measured specific conductance value. The final two inputs of the ANN models are wind-speed and wind-direction at the breach. During low wind speeds, records show that wind does not have a substantial effect on the dynamics of the flow at the breach. During high-wind speeds, it can force all the flow through the breach to move south or north, depending on the wind-direction. Thus, it was important to include the wind data to make the model more robust. After pre-processing and cleaning the data, the final data set had 27,204 entries. This was divided into training (21763 entries), testing (2721 entries) and validation (2720 entries) data set. The final network architecture was developed through a process of refining the attributes of the network. The first stage of development was through the layer structure of the network. The number of layers and the size of those layers were varied through multiple tests to find the best performing network. The networks were evaluated on the training data that they presented. The training and evaluation loss values were compared during model training. With the depth and width of the network determined, the batch size of the data used was changed to find the optimal amount of data required for the network before adjusting the weights. Batch sizes ranging from 12-40 were used to visualize the effect batch size had on the training of the model. Each network was tested at 50 and 150 epoch lengths to view the trend that prolonged training had on the network's performance.

3. RESULTS

3.1. CFD Results

Within the WC Breach and north of the railroad is a submerged berm. The berm is intended for management efforts by serving as a control structure for buoyancy-driven flows moving north-to-south. Depending on current or anticipated conditions, the berm could be modified with construction equipment by removing submerged causeway materials or by adding additional materials, thus increasing or lowering the crest elevation of the berm. The CFD model results indicate that indeed the north-to-south flows are influenced by this submerged berm (see Figs. 3 and 4) and that either a lowering of the lake elevation or a raising of the berm would decrease such flows. As an additional detail the flows in the streamwise or y direction can exceed 1 m/s with a densimetric Froude number of about 0.7. The CFD results clearly show this denser saltwater plunging beneath the less-dense south-to-north flows (see Fig. 4) as was observed in the Utah State University field campaign for a July 2021 condition.

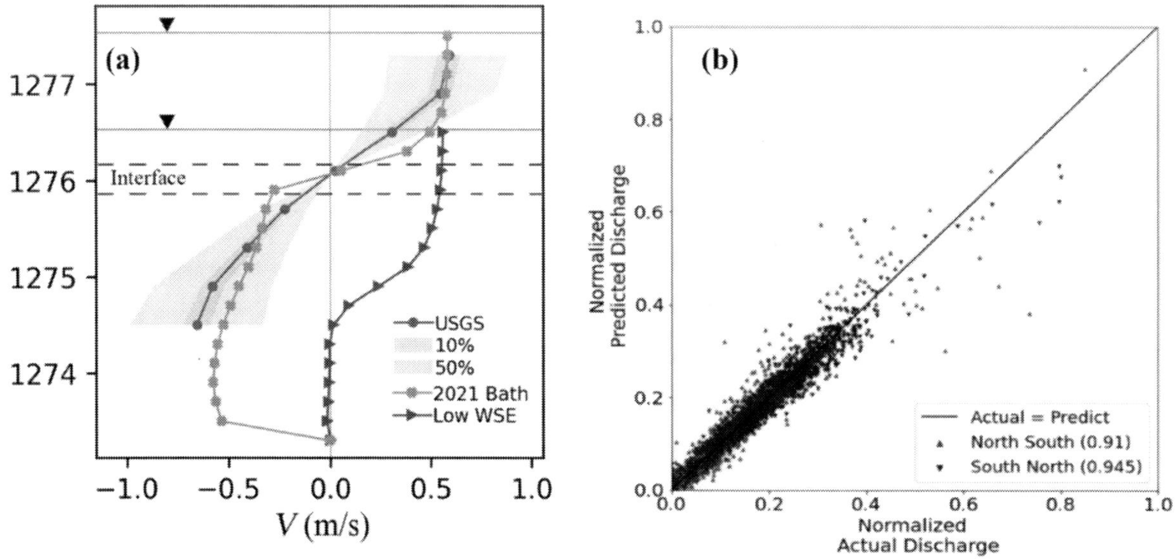

Figure 3. Velocity profile (velocity versus elevation (meters) from USGS compared to CFD results for a specific event in July 2021, and the location of the exchange flow interface via the temperature array (a); and normalized actual versus predicted discharge for the testing data-set using the ANN model (b). Note that if lake levels continue to drop the model predicts a significant decrease in north-to-south flows. The ANN model NSE coefficient is 0.91 for N-S flow and 0.945 for S-N flow.

Such results demonstrate that various combinations of water surfaces, densities, and lake elevations can be used to formulate a head-discharge rating curve through the breach. For additional details of other validation events and results, please see Rasmussen (2022). This study demonstrates that a simple empirical rating curve could be formulated and then incorporated into a hydrologic model of the Great Salt Lake Basin for management efforts. Furthermore, this modeling tool provides two specific advantages:

1. Hypothetical hydraulic scenarios can be simulated
2. Geometric modifications to the breach can be simulated for design purposes.

Specifically, this validated model can be used to consider current drought conditions at the lake and scenarios with even lower lake elevations. Therefore, simulations were performed such as a 0.6 m drop in lake elevation, as shown in Fig. 5. Portions of the lakebed are now exposed and discharges from north-to-south are further reduced. The CFD model also allows exploring any output hydraulic variable at any location within the domain for consideration of potential geometric modifications to the submerged berm. As shown in Fig. 6, sections in the transverse or x direction were taken in the breach at 150 m spacing with (a) located furthest north and (g) furthest south. At the location of the USGS ADV uplooker, CFD results are compared to the USGS field data (see Fig. 3a). Note the significant change in velocity for a reduced water surface elevation.

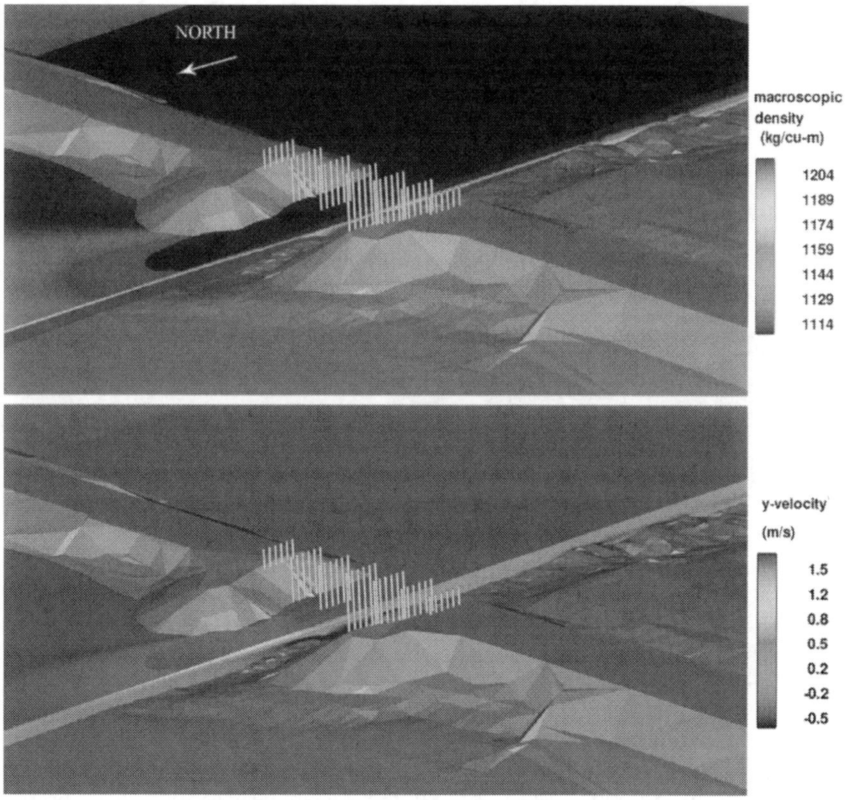

Figure 4. Water density section and velocity gradient in streamwise direction through the WC Breach for the July 2021 event, with surface jet expanding laterally and the lower brine layer traveling south with the velocity section through the WC Breach for July 2021 event.

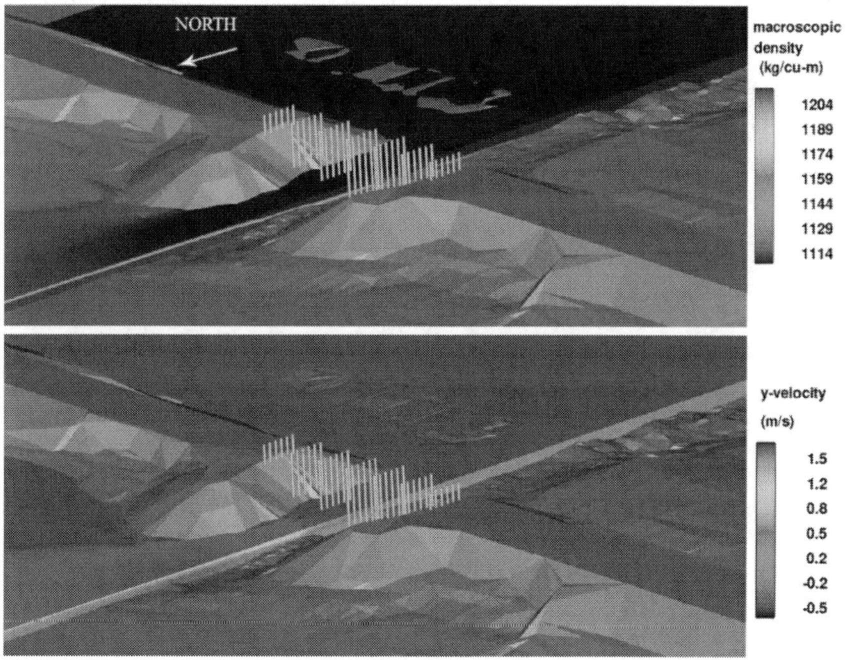

Figure 5. Potential drought scenario if the July 2021 field conditions included lake elevations lowered by 0.6 m.

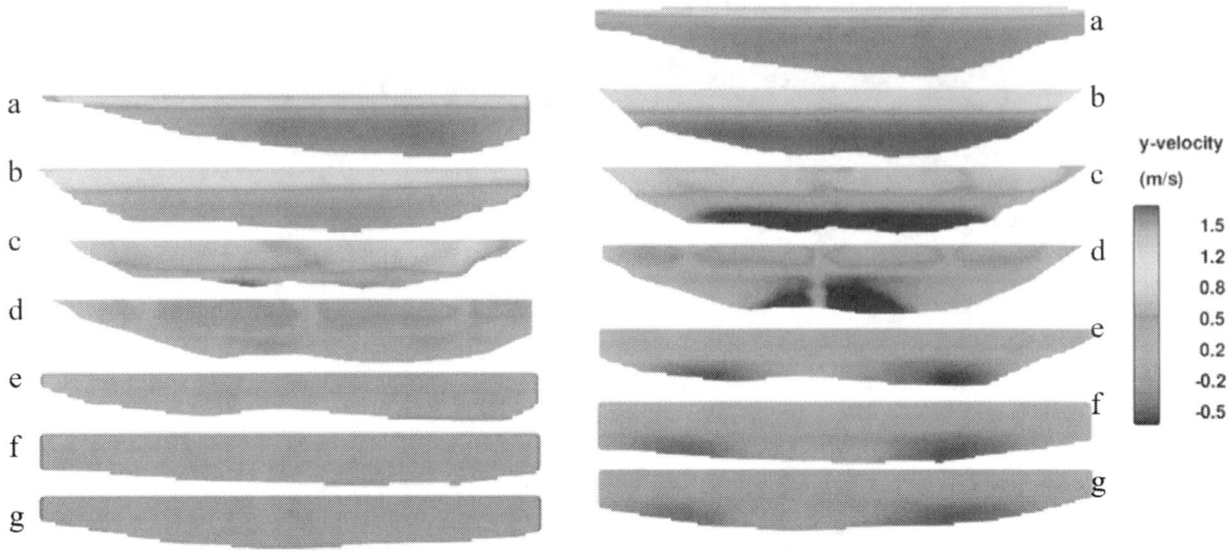

Figure 6. Sections through the WC Breach plotted by streamwise velocity for the July 2021 field event. Worsened lake conditions with a reduced water surface of 0.6 m (left) as compared to actual lake elevations (right). The cross-sections plotted start north of the breach, and go towards the south at a 12 m interval. The streamwise velocity clearly show the weakening of the N-S flow with lowering of the lake elevation.

3.2. ANN Results

Different architectures of the ANN model was trained and the model that performed the best for the testing data-set has been presented here. The ANN with a 7-64-48-32-16-1 architecture (see Fig. 2.) was found to be the most accurate while predicting N-S and S-N flow. The Nash–Sutcliffe efficiency (NSE) coefficient (R^2) for predicting N-S and S-N flow was found to be 0.91 and 0.945, respectively. The corresponding percent bias for the model was –3.65 and 0.84 for N-S and S-N respectively. This means, the ANN model is relatively bias free. Though, the N-S prediction was found to have negative bias, indicating that the ANN model is underestimating the flow compared to observed discharge values. As part of the ANN model development effort, a linear regression model was also developed. In congruence with the expectations, the linear regression model was found to have a substantially lower NSE coefficient ($R^2 \sim 0.8$), compared to the ANN model.

4. CONCLUSIONS

Various modeling tools are available to support water resources management efforts, which includes physics-based models and data-driven models. In this study, CFD and ANN models were used to predict a complex flow behavior at the West Crack Breach in the railroad causeway at the Great Salt Lake. The general flow pattern is an exchange flow or a density-driven bidirectional flow where the denser water from the northern section plunges and flows south, while the less-dense southern water flows at the surface towards the north. Both modeling efforts required field data. The CFD model was able to accurately simulate field conditions and is able to simulate hypothetical scenarios such as different hydraulic conditions or different geometric configurations, both of which are of high interest to the State of Utah. The simulated July 2021 event, had about 1% difference for N-S flow prediction and about 18% for S-N flow prediction, compared to field measurements. The ANN model was also able to accurately predict flows through the breach and is computationally inexpensive, allowing incorporation into a hydrologic model of the basin. It also informs managers what field data is needed for prediction efforts and what field conditions may cause discharges to increase or decrease. However, the ANN model is limited to the range of training data and therefore is not suitable for extrapolation such as predicting performance under future climatic shifts.

5. ACKNOWLEDGMENTS

This research was possible due to funding from Utah Department of Natural Resources (Utah DNR) and ESIP Lab. EL was also supported by EURP fellowship from the College of Engineering at Utah State University.

6. REFERENCES

An, S., P. Y. Julien, and S. K. Venayagamoorthy. (2012). "Numerical simulation of particle-driven gravity currents." *Environ Fluid Mech*, 12 (6): 495–513. https://doi.org/10.1007/s10652-012-9251-6.

Arnow, T., and D. Stephens. (1990). *Hydrologic Characteristics of the Great Salt Lake, Utah: 1847-1986*. 63. U.S. Geological Survey.

Baskin, R. L. (2005). *Calculation of Area and Volume for the South Part of Great Salt Lake, Utah*. Scientific Investigations Map. Scientific Investigations Map. U.S. Geological Survey.

Bedford, D. (2005). "Utah's Great Salt Lake: A Complex Environmental-Societal System." *Geographical Review*, 95 (1): 73–96. https://doi.org/10.1111/j.1931-0846.2005.tb00192.x.

Bink, A. (2021). "Great Salt Lake hits historic low in southern portion." *abc4 News*. Accessed September 27, 2021. https://www.abc4.com/weather/drought/great-salt-lake-hits-historic-low-in-southern-portion/.

Bioeconomics, Inc. (2012). *Economic Significance of the Great Salt Lake to the State of Utah*.

Chanel, P. G., and J. C. Doering. (2008). "Assessment of spillway modeling using computational fluid dynamics." 35: 6.

Crookston, B. M., R. M. Anderson, and B. P. Tullis. (2018). "Free-flow discharge estimation method for Piano Key weir geometries." *Journal of Hydro-environment Research*, 19: 160–167. https://doi.org/10.1016/j.jher.2017.10.003.

Demnati, F., B. Samraoui, F. Allache, A. Sandoz, and L. Ernoul. (2017). "A literature review of Algerian salt lakes: values, threats and implications." *Environ Earth Sci*, 76 (3): 127. https://doi.org/10.1007/s12665-017-6443-x.

Dutta, S., T. E. Tokyay, Y. A. Cataño-Lopera, S. Serafino, and M. H. Garcia. (2014). "Application of computational fluid dynamic modelling to improve flow and grit transport in Terrence J. O'Brien Water Reclamation Plant, Chicago, Illinois." *Journal of Hydraulic Research*, 52 (6): 759–774. https://doi.org/10.1080/00221686.2014.949883.

Dutta, S., B. Crookston, M. Rassmusen, E. Larsen (2021) Predicting Flow through the Causeway of the Great Salt Lake using Hydrodynamic Simulations and Artificial Neural Networks. Scientific report submitted to Utah DNR. https://doi.org/10.26077/a537-fe72

Flow Science Inc. (2022). "FLOW-3D v12.0 User Manual." Flow Science Inc.

Frank, M. G. (2016). "Migratory waterbird ecology at a critical staging area, Great Salt Lake, Utah." Salt Lake City, UT: Utah State University.

Great Salt Lake : a scientific, historical, and economic overview. (1980). Utah Geological Survey.

Gwynn, J. W., and P. A. Sturm. (1987). *Effects of breaching the Southern Pacific Railroad causeway, Great Salt Lake, Utah -- physical and chemical changes, August 1, 1984 - July, 1986*. Utah Geological Survey.

Hahl, D. C., and A. H. Handy. (1969). *Great Salt Lake, Utah: Chemical and Physical Variations of the Brine, 1963-1966*. 36. Salt Lake City, UT: U.S. Geological Survey.

Hassanpour, N., Sadeghfam, S., Abdollahpour, M. and B.M. Crookston (2022). "Predicting hydraulic jump characteristics in a gradually expanding stilling basin with roughness elements by Sugeno Fuzzy Logic." *Journal of Hydroinformatics*. https://doi.org/10.2166/hydro.2022.002

HDR Engineering, Inc. (2016). "Lakeside Sub Bridge As-Built Compliance Set."

HDR Engineering, Inc. (2017). "PERMIT SPK-2011-00755 As-built Drawings."

HDR Engineering, Inc. (2018). *2017 Annual Data Monitoring Report*. Salt Lake City, UT.

Holley, E., R., M. ASCE, and K. M. Waddell. (1976). "Stratified Flow in Great Salt Lake Culvert." *Journal of Hydraulics Division*, 17.

Klotz, E. K., and C. W. Miller. (2010). "Man's Impact on Great Salt Lake Elevations." 54.

Khosravi, K., Khozani, Z.S., Melesse, A. and B.M. Crookston (2022a). "Intelligent flow discharge computations in a rectangular channel with free-overfall conditions." *Neural Computing and Applications.* https://doi.org/10.1007/s00521-022-07112-9

Khosravi, K., Safari, M., Khozani, Z., Crookston B.M., and A. Golkarian. (2022b). "Stacking ensemble-based hybrid algorithms for discharge computation in sharp-crested labyrinth weirs." *Soft Computing.* https://doi.org/10.1007/s00500-022-07073-0(0123456789

Madison, R. J. (1970). "Effects of a Causeway on the Chemistry of the Brine in Great Salt Lake_Utah." 50.

Messager, M. L., B. Lehner, G. Grill, I. Nedeva, and O. Schmitt. (2016). "Estimating the volume and age of water stored in global lakes using a geo-statistical approach." *Nat Commun,* 7 (1): 13603. https://doi.org/10.1038/ncomms13603.

Naftz, D. L., F. J. Millero, B. F. Jones, and W. Reed Green. (2011). "An Equation of State for Hypersaline Water in Great Salt Lake, Utah, USA." *Aquat Geochem*, 17 (6): 809–820. https://doi.org/10.1007/s10498-011-9138-z.

Pendleton, M. C., S. Sedgwick, K. M. Kettenring, and T. B. Atwood. (2020). "Ecosystem Functioning of Great Salt Lake Wetlands." *Wetlands*, 40 (6): 2163 2177. https://doi.org/10.1007/s13157-020-01333-1.

Rasmussen, M. (2022). "Using computational fluid dynamics to predict flow through the West Crack Breach of the Great Salt Lake railroad causeway." MS. Thesis. Utah State University, Logan, Utah.

Salazar, F. and Crookston, B.M. (2019). "A performance comparison of machine learning algorithms for arced labyrinth spillways." *Water by MPDI.* 2019 11(3) 544. https://doi.com/10.3390/w11030544

Stancanelli, L., R. Musumeci, and E. Foti. (2018). "Computational Fluid Dynamics for Modeling Gravity Currents in the Presence of Oscillatory Ambient Flow." *Water*, 10 (5): 635. https://doi.org/10.3390/w10050635.

Torres, C., D. Borman, A. Sleigh, and D. Neeve. (2021). "Application of Three-Dimensional CFD VOF to Characterize Free-Surface Flow over Trapezoidal Labyrinth Weir and Spillway." *J. Hydraul. Eng.*, 147 (3): 04021002. https://doi.org/10.1061/(ASCE)HY.1943-7900.0001852.

Tsakiri, K., Marsellos, A., & Kapetanakis, S. (2018). Artificial neural network and multiple linear regression for flood prediction in Mohawk River, New York. Water, 10(9), 1158.

Waddell, K., W. Gwynn, and HDR Engineering, Inc. (2014). *Bridge Evaluation Report. Union Pacific Railroad Great Salt Lake Causeway Culvert Closure and Bridge Construction Project.* 46.

Waiser, M. J., and R. D. Robarts. (2009). "Saline Inland Waters." *Encyclopedia of Inland Waters*, 634–644. Elsevier.

Williams, W. D. (2002). "Environmental threats to salt lakes and the likely status of inland saline ecosystems in 2025." *Envir. Conserv.*, 29 (2): 154–167. https://doi.org/10.1017/S0376892902000103.

Wine, M. L., S. E. Null, R. J. DeRose, and W. A. Wurtsbaugh. (2019). "Climatization—Negligent Attribution of Great Salt Lake Desiccation: A Comment on Meng (2019)." *Climate*, 7 (5): 67. https://doi.org/10.3390/cli7050067.

Wurtsbaugh, W. A., C. Miller, S. E. Null, R. J. DeRose, P. Wilcock, M. Hahnenberger, F. Howe, and J. Moore. (2017). "Decline of the world's saline lakes." *Nature Geosci*, 10 (11): 816–821. https://doi.org/10.1038/ngeo3052.

Wurtsbaugh, W., C. Miller, S. Null, P. Wilcock, M. Hahnenberger, and F. Howe. (2016). "Impacts of Water Development on Great Salt Lake and the Wasatch Front." 9.

Zadereev, E., O. Lipka, B. Karimov, M. Krylenko, V. Elias, I. S. Pinto, V. Alizade, Y. Anker, A. Feest, D. Kuznetsova, A. Mader, R. Salimov, and M. Fischer. (2020). "Overview of past, current, and future ecosystem and biodiversity trends of inland saline lakes of Europe and Central Asia." *Inland Waters*, 10 (4): 438–452. https://doi.org/10.1080/20442041.2020.1772034.

DOI: https://doi.org/10.26077/9f33-e4bf

Predicting Riverbed Deformation During Floods for a Sustainable Design of Headworks Structures

I. Joshi[1,2], U. Singh[2], M. B. Bishwakarma[2], Y. Kitamura[3] & L. Lia[1]
[1]*Norwegian University of Science and Technology, Trondheim, Norway*
[2]*Hydro Lab, Lalitpur, Nepal*
[3]*Chigasaki Research Institute, Chigasaki, Japan*
E-mail: ishwar.joshi@ntnu.no

***Abstract**: Understanding and predicting riverbed deformation during floods is important for the proper and sustainable design of headworks structures. Field-scale study of floods is challenging, and existing literature shows that mainly qualitative data (Photographic and Videographic) have been used in riverbed deformation studies. On the other hand, laboratory experiments allow studying floods in more detail and under a controlled environment. In this research, bed deformation of river channel during floods was investigated by conducting experiments in a curved flume. Parameters of the experiments were chosen with respect to gravel-cobbled River in Nepal. Four different floods with different magnitudes and durations were investigated in the experiments. Riverbed deformation was quantified by differencing pre and post-flood DEMs obtained using Structure from Motion (SfM) technology. The maximum scour during the floods was measured by scour chains made with light beads in a thread. The results showed that the magnitude and duration of floods are both important factors for riverbed deformation. The volume of channel bed erosion increased with the increase in flood magnitude. However, the total volume of sediment reworked was higher in the case of floods which are of lower magnitude but occur frequently in multiple cycles. The depth of maximum bed scour during the floods was observed much higher than the channel bed resulted at the end of the floods. The difference was more prominent along the bends where the secondary currents are stronger. The understanding could be used to better validate sediment transport models, and thus predict riverbed deformation more accurately for sustainable design of headworks structures.*

***Keywords:** Floods, Maximum scour, Sediment transport, Numerical modeling.*

1. Introduction

The sustainability of headworks structures depends on how it can effectively handle the incoming sediments (Rulot et al. 2012; Joshi et al. 2021). Physical models are constructed to test the performance of headworks structures regarding sediments. Although, physical models give satisfactory results, they will not be economically feasible in predicting future trends of riverbeds which are crucial for sediment management. For e.g., in Peaking Run of River projects (PROR) long term changes in live storage volume are often needed. Changes in live storage can have huge implication on peak power production. Existing literatures (Giri et al. 2016; Giri and Omer 2019; Mool et al. 2017) have used numerical models independently or in conjunction (hybrid) with physical models for long-term prediction of riverbeds in hydropower projects. They have highlighted the importance of sediment transport equations and its impact on the prediction of riverbed.

During high discharge or floods, the riverbed scours during the rising limb and refills during the falling limb. The equilibrium assumption of sediment transport equations may not hold true during the scouring and refilling process (Singh et al. 2019). So, the predictions made with such equilibrium sediment transport equations in numerical model will not be accurate. Thus, it is important to test how different sediment transport equations behave in varying flood conditions.

In this work, novel experiments in laboratory are done, where riverbed deformation by floods of different magnitude and duration are compared. Also, maximum scour during floods using scour chains is measured. The materials, methods and results from the experiments are summarized in this paper.

2. Materials and Methods

2.1. Experimental channel

The experiments were carried out in a 13 m long channel having an average width of 2 m (Figure 1). A 6 m long reach of the channel at the upstream end was curved and also wider than the channel at the downstream end. An inlet channel with a 0.6 m wide rectangular opening expanding to 0.75 m was provided at the upstream end so that sediment could be well distributed at the inlet. Required flow was supplied by pumping water from an underground reservoir to a calibrated V-notch which discharged into the pool. The pool helped to kill turbulence caused by flow from pump and thus ensuring calmer flow into the inlet. A sediment feeder was installed at the inlet channel to supply the sediment at a designated rate. The water level at the downstream end was controlled with the help of an overflow weir. The flow and sediment exiting the channel were collected in an outlet tank at the downstream end. Sediments were allowed to settle in the tank and the water was discharged back to the underground reservoir with the help of a return canal. The weight of the collected sediments was also measured at the end of each experiment.

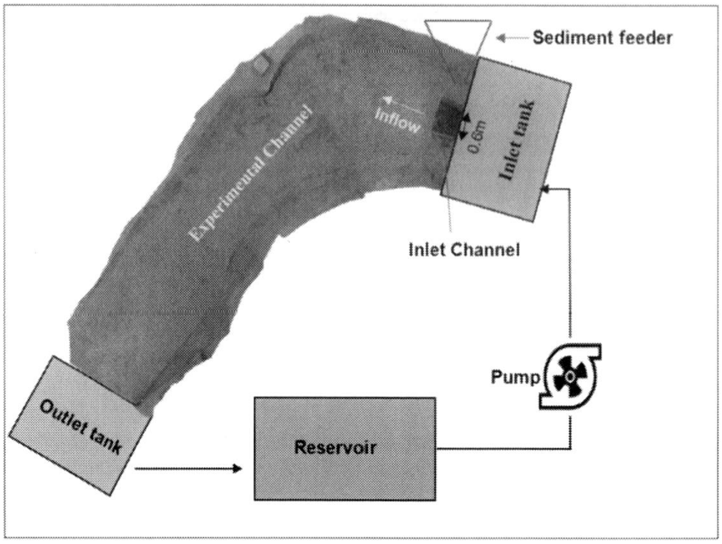

Figure 1. Schematic layout of the experimental setup

2.2. Sediment sizes and composition

The parameters of the experiments were chosen with respect to monitored gravel-cobbled reach at a scale of 1:50. The sediment layer constitutes of non-uniform sediment with a median grain size of 0.9 mm. The sediment used in the experiment was finer and well sorted than the bed material of the monitored reach (Figure 2).

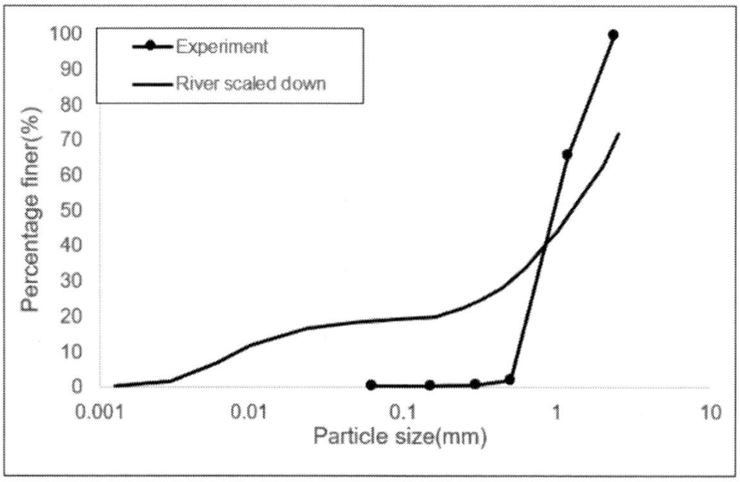

Figure 2. Particle size distribution of the sediment used in the experiment and the bed deposits in the monitored reach

2.3. Instrumentation

2.3.1. Scour chains

Scour chain can measure the maximum scour during a flood and has been used at field scales (Leopold et al. 1995). However, it has not been used at laboratory scales till now, which shows the novelty of this work. Scour chains, 3 at each cross-section, were installed at three different cross-sections, namely X-6, X-9, and X-12 (Figure 3), while constructing initial channel bed. The scour chains were constructed of thread and plastic beads. The beads were spherical and made of plastic (light material) so that local disturbance to bed would not significantly affect the scour depth. Figure 3(b) shows an initial position of installed scour chain before starting experiments. The scour chains were retrieved at the end of the experiments to measure the maximum depth of scour.

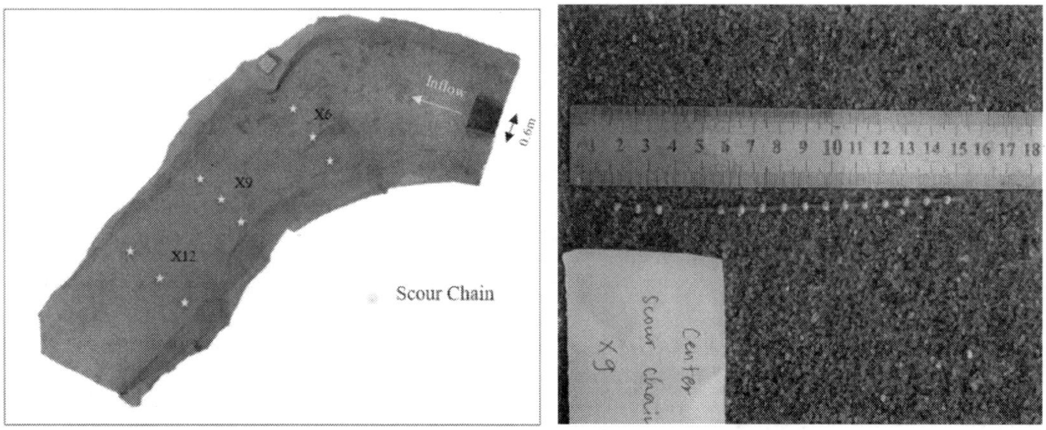

Figure 3: a) Experimental channel with location of scour chain and sampling points, b) initial position of scour chain in the channel bed after the installation.

2.3.2. Structure from Motion

SfM is a photogrammetric method of surveying in which a three-dimensional topography is developed from a series of overlapping two dimensional images (Figure 4). It has been used as tool to acquire bed topography of channels in laboratory experiments (Kasprak et al. 2014; Leduc et al. 2019) as well as in the field scale river (Duró et al. 2018; Javernick et al. 2014). The application of SfM requires establishment of control points along the banks of the experimental channel. These control points should be distributed in such a way that they cover all the observation area. A control point should be visible in at least three images. In this work, the images of the channel bed were acquired using Sony α6300 (model ILCE-6300), mirror-less digital camera with 24-megapixel Exmor RS sensor. A multitude of SfM software are available to process the images, but the general principle remains the same. PhotoScan (developed by Agisoft) was used to process the images in this study. The results from the SfM were validated against the cross-sections measured by dumpy levels. Figure 5 shows that a cross-section extracted from the Digital Elevation Model (DEM) of the channel bed surveyed using SfM is very similar to the cross-section measured by dumpy level. So, data acquired by SfM were used for the analysis.

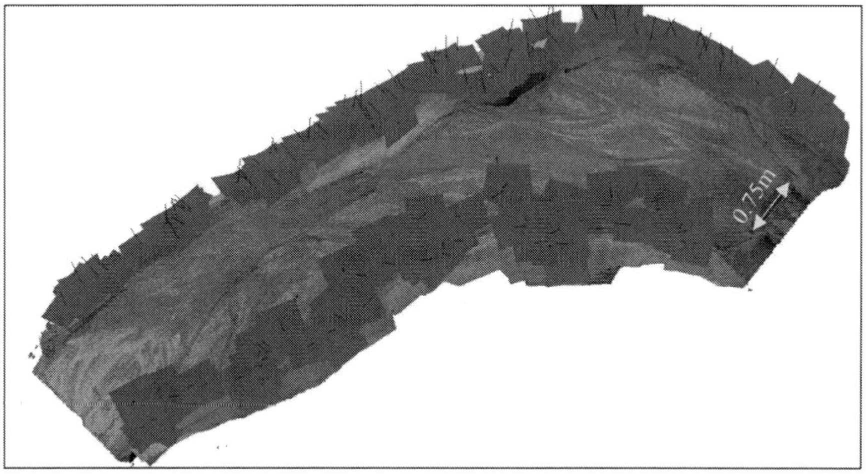

Figure 4: 3D model of the experimental channel developed using Structure from Motion. The blue square shows the camera position.

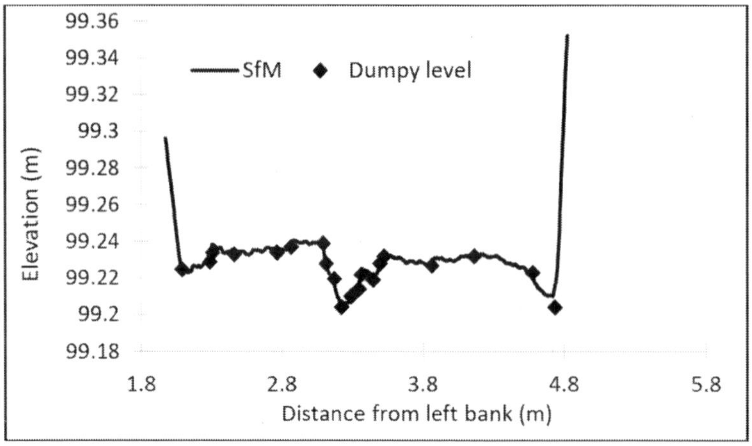

Figure 5: Comparison of a channel cross-section measured from SfM and that surveyed by dumpy level.

2.4. Flow and sediment conditions

The initial flat channel bed was first allowed to develop at a discharge of 4.53 l/s for 15 hours. This discharge corresponds to an average high season discharge observed at the monitored reach, which was also considered as a formative discharge of the channel. Four different flood flows were then simulated in the evolved channel to investigate the riverbed deformation during floods of different magnitudes and duration. The magnitude and duration of simulated flood scenarios are provided in Table 1. The duration of a flood hydrograph for all scenarios were assumed to be 150 minutes, which corresponds to 18 hours in the gravel-cobbled reach. The sediment used to construct the channel bed was also used as the feed material. The sediment feed rates for different flood scenario are show in Figure 6 and Table 1. The sediment represents gravels at prototype. So, the feed rate was estimated using Wong and Parker (2006) sediment transport formula as shown in Equation 1.

$$q^* = 4.93(t_b^* - 0.047)^{1.6} \quad (1)$$

$$q^* = \frac{q_b}{\sqrt{RgD_m}D_m} \quad (2)$$

$$t_b^* = \frac{t_b}{\rho RgD_m} \quad (3)$$

Where, q^* is the dimensionless volume bedload-transport rate per unit width, t_b^* is the dimensionless bed shear stress, q_b is the volumetric bedload transport per unit width ($m^3/s/m$), R is the hydraulic radius (m), D_m is the median particle size (m), g is the acceleration due to gravity (m/s^2), and t_b is the bed shear stress (N/m^2).

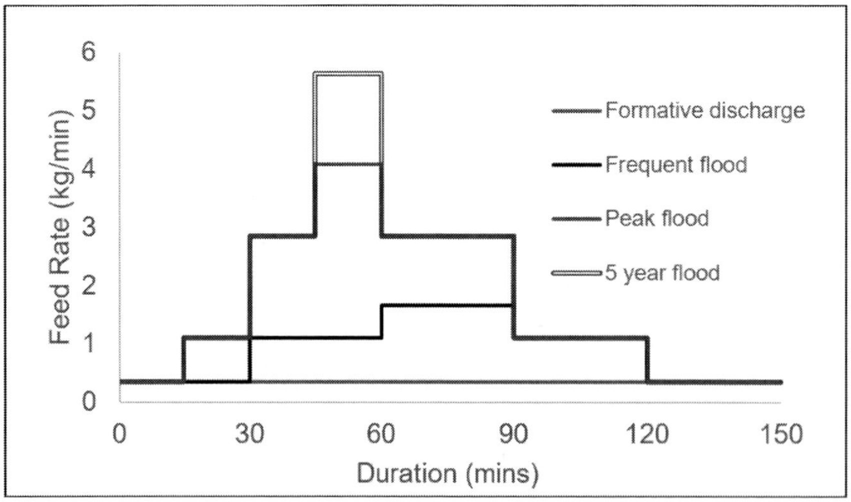

Figure 6: Sediment feed rate corresponding to different flood scenarios

Table 1: Flow and sediment feeding rates applied for different flood scenarios

SN	Flood scenarios	Flow (l/s)		Sediment feed rate (kg/min)		Duration (hours)	Number of hydrographs simulated
		Min	Max	Min	Max		
1	Frequent flood	4.53	11.31	0.35	1.66	2.5	1
2	Peak flood	4.53	22.62	0.35	4.08	2.5	1
3	5 year flood	4.53	29.69	0.35	5.64	2.5	1
4	Frequent flood long	4.53	11.31	0.35	1.66	15	6

3. Results and Discussions

3.1. Channel bed deformation

The channel bed deformation was analyzed by performing the differencing of the DEM's obtained at the different stages of the experiment. The erosional and depositional volumes were also computed. Similarly hypsometric curves were generated for the statistical analysis of the obtained DEMs and their differences.
Hypsometric curves and DEMs of initial beds of all scenarios is presented in Figure 7. Figure 7 shows that the DEMs are more or less identical. However, some discrepancies are observed at the elevation range of 99.26 - 99.22m and 99.34 - 99.30m in the case of peak flood and frequent flood long experiments. This discrepancy is due to some workmanship error during the construction of the initial beds.
Hypsometric curves and DEMs obtained after simulating formative discharge for 15 hours for all scenarios are presented in Figure 8. The planform shown by the DEMs appear different in each of the scenarios. However, hypsometric curve depicts that all of the evolved beds are statistically similar. Thus all flood scenarios were simulated in a statistically similar initial channel bed.
Similarly, Figure 9 shows the difference between the DEMs obtained by formative discharge and the final bed for the respective flood scenarios. The DEM of difference was used to estimate the erosional and depositional volume/area between the final and initial bed (obtained with the formative discharge). Table 2 shows the estimated erosional and depositional volume in each of the experiments.
Frequent flood, peak flood, 5 years flood and frequent flood long duration resulted in 0.173 m^3, 0.263 m^3, 0.335 m^3, and 0.355 m^3 of sediment rework respectively (Table 2). The erosional volumes were 0.074 m^3, 0.134 m^3, 0.251 m^3 and 0.213 m^3 respectively. Similarly, the depositional volumes were 0.099 m^3, 0.129 m^3, 0.084 m^3 and 0.142 m^3 respectively. It can be observed that the amount of bed deformation (i.e. the total reworked volume) increases with the increase in flood magnitude. However, frequently occurring flood with longer simulation resulted in most bed deformation. The magnitude of the erosional volume also increased with the increase in flood magnitude but also depends on the duration of the floods.

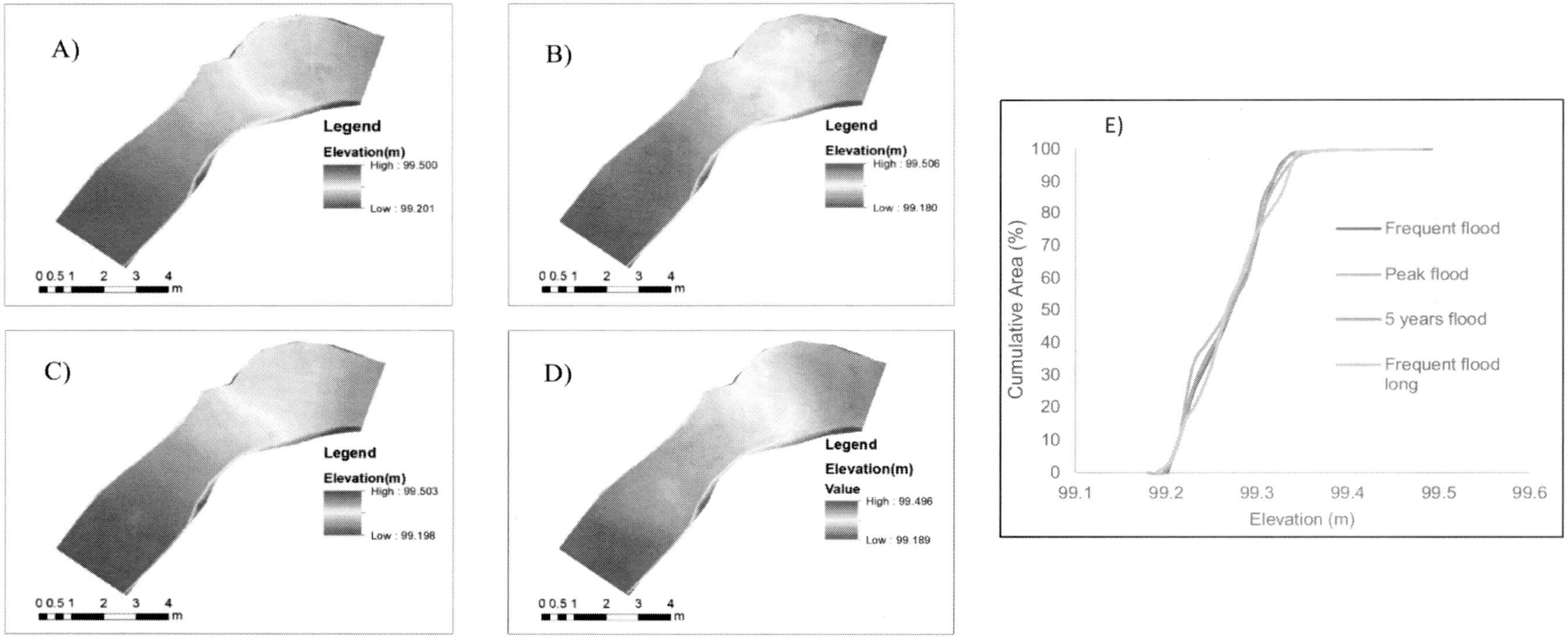

Figure 7: Initial beds of A) Frequent flood; B) Peak flood; C) 5 year flood D) Frequent flood long experiment scenarios. E) The hypsometric curves of DEM's of all the initial beds

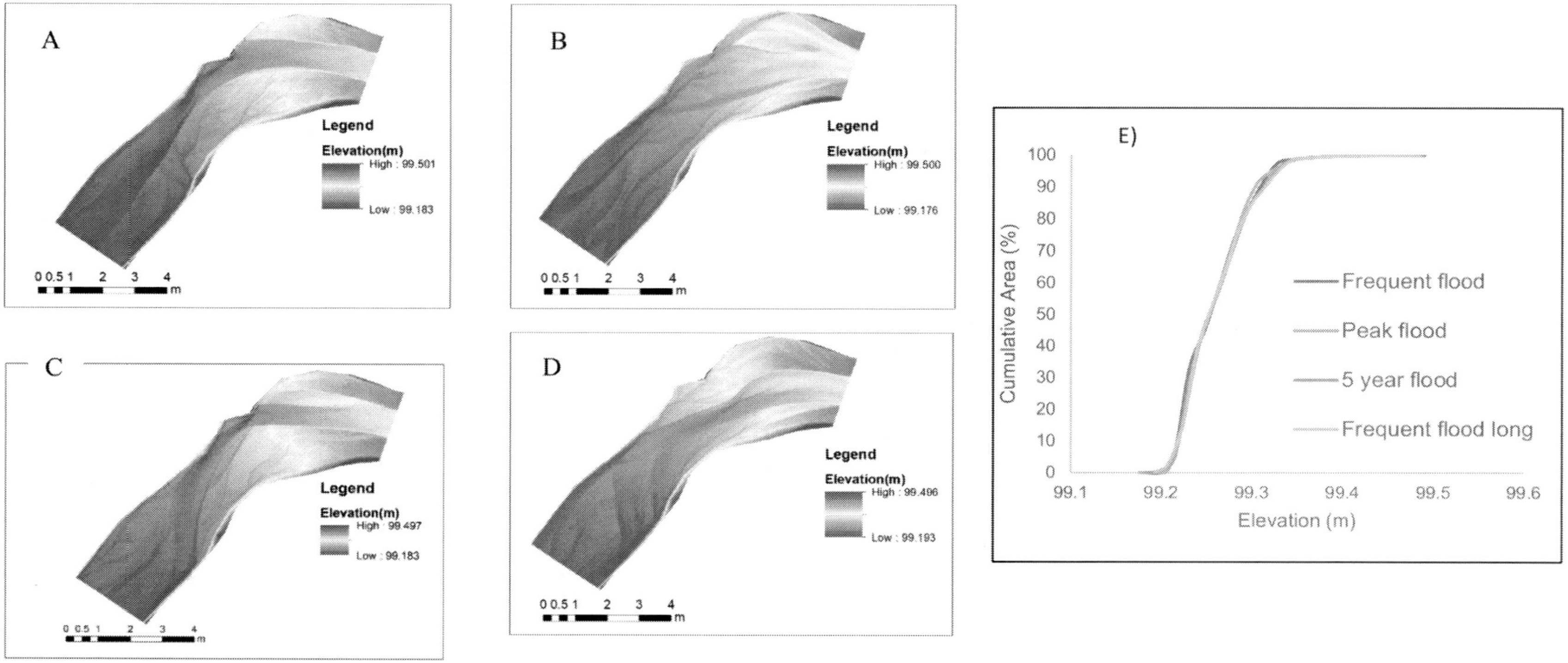

Figure 8: Channel bed resulting from formative discharge during: A) Frequent flood; B) Peak flood; C) 5 year flood and D) Frequent flood long experiment scenarios. E) The hypsometric curves of DEM's of all the beds after formative discharge

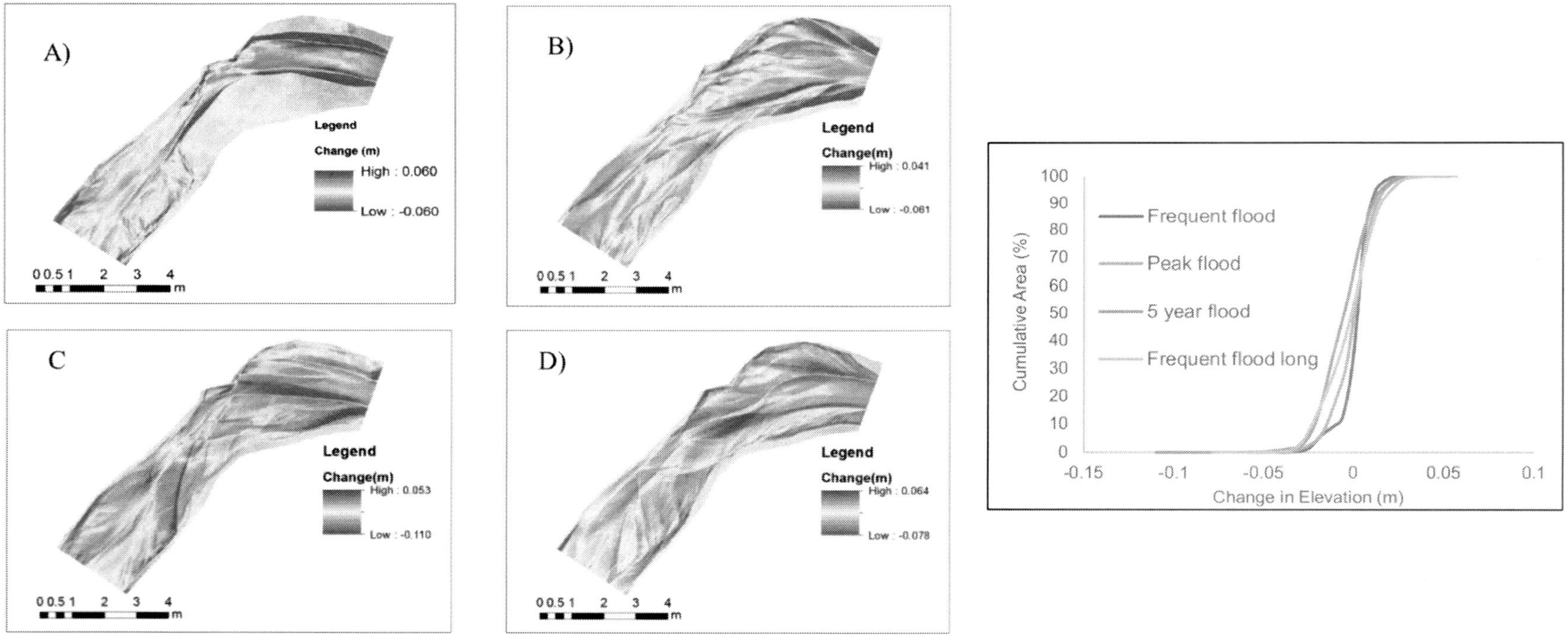

Figure 9: DEM of difference between final bed and bed after formative discharge: A) Frequent flood B) Peak flood C) 5 year flood D) Frequent flood repeated. E) The hypsometric curves of DEM of

Table 2: Erosion and depositional volume for different flood scenarios

Parameters/ Experiments	Frequent flood	Peak flood	5 year flood	Frequent flood long
	(1 cycle)	(1 cycle)	(1 cycle)	(6 cycles)
Erosion Volume (m^3)	0.074	0.134	0.251	0.213
Deposition Volume (m^3)	0.099	0.129	0.084	0.142
Total Reworked Volume (m^3)	0.173	0.263	0.335	0.355

3.2. Maximum scour depth

The maximum scour depth during a flood is higher than the final bed level (Leopold et al. 1995). Figure 10 shows the bed level profile before and after the flood along with final position of the scour chain. The results indicate that maximum scour levels were lower than the final bed levels at the end of the experiments. Table 3 shows the comparison of maximum depth of scour between all flood scenarios. The results indicate that 5 year flood produces maximum scour at cross-section 6, which lie along the bend and peak flood produces maximum scour at cross-section 12. At cross-section 9 both peak flood and 5 year flood produces similar value of maximum scour.

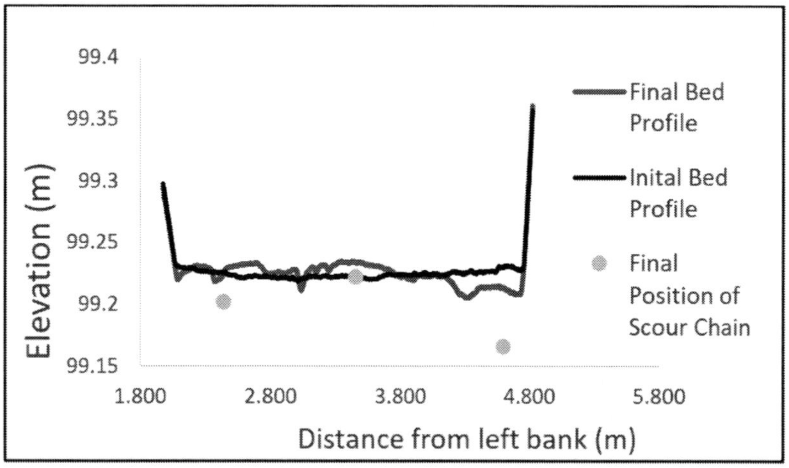

Figure 10: Bed level profiles before and after a flood at X-12. Yellow dot marks the final position of the scour chain.

Table 3: Maximum depth of scour at different cross-sections

SN	Cross-section	Maximum depth of scour(cm)	Flood Scenario	Peak values of flood (lps)
1	X6	12.4	5 year flood	29.69
2	X9	8.8	5 year flood and peak flood	29.69 and 22.62
3	X12	6.4	Peak flood	22.62

4. Conclusions:

Bed deformation of river channel during floods was investigated by conducting experiments in a curved flume. Parameters of the experiments were chosen with respect to a gravel-cobbled river in Nepal, which was monitored in the field. Four different floods namely, frequently observed floods, observed peak flood, 5 year flood and frequently observed flood for longer duration were investigated in the experiments. The experiments allowed studying the phenomena in more detail under controlled environment. Floods scenarios were investigated in an initial channel - resulting from a formative discharge. Maximum scour depths along different cross-sections was measured with the help of scour chains.

The results showed that magnitude and the frequency of floods are both important factors for the riverbed deformation. The volume of channel bed erosion increased with the increase in flood magnitude. However, the total volume of sediment reworked was higher in the case of floods which are of lower magnitude but occur frequently in multiple cycles.

The depth of maximum bed scour during the floods was observed much higher than the channel bed resulted at the end of the floods. The difference was more prominent along the bends where the secondary currents are stronger. This understanding could be used to better validate sediment transport models and design sustainable headworks structures.

5. ACKNOWLEDGMENTS

The funding, for conducting the experiments, from Chigasaki Research Institute (CRI), J-Power, Japan is highly acknowledged. The first author would also like to thank HydroCen and Energize Nepal for providing grant to do research at NTNU, Norway.

6. REFERENCES

Duró, G., A. Crosato, M. G. Kleinhans, and W. S. J. Uijttewaal. 2018. "Bank erosion processes measured with UAV-SfM along complex banklines of a straight mid-sized river reach." *Earth Surf. Dyn.*, 933–953.

Giri, S., and A. Omer. 2019. "Sediment Removal and Beneficial Use (Part-2): Morphological Sediment Removal and Beneficial Use (Part-2): Morphological Modelling of Dakpathar Barrage." (February).

Giri, S., M. Yossef, S. Raman, P. Cleyet-Merle, and B. R. K. Pillai. 2016. "Reservoir sedimentation issue in Pillur reservoir in Nilgiris basin (India): Field reconnaissance and numerical modelling using Delft3D." *River Flow - Proc. Int. Conf. Fluv. Hydraul. RIVER FLOW 2016*, 1463–1470. https://doi.org/10.1201/9781315644479-231.

Javernick, L., J. Brasington, and B. Caruso. 2014. "Modeling the topography of shallow braided rivers using Structure-from-Motion photogrammetry." *Geomorphology*, 213: 166–182. https://doi.org/10.1016/j.geomorph.2014.01.006.

Joshi, I., R. K. Chaudhary, and D. K C. 2021. *Engineering perspective of Melamchi disaster and Mitigation Methods. Zenodo.*

Kasprak, A., J. M. Wheaton, P. E. Ashmore, J. W. Hensleigh, and S. Peirce. 2014. "Journal of Geophysical Research : Earth Surface The relationship between particle travel distance and channel morphology : Results from physical models of braided rivers." *J. Geophys. Res. Earth SurfaceResearch EarthSurface*, 55–74. https://doi.org/10.1002/2014JF003310.Received.

Leduc, P., S. Peirce, and P. Ashmore. 2019. "Short communication: Challenges and applications of structure-from-motion photogrammetry in a physical model of a braided river." *Earth Surf. Dyn.*, 7 (1): 97–106. https://doi.org/10.5194/esurf-7-97-2019.

Leopold, L. B., M. G. Wolman, and J. P. Miller. 1995. *Fluvial processes in geomorphology*. Courier Corporation.

Mool, P., I. Popescu, S. Giri, A. Omer, K. Sloff, and Y. Kitamura. 2017. "Delft3D morphological modelling of sediment management in daily peaking run-of-the-river hydropower (PROR) reservoirs in Nepal." *85th Annu. Meet. Int. Comm. Large Dams*, (i).

Rulot, F., B. J. Dewals, S. Erpicum, P. Archambeau, and M. Pirotton. 2012. *Long-term sediment management for sustainable hydropower. Compr. Renew. Energy*. Elsevier Ltd.

Singh, U., I. Joshi, M. Bishwakarma, and K. Yuichi. 2019. "Experimental Investigation of River Bed Deformation duringLarge Floods." *9th Int. Work. Multimodal Sediment Disasters*. Kyoto, Japan.

Wong, M., and G. Parker. 2006. "Reanalysis and Correction of Bed-Load Relation of Meyer-Peter and Müller Using Their Own Database." *J. Hydraul. Eng.*, 132 (11): 1159–1168.

DOI: https://doi.org/10.26077/34f4-8489

Preliminary Analysis Of The Effect Of Worked Wood Piles In Straight Channels

Deep Roy[1] & Simone Pagliara[2]
[1]*Department of Civil Engineering – National Institute of Technology Tiruchirappalli, Tiruchirappalli, India*
[2]*Laboratory of Hydraulics, Hydrology and Glaciology (VAW) – ETH Zurich, Zurich, Switzerland*
E-mail: pagliara@vaw.baug.ethz.ch

Abstract: *Eco-friendly low-head structures such as blunt nosed chevrons, wood bundles, worked wood piles etc. provide an effective solution to ensure grade and sediment load control in rivers, as well as to introduce morphological variations in the channel bed promoting biodiversity. Worked wood piles are usually installed near the outer bank in order to redirect the flow away from it, thus stabilizing the basin. Moreover, these structures increase the natural organic content in channels and create pronounced scour and dune features in their vicinity, giving shelter to a host of aquatic organisms and fish species. However, till date, a comprehensive analysis of the scour features occurring in the presence of worked wood piles is absent in literature. To shed light on this underexplored topic, the present study provides some initial results on the equilibrium and temporal scour characteristics in the vicinity of worked wood piles. Experiments were conducted with such structures in straight channels under a wide range of hydraulic conditions. Isolated structures as well as series of two structures were tested. Among others, the equilibrium scour morphology due to worked wood piles resulted to be considerably influenced by inflow conditions, tailwater level and number of structures in case of series arrangement. Namely, the scour and dune features in the stilling basin of the structure are more enhanced for higher discharge and tailwater and in case of two worked wood piles in series. Moreover, the development of scour and dune geometry was recorded at regular intervals of time for the entire test duration which revealed that the rate of increase of scour area in case of two worked wood piles in series is higher in specific intervals during the test depending on the hydraulic conditions. Overall, the results offer important observations and a preliminary analysis of the local scour phenomena around worked wood piles, aiming at providing a basis for future design guidelines.*

Keywords: *river restoration structures, scour evolution, equilibrium morphology, inflow conditions.*

1. Introduction

Natural large woody debris elements (diameter > 0.1 m and length > 2 m) form an important component of the riverine ecosystem. They can create localized scour holes and elongated dunes which provides perching places for reptiles as well as nursery for aquatic organisms and fishes (Fischenich and Morrow 2000). In some cases, they are periodically removed from the channel bed to provide better navigation and enhancement in channel capacity (Gippel 1995, Shields and Gippel 1995). Moreover, debris accumulation at bridge piers and hydraulic structures and other constricted channel sections leads to increased flood levels and enhanced scour at the foundation and causes failure of such structures during floods (Palermo et al. 2021a). Nevertheless, an optimum quantity of wood debris is essential in the channel as it acts as an important source of fine particulate organic matter in natural streams and serves as an active agent to stabilize channel dimensions and promote aquatic ecosystem (Pagliara et al. 2021, Ward and Aumen 1986).

In some cases, woody debris can be used as component of various eco-friendly bank protection structures. Traditionally, submerged vanes have been used to protect the outer bank of a meandering river from excessive erosion and provide efficient sediment load control (Odgaard and Mosconi 1987, Odgaard and Spoljaric 1986). Moreover, Bhuiyan et al. (2010) conducted an experimental study on low angle bank attached vanes and estimated the bed morphology, flow and turbulence characteristics in correspondence with such structures. Following this, the focus shifted to the use of structures made from woody debris and stones to stabilize the riverbed morphology as they can be easily integrated into the surrounding environment. Scurlock et al. (2012) studied the scour phenomena occurring downstream of various kinds of rock weirs (A-, U-, and W-shaped) in straight channels. Subsequently, Pagliara and Kurdistani (2015) studied the scour characteristics due to J-Hook vanes which are used in correspondence with outer banks in river bends to provide better protection against bank failure. Furthermore, structures like block ramps and blunt nosed chevrons were studied in straight and curved channels which helps to dissipate upstream flow energy, improve navigability and enhance biodiversity (Palermo et al. 2021b, Roy and Pagliara 2021). Focusing on wood structures, Pagliara et al. (2014, 2015) investigated the scour geometry parameters and equilibrium morphology due to log-vanes and log deflectors, respectively. Triangular shaped log-frame deflectors were studied by Pagliara and Kurdistani (2016) which were successful in driving the scour region away from the outer bank vicinity. Pagliara et al. (2020) conducted experiments on double winged log-frames in straight channels which created localized scour formations optimum for fish nursery. Finally, Pagliara et al. (2021) analyzed scour features due to wood bundles (mainly composed of woody debris) which were used to redirect high velocity gradients away from the outer bank and generate eco-friendly scour pools in the center of the channel. Although some studies can be found in literature pertaining to equilibrium scour characteristics due to eco-friendly low-head hydraulic structures composed from woody debris in the channel, the temporal aspect of scour development is still missing. Therefore, in this paper, we present a new structure called worked wood piles which belongs to the above discussed structure typology and analyze both the temporal and equilibrium scour formation characteristics due to such structures. Experiments were conducted on worked wood piles in straight channels by varying the flow parameters and number of structures in series arrangement. The time dependent development of the scour and dune features in the vicinity of the structure were traced throughout the duration of the test which provided key understanding of the scour evolution process. The various bed morphologies occurring at equilibrium conditions in the stilling basin of such structures have been analyzed and compared to highlight the role played by various hydraulic and structural parameters in characterizing the scour process.

2. Experimental Set-up

Tests were conducted in a straight flume at the hydraulics laboratory of University of Pisa characterized by length = 11 m, width (B) = 0.5 m and height = 0.5 m. The worked wood pile was placed at a distance of s from the channel entrance where s/B =15.8. This distance enabled the formation of fully developed flow at the test section. Initially the test was conducted by placing a single worked wood pile in correspondence with the channel bank. Following this, a second worked wood pile was placed in series combination downstream of the first structure separated by a distance d (d/B = 0.6) and the test was repeated. Figure 1a shows the plan view of the entire stilling basin along with the series arrangement of multiple worked wood piles highlighting the scour and dune regions in the vicinity of the structures. An individual worked wood pile structure consists of three vertical trunks having a diameter of 0.015 m fastened together with metallic strings. Secondary branches having a diameter around 2-3 mm were used as horizontal members to reinforce the main trunks and lock them into position. The horizontal members were placed either parallel to the direction of the flow or making an angle of 60° with it and had an overall length of 0.1 m. The cross-section of the stilling basin along with the series arrangement of multiple worked wood piles has been shown in Fig. 1b. The height of the structure above the channel bed is denoted by h_{st} and in the current tested range of parameters h_{st}/B = 0.17. Notably, for an individual worked wood pile, horizontal reinforcement members were provided at two distinct depths along the main trunk, i.e., $0.25h_{st}$ and $0.75h_{st}$ from the top of the structure. The vertical trunk of the structure was extended well below the channel bed level beyond the maximum scour depth and attached to a firm weight to keep the entire arrangement in position during the test. The maximum scour depth in the vicinity of the first and the second structure is denoted by z_{m-1} and z_{m-2}, respectively. The overall maximum height of the dune is denoted by z_M. The maximum transversal width of the scour and dune is denoted by w_s and w_d, respectively (Fig. 1a). The discharge and tailwater level are denoted by Q and h_{tw}, respectively. For tests in case of both isolated worked wood piles and multiple piles in series, additional protection against scour was given by inserting a long longitudinal wooden member along the channel wall characterized by diameter = 0.015 m and length l_p given by l_p/B = 0.6 (in case of isolated structure) and l_p/B = 1.2 (in case of structures in series). The top level of the longitudinal member coincided with that of the channel bed and the depth was ensured to be sufficient to provide effective cover against excessive scour magnitudes at the bank foundation.

Prior to the commencement of each test the channel bed was levelled. Initially, water was filled up slowly in the channel to a predetermined level and subsequently the discharge was increased to the desired value. The tailwater level was kept constant throughout the test using a sluice gate located at the end of the channel. For each test, clear water conditions were maintained (Oliveto and Hager 2002) and the flow intensity term varied as $0.83 \leq U/U_c \leq 0.91$ where U is the average flow velocity and U_c is the average threshold velocity calculated according to the methodology proposed by Neill (1968). The average duration of the tests was 240 minutes. The test duration was sufficient to obtain the equilibrium scour configuration for this structure typology. During the test, the evolution of the of the scour and dune features in the vicinity of the worked wood piles were recorded at predefined intervals of time, i.e., 1, 5, 10, 30, 60, 90, 120, 180 and 240 minutes with help of multiple high-definition video cameras and measuring scales attached to both sides of the channel. In addition, a 0.1 mm precise point gauge was also used to record the water level depths, scour depths and other significant points of the bed morphology at various time intervals. At the end of test duration, the water was emptied from the channel slowly so as not to disturb the equilibrium scour morphology and the bed was dried. A "Leica Geosystems" laser scanner was used to survey the deformed bed of the channel and the obtained data cloud was subsequently processed through "Cyclone" software to obtain detailed scour maps of the stilling basin having a precision of ±1 mm. The accuracy of the scour maps was crosschecked with point gauge readings and the difference was always negligible.

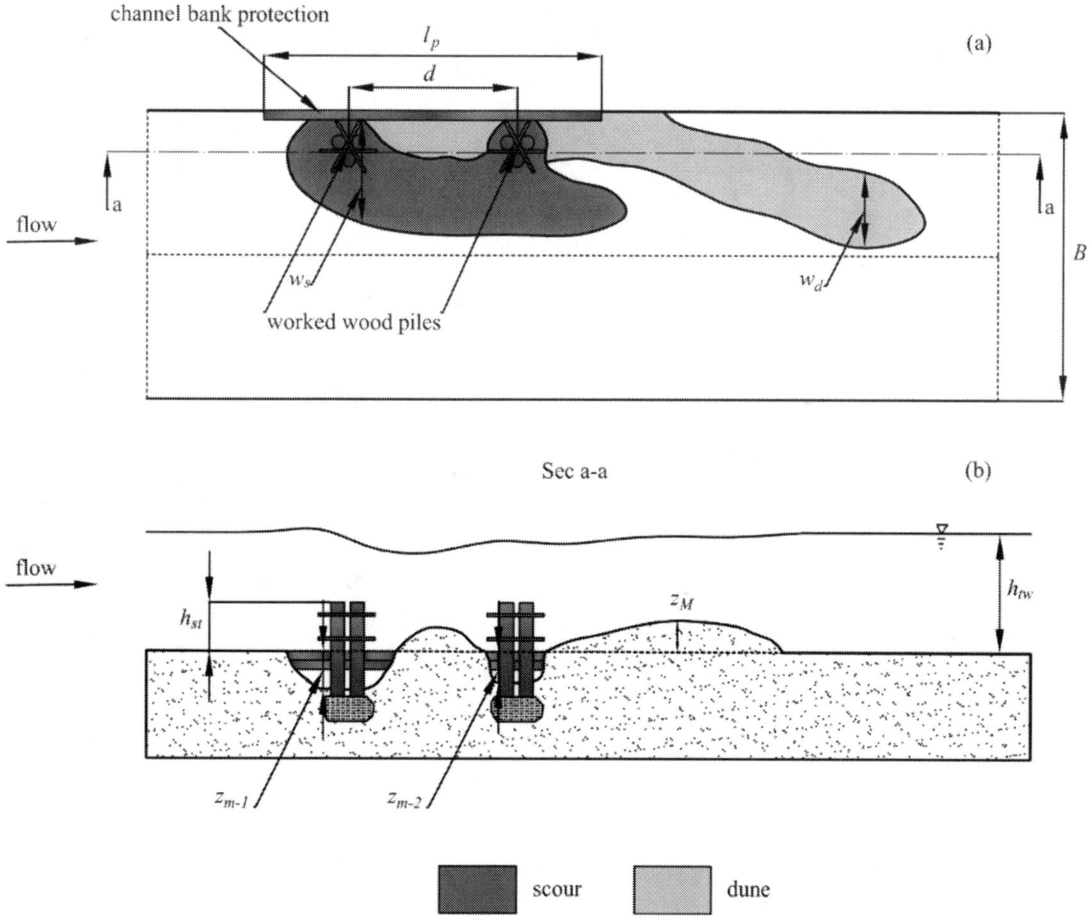

Figure 1. a) Plan and b) cross-section of the test section in case of series arrangement of two worked wood piles highlighting important hydraulic and scour parameters.

The tests were conducted at predefined sets of discharge and tailwater, i.e., $Q = 0.005$ m³/s, $h_{tw} = 0.036$ m; $Q = 0.008$ m³/s, $h_{tw} = 0.056$ m; $Q = 0.011$ m³/s, $h_{tw} = 0.07$ m; $Q = 0.014$ m³/s, $h_{tw} = 0.094$ m. The bed material was uniform having $d_{50} = 1$ mm and $\sigma = (d_{84}/d_{16})^{0.5} = 1.15$ where d_{xx} is the material size for which xx% of the material is finer. Moreover, the density of the bed material and water are 2467 kg/m³ and 1000 kg/m³, respectively. Figs. 2a and b show the equilibrium scour characteristics due to an isolated worked wood pile and series arrangement of two worked wood piles, respectively, in straight channels.

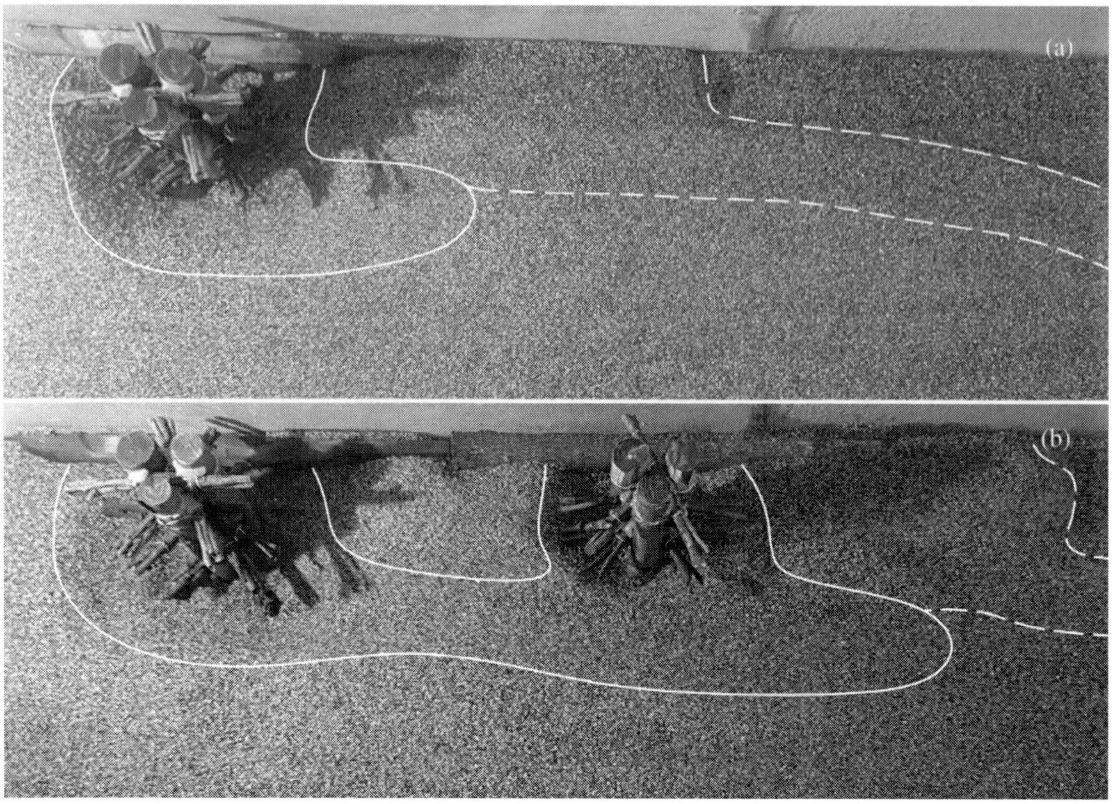

Figure 2. Equilibrium scour morphology due to a) an isolated worked wood pile and b) series arrangement of two worked wood pile structures in straight channels. The scour and dune regions are shown by solid and dashed lines, respectively. The flow is from left to right.

3. Results and discussion

3.1. Equilibrium scour morphology

From a design perspective, the knowledge of the equilibrium scour morphology due to worked wood piles under various hydraulic and structural conditions is of paramount importance. Therefore, in this section, we provide a preliminary analysis of the morphological features due to such structures to identify the key variables governing the scour mechanism. Figs. 3a and b show the scour maps pertaining to an isolated worked wood pile for Q = 0.008 m^3/s, h_{tw} = 0.056 m and Q = 0.011 m^3/s, h_{tw} = 0.07 m, respectively. As shown in Fig. 3a, there is the formation of a distinct scour hole around the worked wood pile which can be used as shelter by fishes and other aquatic organisms. The scour region is followed by an elongated dune region along the channel bank which can serve as excellent spawning site for fishes. Notably, additional protection has been provided along the channel bank within the scour zone by inserting a longitudinal wooden member which reduced the scour magnitude at the base of the bank. Moreover, the elongated dune formed in the downstream stilling basin of the structure provides additional stability to the channel bank on that side and drives the scour towards the central part of the channel. Fig. 3b shows the scour morphology due to the same structure condition but having a higher magnitude of discharge and tailwater, i.e., Q = 0.011 m^3/s, h_{tw} = 0.07 m. In this case, both the scour and dune regions become more pronounced and elongated in nature with otherwise similar characteristics as that of the previous case. This introduces flow variations and helps to create and preserve a variety of aquatic habitats in the channel which are conducive to the health of the river.

The following set of scour maps shown in Figs. 3c and d, are corresponding to series combination of two worked wood pile structures having Q = 0.008 m^3/s, h_{tw} = 0.056 m and Q = 0.014 m^3/s, h_{tw} = 0.094 m, respectively. Like that in case of isolated structure, Fig. 3c shows that there are scour formations in the vicinity of both the worked wood piles in series. However, the scour in the proximity of the first structure is deeper and larger than its counterpart near the second structure. Both the scour holes are connected by an outer shallow scour contour line. In addition, there are dune formations between the two structures and also in the downstream stilling basin after

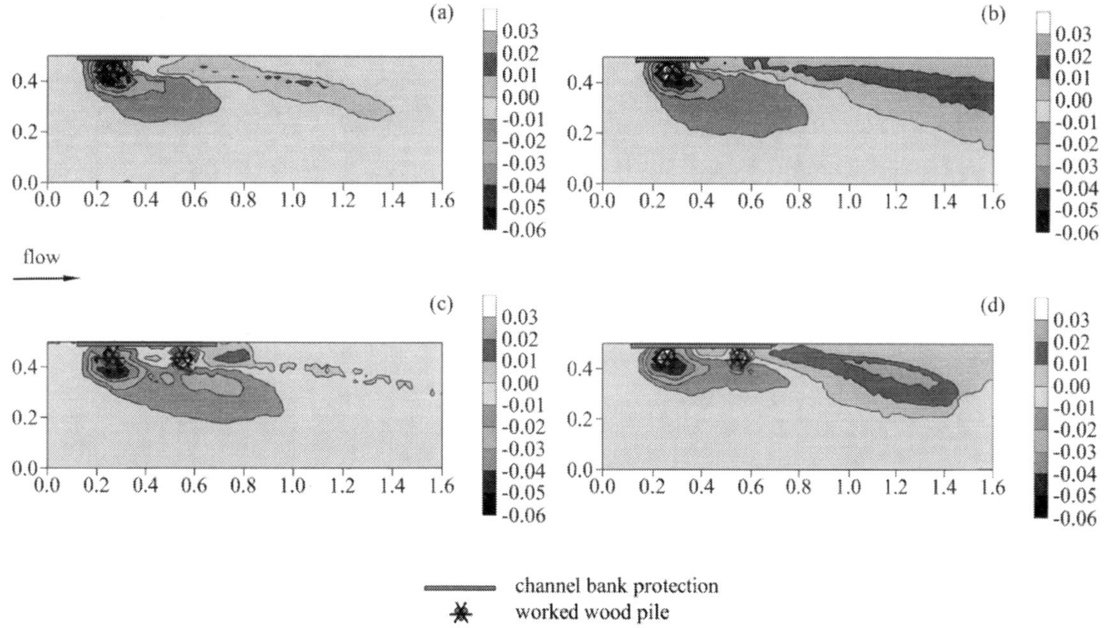

Figure 3. Equilibrium scour maps in case of (a, b) isolated worked wood pile for $Q = 0.008$ m^3/s, $h_{tw} = 0.056$ m and $Q = 0.011$ m^3/s, $h_{tw} = 0.07$ m, respectively, and (c, d) series arrangement of two worked wood piles for $Q = 0.008$ m^3/s, $h_{tw} = 0.056$ m and $Q = 0.014$ m^3/s, $h_{tw} = 0.094$ m, respectively. All dimensions are in meter.

the series arrangement which decreases the risks of bank failure by erosion. Moreover, in Fig. 3d, for a higher inflow condition and tailwater, the resulting equilibrium morphology is characterised by the same features and the dune region becomes more pronounced and extended in nature which partially confines the scour towards the channel center, thereby, providing additional stability. Therefore, for both isolated worked wood pile and series arrangement of worked wood piles, the equilibrium scour morphology is considerably affected by the inflow condition and tailwater which are crucial variables to be considered in the design of such structure typology (Pagliara et al. 2020, Palermo et al. 2019).

On comparing Figs. 3a and c, scour modifications caused by the series combination of worked wood piles can be analyzed for otherwise identical hydraulic conditions. The presence of multiple structures in series combination in the channel increases the complexity of the flow pattern across such an arrangement and there is interaction between the two structures in terms of bed morphology. As shown in Fig. 3c, this leads to the formation of more enhanced scour and dune regions compared to that in case of an isolated structure (Roy and Pagliara 2021) which can have positive impact in terms of promoting biodiversity and fish refuge.

A further discussion on the shape and dimension of scour and dune features are in order. At equilibrium stage, within the tested range of parameters, w_s/B and w_d/B vary as $0.28 \leq w_s/B \leq 0.48$ and $0.26 \leq w_d/B \leq 0.70$, respectively. On the other hand, the maximum magnitude of scour and dune varies as $0.49 \leq z_m/h_{st} \leq 0.63$ and $0.22 \leq z_M/h_{st} \leq 0.28$, respectively ($z_m$ = global maximum scour depth).

3.2. Temporal evolution of scour features

In the preceding section, we have discussed the salient features of the bed morphology when the equilibrium scour configuration has been reached under clear water conditions. However, the temporal evolution of the scour geometry features is equally important for proper design of stilling basin and scour countermeasure systems. Particularly, the high discharge occurring during floods may not continue for sufficient duration to achieve the equilibrium scour morphology as well as maximum scour depth and dune height. In such a case the characteristics of the scour geometry at various stages of time is essential for design purposes. Therefore Figs. 4a and b present temporal scour evolution maps showing the bed morphology characteristics at various time intervals from the start of the experiment till the end of test duration (5, 10, 30, 60, 120 and 240 minutes) for an isolated worked wood pile in case of $Q = 0.008$ m^3/s, $h_{tw} = 0.056$ m and $Q = 0.011$ m^3/s, $h_{tw} = 0.07$ m, respectively. As shown in Fig. 4a, the scour hole first emerges surrounding the wood pile and thereupon steadily expands with time. From 60 minutes onwards, the scour hole takes on an elongated shape and extends alongside the dune in the downstream stilling basin till it reaches the equilibrium configuration at 240 minutes. The dune starts forming behind the worked wood pile at the initial stages and becomes a wing like formation along the channel bank which elongates steadily with

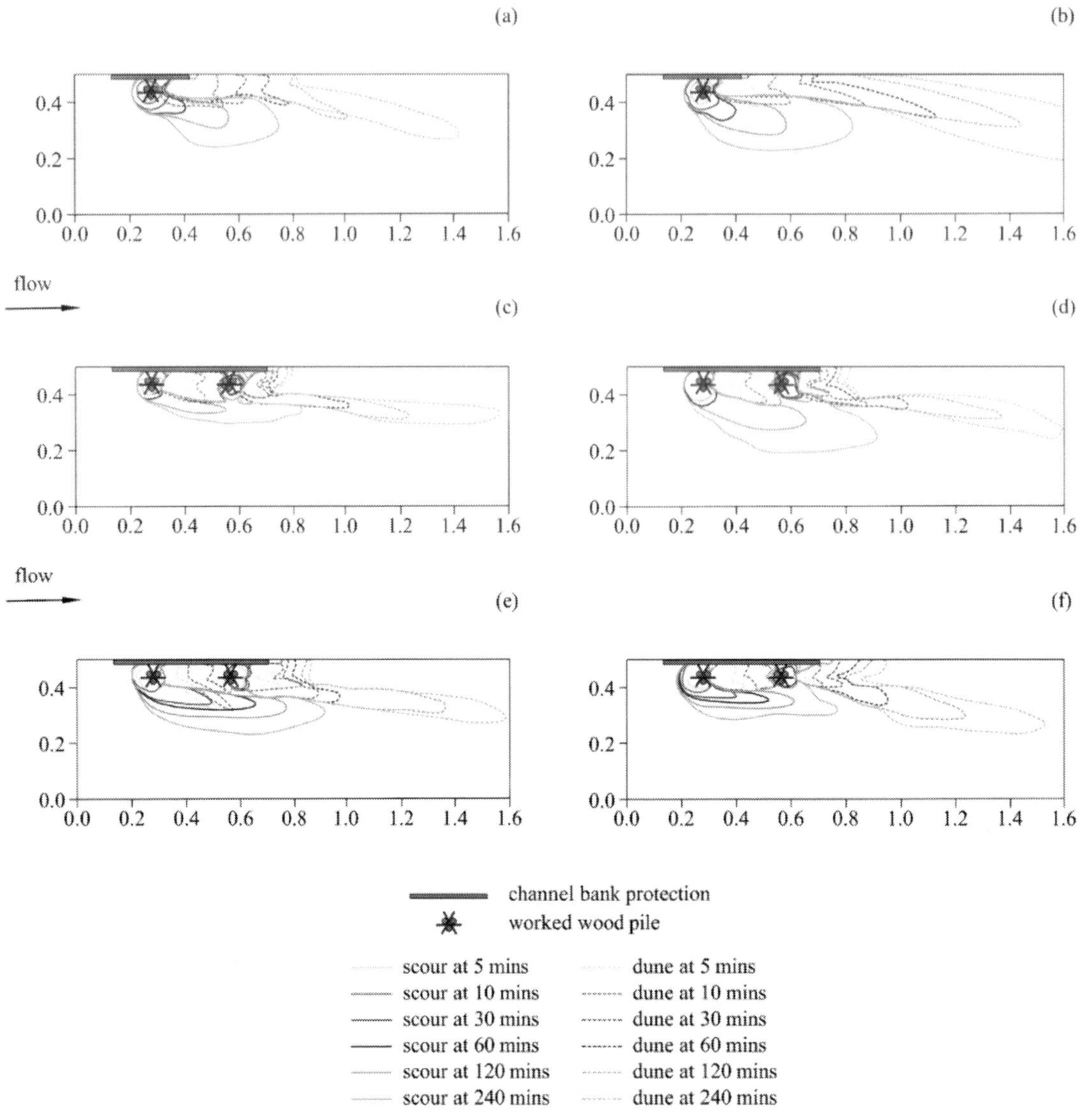

Figure 4. Temporal scour evolution maps for isolated worked wood pile in case of a) $Q = 0.008$ m^3/s, $h_{tw} = 0.056$ m, b) $Q = 0.011$ m^3/s, $h_{tw} = 0.07$ m and for series arrangement of two worked wood piles in case of c) $Q = 0.005$ m^3/s, $h_{tw} = 0.036$ m, d) $Q = 0.008$ m^3/s, $h_{tw} = 0.056$ m, e) $Q = 0.011$ m^3/s, $h_{tw} = 0.075$ m and f) $Q = 0.014$ m^3/s, $h_{tw} = 0.094$ m. All dimensions are in meter.

time. Overall, it can be said that both the scour and dune area enlarge at a fairly uniform rate as the scour and dune contours at various time intervals are evenly spaced. Upon increasing the inflow conditions in Fig. 4b, keeping all other parameters constant, we obtained a similar temporal scour evolution map as that of Fig. 4a. The scour and dune have usual shapes and enlarge at a constant rate. Therefore, the inflow conditions play a limited role in influencing the scour and dune evolution in case of isolated worked wood piles.

Following this, in Figs. 4c, d, e and f, we compared the scour and dune evolution trend due to series arrangement of two worked wood piles for various magnitudes of hydraulic parameters such as $Q = 0.005$ m^3/s, $h_{tw} = 0.036$ m; $Q = 0.008$ m^3/s, $h_{tw} = 0.056$ m; $Q = 0.011$ m^3/s, $h_{tw} = 0.075$ m and $Q = 0.014$ m^3/s, $h_{tw} = 0.094$ m, respectively. In Fig. 4c, the scour holes start developing separately around the individual worked wood piles and combine at a later stage to form a single elongated scour zone. Notably, contrary to that in case of single worked wood pile, the scour enlargement rate is rather slow up till around 60 minutes following which the scour expands rapidly to achieve equilibrium configuration (note the widely spaced scour contours at 60 minutes and 120 minutes). Moreover, there are dune formations in the region between the structures as well as in the downstream stilling basin. The dune starts forming at a uniform rate behind the upstream structure and covers the entire area between the two structures within 30 minutes from the beginning of the test. In addition, the dune behind the downstream structure in the stilling basin starts growing in a wing like formation and rapidly becomes elongated at later stages (30 to 60 minutes onwards) which restricts the scour towards the central part of the channel and enhances basin stability. Upon increasing the inflow conditions in Fig. 4d (Q = 0.008 m^3/s, h_{tw} = 0.056 m), a similar temporal evolution of scour and dune regions can be observed but with more pronounced dune and scour features at equilibrium stage as discussed in the previous section. Subsequently, the discharge and tailwater were further increased to Q = 0.011 m^3/s, h_{tw} = 0.075 m (Fig. 4e) and Q = 0.014 m^3/s, h_{tw} = 0.094 m (Fig. 4f) which resulted in some modification of the scour evolution characteristics. In Fig. 4e, like the previous cases, the scour formation begins separately around each structure and eventually joins together to form a single scour zone around both structures towards the final stages of the test. However, the scour forms in a small circular shape around the structures up to 10 minutes following which it rapidly attains an elongated shape which enlarges with time and attains the equilibrium configuration (note the widely spaced scour contours at 10 and 30 minutes). The formation of dunes between the two structures follows the same pattern as that in previous cases (Fig. 4c and d). In the downstream stilling basin, the dune starts forming into an initial wing like shape and from around 60 minutes expands rapidly into an elongated formation which partially confine the scour region. The scour and dune evolution characteristics in Fig. 4f show substantial similarity with that of the previous case. Therefore, it can be concluded that unlike that in case of isolated worked wood piles, the inflow conditions influence the scour evolution rate for series arrangement of multiple worked wood piles in the tested range of parameters.

In order to highlight the effect of the series combination of multiple worked wood piles on the scour evolution process, Figs. 4a and d have been compared having identical hydraulic conditions but in case of isolated structure and structure series, respectively. For isolated worked wood piles, in Fig. 4a, the rate of enlargement of the scour region is uniform with time and the contour lines are evenly spaced. Conversely, for series arrangement of worked wood piles in Fig. 4d, the scour evolution is slow in the initial stages up to around 60 minutes and thereafter, expands rapidly to form an elongated shape at equilibrium condition which is reached at 240 minutes. As for the dune, the rate of expansion is fairly uniform in case of isolated worked wood pile when compared to series arrangement. In case of the latter, the dune grows slowly in the initial stages as a small wing like shape and from around 30 to 60 minutes, expands rapidly to form an oblong shape. On comparing Figs. 4b and e for a higher set of discharge and tailwater, similar observations can be made. In this case, for series arrangement of two worked piles, the scour growth remains limited up to 10 minutes following which it enlarges rapidly to achieve the equilibrium configuration.

4. Conclusion

Tests were conducted with a new type of low-head eco-friendly hydraulic structure, i.e., worked wood pile in straight channels by varying the flow discharge, tailwater level and number of structures in series arrangement. The equilibrium scour morphology due to worked wood piles is characterized by a distinct scour region in the vicinity of the structure. The dune is formed downstream of the structure along the channel bank in the stilling basin which partially restricts the scour to the central part of the channel. Conversely, for series arrangement of two worked wood piles, the scour and dune become more pronounced and elongated. Moreover, higher inflow conditions usually enhance scour and dune features under otherwise similar conditions thereby playing an important role in the scour process. The scour zone in the vicinity of the structure serves as a low-velocity deep-flow region optimum for growth of aquatic organisms and fishes while the dune region stabilizes the channel bank and acts as spawning sites. Moreover, comparison and analysis of the temporal scour evolution maps pertaining to different cases show that for isolated worked wood piles the rate of expansion of the scour and dune region is uniform with time. Conversely, in case of series arrangement of two worked wood piles, the inflow condition and tailwater affects the growth rate of scour and dune features. The rate of scour expansion is higher in specific time intervals during the test depending on the magnitude of discharge and tailwater level. The dune initially has a wing like shape and starts expanding at a faster rate from around 60 minutes to evolve into an extended oblong shape further downstream in the stilling basin. The knowledge of the evolving scour morphology at various time stages can be valuable for design of scour countermeasure systems particularly during flood flow in the river which may not occur long enough to obtain equilibrium scour configuration. Generally speaking, the inflow conditions, number and spatial arrangement of structures are important variables which needs to be examined under a variety of hydraulic conditions and channel geometry to develop effective design guidelines for worked wood piles.

5. Acknowledgments

This work was conducted in the hydraulics laboratory of the Department of Energy, Systems, Territory and Construction Engineering, University of Pisa, Italy. Moreover, this research did not receive any specific grant from funding agencies in the public, commercial, or not-for-profit sectors.

6. References

Bhuiyan, F., Hey, R.D., and Wormleaton, P.R. (2010). "Bank-attached vanes for bank erosion control and restoration of river meanders." *J. Hydraul. Eng.*, 136(9), 583–596.

Fischenich, J.C., and Morrow, J.V. Jr. (2000). "Streambank habitat enhancement with large woody debris." Army Engineer Waterways Experiment Station Vicksburg MS Engineer Research and Development Center (ERDC TN-EMRRP-SR-13).

Gippel, C.J. (1995). "Environmental hydraulics of large woody debris in streams and rivers." *J. Environ. Eng.*, 121(5), 388-395.

Neill, C. R. (1968). "Note on initial movement of coarse uniform bed-material." *J. Hydraul. Res.*, 6(2), 173–176.

Odgaard, A.J., and Mosconi, C.E. (1987). "Stream bank protection by submerged vanes." *J. Hydraul. Eng.*, 113(4), 520-536.

Odgaard, A.J., and Spoljaric, A. (1986). "Sediment control by submerged vanes." *J. Hydraul. Eng.*, 112(12), 1164-1181.

Oliveto, G., and Hager, W. H. (2002). "Temporal evolution of clear-water pier and abutment scour." *J. Hydraul. Eng.*, 128(9), 811–820.

Pagliara, S., Hassanabadi, L.S., and Kurdistani, S.M. (2014). "Log-vane scour in clear water condition." *River. Res. Appl.*, 31(9), 1176-1182.

Pagliara, S., Hassanabadi, L., and Kurdistani, S.M. (2015). "Clear water scour downstream of log deflectors in horizontal channels." *J. Irrig. Drain. Eng.*, 141(9), 1-8, 04015007.

Pagliara, S., and Kurdistani, S.M. (2015). "Clear water scour at J-Hook Vanes in channel bends for stream restorations." *Ecol. Eng.*, 83, 386–393.

Pagliara, S., and Kurdistani, S.M. (2016). "Flume experiments on scour downstream of wood stream restoration structures." *Geomorphology*, 279, 141-149.

Pagliara, S., Palermo, M., and Roy, D. (2020). "Scour around double-winged log frames under clear water condition." *J. Irrig. Drain. Eng.*, 146(12), 1-11, 04020038.

Pagliara, S., Roy, D., and Palermo, M. (2021). "Scour features at wood bundles." *Water*, 13, 2118.

Palermo, M., Pagliara, S., and Roy, D. (2019). "Erosive processes downstream of arch shaped sills in vegetated channels." *Proc., 38th International Association for Hydro-environment Engineering and Research World Congress*, IAHR, Panama City, Panama, 1560-1569.

Palermo, M., Pagliara, S., and Roy, D. (2021a). "Effect of debris accumulation on scour evolution at bridge pier in bank proximity." *J. Hydrol. Hydromech.*, 69(1), 108–118.

Palermo, M., Roy, D., and Pagliara, S. (2021b). "Morphological characteristics of eco-friendly protected basins downstream of block ramps in river bends." *Geomorphology*, 377, 107587.

Roy, D., and Pagliara, S. (2021). "Equilibrium morphology and scour evolution at blunt nosed chevrons." *River. Res. Appl.*, 38(3), 499-512.

Scurlock, S.M., Thornton, C.I., and Abt, S.R. (2012). "Equilibrium scour downstream of three-dimensional grade-control structures." *J. Hydraul. Eng.*, 138(2), 167-176.

Shields, F.D. Jr., and Gippel, C.J. (1995) "Prediction of effects of woody debris removal on flow resistance." *J. Hydraul. Eng.*, 121(4), 341-354.

Ward, G.M., and Aumen, N.G. (1986) "Woody debris as a source of fine particulate organic matter in coniferous forest stream ecosystems." *Can. J. Fish. Aquat. Sci.*, 43(8), 1635–1642.

DOI: https://doi.org/10.26077/5e0b-783d

Reynolds Stress Modelling of Supercritical Flow in a Narrow Channel

S. Kadia[1], N. Rüther[2] & E. Pummer[1]
[1]*Department of Civil and Environmental Engineering, Norwegian University of Science and Technology, 7491 Trondheim, Norway*
[2]*Chair of Hydraulic Engineering, Technical University of Munich, Arcisstr. 21 80333 Munich, Germany*
E-mail: subhojit.kadia@ntnu.no; subhojitkadia@gmail.com

Abstract: *One of the most effective techniques to combat reservoir sedimentation, especially for small to medium-sized reservoirs, is the installation of Sediment Bypass Tunnels (SBTs). SBTs are designed for supercritical narrow open channel flow conditions. In such narrow channels, the walls and the free surface boundaries can influence the flow characteristics and form secondary currents. These vortex structures can alter the longitudinal velocity distribution and form velocity dip. Besides, they can also alter the bed shear stress distribution and influence the sediment transport. Such complex phenomenon can be modelled using Reynolds Stress Models (RSMs). In this study, the full Launder Reece and Rodi (LRR) pressure-strain model was implemented in OpenFOAM® and the simulated results were compared with former experimental results for flow depth = 0.16 m, Reynolds number = 5.7×10^5, Froude number = 1.84, and channel aspect ratio = 1.25. In addition, the dissipation rate of turbulent kinetic energy at the free surface was modelled following literature. The simpleFoam solver, which is based on the SIMPLE algorithm, was used for the steady state simulation. The predicted velocity distributions, secondary currents, velocity dip, turbulence intensity distributions, and lateral variation in bed shear stress are consistent with literature. The results show that the simulated maximum longitudinal velocity and the cross-sectional average bed shear stress are 0.34% and 0.74% lower than the observed values. An intermediate vortex was developing between the free surface vortex and the bottom vortex for the tested case. However, it remained as a part of the free surface vortex.*

Keywords: *RSM, secondary currents, sediment bypass tunnel, supercritical flow, velocity dip.*

1. Introduction

Sediment bypass tunnels (SBTs) are hydraulic structures with supercritical flows. They route incoming sediments around reservoirs and help to combat the reservoir sedimentation. Small to medium-sized reservoirs can be effectively benefited from the installation of SBTs if the hydro abrasion due to the sediment impacts is resisted by suitable hydraulic design optimisation and invert material use (Boes et al. 2014). SBTs are designed for narrow open channel flows, mostly for aspect ratios ($Ar = b/h$, where b = channel width and h = flow depth) between 1.1 and 2.3 and for Froude numbers Fr between 1.4 and 3.0 (obtained from the data provided in Auel (2014); Demiral (2021); Müller-Hagmann (2018)). For such narrow open channel flows with $Ar \leq 5$, the solid and the free surface boundaries can alter the flow and turbulence characteristics, form 'secondary currents of Prandtl's second kind', and cause velocity dip throughout the channel width (Nezu and Nakagawa 1993; Yang et al. 2004). Even though the amounts of secondary currents generally extend only up to 1.5% to 3% of the bulk velocity $\overline{U}$ or of the maximum longitudinal velocity U_{max} (Broglia et al. 2003; Naot and Rodi 1982; Nezu and Nakagawa 1993; Tominaga et al. 1989), they can yet impact the longitudinal velocity contours, bed shear stress distribution, and sediment transport (Auel et al. 2014; Einstein and Li 1958; Kang and Choi 2006; Nezu and Nakagawa 1993). Such amount can be even up to 5% of $\overline{U}$ as found by Nikitin (2021) in a direct numerical simulation (DNS) study. Therefore, the estimations of secondary currents and their effects become crucial to obtain accurate bed shear stress and sediment transport for narrow channels including SBTs.

The development of secondary currents or vortex structures depends on Ar and their estimations can be affected by the data resolution. Nezu and Nakagawa (1993); Nezu and Rodi (1985) found a stronger 'free surface vortex' and a weaker 'bottom vortex' (marked as I and II in Fig. 1) in each half of narrow open channels. Later, Grega et al. (1995) revealed that a relatively smaller 'inner secondary vortex' develops at the top corner – vortex III in Fig. 1. Broglia et al. (2003); Kang and Choi (2006); Nikitin (2021) also showed the presence of vortex III in their numerical studies. In addition to these three vortices, the numerical studies by Naot and Rodi (1982) – for Ar = 0.6 and by Broglia et al. (2003) – for Ar = 1.0 indicated the presence of an intermediate vortex in between the free surface vortex and the bottom vortex. Basically, this additional vortex is a part of the free surface vortex that gets detached at lower aspect ratios. However, thorough numerical and experimental investigations, especially on the formation of intermediate vortex and on its effect on the flow structure are required. Apparently, Nezu and Nakagawa (1993); Nezu and Rodi (1985) could not observe vortex III due to the unavailability of high resolution measurements of secondary velocities at the corner between sidewall and free surface, which is difficult to acquire.

It was found that most of the previous experimental and computational fluid dynamics (CFD) studies on the 'secondary currents of Prandtl's second kind' in narrow open channel flows were conducted for subcritical flow conditions. However, there are a few experimental and CFD studies available on supercritical narrow channel flows which grew in numbers within the past one decade. Interestingly, Nezu and Nakagawa (1986, 1993) measured all three velocity components using a Laser Doppler Anemometer (LDA) for $Ar \approx 2$ but noticed no significant difference between the tested subcritical flow and the tested supercritical flow (though Fr was not high) – by comparing the non-dimensional flow properties and secondary currents. Previously, Rajaratnam and Muralidhar (1969) measured longitudinal velocity U for a wider range of supercritical flows and found velocity dip throughout the channel width for $Ar \leq 7$. In recent years, Auel et al. (2014) and Demiral et al. (2020) used a 2D LDA to measure U and vertical velocity W in decelerating flows for $2.5 \leq Ar \leq 10.7$ and $0.89 \leq Ar \leq 1.91$, respectively; and Jing et al. (2019) measured these velocity components for $3 \leq Ar \leq 7.5$ using a 2D Particle Image Velocimeter (PIV). Auel et al. (2014) and Jing et al. (2019) noticed velocity dips upto the channel center for $Ar \leq 4 - 5$ and $Ar \leq 3.75$. Both the studies predicted vortices I and II, but Demiral et al. (2020) reported four vortices (see Fig. 2(a)): I, II, III, and an intermediate vortex (for all tested cases in the range $0.89 \leq Ar \leq 1.91$) which were argued to be comparable to the duct flows. However, the predictions made by Demiral et al. (2020) contradict the previous observations (Nezu and Rodi 1985) and CFD simulation results (Broglia et al. 2003; Naot and Rodi 1982) on subcritical flows with comparable aspect ratios. Apparently, the flow non uniformity present in Demiral et al. (2020) led to substantial free surface perturbations which affected the turbulence characteristics and the secondary currents. Furthermore, Auel et al. (2014); Demiral et al. (2020); Jing et al. (2019) had no direct measurement of lateral velocity V and secondary currents. Therefore, further investigation on 3D velocity measurements and turbulence characteristics for uniform and non-uniform supercritical flows is essential to recognise the effect of flow non uniformity on the secondary currents. On the other hand, the CFD researchers modelled the 'secondary currents of Prandtl's second kind' in narrow subcritical open channel flows using different models starting from nonlinear eddy viscosity model to DNS (Broglia et al. 2003; Cokljat 1993; Cokljat and Younis 1995; Kang and Choi 2006; Naot and Rodi 1982; Nikitin 2021; Reece 1977; Shi et al. 1999). Besides, Nasif et al. (2020) and Shinneeb et al. (2021) – used detached-eddy simulations (DES) and Kadia et al. (2021) – used RSM to predict the secondary currents and the related flow features for supercritical narrow channel flows. But these studies did not explore the development of intermediate vortex which is being investigated presently. In fact, the previously used Reynolds stress models (RSMs), LRR (Launder, Reece, and Rodi) based on Launder et al. (1975) and SSG (Speziale, Sarkar, and Gatski) based on Speziale et al. (1991), are a suitable choice because they are computationally less expensive than large eddy simulations (LES) and DNS but still capable to simulate the desired secondary currents and their effects on the flow characteristics. The SSG model involves no wall damping functions which makes it easier to execute than LRR, especially for nonplanar walls. However, noticeable deviations found in the distribution and bulging of the longitudinal velocity around the bottom corner bisector between the simulated flow obtained by Kang and Choi (2006) – using the SSG RSM and what was found by Nezu and Rodi (1985) using experiment and found by Cokljat (1993); Cokljat and Younis (1995) – using the LRR RSM. Furthermore, the implementation of the wall damping functions for a simple rectangular or square geometry, like the present study, is not so complicated. Therefore, the LRR model was used in this study.

Presently, the flow characteristics in narrow open channels and the formation of intermediate vortices are being investigated in detail. This study compares the simulation results obtained for a uniform flow case with $h = 0.16$ m, $b = 0.2$ m, $A_r = 1.25$, Fr = 1.84, and Reynolds number Re = 5.7×10^5 with the experimental results from Demiral et al. (2020). The experiment was performed in a decelerating flow condition but is close to uniform flow. A steady state simulation was performed in OpenFOAM® version dev (The OpenFOAM Foundation 2022a) based on the SIMPLE algorithm (Caretto et al. 1973). The available LRR model in OpenFOAM®, which is based on the simplified pressure-strain model of Launder et al. (1975), was modified for the full pressure-strain model and for the non-linear boundary damping functions used by Cokljat (1993); Cokljat and Younis (1995). Furthermore, the rate of turbulent kinetic energy dissipation ε at the free surface was applied following Naot and Rodi (1982). These modifications helped to overcome the shortcomings of an earlier study by Kadia et al. (2021) for Fr $\approx$ 1.8 and $Ar \approx 2.8$ which was based on the available models in OpenFOAM® and was unable to estimate the near free surface flow correctly.

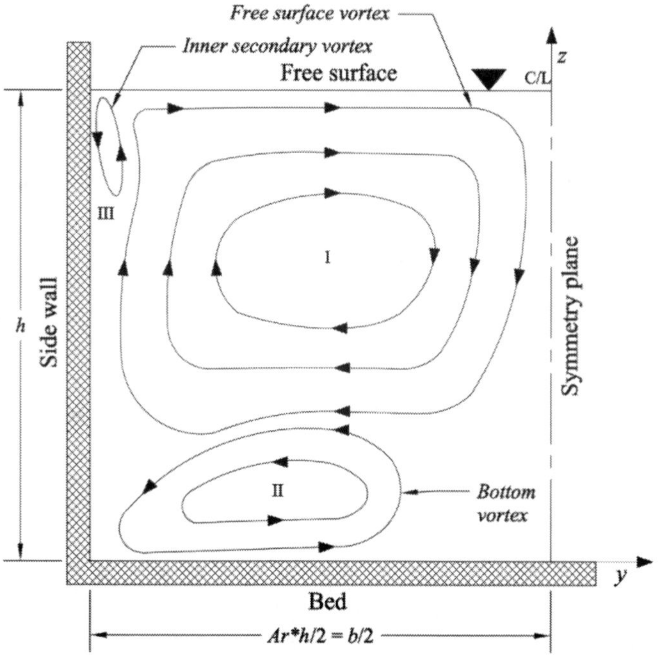

Figure 1. Schematic view of the secondary currents in one half of an open channel for $Ar \approx 2$ (after Kang and Choi (2006)) along with the simulated domain boundaries.

2. Methodology

In LRR RSM, the following transport equation (Alfonsi 2009; Hanjalić and Launder 1972; Launder et al. 1975; Speziale et al. 1991; Wilcox 2006) is solved together with the Reynolds-averaged Navier-Stokes (RANS) equations (see Wilcox (2006) for more details) to obtain the flow properties

$$\frac{D\overline{u_i'u_j'}}{Dt} = \frac{\partial \overline{u_i'u_j'}}{\partial t} + U_k \frac{\partial \overline{u_i'u_j'}}{\partial x_k} = \underbrace{-\left(\overline{u_i'u_k'}\frac{\partial U_j}{\partial x_k} + \overline{u_j'u_k'}\frac{\partial U_i}{\partial x_k}\right)}_{\text{Production}} \underbrace{- 2\nu\overline{\frac{\partial u_i'}{\partial x_k}\frac{\partial u_j'}{\partial x_k}}}_{\text{Dissipation}} \\ \underbrace{+ \overline{\frac{p'}{\rho}\left(\frac{\partial u_i'}{\partial x_j} + \frac{\partial u_j'}{\partial x_i}\right)}}_{\text{Pressure strain}} - \frac{\partial}{\partial x_k}\underbrace{\left(\overline{u_i'u_j'u_k'} + \frac{1}{\rho}\left(\overline{p'u_i'}\delta_{jk} + \overline{p'u_j'}\delta_{ik}\right) - \nu\frac{\partial \overline{u_i'u_j'}}{\partial x_k}\right)}_{\text{Diffusion}} \tag{1}$$

where u_i' is the velocity fluctuation along the direction i, $R_{ij} = -\overline{u_i'u_j'}$ is the specific Reynolds stress tensor, U_i is the mean velocity along i, ρ is the fluid density, ν is the kinematic viscosity of fluid, p' is the pressure fluctuation, and δ_{ik} is the Kronecker's delta. Apparently, the viscous diffusion tensor and the production tensor do not need to be modelled separately (Alfonsi 2009). Here, the simple gradient-diffusion model proposed by Daly and Harlow (1970) was used for the diffusion tensor formed by the triple product of the velocity fluctuations and the diffusion terms related to p' were ignored based on Wilcox (2006). Besides, the deviatoric part of the dissipation rate tensor is generally ignored at high Re following the 'Kolmogorov hypothesis of local isotropy' and the tensor is approximated as a function of the scalar quantity ε which is solved using its transport equation (Alfonsi 2009; Wilcox 2006). Furthermore, the slow part of the pressure strain was modelled following Rotta (1951) (as detailed in Launder et al. (1975); Wilcox (2006)) and the full or complete rapid pressure strain part was modelled based on Launder et al. (1975) following Cokljat (1993); Cokljat and Younis (1995). In addition, the boundary reflection corrections were modelled following Gibson and Launder (1978); Shir (1973). The near boundary damping due to the solid walls and the free surface were incorporated using the nonlinear damping functions used previously (Cokljat 1993; Cokljat and Younis 1995; Gibson and Rodi 1989; Naot and Rodi 1982). It was considered that each boundary acts individually and damps only the normal component (Cokljat 1993; Cokljat and Younis 1995). The model related coefficients were taken from Cokljat (1993); Cokljat and Younis (1995) except the coefficient C_f which is related to the free surface damping. Naot and

Rodi (1982) proposed $C_f = 0.16$, but the same was found unsuitable for the tested simulation – the vortex structure I was unsatisfactory near the free surface. A value of $C_f = 0.25$ was found suitable after trials. Lastly, the turbulence parameters near the walls $\left|\overline{u'w'}/k\right|_{wall} = \left|\overline{u'v'}/k\right|_{wall} = 0.234$ was used following Cokljat (1993), where u', v', and w' are the velocity fluctuations along the longitudinal, lateral, and vertical directions, k is the turbulent kinetic energy. This value is comparable to other studies like – Gibson and Launder (1978): 0.26; Launder et al. (1975): 0.24.

The simulated rectangular domain was 0.16 m high and 0.1 m wide (lateral distance from the channel centre y being from -0.1 to 0 m) which is one half of the channel width. Figure 1 shows the domain boundaries: bed, sidewall, free surface, and symmetry plane. The domain contained only water, i.e., single phase solution. Besides, the effect of the free surface was introduced using a free surface damping function for the Reynolds stress and using the ε boundary condition proposed by Naot and Rodi (1982) which were used in previous RSM studies as discussed earlier. A symmetry plane boundary was used at the channel centre $y = 0$ to reduce the domain size to half and to save the computational cost. One-layer of hexahedral mesh of size ≈ 0.75 mm (satisfying the log-law criteria) was generated after performing a grid convergence study for the U profile and for the bed shear stress τ_b distribution. A total of 28676 cells were created. Besides, a dynamically adjusted longitudinal pressure gradient drove the flow (Kadia et al. 2021; Talebpour 2016; Talebpour and Liu 2019). In addition, the boundary conditions were: inlet and outlet – periodic or cyclic; bed and sidewall – no slip velocity conditions with *fixedFluxExtrapolatedPressure* for the pressure p and with standard wall functions for R_{ij}, ε, and eddy-viscosity ν_t; free surface – slip velocity condition with *fixedFluxExtrapolatedPressure* for p, zero normal gradient conditions for R_{ij}, ν_t being calculated from k and ε, and with the free surface ε boundary condition proposed by Naot and Rodi (1982). Furthermore, second order *limitedLinear* numerical schemes were used for the convection terms. Additionally, the convergence criterion was 5×10^{-5} whereas the relaxation factor was 0.7 for U_i, ε, and R_{ij} and was 0.3 for p. The initial simulation showed pronounced oscillation or checkerboard kind of scenario for the near wall p, W, and V which appeared to be caused by the decoupling of Reynolds stress and velocity that occurs while using collocated grid and cell-centred scheme for RSMs (Kim 2001; The OpenFOAM Foundation 2022b). In May 2022, the OpenFOAM® support team – CFD Direct modified the coupling method in the dev version to eliminate or to reduce the staggering patterns in the results (The OpenFOAM Foundation 2022b). Furthermore, a cell-limited velocity gradient scheme improved the near wall results. The steady state simulation was performed using the *simpleFoam* solver in OpenFOAM®. Finally, the cell centre data were extracted from ParaView (version 5.9.1) and then plotted in Python software – using Turbulucid package (Mukha 2018) to get the contour plots and the secondary velocity diagram.

3. Results and discussion

The simulated U_{max} was 2.659 m/s which is 0.34% less than the observed value. Whereas the simulated maximum W was approximately 1.0% of U_{max} which is lower than what was found by Demiral et al. (2020). Besides, the maximum V and the maximum U_{WV} (where U_{WV} is the resultant secondary velocity) were about 2% of U_{max} which are comparable to the same found earlier by Broglia et al. (2003); Naot and Rodi (1982); Nezu and Nakagawa (1986); Nezu and Rodi (1985); Tominaga et al. (1989), but smaller as compared to what was found by Nikitin (2021). The simulated normalised longitudinal velocity U/U_{max} distribution and normalised vertical velocity W/U_{max} distribution were compared with the experimental results from Demiral et al. (2020) as shown in Figure 2. Furthermore, the comparison of U/U_{max} profiles at three lateral locations, $y/h = -0.563$, $y/h = -0.313$, and $y/h = 0$ is shown in Fig. 3. The results show that the distribution of U/U_{max} is influenced by the secondary currents. Significant differences observed between the simulated vortices (Fig. 4) and the vortices reported by Demiral et al. (2020) (which were based on the measurement of 2D velocities) as shown in Figs. 2(a – b). Even for a narrow channel with $Ar = 1.25$, the free surface vortex dominated the flow structure. Seemingly, the simulated free surface vortex is larger than what was expected by Demiral et al. (2020). Whereas the simulated inner secondary vortex (in Fig. 4) is comparable to the findings of Broglia et al. (2003); Grega et al. (1995); Kang and Choi (2006); Nikitin (2021), but is smaller (upto z/h about 0.8, where z is the vertical distance from the bed) than what was expected by Demiral et al. (2020) in Figs. 2(a – b). Figure 4 and Fig. 2(c) indicate that the parts of the free surface vortex and the bottom vortex which moved the faster water toward the bottom corner caused the bulging of the isovel lines of U/U_{max} toward that corner. Similarly, the parts of the free surface vortex and the inner secondary vortex which carried the quicker water toward the top corner influenced the bulging of the isovel lines toward the top corner. Likewise, the inward component of the free surface vortex near the free surface carried the slower water from the sidewall toward the channel centre, moved the isovel lines toward the centre of the flow area, and eventually formed the velocity dip. However, the bulging of the isovel lines near the free surface is more pronounced in the experimental data than the simulation result. The velocity dip was found throughout the channel width which is expected for such narrow channel with $Ar = 1.25$. At the channel centre, the simulated U_{max} was located at $z/h = 0.62$ which is very close to the observed location of $z/h = 0.59$ (Demiral et al. 2020). In addition, it was found that an intermediate vortex is developing between the free surface vortex and the bottom vortex at around $z/h \approx 0.45$ and $y/h \approx -0.45$. The inward flow of the developing intermediate vortex at $z/h \approx 0.55$ pushed the isovel lines of U/U_{max} toward the channel centre. The effect is more noticeable in the measured data as compared to the simulation result as shown in Figs. 2(a, c) and Fig. 3. However, the developing vortex is still attached to the free surface vortex. It is expected that a further narrow channel will eventually have separate intermediate vortex and free surface vortex. Presently, this aspect is being investigated in detail. Lastly, the upward component of the bottom vortex shown in Fig. 2(d) and Fig. 4 pushed the isovel lines of U/U_{max} away from the bed.

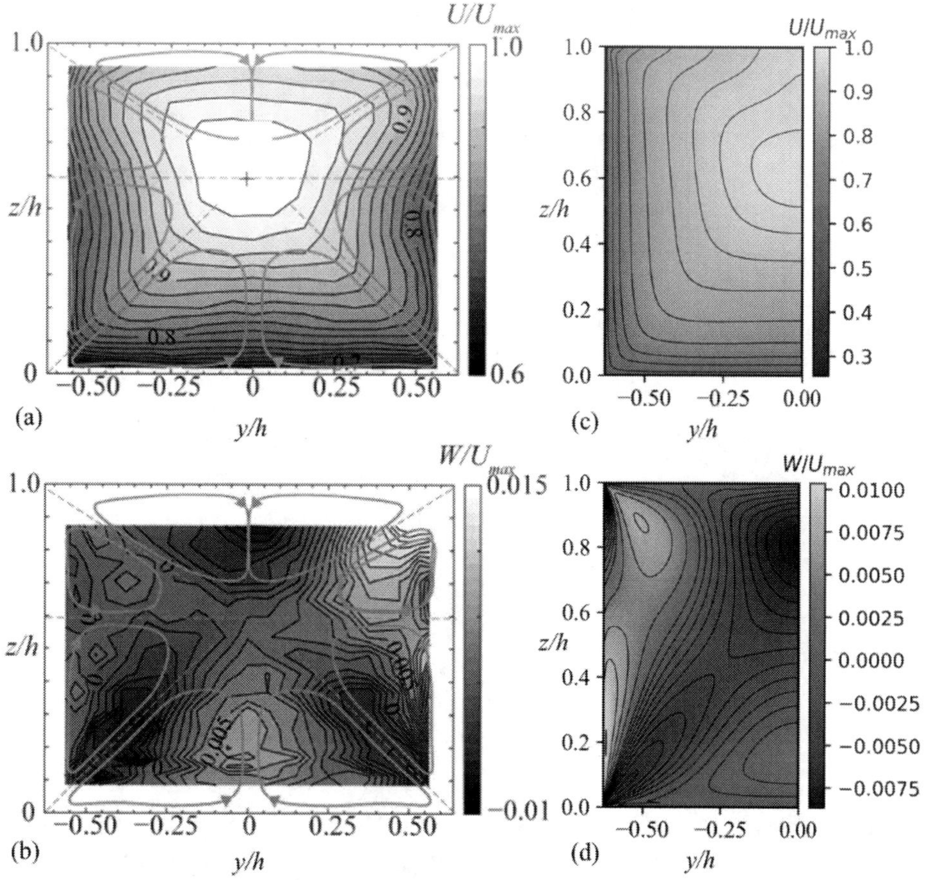

Figure 2(a-d). Comparison between the experimental (Demiral et al. 2020) and the simulated normalised velocity contours for $h = 0.16$ m, $Ar = 1.25$: (a) U/U_{max} – exp., (b) U/U_{max} – RSM, (c) W/U_{max} – exp., and (d) W/U_{max} – RSM.

Figure 5 shows the simulated normalised longitudinal turbulence intensity u_{rms}/U_{*l} distribution (where u_{rms} = longitudinal turbulence intensity and U_{*l} = cross-sectional average shear velocity) and normalised vertical turbulence intensity w_{rms}/U_{*l} distribution (where w_{rms} = vertical turbulence intensity) in comparison with the experimental results from Demiral et al. (2020). Moreover, the simulated u_{rms}/U_{*l} distribution shown in Fig. 5(c) is comparable to the observed distribution shown in Fig. 5(a). The simulated U_{*l} was 0.0897 m/s which is 0.33% lower than the experimental value. It was found that u_{rms} is highest toward the walls where the velocity gradient is steeper in the normal direction. Whereas the gradient and eventually u_{rms} decreased toward the centre of the flow area where the effect of the boundaries reduced. Along the corners, the contour lines of u_{rms}/U_{*l} bulged due to the secondary currents. Similarly, the contour lines bulged inward at z/h around 0.6 due to the developing intermediate vortex. These observations are consistent with the previous observations (Broglia et al. 2003; Kang and Choi 2006; Nezu and Nakagawa 1986, 1993; Shi et al. 1999). However, the simulated values of the maximum u_{rms}/U_{*l} near the walls are lower than the observed values. In addition, w_{rms}/U_{*l} increased toward the sidewall where the lateral turbulence intensity damped and redistributed along the other two directions. Similarly, Fig. 5(d) shows that w_{rms}/U_{*l} remained almost constant near the bed where it got damped. Furthermore, the simulated w_{rms}/U_{*l} reduced toward the free surface which contradicts the observation of Demiral et al. (2020), but is consistent with the earlier studies (Broglia et al. 2003; Kang and Choi 2006; Nezu and Nakagawa 1986, 1993; Shi et al. 1999). Apparently, the flow non uniformity which was present in the experimental work of Demiral et al. (2020) has influenced the free surface perturbation and increased the vertical velocity fluctuations (can be comparable to Song and Chiew (2001); Zhang et al. (2019)). However, the same did not influence u_{rms}/U_{*l} noticeably.

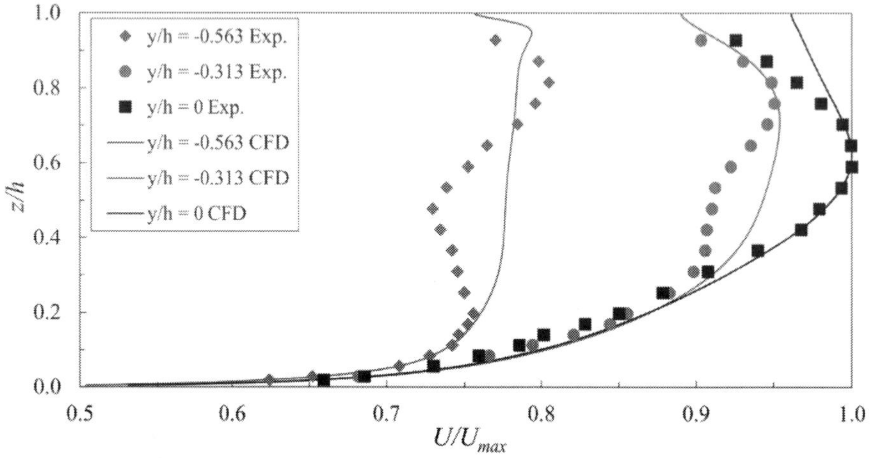

Figure 3. Comparison between the experimental (Demiral et al. 2020) and the simulated normalised longitudinal velocity U/U_{max} profiles at three different lateral locations for h = 0.16 m, Ar = 1.25.

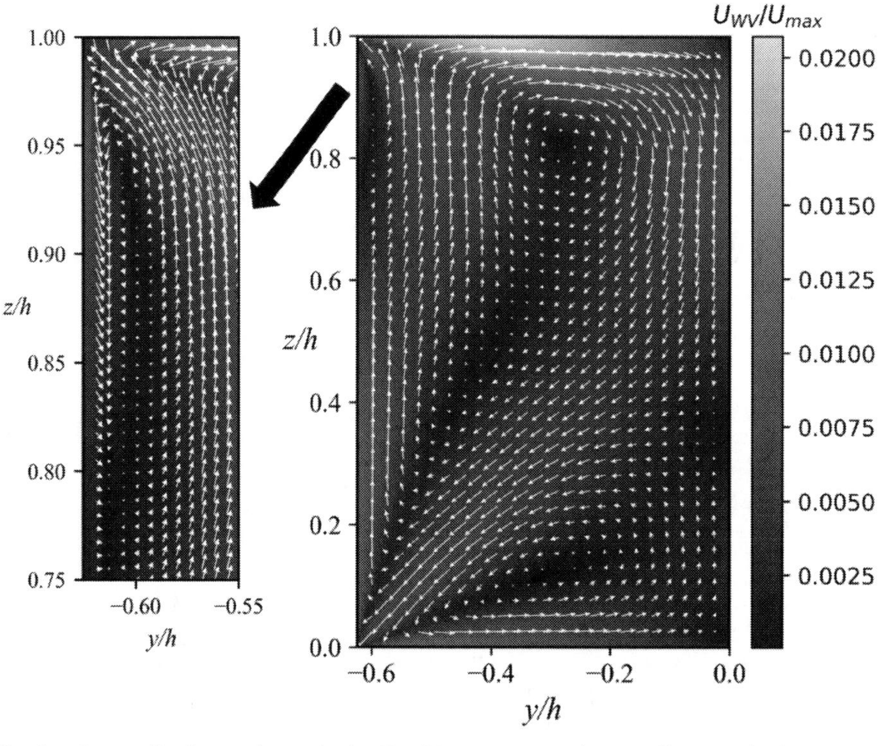

Figure 4. Simulated normalised secondary velocity U_{WV}/U_{max} contour and vector diagrams for h = 0.16 m, Ar = 1.25.

A comparison of the simulated normalised specific Reynolds shear stress $-\overline{u'w'}/U_{*l}^2$ (where $-\overline{u'w'}$ is the primary specific Reynolds shear stress) with the observed results is shown in Fig. 6. It was understood that the variation in gradient $\partial U/\partial z$ affects the distribution of $-\overline{u'w'}$. Both $\partial U/\partial z$ (observed from Fig. 2(c)) and $-\overline{u'w'}$ were higher near the bed. Whereas above the top corner bisector, $-\overline{u'w'} < 0$ was found which is related to $\partial U/\partial z < 0$ caused by the bulging of isovel lines due to free surface vortex as can be observed from Fig. 2(c) and Fig. 4.

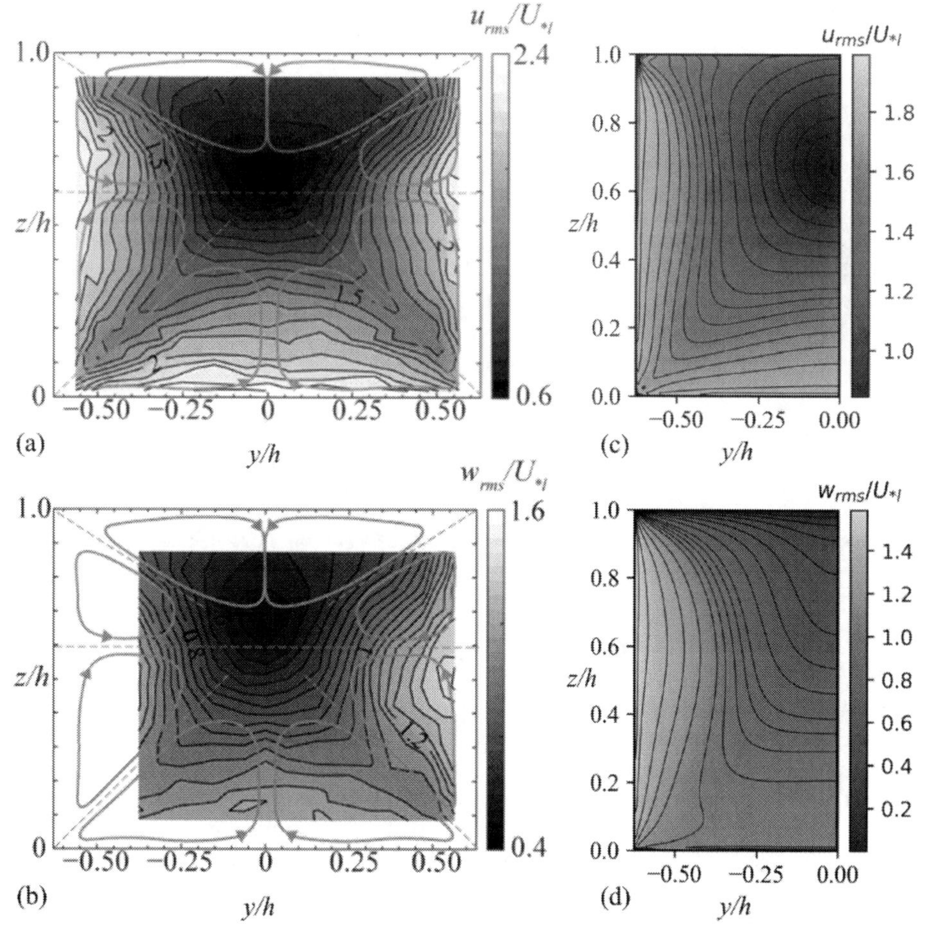

Figure 5(a-d). Comparison between the experimental (Demiral et al. 2020) and the simulated turbulence intensity contours for h = 0.16 m, Ar = 1.25: (a) u_{rms}/U_{*l} – exp., (b) u_{rms}/U_{*l} – RSM, (c) w_{rms}/U_{*l} – exp., and (d) w_{rms}/U_{*l} – RSM.

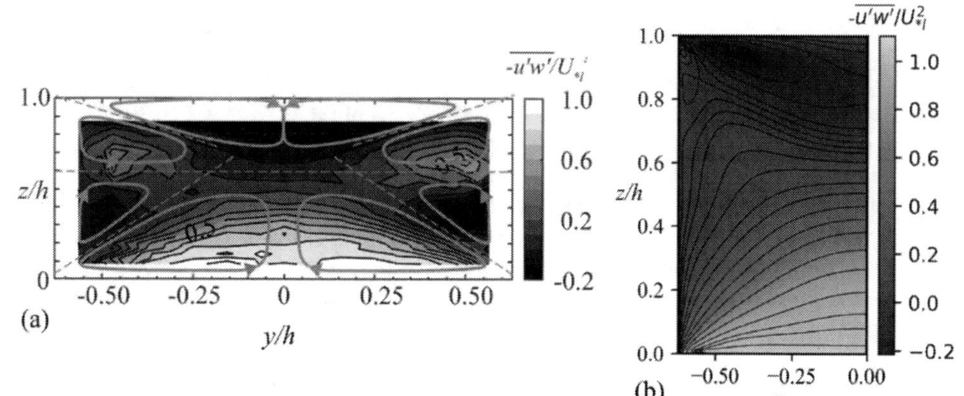

Figure 6(a-b). (a) Experimental specific Reynolds stress contours (Demiral et al. 2020) versus (b) Simulated specific Reynolds stress contours for h = 0.16 m, Ar = 1.25.

The lateral variation in the simulated normalised bed shear stress $\tau_b/\tau_{b\ mean}$ was compared with the experimental results as shown in Fig. 7, where $\tau_{b\ mean}$ is the cross-sectional average bed shear stress. The simulated $\tau_{b\ mean}$ = 8.04 N/m^2

which is 0.74% lower than the observed value. It was found that the simulated τ_b was lowest toward the sidewall and it continued to increase till y/h around -0.3 due to the downward flow of the bottom vortex shown in Fig. 4. Similarly, the upward flow of the bottom vortex bulged U contour lines upward, flattened the near wall gradient, and reduced τ_b toward the channel centre beyond y/h around -0.3. Figure 7 also indicates that the estimated τ_b is marginally higher than the experimental value for $-0.2 \leq y/h \leq 0$. Overall, it appeared that the RSM predicted the bed shear stress distribution suitably which is important for the sediment transport analysis in narrow channels like SBTs.

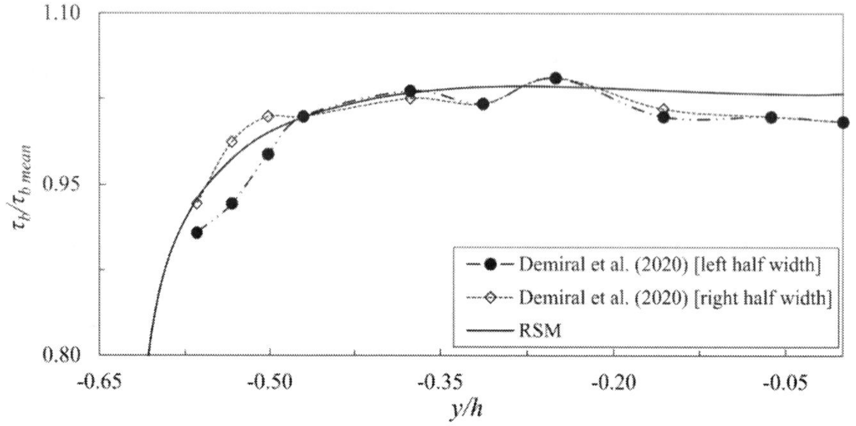

Figure 7. Experimental normalised bed shear stress (Demiral et al. 2020) versus simulated normalised bed shear stress for h = 0.16 m, Ar = 1.25.

4. Concluding remarks

A supercritical uniform narrow open channel flow scenario, like what is observed in SBTs, was simulated in OpenFOAM® using the steady state solver *simpleFoam* to study the secondary flow structures and their influence on the mean and turbulent flow characteristics, bed shear stress distribution and indirectly sediment transport. The existing simplified LRR model in OpenFOAM® was modified by implementing the full LRR pressure-strain model, the quadratic boundary reflection functions, and the free surface ε boundary condition proposed by Naot and Rodi (1982). It was found that the simulated maximum longitudinal velocity, mean shear velocity, and mean bed shear stress are within ± 1% away from the observed values of Demiral et al. (2020) which could be partially due to the measurement uncertainties. As expected, the secondary currents bulged the contour lines of longitudinal velocity and longitudinal turbulence intensity; and the bottom vortex affected the bed shear stress distribution. Similarities observed between the simulated vortex structures and the same found in previous numerical and experimental studies, but significant disparities found when compared them with the vortex structures anticipated by Demiral et al. (2020) which were based on 2D velocity data. The simulated free surface vortex is larger than the anticipated vortex whereas the simulated inner secondary vortex is smaller than the anticipated vortex. A noticeable deviation observed between the simulated and the observed vertical turbulence intensity distributions, especially towards the free surface. Apparently, the flow non uniformity present in the experimental study by Demiral et al. (2020) influenced the free surface perturbation and increased the vertical velocity fluctuations near the free surface. Such effect is absent in the used model since it is based on fully developed flows. Furthermore, the longitudinal turbulence intensities near the walls were lower than the expected value which may be explored more in the future. However, the estimated maximum secondary velocity was about 2% of the maximum longitudinal velocity which is comparable to most of the previous studies. Moreover, the comparison study indicates that the used RSM can efficiently estimate the mean bed shear stress and the bed shear stress distribution (affected by the secondary currents) which are important with reference to the sediment transport process in narrow channels including SBTs. Lastly, the simulated developing intermediate vortex remained connected to the free surface vortex for the tested case. Presently, the study is being extended for more narrow channel flows with different aspect ratios to investigate the development of intermediate vortex in detail.

5. ACKNOWLEDGMENTS

This work is supported by NTNU – project number 81772024. The authors would like to thank Prof. Robert M. Boes (ETH Zurich) for valuable inputs and suggestions in this study. They are thankful to Dr. Ismail Albayrak (ETH Zurich) and Dr. Dila Demiral (ETH Zurich) for sharing the experimental data and the related plots. They would also like to thank Henry Weller and the team at CFD Direct who implemented the improved Reynolds stress – velocity coupling method in OpenFOAM® which was funded by NTNU.

6. REFERENCES

Alfonsi, G. (2009). "Reynolds-averaged Navier-Stokes equations for turbulence modeling." *Appl. Mech. Rev.*, 62(4), 1–20.

Auel, C. (2014). "Flow Characteristics, Particle Motion And Invert Abrasion In Sediment Bypass Tunnels." PhD thesis, ETH Zurich, Switzerland.

Auel, C., Albayrak, I., and Boes, R.M. (2014). "Turbulence Characteristics in Supercritical Open Channel Flows: Effects of Froude Number and Aspect Ratio." *J. Hydraul. Eng.*, 140(4), 04014004.

Boes, R.M., Auel, C., Hagmann, M., and Albayrak, I. (2014). "Sediment bypass tunnels to mitigate reservoir sedimentation and restore sediment continuity." *Proc., Reservoir Sedimentation - River Flow 2014*, CRC Press, London, 221–228.

Broglia, R., Pascarelli, A., and Piomelli, U. (2003). "Large-eddy simulations of ducts with a free surface." *J. Fluid Mech.*, 484, 223–253.

Caretto, L.S., Gosman, A.D., Patankar, S.V., and Spalding, D.B. (1973). "Two calculation procedures for steady, three-dimensional flows with recirculation." *Proc., Third International Conference on Numerical Methods in Fluid Mechanics*, Springer, Berlin, 60–68.

Cokljat, D. (1993). "Turbulence Models for Non-circular Ducts and Channels." Doctoral Thesis, City University of London, England.

Cokljat, D., and Younis, B.A. (1995). "Second-Order Closure Study of Open-Channel Flows." *J. Hydraul. Eng.*, 121(2), 94–107.

Daly, B.J., and Harlow, F.H. (1970). "Transport equations in turbulence." *Phys. Fluids*, 13(11), 2634–2649.

Demiral, D. (2021). "Hydro-abrasion processes and modelling at hydraulic structures and steep bedrock rivers." PhD thesis, ETH Zurich, Switzerland.

Demiral, D., Boes, R.M., and Albayrak, I. (2020). "Effects of Secondary Currents on Turbulence Characteristics of Supercritical Open Channel Flows at Low Aspect Ratios." *Water*, 12(11), 3233.

Einstein, H.A., and Li, H. (1958). "Secondary currents in straight channels." *Trans. Am. Geophys. Union*, 39(6), 1085–1088.

Gibson, M.M., and Launder, B.E. (1978). "Ground effects on pressure fluctuations in the atmospheric boundary layer." *J. Fluid Mech.*, 86(3), 491–511.

Gibson, M.M., and Rodi, W. (1989). "Simulation of free surface effects on turbulence with a Reynolds stress model." *J. Hydraul. Res.*, 27(2), 233–244.

Grega, L.M., Wei, T., Leighton, R.I., and Neves, J.C. (1995). "Turbulent mixed-boundary flow in a corner formed by a solid wall and a free surface." *J. Fluid Mech.*, 294, 17–46.

Hanjalić, K., and Launder, B.E. (1972). "A Reynolds stress model of turbulence and its application to thin shear flows." *J. Fluid Mech.*, 52(4), 609–638.

Jing, S., Yang, W., and Chen, Y. (2019). "Smooth open channel with increasing aspect ratio: Influence on secondary flow." *Water*, 11(9), 1872.

Kadia, S., Pummer, E., and Rüther, N. (2021). "CFD Simulation of Supercritical Flow in Narrow Channels Including Sediment Bypass Tunnels." *Proc., 8th International Junior Researcher and Engineer Workshop on Hydraulic Structures*, Utah State University, Logan, 7.

Kang, H., and Choi, S.U. (2006). "Reynolds stress modelling of rectangular open-channel flow." *Int. J. Numer. Methods Fluids*, 51(11), 1319–1334.

Kim, S.E. (2001). "Unstructured mesh based reynolds stress transport modeling of complex turbulent shear flows." *Proc., 39th Aerospace Sciences Meeting and Exhibit*, AIAA, Reno, NV.

Launder, B.E., Reece, G.J., and Rodi, W. (1975). "Progress in the development of a Reynolds-stress turbulence closure." *J. Fluid Mech.*, 68(3), 537–566.

Mukha, T. (2018). "Turbulucid: A Python Package for Post-Processing of Fluid Flow Simulations." *J. Open Res. Softw.*, 6, 23.

Müller-Hagmann, M. (2018). "Hydroabrasion by High-Speed Sediment-Laden Flows in Sediment Bypass Tunnels." PhD thesis, ETH Zurich, Switzerland.

Naot, D., and Rodi, W. (1982). "Calculation of secondary currents in channel flow." *J. Hydraul. Div.*, 108(HY8), 948–968.

Nasif, G., Balachandar, R., and Barron, R.M. (2020). "Supercritical flow characteristics in smooth open channels with different aspect ratios." *Phys. Fluids*, 32(10), 105102.

Nezu, I., and Nakagawa, H. (1986). "Investigation on three-dimensional turbulent structure in uniform open-channel and closed duct flows." *Doboku Gakkai Ronbunshu, Japan Soc. Civil Engrs.*, 369, 89-98 (in Japanese).

Nezu, I., and Nakagawa, H. (1993). Turbulence in Open Channel Flows (IAHR Monographs), Balkema, Rotterdam.

Nezu, I., and Rodi, W. (1985). "Experimental study on secondary currents in open channel flows." *Proc., 21st IAHR Congress*, Melbourne, vol. 2, 115-119.

Nikitin, N. (2021). "Turbulent secondary flows in channels with no-slip and shear-free boundaries." *J. Fluid Mech.*, 917, A24.

Rajaratnam, N., and Muralidhar, D. (1969). "Boundary shear stress distribution in rectangular open channels." *La Houille Blanche*, 55(6), 603–610.

Reece, G.J. (1977). "A generalized Reynolds-stress model of turbulence." Doctoral thesis, University of London, England.

Rotta, J. (1951). "Statistische Theorie nichthomogener Turbulenz." *Zeitschrift für Phys.*, 129(6), 547-572 (In German).

Shi, J., Thomas, T.G., and Williams, J.J.R. (1999). "Large-eddy simulation of flow in a rectangular open channel." *J. Hydraul. Res.*, 37(3), 345–361.

Shinneeb, A.-M., Nasif, G., and Balachandar, R. (2021). "Effect of the aspect ratio on the velocity field of a straight open-channel flow." *Phys. Fluids*, 33(8), 085110.

Shir, C.C. (1973). "A Preliminary Numerical Study of Atmospheric Turbulent Flows in the Idealized Planetary Boundary Layer." *J. Atmos. Sci.*, 30(7), 1327–1339.

Song, T., and Chiew, Y.M. (2001). "Turbulence Measurement in Nonuniform Open-Channel Flow Using Acoustic Doppler Velocimeter (ADV)." *J. Eng. Mech.*, 127(3), 219–232.

Speziale, C.G., Sarkar, S., and Gatski, T.B. (1991). "Modelling the pressure-strain correlation of turbulence : An invariant dynamical systems approach." *J. Fluid Mech.*, 227, 245–272.

Talebpour, M. (2016). "Numerical Investigation of Turbulent-Driven Secondary Flow." Masters thesis, The Pennsylvania State University, USA.

Talebpour, M., and Liu, X. (2019). "Numerical investigation on the suitability of a fourth-order nonlinear k-ω model for secondary current of second type in open-channels." *J. Hydraul. Res.*, 57(1), 1–12.

The OpenFOAM Foundation. (2022a). "OpenFOAM dev: C++ Source Code Guide." <https://cpp.openfoam.org/dev/> (June 2022).

The OpenFOAM Foundation. (2022b). "GitHub - OpenFOAM-dev." <https://github.com/OpenFOAM/OpenFOAM-dev> (June 2022).

Tominaga, A., Nezu, I., Ezaki, K., and Nakagawa, H. (1989). "Three-dimensional turbulent structure in straight open channel flows." *J. Hydraul. Res.*, 27(1), 149–173.

Wilcox, D.C. (2006). Turbulence Modelling for CFD, DCW Industries, Inc., La Canada, California.

Yang, S.-Q., Tan, S.-K., and Lim, S.-Y. (2004). "Velocity Distribution and Dip-Phenomenon in Smooth Uniform Open Channel Flows." *J. Hydraul. Eng.*, 130(12), 1179–1186.

Zhang, P., Duan, Y., Li, D., Hu, J., Li, W., and Yang, S. (2019). "Turbulence statistics and very-large-scale motions in decelerating open-channel flow." *Phys. Fluids*, 31(12), 125106.

DOI: https://doi.org/10.26077/2c0b-f314

River aggradation-degradation under sand mining: experimental and numerical studies

Nejib H. Abdullahi[1] and Zulfequar Ahmad[1]
[1]*Indian Institute of Technology Roorkee, Roorkee, India*
E-mail: nejibhassen1@gmail.com

Abstract: *Sand mining from the river for different purposes is a common cause of the failure of hydraulic structures. Its presence in a river causes intense bedload transport, which is crucial for designing and operating hydraulic structures. The flow characteristics and aggradation-degradation phenomenon in the vicinity of the sand mining pit have been analyzed with experimental and numerical approaches. The bed level evolution in the vicinity of the pit was measured using a pointer gauge for different temporal and equilibrium stage conditions. A numerical simulation was done using a three-dimensional computational fluid dynamics model, Flow-3D, which was used to compute the bedload transport deposited in the pit and to estimate the spacetime dynamics of the bed evolution of the pit. The volume of fluid (VOF) equation is used to confirm the application of boundary conditions at the free surface in a numerical method. The turbulence formed in the vicinity of the pit was predicted using the turbulence model k-ε with re-normalization group (RNG) extensions. Different bedload transport equations and mesh grid sizes were tried to identify the best fit for aggradation-degradation simulation around sand mining pit. The total volume of bedload transported from the upstream section was deposited in a pit since a pit was used as a bedload trap. The experimental and numerical results showed a close agreement which verified the efficiency of the Flow-3D model.*

Keywords: *Sand mining, bedload transport, pit morphology, aggradation, degradation, bed evolution*

1. Introduction

River beds and banks are frequently exposed to morphological evolutions in different phases through natural or artificial modes. The morphological evolution of a river in the form of aggradation and degradation in the vicinity of the sand mining pit has been analyzed with experimental and numerical methods. The river morphological evolutionary processes are the methods of searching for the following equilibrium conditions. The in-stream mining of sand, gravel, and aggregate damages the river morphology and increases the failure of the hydraulic structures (James 1991; Yanmaz and Cicekdag 2004; Kim 2005; Rovira et al. 2005; Ashraf et al. 2011; Padmalal and Maya 2014). Lee et al. (1993) pioneered an experimental study on the migration speed of a rectangular sand mining pit and reported significant sand exchange in the vicinity of the mining pit. They also classified the process in the sand mining pit into two periods, namely the convection and diffusion periods. The convection period starts from the early deformation of the sand mining pit until the upstream nick point reaches the initial downstream nick. The diffusion period starts from that point onward; similar results were reported by (Lee and Chen 1996; Neyshabouri et al. 2002). In-stream sand mining effectively prevents and controls the river channel overflow in rapidly depositing rivers (Mathias Kondolf 1994).
The sand mining pit bank is eroded, and the erosion expands to the whole channel width and aggravates toward the downstream section of a pit over time (Barman et al. 2018). The mining pit depth can be classified into the mixing layer at the top, the subsurface layer, and the replenishing layer (López 2004). In river channels, incision mostly originates from sand mining pits that spread upstream and downstream of the pit and induce channel instability, hydraulic structure failure, and groundwater table lowering (López 2004; Rinaldi et al. 2005). Azar et al. (2012) investigated the evolution of the riverbed profile due to a sand mining pit using *HEC-RAS*, a one-dimensional hydraulic engineering software. Wu and Wang (2008) studied the morphological evolutions through initially dry sand mining pits using non-equilibrium sand transport and shallow water equations. They reported an intensive degradation of the upstream and downstream nicks, with higher degradation obtained at the upstream nick point. The hydraulic and flood characteristics of the reach of the Tapi River length of 7.75 *km* located between Sardar Bridge and Magdalla Bridge, Surat India were studied by Mehta et al. (2020) using hydrodynamic software. They designed the stable channel using the Copeland method, which is a part of Army Corp of Engineers (*HEC-RAS*) software.

The capability of a two-dimensional numerical model (*CCHE2D*) to evaluate the sediment mining pit-infilling process was studied by Haghnazar et al. (2020). They carried out the comparisons of the experimental and numerical results and found acceptable agreements. They also described the two-dimensional *CCHE2D* model overestimated the aggradation process in the mining pit. Samadi et al. (2018) reported significant river bed level changes after studying the migration of sand and gravel mining on the Sarbaz River in Sistan and Baluchestan, Iran using two-dimensional *CCHE2D* and one-dimensional *HEC-RAS* models. The Geographic Information System-based study in the Rambla de la Viuda stream, Eastern Spain, was undertaken to assess morpho-sedimentary changes due to in-stream gravel mining (Calle et al. 2017). They used *LiDAR* (Light Detection and Ranging), *DTM* (Digital Terrain Model), and *RTK-GPS* (Real-Time Kinematic-Global Positioning System measurements to quantify the river bed degradation and observed a river bed degradation of 3.5 *m*. The consequence of sand and gravel mining on a bridge in Helleh River in Bushehr province, Iran, was studied using three approaches experimental, numerical, and case study (Zolghadr et al. 2021). They reported a bed level change of 1.27 *m* at the bridge foundation for a flood event with a return period of 25 years and initial signs of river meandering.

There are a lot of authorized and non-authorized sand and gravel minings in many active and non-active rivers. It is crucial to predict the impact of sand mining activities and effective management to maintain the river ecosystem's stability and provide economical sand resources. This study investigates the impacts of sand mining on the flow characteristics and morphology of riverbeds and banks using a trapezoidal sand mining pit under non-equilibrium conditions with experimental and numerical methods. A *CFD* numerical simulation software package, flow-3D, was used for numerical simulations, and the simulation was validated with laboratory experiments.

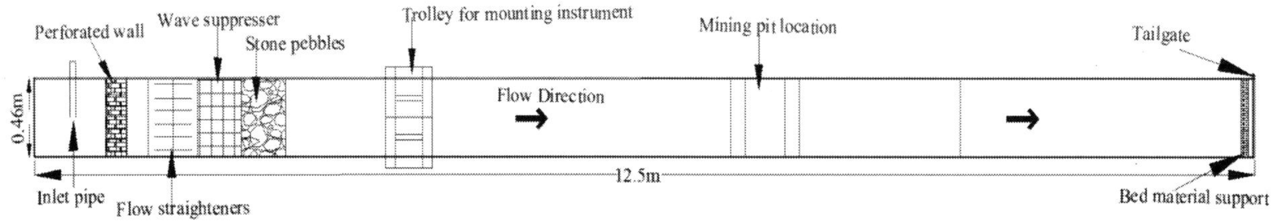

Figure1. Experimental flume plan

2. Experimental Work and Methodology

The experiments in this study were conducted in a tilting flume of 12.5 *m* in length, 0.46 *m* in width, and 0.7 *m* in depth. The flume had a working section of 5 *m* in length, 0.46 *m* in width, and 0.7 *m* in depth and was located at the center of the flume. A perforated wall, flow straighteners, and wave suppresser were provided at the upstream end of the flume to stabilize the flow. Gravels were provided at the flume entrance for the dissipation of energy. The sides of the flumes were constructed of transparent glasses to visualize the flow. The bed materials d_{15}, d_{50}, d_{85}, and σ_g were 0.16 *mm*, 0.24 *mm*, 0.58 *mm*, and 1.2, respectively, filled in a thickness of 0.25 *m* in the flume. The grain size distribution of sand can be considered uniform if the geometric standard deviation is less than 1.4, as Marsh et al. (2004) suggested. The computed geometric standard deviation of the bed material confirms the uniformity of the bed material. By connecting the two containers with a plastic pipe, the longitudinal bed slope was measured and set to 0.002.

The bed material was laid over the channel bed and compacted, leveled, and then flooded with clear water flows for the compaction of the sand particles. After draining the water, a trapezoidal sand mining pit of specified dimensions was constructed. The discharge was kept constant, and the elevations of the channel bed and water surface were measured at various temporal and spatial levels. The flow was in subcritical conditions, and the sand mining pit served as a bedload trap. The plan of the experimental flume is shown in Figure 1. At one-hour intervals, the topography of the mining pit bed, the water surface profile, and other data were recorded. The experimental runs continued until the equilibrium stage, or the mining pit disappeared. The water was drained after reaching equilibrium in order to analyze the morphology of the mining pit bed. The characteristics of the pit, flow conditions, and sand conditions used in this study are given in table 1, where S_o is the longitudinal flume bed slope; Q is the inflow discharge; y is the upstream flow depth; u_* is the shear velocity; u_{*c} is the critical shear velocity of sand; Fr is the Froude number; Re is the Reynolds number; L is the pit length; H is the pit depth, and V is the extracted volume of sand.

Table 1. Characteristics of flow, pit, and sand used in experiments

Test No.	S_o	Q (m^3/s)	y (m)	u_*/u_{*c}	Fr	Re	L (m)	H (m)	V (m^3)
T-1	1/500	0.025	0.142	1.42	0.33	31578	0.40	0.15	0.038

The topography of the mining pit bed and the water surface level was measured with a pointer gauge of the least count of 0.10 *mm*. Initially, the degradation at the upstream nick of the sand mining pit was higher than the downstream nick, and later the aggradation at the upstream nick was higher. The aggradation of the pit was continuous until the pit depth was low to deposit sand. The temporal elevation at the vicinity of the pit was measured at different times, and its volume of aggradation degradation was evaluated. The flow velocity in a sand bed as a solid bed (immobile bed) and sand bed (mobile bed) around the sand mining pit was simulated using the flow 3D and compared with the instantaneous velocity measured using an Acoustic Doppler Velocimeter (*ADV*) to calibrate the results of the simulation. The instantaneous velocities were measured using a Nortek Vectrino plus Acoustic Doppler Velocimeter (*ADV*) with a maximum sampling rate of 200 *Hz*. Its sampling volume is approximately 0.08 *m* below the down-looking probe, with a minimum distance of 0.05 *m* between the probe and sampling volume to minimize flow interference. The spike removal was done using a minimum correlation of 70%, a minimum signal-to-noise (*SNR*) ratio of 15, and an acceleration threshold of 1 to 1.5. As the bedforms significantly impact the Acoustic Doppler Velocimeter probe locations and to simplify the flow complexity due to the pit, the lateral velocities were taken at z = 0.075 *m* above the bed.

3. Numerical Model Setup

3.1. Governing Equations

Flow 3D will be used to solve the Reynolds Average Navier-Stokes equations (*RANS*) with a κ-ε turbulent closure model to evaluate the flow hydraulics (Equation 2-4). The volume of fluid (*VOF*) technique will be used to follow the free surface (Equation 5). The Fractional Area Volume Obstacle Representation (*FAVOR*) is utilized to simulate the surfaces and boundaries of the complex geometry. During the numerical simulation, the mass balance of the system was also controlled (Equation 1).

$$V_F\frac{\partial\rho}{\partial t}+\frac{\partial}{\partial x}(\rho uA_x)+R\frac{\partial}{\partial y}(\rho vA_y)+\frac{\partial}{\partial z}(\rho wA_z)=R_{SOR} \tag{1}$$

$$\frac{\partial u}{\partial t}+\frac{1}{V_F}\left(uA_x\frac{\partial u}{\partial x}+vA_yR\frac{\partial u}{\partial y}+wA_z\frac{\partial u}{\partial z}\right)-\xi\frac{A_yv^2}{xV_F}=-\frac{1}{\rho}\frac{\partial p}{\partial x}+G_x+f_x \tag{2}$$

$$\frac{\partial v}{\partial t}+\frac{1}{V_F}\left(uA_x\frac{\partial v}{\partial x}+vA_yR\frac{\partial v}{\partial y}+wA_z\frac{\partial v}{\partial z}\right)-\xi\frac{A_yuv}{xV_F}=-\frac{R}{\rho}\frac{\partial p}{\partial y}+G_y+f_y \tag{3}$$

$$\frac{\partial w}{\partial t}+\frac{1}{V_F}\left(uA_x\frac{\partial w}{\partial x}+vA_yR\frac{\partial w}{\partial y}+wA_z\frac{\partial w}{\partial z}\right)=-\frac{1}{\rho}\frac{\partial p}{\partial z}+G_z+f_z \tag{4}$$

$$V_F\frac{\partial F}{\partial t}+\nabla.(AUF)=0 \tag{5}$$

where ρ, V_F, R_{SOR}, A, U, and F are fluid density, open volume ratio to flow, mass source, average flow area in (x, y, z) direction, average velocity in (x, y, z) direction, and volume flow function, respectively, and u, v, w are velocity components in (x, y, z), A_x, A_y, A_z are fractional areas open to flow, G_x, G_y, G_z are gravitational forces in each cartesian direction, f_x, f_y, f_z are body force per unit mass in (x, y, z) directions, respectively.

3.2. Sediment Transport

In this study, only bedload transport was considered, and sediment concentration (C_{bed}) and the bedload (q_b) were computed by Van Rijn's (1987) formula given below:

$$C_{bed}=0.015\frac{d_{50}^{0.3}\left(\frac{\tau-\tau_c}{\tau_c}\right)^{1.5}}{g\left(\frac{\rho_s-\rho_w}{\rho_w v^2}\right)^{0.1}} \tag{6}$$

$$\frac{q_b}{d_{50}^{1.5}\sqrt{\frac{\rho_s-\rho_w}{\rho_w}g}}=0.053\frac{\left(\frac{\tau-\tau_c}{\tau_c}\right)^{1.5}}{d_{50}^{0.3}\left(\frac{(\rho_s-\rho_w)g}{\rho_w v^2}\right)^{0.1}} \tag{7}$$

where d_{50}, τ, τ_c, ρ_s, v, g are the sediment particle diameter, the bed shear stress, the critical bed shear stress for movement of sediment particles according to Shields' diagram, the density of sediment, the kinematic viscosity of water and the acceleration of gravity, respectively. The parameter a indicates the reference level equal to half the bed form height.

3.3. Model Development

A 3D computational was considered for the *CFD* simulation. The modeling considers a domain length of 5 *m*, 3 *m* upstream, and 2 *m* downstream from the center of the pit. For the initialization of the model, the flow depth was fixed at 0.25 *m*. The computational domain boundaries at the inlet and outlet are discharge of 0.025 m^3/s and a continuative boundary, respectively. As defined in the flow 3D manual, the continuative boundary condition consists of zero normal derivatives at the boundary for all quantities used to preserve the smooth flow continuity. A symmetry boundary condition was assigned at the top surface of the computational domain, indicating zero-gradient conditions. Likewise, the Wall boundary conditions were assigned on the sides of the computational domain. The modeled geometry with utilized boundary conditions is shown in Figure 2*a*. Different mesh grid sizes were tried to identify the best fit for the simulation of aggradation-degradation around sand mining pit. After many trials, the nested mesh was considered the most suitable for the simulation. Combinations of two mesh grid sizes were used in the model, the finer mesh (143 x 66 x 100) with a grid size of 0.007 *m* in all directions near the mining pit and the coarser mesh (510 x 46 x 75) with a grid size of 0.01 *m* in all directions in other regions, respectively (Figure 2*b*). The computational mesh duration was 720 minutes, and the numerical simulation was done using a 2.1 *GHz* processor computer.

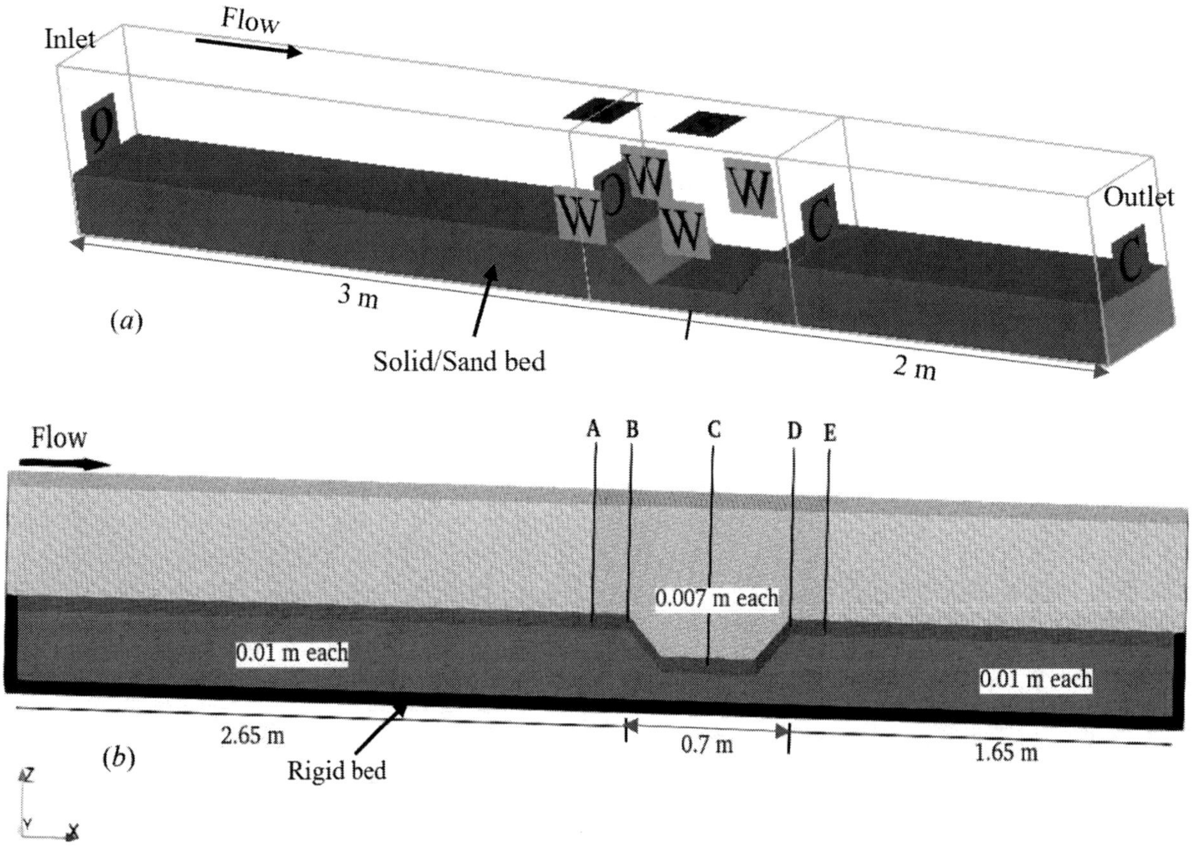

Figure 2. (*a*) Modeled geometry with the utilized boundary conditions (*b*) mesh sizes and selected sections for velocity analysis, here *Q*, *C*, *W*, and *S* are the inflow discharge, Continuative, Wall and Symmetry

The numerical simulation consists of two scenarios. The first scenario (scenario 1) was achieved by simulating the sand bed as an immobile bed (solid bed). Scenario 1 aimed to understand the velocity characteristics induced by a solid bed (immobile) pit as well as to compare the results with the sand bed (mobile) pit velocity distributions. The velocity simulated in scenario 1 is low due to the significant flow depth in the pit and the flow separation induced by the pit. The second simulation scenario (scenario 2) was achieved by simulating the sand bed as a mobile bed (sand bed). The flow characteristics in scenarios one and two were evaluated and compared using similar boundary and initial conditions to understand better the impact of mobile and immobile sand mining pits. Scenario 2 aimed to comprehend the aggradation-degradation characteristics in the vicinity of the sand mining pit were investigated in order to fully understand the impact of the sand mining pit composed of a mobile bed.

4. Results and Discussions

4.1. Velocity Distributions

Five different sections in the vicinity of the sand mining pit are designated for the flow velocity analysis. Section *A* and *B* are located 0.5 *m* and 0.35 *m* upstream of the pit center, section *C* is located at the center of the pit, and sections *D* and *E* are located 0.35 *m* and 0.5 *m* downstream of the pit center, respectively (Figure 2*b*). The sand bed was modeled as a solid bed (immobile) and a sand bed (mobile) using flow 3D to obtain velocity distributions in the longitudinal and cross-sectional directions. Figures 3 and 4 show a comparison of simulated and measured longitudinal and lateral velocity distributions for five different sections in the vicinity of the sand mining pit. The results show that the velocity in the solid bed with a mining pit is lower than that of the sand bed with a mining pit. The flow 3D and experimental results are nearly similar for all five different sections of the sand bed. The differences between the results of the flow 3D for the solid bed and sand bed are due to the complexities of the flow pattern in the sand bed, increasing the turbulence and flow characteristics.

The experimental and flow 3D velocity results for the sand bed condition mostly have a similar pattern in all sections except inside the pit, which may be due to the complexity of the interaction between the turbulent flow and sediment particles not captured by the model. Nevertheless, in general, it can be concluded from these figures that the flow 3D model simulated the flow velocity with reasonable precision. Various trials were done on the mesh sizing, turbulence models, and sediment transport formula to fit best the experimental observations in modeling the aggradation-degradation near the sand mining pit. Thereafter, the *k-ε* with *RNG* extension for turbulence modeling and Van Rijn formula for sediment transport model was adopted for the numerical analysis. Some of the results of the numerical analysis have been presented here due to the paper length limit.

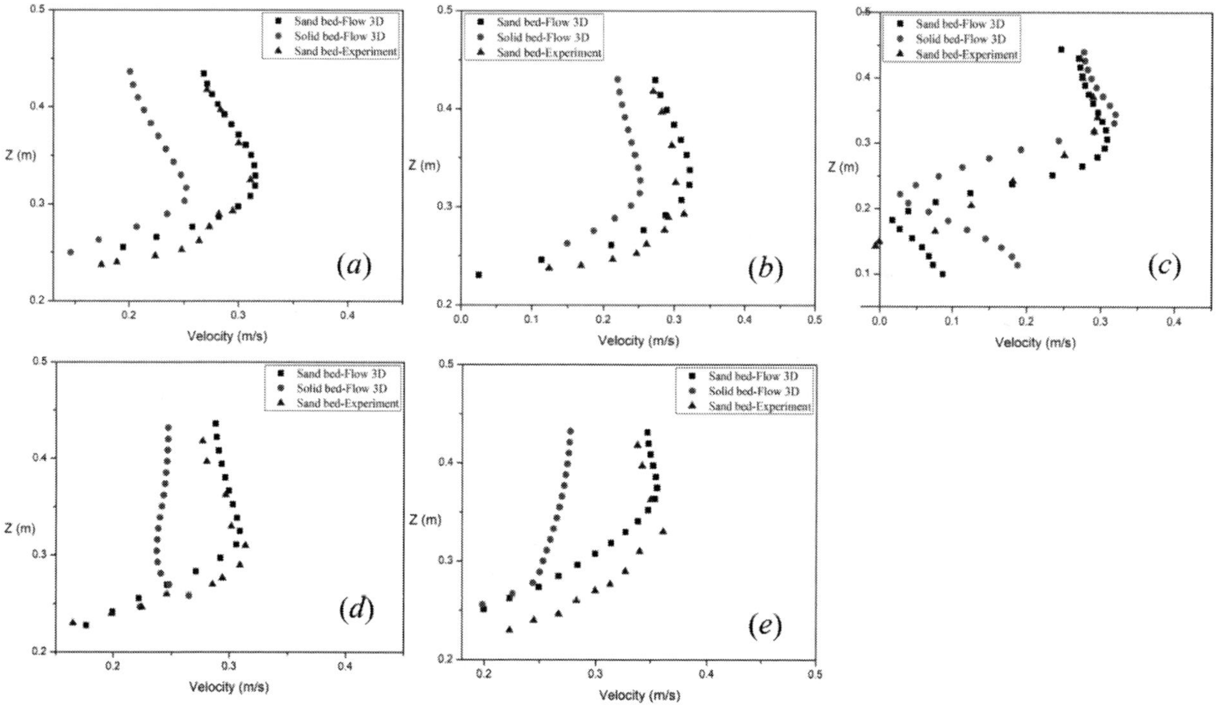

Figure 3. Vertical velocity distributions at different setions (shown in Figure 2*b*) for *a*) *A*, *b*) *B*, *c*) *C*, *d*) *D*, *e*) *E*

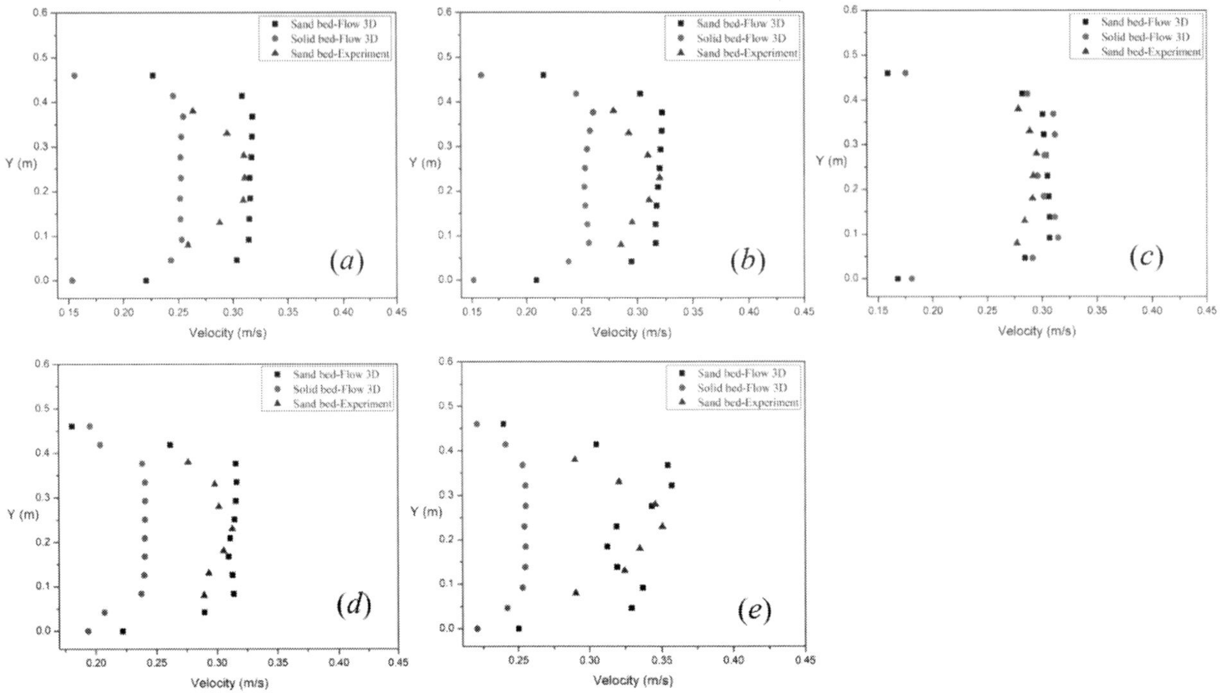

Figure 4. Lateral velocity distributions at different sections (shown in Figure 2*b*) for *a*) *A*, *b*) *B*, *c*) *C*, *d*) *D*, *e*) *E*

4.2. Morphological Sand Bed Changes

The time-spatial aggradation-degradation development in the vicinity of the pit using the experimental and flow 3D for Test *T*-1 is shown in Figure 5 (*a-d*). The intensive degradation and detachment of the upstream nick were observed in the initial phase. In contrast, there is an insignificant change in the downstream nick level in the same phase. The downstream nick side slope was nearly similar to the angle of repose of the bed material (32°) in the initial phase, while this angle was not preserved at the upstream nick side. Afterward, the bedload transport started diffusively, and the upstream nick started aggradation and migrating in the flow directions, whereas the downstream nick started degradation intensively. In the migration phase of the sand mining pit, the upstream nick develops a steep slope nearly equal to the angle of repose of the bed material (32°) and migrates in the flow direction. The downstream nick becomes flattened, and no such angle is preserved. Since the pit served as a bedload trap, the water flowing out of the pit is clear water and caused an intensive degradation to the downstream section of the sand mining pit. The aggradation-degradation near the sand mining pit was simulated using the flow 3D shows aggradation inside a pit and degradation in both the upstream and downstream nick points, with high degradation at the upstream nick (Figure 5 *a-d*). The process of aggradation-degradation was analyzed using flow 3D both in longitudinal and cross-sectional directions, and it shows similar patterns to the experimental results.

The *CFD*-simulated sand bed (mobile bed) variations reveal aggradation inside the pit in a pattern similar to the experimental results. However, it shows more specific degradation at the upstream section of the pit than at the downstream nick; this may be due to the influence of the upstream nick side slope in the initial phase. Sand mining causes aggradation-degradation on the river bed, impacting the river ecosystem and hydraulic structures, which can be determined using numerical methods (flow 3D). The downstream degradation is mainly due to the clear water from the sand mining pit since most of the bedload was settled in the pit (Abdullahi and Ahmad 2022a). This degradation of the top layer of the sand bed also causes the failure of river ecology and can harm flora and fauna near the river bed. The average degradation in the upstream section of the sand mining pit was 9.8% and 5.1% for the flow 3D and experimental results, while it was 6.7% and 14% in the downstream section, respectively. The maximum aggradation depth inside the sand mining pit was 86% and 98% for the flow 3D and experimental results, respectively (Figure 5*d*).

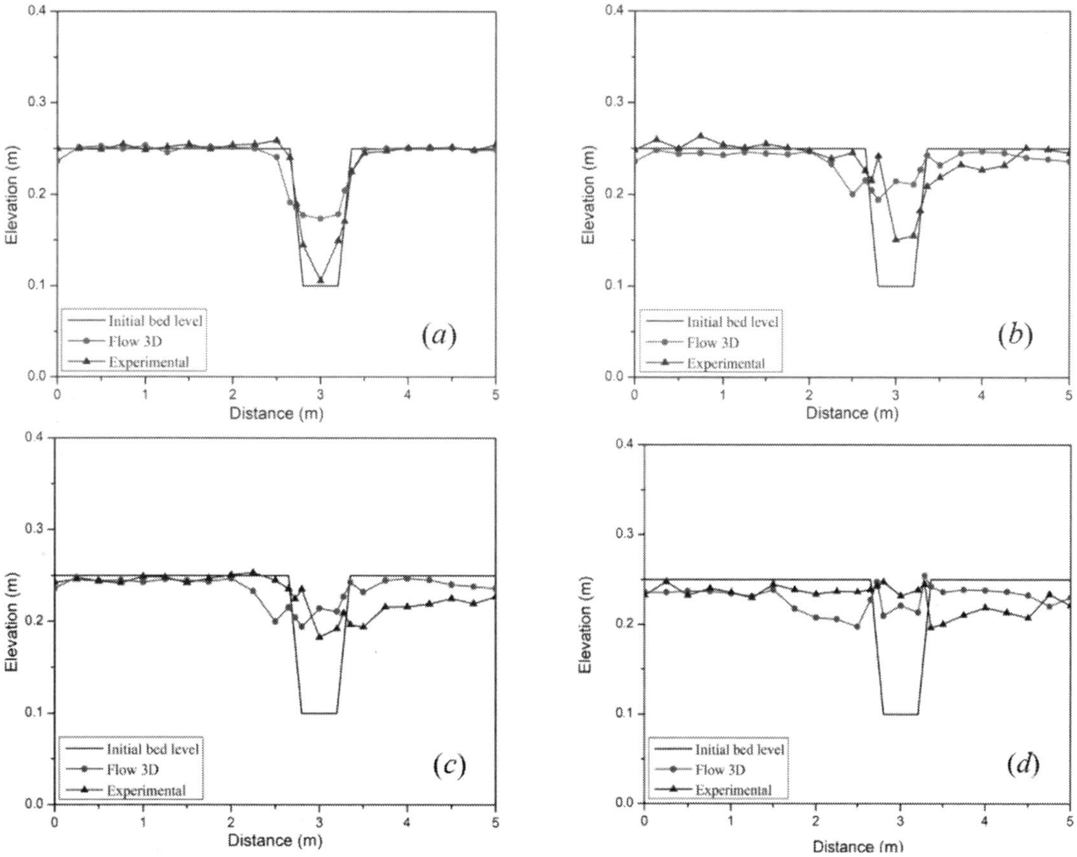

Figure 5. Comparisons of the longitudinal temporal aggradation-degradation profile at the center of the flume for *a*) 1 *hr*, *b*) 3 *hr*, *c*) 7 *hr*, *d*) 12 *hr*

The initial extracted volume of sand from the sand mining pit was 0.038 m^3. After the experimental work and numerical simulations for a similar computational duration, the maximum aggradation volume inside the pit was 0.037 m^3 (97.4%) and 0.033 m^3 (86.8%) for the experimental and numerical simulation (flow 3D), respectively. It can be concluded that 0.001 m^3 and 0.005 m^3 volumes of sand were removed out of the channel as bedload for the experimental and flow 3D model, respectively, after the sand mining pit reached the equilibrium conditions.

Figure 6 shows the numerical simulation results of the aggradation-degradation in the longitudinal and cross-sectional directions and sand bed elevation change. The longitudinal bed change with the span of one-meter length (refer to Figure 2*b* for the locations of sections *A* to *E*) near the pit shows the aggradation inside the pit and degradation in the upstream and downstream nicks with the maximum degradation in the upstream nick. The degradation at the upstream nick was 4.6% and 9% for the experimental and the flow 3D, while it was 21.3% and 3.1% for the downstream nick, respectively. The cross-sectional bed change was also done for the five different sections near the pit. Aggradation was observed inside the sand mining pit, whereas degradation was observed in another four sections *A*, *B*, *D*, and *E* (Figure 6). The locomotion and spatiotemporal morphological evolution of a sediment mining pit were studied by Abdullahi and Ahmad (2022b). They classified the evolution of the sediment mining pit into expansion and migration phases, with the dividing point occurring when the maximum residual depth of the pit flips from the upstream edge to the downstream edge. The expansion phase begins with the initial pit deformation and lasts when the maximum top width prevails. The migration phase begins after the expansion phase is complete and lasts until the pit is filled or completely vanishes.

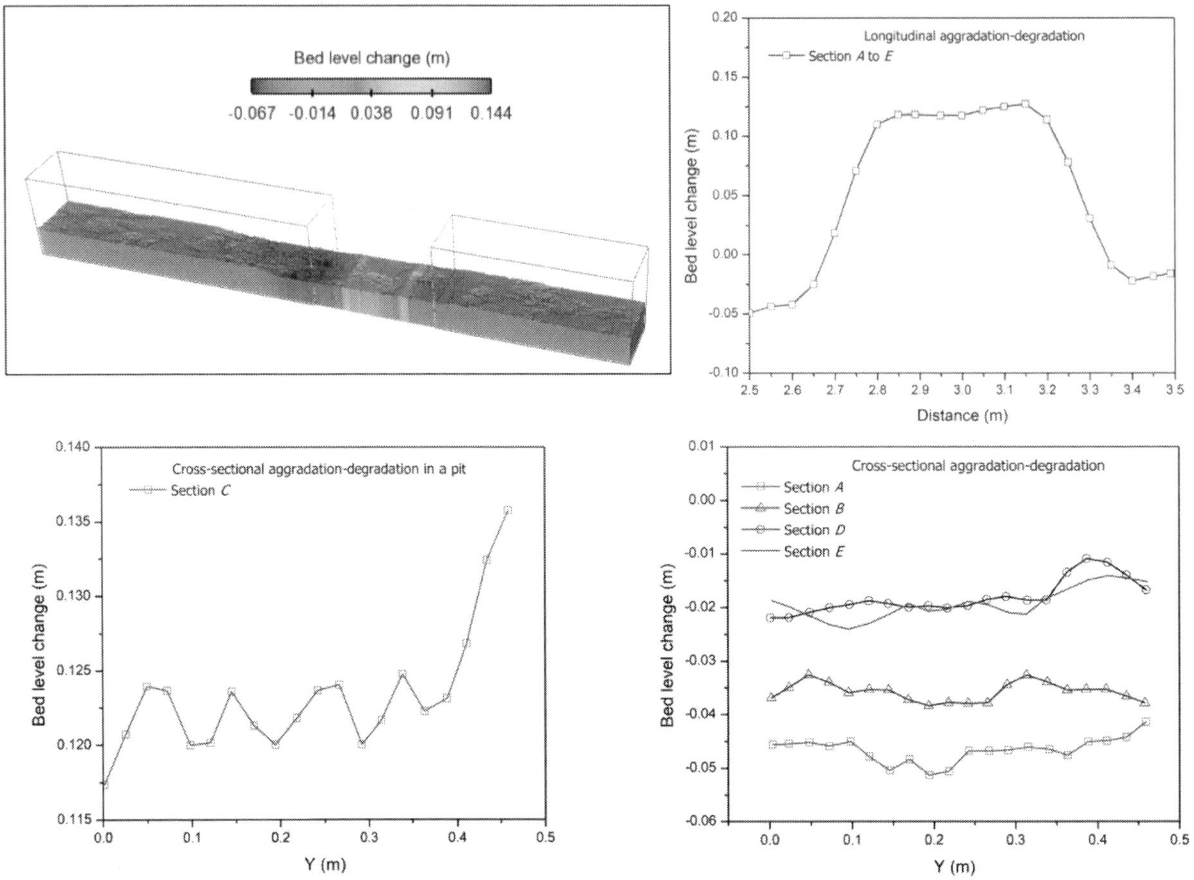

Figure 6. The aggradation-degradation process at the vicinity of sand mining pit using flow 3D model after 12 *hr*

As shown in Figure 6, the cross-sectional degradation was observed in Sections *A*, *B*, *D*, and *E*, with the higher degradation observed in sections *A* and *B* and lower degradation in sections *D* and *E*. The comparisons of the experimental and flow 3D results for the equivalent analysis duration are shown in Figure 7*a*. It is visible from the Figure that the patterns in both the computational methods are similar, except in some areas like the downstream nick point; this is maybe due to the sand mining pit, which creates complex sediment-flow interactions inside the pit and during the evacuation of flow from inside the pit.

There is also an insignificant difference between the flow 3D and experimental work results in the magnitude of degradation in the upstream and downstream nicks caused due to the complexity of flow and bedload transport characteristics near the sand mining pit. In experimental results, the maximum aggradation depth observed inside the pit was 0.147 *m* (Figure 7*a*). In comparison, the flow 3D underestimated the value of aggradation depth to 0.129 *m*. This underestimating can be because of the sediment-flow interaction behavior inside the sand mining pit. In the study of the scour around spur dikes and groynes, a similar difference in results in the experimental and flow 3D of 0.03 *m* and 0.023 *m* was reported by (Karami et al. 2014) and (Choufu et al. 2019), respectively. After analyzing the aggradation-degradation near the pit, a maximum aggradation depth of 98% and 86% inside the sand mining pit was observed under the experimental analysis and a flow 3D model. These results indicate the accuracy of the flow 3D model in calculating the aggradation-degradation near the sand mining pit. The temporal variations of the rate of the aggradation depth in the pit center for both the experimental and flow 3D methods are shown in Figure 7*b*. It can be concluded from this Figure that about 75% of aggradation depth occurred during the first 9% of the computational time for the flow 3D model. The aggradation rate increased with a steep slope until it filled the 75% pit depth and gently increased until the equilibrium conditions. In the experimental work, 38% of aggradation depth occurred during the first 25% of the experimental durations. The aggradation rate continuously increased with a steeper slope until the equilibrium conditions for the next 75% of the experimental durations. From this Figure, it can be seen that the pattern between the flow 3D and experimental results are satisfactory.

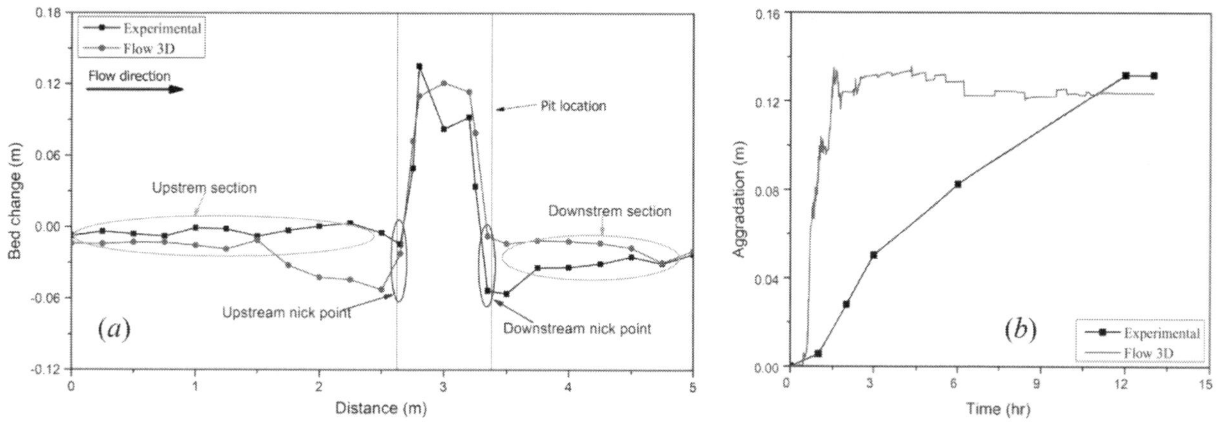

Figure 7. *a*) Comparisons of the aggradation-degradation in the sand mining pit centerline, *b*) time variations of aggradation at the center of the pit

5. Conclusions

The aggradation-degradation in the river due to sand mining was studied with experimental and numerical approaches. The numerical approach shows satisfactory results for the longitudinal and lateral flow velocity distributions with experimental results. The numerical and experimental results of the velocity distributions are nearly similar for all five different sections of the sand bed. The differences between the results of the numerical approach for the solid bed and sand bed are due to the complexities of the flow pattern in the sand bed, increasing the turbulence and flow characteristics. The numerical approach also shows satisfactory results for the aggradation-degradation around the sand mining pit with experimental results. Approximately 75% of the aggradation depth in the pit occurs during the first 9% of the computational time using a numerical approach. The aggradation rate increased within a constant steep slope until it filled 75% of the pit depth and then gradually increased until the equilibrium conditions were reached. In the experimental work, 38% of aggradation depth in the pit occurs during the first 25% of the experimental durations. After analyzing the aggradation-degradation near the pit, a maximum aggradation depth of 98% and 86% inside the sand mining pit was observed using experimental analysis and a flow 3D model. These results indicate the accuracy of the flow 3D model in calculating the aggradation-degradation near the sand mining pit.

6. References

Abdullahi, N. H., & Z. Ahmad. (2022a). "Evaluation of migration speed equations of the upstream nick of the sediment mining pit." *ISH J. Hydraul. Eng.*, 1–8. https://doi.org/10.1080/09715010.2022.2058332

Abdullahi, N. H., & Z. Ahmad. (2022b). "Experimental study of sediment mining pit locomotion and morphological evolution." *Arab. J. Geosci.*, *15*(14), 1–15. https://doi.org/10.1007/s12517-022-10544-1

Ashraf, M. A., M. J. Maah, I. Yusoff, A. Wajid, & K. Mahmood. (2011). "Sand mining effects, causes and concerns: A case study from bestari jaya, Selangor, Peninsular Malaysia." *Sci. Res. Essays*, *6*(6), 1216–1231. https://doi.org/10.5897/SRE10.690

Azar, M. G., M. R. Namaee, & M. Rostami. (2012). "Evaluating a Numerical Model to Simulate the Variation of River Bed Due to a Mining Pit Based on Experimental Data." *Asian J. Appl. Sci.* https://doi.org/10.3923/ajaps.2012.154.163

Barman, B., B. Kumar, & A. K. Sarma. (2018). "Dynamic characterization of the migration of a mining pit in an alluvial channel." *Int. J. Sediment Res.*, *34*(2), 155–165. https://doi.org/10.1016/j.ijsrc.2018.10.009

Calle, M., P. Alho, & G. Benito. (2017). "Channel dynamics and geomorphic resilience in an ephemeral Mediterranean river affected by gravel mining." *Geomorphology*, *285*, 333–346. https://doi.org/10.1016/j.geomorph.2017.02.026

Choufu, L., S. Abbasi, H. Pourshahbaz, P. Taghvaei, & S. Tfwala. (2019). "Investigation of flow, erosion, and sedimentation pattern around varied groynes under different hydraulic and geometric conditions: A numerical study." *Water (Switzerland)*, *11*(2). https://doi.org/10.3390/w11020235

Haghnazar, H., Y. Sangsefidi, M. Mehraein, & H. Tavakol-Davani. (2020). "Evaluation of infilling and replenishment of river sand mining pits." *Environ. Earth Sci.*, *79*(14), 1–18. https://doi.org/10.1007/s12665-020-09106-z

James, L. A. (1991). "Incision and morphologic evolution of an alluvial channel recovering from hydraulic mining sediment." *Geol. Soc. Am. Bull.*, *103*(6), 723–736. https://doi.org/10.1130/0016-7606(1991)103<0723:IAMEOA>2.3.CO;2

Karami, H., H. Basser, A. Ardeshir, & S. H. Hosseini. (2014). "Verification of numerical study of scour around spur dikes using experimental data." *Water Environ. J.*, *28*(1), 124–134. https://doi.org/10.1111/wej.12019

Kim, C. (2005). "Impact analysis of river aggregate mining on river environment." *KSCE J. Civ. Eng.*, *9*(1), 45–48. https://doi.org/10.1007/bf02829096

Lee, H. Y., & S. C. Chen. (1996). "Migration of rectangular mining pit composed of nonuniform sediments." *J. Chinese Inst. Eng. Ser. A/Chung-Kuo K. Ch'eng Hsuch K'an*, *19*(2), 255–264. https://doi.org/10.1080/02533839.1996.9677785

Lee, H. Y., D.-T. Fu, & M.-H. Song. (1993). "Migration of rectangular mining pit composed of uniform sediments." *J. Hydraul. Eng.*, *119*(1), 64–80.

López, J. L. (2004). "Channel response to gravel mining activities in mountain rivers." *J. Mt. Sci.*, *1*(3), 264–269. https://doi.org/10.1007/bf02919330

Marsh, N. A., A. W. Western, & R. B. Grayson. (2004). "Comparison of Methods for Predicting Incipient Motion for Sand Beds." *J. Hydraul. Eng.*, *130*(7), 616–621. https://doi.org/10.1061/(asce)0733-9429(2004)130:7(616)

Mathias Kondolf, G. (1994). "Geomorphic and environmental effects of instream gravel mining." *Landsc. Urban Plan.*, *28*(2–3), 225–243. https://doi.org/10.1016/0169-2046(94)90010-8

Mehta, D., S. M. Yadav, S. Waikhom, & K. Prajapati. (2020). *Stable Channel Design of Tapi River Using HEC-RAS for Surat Region*. 25–36. https://doi.org/10.1007/978-3-030-38152-3_2

Neyshabouri, S. A. A. S., A. Farhadzadeh, & A. Amini. (2002). "Experimental and field study on mining-pit migration." *Int. J. Sediment Res.*, *17*(4), 323–331.

Padmalal, D., & K. Maya. (2014). "Sand Mining Chapter 2 Rivers Structure and Functions." *Environ. Sci. Eng.*, 9–21. https://doi.org/10.1007/978-94-017-9144-1

Rinaldi, M., B. Wyzga, & N. Surian. (2005). "Sediment mining in alluvial channels: Physical effects and management perspectives." *River Res. Appl.*, *21*(7), 805–828. https://doi.org/10.1002/rra.884

Rovira, A., R. J. Batalla, & M. Sala. (2005). "Response of a river sediment budget after historical gravel mining (the Lower Tordera, NE Spain)." *River Res. Appl.*, *21*(7), 829–847. https://doi.org/10.1002/rra.885

Samadi, A., A. Azizian, & J. Mozaffari. (2018). "A Numerical Study of Movement Pattern of Sand and Gravel Mining Using 1D/2D Models." *J. Quant. Geomorphol. Res. Vol*, *6*(3). http://geomorphologyjournal.ir/article-1-901-en.html%0AA

van Rijn, L. C. (1987). "Mathematical modelling of morphological processes in the case of suspended sediment transport." *Delft, Netherlands, Delft Hydraul. Lab., Jun. 1987*, *382*.

Wu, W., & S. S. Y. Wang. (2008). "Simulation of Morphological Evolution near Sediment Mining Pits Using a 1-D Mixed-Regime Flow and Sediment Transport Model." *World Environ. Water Resour. Congr.*, 1–10.

Yanmaz, A. M., & O. Cicekdag. (2004). "Channel mining induced stream bed instability around bridges." *Watershed Manag. Oper. Manag. 2000*, *105*, 1–8. https://doi.org/10.1061/40499(2000)15

Zolghadr, M., S. M. Ali Zomorodian, R. Sha'bani, & H. M. Azamatulla. (2021). "Migration of sand mining pit in rivers: An experimental, numerical and case study." *Meas. J. Int. Meas. Confed.*, *172*(December 2020), 108944. https://doi.org/10.1016/j.measurement.2020.108944

DOI: https://doi.org/10.26077/8226-d4fe

Scour downstream of a Broad Crested drowned Weir

Rajat Chauhan[1], Rakesh Kumar Chaudhary[2] & Zulfequar Ahmad[1]
[1]*Department of Civil Engineering, Indian Institute of Technology Roorkee, Roorkee, India*
[2] *Institute for Integrated Hydro Environment Research, Kathmandu, Nepal*
E-mail: corresponding, logtokumarrakesh063@gmail.com

***Abstract**: A detrimental problem of scour downstream of a Bagmari Syphon in the feeder canal off taking from Farakka Barrage on the Ganga River, West Bengal, India has kindled our interest in understanding the scour process downstream of a broad crested drowned weir. A river was syphoned across the Feeder canal at Bagmari, flushing the top of the syphon with canal bed. However, the canal bed degraded over the years and resulted in exposure and protrusion of the syphon barrel above the canal bed and acting as a broad crested drowned weir. Due to the acceleration of flow over the barrel, scour has developed downstream of the barrel. This problem has been formulated and studied in the laboratory to study the scour process and its mitigation measures using boulders. Experiments have been carried out in the Hydraulics Laboratory of IIT Roorkee, India in a flume of 12.5 m length, 0.46 m width, 0.70 m depth for the different protruding height of weir and flow intensity, keeping ratio of shear stress of approach flow to critical shear stress less than unity. Study on factors affecting maximum scour depth and its temporal variation were analyzed graphically. The scour depth increases with an increase in protruding height of weir and flow intensity. It has been found that maximum scour reduces significantly using boulder apron. The maximum reduction in scour depth with boulder apron lies in the 44% - 68% range and the range of shift in maximum scour depth from weir with boulder apron was 1.6-1.8 times from the original location of maximum scour depth without boulder apron.*

***Keywords:** Broad crested weir, drowned, flow intensity, riprap, scour*

1. Introduction

Hydraulic structures constructed across a channel is always under a risk of failure due to uncertain hydraulic load. One of the risks associated with such structures is failure due to severe scour in its vicinity. Broad crested weir is one of the widely used hydraulic structures found across a channel as a flow regulating structure. Besides, it is also used for flow measurement. Severe local scour downstream of broad crested weir may become a reason for its failure. It is, therefore, important to have a thorough understanding of the mechanism of local scour downstream of a weir.

Flow over a weir is normally free flow but during extreme flood conditions flow depth downstream of weir may rise above the crest causing the weir to be drowned and resulting in submerged flow conditions. The local scour downstream of a broad crested weir in free flow condition is much larger than submerged flow condition since the flow intensity for free flow condition is higher than submerged flow condition. The local scour downstream of a weir like structures such as bed sills, check dam, grade control structures and submerged weirs have been extensively studied and several empirical equations have been proposed (Bormann and Julien 1991; Gaudio et al. 2000; Lenzi et al. 2002, 2003a, b; D'Agostino and Ferro 2004; Marion et al. 2004; Ben Meftah and Mossa 2006; Marion et al. 2006; Scurlock et al. 2012; Lu et al. 2013; Pagliara and Kurdistani 2013; Guan et al. 2014, 2015, 2016).

Sometimes, water conveying structures have to cross a river, such as pipelines, canal syphons etc. may behave as submerged weir if it gets exposed or flow becomes obstructed due to improper level of the river bed and crossing structures. A similar case is found downstream of Bagmari Syphon in the feeder canal off taking from Farakka Barrage on the Ganga River, West Bengal, India. A river was syphoned across the Feeder canal at Bagmari, flushing the top of the syphon with the canal bed. However, the canal bed degraded over the years and resulted in exposure and protrusion of the syphon barrel above the canal bed and acting as a broad crested drowned weir. The flow when reaches the top of the syphon barrel acting as a broad crested weir, gets retarded and as it passes over the weir, the flow gets accelerated and modified, which results in local scour. If precautions such as protection of the syphon barrel behaving as broad crested weir are not undertaken, the structures will no longer remain stable and leads to failure. Thus, it becomes imperative to have a thorough understanding of the local scour mechanism and its mitigation techniques.

The aforementioned problem was formulated in an experimental flume in order to understand the scour process, factors influencing the scour and its mitigation using boulder riprap under different flow intensities maintaining clear water scour. The results of this study are useful in the mitigation of local scour and the development of a numerical model for estimation of local scour downstream of broad crested weir like structures in the water resources projects.

2. Experimental Set-up and Procedure

The experiments were conducted in a tilting flume with a dimension of 12.5 m length, 0.46 m width, 0.70 m depth with a slope of 3/8000 in the Hydraulics laboratory of Department of Civil Engineering, Indian Institute of Technology Roorkee, India. About 35 cm depth of fine sediment was filled to prepare the bed. The working section of the flume was made up of glass on side walls to facilitate visual observations as shown in Figure 1 (a). At the upstream end of the flume, water was fed into a tank to stabilize the water head through a pipe of 150 mm diameter. A flow straightener made of wood was installed at the inlet of the flume to suppress the turbulence in the flow. A small stretch of channel bed downstream of the flow straightener was filled with stones up to 15 cm thickness for stabilizing the flow. Flow passes over a weir model made of steel sheet and rectangular in shape with certain protrusions as shown in Figures 1(a)-(b). At the downstream end of the flume, a tailgate was installed to maintain desired tailwater level in the flume. There was a sharp-crested weir at the outlet of the flume to measure the discharge. A pointer gauge was used to measure initial and final water level to obtain head (H) over the weir. By measuring the head over the weir and knowing the C_d and width of weir, discharge was calculated using the general formula. The dimension of the broad-crested weir model was width, X=0.45 m, length, B=0.25 m and height, Y=0.35 m height as shown in Figure 1(a)-(b) and Figure 2.

The sand characteristics were: median size d_{50} = 0.30 mm; d_{60} = 0.31 mm, d_{10} = 0.23 mm, geometric standard deviation σ_g = 1.23 and coefficient of uniformity C_u = d_{60}/d_{10} = 1.34 respectively. Figure 2 shows a sketch of experimental parameters in the testing flume.

Figure 1. (a) Photometric view of weir embedded in sand river bed in experimental flume & (b) Schematic diagram of the experimental setup in a flume.

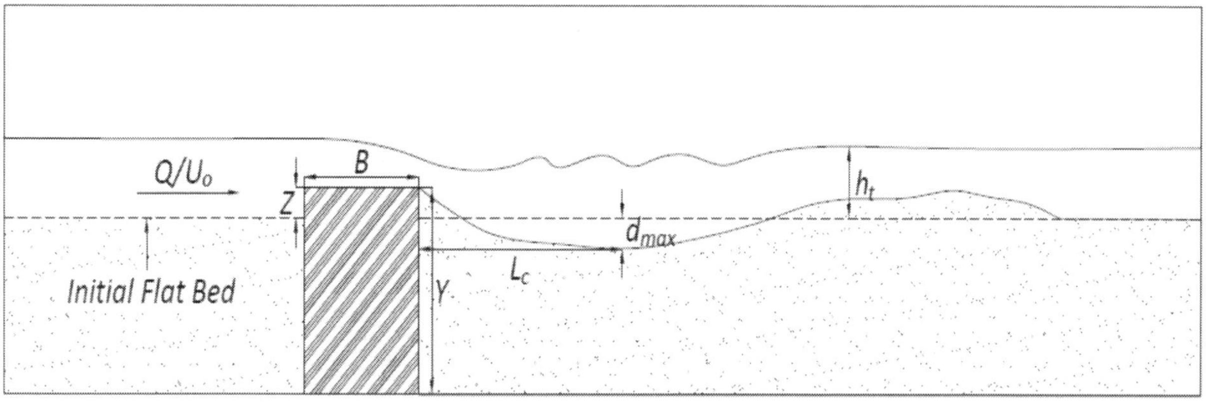

Figure 2. Definition sketch of experimental parameters in a testing flume

Taking all the basic parameters into account, scour depth d_s downstream of the broad-crested drowned weir can be written in a functional form as,

$$d_s = f\ (\rho, \rho_s, \mu, g, h_o, h_t, U_o, d_{50}, \sigma_g, U_c, B, z) \tag{1}$$

where, d_s = scour depth, ρ, ρ_s = densities of water and sediment respectively, μ = dynamic viscosity of water, g = acceleration of gravity, h_o = average approach flow depth, h_t = tailwater depth, U_o = average approach flow velocity, U_c = critical average flow velocity, (d_{50}, σ_g) are the sediment median size and geometric standard deviation of sediment material, B = weir crest width and z = protrusion height of weir.

In this study, sediment size, weir crest width B and fluid properties were kept constant. The approach flow depth h_o depends on weir protrusion height z, tailwater depth h_t and flow intensity U_o/U_c. Since, increase in weir height and approach flow velocity U_o increases flow resistance and discharge, respectively, the approach flow depth h_o also increases, i.e., h_o depends on U_o for constant channel slope. Thus, it is not considered in dimensional analysis. The tailwater depth was in controlled condition. Finally, dimensional analysis results (Guan et al. 2015) as,

$$\frac{d_s}{h_t} = f\ \left(\frac{U_o}{U_c}, \frac{z}{h_t}\right) \tag{2}$$

To analyze the local scour downstream of the weir in clear water conditions, U_o/U_c was kept < 1.0 throughout the experiment. Due to clear water conditions, there was no or minimal scour at the upstream of the structure, hence readings were not taken. The scour development at downstream of the weir was measured with a flat gauge, and readings were taken at center and two edges of the flume at 20 mm intervals along the flume. It usually took 2 hours for scour to reach at maximum scour stage. To analyze the temporal variance, readings were taken at 5 mins, 15 mins, 30 mins, 60 mins and 120 mins. It was observed that scour process attained maximum scour after 2 hours. Approach fow depth h_o, tail water depth h_t, head over weir, H_d were measured with pointer gauge.

For each protruding height of weir (z), discharge was controlled using the valve fitted at the inlet of the flume. Discharge was varied keeping the tail depth constant and the tail depth was controlled with the help of a tail gate provided at the end of the flume. Discharge was controlled in such a way that flow velocity at downstream of the structure, U_o should always be less than critical velocity, U_c corresponding to the approach flow depth, h_o, for this study, flow intensity ranges between $0.25 < U_o/U_c < 0.62$. Critical velocity, U_c is calculated from the equation $U_c/u_{*c} = 5.75\log(5.53\ h_o/d_{50})$, in which the average flow critical shear velocity, u_{*c} = 0.013 m/s was calculated using the Shields diagram for the particle size d_{50} = 0.30 mm. In this study, three different protruding heights of weir were used. i.e., z = 20, 30 and 50 mm and a total of 18 clear water scour tests were carried out. For each weir height, total 6 tests were carried out, 5 on sand bed and one with apron for discharge at which maximum scour was observed.

Sand bed was covered with wooden ply and plastic sheet till tail water depth reached the desired level. Ply and plastic sheet were removed safely to expose sand bed to flow. For each weir height and discharge, the sand bed was leveled to study the scour length and maximum scour depth corresponding to the flow. To analyze the effect of boulder raft/apron on the maximum scour depth and length, boulders of d_{50} = 10 mm were provided up to the 125% of length at which maximum scour is observed. For each weir height., flow conditions were replicated for the discharge value at which maximum scour depth was observed to determine the amount of reduction in scour depth and length in the presence of the apron.

3. Results and Discussion

3.1. Scouring Process

Scour downstream of drowned weir depends upon approach flow intensity and flow regimes (surface flow or impinging jet) (Guan et.al.,2015). Initially, just after the removal of plyboard cover, scour starts rapidly in a vertical and horizontal direction downstream of the submerged weir. The scour hole formed downstream of broad-crested weir exhibits 3D characteristics. The maximum scour hole occurred close to the flume sidewall rather than the center of the flume. This is due to the secondary flow to the transverse section of the flume. This secondary flow is a paired circular flow cells, which are quasi-symmetrically located on both sides of the centerline sand ridge (Guan et.al.,2015). At the beginning, the scour rate was higher for horizontal and vertical directions i.e., scour length and scour depth. Later on, it decreases asymptotically as depicted from Figure 4. Here the flow separation occurs at the end of crest of the weir, later on it rejoins at deepest point of scour hole and finally emerges from scour hole. The mode of sediment transport was suspended load and bed load in the scour hole. Figure 3 depicts two distinct parts, scour hole and dune. The digged material of scour hole gets transported downstream at such a location where there is no sufficient shear stress to carry the bed material further downstream. And this continuous deposition of bed material from scour hole gets piled up and forms dune. Initially, the dune's crest slope was steeper, later with continuous erosion process up to maximum scour depth, it becomes flatter, that can be clearly seen from Figure 3.

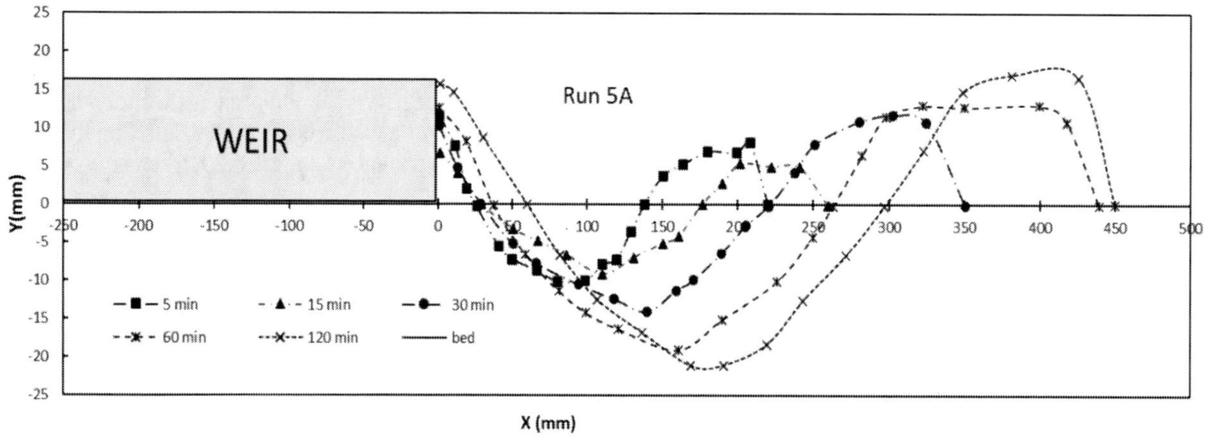

Figure 3. Temporal variation of scour profiles along longitudinal direction for Run 5A

3.2. Temporal variation of maximum scour depth

The temporal evolution of scour depth for all the three protrusions height of weir is shown in Figure 4. This shows that maximum scour occurs rapidly within short time. Later on the rate of scour decreases asymptotically. Here the time versus scour depth curve is divided in to three stages. For first 30 minutes the stage is initial fast stage (stage I) where scour depth increases instantly. It can be observed that at least 75% of maximum scour depth is achieved in Stage-I only. In stage stage-II ≈95% of maximum scour depth is achieved and remaining scour depth is achieved in Stage-III. The maximum scour depth is obtained in 2 hrs.

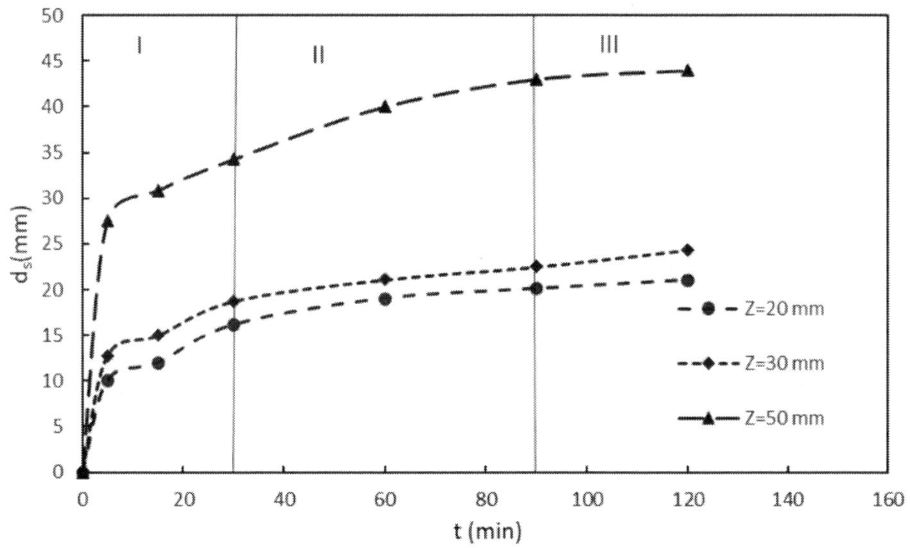

Figure 4. Various stages of scour with respect to time

3.3. Factors affecting maximum scour depth

3.3.1. Flow-Intensity

As in clear-water condition, flow does not bring sediment with it to replenish the scour hole at downstream of the weir, therefore, the relationship between flow intensity and maximum scour depth comes out to be a linear relation. Figure 5 represents a linear relation between d_s/h_t and U_o/U_c. This figure indicates that, for each weir height, the downstream scour depth increases linearly with velocity when $0<U_o/U_c<1$. For clear-water scour, there is no sediment transport upstream of the submerged weir to replenish the downstream scour hole, and the maximum downstream scour depth is determined only by the strength of the overflow. Therefore, for a given weir height and sediment size, an increase in U_o/U_c, which leads to a stronger overflow without bringing more sediment replenishment, results in a deeper downstream scour hole.

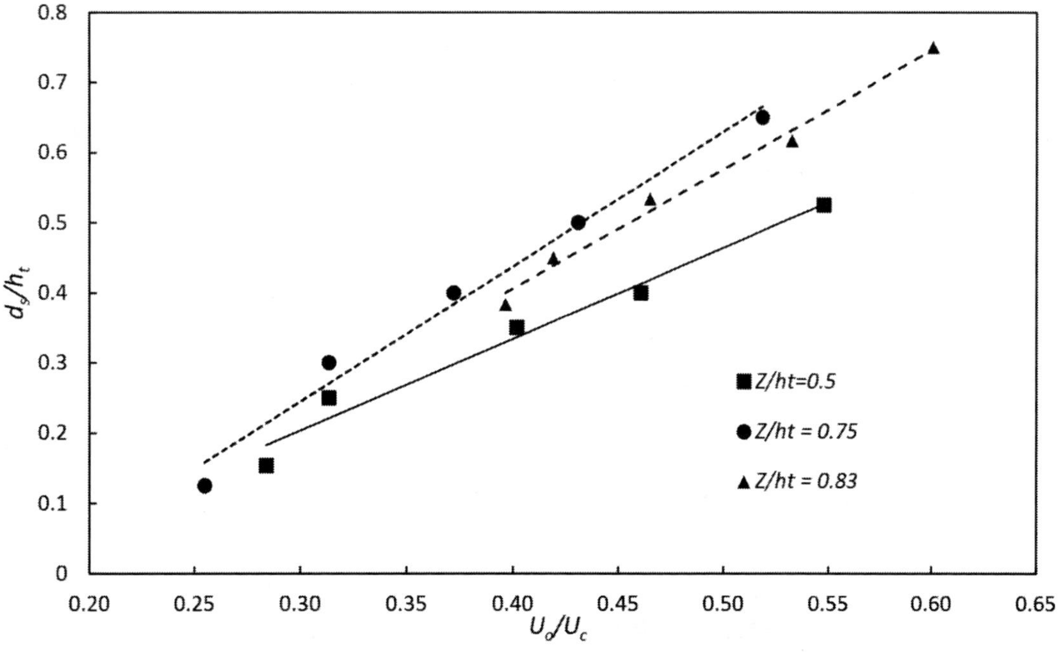

Figure 5. Variation of dimensionless maximum scour depth with flow intensity

3.3.2. Protrusion height of the weir

Figure 6 shows that the maximum downstream clear-water scour depth increases with z/h_t while h_t and U_o/U_c are held constant. For clear-water scour, the upstream sediment bed does not move. With an installed submerged weir, the flow over the weir crest is accelerated, destabilize the downstream sediment bed and generates a scour hole. As the protruding height of weir increases, the velocity head over the weir crest also increases resulting in higher value of approach Froude number. This causes the higher overflow impact on the downstream bed resulting in a deeper downstream scour hole. For zero weir height, no constriction of flow occurs, and the uniform approach flow is not able to move the sediment when $U_o/U_c<1$, resulting in zero scour depth.

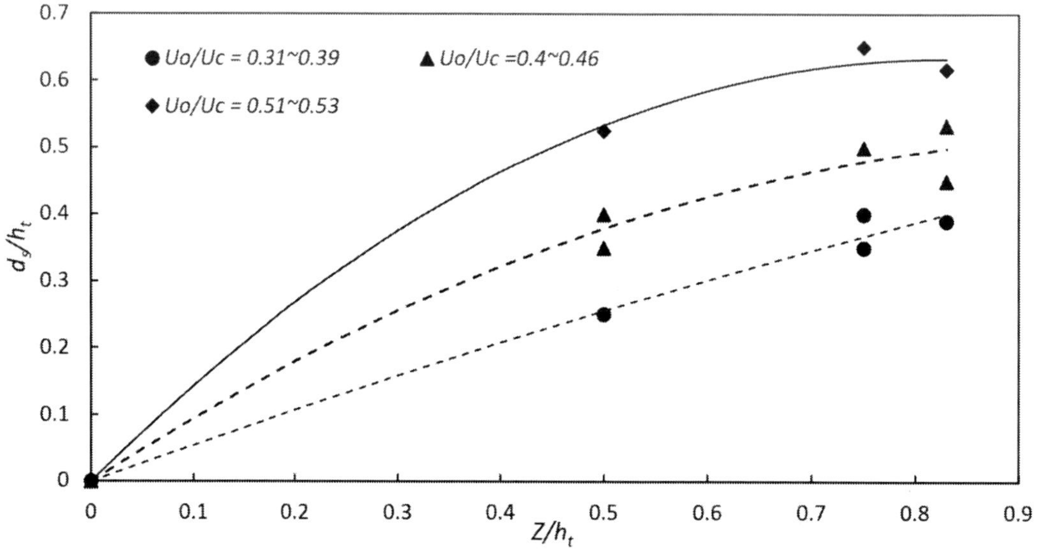

Figure 6. Variation of dimensionless maximum scour depth with dimensionless height of weir

3.4. Scour reduction using boulder apron

An apron of boulders of d_{50} = 10 mm was provided up to 125% of length where maximum scour depth was observed. Test runs were done for same discharge at which maximum scour depths were observed. Figure 7(a) shows the boulder apron placed downstream of the broad crested weir and Figure 7(b) shows scour profile through sidewall at different time interval by using boulder apron. Similarly, Figure 8 represents the comparison of scour pattern with and without apron at same discharge for z = 20mm, 30mm and 50mm, respectively.

Figure 7. (a) boulder apron placed downstream of weir and (b) Maximum scour profile obtained after using apron.

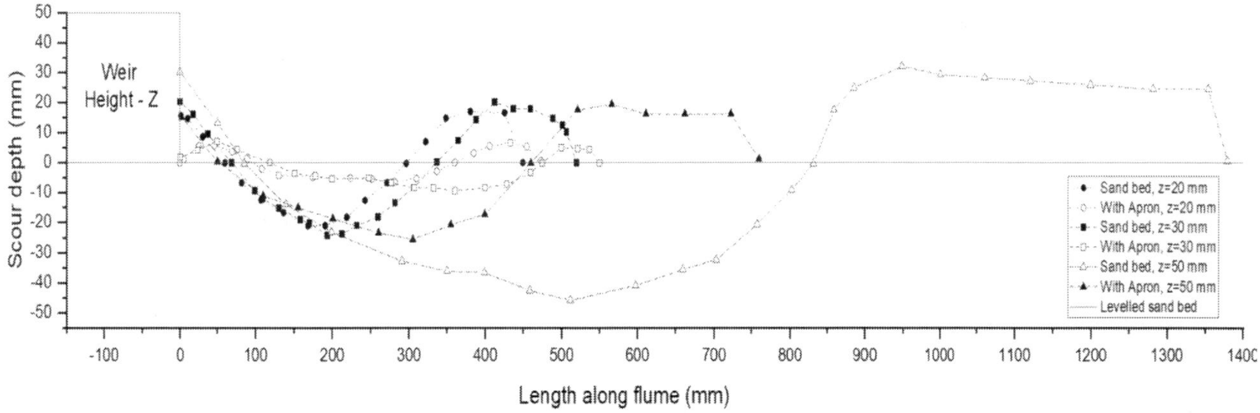

Figure 8. Comparison of scour pattern with and without apron for z =20mm, 30mm and 50mm

3.4.1. Maximum scour depth

With the use of apron excess resistance can be added to prevent scouring of bed sediment. Figure 9 indicates that with use of apron, a significant reduction in maximum scour depth was observed. Maximum scour reduced by 44% in case of z =50 mm and 68% in case of z =20 mm, for z =30 mm it was reduced by 60%.

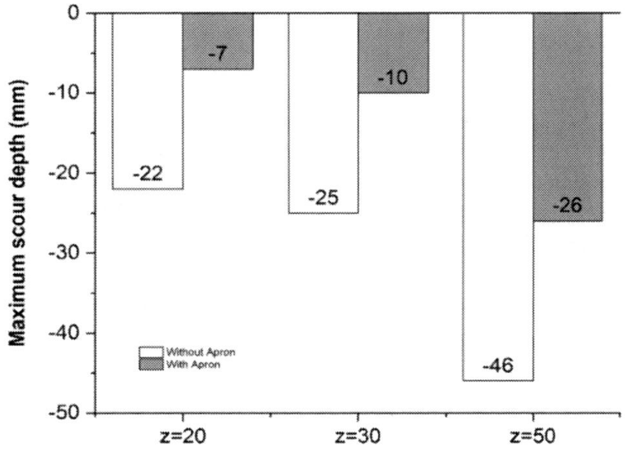

Figure 9. Comparison of reduction in maximum scour depth with use of boulder's apron

3.4.2. Location of maximum scour

The use of boulder apron shifted point of maximum scour depth downstream form toe of the weir. Since the apron roughness placed up to 125 % length of scour moderated the flow intensity resulting reduction in the scour depth. As far as maximum scour depth is concerned from the structure, safer the structure will be from the detrimental problem of scour. For z = 20mm, maximum scour depth was shifted by 1.6 times away from the original location of maximum scour depth without apron and similarly it was 1.8 times for z =30 mm. But it was reduced by 0.6 times for z = 50mm. Figure 10 clearly illustrates the above discussed matter in bar chart format. Thus, the application of boulder apron minimizes the local scour. It can be considered as a good preventive measure since it is successful to reduce scour depth near the structure of weir.

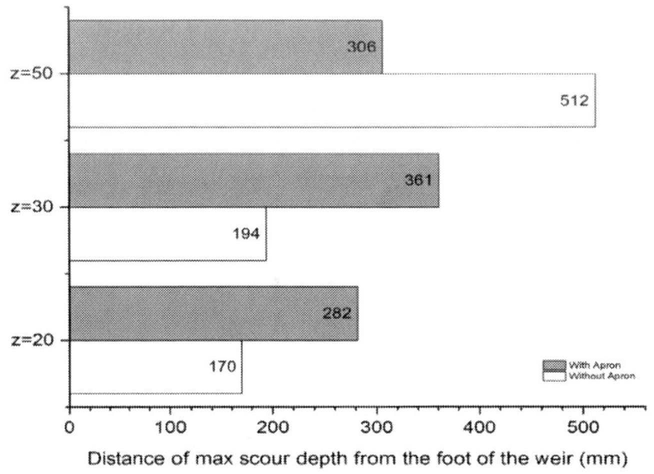

Figure 10. Comparison of location of max scour depth with and without apron

4. Conclusion

Following observations were made from experimental study.

- Maximum scour depth increases proportionally with flow intensity. i.e., maximum scour depth increases with increase of flow intensity (U_o/U_c) for tested range. Similarly, Maximum scour depth increases with increase in protruding height of weir z for three different range of flow intensity (U_o/U_c). At lower range of flow intensity, maximum scour depth increases slowly with increase in protruding height of weir. The rate of maximum scour depth increases with an increase in protruding height of weir at higher range of flow intensity (U_o/U_c).
- The maximum scour reduces significantly using boulder apron. The maximum reduction in scour depth with boulder apron lies in the 44% - 68% range and the range of shift in maximum scour depth from weir with boulder apron was 1.6-1.8 times from the original location of maximum scour depth without boulder apron.
- The results obtained from this study can be used for analyzing the proper length of apron required for protection works of Baghmari syphon of Feeder canal, Farraka Barrage having same tested range of dimensionless parameters.

Further, experimental findings can be substantiated with analytical methods, and two-phase 3D CFD modelling could be done to visualize flow behavior and shear stresses to better understand the phenomenon.

5. REFERENCES

Bormann, N. E., and Julien, P. Y. (1991). "Scour downstream of grade control structures." *J. Hydraul. Eng.*, 117(5), 579–594.

D'Agostino, V., and Ferro, V. (2004). "Scour on alluvial bed downstream of grade-control structures." *J. Hydraul. Eng.*, 131(1), 24–37.

Gaudio, R., Marion, A., and Bovolin, V. (2000). "Morphological effects of bed sills in degrading rivers." *J. Hydraul. Res.*, 38(2), 89–96.

Guan, D., Melville, B. W., and Friedrich, H. (2014). "Flow patterns and turbulence structures in a scour hole downstream of a submerged weir." *J. Hydraul. Eng.*, 140(1), 68–76.

Guan, D., Melville, B. W., and Friedrich, H. (2015). "Live-bed scour at submerged weirs." *J. Hydraul. Eng.*, 04014071.1-11.

Guan, D., Melville, B. W., and Friedrich, H. (2016). "Local scour at submerged weirs in sand-bed channels." *J. Hydraul. Res.*, 54(2), 172-184.

Lenzi, M. A., Marion, A., and Comiti, F. (2003b). "Local scouring at grade-control structures in alluvial mountain rivers." *Water Resour. Res.*, 39(7), 1176.

Lu, J.-Y., Hong, J.-H., Chang, K.-P., and Lu, T.-F. (2013). "Evolution of scouring process downstream of grade-control structures under steady and unsteady flows." *Hydrol. Process.*, 27(19), 2699–2709.

Marion, A., Tregnaghi, M., and Tait, S. (2006). "Sediment supply and local scouring at bed sills in high-gradient streams." *Water Resour. Res.*, 42(6).

Melville, B. W., and Sutherland, A. J., (1988). "Design Methods for Local Scour at Bridge Piers." *J. Hydraul. Eng.*, 114(10)1210-1226

Scurlock, S. M., Thornton, C. I., and Abt, S. R. (2012). "Equilibrium scour downstream of three-dimensional grade-control structures." *J. Hydraul. Eng.*,138(2)167–176.

Pagliara, S., and Kurdistani, S. M. (2013). "Scour downstream of cross-vane structures." *J. Hydro-environ. Res.*, 7(4), 236–242.

DOI: https://doi.org/10.26077/54c1-fd55

Scour Geometry and Dune Formation Characteristics around Customized Structure in Channels

Rajib Das[1], Subhasish Das[1], Michele Palermo[2], Deep Roy[3] & Simone Pagliara[4]
[1]*School of Water Resources Engineering, Jadavpur University, Kolkata, India*
[2]*Department of Energy, Systems, Territory and Construction Engineering, University of Pisa, Pisa, Italy*
[3]*Department of Civil Engineering, National Institute of Technology Tiruchirappally, Tiruchirappally, India*
[4]*Laboratory of Hydraulics, Hydrology and Glaciology (VAW), ETH Zurich, Zurich, Switzerland*
E-mail: rajib.das@jadavpuruniversity.in

Abstract: *In this study, an attempt is made to verify the scour geometry and location of the dunes at a channel downstream where various structures are used as flow barriers. Various structures such as a square structure with a vane attached to one side, a rhombus and two similar structures positioning eccentrically, are used to examine scour geometry and dune formation. Overall, six Clearwater tests are performed, three of which are of single structures and three are a combination of two similar structures. All tests are performed keeping approach flow depth, discharge and bed material fixed. The main focus of the study is to assess the scour on both sides of such structures, relocating sand downstream and their volume. The scour depth, length, planar area and volume are measured and calculated. It is observed that the scour volume and relocating of sediments are significantly more around double structure arrangement. Also in double structure tests, sediment relocation is much greater on the eccentric structure side. As a result, on the eccentric structure side, the scour planner area and the angle of sediment relocation are significantly higher compared to the non-eccentric side. To examine further, the scour hole characteristics and the volumes of sediment relocation are computed from observed data.*

Keywords: *eccentricity; scour depth; sediment relocation; square structure.*

1. Introduction

Sediment transport in rivers is a very important phenomenon and attracts engineers, researchers, and scientists all over the world. Silt deposition and scouring phenomenon lead to the interrelationship between flowing water and bed sediment materials. The channel depth reduces significantly due to sediment deposition at the bed of waterways, which makes navigation difficult and has become a subject of more attention to hydraulic engineers. Vortices developed in front of an unsubmerged obstruction or structures, exert shear stress on bed material. This phenomenon leads to scouring around structures.

Many laboratory-based clear-water scour tests were carried out for forecasting scour geometry. Most of the previous studies have focused on the lowering of scouring volume. Due to the complexity of the nature of scour process, many formulae and mathematical models were derived/developed for obtaining the equilibrium depth of scouring for a certain range of testing conditions (e.g., structure sizes, flow characteristics, scouring duration, and sand material properties). The presence of isolated structures induces a significant increase in local scour around their bases and sides (Chang et al. 2013). Under clear-water condition, the scour depth is reached asymptotically with respect to time and usually takes a few days to attain an equilibrium state. Das et al. (2017) explained that channel depths may reduce considerably for sand deposition and carried out tests on the scour geometry study.

Several studies were performed around multiple structures focusing on predicting equilibrium scour depth (Ashtiani et al. 2010) and the subsequent influence of intermediate gap between structures. Sheppard et al. (1995) conducted two structure arrangements positioning structures flow-wise and described two additional parameters (a) reinforcement and (b) sheltering effect scouring. Few researchers carried out some measurements around double structure arrangements focusing on scour geometry. Das and Mazumdar (2015a-b, 2018), Das et al. (2014a, 2016, 2018), Jaman et al. (2017a,b), and Narayana et al. (2020) observed scour depth variations using multiple, identical structures. The consequences of the longitudinal pier-like structure gap on flow characteristics and scour depth were described by Michael et al. (1991). Das et al. (2014b) explained how inline eccentric pier-like structures may enhance sand transportation towards the channel bank and structure downstream as well.

It is a known fact that removing the sediments and debris by mechanical means from the bed of an alluvial channel is very expensive. Formerly, several studies have been performed to reduce the scouring. But very few researchers have designated the insightful prospect of shifting the bed sediment load towards the channel bank by increasing the scouring process. It ultimately helps in dredging out the deposited sediments without the help of the traditional dredging mechanism.

In this study, it has been checked whether sand load could be diverted and transported to a particular side of flow obstructing structures. The aim is of increasing non-symmetrical scouring and shifting more sand downstream toward the bank. It may help to dredge out the materials from the bed or the bank of the waterway to deepen or widen navigation channels. So, if sand could be diverted and transported towards the bank, it will help in maintaining the constant depth of that channel. As the main aim is to shift the sediments towards the bank, hence different structures which are not commonly used are tested here. The purpose of these structures is different from bridge piers. It is investigated to find the most suitable structures or structure combinations to get maximum sediment shifting towards the bank. In this work, the researchers have focused on different shaped single and double structures. In total six tests were performed, out of which three tests are single structures and three are double structures.

2. Test Setup

In the present work, six tests have been carried out to examine the nature of the scour, and dune formation downstream of structures and to check how far the sediments are shifted towards the bank. The nature of scour geometry, and the nature of sediment shift towards the bank is compared among different structures and structure combinations. Different structures used are a rhombus and a square with a vane attached on one side.

Test 1 was carried out with a rhombus. A single structure was used in this test. A picture after the occurrence of scour is shown in Figure 1(a).

Test 2 was carried out with a square structure with a vane attached to one side. The vane is attached to the upstream face of the square, at an angle of 45^0 and pointing to a direction opposite to the flow. A picture after the occurrence of scour is shown in Figure 1(b).

Test 3 was also carried out with a square structure with a vane attached to one side. In this case, the vane is attached to the downstream face of the square, at an angle of 45^0 and pointing towards the direction of flow. A picture after the occurrence of scour is shown in Figure 1(c).

Test 4 was carried out with two rhombus structures, with the downstream rhombus placed in an eccentric position. A picture after the occurrence of scour is shown in Figure 1(d).

Test 5 was also carried out with two rhombus structures, but in this case, the alignment of the downstream rhombus was made opposite. A picture after the occurrence of scour is shown in Figure 1(e).

Test 6 was carried out with a combination of two square structures with a vane attached to one side. The vane in both structures is attached at an angle of 45^0 and points towards the direction of flow. In the upstream structure, the vane is attached to the upstream face of the square and in the downstream eccentric structure, the vane is attached to the downstream face of the square. A picture after the occurrence of scour is shown in Figure 1(f).

For all tests, the obstructing structures were considered to have a constant side width (b) of 0.070 m. The thickness of the vanes attached to the square structure was 0.005 m and the length of the vanes was taken 0.07 m. In the case of two structure tests, the longitudinal spacing between the structures was taken as $0.75L_{ss}$, where L_{ss} is the combined scour and dune length for single structure tests. The eccentricity in the case of two structure tests was kept as $3b$, with b indicating the side width of the structure. Earlier rectangular vanes were used by Ouyang and Lin (2016). A triangular shape was used by Fouli and Elsebaie (2016).

Each test was started with a flattened bed (Figure 2a). The geometric standard deviation of the selected sand was set at 1.8. The slope of the prepared sand-bed was fixed at 0.00042 for all six tests. The threshold condition was fixed at 68.2% following Das et al. (2013a-b). Incoming flow depth (h) was set constant at 0.125 m for all tests by operating the tailgate, and the discharge of the approaching flow (Q) of 0.025 m^3/s. The ratio of approaching flow depth with the side width of structures (h/b) was kept equal to 1.8 to obtain restricted flow in the rectangular channel. Usually, it is known that inflow Froude number and sediment particle size are significant parameters for evaluating scour around structures. In the present study, the discharge, inflow depth and sediment particle size were kept the same for all six tests, hence the effect of Froude number change and sediment particle was not determined. Before starting the test, the outlet valve was closed, and the flume was filled with water very slowly without disturbing the bed material until the desired flow depth was reached. Then the inflow valve was adjusted to the desired discharge in the flume and the outlet valve was partially opened. The rate of scour was found at its peak in the period of the first three hours and then the scour rate was less. The test conditions of all six tests are noted in table 1. A picture of a test is shown in Figure. 2.

Table 1. Testing conditions.

Test number	Structure shape and number of structure	Flow (Q) (m^3/s)	Flow depth (h) (m)	Structure width (b) (m)	Vanes	Run time (h)
Test 1 (S1)	Rhombus, single	0.025	0.125	0.07	No vane	29
Test 2 (S2)	Square with vane, single	0.025	0.125	0.07	Vane, upstream face	36
Test 3 (S3)	Square with vane, single	0.025	0.125	0.07	Vane, downstream face	35
Test 4 (DS1)	Rhombus, double	0.025	0.125	0.07	No vane	40
Test 5 (DS2)	Rhombus, double	0.025	0.125	0.07	No vane	40
Test 6 (DS3)	Square with vane, double	0.025	0.125	0.07	With vane	40

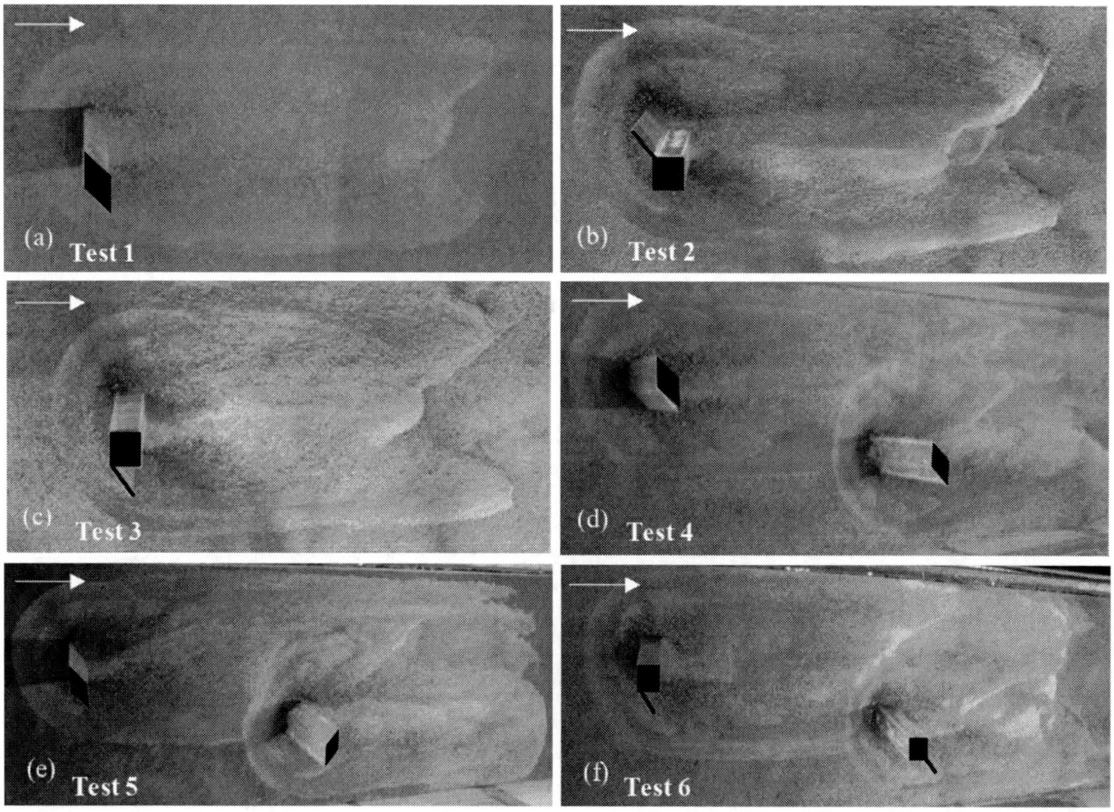

Figure 1. Pictures of six structural arrangements showing scour bed after the test (arrows indicate the flow direction)

Each test run duration was varied from 29 to 42 hours. After the test run, the pump was allowed to stop draining out the water from the working section without modifying the scour topography. After that, testing flume bed sand was allowed for drying, and then the photo of scouring topography around the structures was taken. Then bed morphology was measured using electronic gauges (laser type) fixed with the moving trolley.

It was further investigated whether the scour is symmetric or non-symmetric due to the addition of the vane at the side of the structure, and in addition to it, the formation of the dune and its alignment downstream of the structure was also studied. If it is non-symmetric then it has to be found out on which side of the structure the scour is higher. Comparison of the different scour parameters (equilibrium depth, length of scouring, maximum width of scouring, total scour area, total scour volume and both side volume, angle of shift) is to be carried out with different structures and combinations.

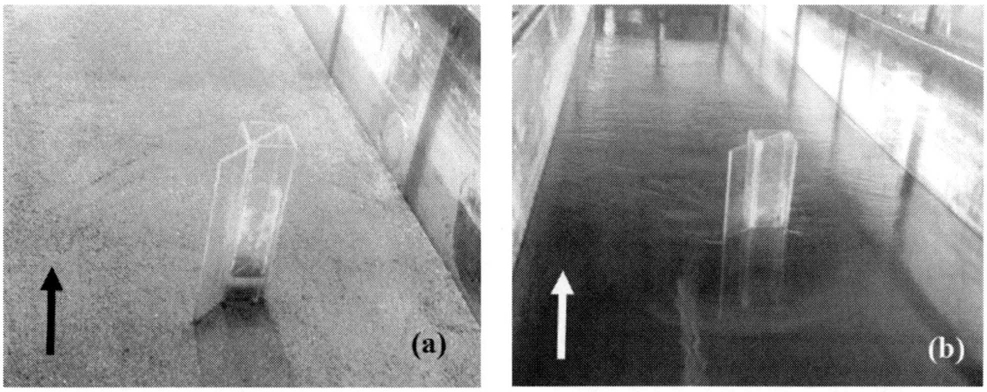

Figure 2. Pictures of a test (a) with original flat-bed and (b) during the experimental run.

(arrows indicate the flow direction)

3. Results and Discussions

A study of local scour around structures in clear-water conditions was carried out. Figures 3-8 refer to contours at locations of scouring, shifting and depositions. Here, the negative contour values indicate the scouring area and the positive contour values indicate dune formation (sand deposition). For comparison of contours here contour plots of all six tests have been shown. It can be observed that the scour and the dune profile is non-symmetric. From beginning to end of six tests, scouring patterns and sand shifting was observed for each arrangement. It is evident from contour plots that the scour and dune profile on the left and right sides of the structure is different. Both the scour and dune on the right side of the structure are more prominent. It shows that the sand deposition is happening towards one side of the structure.

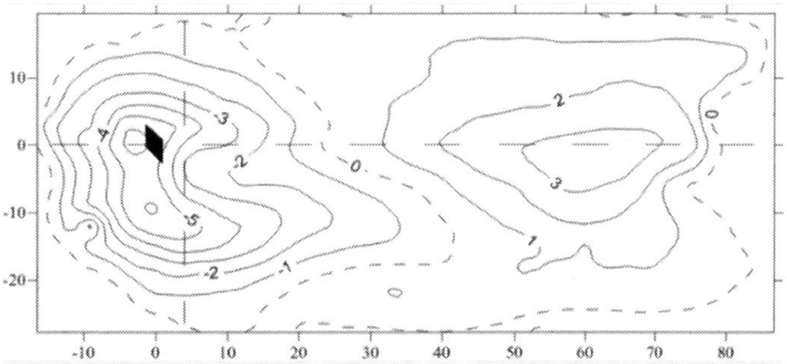

Figure 3. Contour map showing scour for Test 1 (S1), (units are in cm)

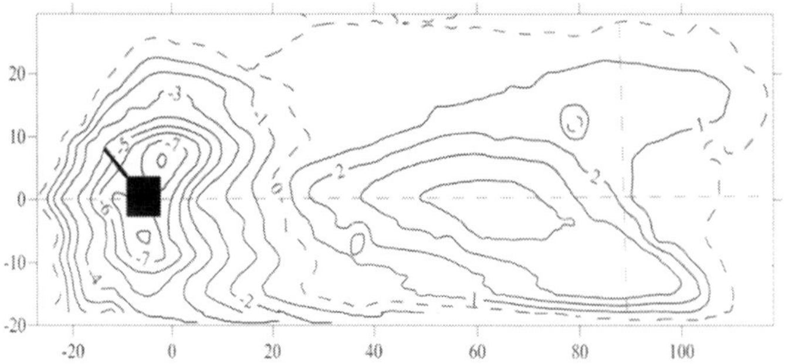

Figure 4. Contour map showing scour for Test 2 (S2), (units are in cm)

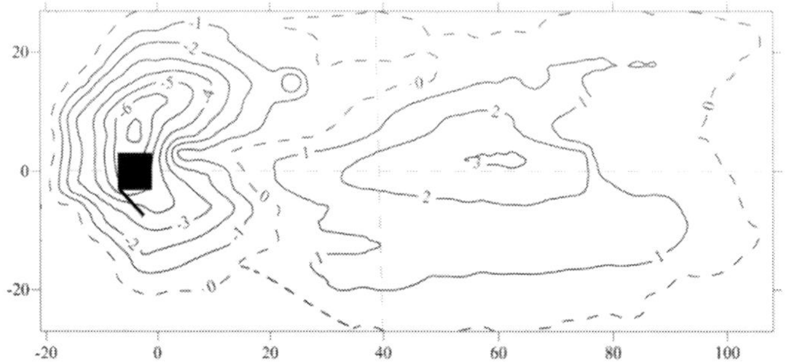

Figure 5. Contour map showing scour for Test 3 (S3), (units are in cm)

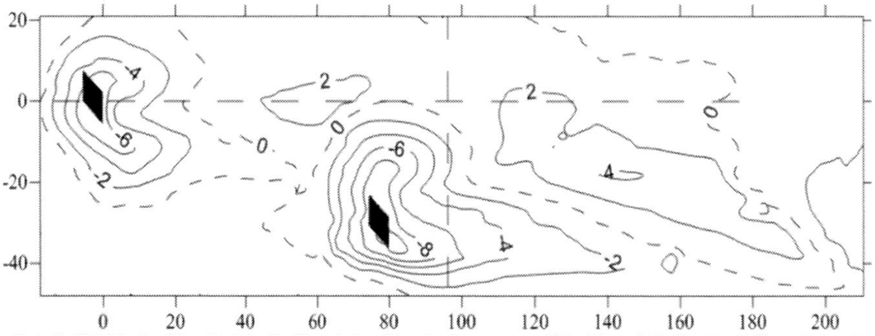

Figure 6. Contour map showing scour for Test 4 (DS1), (units are in cm)

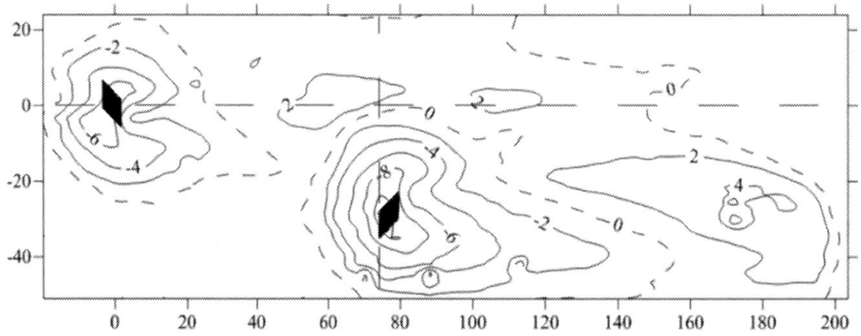

Figure 7. Contour map showing scour for Test 5 (DS2), (units are in cm)

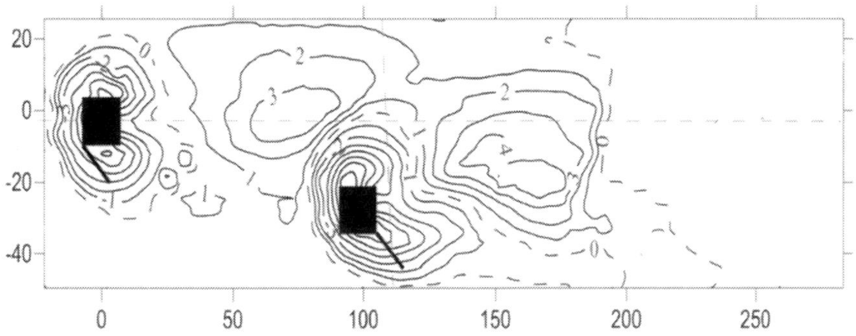

Figure 8. Contour map showing scour for Test 6 (DS3), (units are in cm)

Figure. 9 shows a picture of the equilibrium bed morphology after the second test run. A symmetry line (shown by the dashed line in Figure. 9) perpendicular to the face of the square structure is included. It separates the left and right zones. According to Figure. 9, the left-side zone is above the symmetry line and the left side with respect to flow direction as well. Similarly, the right-side zone is below the symmetry line and the right side with respect to flow direction. The symmetry line also shows the angle of the shift towards right and left banks respecting flow direction. Here it is evident that the angle of shift on the right side is more compared to the left side of the structure.

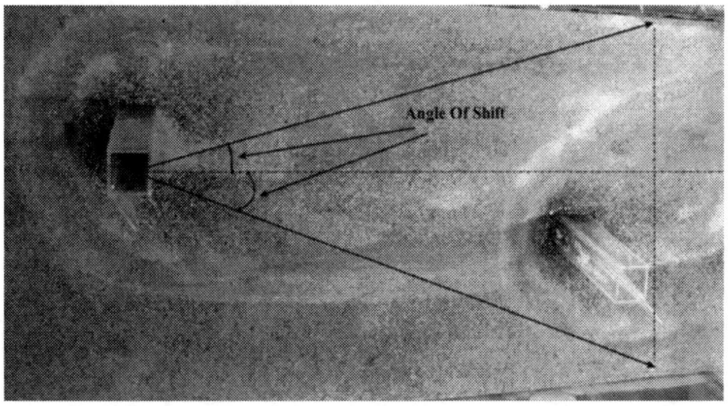

Figure 9. Scour affected zone around the structure for Test 6 (DS3). Angle of shift of scour and dune is shown

After conducting all six tests, geometrical parameters like equilibrium scour depth, length of scour hole, maximum width of scouring, scour hole surface area, the total volume of scouring, scour volume of both sides (left & right) and shifting angle on both sides of the sediment on the left and right side of the channel were analyzed. The morphometric consequences of different shaped structural arrangements on scouring were investigated here. Shifting and deposition patterns were studied. The test results of parameters left and right sides of structures are furnished in Table 2.It can be seen from Table 2 that the angle of the shift of scour and dune is significantly different on the left and right sides of the structure used for tests in this study. Except for tests 2 and 3, in the case of all other tests, the angle of shift is more on the right side of the structure. The same phenomena is observed for scour area and also scour volume when compared between the left and right sides of the structure. The shift on the right side for all parameters like scour area and volume is significantly more for double structure tests.

The results of Table 2 are represented in graphical form in Figure. 10, and it is observed that the sand scour around the double structure was significantly increased. From the test results furnished, the morphology of the scour was non-symmetric at the left and right end of the structure considering from upstream end and both sides of the structure. From the contours, Table 2 and graphical representation (Figure 10) it is evident that the sediment is deposited more on one side of the channel i.e. the dune formed downstream of the structures is not symmetrical. The diversion angle of the scour and dune is more on the right side of the structure. This is significant as it proves that the transported sediment load in a channel can be shifted towards the bank with the help of a suitable structure combination.

Table 2. Scour symmetry analysis (left and right of the structure)

Test No.	Angle of shift (left side) α	Angle of shift (right side) β	Scour area, left side (A_{sl})	Scour area, right side (A_{sr})	Scour volume, left side (V_{sl})	Scour volume, right side (V_{sr})
	Degree	Degree	m^2	m^2	m^3	m^3
Test 1 (S1)	10.53	15.79	0.07316	0.133287	0.0013097	0.00227349
Test 2 (S2)	15.79	12.53	0.160542	0.132173	0.0026602	0.00343308
Test 3 (S3)	14.7	11.5	0.16266	0.077115	0.0027536	0.00126582
Test 4 (DS1)	12.43	25.94	0.19416	0.633199	0.0022374	0.0152374
Test 5 (DS2)	13.61	27.63	0.224943	0.624306	0.00291588	0.01533337
Test 6 (DS3)	14.28	22.69	0.129468	0.766747	0.002656854	0.01342381

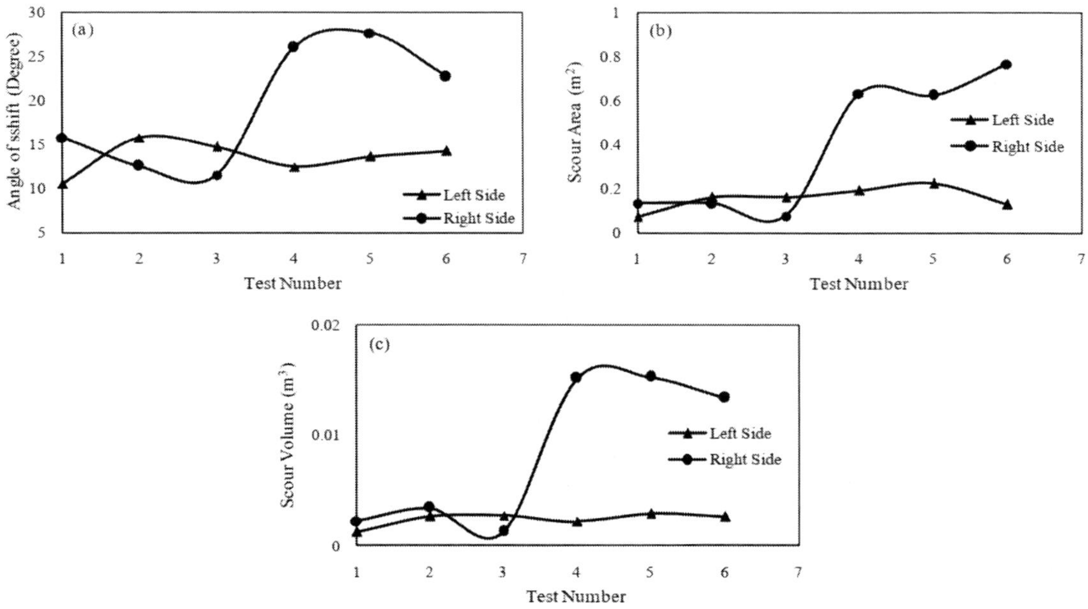

Figure 10. Scour affected zone around the structure for Test 2 (S2). Angle of shift of scour and dune is shown

The geometrical parameters after the test run like equilibrium scour depth, length of scour hole, maximum width of scouring, scour hole surface area, and the total volume of scouring are shown in Table 3. From the test results shown in Table 3, we can see that no significant variation of scour depth is noted for different structures only thing that can be noted is that scour depth for double structure is little more than the single structure. More difference is noted in the case of other parameters such as scour length, width, area and volume. It is noted that in the case of double structure the scour length, area and volume are significantly more for double structure compared to the single structure. Figure 11 (a-d) shows the changes in total scour depth, width, length, area and volume around the structure for different tests.

The sand is removed from the channel bed from around the structures and is deposited at the channel downstream. The determined results clearly show more scouring in the double structure than that for a single structure. From Tables 2, 3 and Figures 9, 10 and 11, a clear and interesting phenomenon is observed. It is observed that scour and dune formation is more on one side of the structure and in the case of the double structure arrangement, the quantity of scour and dune is significantly more and especially on the side where the downstream structure is located. This phenomenon shows how by using structures sediments can be deposited or removed to the side of the channel.

Table 3. Test Results.

Test No.	Maximum Scour Depth	Maximum Scour length	Maximum Scour width	Maximum Scour area	Maximum Scour volume
	m	m	m	m^2	m^3
Test 1 (S1)	0.04428	0.565	0.475	0.207986	0.00362511
Test 2 (S2)	0.04482	0.738	0.556	0.316396	0.0056935
Test 3 (S3)	0.04215	0.698	0.535	0.299244	0.00451944
Test 4 (DS1)	0.04655	1.835	0.756	0.782512	0.01737519
Test 5 (DS2)	0.04638	1.784	0.782	0.854833	0.01830355
Test 6 (DS3)	0.04732	1.956	0.827	1.287618	0.016079868

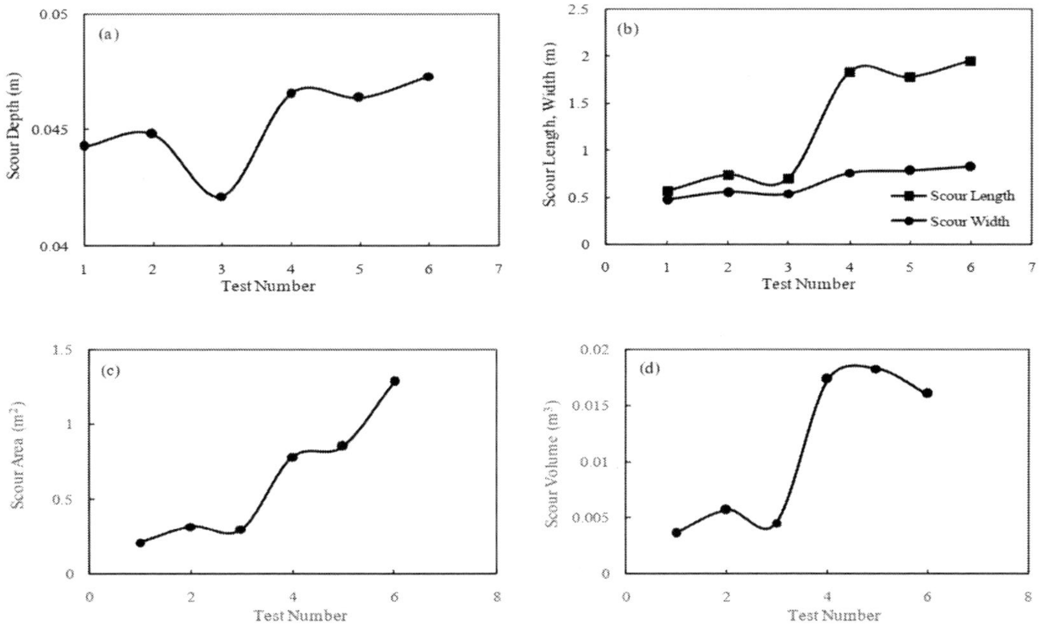

Figure 11. Scour depth, width, length, area and volume variation for different tests

5. Conclusions

In the present study, scour hole characteristics on clear water scour tests have been performed on various structures such as rhombus, square with a vane attached to one side and double structure combination of two rhombi and two squares with a vane attached. Six tests were conducted with constant inflow depth, discharge and uniformly graded sand-bed material maintained in a rectangular flume. In all cases, equilibrium scour depths around such structures were gauged to prepare contour maps, scale measurements of stable beds, and quantify sand scour. From the present laboratory investigation, the following conclusions are drawn:

(i) Using the single rhombus structure, the shifting angle on the right side is found more compared to the left side. However, using the square structure with the vane, the shifting angle on the right side compared to the left side is not found prominent.

(ii) Although, the diversion angle of the scour and the dune is more for the single rhombus structure but for the double structure arrangement this phenomenon is significant and it shows that a large amount of transported sand is shifted towards the right bank.

(iii) Shifting sediment volume is higher for the double structure and also the angle of shift is larger as compared to the single structure.

These outcomes are significant because they provide a suitable structural arrangement for self-dredging of the transported sand load shifted towards the sidewall or bank. It can also be extended into various shapes and geometry of structures with various arrangements and dimensions that can be useful compared to real field problems.

6. REFERENCES

Ashtiani, B.A., Ghorghi, Z.B. and Beheshti, A.A. (2010). "Experimental investigation of clear-water local scour ofcompound piers." *J.Hydraul. Eng.*, 136(6), 343–351.

Chang, W.Y., Constantinescu, G., Lien, H.C., Tsai, W.F., Lai, J.S. and Loh, C.H. (2013). "Flow structure around bridge piers of varying geometrical complexity." *J.Hydraul. Eng.*, 139(8), 812-826.

Das, R., Das, S., Jaman, H. and Mazumdar, A. (2018). "Impact of Upstream Bridge Pier on the Scouring Around Adjacent Downstream Bridge Pier." *Arab. J. Sci. Eng.*, 44(5), 4359–4372.

Das, R., Khwairakpam, P., Das, S. and Mazumdar, A. (2014b). "Clear-Water Local Scour around Eccentric Multiple Piers to shift the Line of Sediment Deposition." *Asian J. Water, Environ.Pollut.*, 11(3), 47-54.

Das, S., Das, R. and Mazumdar, A. (2013b). "Circulation characteristics of horseshoe vortex in the scour region around circular piers." *Water Sci. Eng.*, 6(1), 59-77.

Das, S., Das, R. and Mazumdar, A. (2014a). "Variations of clear water scour geometry at piers of different effective widths." *Turkish J. Eng. Environ. Sci.*, 38(1), 97-111.

Das, S. and Mazumdar, A. (2015a). "Comparison of Kinematics of Horseshoe Vortex at a Flat Plate and Different Shaped Piers." *Int. J. Fluid Mech. Res.*, 42(5), 418-448.

Das, S. and Mazumdar, A. (2015b). "Turbulence flow field around two eccentric circular piers in scour hole." *Int. J. River Basin Manag.*, 13(3), 343–361.

Das, S. and Mazumdar, A. (2018). "Evaluation of Hydrodynamic Consequences for Horseshoe Vortex System developing around two eccentrically arranged Identical Piers of Diverse Shapes." *KSCE J. Civ. Eng.*, 22(7), 2300-2314.

Das, S., Das, R. and Mazumdar, A. (2016). "Comparison of Local Scour Characteristics around Two Eccentric Piers of Different Shapes." *Arab. J. Sci. Eng.*, 41(4), 1199-1213.

Das, S., Das, R.,Mukherjee, R. and Mazumdar, A. (2017). "Enhancement of sediment transportation by increasing scour around a square pier with vane attached on one side." *Proc., 37th IAHR World Congress*, Kuala Lumpur, Malaysia.

Das, S., Midya, R., Das, R. and Mazumdar, A. (2013a), "A Study of Wake Vortex in the Scour Region around a Circular Pier." *Int. J. Fluid Mech. Res.*, 40, (1), 42-59.

Fouli, H. and Elsebaie, I.H. (2016). "Reducing local scour at bridge piers using an upstream subsidiary triangular pillar." *Arab J. of Geosci.*, 9, 598.

Jaman, H., Das, S., Das, R. and Mazumdar, A. (2017a). "Hydrodynamics of Flow Obstructed by Inline and Eccentrically-Arranged Circular Piers on a Horizontal Bed Surface." *J. Inst. Eng. (India): Ser. A*,98(1-2), 77-83.

Jaman, H., Das, S., Kuila, A. and Mazumdar, A. (2017b). "Hydrodynamic flow patterns around three inline eccentrically arranged circular piers." *Arab. J. Sci. Eng.*,42(9), 3973-3990.

Michael, S.A., Mohamed, G.M., Mohamed, S.B.A.M. (1991). "Wake vortex scour at bridge piers." *J.Hydraul. Eng.*, 117(7), 891–904.

Narayana, P.L., Timbadiya P.V. and Patel, P.L. (2020). "Bed level variations around submerged tandem bridge piers in sand beds." *ISH J.Hydraul. Eng.*, 28(1), 149-157.

Ouyang, H.T. and Lin, C.P. (2016). "Characteristics of interactions among a row of submerged vanes in various shapes." *J. Hydro-Environ. Res.*, 13: 14-25.

Sheppard, D.M., Zhao, G. and Copps, T.H. (1995). "Local scour near multiple pile piers in steady currents." *Proc.,International Conference on Water Resources Engineering*, ASCE, San Antonio, USA.

DOI: https://doi.org/10.26077/6b84-106d

SPH Simulation of Non-Aerated Flow over Smooth Invert, Converging Spillways

J.D. Nóbrega[1], J. Matos[2], H.E. Schulz[3] & R.B. Canelas[4]
[1]*Federal University of Goiás, Goiânia, Brazil*
[2]*CERIS, Instituto Superior Técnico, Universidade de Lisboa, Lisbon, Portugal*
[3]*School of Engineering of São Carlos, University of São Paulo, São Paulo, Brazil. Hydro Engineering Solutions, Auburn, Alabama, United States*
[4]*Bentley Systems, Lisbon, Portugal*
E-mail: juliana.nobrega@ufg.br

Abstract: *Spillways may require converging sidewalls because of site or economy constraints. The adverse effects of the converging walls on supercritical flows include the formation of standing waves. In this context, numerical simulations were carried out using the Smoothed Particle Hydrodynamics (SPH) method. The numerical results were compared with experimental data acquired on a spillway model composed of a broad crested weir followed by a 1V:2H sloping chute with wall convergence angle of 9.9° or 19.3°, for a range of discharges. The flow depths at the spillway centerline were reasonably well predicted by the numerical simulations, except on the transition from the weir to the chute. Experimental data and numerical results showed that the sidewall flow depth normalized by the centerline flow depth (uninfluenced by the converging wall), tended to increase along the spillway, reaching maximum values of approximately 2 and 5, for wall convergence angles of 9.9° and 19.3°, respectively. Numerical cross-sectional flow depth profiles were also obtained, showing distinct shapes of the standing waves, depending on the wall convergence angle and position along the spillway. Overall, the simulated development of the standing wave width compared well with the experimental counterparts. The experimental velocity profiles obtained at the spillway centerline with wall convergence angle of 9.9° were also compared with those obtained numerically and the results revealed a fairly good prediction, except near the upstream end of the chute and close to the invert.*

Keywords: *Smoothed Particle Hydrodynamics, spillway, standing wave, supercritical flow, wall convergence.*

1. Introduction

A wall deflection in supercritical flows causes an oblique hydraulic jump, resulting in the abrupt rise of the flow depth in the shock front and its propagation along the channel (Ippen and Harleman 1956). The oblique hydraulic jump or standing wave is also known as shockwave, by analogy to supersonic flows of gases (Chow 1959). Oblique standing waves are typically found on chute spillways with converging transitions. Converging walls might be required because of site and economy constraints when designing new dam spillways, or in dam rehabilitation projects by increasing its crest width (USBR 1987, Hunt et al. 2008, Moran et al. 2021). However, concerns arise because of the possible flow overtopping on the training walls, generation of cross-waves and nonuniformity of the flow approaching the stilling basin.

A number of studies has been focused on the standing waves imposed by wall deflections, to avoid the flow overtopping of the training walls, and to reduce or eliminate the standing waves. Theoretical and experimental studies on horizontal channels with converging walls have been developed to derive fundamental relationships and to evaluate the influence of the various parameters on the standing waves features, such as the approach Froude number, wall convergence angles and types of channel contractions (Dawson 1943, Stone 1946, Ippen and Dawson 1951, Ippen and Harleman 1956, Hager 1992, Hager et al. 1994, Reinauer and Hager 1996, 1998). Investigations addressing inclined chute contractions were also carried out, as presented by Neilson (1976), Reinauer and Hager (1998), and Jan et al. (2009), determining also the role of the bottom slope on the flow patterns of the standing waves.

Numerical approaches were carried out using various numerical methods for horizontal channels with converging walls (Stone 1946, Hager et al. 1994, Causon et al. 1999, Song et al. 1999, Abdo et al. 2018, Salazar et al. 2020). However, fewer numerical studies involving sloping converging channels are found in the literature (Nunes et al. 2018, Salazar et al. 2020). Recently, Salazar et al. (2020) applied the Lagrangian particle finite element method (PFEM) to simulate the flow over two dam spillways, showing its ability to address problems involving high irregularities in free surface. Applications using the Smoothed Particle Hydrodynamics (SPH) method to study the influence of chute contractions in the flow features were not found in the literature. Thus, the present study aimed to perform SPH simulations of the main flow properties over converging spillways on a typical slope of embankment dams, preceded by a broad crested weir, and to compare the results with experimental data.

2. Methodology

2.1. Experimental setup

Experimental tests were carried out in an experimental facility assembled at the Hydraulic and Water Resources Laboratory of the Instituto Superior Técnico, University of Lisbon, in the framework of André and Ramos (2003) and Cabrita (2007). The facility consisted of an 8.0 m long and 0.7 m wide rectangular flume with Plexiglas walls, with a water recirculation system. The spillway model comprised a 0.5 m high and 0.5 m long uncontrolled broad crested weir, followed by a smooth chute with a slope of 1V:2H (chute angle from the horizontal θ = 26.6°) and a stilling basin. The upstream width of the chute was W_0 = 0.7 m, linearly reducing to the downstream width of W_1 = 0.35 m (width ratio of W_r = 0.5), by means of: i) one converging wall of ϕ = 19.3° (André and Ramos 2003); ii) two converging walls of ϕ = 9.9° (Cabrita 2007).

Experiments were carried out for the discharges (Q) of 35, 42, 49 and 56 L/s. These values were selected in order to obtain approach unit discharges (q) of 50, 60, 70 and 80 L/(s·m), relative to the upstream chute width W_0 = 0.7 m. The flow depths and velocities were determined at twenty equally spaced cross-sections perpendicular to the chute invert. The flow depths at the spillway centerline or near the non-converging wall were measured using a point gauge, with 0.05 mm resolution, whereas the flow depths at the flume walls were obtained using graduated rulers positioned perpendicularly to the bottom. To determine the flow velocities, measurements were taken using a Prandtl-Pitot tube with an external diameter of 8 mm, connected to a manometer. The standing wave width was determined by visual observations with the aid of a graduated ruler and photography. Additionally, in the 9.9° converging spillway, a micro-propeller current meter connected to a displacement structure movable in the transverse direction was also used (André and Ramos 2003, Cabrita 2007). The flow depths at the converging wall presented herein are those obtained on the right sidewall of the chute, according to the flow direction. The presented experimental standing wave widths are the mean values obtained on the left and right flume sidewalls for ϕ = 9.9°, or the right flume sidewall, for ϕ = 19.3°.

For the tested discharges, the approach shock number ($S_0 = F_0 \cdot \phi$) ranged from 0.31 to 0.32, for a wall convergence angle ϕ = 9.9°, and from 0.60 to 0.63, for ϕ = 19.3°. In the above formula, ϕ is given in radians, and F_0 is the approach Froude number, which depends on the approach flow depth immediately upstream of the wall deflection (d_0). The value of d_0 was here defined as the experimental flow depth at the position x = 0 (x is the streamwise coordinate) at the chute centerline.

2.2. Numerical model

A scale factor of 10:1 (numerical model: experimental model) and Froude similitude was considered to develop the numerical model, similarly to previous SPH numerical simulations (Nóbrega et al. 2020). Scale effects are expected to be negligible given the experimental flow conditions and occurrence of non-aerated flow for the tested discharges. In fact, for the experimental conditions, the flow depths over the broad crested weir were generally greater than 0.06 m and Reynolds numbers ($Re=UR/\nu$, where U is the mean velocity, R is the hydraulic radius and ν is the kinematic viscosity of water) satisfied the condition $Re > 10^{4.5}$ (Novak et al. 2007).

Continuous inflow fluid layers of uniform velocity were configured in the initial domain on the broad crested weir, for similar flow depth and maximum velocity measured at 0.125 m from the upstream end of the weir. The spillway chute was also initially filled with fluid particles with null velocity. The initial particle spacing (dp) used in simulations was 20 mm, at prototype scale.

The simulations were carried out during twenty seconds of computational time, with output intervals of 0.01 seconds. Considering that the flow depths on the broad crested weir stabilized after twelve seconds of simulation time, the flow depths and velocity profiles were determined as time-averaged values between twelve and twenty seconds of simulation. From the numerical cross-sectional profiles, the standing wave width was defined as the transverse location corresponding to the half value of the wave height. The height of the standing wave was calculated as the difference between the flow depth at the converging wall and the mean flow depth at the region unaffected by the standing wave.

2.3. SPH method

Simulations were performed using the DualSPHysics software, v.4.2. (Crespo et al. 2015, Domínguez et al. 2021), applying the parameters briefly described in the present session. In the SPH method, the properties of the discrete particles representing the fluid are calculated by the convolution between a smoothing kernel function (W) and a function f over an influenced domain, given by the kernel function length (h_k).

The continuity equation in the discrete form is given by Eq. (1), which considers an additional density diffusion term, artificially introduced to reduce the fluid density fluctuations. The chosen density diffusion term was the Delta-SPH model proposed by Molteni and Colagrossi (2009), represented by Eq. (2).

$$\frac{d\rho_a}{dt} = -\sum_b m_b \boldsymbol{v}_{ab} \cdot \nabla_a W_{ab} + D_a \tag{1}$$

$$D_a = 2\delta h_k \sum_b m_b \overline{c_{ab}} \times \left(\frac{\rho_a}{\rho_b} - 1\right) \frac{1}{r_{ab}^2 + 0.01\, {h_k}^2} \cdot \nabla_a W_{ab} \tag{2}$$

where m is the mass of the particle, ρ is the density of the particle, t is the time, $\mathbf{v}$ is the velocity vector, $\mathbf{r}$ is the position vector, $\mathbf{v}_{ab} = \mathbf{v}_a - \mathbf{v}_b$, $\mathbf{r}_{ab} = \mathbf{r}_a - \mathbf{r}_b$, ∇_a is the gradient of the kernel function, $W_{ab} = W(\mathbf{r}_a - \mathbf{r}_b, h_k)$, D_a is the numerical density diffusion term, δ is a magnitude coefficient, and $\overline{c_{ab}} = (c_a + c_b)/2$ is the average speed of sound. The subscripts a and b refer to the generic particle a and its neighboring particle b.

The discrete form of the momentum equation is given by Eq. (3). The dissipation term of this equation was modeled using the sub-particle scale (SPS) model (Eq. 4), which addresses the effects of laminar and turbulent viscosity (Crespo et al. 2015).

$$\frac{d\boldsymbol{v}_a}{dt} = -\sum_b m_b \left(\frac{p_a}{\rho_a^2} + \frac{p_b}{\rho_b^2}\right) \nabla_a W_{ab} + \boldsymbol{g} + \langle \Gamma \rangle_a \tag{3}$$

$$\langle \Gamma \rangle_a = \sum_b m_b \left(\frac{4\nu \boldsymbol{r}_{ab} \cdot \nabla_a W_{ab}}{(\rho_a + \rho_b)(r_{ab}^2 + 0.01 {h_k}^2)}\right) \boldsymbol{v}_{ab} + \sum_b m_b \left(\frac{\tau_a^{ij}}{\rho_a^2} + \frac{\tau_b^{ij}}{\rho_b^2}\right) \nabla_a W_{ab} \tag{4}$$

where p is the pressure, $\mathbf{g}$ is the gravity acceleration vector, $\langle \Gamma \rangle_a$ is the dissipation term, ν is the kinematic viscosity, and $\boldsymbol{\tau}$ is the SPS stress tensor.

The governing equations is completed with an equation of state for weak compressible fluids (Eq. 5), representing explicitly the variation of the pressure with the fluid density. This relation depends upon the reference density (ρ_0 = 1000 kg/m^3), the numerical speed of sound defined as $c_s = \sqrt{\partial p / \partial \rho}$, and the exponent of the equation of state for water (γ = 7).

$$p = \frac{\rho_0 c_s}{\gamma} \left[\left(\frac{\rho}{\rho_0}\right)^{\gamma} - 1\right] \tag{5}$$

3. Results

3.1. Flow depths

The flow depths at the spillway centerline for ϕ = 9.9° and Q = 35 L/s, or at the non-converging wall for ϕ = 19.3° (d_1), for the range of discharges, gradually decreased along the streamwise direction (x), as observed in Fig. 1a,b. For ϕ = 9.9° and discharges ranging from 42 L/s to 56 L/s, the flow depth presented nearly constant or gradually increasing values along the chute, downstream of $x \cong 0.5$ m. Even though the standing wave did not appear to reach the centerline of the spillway, a slight increase in the centerline flow depth is noticeable near the downstream end of the chute, similarly to the results of Adbo et al. (2018), for both 2D depth-averaged and 3D numerical models.

The overall behavior of the flow depth profiles for d_1 was reasonably well predicted by the SPH simulations, although the flow depths at the centerline for ϕ = 9.9° were in general overestimated (Fig. 1a). The mean absolute percentage differences for d_1 at the spillway with ϕ = 9.9° varied from 3.9% to 12.1% for $0.22 \leq x$ (m) ≤ 0.89, and the highest absolute percentage differences were found at x = 0.11 m because of the SPH particles detachment from the bottom on the initial reach of the chute, which was not observed in the experiments. On the other hand, the numerical values for d_1 at the spillway with ϕ = 19.3° were in general lower than the experimental data from André and Ramos (2003), as seen in Fig. 1b. The mean absolute percentage differences for ϕ = 19.3° varied from 8.1 to 12.1% for the range of discharges and $0.17 \leq x$ (m) ≤ 0.89.

The development of the flow depth at the converging wall (d_2) changed with the wall convergence angle. For ϕ = 9.9°, the experimental flow depths gradually reduced up to $x \cong 0.4$ m and were almost constant downstream of this location (Fig. 1c). For ϕ = 19.3°, the flow depths decreased until $x \cong 0.280$ to 0.4 m, depending on the discharge, and gradually increased further downstream (Fig. 1d).

The flow depth profile for d_2 was satisfactorily simulated using the SPH method, the relative differences being within 10%. For ϕ = 9.9°, the highest relative differences between the experimental and numerical flow depths were obtained at x = 0.11 m. For ϕ = 19.3°, the flow depths were overestimated by the SPH method up to x = 0.45 m, whereas they were underestimated downstream of this location, with mean absolute percentage differences varying from 5.4% to 8.5%, for $0.06 \leq x$ (m) ≤ 0.95.

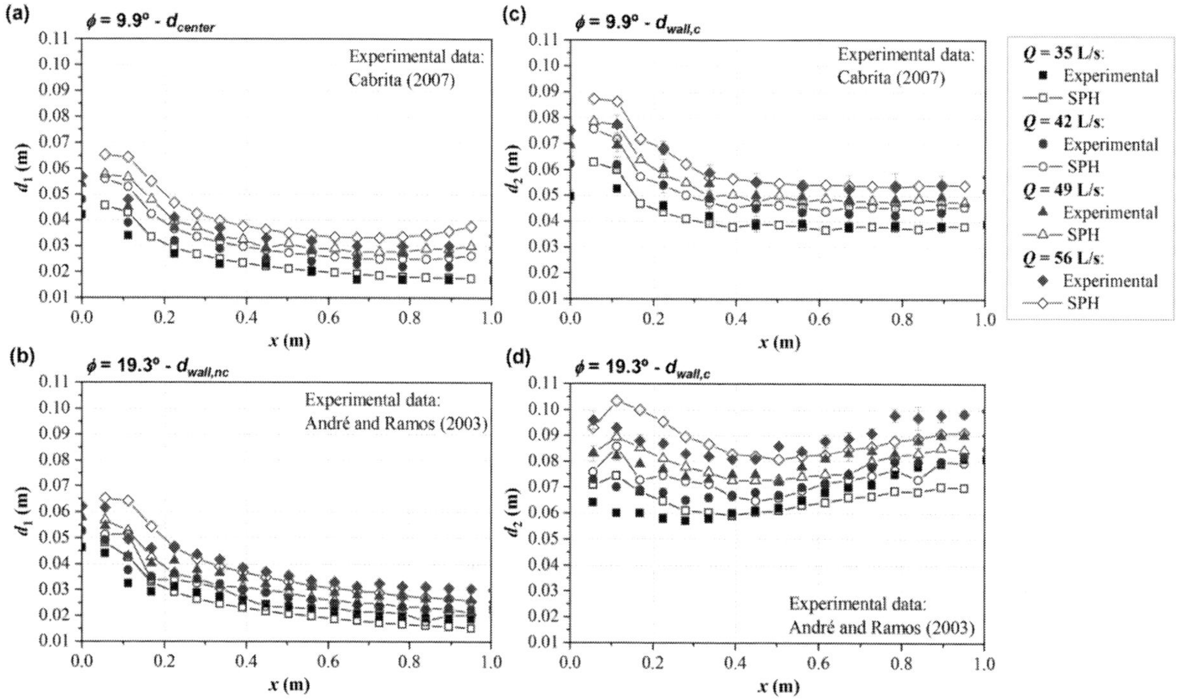

Figure 1. Flow depth along the spillway: (a) centerline, for ϕ = 9.9°; (b) non-converging wall, for ϕ = 19.3°; (c) converging wall, for ϕ = 9.9°; (d) converging wall, for ϕ = 19.3°.

The flow depth contours for Q = 56 L/s are shown in Fig. 2, where z is the transverse coordinate from the right sidewall (in the flow direction) of the broad crested weir. The extension of the standing wave is clearly seen in this figure near the converging wall. From Fig. 2a, one may notice the influence of the converging walls on the gradual increase of the flow depths towards the end of the chute. Another notable aspect is the nonuniformity of the flow depth within the standing wave, similar to an undular shape (Fig. 2a,b).

Normalized experimental and numerical flow depths at the spillway centerline for ϕ = 9.9°, or at the non-converging wall for ϕ = 19.3° (d_1), relative to the critical flow depth at the broad crested weir (d_c), are presented in Fig. 3. The results obtained using the explicit simplified equations from Castro-Orgaz and Hager (2010), for prismatic chutes, are also presented. For the present study, the dimensionless roughness ($r = k_s/d_c$, where k_s is the roughness height) varied between 0.0047 and 0.0035, for discharges of 35 L/s and 56 L/s, respectively. Overall, Castro-Orgaz and Hager (2010) equation, valid for $x/(d_c r) > 2000$, fits well to the experimental data. The largest differences were obtained near the end of the chute contraction for $\phi = 9.9°$ and discharges between 42 and 56 L/s, which would be expected, due to the increase in the centerline flow depth in the converging chute.

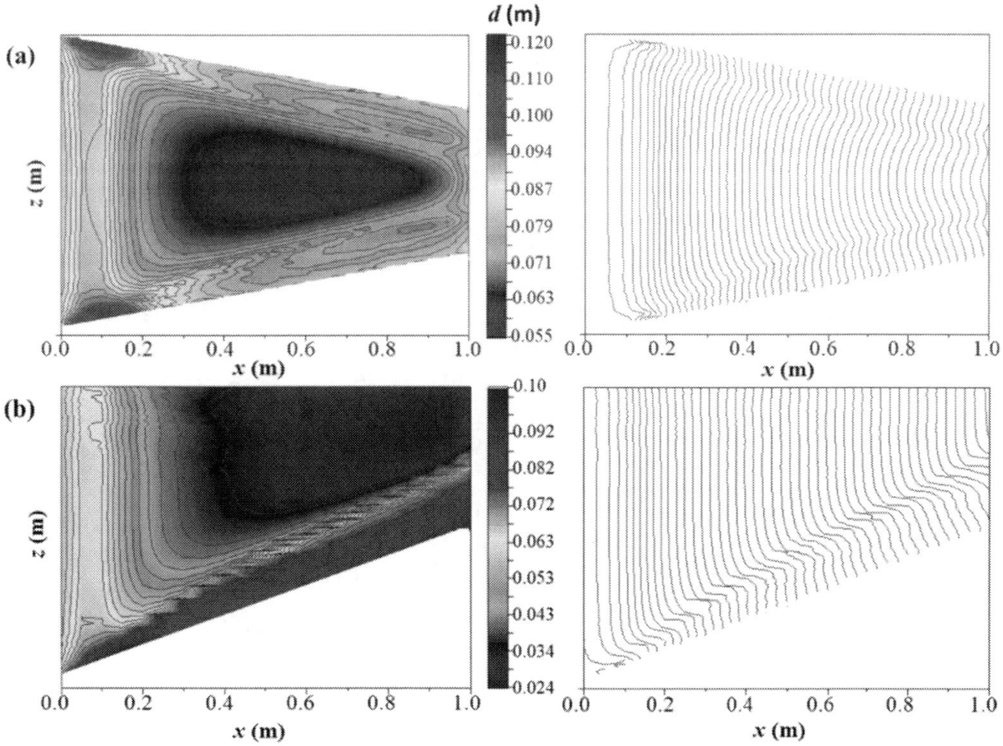

Figure 2. Flow depth contours for Q=56 L/s: (a) $\phi = 9.9°$; (b) $\phi = 19.3°$.

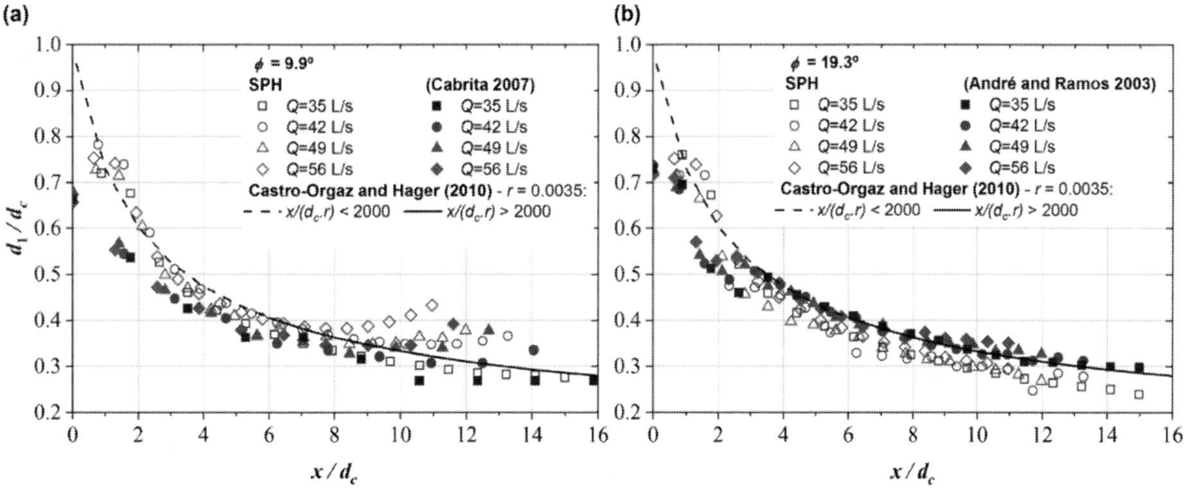

Figure 3. Normalized flow depth development at the spillway centerline or at the non-converging wall: (a) $\phi = 9.9°$; (b) $\phi = 19.3°$

The relative flow depth at the converging wall in respect to the flow depth at the centerline or at the non-converging wall (d_2/d_1) was found to vary significantly with the wall convergence angle (Fig. 4). The relative flow depths for $\phi = 9.9°$ attained maximum values of $d_2/d_1 \cong 1.5$ to 2.3 at the downstream end of the chute, depending on the discharge. In turn, at the 19.3° converging spillway, the relative flow depths continuously increased along the chute, reaching maximum values of $d_2/d_1 \cong 3.5$ to 5.0 near its downstream end.

The formulations for horizontal channels proposed by Ippen and Harleman (1956) and Hager (1989) were used to predict d_2/d_1 as a function of ϕ, the standing wave angle β (in respect to the approach flow direction), and the Froude number. To apply their equations, the local Froude number (F_1) at each cross-section was calculated using the flow depth derived from Castro-Orgaz and Hager (2010) formulation, for identical approach unit discharge. Although the flow depths at the centerline for ϕ = 9.9° were greater than those expected on prismatic chutes near the contraction end (see Fig. 3a), it was assumed, for simplicity, that the flow depths at the centerline were not affected by the converging walls. A reasonable prediction of the relative flow depths was observed by applying Ippen and Harleman (1956) and Hager (1989) equations on a sloping converging chute. The differences among the Ippen and Harleman (1956) formulation and the experimental values for ϕ = 9.9°, for discharges equal or greater than Q = 42 L/s, may be mainly justified by the flow depth increase near the downstream end of the converging chute. In addition, the hypothesis of constant flow depth and velocity at each cross-section, as implicitly assumed when using both formulations on the sloping chute, loses validity near its downstream end, due to the increasing influence of the standing wave.

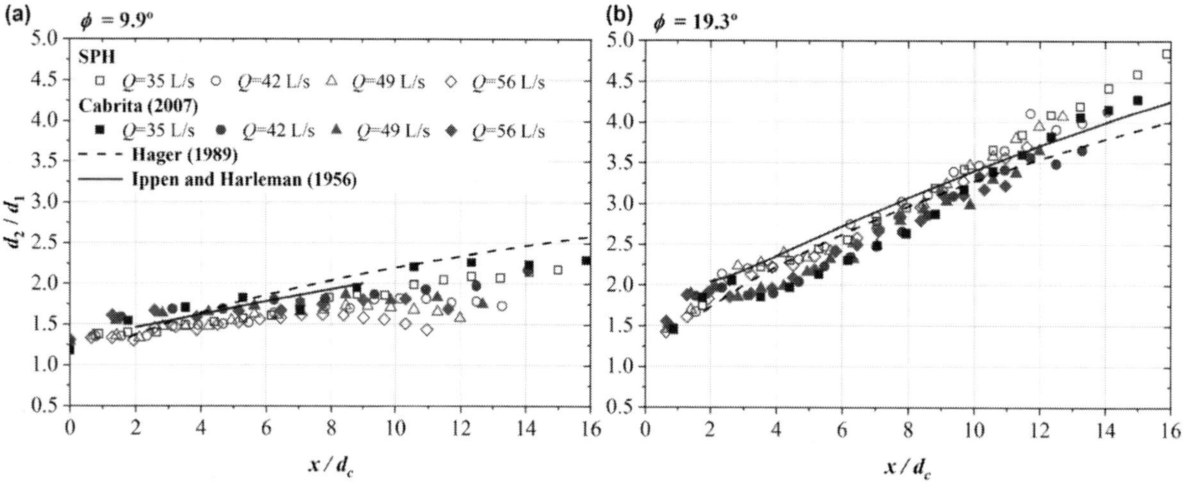

Figure 4. Ratio between the flow depth at the converging wall and at the centerline, or at the non-converging wall, along the spillway: (a) ϕ = 9.9°; (b) ϕ = 19.3°.

3.2. Standing wave characteristics

The shape and width of the standing wave was obtained from the numerical cross-sectional flow depth profiles along the spillway. Considering the range of discharges and the wall convergence angles tested, the standing waves did not reach the spillway centerline for ϕ = 9.9°, or the non-converging wall for ϕ = 19.3°, either experimental or numerically.

Taking into account that the experimental measurements were only taken at the sidewall or at the centerline of the spillway, the discussion presented herein about the shape of the standing wave is based only on the numerical results. Normalized graphs for the standing wave are presented in Fig. 5, for the discharge of 56 L/s. In this figure, d is the flow depth; d_1 is the flow depth at the centerline or at the non-converging wall; d_2 is the flow depth at the converging wall; z is the transverse coordinate from the right sidewall of the broad-crested weir; z_0 is the transverse position in respect to the right sidewall of the converging chute; and F_1 is the local Froude number from numerical calculations; x is the streamwise coordinate; L is the chute length. The local Froude number was calculated as $F_1 = U_1/\sqrt{gd_1}$, where U_1 is the mean velocity given by the ratio between the approach unit discharge and d_1, obtained from the numerical results.

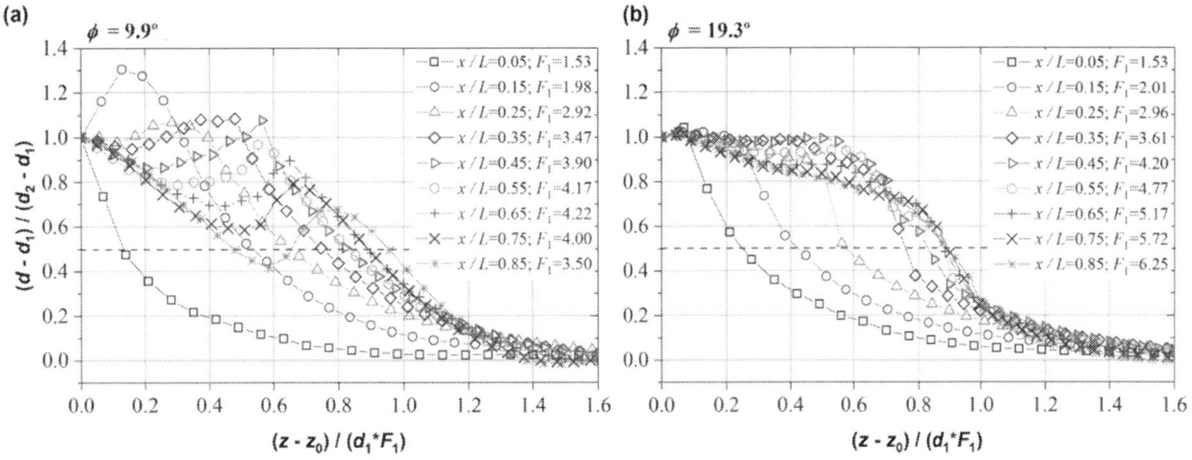

Figure 5. Normalized flow depths at the transverse direction for Q = 56 L/s: (a) ϕ = 9.9°; (b) ϕ = 19.3°.

The characteristics of the standing waves were found to vary with the wall convergence angle and the position along the spillway, as observed in Fig. 5. In the cross-sections near the upstream end of the chute (i.e., $x/L \leq 0.15$), the cross-sectional flow depth profile is markedly distinct from those found further downstream, as would be expected due to the inaccuracy of the SPH simulations in this initial reach of the chute (Nóbrega et al. 2020). In general, the standing wave in the vicinity of the 9.9° converging wall was undular, presenting a wavy shape with a crest of maximum height around 1.3 for x/L = 0.15. This result is reasonably in agreement with those found by Ippen and Harleman (1956) and Hager et al. (1994) on horizontal chutes. The shape of the standing wave in the vicinity of the 19.3° converging wall was similar to a roller type jump, presenting a constant level or gradually decreasing flow depths near the wall, followed by an abrupt change in the flow depth with the transverse coordinate. These observations are consistent with the definitions proposed by Ippen and Harleman (1956) to classify the standing waves on horizontal channels, as undular for $1 < d_2/d_1 < 2$, or direct standing waves with a sharp front for $d_2/d_1 \geq 2$, in analogy to hydraulic jumps (see the relative flow depths in Fig. 4).

Reinauer and Hager (1998) used the shock number ($S_0 = F_0\phi$) as the governing parameter to classify the standing wave shape observed on supercritical flow across a wall deflection. According to these authors, if $S_0 < 0.5$, the standing waves generate surface undulations comparable to undular hydraulic jumps. If $S_0 >$ 1.8-2.0, the standing waves are over forced, occurring the breaking of the shock front and air entrainment, and the flow depths near the wall become excessively high. In the experiments developed by André and Ramos (2003) and Cabrita (2007), the approach shock number varied from 0.31 to 0.32 for ϕ = 9.9°, and from 0.60 to 0.63 for ϕ = 19.3°, which is in conformity with the criterion proposed by Reinauer and Hager (1998).

The numerical values of the standing wave width were determined considering a simplistic approximation to the standing wave front as the transverse position corresponding to the half wave height, as represented by the dashed line in Fig. 5 for $(d - d_1)/(d_2 - d_1) = 0.5$. This approximation is better suited for ϕ = 19.3° because the wave fronts are more sharpened than those observed for ϕ = 9.9°.

Typical development of the normalized standing wave width (w/d_c) along the chute is shown in Fig. 6. The experimental values gradually increased along the chute and tended to nearly constant values, and they were found to be practically independent of the discharge. The results from visual observations are in general lower than those determined using the micro-propeller current meter, which may be due to experimental measurement inaccuracies, namely those related to the subjectivity of visual measurements and the unsteadiness of the turbulent free surface. The maximum experimental standing wave widths normalized by the critical flow depth were w/d_c = 1.9 to 2.2 for x/d_c = 15.9 and Q = 35 L/s, and w/d_c = 1.7 to 1.8 for x/d_c = 11.6 and Q = 56 L/s, for measurements from visual observations or using the micro-propeller current meter. The overall development of the numerical standing wave width along the chute is in reasonably good agreement with the experimental data, except near the chute ends. The mean absolute percentage differences between the numerical results and those from the micro-propeller current meter, for ϕ = 9.9° and $0.22 \leq x$ (m) ≤ 0.89, are limited to 11% for the range of discharges, while the differences are in general greater when compared to visual observation data, namely in the upstream end of the chute.

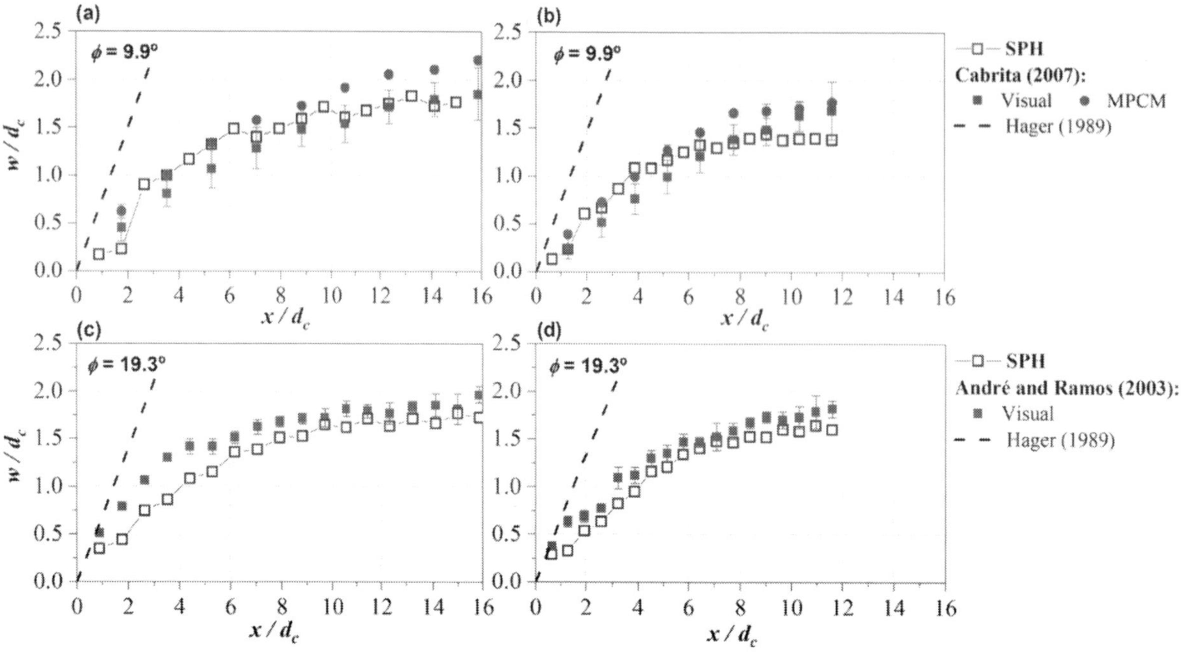

Figure 6. Normalized standing wave width: (a) Q = 35 L/s and ϕ = 9.9°; (b) Q = 56 L/s and ϕ = 9.9°; (c) Q = 35 L/s and ϕ = 19.3°; (d) Q = 56 L/s and ϕ = 19.3°. MPCM - micro-propeller current meter.

In Fig. 6, the dashed line was obtained from the standing wave angle estimated from Hager (1989), at the origin of the chute contraction, as a function of the approach Froude number. This value is considerably larger than those estimated from the experimental data or numerical simulations near the chute origin, for ϕ = 9.9°. This may be explained, at least in part, by the absence of experimental data in the upstream end of the chute (Fig. 6a,b). However, a good approximation of the wave front near the upstream end of the chute is observed for the experimental data corresponding to ϕ = 19.3° (Fig. 6c,d).

3.3. Velocity profiles

The experimental and numerical velocity profiles at the chute centerline are presented in Fig. 7 for the range of discharges for ϕ = 9.9°, where y represents the direction perpendicular to the spillway invert, and the dashed lines represent the free-stream velocities calculated using the ideal-flow theory given by Eq. (6):

$$V_{max} = \sqrt{2g(H_{max} - d_1 cos\theta)} \tag{6}$$

where V_{max} is the free-stream velocity, g is the gravity acceleration constant, H_{max} is the upstream total head relative to the invert, d_1 is the flow depth measured at the flume centerline, and θ chute angle from the horizontal.

The largest percentage differences between the experimental and the numerical velocity profiles were found up to x/L = 0.15, because of the SPH particles detachment at the transition from the broad crested weir to the spillway chute. Inaccuracies on the experimental velocity measurements close to the upstream end of the chute might be also expected, because of the non-hydrostatic pressure distribution due to the flow curvature. For the cross-sections downstream of x/L = 0.20, the mean percentage differences decreased with the streamwise coordinate and were limited to 4.6%, considering the range of discharges tested.

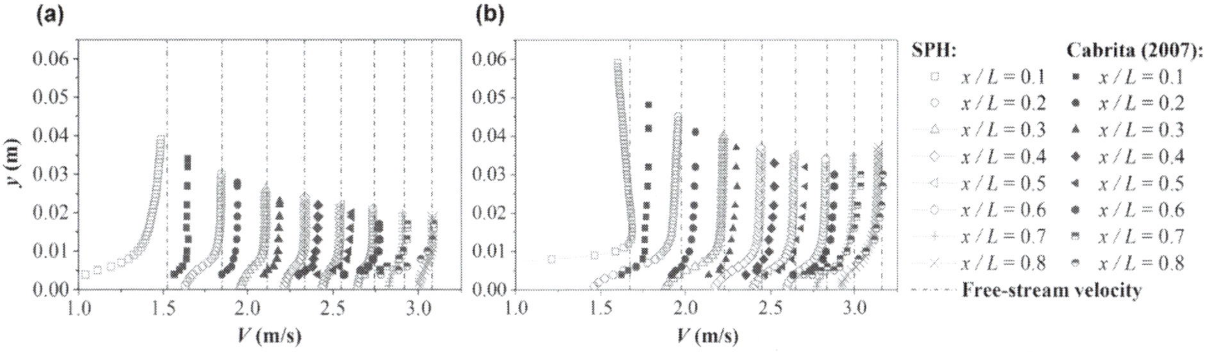

Figure 7. Velocity profiles at the chute centerline of the converging spillway with $\phi = 9.9°$: (a) $Q = 35$ L/s; (b) $Q = 56$ L/s.

4. Conclusions

The comprehension of the complex flow induced by wall convergence on supercritical flows is of interest for designing dam spillways. For this purpose, numerical simulations of the flow over smooth chutes with converging walls were performed using the SPH method. The flow depth profile at the converging wall or at the spillway centerline (or non-converging wall) was found to vary with the wall convergence angle. The general pattern of the profile and the values of the flow depths were reasonably well predicted by the numerical simulations, in particular the flow depth near the converging wall, which is an important parameter to determine the height of the spillway training walls. The relative flow depth at the converging wall in respect to the flow depth at the centerline attained maximum values around 2 and 5 at the downstream chute end for wall convergence angles of 9.9° and 19.3°, respectively. The standing waves characteristics were also analyzed from the numerical flow depth profiles at the transverse direction. An undular shape of the standing wave was observed on the 9.9° converging spillway, for approach shock numbers lower than 0.5, whereas a roller type shape was noted for the wall convergence angle of 19.3°, for approach shock numbers slightly greater than 0.5. The experimental and numerical results for the standing wave width showed a gradually increasing trend along the chute, tending to nearly constant values. The velocity profiles at the chute centerline are also reasonably in agreement with the experimental counterparts, except near the upstream end of the chute and close to the invert.

5. ACKNOWLEDGMENTS

The authors acknowledge the support of the Portuguese Foundation for Science and Technology (FCT), through the RECI/ECM-HID/0371/2012 Project. Thanks are also extended to Prof. Rui Ferreira, for providing the computational server to run the simulations, and to Luís Palma Mendes, for his assistance.

6. REFERENCES

Abdo, K., Riahi-Nezhad, C.K., and Imran, J. (2018). "Steady supercritical flow in a straight-wall open channel contraction." *J. Hydraul. Res.*, 57(5), 647-661.

André, M., and Ramos, P. (2003). Hidráulica de descarregadores de cheia em degraus: Aplicação a descarregadores com paredes convergentes, Graduate Research Rep., Instituto Superior Técnico, Universidade de Lisboa, Lisbon, Portugal.

Cabrita, J. (2007). Caracterização do escoamento deslizante sobre turbilhões em descarregadores de cheias em degraus com paredes convergentes, M.Sc. thesis, Instituto Superior Técnico, Universidade de Lisboa, Lisbon, Portugal.

Castro-Orgaz, O., and Hager, W.H. (2010). "Drawdown curve and turbulent boundary layer development for chute flow". *J. Hydraul. Res.*, 48(5), 591-602.

Causon, D.M., Mingham, C.G., and Ingram, D.M. (1999). "Advances in calculation methods for supercritical flow in spillway channels." *J. Hydraul. Eng.*, 125(10), 1039-1050.

Chow, V.T. (1959). Open-channel Hydraulics, McGraw-Hill, New York.

Crespo, A.J.C., Domínguez, J.M., Rogers, B.D., Gómez-Gesteira, M., Longshaw, S., Canelas, R., Vacondio, R., Barreiro, A., and García-Feal, O. (2015). "DualSPHysics: Open-source parallel CFD solver based on Smoothed Particle Hydrodynamics (SPH)." *Comput. Phys. Comm.*, 187, 204-216.

Dawson, J.H. (1943). The effect of lateral contractions on super-critical flow in open channels, PhD Thesis, Lehigh University, Pennsylvania, United States.
Domínguez, J.M., Fourtakas, G., Altomare, C., Canelas, R.B., Tafuni, A., García-Feal, O., Martínez-Estévez, I., Mokos, A., Vacondio, R., Crespo, A.J.C., Rogers, B.D., Stansby, P.K., and Gómez-Gesteira, M. (2021). "DualSPHysics: From fluid dynamics to multiphysics problems." *Comput. Part. Mech.*, 1-29. doi:10.1007/s40571-021-00404-2.
Hager, W.H. (1989). "Supercritical flow in channel junctions." *J. Hydraul. Eng.*, 115(5), 595-616.
Hager, W.H. (1992). "Spillways – shockwaves and air entrainment." ICOLD Bulletin 81. International Commission for Large Dams, Paris.
Hager, W.H., Schwalt, M., Jimenez, O., and Chaudhry, M.H. (1994). "Supercritical flow near an abrupt wall deflection." *J. Hydraul. Res.*, 32(1), 103-118.
Hunt, S.L., Kadavy, K.C., Abt, S.R., and Temple, D.M. (2008). "Impact of converging chute walls for roller compacted concrete stepped spillways". *J. Hydraul. Eng.*, 134(7), 1000-1003.
Ippen, A.T., and Dawson, J.H. (1951). "Design of channel contractions - high velocity flows in open channels." *Trans. Am. Soc. Civil Eng.*, 116(1), 326-346.
Ippen, A.T., and Harleman, D.R.F. (1956). "Verification of theory for oblique standing waves." *Trans. Am. Soc. Civil Eng.*, 121(1), 678-694.
Jan, C.-D., Chang, C.-J., Lai, J.-S., and Guo, W.-D. (2009). "Characteristics of hydraulic shock waves in an inclined chute contraction – Experiments." *J. Mech.*, 25(2), 129-136.
Molteni, D., and Colagrossi, A. (2009). "A simple procedure to improve the pressure evaluation in hydrodynamic context using the SPH." *Comput. Phys. Commun.*, 180(6): 861-872.
Moran, R., Toledo, M.A., Peraita, J., and Pellegrino, R. (2021). "Energy dissipation in stilling basins with side jets from highly convergent chutes." *Water,* 13(10), 1343.
Neilson, F.M. (1976). Convex chutes in converging supercritical flow, Report, U.S. Army Engineer Waterways Experimentation Station Hydraulics Laboratory, Mississippi, United States.
Nóbrega, J.D., Matos, J., Schulz, H.E., and Canelas, R.B. (2020). "Smooth and stepped spillway modelling using the SPH method". *J. Hydraul. Eng.,* 146(8), 04020054.
Novak, P., Moffat, A.I.B., Nalluri, C., and Narayanan, R. (2007). Hydraulic structures, 4 ed., Taylor & Francis, New York.
Nunes, F., Matos, J., and Meireles, I. (2018). "Numerical modelling of skimming flow over small converging spillways". *Proc., Protections 2018: 3rd International Conference on Protection against Overtopping*, 6-8 June 2018, Grange-over-Sands, UK.
Reinauer, R., and Hager, W.H. (1996). "Shockwave reduction by chute diffractor." *Exp. Fluids*, 21(3), 209-217.
Reinauer, R., and Hager, W.H. (1998). "Supercritical flow in chute contraction." *J. Hydraul. Eng.*, 124(1), 55-64.
Salazar, F., San-Mauro, J., Celigueta, M.A., and Oñate, E. (2020). "Shockwaves in spillways with the particle finite element method." *Comput. Part. Mech.*, 7, 87-99.
Song, C.S., and Zhou, F. (1999). "Simulation of free surface flow over spillway". *J. Hydraul. Eng.*, 125(9), 959-967.
Stone, A.A. (1946). Theoretical investigation of standing waves in hydraulic structures, M.Sc. thesis, Massachusetts Institute of Technology, Boston, United States.
USBR (1987). Spillways. In Design of small dams, 3 ed., United States Department of the Interior, Washington, US, 339–434.